A Specialist Periodical Report

Inorganic Chemistry of the Transition Elements
Volume 4

A Review of the Literature Published between October 1973 and September 1974

Senior Reporter
B. F. G. Johnson, *University Chemical Laboratory, University of Cambridge*

Reporters
F. L. Bowden, *University of Manchester Institute of Science and Technology*
D. W. Clack, *University College, Cardiff*
R. Davis, *Kingston Polytechnic, Kingston-upon-Thames*
C. D. Garner, *University of Manchester*
J. Howell, *Keele University*
M. Hughes, *University of Cambridge*
L. A. P. Kane-Maguire, *University College, Cardiff*
W. Levason, *University of Manchester Institute of Science and Technology*
C. A. McAuliffe, *University of Manchester Institute of Science and Technology*
J. A. McCleverty, *University of Sheffield*

The Chemical Society
Burlington House, London W1V 0BN

ISBN: 0 85186 530 5
ISSN: 0 305-9774
Library of Congress Catalog Card No: 72-83458

Set and printed in Great Britain by
Page Bros (Norwich) Ltd, Norwich

Foreword

This is the fourth volume of the Specialist Periodical Report covering Transition Metal Chemistry, including the lanthanides and actinides. It is a review of the literature published between October 1973 and September 1974. The layout is similar to that of the earlier reports. Chapter 1 deals with the chemistry of the Early Transition Metals, excluding Scandium, Yttrium, and the Lanthanides. The chemistry of the First Transition Period from Manganese to Copper is given in Chapter 2. The chemistry of the Noble Metals is reported in Chapter 3 and Chapter 5 is concerned with Scandium, Yttrium, the Lanthanides and Actinides. This year we have included an additional Chapter devoted to the chemistry of Zinc, Cadmium, and Mercury, which carries a sub-section concerned with biological aspects of these metals. Again, we have attempted to be comprehensive and have chosen to place much of the data in tabulated form.

B. F. G. JOHNSON

Contents

Chapter 1 The Early Transition Metals   1
    I: Titanium, Zirconium, Hafnium, Vanadium, Niobium, and
    Tantalum
    By. F. L. Bowden

1 **Titanium**   1
  Introduction   1
  Binary Compounds and Related Systems   3
    Halides and Oxyhalides   3
    Oxides   4
    Chalcogenides   5
    Carbides, Silicides, and Germanides   7
    Nitrides, Borides, *etc.*   7
  Titanium (II)   8
  Titanium(III)   8
    Halides and Oxyhalides   8
    O-Donor Ligands   10
    N-Donor Ligands   12
    Mixed N- and O-Donor Ligands   12
    Cyclopentadienyl Complexes   14
    Miscellaneous   16
  Titanium(IV)   16
    Halides, Oxyhalides, and Pseudohalides   16
    O-Donor Ligands   18
    S-Donor Ligands   24
    N-Donor Ligands   24
    Mixed N- and O-Donor Ligands   26
  Organometallic Titanium(IV) Compounds   28
    Titanium–Carbon σ-Bonded Complexes   28
    Cyclopentadienyl Complexes   30

2 **Zirconium and Hafnium**   31
  Introduction   31
  Binary Compounds and Related Species   31
    Halides and Oxyhalides   31
    Oxides and Chalcogenides   33
    Carbides, Silicides, and Related Compounds   34

Nitrides   34
Borides   34
Zirconium(II), Zirconium(III), Hafnium(II), and
    Hafnium(III)   34
Zirconium(IV) and Hafnium(IV)   34
    Halides and Oxyhalides   34
    O-Donor Ligands   35
    S-Donor Ligands   42
    N-Donor Ligands   42
    Mixed N- and O-Donor Ligands   43
Organometallic Zirconium(IV) and Hafnium(IV)
    Compounds   44
    σ-Bonded Complexes   44

3 Vanadium   45
    Introduction   45
    Carbonyl and other Low Oxidation State Compounds   45
    Binary Compounds and Related Species   46
        Halides and Oxyhalides   46
        Oxides   47
        Nitrides   47
        Borides   47
        Silicides   47
        Hydrides   48
    Vanadium(II)   48
    Vanadium(III)   49
        Halides and Oxyhalides   49
        O-Donor Ligands   49
        S-Donor Ligands   50
        N-Donor Ligands   50
        Mixed N- and O-Donor Ligands   51
    Compounds with Metal–Carbon Bonds   51
    Vanadium(IV)   52
        Halides and Oxyhalides   52
        O-Donor and Mixed N- and O-Donor Ligands   52
        Schiff Base Complexes   56
        S-Donor Ligands   58
        N-Donor Ligands   61
        Compounds with Metal–Carbon Bonds   62
        Miscellaneous   62
    Vanadium(V)   63
        Halides and Oxyhalides   63
        O-Donor Ligands   64
        S- and Se-Donor Ligands   69
        N- and O-Donor Ligands   69

4 Niobium and Tantalum   70
  Introduction   70
  Carbonyl and other Low Oxidation State Compounds   70
  Binary Compounds   72
    Halides   72
    Oxides   72
    Chalcogenides   72
    Nitrides and Related Compounds   73
    Silicides and Borides   73
    Hydrides   73
  Compounds with Nb–Nb or Ta–Ta Bonds   73
  Niobium(III) and Tantalum(III)   74
  Niobium(IV) and Tantalum(IV)   75
  Niobium(V) and Tantalum(V)   77
    Halides and Oxyhalides   77
    O-, S-, Se-, and Te-Donor Ligands   79
    N-Donor Ligands   84
    Mixed N- and S-Donor Ligands   85
    Organometallic Niobium(V) and Tantalum(V)
      Compounds   85

*II: Chromium, Molybdenum, Tungsten, Technetium, and
Rhenium*   87
By C. D. Garner

1 Chromium   87
  Introduction   87
  Carbonyl and Thiocarbonyl Complexes   87
  η-Complexes   89
  Trifluorophosphino- and Trichlorostannano-
    complexes   89
  Dinitrogenyl Complexes   89
  Nitrosyl Complexes   89
  Compounds with Chromium–Metal Bonds   90
  Binary Compounds and Related Systems   90
    Halides   90
    Oxides   90
    Chalcogenides   90
    Compounds with Elements of Group V   91
    Carbides, Silicides, and Germanides   92
    Borides   92
  Chromium(II)   92
    Halides   92
    O-Donor Ligands   93
    N-Donor and P-Donor Ligands   93
  Chromium(III)   93

Optically Active Complexes 94
Halide and Oxyhalide Complexes 94
O-Donor Ligands 96
Polymeric Complexes containing Bridging
    O-Donor Ligands 98
Complexes with S-Donor, Se-Donor, or Te-Donor
    Ligands 100
N-Donor Ligands 100
Mixed N-Donor and Other Donor Ligands 103
P-Donor Ligands 104
Cyano-complexes 104
Organometallic Complexes 105
Chromium(IV) 105
Chromium(V) 106
Chromium(VI) 107
    Halides and Oxyhalides 107
    O-Donor Ligands 107
    S-Donor Ligands 109
    N-Donor Ligands 109

2 Molybdenum and Tungsten 109
Introduction 109
Carbonyl and Thiocarbonyl Complexes 113
η-Complexes 114
Trifluorophosphino- and Trichlorostannano-
    complexes 115
Dinitrogenyl and Related Complexes: Nitrogen
    Fixation 115
Nitrosyl and Thionitrosyl Complexes 117
Cyano- and Isocyano-complexes 118
Pyrazolylborato-complexes 120
Hydrido-complexes 120
Binary Compounds and Related Systems 121
    Halides 121
    Oxides 123
    Chalcogenides 123
    Nitrides 124
    Carbides 124
    Silicides 124
    Borides and Gallides 125
Compounds containing Mo—Mo, W—W, or
    Related Bonds 125
Molybdenum(II) and Tungsten(II) 128
Molybdenum(III) and Tungsten(III) 129
Molybdenum (IV) and Tungsten(IV) 130
Molybdenum(V) and Tungsten(V) 132
    Mononuclear Complexes 132

Dinuclear and Polynuclear Complexes 135
Molybdenum and Tungsten Bronzes 136
Molybdenum(VI) and Tungsten(VI) 138
Halide, Oxyhalide, and Related Complexes 138
O-Donor Ligands 139
Iso- and Hetero-polyanions 146
S- and Se-Donor Ligands 149
N-Donor Ligands 150
Organometallic Compounds 151

3 Technetium and Rhenium 151
Introduction 151
Carbonyl Complexes 152
Dinitrogenyl Complexes 154
Nitrosyl Complexes 155
Binary Compounds and Related Systems 155
Oxides 155
Carbides, Silicides, and Borides 156
Compounds containing Tc—Tc or Re—Re Bonds 156
Rhenium(III) 158
Technetium(IV) and Rhenium(IV) 158
Rhenium(V) 159
Technetium(VI) and Rhenium(VI) 160
Technetium(VII) and Rhenium(VII) 161

Chapter 2 Elements of the First Transition Period 162
I: Manganese and Iron
By R. Davis

1 Manganese 162
Carbonyl Compounds 162
Nitrosyl and Nitrogenyl Compounds 171
Manganese(II) 172
Halides 172
Complexes 173
N-Donor ligands 173
O-Donor ligands 174
S-Donor ligands 176
Mixed donor ligands 176
Manganese(III) 177
Complexes 177
Manganese(IV) 178
Higher Oxidation States of Manganese 179
Oxides and Sulphides 179

2 Iron 180
Carbonyl Compounds 180
Nitrosyl and Aryldiazo Complexes 191

Other Iron(0) Compounds   194
Iron(II)   194
  Halides, Cyanides, and Hydrides   194
  Complexes   195
    N-Donor ligands   195
    O-Donor ligands   195
    S-Donor and P-donor ligands   199
    Mixed donor ligands   200
    Isocyanide complexes   201
Iron(III)   201
  Halides and Cyanides   201
  Complexes   203
    N-Donor ligands   203
    O-Donor ligands   204
    S-Donor ligands   206
    Mixed donor ligands   208
Iron(IV)   209
Model Compounds for Iron in Biological Systems   210
Oxides and Sulphides   214
Other Papers   215

3 Formation and Stability Constants (Manganese and Iron)   215

4 Bibliography   216

II: Cobalt, Nickel, and Copper   218
By C. A. McAuliffe and W. Levason
1 Cobalt   218
  Carbonyls   218
    Group IVB Donors   219
    Group VB Donors   220
  Nitrosyls   221
  Cobalt(I)   222
  Cobalt(II) Complexes   224
    Halides and Pseudohalides   224
    Amine Complexes   225
    Pyridine and Related Ligands   225
    Imidazole and Related Ligands   229
    Macrocyclic N-Donors   229
    Other N-Donors Ligands   232
    P-Donor and As-Donor Ligands   233
    O-Donor Ligands   234
    S-Donor Ligands   237
    Mixed Donor Ligands   239
    Other Compounds   243
    Interaction with Dioxygen   243
    Carbaboranes   246

Oxides and Other Simple Anions 248
Cobalt(III) Complexes 248
    Halides 248
    Ammine Complexes 248
    Diamine Complexes 249
    Polyamine Donors 252
    Macrocylic N-Donors 253
    Other N-Donors 255
    Oximate Compounds 255
    Schiff Bases 257
    P-Donors, As-Donors, and Sb-Donors 259
    O-Donors 259
    S-Donors 260
    Amino-acid Complexes 261
    Other Mixed Donor Ligands 263
    Polynuclear Bridged Complexes 266
Cobalt(IV) 267

2 Nickel 267
Carbonyls and Nitrosyls 267
Nickel(0) 269
Nickel(I) 270
Nickel(II) 271
    Halides and Pseudohalides 271
    N-Donors 271
        Pyridine and related ligands 271
        Amine donors 272
        Macrocyclic N-donors 274
        Other N-donors 278
    P-Donors, As-Donors, and Sb-Donors 281
    O-Donors 284
    S-Donors 287
    Carbaboranes 290
    Mixed Donor Ligands 291
    Other Compounds 302
Nickel(III) 303
Nickel(IV) 303
ESCA Measurements 304

3 Copper 304
Copper(I) 304
    Halides and Pseudohalides 304
    N-Donors 305
    P-Donors and As-Donors 306
    O-Donors and S-Donors 307
    Mixed Donors and other Components 308

Copper(II)   309
  Halides and Pseudohalides   309
  N-Donors   310
  Macrocycles   311
  O-Donors and S-Donors   312
  Mixed Donors   313
  Copper(III)   318

**4 X-Ray Data (Copper), Physical Data (Copper), and
Formation and Stability Constants (Cobalt, Nickel, and
Copper)**   319

5 Bibliography   327

Chapter 3 The Noble Metals   329
           I: Ruthenium, Osmium, Rhodium, and Iridium
           By L. A. P. Kane-Maguire

1 Ruthenium   329
  Cluster Compounds   329
  Ruthenium-(−II), -(0), and -(I)   329
  Ruthenium(II)   330
    Group VII Donors   330
      Hydride-phosphine complexes   330
      Halogeno-carbonyl and -phosphine complexes   330
    Group VI Donors   331
      O-Donor ligands   332
      S-Donor ligands   333
    Group V Donors   333
      Molecular nitrogen complexes   333
      Nitrosyl complexes   334
      Other N-donor ligands   334
      P- and As-donor ligands   338
    Group IV Donors   339
      C-Donor ligands   339
  Mixed Ruthenium-(II) and -(III) Complexes   340
  Ruthenium(III)   341
    Group VII Donors   341
    Group VI Donors   341
    Group V Donors   341
      N-Donor ligands   341
      As-Donor ligands   342
  Ruthenium(IV)   342
    Group VII Donors   342
    Group VI Donors   343
    Group V Donors   343
  Ruthenium(V) and Higher Oxidation States   344

2 Osmium 344
Cluster Compounds 344
Osmium(II) 346
Group VII Donors 346
Hydrido- and halogeno-carbonyl complexes 346
Group VI Donors 347
Group V Donors 347
N-Donor ligands 347
P-Donor ligands 348
Group IV Donors 348
Osmium(III) 349
Osmium(IV) 350
Osmium(VI) 350
Osmium (VIII) 351

3 Rhodium 352
Cluster Compounds 352
Rhodium(0) 353
Rhodium(I) 354
Group VII Donors 354
Hydrido-phosphine complexes 354
Halogeno-carbonyl and -phosphine complexes 354
Group VI Donors 356
O-Donor ligands 356
S-Donor ligands 358
Group V Donors 358
N-Donor ligands 358
P- and As-Donor ligands 360
Group IV Donors 361
Group III Donors 362
B-Donor ligands 362
Rhodium(II) 362
Group VI Donors 362
Other Donor Ligands 363
Rhodium(III) 363
Group VII Donors 363
Group VI Donors 364
O-Donor ligands 364
S-Donor ligands 365
Group V Donors 366
N-Donor ligands 366
P- and As-Donor ligands 368
Group IV Donors 369
Rhodium-(IV) and -(V) 369

4 Iridium 371
Cluster Compounds 371

Iridium(I)   371
  Hydrido- and Halogeno-carbonyl or -phosphine
  Complexes   371
  Group VI Donors   372
  Group V Donors   373
    N-Donor ligands   373
    P- and As-donor ligands   374
  Group IV Donors   375
  Group III Donors   376
    B-Donor ligands   376
Iridium(III)   376
  Group VII Donors   376
  Group VI Donors   376
    O-Donor ligands   376
    S- and Se-donor ligands   376
  Group V Donors   377
    N-Donor ligands   377
    P- and As-donor ligands   378
  Group IV Donors   379
Iridium(IV)   379
  Group VII Donors   379
  Group VI Donors   379
    O-Donor ligands   379

II: Palladium, Platinum, Silver, and Gold   381
By D. W. Clack

1 Palladium   381
  Palladium(0)   381
  Palladium(I)   383
  Palladium(II)   384
  Group VII and Hydride Donors   384
  Group VI   386
    O-Donor ligands   386
    Mixed O- and N-donor ligands   387
    S-Donor ligands   388
    Se- and Te-donor ligands   393
  Group V   393
    N-Donor ligands   393
    P-, As-, and Sb-donor ligands   396
  Group IV Donors   397

2 Platinum   401
  Cluster Compounds   401
  Platinum(0) and Platinum(I)   402

Platinum(II) 403
  Group VII and Hydride Donors 403
  Group VI Donors 405
  Group V 409
    N-Donor ligands 409
    P-Donor ligands 413
    Sb-Donor ligands 415
  Group IV 416
    C-Donor ligands 416
    Si-Donor ligands 417
  Group III 418
    Boron donor ligands 418
Platinum(IV) 418
  Group VII Donors 418
  Oxygen Donors 419
  Nitrogen Donors 419
Platinum(V) 422

3 Silver 424
  Silver (I) 424
    Group VII Donors 424
    Group VI 425
      O-Donor ligands 425
      S-Donor ligands 425
    Group V 426
      N-Donor ligands 426
      P-Donor ligands 427
    Group IV 427
      C-Donor ligands 427
  Silver(II) and Silver(III) 428

4 Gold 431
  Gold(I) 431
    Group VII Donors 431
    Group VI Donors 431
    Group V 431
      N-Donor ligands 431
      P-Donor ligands 432
    Group VI 432
      C-Donor ligands 432
  Gold(II) 432
  Gold(III) 432
    Group VII Donors 432
    Group VI 433
      S-Donor ligands 433
  Gold(V) 434

Chapter 4 *Zinc, Cadmium, and Mercury*   435
          By J. Howell and M. Hughes
          1 **Introduction**   435

          2 **Zinc**   435

          3 **Cadmium**   446

          4 **Mercury**   451

          5 **The Bio-inorganic Chemistry of Zinc, Cadmium, and Mercury** 462
            Enzymic Reactions involving Zinc   462
            Metal–Protein Complex Formation   465
            Studies involving Reaction Mechanisms   468

Chapter 5 *Scandium, Yttrium, the Lanthanides, and the Actinides* 471
          By J. A. McCleverty
          1 **Scandium and Yttrium**   471
            Structural Studies   471
            Chemical Studies   471

          2 **The Lanthanides**   473
            Structural Studies   473
            Chemical Studies   474
            Lanthanide Shift Reagents   486

          3 **The Actinides, Uranyl and Related Species**   487
            Structural Studies   487
            Chemical Studies   488
            Uranyl and Related Compounds   493
            Structural Studies   493
            Chemical Studies   494

          Author Index   500

1

The Early Transition Metals

BY F. L. BOWDEN AND C. D. GARNER

PART I : Titanium, Zirconium, Hafnium, Vanadium, Niobium, and Tantalum *by F. L. Bowden*

1 Titanium

Introduction.—There have been several reviews of the chemical technology of titanium and its compounds.[1-3a] The organometallic[4] and structural[4a] chemistry of titanium reported during 1972 has been reviewed as have some aspects of its synthetic chemistry.[5] According to *ab initio* calculations the Ti—C bond in TiCO and TiCO$^+$ can be described as a CO non-bonding pair that is shifted slightly onto the Ti and a shift of Ti $3d$ π-orbitals back onto the CO.[6]

Vibrational assignments have been made for bis(cyclo-octatetraene)titanium. The intensities of the skeletal vibrations indicate that the two C_8H_8 rings are more electrostatically bonded than in the corresponding Th and U compounds.[7] X-Ray photoelectron spectra of $(C_5H_5)M(C_7H_7)$ (M = Ti, V or Cr) show that the oxidation state of the metal increases in the sequence M = Cr < V < Ti, resulting in an increased electron density on the ligands in the same sequence.[8] The highest negative charge is found on the C_7H_7 ring for the Ti compound, in accord with other results.[9]

The magnetic behaviour of $Cp_2Ti^+pic^-$ has been discussed in terms of the ligand-field model.[10]

The blue, diamagnetic mono- and di-methyl substituted complexes $(h^5\text{-}C_5H_4R)$ $(h^7\text{-}C_7H_6R)$Ti have been prepared by the reaction of $(h^5\text{-}C_5H_4R)TiCl_3$ with Pr^iMgBr in the presence of an excess of C_7H_7R. They are stable up to 300°C but are decomposed by air and water. Their mass spectra indicate substantial rearrangement involving the transfer of a CH or CCH$_3$ fragment from the seven- to the five-membered ring leading to ions of dibenzenetitanium derivatives.[11]

[1] H. Rechmann, *Chem.-Ztg.*, 1973, **97**, 320.

[2] J. Baratossy, *Magyar Alum.*, 1973, **10**, 327 (*Chem. Abs.*, 1974, **81**, 28 426).

[3] U. V. Koppikar. *Chem. Ind. Develop.*, 1973, **7**, 19.

[3a] I. Sugiyama, *Chitaniamu Jirokoniumu*, 1973, **21**, 136 (*Chem. Abs.*, 1974, **80**, 70 871).

[4] F. Calderazzo, *J. Organometallic Chem.*, 1973, **53**, 179.

[4a] M. B. Hursthouse, in 'Molecular Structure by Diffraction Methods', ed. G. A. Sim and L. E. Sutton. (Specialist Periodical Reports), The Chemical Society, London, 1973, Vol. 1, p. 443.

[5] K. S. Mazdiyasui, *Ann. Reports Inorg. Gen. Synthesis*, 1973, **1**, 55.

[6] A. P. Morton and W. A. Goddard, *J. Amer. Chem. Soc.*, 1974, **96**, 1.

[7] L. Hocks, J. Goffart, and G. Duyckaerts, *Spectrochim. Acta*, 1974, **30A**, 907.

[8] C. J. Gruenenboom, G. Sawatzky, H. J. De Liefde Meijer, and F. Jellinek, *J. Organometallic Chem.*, 1974, **76**, C4.

[9] S. Evans, J. C. Green, and S. E. Jackson, *J.C.S. Dalton* 1974, 304.

[10] K. D. Warren, *Inorg. Chem.*, 1974, **13**, 1317.

[11] H. T. Verkouw and H. O. Van Oven, *J. Organometallic Chem.*, 1973, **59**, 259.

The full report of the crystal structure of Cp_3Ti has appeared.[11a] Diphenylacetylene reacts with $CpTi(CO)_2$ in the cold to give the yellow monoacetylene complex $Cp_2Ti(CO)PhC_2Ph$; the acetylene is readily displaced by CO and the complex is an active catalyst for the hydrogenation of olefins and acetylenes. Under more vigorous reaction conditions the metallocycle (1) is formed.[12]

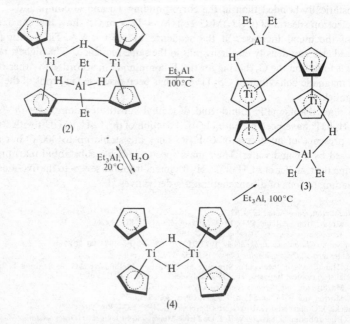

(1)

X-Ray diffraction and chemical studies[13] on $[(C_5H_5)Ti]_2(H)(H_2AlEt_2)(C_{10}H_8)$ (2) and $[(C_5H_4)TiHAlEt_2]_2(C_{10}H_8)$ (3), and ^{13}C n.m.r. studies of 'titanocene' (4)[14] have finally established the structure of this unusual molecule. In (2) the titanium atoms are bridged by a μ-(h^5-C_5H_4–C_5H_4), fulvalene, ligand and by H and H_2AlEt_2 ligands, whereas in (3) the bridging ligands include both a fulvalene ligand and an h^1-C_5H_4–h^5-C_5H_4 ligand. The chemical relationships between 'titanocene', (2), and (3) indicate strongly that 'titanocene' has the structure shown[13] and this has been supported by ^{13}C n.m.r. spectroscopy.[14]

[11a] R. A. Forder and K. Prout, *Acta Cryst.*, 1974, **30B**, 491.
[12] G. Fachinetti and C. Floriani, *J.C.S. Chem. Com.*, 1974, 66.
[13] L. J. Guggenberger and F. N. Tebbe, *J. Amer. Chem. Soc.*, 1973, **95**, 7870.
[14] A. Davisson and S. S. Wreford, *J. Amer. Chem. Soc.*, 1974, **96**, 3017.

'Titanocene' has been used to reduce azo-compounds, ketones, alkynes, and organic halides. Results of deuterium-labelling experiments indicate that the hydrogen in the reduced products is derived from the h^5-C_5H_5 rings of the 'titanocene'.[15] In view of the dihydride structure of (4) it seems more likely that deuterium labelling of titanocene has produced its deuterium analogue and that reduction occurs *via* the hydride hydrogens.

E.s.r. evidence has been presented to support the proposal that hydrogen and hydrogen–alkyl complexes are formed by treating Cp_2M–$AlEt_3$ complexes with H_2.[16] The involvement of titanocene and related species in the reduction of dinitrogen has been further investigated.[17] Permethyltitanocene undergoes a rapid and reversible reaction with N_2 forming $C_5Me_5TiN_2TiC_5Me_5$. Solutions of this compound in toluene absorb more nitrogen at low temperatures with an accompanying colour change from dark blue to intense purple-blue. At $-80°C$ the absorbed N_2 is entirely retained, even under reduced pressure. The amount of nitrogen released when the low-temperature N_2 complex reverts to $(C_5Me_5)_2Ti$ at room temperature is consistent with the stoicheiometry $(C_5Me_5)_2TiN_2$. Furthermore, the ^1H and ^{13}C n.m.r. spectra of solutions of the complex below $-62°C$ indicate that it exists in two forms. The ^{15}N n.m.r. spectrum of $(C_5Me_5)_2Ti^{15}N^{15}N$ at $-61°C$ showed two doublets, and a singlet at lower field; the equilibrium (5) $\rightleftharpoons$ (6) was proposed. The doublet structure of the higher-field signals was attributed to ^{15}N–^{15}N coupling $[J(^{15}N-^{15}N) = 7 \pm 2\ Hz]$ between the non-equivalent nuclei of (5). The i.r. spectrum provided further evidence for two forms of the complex and showed that the $N\equiv N$ bond orders are affected to similar extents by co-ordination in both complexes.

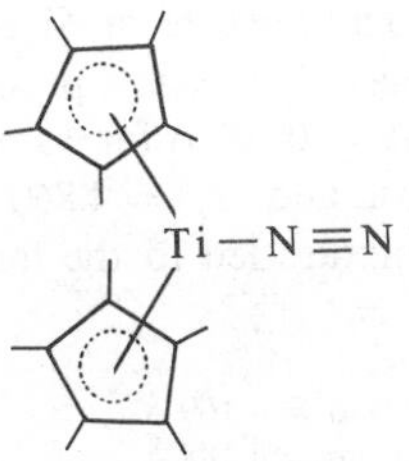

(5) 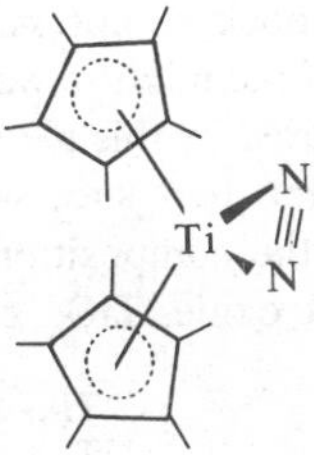   (6)

Binary Compounds and Related Systems.—*Halides and Oxyhalides.* At 1100—$1450°C$ $TiCl_4$ reacts with metallic Ti to give lower Ti chlorides with a $Cl:Ti$ ratio in the gas phase of 2.00—2.50. An increase in temperature does not affect the degree of reaction of $TiCl_4$. Reduction of the partial pressure of $TiCl_4$ vapour in the original mixture with Ar reduces the $Cl:Ti$ ratio from 2.60 to 2.00.[18] The fraction of Ti^{2+} ions in a NaCl melt at a concentration of Ti of 0.83—0.4 wt. % and at 950—$1100°C$ is close to unity $(0.80$—$0.97)$. The amount of Ti^{2+} increases with decreasing temperature. At

[15] T. R. Nelson, *Diss. Abs.* (*B*), 1973, **34**, 1044.
[16] F. K. Shmidt. V. V. Suraer, S. M. Krasnopol'skaya, and V. G. Lipovich, *Kinet. i Katalitz*, 1973, **14**, 617 (*Chem. Abs.*, 1973, **79**, 91 333).
[17] J. E. Bercaw, E. H. Rosenberg, and J. D. Roberts, *J. Amer. Chem. Soc.*, 1974, **96**, 612.
[18] I. Ya. Pishchai and Ya. G. Goroshchenko, *Zhur. priklad Khim.*, 1973, **46**, 1180 (*Chem. Abs.*, 1973, **79**, 60 952).

the temperature of the electrolytic processes (850—900 °C) this value is still lower, as indicated by the isotherms of equilibrium potentials of Ti relative to a Cl comparison electrode as a function of the amount of Ti dissolved in the melt. With an increase in temperature the equilibrium constant of the reaction

$$2TiCl_3 + Ti \rightarrow 3TiCl_2$$

decreases. For low concentrations of Ti and relatively low temperatures (800—900 °C) it may be considered that all the Ti dissolved in the melt is in the form of Ti^{2+}.[19] Disproportionation of $TiCl_3$ begins at 550 °C in an inert atmosphere and at 660—80 °C in a $TiCl_4$ atmosphere. At 600 °C, 95—96 % $TiCl_3$ is still present.[20] The equilibrium diagram, electrical conductivity, and density of fused $TiCl_4$—$TiBr_4$ mixtures indicate almost ideal solution behaviour without appreciable chemical interaction in either the liquid or solid phase.[21] This is at variance with an earlier report.[22] $TiCl_4$ has been investigated as an n.m.r. shift reagent.[23]

Titanium(III) oxychloride has been obtained as a hydrate from the hydrolysis of $TiCl_2$–$TiCl_3$ mixtures in air,[24] and as the anhydrous material from the reaction between TiO_2 and $TiCl_2$ at 650—700 °C.[25] Further heating to 700—750 °C eliminates unchanged $TiCl_2$ and impurities such as $TiCl_3$. TiOCl free of impurities is stable in air.

An electron diffraction study of $TiBr_4$ and TiI_4 has shown them to have regular tetrahedral structures.[26] Pure TiF_4 has been obtained in high yield (91 %) by treating preoxidized ilmenite with FeF_3 at 850 °C.[27]

Oxides. Experimental work concerning the equilibrium thermodynamic properties of Ti_2O, Ti_3O, and Ti_6O has been reviewed.[28] Cubic (NaCl type) TiO has been prepared by shock compression of an equimolar mixture of Ti and TiO_2.[29] The temperature of the mixture was estimated to reach 3000 K at a pressure of 850 kbar. The TiO prepared by this method had $a = 4.179 \pm 0.002$ Å for a single compression and 4.177 Å for two successive compressions, and $d_4^{25} = 4.860 \pm 0.005$ g cm^{-3}. Variations in the composition of the initial mixture led to the formation of non-stoicheiometric oxides TiO_x, e.g.

$$2Ti + 3TiO_2 \rightarrow 5TiO_{1.2} \qquad a = 4.169 \text{ Å}$$
$$3Ti + 2TiO_2 \rightarrow 5TiO_{0.8} \qquad a = 4.188 \text{ Å}$$

[19] V. A. Reznichenko and F. B. Khalimov, *Protessy. Proizvod. Titana Ego Dvuokisi*, 1973, 188 (*Chem. Abs.*, 1973, **79**, 26 458).

[20] A. A. Krasneninnikova, A. A. Furman, and Z. I. Latina, *Zhur. priklad Khim.*, 1973, **46**, 2305 (*Chem. Abs*, 1974, **80**, 33354).

[21] R. V. Chernov and V. M. Moshnenko, *Zhur. neorg. Khim.*, 1973, **18**, 1964.

[22] G. P. Luchinskii, *Zhur. fiz. Khim.*, 1972, **46**, 2959.

[23] A. K. Bose, P. R. Srinvasan, and G. Trainor, *J. Amer. Chem. Soc.*, 1974, **96**, 3670.

[24] E. F. Klyuchnikova, V. G. Gopienko, and O. I. Arakelyan, *Protessy Proizvod. Titana Ego Dvuokisi*, 1973, 204 (*Chem. Abs.*, 1974, **80**, 147 792).

[25] L. D. Polyachenok, G. I. Novikov, and O. G. Polyachenok, *Obshchei Priklad Khim.*, 1972, 31 (*Chem. Abs.*, 1973, **79**, 48 762).

[26] G. V. Girichev, E. Z. Zasorin, N. I. Giricheva, K. S. Krasnov, and V. P. Spiridanov, *Izvest. V.U.Z., Khim. i Khim. Tekhnol.*, 1974, **17**, 468 (*Chem. Abs.*, 1974, **81**, 17 891).

[27] J. H. Moss and B. Leng, Ger. Offen., 2 314 923,

[28] I. I. Kornilov (ed.), *Nov. Konstr. Mater.-Titan, Tr. Nauch.-Tekh, Soveshch. Met., Metalloved. Primen. Titana 8th.* 1970, 24 (*Chem. Abs.*, 1973, **79**, 10 379).

[29] E. D. Ruchkin, A. E. Strelyaev, V. A. Lazareva, and S. S. Batsanov, *Zhur. neorg. Khim.* 1973, **18**, 2855.

The electronic structure of titanium monoxide has been studied by X-ray spectro-scopy and ESCA.[30] The dissociation constant and heat capacity of TiO at 3000—10000 K have been reported as $K_p = 7.77 \times 10^{-6}$—$3.39 \times 10^3$ atm and $C_p = 9.65$—11.81 cal mol^{-1} K^{-1}.[31] Analyses of the rotational spectra of the TiO molecule have been carried out.[32, 33]

A comparison of the intensities of the M—O absorption band for TiO_{1+x} and $VO_{1\pm x}$ shows the former to diminish more rapidly with increasing x; this effect has been ascribed to the lower degree of polarity in the Ti—O bond.[34] The same explana-tion has been offered to account for the presence of a deformation absorption in the spectra of TiO, Ti_2O_3, and TiO_2. This absorption band is absent from the spectra of HfO_2 and ZrO_2.[35] The magnetic susceptibilities and the e.p.r. and optical spectral parameters of $Ti_xO_y (x = 3$—$10; y = 5, 7, 9, 11, 13, 15, 17$ or 19) have been interpreted in terms of the presence of non-interacting, localized Ti^{3+} ions. These range from *ca.* 0.01 % of the total Ti^{3+} in Ti_3O_5 to *ca.* 40 % of the total in $Ti_{10}O_{19}$. The remaining Ti^{3+} ions are thought to be involved in homopolar bonds within specific groups of Ti^{3+} ions.[36]

Ti_2O_3 and $(Ti_{0.900}V_{0.100})_2O_3$ have been shown to be isostructural with Al_2O_3, having rhombohedral unit cells with dimensions Ti_2O_3, $a = 5.4325(8)$ Å, $\alpha = 56.75$ (1)°, $(Ti_{0.900}V_{0.100})_2O_3$, $a = 5.4692(8)$ Å, $\alpha = 55.63(1)°$. The effect of substitution by V^{3+} is to increase the metal–metal distance across shared octahedral faces from 2.579 Å in Ti_2O_3 to 2.658 Å in the Ti—V compound while decreasing the metal–metal distance across the shared octahedral edge from 2.997 Å to 2.968 Å.[37]

TiO_2 has been prepared by the injection of compressed air into molten $TiBr_4$[38] and by the vapour-phase hydrolysis of $TiCl_4$.[39] A thermodynamic analysis of the latter system indicated that the formation of Ti_2O_3 and Ti_3O_5 byproducts is favoured at higher temperatures. Reduction of the titanium oxides Ti_xO ($x = 19, 6, 5, 4, 3$ or 2) with hydrogen at 600 °C, 100 kg cm^{-2} over 48 h gave mixtures of Ti_2O and TiH_2. No reaction was observed with TiO or TiO_2. TiO_2 has the maximum stability towards hydrogen.[40]

Chalcogenides. The crystal structures of titanium sulphides have been reviewed.[41] Ti_3S_8 has several structural features in common with TiS_2 including trigonal-prismatic co-ordination polyhedra about S atoms and cubic co-ordination polyhedra about Ti atoms. Whereas six cube faces are shared with trigonal prisms in TiS_2 only,

[30] E. A. Zhurakovskii, T. N. Bondarenko, and V. P. Dzeganovskii, *Doklady Akad. Nauk. S.S.S.R.*, 1974, **215**, 1360.
[31] W. Zyrnicki, *Roczniki Chem.*, 1973, **47**, 239, (*Chem. Abs.*, 1973, **79**, 108 264).
[32] C. Linton, *J. Mol. Spectroscopy*, 1974, **50**, 235.
[33] C. Linton, *J. Mol. Spectroscopy*, 1974, **51**, 194.
[34] M. I. Aivazov and A. Kh. Muranovich, *Izvest. Akad. Nauk S.S.S.R.*, *neorg. Materialy*, 1973, **9**, 2156.
[35] B. T. Kaminskii, A. S. Plygunov, and G. N. Prokof'eva, *Ukrain. khim. Zhur.*, 1973, **39**, 946, (*Chem. Abs.*, 1974, **80**, 21 116).
[36] J. F. Houlihan and L. N. Mulay, *Inorg. Chem.*, 1974, **13**, 745.
[37] W. R. Robinson, *J. Solid State Chem.*, 1974, **9**, 255.
[38] G. Holland, Ger. Offen., 2 232 579.
[39] Yu. Kh. Shaulov, Yu. M. Shtrambrand, A. N. Shinkarev, N. I. Andreeva, and E. A. Ryabenko, *Zhur. fiz. Khim.*, 1973, **47**, 2714.
[40] A. M. Sukhotin, E. I. Antonovskaya, E. V. Sgibnev, I. I. Kornilov, and V. V. Vavilova, *Doklady Akad. Nauk S.S.S.R.*, 1973, **209**, 1121.
[41] O. Watenabe, *Chitaniumu Jurukoniumu*, 1973, **21**, 129 (*Chem. Abs.*, 1973, **78**, 49774).

three are shared in Ti_3S_8.[42] Carbon disulphide has been used as a sulphurizing agent in the preparation of Ti_3S_4.[43] A material $Ti_{1.2}S_2$ resembling Ti_3S_4 in its X-ray powder diffraction pattern has been obtained from TiO_2 carbon mixtures and H_2S. Intermediate oxygen-containing sulphides are formed and these must be mixed with more carbon and reheated to remove all the oxygen.[44]

There is some conflict concerning the electronic nature of TiS_2.[45–48] A band gap between the highest occupied and lowest unoccupied orbitals of *ca.* 1 eV has been widely accepted. However, an MO energy level diagram derived on the basis of octahedral symmetry for TiS_2 from X-ray band spectra showed no evidence of a gap wider than 0.1 eV, the limit of error; that is, TiS_2 is a metal or semi-metal rather than a small-gap semiconductor. A similar conclusion was drawn from X-ray photoelectron data on TiS_2. Both these results are very different from the 2.0—2.7 eV gap derived from optical studies of TiS_2. M_xTiS_2 phases (M = Li, Na, or K; $x = 0.3$—0.5) have been prepared by the reaction of TiS_2 with alkali-metal halide melts in an H_2S–CS_2 atmosphere. The structures of these compounds consist of alkali-metal cations located between negatively charged $(TiS_2)^{x-}$ layers of the CdI_2 type. They form hydrates with water, the interlayer distances depending on the radius of the alkali-metal cation and on the ambient water-vapour pressure. Water can be replaced partially or totally by polar molecules, which solvate the titanium sulphide layers and the alkali-metal cations. The cations can be replaced by NH_4^+, Ca^{2+}, and R_4N^+ ions.[49] Single crystals of $FeTiX_4$ grown by isothermal vapour growth exhibit metallic behaviour.[50] Lattice parameters have been determined for some ternary titanium-containing chalcogenides: TiS_xSe_{2-x}, TiS_xTe_{2-x}, $TiSe_xTe_{2-x}$ $(0 < x < 2)$,[51] $Ti_{0.50}NbSe_2$.[52] Layer intercalation compounds of TiS_2 with a variety of amines have been prepared.[53–56] The limiting composition $(C_nH_{2n+1}NH_2)_x$ TiS_2($x = 0.95$—1.02 for $n = 4$—10 and $x = 1.05$ for $n \geqslant 16$) was established for n-alkylamine adducts which were obtained by displacement of N_2H_4 from the adduct TiS_2,N_2H_4. The guest molecules are arranged in bimolecular layers between layers of the host lattice.[53] With substituted imidazoles as guest molecules no increase in interlayer spacing occurs on substitution of H by CH_3 on nitrogen, while CH_3 substitution on C-2 causes an increase of *ca.* 4 Å.[54] At low values of n (3—6), both alkyl chains of the secondary amines $(C_nH_{2n+1})NH$ ($n = 3$—10) are parallel to the TiS_2 layers,[55] but the angle of inclination of the longitudinal axis of the alkyl chain towards the layers increases sharply at higher n values.

[42] J. P. Owens and H. F. Franzen, *Acta Cryst.*, 1974, **B30**, 427.
[43] N. V. Vekshina and V. M. Lantratov, *Zhur. Priklad Khim.*, 1973, **46**, 917 (*Chem. Abs.*, 1973, **79**, 13 043).
[44] B. T. Kaminskii and A. S. Plygunov, *Ukrain. khim. Zhur.*, 1973, **39**, 1224 (*Chem. Abs.*, 1974, **80**, 77 814).
[45] D. E. Ellis and A. Seth, *Internat. J. Quantum Chem.* (*Symp.*), 1973, **7**, 223.
[46] H. W. Myron and A. J. Freeman, *Phys. Rev.* (*B*), 1974, **9**, 481.
[47] D. W. Fischer, *Phys. Rev.* (*B*), 1973, **8**, 3576.
[48] G. K. Wertheim, F. J. DiSalvo, and D. N. E. Buchanan, *Solid State Com.*, 1973, **13**, 1225.
[49] R. Schoellhorn and A. Weiss, *Z. Naturforsch.*, 1973, **28b**, 711.
[50] S. Muranaka and T. Takada, *Bull. Inst. Chem., Res., Kyoto Univ.*, 1973, **51**, 287 (*Chem. Abs.*, 1974, **81**, 30 671).
[51] H. P. Rimington and A. A. Balchin, *J. Crystal Growth*, 1974, **21**, 171.
[52] A. Meerschaut and J. Rouxel, *Compt. rend.*, 1973, **277**, C, 163.
[53] A. Weiss and R. Ruthardt, *Z. Naturforsch.*, 1973, **28b**, 249.
[54] A. Weiss, R. Ruthardt, and H. Orth, *Z. Naturforsch.*, 1973, **28b**, 446.
[55] A. Weiss and R. Ruthardt, *Z. Naturforsch.*, 1973, **28b**, 522.

Carbides, Silicides, and Germanides. Single crystals of TiC were obtained from the reaction

$$TiCl_4(l) + CCl_4(l) + H_2 \rightarrow TiC(s) + HCl(g)$$

in a flow system. Below 1200 °C TiC condensed as a polycrystalline film; octahedral crystals formed above 1600 °C.[57] The titanium–carbon phase diagram has been analysed by a statistical thermodynamic method.[58] Within the homogeneity region, titanium carbide acquired two crystal structures: (i) a NaCl-type structure, at carbon concentrations >40 atom % and (ii) a structure belonging to the space group $Fd3m$ at carbon concentrations <40 atom %.[59] Titanium silicides are included in a review of the electronic structures of metal silicides.[60] The enthalpy and heat capacity of TiS_2 have been reported,[61] as have the heats of formation of TiSi, Ti_5Si_3, and $TiSi_2$.[62] TiS_2 resists oxidation owing to the formation of thin protective films of oxides and silicate glasses; at 500 °C the layer contains α-quartz and TiO_2, at 700 °C anatase, and at higher temperatures still, rutile containing traces of metallic titanium.[63] The resistance of titanium germanides to oxygen decreases in the series TiGe $>$ TiGe$_2$ $>$ Ti_5Ge_3; metal oxides and GeO_2 are the reaction products.[64]

Nitrides, Borides, etc. X-Ray diffraction studies on annealed TiN_x show the homogeneity region to extend from $TiN_{0.63}$ to $TiN_{1.0}$ whereas δ and ε phases coexist at $TiN_{0.40}$—$TiN_{0.60}$. Ti and N atoms form two sublattices in the δ phase which are of the NaCl type. Both sublattices are less occupied in quenched specimens than in those that have been annealed indicating the formation of thermally induced vacancies.[65]

Titanium borides have been reviewed[66, 67] and optimum conditions given for their preparation from boron and titanium oxides.[68] Electrolysis of a solution containing cryolite, NaOH, sodium borate, NaCl, and rutile at 1100 °C results in the crystallization of TiB_2 at the cathode.[69] Mixtures of boride and nitride phases are formed during the interaction of BN with TiB_2 at 1200—2000 °C.[70]

FeTi reacts directly with hydrogen to form successively the hydrides of approximate compositions FeTiH and $FeTiH_2$. Both have dissociation pressures >1 atm at 0 °C

[56] A. R. Beal and W. Y. Liang, *Phil. Mag.*, 1973, **27**, 1397.
[57] V. S. Sinel'nikova and I. I. Timofeeva, *Vyrashchivanie Monokrist. Tugoplakikh Redk. Metal.*, 1973, 33 (*Chem. Abs.*, 1974, **81**, 6962).
[58] V. N. Zagryazkin, E. V. F. Fiveiskii, and A. S. Panov, *Zhur. fiz. Khim*, 1973, **47**, 1951.
[59] M. P. Arbuzov, B. V. Khaenko, E. T. Kachkovskaya, and Ya. S. Golub, *Ukrain fiz. Zhur.*, 1974, **19**, 497 (*Chem. Abs.*, 1974, **81**, 42 615).
[60] E. A. Zhurakovskii, P. V. Gel. and V. I. Sokolenko. *Visn. Akad. Nauk Ukrain. S.S.S.R.*, 1973, 8 (*Chem. Abs.*, 1974, **80**, 137 243).
[61] V. P. Bondarenko, L. A. Dvorina, V. I. Zmii, E. N. Fornichev, N. P. Slyusar, and A. D. Krivorotenko, *Porosh. Met.*, 1973, 47 (*Chem. Abs.*, 1974, **80**, 52 995).
[62] V. D. Savin, *Zhur. fiz. Khim.*, 1973, **47**, 2527.
[63] R. F. Voitovich and E. A. Pugach, *Porosh. Met.*, 1974, 63 (*Chem. Abs.*, 1974, **80**, 90 591).
[64] V. S. Birayava, *Ref. Zhur. Khim.*, 1973, Abs. No. 3V26 (*Chem. Abs.*, 1973, **79**, 12 981).
[65] M. P. Arbuzov, B. V. Khaenko, and E. T. Kachkovskaya, *Porosh. Met.*, 1973, **13**, 69 (*Chem. Abs.*, 1973, **79**, 71 069).
[66] T. Atohda, *Chitaniumu Jurukiumu*, 1973, **21**, 145 (*Chem. Abs.*, 1974, **81**, 29 744).
[67] H. Kunst, *Z. Wirt. Fertigung*, 1973, **68**, 501 (*Chem. Abs.*, 1974, **80**, 99 010).
[68] Yu. B. Paderno and V. V. Odintsov, *Metalloterm. Protessy Khim. Met.*, *Mater. Konf.* 1971, 39 (*Chem. Abs.*, 1974, **80**, 66 230).
[69] J. Gomes, U.S.P., 3 775 271.
[70] G. V. Samsonov, A. L. Burykina, O. A. Medvedeva, and V. P. Kosteruk, *Porosh. Met.*, 1973, 50 (*Chem. Abs.*, 1974, **80**, 73 297).

unlike the very stable TiH_2. The lower hydride FeTiH has tetragonal symmetry and $d = 5.88 \, g \, cm^{-3}$ ($FeTiH_{1.06}$), whilst the higher hydride has a cubic structure and $d = 5.47 \, g \, cm^{-3}$ ($FeTiH_{1.93}$).[71] Titanium carbonitride has been prepared by nitriding the carbon-deficient carbide. The activation energy of the nitriding reaction decreases with decreasing proportion of carbon in the carbide[72] and the stability of the carbonitride to oxidizing acid mixtures increases with increasing carbon content.[73] The homogeneity region for TiC_xN_z has been established as $x = 0.49$—1.0, $z = 0.42$—1.0. The crystal lattice of samples with $x + z < 1$ has vacancies only in the non-metal sublattice.[74]

Titanium(II).—Very few investigations of this oxidation state have been reported this year. The versatility of reduced titanium species in reductive fixation reactions has been further exemplified by the fixation of CO[75] and CO_2.[76] The $TiCl_3$,3THF–Mg–THF system reacts with CO changing from blue to black; with a 10-fold excess of Mg, three moles of CO per g atom of Ti are fixed. Acetylene appears in the hydrolysis products of the black fixation product indicating the reduction of CO to carbide.[75] $TiCl_4$,2THF is reduced by metallic magnesium according to

$$2TiCl_4,2THF + 6Mg + 2THF \rightarrow (THF,Cl_2Mg_2Ti)_2 + 2MgCl_2(THF)_2$$

The titanium complex is diamagnetic which, if the titanium is present as Ti^{II}, indicates a strong exchange interaction. It reacts with CO_2 forming $THF.Ti(OOCH)_2MgCl_{1.5}$, which yields formic acid on hydrolysis and ethyl formate with C_2H_5I. This indicates that the hydrogen is bound to the carbon atom of the co-ordinated CO_2 molecule as in (7) or (8). The formation of C—H bonds implies the presence of reactive metal–hydride intermediates.[76]

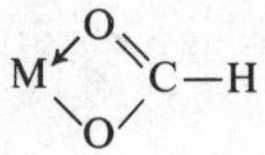

(7) 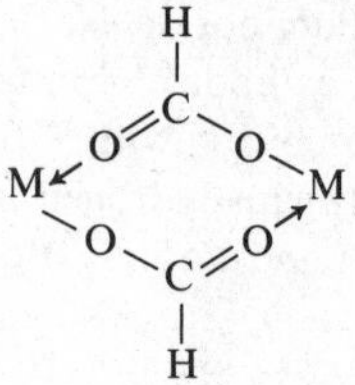   (8)

Titanium(III).—*Halides and Oxyhalides.* Heating the binary fluorides TiF_3 and MF (M = Na, K, Rb, Cs, or Tl) in a closed system at 450—600 °C results in the formation of the light-blue cubic Elpasolithes Cs_2KTiF_6 ($a = 9.12$ Å), Rb_2KTiF_6 ($a = 8.93$ Å), and Rb_2NaTiF_6 ($a = 8.53$ Å) as well as Cs_2NaTiF_6 (hexagonal $a = 6.27$, $c = 30.91$ Å, isotypic with Cs_2NaCrF_6) and Tl_3TiF_6 ($a = 8.88$ Å). The reflectance spectra of the

[71] J. J. Reilly and R. H. Wiswall, *Inorg. Chem.*, 1974, **13**, 218.

[72] G. V. Samsonov, I. I. Bilyk and V. V. Morozov, *Izvest. Akad. Nauk S.S.S.R., neorg. Materialy*, 1973, **9**, 1724.

[73] G. D. Bogomolov and G. D. Lyubimov, *Tr. Inst. Khim., Ural. Nauch. Tsentr., Akad. Nauk S.S.S.R.*, 1973, **25**, 19 (*Chem. Abs.*, 1974, **81**, 20 366).

[74] B. F. Mitrofanov, Yu. G. Zainulin, S. I. Alyamovskii, and G. P. Shveikin, *Izvest. Akad. Nauk S.S.S.R., neorg. Materialy*, 1974, **10**, 745.

[75] B. Jezowska-Trzebiatowska and P. Sobota, *J. Organometallic Chem.*, 1973, **61**, C43.

[76] B. Jezowska-Trzebiatowska and P. Sobota, *J. Organometallic Chem.*, 1974, **76**, 43.

quaternary fluorides are similar to the spectrum of TiF_3, the absorption maxima being shifted to higher energies by 1500—1800 cm^{-1}. TiF_3 obeys the Curie–Weiss law between 150—295 K, with $\theta = -75$ K; this is substantially larger than that found previously (-10 K). Below 150 K, TiF_3 deviates from Curie–Weiss behaviour and exhibits a Néel point at 50 K. Cs_2KTiF_6 obeys the Curie–Weiss law over the temperature range 75—298 K, $\theta = 42$ K, $\mu = 1.8$ BM.[77] The quaternary oxyfluoride Ti_2NbO_5F has been prepared from Ti_2O_3 and NbO_2F at 900 °C. It has a rutile-type structure.[78] $KTiF_4$ is formed as an intermediate in the reaction of K_3TiF_6 with water. Subsequent hydrolysis, accompanied by oxidation to Ti^{4+}, affords H_2TiO_3 and K_2TiOF_4.[79] These hydrolysis products are present as impurities in metallic titanium produced by a hydrometallurgical method.

Glycerine solutions of $TiCl_3$ have a pH of *ca.* 1 and a redox potential *ca.* 250 mV (*vs.* SCE). When the solutions were made alkaline with Na_2CO_3 or NaOH the redox potential fell to -1250 and -1435 mV respectively. These solutions reduce many ions, some to the elemental state; Ti^{III} was proposed as a titrant for U^{VI}, Hg^{II}, Cd^{II}, and and Pb^{II}.[80]

Modification of the reactivity of metal-containing fragments by polymeric ligands continues to be a subject of great interest. Two papers have appeared which illustrate this effect in the reaction of $TiCl_3$ with oxygen.[81,82] In pyridine solution, $TiCl_3$ undergoes an initial rapid reaction accompanied by a colour change from brown to green. This is followed by a much slower uptake of oxygen which is first order in $TiCl_3$ and O_2. A dimeric Ti^{IV}–O–Ti^{IV} structure for the final product is supported by the stoicheiometry of the overall reaction; the absence of peroxide activity in the product, and the appearance of $v(Ti$–O–$Ti)$ at 830 cm^{-1} in the i.r. spectrum. Evidence was obtained that the solute species responsible for the initial rapid uptake of oxygen is a Ti^{III} dimer. A weak, 'half-field', $\Delta M_s = \pm 2$ signal centred on $g = 4.03$ appears in the e.p.r. spectrum of frozen solutions of $TiCl_3$ in pyridine; comparison of experimental and computer-simulated spectra established a Ti—Ti distance of *ca.* 3.0 Å, consistent with the presence of the $Ti_2Cl_9^{3-}$ ion in solution. This was further supported by the comparison between the electronic spectrum of the brown solution and the reflectance spectrum of $CsTi_2Cl_9$, by the strong chloride ion dependence of the colour of the Ti^{III} solutions, and by the accelerating effect of added chloride ion on the initial rate of oxygen uptake.[81] The first step proposed for the solution reaction of $TiCl_3$ with O_2 is direct electron transfer from Ti^{III} to O_2. Support for this proposal comes from observations on the reaction of O_2 with $TiCl_3$ bound to a divinylbenzene–4-vinylpyridine copolymer.[82] Admission of oxygen to the Ti^{III}–polymer results in a diminution of the Ti^{III} e.p.r. signal ($g = 1.960$) and the growth of a new signal ($g = 2.004$, 2.010, or 2.022) characteristic of the $O_2^-\cdot$ species. This signal diminishes on evacuation of the system and is restored when oxygen is readmitted but slowly weakens under any atmosphere. The formation of a polymer-bound Ti^{IV}–$O_2^-\cdot$ species

[77] E. Alter and R. Hopper, *Z. anorg. Chem.*, 1974, **403**, 127.
[78] J. Senegas and J. Galy, *Compt. rend.*, 1973, **277**, C, 1243.
[79] T. M. Burmistrova, V. A. Reznichenko, and G. A. Menyailova, *Protessy Proizvod. Titana Ego Dvuoskii*, 1973, 198 (*Chem. Abs.*, 1973, **79**, 147 791).
[80] M. T. Dopico Vivero, *Rev. Fac. Cienc., Univ. Oviedo*, 1973, **14**, 3 (*Chem. Abs.*, 1974, **80**, 127 751).
[81] C. D. Schmulbach, C. C. Hinkley, C. Kolich, and T. A. Ballentine, *Inorg. Chem.*, 1974, **13**, 2026.
[82] Y. Chimura, *Chem. Lett*, 1974, 393.

was proposed. Apparently the isolation of Ti centres and the relative rigidity of the polymeric ligand militate against the rapid formation of $Ti^{IV}-O^{2-}-Ti^{IV}$ dimers.

The phase diagram of the ternary system $K_3TiCl_6-K_2ZrCl_6-TiCl_3$ has been investigated;[83] three subsystems were formed:
(i) $KCl-K_3TiCl_6-K_2ZrCl_6$, (ii) $K_3TiCl_6-K_2ZrCl_6-TiCl_3$, and
(iii) $K_2ZrCl_6-TiCl_3-ZrCl_4$.

A series of bromoaquatitanium(III) complexes $M_2TiBr_5(H_2O)_n$ (M = Cs, Rb, or pyH; n = 1, 3 or 5) and $M_2(H_2(H_3O)TiBr_6(H_2O)_5$ (M = pyH or quinH) has been prepared.[84] Their electronic spectra obey Jørgensen's law indicating that they contain species of the type $[TiBr_n(H_2O)_{6-n}]^{3-n}$. Bromination of $C_5Me_5Ti(OEt)_3$ with acetylbromide in the dark gave $C_5Me_5TiBr_3$ which was converted into $C_5Me_4CH_2BrTiBr_3$ by bromine or N-bromosuccinimide.[85] Previous claims that $K_3Ti(CN)_6$ and $K_3Ti(CN)_6,KCN$ are obtained from aqueous solutions of $TiCl_3$ and KCN have not been confirmed;[86] only the dark blue titanium(III) hydroxide could be obtained. The reaction between $TiBr_3$ and KCN in liquid ammonia however, gave a grey-green product of stoicheiometry $K_5Ti(CN)_8$. This has a μ_{eff} = 1.71 BM close to the spin-only value for Ti^{III}, it contains free KCN thus eliminating the eight-co-ordinate $[Ti(CN)_8]^{4-}$ structure, and its electronic spectrum with absorption bands at 22400, 16800, and 9700 cm^{-1} rules out six-co-ordination but is consistent with a mono-capped trigonal-prismatic C_{2v} structure, $K_4Ti(CN)_7,KCN$.

Titanium tetrahalides are reduced by LiBH_4 in pentane to $(BH_4)_2TiCl_2Ti(BH_4)_2$ and $Ti(BH_4)_3$. The chlorine-bridged dimer reacted with THF to give $Ti(BH_4)_3,2THF$ and $TiCl_3,nTHF$.[87]

O-Donor Ligands. Aqueous solutions of titanium(III) between pH 1.0 and 3.5 exhibit a single-line e.p.r. signal with g = 1.9408 and a linewidth of 95 $\pm$ 3 G which accounts for only 2% of the total titanium(III).[88] The g-value, linewidth, and pH dependence of the intensity of the signal are independent of added Cl^- and I^- ions up to 2 × 10^3 molar excess, indicating that the anion is not required in the primary co-ordination sphere; at this level, SO_4^{2-} destroys the signal. Since the predominant species in aqueous solutions of titanium(III) in the pH range 0.0—3.0 are $Ti(H_2O)_6^{3+}$ and $Ti(OH)^{2+}$ the former of which is e.p.r. inactive, it was suggested that the $[Ti(H_2O)_4-(OH)_2]^+$ is the e.p.r. active ion. Possibilities such as (9), (10), and $[TiO(H_2O)_5]^+$ were considered less likely candidates on the grounds that the e.p.r. signal was not enhanced by the deliberate addition of titanium(IV) or oxygen, and that there is no evidence of a very strong axial component due to Ti=O in the ligand field of the titanium(IV) ion. The third paper of a series on the kinetics of redox reactions of

$$(H_2O)_4Ti^{III} \overset{O}{\underset{O}{\diagdown\diagup}} Ti^{IV}(H_2O)_4 \qquad\qquad (H_2O)_5Ti^{III}-O-Ti^{IV}(H_2O)_5$$

(9)   (10)

[83] E. N. Ryabov, A. A. Golubev, S. V. Aleksandrovskii, and R. A. Sandler, *Zhur. neorg. Khim.*, 1974, **19**, 835.
[84] S. E. Adnitt, D. W. Barr, D. Nicholls, and K. R. Seddon, *J.C.S. Dalton*, 1974, 644.
[85] A. N. Nesmeyanov, O. V. Nogina, G. I. Drogunova, and B. V. Lokshin, *Izvest. Akad. Nauk S.S.S.R.*, *Ser. khim.*, 1973, 406.
[86] D. Nicholls, T. A. Ryan, and K. R. Seddon, *J.C.S. Chem. Comm.*, 1974, 635.
[87] K. Franz and H. Noeth, *Z. anorg. Chem.*, 1973, **397**, 247.
[88] P. I. Premovic and P. R. West, *Canad. J. Chem.*, 1974, **52**, 2919.

titanium ions in aqueous solution has appeared,[89] concerning the reaction between titanium(III) and vanadium(V). The $[H^+]^{-1}$ dependence of the rate was taken to indicate the need for one of the reactants to change its degree of hydrolysis prior to the formation of the activated complex. An inner-sphere mechanism with oxo-bridged activated complexes was suggested, although it was pointed out that outer-sphere paths cannot be excluded. In an independent study of this reaction,[90] evidence was presented for an initial equilibrium step followed by the formation of the hydroxide-bridged intermediate $TiOHVO_2^{3+}$. Anhydrous $TiCl_3$ reacted with $CO(NH_2)_2$ in EtOH, $MeNCONH_2$ in MeCN, $CO(NHMe)_2$ in THF, $PhNHCONH_2$ in MeCN, biuret (L^1) in dioxan (diox) and biurea (HL^2) in H_2O to give: $[Ti\{CO(NH_2)_2\}_6]Cl_3$, $TiCl_3,5MeNHCONH_2$, $TiCl_3,4CO(NHMe)_2$, $TiCl_3,4PhNH\text{-}CONH_2$, $TiCl_3,2diox,L^1$, and $TiL_3^2,12HL^2$, respectively.[91] The reflectance spectra of the complexes and their magnetic susceptibilities were characteristic of titanium(III) in a tetragonally distorted octahedral ligand field, the splitting of the $^2T_{2g}$ term ranging from 225 to 855 cm^{-1} and the spin–orbit coupling constants (λ^1) from 85—110 cm^{-1} ($\lambda = 154$ cm^{-1} for the free Ti^{3+} ion).

Titanium(III) reduces maleic and fumaric acids to succinic acid. A rapid change in the absorption spectrum of the titanium solution when the acid is added is not accompanied by the consumption of titanium(III) indicating that titanium(III)-carboxylate complexes are involved as intermediates in the redox reaction.[92] Carboxylate complexes are also formed with glycolic[93] and lactic[94] acids. The i.r. spectrum of the 1:3 lactate complex is reported to indicate co-ordination through the ligand's hydroxy-groups. 5,5'-Thiodisalicylic acid forms an insoluble 1:1 titanium(III) complex.[95] Octahedral co-ordination of the metal atom, metal–metal interaction in a polymeric structure and unidentate co-ordination by the carboxylate group are indicated by a single absorption band in the electronic spectrum at 20800 cm^{-1}, a magnetic moment of $\mu = 1.56$ BM and the i.r. spectrum $[\nu_{sym}(COO)$ 1600 and $\nu_{asym}(COO)$ 1500 cm$^{-1}]$ respectively. The polymeric structure (11) has also been

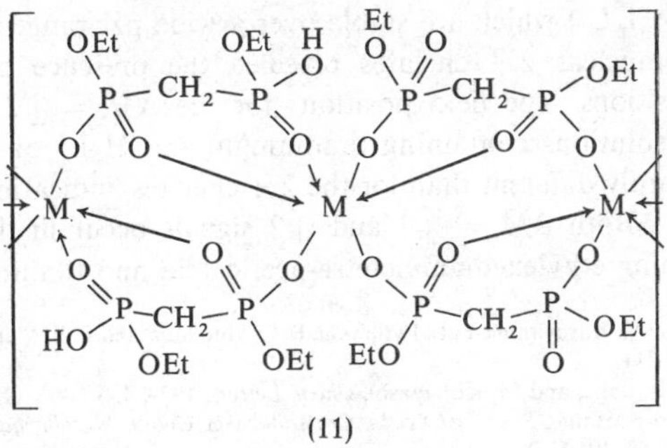

(11)

[89] J. D. Ellis and A. G. Sykes, *J.C.S. Dalton*, 1973, 2553.
[90] P. T. Logan, *Diss. Abs. (B)*, 1973, **34**, 1415.
[91] M. Geis-Blazekova, *Z. anorg. Chem.*, 1973, **402**, 1.
[92] R. D. Foust and P. C. Ford, *J. Inorg. Nuclear Chem.*, 1974, **36**, 930.
[93] P. B. Chakrawarti and H. N. Sharma, *Sci. Cult.*, 1973, **39**, 344 (*Chem. Abs.*, 1973, **79**, 142 433).
[94] P. B. Chakrawarti and H. N. Sharma, *Sci. Cult.*, 1973, **40**, 114 (*Chem. Abs.*, 1974, **80**, 152 259).
[95] P. C. Srivasta, K. B. Pandeya, and H. L. Nigam, *J. Inorg. Nuclear Chem.*, 1973, **35**, 3613.

suggested to account for the subnormal magnetic moment $\mu_{\text{eff}} = 1.54$ BM and insolubility of the yellow-brown complex of stoicheiometry TiL(LH) [L = CH_2-$\{O(O)P(OR)\}_2$] formed in the reaction between tetraethylmethylenediphosphonate and $TiCl_3$.[96] Pure titanium tetrametaphosphate, $Ti_4(P_4O_{12})_3$, has been prepared by heating together titanium metal and phosphoric acid in molar ratios of between 1:1 and 1:5 under nitrogen or *in vacuo*.[97] In the presence of air contamination by TiP_2O_7 occurs. $Ti_4(P_4O_{12})_3$ is isomorphous with the A type of aluminium tetrametaphosphate. The crystal structure of $[TiCl_2(HOCHMe_2)_4]Cl$ has been reported.[98] The Cl atoms are *cis*, $d(Ti—Cl) = 2.32$ Å, $d(Ti—O) = 2.10$ Å. The Me_2CHOH groups are oriented so that the OH groups can form intermolecular H bridges of the type O—H$\cdots$Cl. A polarographic study of titanium-(III) and -(IV) in the presence of excess diethanolaminophosphonic acid (Questex PF), over the pH range 1.4—11 has been carried out.[99] The polarographic wave is due to a reversible, diffusion-controlled, one-electron reduction of Ti^{IV} to Ti^{III}. Above pH 2 the half-wave potential shows a dependence on pH which is consistent with the involvement of $2H^+$ in the electrode reaction. In an i.r. investigation of donor–acceptor interactions, β-$TiCl_3$ was produced in the i.r. cell by the hydrogen reduction of $TiCl_4$. The interaction with the β-$TiCl_3$ caused $v(C=O)$ to be lowered by 70 cm^{-1} for acetone and 130 cm^{-1} for ethylacetate.[100]

N-*Donor Ligands*. Cinchonine forms a 1:2 complex with $TiCl_3$ in which the ligand co-ordinates through the N atom of the heterocyclic ring. The presence of two lattice-held H_2O molecules in the complex is indicated by thermogravimetric analysis and i.r. spectroscopy.[101]

Mixed N- *and* O-*Donor Ligands*. Two studies of aminocarboxylic acid complexes of titanium(III) have shown that the metal is stabilized to oxidation by oxygen through complex formation.[102, 103] Potentiometric titration data have shown that titanium(III) will form 2:1 and 1:1 chelates with 3,6-dioxaoctane-1,8-diamine-NNN'N'-tetra-acetic acid (H_4L^1), diethylenetriamine-penta-acetic acid (H_5L^2), and triethylenetetra-amino-hexa-acetic acid (H_6L^3) which are stable over a wide pH range.[104] E.p.r. measurements on the dinuclear 2:1 chelates revealed the presence of $\Delta M_s = \pm 1$ and $\Delta M_s = \pm 2$ transitions. The field positions for the $\Delta M_s = \pm 2$ transitions in the e.p.r. spectra of solutions containing titanium(III) and H_4L^1 or H_6L^3 in 1:1 mole ratio are significantly different than for the 2:1 chelates, indicating the formation of a dimeric species. Both $\Delta M_s = \pm 1$ and ± 2 signals occur in the 77 K spectra of solutions containing ethylenediaminetetra-acetic acid and titanium(III) in 1:1 mole

[96] C. M. Mikulski, N. M. Karayannis, L. L. Pytlewski, R. O. Hutchins, and B. E. Maryanoff, *J. Inorg. Nuclear Chem.*, 1973, **35**, 4011.

[97] M. Tsuhako, I. Motooka, and M. Kobayashi, *Chem. Letters*, 1974, 435.

[98] M. Handlovic and F. Hanic, *Proc. 2nd Semin. Crystallochem. Coord. Metallorganic Compounds*, 1973, 32 (*Chem. Abs.*, 1974, **80**, 101 135).

[99] R. L. Grassi, L. M. Fachinetti, and S. H. Tarulli, *Anales de Quim.*, 1973, **69**, 991 (*Chem. Abs.*, 1974, **80**, 66 140).

[100] I. V. Ikonitskii, K. V. Nel'son, and V. O. Reikhofel'd, *Zhur. Priklad Spektroskopii*, 1973, **19**, 156 (*Chem. Abs.*, 1973, **79**, 85 112).

[101] M. M. Khan, S. M. F. Rahman, and N. Ahmad, *Indian J. Chem.*, 1974, **12**, 111.

[102] E. G. Yakovleva, N. I. Pechurova, and L. I. Martynenko, *Izvest. Akad. Nauk S.S.S.R., Ser. khim.*, 1973, 702, 1699.

[103] E. G. Yakovleva, N. I. Pechurova, L. I. Martynenko, and V. I. Spitsyn, *Zhur. neorg. Khim.*, 1973, **18**, 1519.

[104] D. J. Cookson, T. D. Smith, and J. R. Pilbrow, *J.C.S. Dalton*, 1974, 1396.

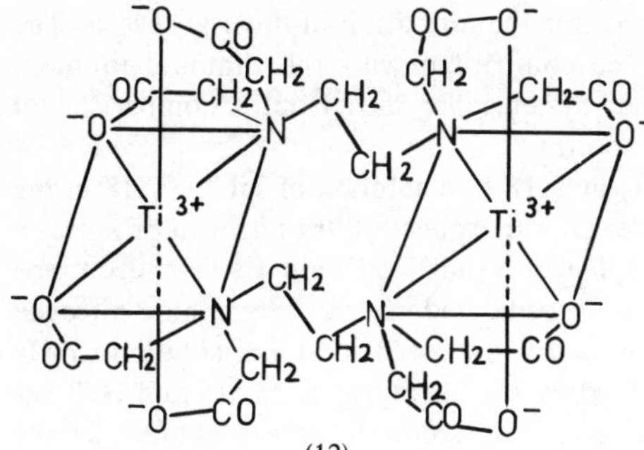

(12)

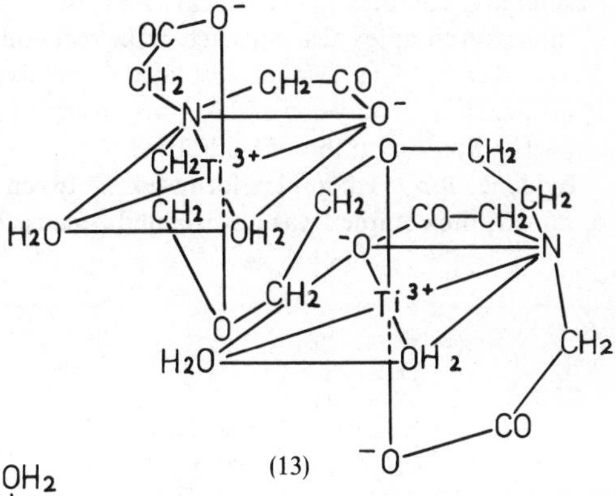

(13)

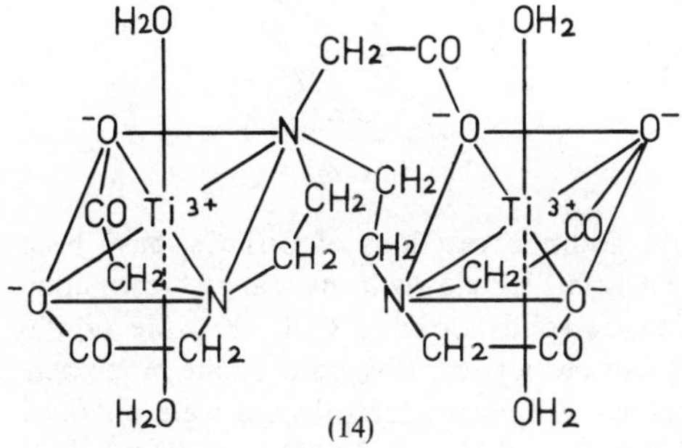

(14)

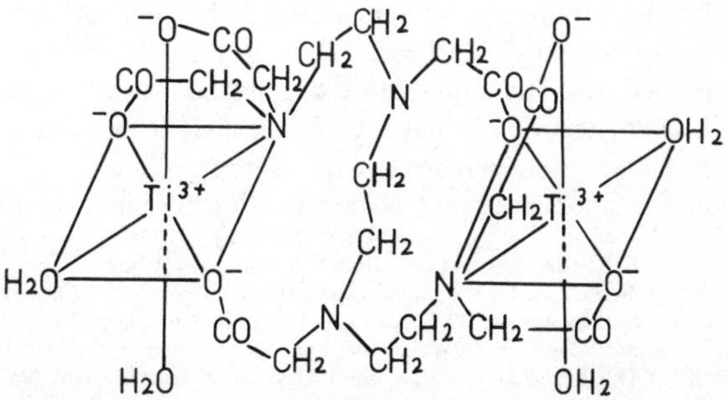

(15)

ratio; this provides unequivocal evidence for the formation of dimeric species. The structures (12)—(15) were chosen to be compatible with the titanium–titanium separations and symmetries of the pair systems obtained from a comparison of experimental and computer-simulated spectra.

The addition of a solution of salenH$_2$ in THF to a solution of TiCl$_3$,3THF in the same solvent at room temperature under an atmosphere of dry nitrogen produces a green precipitate of trichloro[NN'-ethylenebis(salicylideneimine)]tetrahydrofuran-titanium(III) (16) in which the ligand is co-ordinated *via* the azomethine nitrogen atoms only.[105] Over a period of several hours, the green solid and solution slowly turned red. This change was accompanied by the liberation of *ca.* 0.5 mol HCl per titanium atom. Red crystals isolated from this solution were identified as the titanium(IV) complex [TiCl$_2$(salen)],THF by X-ray crystallography. Oxidation to a titanium(IV) complex also occurred in the reaction between salpnH$_2$ and TiCl$_3$,3THF. A second complex (17) was also isolated from this reaction. Formation of (17) involves hydrogenation of only one of the C=N groups of the Schiff base. It is the first example of partial hydrogenation of Schiff base.

Evidence for 2:1 complex formation between titanium(III) and L-lysine has been obtained; the complex was not isolated because of its instability to air.[106]

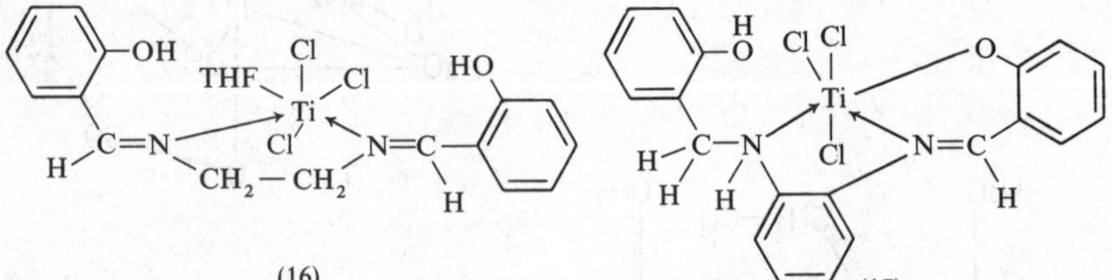

(16)   (17)

Cyclopentadienyl Complexes. Oxidative addition reactions of CpTiCl$_2$ have been reported. A blue-green solution of CpTiCl$_2$ in THF slowly turns orange-brown upon addition of an organic disulphide. Orange solids CpTi^{IV}(SR)Cl$_2$ (R = Me or Ph) which are very sensitive to hydrolysis can be isolated. Five-co-ordinate monomers CpTiX$_2$(S$_2$CNR$_2$) containing the bidentate dithiocarbamato ligand were obtained from the analogous reaction with thiuram disulphides [R$_2$NC(S)S—SC(S)NR$_2$].[107] Further studies of the magnetic behaviour of the cyclopentadienyltitanium(III) derivatives Cp$_2$TiX (X = Cl, Br, I, or OBz)[108,109] and Cp$_2$TiN$_2$TiCp$_2$[110] have been carried out. The temperature dependence of the magnetic susceptibility for Cp$_2$TiX is characteristic of singlet–triplet behaviour in a dimeric X-bridged system with the d–d interaction increasing in the series F < Cl ≈ I < Br; it was noted that the position of I is anomalous but no explanation for this was offered. The normal magnetic

[105] F. L. Bowden and D. Ferguson, *J.C.S. Dalton*, 1974, 460.
[106] O. Farooq, A. U. Malik, and N. Ahmad, *J. Electroanalyt. Chem. Interfacial Electrochem.*, 1973, **46**, 136.
[107] R. S. P. Coutts and P. C. Wailes, *J. Organometallic Chem.*, 1974, 73.
[108] R. S. P. Coutts, P. C. Wailes, and R. L. Martin, *Austral. J. Chem.*, 1973, **26**, 2101.
[109] O. D. Ubozhenko and A. A. Zharkikh, *Zhur. Khim.*, 1973, Abs. No. 8B467 (*Chem. Abs.*, 1973, **79**, 136305).
[110] I. N. Ivleva, A. K. Shilova, S. I. Salienko, and Yu. G. Borod'ko, *Doklady Akad. Nauk S.S.S.R.*, 1973, **213**, 116.

susceptibility, the absence of $\Delta M_s = \pm 2$ transitions from the e.p.r. spectrum, and the high solubility of $Cp_2Ti(OBz)$ in non-polar solvents support the proposal of a monomeric structure.[109] The —N=N— bridge provides a path for the antiferromagnetic exchange observed for $Cp_2TiN_2TiCp_2$.[110] Two conflicting reports of attempts to prepare the elimination-stabilized titanium(III) alkyl $Cp_2TiCH_2SiMe_3$ have appeared. Treatment of $(Cp_2TiCl)_2$ or $(CpTiCl_2)_n$ with stoicheiometric amounts of Me_3SiCH_2-$MgCl$ or Me_3SiCH_2Li has been reported to give the titanium(IV) products of disproportionation, Cp_2TiR_2 and $CpTiR_3$ ($R = Me_3SiCH_2$). Black precipitates formed in these reactions were suggested to be bivalent titanium species.[111] According to another report,[112] green, thermally stable (at $25\,°C$) crystals of Cp_2TiR ($R = Me_3SiCH_2$ or C_6F_5) can be obtained from the reaction of $(Cp_2TiCl)_2$ and the appropriate organolithium reagent at $-78\,°C$. These organotitanium(III) derivatives are analogues of the aryl complexes Cp_2TiAr prepared from Cp_2TiCl and $ArMgCl$. Like these, the Me_3SiCH_2 complex reacts with N_2 at $-80\,°C$ affording a deep-blue solution. No reaction occurs with the C_6F_5 derivative which remains green; previously, this compound has been reported to be purple.[113] Nitrogen complex formation is also indicated by low-temperature studies of the systems N_2-$Cp_2TiCHMe_2$[114] and N_2-$CpTiCl_2$-Pr^iMgCl.[115] The e.p.r. spectrum of the dicyclopentadienyl complex in the presence of N_2 at -60 to $-100\,°C$ confirmed the formation of a paramagnetic molecule which dimerized to $(Cp_2TiCHMe_2)_2$; this then reacted with N_2 to give the diamagnetic complex $(Cp_2TiCHMe_2)N_2$. A blue complex is formed in the reaction between $CpTiCl_2$, Pr^iMgCl, and N_2 at $-60\,°C$. The reaction is second-order with respect to $(CpTiCl_2)$. D.T.A. of the compounds Cp_2TiR showed[116] the thermal stability sequence to be

$$R = Ph \approx m\text{-}MeC_6H_4 \approx p\text{-}MeC_6H_4 < CH_2Ph < o\text{-}MeC_6H_4 \approx C_6F_5 < 2,6\text{-}Me_2C_6H_3 \approx$$
$$2,4,6\text{-}Me_3C_6H_2;$$

quantitative yields of RH were obtained in each case. Temperatures at which rapid decomposition occurred ranged from $29\,°C$ ($R = Ph$) to $230\,°C$ ($R = 2,4,6\text{-}Me_3C_6H_2$) indicating the marked dependence of thermal stability on the nature of R. The source of hydrogen for the formation of RH appears to be the cyclopentadienyl ligand. It was suggested that thermal decomposition occurs by an intermolecular reaction in which the 15-electron Cp_2TiR molecule, acting as an electrophile, attacks the cyclopentadienyl group of another molecule with the elimination of RH and the formation of bridging Ti—C_5H_4—Ti structures. For the compounds $(Cp_2TiR)_2N_2$ the thermal stability sequence is:

$$R = o\text{-}MeC_6H_4 < C_6F_5 < CH_2Ph < m\text{-}MeC_6H_4 \approx p\text{-}MeC_6H_4 \approx Ph$$

surprisingly opposite to that of the Cp_2TiR series. The co-ordinated nitrogen occupies the reaction site of thermal decomposition and at the same time reduces the electrophilic character of the Cp_2TiR moiety. It was also suggested that stabilization of

[111] M. L. H. Green and C. R. Lucas, *J. Organometallic Chem.*, 1974, **73**, 259.

[112] T. Chivers and E. D. Ibrahim, *J. Organometallic Chem.*, 1974, **77**, 241.

[113] J. H. Tenben and H. J. De Liefde Meijer, *J. Organometallic Chem.*, 1972, **46**, 313.

[114] V. B. Panov, Yu. M. Shul'ga, E. F. Kvashina, and Yu. G. Borod'ko, *Kinetika i Kataliz*, 1974, **15**, 518 (*Chem. Abs.*, 1974, **80**, 49 777).

[115] Yu. G. Borodko, E. F. Kvashina, V. B. Panov, and A. E. Shilov, *Kinetika i Kataliz*, 1973, **14**, 255 (*Chem. Abs.*, 1974, **81**, 49 777).

[116] J. H. Teuben, *J. Organometallic Chem.*, 1974, **69**, 241.

Cp_2TiR compounds might be achieved by blocking the empty titanium orbital sterically or by complexation with a donor molecule. This suggestion is borne out by the isolation and high thermal stability ($d \sim 225\ °C$) of the compound $Cp_2Ti[2-(Me_2NCH_2)C_6H_4]$ (18) in which there is intramolecular co-ordination of the dimethyl-amino-group.[117] In a related attempt to stabilize cyclopentadienyltitanium(III) alkyls, a series of complexes Cp_2TiClL ($L = R^1NH_2$ or R_3^2P and $R^1 = H$, Me, Et, Ph, or C_3H_5; $R^1NH_2 = py$; $R_3^2 = Me_2Ph$ or $MePh_2$) and $CpTiCl_2,L_2$ [$L = Me_2PhP$, $MePh_2P$, or $L_2 = (MePCH_2)_2, (Ph_2PCH_2)_2$] were prepared;[111] they reacted rapidly

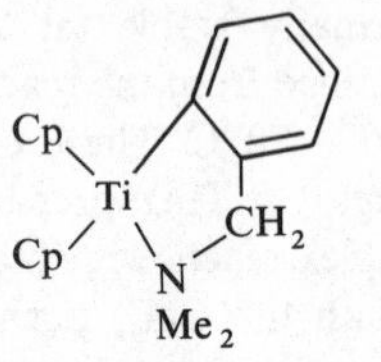

(18)

with ethyl-, allyl-, and phenyl-magnesium bromides or with Bu^nLi and $PhLi$ even at $-90\ °C$ to give dark mixtures from which the only tractable products to be obtained were the ligands L.

The reaction between $TiCl_3,3THF$ and $CH_2{=}CHSiMe_2CH_2MgCl$ yields $(CH_2{=}CHSiMe_2CH_2)_3Ti^{III}$ which has been used as a polymerization catalyst.[118] *Miscellaneous.* Two previously reported products from $[ReCl(N_2)(PMe_2Ph)_4]$ and $TiCl_3$ have been shown to be derived from Ti^{IV}.[119]

Hydrazine is an intermediate in the formation of ammonia from reduction of N_2 by titanium(III) in the presence of molybdenum compounds. Hydrogen is also produced by reaction of the titanium(IV) with the methanol solvent. Activation energies were determined for the formation of these three products.[120] The preparation and n.m.r. spectra of Mg, Zn, and Cd derivatives of nonahydrohexaboratobis(cyclopentadienyl)titanium(III) have been reported.[121]

Titanium(IV).—*Halides, Oxyhalides, and Pseudohalides.* Uncertainties surrounding the occurrence and structure of polymeric fluorotitanium(IV) anions have been dispelled by an ^{19}F n.m.r. study of the system $TiF_4{-}(Pr_2^nNH)_2TiF_6$ in liquid SO_2.[122] TiF_4 is insoluble in liquid SO_2 but dissolves in the presence of TiF_6^{2-}. At a $TiF_4:TiF_6^{2-}$ ratio of *ca.* 1:3 the principal species in solution is the $Ti_2F_{11}^{3-}$ ion (19). As the ratio increases new signals appear which reach a maximum intensity at 1:1 and are therefore attributable to $(TiF_5)_n^{n-}$. A detailed analysis of the spectrum indicates the probable formulation of this ion as $(Ti_2F_{10})^{2-}$ (20). The spectra of mixtures with $TiF_4:TiF_6^{2-}$ ratios > 1 are complex; at a ratio of 3:1 there are two major signals in the intensity

[117] D. Ytsma, J. G. Hartsuiker, and J. H. Teuben, *J. Organometallic Chem.,* 1974, **74**, 239.
[118] Fr. P., 2 153 469, (*Chem. Abs.,* 1973, **79**, 79 536).
[119] R. Robson, *Inorg. Chem.,* 1974, **13**, 475.
[120] N. T. Denisov, G. G. Terekhina, N. I. Shuvalova, and A. E. Shilov, *Kinetika i Kataliz,* 1973, **14**, 939 (*Chem. Abs.,* 1974, **81**, 52 748)
[121] D. L. Denton, *Diss. Abs. (B),* 1974, **34**, 3688.
[122] P. A. W. Dean, *Canad., J. Chem.,* 1973, **51**, 4024.

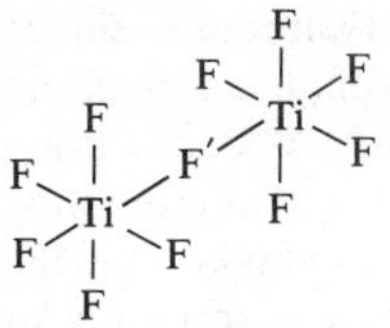

    (19) 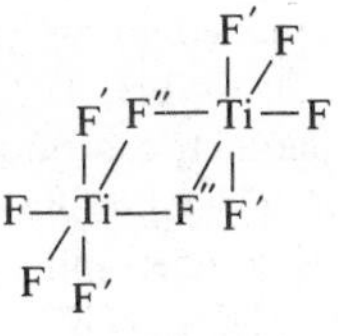   (20) 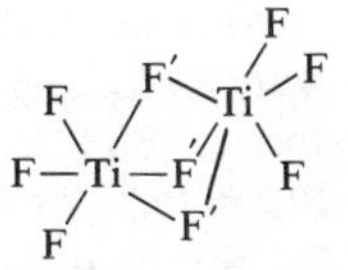   (21)

ratio 2:1. It is suggested that these are due to the $Ti_2F_9^-$ ion (21) having a face-joined bioctahedral structure. This is consistent with the 3:1 $TiF_4:TiF_6^{2-}$ ratio necessary to generate these signals.

The low-temperature ^{19}F n.m.r. spectra of the mixtures $TiF_6^{2-}-TiX_6^{2-}$ (X = Cl, Br, NCS, or NCO) in liquid SO_2 show that rapid fluorine exchange is occurring even at the lowest accessible temperatures for X = Cl or Br. Spectra of the mixed species $(TiF_{6-x}X_x)^{2-}$ are observed for X = NCS or NCO.[123] In these cases fluorine-bridged dimers also occur in stable equilibrium. The existence of these dimers was taken to indicate the following mechanism of redistribution:

$$TiF_6^{2-} + TiX_6^{2-} \rightleftharpoons F_5Ti-F^*-TiX_5^{3-} + X^-$$
$$X^- + F_5Ti-F^*-TiX_5^{3-} \rightleftharpoons XF_5Ti^{2-} + F^*TiX_5^{2-}$$

The ^{19}F chemical shifts of $TiF_{6-x}X_x^{2-}$ (X = NCS or NCO) and of some of the related mixed complexes are reasonably represented by the equation $\delta = pC + qT$, where C and T are constants characteristic of the substituent X and p and q are the number of ligands X cis and trans respectively to the resonating fluorine. The constants T appear to be a measure of the π-donor ability of the ligand X and decrease in the series F > NCO > NCS. Structural assignments have been made on the basis of the ^{19}F n.m.r. spectra of complexes formed in the following systems:[124] TiF_4–$TiCl_4$–dioxan; TiF_4–$TiBr_4$–2,6-dimethylpyridine N-oxide; TiF_4–$TiCl_4$–thioxan; TiF_4–$TiCl_4$–dithiane; TiF_4–H_2O–$ClCH_2CN$; TiF_4–H_2O–methylal; TiF_4–$TiCl_4$–H_2O–1,2-dimethoxyethane; TiF_4–$TiBr_4$–H_2O–1,2-dimethoxyethane. The phase diagrams of the following systems have been analysed: KF–TiF_4, NaF–TiF_4, NaF–KF–TiF_4[125] and LiF–NaF–Li_2TiF_6–Na_2TiF_6, KF–LiF–K_2TiF_6–Li_2TiF_6.[126] The hexafluorotitanates $(N_2H_5)_2TiF_6$, $(N_2H_5)_2TiF_6,2HF$, and $(N_2H_6)TiF_6$ have been prepared and characterized by thermal analysis and i.r. spectroscopy.[127] Ammonia and nitrogen are released in the thermal decomposition of $(N_2H_5)_2TiF_6$. Barium hydroxyfluorotitanate ($BaTiOF_4,0.5H_2O$) is precipitated by the addition of CO_3^{2-} ions to solutions containing K_2TiF_6 and $BaCl_2$; it decomposes to BaF_2 and titanium(IV) hydroxide upon further addition of CO_3^{2-}.[128] Mixed $M_2^1M^2OF_4$-type compounds (M^1 = K or Na; M^2 = Ti or Zr) have been prepared,[129] as have the compounds $Cs_xM_y^1M_{2-y}^2X_6$ (M^1 = Ti, M^2 = V, Cr, Fe, Ga, Ti, Nb, or W; X = F or F and O, x = 0.4 or 0.5).[130]

[123] P. A. W. Dean and B. J. Ferguson, *Canad. J. Chem.*, 1974, **52**, 667.
[124] S. R. Borden, *U.S. Nat. Tech. Inform. Serv., A.D. Reports*, 1973, No. 766336 (*Chem. Abs.*, 1974, **80**, 52 939).
[125] R. V. Chernov and I. M. Ermolenko, *Zhur. neorg. Khim.*, 1973, **18**, 2238.
[126] I. M. Ermolenko and R. V. Chernov, *Zhur. neorg. Khim.*, 1973, **18**, 2528.
[127] J. Slivnik, J. Macek, B. Orel, and B. Sedej, *Monatsh.*, 1973, **104**, 624.
[128] A. M. Golub, I. I. Voitko, and S. P. Rozhenko, *Izvest. Akad. Nauk S.S.S.R., neorg. Materialy*, 1974, **10**, 287.
[129] A. A. Kazain and T. V. Afana'ev, *Ref. Zhur., Met.*, 1973, Abs. No. 2G245. (*Chem. Abs.*, 1973, **79**, 147 794)
[130] D. Babel, F. Biner, and G. Paussewang., *Z. Naturforsch.*, 1973, **28b**, 213.

Whereas all the metal halides examined so far yield mononuclear halide anions with RPX_2 and Bu^iX (R = alkyl, Cl, or Br), $TiCl_4$ forms the dinuclear titanium species $(Bu^tPBr_3)(Ti_2Cl_9)$. A stoicheiometric product could not be obtained using Bu^tCl, PCl_3, or $MePCl_2$ with $TiCl_4$.[131] $TiCl_4$ behaves as a Lewis acid in promoting the disproportionation of monochloramine, and subsequently combines with the ammonium chloride formed to give $(NH_4)_2(TiCl_6)$; this decomposes thermally to $TiN_{0.98}$.[132] Force constants and mean amplitudes of vibration of the $TiCl_6^{2-}$ ion have been calculated.[133, 134] Chloride–thiocyanate exchange reactions between $TiCl_4$ and NH_4CNS in dry THF or MeCN lead to the mixed complexes $TiCl_n$-$(NCS)_{4-n}(sol)_2$ (n = 1, 2, or 3; sol = THF or MeCN).[135] Complexes are also formed in acetone and DMF.[136] Their i.r. spectra indicate that the NCS ion is bound to titanium *via* the N atom. The stepwise formation constants of the six complexes formed in acetone and DMF have been reported.[136]

O-Donor Ligands. A large number of studies have been concerned with the preparation and properties of titanates Ti^{IV} mixed oxide compounds. Li_4TiO_4 has been found to be isostructural with Li_4GeO_4; it is only the second titanate known to contain tetrahedrally co-ordinated Ti^{4+}.[137] Potassium tetratitanate $K_2Ti_4O_9$ has previously been reported as the product of the solid-state reaction between TiO_2 and K_2CO_3 at 900 °C and as the product of the thermal decomposition of K_2TiF_6 at 820–1000 °C. It appears that the thermal decomposition product is in fact an oxy-fluorotetratitanate of probable stoicheiometry $K_2Ti_4O_8F_2$. Its X-ray powder diffraction pattern is significantly different from that of an authentic sample of $K_2Ti_4O_9$. The lattice parameters of K_2TiO_3 and $K_2Ti_4O_9$ have been determined.[138] A new method for the preparation of the metatitanates M_2TiO_3 (M = Ag or K) and $MTiO_3$ (M = Pb or Ba) involves the hydrolysis of titanium hexahalides and oxyhalides in water, non-aqueous solvents or melts *e.g.*[139]

$$H_2TiCl_6 \; + \; Ag_2O \xrightarrow[\text{5 days}]{\text{EtOH, 36°C}} Ag_2TiO_3$$

$$(NH_4)_2TiCl_6 + Pb(NO_3)_2 \xrightarrow{300°C} PbTiO_3$$

The unit cell of β-Ba_2TiO_4 contains four isolated TiO_4 tetrahedra linked by Ba atoms;[140] Ti—O distances range from 1.766—1.836 Å. One Ti—O bond is significantly shorter than the others as a result of the O being bonded to only three Ba atoms while each of the others is bonded to four. A new relationship between bond strength (S) and bond length (R) has been proposed, *viz* $S = 0.666 \; (R/1.953)^{-5.25}$. Since Pb is volatile at high temperatures, a large excess of PbO is needed to explore the entire triangle $PbTiO_3$–$PbNb_2O_6$–$3PbONb_2O_5$.[141] White to orange crystalline products

[131] J. I. Bullock, F. W. Parrett, and N. J. Taylor, *Canad. J. Chem.*, 1974, **52**, 2880.
[132] H. P. Fritz and W. Treptow, *Z. Naturforsch.*, 1973, **28b**, 584.
[133] A. N. Pandey, D. K. Sharma, H. S. Singh and B. P. Singh, *Z. Naturforsch.*, 1973, **28a**, 1155.
[134] V. K. Gupta, R. Babu, A. N. Pandey, and Z. H. Zaidi, *Indian J. Pure Appl. Phys.*, 1973, **11**, 88.
[135] H. Bochland and H. Steinecke, *Z. Chem.*, 1974, **14**, 65.
[136] A. M. Sych and R. I. Mrikhina, *Ukrain khim. Zhur.*, 1973, **39**, 558 (*Chem. Abs.*, 1973, **79**, 150 028).
[137] B. L. Dubey and A. R. West, *J. Inorg. Nuclear Chem.*, 1973, **35**, 3713.
[138] A. J. Eeasteal and D. J. Udy, *J. Inorg. Nuclear Chem.*, 1973, **35**, 3956.
[139] V. Marecek, J. Novak, and J. Strajblova, *Konf. Keram. Elektron, 4th*, 1971, XLI, 1 (*Chem. Abs.*, 1973, **79**, 26 607).
[140] Kang Kun Wu and I. D. Brown, *Acta Cryst.*, 1973, **29B**, 2009.
[141] J. Bachelier, N. Hervieu, and E. Quemeneur, *Bull. Soc. chim. France*, 1973, 2593.

were obtained from mixtures heated to 1000 °C. The compounds formed confirm the existence of lacuniary pyrochlore types, with simultaneous anionic and cationic gaps in the triangular structure. A new pyrochlore $Pb(Ti_xSn_{1-x})O_3$ has been prepared and its transition to a perovskite structure studied.[142] Table 1 summarizes other work on titanates and Ti^{IV} mixed oxides.

The structure of a new mineral $Sr_3TiSi_4O_{12}(OH),2H_2O$ has been reported.[143] Single silica chains similar to those in haradaite run along the b axis direction and the Ti is surrounded by four O atoms of silica chains and two OH radicals to form octahedral chains along the silica chains. Strontium atoms are surrounded by seven or nine O atoms, OH, and H_2O. A complex titanium-containing polysiloxane has been prepared from $PhSi(OH)_2ONa$ and $TiCl_4$; it has catalytic activity in cumene cracking and the dehydration of EtOH.[144] The insertion of ketones and keten derivatives into the Ti—O bond of titanium alkoxides leads to complex alkoxy-titanates containing terminal carboxylate groups.[145] Molecular peroxo complexes of the type $Ti(O_2)F_2(H_2O)L$ (L = o-phen or bipy), $Ti(O_2)F_2(H_2O)L$, and $Ti(O_2)$-$SO_4L_3,3H_2O$ (L = picoline N-oxide) have been prepared. Medium-to-strong bands in the i.r. spectra in the region 520—620 cm^{-1} have been assigned to $v_{asym}Ti(O_2)$ and $v_{sym}Ti(O_2)$.[146] The e.p.r. spectra of frozen solutions containing titanium peroxo complexes have been analysed.[147] Successive treatment of $TiCl_4$ with H_2O_2, NH_4OH, and $In(NO_3)_3$ gave $[In(OH)_2]_2[O_2TiO(OH)_2],2H_2O$; with excess oxalic acid and $In(NO_3)_3$, $In_2[Ti(OH)_2(C_2O_4)_2]_3,3H_2O$ was formed. Thermal decomposition of the peroxo complex gave In_2O_3, TiO_2, H_2O, and oxygen.[148]

X-Ray diffraction studies have shown $CsTi(SO_4)_2,12H_2O$ to be a β-alum iso-morphous with $CsAl(SO_4)_2,12H_2O$.[149] Tensimetric studies of the composition of $TiOSO_4,nH_2O$ formed during the evaporation of an aqueous solution indicate that, at < 0.5 mmHg water vapour pressure, the solid glassy product contains two or three molecules of H_2O. Solid $TiO(SO_4)(H_2O)_2$ contains a bidentate sulphate group. De-hydration at 300 °C gives an anhydrous amorphous product containing Ti—O—Ti—O—Ti chains with multiple titanium–oxygen bonds.[150] Thermal and X-ray analytical data indicate that an exchange reaction occurs in the $PbSO_4[Pb_3(PO_4)_2]$–$MTiO_3$ system (M = Ca, Ba, Zn, or Cd) to give $PbTiO_3$ and MSO_4 or $M_3(PO_4)_2$. The exchange reaction corresponds in all cases to two exothermic effects, the first of which corre-sponds to the main reaction phase and the second to crystallization.[151] Bands below 800 cm^{-1} in the Raman spectrum of titanium(IV) nitrates have been reassigned.[152] The intensities of bands at 1600, 1220, and 1000 cm^{-1} due to N—O vibrational modes vary according to the mode of co-ordination of the NO_3 group.

[142] S. Yoshida, T. Matsuzaki, T. Kashivazaki, K. Mori, and K. Tarama, *Bull. Chem. Soc. Japan*, 1974, **47**, 1568.
[143] T. Mizota, M. Komatsu, and K. Chihara, *Mineral J.*, 1972, 302 (*Chem. Abs.*, 1974, **81**, 42 709).
[144] I. M. Kolesnikov, G. M. Panchenkov, K. A. Andrianov, A. A. Zhdanov, M. M. Levitskii, and N. N. Belov, *Izvest. Akad. Nauk S.S.S.R., Ser. khim*, 1974, 488.
[145] L. Vuitel, Swiss P., 536 858 (*Chem. Abs.*, 1973, **79**, 78 093).
[146] J. Bala-Pala and J. E. Guerchais, *Bull. Soc. chim. France.*, 1973, 1545.
[147] V. M. Berdnikov, P. V. Schastnev, A. A. Merkulov, A. V. Balaev, N. M. Bazhin, and Yu. V. Gatillov, *Zhur. strukt. Khim.*, 1973, **14**, 634.
[148] Gh. Macarovici, E. Perte, M. Birou, and M. Strajescu, *Stud. Univ. Babes-Bolyai Ser. Chem.*, 1973, **18**, 123 (*Chem. Abs.*, 1974, **81**, 32 704).
[149] J. Sygusch, *Acta Cryst.*, 1974, **30B**, 662.
[150] Ya. G. Goroshchenko, R. V. Kuprina, and A. M. Kalinichenko, *Zhur. neorg. Khim.*, 1974, **19**, 370.
[151] I. N. Belyaev, I. I. Belyaeva, and L. N. Aver'yanova, *Zhur. neorg. Khim.*, 1974, **19**, 224.
[152] D. W. Arnos and G. W. Plewett, *Spectrochim. Acta.*, 1974, **30A**, 453.

Table 1

Compound	Source	Properties reported	Ref.
K_2O, TiO_2			
$2K_2O$, $3TiO_2$			
K_2O, $2TiO_2$	K_2CO_3–TiO_2	i.r., t.d, X	a
K_2O, $4TiO_2$			
K_2O, $6TiO_2$			
$MgTiO_3$	$MgCO_3 + H_2TiF_6$	X	b
$CaTiO_3$	—	homogeneity range	c
$BaTiO_3$	$BaCO_3 + TiO_2$	—	d
$Ba_2Ge_2TiO_8$	—	X	e
$BaTi_2Fe_4O_{11}$	$BaCO_3$ + d-Fe_2O_3 + TiO_2 (anatase)	X	f
Al_2TiO_5	—	ΔH	g
Ga_2TiO_5	—	X	h
$Ga_2Ti_2O_7$	—	X	h
Ga_2TiO_5	$[Ga(OH)_2]_2[O_2TiO(OH)_2]$, $3H_2O$ at 560 °C		i
In_2TiO_5	—	X	h
$Tl_2Ti_2O_7$	—	X	h
$PbTi_3O_7$	$PbO_2 + TiO_2$	X	j
$TiNb_2O_7$	—	X	k
$Ti_2Nb_{10}O_{29}$	—	X	k
$Ti_xV_{1-x}O_{1.5}$	$(NH_4)_2[(Ti_xV_{1-x})O(C_2O_4)_2]$ at 600—1300 °C	X	l
$Ti_8Cr_2O_{19}$	—	X, t	m
$Ti_7Cr_2O_{17}$			
$Ti_6Cr_2O_{15}$			
$Ti_5Cr_2O_{13}$	—	t.d.	m
$Ti_4Cr_2O_{11}$			
$FeTiO_3$	$FeCl_2, 4H_2O$ + aq. H_2TiF_6	X	b
$FeMg_{1-x}TiO_3$	$FeCl_2,4H_2O + MgCO_3$ + aq. H_2TiF_6	X	b
$x = 0.015—0.2$			
$2CoO.3TiO_2.0.2O_2$	Cobalt(III)ammine		
$2Co,3TiO_2$	titanyloxalate	i.r.	n
$2/3Co_3O_4,3TiO_2,5/6O_2$	complexes at 300—1000 °C		
$Ni_5TiB_2O_{10}$	—	X	o
$La_4Ca_{0.5}Ti_{4.5}O_{15.5}$			
$La_4CaTi_5O_{17}$	$La_2Ti_2O_7$—$CaTiO_3$	X	p
$La_4Ca_2Ti_6O_{20}$			
$LnTiTaO_6$	$Ln_2O_3 + TiO_2 + Ta_2O_5$	X, d	q
(Ln = La—Sm, Ga—Yb)			
$M_2O_3,2TiO_2$	thermal d. of nitrate salts	—	r
(M = La—Lu, Y)			
$LnTiNbO_6$	calcination of hydroxide mixtures	—	s
(Ln = La, Ce, Pr, or Nd)			
$EuTiNbO_6$	$Eu(OH)_3 + Ti(OH)_4 + Nb(OH)_5$ 970 °C	X, d	t
$SmTiNbO_6$	$Sm(OH)_3 + Ti(OH)_4 + Nb(OH)_5$ 960 °C	X, d	t

(a) E. K. Belyaev, N. M. Panasenko, and V. M. Tomenko. *Izvest. Akad. Nauk S.S.S.R., neorg. Materialy*, 1974, **10**, 460; (b) P. J. Gellings and G. M. H. Van de Velde, *Z. Naturforsch.*, 1973. **28b**, 405; (c) N. G. Kisel. T. F. Limar, L. P. Mudrolyubova, and I. F. Cherednichenko, *Izvest. Akad. Nauk. S.S.S.R., neorg. Materialy*, 1974, **10**, 465; (d) S. A. Kutolin, A. I. Vulikh, and A. E. Shammasova, U.S.S.R. P., 392,001 (Cl. C o 1f) (*Chem. Abs.*, 1974, **81**, 15245); (e) M. Kimura, K. Doi, S. Nanamatsu, and T. Kawamura, *Appl. Phys. Letters* 1973, **23**, 531; (f) F. Haberey and M. Velicescu, *Acta Cryst.*, 1974, **30B**, 1507; (g) M. S. J. Garni and R. McPherson, *Thermochim. Acta*, 1973, **7**, 251; (h) V. E. Plyushchev and I. A. Pozdin, *Khim. Khim. Tekhnol., Trudy Yubileinoi Konf., Posvyashch. 70-Letlyu. Inst.*, 1970. 351 (*Chem. Abs.*, 1974, **81**, 42608)); (i) E. Perte and M. Strajescu,

The acceptor strengths of $Ti(OPh)_4$ and $TiCl_4$ have been investigated by e.p.r. studies of their complexes with aliphatic nitroxides.[153] Increased acceptor character increases the O character of the nitroxide's π-bonding electrons and at the same time increases the N character of the unpaired π^*-electron; therefore, there should be an increases in N h.f.s. and a decrease in O h.f.s. According to these criteria, $TiCl_4$ is a better Lewis acid than $Ti(OPh)_4$. The strong monobasic acids HY ($Y = SO_3F$, SO_3Me, PO_2F, or CO_2CF_3) react with $TiCl_4$ to give $TiCl_2Y_2$ when they are in excess.[154] With excess $TiCl_4$, HSO_3F forms $TiCl_{3.33}(SO_3F)_{0.66}$, HSO_3Me forms $TiCl_3SO_3Me$, but HCO_2CF_3 and HPO_2F again give $TiCl_2Y_2$. The vibrational spectra of the complexes $TiCl_2Y_2$ and $TiCl_3SO_3Me$ are interpreted in forms of hexaco-ordinated titanium and bi- or ter-dentate oxyacid groups bridging metal atoms to give polymeric structures. $TiCl_{3.33}(SO_3F)_{0.66}$, which can also be formed from $Sn(SO_3F)_4$ and excess $TiCl_4$, has $\nu(S-F)$ *ca.* 200 cm^{-1} lower than in KSO_3F and $\nu_{asym}SO_3$ 30 cm^{-1} lower. It was suggested that in $TiCl_{3.33}(SO_3F)_{0.66}$ the SO_3F group is quadridentate, binding through fluorine and the three oxygens.[154] The $Ti(SO_3Me)_nCl_{4-n}$ complexes are insoluble in $MeSO_3H$ but form adducts with bases (B) such as py, butylamine, or quin which dissolve to give colourless highly conducting solutions.[155] The conductivity values are consistent with:

$$Ti(SO_3Me)_nCl_{4-n},2B + 4MeSO_3H \rightarrow H_2[\overset{\frown}{Ti}(SO_3Me)_{n+2}Cl_{4-n}] + 2BH^+ + 2MeSO_3^-$$

Solvolytic reactions of $TiCl_2(OAc)_2$ with alcohols yield $TiCl_2(OAc)OR$ and $TiCl_2$-$(OR)_2$ except with Bu^tOH which substitutes OH for Cl yielding $TiCl(OH)(OAc)_2$.[156] (Organoxy)chlorotitanium(IV) compounds of cyclohexanol, catechol, resorcinol, thymol, vanillin, and cyclohexane-1,4-diol, and chlorotitanium acylates derived from esters of sulphuric, oxalic, and phenoxyacetic acids have been prepared.[157] The acylates are good intermediates for the synthesis of other (organoxy)titaniums by substitution of Cl with organic hydroxy and ester compounds. Reactions between orthotitanic esters and sulphuryl/thionyl chloride were also studied as an alternative route for the preparation of chloro-organotitanium compounds. Titanium alkoxides have been prepared from aliphatic alcohols and $TiCl_4$ in refluxing *sym*-tetrachloroethane.[158] However, salicylaldehyde undergoes Meerwein–Ponndorf reduction with $TiCl_4$[159] rather than the complex formation reported previously. Thermal decomposition of the adducts $TiCl_4,L^1$ and $TiCl_4,2L^2$ ($L^1 = HCO_2CHMe_2$, $AcOCHMe_2$,

[153] A. H. Cohen and B. M. Hoffmann, *Inorg. Chem.*, 1974, **13**, 1484.

[154] J. R. Dalziel, R. D. Klett, P. A. Yeats, and F. Aubke, *Canad. J. Chem.*, 1974, **52**, 231.

[155] R. C. Paul, V. P. Kapila, and S. K. Sharma, *J. Inorg. Nucl. Chem.*, 1974, **36**, 1933.

[156] H. Lee, Y. S. Uh, Y. S. Sohn, and Q. W. Choi, *Daehan Hwahak Hwoejee*, 1973, **17**, 174 (*Chem. Abs.*, 1973, **79**, 38116).

[157] C. Gopinathan and J. Gupta, *Indian J. Chem.*, 1973, **11**, 948.

[158] P. J. Joseph, *Indian Chem. J., Annu.*, 1972, 135.

[159] R. C. Mehrota, V. D. Gupta, and P. C. Bharaca, *Indian J. Chem.*, 1973, **11**, 814.

Rev. Roumaine Chim., 1974, **19**, 395; (*j*) K. Kato, I. Kawada, and K. Muramatsu, *Acta Cryst.*, 1974, **30B**, 1634; (*k*) R. B. Von Dreele and A. K. Cheetham, *Proc. Roy. Soc.*, 1974, **A338**, 311; (*l*) F. Maillot and R. A. Paris, *Compt. rend.*, 1973, **277**, *C*. 207; (*m*) C. W. Lee, *Acta Geol. Taiwan*, 1973, 39 (*Chem. Abs.* 1974, **80**, 55381); (*n*) C. G. Macarovici and M. Strajescu, *Stud. Univ. Babes-Bolyai, Ser. Chim.*, 1973, **18**, 67. (*Chem. Abs.* 1973, **79**, 99911); (*o*) C. G. F. Stenger, C. G. Verschoor, and D. J. W. Ijdo, *Mater. Res. Bull.*, 1973, **8**, 1285; (*p*) M. Nanot, F. Queyroux, and J. C. Gilles, *Compt. rend.*, 1973, **227**, *C*, 505; (*q*) V. V. Kazantsev, E. I. Krylov, A. K. Borisov, and A. I. Chupin, *Zhur. neorg. Khim.*, 1974, **19**, 930; (*r*) N. I. Timofeeva and Z. I. Krainova, *Izvest. Akad. Nauk S.S.S.R., neorg. Materialy*, 1973, **9**, 1756; (*s*) A. M. Sych and V. G. Klenus, *Izvest. Akad. Nauk S.S.S.R., neorg. Materialy*, 1973, **9**, 1219; (*t*) E. I. Krylov, A. K. Borisov, Yu. N. Makurin, and G. G. Kasimov, *Izvest. Akad. Nauk S.S.S.R., neorg. Materialy*, 1974, **10**, 74.

$CH_2ClCO_2CHMe_2$, HCO_2Pr, HCO_2Bu, $AcOBu$, or $PrCO_2Et$; $L^2 = HCO_2CHMe_2$, $AcOCHMe_2$, $CCl_3CO_2CHMe_2$, $PrCO_2Et$, or $HCO_2C_8H_{17}$) gives $TiCl_2(O_2CR)$, $TiOCl_2$, and TiO_2.[160] Carboxylate complexes have also been obtained directly from $TiCl_4$ and RCO_2H; dimeric structures based on two bridging carboxylate groups have been proposed on the basis of i.r. and n.m.r. spectroscopic evidence.[161] N.m.r. studies of solutions of $TiCl_4$ in RCO_2CHMe_2 (R = H, Me, $ClCH_2$, or CCl_3) showed that carbonium ions are generated only when R = CCl_3.[162] A correlation between the composition of ionic complexes formed between $TiCl_4$ and esters, and Taft induction constants has been proposed.[163] Copolymerization of the disodium salt of 1,1-dicarboxycobalticinium hexafluorophosphate and $Cp_2Ti(PF_6)_2$ yielded the

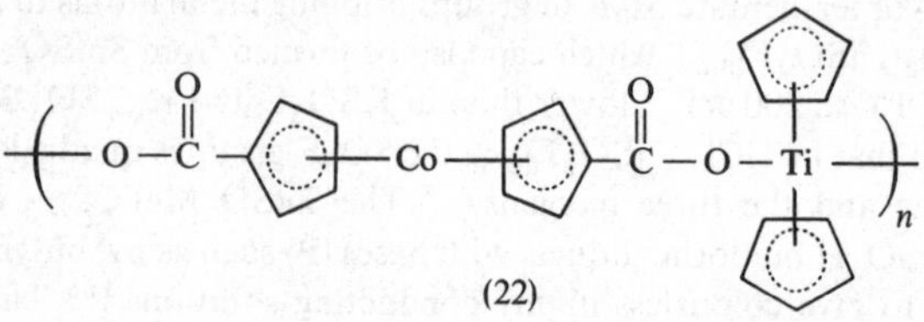

(22)

polymer (22), n = *ca.* 180. Thermal degradation of the polymer sets in at *ca.* 150 °C, at 600 °C the weight losses are *ca.* 40% (in air) and *ca.* 20% (in N_2) showing that oxidation is primarily responsible for the degradation.[163a] The reaction between ethylene glycol and titanium(IV) is second order, the rate increasing rapidly with increasing pH. Complex formation accounts for the decrease in the titanium(IV) polarographic wave height when ethylene glycol is added and the negative slope of the diffusion current as a function of ethylene glycol concentration.[164] Ligand-exchange equilibria for some dihalogenobis(β-diketonato)titanium(IV) and diethoxy-bis(β-diketonato)titanium(IV) complexes have been studied by n.m.r. spectroscopy and diketonate exchange was examined.[165] Deviations from random scrambling were rationalized on the basis of an electrostatic model.

Attempts to prepare metallorganic polymers *via* thermal condensation between aromatic diols and Ti(L–L)$_2$(alkoxy)$_2$(L–L = chelate ligand) complexes gave insoluble powders and low molecular weight products. It was suggested that the *cis* stereochemistry of the titanium monomers was responsible for the failure to obtain polymers.[166] Various Ti(L—L)$_2$(phenoxy)$_2$(L—L = acac, oxalate, or quin) were synthesized but these too had *cis* octahedral structures. In solution rapid Me group exchange occurs for Ti(acac)$_2$(phenoxy)$_2$. A dissociative acac bond rupture mechanism was favoured in preference to a twist-type mechanism. Isopropyl group exchange in complexes containing this group appears to occur *via* a twist-type mechanism with concomitant inversion of the molecule. The chemistry of titanium(IV) chelates has been extended by the synthesis of a large number of new complexes,[167] mostly either

[160] Yu. A. Lysenko and L. I. Khoklova, *Zhur. neorg. Khim.*, 1974, **19**, 1268.
[161] Y. S. Uh, H. S. Lee, and Y. S. Sohn, *Daehan Hwahak Hwoejee*, 1973, **17**, 115.
[162] Yu. A. Lysenko and L. I. Khokhlova, *Zhur. obshchei Khim.*, 1974, **44**, 403.
[163] Yu. A. Lysenko and L. I. Khokhlova, *Zh. obshchei Khim.*, 1974, **44**, 487.
[163a] C. E. Carraher and J. E. Sheats, *Makromol. Chem.*, 1973, **166**, 23.
[164] C. Liegois, *Bull. Soc. chim. France*, 1973, 2177.
[165] R. C. Fay and R. N. Lowry, *Inorg. Chem.*, 1974, **13**, 1309.
[166] K. R. Taylor, *Diss. Abs. (B)*, 1974, **34**, 4856.
[167] C. Gopinathan and J. Gupta, *Indian J. Chem.*, 1974, **12**, 103.

by reaction between chlorotitanium chelates and aliphatic or aromatic hydroxy-compounds or dialkyl sulphates or by treatment of sulphato-, oxalato-, or chloro-titanium chelates with chelating ligands. Ammonia adducts with up to four molecules of ammonia per titanium atom were obtained by passing gaseous ammonia through a suspension of the complex in ether at 0 °C. These adducts were more resistant to hydrolysis than the parent complexes. The detailed structures of several of the new complexes are under study. Replacement of the CH hydrogen of acetylacetone by the cyanide group produces a terdentate ligand which reacts with $TiCl_4$ in a 1:1 molar ratio to give the tetrameric complex (23).[168] With a 1:2.5 molar ratio $TiCl_2L_2$ (LH = 3-cyanopenta-2,4-dienone) is formed. The complexes $TiO(C_6H_7O_7)^+$, TiO-

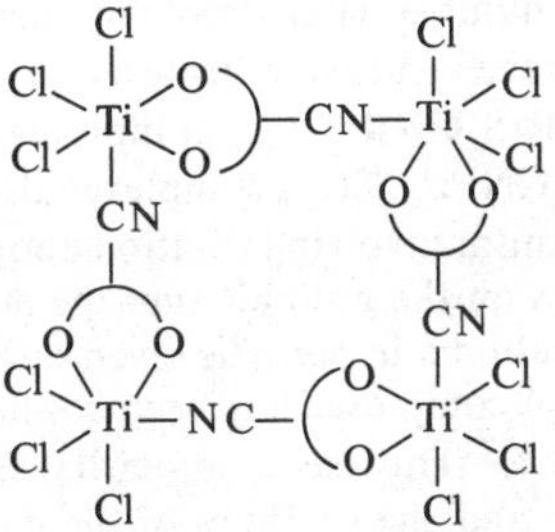

(23)

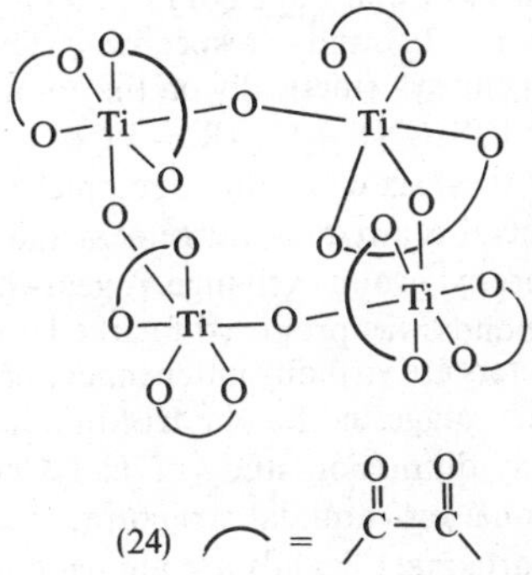

(24) $\frown$ = $\underset{/}{\overset{O}{\underset{\|}{C}}} - \underset{\backslash}{\overset{O}{\underset{\|}{C}}}$

$(C_6H_7O_7)_2$, $TiOC_4H_5O_6^+$, and $TiO(C_4H_5O_6)_2$ form in Ti^{IV} citrate and tartrate solutions, respectively, in 0.2 M-HCl.[169]

Most so-called titanyl complexes do not contain the (—Ti=O) unit and X-ray diffraction studies have shown that ammonium titanyl oxalate is no exception. It contains the tetrameric unit cyclo-tetra-di-μ-oxo-*cis*-dioxalatotitanate(IV) (24).[170] Each titanium atom has an approximately octahedral stereochemistry provided by two *cis* oxalate groups and two bridging oxygen atoms. Titanium–oxygen bonds *trans* to these oxygens are longer (*ca.* 2.0 Å) than the others (*ca.* 1.8 Å). The Ti_4O_4 ring has alternate Ti—O bond lengths of 1.84 and 1.78 Å. Adamantyl derivatives of pyrogallol, β-resorcylic acid, and protocatechuic acid form the complexes $Ti(OH)_3$-(H_2L) and $[Ti(OH)_3(H_2L)_2]^-$ (H_3L = ligand) which have absorption peaks at 426, 330—360, and 380—420 nm, respectively. Additionally, the pyrogallol derivative forms $[Ti(OH)_3(H_3L)]^+$.[171, 172]

$TiCl_4$ has been reported to react with $OP(OBu)_3$ to give $(TiCl_2L_2)Cl_2$ and $(TiCl_2L_2)$-$(TiCl_6)$ [L = $OP(OBu)_3$] in which the $(BuO)_3PO$ is bidentate, binding through P—O and the O atom of the POBu moiety.[173] The adducts $[(BuO)_3P]_nTiCl_4$ (n = 1 or 2) are formed with $(BuO)_3P$; they are converted into $(BuO)BuP(O)OTiCl_2(OH)$ and

[168] D. W. Thompson, P. B. Barrett, J. P. Lefelholz, and G. A. Lock, *J. Coord. Chem.*, 1973, **3**, 119.
[169] V. K. Zolotukhin and O. M. Gnatyshin, *Zhur. neorg. Khim.*, 1973, **18**, 2761.
[170] G. M. H. Van de Velde, S. Harkema, and P. J. Gellings, *Inorg. Nuclear Chem., Letters*, 1973, **9**, 1169.
[171] G. Yu. Stepanova, S. Yu. Shnaiderman, and I. Ya. Bleikher, *Zhur. obshchei Khim.*, 1973, **43**, 1552.
[172] G. Yu. Stepanova and Ya. I. Bleikher, *Ukrain khim. Zhur*, 1973, **39**, 1176.
[173] T. N. Sumarokova, Yu. A. Nevskaya, and T. D. Ibraeva, *Izvest. Akad. Nauk Kaz. S.S.R., Ser. khim.*, 1973, **23**, 28 (*Chem. Abs.*, 1973, **79**, 111 274).

(BuO)BuP(O)OTiCl$_2$OP(O)Bu(OBu), respectively, on being heated to 130—140°C.[174] The standard enthalpy and entropy of adduct formation between TiBr$_4$ and POCl$_3$ have been determined as $\Delta H^\circ_{526} = 13.5 \pm 0.5$ kcal mol^{-1} and $\Delta S^\circ_{526} = 30.3 \pm 1$ e.u.[175] The adducts TiCl$_4$,L (L = O-methoxycarbonylphenyldiphenylarsine oxide or O-methoxycarbonylphenyldi-p-tolylarsine oxide) have been assigned six-co-ordinate structures involving co-ordination through the OMe oxygen atom of the ester group and the arsenyl oxygen on the basis of i.r. spectral data.[175a]

S-*Donor Ligands.* The full report of the preparation and characterization of the titanium(IV) dithiocarbamato-complexes Ti(S$_2$CNR$_2$)$_n$Cl$_{4-n}$ (n = 2, 3, or 4; R = Me, Pr^i, Bu^i, or, when n = 3, Et) has appeared.[176] Molecular weight, conductance, and i.r. data show that these complexes may be assigned co-ordination numbers 6, 7, and 8 when n = 2, 3, and 4, respectively. The C$\dddot{=}$N, C$\dddot{=}$S, and Ti—Cl stretching frequencies depend systematically on the co-ordination number. High dipole moments of ca. 9 D for Ti(S$_2$CNR$_2$)$_2$Cl$_2$ (R = Pr^i or Bu^i) in benzene solution indicate a *cis* configuration for these six-co-ordinate complexes. All the Ti(S$_2$CNR)$_n$Cl$_{4-n}$ complexes undergo a fast metal-centred rearrangement and the Ti(S$_2$CNPr$^i_2$)$_n$Cl$_{4-n}$ complexes also undergo isopropyl group exchange. A gear-like mechanism involving rotation about C—N single bonds was proposed for the latter process on the grounds that the activation parameters are virtually independent of co-ordination number. The seven-co-ordinate geometry suggested for the tris(dithiocarbamato) complexes has been confirmed by an X-ray diffraction study of Ti(S$_2$CNMe$_2$)$_3$Cl.[177] This has a somewhat distorted pentagonal-bipyramidal structure; the titanium and the sulphurs of the equatorial dithiocarbamate ligands are almost coplanar with Ti—S bond lengths of 2.473(3)—2.577(3) Å. The chlorine and the axial sulphur are displaced from the quasi fivefold axis by 10 and 5°, respectively, while the remaining sulphur is appreciably below the equatorial plane (0.61 Å) because of the restricted 'bite' (ca. 69°) of the dithiocarbamate ligand. There is considerable crowding in the equatorial plane where S···S interligand distances are as much as 0.5 Å less than the sum of the S atom van der Waals radii. The seven-co-ordinate geometry exemplified here appears to be preferred for complexes of the type M(chelate)$_3$X where X is a ligand which forms a relatively strong covalent M—X bond.

Differences between the heats of formation of titanium halide complexes with aliphatic mono- and di-sulphides (n-C$_3$H$_7$)$_x$TiX$_4$ (x = 1 or 2; X = Cl or Br) and Bu^{n}S(CH$_2$)$_3$SBu^{n}TiBr$_4$, in aliphatic and aromatic solvents have been attributed to weak molecular complex formation between the titanium sulphide adduct and the aromatic molecule.[178] Variations in the donor ability of the aromatic solvent did not produce any corresponding variation in ΔH_f.

N-*Donor Ligands.* The orange crystalline product of the reaction between TiCl$_4$ and N(SiMe$_3$)$_3$ in benzene which was originally thought to be a tetramer has been shown to be *catena*-di-μ-chloro-bis-μ-(trimethylsilylamino)-di[chlorotitanium(IV)] (25) by

[174] A. G. Akhmadullina, E. G. Yarkova, N. A. Mukmeneva, P. A. Kirpichinkov, A. A. Muratova, and A. N. Pudovik, *Zhur. Priklad Khim.,* 1973, **46**, 1291 (*Chem. Abs.,* 1973, **79**, 61 085).
[175] A. V. Suvorov and D. Sharipov, *Zhur. neorg. Khim.,* 1974, **19**, 572.
[175a] S. S. Sandhu and H. Singh, *J. Indian Chem. Soc.,* 1973, **50**, 373.
[176] A. N. Bhat, R. C. Fay, D. F. Lewis, and A. F. Lindmark, *Inorg. Chem.,* 1974, **13**, 886.
[177] D. F. Lewis and R. C. Fay, *J. Amer. Chem. Soc.,* 1974, **96**, 3843.
[178] I. P. Romm, E. N. Kharlamova, and E. N. Gur'yanova, *Zhur. obshchei Khim.,* 1973, **43**, 891.

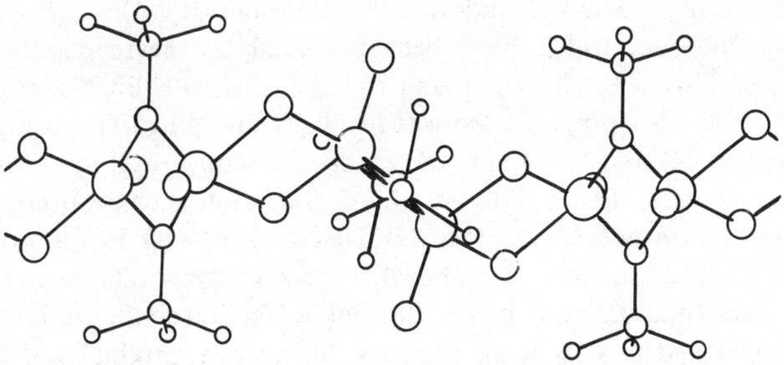

(25)

X-ray crystallography.[179] The structure consists of planar four-membered $(Ti-N)_2$ rings with planar geometry at N which are linked by Cl-bridges, giving five-co-ordinate Ti atoms in approximately trigonal-bipyramidal geometry. The Si atoms lie in the ring planes and thus there is the possibility of π-interaction between the $N(p)$, $Ti(d)$, and $Si(d)$ orbitals. Ti—N distances in (25) average 1.89 Å; this is substantially less than the Ti—N bond lengths in bis(dimethylamino)titanium difluoride tetramer (26). N.m.r., i.r., and electronic spectroscopic data indicate that octamethyl-cyclotetrasilazane (OMT) and hexamethylcyclotrisilazane (HMT) utilize only two of their available N-atom donors in the adducts $(TiCl_4)_2$,OMT (27) and $(TiCl_4)_2$HMT (28), apparently because of the strong preference of the early transition elements for

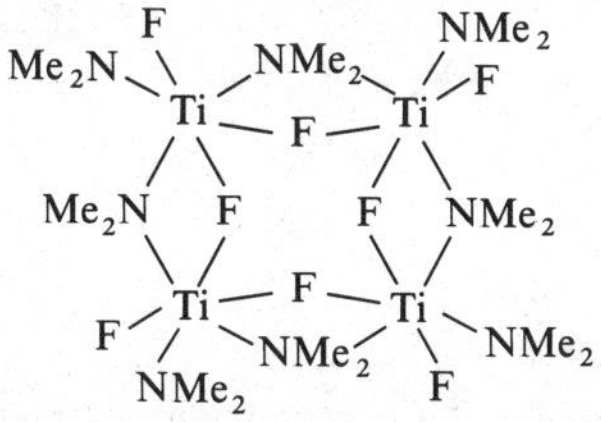

(26)

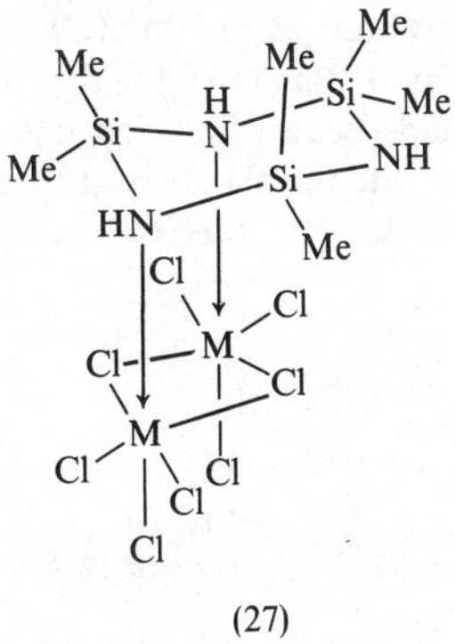

(27)

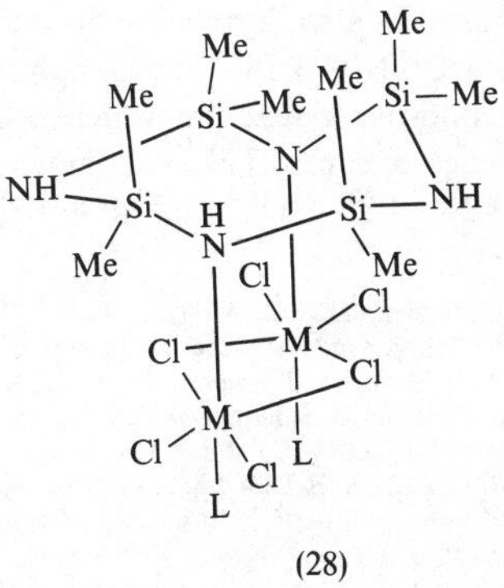

(28)

[179] N. W. Alcock, M. Pierce-Butler, and G. R. Willey, *J.C.S. Chem. Comm.*, 1974, 627.

hexaco-ordination.[180] Dialkylamidotitanium fluorides $(R_2N)TiF_3$ (R = Me and Et), $(R_2N)_2TiF_2$, and $(Et_2N)_3TiF$ have been prepared by the reaction of TiF_4 with Me_3SiNR_2, of TiF_4 with $(R_2N)_4Ti$, and of $(Et_2N)_4Ti$ with $PhPF_4$, respectively.[181] $(Me_2N)TiF_3$ was also prepared from $(Me_2N)_4Ti$ and PF_5. The compounds were characterized by ^1H and ^{19}F n.m.r. and i.r. spectroscopy. $(Et_2N)_3TiF$ is monomeric in solution, whereas $(Et_2N)_2TiF_2$ is oligomeric with a concentration-dependent degree of association. Me_2NTiF_3, $(Me_2N)_2TiF_2$, and $(Et_2N)TiF_3$ form co-ordination polymers. The difluoride has been shown to be a tetramer (26) by *X*-ray crystallography.[182] Each titanium atom has a distorted octahedral configuration and is linked to one terminal and two bridging F atoms and to one terminal and two bridging NMe_2 groups. The Ti—N distances are: 1.99 Å (terminal N), 2.14—2.19 Å (bridging N) and the Ti—F distances are 1.77 Å (terminal F), 2.00—2.06 Å (bridging F). Phthalonitrile reacts with titanium to form a metal–phthalocyanine dimer, for which a bridged structure was proposed.[183] I.r. spectroscopic data have also been cited as evidence that 2-aminothiazole (L) co-ordinates to titanium(IV) in $TiCl_4,2L$ only *via* the NH_2 nitrogen, whereas the N-acylated ligand uses its O donor atom.[184] The heats of neutralization of $TiCl_4$ with a series of bases have been determined as $-\Delta H$ = 47.6 (Et_3N), 40.0 (α-pic), 32.5 (py), and 28.9 ($PhNMe_2$) kcal mol^{-1}.[185] A study has been made of the thermal decomposition of the biguanide complexes of the first transition series metals including titanium.[186] Tetrakis(dimethylamino)titanium reacts with the metal carbonyls $M(CO)_x$ (M = Mo, W, Fe, or Ni) without CO evolution.[186a] N.m.r. and i.r. spectroscopic data are consistent with the carbenoid structure (29) for the product from $Fe(CO)_5$.

$$(Me_2N)_3Ti-O$$
$$C\!\!\!-\!\!\!-Fe(CO)_4$$
$$Me_2N$$

(29)

Mixed N- *and* O-*Donor Ligands.* TiO_2 reacts with N-methylsalicylhydroxamic acid (LH_2) in a 1:1 aqueous ethanol solution containing HCl to form a 1:2 complex; yellow $TiO(LH)_2$ separated from pH 6—7 solutions containing edta.[187] A 1:2 complex is also formed with the bis-(N-phenylhydroxamic acid), $PhN(OH)CO-(CH_2)_5CON(OH)Ph$.[188] The complexing of Methylthymol Blue (LH_6) with Ti^{IV} is a second-order reaction which becomes irreversible and pseudo first order in the presence of excess Ti^{IV}. The dependence of the reaction rate on pH showed that, at pH 1.10—1.85, TiO^{2+} and $TiO(OH)^+$ react with LH_4^{2-} to give $TiL + H_2O + nH^+$.[189]

[180] J. Hughes and G. L. Willey, *J. Amer. Chem. Soc.*, 1973, **95**, 8758.
[181] H. Buerger and K. Wiegel, *Z. anorg. Chem.*, 1973, **398**, 257.
[182] W. S. Sheldrick, *J. Fluorine Chem.*, 1974, **4**, 415.
[183] C. Gall and D. Simkin, *Canad. J. Spectroscopy*, 1973, **18**, 130.
[184] P. P. Singh and U. P. Shukla, *Austral. J. Chem.*, 1974, **27**, 1827.
[185] R. C. Paul, K. S. Dhindsa, S. C. Ahluvalia, and S. P. Narula, *Indian J. Chem.*, 1973, **11**, 277.
[186] P. V. Babykutty, P. Indrasenan, R. Anantaraman, and C. G. R. Nair, *Thermochimica Acta*, 1974, **8**, 271.
[186a] W. Petz, *J. Organometallic Chem.*, 1974, **72**, 369.
[187] E. A. Zhurakovskii, E. A. Ryabenko, and V. A. Reznichenko, *Zhur. neorg. Khim.*, 1974, **19**, 9.
[188] N. Ghosh, *J. Indian Chem. Soc.*, 1973, **50**, 415.
[189] T. B. Mal'kova and L. A. Fomina, *Zhur. neorg. Khim.*, 1974, **19**, 697.

Up to pH 2, edta (LH_4) forms $(TiOHL)^-$, and $(TiOL)^{2-}$ above this pH.[190] Ternary complex formation in the titanium–polyphenol–aminopolycarboxylic acid system has been studied.[191] Tiron forms ternary chelates with nitrilotriacetic acid, edta, and $[(HO_2CCH_2)_2NCH_2CH_2]_2NCH_2CO_2H$. Pyrocatechol, dibromogallic acid, and gallic acid form mixed complexes only with nitrilotriacetic acid. Chromotropic acid forms a complex[192] with Ti^{IV} but not with the Ti^{IV}–polyphenol system.

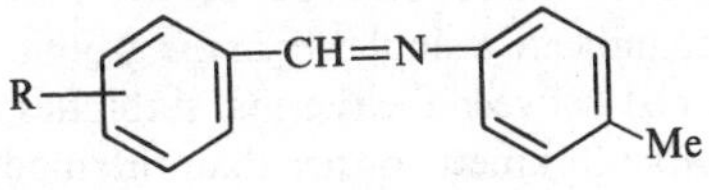

(30)

Studies of titanium(IV)–Schiff base interactions continue. Attempts[193] to establish a linear correlation between the heat of mixing (Q) of $TiCl_4$ and the Schiff bases (30); R = H, *p*-tolyl, *p*-Me_2CH, *p*-MeO, *p*-Cl, *p*-Br, *p*-NO_2, or *m*-NO_2) and the change in the C=N stretching frequency, $\Delta v(C=N)$, which occurs on co-ordination and is a measure of the metal–ligand bond strength were unsuccessful, although such a correlation exists for the $SnCl_4$ adducts. However, a linear correlation was established between Q and the Hammett substituent constants of R. Complex formation between $TiCl_4$ and the Schiff bases PhCH=NR (R = Ph or Bu) and $MeEtC=NC_8H_{17}$ has also been investigated.[194, 195] Structure (31) has been proposed for the orange complex derived from salicylaldehydesalicylhydrazone.[196] An interesting ligand exchange and isomerization occurs in the reaction between the *NN'*-ethylenebis(salicylideneimine)-cobalt(II) and $TiCl_4$.[197] Two chlorine atoms are transferred to cobalt while the phenolic oxygens become co-ordinated to the titanium as in (32). According to electronic spectral and magnetic data the Co^{II} ion has acquired tetrahedral co-ordination. $TiCl_4,2HL$ complexes (HL = Schiff base derived from salicylaldehyde and *o*-, *p*-, *m*-toluidine, *p*-anisidine, $EtNH_2$, $PrNH_2$, and $BuNH_2$) decompose thermally under oxygen in two stages. The intermediates in the first stage are oxidized in the second; no $TiCl_2L_2$ complexes were observed.[198] Vanillin thiosemicarbazone forms a $TiCl_4$ adduct.[199]

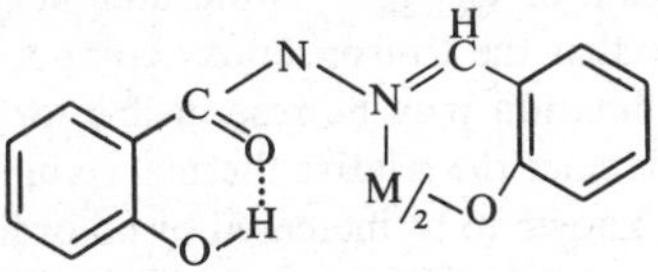

(31)

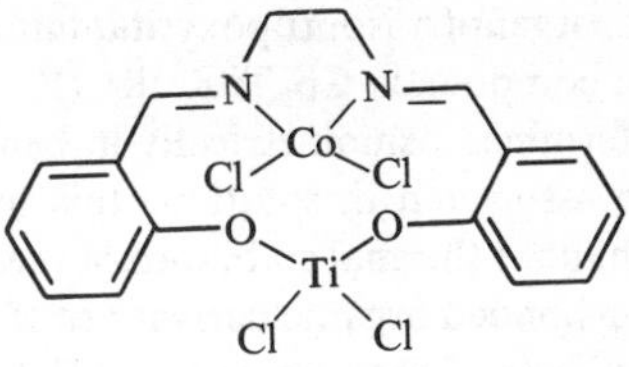

(32)

[190] B. Karadakov and P. Nenova, *God. Vissh. Khimikotekhnol. Inst. Sofia.*, 1968, **15**, 287 (*Chem. Abs.*, 1973, **79**, 24 067).

[191] S. Koch and G. Ackermann, *Z. Chem.*, 1974, **14**, 103

[192] J. Duda and J. Maslowska, *Roczniki Chem.*, 1973, **47**, 899 (*Chem. Abs.*, 1973, **79**, 11 259).

[193] V. A. Kogan, A. S. Egorov, and O. A. Osipov, *Zhur. neorg. Khim.*, 1973, **18**, 2091.

[194] A. S. Egorov, V. A. Kogan, and O. A. Osipov, *Zhur. obshchei Khim.*, 1973, **43**, 2722.

[195] V. A. Kogan, A. S. Egorov, O. A. Osipov, and V. P. Sokolov, *Zhur. obshchei Khim.*, 1973, **43**, 2719.

[196] K. K. Narang and A. A. Aggarwal, *Inorg. Chim. Acta*, 1974, **9**, 137.

[197] N. S. Biradar, V. B. Mahale, and V. H. Kulkarni. *Rev. Roumaine Chem.*, 1973. **18**. 809.

[198] D. Negoiu, A. Kriza, and M. Merehes, *An. Univ. Bucurest Chem.*, 1971, **20**, 25 (*Chem. Abs.*, 1973, **79**, 38 110).

[199] S. A. Kudritskaya, *Mater. Dokl. Nauch-Tekh. Konf., Kishinev. Politekh, Inst.*, 9th, 1972, 297 (*Chem. Abs.*, 1974, **81**, 25 264).

Organometallic Titanium(IV) Compounds.—As in earlier volumes only selected aspects of this chemistry are presented.

Titanium–Carbon σ-Bonded Complexes. The vertical ionization potentials of $Ti(CH_2SiMe_3)_4$ are consistently slightly higher than those of $Ti(CH_2CMe)_4$, in accord with the greater electron-releasing properties of the neopentyl group.[200] The closeness of the three upper photoelectron bands for the titanium complexes and their tin analogues was taken to indicate a similarity in the ground-state electronic properties, and the marked contrast between the thermal stabilities of the isoleptic titanium and tin alkyls is attributable to kinetic rather than thermodynamic effects. Methyl-group exchange between $TiMe_4$ and Me_3Al in ether–hexane solutions has been studied by ^2H n.m.r. spectroscopy.[201] At a molar ratio $TiMe_4:Et_2O > 1$, rapid exchange occurs between unsolvated metal alkyls. Tetramethyltitanium etherate reacts with Me_3Al in two steps, an initial fast complexation to $Me_3Ti(AlMe_4)$ which accounts for the transfer of a single Me group from the Ti to the Al atom and a subsequent slow reverse process which accounts for the ultimate random equilibrium distribution of the labelled Me groups amongst both metal atoms. Mixing the etherates of both Ti and Al methyl compounds leads to the appearance of an extra signal at τ 5.6 which is due to the Me resonance of the species $(TiMe_3)^+$ in the solvent separated ion-pair form of the complex $(TiMe_3)^+(AlMe_4)^-$. The orange product of the reaction between $TiCl_4,2py$ and $PhCH_2MgCl$ in Et_2O, previously thought to be the ether adduct of $Ti(CH_2Ph)_4$, has shown to be μ-oxobis[tribenzyltitanium(IV) by X-ray crystallography;[202] the same compound forms on refluxing $Ti(CH_2Ph)_4$ in Et_2O. The distance Ti—O (1.80 Å) is long compared with 1.73 Å in $(CpTiCl_2)_2O$ but short compared with 1.99 and 1.95 Å in rutile. Both titanium atoms are tetrahedrally co-ordinated, with Ti—CH_2 distances of 2.08 Å the same as in $(PhCH_2)_4Ti$, and with O—Ti—CH_2 angles of 113° and CH_2—Ti—CH_2 angles of 105°. The slight opening of the former angle and closing of the latter compared to 109° evidently being required to minimize repulsion between the benzyl groups.

Several lithium alkyltitanates $Li(TiR_n^1R_{5-n}).xSol$ have been prepared,[203] as have the organotri-isopropoxytitanium compounds $(OPr^i)_3TiR$ (R = Me or C_6F_5) and the compounds $Cp_2Ti(R^1)R^2$ (R^1 = Me, R^2 = Ph or C_6F_5).[204] Molecular weights determined osmometrically in benzene indicate that the tri-isopropoxy compounds are associated in solution; this internal co-ordination may be responsible for the enhanced thermal stabilities of these compounds since the relative thermal stabilities of σ-bonded organic derivatives of titanium are known to be increased by co-ordination with donor ligands. Cp_2TiCl_2 reacts with $LiCH_2SPh$ at $-20\,°C$ in THF–toluene to give yellow-orange $Cp_2Ti(CH_2SPh)_2$; this formed $Cp_2Ti(SPh)_2$ by elimination of CH_2.[205] Studies of selectivity and dynamic stereochemistry in titanocene complexes have centred round the preparation of complexes with asymmetric metal centres. $Cp_2TiPhCl$ treated with *o*-cresol or NaOMe gave $Cp_2Ti(Ph)OR$ (R = *o*-MeC_6H_4O or OMe). $CpCp'TiXY(X = Y = Cl)$ (33) and $PhMgCl$ or C_6F_5-MgBr gave the diastereoisomers of (33; X = Cl, Y = Ph or C_6F_5). Replacement of

[200] M. F. Lappert, J. B. Pedley, and G. Sharp, *J. Organometallic Chem.*, 1974, **66**, 271.
[201] L. S. Bresler, A. S. Khachaturov, and I. Ya. Poddubnyi, *J. Organometallic Chem.*, 1974, **64**, 335
[202] H. Stoeckli-Evans, *Helv. Chim. Acta*, 1974, **57**, 684.
[203] J. Mueller, H. Rau, P. Zdunneck, and K. H. Thiele, *Z. anorg. Chem.*, 1973, **401**, 113.
[204] M. D. Rausch and H. B. Gordon, *J. Organometallic Chem.*, 1974, **74**, 85.
[205] R. Taube and D. Steinborn, *J. Organometallic Chem.*, 1974, **65**, C9.

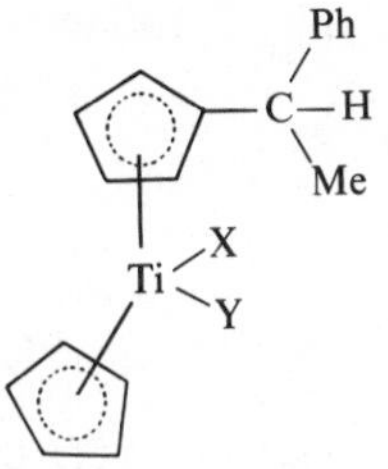

(33) 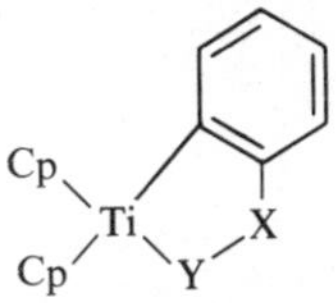   (34)

the Cl in (33) by OMe gave the diastereoisomers (33; X = OMe, Y = C_6F_5).[206] Insertion of CO into the Ti—C bond of $Cp_2TiR(X)$ (R = Me or CH_2Ph, X = Cl; R = Et, X = Cl) produces the corresponding acyl complex. The reaction is reversible in the case of R = CH_2Ph, CO being eliminated when the solution is saturated with N_2 or evacuated.[207] An O-sulphinate is formed from $Cp_2TiClEt$ and SO_2.[208] Analogues of (34; X = CO, Y = O) with X = Ph_2C or PhN=O have been prepared by treating benzophenone and PhNCO, respectively, with diphenyltitanocene.[209] Attempts to effect similar reactions using COS, CS_2, or PhCHO were unsuccessful. In a continuing series of studies on model compounds for the proposed intermediates in the Ziegler–Natta polymerization of olefins, attempts have been made to prepare titanium complexes containing a co-ordinated olefin.[210] The stabilizing effect of cyclic chelate formation on complexes containing both a donor atom and a C=C bond observed previously prompted an examination of complex formation between some titanium(IV) halides and derivatives of the CH_2=CHCH$_2$CH$_2$O group. The results are summarized below:

$$OMeCH_2CH_2CH=CH_2 + MeTiCl_3 \rightarrow \text{violet compound} \xrightarrow{-78\,^{\circ}C} Ti^{III} \text{ species}$$
$$OMeCH_2CH_2CH=CH_2 + TiCl_4 \rightarrow [TiCl_4(MeOCH_2CH_2CH=CH_2)]_2$$
$$Li(OCH_2CH_2CH=CH_2) + TiCl_4 \rightarrow TiCl_3(OCH_2CH_2CH=CH_2)$$
$$CH_2=CHCH_2CH_2OH + Cp_2TiCl_2 \xrightarrow{Et_3N} \underset{A}{Cp_2TiCR(OCH_2CH_2CH=CH_2)}$$
$$Cp_2TiCl(OCH_2CH_2CH=CH_2) + MeLi \rightarrow \underset{B}{Cp_2TiMe(OCH_2CH_2CH=CH_2)}$$

Treatment of A or B with $MeAlCl_2$ produced intractable highly air-sensitive mixtures. In no case was evidence for C=C co-ordination obtained. Deprotonation of one of the methyl groups of $LiMe_3SiNSiMe_3$ occurs in the formation of the elimination-stabilized alkyl (35).[211] Protonic acids react with bis-(π-allyl)nickel to yield products in which one allyl group is replaced by the anion of the acid; however, the reaction with Lewis acids is more complex. Interest in these systems stems from the ability of the solid product from the reaction between (π-C_3H_5)$_2$Ni and $TiCl_4$ to catalyse the *cis*-polymerization of butadiene. The cryoscopic titration curve of $TiCl_4$ with (π-C_3H_5)$_2$Ni in heptane solution showed a sharp break at a Ti:Ni ratio close to unity. I.r. data on

[206] J. C. Leblanc, C. Moise, and T. Bounthakna, *Compt. rend.*, 1974, **278**, C, 973.
[207] G. Fachinet and C. Floriana, *J. Organometallic Chem.*, 1974, **71**, C5.
[208] U.S.P. 3 728 365 (*Chem. Abs.*, 1973, **79**, 5462).
[209] I. S. Kolomnikov, T. S. Lobeeva, L. A. Pulova, and M. E. Vol'pin, *Izvest. Akad. Nauk S.S.S.R., Ser. khim.*, 1973, 1379.
[210] R. J. H. Clark and M. A. Coles, *J.C.S. Dalton*, 1974, 1462.
[211] C. R. Bennett and D. C. Bradley, *J.C.S. Chem. Comm.*, 1974, 29.

$$\begin{array}{c}
\text{H}_2 \\
\text{C} \\
\text{Cp} \diagdown \diagup \diagdown \\
\qquad \text{Ti} \qquad \text{SiMe}_2 \\
\text{Cp} \diagup \diagdown \diagup \\
\text{N} \\
| \\
\text{SiMe}_3
\end{array}$$

(35)

the solid product indicate the preservation of the π-allyl groups and furthermore that at least one of them is bonded to the titanium atom. On this basis, the complex was formulated as π-$C_3H_5NiCl\cdot C_3H_5$—$TiCl_3$.[212]

Cyclopentadienyl Complexes. X-Ray diffraction studies of several cyclopentadienyl complexes have been reported this year. The sulphur atoms and the h^5-C_5H_5 ligands of $Cp_2TiS_2(CH)_2$ adopt a distorted tetrahedral arrangement around the titanium atom with two staggered five-membered rings tilted at an angle of 51.2°. The TiS_2 plane is folded out of the $S_2C_2H_2$ plane at an angle of 46.1°.[213] Silylation of Cp_2TiCl_2 with SiH_3K yielded the olive green dinuclear $(Cp_2TiSiH_2)_2$. There is a substantial distortion from ideal tetrahedral geometry about the titanium atoms with Cp—Ti—Cp and Si—Ti—Si angles of 126 and 79.2° respectively, whereas the Ti—Si—Ti angle of 102.8° is close to the tetrahedral angle.[214] An X-ray diffraction study of $Cp_2Ti(NCO)_2$ has established that the NCO ligand is bound to the metal atom *via* its N atom.[215] The central distorted $ZnCl_4^{2-}$ tetrahedron of $Zn(Cl_2TiCp_2)_2,2C_6H_6$ is linked along the Cl—Cl edges to two distorted Cp_2TiCl_2 tetrahedra in such a way that their centres are nearly colinear. The two benzene molecules are to be regarded as solvent of crystallization.[216] A mass spectroscopic investigation of $Cp_2TiX_2(X = F, Cl, Br, or I)$, $CpTiX_3$, $MeC_5H_4TiX_3$, and $Me_5C_5TiX_3$ (X = Cl, Br, or OEt) has been carried out.[217] The intensity of the $(M - 2X)^+$ ion peak in the Cp_2TiX_2 series is greatest for X = I and least for X = F, in accord with previous observations that overlap between the lone-pair electrons of the halogen and the vacant Ti orbitals is greatest for the fluoro compounds. Other degradation pathways for Cp_2TiX_2 include loss of HX and the C_5H_5 ligand. Cp_2TiX_3 readily eliminates one halogen atom from the parent ion but the remaining two are much more difficult to remove. This compares with the result of treating $CpTiX_3$ with EtOH when only one halogen enters into the exchange reaction. Unlike the spectra of the unsubstituted Cp complexes, the spectra of $[Me_3C_5]TiX_3$ contain a very intense peak corresponding to $(M - HX)^+$. One of the methyl groups must be the source of the hydrogen leading to the formation of a fulvene type structure. Asymmetric titanocene dichlorides with two planes of chirality have been prepared; diastereoisomeric derivatives have been obtained by the replacement of one Cl by OPh.[218] Asymmetric decomposition of titanocene derivatives in optically active mandelic acid led to the optically active titanocenes [36; Y = OPh,

[212] E. V. Kristal'nyi, N. V. Kozlova, E. V. Zabolotskaya, A. R. Gantmakher, and B. A. Dolozoplosk, *Doklady Akad. Nauk S.S.S.R.*, 1973, **211**, 1122.
[213] A. Kutoglu, *Acta Cryst.*, 1973, **29B**, 2891.
[214] G. Henchen and E. Weiss, *Chem. Ber.*, 1973, **106**, 1747.
[215] A. A. Kossiakoff, *Diss. Abs. (B)*, 1973, **33**, 4744.
[216] C. G. Vauk, *J. Cryst. Mol. Structure*, 1973, **3**, 201.
[217] A. N. Nesmeyanov, Yu. S. Nekrasov, V. F. Sizoi, O. V. Nogina, and V. A. Dubovitsky, *J. Organometallic Chem.*, 1973, **61**, 225.
[218] A. Dormond, D. K. Tirouflet, and J. Tirouflet, *Compt. rend.*, 1974, **278**, C, 1207.

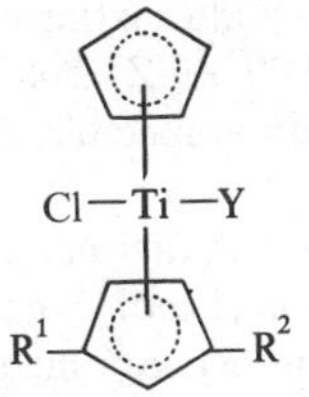

(36) 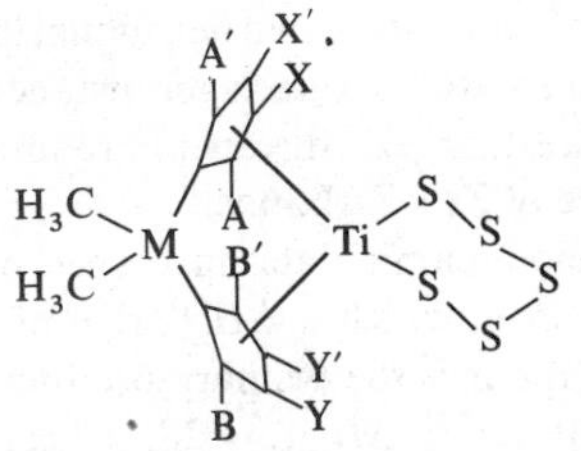   (37)

3,6-Me(Me$_2$CH)C$_6$H$_3$O, R^1 = H, R^2 = CHMe$_2$; Y = o-CH$_3$C$_6$H$_4$O, R^1 = H, R^2 = CHPh$_2$; Y = 3,6-Me(Me$_2$CH)C$_6$H$_3$O, 2,6-Me$_2$C$_6$H$_3$O; R^1 = Me, R^2 = CHMe$_2$; Y = Cl, R^1 = Me, R^2 = CHMe$_2$). These compounds are chiral only around the Ti atom.[219] The relative configuration of the two chiral moieties in the racemic form (m.p. 164°) of (h^5-3-MeC$_5$H$_3$CMe$_2$Ph) (h^5-C$_5$H$_5$)Ti(2,6-Me$_2$C$_6$H$_3$O)Cl has been determined by X-ray diffraction.[220]

Cp$_2$TiX$_2$ derivatives of the non-linear pseudohalides X = N(CN)$_2$, C(CN)$_3$, and ONC(CN)$_2$, have been prepared from Cp$_2$TiCl$_2$ and the appropriate silver salt in MeCN. The ONC(CN)$_2$ is co-ordinated *via* its O atom.[221] Complex (37) exhibits an AA'BB'XX'YY' pattern in its ^{1}H n.m.r. spectrum; owing to the asymmetry of the S$_5$ chelate ring the two rings of the corresponding dichloro-complex are equivalent.[222] Cp$_2$TiL$_2$(L = SeMe, SeCH$_2$Ph, SePh, SeC$_6$H$_4$Me-p, TePh, or TeC$_6$H$_4$Me-p) complexes have been prepared as part of an examination of the donor properties of the Group VI ligand atoms.[223] The Cp proton resonances indicated that the SeMe and SeCH$_2$C$_6$H$_5$ groups are more able to enter into titanium-selenium π-bond formation. Interfacial and solution polymerization techniques have been used in the preparation of titanium polyesters from the sodium salts of adipic, azelic, itaconic, thiosuccinic, terephthalic, and thiodiproprionic acids.[224] The i.r. spectra of Cp$_2$TiCl$_2$, (C$_5$D$_5$)$_2$TiCl$_2$, and Cp$_2$Ti(OCH$_2$Ph)$_2$ have been reported.[225]

2 Zirconium and Hafnium

Introduction.—A text describing the chemistry of hafnium has been published.[226] The structural,[3a, 227] synthetic,[5] and organometallic[4] chemistry of zirconium and hafnium have been reviewed, and the compilation of the chemistry of organozirconium compounds has been revised.[228] An account of the controversy following Urbain's claim, in 1911, to have isolated and identified element 72 has appeared.[229]

Binary Compounds and Related Species.—*Halides and Oxyhalides.* The i.r. spectra of matrix-isolated ZrF$_2$, ZrF$_3$, and ZrF$_4$ have been obtained.[230] The crystal structure

[219] A. Dormond, J. Tirouflet, and F. Moigue, *J. Organometallic chem.*, 1974, **69**, C7.
[220] C. Lecomte, Y. Duvansoy, and J. Protas, *J. Organometallic Chem.*, 1974, **73**, 67.
[221] K. Issleib, H. Kochler, and G. Wille, *Z. Chem.*, 1973, **13**, 347.
[222] H, Kopf and W. Kahl, *J. Organometallic Chem.*, 1974, **64**, C37.
[223] M. Sato and T. Yoshida, *J. Organometallic Chem.*, 1974, **67**, 395.
[224] C. E. Carraher, *Makromol. Chem.*, 1973, **166**, 31.
[225] N. N. Vyshinskii and T. I. Ermolaeva, *Trudy Khim. i khim. Tekhnol.*, 1973, 64 (*Chem. Abs.*, 1974, **80**, 101835).
[226] I. A. Sheka and F. F. Karlysheva, *Chem. Abs.*, 1973, **79**, 132497).
[227] T. E. Dermott, *Coord. Chem. Revs.*, 1973, **11**, 1.
[228] See ref. 5.
[229] R. T. Allsop, *Educ. Chem.*, 1973, **10**, 222.
[230] R. H. Hauge, J. L. Margrave, and J. H. Hastie, *High. Temp. Sci.*, 1973, **5**, 89.

of $ZrCl^{231}$ consists of two hexagonal layers of Cl atoms alternating with two hexagonal layers of Zr atoms. Interlayer distances are Cl–Cl 2.00, Zr–Zr 2.08, and Zr–Cl 2.40 Å. The nearest Zr—Zr distances are much smaller than in metallic Zr and indicate the existence of Zr—Zr bonds.

Thermodynamic data have been obtained[232] for the thermal disproportionation of ZrCl from which a $\Delta H_f^0(ZrCl)$ of $- 68.5 \pm 1.5$ kcal mol^{-1} has been calculated. $ZrCl_2$ vapour is the primary product of the reaction between metallic Zr and $ZrCl_4$ at $> 900\,^\circ C$;[233] $\Delta H_{subl} = 28.45$ kcal mol^{-1} and $\Delta S_{subl} = 21.8$ e.u. were calculated from the experimental data. Chemical and X-ray analysis[234] of $ZrCl_2$ at 500—618 °C showed that disproportionation in an equilibrium state proceeds continuously with a solid-phase composition change from $ZrCl_2$ to Zr. The low-temperature (230—310 °C) reduction of an $AlCl_3$–$ZrCl_4$ eutectic containing 76% $AlCl_3$ with metallic Zr provides a convenient route to $ZrCl_3$ free of Zr or Al.[235] A soluble blue species, formed as the liquid eutectic is raised to the reaction temperature, may be a Zr^{III} species. It appears that the crystal growth of the $ZrCl_3$ product results from decomposition of the blue species *via* a solvated Zr^{III} halide rather than *via* the gas phase, it is limited by the ability of the system to transport the soluble Zr species to the growth site and by a competing reaction which results in the deposition of a brown aluminium-containing Zr^{II} compound. $ZrCl_3$ prepared by the direct reduction of $ZrCl_4$ with Zr is contaminated with finely dispersed metallic Zr.[236] A standard heat of sublimation for $ZrCl_4$ of 27.06 ± 1 kcal mol^{-1} was obtained [237] from a mass spectrometric study of its vapour over the temperature range 100—175 °C. The force constants have been evaluated for ZrX_4 (X = F, Cl, Br, or I).[238] $ZrBr_3$ disproportionates in two stages.[239] The first, which occurs up to 500—550 K has the stoicheiometry

$$6ZrBr_3(s) \rightarrow 5ZrBr_{2.8}(s) + ZrBr_4(g)$$

The final residue from the second stage has Zr:Br ratios ranging from $\sim 1:1.4$ to $1:0.8$. It was suggested that this may be the monobromide phase formed according to:

$$\tfrac{5}{3}ZrBr_{2.8}(s) \rightarrow ZrBr_4(g) + \tfrac{2}{3}ZrBr(s)$$

Eleven different species were distinguished during an effusion equilibrium study of the non-stoicheiometric zirconium iodides.[240] Enthalpy and entropy changes for the disproportionation reactions leading to these species were evaluated as were the heats of formation of $ZrI_n(n = 4, 3.2, 3.17, 3, 2.82,$ or $2.5)$. A distorted tetrahedral co-ordination of the fluorines around hafnium in HfF_4 appears likely according to

[231] S. I. Troyanov. *Vestnik Moskov Univ., Khim.*, 1973, **14**, 369 (*Chem. Abs.*, 1973, **79**, 97963).
[232] V. I. Tsirel'nikov, Yu. S. Khodeev, and S. I. Troyanov, *Zhur. fiz. Khim.*, 1973, **47**, 1567 (*Chem. Abs.*, 1973, **79**, 108831).
[233] L. N. Shelest and E. K. Safronov, *Ref. Zhur. Khim.*, 1973, Abs. No. 12B714 (*Chem. Abs.*, 1974, **81**, 31075).
[234] L. N. Shelest, T. Z. Maiskaya, E. K. Safronov, and A. S. Mikhailova, *Ref. Zhur. Met.*, 1973, Abs. No. 5G322 (*Chem. Abs.*, 1974, **81**, 74922).
[235] E. M. Larsen, J. W. Moyer, F. Gil-Arnao, and M. J. Camp, *Inorg. Chem.*, 1974, **13**, 574.
[236] A. N. Naumchik and B. G. Vorotinova, *Ref. Zhur. Met.*, 1973, Abs. No. 5G330 (*Chem. Abs.*, 1974, **80**, 72523).
[237] Yu. S. Khodeev and V. I. Tsirel'nikov, *Vestnik Moskov. Univ., Khim.*, 1973, **14**, 611 (*Chem. Abs.*, 1974, **80**, 53602).
[238] N. K. Sanyal and L. Dixit, *Proc. Nat. Acad. Sci. India (A)*, 1972, **42**, 75 (*Chem. Abs.*, 1974, **80**, 65112).
[239] A. S. Normanton and R. A. Shelton, *J. Less-Common Metals*, 1973, **32**, 111.
[240] A. K. Baev and R. A. Shelton, *Zhur. fiz. Khim.*, 1973, **47**, 2382 (*Chem. Abs.*, 1974, **80**, 20154).

the results of a Mössbauer study.[241] $HfCl_2$ has been prepared[242] from Hf and $HfCl_4$ at 400 °C; its disproportionation over the range 673—713 K has been examined and $\Delta H_f^0(298\ K) = 132 \pm 2.6\ \text{kcal mol}^{-1}$ and $\Delta S° = 29.2 \pm 0.8$ e.u. Thermodynamic data have also been calculated for HfX, HfX_2 and HfX_3 (X = Cl, Br, or I) over the temperature range 298—2000 K.[243, 244]

Oxides and Chalcogenides. The equilibrium constants of the exchange reaction

$$M^1(g) + M^2O(g) = M^1O(g) + M^2(g)\ (M^1 = \text{Hf or Zr}, M^2 = \text{Th or Y})$$

have been determined by a mass spectrometric method[245] and the data used to calculate the standard free energies of formation of gaseous ZrO. Thermal methods do not appear to be suitable for the preparation of solid ZrO. A 15 h sintering of Zr and ZrO_2 at 1400 °C and 10^{-4} mmHg gave an α-Zr solid solution and monoclinic ZrO_2. Mixtures of Zr, ZrO_2, and MO (M = Fe, Co, Ni, Ti, V, or Nb) gave phases such as $(Zr_{0.95}M_{0.05})O$ and $(Zr_{0.8}M_{0.2})O$ whose diffraction patterns showed only weak lines attributable to ZrO.[246]

Neither ZrO_2 nor HfO_2 exhibits an i.r. deformation mode in the region of $1080\ \text{cm}^{-1}$; this has been attributed to an increase in ionic bonding in these compounds compared with the titanium oxides where the deformation mode is prominent.[35] Single- and poly-domain particles of tetragonal ZrO_2 have been obtained from the dehydration of ZrO_2, nH_2O.[247] The difference Raman spectrum of tetragonal ZrO_2 obtained by graphical subtraction of the spectrum of the monoclinic phase from the spectrum of a mixture of the two phases has all six Raman bands predicted by group theory;[248] it appears that an earlier report in which 12 bands were listed must also have concerned the two-phase mixture.

Single crystals of $Zr_{2.00}As_{2.80}Te_{0.92}$[249] and HfS_xSe_{2-x}[250] (x = 0, 0.5, 1, 1.5, or 2) have been grown and their lattice parameters determined. Non-stoichiometry in the $Zr(Se_xTe_{1-x})_2$ system has been studied[251] and the limits of the homogeneous domains of this system and of Zr_xSe_2, Zr_xTe_2, $HfSe_2$, and Hf_xTe_2[252] have been determined. X-Ray diffraction studies have shown $HfTe_5$ to be isostructural with $ZrTe_5$.[253]

Intercalation of alkali-metal atoms into zirconium disulphide has been reviewed.[254] Only alternate layers of vacant sites are occupied by alkali-metal atoms when ZrS_2 is treated with a dilute solution of alkali metal in liquid ammonia.[255] At higher con-

[241] N. A. Filatova, V. A. Kozhemyakin, and B. G. Korshunov, *Ref. Zhur. Khim.*, 1973, Abs. No. 12B715 (*Chem. Abs.*, 1974, **81**, 31 074).
[242] Lin Ching Lu, *Diss. Abs.* (*B*), 1973, **34**, 1689.
[243] V. I. Tsirel'nikov, *Zhur. fiz. Khim.*, 1973, **47**, 1608 (*Chem. Abs.*, 1973, **79**, 97 723).
[244] K. S. Krasnov, E. V. Morozov, N. V. Filippenke, and N. I. Giricheva, *Izvest. V.U.Z., Khim., khim. Tekhnol.*, 1973, **16**, 1500 (*Chem. Abs.*, 1974, **80**, 41 640).
[245] R. J. Ackermann and E. G. Rauh. *J. Chem. Phys.*, 1974, **60**, 2266.
[246] S. I. Alyamovskii, Yu. G. Zainulin, G. P. Shveikin, and P. V. Gel'd, *Izvest. Akad. Nauk S.S.S.R., neorg. Materialy*, 1973, **9**, 1837.
[247] T. Mitsuhashi, M. Ichihara, and U. Tatsuke, *J. Amer. Ceram. Soc.*, 1974, **57**, 91.
[248] D. P. C. Thackeray, *Spectrochim. Acta*, 1974, **30A**, 549.
[249] A. Masset and Y. Jeannin, *J. Solid State Chem.*, 1973, **7**, 124.
[250] H. P. B. Rimmington and A. A. Balchin, *J. Mater. Sci.*, 1974, **9**, 343.
[251] A. Gleizes and Y. Jeannin, *J. Less-Common Metals*, 1974, **34**, 165.
[252] A. Kjekshus and L. Brattas, *Acta Chem. Scand.*, 1973, **27**, 1290.
[253] S. Furuseth, L. Brattas, and A. Kjekshus, *Acta Chem. Scand.*, 1973, **27**, 2367.
[254] A. Leblanc-Soreau, M. Danot, L. Trichet, and J. Rouxel, *Mater. Res. Bull.*, 1972, **9**, 191.
[255] J. Cousseau, L. Trichet, and J. Rouxel, *Bull. Soc. chim. France*, 1973, 872.

centrations further occupation occurs to give $NaZrS_2$ or $KZrS_2$ with Na occupying only octahedral sites whilst K occupies both octahedral and prismatic sites.

Carbides, Silicides, and Related   Compounds. Thermodynamic data have been obtained[58] for Zr and Hf carbides. $ZrSi_2$ is more resistant to oxidation than $TiSi_2$,[63] it forms lower silicides and ZrO_2 at 600 °C and $ZrSiO_4$ at higher temperatures. At temperatures below 700 °C the oxidized layer of $HfSi_2$ contains HfSi, HfO_2, and α-quartz; above 770 °C there is an abrupt change, $HfSiO_4$ and an unknown phase, possibly $(HfO_2)_x(SiO_2)_y$, being produced. ZrVSi and HfVSi have a PbFCl type structure with the lattice parameters $a = 3.695$, $c = 7.212$ Å and $a = 3.662$, $c = 7.135$ Å, respectively.[256] The co-ordination of Si atoms in Zr_2Si, Zr_3Si, Zr_5Si_3, Zr_3Si_2, Zr_5Si_4, and α- and β-ZrSi has been studied.[257] The chemical reactivity of some zirconium germanides has been examined, reactivity to oxygen and acids increased in the series $ZrGe > ZrGe_2 > Zr_2Ge_3 > Zr_3Ge$.[64]

Nitrides. The hafnium nitride molecule HfN has been identified by molecular beam mass spectrometry of the vapour phase above $HfN_{1.0}$ at > 2800 K.[257a] The enthalpy of the reaction $HfN(g) = Hf(g) + 0.5N_2(g)$ was determined as $\Delta H°_0 = 60.5 \pm 30$ kJ mol^{-1}, and the dissociation energy of HfN(g) as $D°_0 = 531 \pm 30$ kJ mol^{-1}.

The formation of zirconium carbonitrides from carbon-deficient carbides has been studied.[72]

Borides. The preparation of ZrB_{12} has been reviewed[67,68] and an electrochemical method for its preparation has been patented.[69]

Zirconium(II), Zirconium(III), Hafnium(II), and Hafnium(III).—The white soluble dimer$[(\pi\text{-tetrahydroindenyl})_2ZrH_2]_2$ produced by the action of hydrogen gas on dimethyldi-π-indenylzirconium exhibits ^1H n.m.r. signals at $\delta - 1.96$ and 4.59.[258] The former is in the region expected for hydrido protons but the latter is at an unusually low field position. It has been tentatively assigned to a pair of bridging hydrogens by analogy with $(CpZrH_2,AlMe_3)_2$ which has a pair of bridging Zr—H—Zr groups and exhibits ^1H resonances in the range $\delta - 0.9$ to -3.0.

Steric effects appear to control the outcome of mixing benzene solutions of the three isomeric lutidines with solid zirconium trichloride.[259] 2,6-Lutidine showed no sign of reaction whereas 2,4- and 3,5-lutidine undergo two reactions simultaneously, the incorporation of ligand into the solid phase, and the entry of some Zr^{III} into solution. The solid adducts have the stoicheiometries $ZrCl_3,2(3,5$-lutidine) and $ZrCl_3,1.3(2,4$-lutidine). The products recovered from the solution phase have X-ray powder patterns characteristic of the $ZrCl_4,2L$ adducts, although they are contaminated by paramagnetic material, possibly Zr^{II} species formed *via* disproportionation of Zr^{III}.

Zirconium(IV) and Hafnium(IV).—*Halides and Oxyhalides.* Raman spectroscopic measurements[260] on $LiF–NaF–ZrF_4$ mixtures at 650 °C have identified a six-co-ordinate environment for zirconium in a 31–36–33 mol % $LiF–NaF–ZrF_4$ melt. The

[256] Ya. P. Yarmolyuk, J. Loub, and E. I. Gladyshevskii, *Dopov. Akad. Nauk Ukrain R.S.R., Ser. A.,* 1973, **35**, 943 (*Chem. Abs.,* 1974, **80**, 20 381).
[257] N. M. Zhaveronkov, *Khim. Metal. Splavov,* 1973, 19 (*Chem. Abs.,* 1973, **79**, 150 369).
[257a] J. Kohl and C. A. Stearus, *J. Phys. Chem.,* 1974, **78**, 273.
[258] H. Weigold, A. P. Bell, and R. I. Willing, *J. Organometallic Chem.,* 1974, **73**, C23.
[259] E. M. Larsen and T. E. Henzler, *Inorg. Chem.,* 1974, **13**, 581.
[260] L. M. Toth, A. S. Quist, and G. E. Boyd, *J. Phys. Chem.,* 1973, **77**, 1384.

occurrence of eight-, seven-, and five-co-ordinated zirconium in molten fluoride was inferred from frequency shifts accompanying changes in 'free' fluoride ion concentration caused by varying the mole % of ZrF_4. High temperature hydrolysis reactions occur between steam and MCl–$MZrF_6$ melts (M = Na or K) affording the oxyfluoride compounds M_2ZrOF_4.[129] Phase equilibria in the system CsF–HfF have been studied.[261] The system forms the compounds Cs_3HfF_7, $CsHfF_5$, and three polymorphic forms of Cs_2HfF_6. This last compound is formed by a peritectic reaction at 630 °C. Its homogeneity region at 630 °C is 24—31 % HfF_4. The lattice parameters of the complex fluorides RbM_4F_{21} (M = Zr or Hf), formed by heating Rb_2MF_6 with $(NH_4)_3MF_7$, have been determined.[262] Force constants have been evaluated for the MCl_6^{2-} species (M = Zr or Hf);[133,134] species containing the $ZrCl_6^{2-}$ ion have been identified in a phase diagram study of the ternary system $KCl–ZrCl_4–TiCl_3$.[83]

Monochloramine reacts with $ZrCl_4$ or $HfCl_4$ in the absence of air to give $(NH_4)_2$-(MCl_6) (M = Zr or Hf).[132] Salts of the $ZrCl_6^{2-}$ ion are also formed from $ZrOCl_2,8H_2O$ and N-benzylquinoline or NN-dimethyldibenzylamine.[263] X-Ray spectra have been recorded for the pyridinium salts L_2ZrCl_6, $L_2Zr(OEt)Cl_5$, L_2ZrOCl_4, and $LZrOCl_3$ (L = pyH) and the results used to calculate the effective charges on Zr and the ligands.[264] The behaviour of M_2ZrCl_6 (M = K, Rb, or Cs) when heated in a stream of O_2 has been studied.[265] A drop calorimetric method has been used to obtain the enthalpy of fusion, heat content, and heat capacity of Cs_2ZrCl_6.[266] Electron-diffraction studies of ZrI_4 and HfI_4 vapours at 250—270 °C have confirmed the tetrahedral structure for these molecules.[267] The metal–iodine distances are the same (2.660 ± 0.005 Å) and the I—I distances are 4.337 (Zr) and 4.341 Å (Hf).

O-*Donor Ligands.* Table 2 summarizes the results of some studies concerning zirconates, hafnates, and related mixed-oxide systems. The results of a detailed thermal and spectroscopic study[268] of $ZrOCl_2,8H_2O$ and $ZrOBr_2,8H_2O$ support the idea that these two isomorphous species are also isostructural; that is, the bromide must also contain the tetrameric cation $[Zr_4(OH)_8(H_2O)_{16}]^{8+}$. Two equivalents of water are lost from the hydrates at < 50 °C; this process is easily reversed and i.r. spectral data indicate that the eight-co-ordinate structure remains unchanged. The remaining one lattice water molecule per zirconium atom is lost along with one of the more weakly co-ordinated water molecules in the second dehydration stage. I.r. data on the $[Zr_4(OH)_8(H_2O)_{12}]X_8$ species indicate that the co-ordination geometry of the Zr^{IV} ions is similar to the distorted seven-co-ordinate square-base, trigonal-top prism structure of ZrO_2 (baddelyite). The third stage of dehydration involves loss of one equivalent of halide. The $Zr_4(OH)_8$ unit is also thought to be the core of the

[261] A. A. Kosorukov, V. Ya. Frenkel, Yu. M. Korenev, and A. V. Novoselova, *Zhur. neorg. Khim.*, 1973, **18**, 1938.
[262] Yu. M. Korenev and A. A. Kosorukov, *Vestnik. Moskov Univ., Khim.*, 1973, **14**, 426 (*Chem. Abs.*, 1973 **79**, 140 565).
[263] A. M. Kostrova and G. S. Markov, *Zhur. neorg. Khim.*, 1973, **18**, 1796.
[264] R. L. Barinskii, I. M. Kulikova, I. B. Barskaya, and G. M. Toptygina, *Izvest. Akad. Nauk S.S.S.R., Ser. fiz.*, 1974, **38**, 516.
[265] T. I. Beresnev, L. A. Denisova, and A. N. Ketov, *Zhur. neorg. Khim.*, 1974, **19**, 702.
[266] A. S. Kucharski and S. N. Flengas, *Canad. J. Chem.*, 1974, **52**, 946.
[267] E. Z. Zasorin, G. V. Girichev, V. P. Spiridonov, K. S. Krasnov, and V. I. Tsirel'nikov, *Izvest. V.U.Z., Khim. i khim. Tekhnol.*, 1973, **16**, 802 (*Chem. Abs.*, 1973, **79**, 7111).
[268] D. A. Powers and H. B. Gray, *Inorg. Chem.*, 1973, **12**, 2721.

Table 2   *Zirconates, hafnates, and related mixed-oxide systems*

Compound	Source	Properties reported	Ref.
Li_2O,ZrO_2	Li salts + ZrO_2	Rate of formation independent of $Li_2O:ZrO_2$ molar ratio	a
Li_2O,HfO_2 $2LiO_2,HfO_2$ $4LiO_2,HfO_2$	HfO_2–Li_2O (Li_2CO_3) at 800—940 °C		b
$SrZrO_3$	$SrCO_3$ + ZrO_2 1000—1200 °C	X Orthorhombic to cubic phase transition at 1300 °C	c
$BaZrO_3$	$BaCO_3$ + ZrO_2		d
$PbZrO_3$	—	X	e
$PbZrO_3$	$PbSO_4$ + $MZrO_3$ (M = Ca, Sr, or Ba)	X, t.d.	f
$Pb(Ti_yZr_{1-y})O_3$	—	Range of intrinsic non-stoicheio-metry determined at 1000 °C	g
$PbHfO_3$	1:1.269 HfO_2–$PbCO_3$ 1—6 h at 750—900 °C	X Identical with that previously reported	h
$ZrO_2.M_2O_3$ (M = La—Lu, Y)	Thermal dehydration of nitrates	—	i
$HfTiO_4$	HfO_2–TiO_2	X Temperatures of polymorphic transformation of the HfO_2-based solid solutions determined	j
$Zr(VO_4)_2,xH_2O$	—	t.d., X Crystallization begins at 450 °C	k
$Zr(MoO_4)_2$	ZrO_2 + MoO_3	Optical properties	m
$Hf(MoO_4)_2$	HfO_2 + MoO_3	Optical properties	
$ZrCoO_x$ (x = 0.35—1)	—	Magnetic properties	n
$ZrNiO_x$ (x = 0.2—1) Hf_3NiO		Magnetic properties	
$CdSn_xZr_{1-x}O_3$	—	X Perovskite structure	o
$ZrSiO_4$	ZrO_2 + SiO_2	X	p
$HfSiO_4$	+ MoO_3, 1100 °C 20 h		
ZrC_xO_y	$ZrC + ZrO_2 + Zr + C$, 1700 °C	X	q, r
ZrN_xO_y	$Zr; ZrO_2, N_2$	X 17 single-phase compounds obtained with NaCl-type cubic lattice.	s

(a) E. K. Belyaev and V. F. Annopolskii, *Zhur. neorg. Khim.*, 1974, **19**, 550; (b) E. K. Belyaev and V. F. Annopolskii, *Zhur. neorg. Khim.*, 1973, **18**, 1806; (c) A. E. Solov'eva, A. M. Gavrish, and E. I. Zoz, *Izvest. Akad. Nauk. S.S.S.R., neorg. Materialy*, 1974, **10**, 469; S. A. Kutolin, A. I. Vulikh, and A. E. Shammasova, *U.S.S.R.P.*, 392,001 (Cl.C.01f) (*Chem. Abs.*, 1974, **81**, 15245); (e) G. E. Shatalova, V. S. Filip'ev, L. M. Katsnel'son, and E. G. Fesenko, *Kristallografiya*, 1974, **19**, 412; (f) I. N. Belyaev, I. I. Belyaeva, and L. N. Aver'yanova, *Zhur. neorg. Khim.*, 1974, **19**, 224; (g) R. L. Holman and R. M. Fulrath, *J. Appl. Phys.*, 1973, **44**, 5227; (h) I. V. Vinarov, A. N. Grinberg, and L. Ya. Filatov, *Zhur. neorg. Khim.*, 1974, **19**, 1123; (i) N. I. Timofeeva, Z. I. Krainova, and V. N. Sakovich, *Izvest. Akad. Nauk S.S.S.R., neorg. Materialy*, 1973, **9**, 1756; (j) S. B. Kocheregin and A. K. Kuznetsov, *Izvest. Akad. Nauk S.S.S.R., neorg. Materialy*, 1974, **10**, 297; (k) V. A. Shichko and E. S. Borchinova, *Zhur. priklad Khim.*, 1974, **47**, 247 (*Chem. Abs.*, 1974, **80**, 115530); (l) R. I. Spinko, V. N. Lebedev, R. V. Kolesova, and E. G. Fensko, *Kristallografiya*, 1973, **18**, 1287; (m) N. V. Gul'ko, A. G. Karaulov, N. M. Taranukha, and A. M. Gavrish, *Izvest. Akad. Nauk S.S.S.R., neorg. Materialy*, 1973, **9**, 1766; (n) R. Sobezak, *Monatsh.*, 1973, **104**, 1526; (o) R. I. Spinko, V. N. Lebedev, R. V. Kolesova, and E. G. Fesenko,

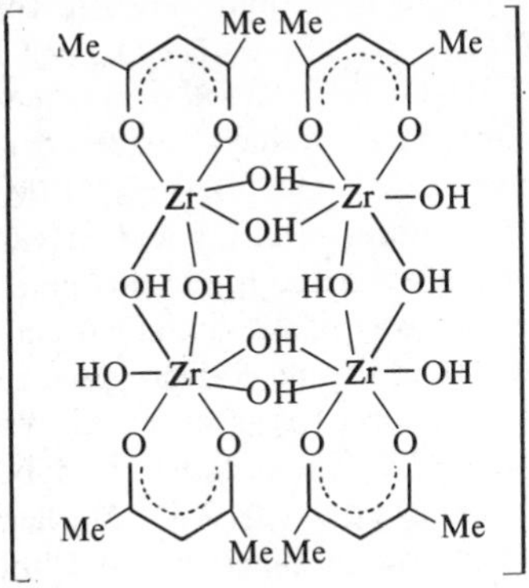

(38)

[Zr(acac)$_4$(OH)$_{11}$]$^+$ ion (38) which it is proposed[269] is formed in the reaction between Zr(acac)$_4$ and Na$_2$(3)-1,2-B$_9$C$_2$H in THF. The ^{11}B n.m.r. spectrum of the product is congruent with that of Me$_4$N$^+$[(3)-1,2-B$_9$C$_2$H$_{12}$]$^-$ and thus the product would appear to be [Zr$_4$(acac)$_4$(OH)$_{11}$][(3)-1,2-B$_9$C$_2$H$_{12}$] rather than the expected zirconium insertion product.

X-Ray diffraction studies have identified[270] a distorted pentagonal-bipyramidal environment about zirconium in Zr(OH)$_2$SO$_4$H$_2$O with Zr—O distances 2.10—2.19 Å; almost planar infinite [Zr(OH)$_2$]$_n^{2n+}$ chains are joined in one direction by sulphate groups, the layers so formed are held together only by hydrogen bonds and van der Waals forces. The [M(OH)$_2$]$_n^{2n+}$ unit is also present in Hf(OH)$_2$SO$_4$[271] and in Hf$_4$(OH)$_8$(CrO$_4$)$_4$,H$_2$O.[272]

Complex formation and the depolymerizing effect of NaCl are suggested as the causes of the increase in solubility of ZrO$_2$ in NaPO$_3$–NaCl melts from 1 mol % at zero NaCl content to 3 mol % at 40 mol % NaCl.[273] Phase transformations in the alumina–silica–zirconium dioxide[274] and zirconium dioxide–cerium dioxide systems[275] have been studied and the phase diagram of the system ZrO$_2$–YO$_{1.5}$ has been revised.[276] Isomorphous substitution of Zr atoms by Ca atoms accounts for the variation of 0—20 mol % Ca in Zr(OH)$_4$ as a function of Ca(OH)$_2$ concentration in

[269] A. R. Siedle, *J. Inorg. Nuclear Chem.*, 1973, **35**, 3429.
[270] M. Hansson, *Acta Chem. Scand.*, 1973, **27**, 2614.
[271] M. Hansson, *Acta Chem. Scand.*, 1973, **27**, 2455.
[272] M. Hansson and W. Mark, *Acta Chem. Scand.*, 1973, **27**, 3467.
[273] F. F. Grigorenko and V. M. Solomakha, *Ref. Zhur. Khim.*, 1972, Abs. No. 8B1582 (*Chem. Abs.*, 1974, **81**, 17368).
[274] F. A. Matveeva and Yu. P. Lyban, *Ref. Zhur. Khim.*, 1973, Abs. No. 11B781 (*Chem. Abs.*, 1974, **80**, 41610).
[275] A. Ono, *Mineral J.*, 1973, **7**, 228 (*Chem. Abs.*, 1974, **80**, 41759).
[276] K. K. Srivasta, R. N. Patil, C. B. Choudhar, K. V. G. Gokhale, and E. C. Subbarao, *Trans. J. Brit. Ceram. Soc.*, 1974, **73**, 85.

Kristallografiya, 1973, **18**, 1287; (*p*) A. N. Tsrigunov, N. N. Shevchenko, and E. N. Shumilin, *Vestik. Moskov. Univ. Khim.* 1973, **14**, 610 (*Chem. Abs.* 1974, **80**, 41700); (*q*) Yu. G. Zainulin and S. I. Alyamovskii, *Tr. Inst. Khim., Ural. Nauch. Tsentr. Akad. Nauk S.S.S.R.*, 1973, **25**, 33 (*Chem. Abs.*, 1974, **81**, 42490); (*r*) Yu. G. Zainulin, *ibid.*, p. 26 (*Chem. Abs.* 1974, **81**, 42488); (*s*) P. V. Gel'd, *ibid.*, p. 21 (*Chem. Abs.* 1974, **81**, 42489).

solution, contact time, temperature, and $Zr(OH)_4$ concentration.[277] Four independent phases $K_xZr_yGe_pO_q$, K_3ZrF_7, $K_2Ge_4O_9$, and ZrO_2 were obtained in the system $ZrO_2-GeO_2-KF-H_2O$; their lattice parameters have been determined.[278] Solid–solid phase equilibria in ZrO_2–metal oxide systems have been reviewed.[279] Zirconyl oxalates of the type $MZrO(C_2O_4)_2,xH_2O$ (M = Ca, $x = 4$;[280] M = Ba, $x = 5$[281]) decompose to mixtures of ZrO_2 and M_2CO_3 at *ca.* 400 °C. At higher temperatures solid-phase reactions lead to the zirconates $MZrO_3$.

Several aspects of the chemistry of the zirconium phosphate ion exchangers have béen investigated. Bands in the i.r. and Raman spectra of crystalline α-$Zr(HPO_4)_2$, H_2O have been assigned on the basis of an examination of the spectral changes caused by dehydration at 383 K. The lattice H_2O appears to be asymmetric with O—H bond lengths differing by up to 0.04 Å.[282] The heats of Na^+-H^+ exchange on samples of Zr phosphate of widely different crystallinites have been determined calorimetrically.[283] For a highly crystalline sample $\Delta H° = -6.90 \pm 0.10$ kcal mol^{-1} for the first 50% of exchange corresponding to the change from $Zr(HPO_4)_2,H_2O$ to $Zr(NaPO_4)(HPO_4),5H_2O$ and $\Delta H° = +6.45 \pm 0.15$ kcal mol^{-1} for the second half of exchange when the half-exchanged form is converted into $Zr(NaPO_4)_2,3H_2O$. The reaction is initially exothermic, passes through a broad maximum, and becomes progressively more endothermic. It was proposed that the second half of the exchange is endothermic because of electrostatic repulsions between two Na^+ ions in a single cavity and because of the dehydration step. The gradual change in enthalpy was explained on the basis of an irregular structure for the exchanger with cavities of different sizes, the largest being occupied first. These proposals are supported by the results of studies on the variation of Na^+-H^+ ion exchange and unit cell dimensions in highly crystalline α-zirconium phosphates.[284] Hysteresis loops occur in the forward and reverse Li^+-K^+ ion-exchange isotherms of zirconium phosphate both in aqueous solution[285] and in KNO_3-$LiNO_3$ melts at 300 °C.[286] These are attributed to the formation of different Li–K phase percentages in the two opposite processes for the aqueous system and to the formation of an immiscible $ZrLiK(PO_4)_2$ phase with an interlayer spacing of 7.6 Å in the forward reaction without a corresponding change in the reverse process for the fused-salt system. The half Na^+ ion-exchanged Zr phosphate has been found to be very much more effective than the H^+ form for the uptake of transition-metal ions; the same is true for the alkaline-earth metal ions and it is suggested that there is a common exchange mechanism.[287] Zirconium

[277] V. V. Sakharov, A. P. Fedosov, S. S. Korovin, L. F. Karmanova, and I. A. Apraksin, *Zhur. neorg. Khim.*, 1974, **19**, 13.
[278] N. A. Nosrev, L. N. Dem'vanets, V. V. Ilyukhin, and N. V. Belov, *Kristallografiva*, 1974, **19**, 432 (*Chem. Abs.*, 1974, **81**, 30 305).
[279] M. Zaharescu, M. Mavrodin-Tarabic, and I. Onaca, *Rev. Roumaine Chim.*, 1974, **24**, 888.
[280] I. Murase and C. Kawashima, *Seikei Daigaku Kogakubu Kogaku Hokoku*, (*Chem. Abs.*, 1974, **80**, 16 975).
[281] A. I. Sheinkinan, N. V. Yunusova, V. P. Osachov, and A. A. Kondrashenkov, *Zhur. neorg. Khim.*, 1974, **19**, 380.
[282] S. E. Horsley, D. V. Nowell, and D. T. Stewart, *Spectrochim. Acta*, 1974, **30A**, 535.
[283] A. Clearfield and L. Kullberg, *J. Phys. Chem.*, 1974, **78**, 152.
[284] A. Clearfield, L. Kullberg, and A. Oskarsso, *J. Phys. Chem.*, 1974, **78**, 1150.
[285] G. Alberti, U. Costantine, S. Allulli, M. A. Massucci, and N. Tamassini, *J. Inorg. Nuclear Chem.*, 1974, **36**, 653.
[286] G. Alberti, U. Costantine, S. Allulli, M. A. Massucci, and N. Tamassini, *J. Inorg. Nuclear Chem.*, 1974, **36**, 661.
[287a] S. Allulli, A. La Ginestra, M. A. Massucci, M. Pelliccioni, and N. Tamassini, *Inorg. Nuclear Chem. Letters*, 1974, **10**, 337.

molybdovanadates with Zr:V:Mo ratios of $1.68:1.00:0.088$; $3.33:1.00:0.285$ and $6.76:1.00:0.815$ are weakly acidic polyfunctional cation exchangers with a preference for Li^+ over Na^+. The weak acidity arises from $H_2VO_4^-$ units bound to the zirconyl groups of the zirconium molybdovanadate framework.[287a] The basic crystalline zirconium tungstate, $Zr_2W_2O_7(OH)_2(H_2O)_2$, has been shown by X-ray diffraction[287b] to be isostructural with basic zirconium molybdate. Chains of H-bonds involving the oxo-hydroxo- and aquo-groups play a secondary binding role.

A new sulphate–zirconate phase has been observed during studies of the ZrO_2–H_2SO_4–Na_2SO_4–H_2O system.[288] A thermogravimetric and X-ray diffraction study of zirconium and hafnium disulphates $M(SO_4)_2,4H_2O$ obtained by different methods has shown[289] that regardless of the nature of the initial compound, water is lost mainly *via* the formation of the monohydrate and then of the anhydrous disulphate. There is also evidence for the formation of $ZrOSO_4$ and $2ZrO_2,SO_3$ *via* loss of SO_3. The behaviour of $Hf(SO_4)_2$ in aqueous solution has been examined.[290] $Zr_2(OH)O_{0.5}$-$(SO_4)_3(H_2O)_6$ has been prepared during a study of the effects of bridging oxo-groups on the stability of zirconium sulphates prepared[291] from oxosulphate solutions.

Breaks in the conductometric titration curves for the aqueous $ZrOCl_2$–K_3-$[MoO(OH)(CN)_4]$ system occur at molar ratio $3:2$ with the potassium salt as titrant and at $1:1$ and $3:2$ with $ZrOCl_2$ as titrant corresponding to the formation of $KZrO[MoO(OH)(CN)_4]$ and $(ZrO)_3[MoO(OH)(CN)_4]_2$.[292] The e.p.r. spectra of frozen solutions of Zr and Hf peroxo complexes have been analysed.[147] The increase in conductivity with decreasing concentration of solutions of ZrO_2L_2 (L = quinoline N-oxide) has been interpreted[293] in terms of the equilibrium:

$$ZrOCl_2L_2 + 2MeOH \rightleftharpoons [ZrO(OMe)_2L_2] + 2HCl$$

Extrapolation to zero concentration gives a conductance value very close to that reported previously for HCl in MeOH.

In view of the high solubility of $ZrCl_4$ and $HfCl_4$ in $MeNO_2$, MeCN, DMF, and DMSO, heats of solvation of the tetrahalides in these solvents have been determined by a calorimetric method as the differences between the heats of solution of the tetrachloride and its solvate;[294] the heats of solvation fall in the series

$$DMSO > DMF > MeCN > MeNO_2 \text{ and } Hf > Zr$$

These results explain previous observations of the greater stability of zirconium thiocyanate and selenocyanate complexes compared with their hafnium analogues, and the greater stability of zirconium and hafnium complexes in MeCN compared with DMF in terms of competition between the ligand and solvent molecules for co-ordination sites on the metal. Zirconium alkoxides have been prepared from $ZrCl_4$ and aliphatic alcohols[158] but with salicylaldehyde a Meerwein–Ponndorf

[287b] R. G. Safina, N. E. Denisova, and E. S. Boichinova, *Zhur. priklad Khim.*, 1973, **46**, 2432 (*Chem. Abs.*, 1974, **80**, 52641).
[288] L. G. Nekhamkin, L. G. Laube, V. K. Kozlova, L. M. Zaitsev, P. K. Eroshin, I. A. Kukushkina, and A. P. Pokhodenko, *Ref. Zhur. Khim.*, 1973, Abs. No. 12B829 (*Chem. Abs.*, 1974, **80**, 20000).
[289] Ya. G. Barenkova, L. G. Ashaev, and A. E. Potsov, *Zhur. neorg. Khim.*, 1973, **18**, 131.
[290] M. M. Grodneva, D. L. Motov, and R. F. Ohrimenko. *Zhur. neorg. Khim.*, 1973, **18**, 1450.
[291] T. Z. Maiskaya, L. G. Nekhamkin, and L. M. Zaitsev, *Zhur. neorg. Khim.*, 1974, **19**, 951.
[292] Kabir-Ud-Din, A. A. Khan, and M. Aijaz Beg, *J. Inorg. Nuclear Chem.*, 1974, **36**, 468.
[293] A. K. Majumdar and R. G. Bhattacharyya, *J. Inorg. Nuclear Chem.*, 1973, **35**, 4296.
[294] A. M. Golub, T. P. Lishko, and P. F. Lozovskaya, *Zhur. neorg. Khim.*, 1974, **19**, 23.

reaction occurs[159] contrary to a recent report of salicylaldehydato complex formation. Cp_2MCl_2 (M = Zr or Hf) give aqueous solutions containing the Cp_2M^{2+} ion which have been used in the preparation of metal-containing polyesters by interfacial and solution techniques.[224] Extensive hydrolysis occurs in solutions of the Hf compound aged for >15 min. Acylates of zirconium and hafnium have been detected, during the thermal decomposition of various MCl_4,L^1 and $MCl_4,2L^2$ adducts.[160]

Electrophoresis and˙ ion-exchange experiments have established the existence of neutral 1:1 complexes of Zr^{IV} and ˌHf^{IV} with glycolic acid,[295] and of Zr^{IV} with trihydroxyglutaric acid.[296] The stability constants of these complexes, the related cationic complexes $[M(OH_2)(OH)CH_2COO]^+$ (M = Zr or Hf), the Zr^{IV} and Hf^{IV} phenylacetic acid complexes,[297] and Zr^{IV} complexes with trihydroxyglutaric[296] and tartaric acids[298-300] have been reported. Thermodynamic parameters for the formation of $[Zr(OH)_3A^-]$ (AH_2 = tartaric acid) are $\Delta G° = -8.28\,kcal\,mol^{-1}$, $\Delta H° = -10.3\,kcal\,mol^{-1}$ and $\Delta S° = 6.79\,e.u.$[298] According to their i.r. spectra,[301] the hydroxyglutarates $M_4(C_5H_4O_6)_3(OH)_{10}.8H_2O$ (M = Zr^{IV}, or Hf^{IV}) have one bidentate and one unidentate carboxylate group. Partial decomposition of the organic ligand occurred at 120—300°C leading to the formation of $M_4O_6(C_5H_4O_6)$-$(OH)_2$. The thermogravimetric analysis of the Zr^{IV} and Hf^{IV} basic succinates $M_4L_5(OH)_6$ indicates loss of water up to *ca.* 220°C and then loss of ligand up to *ca.* 400°C, the MOL complexes being formed at this temperature.[302] $ZrOCl_2$ forms the 1:1 complexes $Zr(OH)_3L$ (LH = 3- or 5-nitrosalicylic acid, 3,5-dinitrosalicylic acid, or 5-bromosalicylic acid).[303]

Both unidentate and bidentate co-ordination of the halogenoacetate groups $O_2CCH_nX_{3-n}$ (n = 1 or 2; X = Cl or Br) are indicated by the i.r. spectra of the bis(cyclopentadienyl)Mbis(halogenoacetates) (M = Zr^{IV} or Hf^{IV}).[304] Bands in the spectra of the M^{IV} trihalogenoacetate complexes indicate the likely involvement of only unidentate co-ordination. The existence of both bidentate and unidentate co-ordination in the Cp_2MX_2 series raises the question of the co-ordination geometry of the metal atom, which is commonly distorted tetrahedral; presumably in the halogenoacetate series five-co-ordination is involved, and it is interesting that the four carboxylate bands are retained in the i.r. spectrum of $Cp_2Zr(O_2CCH_2Cl)_2$ in benzene solution.

Preparations of the zirconium chelates $Zr(acac)_4$ and $Zr(Pr^i)_2(acac)_2$ have been patented.[305, 306] Mass spectrometry has been applied to the analysis of volatile Zr

[295] M. S. Popov, *Poverkh. Yavleniya Adsorbtsiya, Koord. Vzaimodeistvie,* 1972, 30 (*Chem. Abs.,* 1973, **79**, 118972).
[296] V. K. Zolotukhin, O. M. Gnatyshin, and V. V. Gaiduk, *Ukrain khim., Zhur.,* 1973, **39**, 1099 (*Chem. Abs.,* 1974, **80**, 41542).
[297] S. S. Kalinina, Z. N. Prozorovskaya, L. N. Komissarova, L. I. Yuranova, and V. I. Spitsyn, *Zhur. neorg. Khim.,* 1973, **18**, 1470.
[298] Ts. B. Konunova and L. S. Kachkar, *Zhur. neorg. Khim.,* 1973, **18**, 1527.
[299] L. S. Kachkar, *Ref. Zhur. Khim.,* 1973, Abs. No. 2V107 (*Chem. Abs.,* 1973, **79**, 84013).
[300] V. K. Zolotukhin and O. M. Gnatyshin, *Ukrain khim. Zhur.,* 1973, **39**, 1243 (*Chem. Abs.,* 1974, **81**, 4838).
[301] L. A. Purik and V. V. Serebrennikov, *Tr. Tomsk. Gos. Univ.,* 1973, **240**, 115 (*Chem. Abs.,* 1974, **80**, 70253).
[302] L. A. Tsurik and V. V. Serebrennikov, *Tr. Tomsk. Univ.,* 1973, **240**, 124 (*Chem. Abs.,* 1974, **80**, 140719).
[303] C. S. Pande and G. N. Misra, *Indian J. Chem.,* 1973, **11**, 292.
[304] E. M. Brainina, E. V. Bryukhova, B. V. Lokshin, and N. S. Alimov, *Izvest. Akad. Nauk S.S.S.R., Ser. khim.,* 1973, 891.
[305] Jap. P. 7304771 (*Chem. Abs.,* 1973, **79**, 21119).
[306] Y. Takaoka, Jap. P. 7399122.

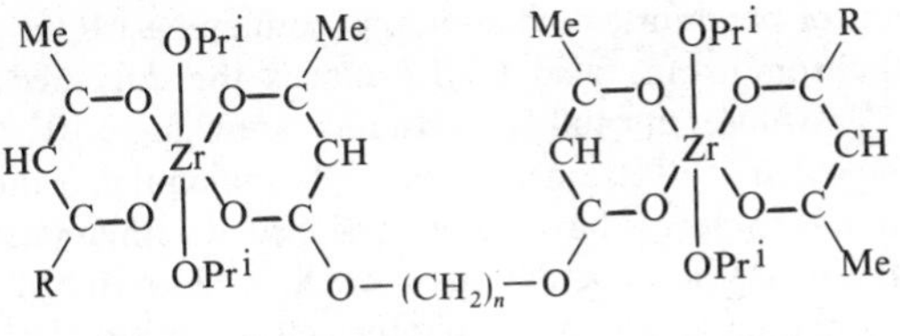

(39)

and Hf diketonates derived from geological and lunar samples; the method has been used to estimate the Zr:Hf ratios in these samples.[307] Replacement of one Pr^iO group per zirconium atom of the alkylenebis(acetoacetato)hexa-alkoxydizirconium complexes (39) by acac leads to six-co-ordinate complexes, but further substitution could not be achieved.[308] Heating $CpHfR_2^1Cl$ (R^1 = acac or dbm) with β-diketones (R^2H) in the presence of Et_3N gave the mixed diketonate complexes $CpHfR_2^1R^2$.[309] These complexes are monomeric and contain bidentate diketonate ligands. Evidence has been presented[310] which favours a twisting mechanism rather than a bond-rupture mechanism for the metal-centred rearrangements of zirconium β-diketonate complexes. The activation parameters for interchange of the diketonate rings in $CpZr(acac)_2X$ (X = Cl, Br, or I) are independent of the nature of X. $CpZr(dik)_3$ (dik = $Bu^tCOCHCOBu^t$ or acac) undergoes a low-temperature interchange of alkyl groups on ligands spanning the edges of the pentagonal-bipyramidal structure (40)

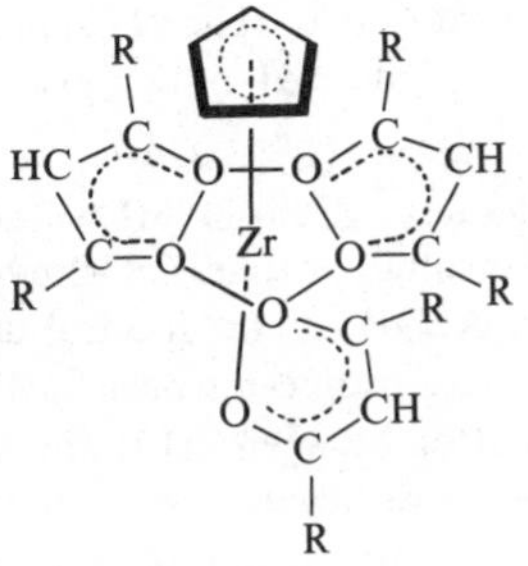

(40)

and a higher temperature process in which the equatorial ligands and the unique ligand spanning the axial position and an equatorial position are interchanged. Neither process is influenced by the addition of diketone ligand. Rates of ligand interchange are similar for $CpZr(dik)_2X$ and $CpZr(dik)_3$, but are larger by a factor of *ca.* 10^6 for the cationic $CpZr(dik)_2^+$ species. Replacing Cp by Cl causes an increase in rate. These observations are inconsistent with a Zr—O bond-breaking process but can be accommodated by a mechanism involving twisting of the ligands. Lineshape analysis of the n.m.r. spectra indicates a combination of facial and digonal twists.

[307] J. J. Leary, *Diss. Abs. (B)*, 1974, **34**, 4253.
[308] U. B. Saxena, A. K. Rai, and R. C. Mehrotra, *Indian J. Chem.*, 1973, **11**, 178.
[309] M. Kh. Minacheva, E. M. Brainina, and L. I. Terekhova, *Izvest. Akad. Nauk. S.S.S.R., Ser. khim.*, 1973, 2359.
[310] T. J. Pinnavaia, J. J. Howe, and R. E. Teets, *Inorg. Chem.*, 1974, **13**, 1074.

U.v. irradiation of perchloric acid solutions containing $MOCl_2$ (M = Zr or Hf) and 2,7-dichlorochromotropic acid (H_4L) affords the dark red complex anions $[M(OH)_2(HL)_2]^{4-}$ whose instability constants are 2.83×10^{-14} (M = Zr) and 2.30×10^{-13} (M = Hf).[311] Tetrameric structures analogous to those of the corresponding glycinato-complexes have been assigned to the α-alanine complexes $M(OH)_2A_2X_23H_2O$ (AH = α-alanine, M = Zr, X = Cl or Br; M = Hf, X = Cl or NO_3) on the basis of their i.r. and X-ray photoelectron spectra.[312,313] Co-ordination through the carboxylate group only of the amino-acid is also indicated for the complexes $Zr_3Cr_2(SO_4)_3$(amino-acid)(OH)$_{12}$,$9H_2O$ which are being studied as models for the binding of tanning mixtures containing chromium and zirconium to proteins.[314]

S-*Donor Ligands.* The complexes Cp_2ZrL_2 (L = SePh, SeC_6H_4Me-*p*, or TlPh) have been prepared.[223] Oligomeric zirconium polythioethers $HO[Zr(Cp)_2—S(CH_2)_nS—(Cp)_2Zr]_nOH$ have been obtained[315] from the copolymerization of Cp_2ZrCl_2 and aliphatic dithiols in an aqueous medium. Preliminary X-ray diffraction data for the tris(benzenedithiolato)zirconium dianion[316] indicate a geometry intermediate between the trigonal prism and antiprismatic limits although it is closer to the latter. The M—S bond lengths 2.544 Å are the longest for the isoelectronic series $M(S_2C_6H_4)_3^{n-}$ (M = Mo, n = 0; M = Nb, n = 1; M = Zr, n = 2). Eight-co-ordinate chelates exhibit a high degree of stereochemical non-rigidity in solution, and tetrakis[N-methyl-N-(pentafluorophenyl)(dithiocarbamato)]zirconium(IV) is no exception.[317] Its Me—N ^{13}C resonance remains a singlet down to −130 °C. The ^{19}F signals begin to broaden at <0 °C. The ^1H n.m.r. spectrum of $(EtNCS_2)_xZr$-$(S_2CNPr)_y$ (x = 1—3, y = 1—3) showed only signals due to one ethyl and one propyl group.

N-*Donor Ligands.* The adducts $ZrCl_4,2L$ (L = 2,6-lutidine, 2-chloropyridine, 4-aminopyridine, 2-amino-5-bromopyridine, 2,2'-bipyridyl, and 4,4'-bipyridyl) are formed from $ZrCl_4$ and L in AcOH.[318] I.r. spectral data indicate co-ordination of both the amino-group and the pyridine nitrogen in the 2-amino-5-bromopyridine adduct. Heats of adduct formation between $ZrCl_4$ and a series of pyridine derivatives have been found to become more negative with increasing pK_a of the amine.[319] Ammonolysis of ZrF_4 in a current of ammonia at 580 °C affords the zirconium nitride fluorides ZrN_xF_{4-3x} ($0.906 \leqslant x \leqslant 0.936$) which are stable to air and moisture and are hardly attacked even by concentrated acid or alkali.[320] The structure of $ZrN_{0.906}F_{1.28}$ is of the fluorite type. ZrF_4,L, $ZrF_4,2L$, $HfF_4,2L$ and $HfF_4,3L$ (L = N_2H_4) have been prepared from the appropriate metal salt and anhydrous N_2H_4 and their thermogravimetric and spectral data reported.[321]

[311] J. Duda and J. Maslowska, *Roczniki. Chem.*, 1973, **47**, 1337 (*Chem. Abs.*, 1974, **80**, 20097).
[312] G. S. Kharitonova and L. N. Pankratova, *Zhur. neorg. Khim.*, 1974, **19**, 570.
[313] G. S. Kharitonova, V. I. Nefedov, L. N. Pankratova, and V. Pershin, *Zhur. neorg. Khim.*, 1974, **19**, 860.
[314] M. H. Davis and J. G. Scroggie, *Austral. J. Chem.*, 1974, **27**, 279.
[315] C. E. Carraher and R. J. Nordin, *J. Appl. Polymer Sci.*, 1974, **18**, 53.
[316] M. J. Bennett, M. Cowie, J. L. Martin, and J. Takats, *J. Amer. Chem. Soc.*, 1973, **95**, 7504.
[317] E. L. Muetterties, *Inorg. Chem.*, 1974, **13**, 1011.
[318] Ts. B. Konunova and M. F. Frunze, *Ref. Zhur. Khim.*, 1973, Abs. No. 2V106 (*Chem. Abs.*, 1973, **79**, 61056).
[319] Ts. B. Konunova and M. F. Frunze, *Zhur. neorg. Khim.*, 1973, **18**, 1800.
[320] W. Jung and R. Juza, *Z. anorg. Chem.*, 1973, **399**, 129.
[321] P. Glavic and J. Slivnik, *Inst. Josef. Stefan, IJS Reports*, 1972, R-613 (*Chem. Abs.*, 1973, **79**, 12982).

Analytical data and the close similarity between the i.r. spectra of phenylhydrazine hydrochloride and complexes derived from $ZrCl_4$ and phenyl- or o-substituted phenyl-hydrazines support the view that these complexes should be formulated as $(RH)_2(ZrCl_{6-x}R_x),mH_2O$.[322]

Haematin acts as a sensitive reagent for hafnium.[323] It forms a red-brown 2:1 chelate which is soluble at pH 2.0, unlike complexes of haematin with other metals; the formation constant of the chelate at this pH is 1.6×10^6. I.r., electronic, and X-ray diffraction studies indicate nitrogen bonding of the ambidentate NCS,NCSe, and NCO ligands in the complexes $M_2^I(M^2L_6)$ (M^1 = K, M^2 = Zr or Hf, L = NCS or NCSe; M^1 = Et_4N, M^2 = Zr or Hf, L = NCO)[324] and Cp_2ZrL_2 (L = NCO or NCS);[215] the $[M(NCO)_6]^{2-}$ and $[M(NCS)_6]^{2-}$ complexes are isostructural with their niobium and tantalum analogues. The cyclopentadienyl pseudohalide complexes Cp_2ZrL_2 [L = $N(CN)_2$ or $C(CN)_3$] have been prepared from Cp_2ZrBr_2 and the appropriate silver salt in MeCN or THF.[221]

Mixed N- and O-Donor Ligands. A three-stage mechanism involving preliminary partial dissociation of the Xylenol Orange complex MX_2(M = Zr or Hf) to MX and X, subsequent addition of the anion (Y^{n-}) of the attacking chelating species to form MXY^{m-} and final expulsion of the second X to give MY^{x-} and X has been proposed on the basis of stopped-flow kinetic investigations of the substitution of Xylenol Orange by some aminopolycarboxylic acids.[325,326] A further example of eight-co-ordination in zirconium compounds has been established by an X-ray diffraction study[327] of Zr(edta),4H$_2$O. Co-ordination of both the NH$_2$ and OH groups of ethanolamine (LH) in $ZrCl_2,L_2$ is indicated by changes in the i.r. spectrum of the ligand which occur on complex formation.[328]

Several zirconium Schiff base complexes have been prepared: $Zr(Pr^iO)_4$,Pr^iOH reacts with terdentate and quadridentate Schiff bases to give yellow or brown $Zr(Pr^iO)_2L$ and ZrL_2.[329] Structures containing eight-co-ordinate zirconium have been suggested for the ZrL_2 complexes. Monofunctional bidentate Schiff bases (HL = benzylidene-2-hydroxyethylamine, benzylidene-2-hydroxypropylamine, salicylideneaniline, β-hydroxynaphthylideneaniline, and benzylidene-o-aminophenol) give $Zr(Pr^iO)_{4-x}L_x$, x = 1—4.[330] Bands at 3280, 2650, and 1070—1050 attributable to $v(NH)$, $v(SH)$, and $v(C=S)$ in the O,N,S Schiff bases (41) are absent from the

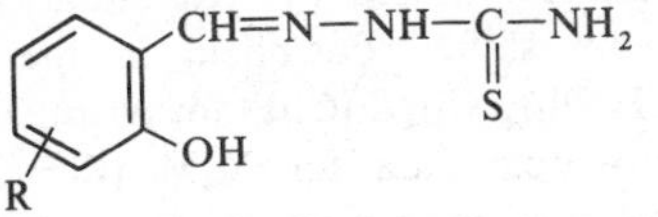

(41)   R = H, 3-Me, 4-Me, 5-Me, 5-Cl

[322] V. R. Timofeeva and M. S. Novakovskii, *Zhur. obshchei Khim.*, 1973, **43**, 1417.

[323] Y. P. Dick and N. Vesely, *Analyt. Chem. Acta*, 1973, **66**, 377

[324] G. Galliart, *Diss. Abs.*, 1974, **34**, 3152.

[325] R. V. Suchkova, L. I. Budarin, and K. B. Yatsimirskii, *Teor. i eksp. Khim.*, 1973, **9**, 614 (*Chem. Abs.*, 1974, **81**, 41 262).

[326] R. V. Suchkova, L. I. Budarin, and K.B. Yatsimirskii, *Zhur. neorg. Khim.*, 1973, **18**, 2748.

[327] A. I. Pozhidaev, T. N. Polynova, M. A. Brai-Koshits, and L. I. Martynenko, *Zhur. strukt. Khim.*, 1973, **14**, 947.

[328] E. I. Toma, *Ref. Zhur. Khim.*, 1973, Abs. No. 2V105 (*Chem. Abs.*, 1973, **79**, 61 058).

[329] S. R. Gupta and J. P. Tandon, *Monatsh.*, 1973, **104**, 1283.

[330] S. R. Gupta and J. P. Tandon, *Bull. Acad. polon. Sci.*, Ser. Sci. chim., 1973, **21**, 911 (*Chem. Abs.*, 1974, **80**, 55 432).

spectra of their zirconyl complexes.[331] These complexes are non-electrolytes in DMF and 1:1 electrolytes in DMSO[332] and have been assigned six-co-ordinate structures, whereas the i.r. spectral data for $ZrOCl_2(RN=CHC_6H_4OH\text{-}o)_2$ (R = Ph, $p\text{-MeC}_6H_4$, $p\text{-ClC}_6H_4$, or $p\text{-MeOC}_6H_4$) indicate N-only co-ordination of the Schiff base and hence five-co-ordinate structures.[333]

Organometallic Zirconium(IV) and Hafnium(IV) Compounds.—σ-Bonded Complexes. Above 20 °C, the ^{1}H n.m.r. spectrum of a 3:2 mixture of $(PhCH_2)_4Hf$ and 3,5-lutidine (L) shows a single methylene resonance due to L exchange between $(PhCH_2)_4Hf$ and $(PhCH_2)_4Hf,L$.[334] The results of linewidth measurements are consistent with a dissociative mechanism for the exchange, and this is supported by the large positive entropy of activation. Whereas there is no detectable dissociation of $(PhCH_2)_4Hf,L$, dissociation of $(PhCH_2)_4Zr,L$ is extensive.[334, 335] Yellow-brown $PhZrCl_3,3THF$ is formed in the reaction between $ZrCl_4$ and $PhMgCl$ at 0 °C in toluene.[336] It is very air- and moisture-sensitive and neither analytical nor molecular weight data could be obtained. However, several adducts with donor ligands have been isolated. Prolonged reaction of $PhZrCl_3,3THF$ in neat MeCN afforded $PhZrCl_3,2MeCN$ which in turn gave $PhZrCl_3,L$ (L = 2,2'-bipyridyl, 4,4'-bipyridyl, and diphos) with bidentate donors. The co-ordination compounds of $PhZrCl_3$ are kinetically more inert than their alkyl analogues being more thermally stable and failing to undergo insertion reactions with e.g. SO_2 and PhNCO. It was suggested that this stability may arise from steric hindrance by the bulky phenyl group or from the lack of a low-energy β-elimination pathway for decomposition or by a combination of the two. In contrast to the $PhZrCl_3$ adducts, $(PhCH_2)_4Zr$ readily undergoes insertion reactions with SO_2, PhNCO, and MeNCS yielding $Zr(CH_2Ph)(SO_2CH_2Ph)_3$, $Zr[NPhC(=O)\text{-}CH_2Ph]_4$ and $Zr[MeNC(=S)CH_2Ph]_4$, respectively.[337] Bands at 900 and 1040 cm^{-1} in the i.r. spectrum of the SO_2 product indicate that it is an O-sulphinate.

Despite its poor Lewis acceptor properties compared with $ZrCl_4$, $Zr(CH_2Ph)_4$ also forms a series of adducts[337] $Zr(CH_2Ph)_4,L$ (L = 2,2-bipyridyl, 4,4-bipyridyl, py, 2-methylpyridine, thioxan, THF, 2-cyanopyridine, pyrazine, or p-phenylenediamine) and $Zr(CH_2Ph)_4,2L$ (L = 2-methylpyridine, quinoline, 3-cyanopyridine, 4-cyano-pyridine, or pyridazine). Solid products could not be obtained with PhCN, MeCN, py or dioxan, in accord with previous observations that the benzylmethylene resonances of $Zr(CH_2Ph)_4$ are not shifted by weak donors. Surprisingly, the benzyl-methylene resonances are shifted downfield in the adducts compared with $Zr(CH_2Ph)_4$. Like $Zr(CH_2Ph)_4$, $Cp_2Zr(Cl)Et$ forms an O-sulphinate with SO_2.[208]

Photoelectron spectroscopic data for R_4M (R = Me_3CCH_2 or Me_3SiCH_2; M = Zr or Hf) indicate a constancy of central atom parameters in keeping with the known similarities between Zr and Hf.[200] The first example of a zirconium-containing metallofluorene, $Cp_2Zr(C_{12}F_8)$ (42) has been obtained[338] from the reaction between Cp_2ZrCl_2 and 2,2'-dilithio-octafluorobiphenyl. It is more susceptible to hydrolysis

[331] N. S. Biradar and A. L. Locker, *J. Inorg. Nuclear Chem.*, 1974, **36**, 1915.
[332] N. S. Biradar and A. L. Locker, *Indian J. Chem.*, 1973, **11**, 833.
[333] N. S. Biradar, A. L. Locker, and V. H. Kulkarni, *Rev. Roumaine Chim.*, 1974, **19**, 45.
[334] J. J. Felten and W. P. Anderson, *Inorg. Chem.*, 1973, **12**, 2334.
[335] J. D. Felden, *Diss. Abs. (B)*, 1974, **34**, 4864.
[336] J. F. Clarke, G. W. A. Fowles and D. A. Rice. *J. Organometallic Chem.*, 1974, **76**, 349.
[337] J. F. Clarke, G. W. A. Fowles, and D. A. Rice, *J. Organometallic Chem.*, 1974, **74**, 417.
[338] S. A. Gardner, H. B. Gordon and M. D. Rausch, *J. Organometallic Chem.*, 1974, **60**, 179.

than its hafnium analogue but less so than its σ-bonded relative $Cp_2Zr(C_6F_5)_2$. Over a period of months even under nitrogen (42) slowly releases octafluorobiphenyl.

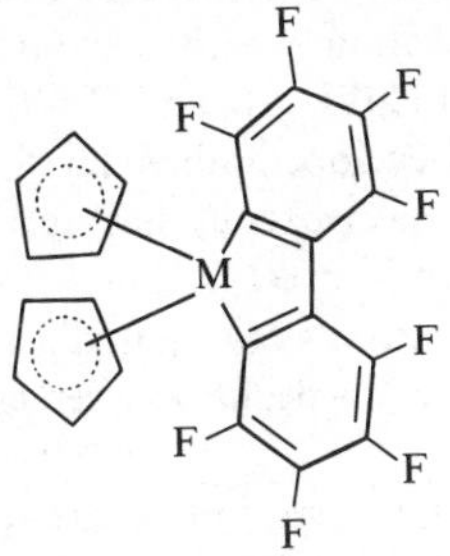

(42)

3 Vanadium

Introduction.—The structural[339] and organometallic chemistry[340] of vanadium has been reviewed and the compilation of the inorganic chemistry of vanadium has been revised.[341]

Carbonyl and other Low Oxidation State Compounds.—Tris-(2,2'-bipyridyl)vanadium(0) solutions in DMF exhibit three one-electron reduction waves corresponding to $VL_3 \rightarrow VL_3^-$ (L = 2,2'-bipyridyl) $VL_3^- \rightarrow VL_3^{2-}$ and $VL_3^{2-} \rightarrow VL_3^{3-}$, and two oxidation waves corresponding to $VL_3 \rightarrow VL_3^+$ and $VL_3^+ \rightarrow VL_3^{2+}$ whose half-wave potentials differ by only 0.10 V.[342] The difference for the isoelectronic chromium complexes is 0.49 V. An explanation has been offered for the small difference between the vanadium potentials in terms of the possible diamagnetism of the $d^4VL_3^+$ species arising from a substantial splitting of the t_{2g} orbitals.

The spin-only value, 2.80 ± 0.17 BM, for the magnetic moment of the V^I cation of V(mesitylene)$^+$ cannot be reconciled with a ligand-field model, even allowing for very substantial distortions from $C_{\infty v}$ symmetry.[10]

A normal-co-ordinate analysis of the vibrational spectrum of bis(cyclo-octatetraene)vanadium has been carried out on the basis of a structure with C_{8v} symmetry.[7] Only weak skeletal vibrations are observed; these absorption bands are absent from the spectrum of $(C_8H_8)^{2-}K_2^+$ but they are strong in the Th and U complexes, indicating that the vanadium complex is more ionically bonded. Core-level binding energies have been used[8] to estimate the oxidation state of vanadium in Cp_2VCl (+1.9) Cp_2V (+0.7) and $CpVC_7H_7$ (+0.8) on the assumption that vanadium is V^0 in the metal and V^{4+} in VO_2. This assumption leads to a slight over-estimation of the charge on the metal atom. Interestingly, Cp_2V and $CpVC_7H_7$ show approximately equal metal oxidation states, in good agreement with the results of MO calculations which gave oxidation states of +0.609 and +0.648, respectively. The He^I photoelectron spectra of $(\pi\text{-}C_6H_3Me_3)_2V$ and $CpVC_7H_7$ have been assigned[9] using a simple MO model.

339 M. B. Hursthouse, ref. 4a, p. 451.
340 P. C. Wailes, *J. Organometallic Chem.*, 1974, **75**, 325.
341 Gmelin's Handbook of Inorganic Chemistry System No. 48, 49, 50, Index 8th edn.
342 T. Saji and A. Shigeru, *Chem. Letters*, 1974, 203.

Mass spectroscopic evidence has been presented[343] to support the proposal that the new compound $CpV(\pi\text{-}C_6H_6)$ is formed along with Cp_2V and $(\pi\text{-}C_6H_6)_2V$ in the reaction between $CpVCl_3$, cyclohexa-1,3-diene, and Pr^iMgBr. Dehydrogenation of the olefin also occurs in the reaction where cyclo-octa-1,3,5-triene replaces the cyclohexadiene; here, a mixture of $CpVC_8H_9$ and $CpVC_8H_8$ is formed which can be converted into pure olive-brown cyclopentadienylcyclo-octatrienylvanadium(0) by catalytic dehydrogenation. Mass spectroscopy has also been used[344] to obtain an average dissociation energy for Cp_2V of 91 kcal mol^{-1} and to analyse the product mixture from the Friedel–Crafts synthesis of bis(ethylbenzene)vanadium.[345] A major feature of the spectrum is six intense ion-peaks between m/e 235 and m/e 375 differing by m/e 28.

The composition and i.r. and n.m.r. spectra of the red complex $\eta\text{-}C_3H_5V(CO)_5$ from allyl chloride and $Na[V(CO)_6]$ are in accordance with a π-bonded C_3H_5 moiety;[346] analogous complexes with substituted allyl groups can be obtained by the same route or by the photochemical addition of $HV(CO)_6$ to substituted butadienes. U.v. irradiation of a solution of $[Na(diglyme)_2]$ $[V(CO)_6]$ in diglyme in the presence of PF_3 results in the complete exchange of CO for PF_3,[347] the first time this has been achieved with a carbonylmetalate(-I) species. Treatment of the $[V(PF_3)_6]^-$ salt with phosphoric acid affords $HV(PF_3)_6$ which is obtained as pale yellow crystals by sublimation. Irradiation of solutions containing equimolar amounts of the $[V(CO)_6]^-$ anion and phosphorus ligand (L) gives $[V(CO)_5L]^-$ in high yield.[348] Attempts have been made to correlate the ^{51}V n.m.r. chemical shifts and the carbonyl stretching frequencies of these complexes with ligand σ- and π-bonding capacities for a wide variety of phosphorus and arsenic ligands.

Binary Compounds and Related Species.—*Halides and Oxyhalides.* A study of the reduction of VF_5 by hydrogen in the gas phase has shown[349] that a quantitative yield of crystalline vanadium can only be obtained by using a two- to five-fold excess of reductant. The enthalpy of formation of $VF_5(g)$ has been determined[350] as $\Delta H_f^\circ(VF(g), 298.15\,K) = -343.25 \pm 0.20$ kcal mol^{-1}. An analysis[351] of the hot band structure for the ν_7 fundamental of VF_5 leads to a calculated barrier to intramolecular ligand exchange of 1.8 kcal mol^{-1}. A d-orbital ordering of $d_{z^2} < d_{xy} < d_{yz} < d_{x^2-y^2} < d_{xz}$ has been obtained[352] for the pseudotetrahedral molecule VF_2Cl_2, and the intensities of its electronic spectral bands have been interpreted in terms of d–p mixing and covalency effects. Vaporization thermodynamic data for $VOCl_3$ and VCl_4 have been determined[353] using a quartz tensimeter. The heats of evaporation

[343] J. Muller and W. Goll, *J. Organometallic Chem.*, 1974, **71**, 257.
[344] P. E. Gaivoronskii and N. V. Larin, *Tr. Khim. i Khim. Tekhnol.*, 1973, 74 (*Chem. Abs.*, 1974, **80**, 101 432).
[345] G. G. Devyatykh, P. E. Gaivoronskii, N. V. Larin, V. A. Umilin, and V. K. Vanchagova, *Zhur. obshchei Khim.*, 1974, **44**, 590.
[346] M. Schneider and E. Weiss, *J. Organometallic Chem.*, 1974, **73**, C7.
[347] T. Kruck and H. Hempel, *Angew. Chem.*, 1974, **86**, 233.
[347a] F. Calderazzo, G. Fachinetti, and C. Floriani, *J. Amer. Chem. Soc.*, 1974, **96**, 3695.
[348] D. Rehder and J. Schmidt, *J. Inorg. Nuclear Chem.*, 1974, **36**, 333.
[349] V. V. Mikulenok, E. G. Rakov, and B. N. Sudarikov, *Trudy Moskov Khim. Tekhnol. Inst.*, 1972, **71**, 83 (*Chem. Abs.*, 1974, **80**, 113 233).
[350] K. Johnson and W. N. Hubbard, *J. Chem. Thermodyn.*, 1974, **6**, 59.
[351] R. R. Holmes, L. S. Couch, and C. J. Hora, *J.C.S. Chem. Comm.*, 1974, 175.
[352] D. K. Johnson, H. J. Stoklosa, and J. R. Wasson, *J. Inorg. Nuclear Chem.*, 1974, **36**, 525.
[353] N. V. Galitskii and G. N. Lebedev, *Ref. Zhur. Met.*, 1973. Abs. No. 2G256 (*Chem. Abs.*, 1973, **79**, 129 415).

are $\Delta H_v^{\circ}(298 \text{ K})$ 9.2 $\pm$ 0.1 and 10.6 $\pm$ 1 kcal mol^{-1}; the entropy changes $\Delta S_v^{\circ}(298 \text{ K})$ 23.9 $\pm$ 0.2 and 26.6 $\pm$ 0.2 e.u.; and the boiling points at 760 mmHg = 126.4 $\pm$ 0.8 and 149.5 $\pm$ 0.9°C.

Oxides. X-Ray and ESCA spectroscopy have been used to study the electronic structure of VO.[30] Oxidation to V_2O_5(s) in a high-temperature microcalorimeter has been used[354] to determine the enthalpies of formation of V_2O_3, V_3O_5, V_2O_4, and V_6O_{13}. Enthalpy data have also been obtained independently for V_2O_3 and V_2O_5.[355] The effect of substitution by V^{III} ions into Ti_2O_3 is to increase metal–metal distances across shared octahedral faces from 2.579 Å in Ti_2O_3 to 2.658 Å in $(Ti_{0.900}V_{0.100})_2O_3$, while decreasing metal–metal distances across the shared octahedral edge from 2.997 to 2.968 Å. Metal–oxygen distances exhibit only very small changes.[37] The valence vibrational bands of MO (M = Ti or V) occur in the same position indicating that the bond strengths are very similar.[34] However, the intensity of the band in $VO_{1\pm x}$ weakens more slowly with increasing x than that of $TiO_{1\pm x}$; this has been attributed to the greater degree of polarity in the V—O bond. The oxides V_2O_3, V_4O_7, V_5O_9, V_6O_{11}, and V_7O_{13} were prepared from V_2O_3–VO_2 mixtures at 800 °C. They give e.p.r. signals, due to V^{IV} in the V_xO_{2x-1} lattice, whose g-values, linewidths, and relative intensities have been tabulated.[356]

Nitrides. At 500—1500 °C hexagonal V_3N is in thermodynamic equilibrium with α-VN solid solution;[357] ageing of V–N alloys containing 8 and 10 atom % N at 500 °C results in the formation of an intermediate cubic phase which has a lattice parameter 3.105 Å some 2 % larger than that of the corresponding α-solid solution.

Borides. Vanadium borides have been reviewed.[67,68] The results of a ^{51}V n.m.r. study[358] of VB are consistent with the boron acting as an acceptor of electrons from the metallic α-band. However, there are surprising differences between the n.m.r. parameters of VB and those of the isostructural NbB which indicate that the details of the band structure of even very closely related borides are probably quite dissimilar.

Silicides. There has been a review of the electronic structures of vanadium silicides.[60] Thermodynamic data have been obtained for VSi_2 prepared by the diffusion saturation of vanadium with silicon under vacuum.[61] Pure silicon reacted with vanadium at 600—1000 °C forming VSi_2 as the silicon-rich phase.[359] Higher temperatures were required to effect reaction between partially oxidized silicon and vanadium; here the products were V_3Si, V_5Si_3, and V_2O_5. ^{51}V n.m.r. spectroscopy[360] of V_3Si indicates the electric field gradient tensor to be axially symmetric at the V nuclei with its symmetry axis parallel to the direction of the chain of transition element atoms. Vanadium disilicide oxidizes to mixtures of α-quartz, lower vanadium oxides, VN, V_5Si_3, and V_3Si at temperatures >500 °C. Its exceptionally high resistance to oxidation compared with VC and VB_2 is ascribed to the formation of a SiO_2

[354] T. V. Charlu and O. J. Kleppa, *High Temp. Sci.*, 1973, **5**, 260 (*Chem. Abs.*, 1973, **79**, 119 050).

[355] N. P. Slyusar, A. D. Krivorotenko, E. N. Fomichev, A. A. Kalashnik, and V. P. Bondarenko, *Zhur. fiz. Khim.*, 1973, **47**, 2706.

[356] Yu. M. Belyakov, N. A. Perelyaev, A. K. Chirkov, and G. P. Shveikin, *Zhur. neorg. Khim.*, 1973, **18**, 3360.

[357] G. Hoerz, *J. Less-Common Metals*, 1974, **35**, 207.

[358] R. B. Creel, S. L. Segel, R. J. Schoenberger, R. G. Barnes, and D. R. Jorgesen, *J. Chem. Phys.*, 1974, **60**, 2310.

[358] K. N. Tu, J. F. Ziegler, and C. J. Kircher, *Appl. Phys. Letters*, 1973, **23**, 493.

[360] B. N. Tret'yakov, V. A. Marchenko, V. B. Kurtisin, and B. N. Kodess, *Zhur. Eksp. i Teor. Fiz.*, 1973, **65**, 1551 (*Chem. Abs.*, 1974, **80**, 8752).

coating.[361] The crystal structures of the ternary silicides ZrVSi and HfVSi have been determined.[256]

Hydrides. Hydrogenation of degassed activated vanadium occurs at -20 to $-60\,°C$ and is complete in 8—10 days forming the cubic hydride $VH_{1.93}$ whose lattice parameter is $a = 4.26\,\text{Å}$.[362] Two quadrupole-split ^2H spectra have been observed[363] for the interstitial vanadium deuterides VD_x $0.2 \leqslant x \leqslant 0.9$.

Vanadium(II).—The formation enthalpies of $RbVCl_3$, Rb_2VCl_4, and $CsVCl_3$ have been determined.[364]

The co-ordination chemistry of vanadium(II) has been extended by the preparation and characterization of complexes with aliphatic and heterocyclic amines.[365] Intense charge-transfer bands dominate the reflectance spectrum of the dark-red or red-brown $V(amine)_4X_2$ complexes (amine = py, β- and γ-picoline; X = Cl, Br, or I) but assignments of some of the d–d transitions indicate a *trans*-dihalogeno structure; this is consistent with the spin-only magnetic moments *ca.* 3.87 BM. In contrast, the yellow $V(amine)_2Br_2$ complexes obtained with β- and γ-picoline exhibit considerable antiferromagnetic interaction indicative of polymeric bromine-bridged structures.

A single-crystal X-ray diffraction study[366] of tetrapyridinodichlorovanadium(II) shows the metal to have a tetragonal stereochemistry, with *trans* V–Cl distances of $2.462\,\text{Å}$ and V—N distances of $2.189\,\text{Å}$. These are comparable to those in vanadyl complexes containing an unsaturated nitrogen ligand *trans* to oxygen and to those in V^{III} and V^{IV} adducts of trimethylamine.

Magnetically dilute complexes with electronic spectra characteristic of octahedral V^{II} have been obtained with the chelating ligands ethylenediamine, 1,2- and 1,3-propanediamine, diethylenetriamine, bipy, and 1,10-phen,[365] Distortions from octahedral symmetry seem unlikely to account either for the high intensity of the absorption bands or the complicated nature of the electronic spectra of the blue 8-aminoquinoline complexes $V(amq)_2Cl_2,2H_2O$, $V(amq)_2Br_2,2H_2O$ and $V(amq)_3I_2$, $2H_2O$. The spectra of the chloride and bromide complexes are very similar, suggesting that the halide is not co-ordinated. Extensive mixing of charge-transfer and d–d bands may account for the intensities, and vibrational interactions for the band shoulders. Vanadium(II) gives 1:3 and 2:1 complexes with 4,7-diphenyl-1,10-phenanthroline;[367] blue VL_3^{2+} ($\lambda_{max} = 680$ nm) forms optimally at pH 3—6 and the less intense V_2OL^{2+} ($\lambda_{max} = 600$ nm) at pH 0.3—2.0. The stability constant of the 1:1 V^{II}–picolinic acid complex decreases with increasing pH and levels out at pH 1—2; with nicotinic and isonicotinic acids oxidation to V^{III} occurs.[368] New methods for the preparation of complex vanadium(II) cyanides have been reported.[369,370] Nitrogen is reduced to ammonia and hydrazine, and hydrazine is reduced to ammonia by V^{II} in aqueous

[361] R. F. Voitovich and E. A. Pugach, *Porosh. Met.,* 1974, 43 (*Chem. Abs.,* 1974, **80**, 103 350).
[362] A. A. Chertkov, L. N. Padurets, and V. I. Mikheeva, *Doklady Akad. Nauk. S.S.S.R.,* 1974, **215**, 610.
[363] K. P. Roenker, *Nuclear Sci. Abs.,* 1973, **28**, (10), 24 935 (*Chem. Abs.,* 1974, **80**, 65 314).
[364] E. Lupenko and A. I. Efimov, *Vestnik Leningrad. Univ. Fiz. Khim.,* 1973, 100 (*Chem. Abs.,* 1974, **81**, 17 526).
[365] M. M. Khamar, L. F. Larkworthy, K. C. Patel, D. J. Phillips, and G. Beech, *Austral. J. Chem.,* 1974, **27**, 41.
[366] D. J. Brauer and C. Krueger, *Cryst. Struct Comm.,* 1973, **2**, 421.
[367] N. L. Babenko, A. I. Busev, and M. Sh. Blokh, *Zhur. neorg. Khim.,* 1973, **18**, 2108.
[368] D. Katakis and E. Vrachnou-Astra. *Chem. Chron.,* 1972, 225. (*Chem. Abs.,* 1973, **79**, 58 308.)
[369] V. V. Dovgei, K. N. Mikhalevich, and A. N. Sergeeva, *Zhur. neorg. Khim.,* 1974, **19**, 943.
[370] V. V. Dovgei, *Ref. Zhur. Khim.,* 1973, Abs. No. 16V101 (*Chem. Abs.,* 1974, **80**, 43 577).

solution.[371] The yield of hydrazine diminishes with increasing temperature due to disproportionation.[372] Vanadium(III) inhibits the reduction to hydrazine[373] consonant with the view that V^{II} is the active centre at which N_2H_4 forms; a mechanism of reduction involving a V^{II}—N≡N—V^{II} intermediate has been proposed. Spectroscopic and polarographic studies indicate only weak complex formation between acrylonitrile and V^{II}.[374] Slow irreversible oxidation to V^{III} occurs and it was suggested that it is this oxidation which accounts for an earlier report of strong complex formation in this system.

Carbonylation of bis(cyclopentadienyl)vanadium(II) gives deep-brown paramagnetic Cp_2VCO^{347a} whose low $\nu(CO)$ of 1881 cm^{-1} is quite common for bent bis(cyclopentadienyl)carbonyls of d^2-d^4 metals. $Cp_2V(I)CO$ has $\nu(CO)$ 1953 cm^{-1} in accord with the metal's increased oxidation state.

Vanadium(III).—*Halides and Oxyhalides.* Solid-state reactions[275] in the systems VF_3-M^1F and VF_3-M^2F_2 (M^1 = Tl^I, M^2 = Ca, Sr, Ba, or Pb) have yielded 17 double fluorides of five different structural types depending on the VF_3:MF ratio. The chief structural type is the cryolite-like M_3VF_6.

Halogenovanadium(III) complexes, $H(VCl_4),5Et_2O$, $H_2(VCl_5),7Et_2O$, and $H(VCl_3Br),5Et_2O$, have been prepared[376] by treating VCl_3 in anhydrous ether with gaseous HCl or HBr; also prepared were $H(VCl_4),2L$ and $(HVCl_3Br),4L$ (L = pyrimidine).

V_2O_3 reacts with NbO_2F in a sealed tube at 900 °C to form V_2NbO_5F;[78] this has a rutile-type structure and its magnetic susceptibility follows the Curie–Weiss law with a Curie temperature of -17 K.

VOCl absorbs 0.5 mole of ammonia in 6 h then the reaction slows down and takes about one month for completion.[376a] An i.r. examination of samples taken at various stages indicated that the reaction proceeds *via* $VOCl,nNH_3$ which then undergoes ammonolysis to $VONH_2$. The low value of 2.22 BM for the magnetic moment of $VONH_2$ was attributed to exchange interactions through polymeric —O—V—O— units.

O-Donor Ligands. The formulation of $V(OH)SO_4$ as a hydroxosulphate has been confirmed by i.r. and crystallographic methods;[377] on hydration it forms a new type of jarosite, $(H_3O)V_3(OH)_6(SO_4)_2$. Whereas both V_2O_5 and VO_2 react readily with potassium at its melting point, V_2O_3 only reacts at temperatures >180 °C. However, it resembles V_2O_5 in giving a mixture of VO and KVO_2 as the initial reaction products these are oxidized subsequently to K_3VO_4.[377a] The rate of reaction of V_2O_3 with Na_2CO_3 is very low[377b] as a result of the formation of a solid solution of $Na_{0.33}V_2O_{3.17}$. V_2O_3 has an extensive range of solid solutions in VO_2; V_2O_3, V_3O_5,

[371] L. A. Nikonova, N. I. Pershikova, M. V. Bodeiko, L. G. Oliinyk, D. N. Sokolov, and A. E. Shilov, *Doklady Akad. Nauk. S.S.S.R.*, 1974, **216**, 140.

[372] D. V. Sokol'skii, Ya. A. Dorfmann, and Yu. M. Shindler, *Zhur. obshchei Khim.*, 1974, **44**, 20.

[373] N. T. Denisov, N. I. Shuvalova, and A. E. Shilov, *Kinetika i Kataliz.*, 1973, **14**, 1325.

[374] S. Tidwell, D. A. Zatko, J. W. Prather, and H. M. Carney, *J. Coord. Chem.*, 1973, **2**, 317.

[375] J. C. Cretenet, *Rev. Chim. Minerale*, 1973, **10**, 399.

[376] A. G. Galinos and E. D. Manesis, *Ann. Chim. France*, 1973, **8**, 369.

[376a] N. I. Vorob'ev, V. V. Pechkovskii, and L. V. Kobets, *Zhur. neorg. Khim.*, 1974, **19**, 3.

[377] J. Tudo, G. Laplace, M. Tachez, and F. Theobald, *Compt. rend.*, 1973, **277**, C, 767.

[377a] M. G. Barker, A. J. Hooper, and R. Lintonbon, *J.C.S. Dalton*, 1973, 2618.

[377b] V. L. Volkov, N. Kh. Valikhanova, and A. A. Fotiev, *Zhur. neorg. Khim.*, 1973, **18**, 3208.

C

V_4O_7, V_5O_9, V_6O_{11}, V_7O_{13}, V_8O_{15}, and VO_2 were observed in the VO_2–V_2O_3 system at 1307 K and 10^{-4}—10^{-12} atm oxygen partial pressure. The standard free energies of stepwise formation of the vanadium oxides under these conditions were determined.[377c]

Superexchange mechanisms acting *via* —O—P—O— or —O—C—O— bridges have been proposed to account for the subnormal magnetic moments of the polymeric diethylmethylenediphosphonate complex (11; M = V)[96] and of the dimeric carboxylates $[CpV(CO_2R)_2]$ (R = CF_3, Ac, Bz, m-$O_2NC_6H_4$, or p-$O_2NC_6H_4$).[109]

A change in colour from dark blue to light yellow accompanies the electroreduction of V^{IV} pyrocatechol and pyrogallol chelates.[378] An analysis of the polarographic data indicates that the reduction of the V^{IV} chelates at various pH's can be described as follows:

$$V^{IV}OL + 2H^+ + e^- = V^{III}L^+ + H_2O \qquad\qquad pH = 3.4—4.2$$
$$V^{IV}OL_2^{2-} + 2H^+ + H_2O + e^- = V^{III}(OH)_2L^- + H_2L \qquad pH = 4.9—5.9$$
$$V^{IV}OL_2^{2-} + H_2O + e^- = V^{III}(OH)_2L_2^{3-} \qquad\qquad pH = 7—8.5$$
$$V^{IV}OL_2^{2-} + 4H^+ + e^- = V^{III}L^+ + H_2L + H_2O \qquad pH = 4.2—4.8$$

There are clear similarities in the negative-ion mass spectra of the M(trifluoroacetylacetonate)$_3$ complexes (M = V^{III} or Al^{III}).[379] The major ion peak corresponds to $(ML_3)^-$ and this is followed in intensity by one for $(L)^-$; changes in oxidation state are evident for the metals in the fragment ions.

S-Donor Ligands. V^{III} complexes of the 3- and 5-phenylpyrazolinedithiocarbamato ligands have been prepared and characterized by elemental analysis and i.r. and electronic spectroscopy.[380]

N-Donor Ligands. Solutions of VCl_3,$6H_2O$ in pyridine–isoamyl alcohol do not show the charge-transfer band due to the $V_2(OH)_2^{4+}$ species,[381] indicating that under these conditions hydrolysis of the V^{III} ion does not occur. Fresh solutions of VCl_3,$6H_2O$ in pyridine–isoamyl alcohol are violet with a maximum colour intensity at V:py of *ca.* 1:6; however, the colour changes rapidly to blue and then green in air. Extraction of the violet solution with CCl_4 affords a violet organic phase from which violet crystals can be isolated; the crystals decompose in air with the liberation of pyridine. These observations point to the formation of unstable V^{III} pyridine complexes. Changes in the pH and conductivity of the VCl_3,$6H_2O$–isoamyl alcohol–pyridine solutions have been interpreted in terms of replacement of py by H_2O on vanadium and hydrolysis of the hydrated vanadium pyridine complex so formed.

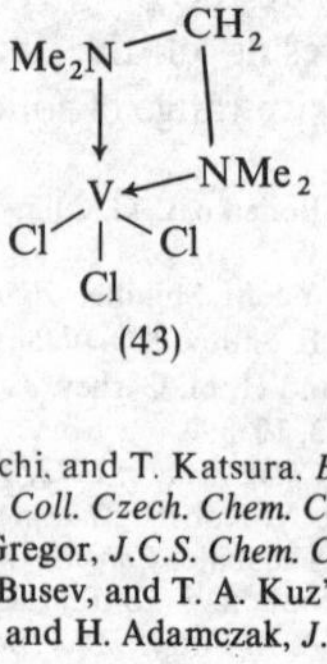

$$Me_2N \underset{Cl}{\overset{CH_2}{\underset{|}{\overset{|}{V}}}} \overset{NMe_2}{\underset{Cl}{\nearrow}}$$

(43)

[377] (c) H. Endo, M. Wakihara, M. Taniguchi, and T. Katsura. *Bull. Chem. Soc. Japan*, 1973, **46**, 2087.
[378] J. Zelinka, M. Bartusek, and A. Okac, *Coll. Czech. Chem. Comm.*, 1974, **39**, 83.
[379] I. W. Fraser, J. L. Garnett, and I. K. Gregor, *J.C.S. Chem. Comm.*, 1974, 365.
[380] V. M. Byr'ko, M. B. Polinskaya, A. I. Busev, and T. A. Kuz'mina, *Zhur. neorg. Khim.*, 1973, **18**, 1557.
[381] L. Pajdowskii, Z. Zarwecka, K. Fried, and H. Adamczak, *J. Inorg. Nuclear Chem.*, 1974, **36**, 585.

The formation of bis(dimethylaminomethane)trichlorovanadium(III) (43) by the reduction of VCl_4 with tetrakis(dimethylamino)diborane[382] requires that oxidation of a methyl group, dimethylamino-group transfer, and transfer of a methyl group to nitrogen must have occurred. The $CH_2(NMe_2)_2$ ligand undergoes quantative displacement by py affording VCl_3,2py, and a protonation-cleavage reaction with HCl in which $Me_2NH_2^+Cl^-$ and $Me_2NCH_2^+Cl^-$ are liberated along with VCl_3. Electronic and i.r. spectral data are respectively, consistent with five-co-ordinate V^{III} and a chelate ligand structure.

Vanadium(III) forms more stable complexes with $NCSe^-$ ion in acetone and methyl cyanide than in DMF.[383] Spectroscopic detection of the 1:3 and 1:2 complexes in acetone, the 1:2 complex in methyl cyanide, and the 1:1 complex in all three solvents has been achieved. Stepwise stability constants of the complexes are reported as are their electronic spectral parameters. $V(antipyrene)_3(NCSe)_3$ and $V(diantipyrrylmethane)(NCSe)_3$ have been isolated from ethanol solutions.

A green solvate formulated as $Et_4NVCl_3N_3(MeCN)_2$ is produced in the reaction between Et_4NN_3 and VCl_3 in methyl cyanide at room temperature.[384] The solvent is lost at 50—70 °C *in vacuo* and at 100 °C nitrogen is evolved with the formation of the grey-green vanadium(v) complex Et_4NVCl_3N. It has been suggested that back-bonding from the vanadium's d-orbitals into the π^*-antibonding orbitals of the N_3^- ligand promotes expulsion of nitrogen by weakening the N—N bond. The remaining co-ordinated N^- species can oxidize either the metal or the chloride ligand.

Mixed N- and O-Donor Ligands. Spectrophotometric titration methods have been used to establish the existence and stoicheiometry of V^{III} complexes with alanine, α-aminobutyric acid, lysine, and glycylglycine; pH measurements showed that the amino-acid ligand is co-ordinated in the anion form.[385]

Compounds with Metal–Carbon Bonds. A magnetic moment of 2.6 BM for Cp_2V-$[2-(Me_2NCH_2)C_6H_4]$ indicates two unpaired electrons per vanadium atom and therefore that internal co-ordination through the nitrogen atom is unlikely since this should lead to electron pairing;[117] this inference is further supported by the close similarity between the electronic spectra of the complex and Cp_2VR (R = aryl) where such co-ordination is impossible.

Trimesitylvanadium has been prepared in 45% yield by the Grignard reaction of mesitylmagnesium bromide with VCl_3,3THF.[386]

A full account of the crystal structure determination of potassium heptacyano-vanadate(III) dihydrate has appeared.[387] Apart from clearing up the uncertainty surrounding the nature of the vanadium(III) cyanide complex, this study has established the $V(CN)_7^{4-}$ ion as one of the very few examples of a seven-co-ordinate complex containing only simple unidentate ligands. The choice of this co-ordination number has been rationalized on the basis of the nine-orbital rule. The two electrons of the paramagnetic $V(CN)_7^{4-}$ ion ($\mu_{eff} = 2.8$ BM) occupy two separate orbitals,

[382] R. F. Kiesel and E. P. Schram, *Inorg. Chem.*, 1974, **13**, 1313.

[383] V. V. Shopenko, E. I. Ivanova, and A. S. Grigor'eva, *Ukrain. Khim. Zhur.*, 1973, **39**, 754 (*Chem. Abs.*, 1973, **79**, 132413).

[384] M. Kasper and R. D. Bereman, *Inorg. Nuclear Chem. Letters*, 1974, **10**, 443.

[385] Z. Karwecka and L. Pajdowski, *Advan. Mol. Relaxation Processes*, 1973, **5**, 45

[386] W. Siedel and G. Kreisel, *Z. Chem.*, 1974, **14**, 25.

[387] R. Levenson and L. R. Towns, *Inorg. Chem.*, 1974, **13**, 105.

allowing seven ligands to donate electron pairs and the V^{III} thereby to achieve the number of orbitals of krypton. Crystal-field parameters for the pseudo-D_{5h} $V(CN)_7^{4-}$ ion have been calculated and used as the basis of an analysis of its electronic spectrum.[388]

Vanadium(IV).—Accumulating structural data have made it clear that the Ballhausen–Dahl model developed to describe the bonding in the complexes Cp_2MH_2 (M = Mo or W), in which a sterically active non-bonding orbital is directed between the hydride ligands, cannot be extended to d^1 and d^2 M^{IV} complexes of the type Cp_2ML_2.

Single-crystal e.p.r. studies and MO calculations on several vanadium complexes[389] indicate that the highest occupied MO has substantial (50—70%) $3d_{z^2}$ character and very little $3d_{x^2-y^2}$ or $4s$. This is entirely compatible with the observed decrease of *ca.* 6° in the L—M—L bond angle which occurs on occupation of this orbital *i.e.* on going from Ti^{IV}, d^0 to V^{IV}, d^1.

Halides and Oxyhalides. VF_4 is formed in the oxidizing reaction of VF_5 on UF_4; VF_4 reacts with UF_6 to give UF_5 and VF_5.[390]

Reduction to V^{IV} occurs in the reaction between $VOBr_3$ and ionic bromides, ferrocene, MeCN and a range of uni- or bi-dentate ligands.[391, 392] The oxovanadium (IV) complexes $(Et_4N)_2(VOBr_4)$, $(pyH)[VOBr_3(MeCN)_2]$, $(pyH)[VOBr_3]$, $(quinH_3)$-$[VOBr_5]$, $(FeCp_2)(VOBr_3)$, and $VOBr_2(MeCN)_3$ have been isolated, as have several $VOBr_2$ complexes: $VOBr_2,L$ (L = THF or 1,4-dioxan), $VOBr_2,(L–L)$, (L–L = 1,2-dimethoxyethane), $VOBr_2L_3$ (L = Ph_2SO), and $VOBr_2L_5$ (L = Me_2SO). The last-named was formulated as $[VO(Me_2SO)_5]Br_2$. The others gave evidence of either a trigonal-bipyramidal or a square-pyramidal structure. A change between these two structural types was proposed to account for the complete and reversible change which occurs in the far-i.r. spectrum of the $VOBr_4^{2-}$ ion in the temperature range 25—30 °C.

O-Donor and Mixed N- and O-Donor Ligands. Corner-sharing between two VO_6 octahedra and two SO_4 tetrahedra builds up the cyclic molecular unit of $VOSO_4,3H_2O$; the blocks which resemble those in $MgSO_4,6H_2O$ and $FeSO_4,4H_2O$ are linked by hydrogen bonds between the water molecules.[393]

An examination of the magnetic susceptibility, thermal analysis characteristics, and X-ray powder patterns of samples of $VO_{2+x}(-0.02 \leqslant x \leqslant 0.03)$ prepared by heating V_2O_5–V_2O_3 mixtures in silica tubes has shown that the homogeneity range for VO_2 is very narrow.[393a]

Fe_2VO_4 is a mixed valence oxide, $Fe^{III}(Fe^{II}V^{III})O_4$.[394] Progressive substitution of iron by cobalt drives Fe^{III} into the octahedral sites in the oxide lattice then as Co^{II} replaces Fe^{III} the V^{III} is oxidized to V^{IV} finally yielding $Co^{II}(Co^{II}V^{IV})O_4$ at complete substitution. Ion replacement has also been studied[394a] in the molybdovanado-

[388] R. A. Levenson and R. J. G. Dominguez, *Inorg. Chem.*, 1973, **12**, 2342.
[389] J. L. Petersen and L. F. Dahl, *J. Amer. Chem. Soc.*, 1974, **96**, 2248.
[390] J. Reynes, L. Bethuel, and J. Aubert, *Nuclear Sci. Abs.*, 1973, **27**, 446 (*Chem. Abs.*, 1973, **79**, 86982).
[391] D. Nicholls and K. R. Seddon, *J.C.S. Dalton*, 1973, 2747.
[392] D. Nicholls and K. R. Seddon, *J.C.S. Dalton*, 1973, 2751.
[393] F. Theobald and J. Galy, *Acta. Cryst.*, 1973, **29B**, 2732.
[393a] T. Horlin, T. Niklewski and M. Nygren, *Chem. Comm. Univ. Stockholm*, 1973, No. 1. (*Chem. Abs.*, 1973, **79**, 26 599).
[394] A. M. Perry, J. C. Bernier, and A. Michel, *Rev. Chim. Minerale*, 1973, **10**, 543.
[394a] M. Otake, Y. Komiyama, and T. Otaki, *J. Phys. Chem.*, 1973, **77**, 2896.

phosphoric heteropolyacids $H_{3+x}Mo_{12-x}V_xP_{40}, nH_2O$ ($x = 0$, 1, and 2) where replacements of Mo by V ions affords a promising way of clarifying the nature of the reduced form of the heteropoly compound. The hyperfine coupling constants of the V^{IV} observed for $x = 1,2$ are almost the same as those of organic vanadyl complexes, which indicates localization of one electron on the vanadium ion. A mechanism for this localization is proposed. Several narrow lines split by *ca.* 15 gauss are observed for $x = 1$ and 2; the splitting is presumed to be due to interactions with neighbouring hydroxylated Mo anions and the narrowness of the lines to the separation of the paramagnetic ions in the heteropoly cage which is sufficiently large to remove dipolar broadening.

Interest in the interpretation of the spectral and magnetic properties of oxo-vanadium(IV) complexes has grown as it has become apparent that the general assumption of overall C_{4v} symmetry for these complexes is unjustified. Bis-(2-methyl-8-quinolato)oxovanadium(IV), VO(quin)$_2$,[395] and bis(tetramethylurea)dichloro-oxo-vanadium(IV), VO(tmu)$_2$Cl$_2$,[396] for example, have been found to be five-co-ordinate with a trigonal-bipyramidal co-ordination polyhedron about the vanadium atom. A crystal-field model has been developed which gives a good account of the electronic and e.p.r. spectra of VO(quin)$_2$.

Except for the splitting of the d_{xz} and d_{yz} levels, the Ballhausen–Gray energy-level scheme for vanadyl complexes of C_{4v} microsymmetry is only slightly modified for complexes of lower symmetry. It was suggested that this is due to the dominance of the strong axial field of the vanadyl oxygen atom. E.p.r. data and MO calculations on VO(tmu)$_2$Cl$_2$[396] accord with these findings and indicate that the overall average environment of the d_{xy} unpaired electron is very similar for this complex and for the C_{4v} VOCl$_4^{2-}$ ion.

A crystal-field model has also been shown[397] to give reasonable agreement between the calculated and observed positions of the first two electronic absorption bands of some C_{4v} oxovanadium(IV) complexes.

The implication of preliminary magnetic c.d. measurements[398] on the $[VO(H_2O)_5]^{2+}$ ion is that out-of-plane π-bonding to the equatorial ligands is more important in excited states than has hitherto been assumed.

The e.p.r. spectra of frozen 1:1 aqueous ethylene glycol solutions containing vanadyl DL- and DD-tartrates are consistent with the formation of dimeric species.[399] Computer simulation of the spectra assuming a monoclinic dimer model with the V—V separation d(V—V) and the angle (α) between the V—V internuclear line and the V=O bond as structural parameters gave best fits for d(V—V) = 4.08, 4.18 Å; $\alpha = 28 \pm 2$, 0—10° for the DL- and DD-tartrates, respectively. For the DL-tartrate dimer, the excellent agreement between the structural parameters for the solid complex [d(V—V) = 4.082 Å, $\alpha = 28°$] and the frozen solution indicates that the same structure exists in both states. For the DD-tartrate however, although the overall symmetry is maintained, there is a slight reduction in d(V—V) from 4.35 Å in the solid which may arise from a compression due to the host lattice.

E.p.r. data indicate that in the reaction product formed from VCl$_4$ and the surface

[395] H. J. Stoklosa, J. R. Wasson, and J. McCormick, *Inorg. Chem.*, 1974, **13**, 592.
[396] H. A. Kuska and P. H. Yang, *Inorg. Chem.*, 1974, **13**, 1090.
[397] J. R. Wasson and H. J. Soklosa, *J. Inorg. Nuclear Chem.*, 1974, **36**, 227.
[398] D. J. Robbins, M. J. Stillman, and A. J. Thomson, *J.C.S. Dalton*, 1974, 813.
[399] A. D. Toy, T. D. Smith, and J. R. Pilbrow, *Austral. J. Chem.*, 1974, **27**, 1.

hydroxy-groups of high-area thorium oxide, each vanadium is bound to the surface by two oxygens[400] and that the preferred orientation of $VOCl_2$ in $VOCl_2$-loaded cellulose is that with the V—O internuclear direction perpendicular to the cellulose chain.[401]

Entry of VO^{2+} ions into the lattice of magnesium tetrahydrate single crystals results in the formation of the $[VO(H_2O)_3O_2]^{2+}$ species where both oxygen atoms belong to one co-ordinating acetate group.[402] It is suggested that the approximately axial symmetry indicated by the e.p.r. spectrum is achieved by a displacement of the co-ordinating water molecules. Analysis[403] of the single-crystal e.p.r. spectrum of the $V^{IV}W_5O_{19}^{4-}$ ion in a $(MeNH_3)_2Na_2(V_2^VW_4O_{19})$,$6H_2O$ host showed three mutually perpendicular g directions consistent with a structure for the anion analogous to that found for $Nb_6O_{19}^{8-}$. It was only possible to reconcile the g-values with previously reported optical data by assuming an unusually large value *ca.* 240 cm^{-1} for the spin–orbit coupling constant of vanadium.

Two kinds of exchanging ligand are evident from studies[404] of the temperature dependence of the n.m.r. linewidth of the solvent formyl resonance in solutions of VO^{2+} in DMF. Axial exchange is *ca.* 230 times faster than equatorial exchange. The rate of equatorial exchange in $VO(dimethylacetamide)_5^{2+}$ is 25 times faster than that in $VO(DMF)_5^{2+}$ and has a lower enthalpy of activation which is consistent with a weaker vanadium–ligand bond. Axial exchange does not produce broadening of the Me resonances in either DMF or DMA which could be distinguished from outer-sphere broadening. Increases in the exchange rate of water protons in the hydration shell of the VO_{aq}^{2+} ion caused by the addition of sodium dodecylsulphate to an aqueous solution of $VOSO_4$ or by the emulsification of the solution with toluene have been attributed[405] to the absorption of the VO^{2+} ion at the micelle and emulsion drop surfaces.

$VOSO_4$,$2L$,$3H_2O$ (L = urea)[406] and VOL_2,mB (LH = acacH, dbm, bza, or CF_3COCH_2COPh; B = py, 3-Mepy, 4-Mepy, bipy, or *o*-phen; $m = 1$ or 2)[407] have been prepared and subjected to thermogravimetric analysis. VOL^1L^2 containing bidentate ligands is reported to be formed from pyrocatechol (H_2L^1), diantipyryl-methane (L^2), and either NH_4VO_3 or $VOSO_4$.[408]

Studies of vanadyl diketonate complexes $VO(dik)_2L$ (L = heterocyclic amine) continue. The stability constants of $VO(dik)_2L^1$ ($L^1 = 4$-methylpyridineN-oxide) fall in the series dik = thenoyltrifluoroacetone (tta) > trifluoroacetone (tfa) > dbm > bzac > acac > dpm whereas the pK_a's of the diketones fall in the series dbm > bzac > acac > tta > tfa.[409] The i.r. spectra of the complexes in solution show two

[400] B. W. Martin, *Diss. Abs.* (*B*), 1974, **34**, 4900.

[401] P. H. Kasai and D. McLeod, *J. Phys. Chem.*, 1974, **78**, 308.

[402] M. V. Krishnamurthy, *Z. phys. Chem.*, 1974, **254**, 17.

[403] H. Junsoo So, C. M. Flynn, and M. T. Pope, *J. Inorg. Nuclear Chem.*, 1974, **36**, 329.

[404] G. A. Miller and R. E. D. McClung, *J. Chem. Phys.*, 1973, **58**, 4358.

[405] E. E. Zaev and L. M. Khalilov, *Kolloid Zh.*, 1974, **36**, 147.

[406] T. A. Azizov, O. F. Khodzhaev, M. T. Saibova, and N. A. Parpiev, *Uzbek. Khim. Zhur.*, 1973, **17**, 26 (*Chem. Abs.*, 1973, **79**, 61013).

[407] D. G. Batyr and V. T. Balan, *Izvest. Akad. Nauk Mold S.S.R.*, Ser. Biol. Khim. Nauk, 1973, 61 (*Chem. Abs.*, 1973, **79**, 13002).

[408] E. P. Klimenko and S. Ya. Shnaiderman, *Izvest. V.U.Z. Khim. i khim. Tekhnol.*, 1973, **16**, 867 (*Chem. Abs.*, 1973, **79**, 86922).

[409] N. S. Al-Niaimi, A. R. Al-Karaghouli, S. M. Aliwi, and M. G. Jalhoom, *J. Inorg. Nuclear Chem.*, 1974, **36**, 283.

shifted $v(V=O)$ bands, indicating a *cis–trans* equilibrium; relative intensities suggest that the *trans*-isomer predominates for dik = acac and the *cis* for dik = bzac and dbm, and that the latter is the exclusive isomer for dik = dpm, tta, and tfa. A polymeric structure was suggested for the solid tta and tfa complexes.

Solution equilibrium studies[410] have identified both 1:1 and 1:2 complexes found between VO^{2+} ions and oxalic acid, but only 1:1 complexes for phthalic, maleic, succinic, and adipic acids. Equilibrium constants for the formation and subsequent hydrolysis of these complexes have been determined. $VO(H_2L)^-$ (H_5L = trihydroxyglutaric acid) is the principal complex species in mixed aqueous acetone solutions but minor amounts of $(VO)_3(H_2L)_2$ are also formed;[411] the stability of $VO(H_2L)^-$ increases with increasing acetone content. Solvent effects are also evident in vanadium(IV)–quercetin system;[412] the protonated ligand RH_2^+ forms VOR^+ and VR_2^{2+} in methanol at pH 3 but only VOR^+ in aqueous methanol.

The product of the reaction between VCl_4 and $P_2O_3Cl_4$ in CCl_4 at $0\,°C$ is insoluble in all common solvents, softens rather than melts and exhibits neither terminal $v(P—O)$ nor $v(P=O)$ absorption bands in its i.r. spectra.[413] On these grounds it has been assigned the polymeric structure (44).

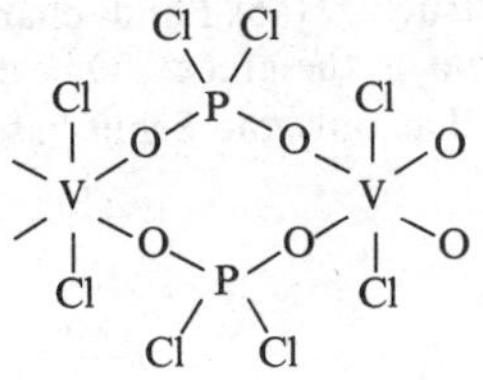

(44)

A linear relationship has been established[414] between the pK_a's and the metal–ligand stability constants for a series of VO^{2+} complexes with substituted acetophenones.

Whereas the VO^{2+} ion forms only a 1:2 complex with L-lysine,[414] there is evidence for both 1:1 and 1:2 complexes with glycine[415] and with NN'-dihydroxyethylglycine.[416] In the 1:1 glycine complexes, the amino-acid can function as a unidentate

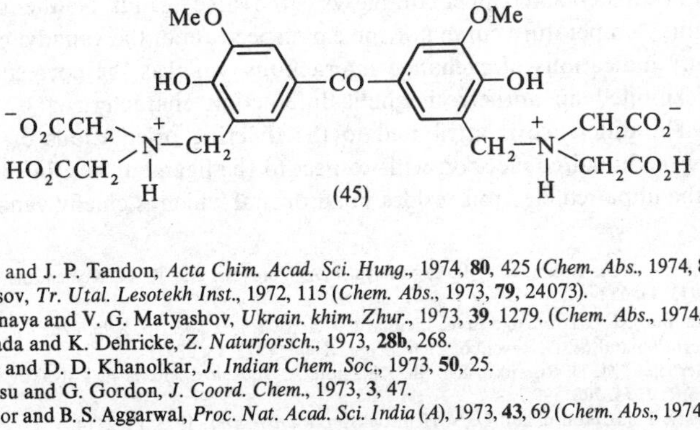

(45)

[410] S. P. Singh and J. P. Tandon, *Acta Chim. Acad. Sci. Hung.*, 1974, **80**, 425 (*Chem. Abs.*, 1974, **81**, 30398).
[411] A. Mudretsov, *Tr. Ural. Lesotekh Inst.*, 1972, 115 (*Chem. Abs.*, 1973, **79**, 24073).
[412] N. V. Chernaya and V. G. Matyashov, *Ukrain. khim. Zhur.*, 1973, **39**, 1279. (*Chem. Abs.*, 1974, **80**, 14367).
[413] A. F. Shihada and K. Dehricke, *Z. Naturforsch.*, 1973, **28b**, 268.
[414] D. B. Ingle and D. D. Khanolkar, *J. Indian Chem. Soc.*, 1973, **50**, 25.
[415] H. Tomiyasu and G. Gordon, *J. Coord. Chem.*, 1973, **3**, 47.
[416] R. C. Kapoor and B. S. Aggarwal, *Proc. Nat. Acad. Sci. India (A)*, 1973, **43**, 69 (*Chem. Abs.*, 1974, **81**, 42219).

or as a bidentate ligand, and the unidentate ligand complex has a protonated analogue.[415] Ligand (45 = H_6L) forms a 1:1 V^{IV} complex $(VOHL)^{3-}$.[417] Complexes of the nucleoside uridine (u) with oxovanadium(IV) and vanadium(V) are potent inhibitors of the hydrolysis of uridine-2′,3′-phosphate catalysed by the enzyme ribonuclease.[417a] That this is a competitive inhibition is indicated by the fact that the association constant for binding of the V^{IV}–u complex is 1000 times larger than that for binding of the substrate and 40 times larger than that for binding of the product uridine-3′-phosphate.

It has been suggested either that the metal ion binds to co-ordinating groups on the enzyme, *e.g.* the imidazoles of histidine residues which form part of the active site, or that the metal complex has or can easily adopt a substrate-like structure.

Schiff Base Complexes. Condensation of salicylaldehyde and salicyladehyde thiosemicarbazone in basic solution occurs only in the presence of transition-metal ions affording complexes of the type $M^1(M^2L),nH_2O$ (M^1 = K, Na, or NH_4; M^2 = VO, Cu, or Ni; H_3L = o-$OCH_6H_4CH=N-NHC(S)N=CHC_6H_4OH-o$).[418,419] Replacement of M^1 by the methyl group on treatment with methyl iodide produces volatile thioether derivatives from which mass spectrometric evidence has been obtained that (M^2L) has the structure (46). Bands characteristic of the salicylaldehyde unit appear in the i.r. spectrum of the green VO^{2+} complex (31) of salicylaldehyde salicylhydrazone, indicating that only the Schiff base part of the ligand is co-ordinated.[196]

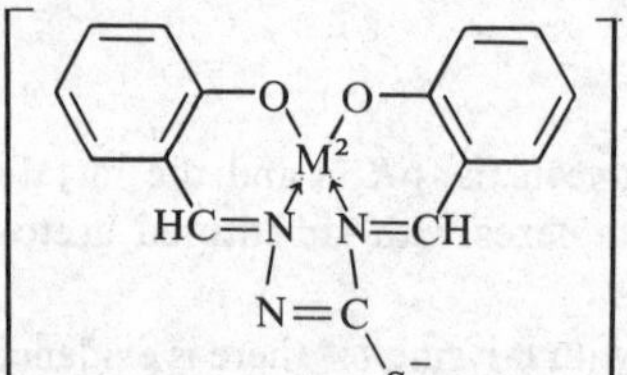

(46) 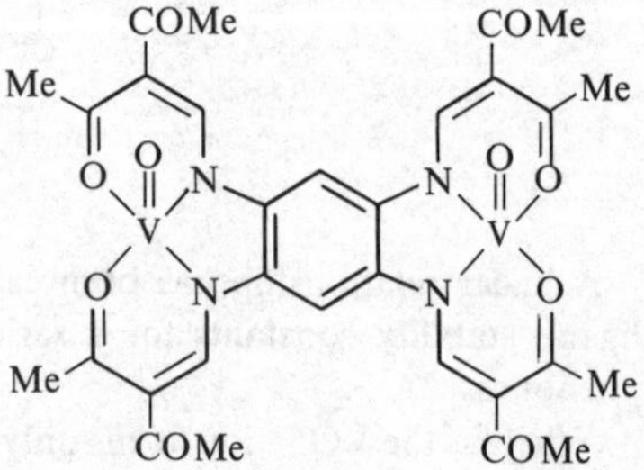　　　　　　　　　　　　　　　(47)

Several investigations have been concerned with an examination of the magnetic behaviour of vanadyl and copper complexes of the same ligands. Neither the magnetic susceptibility–temperature curve nor the e.p.r. spectrum of the vanadyl complex (47) showed any indications of exchange interactions, whereas the corresponding Cu^{II} complex exhibited an antiferromagnetic interaction characterized by $J = -12.2$ cm^{-1}.[420] The difference is attributed to the differing orientations of the orbitals containing the unpaired electron with respect to the ligand atoms. In the vanadium complex the unpaired electron resides in an orbital which is chiefly vanadium d_{xy} in

[417] N. Voronina, A. A. Ivakin, I. V. Podogomaya, I. A. Yolovets, and K. N. Klyachkina, *Zhur. obshchei khim.*, 1973, **43**, 632.

[417a] R. N. Lindquist, J. L. Lynn, and G. E. Lienhard, *J. Amer. Chem. Soc.*, 1973, **95**, 8762.

[418] N. V. Gerbeleu and M. D. Revenko, *Zhur. neorg. Khim.*, 1973, **18**, 2397.

[419] N. V. Gerbeleu, M. D. Revenko, Kh. Sh. Khariton, A. V. Ablov, and V. A. Polyakov, *Doklady Akad. Nauk S.S.R.*, 1973, **208**, 599.

[420] E. F. Hasty, T. J. Colburn, and D. N. Hendrickson, *Inorg. Chem.*, 1973, **12**, 2414.

character, lying in the molecular plane of the complex; thus there is no opportunity for spin polarization between the unpaired electron and the electrons of the bridging ligand.

Other systems where substantial exchange interactions were anticipated by analogy with the copper complexes included the 5-substituted N-(2-hydroxyphenyl)-salicylaldimine vanadyl complexes[421] and the oxinate complexes VO(oxine)X (X = Cl, Br, or OH)[422]. The room temperature magnetic moments are on the whole close to the spin–only value for one unpaired electron and decrease only slightly with temperature. The susceptibility–temperature curve of the salicylaldimine complexes did not fit the Bleaney–Bowers equation for two interacting vanadium atoms. Such exchange as there is in the oxinate complexes is thought to occur *via* the halogen bridges of a polymeric structure. The bis(thioxinate)vanadyl complex has an almost temperature-invariant magnetic moment which probably indicates a monomeric five-co-ordinate structure.

Spectroscopic and magnetic evidence has been presented[423] in favour of a square-pyramidal structure for vanadyl complexes of the terdentate Schiff bases derived from 3-aldehydosalicylic acid and glycine or anthranillic acid, and an octahedral structure for the mixed complexes $VOL^1L^2(L^1$ = terdentate dibasic hydrazone Schiff bases, L^2 = bipy or o-phen).[424]

Vanadyl complexes of Schiff bases derived from salicylaldehyde and isopropanol-amine and 2-amino-2-methylpropanol have been prepared.[425] Schiff bases derived from 2-hydroxy-1-naphthaldehyde and ethanolamine or propanolamine co-ordinate to vanadium (as VO^{2+}) through O,N, and O as terdentate dibasic ligands.[426] The vanadyl complexes have subnormal magnetic moments (μ_{eff} = 1.34—1.51 BM at 25 °C) and a temperature variation of the magnetic susceptibility characteristic of antiferromagnetic exchange between V atoms in a dinuclear species. J values indicate that exchange is more pronounced in $VOL^1[H_2L^1$ = N-(2-hydroxynaphthylidene)-2-hydroxyethanamine] (J = -281 cm^{-1}) than $VOL^2[H_2L^2$ = N-(2-hydroxynaphthylidene)-3-hydroxypropanamine] (J = -147 cm^{-1}), the difference being attributed to the chelate ring effect.

The VO^{2+} complex of the ligand NN'-bis(salicylidene)-1,1-(dimethyl)ethylene-diamine has μ_{eff} = 1.76 BM, which is almost temperature invariant; there is much evidence that the complex is monomeric with a trigonal-bipyramidal structure.[427]

Low values of v(V=O) indicate polymeric structures in the solid state for oxo-vanadium(IV) chelates with the quadridentate Schiff bases NN-bis(salicylidene)-propane-1,3-diamine, its 3-methyl and 3-isopropyl derivatives, and bis(salicylidene)-(S)-$(+)$-butane-1.3-diamine. $(Sal)_2(+)$tnMe. Dissociation occurs in non-donor solvents and in donor solvents to give five- and six-co-ordinate monomers.[428] The c.d. spectrum of VO(Sal)$_2(+)$tnMe shows that a configurational effect contributes to

[421] G. M. Klesova, L. V. Mosina, V. V. Zelentsov, Yu. V. Yablokov, and V. I. Spitsyn, *Zhur. neorg. Khim.*, 1974, **19**, 1155.

[422] G. M. Klesova, V. V. Zelentsov, and V. I. Spitsyn, *Zhur. obshchei Khim.*, 1973, **43**, 454.

[423] K. Dey, K. K. Chatterjee, and S. K. Sen, *J. Indian Chem. Soc.*, 1973, **50**, 167.

[424] R. L. Dutta and G. P. Sengupta, *J. Indian Chem. Soc.*, 1973, **50**, 640.

[425] G. N. Rao and S. C. Rustagi, *Indian J. Chem.*, 1973, **11**, 1181.

[426] A. Syamal, *Indian J. Chem.*, 1973, **11**, 363.

[427] K. S. Patel and J. C. Bailar, *J. Coord. Chem.*, 1973, **3**, 113.

[428] R. L. Farmer and F. L. Urbach, *Inorg. Chem.*, 1974, **13**, 587.

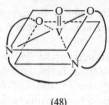

(48)

the disymmetry of the molecule. Two oppositely signed c.d. bands indicate that the two azomethine bonds are not coplanar and that the terminal chelate rings have a skew relationship to one another. The proposed dominant configuration of $VO(Sal)_2$-$(+)$tnMe is (48). $\Delta\varepsilon$ values of the compound are significantly lower in donor solvents than in non-donor solvents suggesting that the ligand may adopt a more planar geometry in the former to minimize steric interactions with the axial donor; this also accounts for the loss of the c.d. couplet in donor solvents.

Vanadyl and copper(II) ions catalyse the β-elimination reaction of O-phospho-threonine in the presence of pyridoxal.[429] Equilibrium spectroscopic studies of the threonine–metal ion–pyridoxal system have identified a metal-ion complex of the amino-acid-pyridoxal Schiff base. The catalytic effect of the metal is ascribed to its electron-withdrawing effect. It was suggested that the specific catalytic effect of Cu^{2+} and VO^{2+} arises from their reluctance to co-ordinate the phosphate in an axial position. Other metal ions such as nickel can also form the Schiff base complex but probably stabilize the phosphothreonine system by chelate formation.

S-Donor Ligands. Superhyperfine splittings in the e.p.r. spectra of C_{4v} vanadyl chelates containing the OVS_2X unit (X = ^{31}P or ^{75}As) have been attributed to direct $V(3d_{x^2-y^2}) - X(ns)$ ($n = 3$, X = P; $n = 4$, X = As) transannular interactions.[430,431] Of the 24 lines anticipated for coupling between $^{51}V(I = \frac{7}{2})$ and two equivalent $^{13}C(I = \frac{1}{2})$ nuclei, 18 were resolved in the e.p.r. spectrum of $VO(S_2{}^{13}CNEt_2)_2$,[432] showing that the $C(2s)$ orbital can also participate in transannular interactions. The ^{13}C superhyperfine coupling constant of $6.1 \times 10^{-4}\,cm^{-1}$ is substantially lower than those for ^{31}P or ^{75}As, which lie in the range 28—$50 \times 10^{-4}\,cm^{-1}$. The ^{51}V hyperfine coupling constants for a series of oxovanadium(IV) alkoxyethyl and alkoxyphenyl dithiophosphinates $[(R^1O)R^2PS_2]VO$ lie in the ranges $^VA_0$ 82.5—88.0 and $^PA_0$ 25.9—$45.8 \times 10^{-4}\,cm^{-1}$ for a wide range of substituents, indicating a relatively constant metal–ligand covalency.[433,434]

MO calculations[430] on the model compounds $VO(S_2PX_2)_2$ (X = H or F) show that the V—P interaction is antibonding for X = F and bonding for X = H, yet ^PA for $VO(S_2PF_2)_2$ is larger than any value reported for other vanadyl dithiophosphate or dithiophosphinate complexes. It seems that the type of interaction across the ring is much less important than the magnitude (C_s^2) of the phosphorus $3s$ coefficient in the ground state.

[429] Y. Murakami, H. Knodo, and A. E. Martell, *J. Amer. Chem. Soc.*, 1973, **95**, 7138.
[430] H. J. Stoklosa, G. L. Seebach, and J. R. Wasson, *J. Phys. Chem.*, 1974, **78**, 962.
[431] R. N. Mukherjee, S. Shanbhag, M. S. Venkateshan, and M. D. Zingde, *Indian J. Chem.*, 1973, **11**, 1066.
[432] H. J. Stoklosa and J. R. Wasson, *Inorg. Nuclear Chem. Letters*, 1974, **10**, 377.
[433] D. R. Lorenz, D. K. Johnson, H. J. Stoklosa, and J. R. Wasson, *J. Inorg. Nuclear Chem.*, 1974, **36**, 1184.
[434] B. M. Kozynev, *Radio Spektroskopiya*, 1973, 76 (*Chem. Abs.*, 1973, **79**, 25384).

Bis(cyclopentadienyl)vanadium(IV) NN'-dialkyldithiocarbamate and OO'-dialkyl-dithiophosphate complexes have been prepared[435] from Cp_2VCl_2 in aqueous solution. Salts of the dithiophosphonic acids did not produce the desired complexes but these could be obtained from the free acids in their corresponding alcohols or with the salts in the presence of a small amount of concentrated hydrochloric acid. It was suggested that in aqueous solution Cp_2VCl_2 hydrolyses to $Cp_2V(OH)^+$ and the dithiophosphonate ion cannot displace OH from the vanadium; in the presence of acid Cp_2V^{2+} is formed and reacts readily. The complexes $(Cp_2VS_2NEt_2)BF_4$ and $[Cp_2VS_2P(C_6H_{11})_2]Cl$ have been used[436] in an extension of studies on the C_{4v} oxovanadium(IV) chelates to C_{2v} biscyclopentadienyl complexes. These studies were intended to test the generality of the electron delocalization mechanism for trans-annular interactions. Such an interaction was indeed found but reduced by *ca.* 10—15% for the ligand superhyperfine splitting and 30—35% for the vanadium hyperfine splitting. Structural data are required before the causes of these differences can be ascertained with any certainty.

Independent studies[437-439] of base adduct formation by $VO(S_2PR_2)_2$ (R = Me, Ph, OMe, or OEt) chelates provide evidence, in the shape of a gradual change from 24- to 16- to eight-line spectra with increasing base concentration, for the progressive destruction of the chelate rings during the formation of 1:1 and 1:2 adducts. This is in sharp contrast to observations on vanadyl-O_4 and vanadyl-O_2N_2 chelates where co-ordination of base *trans* to the vanadyl oxygen completes a pseudo-octahedral stereochemistry. I.r. and e.p.r. data[437] on solutions of the S_4 chelates containing low concentrations of base are consistent with the presence of the two species (49) and (50) which are undergoing fast exchange. Coupling between the P atoms in the cross-chelate (49) is probably too small to be detected and the doublet (16 lines) spectrum is due only to (50). Further attack of base produces the 2:1 adduct (51).

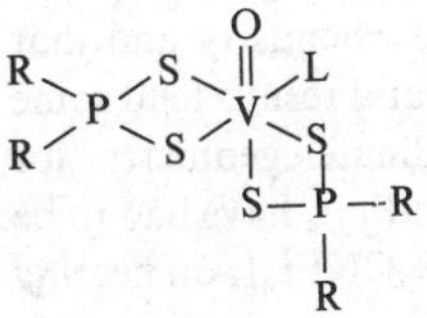

(49) 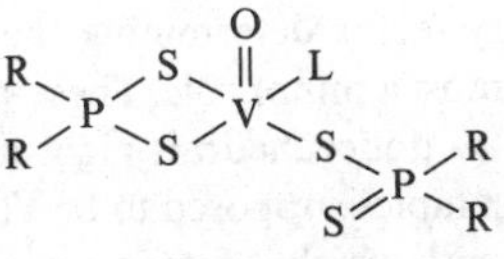   (50) 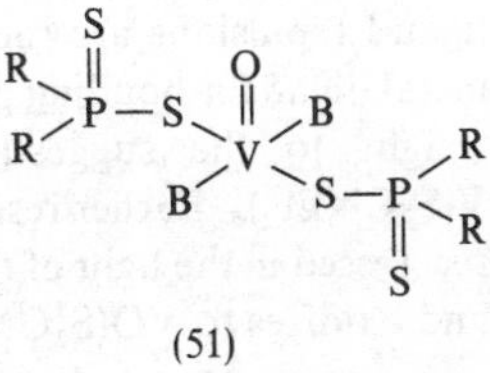   (51)

Two vanadium(IV) ethylxanthate complexes have been prepared;[440] their e.p.r. parameters are practically identical with those of the corresponding diethyldithio-carbamate complexes and on this basis analogous structures were proposed, *i.e.* a dodecahedral eight-co-ordinate structure with four bidentate ethylxanthate ligands and a distorted tetrahedral structure with four unidentate ligands. Addition of phosphorus(III) ligands (L) such as trialkylphosphites to solution of the ethyl xanthate

[435] A. T. Casey and J. R. Thackeray, *Austral. J. Chem.*, 1974, **27**, 757.
[436] H. J. Stoklasa and J. R. Wasson, *Inorg. Nuclear Chem. Letters*, 1974, **10**, 401.
[437] G. A. Miller and R. E. D. McClung, *Inorg. Chem.*, 1973, **12**, 2552.
[438] M. Sato, Y. Fujita, and T. Kwan, *Bull. Chem. Soc. Japan*, 1973, **46**, 3007.
[439] M. Sato and T. Kwan, *Bull. Chem. Soc. Japan*, 1973, **46**, 3745.
[440] G. N. Koshkina, I. V. Ovchinnikov, and A. D. Troitskaya, *Zhur. obshchei Khim.*, 1973, **43**, 956.

complexes results in the appearance of new lines in the e.p.r. spectra which are attributed to a five-co-ordinate (ethylxanthate)$V^{IV}L$ species. The e.p.r. parameters of these species, especially $^V A_0$, are insensitive to changes in the substituents on phosphorus, this is taken to indicate a low degree of covalence in the V—P bond. Differences in the pH of the starting diethyldithiocarbamate solution are thought to be responsible for the formation of the four different $VO(S_2CNEt_2)_2$ complexes detected by e.p.r. spectroscopy.[441] VOL,H_2O (HL = 3- and 5-phenyl and 3,5-diphenyl pyrazoline-dithiocarbamic acid) have been prepared and studied by i.r. and e.p.r. spectroscopy.[380]

Thiosalicylic acid forms a green 2:1 complex with VO^{2+} in aqueous ethanol[442] (log K_1 = 10.24, log K_2 = 8.39). Addition of pyridine causes a more than 10-fold increase in the molar extinction coefficient and a yellow shift affording a chloroform extractable complex which can be used for the analysis of vanadium. The thermodynamic parameters for the 2:1 complex of furfurylmercaptan and VO^{2+} are (35 °C) log K, 8.18; $\Delta H°$ = − 8.18 kcal mol^{-1}, $\Delta G°$, − 23.31 kcal mol^{-1} $\Delta S°$, + 49.12 e.u.[443] Stepwise formation constants of the VO^{2+} complex of p-(mercaptoacetamide)chloro-benzene have been determined.[444]

An earlier reassignment of the ground state of the $V(mnt)_3^{2-}$ ion (mnt = maleoni-triledithiolate) to $3d_{xy}$ has been criticized, and the results of a single-crystal e.p.r. study of the $V(mnt)_3^{2-}$ ion doped into the isomorphous $(Ph_4As)_2Mo(mnt)_3$ lattice have been offered in support of substantial $3d_{z^2}$ character for the orbital containing the unpaired electron, although it was pointed out that metal–ligand covalency is quite high.[445]

X-Ray crystallographic analyses[446] of the structures of tetrakis(phenyldithio-acetato)vanadium(IV) and tetrakis(dithiobenzoato)vanadium(IV) provide the first structural proof of the existence of the (VS_8) co-ordination polyhedron; V—S distances are 2.46—2.55 and 2.44—2.58 Å, respectively. In both complexes the shape of the VS_8 unit approximates closely to that of a dodecahedron, indicating that ligand–ligand repulsions are the primary factor determining the stereochemistry and that metal–ligand π-bonding plays at most a minor role. These structural results lend some weight to the suggestion of a dodecahedral eight-co-ordinate geometry for $V(S_2CNEt_2)_4$. Earlier results on samples supposed to be $V(S_2CNEt_2)_4$ have had to be reassessed in the light of the ease with which it dissociates to $V(S_2CNEt_2)_3$ on heating and oxidizes to $VO(S_2CNEt)_2$.[447]

A one-electron reduction to a V^{III} species occurs as the first step in the electro-chemical reduction of Cp_2V^{IV} complexes with a wide range of chelating ligands.[448, 449] A subsequent dissociation step which is markedly catalysed by oxygen occurs with some of the complexes with S donors; this behaviour may be responsible for some deviations from reversible behaviour. Half-wave potentials for the first reduction step showed a reasonable correlation with the amount of charge available for donation

[441] N. B. Kalinichenko, I. Marov, A. N. Ermakov, and Yu. I. Dubrov, *Zhur. neorg. Khim.*, 1973, **18**, 2681.
[442] R. S. Ramakrishna and S. Pathmanaban, *J. Inorg. Nuclear Chem.*, 1974, **36**, 741.
[443] R. S. Saxena and S. S. Sheelwant, *J. Inst. Chem., Calcutta*, 1973, **45**, 164.
[444] S. N. Kakkar and P. V. Shadikar, *Indian J. Chem.*, 1973, **11**, 1325.
[445] Whei-Lu Kwik and E. I. Stieffel, *Inorg. Chem.*, 1973, **12**, 2337.
[446] M. Bonamico, G. Dessy, V. Fares, and L. Scaramuzza, *J.C.S. Dalton*, 1974, 1258.
[447] D. C. Bradley, I. F. Rendall, and K. D. Sales, *J.C.S. Dalton*, 1973, 2228.
[448] A. M. Bond, A. T. Casey and J. R. Thackeray, *J.C.S. Dalton*, 1974, 773.
[449] A. M. Bond, A. T. Casey, and J. R. Thackeray, *Inorg. Chem.*, 1974, **13**, 84.

by the ligand. Attempts to prepare the V^{III} complexes electrochemically were unsuccessful due to the occurrence of second and subsequent reduction steps and the extreme air sensitivity of the low-valent vanadium species.

N-*Donor Ligands*. Stable products of nitrogen reaction with V^{IV} have been established by a combination of e.p.r. and i.r. spectroscopy and polarography in the system nitrogen–sodium vanadate–magnesium–water.[450]

Anhydrous HCN reacts with VCl_4 at room temperature to give $VCl_4(NCH)_2$.[451] X-Ray crystallography has established a slightly distorted octahedral stereochemistry for the vanadium atom with two *cis* nitrogen atoms. Hydrogen bonding occurs between the hydrogens and chlorines of different octahedral units. Disproportionation of NH_2Cl occurs in its reaction with VCl_4 which affords $(NH_4)_2(VCl_6)$.[132] $VOCl_2,5NH_3$ is formed from $VOCl_2$ and liquid ammonia;[452] at temperatures $>200\,°C$ the principal products are VN and V_2O_3. When $VOCl_3$ is saturated with dry HBr at $0\,°C$ halogen exchange occurs and $VOBr_3$ is formed; on being heated to $160\,°C$ in a stream of HBr, $VOBr_3$ affords pure $VOBr_2$.[453] Pure $VOBr_2$ cannot be obtained directly from $VOCl_3$ and HBr at $160\,°C$ because contamination by chloride ions occurs. $VOBr_2$ dissolves in liquid ammonia at $-37\,°C$, evaporation of the brown solution yields grey-brown $[VO(NH_3)_5]Br_2$ which decomposes at $120\,°C$ to $VOBr_2(NH_3)_2$. This has been assigned a six-co-ordinate polymeric structure with V—O—V— bridges on the basis of the low values for v(V—Br) $300\ cm^{-1}$ and v(V—O) $871\ cm^{-1}$.

The complexes $[VO(H_2O)_{5-x}(NCS)_x]^{(2-x)}$ $(x = 1—5)$ have been identified by e.p.r. studies[454] of aqueous ethanol and acetone solutions containing V^{IV} and SCN^- ions. An i.r. spectroscopic examination[455] of $M_2[VO(CNS)_4,MeCN]$ and $M_2[VO(CNS)_4]$ ($M^+ = Et_4N^+$) has increased the amount of spectroscopic information on vanadyl thiocyanate complexes. Two types of V—N bond are evident; the V—N(NCS) bond absorbing in the region $360—390\ cm^{-1}$ and the V—N(MeCN) bond adsorbing at lower frequencies *ca.* $210\ cm^{-1}$. The co-ordination of MeCN to the $VO(CN)_4^{2-}$ ion causes the expected reduction in v(V—O).

The pK_a and bulk of the ligand and the reaction conditions all influence the stoicheiometry of the products from the reactions of $VOCl_2,3MeOH$ with aliphatic and heterocyclic amines.[456] Methanolysis is favoured by bulky ligands with a high pK_a. At high ligand concentrations there is a greater tendency to adduct formation but even here sterically hindered ligands promote methanolysis. When an excess of $VOCl_2$, $3MeOH$ is employed, the product is the protonated base salt of the $[VOCl_2(OMe)]^-$ ligand; monobasic ligands yield 1:1 adducts, 4,4′-bipyridyl yields both 1:1 and 1:2 adducts, and *NNN′N′*-tetramethylethylenediamine only a 1:2 adduct.

At ligand to $VOCl_2,3MeOH$ ratios of 4:1 two types of complex are formed. Simple adducts $VOCl_2,xL$ $(x = 1.5, 2, or 3)$ and the products of methanolysis, $VO(OMe)_2$

[450] B. Jezowska-Trzebiatowska, E. Kaczurba, and P. Sobota, *Roczniki Chem.*, 1974, **48**, 349 (*Chem. Abs.*, 1974, **80**, 152300).
[451] G. Constant, J. C. Daran, Y. Jeannin, and R. Morancho, *J. Coord. Chem.*, 1973, **2**, 303.
[452] N. I. Vorob'ev, V. V. Pechkovskii, and L. V. Kobets, *Zhur. neorg. Khim.*, 1973, **18**, 2171.
[453] A. Anagnostopoulos, D. Nicholls, and M. E. Pettifer, *J.C.S. Dalton*, 1974, 569.
[454] N. B. Kalinichenko, I. N. Marov and A. N. Ermakov, *Zhur. neorg. Khim.*, 1873, **18**, 1260.
[455] J. P. Brunette and M. J. F. Leroy, *Compt. rend.*, 1974, **278**, C, 347.
[456] G. W. A. Fowles, D. A. Rice, and J. D. Wilkins, *Inorg. Chim. Acta*, 1973, **7**, 642.

and amine hydrochloride. 1:3 Complexes result from the reactions of neat pyridine, 4-methylpyridine, and isoquinoline with $VOCl_2,3MeOH$ whereas, with 2- and 3-methylpyridine partial methanolysis occurs to give $VOCl(OMe)2$-Mepy and VOCl-$(OMe)2(3$-Mepy$)$.

The extraction of V^{IV} into n-butanol in the presence of salicylic acid is enhanced by neutral donor molecules which act by displacing water from the co-ordination sphere of the vanadium thus rendering the complex less hydrophilic.[457] Extraction efficiency increases in the order *o*-phen > isoquin > quin > 4-pic > 3-pic > py. Steric hindrance prevents co-ordination of 2-pic.

The heat of sublimation of vanadyl phthalocyanine has been determined as 223 ± 8 kJ mol^{-1}.[458]

E.p.r. studies[459] indicate the formation of sandwich-type dimer complexes of VO^{2+} and etioporphyrin (EP) in light petroleum at 77 K; the e.p.r. data are consistent with an EP–EP plane distance of *ca.* 3.5 Å. A planar ligand local environment for V^{IV}, possibly provided by a phthalocyanine- or porphyrin-type ligand, has been proposed[460] to account for the axial microsymmetry of the V^{IV} centre and the absence of zero-field splitting indicated by e.p.r. studies of vanadium in mineral oil.

Compounds with Metal–Carbon Bonds. Treatment of $PhCH_2MgCl$ with VCl_4 in a 6:1 ratio at $-18\,°C$ in Et_2O gave unstable benzyl complexes $(PhCH_2)_nVCl_{4-n}$ where $n < 4$. Pyrolysis at 150 °C gave $(PhCH_2)_2V,MgCl_2,Et_2O$.[461] The i.r. spectra of $Cp_2V(OBz)_2$ and $(C_5D_5)_2V(OBz)_2$ have been analysed.[225] Tetrakis-(1-norbornyl)-vanadium,$(nor)_4V$, only gives an e.p.r. spectrum below 143 K because of its very short spin–lattice relaxation time.[462]

An X-ray diffraction study of potassium oxopentacyanovanadate(IV) has shown it to contain discrete $[VO(CN)_5]^{3-}$ ions in which the V=O bond distance is 1.64 Å,[463] the mean V—C(eq) bond distance is 2.14 Å and the greater V—C(ax) bond distance of 2.31 Å is characteristic of a vanadium–ligand bond *trans* to the V=O^{2+} entity.

Miscellaneous. The stable dark-green dinuclear complex $[VORu(NH_3)_5(H_2O)_n]^{4+}$, containing the VO^{2+} ligand, was obtained from an attempted oxidation of $Ru(NH_3)_5$-H_2O^{2+} with VO^{2+}.[464] It can also be obtained from VO^{2+} and $Ru(NH_3)_6^{2+}$, but here the acid-catalysed formation of $Ru(NH_3)_5H_2O^{2+}$ is the rate-determining step; the pentammine complex lies on the route to the VO^{2+} complex from $Ru(NH_3)_5Cl^{2+}$ and $V(H_2O)_6^{3+}$, being formed in a redox electron-transfer reaction between VO^+ or $V(OH)_2^+$ and the Ru^{III} species. The stability of the dinuclear complex and the decrease of 30 cm^{-1} in $v(V=O)$ which occurs on co-ordination of the VO^{2+} species are interpreted in terms of the VO^{2+} ligand acting as an analogue of CO or N_2 and using its vacant antibonding π-orbitals for back donation from the d_{xz} and d_{yz} orbitals of the ruthenium atom.

[457] V. P. Ranga and V. V. R. Sastry, *J. Inorg. Nuclear Chem.*, 1974, **36**, 415.
[458] A. G. Kay, *Austral. J. Chem.*, 1973, **26**, 2425.
[459] M. P. Tsvirko and K. N. Solov'ev, *Zhur. Priklad Spektroskopii*, 1974, **20**, 115 (*Chem. Abs.*, 1974, **80**, 138 848).
[460] G. Galambos, G. Korosi, P. Siklos, and F. Judos, *Magyar Kém. Folyóirat*, 1973, **79**, 364 (*Chem. Abs.*, 1973, **79**, 110040).
[461] V. N. Latyaeva, A. N. Lineva, and V. V. Drobotenko, *Trudy Khim. i khim. Tekhnol.*, 1973, 35 (*Chem. Abs.*, 1974, **80**, 83176).
[462] B. K. Bower and C. W. James, *Inorg. Chem.*, 1974, **13**, 759.
[463] S. Jagner and N. G. Vannerberg, *Acta Chem. Scand.*, 1973, **27**, 3482
[464] H. De Smedt, A. Persoons, and L. De Maeyer, *Inorg. Chem.*, 1974, **13**, 90.

Vanadium(v).—*Halides and Oxyhalides.* Complex fluoroanion formation by VF_5 has been studied by ^{19}F n.m.r. spectroscopy[465] and by i.r. and Raman spectroscopy and X-ray diffraction.[466] VF_5 is soluble in sulphuryl chlorotrifluoride; the solution gives a single ^{19}F n.m.r. signal at -75 °C but three lines in the intensity ratio 2:2:1 at -135 °C.[455] Such a spectrum is indicative of linear or cyclic chains of VF_5 molecules in an octahedral configuration with *cis*-fluorine bridges, and shows that the structure of VF_5 in solution at low temperature is similar to that in the solid state. $AgVF_6$ hydrolyses very readily in moist air and it now seems clear that the n.m.r. spectrum previously attributed to the VF_6^- ion is that of the hydrolysis product $VF_4(OH)_2$ with the OH groups axial. The fluorine resonance spectrum of an equimolar solution of $Bu_4^nNTaF_6$ and VF_5 in sulphuryl chlorotrifluoride at -130 °C is assigned to structure (52). One feature of the spectrum of $VTaF_{11}^-$ which is unique for the Group VB

(52)

elements is the absence of observable spin coupling for J_{c-d} and J_{d-e}. The chemical shifts of the fluorines bound to Ta indicate that VF_5 is a much weaker electron-withdrawing group than the other Group V pentafluorides. $ClOF_3$ combines with VF_5 to give a white moisture-sensitive solid adduct $ClOF_2^+VF_6^-$ which has a dissociation pressure of 2.5 mmHg at room temperature.[466] Both orbital valence and Urey–Bradley force fields were calculated for the VF_6^- ion.

The photolytic and electrolytic dissociation of $VOCl_3$ and VF_5 have been investigated[467] and evidence presented for the formation of a V_2 molecule with a lifetime of *ca.* 150 μs and an upper dissociation limit of 60 kcal mol^{-1}.

Ammonium bifluoride reacts with ammonium vanadate at 75 °C to give $(NH_4)_3$-VO_2F_4.[468] This undergoes thermal decomposition to a mixture of $(NH_4)_2VO_2F_3$, $(NH_4)_3V_2O_4F_3$, and $NH_4VO_2F_2$. X-Ray diffraction studies have established the polymeric nature of the $VO_2F_2^-$ ion, the structure (53) consists of infinite chains of strongly distorted octahedra with common edges. Two adjacent polyhedra are linked by asymmetrical oxygen and fluoride bridges. Oxotetrafluorovanadates prepared by the reaction of V_2O_5 with MF (M = Na, K, Rb, Cs, Tl, or Me_4N) in 40% HF also have infinite chain structures but with only fluorine bridges.[470] The V—F bridge bond lengths are 1.875 and 2.333 Å and the terminal V—O and V—F bond lengths are 1.572 and 1.793 Å, respectively. Compounds of the type $VOF_3(LL)$ (LL = *o*-phen or bipyO$_2$), $VO_2F(LL)$ (LL = *o*-phen or bipy) and VOF_3L_2 (L = pyO or picO) are the first examples of molecular derivatives of oxofluorovanadium(v) with organic

[465] S. Brownstein and G. Latremouille, *Canad. J. Chem.*, 1974, **52**, 2236.
[466] R. Bougon, T. Bui Huy, A. Cadet, P. Charpin, and R. Rousson, *Inorg. Chem.*, 1974, **13**, 690.
[467] P. I. Stepanov, N. N. Kabankova, E. N. Moskvitina, and Yu. Ya. Kuzyakov, *Vestnik Moskov Univ., Khim.*, 1973, **14**, 645 (*Chem. Abs.*, 1974, **80**, 101940).
[468] E. G. Rakov, L. K. Marina, B. V. Gromov, and M. M. Ol'Shevskaya, *Zhur. neorg. Khim.*, 1973, **18**, 1220.
[469] R. Mattes and H. Rieskamp, *Z. anorg. Chem.*, 1973, **399**, 205.
[470] (*a*) H. Rieskamp and R. Mattes, *Z. anorg. Chem.*, 1973, **401**, 158; (*b*) A. J. Edwards, D. R. Slim, J. Sala-Pala, and J. E. Guerchais, *Compt. rend.*, 1973, **276**, C, 1377.

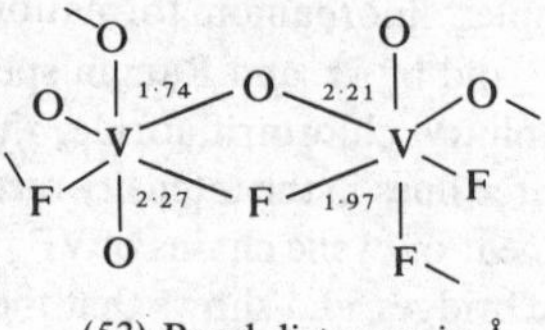

(53) Bond distances in Å

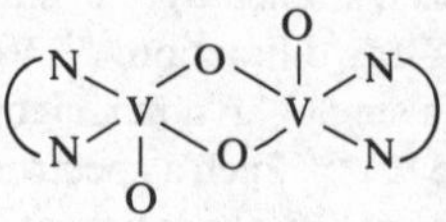

(54)

ligands.[470a] Preliminary *X*-ray diffraction results on $VO_2F(bipy)$[470b] indicate a distorted octahedral stereochemistry for the vanadium achieved by formation of an asymmetric di-μ-oxo dimeric structure (54) with V—O (bridge) distances of 1.69 and 2.35 Å and a V—O (terminal) distance of 1.60 Å. Thus the VO_2F derivative differs from corresponding derivatives of VOF_3 and CrO_2F_2 where the metal's octahedral stereochemistry is obtained by means of fluorine bridging. The asymmetry of $VO_2F(bipy)$ is also evident in the V—N(bipy) bond lengths of 2.14 and 2.19 Å. Liquid vapour equilibria have been studied for the binary system $VOCl_3$–$SiCl_4$.[471] Halogen exchange occurs between $VOCl_3$ and HBr at 0 °C affording a convenient route to pure $VOBr_3$.[453]

O-Donor Ligands. Table 3 summarizes much of the work on vanadates and mixed oxides containing vanadium(v) which has been reported this year.

Table 3 *Vanadates and mixed-oxide compounds containing vanadium(V)*

Compound	Properties Reported	Ref.
V_2O_5	O exchange between O and V_2O_5	a
V_6O_{13}	X	b
V_nO_{2n-u} $n = 5$ and 8	Thermodynamic properties	c
$LiVO_3$	X	d
$LiVO_3$-	X	e
KVO_4	i.r., t.d.	f
$Mg_2V_2O_7$	X	g
$Ca_2V_2O_7$	X	g
BaV_2O_6,H_2O	X, t.d.	h
$SrV_2O_6,4H_2O$	X, t.d.	h
BaV_2O_6	X	i
M_2O_3–V_2O_5 (M = B, Al, Ga, In, or Tl)	X	j
PbV_2O_6	X	k
α-$Cu_2V_2O_7$	X (low-temperature form)	l
β-$Cu_2V_2O_7$	X (high-temperature form)	m
$Ni_2V_2O_7$	n.m.r.	n
$Ag_{2-x}V_4O_{10}$, $x = 0.57$	X	o
$Ag_{4-x}V_4O_{12}$, $x = 1.05$	X	p
MVO_4 (M = La, Nd, or Y), Nd_5VO_{10}	Prep, optical props.	q
MVO_4 (M = Pr, Sm, Nd, Eu, or Lu)	^{51}V n.m.r.	r
MVO_4, $M_4(V_2O_7)_3$, $M(VO_3)_3$, (M = Ce or Pr)	Potentiometric titration of formation reaction	s
$SmVO_4$, $Sm_4(V_2O_7)_3$, $Sm_4(V_4O_{16})$, $Sm_2V_{10}O_{28}$	Potentiometric titration of formation reaction	t

[471] K. V. Tret'yakova and L. A. Nisel'son, *Izvest. V.U.Z. Tsvet. Met.*, 1974, **17**, 106 (*Chem. Abs.*, 1974, **81**, 17370).

Table 3 *(cont.)*

Compound	Properties Reported	Ref.
$LiMgVO_4$	X	u
$(NH_4)_2Ca_2V_{10}O_{28},17H_2O$	i.r., X t.g.a.	v
$NH_4Ca(VO_3)_3,4H_2O$		
$KCa(VO_3)_3,5H_2O$	Preparation	w
$K_2Zn_2V_{10}O_{28},16H_2O$		
$K_2Cd_2V_{10}O_{28},16H_2O$		
$K_2Cd_2V_{10}O_{28}$	t.g.a., i.r.	x
$K_4CdV_{10}O_{28},10H_2O$	X	
$Cs_4CdV_{10}O_{28},11H_2O$		
$AlIn(VO_4)_2$		
$GaIn(VO_4)_2$		
α-, β-, and γ-TlIn$(VO_4)_2$		
α- and β-GaTl$(VO_4)_2$	X	j
α-, β-AlTl$(VO_4)_2$		
Tl_3VO_4		
$TlV_3O_9, Tl_2V_8O_{23}$		
$M_xV_yMo_{1-x}O_3 (x = 0.13)(M = K, Rb, or Cs)$	X	y
$Cs_4Mo_2V_6O_{23}$		
$Cs_4Mo_2V_{10}O_{33}$	t.g.a.	z
$Cs_2MoV_{10}O_{28-75}$ (a bronze)		
$Na_5V_3W_3O_{19},18H_2O$		
$Na_4V_2W_5O_{19},14H_2O$		
$Na_5V_4W_8O_{37.5},26H_2O$	t.g.a., i.r.	aa
$Na_5V_2W_{10}O_{37.5},8H_2O$		
$Li_5V_3W_3O_{19},26H_2O$		
$V_{1-x}M_xO_2$ (M = Cr, Mo, or W)	X, mag., t.g.a.	bb
$(x = 0—1)$		
$V_{1-x}Ti_xO_2$	Raman, conduction	cc
$Mg(VO_3)_2 Mg(V_4O_7)$	X	dd
TiO_2–VO_2 Solid solutions	X	ee
$Ca(VO_3)_2$–KVO_3	phase diagram	ff

(a) N. D. Gol'dshtein, O. V. Zamyantia, Yu. A. Mishchenko, and A. I. Gel'bshtein, *Zhur. fiz. Khim.*, 1973, **47**, 1549; (b) I. Kawada, N. Mitsuko, M. Saeki, M. Ishii, N. Kimizuka, and M. Nakihara, *J. Less-Common Metals*, 1973, **32**, 171; (c) I. A. Vasil'eva, I. S. Sukhushina, and R. F. Balabaeva *Zhur. fiz. Khim.*, 1973, **47**, 2162; (d) P. Mahe and M. R. Lee, *Compt. rend.*, 1973, **277**, 307; (e) R. Swanson, G. W. Martin, and R. S. Feigleson, *J. Cryst. Growth*, 1973, **20**, 306; (f) N. A. Vorob'eva and G. A. Bogdanov, *Izvest. V.U.Z., Khim. i khim. Tekhnol.*, 1973, **16**, 1623 (*Chem. Abs.*, 1973, **79**, 55401); (g) J. C. Pedregosa, E. J. Baran, and P. J. Arymonino, *Z. Kristallogr. Kristallgeom. Kristallphys. Kristallchem.*, 1973, **137**, 221 (*Chem. Abs.*, 1974, **80**, 113 561 d); (h) S. V. Alchangyan and I. P. Kislyakov, *Ref. Zhur. Khim.* 1973, Abs. No. 3B811 (*Chem. Abs.*, 1973, **79**, 12983 s); (i) S. Launay and J. Thoret, *Compt. rend.*, 1973, **277**, D, 541; (j) B. G. Golovkin and A. A. Fotiev, *Zhur. neorg. Khim.*, 1973, **18**, 2574; (k) B. D. Jordan and C. Calvo, *Canad. J. Chem.*, 1974, **52**, 2701; (l) D. Mercurio-Lavaud and B. Frit, *Acta Cryst.*, 1973, **29B**, 2737; (m) D. Mercurio-Lavaud and B. Frit, *Compt. rend.*, 1973, **277**, C, 1101; (n) V. N. Lisson, R. N. Pletnev, and V. A. Grubanov, *Zhur. fiz. Khim.*, 1973, **47**, 1313; (o) Yu. N. Drozdov, E. A. Kuz'min, and N. V. Belov, *Doklady Akad. Nauk S.S.S.R.*, 1973, **210**, 339; (p) Yu. N. Drozdov, E. A. Kuz'min, and N. V. Belov, *Kristallografiya*, 1974, **19**, 65 (*Chem. Abs.*, 1974, **80**, 101 092); (q) E. N. Isupova, N. A. Godina, and E. K. Keler, *Izvest. Akad. Nauk S.S.S.R., neorg. Materialy*, 1973, **9**, 1214; (r) R. N. Pletnov, V. N. Lisson, V. A. Gubanov, and A. K. Chirkov, *Fiz. Tverd. Tela*, 1974, **16**, 289 (*Chem. Abs.* 1974, **80**, 102 048); (s) R. S. Saxena and M. C. Jain, *J. Indian. Chem. Soc.*, 1973, **50**, 77; (t) G. S. Shivahare and N. D. Joshi, *ibid.*, p. 429; (u) M. Th. Paques-Ledent, *Chem. Phys. Letters*, 1974, **24**, 231; (v) M. Sivak and L. Zurkova, *Chem. Zvesti*, 1973, **27**, 756; (w) A. A. Ivakin, A. P. Yatsenko, and M. P. Glazyrin, *U.S.S.R.P.* 391 062 (*Chem. Abs.*, 1974, **81**, 5150); (x) L. Ulicka and C. Vargova, *Chem. Zvesti*, 1973, **27**, 152; (y) B. Darriet and J. Galy, *J. Solid State Chem.*, 1973, **8**, 183; (z) B. V. Slobodin, M. V. Mokhosoev, and P. T. Zvoleiko, *Zhur. neorg. Khim.*, 1974, **19**, 388; (aa) D. F. Takezhanova, D. U. Begalieva, A. K. Il'yasova, and A. Bekturov, *Doklady Akad. Nauk S.S.S.R.*, 1974, **215**, 1139; (bb) M. Nygren, *Chem. Comm. Univ. Stockholm*, 1973; (cc) L. L. Chase, *Phys. Letters*, 1973, **46**, 215; (dd) E. Pollert, *Silikaty*, 1973, **17**, 337; (ee) F. Mailot and R. A. Paris, *Compt. rend.*, 1973, **277**, C, 1361; (ff) M. P. Glazyrin, A. A. Ivakin, S. I. Alyamovskii, and A. P. Yatsenko, *Zhur. neorg. Khim.*, 1974, **19**, 840.

An oxoperoxovanadium(v) intermediate $[VO(O_2)]^+$ has been proposed[471a] to occur in the oxygen-dependent part of the reaction between V^V and I^- ions in the presence of oxygen. This intermediate breaks down in the presence of an excess of I^- ions according to the equation

$$[VO(O_2)]^+ + 2I^- + 2H^+ \rightarrow VO_2^+ + I_2 + H_2O$$

The main features of the single-crystal Raman spectrum of vanadinite $Pb_5(VO_4)_3Cl$ have been accounted for.[472] New data on the isomorphism and synthesis of vanadium garnets have been presented.[473] Uranite forms the intercalation compounds C_nH_{2n+1}-$NH_3[VO_2V_3O_9]$ $(n = 6—18)$ by cation exchange at pH 5—6.[474] The alkyl chains are poorly ordered in monolayers and orientated nearly perpendicular to the μO_2 trimetavanadate layers. On cation exchange at pH 7—9 intercalation compounds containing up to one additional alkylamine are formed, which contain close-packed, well-ordered alkyl monolayers. With excess amines or with alkanols, 1:3 intercalation compounds are formed with the alkyl chains ordered to bimolecular layers.

According to thermodynamic data for 400 °C, lower oxides of vanadium are the likely products of the reaction between V_2O_5 and potassium. V_2O_5 reacts readily with liquid potassium to give VO and KVO_2 but subsequent oxidation by oxygen dissolved in the liquid metal results in the formation of K_3VO_4.[377a] In a detailed investigation[475] of the alkali-rich portion of the K_2O–V_2O_5 system several potassium oxo-compounds were allowed to react with V_2O_5. The compounds KVO_3 (pale orange) $K_{32}V_{18}O_{61}$ (very pale green) and $K_4V_2O_7$ (grey-green) were obtained from K_2CO_3 and V_2O_5 in the molar ratios 1:1, 16:9, and 2:1, respectively, at 700 °C in oxygen. The colours of the compounds and their paramagnetism, which was in excess of that predicted on the basis of the combined V_2O_5, were attributed to reduced vanadium centres created by O_2 loss. KVO_3 with only a very slight colour was obtained from KOH and V_2O_5 and from KOH and NH_4VO_3. KVO_3, $K_{32}V_{18}O_{61}$, $K_4V_2O_7$, and K_3VO_4 were shown to lose oxygen on heating and three forms of K_3VO_4 were identified. A study of the V_2O_5–Na_2CO_3 system has been carried out.[377b]

Three papers from the same group are devoted to reports on the acidification of vanadate solutions.[476a-c] Above pH 13 the major species in equilibrated solutions of vanadate is the VO_4^{3-} ion, at lower pHs protonated decavanadates form. The VO_4^{3-} ion reacts with an excess of acid in at least two ways.[476a] Rapid reactions lead to the formation of an unknown intermediate and decavanadate, both species then degrading to the final product VO_2^+. The results of a kinetic study[476b] of the breakdown of decavanadate are interpreted in terms of an equilibrium between $H_2V_{10}O_{28}^{4-}$, $H_3V_{10}O_{28}^{3-}$, and $H_4V_{10}O_{28}^{2-}$ with the decomposition occurring through $H_4V_{10}O_{28}^{2-}$. Large cations such as R_4N^+ and R_3S^+ markedly decrease the rate of the reaction between $V_{10}O_{28}^{4-}$ and acid due to the formation of stable ion pairs.[476c] A depolarized light-scattering study[477] of vanadate solutions in the pH range 14—8.5 indicates the

[471a] S. Celsi, F. Secco, and M. Venturini, *J.C.S. Dalton*, 1974, 793.

[472] D. M. Adams and I. R. Gardner, *J.C.S. Dalton*, 1974, 1505.

[473] B. V. Mill and G. Ronniger, *Fiz. Khim. Ferritov*, 1973, 98 (*Chem. Abs.*, 1973, **79**, 99913).

[474] K. Beneke, U. Grosse-Brauckmann, and G. Lagaly, *Z. Naturforsch.*, 1973, **28b**, 408.

[475] M. G. Barker and A. J. Hooper, *J.C.S. Dalton*, 1973, 2614.

[476] (a) B. W. Clare, D. L. Keppert, and D. W. Watts, *J.C.S. Dalton*, 1973, 2476;
(b) B. W. Clare, D. L. Keppert, and D. W. Watts, *ibid.*, p. 2479;
(c) B. W. Clare, D. L. Keppert, and D. W. Watts, *ibid.*, p. 2481.

[477] A. Gaglani, N. Asting, and W. H. Nelson, *Inorg. Chem.*, 1974, **13**, 1715.

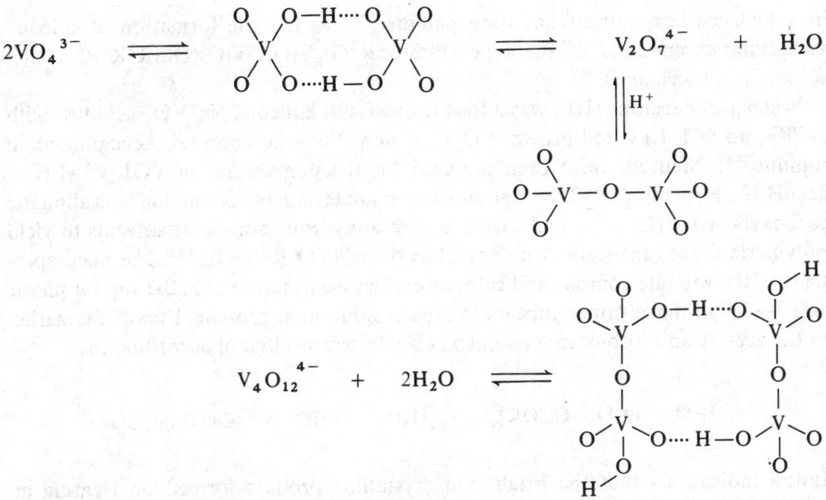

Scheme 1

formation of isopolyvanadate species as the pH is lowered (Scheme 1). These results are consistent with ^{51}V n.m.r. and potentiometric titration studies. Molecular sieving of isopolyvanadates on dextran and polyacrylamide gels occurs at pH 6 but at lower pHs chelate formation takes place with the gel matrix.[478]

Three complexes $VO_2H_2PO_4$, $VO_2HPO_4^-$, and $VO_2(HPO_4)_2^{3-}$ have been identified[479] in solutions containing VO_2^+ and orthophosphate ions. The preparation and characterization of 11 tungstovanadate(v) complexes have been reported;[480] their interconversions are summarised in Scheme 2. Attempts to reduce the ions $V_3W_{10}O_4O^{5-}$ or $V_4W_9O_4O^{6-}$ lead to products of variable composition containing V^V and V^{IV}. The preparation of triethyl orthovanadate from $VOCl_3$, EtOH, and NH_3

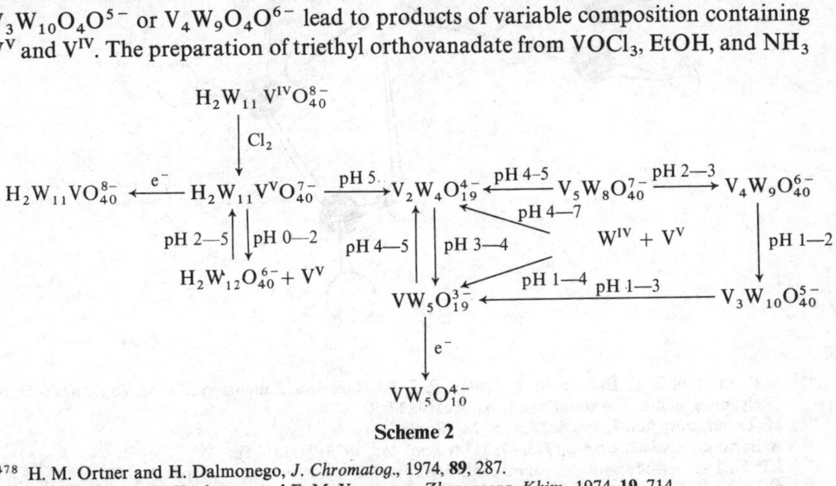

Scheme 2

[478] H. M. Ortner and H. Dalmonego, *J. Chromatog.*, 1974, **89**, 287.
[479] A. A. Ivakin, L. D. Kurbatov, and E. M. Voronova, *Zhur. neorg. Khim.*, 1974, **19**, 714.
[480] C. M. Flynn, M. T. Pope, and S. O'Donnell, *Inorg. Chem.*, 1974, **13**, 831.

in a hydrocarbon solvent has been patented[481] as has the formation of silicon-vanadium compounds of the type $(PhSiR^1R^2O)_2V(O)OV(O)(OSiR^1R^2Ph)_2$ (R^1, $R^2 = C_{1-4}$ alkyl, aryl).[482]

When acetylacetone (HL) was added dropwise to aqueous Na_3VO_4 acidified with $HClO_4$ to pH 1.14, red-brown VO_2L, a new V^V acac complex, precipitated in minutes.[483] Methods have been reported for the preparation of VOL_2Cl (LH = acacH, tfacH, or dbm).[484] Tris-t-butylorthovanadate undergoes reaction with aliphatic carboxylic acids (C_1—C_5) or benzoic acid in anhydrous non-polar solvents to yield polymeric dioxovanadium(v) monocarboxylates, $(VO_2RCOO)_n$.[485] The mass spectrum of the valerate complex indicates its existence as a tetramer in the vapour phase, and 2—3 % of the tetramer appears to exist in solution in benzene. Depolymerization of the acetate and isobutyrate complexes is effected by ethanol according to:

$$\text{\textlbrackdbl}O-\overset{|}{V}(=O)-OCOR]_n + nC_2H_5OH \rightarrow nHO-V(=O)(OEt)(OCOR)$$

Initial indications that the bright red crystalline product formed on treating an alkaline vanadate solution containing oxalate ions with H_2O_2 was the monoperoxo-vanadate, $K_3[VO(O_2)(C_2O_4)_2],2H_2O$, have been shown to be incorrect.[486] An X-ray diffraction study has shown the compound to be the potassium salt of the [bis(oxalato)-dioxovanadium(v)]$^{3-}$ ion. The anion has irregular octahedral geometry (55) with

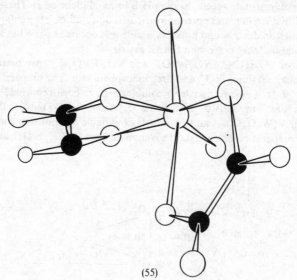

(55)

[481] A. F. Zhigach, B. M. Esel'sen, A. F. Popov, P. Z. Sorokin, A. S. Zahhartseva, V. M. Zapletnyak, D. P. Chelyanova, and R. E. Propof'eva, U.S.S.R.P., 422739.

[482] M. Takamizawa and T. Fujino, *Jap. P.* 73, **48**, 423.

[483] A. Bartecki, *Roczniki Chem.*, 1973, **47**, 217 (*Chem. Abs.*, 1973, **79**, 13025).

[484] J. P. Fackler, S. Anderson, J. P. Jones, and S. J. Kopperl, *Synth. React. Inorg. Metal-Org. Chem.*, 1974, **4**, 49.

[485] F. Preuss, J. Woitschach, and H. Schug, *J. Inorg. Nuclear Chem.*, 1973, **35**, 3723.

[486] R. E. Drew, F. W. B. Einstein, and S. E. Gransden, *Canad. J. Chem.*, 1974, **52**, 2184.

the two double-bonded oxygen atoms *cis* to each other. The V—O (oxalate) bonds are of two significantly different types: those that are *trans* to each other (2.009, 1.995 Å) and those that are *trans* to oxo ligands (2.158, 2.230 Å) indicating the marked *trans*-effect of the oxo ligand. The preparation[487] of $M_2[VO_2F(C_2O_4)]$ (M = Na, K, Rb, Cs, NH_4, or Me_4N) has extended the range of complexes containing fluoride, oxygen, and oxalate together as ligands, only Mo and W complexes being known previously. Since a co-ordination number of six is most probable for the vanadium atom, fluorine bridging probably occurs; the V=O and V—O regions of the i.r. spectra are identical with those of VO_2(oxalate)$_2$. A brief note of the preparation of $(NH_4)_3V(O_2)_2F_4$ has appeared.[488]

S-*Donor and Se-Donor Ligands.* Dark violet crystals of the cubic thiometallate Tl_3VS_4 are formed in the reaction between VS_2 and Tl_2S at 500 °C;[489] this reaction involves disproportionation of the VS_2 ion, stoicheiometric $VS_{0.95}$ being the other vanadium-containing product. Single crystals of $Fe_2V_2S_4$ and $Fe_2V_2Se_4$ have been produced by isothermal vapour growth in closed tubes[50] and the crystallographic parameters of the ternary sulphide VUS_3 have been determined.[490] The possibility that the complex originally formulated as $VO(S_2CNEt)_3$ could contain co-ordinated peroxide has been ruled out by an *X*-ray crystallographic study of this compound.[491] The metal atom is seven co-ordinate, lying at the centre of a pentagonal bipyramid with the oxygen at one of the axial apices and sulphur at the other six sites. In common with other vanadium complexes containing the VO unit, the axial M–S distance is appreciably greater (2.75 Å) than the equatorial (*av.* 2.55—2.60 Å).

N-*Donor and O-Donor Ligands.* Quinolinol complexes of vanadium(v) have been investigated as potential models for vanadium compounds of biological interest.[492, 493] Strong covalent metal–ligand oxygen bonding but only weak labile metal–nitrogen interactions are indicated by the n.m.r. parameters of the 8-quinolinol and 2-methyl-8-quinolinol complexes $VO(OH)(H_2O)L_2$ in DMSO. The complex VOL_2I (L = 8-quinolinol) precipitates when an aqueous solution of KI and the ligand is added in excess to a vanadium(v) solution at pH 4.[494] Rate constant and activation energy data have been obtained[495] for the oxidation of edta and $HO_2CCH_2N-[CH_2CH_2N(CH_2CO_2H)_2]_2$ by V^V which proceeds *via* the intermediate complexes $VO_2H_2A^-$. Oxidation of the tetracarboxylic acid is slower than that of the penta-carboxylic analogue.

Equilibrium[496] and kinetic[497] studies have also been reported for the vanadium(v)–hydroxylamine system. The complex VO_2NH_3OH is suggested to be the major

[487] H. Reiskamp and R. Mattes, *Z. Naturforsch.*, 1974, **29b**, 271.
[488] G. Pausewang, *Z. Naturforsch.*, 1974, **29b**, 272.
[489] V. Schmidt and W. Ruedorff, *Z. Naturforsch.*, 1973, **28b**, 25.
[490] H. Noel, *Compt. rend.*, 1973, **277**, C, 463.
[491] J. C. Dewan, D. L. Keppert, C. L. Raston, D. Taylor, A. H. White, and E. N. Maslen, *J.C.S. Dalton*, 1973, 2082.
[492] L. W. Amos and D. T. Sawyer, *Inorg. Chem.*, 1974, **13**, 78.
[493] L. W. Amos, *Diss. Abs. (B)*, 1974, **34**, 3135.
[494] Y. Anjaneyulu, A. S. R. Murty, R. V. Krishna, and R. Pandu, *Proc. Nat. Acad. Sci. India (A)*, 1973, **43**, 84 (*Chem. Abs.*, 1974, **84**, 20316).
[495] R. P. Tishchenko, N. I. Nechurova, L. Martynenko, and I. Spitsyn, *Izvest. Akad. Nauk S.S.S.R.*, *Ser. khim.* 1973, 1212.
[496] G. Bengtsson, *Acta Chem. Scand.*, 1973, **27**, 2554.
[497] G. Bengtsson, *Acta Chem. Scand.*, 1973, **27**, 3053.

vanadium-containing species. Kinetic data are consistent with a two-stage decomposition process for the complex:

$$VO_2NH_3OH^{2+} \rightarrow VO^{2+} + NH_2O + H_2O$$
$$2NH_2O \rightarrow N_2 + 2H_2O$$

4 Niobium and Tantalum

Introduction.—The compilation of the inorganic chemistry of niobium and tantalum has been revised.[341] Reviews have appeared on the physicochemical properties of tantalum compounds and alloys,[498, 499] on the extraction, properties, and uses of niobium and tantalum,[500] and on the structures of their compounds determined by diffraction methods.[339] The half-life of [182]Ta has been determined as 114.74 ± 0.08 days.[501]

Chemical shift data have been obtained by photoelectron spectroscopy[501a] for the core electrons of niobium and tantalum compounds. The similarity in chemical shifts between structurally similar compounds of these two elements is in accord with the similarity in their ionic radii.

Carbonyl and other Low Oxidation State Compounds.—An attempt to form a cyclopentadiene complex by displacement of the phosphine ligand from $CpNb(Ph_3P)_2(CO)_2$ resulted instead in the formation of a complex formulated as the dihydride $CpNb(Ph_3P)_2CO(H)_2$ (56).[502] In (56) the plane of the Cp ligand is almost parallel with the plane defined by the two phosphorus atoms and the carbon of the CO ligand. The P ligands are approximately *trans* and displaced slightly towards the carbonyl ligand indicating that hydride ligands could be co-ordinated in the 'vacant' co-ordination sites. A similar type of distortion, this time of a carbonyl ligand, indicates a probable location for the hydride ligand of $Ta(CO)_2H(Me_2PCH_2CH_2PMe_2)_2$ (57).[503] The phosphorus and carbonyl carbon atoms define a distorted octahedron about the tantalum atom and the location of the hydride atom in (57) gives a seven-co-ordinate

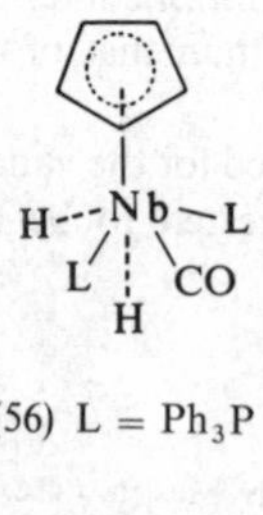

(56) L = Ph$_3$P

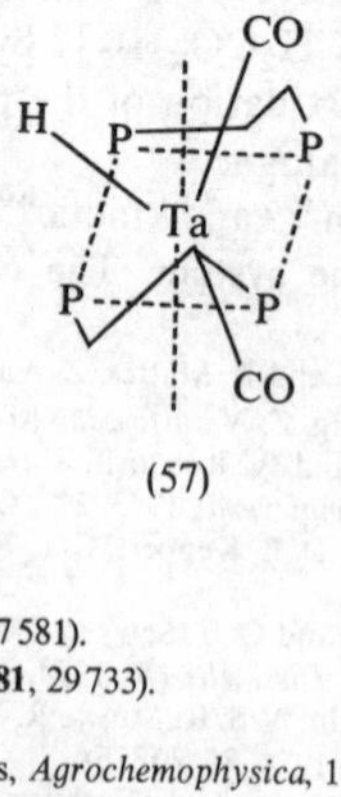

(57)

[498] R. Ferro, *Atomic Energy Rev.*, 1973, 67 (*Chem. Abs.*, 1973, **79**, 117581).

[499] A. L. Dragoo, *Atomic Energy Rev.*, 1972, 131 (*Chem. Abs.*, 1974, **81**, 29733).

[500] J. Van Der Planken, *Chem.–Ztg.*, 1973, **97**, 331.

[501] C. J. Visser, J. H. M. Karsten, F. J. Hassbrock, and P. G. Marais, *Agrochemophysica*, 1973, **5**, 15 (*Chem. Abs.*, 1974, **80**, 89692).

[501a] G. E. McGuire, G. K. Schweitzer, and T. A. Carlson, *Inorg. Chem.*, 1973, **12**, 2450.

[502] N. I. Kirillova, A. I. Gusev, A. A. Pasynskii, and Yu. T. Struchkov, *Zhur. strukt. Khim.*, 1974, **15**, 288.

[503] P. Meakin, L. J. Guggenberger, F. N. Tebbe, and J. P. Jesson, *Inorg. Chem.*, 1974, **13**, 1025.

structure which is a distorted capped octahedron. At low temperatures (8 °C) the continuous-wave mode hydride region ^{1}H n.m.r. spectrum of (57) is a triplet of triplets which can be fitted by an AA'BB' model and is consistent with the solution structure being the same as the crystal structure. The high-temperature (75 °C) limit spectrum is a binomial quintet indicating an exchange process which averages the phosphorus atoms. Lineshape analysis indicates two possible mechanisms of H exchange, one involving a bridging hydride intermediate present in low concentrations and another involving dissociation of one end of the diphosphine ligand. $Cp_2Nb(CO)SH$ adopts the wedge-shaped configuration of the Cp rings characteristic of molecules of the type Cp_2ML_2 (M = Metal of Group IVA or VA).[504] The large SH group causes an increase in the angle between the Cp ligands (45°) relative to that in $Cp_2NbCO(H)$.

A continuation of studies on the interaction of multiply-bonded molecules with alkyne complexes of niobium has resulted in the producton of the novel complex $Cp(PH_3)Nb[HCPh(CPh)_3C(=NH)Me]$ (58), with an unusual *hexahapto*-aza-allyl-allylic system, from $CpNb(CO)[PhC_2Ph]_2$ and MeCN.[505] Five carbon atoms and one nitrogen are arranged in a chain and separated from the central Nb atom by 2.1—2.5 Å. The phosphine ligand is apparently derived from the phosphorus pentoxide drying agent.

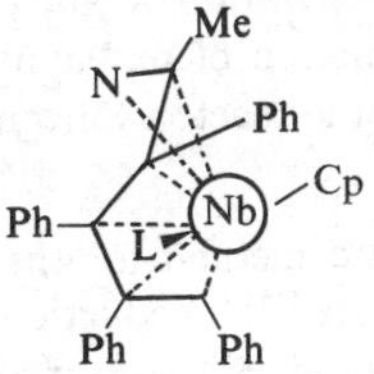

(58) L = PPh_3

Metal-substituted acetylenes $Ph_3MC{\equiv}CPh$ (M = Si, Sn, or Ge) form co-ordinatively unsaturated complexes $Cp_2Nb(CO)(Ph_3MC{\equiv}CPh)$ on reaction with $Cp_2Nb(CO)_4$; changes in reaction conditions or reactant ratios do not influence the stoicheiometry of the products.[506]

A tensimetric study of the $Ni(CO)_4$–$CpNb(CO)_4$ system has been carried out, solubilities of $CpNb(CO)_4$ in $Ni(CO)_4$ being given.[507] The e.p.r. spectrum of Cp_4Nb has been analysed.[508]

An energy-level diagram for $HTa(CO)_6$ has been constructed on the basis of MO calculations; the compound was prepared by treating Ta_2O_5 with a mixture of H_2 and CO.[509]

504 N. I. Kirillova, A. I. Gusev, A. A. Pasynskii, and Yu. T. Struchkov, *Zhur. strukt. Khim.*, 1973, **14**, 868.
505 N. I. Kirillova, A. I. Gusev, A. A. Pasynskii, and Yu. T. Struchkov, *J. Organometallic Chem.*, 1973, **63**, 311.
506 A. N. Nesmeyanov, N. E. Kolobova, A. B. Antonova, K. N. Anisomov, and O. M. Khitrova, *Izvest. Akad. Nauk S.S.S.R., Ser. khim.*, 1974, 859.
507 A. K. Baev and L. G. Fedulova, *Zhur. fiz. Khim.*, 1973, **47**, 2523.
508 Yu. A. Bobrov, A. D. Krivospitskii, and G. K. Chirkin, *Zhur. strukt. Khim.*, 1973, **14**, 813.
509 P. U. Torgashev, *Ref. Zhur. Khim.*, 1973, Abs. No. 18B34 (*Chem. Abs.*, 1974, **81**, 32702).

Binary Compounds.—*Halides.* Thermodynamic properties of $NbCl_5$[510] and $TaCl_5$[511] have been calculated from vapour pressure data obtained for the solid and liquid phases. Normal-co-ordinate analyses have been carried out[512] on the Nb and Ta pentahalides MX_5 (X = F, Cl, or Br); the results are in agreement with these published for X = Cl or Br but not for X = F. A Raman spectroscopic study[513] of molten $NbCl_5$ in the temperature range 220—320 °C indicates the existence of a monomer-dimer equilibrium. According to computer calculations on the system MF_5–H_2–HF–M (M = Nb or Ta) quantitative yields of crystalline M are not to be expected with stoicheiometric concentration ratios of MF_5 and H_2 at temperatures < 3000 K.[349] However, the theoretical yield can be obtained by using a two- to five-fold excess of H_2 and 7.4—16.4 vol% of MF_5 at 1600—1800 K for TaF_5 and 1300—1500 K for NbF_5.

A heat of fusion of 3.0 ± 0.05 kcal mol^{-1} has been obtained[514] for TaF_5; heats of evaporation are 12.92 kcal mol^{-1} for NbF_5 and 12.97 kcal mol^{-1} for TaF_5.

Published data have been used[515] to calculate the enthalpies and free energies of reaction for the chlorination of Nb_2O_5 and Ta_2O_5 with Cl_2 and S_2Cl_2 at 298 or 600 K. The Nb_2O_5 reaction goes in 100% yield at 260—270 K, *ca.* 100 K lower than for Ta_2O_5.

Convenient methods for the preparation of NbF_5 and NbF_4 have been reported.[516] NbF_5 is prepared by the fluorination of niobium metal with SnF_2, volatile NbF_5 condensing on the cooled lid of the reactor. Silicon is used as the reducing agent in the preparation of NbF_4.

Oxides. Details of an electric arc method for the preparation of pure NbO from Nb_2O_5 and Nb have been reported.[517] Reduction of Nb_2O_5 with ammonia in the temperature range 600—1300 °C gives the δ-oxynitride at 800 °C, NbO_2 between 800 and 820 °C and hexagonal NbN above 820 °C.[518]

Chalcogenides. NbS_2 and TaS_2 have been prepared by the calcination of the metal oxide with CS_2 under argon at 1000 °C.[43] Interest in the superconductivity of transition-metal dichalcogenides and their intercalation compounds has prompted a refinement of the structural parameters of NbS_2.[519] The structure consists of lamina or 'sandwiches' formed by a layer of niobium between two layers of sulphur atoms stacked upon each other, with three lamina in a repeat unit. The niobium has a six-fold trigonal prismatic co-ordination environment with the S—S separation along the ends or within the sulphur layer of 3.330 Å slightly longer than along the sides or between the sulphur layers (3.114 Å).

The preparation, supercondictivity, electrical conductivity, and electronic structure of $NbSe_2$ have been reviewed.[520] Electron diffraction and *X*-ray measurements on

[510] L. I. Staffansson and P. Enhag. *Acta Chem. Scand.*, 1973, **27**, 2733.

[511] L. I. Staffansson and P. Enhag, *Scand. J. Met.*, 1973, **2**, 305 (*Chem. Abs.*, 1974, **80**, 149396).

[512] S. P. So, *J. Mol. Structure*, 1973, **16**, 311.

[513] W. Bues, F. Demiray, and H. A. Oeye, *Z. phys. Chem.*, 1973, **84**, 1 (*Chem. Abs.*, 1973, **79**, 24088).

[514] G. C. Fedorov, A. A. Tsvetkov, E. G. Rakov, and B. N. Sudarikov, *Zhur. neorg. Khim.*, 1973, **18**, 2003.

[515] I. A. Glukhov, L. M. Shalukhina, and A. Sharipov, *Doklady Akad. Nauk. Tadzh. S.S.R.*, 1973, **16**, 28 (*Chem. Abs.*, 1973, **79**, 24138).

[516] F. P. Govtsema, *Inorg. Synth.*, 1973, **14**, 105.

[517] T. B. Reed and E. R. Pollard, *Inorg. Synth.*, 1973, **14**, 131.

[518] R. A. Guidotti and D. G. Kesterke, *Met. Trans.*, 1973, **4**, 1233 (*Chem. Abs.*, 1973, **79**, 56270).

[519] B. Morosin, *Acta Cryst.*, 1974, **30B**, 551.

[520] Y. Ishizawa, *Kotai Butsuri*, 1974, **9**, 209 (*Chem. Abs.*, 1974, **81**, 30761).

1T–TaS$_2$ have been shown[521] the existence of a superlattice containing 26 formula units per unit cell, consistent with the compound's observed physical properties.

Nitrides and Related Compounds. Ta$_3$N$_5$ obtained from the reaction of TaCl$_5$ with ammonia at temperatures < 800 °C is isostructural with anosovite (Ti$_3$O$_5$).[522] The Ta atoms are surrounded by six nitrogen atoms in a distorted octahedral arrangement with Ta–N distances ranging from 1.96 to 2.26 Å; TaN$_6$ octahedra are linked by edge- and corner-sharing. In the green polymorph of TaON,[523] four nitrogens are located on one side of the metal and three oxygens on the other; an ordered arrangement is expected on simple electrostatic grounds. Ta—Ta distances across the planes of the oxygens are substantially larger (3.90 Å) than those (3.30 Å) across the planes of the nitrogens, indicating a strong Ta–N interaction; this is further supported by the most striking feature of the TaON structure, the very short N–N distances 2.51—2.64 Å (compared with 3.65 Å in Li$_3$N and 2.8 Å in Th$_3$N$_4$) which are attributed to a substantial degree of covalency in the Ta—N interaction.

Silicides and Borides. NbSi$_2$ and TaSi$_2$ are much less resistant to oxidation than their vanadium analogue;[361] between 600 and 900 °C oxidation is so vigorous that samples are converted completely into a scale which, for niobium, consists of α-Nb$_2$O$_5$, β-Nb$_2$O$_5$, and Nb$_5$Si$_3$. At 1000 °C TaSi$_2$ is oxidized to β-Ta$_2$O$_5$, Ta$_2$N, and Si$_3$N$_4$.

Borides of Nb and Ta have been reviewed[67] and a method for their preparation from boron and metal oxide has been reported.[68]

Hydrides. According to X-ray and neutron diffraction and metallographic studies of the Nb–H system,[524] the H may be considered a lattice gas with phase transitions. In the α-, α′-, β-, and γ-phases of the system, H occupies tetrahedral interlattice positions. Whereas direct reaction between niobium metal and hydrogen occurs only after repeated 'activation' of the metal by hydrogen absorption at *ca.* 7 atm and 350 °C, NbH$_2$ is formed at temperatures as low as 22 °C in mixtures of LaNi$_5$H$_{6.7}$ and Nb.[525] The extraordinary catalytic effect of the lanthanum–nickel complex is attributed to the presence of surface-absorbed atomic hydrogen species which are able to diffuse into the niobium lattice. There has been a review of the Ta–H system.[526]

Compounds with Nb—Nb or Ta—Ta Bonds.—Further interest in the (M$_6$X$_{12}$)$^{n+}$ cluster cations (M = Nb or Ta; X = Cl or Br) prompted a search for a rapid and efficient method for the preparation of the hydrates M$_6$X$_{14}$,8H$_2$O. An improved synthesis involving the conproportionation of metal, metal halide, and alkali-metal halide has been developed.[527] This has the advantage of employing readily available starting materials, and since both oxidizing and reducing agents are transformed into product large amounts of product can be obtained even at low yields; furthermore, the metal is the reductant and, therefore, over-reduction of the pentahalide to the metal cannot occur. An X-ray diffraction study[528] of [Me$_4$N]$_3$[(Nb$_6$Cl$_{12}$)Cl$_6$] has shown the Nb—Nb distances to be 2.97 Å (av.) and the Nb—Cl distances to be 2.43 Å

521 F. J. Di Salvo, R. G. Maines, J. V. Waszczak, and R. Schwall, *Solid State Comm.*, 1974, **14**, 497.
522 J. Straehle, *Z. anorg. Chem.*, 1973, **40**, 47.
523 D. Armytage and B. E. F. Fender, *Acta Cryst.*, 1974, **30B**, 809.
524 M. A. Pick, *Ber. Kernforschungsanlage Juelich*, 1973, No. 951-FF. (*Chem. Abs.*, 1974, **80**, 41339).
525 D. H. W. Carstons and J. D. Farr, *J. Inorg. Nuclear Chem.*, 1974, **36**, 461.
526 R. Hanada, *Tetsu To Hagane*, 1973, **59**, A157 (*Chem. Abs.*, 1974, **81**, 29734).
527 F. W. Koknat, J. A. Parsons, and A. Vongvusharintra, *Inorg. Chem.*, 1974, **13**, 1699.
528 F. W. Koknat and R. E. McCarley, *Inorg. Chem.*, 1974, **13**, 295.

(bridging) and 2.52 Å (terminal), compared with Nb—Nb distances of 2.92 and 3.02 Å, Nb—Cl (bridging) distances of 2.48 and 2.42 Å and Nb—Cl (terminal) distances of 2.61 and 2.46 Å in $K_4(Nb_6Cl_{18})$ and $(Me_4N)_2Nb_6Cl_{18}$, respectively. Thus the Nb—Nb distances increase gradually and the Nb—Cl(terminal) distances decrease gradually upon stepwise removal of two electrons from an $(Nb_6Cl_{18})^{4-}$ anion. These results are in accord with an MO scheme for metal–metal bonding in the $(M_6X_{12})^{n+}$ cluster cations which has been described in terms of a combination of the d_{xy}, d_{xz}, d_{yz} and d_{z^2} orbitals of all six metal atoms leading to eight bonding and six anti-bonding MO's. In the $(M_6X_{12})^{2+}$ cation all eight bonding orbitals should be filled and the anti-bonding orbitals should be vacant. Oxidation of the cluster should, therefore, lead to an increase in Nb—Nb distances since electrons have been removed from bonding orbitals.

The possibility that one of the products of the reaction between $NbCl_5$, C_6Me_6, Al, and $AlCl_3$ at 130 °C has a hexanuclear structure analogous to $(Nb_6Cl_{12})Cl_6^{n-}$ ($n = 2$, 3, or 4) has been eliminated, at least for the solid state, by a single-crystal X-ray diffraction study[529] which has shown it to have the trinuclear structure (59).

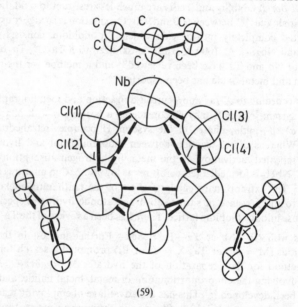

(59)

Niobium(III) and Tantalum(III).—Phase diagrams of the K_2NbCl_5–LiCl–LiF, K_2NbCl_5–KCl–LiF, and K_2NbCl_5–KF–LiF subsystems of the K^+, Li^+, Nb^{3+}, Cl^-, F^- system have been determined.[530] Zinc powder reduction of an aqueous niobium(v) solution affords diamagnetic red-violet crystals of $K[Nb^{III}(SO_4)_2,4H_2O]$ one of the few examples of inorganic Nb^{III} compounds.[531] As expected, the Nb^{III} species is readily oxidized, turning white on exposure to air owing to the formation of Nb^V,

[529] M. R. Churchill and S. W. Y. Chang, *J.C.S. Chem. Comm.*, 1974, 248.
[530] V. V. Safonov, G. M. Chumakova, and R. B. Ivnitskaya, *Izvest. V.U.Z. Tsvet. Met.*, 1974, **17**, 113 (*Chem. Abs.*, 1974, **81**, 17371).
[531] R. N. Gupta and B. K. Sen, *Z. anorg. Chem.*, 1973, **398**, 312.

and reducing Ag^I and Cu^{II} to the metals. Its strongly acidic solutions are red; they become blue on dilution and finally precipitate $Nb^{III}(OH)_3$. Complexes of the type $(MeCp)_2TaX(R^1C_2R^2)$ $(X = I \text{ or } H)$ $(R^1 = R^2 = Pr^i; R^1 = Me, R^2 = Bu^n, Pr^i, \text{ or } Bu^t)$ show remarkable stability for transition-metal complexes of dialkylacetylenes.[532] They have been prepared by the reactions:

$$(MeCp)_2TaH_3 + PhI + R^1C_2R^2 \rightarrow (MeCp)_2Ta(I)R^1C_2R^2$$
$$(MeCp)_2Ta(I)R^1C_2R^2 + LiAlH_4 \rightarrow (MeCp_2)Ta(H)R^1C_2R^2$$

Treatment of a solution of the hydride complex $(R^1 = R^2 = Pr^i)$ with HBF_4 causes the liberation of an almost quantitative yield of *cis*-oct-4-ene, probably *via* initial protonation of the metal. Diamagnetic allyl complexes Cp_2MR $(M = Nb^{III} \text{ or } Ta^{III}; R = \pi\text{-allyl}, \pi\text{-1-methylallyl}, \text{ or } \pi\text{-2-methylallyl})$ have been prepared[533] by the reductive alkylation of paramagnetic Cp_2MCl_2. There is a steady decrease in $v(C\equiv C)$ for the series $M = Ti$ (1509), Nb (1480), Ta (1450) cm^{-1} indicative of the increasing strength of the metal–olefin bond. Stepwise elimination of the allyl group for the Nb and Ta complexes compared with a single-step elimination for Ti is also consistent with a stronger metal–allyl bond for the heavier metals, whilst the instability of Nb and Ta complexes in solution is attributed to a greater polarity in the metal–carbon bond.

Niobium(IV) and Tantalum(IV).—Studies of magnetic exchange in paramagnetic cluster compounds have been extended to the niobium complex $(CpNbCO_2H)_3$-$(OH)_2O_2$.[534] The variation of magnetic susceptibility with temperature for this compound has been interpreted in terms of antiferromagnetic exchange $[(-J) \sim 225 \text{ K } (-156 \text{ cm}^{-1})]$ between three d^1Nb^{IV} ions located at the vertices of a regular triangle, and the equivalence of the niobium atoms has been confirmed by X-ray analysis. Each Nb atom is linked by two bridging ligands, formate and hydroxyl or bidentate oxygen. A distinction between these latter two ligands could not be made as hydrogen atoms were not located. One oxygen atom is linked to all three Nb atoms and lies 0.96 Å above the Nb_3 plane. Nb—Nb bond formation is considered unlikely on the grounds of the paramagnetism of the complex, the requirement for an increase in co-ordination number of two, and the large Nb—O—Nb bridging angles. $(dpm)_4Nb^{IV}$ has been shown[534a] to have a D_4 square-antiprismatic co-ordination polyhedron. This is the first report of this structural type for eight co-ordination.

Reduction of a methanolic solution of $NbCl_5$ at a mercury pool cathode followed by reaction with concentrated aqueous KCN afforded orange crystals of $K_4[Nb(CN)_8]$, $2H_2O$, the first example of an unsubstituted Nb^{IV}cyano-complex.[535] Further reduction gave red unstable $K_5[Nb(CN)_8]$. The Nb^{IV} complex is isomorphous with $K_4[Mo(CN)_8],2H_2O$, which is known to have a dodecahedral anion; the e.p.r. parameters of the solid are also consistent with a dodecahedral structure. A small but significant difference between the solid and solution e.p.r. parameters for $[Nb(CN)_8]^{4-}$ indicates a structural change, probably to antiprismatic. Tantalum(IV) compounds

[532] J. A. Labinger, J. Schwartz, and J. M. Townsend, *J. Amer. Chem. Soc.*, 1974, **96**, 4009.
[533] A. Van Baalen, C. J. Groenenboom, and H. J. De Liefde Meijer, *J. Organometallic Chem.*, 1974, **74**, 245.
[534] V. T. Kalinnikov, A. A. Pasynskii, G. M. Larin, V. M. Novolortsev, Yu. T. Struchkov, A. I. Gusev, and N. I. Kirillova, *J. Organometallic Chem.*, 1974, **74**, 91.
[534a] G. Podolsky, *Diss. Abs. (B)*, 1973, **33**, 5199.
[535] P. M. Kiernan, J. F. Gibson, and W. P. Griffith, *J.C.S. Chem. Comm.*, 1973, 816.

containing H^- and AlH_4^- ligands have been obtained[536] from the reactions of tantalum(v) halides and $LiAlH_4$. The initial product is $Ta(AlH_4)_4$ which loses AlH_3 to give $Ta(AlH_4)_{4-n}H_n$

Intercalation of guest species into the layer-structured disulphides of the Group IVB and VB transition metals often requires high reaction temperatures thus precluding the formation of some thermally unstable intercalation compounds. The development of a new electrochemical intercalation technique operating at normal temperatures is, therefore, especially welcome.[537] The host species acts as the cathode of an electrolytic cell and the guest species is generated by electrolysis of a solution of one of its salts. Ammonium or iminium salts were used as sources of ammonia and amines and hydrogen was evolved as the neutral species was intercalated into the crystal lattice. The lattice constants of the intercalation compounds were the same as those of compounds produced by thermal methods.

Another method[538] which avoids the use of high temperatures is 'cavitation' of a dichalcogenide such as TaS_2 with guest molecules by means of the intense agitation provided by an ultrasonic microprobe. A slightly higher degree of intercalation ($x = 0.26$) was achieved for $TaS_2(4\text{-vinylpyridine})_x$ with this method in 7 h compared with that ($x = 0.23$) obtained by heating the components at 140 °C for 72 h. 4-Vinyl-pyridine was used as the guest molecule in order to study polymerization processes in a preorientated medium. The greater thermal stability of the intercalate prepared by heating was attributed to the presence of poly-4-vinylpyridine as the guest species. Intercalation into NbS_2, TaS_2, and Ta_2S_2C has been studied in detail. Measurements of the interlayer distances in the host lattices for a series of amines as guest molecules showed[539] that the arrangement of the guest molecules in the interlayer space is dominated by the host lattice–amine interaction for amines with C_4—C_9 alkyl chains, and by van der Waals interaction between the alkyl chains for amines with C_{12}—C_{18} alkyl chains. Phase transitions corresponding both to conformational changes of the guest molecules and to changes in the amine–NbS_2 ratio were observed with NbS_2 and n-octadecylamine and oleyl amine in the temperature range 20—120 °C. Amides, N-substituted amides, phosphines, phosphine oxides, phosphoric amides, and metal ions have also been used as guest molecules.[540 – 543] Bright-field electron micrographs have been recorded[544] at 140 K for $NbSe_2$, $PhNH_2$–$NbSe_2$ and py–NbS_2. Selected area diffraction patterns from both intercalation compounds showed the presence of a 2×2 superlattice.

Redetermined molecular parameters[545] of $Cp_2NbX(S_2)$ ($X = Cl$ or S) agree with those obtained originally.

[536] A. I. Golovanova, M. Kost, and V. I. Mikheeva, *Izvest. Akad. Nauk S.S.S.R., Ser. khim.*, 1973, 1448.

[537] M. S. Whittingham, *J.C.S. Chem. Comm.* 1974, 328.

[538] J. J. Winter, A. A. Leupold, F. Rothwarf, Hsu, Che-Hsiung, M. M. Labes, J. T. Breslin, and D. J. Edmiston, *Nature Phys. Sci.*. 1973, **246**, 122.

[539] R. Schoellhorn and E. Sick, *Z. Naturforsch.*, 1973, **28b**, 168.

[540] R. Schoellhorn and A. Weiss, *Z. Naturforsch.*, 1973, **28b**, 172.

[541] R. Schoellhorn and A. Weiss, *Z. Naturforsch.*, 1973, **28b**, 716.

[542] G. V. S. Rao, M. W. Shafer and L. J. Tao, *Mater. Res. Bull.*, 1973, **8**, 1231 (*Chem. Abs.*, 1973, **79**, 150847).

[543] F. J. Di Salvo, G. W. Hull, L. H. Schwartz, J. M. Voorhoeve, and J. V. Waszczak, *J. Chem. Phys.*, 1973, **59**, 1922.

[544] P. M. Williams and B. A. Robinson, *Nature Phys. Sci.*, 1973, **245**, 79.

[545] R. Roder, *Diss. Abs.* (*B*), 1974, **34**, 4866.

Niobium(V) and Tantalum(V).—*Halides and Oxyhalides.* Halide ion exchange and solvent substitution have been observed[546] in a ^{19}F n.m.r. spectroscopic study of the TaF_5–$TaCl_5$–MeCN system. The species formed include (TaF_xCl_{5-x}) $(x = 1$—$5)$, $[TaF_2Cl_2(MeCN)_2]^+$, $[TaFCl_2(MeCN)_3]^{2+}$, and $[TaFCl_3(MeCN)_2]$. In the system $TaCl_5$–SnF_4–MeCN, TaF_4Cl and $[SnCl_4(MeCN)_2]$ are the major observed species, ligand redistribution occurring with TaF_4Cl. Tantalum(V) fluoro-complexes are more stable than those of Nb^V or Sn^{IV}, as evidenced by the lack of reaction between TaF_5 and $SnCl_4$ in MeCN and by the formation of $[NbCl_4(MeCN)_2]^+$ and $(TaF_5Cl)^-$ from either TaF_5 and $NbCl_5$ or $TaCl_5$ and NbF_5 in MeCN. $[NbX_4(MeCN)_2]^+$ cations (X = Cl or Br) have also been detected in a ^{93}Nb and ^{19}F n.m.r. study of the pentahalides in MeCN solution.[547] It was also observed that $NbOCl_3$ reacts with HF or HCl in MeCN to give the NbX_6^- anion. TaF_5 forms 1:1 and 1:2 complexes with Me_2O and Me_3N.[548] At $-105\,^{\circ}C$ the ^{19}F n.m.r. spectrum of the 1:1 Me_2O complex is typical of an AB_4 system, but as the temperature is raised the spectrum collapses to a singlet. A comparison of the experimental and computer-simulated spectra gives the best fit for an intermolecular exchange process involving a small amount of free TaF_5. Analysis of the intensities of the fluorine resonances for the 2:1 $Me_2O:TaF_5$ complex as a function of temperature indicated a series of equilibria between TaF_5, TaF_6, and $TaF_4(Me_2O)_n^+$ ($n = 1$—4). Fluorine exchange is fast even at $-105\,^{\circ}C$ for the 2:1 $Me_3N:TaF_5$ complex.

Whereas there is no evidence for complex formation between BF_3 and the Group VA hexafluoroanions, it is strongly absorbed by methylene chloride solutions of the Group VB hexafluoro-anions.[549] Absorption occurs to a maximum of one equivalent of BF_3 for solutions of $MNbF_6$ (M = Ag or Bu_4N) but this is exceeded for $MTaF_6$ solutions. It has been suggested that complexing of one equivalent of BF_3 can be represented, at least in part, by the reactions:

$$BF_3 + MF_6^- \rightleftharpoons BF_4^- + MF_5$$
$$MF_5 + MF_6^- \rightleftharpoons M_2F_{11}^-$$
$$BF_3 + BF_4^- \rightleftharpoons B_2F_7^-$$

^{19}F data on solutions initially equimolar in $Ta_2F_{11}^-$ and BF_3 are consistent with the further absorption of BF_3 by TaF_6^- solutions being due to the formation of the $BTa_2F_{14}^-$ ion. Only weak complex formation was observed between PF_5 and the Group VB hexafluoroanions. The known propensity of the Group VB elements for complex fluoroanion formation in solution, prompted an attempt to produce hetero-dinuclear complex fluoro-anions containing tungsten.[550] No interaction could be observed between WF_6 and BF_3, AsF_5, VF_5, TaF_5, $Bu_4^nNNbF_6$, AgF_6, $Bu_4^nNSbF_6$, or ethyl fluoride, indicating the weak fluorine donor and acceptor capacity of the hexa-fluoride. Reactions of WF_6^- with binary fluorides could be interpreted in terms of disproportionation to WF_6 and WF_4. There was evidence of complex formation between WF_4 and the fluoroanions derived from BF_3, TaF_5, and NbF_5.

[546] Yu. A. Buslaev, E. G. Il'in, and M. N. Scherbakova, *Doklady Akad. Nauk. S.S.S.R.,* 1974, **214**, 842.
[547] Yu. A. Buslaev, V. D. Kopanv, S. M. Sinitsyna, and V. G. Khlebodarov, *Zhur. neorg. Khim.,* 1973, **18**, 2567.
[548] S. Brownstein and M. J. Farrall, *Canad. J. Chem.,* 1974, **52**, 1958.
[549] S. Brownstein, *J. Inorg. Nuclear Chem.,* 1973, **35**, 3567.
[550] S. Brownstein, *J. Inorg. Nuclear Chem.,* 1973, **35**, 3575.

^{93}Nb n.q.r. spectra have been recorded[551] for NbF$_5$, XeF$_2$,NbF$_5$, XeF$_2$,2NbF$_5$, NbF$_5$L (L = Me$_2$O or CH$_2$ClCN), and NbF$_5$,2py. A quadrupole coupling constant of 34.6 MHz at 77 K for XeF$_2$,NbF$_5$ is much greater than would be expected for a structure involving NbF$_6$ ions; it has been interpreted in terms of the participation of one of the Nb—F bonds in fluorine bridging in a structure such as (60) which is analogous to that of XeF$_2$,2SbF$_5$. Compounds with NbF$_5$:PbF$_2$ ratios of 4:1, 5:2, 2:1, 1:1, and 1:2 have been detected in a study of the phase diagram of the NbF$_5$–PbF$_2$ system.[552]

(60)

Several hydrazinium-(2+) and -(1+) fluoroniobates(v) were isolated from solutions of NbF$_5$ and N$_2$H$_4$,HF in anhydrous HF or solutions of Nb in HF of various concentrations in the presence of N$_2$H$_4$,HF.[553] N$_2$H$_6$(NbF$_6$)$_2$ and N$_2$H$_6$NbF$_7$ were isolated from anhydrous HF, whereas N$_2$H$_6$NbOF$_5$ and some ill-defined products or mixtures were obtained from HF solutions. The mixed oxofluorometalates Cs$_2$TaOF$_5$[553a] and M$_2$(LiNb)O$_2$F$_4$[553b] (M = K, Rb, or Tl) have been prepared.

Niobium and tantalum are extracted by n-trioctylamine (B) from HF solution in the form (BH)HNb(OH)$_2$F$_5$, (BH)HNb(OH)F$_6$, and (BH) HTaF$_7$,[554] Symmetrics and space groups have been determined[555] for Na$_2$Ta$_2$O$_5$F$_2$ prepared by heating stoicheiometric amounts of NaF and Ta$_2$O$_5$ in a sealed tube at 700—950 °C and Ag$_2$M$_2$O$_5$F$_2$ (M = Nb or Ta) obtained from Ag$_2$O and MO$_2$F at 650 °C under oxygen. Covalency effects in oxyhalide complexes of the Group VB transition metals are difficult to study because only vanadium forms stable oxyhalide complexes. An e.p.r. study of the (NbOF$_5$)$^{3-}$ radical has been carried out,[556] this species being produced by γ-irradiation of single crystals of K$_2$NbOF$_5$,H$_2$O at 77 K. The most significant evidence of covalency in the (NbOF$_5$)$^{3-}$ complex is the presence of ligand superhyperfine interaction the magnitude of which indicates strong in-plane π-bonding between the niobium 4$^d_{xy}$ orbital and the 2p_y orbitals of fluorine. On warming, the (NbOF$_5$)$^{3-}$ radical decays at -80 °C forming a second radical; the e.p.r. spectrum of this is consistent with its formulation as a hole species formed by electron loss from (NbOF$_5$)$^{3-}$ NH$_4$NbOF$_4$ crystallizes from solutions containing Nb$_2$O$_5$, NH$_4$F, and HF. The NbO$_2$F$_4$ species has a tetragonally distorted octahedral structure with Nb—O—Nb chains.[557] Disproportionation of NH$_2$Cl in the presence of the Group VB pentachlorides affords the ammonium salts of the MX$_6^-$ anions (M = Nb or Ta); they were

[551] J. C. Fuggle, J. M. Holloway, D. A. Tong, D. W. A. Sharp, and J. M. Winfield, *J.C.S. Dalton*, 1974, 205.
[552] D. Bizot, *Rev. Chem. Minerale*, 1973, **10**, 579.
[553] B. Friec and M. Bertoncelj, *Inst. Jozef Stefan. IJS Reports*, 1971, R-603 (*Chem. Abs.*, 1973, **79**: 111 270).
[553a] G. Pausewang, *Z. Naturforsch.*, 1974, **29b**, 49.
[553b] M. Bolte and J. P. Besse, *Compt. rend.*, 1973, **277**, C, 1359.
[554] E. S. Pal'shin and L. A. Ivanova, *Zhur. Analit. Khim.*, 1973, **28**, 1741 (*Chem. Abs.*, 1974, **81**, 41 346).
[555] J. P. Chaminade, M. Vlasse, M. Pouchard, and P. Hagmenuller, *Compt. rend.*, 1973, **277**, C, 1141.
[556] J. R. Shock and M. T. Rogers, *J. Chem. Phys.*, 1973, **58**, 3356.
[557] V. I. Pakhomov, R. L. Davidovich, T. A. Kaidalova, and T. F. Levchishina, *Zhur. neorg. Khim.*, 1973, **18**, 1240.

characterized by d.t.a. and by i.r. and Raman spectroscopy[132] and their force constants and mean amplitudes of vibration have been evaluated.[133] Effective charges on the metal and the ligands have been calculated[264] from spectroscopic data on $RbNbCl_6$, Et_4NNbF_6, K_2NbOF_5, and Rb_2NbOCl_5. The donor ability of solvents towards $NbCl_5$ as measured by calorimetry and i.r. spectroscopy has been found to be: $DMF \gg MeCN > AcOPr \sim AcOBu > MeCOEt > diox > C_6H_6$.[558]

Ketones form coloured solutions of $1:1$ complexes which hydrolyse on contact with air; i.r. spectra indicate solvation by a second molecule of ketone. Heats of solution of $NbCl_5$ in MeCOR (R = Et, Pr, or Bu^i) and cyclohexanone are reported.[559]

O-, S-, Se-, and Te-Donor Ligands. Table 4 summarizes much of the data reported on niobates, tantalates, and mixed oxides containing niobium(V) and tantalum(V) published this year.

Compound formation between Nb_2O_5 and some oxides of bivalent and tervalent elements, in particular the lanthanide oxides, has been reviewed.[559a] Phases isostructural with the niobium oxide and titanium–niobium oxide block structures have been detected[559b] in the system MgF_2–Nb_2O_5; this is in agreement with a prediction based on the close similarity between the ionic radii of Mg^{2+} (0.66 Å), Ti^{4+} (0.68 Å), and Nb^{5+} (0.69 Å). The topochemical formation of niobium and tantalum hydroxides[559c] and their solubilities in citric acid[560] and HSCN and HCl[561] have been investigated. Thermodynamic calculations show that the reduction of Nb_2O_5, NbO_2, NbO, or Ta_2O_5 to the metal and Na_2O by sodium has a $\Delta G°$ in the range 2.4—5.0 kcal mol^{-1}. Indeed, the Group VB oxides reacted with sodium at 400 or 600 °C to give the metal and cubic Na_3MO_4 (M = Nb or Ta) as equilibrium products.[562] The Na_3MO_4 compounds represent the maximum amount of sodium oxide that can exist with regular MO_6 octahedra.

The peroxyniobium compounds $M^1M^2_2[Nb(O_2)_4]$ ($M^1 = M^2 = $ Li, Cs, Rb; $M^1 = NH_4$, $M^2 = Cs$; $M^1 = Cs$, $M^2 = NH_4$; $M^1 = Rb$, $M^2 = NH_4$) and Rb_3-$[NbO(O_2)_3]$ have been characterized[563] by X-ray diffraction, electrical conductivity, and thermal decomposition studies. Thermal methods have also been used[563a] in a study of the oxonium and ammonium pyrochlores H_3OMWO_6, $H_4OM_2O_6$, and NH_4MWO_6 (M = Nb or Ta).

Tetraperoxotantalates $M_3Ta(O_2)_4$ with organic cations as well as oligomeric peroxypolytantalates have been prepared[564] for the first time by perhydrolysis of $Ta(OEt)_5$ in the presence of bases. A medium intensity band in the region of 840 cm^{-1} is assigned to the O—O stretching vibration. Alkali-metal and mixed alkali-metal–ammonium tetraperoxoniobates have also been prepared.[564a] A peroxo group co-

558 L. V. Surpina and Yu. V. Kolodyazhnyi, *Zhur. neorg. Khim.*, 1973, **18**, 2853.

559 L. V. Surpina and Yu. V. Kolodyazhnyi, and O. A. Osipov, *Zhur. obshchei Khim.*, 1973, **43**, 1165.

559a G. Tilloca, *Rev. Internat. Hautes Temp. Refract.*, 1973, **10**, 183 (*Chem. Abs.*, 1974, **80**, 9882).

559b F. J. Lincoln, J. L. Hutchinson, and J. S. Anderson, *J.C.S. Dalton*, 1974, 115·

559c V. V. Sakharov, N. E. Ivanova, S. S. Korovin, and M. A. Zakharov, *Zhur. neorg. Khim.*, 1974, **19**, 579.

560 L. Pets, *Tr. Tallin. Politekh. Inst. Ser. A.* 1971, No. 303, 105 (*Chem. Abs.*, 1974, **80**, 74696).

561 V. V. Nakrasova, E. S. Pal'shin, and B. F. Myasoedov, *Radiochem. Radioanalyt. Letters*, 1974, **16**, 257.

562 M. G. Barker, A. J. Hopper, and D. J. Wood, *J.C.S. Dalton*, 1974, 55.

563 E. N. Traggeim, R. N. Shchelokov, and M. A. Michnik, *Khim. i Khim. Tekhnol. Tr. Yubileinoi Konf. Posurashch. 70-Letiyu Inst.* 1970. 294 (*Chem. Abs.*, 1974, **81**, 20308).

563a D. Groult, C. Michel, and B. Raveau, *J. Inorg. Nuclear Chem.*, 1974, **36**, 61.

564 J. Fuchs and D. Lubkoll, *Z. Naturforsch.*, 1973, **28b**, 590.

564a R. N. Shchelokov, E. N. Traggeim, M. A. Michnik, and S. V. Morozova, *Zhur. neorg. Khim.*, 1973, **18**, 2104.

Table 4 *Niobates, tantalates, and oxide compounds containing niobium(V) or tantalum(V)*

Compound	Properties reported	Ref.
$Nb_{12}O_{29}$	Electron microscopy	a
Nb_2O_5	Isotope exchange with ^{18}O	b
Nb_2O_5	Temperature–pressure phase relations	c
Nb_2O_5	T.g.a. and elec. cond.	d
Nb_2O_5	X	e
$Nb_{22}O_{54}$	X	f
Ta_2O_5	Elec. cond, thermoelec. power, and wt. change	g
$Nb_xTa_yO_2$ $(0.8 \leqslant (x + y) \leqslant 1.0)$	X	h
$NaNbO_3$	X	i
Na_3MO_4 (M = Nb or Ta)	X	j
$K_{5.9}Li_{3.8}Nb_{10}O_{30}$ $K_{5.0}Li_{2.4}Nb_{10}O_{30}$	Polarized Raman spectra	k
$Na_{14}Nb_{12}O_{37},32H_2O$ $K_{14}Nb_{12}O_{37},27H_2O$	Formation	l
$KTa_{0.65}Nb_{0.35}O_3$	Magnetic properties	m
$NaCa_4Nb_5O_{17}$ $Na_2Ca_4Nb_6O_{20}$	X	n
$K_2CaNb_5O_{14}F$ $K_2SrNb_5O_{14}F$ $KNaSrNb_5O_{14}F$ $K_2BaNb_5O_{14}F$ $KNaBaNb_5O_{14}F$	Magnetic properties	o
$Ca_2Nb_2O_7$	X	p
Sr_2NbO_5	Prep., X	q
$Sr_xBa_{1-x}Nb_2O_6$	Polarized Raman spectrum	r
$Ba_4Nb_3LiO_{12}$	X	s
$TlNbB_2O_6$	X	t
$PbNb_2O_6$	Magnetic properties	u
$Pb(Li_{0.25}Nb_{0.75})O_3$	X	v
$Nb_2O_5-PbTiO_3$	D.t.a., electron microprobe analysis. Knudsen effusion measurements	w
$SbNbO_4$ $SbTaO_4$	Single-crystal growth	x
$(Sr, Ba)Bi_2Ta_2O_9$	X	y
M_2NbO_5F(M = Ti, V, or Cr)	X, magnetic properties	z
$FeNbO_4$	Neutron diffraction, d.	aa
$CuNb_2O_6-CuTa_2O_6$ $CuTa_2O_6-NiTa_2O_6$ $CuTa_2O_6-MnTa_2O_6$	Phase relationships	bb
$Ln_{0.22}(M_{0.66}W_{0.33})O_3$ $Ln_{0.05}(M_{0.15}W_{0.85})O_3$ (M = Nb or Ta; Ln = La, Nd, Sm, Eu, Gd, Dy, Ho, Er or Yb)	Prep., X	cc
$LaMO_4$ (M = Nb or Ta)	Prep.	dd
$LnTiNbO_6$ (Ln = Gd, Tb or Dy)	X, i.r.	ee
$Ln(NbO_3)_3$ (Ln = La, Pr or Nd)	I.r.	ff
Ln_3NbO_7 (Ln = La, Pr, Nd or Sm)	Prep.	gg
$NdNbO_4$ $Nd_{0.33}NbO_3$	Elec. cond., dielec. permeability	hh

Table 4 *(cont.)*

Compound	Properties reported	Ref.
Eu_2TaAlO_6 $Eu_3MgTa_2O_9$ $Eu_3MnTa_2O_9$	X, magnetic properties	ii
A^1MO_3, $A^2M_2O_6$, and $A^2M_2O_7$ (A^1 = Li, Na, K, Ag or Tl; A^2 = Ca, Sr, Ba, Pb, Cd, Mg, Ni, Co, Cu or Zn; M = Ta, Nb or Sb)	I.r.	jj
$A^+B^{5+}O_3$ (B = Nb or Ta)	X	kk

(a) S. Iijima, S. Kimura, and M. Goto, *Acta Cryst.*, 1973, **29A**, 632; (b) R. A. Ross, D. G. Klissurski, and T. J. Griffith, *Z. phys. Chem.*, 1973, **86**, 50; (c) J. L. Waring, R. S. Roth, and H. S. Parker, *J. Res. Nat. Bur. Stand.*, (A), 1973, **77**, 705 (*Chem. Abs.*, 1974, **80**, 31 256); (d) W. C. Tripp and J. M. Wimmer, *U.S. Nat. Tech. Inform. Serv. AD Reports* 1973, No. 773175/5GA (*Chem. Abs.*, 1974, **81**, 7043); (e) L. S. Palatnik, Yu. I. Malyuk, and V. V. Belozerov. *Doklady Akad. Nauk S.S.S.R.*, 1974, **215**, 1182; (f) S. Iijima, S. Kimura, and M. Goto, *Acta Cryst.*, 1974, **30A**, 251; (g) J. E. Stroud, W. C. Tripp, and J. M. Wimmer, *J. Amer. Ceram. Soc.*, 1974, **57**, 172; (h) A. V. Anan'in, O. N. Breusov, A. N. Dremin, V. N. Drobyshev, and S. V. Pershin, *Zhur. neorg. Khim.*, 1974, **19**, 710; (i) C. N. W. Darlington and H. D. Megaw, *Acta Cryst.*, 1973, **29B**, 2171; (j) J. Darriet and A. Vidal, *Compt. rend.*, 1973, **277**, C, 1235; (k) D. Siapkas and R. Clarke, *Phys. Stat. Sol. (B)*, 1974, **62**, 43; (l) Z. M. Titova and A. K. Sharova, *Tr. Inst. Khim. Ural Nauch. Tsentr. Akad. Nauk S.S.S.R.*, 1973, **25**, 145 (*Chem. Abs.*, 1974, **81**, 20 265); (m) J. A. Rubin, *U.S. Nat. Tech. Inform. Serv.*, AD Reports, 1973, 764 902/3 (*Chem. Abs.*, 1974, **80**, 31 167); (n) A. Carpy, P. Amestoy, and J. Galy, *Compt. rend.*, 1973, **277**, C. 501; (o) J. Ravez, D. Tourneur, J. Grannec, and P. Hgennauller, *Z. anorg. Chem.*, 1973, **399**, 34; (p) K. Scheunemann and H. K. Muller-Buschbaum, *J. Inorg. Nuclear Chem.*, 1974, **36**, 1965; (q) G. G. Kasimov, E. G. Vovkotrub, and E. I. Krylov, *Zhur. neorg. Khim.*, 1974, **19**, 269; (r) K. G. Bartlett and L. S. Wall, *J. Appl. Phys.*, 1973, **44**, 5192; (s) E. F. Jendrek, A. D. Potoff, and L. Katz, *J. Solid State Chem.*, 1974, **9**, 375; (t) M. Gasperin, *Acta Cryst.*, 1974, **30B**, 1181; (u) T. Yamada, *Appl. Phys. Letters*, 1973, **23**, 213; (v) S. Udagawa, S. Shin, and K. Kamata, *Mater. Res. Bull.*, 1973, **8**, 1165 (*Chem. Abs.*, 1973, **79**, 150 390); (w) D. Hennings, *Proc. Internat. Symp. on Reactivity of Solids*, 1972, 149 (*Chem. Abs.*, 1973, **79**, 149 872); (x) L. A. Ivanova, V. I. Popolita, S. Y. Stefanov, A. N. Lobachev, and Y. N. Venevtse, *Kristallografia* 1974, **19**, 573; (y) R. E. Newnham, R. W. Wolfe, R. S. Jorsey, and F. A. Diaz-Colon, *Mater. Res. Bull.*, 1973, **8**, 1183 (*Chem. Abs.*, 1973, **79**, 150 387); (z) J. Senegas and J. Galy, *Compt. rend.*, 1973, **277**, C, 1243; (aa) H. Weitzel and H. Schroecke, *Neues Jahrb. Mineral. Abh.*, 1973, **119**, 285 (*Chem. Abs.*, 1974, **80**, 7857); (bb) G. V. Buzuev, *Zhur. neorg. Khim.*, 1973, **18**, 2489; (cc) G. Desgardin and B. Raveau. *J. Inorg. Nuclear Chem.*, 1973, **35**, 2295; (dd) P. D. Peshev, G. M. Bliznakov, and M. S. Ivanova, *Doklady Bolg. Akad. Nauk*, 1973, **26**, 891 (*Chem. Abs.*, 1974, **80**, 85 190); (ee) A. M. Sych and V. G. Klenus, *Izvest. Akad. Nauk S.S.S.R., neorg. Materialy*, 1974, **10**, 634; (ff) A. M. Sych, A. T. Belokon, V. P. Dem'yanenko, and L. A. Ermenko, *Ukrain. fiz. Khim.*, 1973, **18**, 787 (*Chem. Abs.*, 1973, **79**, 47 418); (gg) T. I. Panova, N. A. Godina, and E. K. Keler, *Izvest. Akad. Nauk. S.S.S.R., neorg. Materialy*, 1973, **9**, 628; (hh) A. M. Sych, L. A. Eremenko, L. A. Zastavker, and M. M. Nekrasov, *Izvest. Akad. Nauk S.S.S.R., neorg. Materialy*, 1974, **10**, 496; (ii) H. Parent, J. C. Bernier, and P. Poix, *Compt. rend.*, 1974, **278**, C, 49; (jj) C. Rocchiccioli-Deltcheff and R. Franck, *Ann. Chim. (Italy)*, 1974, **9**, 43; (kk) B. Goodenough and A. Kafalas, *J. Solid State Chem.*, 1973, **9**, 493.

ordinated to niobium stabilizes Nb_2O_7,xH_2O. This acid consists of two six-co-ordinate niobium atoms connected by bridging hydroxy-groups; it has been formulated[565] as $Nb_2O_2(O_2)_2(OH)_2,xH_2O$, and yields $(Me_4N)_3Nb(O_2)_4$ on treatment with $(Me_4N)_3Nb(O_2)_4$. The new peroxotetrafluorotantalate(v), $[Ta(O_2)F_4(H_2O)]^-$, has been isolated[565a] as its Et_4N salt. Diketonate ligands displace the water molecule and a fluoride ion affording the crystalline complexes $Et_4N[Ta(O_2)F_3dik]$ (dik = acac, dbm, bztfacac, ttfa, and hfacac). The ^{19}F n.m.r. spectra of MeCN solutions of the complexes containing unsymmetrical diketonate ligands show two AX_2 patterns, indicating the existence of geometrical isomers in these solutions. A structure composed of parallel layers of NbO_6 octahedra and SO_4 tetrahedra connected by Cs^+ ions

[565] D. Driss, T. Abdelaziz, and S. Bernard, *Compt. rend.*, 1973, **277**, C, 945.
[565a] J. Y. Calves and J. E. Gerchais, *J. Fluorine Chem.*, 1974, **4**, 47.

D

has been established[566] for $CsNbO(SO_4)_2$ by X-ray crystallography; the Rb^+ and NH_4^+ salts have analogous structures.[567] On heating, the Rb^+ salt forms $9Rb_2O$, $10Nb_2O_5,8SO_3$ *via* $[Rb_2SO_4,Nb_2O_3(SO_4)_3]$.[568]

Only ligands having a high charge density on the donor atom, such as amines and amine or phosphine oxides are capable of combating the tendency of the Group VB penta-alkoxides to dimerize; with other ligands the equilibrium lies well on the side of the dimer.[569] Fast exchange between free and co-ordinated ligand[570] precludes isolation of co-ordination complexes by evaporation of solvent. Solvent, concentration, and temperature conditions for precipitation of crystalline complexes are critical and highly dependent on the ligand. Complexes $Nb(OMe)_4L$ (L = py, γ-pic, Me_3NO, or HMPA) were obtained by slow crystallization at $-35 °C$ from toluene or MeCN solutions of the pentamethoxide containing ligand in several-fold excess. Hydrazine adducts of $Ta(OMe)_5$ and $Nb(OMe)_5$ precipitated immediately on addition of ligand to toluene solutions of the alkoxides. Solutions of the niobium adducts in polar solvents have n.m.r. spectra which are identical with those of solutions of equimolar mixtures of the penta-alkoxide and of the ligand in the same solvent and at the same concentration as expected from the dimer $\rightleftharpoons$ adduct equilibrium. Tantalum pentaethoxide reacts with diethylhydroxylamine and with oximes eliminating ethanol to form a new series of compounds with Ta—O—N bonds $(OEt)_nTa(ONEt_2)_{5-n}$ and $(OEt)_nTa(ON{=}CR^1R^2)_{5-n}$ ($n = 1$—5; $R^1 = R^2 = $ Me or Et;[570a] $R^1 = $ Me, $R^2 = $ Et). Estimation of the ethanol removed azeotropically with benzene provided a convenient way of following the progress of the reaction. Higher alkoxides could be obtained by heating the ethoxides in refluxing alcohols ROH ($R = Bu^t, Bu^n$, or Pr^i) but only three OEt groups could be replaced to afford products of the type $(OEt)(OR)_3Ta(ON{=}CR^1R^2)$. The i.r. spectra of the tantalum oxinates had two absorption bands in the region 1600—1650 cm^{-1}; these were assigned to C$\equiv$N stretching vibrations indicating the presence of oximato-groups in two different environments. Conductivity and dipole moment measurements in solution and i.r. spectra of the solid adducts NbX_5,nHMPA (X = F, $n = 2$; X = Cl or Br, $n = 1$) and TaX_5,HMPA (X = F, Cl, or Br) indicate[571] that the HMPA is bound to the metal through its oxygen atom and that ionic dissociation does not occur in solution. Co-ordination through oxygen is also indicated[572] by i.r. spectroscopy for $NbO(OH)$-Cl_2,tbp, the form in which $NbCl_5$ is extracted into tributylphosphate.

Substitution reactions of octahedral complexes of transition metals with few d-electrons are expected to proceed *via* associative mechanisms. Kinetic evidence has been presented[573] in support of such a mechanism for ligand exchange in the adducts $TaCl_5$,L (L = Me_2S, Me_2Se, or Me_2Te) which has large negative entropies

[566] A. A. Kashaev and G. V. Sokolova, *Kristallografiya*, 1973, **18**, 620.

[567] A. A. Kashaev, G. E. Postoenko, and E. A. Zelbst, *Kristallografiya*, 1973, **18**, 1278 (*Chem. Abs.*, 1974, **80**, 101168).

[568] M. I. Andreeva, E. A. Podozerskaya, V. Ya. Kuznetsov, and R. A. Popova, *Zhur. neorg. Khim.*, 1973, **18**, 1215.

[569] L. G. Hubert-Pfalzgraf and J. G. Riess, *J.C.S. Dalton*, 1974, 585.

[570] L. G. Hubert-Pfalzgraf and J. G. Reiss, *Bull. Soc. Chim. France*, 1973, 1201.

[570a] R. C. Mehrotra, A. K. Rai, and R. Bohra, *Z. anorg. Chem.*, 1973, **399**, 338.

[571] J. R. Masaguer and J. Sordo, *Anales de Qhim.*, 1973, **69**, 1263 (*Chem. Abs.*, 1974, **80**, 77784).

[572] M. A. Zahharov, V. M. Klyuchinkov, S. S. Korovin, and I. A. Apraksin, *Zhur. neorg. Khim.*, 1973, **18**, 1331.

[573] R. Good and A. E. Merbach, *J.C.S. Chem. Comm.*, 1974, 163.

of activation. Surprisingly, however, exchange of Me_2O has a positive ΔS^* indicating the occurrence of a different, presumably dissociative mechanism; this is supported by the fact that the exchange rate is independent of free ligand concentration.

Information on stereolability and dissociation processes for molecular early transition-metal derivatives has been obtained[574] by a low-temperature ^1H n.m.r. study of $NbOCl_3,2OSMe_2$ in CD_3CN solution. At $-40\,°C$ the spectrum of the solution consists of five signals; the effects of dilution and the addition of free ligand on their intensities show them to be due to geometrical isomers of octahedral $NbOCl_3,2OSMe_2$ and to free ligand released on dissociation of the adducts. Above $60\,°C$ there is only one signal, indicating that all the solution species are in rapid equilibrium. Difficulties encountered in distinguishing between free and co-ordinated water in a series of oxobis(oxalato)aquoniobate(v) salts by vibrational spectroscopy prompted an examination[575] of the use of dimethyl sulphoxide (DMSO) and tetra-methylene sulphoxide (TMSO) as replacement ligands, since there are marked differences between the vibrational spectra of the free and co-ordinated sulphoxide molecules. Several oxobis(oxalato)(sulphoxide)niobate(v) complexes were prepared: $M[NbO(C_2O_4)(DMSO)_2],nDMSO$ ($n = 1$, $M = NH_4$, K, or Rb; $n = 0$, $M = NH_4$, K, Rb, or Cs), $M[NbO(C_2O_4)_2 (TMSO)_2]$, $nTMSO$ ($n = 1$, $M = NH_4$, K, or Rb; $n = 0$, $M = NH_4$ or K). Their i.r. spectra are in agreement with thermal analyses which indicate the presence of two types of sulphoxide in the complexes containing three sulphoxide molecules. Strong absorption at *ca.* $1700\,cm^{-1}$ is assigned to the co-ordinated oxalate groups. The similarity in position, intensity, and multiplicity of this absorption in the spectra of the aquo- and sulphoxide complexes implies similar environments for the oxalate ligands in the two series, which are thus assumed to be structurally analogous.

Further crystallographic studies on oxobis(oxalato)aquoniobate(v) complexes have established[576] an isostructural relationship between $M[NbO(C_2O_4)_2(H_2O)_2]$, nH_2O ($n = 2$, $M = NH_4$, Rb, or Cs), and between the salts with ($n = 3$, $M = NH_4$ or K). An examination of the Nb^V–dibromogallic acid–oxalate system showed that only binary oxalate complexes are formed.[577] According to a thermal, X-ray, and neutron-activation investigation of the dichelate of niobium with propyl-3,4,5-trihydroxybenzoate (PTB), it has the form $K[NbO\{C_6H_2(OH)(O_2)(CO_2)Pr\}_2]$, $2PTB,3H_2O$.[578]

Continuing studies of potential seven co-ordination in trihalogeno-niobium and -tantalum complexes with bidentate ligands have involved a crystal-structure determination on $TaCl_3(OCH_2CH_2OMe)_2$.[579] This revealed the metal in a distorted octahedral environment with one of the (2-methoxy)ethanol ligands acting as a bidentate ligand and the other as a unidentate, as in (61). It appears that seven co-ordination is forbidden by the bulk that would arise in the pentagonal girdle. Evidently the preference for two apical chlorines and one chlorine equatorial is sufficient to prevent the bidentate apical–equatorial co-ordination of the oxygen ligand. Ta—Cl bond lengths (2.36—2.37 Å) are normal for six-co-ordinate tantalum.

[574] J. G. Riess, R. C. Muller, and M. Postel, *Inorg. Chem.*, 1974, **8**, 1802.
[575] N. Brnicevic and C. Djordjevic, *J.C.S. Dalton*, 1974, 165.
[576] N. Galesic, S. Popovic and M. Sljukic, *Croat. Chem. Acta*, 1973, **45**, 437 (*Chem. Abs.*, 1973, **79**, 71145).
[577] S. Koch and G. Ackermann, *Chem. Anal. (Warsaw)* 1973, **18**, 49 (*Chem. Abs.*, 1973, **79**, 24043).
[578] F. Jasim, *Thermochimica Acta*, 1973, **6**, 439.
[579] M. G. B. Drew and J. D. Wilkins, *Inorg. Nuclear Chem. Letters*, 1974, **10**, 549.

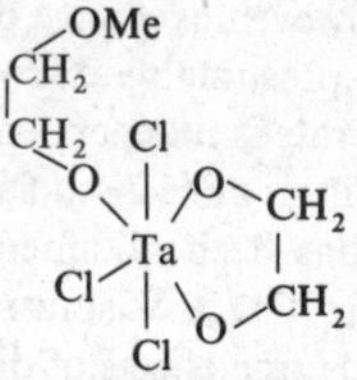

$$\begin{array}{c}
\text{OMe} \\
| \\
\text{CH}_2 \\
| \\
\text{CH}_2 \quad \text{Cl} \\
\text{O} \diagdown \ | \ \diagup \text{O} \diagdown \text{CH}_2 \\
\text{Ta} \qquad | \\
\text{Cl} \diagup \ | \ \diagdown \text{O} \diagup \text{CH}_2 \\
| \\
\text{Cl}
\end{array}$$

(61)

Reaction of MCl_5 (M = Nb or Ta) with dipivaloylmethane ($dpmH_4$) in dichloro-methane gave $M(dpm)_2Cl_3$, whilst with neat ligand $M(dpm)Cl_4$ was obtained.[534a] Conductivity data indicate that $Ta(dpm)_2Cl_3$ is highly dissociated in nitrobenzene and in nitromethane solution; an equilibrium involving $Ta(dpm)_4^+$ and $TaCl_6$ was suggested, as was the presence of an eight-co-ordinate cation in the solid. An attempt to obtain the $Nb(dpm)_4^+$ cation by chlorine oxidation of $Nb(dpm)_4$ in oxygen-free CCl_4 led to abstraction of ligand oxygen by the metal and the formation of the polymeric oxo-complex $[NbOCl_2\text{-}(dpm)]_x$.

The mass spectra of Nb^V and Ta^V complexes with chloro, alkoxy, acetylacetonato, or salicylaldehydato ligands have been recorded.[580]

N-*Donor Ligands.* Colourless niobium and tantalum tetrachloride azides can be obtained[581] by the reaction of the corresponding pentachlorides with ClN_3 or NaN_3. The tetrachloride azides are explosive and hydrolyse readily in air. $TaCl_4N_3$ is dimeric in solution and in the solid state, the monomer units being linked together by the α-N atoms of the azide groups. Two slightly different forms of the dimers have been revealed by an *X*-ray diffraction study; they differ by *ca.* 6° in the angle between one of the azide groups and the $(Ta-\alpha-N)_2$ ring. Ta—N distances range from 2.14(2) to 2.21(2)Å.

Complex formation between $Nb(OH)_5$ and methylamine has been established[582] by an n.m.r. study. Two signals are observed in the n.m.r. spectrum of an aqueous solution of methylamine containing $Nb(OH)_5$. One of these is due to the methyl protons of the amine and the other to the rapidly exchanging protons of the OH, NH_2, and H_2O groups. The i.r. spectra of solutions and precipitates formed by $Nb(OH)_5$ and methylamine or monoethanolamine have been interpreted[583] in terms of Nb—N co-ordination. Niobium oxohalide complex formation with tri-n-octylamine has been studied by reverse ebullioscopy.[584]

Benzimidazole and 2-substituted benzimidazoles form 4:1 complexes with $NbCl_5$ which precipitate from THF solution and behave as 1:1 electrolytes in DMF.[585] A band at 1639 cm^{-1} in the i.r. spectra of the benzimidazole complex is assigned to ν(C=N); it is 26 cm^{-1} higher than in free benzimidazole, in accord with co-ordination of the ligand through its azomethine nitrogens only; this is supported by the retention

[580] D. Stefanovic, L. Stambolija, and V. Katovic, *Org. Mass Spectrometry*, 1973, **7**, 1357.
[581] J. Strahle, *Z. anorg. Chem.*, 1974, **405**, 139.
[582] N. D. Milovidova, B. M. Moiseev, and N. V. Petrova, *Zhur. neorg. Khim.*, 1974, **19**, 405.
[583] N. V. Petrova, L. S. Solntseva, and N. S. Mikhailova, *Izvest. V.U.Z. Tsvet. Met.*, 1974, **17**, 38 (*Chem. Abs.*, 1974, **81**, 15543).
[584] S. M. Sinetsyna, T. M. Gorlova and V. F. Chistyakov, *Zhur. neorg. Khim.*, 1973, **18**, 2114.
[585] N. S. Birader, T. R. Goudar, and V. H. Kulkarni, *J. Inorg. Nuclear Chem.*, 1974, **36**, 1181.

of a band at $1330\,cm^{-1}$ attributable to the OH in-plane deformation vibration on complex formation.

A new series of niobium derivatives (62) of the *N*-phenylurethanes has been obtained[586] from the insertion reaction between niobium penta-alkoxides and phenyl isocyanate. Most of the products are yellow liquids, one is white and the rest are brown or black solids or semi-solids. They exhibit $v(C=O)$ in the range 1690—$1730\,cm^{-1}$. Insertion of up to three molecules of phenyl isocyanate occurs readily at

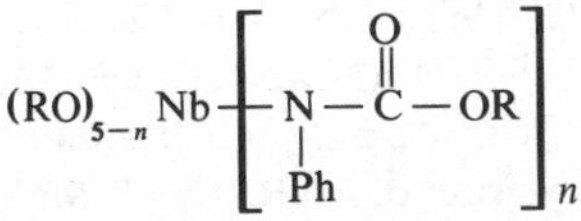

$$(RO)_{5-n} Nb \left[N - \overset{\overset{\displaystyle O}{\|}}{C} - OR \right]_n$$
$$\underset{Ph}{|}$$

(62)

room temperature. The mono-insertion products are thermally stable and can be distilled under reduced pressure but the di- and tri-insertion products tend to disproportionate. A freshly prepared sample of the insertion product from niobium penta-isopropoxide gave isopropyl-*N*-phenylcarbamate on hydrolysis. Although $TaCl_3(NN'$-di-isopropylacetamidinato$)_2$ belongs to the pentagonal-bipyramidal structural class characteristic of (chelate)MCl_3 compounds (M = Nb or Ta), there are considerable distortions from the ideal geometry owing to the steric demands of the bulky *N*-isopropyl groups.[587] The two four-membered rings are planar and are twisted out of the plane of the pentagonal girdle in opposite directions.

Mixed N- *and* S-*Donor Ligands.* Despite difficulties created by rapid exchange and ligand redistribution reactions, it has been possible to observe the ambidentate behaviour of the SCN^- ion in $NbCl_5$ acetonitrile solutions by ^{93}Nb n.m.r. spectroscopy.[588] This technique has also allowed the identification of specific geometrical isomers in labile equilibria for halogen and pseudo-halogen ligands. $NbCl_3[NMeC-(=S)Me]_2$, the product from the insertion of MeNCS into the Nb—C bonds of Me_2NbCl_3, has been shown by *X*-ray diffraction to be another example of the pentagonal-bipyramidal class of seven-co-ordinate molecules.[589] The MMeC(=S)Me ligands and one chlorine atom occupy the pentagonal girdle positions with the ligand atom sequence Cl, N, S, S, N. Distortion of the girdle is caused by the small N—S ligand bites, *ca.* 63° compared with the ideal value of 72°, and by steric interactions between the girdle chlorine and the *N*-methyl groups, which also lengthen the Nb—Cl(girdle) bond compared with the Nb—Cl(axial) bonds.

Organometallic Niobium(v) *and Tantalum*(v) *Compounds.* Reports this year have concerned two main areas of interest: the preparation of alkyl-niobium and -tantalum halides and their reactions with donor ligands, and insertion reactions into methyl-metal bonds. The first examples of monomethyl niobium and tantalum compounds have been prepared by the low-temperature methylation of the appropriate halides

[586] R. C. Mehrotra, A. K. Rai, and R. Bohra, *J. Inorg. Nuclear Chem.*, 1974, **36**, 1887.
[587] M. G. B. Drew and J. D. Wilkins, *J.C.S. Dalton*, 1974, 1579.
[588] A. T. Rake, *Diss. Abs.* (*B*), 1973, **34**, 1418.
[589] M. G. B. Drew and J. D. Wilkins, *J.C.S. Dalton*, 1974, 198.

with HgMe$_2$ or SnMe$_4$.[590] The pure orange-brown product is very soluble in common non-co-ordinating solvents; it is monomeric in the vapour phase and dimeric in benzene. MeNbCl$_4$ forms 1:1 adducts in solution with MeCN, NMe$_3$, SMe$_2$, thioxan, P(NMe$_2$)$_3$, PPh$_3$, AsPh$_3$, γ-pic, and diphos and 2:1 adducts with Me$_2$O and OP(OMe)$_3$; several of these were isolated. More complicated reactions involving chlorine exchange and oxygen abstraction from the ligand occurred with ligands such as Me$_3$PO, Ph$_3$PO, and (Me$_2$N)$_3$PO. Indeed, such reactions have been employed successfully in the preparation[591] of new σ-alkyl oxohalide compounds of niobium and tantalum, Me$_2$MOX$_2$,2L (M = Nb, X = Cl, L = Me$_2$SO, Me$_3$NO, pyO, Me$_3$PO, Ph$_3$PO, (Me$_2$N)$_3$PO, O[OP(NMe$_2$)$_2$]$_2$ or OAsPh$_3$; M = Nb, X = Br, L = Ph$_3$PO; M = Ta, X = Cl, L = Ph$_3$PO). An alternative route to this type of compound involves direct alkylation of MOCl$_3$ with MeMgI in ether–toluene (1:2) followed by the addition of the ligand.

The acceptor properties of Me$_3$MCl$_2$ (M = Nb or Ta) are considerably less than those of Me$_2$MCl$_3$ and MeMCl$_4$. Of a large number of unidentate ligands tested, only py, 3-Mepy, Cl$^-$, and Ph$_3$PO gave isolable products.[592] Complexes of the type Me$_3$MCl$_2$,L–L were obtained with the bidentate ligands L–L = dme, tren, bipy, 4,4'-bipy, diphos, MeSCH$_2$CH$_2$SMe, EtS(CH$_2$)$_2$SEt, and Ph$_2$PCH$_2$PPh$_2$. I.r. and n.m.r. data were consistent with seven co-ordinate structures.

Insertion reactions of the methyl-niobium and -tantalum halides with nitric oxide, organic nitrites, isocyanates, thiocyanates, and isothiocyanates have been reported. Multiple insertions of NO into the M—C bonds of Me$_2$MCl$_3$ and Me$_3$MCl$_2$ (M = Nb or Ta) occurred at low temperatures, yielding the N-methyl-N-nitrosohydroxyl-amine complexes MCl$_3$[ON(Me)NO]$_2$ and MeMCl$_2$[ON(Me)NO]$_2$, respectively.[593] A single-crystal X-ray study of MeTaCl$_2$[O(Me)NO]$_2$ has identified it as a member of the pentagonal-bipyramidal group of seven-co-ordinate Group VB monomers. There are four oxygens and a methyl group in the pentagonal girdle, the co-ordination sphere being completed by two axial chlorine atoms. MeTaCl$_2$[ON(Me)NO]$_2$ is substantially less reactive than other alkyls of tantalum and niobium; it is unaffected by 2 h exposure to dry oxygen at room temperature, whereas Me$_2$NbCl$_3$, for example, reacts explosively with pure oxygen. Cleavage of the methyl group can be effected with Et$_2$NH, EtCO$_2$H, HCl, or MeOH, the last yielding the complex MeOTaCl$_2$-[ON(Me)NO]$_2$. Both the Ta and Nb alkyl complexes are more thermally stable than six-co-ordinate complexes of the type MeMCl$_4$L (L = unidentate ligand). Insertion of RNCS (R = Me or Ph) into the metal–carbon bonds of Me$_x$MCl$_{5-x}$ (M = Nb or Ta, x = 1 or 2) gave the deeply coloured N-substituted thioacetamide complexes XMCl$_3$[NR—C(S)Me] (M = Nb or Ta, R = Me, X = Cl; M = Ta, R = Ph, X = Cl or Me; M = Ta, R = X = Me) and NbCl$_3$[NMe—C(S)Me]$_2$.[594] The rates of insertion followed the sequence MeMCl$_4$ > Me$_2$MCl$_3$ > Me$_3$MCl$_2$: they parallel the established order of acceptor properties of the methyl-niobium and -tantalum chlorides. Only 1:1 donor–acceptor complexes were formed between MeMCl$_4$ or Me$_2$MCl$_3$ (M = Nb or Ta) and MeSCN. The seven-co-ordinate nature

[590] C. Santini-Scampucci and J. G. Riess, *J.C.S. Dalton*, 1973, 2436.
[591] C. Santini-Scampucci and J. G. Riess, *J.C.S. Dalton*, 1974, 1433.
[592] G. W. A. Fowles, D. A. Rice, and J. D. Wilkins, *J.C.S. Dalton*, 1974, 1080.
[593] J. D. Wilkins and M. G. B. Drew, *J. Organometallic Chem.*, 1974, **69**, 111.
[594] J. D. Wilkins, *J. Organometallic Chem.*, 1974, **65**, 383.

of NbCl$_3$[NMe—CS(Me)]$_2$ was confirmed by X-ray diffraction (see previous section). Analogues of the RNCS insertion products MCl$_3$[NR—C(=O)Me]$_2$, MeMCl$_2$[NPh—C(=O)Me]$_2$, MeTaCl$_2$[NMe—C(=O)Me]$_2$, and TaCl$_2$[NMe—C-(=O)Me]$_3$ containing acetamide groups derived from insertion of isocyanates into the metal–carbon have been obtained[595] and the complexes MeTaCl$_2$[C(Me)=NR] (R = cyclohexyl or *p*-tolyl) have been isolated from the reactions of Me$_3$TaCl$_2$ with RNC. Trimethylbis[bis(pyrazolyl)borate]tantalum(v) has been prepared.[596]

PART II Chromium, Molybdenum, Tungsten, Technetium, and Rhenium *by C. D. Garner*

1 Chromium

Introduction.—Reviews have been published describing recent developments in the organometallic[1] and synthetic[2] chemistry of chromium. Interesting new aspects of chromium chemistry reported this year include the electrochemical generation[3] of the new cation [Cr(CO)$_6$]$^+$, the first report of thiocarbonyl complexes of chromium[4, 5] and the characterization of (Et$_4$N)$_6$[Cr(SnCl$_3$)$_6$].[6] Bivalent chromium in anhydrous zeolite *A* has been shown to be a particularly simple, reversible, binder of molecular oxygen.[7] [CrMo(OAc)$_4$] has been reported[8] and oxidation and substitution of [Cr$_2$(OAc)$_4$] in aqueous media have been shown to involve prior dissociation into Cr(OAc)$_2$ units.[9]

Carbonyl and Thiocarbonyl Complexes.—For a comprehensive survey the reader should consult the Organometallic Specialist Periodical Report.

Further thermodynamic data,[10] absolute integrated intensities for the i.r. absorption bands,[11] and more Raman data[12] have been presented for Cr(CO)$_6$. Core[13] and valence region[14] photoelectron spectra have been determined for Cr(CO)$_6$ and discussed on the basis of MO data. Examination of the reaction products of Cr atoms with CO at 4.2—10 K has established the existence of a Cr(CO)$_5$ species having a D_{3h}, trigonal-bipyramidal stereochemistry. The data obtained using pure

595 J. D. Wilkins, *J. Organometallic Chem.*, 1974, **67**, 269.
596 D. H. Williamson, C. Santini-Scampucci, and G. Wilkinson, *J. Organometallic Chem.*, 1974, **77**, C25.
1 D. J. Darensbourg, *J. Organometallic Chem.*, 1973, **62**, 299.
2 G. A. Tsigdinos and F. W. Moore, *Ann. Reports Inorg. Gen. Synthesis*, 1973, **1**, 89.
3 C. J. Pickett and D. Pletcher, *J.C.S. Chem. Comm.*, 1974, 660.
4 B. D. Dombek and R. J. Angelici, *J. Amer. Chem. Soc.*, 1973, **95**, 7516.
5 G. Jaouen, *Tetrahedron Letters*, 1973, 5159.
6 T. Kruck and H. Breür, *Chem. Ber.*, 1974, **107**, 263.
7 R. Kellerman, P. J. Hutta, and K. Klier, *J. Amer. Chem. Soc.*, 1974, **96**, 5946.
8 C. D. Garner and R. G. Senior, *J.C.S. Chem. Comm.*, 1974, 580.
9 R. D. Cannon and J. S. Lund, *J.C.S. Chem. Comm.*, 1973, 904.
10 G. V. Burchalova, M. S. Sheiman, V. G. Syrkin, and I. B. Rabinovich, *Trudy Khim. i khim. Tekhnol.*, 1973, 47 (*Chem. Abs.*, 1974, **80**, 149416).
11 S. Kh. Samvelyan, V. T. Aleksanyan, and B. V. Lokshin, *J. Mol. Spectroscopy*, 1973, **48**, 47.
12 D. M. Adams, W. S. Fernando, and M. A. Hooper, *J.C.S. Dalton*, 1973, 2264.
13 J. A. Connor, M. B. Hall, I. H. Hillier, W. N. E. Meredith, M. Barber, and Q. Herd, *J.C.S. Faraday II*, 1973, **69**, 1677.
14 B. R. Higginson, D. R. Lloyd, P. Burroughs, D. M. Gibson, and A. F. Orchard, *J.C.S. Faraday II*, 1973, **69**, 1659.

$^{12}C^{16}O$ are very different from those in $^{12}C^{16}O$–Ar matrices and indicate a number of species which strongly suggest that the square-pyramidal $Cr(CO)_5$ moiety, originally considered to be the primary product of $Cr(CO)_6$–Ar matrix photolysis, is actually $Cr(CO)_5L$ (where L is either an O-bonded isocarbonyl or a sideways-bonded carbonyl ligand). The implication of these results for the structure and bonding of metal pentacarbonyls and their anions have been discussed.[15] $[Cr(CO)_5]^-$ has been produced[16] by the vacuum-u.v. photolysis of matrix-isolated $Cr(CO)_6$. Optical absorption maxima have been reported[17] for this ion and $[Cr(CO)_4]^-$ following their generation by an ion cyclotron resonance technique. This method has also been used in other studies, the results of which suggest that the new dianions $[Cr(CO)_8]^{2-}$ and $[Cr(CO)_6]^{2-}$ may have been prepared.[18] $Cr(CO)_6$ in MeCN solution containing Bu_4NBF_4 may be oxidized at a Pt electrode to the 17-electron species $[Cr(CO)_6]^+$ whose e.s.r. spectrum has been determined.[3] New, accurate positive-ion mass spectral data have been presented[19] for $Cr(CO)_6$.

Changes in the electronic spectrum of $Cr(CO)_6$ produced by substituting CO by an amine have been studied experimentally and theoretically and discussed in terms of σ- and π-bonding effects.[20] ESCA data for $[(OC)_5CrC(OMe)Me]$ have been interpreted to suggest that none of the carbene carbon atoms has a charge more positive than that of the carbonyl carbons. Therefore, the carbene carbon does not appear to be best described as a carboxonium atom.[21] ^{119}Sn Mössbauer, $^{119}Sn-^1H$ three-bond n.m.r. coupling constants, and ESCA data for the $[(OC)_5CrSn(Bu^t)_2L]$ (L = THF, DMSO, or py) 'stannylene' complexes favour[22] Sn^{IV}. $[(OC)_5CrSnX_2]$ (X = Cl, Br, or I) and $[(OC)_5CrGeCl_2]$ have been prepared by the photochemical reaction of $Cr(CO)_6$ in SnX_2 or $CsGeCl_3$ in THF. Further reaction of these compounds with Me_4NX in these media produced $[(OC)_5CrSnX_3]^-$ or $[(OC)_5CrGeCl_3]^-$, respectively.[23] Dimethylketimine has been stabilized in the $[(OC)_5CrNH=CMe_2]$ complex.[24] $[Cr(CO)_5F]^-$ has been prepared by the reaction of $Cr(CO)_6$ with $(Ph_3P)_2NF$, or $[Cr_2(CO)_{10}]^{2-}$ with AgF, and characterized by i.r. spectroscopy.[25] The anions $[Cr(CO)_5X]^-$ (X = Cl, Br, or I) undergo two one-electron electrochemical oxidation steps in acetone solution. The first is reversible and produces $[Cr(CO)_5X]$, the second forms $[Cr(CO)_5X]^+$ which may be stabilized at $-75\,°C$, thus supporting earlier suggestions that chromium(II) halogenocarbonyls may be obtained.[26]

$Na_2[Cr(CO)_5]$ reacts with Cl_2CS in THF to produce $Cr(CO)_5(CS)$ in *ca.* 5% yield. This thiocarbonyl complex reacts with Ph_3P to produce *trans*-$[Cr(CO)_4$-$(CS)PPh_3]$ and with primary amines (RNH_2) to produce a thiocarbene, which

[15] E. P. Kündig and G. A. Ozin, *J. Amer. Chem. Soc.*, 1974, **96**, 3820.
[16] J. K. Burdett, *J.C.S. Chem. Comm.*, 1973, 763.
[17] R. C. Dunbar and B. B. Hutchinson, *J. Amer. Chem. Soc.*, 1974, **96**, 3816.
[18] R. C. Dunbar, J. F. Ennever, and J. P. Fackler, jun., *Inorg. Chem.*, 1973, **12**, 2734.
[19] P. E. Gaivoronskii, *Trudy Khim. i khim. Tekhnol.*, 1973, 72 (*Chem. Abs.*, 1974, **80**, 101434).
[20] F. A. Cotton, W. T. Edwards, F. C. Rauch, M. A. Graham, R. N. Perutz, and J. J. Turner, *J. Coordination Chem.*, 1973, **2**, 247.
[21] W. B. Perry, T. F. Sehaaf, W. L. Jolly, L. J. Todd, and D. L. Cronin, *Inorg. Chem.*, 1974, **13**, 2038.
[22] G. W. Grynkewich, B. Y. K. Ho, T. J. Marks, D. L. Tomaja, and J. J. Zuckerman, *Inorg. Chem.*, 1973, **12**, 2522.
[23] D. Uhlig, H. Berens, and E. Lindner, *Z. anorg. Chem.*, 1973, **401**, 233.
[24] R. B. King and W. M. Douglas, *J. Amer. Chem. Soc.*, 1973, **95**, 7528.
[25] W. Douglas and J. K. Ruff, *J. Organometallic Chem.*, 1974, **65**, 65.
[26] A. M. Bond, J. A. Bowden, and R. Colton, *Inorg. Chem.*, 1974, **13**, 602.

readily loses H_2S, affording the corresponding isocyanide complex, $[(OC)_5CrC\equiv NR]$.[4] Irradiation of $[(\eta^6\text{-}C_6H_5CO_2Me)Cr(CO)_3]$ with *cis*-cyclo-octene and treatment of the product with CS_2 yields $[(\eta^6\text{-}C_6H_5CO_2Me)Cr(CO)_2(CS)]$ which produces $[(\eta^6\text{-}C_6H_5CO_2Me)Cr(CO)(CS)\{P(OEt)_3\}]$ on irradiation with $(EtO)_3P$. The corresponding benzoic acid derivative of this latter compound has a pK_a of 5.20 which, when compared with that for the corresponding biscarbonyl complex, suggests that π-back-bonding to the CS group is greater than to the CO one in these compounds.[5]

η-Complexes.—[13]C n.m.r. chemical shift data for a series of closely related $[(\eta^6\text{-}arene)Cr(CO)_3]$ complexes have been shown to afford a linear measure of Cr to CO π-back donation.[27] Thermodynamic heats of formation for several bis-$(\eta^6\text{-arene})$ chromium halides, $(\eta^6\text{-}C_6H_6)_2Cr$, and some $[(\eta^6\text{-arene})Cr(CO)_3]$ derivatives have been reported.[28]

The molecular structure of $(\eta^6\text{-}C_5H_5)_2Cr$ has been determined by gas-phase electron diffraction; $Cr-C = 216.9(4)$ pm and the C—H bonds are bent $2.9(1.1)°$ out of the plane of the C_5 ring towards the metal atom.[29] Ligand-field theory has been applied[30] to discuss the electron structure of $(\eta^5\text{-}C_5H_5)_2Cr$, $[(\eta^5\text{-}C_5H_5)_2Cr]^+$, and $[(\eta^6\text{-}C_6H_6)_2Cr]^+$. New positive-ion mass spectral data have been presented for $(\eta^5\text{-}C_5H_5)_2Cr$.[31]

Trifluorophosphino- and Trichlorostannano-complexes.—The photoelectron spectrum of $Cr(PF_3)_6$ has been determined and compared with other available data for transition-metal trifluorophosphino- and carbonyl-complexes.[32]

$(Et_4N)SnCl_3$ reacts with $(\eta^6\text{-}C_6H_6)_2Cr$ in refluxing THF or acetone to produce $(Et_4N)_6[Cr(SnCl_3)_6]$, the existence of which suggests that $SnCl_3$ is a good π-back-bonding ligand. $Me_4N[Cr(CO)_5SnCl_3]$ and *fac*-$(Me_4N)_3[Cr(CO)_3(SnCl_3)_3]$ were also reported in this study.[6]

Dinitrogenyl Complexes.—A theoretical study of $[(\eta^6\text{-}C_6H_6)Cr(CO)_{3-n}(N_2)_n]$ ($n = 0$ or 1) has shown that dinitrogen is a weaker π-acceptor and a stronger σ- and π-donor ligand than carbonyl in these complexes.[33] The electronic structure of linear Cr—N—N fragments have also been investigated using SCF MO calculations.[34]

Nitrosyl Complexes.—I.r. and e.s.r. spectral data of several chromium nitrosyl complexes have been determined and discussed with respect to the metal–ligand bonding interactions.[35] $[Cr(NO)(AsPhR_2)(S_2CX)(H_2O)_2]$ (R = Me, Et, Ph, or Bu; X = OEt or NEt_2) have been prepared by treating the corresponding $[Cr(NO)(S_2CX)(H_2O)_3]$ complex with PhR_2As in acetone and characterized by e.s.r. spectroscopy.[36] Denitrosylation of $[Cr(NO)_2(NCMe)_4](PF_6)_2$ with RNC (R = Me, Bu^t, or p-ClC_6H_4) has

[27] G. M. Bodner and L. J. Todd, *Inorg. Chem.*, 1974, **13**, 1335.

[28] J. A. Connor, H. A. Skinner, and Y. Virmani, *J.C.S. Faraday I*, 1973, **69**, 1218; P. N. Nikolaev, V. A. Safonov, E. M. Moseeva, and I. B. Rabinovich, *Trudy Khim. i khim. Tekhnol.*, 1972, 54 (*Chem. Abs.*, 1973, **79**, 77869).

[29] A. Haaland, J. Lusztyk, D. P. Novak, J. Brunvoll, and K. B. Starowieyski, *J.C.S. Chem. Comm.*, 1974, 54.

[30] K. D. Warren, *Inorg. Chem.*, 1974, **13**, 1243, 1317.

[31] P. E. Gaivoronskii and N. V. Larin, *Trudy Khim. i khim. Tekhnol.*, 1973, 74 (*Chem. Abs.*, 1974, **80**, 101432).

[32] J. F. Nixon, *J.C.S. Dalton*, 1974, 2226.

[33] N. J. Fitzpatrick and N. J. Mathews, *J. Organometallic Chem.*, 1973, **61**, C45.

[34] S. M. Vinogradova and Yu. G. Borod'ko, *Zhur. fiz. Khim.*, 1973, **47**, 789.

[35] T. V. Ovchinnikov, *Izvest. Akad. Nauk S.S.S.R., Ser. khim.*, 1973, 975.

[36] O. I. Kondrat'eva, A. D. Troitskaya, N. A. Chadaeva, A. I. Chiukova, G. M. Usucheva, and A. E. Ivantsov, *Zhur. obshchei Khim.*, 1973, **43**, 2087.

afforded the diamagnetic compounds $[Cr(NO)(CNR)_5](PF_6)$. which can be oxidized either voltametrically or chemically to the corresponding paramagnetic ($S = \frac{1}{2}$) $[Cr(NO)(CNR)_5](PF_6)_2$ derivative. Voltametric evidence for the $[Cr(NO)(CNR)_5]^{2+}$ cations has also been obtained.[37]

Compounds with Chromium–Metal Bonds.—The crystal structure of $[(\eta^5\text{-}C_5H_5)_2\text{-}Cr_2(CO)_6]$ shows a crystallographic centre of symmetry with the two $(\eta^5\text{-}C_5H_5)(OC)_2Cr$ moieties linked by a Cr—Cr bond (328.1 pm) which is far longer than expected and too long for an unstrained single bond. ^1H N.m.r. data show that this complex exists as a solvent- and temperature-dependent mixture of *anti*- and *gauche*-rotamers. The barrier to internal rotation, $\Delta G^{\ddagger}_{298} = 50 \pm 3$ kJ mol^{-1}, is substantially lower than in the corresponding Mo compound. Marked line-broadening in the ^1H n.m.r. spectrum at $-10\,°C$ suggests dissociation into $(\eta^5\text{-}C_5H_5)Cr(CO)_2$ radicals.[38] Similar fluxional behaviour for $[(\eta^5\text{-}C_5H_5)_2Cr_2(NO)_2]$ has also been studied by ^1H n.m.r. and i.r. spectroscopy.[39] A vibrational analysis of the i.r. spectroscopic data for $[(OC)_5MnM(CO)_5]^-$ (M = Cr, Mo, or W) has suggested that the force constant for the metal–metal bonds increases as M = W > Mo > Cr.[40]

$[CrMo(OAc)_4]$. a metal(II) carboxylate with a heteronuclear metal–metal bond, has been prepared by gradual addition of $Mo(CO)_6$ in AcOH–CH$_2$Cl$_2$ (5:1) to a refluxing solution of $[Cr_2(OAc)_4(OH_2)_2]$ in AcOH–Ac$_2$O (10:1). Vibrational spectroscopic data suggest that the force constant of the Cr—Mo bond is *ca.* 0.66 that of the corresponding Mo—Mo one. Consistent with this conclusion, this heteronuclear bond appears to be quite labile.[8] A study of the oxidation and substitution of $[Cr_2(OAc)_4]$ in aqueous media has shown that these are preceded by unimolecular dissociation into $Cr(OAc)_2$ species, the rate-determining step probably being the cleavage of a Cr—O bond.[9] Extended Hückel MO calculations have been accomplished for $[Cr_2Cl_9]^{3-}$ and the Cr—Cr interaction in this ion assessed.[41]

Binary Compounds and Related Systems.—*Halides*. The preparation and properties of the higher valent chromium halides have been reviewed.[42] Fluorination of chromium by XeF$_2$ and XeF$_6$ has been studied under a variety of conditions leading to the formation of CrF$_2$, CrF$_3$, and CrF$_4$,XeF$_6$.[43] A neutron-diffraction study of the charge distribution within CrF$_3$ has been reported[44a] and the Cl K-absorption spectrum of CrCl$_2$ determined.[44b] The complex heterogeneous equilibria between chromium-(II),-(III), and -(IV) iodides and iodine have been studied between 519 and 591 °C. Evidence for the formation of CrI$_4$(g) was obtained and a value for its heat of formation calculated.[45]

Oxides. The preparation and properties of the higher valent oxides of chromium have been reviewed.[42] $\Delta\text{-}Cr_2O_3$ has been obtained by heating Cr_2O_3 at 1550 °C and

[37] M. K. Lloyd and J. A. McCleverty, *J. Organometallic Chem.*, 1973, **61**, 261.
[38] R. D. Adams, D. E. Collins, and F. A. Cotton, *J. Amer. Chem. Soc.*, 1974, **96**, 749.
[39] R. M. Kirchner, T. J. Marks, J. S. Kristoff, and J. A. Ibers, *J. Amer. Chem. Soc.*, 1973, **95**, 6602; J. L. Calderon, S. Fontana, E. Frauendorfer, and V. W. Day, *J. Organometallic Chem.*, 1974, **64**, C10.
[40] J. R. Johnson, R. J. Ziegler, and W. M. Risen, *Inorg. Chem.*, 1973, **12**, 2349.
[41] D. V. Korol'kov, and Kh. Missner, *Teor. i eksp. Khim.*, 1973, **9**, 336 (*Chem. Abs.*, 1973, **79**, 97135).
[42] C. Rosenblum and S. L. Holt, *Transition Metal Chem.*, 1972, **7**, 87.
[43] B. Zemva and J. Slivnik. *Inst. Jozef Stefan, IJS Rep.*, 1972, R-609 (*Chem. Abs.*, 1973, **79**, 26693).
[44] (a) A. J. Jacobson, L. McBride and B. E. F. Fender. *J. Phys. (C)*, 1974, **7**, 783; (b) C. Sugiura, *J. Chem. Phys.*, 1973, **58**, 5444.
[45] C.-F. Shieh and N. W. Gregory. *J. Phys. Chem.*, 1973, **77**, 2346; C. Shieh, *Diss. Abs. (B)*, 1973, **34**, 1456.

characterized by X-ray powder diffraction and i.r. spectroscopy.[46] Study of the interconversion of CrOOH and CrO_2 has indicated that the solids are linked by a topotactic reaction with the former compound possessing a symmetrical hydrogen-bond.[47] The Cr_2O_3–CrO_2 phase boundary has been investigated under a high pressure of O_2, and thermodynamic data for the interconversion of these oxides have been determined.[48] The formation and stability region of pure ferromagnetic CrO_2, originating from the thermal decomposition of CrO_3 under pressure, has been studied.[49] X-Ray and electron diffraction spectroscopy and electron microscopy have been used[50] to characterize the rutile-derived shear phases of composition Cr_nO_{2n-1} (n = 4 or 6), $V_{1-x}Cr_xO_2$ ($0 \leqslant x \leqslant 0.15$) and $(Ti,Cr)O_x$ ($1.875 \leqslant x \leqslant 1.93$) and the phase relationships and physical properties of these vanadium systems have been reviewed.[51]

Chalcogenides. The kinetic and thermodynamic properties of chromium sulphides have been investigated under various conditions of temperature and pressure of S_2,[52] and the non-stoicheiometry, structure, and thermodynamic relationships for CrS_x ($1.00 \leqslant x \leqslant 1.54$) further studied.[53] Cr_3S_4, $NiCr_2S_4$, and $Ni_{1-x}Cr_{2+x}S$ (x = 0.20—0.74) crystals have been prepared by chemical transport reactions using $AlCl_3$–$CrCl_3$ mixtures as the transporting agents,[54] as have the phases Cr_2S_3, Cr_2Se_3, and $Cr_2S_{3-x}Se_{3-x}$ (x = 0.75—2.70). The transport rate data were used to calculate the heats of formation of $Cr_2S_3(s)$ and $Cr_2Se_3(s)$ as 357 ± 30 and 268 ± 30 kJ mol^{-1}, respectively.[55] The magnetic properties of compounds in the Cr–Se system have been investigated between 4.2 and 230 K.[56] X-Ray powder diffraction data have been reported[57] for the new sulphides $CrUS_3$ and CrU_8S_{17} and the crystal structure of $CrNb_2Se_4$ has been determined.[58]

Compounds with Elements of Group V. An effusion–mass spectrometric study has afforded values of 380 ± 20 and 450 ± 20 kJ mol^{-1}, respectively, for the dissociation energy and heat of formation at 298 K of CrN(g).[59] X-Ray crystallographic data have been determined for TiCrAs and the metal–metal bonding in this and other NiAs-related structures has been discussed.[60] The thermodynamic properties of

[46] S. Ahmed, *Sci. Res. (Dacca),* 1970, **7**, 145 (*Chem. Abs.,* 1973, **79**, 111229).

[47] F. P. Temme and T. C. Waddington, *J. Chem. Phys.,* 1973, **59**, 817; Y. Shibasaki, F. Kanamura, and M. Koizumi, *Mater. Res. Bull.,* 1973, **8**, 559; M. A. Alario-Franco, J. Fenerty, and K. S. W. Sing, 'Proceedings of the Seventh International Symposium on the Reactivity of Solids', ed. J. S. Anderson, Chapman and Hall, London, 1972, p. 327.

[48] Y. Shibasaki, F. Kanamura, M. Koizumi, and S. Kume, *J. Amer. Ceram. Soc.,* 1973, **56**, 248.

[49] Y. Mihara, E. Hirota, and Y. Terada, *Jap. P.,* 7318319 (*Chem. Abs.,* 1974, **80**, 75919); Z. Drbalek, D. Rykl, V. Seidl, and V. Sykora. *Sb. Vys. Sk. Chem.-Technol. Praze. Mineral.,* 1973, **G15**, 43 (*Chem. Abs.,* 1973, **79**, 142394).

[50] M. A. Alario-Franco, J. M. Thomas, and R. D. Shannon, *J. Solid State Chem.,* 1974, **9**, 261; G. Villeneuve, M. Drillon, and P. Hagenmuller, *Mater. Res. Bull.,* 1973, **8**, 1111; D. K. Philp and L. A. Bursill, *Acta Cryst.,* 1974, **A30**, 265.

[51] M. Nygren, *Chem. Commun. Univ. Stockholm,* 1973, No. 2 (*Chem. Abs.* 1973, **79**, 24168).

[52] K. Nishida, K. Nakayama, and T. Narita, *Corros. Sci.,* 1973, **13**, 759 (*Chem. Abs.,* 1974, **80**, 62678).

[53] K. N. Strafford and A. F. Hampton, *J. Mater. Sci.,* 1973, **8**, 1534; D. J. Young, W. W. Smeltzer, and J. S. Kirkaldy, *J. Electrochem. Soc.,* 1973, **120**, 1221.

[54] H. D. Lutz and K. H. Bertram, *Z. anorg. Chem.,* 1973, **401**, 185.

[55] H. D. Lutz, K. H. Bertram, M. Sreckovic, and W. Molls, *Z. Naturforsch.,* 1973, **28B**, 685.

[56] M. Yuzuri, *J. Phys. Soc. Japan,* 1973, **35**, 1252.

[57] H. Noel, *Compt. rend.,* 1973, **277**, C, 463.

[58] A. Meerschaut and J. Rouxel, *Compt. rend.,* 1973, **277**, C, 163.

[59] R. D. Srivastava and M. Farber, *High Temp. Sci.,* 1973, **5**, 489 (*Chem. Abs.,* 1974, **80**, 137632).

[60] V. Johnson. *Mater. Res. Bull.,* 1973, **8**, 1067.

CrSb and $CrSb_2$ have been examined in e.m.f. studies and their free energies of formation calculated as 7.1 and 8.6 kJ(g atom)$^{-1}$, respectively.[61]

Carbides, Silicides, and Germanides. The preparation of chromium carbides and related compounds has been reviewed.[62] The thermodynamic parameters for Cr_3C_2, Cr_7C_3, and $Cr_{23}C_6$ formation between 690 and 1133 K have been calculated from e.m.f. and other data.[63] The high-temperature phase in the Mo—Cr—C system, ζ-$(MoCr)_4C_3$, has been shown to be isomorphous with ζ-Hf_4N_3 by X-ray powder photography.[64] The electronic structures of Cr_3Si, CrSi, and Cr_5Si_3 have been investigated by X-ray emission and absorption spectroscopy,[65] and the magnetic structure of CrGe has been probed by neutron diffraction techniques at 4.2 K.[66]

Borides. Cr_2B, Cr_5B_3, and CrB have been characterized in X-ray diffraction studies of Cr–B system[67] and $CeBr_2B_6$, $CeCrB_4$, and $Ir_{10}Cr_{13}B_6$ have been similarly identified in their respective M–Cr–B systems.[68]

Chromium(II).—The applications of chromium(II) salts in preparative organic chemistry have been reviewed.[69] The kinetics of oxidation of Cr^{II} to Cr^{III} by halogen radical anions, a particularly simple one-electron oxidation scheme, have been determined.[70] Hydrated electrons are formed[71] during the photochemical oxidation of aqueous chromium(II).

Halides. The interaction between CrF_2 and CO in Ar matrices at low temperature has been studied by i.r. spectroscopy.[72] The new compounds $CrMF_5$ (M = Al, Ga, Ti, or V) have been prepared, they are isostructural with Cr_2F_5 and their magnetic properties have been investigated from 298 to 4.2 K.[73] The preparations of $MCr^{II}Cr^{III}F_6$ (M = Rb, Cs, or Tl) have been described and the magnetic and Mössbauer characteristics of these compounds have been discussed in relation to their crystal structures.[74] Similarly, the variable-temperature magnetic properties of K_2CrCl_4 have been interpreted in terms of a tetragonally distorted octahedral $Cr^{II}Cl_6$ unit.[75] The phase composition of the $NaCl$–$CrCl_2$ system has been studied and Na_2CrCl_4 further characterized.[76] The physical properties of the compounds

[61] L. V. Goncharuk and G. M. Lukashenko. *Porosh. Met.*, 1973, **13**, 55 (*Chem. Abs.*, 1973, **79**, 35738).
[62] S. Windisch and H. Nowotny. 'Preparative Methods in Solid State Chemistry, ed. P. Hagenmuller, Academic, New York, 1972, p. 533.
[63] V. N. Eremenko and V. R. Sidorko, *Porosh. Met.*, 1973, **13**, 51 (*Chem. Abs.*, 1973, **79**, 108825).
[64] E. Rudy, *J. Less-Common Metals*, 1973, **33**, 327.
[65] E. A. Zhurakovskii, P. V. Gel, V. V. Sokolenko, and A. I. Sokolenko, *Ukrain. fiz. Zhur.*, 1973, **18**, 843 (*Chem. Abs.*, 1973, **79**, 47415).
[66] M. Kolenda, J. Leciejewicz, and A. Szytula, *Phys. Status Solidi (B)*, 1973, **57**, K107 (*Chem. Abs.*, 1973, **79**, 59309).
[67] M. Lucco Borlera and G. Pradelli, *Met. Ital.*, 1973, **65**, 421 (*Chem. Abs.*, 1974, **80**, 41351).
[68] Yu. B. Kuz'ma, S. I. Svarichevskaya, and V. N. Fomenko, *Izvest. Akad. Nauk S.S.S.R., neorg. Materialy*, 1973, **9**, 1542.
[69] P. Rogl and H. Nowotny, *Monatsh.*, 1973, **104**, 1325.
[69] J. R. Hanson, *Synthesis*, 1974, 1.
[70] G. S. Laurence and A. T. Thornton, *J.C.S. Dalton*, 1974, 1142.
[71] H. Hartmann, J. Müller, and H. Kelm, *Naturwissen.*, 1973, **60**, 256.
[72] D. A. Van Liersburg and C. W. De Kock, *Nuclear Sci. Abs.*, 1973. **28**. 26905 (*Chem. Abs.*, 1974. **80**. 65255).
[⋅] A. Tressaud, J. M. Dance, J. Ravez, J. Portier, P. Hagenmuller, and J. B. Goodenough, *Mater. Res. Bull.*, 1973, **8**, 1467; J. M. Dance and A. Tressaud, *Compt. rend.*, 1973, **277**, C, 379.
[74] E. Banks, J. A. Deluca, and O. Berkooz, *J. Solid State Chem.*, 1973, **6**, 569.
[75] W. E. Gardner and A. K. Gregson, *Progr. Vac. Microbalance Tech.*, 1973, **2**, 183 (*Chem. Abs.*, 1974, **80**, 114077).
[76] V. G. Gopienko, S. N. Shkol'nikov, and E. F. Klyuchnikova, *Zhur. priklad Khim. (Leningrad)*, 1973, **46**, 1123.

MCrX$_3$ (M = Li—Cs; X = halide) have been reviewed[77a] and mass spectrometric studies of the compositions of MCrCl$_3$ (M = K or Cs) vapours have been reported and the thermodynamic characteristics for the evaporation of these compounds determined.[77b] Electrode potential data for Cr^{II} in KCl–LiCl melts have been reported.[78] CsCrI$_3$ has been obtained by melting CsI and CrI$_2$ (1:1) in a sealed, evacuated quartz tube. The lattice parameters of this compound have been determined and compared with those of other compounds of this formula type.[79]

O-*Donor Ligands.* The ion-exchange of Cr^{II} into *A*-type zeolites under oxygen-free conditions yields a pale-blue air-stable material containing 1.5 Cr^{II} atoms per unit cell. The electronic spectral and magnetic properties of this material support the idea that the Cr^{II} is co-ordinated by three zeolitic oxygen atoms arranged to give D_{3h} symmetry about the metal. Exposure of this Cr^{II}-containing zeolite to dry oxygen at 1 atm results in an instant colour change to grey, a change which is reversed *in vacuo*. It is suggested that the Cr^{II} in this zeolite is able to bind molecular oxygen as Cr^{III}–O$_2^-$, and the room temperature value of μ_{eff}, 3.7 BM, is consistent with spin-pairing between these two ions. Thus Cr^{II} on anhydrous zeolite *A* appears to be a particularly simple and reversible binder of O$_2$.[7]

The crystal structure of CrCl$_2$,4H$_2$O has been determined, the Cr^{II} being co-ordinated to four H$_2$O molecules arranged in a square plane (Cr—O = 208 pm) and two Cl atoms mutually *trans* (Cr—Cl = 276 pm). The electronic spectrum and thermal decomposition of this compound were also reported in this study.[80] The thermal decomposition of M$_2$[Cr(CO$_3$)$_2$],xH$_2$O (M = Na or K) has been followed by t.g.a.[81a] The dioctylphosphonate [Cr{OP(C$_8$H$_7$)$_2$O}$_2$] has been prepared, and its i.r. and electronic spectra suggest that it possesses a polymeric structure which involves several different types of metal–phosphinate linkage. This blue material undergoes a reversible colour change to pink above 185 °C.[81b]

N-*Donor and* P-*Donor Ligands.* Chromium(II) interacts with dinitrogen and ammonia on the surface of Cr^{II}-doped silica under atmospheres of N$_2$ and NH$_3$, respectively.[82] The dissolution of chromium metal in KSCN at 180—220 °C has been followed[83] by electronic spectroscopic and polarographic studies, the results of which indicated the formation of [Cr(NCS)$_6$]$^{4-}$.

The electrochemical reduction of [CrCl$_2$(diphos)$_2$]$^+$ has afforded the strongly reducing complex [CrCl$_2$(diphos)$_2$].[84]

Chromium(III).—Electronic energy transfer between Cr^{III} and Co^{III} complexes has been suggested to occur in certain instances from electronic and luminescence spectral studies.[85] An investigation of the quenching of [Ru(bipy)$_3$]$^{3+}$ phosphorescence by Cr^{III} complexes has shown that the nature of the ligands and the geometrical

[77] (a) J. F. Ackerman. G. M. Cole, and S. L. Holt, *Inorg. Chim. Acta*, 1974, **8**, 323; (b) I. A. Rat'kovskii. T. A. Pribytkova. and G. I. Novikov, *Zhur. fiz. Khim.*, 1973, **47**, 2950; 1974, **48**, 231.
[78] A. F. Yarovoi, V. N. Danilin, I. T. Sryvalin. and B. P. Burylev, *Zhur. fiz. Khim.*, 1973, **47**, 2416.
[79] T. Li, G. D. Stucky, and G. L. McPherson, *Acta Cryst.*, 1973, **B29**, 1330.
[80] H. G. Von Schnering and B. H. Brand. *Z. anorg. Chem.*, 1973, **402**, 159.
[81] (a) C. Baour, R. Ouahes, and H. Suquet. *Compt. rend.*, 1973, **276**, C, 1795; (b) H. D. Gillman, *Inorg. Chem.*, 1974, **13**, 1921.
[82] H. L. Krauss and B. Rebenstorf. *Z. anorg. Chem.*, 1973, **402**, 113; C. Hierl and H. L. Krauss, *Z. anorg. Chem.*, 1973, **401**, 263.
[83] S. V. Volkov. N. Kh. Tumanova. and N. I. Buryak, *Zhur. neorg. Khim.*, 1973, **18**, 1536.
[84] L. F. Warren and M. A. Bennett, *J. Amer. Chem. Soc.*, 1974, **96**, 3340.
[85] E. Cervona. C. Conti. and G. Sartori. *Gazzetta*, 1973, **103**, 923.

configuration of the complex are very important in determining the quenching ability of a particular complex.[86] A correlation has been found between the energy of the first d–d band of a number of Cr^{III} complexes and the rate of electron exchange with Cr^{II} complexes. This correlation has been interpreted in terms of the ease of electronic rearrangement at the Cr^{III} site in the activated complex prior to electron transfer.[87] A thorough evaluation of the *trans*-effect in substitution reactions of Cr^{III} octahedral complexes has been presented,[88] and further studies of *cis*-activation by oxyanions co-ordinated to Cr^{III} have been reported.[89] A comprehensive survey of the studies concerned with the kinetics and mechanism of Cr^{III} complexes appears in the Inorganic Reaction Mechanisms Specialist Periodical Report.

Optically Active Complexes. Irradiation of octahedral $[Cr^{III}(LL)_3]$ complexes with a laser beam at a wavelength close to the c.d. band maximum has been shown to induce optical activity, the extent of photoresolution amounting to a few per cent.[90] The natural optical activity of $[Cr(en)_3]^{3+}$ has been well interpreted using an SCC MO model.[91] A correlation of the absolute configuration of tris(*trans*-1,2-cyclohexanediamino)-complexes of Cr^{III} and M^{III} (M = Co, Rh, or Ir) using X-ray powder photography has led to the conclusion that all these complexes have the same configuration.[92] The four possible diastereoisomers of tris{(+)-3-acetylcamphorato}-chromium(III) have been separated by preparative t.l.c. on silica gel and an absolute configuration assigned to each complex from n.m.r., c.d., and X-ray diffraction spectral data.[93]

Halide and Oxyhalide Complexes. The charge distribution in $[CrF_6]^{3-}$ has been estimated by a semi-empirical MO approach.[94] The crystal structure of $CaCrF_5$ has been reinvestigated and the CrF_6 unit found to be more nearly regular than previously reported; Cr—F distances range from 185 to 194 pm.[95] The compounds $MCr^{II}Cr^{III}F_6$ (M = Rb, Cs, or Tl) have been prepared and their magnetic and Mössbauer characteristics determined.[74] Lattice constants and magnetic data have been reported for $M_2^1M^2CrF_6$ (M^1,M^2 = Li—Cs, or Tl).[96] Cubic $RbNiCrF_6$-type phases $LiM_{0.5}Cr_{1.5}F_6$ (M = Rb or Cs) have been obtained and shown to involve a statistical distribution of M^I and $\frac{1}{3}Cr^{III}$ ions over the 'Ni' sites.[97]

$[CrCl_6]^{3-}$ has been shown by e.s.r. spectroscopy to maintain an undistorted octahedral geometry in $[Co(NH_3)_6][InCl_6]$, even though the host lattice symmetry is rhombic.[98] The reactions of $CrCl_3$ with the chlorides of a variety of metals have been studied by X-ray diffraction and t.g.a. methods and a variety of compounds

[86] F. Bolletta, M. Maestri, L. Moggi, and V. Balzani, *J. Amer. Chem. Soc.*, 1973, **95**, 7864.

[87] G. St. Nikolov, *J. Inorg. Nuclear Chem.*, 1974, **36**, 1841.

[88] D. E. Bracken and H. W. Baldwin, *Inorg. Chem.*, 1974, **13**, 1325.

[89] G. Guastalla and T. W. Swaddle, *Inorg. Chem.*, 1974, **13**, 61; *Canad. J. Chem.*, 1974, **52**, 527; S. N. Choi and D. W. Carlyle, *Inorg. Chem.*, 1974, **13**, 1818.

[90] V. S. Sastri, *Inorg. Chim. Acta*, 1973, **7**, 381.

[91] R. S. Evans, A. F. Schreiner, and P. J. Hauser, *Inorg. Chem.*, 1974, **13**, 2185.

[92] P. Anderson, F. Galsbol, S. E. Harnung, and T. Laier, *Acta Chem. Scand.*, 1973, **27**, 3973.

[93] R. M. King, *Diss. Abs. (B)*, 1973, **33**, 5196.

[94] L. E. Harris and E. A. Boudraux, *Chem. Phys. Letters*, 1973, **23**, 434.

[95] K. K. Wu, and I. D. Brown, *Mater. Res. Bull.*, 1973, **8**, 593.

[96] D. Babel, R. Haegele, G. Pausewang, and F. Wall, *Mater. Res. Bull.*, 1973, **8**, 1371.

[97] G. Courbion, C. Jacoboni and R. De Pape, *Mater. Res. Bull.*, 1974, **9**, 425.

[98] E. W. Stout, jun., and B. B. Garrett, *Inorg. Chem.*, 1973, **12**, 2565.

Table 1 *Mixed-oxide compounds containing chromium*(III)

Compound	Source	Properties reported*	Ref.
$NaCrO_2$	$Na + Cr_2O_3$ or CrO_2	X	a
$MgCr_2O_4$	$MgF_2 + Cr_2WO_6$ at 1400 °C	μ	b
$CrBO_3$	oxidation of CrB_2	X	c
$LiAl_{5-x}Cr_xO_8$ ($x \leqslant 0.5$)	$LiO–Al_2O_3–Cr_2O_3$ system	X	d
$Cr(PO_3)_3$	$Cr_2O_3 + (NH_4)_3PO_4$ at 1100—1200 °C	i.r., X	e
$MgCr_2P_2O_7$	$MgO–Cr_2O_3–P_2O_5$ system	X	f
$CrAsO_4, H_2O$		X, scorodite structure	g
$Cr(IO_3)_3$		e, μ, t.d., X	h
$Ti_8Cr_2O_{19}$ $Ti_7Cr_2O_{17}$ $Ti_5Cr_2O_{11}$	$TiO_2–Cr_2O_3$ system	t.d., X	i
Cr_2NbO_5F	$NbO_2F + Cr_2O_3$ at 900 °C	X, trirutile structure	j
$M^1CrM^2O_4$ ($M^1 = Co, M^2 = Mn$; $M^1 = Mn, M^2 = Ga$)	$M^1O–Cr_2O_3–M_2^2O_3$ system	X, cation distribution	k
MCr_2O_4 ($M = Fe$ or Ni)	$MO + Cr_2O_3$	ΔH	l
MCr_2O_4 $MCr_{2-x}Al_xO_4$ ($M = Ca$ or Cd)	$MO–Cr_2O_3–Al_2O_3$ systems	e, X	m
Eu_2TaCrO_6	$Eu_2O_3–Ta_2O_5–Cr_2O_3$ system	μ, X, perovskite lattice	n
$LnCrO_3$ ($Ln = lanthanide$)		X	o

* For details of symbols and abbreviations, see Vol. 2, p. 178.

(a) L. N. Yannopoulos, *J. Inorg. Nuclear Chem.*, 1974, **36**, 214. (b) W. Kunnamann, *Inorg. Syn.*, 1973, **14**, 134; (c) V. A. Lavrenko, L. A. Glebov, and E. S. Lugovskaya, *Zasch. Metal*, 1973, **9**, 291 (*Chem. Abs.*, 1973, **79**, 26 708)); (d) T. Truchanowicz and R. Wadas, *Roczniki Chem.*, 1973, **47**, 1079 (*Chem. Abs.* 1973, **79**, 130 106); (e) A. Dimante, Z. Konstants, A. A. Myagkova, J. Bite, and A. Vaivads, *Latv. P.S.R. Zinat. Akad. Vestis., Khim. Ser.*, 1973, 496 (*Chem. Abs.*, 1973, **79**, 130 196); (f) V. M. Ust'yantsev, and M. G. Tretnikova, *Izvest. Akad. Nauk. S.S.S.R., neorg. Materialy*, 1973, **9**, 1652; (g) M. Ronis and F. D'Yvoire, *Bull. Soc. chim. France*, 1974, 78. (h) S. C. Abrahams, R. C. Sherwood, J. L. Bernstein, and K. Nassau, *J. Solid State Chem.*, 1973, **7**, 205; K. Nassau, J. W. Shiever, and B. E. Prescott, *J. Solid State Chem.*, 1973, **7**, 186; (i) C. W. Lee, *Acta Geol. Taiwan*, 1973. **16**, 39, (*Chem. Abs.*, 1974, **80**, 55 381); (j) J. Senegas and J. Galy, *Compt. rend.*, 1973, **277**, C, 1243; (k) P. D. Bhalerao, D. K. Kulkarni, and V. G. Kher, *Pramana*, 1973, **1**, 230 (*Chem. Abs.*, 1974, **80**, 53 299); B. Mansour, N. Baffier, and M. Huber, *Compt. rend.*, 1973. **277**, C, 867; (l) V. A. Levitskii, S. G. Popov, D. D. Ratiani, and B. G. Lebedev, *Termodin. Kinet. Protsessov Vosstanov. Metal. Mater. Konf.*, 1969, 36 (*Chem. Abs.*, 1973, **79**, 10 582); A. Kozlowska-Rog and G. Rog, *Roczniki Chem.*, 1973, **47**, 869 (*Chem. Abs.*, 1973, **79**, 97 700); (m) H. P. Fritzer, 'Proceedings of the Seventh International Symposium on Reactive Solids', ed. J. S. Anderson, Chapman and Hall, London, 1972, p. 354; (n) H. Parant, J. C. Bernier, and P. Poix, *Compt. rend.*, 1974, **278**, C, 49; (o) Z. A. Zaitseva, A. L. Litvin, and A. V. Shevchenko, *Dopov. Akad. Nauk Ukrain RSR, Ser. B*, 1973, **35**, 1099 (*Chem. Abs.*, 1974, **80**, 101 142).

characterized.[99] The synthetically useful complexes CrX_3,nL (X = Cl, Br, or I; L = py, or THF; *n* usually = 3) have been isolated from the reaction of powdered chromium with HgX_2 in the presence of the donor solvent.[100] The dimerization of olefins has been shown to be catalysed by a variety of Cr^{III} chloro-complexes.[101]

[99] N. V. Galitskii, V. I. Kudryavtsev, and F. F. Klyuchnikova, *Ref. Zhur. Met.*, 1973, Abs. 2G412 (*Chem. Abs.*, 1973, **79**, 140 164).
[100] G. E. Parris, *Syn. Inorg. Metal-Org. Chem.*, 1973, **3**, 245.
[101] E. A. Zuech, U.S. P., 3 726 939 (*Chem. Abs.*, 1973. **79**, 18 035).

O-*Donor Ligands.* SnO_2 containing a small amount of Cr_2O_3 is a very active catalyst for the adsorption and reduction of NO with CO, H_2, or C_2H_4 at low temperatures.[102] The formation of $ZnCr_2O_4$ when K_2CrO_4 dissolves in fused $ZnCl_2$ at temperatures $>350\,°C$ has been followed by electronic spectroscopy.[103] The compounds formed in the systems $Na_2O–Cr_2O_3–V_2O_5$[104] and $Cr_2O_3–Fe_2O_3–TiO_2–ZrO_2$[105] have been studied by X-ray diffraction spectroscopy. $[SiMo_{11}CrO_{39}.H_2O]^{5-}$ has been prepared by the addition of $Cr(NO_3)_3$ to a solution of α-molybdosilicic acid[106] and tungstochromates have been prepared in a similar manner.[107] The other mixed oxide compounds containing Cr^{III} which have been reported this year are listed in Table 1.

The thermal dehydration of $Cr(OH)_3.3H_2O$ to the amorphous hydroxide has been studied.[108] X-Ray diffraction data recorded for aqueous solutions of $[Cr(H_2O)_6]Cl_3$ show that the $[Cr(H_2O)_6]^{3+}$ ions pass relatively undisturbed from the solid into solution.[109] A deuterium magnetic resonance study of $[Cr(H_2O)_6]^{3+}$ has confirmed[110] the linear trend of the spin densities at the hydrogen atoms in $[M(H_2O)_6]^{3+}$ (M = Ti, V, or Cr) ions predicted earlier.

The solubility isotherm for the $TiO_2–Cr_2O_3–SO_3–H_2O$ system at 150°C has been presented and $H[Cr(SO_4)_2(H_2O)]$ identified as one of the solid phases.[111] The reactions of the alkali-metal sulphates M_2SO_4 (M = Na, K, or Cs) with $Cr_2(MoO_4)_3$ have been shown to produce $M_2Cr_2(MoO_4)_3$, the X-ray diffraction characteristics of which were reported.[112] A series of salts containing the pentamminodithionato-chromium(III) cation, $[Cr(NH_3)_5(S_2O_6)]^-$ have been prepared and characterized by i.r. and u.v.–visible spectroscopy.[113]

E.s.r. spectra have been reported[114] for several oxalato-chromium(III) complexes and the thermal decomposition products of $Cr_2(C_2O_4)_3.6H_2O$ characterized by X-ray powder photography.[115] The stability constants for 1:1, 1:2, and 1:3 chromium(III) formato-[116] and tartrato-[117] complexes have been determined, and the constitution and charge of 1:1 and 1:2 chromium(III) complexes with aspartic acid and asparagine reported.[118]

$Cr(HL),4H_2O$ [$H_4L = 5,5'$-dithiosalicylic acid (1)] has been prepared and characterized[119] as have 1:1, 1:2, and 1:3 complexes of chromium(III) with pyrocatechol

[102] F. Solymosi and J. Kiss. *J.C.S. Chem. Comm.* 1974. 509.
[103] D. H. Kerridge and I. A. Sturton, *Inorg. Chim. Acta*, 1974, **8**, 37.
[104] R. C. Kerby and J. W. Wilson, *Canad. J Chem.*, 1973, **51**, 1032.
[105] I. E. Grey, A. F. Reid, and J. G. Allpress, *J. Solid State Chem.*, 1973, **8**, 86.
[106] M. Fournier and R. Massart. *Compt. rend.*, 1973. **276**. C. 1517.
[107] V. I. Spitsyn, I. D. Kolli. T. I. Evchenko. and Kh. I. Lunck, *Zhur. neorg. Khim.*, 1973, **18**, 2176; V. I. Spitsyn, I. D. Kolli, and I. D. Bogatyreva, *Zhur. neorg. Khim.*, 1974, **19**, 564.
[108] R. Giovanoli and W. Stadelman, *Thermochim. Acta*, 1973, **7**, 41.
[109] A. Cristini, G. Licheri, G. Piccaluga, and G. Pinna, *Chem. Phys. Letters*. 1974, **24**, 289.
[110] A. M. Achlama and D. Fiat, *J. Chem. Phys.*, 1973, **59**, 5197.
[111] E. K. Sikorskaya and Ya. G. Goroshchenko. *Ukrain. Khim. Zhur.*. 1973, **39**, 322 (*Chem. Abs.*, 1973. **79**, 10452).
[112] E. I. Get'man, V. L. Buhikhanov, and M. V. Mokhosoev, *Zhur. neorg. Khim.*, 1974, **19**, 719.
[113] J. M. Coronas and J. Casabo, *An. Quim.*, 1974, **70**, 330 (*Chem. Abs.*, 1974, **80**, 152284).
[114] W. T. M. Andriessen, *Rec. Trav. chim.*, 1973, **92**, 1389 (*Chem. Abs.*, 1974, **80**, 89236).
[115] M. G. Lyapilina, E. I. Krylov, and V. A. Sharov, *Zhur. neorg. Khim.*, 1974, **19**, 400.
[116] P. H. Tedesco and V. B. De Rumi, *An. Soc. Cient. Argentina*, 1973, **195**, 63 (*Chem. Abs.*, 1973. **79**, 108678).
[117] R. Pastorek, *Acta Univ. Palacki Olamuc.*, *Fac. Rerum Natur.*, 1971. **33**. 389 (*Chem. Abs.*, 1973. **79**. 108746).
[118] A. Lassocinska, *Roczniki Chem.*, 1973, **47**, 889 (*Chem. Abs.*, 1973, **79**, 126757).
[119] P. C. Srivastava, K. B. Pandeya, and H. L. Nigam, *J. Inorg. Nuclear. Chem.*, 1973. **35**, 3613.

violet.[120] Chromium(III)–sucrose interactions have been monitored by pH in alkaline solutions.[121] Spectral studies have shown that in $[CrL_3Cl_3]$ (L = β-furfuraldoxine) the ligands are O-bonded and unidentate.[122] The etherates $H[CrX_4(OEt_2)_2],2Et_2O$ (X = Cl or Br) have been prepared by addition of Cr metal to a saturated solution of HX in anhydrous Et_2O.[123]

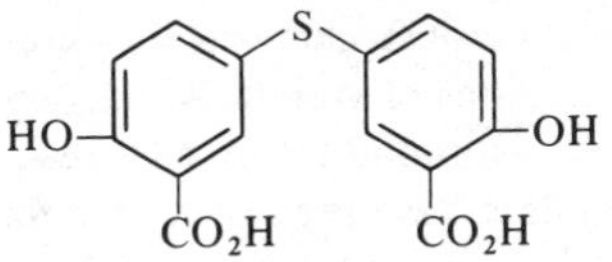

(1)

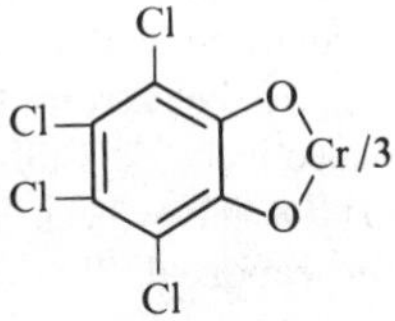

(2)

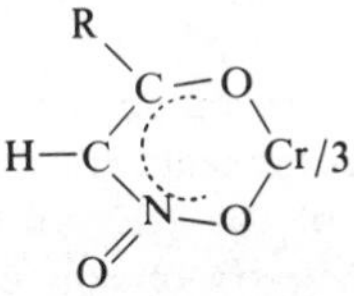

(3) R = H or Ph

The neutral tris-(o-benzoquinone) complex (2) has been prepared by refluxing $Cr(CO)_6$ with o-tetrachlorobenzoquinone. The i.r. spectrum of this complex is consistent with the exclusive presence of O-bonded reduced o-benzoquinone ligands. The complex undergoes three reversible oxidation steps, probably involving oxidation of the reduced ligands.[124] The complexes $Cr(LL)_3$ (LL = 2-nitroacetophenone or 2-nitroacetone) have been prepared and visible spectral and X-ray powder diffraction data used to suggest the structure (3). Two crystalline forms of tris-(2-nitroaceto-phenonato)chromium(III) were obtained which are probably *cis–trans*-isomers arising from the unsymmetrical nature of the chelate.[125] The crystal structures of tris(acetylacetonato)chromium(III)bis(thiourea)[126] and tris-(1,3-propanedionato) chromium(III)[127] have been determined and the molecular dimensions found to be similar to the corresponding ones of $Cr(acac)_3$; the former compound involves thiourea molecules hydrogen-bonded to $Cr(acac)_3$ molecules. Far-i.r.[128a] and X-ray photoelectron spectra[128b] have been recorded for a series of 3-substituted pentane-2,4-dionatochromium(III) complexes and the implications of these data to bonding within the chelate ring discussed. Some 20 tris-(1,3-diketonato)chromium(III) complexes have been studied polarographically and the differences in the half-wave potentials for one-electron reduction interpreted as reflecting a difference in the

[120] E. Chiacchierini, G. De Angelis, and P. Coccanari, *Gazzetta*, 1973, **103**, 413.
[121] S. S. M. A. Khorasani and A. M. S. Alam, *Dacca Univ. Stud.*, 1973, **21**, 37 (*Chem. Abs.*, 1974, **80**, 76234).
[122] B. Sen and M. E. Pickerell, *J. Inorg. Nuclear Chem.*, 1973, **35**, 2573.
[123] A. Galinos, D. Kaminaris and I. Triantafillopoulou, *Chem. Chron.*, 1972, **1**, 243 (*Chem. Abs.*, 1973, **79**, 61060).
[124] C. G. Pierpont, H. H. Downs, and T. G. Rukavina, *J. Amer. Chem. Soc.*, 1974, **96**, 5573.
[125] R. Astolfi, I. Collamati, and C. Ercolani, *J.C.S. Dalton*, 1973, 2238.
[126] W. B. Wright and E. A. Meyers, *Cryst. Struct. Commun.*, 1973, **2**, 477.
[127] B. Andrelczyk, *Diss. Abs. (B)*, 1973, **33**, 5190.
[128] (a) C. A. Fleming and D. A. Thornton, *J. Mol. Struct.*, 1973, **17**, 79; (b) R. Larsson and B. Folkesson, *Chem. Scripta*, 1973, **4**, 189 (*Chem. Abs.*, 1974, **80**, 7285).

electron densities at the co-ordinated oxygen atoms.[129] The enthalpies of solution of $Cr(acac)_3$ in acetone, benzene, and chloroform have been detected calorimetrically and a specific solvent–solute interaction identified in the case of chloroform, the value obtained[130] being *ca.* 25 kJ mol^{-1}. $[Cr\{(OR)_2PS_2\}_3]$ (R = Et or Pr^i) have been successfully eluted by gas chromatography.[131] The *cis–trans* isomerization of $[Cr(tfa)_3]$ (Htfa = 1,1,1-trifluoropentane-2,4-dionate) has been studied[132] in the gas phase and it appears that a twist mechanism is operative. The similarity of these results to those for isomerizations of various $M(tfa)_3$ complexes in solution suggest that the solvent may not be as important in such reactions as previously thought. Negative-ion mass spectral data have been presented for tris-(1,1,1,5,5,5-hexafluoro-pentane-2,4-dionato)chromium(III) and other metal complexes and this technique has potential value for the characterization of inorganic complexes.[133]

Polymeric Complexes containing Bridging O-Donor Ligands. The dinuclear μ-oxo-complexes $[(H_2O)L_2CrOCrL_2(OH_2)]$, derived from methimine and α-aminobutyric acid (HL), have been prepared and characterized by i.r. and electronic spectroscopy; their magnetic moments are 1.69 and 2.02 BM, respectively.[134] Full details of the crystal structure of $[(H_3N)CrOHCr(NH_3)_5]Cl_5.H_2O$ have been published,[135] luminescence spectra of salts containing this cation reported,[136] and salts of the cation $[(H_3N)_5CrOHCr(NH_3)_4Cl]^{4+}$ prepared and their electron transfer and hydrolysis reactions studied.[137] The magnetic properties of $[(en)_2Cr(OH)_2Cr(en)_2]X_4$ (X = Cl, Br, or I) have been determined from 1.6 to 300 K and the occurrence of super-exchange coupling between the Cr^{III} centres has been demonstrated.[138] μ,μ'-Dialkoxytetrakis-(3-bromopentane-2,4-dionato)dichromium(III)   (alkoxy = MeO, EtO, PrO, or BuO) complexes have been prepared by refluxing tris-(3-bromo-pentane-2,4-dionato)chromium(III) in the appropriate alcohol.[139] A series of carbonato-complexes of chromium(III) has been prepared and shown to involve the anions $[Cr_2(OH)_n(CO_3)_2]^{(n-2)-}$ (n = 4 or 5), $[Cr_2(OH)_4(CO_3)_3]^{4-}$, $[Cr_3(OH)_7-(CO_3)_3]^{4-}$, and $[Cr_4(OH)_9(CO_3)_5]^{7-}$.[140]

Exchange interactions in $[Cr_3O(OAc)_6,3H_2O]Cl,nH_2O$ have been re-evaluated, the contribution of dipole–dipole interactions being discussed,[141] and other studies of such interactions in these trinuclear chromium(III) and mixed chromium(III)–iron(III) acetato-clusters have been reported.[142]

[129] R. F. Handy and R. L. Lintvedt, *Inorg. Chem.*, 1974, **13**, 893.
[130] R. J. Irving and R. A. Schulz, *J.C.S. Dalton*, 1973, 2414.
[131] T. J. Cardwell and P. S. McDonough, *Inorg. Nuclear Chem. Letters*, 1974, **10**, 283.
[132] C. Kutal and R. Sievers, *Inorg. Chem.*, 1974, **13**, 897.
[133] I. W. Fraser, J. L. Garnett, and I. K. Gregor, *J.C.S. Chem. Comm.*, 1974, 365.
[134] C. A. McAuliffe and W. D. Perry, *Inorg. Nuclear Chem. Letters*, 1974, **10**, 367.
[135] J. T. Veal, D. Y. Jeter, J. C. Hempel, R. P. Eckberg, W. E. Hatfield, and D. J. Hodgson, *Inorg. Chem.*, 1973, **12**, 2928.
[136] C. D. Flint and P. Greenough, *J.C.S. Faraday II*, 1973, **69**, 1469.
[137] I. E. Gearon, *Diss. Abs. (B)*, 1973, **34**, 1414.
[138] B. Jasiewicz, M. F. Rudolf, and B. Jezowska-Trzebiatowska, *Acta Phys. Polon. (A)*, 1973, **44**, 623 (*Chem. Abs.*, 1974, **80**, 65040).
[139] K. Kasuga, T. Itou, and Y. Yamamoto, *Bull. Chem. Soc. Japan*, 1974, **47**, 1026.
[140] A. K. Sengupta and A. K. Nandi, *Z. anorg. Chem.*, 1974, **404**, 81.
[141] V. A. Gaponenko, Yu. V. Yablokov, and M. V. Eremin. *J. Mol. Structure*, 1973, **19**, 219.
[142] A. V. Ablov, T. N. Zhikhareva, V. M. Novotortseva, V. J. Tsuberblat, M. I. Belinskii, and G. M. Larin, *Zhur. fiz. Khim.*, 1974, **48**, 751; Yu. V. Yablokov, V. A. Gaponenko, M. V. Eremin, V. V. Zelentsov, and T. A. Zhemchuzhnikova, *Zhur. eksp. teor. Fiz.*, 1973, **65**, 1979, (*Chem. Abs.*, 1974, **80**, 31819); M. I. Belinskii, B. S. Tsukerblat, and A. V. Ablov, *Ukrain. fiz. Zhur.*, 1973, **18**, 1568.

The new pyridine and β-picoline (L) complexes $[Cr_3O(OAc)_6L_3]^+$ have been synthesized, their inertness to electrochemical oxidation and reduction being established.[143] $[Cr_3(OH)_{12}(O_2C)_2C_6H_4]$ has been prepared by addition of $K_2[(O_2C)_2-C_6H_4]$ to an aqueous alkaline solution of chromium(III) and the compound characterized by i.r., t.g.a., and d.t.a. studies.[144] The chromium analogue of Werner's brown salt, $[Cr\{(OH)_2Cr(en)_2\}_3](NO_3)_6,xH_2O$, has been isolated from an aqueous ethylenediamine solution of chromium(III) by ion-exchange and precipitation techniques and shown to be isomorphous with the corresponding cobalt(III) salt.[145]

Table 2 *Mixed-metal sulphides, selenides, and tellurides containing chromium(III)*

Compound	Comments and reported properties	Ref.
$KCrS_2$	μ, st.	a
$TlCr_3S_5$	μ, st.	b
$FeCr_2S_4$	X	c
$M_{1-x}Ni_xCr_2S_4$ (M = Mn, Fe, or Co)	phase relationships in these spinels characterized	d
$CuCr_2X_4$ (X = S, Se, or Te)	c, μ	e
$MCrX_2$ (M = Na or Ag; X = S or Se)	μ, structure determined by neutron diffraction at 4 and 300 K	f
$CdCr_2Se_4$	t.d., X, spinel structure	g
$MCr_2(S_{1-x}X_x)$ $(0 \leqslant x \leqslant 1$; M = Cu, X = Se or Te; M = Hg, X = Se)	X, spinel structure	h
$EuCr_2S_{4-x}Se_x$ $(0 \leqslant x \leqslant 4)$	X	i
$MCrS_3$ (M = Y, Gd, Dy, Ho, or Er)	e, X	j

(a) B. Van Laar and F. M. R. Engelsman, *J. Solid State Chem.*, 1973, **6**, 384; (b) C. Platte and H. Sabrowsky, *Naturwiss*, 1973, **60**, 474; (c) K. Nishida and T. Narita, *Hokkaido Daigaku Kogakubu Kenkyu Hokoku*, 1973, 127 (*Chem. Abs.*, 1974, **80**, 66 226); (d) M. Robbins, P. Gibart, D. W. Johnson, R. C. Sherwood, and V. G. Lambrecht, jun., *J. Solid State Chem.*, 1974, **9**, 170; (e) L. I. Valiev, I. G. Kerimov, S. Kh. Babaev, and Z. M. Namazov, *Izvest. Akad. Nauk. Azerb. S.S.R., Ser. Fiz.-Tekh. Mat. Nauk.*, 1973, 44 (*Chem. Abs.*, 1974, **80**, 113 696); (f) F. M. R. Engelsman, G. A. Wiegers, F. Jellinek, and B. Van Laar, *J. Solid State Chem.*, 1973, **6**, 574; (g) I. Kolowos, A. Knobloch, and J. Sieler, *Krist. Tech.* 1974, **9**, 157 (*Chem. Abs.*, 1974, **80**, 149 502); K. G. Barraclough and A. Neyer, *J. Cryst. Growth*, 1973, **20**, 212; (h) E. Riedel and E. Horvath, *Z. anorg. Chem.*, 1973, **399**, 219; D. Konopka, I. Kozlowska, and A. Chelowski, *Phys. Letters*, 1973, **44A**, 289; (i) H. Pink and H. Goebel, *Z. anorg. Chem.*, 1973, **402**, 312; (j) T. Takahashi, S. Osaka, and O. Yamada, *J. Phys. Chem. Solids*, 1973, **34**, 1131.

The polymerization of chromium(III)-containing species in weakly acidic solutions has been studied by dialysis and the formation of polyanions of molecular weight 700—800 identified. These polyanions are suggested to be composed of $\{Cr(OH)_2-(H_2O)_4Cr\}^{4+}$ units arranged in a linear manner.[146]

Chromium(III) in lithium phosphate glass appears to be octahedrally co-ordinated and the value of μ_{eff} (3.2 BM) suggests that some exchange interaction between the metal centres occurs in this medium.[147] Polymeric complexes of chromium(III)

[143] S. Uemara, A. Spencer, and G. Wilkinson, *J.C.S. Dalton*, 1973, 2565.
[144] P. Lumme and J. Tummavuori, *Acta Chem. Scand.*, 1973, **27**, 2287.
[145] P. Andersen and T. Berg, *J.C.S. Chem. Comm.*, 1974, 600.
[146] P. I. Kondratov and T. S. Kondratova, *Izvest. V. U.Z., Khim. i khim. Tekhnol.*, 1973, **16**, 1444 (*Chem. Abs.*, 1974, **80**, 20115).
[147] M. Berretz and S. L. Holt, *J. Inorg. Nuclear Chem.*, 1974, **36**, 49.

have also been identified for phosphinates,[148] diethylenemethylenediphosphonate,[149] and diorgano-phosphate and -phosphonate ligands.[150]

Complexes with S-Donor, Se-Donor, or Te-Donor Ligands. The interatomic distances and valence distributions in the chromium thiospinels have been reviewed.[151] Several mixed-metal sulphides, selenides, and tellurides containing chromium(III) have been reported this year, as detailed in Table 2.

Several chromium(III) complexes of the dithio-oxalate have been prepared which show the versatility of this ligand. $[Cr(S_2C_2O_2)_3,\{M(PPh_3)_2\}_3]$ (M = Cu or Ag) involve the bonding modes $Cr-S_2C_2O_2-Cu$ and $Cr-OSC_2SO-Ag$, respectively, and both of these thermally isomerize to the bonding mode $Cr-O_2C_2S_2-M$.[152] The thiocarbonato-complex. $[Cr(CS_3)_3]^{3-}$. has been isolated and characterized by i.r., electronic, and photoelectron spectroscopy.[153] Temperature-dependent ^1H n.m.r. spectra of a series of tris-NN-dialkyldithiocarbamato-complexes, including $[Cr(S_2CNEt_2)_3]$, in non-co-ordinating media have been recorded. Of all the complexes examined only the chromium one was found to be rigid.[154] The one-electron oxidation and reduction behaviour of chromium(III) and other metal dithiocarbamato-complexes have been studied[155] and the electronic and e.s.r. spectra of tris-(piperidyldithiocarbamato)chromium(III) determined.[156]

A general procedure for the synthesis of thiolato-bisethylenediamine chromium(III) complexes has been developed. These complexes exhibit an intense absorption in the near u.v. characteristic of $Cr^{III}-S$ bonding.[157] The crystal structure of $[Cr(en)_2-(SCH_2CO_2)]ClO_4$ has been determined as part of a structural study of sulphur *trans*-effects. However, in this compound no significant lengthening of the $Cr-N$ bond *trans* to the S atom was found.[158] The co-ordination of $(PhO)_2PS(Se)^-$ to chromium(III) has been monitored by i.r. spectroscopy and $Cr-S$ and $Cr-Se$ stretching frequencies were observed.[159] The e.s.r. characteristics of chromium(III) complexes with a variety of dithiophosphonates have been reported.[160]

N-Donor Ligands. The preparation, i.r. spectra, and X-ray diffraction characteristics of a series of oxo- and thio-anion salts of $[Cr(NH_3)_6]^{3+}$ have been determined,[161] and the ion association constants of $[Cr(NH_3)_6]^{3+}$ and $[Cr(en)_3]^{3+}$ with I^- and

[148] H. D. Gillman, P. Nannelli, and B. P. Block, *J. Inorg. Nuclear Chem.*, 1973, **35**, 4053; H. D. Gillman, J. P. King, and B. P. Block, *U.S. Nat. Tech. Inform. Serv. AD. Reports.* 1971, No. 763692; P. Nannelli, B. P. Block, J. P. King, A. J. Saraceno, O. S. Sprout, jun., N. D. Peschenko, and G. H. Dahl, *J. Polymer Sci., Polymer Chem.*, 1973, **11**, 2691; H. D. Gillman and P. Nannelli, *U.S. Nat. Tech. Inform. Serv. AD Reports*, 1973, No. 766837/9.
[149] C. M. Mikulski, N. M. Karayannis, L. L. Pytlewski, R. O. Hutchins, and B. E. Maryanoff, *J. Inorg. Nuclear Chem.*, 1973, **35**, 4011.
[150] C. M. Mikulski, N. M. Karayannis, and L. L. Pytlewski, *J. Inorg. Nuclear Chem.*, 1974, **36**, 971.
[151] E. Riedel and E. Horvath, *Mater. Res. Bull.*, 1973, **8**, 973.
[152] D. Coucouvanis and D. Piltingsrud, *J. Amer. Chem. Soc.*, 1973, **95**, 5556.
[153] A. Müller, P. Christophliemk, I. Tossidis, and C. K. Jørgensen, *Z. anorg. Chem.*, 1973, **401**, 274.
[154] L. Que, jun. and L. H. Pignolet, *Inorg. Chem.*, 1974, **13**, 351.
[155] R. Chant, A. R. Henrickson, R. L. Martin, and N. M. Rohde, *Austral. J. Chem.*, 1973, **26**, 2533.
[156] E. R. Price and J. R. Wasson, *J. Inorg. Nuclear Chem.*, 1974, **36**, 67.
[157] C. J. Weschler and E. Deutsch, *Inorg. Chem.*, 1973, **12**, 2682.
[158] R. C. Elder, L. R. Florian, R. E. Lake, and A. M. Yacynych, *Inorg. Chem.*, 1973, **12**, 2690.
[159] I. M. Cheremisina, L. A. Il'ina, and S. V. Larionov, *Zhur. neorg. Khim.*, 1973, **18**, 1278.
[160] P. M. Solozhenkin, E. V. Semenov, O. N. Grishina, N. I. Zemlyanskii, and Ya. I. Mel'nik, *Doklady Akad. Nauk Tadzh. S.S.S.R.*, 1973, **16**. 37 (*Chem. Abs.*, 1974, **80**, 54313).
[161] A. Müller, I. Böschen, and E. J. Baran, *Monatsh.*, 1973, **104**, 821.

ClO_4^- obtained.[162] The electronic structures of series of the ions $[Cr(NH_3)_5X]^{n+}$ and *trans*-$[CrL_4XY]^{n+}$ $[L_4 = (NH_3)_4, (py)_4,$ or $(en)_2]$ have been investigated by electronic or e.s.r. spectroscopy.[163] New syntheses of the bis(ethylenediamino)- and tetrammino-chromium(III) complexes $[Cr(en)_2X_2]ClO_4$ (X = NCS, F, Cl, Br, or I), $[Cr(en)_2L]ClO_4$, and $[Cr(NH_3)_4L]ClO_4$ (H_2L = oxalic, malonic, succinic, or *p*-aminosalicylic acid) have been published.[164] The reaction of the previously unreported complex $[Cr(pn)F_2Br(H_2O)]$ (pn = propane-1,2-diamine) with stoicheiometric amounts of en or propane-1,3-diamine (tmd) in refluxing absolute alcohol has been used to prepare the new mixed diamine complexes *trans*-$[Cr(pn)(NN)F_2]Br$ (NN = en or tmd). Reactions using pn under the same conditions gave both *cis*- and *trans*-$[Cr(pn)_2F_2]Br$, whilst *trans*-1,2-diaminocyclohexane (dach) afforded *cis*-$[Cr(pn)(dach)F_2]Br$ as the major product.[165] The related cationic complexes $[Cr(NN)(N'N')FX]^{2+}$ ($NN \neq N'N'$ = en, pn, or tmd; X = H_2O, Cl, or Br) have also been prepared for the first time,[166] and the acid hydrolysis of $[Cr(tmd)_2FX]^+$ (X = F, Cl, or Br) has been investigated.[167] Solvent extraction studies of chromium (III) by tri-n-octylamine[168] and a variety of high molecular weight primary amines[169] have been completed. The preparation and properties of chromium(III) complexes with various protonated multidentate amine ligands exhibiting less than their potential denticity have been reviewed.[170]

The preparation, characterization, and kinetics of the aquation of the $[Cr(H_2O)_5-L]^{3+}$ (L = 3-chloro- or 3-cyano-pyridine) complexes have been reported. Their synthesis involves the reduction of a corresponding pyridine adduct of diperoxochromium(VI) with iron(II), followed by purification using ion-exchange techniques.[171] $[Cr(py)_3X_3]$ (X = Cl or Br) are isostructural with $[Mo(py)_3Cl_3]$ and are therefore presumed to have the *mer*-configuration.[172] Spectral studies have established that $[CrL_3Cl_3]$ (L = pyrazine mono-*N*-oxide) involves each L co-ordinated as N-bonded unidentate ligand.[173] α-*cis*-$[Cr_2L_2Cl_2]Cl$ [L = 2-(aminomethyl)pyridine] reacts with moist AgO and then HNO_3 to produce the all-*trans*-$[CrL_2(H_2O)_2]$ $(NO_3)_3$ complex, which reacts with HX (X = Cl or Br) to afford the corresponding all-*trans*-$[CrL_2X_2]X$ derivatives.[174] The synthesis of $[Cr(NN)_3]^{3+}$ (NN = bipy or phen) ions has been discussed and the optimum yields shown to require oxidation of the corresponding chromium(II) complex, in such a manner that the product is protected from reaction with the constituents of the remaining unoxidized solution.[175]

162 T. Takahashi, T. Koiso, N. Tanaka, *Nippon Kagaku Kaishi*, 1974, 65 (*Chem. Abs.*, 1974, **80**, 87898).
163 L. E. Mohrmann, jun. and B. B. Garrett, *Inorg. Chem.*, 1974, **13**, 357; E. Pederson and H. Toftlund, *Inorg. Chem.*, 1974, **13**, 1603; R. L. Klein, jun., N. C. Miller, and J. R. Perumareddi, *Inorg. Chim. Acta*, 1973, **7**, 685.
164 M. Muto, *Chem. Letters*, 1973, 755.
165 J. W. Vaughn and J. Marzowski, *Inorg. Chem.*, 1973, **12**, 2346.
166 J. W. Vaughn and G. J. Seiler, *Inorg. Chem.*, 1974, **13**, 598.
167 J. M. DeJovine, W. R. Mason, and J. W. Vaughn, *Inorg. Chem.*, 1974, **13**, 66.
168 B. E. McLellan, M. K. Meredith, R. Parmelee, and J. P. Beck, *Analyt. Chem.*, 1974, **46**, 306.
169 T. N. Simonova, I. A. Shevchuk, I. N. Malakha, and L. I. Konovalenko, *Zhur. analit. Khim.*, 1973, **28**, 1825.
170 C. S. Garner, *Nucl. Sci. Abs.*, 1973, **28**, 10621.
171 A. Bakac, R. Marcec, and M. Orthanovic, *Inorg. Chem.*, 1974, **13**, 57.
172 J. V. Brencic, *Z. anorg. Chem.*, 1974, **403**, 218.
173 A. N. Speca, L. L. Pytlewski, and N. M. Karayannis, *J. Inorg. Nuclear Chem.*, 1973, **35**, 4029.
174 K. Michelsen, *Acta Chem. Scand.*, 1973, **27**, 1823.
175 D. M. Soignet and L. G. Hargis, *Syn. Inorg. Metal-Org. Chem.*, 1973, **3**, 167.

The electronic spectrum of the electrochemically generated unstable reduction products of $[Cr(bipy)_3]^{3+}$ has been further reported.[176]

PhOCrL,PhOH (H_2L = octaethylporphyrin) has been prepared by the reaction of H_2L with $Cr(acac)_3$ and its mass, i.r., and ^1H n.m.r. spectra have been recorded.[177] Octamethylcyclotetrasilazane, $(Me_2SiNH)_4$, reacts with a benzene solution of $[CrCl_3(Me_3N)_2]$ to give an immediate purple precipitate of $(CrCl_3)_2(Me_2SiNH)_4$, $2NMe_3$. On the basis of spectral data, the structure (4) has been proposed.[178]

The new Reinecke-salt-like compounds $K[Cr(NCS)_4L_n]$ (n = 1 or 2; L = o- or m-phenylenediamine, respectively) have been prepared by the reaction of $K_3[Cr-(NCS)_6]$ with L and their thermal stabilities investigated.[179] Diquinolino- and diiodoquinolino-silver(I) iodide salts of the anions $[Cr(NCS)_6]^{3-}$ and $[Cr(NCS)_4L_2]^-$ (L = aniline, o- or p-toluidine) have been prepared and characterized.[180] Chromium(III) thiocyanato-complexes with biguanide derivatives, $K[Cr(NCS)_4L]$ [L = phenyl-. p-chlorophenyl-, tolyl-, or 1-(p-chlorophenyl-5-isopropyl)biguanide], have been prepared by the reaction of $K_3[Cr(NCS)_6]$ with L in anhydrous EtOH and their i.r. and electronic spectra reported.[181] *cis*- and *trans*-$[Cr(en)_2(NH_3)(NCS)]^{2+}$ have been prepared and both shown to have a λ_{max} value at 476 nm, but with the expected relative magnitudes of ε values, *cis* > *trans*, as 120 to 80.[182] The thermal decompositions of the salts $[Cr(NH_3)_{6-x}(NCS)_x](NCS)_{3-x}$ (x = 0. 1, or 2) have been characterized. Under d.t.a. conditions $[Cr(NH_3)_6](NCS)_3$ changes stepwise to $[Cr(NH_3)_5(NCS)](NCS)_2$, *trans*-$[Cr(NH_3)_4(NCS)_2](NCS)$, and finally to *mer,fac*-$[Cr(NH_3)_3(NCS)_3]$.[183]

The syntheses and the kinetics of aquation of the new azido-complexes *cis*-$[Cr(C_2O_4)_2(N_3)_2]^{3-}$ and *cis*-$[Cr(C_2O_4)_2(N_3)(OH_2)]^{2-}$ have been reported.[184]

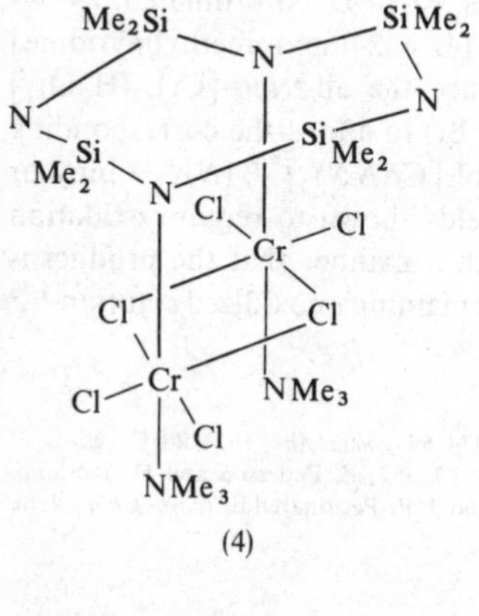

(4)

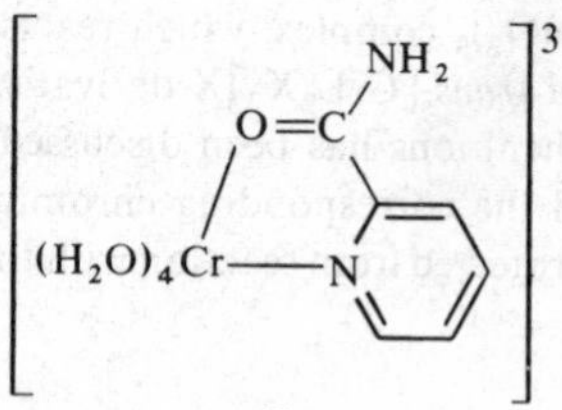

(5)

[176] Y. Sato, *Chem. Letters*, 1973, 1027.
[177] J. W. Buckler, L. Puppe, K. Rohbock, and H. H. Schneehage, *Chem. Ber.*, 1973, **106**, 2710.
[178] J. Hughes and G. R. Willey, *J. Amer. Chem. Soc.*, 1973, **95**, 8758.
[179] I. Ganescu, C. Varhelyi, and M. Proteasa, *Rev. Chim. Minerale*, 1973, **10**, 671.
[180] P. K. Nathur, *J. Inorg. Nuclear Chem.*, 1974, **36**, 943.
[181] C. Gheorghiu and C. Guran, *An. Univ. Bucuresti, Chem.*, 1972, **21**, 83 (*Chem. Abs.*, 1973, **79**, 38084).
[182] A. D. Kirk and T. L. Kelly, *Inorg. Chem.*, 1974, **13**, 1613.
[183] H. Oki, M. Yasuoka, and R. Tsuchiya, *Bull. Chem. Soc. Japan*, 1974, **47**, 652.
[184] K. R. Ashley and R. S. Lamba, *Inorg. Chem.*, 1974, **13**. 2117.

Mixed N-*Donor and Other Donor Ligands.* The chromium-containing product of chromium(II) reduction of 2-, 3-, and 4-carboxamidopyridinopentamminocobalt(III) has been formulated as (5), $\lambda_{max} = 553(29)$, 400(45), and 353(32) nm.[185] The CrL_3X_3 (X = Cl. Br. or I) complexes of thiomorpholin-3-one, thiomorpholin-3-thione, thiazolidine-2-thione, and thiazolidine-2-selenone (L), prepared by treating CrX_3, nH_2O with L, are considered to involve L co-ordinated as a unidentate ligand *via* its N or O atom.[186] The complexes $Cr(acac)_{3-x}L_x[x = 1,2,$ or 3, HL = $Ph_2P(O)$-NHP(O)Ph_2] have been prepared and their i.r. spectral, mol. wt. and t.g.a. characteristics determined.[187] A report of the preparative, i.r. spectral, and magnetic properties of CrLCl [$H_2L = NN'$-o-phenylbis(salidenimine)] has appeared.[188]

Diaquo-μ-triethylenetetraminehexa-acetatodichromium(III) hexahydrate has been shown by X-ray crystallography to consist of discrete centrosymmetric dinuclear units. This chelate functions as a quinquedentate ligand to each of the metal atoms, the octahedral co-ordination of which is completed by a water molecule.[189] Polynuclear compounds of chromium(III) with several alkylenedinitrilotetracetato-ions have been identified in aqueous media and their molecular weights shown to reach 1.9×10^5, the units possibly possessing a linear macromolecular structure.[190] The absorption spectral characteristics of chromium(III) complexed to edta[191] and propylenediaminetetra-acetate[192] have been presented and the polarographic behaviour of chromium(III) complexes with several poly(aminocarboxylate) ligands determined at a dropping mercury electrode.[193]

The formation constant of tris(L-alaninato)chromium(III) has been determined (log $\beta_3 = 25.27$) using c.d. techniques.[194] $Na_2[Zr_3Cr_2(SO_4)_4(OH)_{12}],6H_2O$ reacts with amino-acids at a moderately acidic pH to give either 1:1 or 1:2 complexes in which sulphato-group(s) of the polymeric unit are displaced by one or two amino-acid residues, respectively.[195] The complexing of chromium(III) with various amino-acids has been monitored spectrophotometrically and the formation of various $[CrL_n(H_2O)_{6-n}]^{3+}$ (n = 1—6, L = glycine, dl-α-alanine, or dl-α-leucine) complexes established.[196] Several stable complexes of chromium(III) have been prepared with various nucleotides, including ADP and ATP, by heating the nucleotide with Cr^{3+}_{aq} in acidic solution, and the complexes purified by ion-exchange methods.[197] Chromium(III) complexes of N-methyl-l-methoxy-acethydroxamate and -benzhydroxamate, ligands considered to be models of the siderochromes, have been prepared and the assignments of the conformational isomers described.[198] A ^{51}Cr—

[185] R. J. Balahura, *Inorg. Chem.*, 1974, **13**, 1350.

[186] C. Preti and G. Tosi, *Canad. J. Chem.*, 1974, **52**, 2845.

[187] J. P. King and B. P. Block, *Govt. Rep. Announce.*, (U.S.), 1973, **73**, 55.

[188] C. G. Macarovici and E. Mathe, *Stud. Univ. Babes-Bolyai, Ser. Chem.*, 1973, **18**, 107 (*Chem. Abs.*, 1973, **79**, 99941).

[189] G. D. Fallon and B. M. Gatehouse, *Acta Cryst.*, 1974, **30B**, 1987.

[190] A. Ricard and P. Souchay, *Compt. rend.*, 1973, **277**, C, 973.

[191] J. S. Arribas, R. Moro-Garcia, M. L. Alvarez-Bartolome, and C. Garcia-Bao, *Inform. Quim. Anal.*, 1973, **27**, 201 (*Chem. Abs.*, 1974, **80**, 103332).

[192] J. M. Suarez-Cardeso and S. Gonzalez-Garcia, *An. Quim.*, 1973, **69**, 483 (*Chem. Abs.*, 1973, **79**, 85130).

[193] E. Chiacchierini and R. Cocchieri, *Rev. Roumaine Chim.*, 1973, **18**, 1441 (*Chem. Abs.*, 1974, **80**, 9811).

[194] P. Vielen and A. Bonniol, *Compt. rend.*, 1973, **276**, C, 1769.

[195] M. H. Davis and J. G. Scroggie, *Austral. J. Chem.*, 1974, **27**, 279.

[196] R. I. Burdykina and A. I. Falicheva, *Izvest. V. U.Z. Khim. i khim. Tekhnol.*, 1973, **16**, 476 (*Chem. Abs.*, 1973, **79**, 10488).

[197] M. L. De Pamphilis and W. W. Cleland, *Biochemistry*, 1973, **12**, 3714

[198] J. Leong and K. N. Raymond. *J. Amer. Chem. Soc.*, 1974. **96**, 1757.

peptide complex from brewer's yeast has been isolated and partially characterized; it is probable that the peptide contains at least six amino-acids.[199]

$CrCl_3$ reacts with fluorenone thiosemicarbazone (L) in alcoholic solution to form $CrCl_3L$, the i.r. spectrum of which indicates that L co-ordinates *via* the N and S atoms of its $C=N$ and $C=S$ groups, the Cr^{III} centres being linked by chloride bridges to give a polymeric structure.[200] *o*-Hydroxy-4-benzamidothiosemicarbazide has been shown by electronic and i.r. spectroscopy to function as a terdentate (*ONS*) ligand to chromium(III).[201] Chromium(III) complexes with thiovanol have been prepared and characterized by spectroscopic and magnetic studies,[202] and the relative stabilities of glycolanilide and thioglycolanilide chelates of chromium(III) have been assessed.[203] The acid $H[Cr(L-cysteinate)_2], 2H_2O$ has been obtained from aqueous solutions of chromium(III) and L-cysteine by successive recrystallizations and its c.d. curve, rotatory power, and electronic and i.r. spectra were determined.[204]

P-Donor Ligands. $CrCl_3$ reacts with diphos to produce the red cationic species $[CrCl_2(diphos)_2]^+$ ($\mu_{eff} = 3.9$ BM) and its electrochemical reduction has been characterized.[84]

Cyano-complexes. The crystal structure of $Mn_3[Cr(CN)_6]_2, xH_2O$ has been determined and the cyano-groups shown to bridge the Mn^{II} and Cr^{III} centres with $Cr-C = 206(1)$ pm.[205] The bond lengths in $Cs_2Li[Cr(CN)_6]$ have been compared with those in related compounds in a discussion of cyano–metal bonding in the transition-metal series.[206] Single-crystal neutron diffraction and vibrational spectroscopic data for $CsLi[Cr(CN)_6]$ and related salts have been reported and the high symmetry of their metal centres (O_h) seen to allow for an unambiguous assignment of their vibrational frequencies.[207] The solution and single-crystal Raman spectra of $K_3[Cr(CN)_6]$ have been recorded and a new assignment made of some of the vibrational modes.[208] An *X*-ray photoelectron spectral study of the valence region of $[Cr(CN)_6]^{3-}$ has shown that the energy shifts of the levels originating as cyanide levels are significantly less than anticipated.[209]

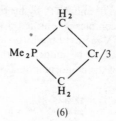

(6)

[199] H. J. Votava, C. J. Hahn, and G. W. Evans, *Biochem. Biophys. Res. Comm.*, 1973, **55**, 320.
[200] F. I. M. Taha and M. K. Hassan, *U.A.R.J. Chem.*, 1971, **14**, 531 (*Chem. Abs.*, 1973, **79**, 86923).
[201] M. P. Swami, P. C. Jain, and A. K. Srivastava, *Roczniki Chem.*, 1973, **47**, 2013 (*Chem. Abs.*, 1974, **80**, 89077).
[202] K. M. Kanth, K. B. Pandeya, and H. L. Nigam, *Indian J. Chem.*, 1973, **11**, 1031.
[203] K. P. Apte and A. K. Bhattacharya, *J. Inorg. Nuclear Chem.*, 1973, **35**, 3924.
[204] A. Bonniol, *Compt. rend.*, 1974, **278**, C, 5.
[205] H. U. Gudel, H. Stucki, and A. Ludi, *Inorg. Chem. Acta*, 1973, **7**, 121.
[206] R. R. Ryan and B. I. Swanson, *Inorg. Chem.*, 1974, **13**, 1681.
[207] B. I. Swanson and L. H. Jones, *Inorg. Chem.*, 1974, **13**, 313; J. R. Armstrong, B. M. Chadwick, D. W. Jones, J. E. Sarneski, H. J. Wilde, and J. Yerkess, *Inorg. Nuclear Chem. Letters*, 1973, **9**, 1025.
[208] P. W. Jenseen, *J. Mol. Structure*, 1973, **17**, 377.
[209] A. Calabrese and R. G. Hayes, *J. Amer. Chem. Soc.*, 1974, **96**, 5054.

Organometallic complexes. The crystal structure of $Li[CrMe_6]$,3dioxan has been determined and shown to contain an octahedral anion ($Cr—C = 230$ pm) which is slightly distorted by weak interaction with the cations ($Li \cdots Me = 217$ pm).[210] Tris(dimethylphosphoniodimethylido)chromium(III) (6) has been prepared by the reaction of $Li_3[CrPh_6]$ with $(Me_4P)Cl$ in THF under anaerobic conditions, or of $CrCl_3$,3THF with $Me_3P=CH_2$ 'metalized' with MeLi or PhLi, and this new compound characterized by i.r. and mass spectrometry.[211] Several new phenylchromium(III) complexes have been prepared. $CrPh_3$,3THF reacts with LiL (HL = pyrrole), $KNPh_2$, or $LiPR_2$ (R = cyclohexyl) to produce $Li[CrPh_3L]$,3THF, $K_2[CrPh_2(NPh_2)]$,2THF or $Li_2[Cr(PR_2)_2Ph_2]$, respectively. $CrCl_3$,3THF reacts with LiL to give $CrLCl_2$,3THF. N.m.r. spectra were recorded for all these compounds.[212] The kinetics and mechanism of the cleavage reactions of Cr^{III}–alkyl bonds have been further investigated.[213]

The isotropic shifts of the ^{11}B and ^{13}C nuclear resonances and the e.s.r. spectrum of $M[Cr^{III}\{2,3\text{-}Me_2\text{-}2,3\text{-}C_2B_9H_2\}_2]$ (M = Cs or Et_4N) salts have been determined and used to suggest that the mode of electron delocalization within this anion is primarily ligand-to-metal charge transfer.[214]

Chromium(IV).—CrF_4,XeF_6 has been prepared by the reaction of CrF_2 with XeF_6.[43, 215] Several members of the $CrO_{2-x}F_x$ series have been prepared by the reaction of CrO_2 with CrF_2 at high pressure. Homogeneous, single-phase products were obtained for $0 \leqslant x \leqslant 0.28$, which possess the rutile structure and are ferromagnetic.[216a] A new series of calcium lanthanide chromates, $Ca_{1+x}M_{1-x}(CrO_4)$ (M = Pr, Nd, Sm, Eu, or Gd) has been prepared and shown to contain both Cr^{III} and Cr^{IV} atoms.[216b]

Na_4CrO_4 has been identified by X-ray powder diffractometry and magnetic susceptibility measurements as a product of the reaction between Na_2O and Cr_2O_3 at high temperatures.[217] MnO reacts[218] with CrO_2 at high pressure and temperature to produce $MnCrO_3$ which is isomorphous with the corresponding vanadium ilmenite.

Treatment of CrO_2Cl_2(bipy) with aqueous HCl has been shown to produce $CrOCl_2$(bipy), which was characterized by i.r. spectral and conductance measurements.[219]

The crystal structure of $[Cr(CH_2CMe_2Ph)_4]$ has been determined and the Cr^{IV} atom shown to be co-ordinated by a slightly distorted tetrahedron of carbon atoms ($Cr—C = 205$ pm).[220] The e.s.r. spectrum of tetrakis-(1-norbornyl)chromium(IV) has been recorded for the compound contained in organic glasses at low temperatures.

210 J. Krause and G. Marx, *J. Organometallic Chem.*, 1974, **65**, 215.

211 E. Kurras, U. Rosenthal, H. Mennenga, and G. Oehme, *Angew. Chem. Internat. Edn.*, 1973, **85**, 913.

212 W. Seidel and W. Reichardt, *Z. anorg. Chem.*, 1974, **404**, 225.

213 J. H. Espenson and D. A. Williams, *J. Amer. Chem. Soc.*, 1974, **96**, 1008. D. A. Williams, *Nuclear Sci. Abs.*, 1973, **28**, 26878.

214 R. J. Wiersema and M. F. Hawthorne, *J. Amer. Chem. Soc.*, 1974, **96**, 761.

215 B. Zemva, J. Zupan, and J. Slivnik, *J. Inorg. Nuclear Chem.*, 1973, **35**, 3941.

216 (a) B. L. Chamberland, C. G. Frederick, and J. L. Gillson, *J. Solid State Chem.*, 1973, **6**, 561; (b) A. Daoudi and G. LeFlem, *Mater. Res. Bull.*, 1973, **8**, 1103.

217 L. Lavielle, H. Kessler, A. Hatterer, *Bull. Soc. chim. France*, 1973, 1918.

218 H. Sawamoto, *Mater. Res. Bull.*, 1973, **8**, 767.

219 M. N. Majumder and A. K. Saha, *J. Indian Chem. Soc.*, 1973, **50**, 162.

220 V. Gramlich and K. Pfefferkorn, *J. Organometallic Chem.*, 1973, **61**, 247.

In most of these glasses the complex apparently assumes several conformations which are distinguishable by their zero-field parameters.[221]

Oxidation of aromatic aldehydes by chromium(IV) has been further investigated and a mechanism proposed consisting of a rate-limiting decomposition of a chromium(IV) complex of the aldehyde hydrate.[222] Several other studies of the role of chromium(IV) in chromium(VI) oxidation have been presented.[223]

Chromium(V).—The electronic structure of $[CrO_4]^{3-}$ has been calculated using an unrestricted CNDO MO approximation,[224a] and the force field appropriate to this ion discussed.[224b] $Ca_3(CrO_4)_2$ has been shown to be isomorphous with the mineral whitlockite by X-ray crystallographic techniques.[225]

The e.s.r. spectrum of Cr^V centres in $K_2Cr_2O_7$ has been reported,[226] as has that for $[CrOF_5]^{2-}$ incorporated into a single crystal of a suitable host lattice. The e.s.r. parameters of this latter complex were discussed using an MO model.[227] A study of the Overhauser effect in ethyleneglycol complexes of chromium(V) has been reported and the results discussed in terms of the electron-density distribution within these species.[228] An e.s.r. study of the photochemical oxidation of diethyleneglycol by $[Cr_2O_7]^{2-}$ has shown that a chromium(V) species is formed following a photochemical redox reaction between the two reactants. This complex can be stabilized at low temperatures, but at room temperature it changes into another chromium(V) species.[229] The formation and disappearance of relatively stable chromium(V) intermediates during the reaction of chromic acid with oxalic acid have been followed by e.s.r. spectroscopy. These studies indicate the existence of two chromium(V) entities which are in rapid equilibrium and whose relative concentrations depend upon the nature of the solvent.[230] A mechanism for the oxidation of cyclobutanol by chromium(V) has been proposed,[231] and the chromic acid oxidation of benzaldehyde in 96% HOAc shown to involve a chromium(V) intermediate.[232]

E.s.r. data have been reported for the complexes $[CrOX_2\{S_2P(OEt)_2\}]$ (X = Cl or NCS) and discussed in terms of the electronic structure of these complexes.[233] Two new and kinetically inert complexes of chromium(V) have been reported. $[CrO(NN)-Cl_3]$ $(NN = $ bipy or phen) have been obtained by the dehydrochlorination of the corresponding $(H_2NN)[CrOCl_5]$ salt in a dry CO_2 atmosphere at 80 °C.[234] CrO_2Cl_2 reacts with hexamethylmelamine (L) in anhydrous EtOAc to produce $[CrO_2ClL]$.[235]

[221] G. A. Ward, B. K. Bower, M. Findlay, and J. C. W. Chien, *Inorg. Chem.*, 1974, **13**, 614.
[222] J. Roček and C. S. Ng, *J. Amer. Chem. Soc.*, 1974, **96**, 2840.
[223] J. Roček and C. S. Ng, *J. Amer. Chem. Soc.*, 1974, **96**, 1522; M. Doyle, R. J. Suredo, and J. Roček, *J. Amer. Chem. Soc.*, 1973, **95**, 8352; K. B. Wiberg and S. K. Mukherjee, *J. Amer. Chem. Soc.*, 1973, **96**, 1884.
[224] (a) D. A. Copeland, *Theor. Chim. Acta*, 1973, **32**, 41; (b) F. Gonzalez-Vilchez and W. P. Griffith, *An. Quim.*, 1973, **69**, 617 (*Chem. Abs.*, 1973, **79**, 47251).
[225] D. P. Sinha and B. C. Srivastava, *Indian J. Phys.*, 1973, **47**, 746.
[226] A. A. Alybakov, Z. Kh. Gubaidullin, and V. A. Gubanova, *Krist. Tech.*, 1974, **9**, 51 (*Chem. Abs.*, 1974, **80**, 65298).
[227] J. T. C. Van Kemenade, *Rec. Trav. Chim.*, 1973, **92**, 1102 (*Chem. Abs.*, 1974, **80**, 32293).
[228] V. N. Fedorov and N. B. Yunusov, *Radiospektoroskopiya*, 1973, 154 (*Chem. Abs.*, 1973, **79**, 25381).
[229] M. Mitewa, A. Malinovski, P. R. Bontchev, and K. Kabassonov, *Inorg. Chim. Acta*, 1974, **8**, 17.
[230] V. Srinivasan and J. Roček, *J. Amer. Chem. Soc.*, 1974, **96**, 127.
[231] F. Hasan and J. Roček, *J. Amer. Chem. Soc.*, 1974, **96**, 534.
[232] K. B. Wibert and G. Szeimies, *J. Amer. Chem. Soc.*, 1974, **96**, 1889.
[233] I. F. Gainulin, *Radiospektroskopiya*, 1973, 76 (*Chem. Abs.*, 1973, **79**, 25384).
[234] S. Sarkar and J. P. Singh, *J.C.S. Chem. Comm.*, 1974, 509.
[235] N. Gunduz, *Comm. Fac. Sci. Univ. Ankara (B)*, 1972, **19**, 109 (*Chem. Abs.*, 1974, **80**, 103343).

Chromium(vi).—*Halides and Oxyhalides.* New preparative routes to $CrOF_4$ and CrO_2F_2 have been developed using the direct fluorination of CrO_3 in a static system. The product formed depends upon the reaction temperature, CrO_2F_2 being obtained at 150 °C and $CrOF_4$ at 220 °C, suggesting that fluorination of CrO_3 proceeds by the stepwise replacement of oxygen by fluorine. Physical constants and thermodynamic data derived from vapour-pressure measurements of $CrOF_4$ show that this compound has some physical properties which differ from those of the oxidetetrafluorides of the second- and third-row transition metals, the most significant difference being the smaller liquid range.[236] Other thermodynamic data have been calculated[237] for $CrOF_4$. Polarized absorption spectra, recorded for $[CrO_3X]^-$ (X = F or Cl) ions in a $KClO_4$ lattice at 4 K, have shown[238] that the lowest energy charge-transfer configuration of these ions is formed by the donation of an electron from the orbitals almost entirely localized on the oxygen atoms and which correlate with the t_1 orbitals in $[CrO_4]^{2-}$. The i.r. and Raman spectra of $[CrO_3Br]^-$ have been recorded and a normal co-ordinate analysis presented for ions of this type.[239] The mean amplitude of vibration of CrO_2Cl_2 has been calculated,[240] the thermal decomposition of this compound investigated,[241] and its oxidation of organic compounds further described[242] and discussed.[243]

O-Donor Ligands. A crystal-structure determination of tris(trimethyltin)chromate hydroxide has shown that this compound is a complicated network polymer in which $[CrO_4]^{2-}$ tetrahedra bridge tin(IV) atoms.[244a] The existence of phase transformations in M_2CrO_4 (M = Tl or Ag) has been confirmed by d.t.a., X-ray diffraction, and electrical conductance measurements.[244b] Further vibrational spectroscopic data have been reported[245] for $[CrO_4]^{2-}$, including a determination of the Raman intensity of the v_1 stretching mode[246] and a discussion of force-field appropriate to this ion.[224b]

The location of the protons in $M_2Cr_2O_7,2H_2O$ (M = Li or Na) has been described using ^1H n.m.r. spectroscopy on single crystals.[247] However, the complete elucidation of the constitution of these salts has required a neutron-diffraction study. For '$Li_2Cr_2O_7,2H_2O$' this has shown the correct formulation to be $Li_2[Cr_2O_6(OH)_2]$,-H_2O.[248] A crystal-structure determination[249] of $MgCr_2O_7,2(CH_2)_6N_4,6H_2O$ has shown the compound to contain dichromate anions in which ∠ CrOCr = 128°. A

[236] A. J. Edwards, W. E. Falconer, and W. A. Sunder, *J.C.S. Dalton*, 1974, 541.
[237] E. G. Rakov, L. G. Koshechko, B. N. Sudarikov, and V. V. Mikulenok, *Trudy Moskov Khim.-Tekhnol. Inst.*, 1972, **71**, 28 (*Chem. Abs.*, 1974, **80**, 100 793).
[238] D. B. Jeans, J. D. Penfield, and P. Day, *J.C.S. Dalton*, 1974, 1777.
[239] A. Müller, K. H. Schmidt, E. Ahlborn, and C. J. L. Lock, *Spectrochim. Acta*, 1973, **29A**, 1773.
[240] E. J. Baran, *Anales Asoc. quim. argentina*, 1973, **61**, 141 (*Chem. Abs.*, 1974, **80**, 54087).
[241] A. Funke and P. Kleinert, *Z. anorg. Chem.*, 1974, **403**, 156.
[242] F. W. Bachelor and U. O. Cheriyan, *Tetrahedron Letters*, 1973, 3291; M. J. Leigh and J. A. Strickson, *J.C.S. Perkin II*, 1973, 1476.
[243] F. Freeman, *Rev. Reactive Species Chem. React.*, 1973, **1**, 37 (*Chem. Abs.*, 1973, **79**, 114637).
[244] (a) A. M. Domingos and G. M. Sheldrick, *J.C.S. Dalton*, 1974, 477; (b) M. Natarajan and E. A. Secco, *Canad. J. Chem.*, 1974, **52**, 712.
[245] A. Müller, I. Böschen, E. J. Baran, and P. J. Aymonino. *Monatsh.*, 1973, **104**, 836; W. B. Grant and S. Radhakrishna, *Solid State Comm.*, 1973, **13**, 109.
[246] H. Schulze and A. Müller, *Adv. Raman Spectroscopy.* 1972, **1**, 546.
[247] G. R. Bulka, V. M. Vinokurov, E. A. Kuz'min, Yu. A. Kharitonov, and N. V. Belov, *Kristallografiya*, 1973, **18**, 984.
[248] I. D. Datt and R. P. Ozerov, *Kristallografiya*, 1974, **19**, 110.
[249] F. Dahan. *Acta Cryst.*, 1974, **B30**, 22.

Table 3   *Mixed oxide compounds containing chromium*(VI)

Compound	Comments and reported properties	Ref.
Na_2CrO_4, $4H_2O$	st	a
$MgCrO_4$, $AcNH_2$, $6H_2O$	t.d.. X	b
$CaCrO_4$, $2H_2O$		
$CaCrO_4$, $4OC(NH_2)_2$	i.r.	c
$Al_2(OH)_4CrO_4$, H_2O		
$Al(OH)CrO_4$, H_2O	prepared in the Al_2O_3–CrO_3–H_2O	
α- and β-$Al_2(CrO_4)_2Cr_2O_7$, $3H_2O$	system at 110—150°C; i.r.. t.d.. X	d
Pb_2CrO_5		
β-Pb_3CrO_6	prepared in the PbO–PbCrO$_4$ system; X	e
Pb_5CrO_8		
$Hf_4(OH)_8(CrO_4)_4$, H_2O	st; isomorphous with corresponding Zr^{IV} compound	f
$KFe(CrO_4)_2$, H_2O	st; composed of $[Fe_2(CrO_4)_4, 2H_2O]^{2-}$ chains	g
β-$RbFe(CrO_4)_2$	st; composed of $\{Fe(CrO_4)_2O_2\}$ chains	h
$MFe(CrO_4)_2$, nH_2O ($M = Na$. K. Rb. Tl. or NH_4)	i.r., X	i
γ-$Cu_2(OH)_2CrO_4$	st	j
$CuCrO_4.xCu(OH)_2,nH_2O$ ($x = 0$, 0.66, 1.5, or 3)	t.d., X	k
$Li_3Ag(CrO_4)_2$	m.p. 740°C	l
$ZnCrO_4$	t.d.	m
$M_2La(CrO_4)_2$, $2H_2O$ ($M = Li$, Na, K, NH_4. or $\frac{1}{2}Mg$)	prepared by the treatment of M_2CrO_4 with $La_2(CrO_4)_2$ in H_2O	n
La_2CrO_6	X	
$La_5Cr_2O_{13}$	prepared by the thermal decomposition	
$La_7Cr_3O_{19}$	of La_2CrO_6	o
$Nd_2(CrO_4)_3$, nH_2O	ΔH	p
$CsM(CrO_4)_2$, H_2O ($M = Nd$ or Sm)	t.d., X	q
$M_2(UO_2)_2(CrO_4)_3$, $6H_2O$ ($M = Na$. K, Rb, Cs. or NH_4)	prepared by the treatment of M_2CrO_4 with $UO_2(NO_3)_2$ in H_2O	r

(a) H. Ruben, I. Olovsson, A. Zalkin, and D. H. Templeton. *Acta Cryst.*, 1973, **B29**, 2963; (b) N. N. Gusto-mesova and A. S. Karnaukhov, *Zhur. neorg. Khim.*, 1974, **19**, 525; (c) A. M. Slobodchikov and N. N. Runov, *Uch. Zap. Yaroslav. Gos. Pedagog. Inst.*, 1972, **103**, 93 (*Chem. Abs.*, 1973, **79**, 71984); (d) Y. Cudennec and A. Bonnin, *J. Inorg. Nuclear Chem.*, 1974, **36**, 273; (e) Y. Wǎtanabe and Y. Otsubo, *Nippon Kagaku Kaishi*, 1973, 1603, (*Chem. Abs.*, 1973, **79**, 97962); (f) M. Hansson and W. Mark, *Acta Chem. Scand.*, 1973, **27**, 3467; (g) V. Debelle, P. Gravereau, and A. Hardy, *Acta Cryst.*, 1974, **30B**, 2185; (h) A. Bonnin, A. Hardy, and E. Garnier, *Compt. rend.*, 1973, **276**, C, 1381; (i) A. Mellier and P. Gravereau, *Spectrochim. Acta*, 1973, **29A**, 2043; (j) A. Rivu, Y. Gerault, and A. Lecerf, *Bull. Soc. Franç. minerale, Cryst.*, 1973, **96**, 25; (k) L. Walter-Levy and M. Goreaud, *Bull. Soc. chim. France*, 1973, 830; (l) N. P. Burmistrova and D. M. Shakirova, *Zhur. neorg. Khim.*, 1973, **18**, 2502; (m) R. P. Clark and F. W. Reinhardt. *Thermochimica Acta*. 1974. **8**. 185; (n) I. J. Martin, *Ion* (*Madrid*), 1973, **33**, 134, (*Chem. Abs.*, 1973, **79**, 26629); (o) R. Berjoan, J. P. Traverse, and J. P. Coutures, *Rev. Chim. minerale*, 1973, **10**, 309; (p) S. B. Tsyrenova, Yu. L. Suponitskii, and M. Kh. Karapet'yants, *Zhur. fiz. Khim.*, 1973, **47**, 2132; (q) T. I. Kuzina, I. V. Shakhno, A. N. Krachak, and V. E. Plyushchev, *Zhur. neorg. Khim.*, 1973, **18**, 2727; (r) I.J. Martin, *Ion* (*Madrid*), 1973, **33**, 136, (*Chem. Abs.*, 1973, **79**, 26606).

similar study[250] of $Rb_2Cr_3O_{10}$ has identified trichromate anions which consist of three CrO_4 tetrahedra joined by shared corners with $\angle$ CrOCr = 136 and 140°. The results of other studies of mixed oxide compounds containing chromium(VI) reported this year are summarized in Table 3.

The relative ease of formation of $LiOBu^n$ on treating $LiBu^n$ with several peroxo-

[250] P. Lofgren, *Chem. Scripta*, 1974, **5**, 91.

metal complexes, including $[CrO(O_2)_2py]$, has been assessed.[251] Ethereal blue peroxychromium(VI) species have been further characterized by investigating their reactions with PbO_2, dimethylaniline, or *o*-aminophenol.[252]

The properties of the ternary layers formed between Cr_{aq}^{VI} and phenylphosphonate diesters in an organic medium have been investigated spectrophotometrically; however, no conclusive evidence in favour of complex formation was obtained.[253] Kinetic studies have shown that oxidation of $PhP(OH)_2$ by an aqueous $HClO_4$ solution of chromium(VI) is preceded by complex formation between the Cr^{VI} and P^{III} centres and a similar mechanism has been proposed for the analogous oxidation of arsenic(III).[254]

A large number of studies of the mechanism of chromium(VI) oxidation of organic compounds have been published and in several cases the prior formation of complexes between alcohols[229, 255] and aldehydes[256] was suggested and good evidence for the generation of chromium-(IV) and/or -(V) species presented.[229, 230, 232] Similarly, oxidation of NH_3OH^+ by chromium(VI) has been found to proceed *via* formation of an O-bonded chromate ester, with chromium-(IV) and -(V) species being involved in the redox reaction sequence.[257]

S-*Donor Ligands.* The rapid formation of the complex $[\{Me_3C(CH_2)_3CS\}CrO_3]^-$ has been established as the first step in the oxidation of methylnonane-6-thiol in HOAc solution.[258]

N-*Donor Ligands.* A wide range of alcohols can be oxidized effectively and conveniently by a complex of chromium(VI) and 3,5-dimethylpyrazole, prepared as a dark red solution by adding one mole of 3,5-dimethylpyrazole to a suspension of CrO_3 in CH_2Cl_2 at room temperature.[259] Oxidation of $(N_2H_5)^+$ by chromium(VI) has been extended to higher acidities than previously studied and the reaction shown to proceed through an N-bonded chromate ester.[260]

Formation and decomposition reactions of CrO_2Cl_2 complexes with 2,2′-bipyridyl[219] and ethylenediamine[261] have been further studied.

2 Molybdenum and Tungsten

Introduction.—The papers presented at the conference on 'The Chemistry and Uses of Molybdenum' have now been published[262] and recent developments in the synthetic[2] and organometallic[1] chemistry of molybdenum and tungsten have been reviewed.

[251] S. L. Regen and G. M. Whitesides, *J. Organometallic Chem.*, 1973, **59**, 293.

[252] O. P. Tomar, R. Singh, and J. Singh, *J. Indian Chem. Soc.*, 1973, **50**, 83, 209.

[253] W. R. Mountcastle, jun, W. H. Martin, P. T. Miller, J. B. Nabors, jun., and J. P. Scarborough, *J. Inorg. Nuclear Chem.*, 1973, **35**, 4175.

[254] K. K. Sen Gupta, and J. K. Chakladar, *J.C.S. Perkin II*, 1973, 929; *J.C.S. Dalton*, 1974, 222.

[255] J.-C. Richer and J.-M. Hackey, *Canad. J. Chem.*, 1974, **52**, 2475; J. Roček and A. E. Radkowsky, *J. Amer. Chem. Soc.*, 1973, **95**, 7123; F. Hasan and J. Roček, *ibid.*, p. 5421.

[256] J. Roček and C.-S. Ng., *J. Org. Chem.*, 1973, **38**, 3348.

[257] R. A. Scott, G. P. Haight, jun., and J. N. Cooper, *J. Amer. Chem. Soc.*, 1974, **96**, 4136.

[258] I. Baldea and S. Schoen, *Stud. Univ. Babes-Bolyai, Ser. Chem.*, 1973, **18**, 47 (*Chem. Abs.*, 1973, **79**, 136207).

[259] E. J. Corey and G. W. J. Fleet, *Tetrahedron Letters*, 1973, 4499.

[260] G. P. Haight, jun., T. J. Huang, and H. Platt, *J. Amer. Chem. Soc.*, 1974, **96**, 3137.

[261] M. N. Majumder and R. K. Mukhopadhyay, *Indian. J. Chem.*, 1973, **11**, 183.

[262] *J. Less-Common Metals*, 1974, **36**, 1–536. The papers presented at the Conference on the Chemistry and Uses of Molybdenum, Reading, 1973.

The dramatic increase in the number of papers concerned with the chemistry of molybdenum has continued unabated throughout the period covered by this report. A major stimulus is the role of molybdenum in the various molybdenum-containing enzymes and evidence for an additional member of this family, carbon dioxide reductase from *Clostridium pasteurianum*, has been presented.[263] A review of the effects of molybdenum on protein synthesis and symbiotic nitrogen fixation has been published.[264] Interesting developments in model systems of the molybdenum-containing enzymes include the catalytic reduction of acetylene by a series of molybdenum(IV) complexes[265] and a detailed appraisal of the role of acids in the reduction of acetylene catalysed by molybdenum thiol complexes.[266] A μ-dinitrogen complex, $[(Me_2PhP)_4ClReN_2MoCl_4(OMe)]$, has been prepared and shown by X-ray crystallography to have the structure (7), in which the N—N separation of 118 pm corresponds to that of a double bond.[267] An X-ray crystallographic study has shown that the di-imino-reduction product of co-ordinated dinitrogen, in the complex $[W(N_2H_2)\text{-}Cl(diphos)_2]^+$, has an essentially linear W—N—N geometry, with W—N and N—N bond lengths of 173 and 137 pm, respectively.[268] Further interesting recent developments in molybdenum–nitrogen chemistry include the novel hydrazido-complexes (8), prepared by the reaction of the corresponding compounds $[(\eta^5\text{-}C_5H_5)Mo(NO)\text{-}X_2]_2$ and $R^1R^2NNH_2$. The compound (8; X = I, $R^1 = R^2 =$ Me) was characterized by X-ray crystallography. $PhNHNH_2$ reacts with $[(\eta^5\text{-}C_5H_5)Mo(NO)I_2]_2$ to afford $[(\eta^5\text{-}C_5H_5)Mo(NO)I(N_2HPh)_2]_2$ which has been suggested to possess the structure

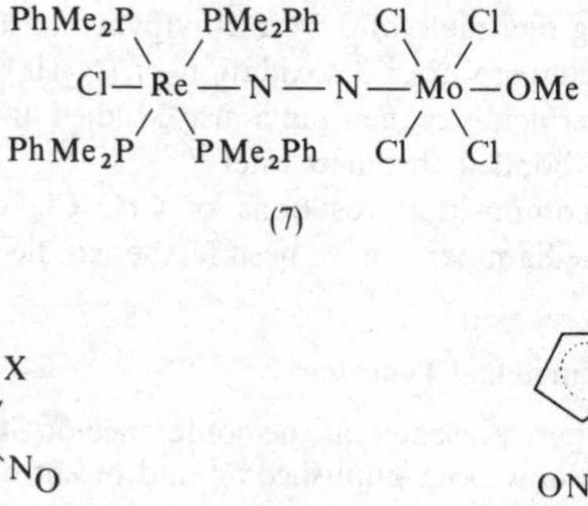

(7)

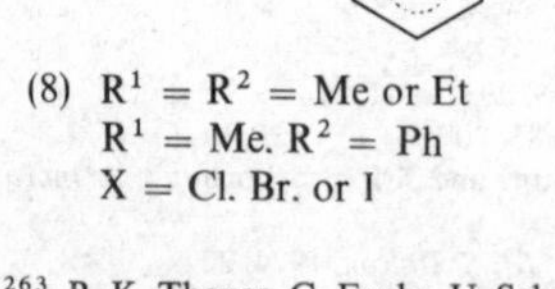

(8) $R^1 = R^2 =$ Me or Et
    $R^1 =$ Me. $R^2 =$ Ph
    X = Cl. Br. or I

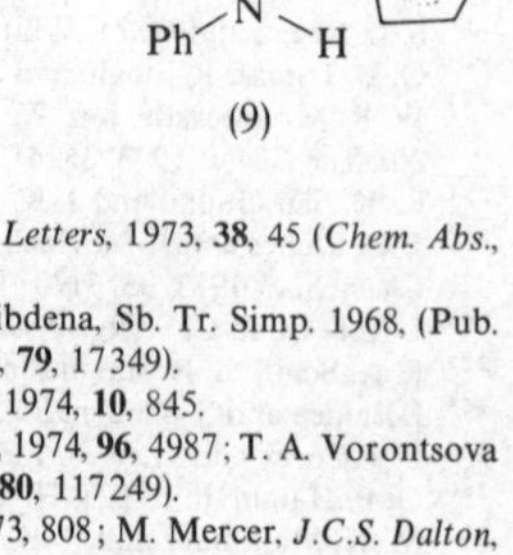

(9)

[263] R. K. Thauer, G. Fuchs, U. Schnitker, and K. Jungermann, *F. E. B. S. Letters*, 1973, **38**, 45 (*Chem. Abs.*, 1974, **80**, 92520).
[264] E. I. Ratner, T. A. Akimochkina, and S. A. Samoilova, *Biol. Rol. Molibdena, Sb. Tr. Simp. 1968.* (Pub. 1972) 24, ed. Ya. V. Peive, Nauka, Moscow, U.S.S.R. (*Chem. Abs.*, 1973, **79**, 17349).
[265] H. J. Sherrill, J. H. Nibert, and J. Selbin, *Inorg. Nuclear Chem. Letters*, 1974, **10**, 845.
[266] A. P. Khrushch, A. E. Shilov, and T. A. Vorontsova, *J. Amer. Chem. Soc.*, 1974, **96**, 4987; T. A. Vorontsova and A. E. Shilov, *Kinetika i Kataliz*, 1973, **14**, 1326 (*Chem. Abs.*, 1974, **80**, 117249).
[267] M. Mercer, R. H. Crabtree, and R. L. Richards, *J.C.S. Chem. Comm.*, 1973, 808; M. Mercer, *J.C.S. Dalton*, 1974, 1637.
[268] G. A. Heath, R. Mason, and K. M. Thomas, *J. Amer. Chem. Soc.*, 1974, **96**, 259.

(9) on the basis of ^{1}H and ^{13}C n.m.r. spectral studies.[269] The reaction of trimethyl-silylmethyl azide with molybdenum-(III) or -(IV) halide complexes in non-aqueous solvents has been shown to afford an easy route to nitrido-complexes of molybdenum-(V) and -(VI).[270] The reactions of the molybdenum(VI) nitrido-complexes $[MoN(S_2CNR_2)_3]$ $[R_2 = Me_2, Et_2,$ or $(CH_2)_5]$ with sulphur have produced the first reported thionitrosyl complexes, $[Mo(NS)(S_2CNR_2)_3]$, in high yield (see p. 118).[271]

The half-life of ^{181}W has been re-investigated and a value of 120.95 ± 0.02 days determined, which differs significantly from the currently accepted value.[272] The u.v. absorption spectra of molybdenum atoms isolated in rare-gas matrices at 14 K have been correlated with similar gas-phase spectral data and assigned in spherical symmetry. Diffusion of the metal atoms in an Ar matrix was also studied and some tentative evidence obtained for dimer formation.[273] The standard heat of vapourization of molybdenum has been determined[274] as 689.3 kJ (g atom)$^{-1}$.

The relative bonding energies of hydrogen chemisorbed at three symmetric sites on a W(100) surface have been computed by an extended Hückel MO approach and the preferred site calculated to be that directly above a surface tungsten atom.[275] ESCA has been used to provide a valuable insight into the chemisorption of CO and O_2 on a molybdenum or a tungsten surface.[276] I.r. spectral studies of chemisorbed CO and O_2 on a tungsten surface have also been reported. The former molecules produce a band maximum at *ca.* 1950 cm^{-1} and the latter produce bands characteristic of terminal (970 cm^{-1}) and bridging (500—1000 cm^{-1}) oxo-groups.[277]

The co-condensation of molybdenum or tungsten atoms in the gas phase with buta-1,3-diene (1 : 100) at liquid nitrogen temperatures, has been shown to afford the corresponding tris(butadiene)metal(0) complex.[278] The treatment of $[Mo(acac)_3]$ and diphos in toluene with $AlEt_3$ under Ar has produced the new ethylene co-ordination compound $[Mo(C_2H_4)(diphos)_2]$. together with the hydrido-species $[MoH(acac)-(diphos)_2]$.[279] The X-ray crystal structure of $[\{Et_2B(pyrazolyl)_2\}(C_7H_7)(CO)_2Mo]$ has been determined and the C_7H_7 ring shown to function as a three-electron donor to the molybdenum. However, the metal attains its complement of 18 electrons in this compound, acquiring two *via* a strong interaction with an α-hydrogen atom of an ethyl group (Mo—H = 193 pm).[280] Evidence for α-elimination of a hydrogen atom has been presented[281] for the methyl group of $[(\eta^5\text{-}C_5H_5)_2W(PR_3)Me]^+$, and

[269] W. G. Kita, J. A. McCleverty, B. E. Mann, D. Seddon, G. A. Sim, and D. I. Woodhouse, *J.C.S. Chem. Comm.*, 1974, 132.

[270] J. Chatt and J. R. Dilworth, *J.C.S. Chem. Comm.*, 1974, 517.

[271] J. Chatt and J. R. Dilworth, *J.C.S. Chem. Comm.*, 1974, 508.

[272] W. A. Myers and R. J. Nagle, jun., *J. Inorg. Nuclear Chem.*, 1973, **35**, 3985.

[273] D. W. Green and D. M. Gruen, *J. Chem. Phys.*, 1974, **60**, 1797.

[274] G. I. Nikolaev, and A. M. Nemets, *Ref. Zhur. Met.*, 1972, Abs. 9A16 (*Chem. Abs.*, 1973, **79**, 70799).

[275] L. W. Ander, R. S. Hansen, and L. S. Bartell, *J. Chem. Phys.*, 1973, **59**, 5277.

[276] S. J. Atkinson, C. R. Brundle, and M. W. Roberts, *J. Electron Spectrosc. Relat. Phenomena*, 1973, **2**, 105; T. E. Madey, J. T. Yates, jun., and N. E. Erickson. *Chem. Phys. Letters*, 1973, **19**, 487.

[277] G. Blyholder and M. Tanaka, *Bull. Chem. Soc. Japan*, 1973, **46**, 1876.

[278] P. S. Skell, E. M. Van Dam, and M. P. Silvon, *J. Amer. Chem. Soc.*, 1974, **96**, 626.

[279] T. Ito, T. Kokubo, K. Yamamoto, A. Yamamoto, and S. Ikeda, *J.C.S. Chem. Comm.*, 1974, 136; *J.C.S. Dalton*, 1974, 1783.

[280] F. A. Cotton and V. W. Day. *J.C.S. Chem. Comm.*, 1974, 415.

[281] N. J. Cooper and M. L. H. Green, *J.C.S. Chem. Comm.*, 1974, 761

the photo-induced insertion of a tungsten atom into a methyl C—H bond in *p*-xylene or mesitylene has been described.[282]

The first metal(II) carboxylate with a heteronuclear metal–metal bond, [CrMo-(OAc)$_4$], has been described.[8] The first examples of dialkylamido- and alkoxido-complexes of molybdenum(III), [Mo$_2$L$_6$] (L = NMe$_2$ or OBu^t) are valuable additions to the class of metal–metal triple-bonded compounds.[283] Considerable progress in the characterization of aquo-species of molybdenum-(III), -(IV), and -(V) has been made this year. The ions (10)–(12), are thought to be the blue green molybdenum(III),[284] the red molybdenum(V),[285] and the yellow molybdenum(V)[286] aquated ions, respectively, formed in acidic media. The di-μ-sulphido-ion Mo$_2$O$_2$S$_2^{2+}$ has been separated from the complex Na$_2$[MoO$_2$S$_2$(cysteinate)$_2$] and shown to be unusually resistant to acid cleavage.[287]

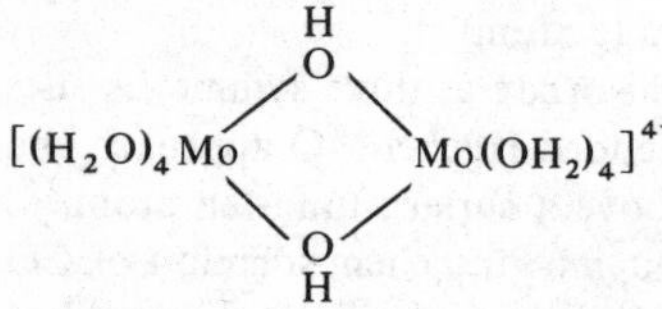

$$[(H_2O)_4Mo \overset{\overset{\displaystyle H}{\displaystyle O}}{\underset{\underset{\displaystyle H}{\displaystyle O}}{}} Mo(OH_2)_4]^{4+}$$

(10)

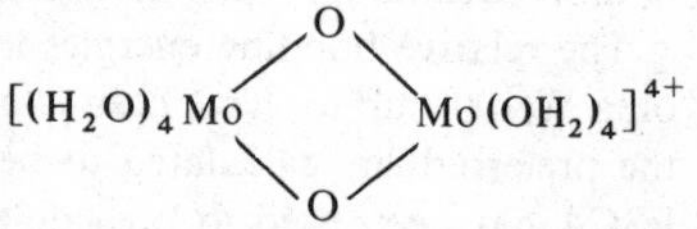

$$[(H_2O)_4Mo \overset{\displaystyle O}{\underset{\displaystyle O}{}} Mo(OH_2)_4]^{4+}$$

(11)

$$[(H_2O)_3Mo \overset{\overset{\displaystyle O \quad O \quad O}{\displaystyle \| \quad \quad \|}}{\underset{\underset{\displaystyle O}{}}{}} Mo(OH_2)_3]^{2+}$$

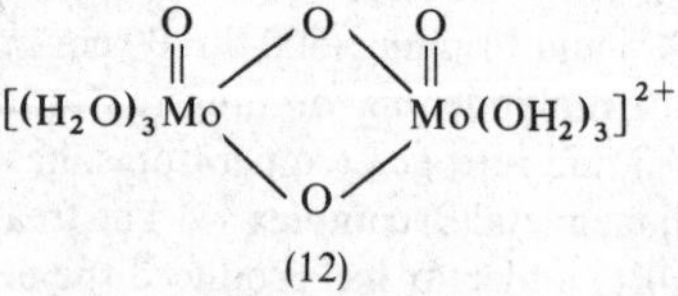

(12)

Organometallic complexes of molybdenum and tungsten have been supported on silica or alumina and activated at temperatures $\leqslant 373$ K prior to investigation of their properties as propane-disproportionation catalysts. It was concluded that molybdenum and tungsten complexes may provide efficient catalysis provided that there is a suitable means of interaction with the support. The original oxidation state of the metal is not important, provided that high oxidation state of the metal precursors can be reduced at some stage during the catalytic preparation.[288] A large number of related studies of olefin-disproportionation[289, 290] and other[291, 292]

[282] K. Elmitt, M. L. H. Green, R. A. Forder, I. Jefferson, and K. Prout, *J.C.S. Chem. Comm.*, 1974, 747.

[283] (a) M. H. Chisholm and W. Reichert, *J. Amer. Chem. Soc.*, 1974, **96**, 1249; (b) F. A. Cotton, B. A. Frenz, and L. Shive, *J.C.S. Chem. Comm.*, 1974, 480.

[284] M. Ardon and A. Pernick, *Inorg. Chem.*, 1974, **13**, 2275.

[285] M. Ardon and A. Pernick, *J. Amer. Chem. Soc.*, 1973, **95**, 6871.

[286] M. Ardon and A. Pernick, *Inorg. Chem.*, 1973, **12**, 2484.

[287] B. Spivack and Z. Dori, *J.C.S. Chem. Comm.*, 1973, 909.

[288] J. Smith, W. Mowat, D. A. Whan, and E. A. V. Ebsworth, *J.C.S. Dalton*, 1974, 1742.

[289] W. Mowat, J. Smith, and D. A. Whan, *J.C.S. Chem. Comm.*, 1974, 34.

[290] A. Ismayel-Milanovic, J. M. Bassett, H. Praliaud, M. Dufaux, and I. De Mourgues, *J. Catalysis*, 1973, **31**, 408; L. Burlamacchi, G. Martino, and F. Trifiro, *J. Catalysis*, 1974, **33**, 1; R. I. Maksimovskaya, *Kinetika i Kataliz*, 1973, **14**, 265 (*Chem. Abs.*, 1973, **79**, 23766).

[291] K. S. Seshadri and L. Petrakis, *J. Catalysis*, 1973, **30**, 195 (*Chem. Abs.*, 1973, **79**, 70580); H. Ueda and N. Toda, *Canad. J. Chem.*, 1973, **51**, 885; M. Akimoto and E. Echigoya, *Bull. Chem. Soc. Japan*, 1973, **46**, 1909; R. F. Howe and I. R. Leith, *J.C.S. Faraday*, 1, 1973, **69**, 1967.

[292] P. Biloen and G. T. Pott. *J. Catalysis*, 1973, **30**, 169 (*Chem. Abs.*, 1973, **79**, 70597).

catalytic systems have been reported, and in several such systems molybdenum(v) signals have been detected by e.s.r. spectroscopy.[290, 291]

An e.s.r. study of $Na_{0.33}WO_3$ at 15 K has provided the first such conclusive evidence of tungsten(v) centres in the tungsten bronzes.[293] Mössbauer spectra of several tungsten compounds have provided much information on the electronic distribution in these materials.[294] Core electron binding energies of several tungsten compounds have been determined and the results of these and other related studies discussed. It is suggested that the importance of crystal potentials must be realized when considering such data.[295]

Carbonyl and Thiocarbonyl Complexes.—As indicated for chromium only a selection of material published this year is reported.

In a cholesteric solvent, $Mo(CO)_6$ and other optically isotropic molecules exhibit Cotton effects which arise due to the large circular birefringence of the medium.[296] Further [13]C n.m.r. spectral data for carbonyl compounds of the Group VI metals have been presented,[297] including the first reported measurement of $^1J(^{95}Mo-^{13}C)$, as 68 Hz, for $Mo(CO)_6$. A direct measurement of the core shifts between free and complexed carbonyl groups has been presented.[13] The valence region photoelectron spectra of $Mo(CO)_6$ and $W(CO)_6$ have been studied and discussed[14] on the basis of published MO calculations. Absolute integrated intensities for the i.r. absorption bands of $Mo(CO)_6$ and $W(CO)_6$ have been measured and the values used, *via* a dipole moment calculation, to assess the effective atomic charges.[11] A solution and single-crystal Raman study of $Mo(CO)_6$ and $W(CO)_6$ has been completed. The solution study identified the $v_{10}(t_{2g})$ vibrational mode for the first time and single-crystal data for $Mo(CO)_6$ at < 120 cm^{-1} contained many previously unreported lines.[12] Further thermodynamic[10] and mass spectral[19] data have been presented for $Mo(CO)_6$ and $W(CO)_6$.[4] A cyclic voltammetric study of the oxidation of these carbonyls, at a Pt electrode in $MeCN-Bu_4NBF_4$, has identified[3] an irreversible two-electron process, which contrasts to the formation of $Cr(CO)_6^+$ under these conditions for $Cr(CO)_6$.

$Na_2[W(CO)_5]$ reacts with Cl_2CS in THF to produce $W(CO)_5CS$ in 12—15% yield. The corresponding molybdenum thiocarbonyl complex did not appear to be sufficiently stable to be detected under the gas chromatographic conditions used. $W(CO)_5CS$ reacts with Ph_3P or py (L) to produce the corresponding *trans*-$[W(CO)_4-(CS)L]$, and with piperidine or Me_2NH in hexane at room temperature to form the corresponding $[(OC)_5WC(SH)NR_2]$ thiocarbene complex; the latter reaction also occurs with primary amines RNH_2. However, the thiocarbenes readily lose H_2S to form the corresponding isocyanide complex $[(OC)_5WC\equiv NR]$.[4]

$[W(CO)_5F]^-$ has been prepared by the reaction of $W(CO)_6$ with $(Ph_3P)_2NF$, and by the reaction of $[W_2(CO)_{10}]^{2-}$ with AgF, and its i.r. spectrum recorded.[25] An electrochemical oxidation of the $[M(CO)_5X]^-$ (M = Mo or W; X = Cl, Br, or I) anions has shown that the $[M(CO)_5X]$ species possess no inherent stability and either disproportionate, or react further to produce $[M(CO)_4X_3]^-$ as the final product.[26] An X-ray crystallographic study of tetraethylammonium monothenoyltri-

[293] H. F. Mollet and B. C. Gerstein. *J. Chem. Phys.*, 1974, **60**, 1440.
[294] R. Gancedo, A. G. Maddock, R. H. Platt, and A. F. Williams, *J.C.S. Dalton*, 1974, 1314.
[295] G. E. McGuire, G. K. Schweitzer, and T. A. Carlson, *Inorg. Chem.*, 1973, **12**, 2450.
[296] S. F. Mason and R. D. Peacock, *J.C.S. Chem. Comm.*, 1973, 712.
[297] B. E. Mann, *J.C.S. Dalton*, 1973, 2012.
[298] G. H. Barnett, M. K. Cooper, M. McPartlin, and G. B. Robertson, *J.C.S. Chem. Comm.*, 1974, 305.

E

fluoroacetonatopentacarbonyltungsten(0) has shown that the anion (13) contains the first example of a *trans*-β-diketonato-ligand.[298] Dimethylketenimine has been stabilized by co-ordination to tungsten in $[(OC)_5WNH=CMe]$.[24] $[Mo_2(CO)_6$-$(Ph_3PNH)_3]$, formed by the photolysis of *cis*-$[Mo(CO)_4(Ph_3PNH)_2]$, has been shown to have a molecular structure in which each metal atom has a distorted octahedral environment. One face of this octahedron is composed of three bridging N-atoms from the triphenylphosphinimino-groups, a new mode of bonding for these ligands.

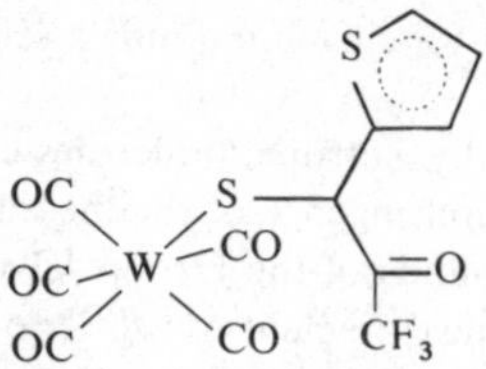

(13) 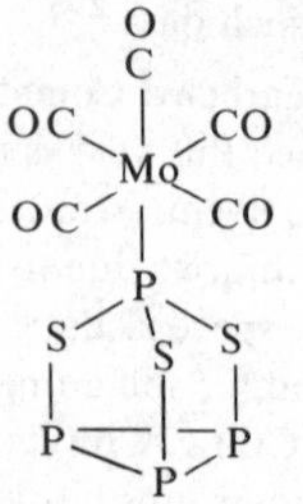　　　　　　　　　　　　　　　　　(14)

The Mo · · · Mo separation is 335.4(6) pm and therefore there is no metal–metal bond in this compound.[299] $[Mo(CO)_4P(C_6H_{11})_3]$ has been generated by the photolysis of $[Mo(CO)_5P(C_6H_{11})_3]$ in a hydrocarbon glass. Two isomeric species were characterized, with a vacant site *cis* or *trans* to the phosphino-group; the latter may be converted into the former by irradiation with visible light.[300] A determination of the crystal structure of $[Mo(CO)_5(P_4S_3)]$ (14) has shown that the tetraphosphorus trisulphide cage has a strong phosphine-like interaction with the $Mo(CO)_5$ group, Mo—P = 247.7(6) pm.[301] The complexes $[M(CO)_5SnX_2]$ (M = Mo or W; X = Cl, Br, or I) and $[M(CO)_5GeCl_2]$ have been prepared by the photochemical reaction of $M(CO)_6$ with SnX_2 or $CsGeCl_3$ in THF. Further reaction of these compounds with Me_4NX produced $[M(CO)_5SnX_3]^-$ or $[M(CO)_5GeCl_3]^-$, respectively.[23] The crystal and molecular structures of $[(\eta^7-C_7H_7)Mo(CO)_2SnPh_nCl_{3-n}]$ (n = 1 or 2) have been determined by X-ray crystallography.[302] X-Ray crystal structures have also been reported for $[(\eta^5-C_5H_5)(CO)_3MoZnCl(OEt_2)]_2$ and $[\{(\eta^5-C_5H_5)(CO)_3Mo\}_2$-Zn]. The former is a centrosymmetric dimer with a $ZnCl_2Zn$ central moiety and $(\eta^5-C_5H_5)(CO)_3Mo$ and Et_2O groups completing the distorted tetrahedral co-ordination of each zinc atom [Mo—Zn = 263.2(1) pm], and the latter has a linear Mo—Zn—Mo arrangement, with Mo—Zn separations of 253.8(1) pm.[303]

η-Complexes.—The compounds $[(\eta^5-C_5H_5)M(CO)_3Cl]$ (M = Mo or W) react with acetylenes (RC_2R) to produce the corresponding 16-electron species $[(\eta^5-C_5H_5)M$-$(RC_2R)Cl]$. The reaction with TlC_5H_5 of the compound with M = Mo and R = CF_3 has been shown to produce $[(\eta^5-C_5H_5)Mo(F_3CC_2CF_3)\{F_3C=C(CF_3)C_5H_5\}]$, following an insertion reaction involving one of the co-ordinated hexafluorobut-2-yne

[299] J. S. Miller, M. O. Visscher, and K. G. Caulton, *Inorg. Chem.*, 1974, **13**, 1632.
[300] J. D. Black and P. S. Braterman, *J. Organometallic Chem.*, 1973, **63**, C19.
[301] A. W. Cordes, R. D. Joyner, R. D. Shores, and E. D. Dill, *Inorg. Chem.*, 1974, **13**, 132.
[302] H. E. Sasse and M. L. Ziegler, *Z. anorg. Chem.*, 1973, **402**, 129.
[303] J. St. Denis. W. Butler. M. D. Glick and J. P. Oliver. *J. Amer. Chem. Soc.*, 1974. **96**, 5427.

ligands.[304] The treatment of $Mo(acac)_3$ with diphos in toluene under Ar produces the π-ethylene complex $[Mo(C_2H_4)(diphos)_2]$ and $[MoH(acac)(diphos)_2]$, in relative amounts which depend upon the quantity of $AlEt_3$ used. The former compound is converted into the latter on reaction with Hacac. N.m.r. studies suggest that the ethylene complex has the square-pyramidal structure (15). This complex reacts with N_2 or tetracyanoethylene (TCNE) to afford *trans*-$[Mo(N_2)_2(dppe)_2]$ or $[(\eta^2\text{-TCNE})\text{-}Mo(diphos)_2]$, respectively.[279] The co-condensation of molybdenum or tungsten atoms in the gas phase with buta-1,3-diene (1:100) at 77 K has been shown to produce a yellow matrix, from which the corresponding tris(butadiene)metal(0) complexes were isolated as pure crystalline materials.[278]

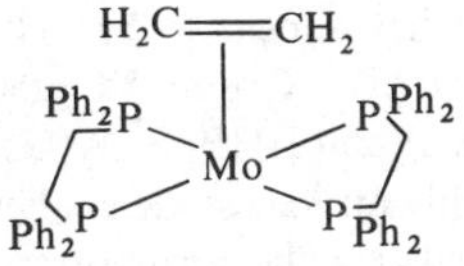

$$H_2C{=\!=}CH_2$$

(structure 15)

(15)

Bis(η^6-aryl) derivatives of molybdenum have been prepared by the co-condensation of the arene and molybdenum atoms at 77 K.[305] Friedel–Crafts reactions catalysed by $[(\eta^6\text{-arene})Mo(CO)_3]$ complexes have been developed further, and such reactions with organic halides shown to proceed *via* an ionic mechanism.[306]

Trifluorophosphino- and Trichlorostannano-complexes.—The reaction of PF_3 with MoO_3 or WO_3 at 4000 atm. and 300°C has been shown to form the corresponding $M(PF_3)_6$ complex.[307]

Et_4NSnCl_3 reacts with $(\eta^6\text{-}C_6H_6)_2Mo$ in refluxing THF or Me_2CO to produce $(Et_4N)_6[Mo(SnCl_3)_6]$, the existence of which suggests that $SnCl_3$ is a good π-backbonding ligand. Related $SnCl_3$ compounds also prepared in this study include $(Me_4N)[M(CO)_5SnCl_3]$ (M = Mo or W) and *fac*-$(Me_4N)_3[Mo(CO)_3(SnCl_3)_3]$.[6]

Dinitrogenyl and Related Complexes; Nitrogen Fixation.—The electronic structure of the linear fragments Mo—N—N have been assessed using an SCF MO model and N_2 again suggested to be predominantly a π-acceptor ligand.[34] Full reports have been published of the preparation, characterization, and reactions of complexes of the type *cis*-$[Mo(N_2)_2(PMe_2Ph)_4]$, *trans*-$[Mo(N_2)_2(PPh_2Me)_4]$, *trans*-$[Mo(N_2)_2(LL)_2]$, and *trans*-$[Mo(N_2)_2(LL)_2]I_3$ (LL = diphos, diars, or $Ph_2PCH_2CH_2AsPh_2$).[308] The oxidation of *trans*-$[Mo(N_2)_2(diphos)]$ has also been studied in an investigation which reported the synthesis of $[Mo(N_2)(diphos)_2Cl]$ by the reduction of $[Mo_2\text{-}(diphos)_2Cl_6],2C_6H_6$ by sodium in a THF solution containing diphos.[309] The complex $[(Me_2PhP)_4ClReN_2MoCl_4(OMe)](7)$ has been prepared (see p. 110).[267]

X-Ray crystallography has shown that the hydrazido-dianion, formed by reduction of a co-ordinated dinitrogenyl group, in $[W(N_2H_2)Cl(diphos)]^+$, has an essentially linear W—N—N arrangement (W—N = 173, N—N = 137 pm and $\angle$WNN = 173°).

[304] J. L. Davidson, M. Green, D. W. A. Sharp, F. G. A. Stone, and A. J. Welch, *J.C.S. Chem. Comm.*, 1974, 706.
[305] F. W. S. Benfield, M. L. H. Green, J. S. Ogden, and D. Young, *J.C.S. Chem. Comm.*, 1973, 866.
[306] J. F. White and M. F. Farona, *J. Organometallic Chem.*, 1973, **63**, 329.
[307] A. P. Hagen and E. A. Elphingstone, *J. Inorg. Nuclear Chem.*, 1973, **35**, 3719.
[308] T. A. George and C. D. Seibold, *Inorg. Chem.*, 1973, **12**, 2544, 2548.
[309] C. Miniscloux, G. Martino, and L. Sajus, *Bull. Soc. chim. France*, 1973, 2183.

The nature of this metal–di-imino bonding has been discussed, as has the possible relevance of such a species to the function of nitrogenase.[268] The hydrazido-complexes $[\{(\eta^5\text{-}C_5H_5)Mo(NO)X\}_2N_2R^1R^2]$(8) have been characterized. The favoured electronic arrangement in such species is that in which the Mo' atom acquires an 18-electron configuration, with the N' atom donating a pair of σ-electrons over the distance 205.4 pm and the N atom donating a pair of 'σ'- and a pair of 'π'-electrons over the distance 192.0 pm.[269]

I.r. and Raman spectra have been measured for a series of arylazo-complexes of the type $[\{RB(pz)_3\}Mo(CO)_2(N_2Ar)]$ and a band in the region of 1530—1580 cm^{-1} assigned to the $v(N-N)$ stretching mode.[310a] Oxidative addition of halogens to various arylazo-derivatives of molybdenum and tungsten has been studied. New complexes prepared include what appear to be first examples of a bridging ArN_2 group and an ArN_2^- complex of a Group VI metal, respectively, in $[(HBpz)_3\text{-}Mo(N_2Ph)X]_n$ (X = Cl, Br, or I) and $[(HBpz)_3W(N_2Ph)X_2]_2$ (X = Cl, Br, or I).[310b] $[(\eta^5\text{-}C_5H_5)_2MoH_2]$ reacts with an excess of azobenzene, methyl, or ethyl azodicarboxylate, or azothiobenzoyl, *via* the formation of the corresponding hydrido-hydrazino-complex $[(\eta^5\text{-}C_5H_5)MoH\{\sigma\text{-}NRNHR\}]$, to form the π-complexes $[(\eta^5\text{-}C_5H_5)_2Mo(PhN=NPh)]$, or the metalloheterocycle, $[(\eta^5\text{-}C_5H_5)_2\text{-}\overline{MoN(COR)N=C(R)O}]$ (R = Ph, OMe, or OEt). 4-Phenyl-1,2,4-triazoline-3,5-dione behaves differently towards $[(\eta^5\text{-}C_5H_5)_2MoH_2]$ and affords the metallated heterocycle, $[(\eta^5\text{-}C_5H_5)_2Mo\{\sigma\text{-}\overline{NN=C(OH)N(Ph)CO}\}_2]$.[310c]

Contrary to optimistic reports,[311a] no NH_3 appears to be formed in the reaction between *trans*-$[Mo(N_2)_2(\text{diphos})_2]$, $[Fe_4S_4(SEt)_4]^{2-}$, and HCl. The main products are $[MoCl_2(\text{diphos})_2]^+$, H_2S, EtSH, H_2, iron chlorides and, in quantitative yield, N_2.[311b] Further developments of the molybdate–thiol/NaBH$_4$ models of nitrogenase have been presented. In particular, these model systems have been extended to include ferredoxin-type complexes as electron-transfer agents. These new systems efficiently catalyse acetylene reduction, even with dithionite as the reductant, and thus duplicate nitrogenase in another important respect.[312] The role of ATP in these molybdate–thiol/NaBH$_4$ catalyses has been critically examined. The stimulating effect of ATP on the molybdate–glutathione/NaBH$_4$ system for the reduction of acetylene has been shown to be very dependent upon the pH of the ATP solution added to the reaction mixture. The stimulation appears to be due to the non-specific effect of protic acid addition, rather than to a 'macroergic' P—O bond function. In the light of these results the H/D substitution studies for acetylene reduction by the model systems, the function of the molybdenum centres in both the model and the enzymatic catalysts has been discussed.[266] The catalytic role of metals and free SH groups in nitrogenase has been studied using specific inhibitors and nitrogenase preparations containing molybdenum and vanadium. The results strongly suggest that the ATP-dependent H_2 formation and the N_2 reduction occur at molybdenum centre(s) in the enzyme.[313]

[310] (a) D. Sutton, *Canad. J. Chem.*, 1974, **52**, 2634; (b) M. E. Deane and F. J. Lalor, *J. Organometallic Chem.*, 1974, **67**, C19; (c) A. Nakamura, M. Aotake, and S. Otsuka, *J. Amer. Chem. Soc.*, 1974, **96**, 3456.

[311] (a) E. E. van Tamelen, J. A. Gladysz, and C. R. Brulet, *J. Amer. Chem. Soc.*, 1974, **96**, 3020; (b) J. Chatt, C. M. Elson, and R. L. Richards, *J.C.S. Chem. Comm.*, 1974, 189.

[312] G. N. Schrauzer, G. W. Kiefer, K. Tano, and P. A. Doemeny, *J. Amer. Chem. Soc.*, 1974, **96**, 641.

[313] R. I. Gvozdev, A. P. Sadkov, A. I. Kotel'nikov, and G. I. Likhenshtein, *Izvest. Akad. Nauk S.S.S.R.*, *Ser. biol.*, 1973, 488 (*Chem. Abs.*, 1974, **80**, 11847).

Nitrosyl and Thionitrosyl Complexes.—A general method for the synthesis of transition-metal nitrosyl complexes has been presented.[314] $[(\eta^5\text{-}C_5H_5)W(NO)_2CO]PF_6$ has been prepared by the reaction of $NOPF_6$ with $[(\eta^5\text{-}C_5H_5)W(NO)(CO)_2]$ in CH_2Cl_2–MeCN at 195 K. This new complex appears to be a useful precursor for the preparation of a variety of new tungsten nitrosyl complexes.[315] *cis*-$[M(NO)_2\text{-}(MeCN)_4]^{2+}$ (M = Mo or W) has been shown by ^1H n.m.r. spectral studies to undergo a stereospecific exchange of the MeCN ligands *via* a dissociative pathway. This stereospecificity appears to be associated with the rigidity of the five-co-ordinate $[M(NO)_2(MeCN)_3]^{2+}$ intermediate.[316] The complexes $[Mo(NO)_2Cl_2L_2]$ [L = EtCN, Me_2CO, Me_2SO, $(Me_2N)_3PO$, py, or DMF] have been prepared by the reaction of $MoCl_5$ with NO in refluxing 1,2-dichloroethane, followed by the addition of L. The position of the $v(N\text{—}O)$ i.r. stretching mode in these compounds has been correlated with the nature of L, according to synergic bonding considerations.[317] A similar route has been used to prepare $[Mo(NO)_2Cl_2(OPPh_3)_2]$.[318] An X-ray photoelectron study of $[Mo(NO)_2Cl_2(PPh_3)_2]$ has been reported.[319]

The full account of the preparation and subsequent reactions of the cationic nitrosyl complexes $[M(NO)(CO)_3(diphos)]PF_6$ (M = Mo or W) has been published. These complexes react with NaX (X = halide, S_2CNMe_2, or S_2CNEt_2) salts to form the corresponding $[M(NO)(CO)_2(diphos)X]$ derivative. With phosphine ligands (L) in $CHCl_3$, $[M(NO)(CO)L(diphos)Cl]$ is obtained, but in acetone, $[M(NO)(CO)_2L_2\text{-}(diphos)](PF_6)$,(acetone) and $[M(NO)(CO)(diphos)_2](PF_6)$ are formed. Decarbonylation occurs when $[Mo(NO)(CO)_3(diphos)](PF_6)$ is heated under reflux in $CHCl_3$ to produce $[Mo(NO)(diphos)Cl_2]_n$ and other polynuclear nitrosyl species.[320]

Treatment of the dimers $[(\eta^5\text{-}C_5H_5)Mo(NO)X_2]_2$ (X = Cl, Br, or I) with the hydrazines $R^1R^2NNH_2$ ($R^1 = R^2$ = Me or Et; R^1 = Me, R^2 = Ph) results in the formation of the novel hydrazido-complexes $[\{(\eta^5\text{-}C_5H_5)Mo(NO)X\}_2N_2R^1R^2]$(8). $[(\eta^5\text{-}C_5H_5)_2Mo(NO)I]$ reacts with Me_2NNH_2 to give a mixture of two compounds, the orange-red analogue of (8) with $R^1 = R^2$ = Me and X = I, and an orange-yellow σ-hydrazido-species, $[(\eta^5\text{-}C_5H_5)Mo(NO)I(NHNMe_2)]$. The dimers $[(\eta^5\text{-}C_5H_5)\text{-}Mo(NO)X_2]_2$ (X = Cl or Br) react with $PhNHNH_2$ to give the R^1 = H, R^2 = Ph, X = Cl or Br analogues of (8). However, $PhNHNH_2$ reacts with $[(\eta^5\text{-}C_5H_5)Mo\text{-}(NO)I_2]_2$ to produce (9).[269] The electrochemical oxidation of thio-bridged dimers of the type $[(\eta^5\text{-}C_5H_5)Mo(NO)(SR)_2]$ has been described and the process shown to produce the corresponding monocations.[321] Several fluxional allylnitrosyl complexes of molybdenum have been studied, as summarized in Scheme 1. These compounds are of interest because the collective ligand environment may confer upon the metal an unusual electronic and/or structural configuration. As part of this study, the crystal structures of the compounds (16) and (17) were determined.[322] Both these compounds contain essentially linear Mo—N—O arrangements and the former (16)

[314] P. Bandyopadhyay and S. Rakshit, *Indian J. Chem.*, 1973, **11**, 496.
[315] R. P. Stewart. jun.. *J. Organometallic Chem.*. 1974. **70**. C8.
[316] B. F. G. Johnson. A. Khair. C. G. Savory. and R. H. Walter. *J.C.S. Chem. Comm.*, 1974. 744.
[317] R. Taube and K. Seyferth. *Z. Chem.*, 1973, **13**. 300.
[318] L. Bencze, J. Kohan, B. Mohai, and L. Marks, *J. Organometallic Chem.*, 1973, **70**, 421.
[319] W. B. Hughes and B. A. Baldwin, *Inorg. Chem.*, 1974, **13**, 1531.
[320] N. G. Connolly, *J.C.S. Dalton*, 1973, 2183.
[321] P. D. Frisch, M. K. Lloyd, J. A. McCleverty, and D. Seddon, *J.C.S. Dalton*, 1973, 2268.
[322] N. A. Bailey, W. G. Kita, J. A. McCleverty, A. J. Murray, B. E. Mann, and N. W. J. Walker. *J.C.S. Chem. Comm.*. 1974. 592.

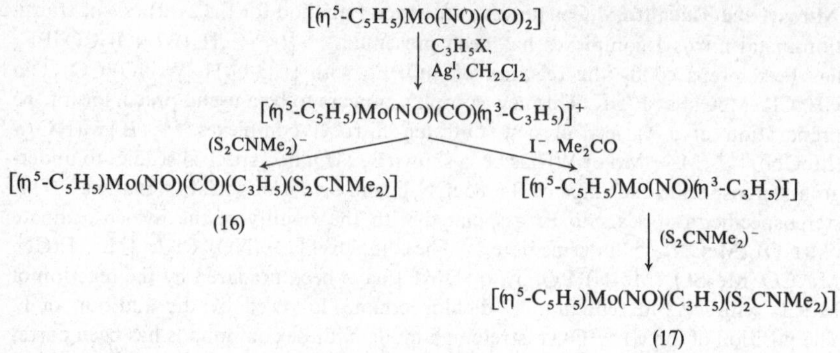
[(η5-C5H5)Mo(NO)(CO)2]
C3H5X,
AgI, CH2Cl2
[(η5-C5H5)Mo(NO)(CO)(η3-C3H5)]+
(S2CNMe2)−
I−, Me2CO
[(η5-C5H5)Mo(NO)(CO)(C3H5)(S2CNMe2)]
(16)
[(η5-C5H5)Mo(NO)(η3-C3H5)I]
(S2CNMe2)−
[(η5-C5H5)Mo(NO)(C3H5)(S2CNMe2)]
(17)

Scheme 1

appears to be derived from $[(\eta^5\text{-}C_5H_5)Mo(NO)(\eta^3\text{-}C_3H_5)I]$ by the nucleophilic attack of the dithiocarbamato-group at a terminal allylic carbon atom.[322]

The nitrido-complexes $[MoN(S_2CNR_2)_3]$ [R_2 = Me_2, Et_2, or $(CH_2)_5$] react with sulphur to give a high yield of the corresponding thionitrosyl species $[Mo(NS)\text{-}(S_2CNR_2)_3]$. They are orange-red, air-stable, crystalline solids, which dissolve as monomeric species in 1,2-dichloroethane. Their ^1H n.m.r. spectra are consistent with a pentagonal-bipyramidal structure having an apical NS group and their ν(N—S) stretching mode occurs at *ca.* 1100 cm^{-1}. The original nitrido-complexes are regenerated by treatment with Bu_3P.[271]

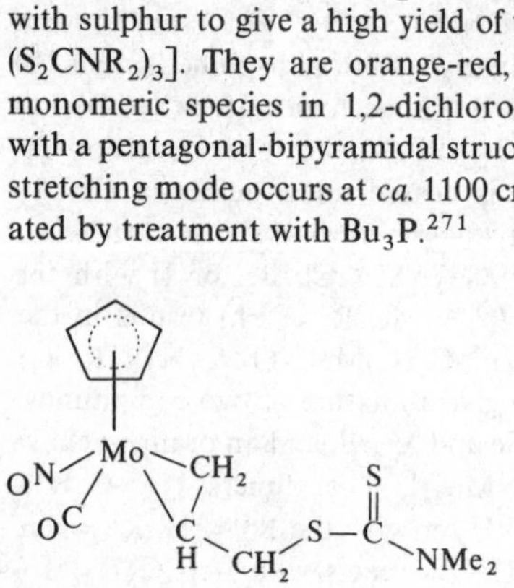
Mo
ON
OC
CH2
C
H
CH2
S
S
C
NMe2

(16)

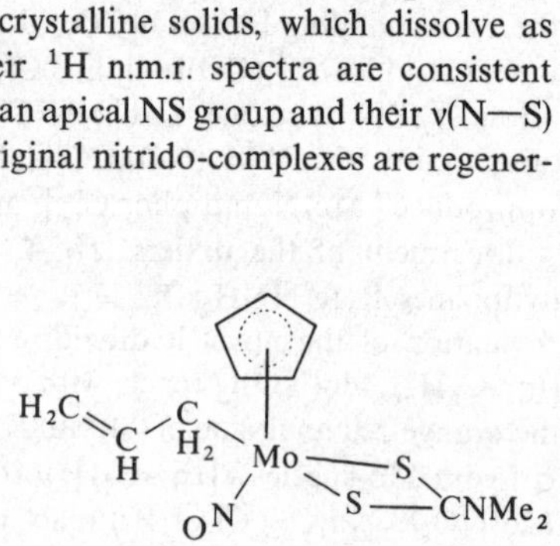
H2C
C
H
C
H2
Mo
S
S
CNMe2
ON

(17)

Cyano- and Isocyano-complexes.—The interaction of ZrO^{2+} with $[MoO(OH)\text{-}(CN)_4]^{3-}$ has been studied and the compounds $KZrO[MoO(OH)(CN)_4]$ and $(ZrO)_3\text{-}[MoO(OH)(CN)_4]_2$ identified.[323] The *X*-ray crystal structure of $Na_3[MoO(OH)\text{-}(CN)_4],4H_2O$ shows the metal atom to occupy a special position on a mirror plane.[324]

The protonation constants of the $[WO_2(CN)_4]^{4-}$ ion have been determined to be $\log K_1 > 12$ and $\log K_2 = 8.84$ at infinite dilution.[325]

The salts $MFe[Mo(CN)_8]$ (M = K, Rb, Cs, or NH_4) have been prepared by treating $FeCl_3$ with $Li_4[Mo(CN)_8]$ and MCl and the compounds shown to possess a cubic unit cell by *X*-ray powder diffractometry. The corresponding $M_2Fe[Mo(CN)_8]$ salts have been prepared in an analogous manner using $FeCl_2$. Their i.r. spectral

323 K. U. Din, A. A. Khan, and M. A. Beg, *J. Inorg. Nuclear Chem.*, 1974, **36**, 468.
324 K. Stadnicka, *Roczniki Chem.*, 1973, **47**, 2021 (*Chem. Abs.*, 1974, **80**, 88191).
325 E. Hejmo, A. Kanas, and A. Samotus, *Bull. Acad. polon. Sci., Ser. Sci. chim.*, 1973, **21**, 311 (*Chem. Abs.*, 1973, **79**, 70708).

properties are consistent with the presence of cyano-bridges between the Mo^{IV} and Fe^{II} centres.[326] Magnetic studies of the salts $M_2[Mo(CN)_8],nH_2O$ (M = Mn—Zn) have, however, emphasized the weakness of the electronic interactions between the Mo^{IV} and M^{II} centres in these compounds.[327] The reactions and the redox characteristics of the system $[Mo(CN)_8]^{3-}$–Hg have been studied polarographically, and the products shown to include $Hg_4[Mo(CN)_8]$ and $Hg(CN)_2$.[328]

Photolysis of a liquid NH_3 solution of $(NH_4)_4[Mo(CN)_8]$ has been reported; two photolytic intermediates were identified. Such photolysis was used to prepare $Mo(CN)_4(NH_3)_2$, which was characterized by spectrophotometric and magnetic studies. Preliminary results concerning the photolysis of $[W(CN)_8]^{4-}$ in liquid NH_3 have also been reported.[329] Improved syntheses of $K_3[MoO(OH)(CN)_4],2H_2O$ and $K_4[MoO_2(CN)_4],8H_2O$ have been developed based on the photolysis of $K_4[Mo(CN)_8]$ in aqueous ammonia.[330] A kinetic study of the photosubstitution of $K_4[Mo(CN)_8]$ in 0.1 M aqueous NaOH has been published,[331] as have the results of several other photolyses of the $[M(CN)_8]^{n-}$ (M = Mo or W; n = 3 or 4) ions.[332]

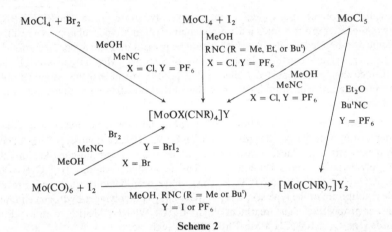

Scheme 2

Synthetic routes to molybdenum isocyanide complexes have been extended and the routes presently available are summarized in Scheme 2. The complexes $[MoOCl(CNR)_4]Y$ are new; they have been extensively characterized and shown to contain predominantly *trans*-cationic units.[333]

[326] D. I. Zubritskaya, D. I. Semenishin, Yu. A. Lyubechenko, and K. N. Mikhalevich, *Zhur. neorg. Khim.*, 1973, **18**, 2289; D. I. Zubritskaya, K. N. Mikhalevich, V. M. Litvinchuk, and Yu. A. Lyubchenko, *ibid.*, 1974, **19**, 706.
[327] G. F. McKnight and G. P. Haight, jun., *Inorg. Chem.*, 1973, **12**, 3007.
[328] T. Jarzyna, Z. Stasicka, and T. Senkowski, *Roczniki Chem.*, 1973, **47**, 1801 (*Chem. Abs.*, 1974, **80**, 55305).
[329] R. D. Archer and D. A. Drum, *J. Inorg. Nuclear Chem.*, 1974, **36**, 1979.
[330] R. L. Marks and S. E. Popielski, *Inorg. Nuclear Chem. Letters*, 1974, **10**, 885.
[331] E. Nagorsnik, P. Thomas, and E. Hoyer, *Inorg. Nuclear Chem. Letters*, 1974, **10**, 353.
[332] Z. Stasicka and H. Bulska, *Roczniki Chem.*, 1973, **47**, 1365 (*Chem. Abs.*, 1974, **80**, 54467); Z. Stasicka, *Roczniki Chem.*, 1973, **47**, 485 (*Chem. Abs.*, 1973, **79**, 13042); *Zesz. Nauk. Uniw. Jagiellon., Pr. Chem.*, 1973, **18**, 39 (*Chem. Abs.*, 1973, **79**, 151494); A. Samotus, *Adv. Mol. Relaxation Processes*, 1973, **5**, 121 (*Chem. Abs.*, 1974, **80**, 32458); *Roczniki Chem.*, 1973, **47**, 251, 265 (*Chem. Abs.*, 1973, **79**, 47765, 47767).
[333] M. Novotny and S. J. Lippard. *Inorg. Chem.*, 1974. **13**, 828.

Pyrazolylborato-complexes.—An X-ray crystallographic study[280] of $[\{Et_2B(pz)_2\}_2$-$(C_7H_7)Mo(CO)_2]$ has shown that the metal atom attains a formal 18-electron configuration and interacts strongly with an α-hydrogen atom of an ethyl group (Mo—H = 193 pm), thus leaving four unused $p\pi$-electrons in the *trihapto*-cyclo-heptatrienyl ring. The C—H—Mo interaction is very similar to, but even stronger than, that previously observed in $[\{Et_2B(pz)_2\}Mo(CO)_2(\eta^3\text{-}PhC_3H_4)]$, the full report of which has now been published.[334] Such α-hydrogen interactions therefore appear to be structurally and thermodynamically competitive with $\eta\text{-}C_nH_n$-metal interactions. N.m.r. studies of $[\{Et_2B(pz)_2\}Mo(CO)_2(\eta^3\text{-allyl})]$ complexes have established that the structure found for the 2-PhC_3H_4 allyl derivative in the solid state is adopted in solution. However, the molecules are fluxional in two respects and the details of their atomic rearrangements have been determined.[335] The crystal structure of $[\{HB(pz)_3\}Mo(CO)_2(\eta^3\text{-}2\text{-}MeC_3H_4)]$ has been solved. The co-ordination sphere of the metal atom consists of one N atom from each of the three pyrazolyl rings, two C atoms from σ-bonded carbonyl groups, and a η^3-2-MeC_3H_4 group.[336] $[\{Et_2B(pz)_2\}$-$(Hpz)Mo(CO)_2(\eta^3\text{-}C_3H_5)]$ has been shown by X-ray crystallography to contain a unidentate pyrazole ligand and a bidentate pyrazolylborato-group. The $\overline{\text{MoNNBNN}}$ ring of the latter group adopts a distorted chair conformation, probably as a result of specific non-bonded repulsions which would be present in the boat conformation.[337] A further crystal-structure determination of $[\{B(pz)_4\}Mo(CO)_2(\eta^5\text{-}C_5H_5)]$ has confirmed the details reported earlier (Vol. 2, p. 119), except that the assignment of a carbon and a nitrogen atom adjacent to one of the co-ordinated nitrogen atoms was reversed.[338]

Hydrido-complexes.—The reaction of $Mo(acac)_3$ and diphos in toluene with $AlEt_3$ under Ar affords $[Mo(C_2H_4)(diphos)_2]$ and $[MoH(acac)(diphos)_2]$. The former complex reacts with Hacac to produce the latter and this hydrido-complex initiates the polymerization of acrylonitrile.[279] The reversible reactions $[(\eta^5\text{-}C_5H_5)_2W$-$(PR_3)CD_3]^+ \rightleftharpoons [(\eta^5\text{-}C_5H_5)W(CD_2PR_3)D]^+$ have been demonstrated, and a reversible equilibrium of the type W—$CH_3 \rightleftharpoons W(=CH_2)H$ involving α-hydrogen elimination was proposed.[281] The treatment of $[(\eta^5\text{-}C_5H_5)_2W(C_2H_4)Me]PF_6$ with Me_2PhP affords $[(\eta^5\text{-}C_5H_5)_2W(CH_2PMe_2Ph)H]PF_6$ in high yield.[339]

Irradiation of a mesitylene solution of $[(\eta^5\text{-}C_5H_5)WH_2]$ with a Hg lamp (366 nm) affords the orange bis-alkyl derivative $[(\eta^5\text{-}C_5H_5)_2W\{CH_2(3,5\text{-}Me_2C_6H_3)_2\}]$. The X-ray crystal structure of this compound has been determined and the geometry of the constituent molecules (18) identified; $H_2CWCH_2 = 75.3°$ is in the region expected for a normal bent d^2-bis-$(\eta^5\text{-}C_5H_5)$ species. The corresponding irradiation process in p-xylene affords the yellow-orange compound, $[(\eta^5\text{-}C_5H_5)_2W\{CH_2C_6H_4Me_2\}_2]$. The formation of these alkyl derivatives appears to be the first example of the direct insertion of a transition-metal atom into an unco-ordinated C—H group.[282]

[334] F. A. Cotton, T. LaCour and A. G. Stanislowski, *J. Amer. Chem. Soc.* 1974, **96**, 754.
[335] F. A. Cotton and A. G. Stanislowski. *J. Amer. Chem. Soc.*, 1974, **96**, 5074.
[336] E. M. Holt. S. L. Holt. and K. J. Watson. *J.C.S. Dalton*, 1973, 2444.
[337] F. A. Cotton, B. A. Frenz, and A. G. Stanislowski, *Inorg. Chim. Acta*, 1973, **7**, 503.
[338] E. M. Holt and S. L. Holt, *J.C.S. Dalton*, 1973, 1893.
[339] N. J. Cooper and M. L. H. Green. *J.C.S. Chem. Comm.*, 1974. 208.

The reactions of $[(\eta^5\text{-}C_5H_5)_2MoH_2]$ with some activated olefins have been studied as model systems for homogeneous hydrogenation catalysis. Stereochemical studies using $[(\eta^5\text{-}C_5H_5)_2MoD_2]$ and dimethyl fumarate or maleate has established that the insertion reactions produce selectively the *cis* adduct, *threo-* or *erythro-σ*-alkyl compound, and that unimolecular alkane elimination from this compound retains the configuration at the σ-bonded α-carbon atom.[340] The full account of the insertion and substitution reactions of $[(\eta^5\text{-}C_5H_5)_2WH_2]$ with $NaO_2CCl_{3-n}F_n$ ($n = 2$ or 0) has now been published.[341] The reactions of $[(\eta^5\text{-}C_5H_5)_2MoH_2]$ with azo- and diazo-compounds are described on p. 116.[310c] White phosphorus reacts with $[(\eta^5\text{-}C_5H_5)_2MoH_2]$ to produce the diphosphorane derivative $[(\eta^5\text{-}C_5H_5)_2\text{-}MoP_2H_2]$.[342] $MeSO_3F$ converts $[(\eta^5\text{-}C_5H_5)_2MoH_2]$ into an unstable product which has been formulated as $[(\eta^5\text{-}C_5H_5)_2MoH_2Me]SO_3F$.[343]

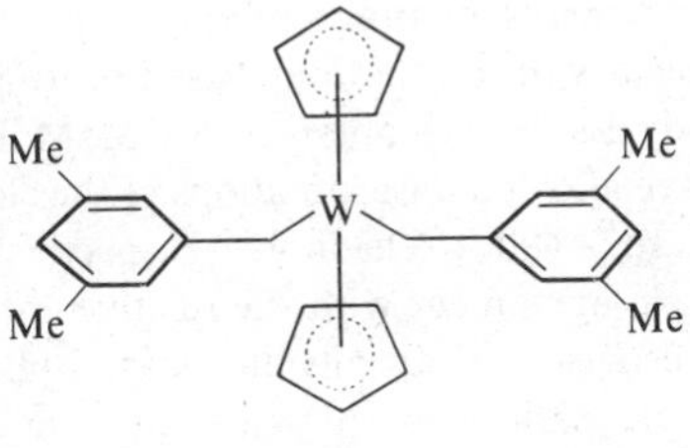

(18)

The full report of the crystal structure of $[\{(\eta^5\text{-}C_5H_5)(\mu\text{-}\eta^5\text{-}C_5H_4)MoH\}_2Al_3Me_5]$ has been published[344] and the crystal structure of a product from the reaction between $[(\eta^5\text{-}C_5H_5)_2MoH_2]$ and $LiBu^n$ determined. This product is $[(\eta^5\text{-}C_5H_5)_2\text{-}MoHLi]_4$ and it consists of eight-membered Mo—Li rings, with Li atoms bridging $(\eta^5\text{-}C_5H_5)_2Mo$ groups. The ring has crystallographic C_2 symmetry and the long Mo—Li bonds involved (270 pm) are presumed to reflect the electron-deficient nature of the compound. The positions of the hydrogen atoms are still unknown, but there is probably one such atom bonded to each molybdenum centre.[345] The reaction between $Li_3[MoPh_6]$ and $MoCl_3,3THF$ in $THF\text{–}Et_2O$ solution at $-20\,°C$ has been shown to produce the dihydro-complex $Li_2MoPh_2H_2,2THF$.[346]

The full account of the crystal structure of $[MoH_4(PMePh_2)_4]$ has been published and the fluxional behaviour of this and related molecules discussed with reference to this structural information.[347]

Binary Compounds and Related Systems.— *Halides.* New simple reductive fluoride elimination routes to MoF_4 and MoF_5 have been reported. MoF_6 is conveniently reduced by elemental silicon (4:1) or by dihydrogen (2:1) in anhydrous HF to produce MoF_5. This compound in turn may be reduced by elemental silicon (4:1)

[340] A. Nakamura and S. Otsuka, *J. Amer. Chem. Soc.*, 1973, **95**, 7262.
[341] K. S. Chen, J. Kleinberg, and J. A. Landgrebe, *Inorg. Chem.*, 1973, **12**, 2826.
[342] J. C. Green, M. L. H. Green, and G. E. Morris, *J.C.S. Chem. Comm.*, 1974, 212.
[343] D. Strope and D. F. Shriver. *J. Amer. Chem. Soc.*, 1973, **95**, 8197.
[344] S. J. Rettig, A. Storr, B. S. Thomas, and J. Trotter, *Acta Cryst.*, 1974, **B30**, 666.
[345] F. W. S. Benfield, R. A. Forder, M. L. H. Green, G. A. Moser, and K. Prout, *J.C.S. Chem. Comm.*, 1973, 759.
[346] B. Heyn and H. Still, *Z. Chem.*, 1973, **13**, 191.
[347] L. J. Guggenberger. *Inorg. Chem.*, 1973, **12**, 2295.

to MoF_4. The direct reaction of MoF_6 with Si (2:1) in anhydrous HF affords a green oil which may be a polymerized mixture of MoF_5 and MoF_4.[348] The i.r.-active stretching fundamental, ν_3, of WF_6 molecules in the gas phase has been studied and a value for the W—F stretching force constant estimated as 5.50 ± 0.07 mdyn Å^{-1}.[349] The vibrational spectrum for matrix-isolated MoF_5 molecules has been presented and assigned assuming a monomeric trigonal-bipyramidal structure.[350] Thermodynamic properties have been calculated for $MoF_5(g)$ from spectroscopic data[351] and new thermochemical data obtained for WF_4, WF_5, and WF_6.[352] The enthalpies of formation of $MoF_5(c)$ and $WF_5(c)$ have been determined and used in a discussion of the thermodynamic relationships within the molybdenum and tungsten fluoride and oxyfluoride systems.[353a] Reaction enthalpies for the tungsten–fluorine system have been determined from transport study measurements. The data differ considerably from those deduced from corresponding tungsten–chlorine values. Furthermore, preliminary measurements of tungsten transport in a chlorine atmosphere have shown little agreement with estimations based on tungsten–chlorine enthalpy values.[353b] New thermodynamic data obtained for $M^1M^2F_6$ (M^1 = K, Rb, or Cs; M^2 = Mo or W) salts have allowed for estimations of the electron affinities of MoF_6 and WF_6, as -517 ± 6 and -490 ± 5 kJ mol^{-1}, respectively.[354] The lower electron affinity $WF_6 < MoF_6$ is in agreement with the relative chemical reactivity of these compounds. MoF_6 has been used to partially fluorinate 1-phenylphosphorinan-4-one 1-oxide and to convert aromatic acid chlorides into the corresponding trifluoromethylated derivatives.[355]

The reduction of WCl_6 by dihydrogen to produce WCl_n (n = 5, 4, 2.6, or 2) has been described.[356] The best conditions for the formation of $MoCl_3$ *via* the reduction of $MoCl_5$ with anhydrous $SnCl_2$ or red phosphorus have been determined and phase equilibria in the $MoCl_3$–$MoCl_5$ system described.[357] WCl_4–WCl_5 equilibria have been examined in a NaCl–KCl melt at temperatures between 1000 and 1250 K. Higher temperatures favour WCl_5 and at >1240 K WCl_4 is unstable under these conditions.[358] β-WCl_6 has been shown[359] by X-ray crystallography to be isostructural with UCl_6, the tungsten atoms occupying octahedral holes in the h.c.p. arrangement of chlorine atoms. α-WCl_6, whilst having a similar h.c.p. arrangement of the chlorine atoms, has a different arrangement of the tungsten atoms.

The enthalpies of the reactions between $Mo(CO)_6$ or $W(CO)_6$ and iodine vapour at temperatures between 520 and 540 K have been determined. The corresponding tri-iodides are formed and the results obtained lead to the values of ΔH_f° for $MoI_3(c)$ and $WI_3(c)$ being calculated as -113 ± 10 and -46 ± 13 kJ mol^{-1}, respectively.[360]

[348] R. T. Paine and L. B. Asprey, *Inorg. Chem.*, 1974, **13**, 1529.

[349] R. S. McDowell and L. B. Asprey, *J. Mol. Spectroscopy*, 1973, **48**, 254.

[350] N. Acquista and S. Abramowitz, *J. Chem. Phys.*, 1973, **58**, 5484.

[351] V. T. Orekhov and A. G. Rybakov, *Zhur. fiz. Khim.*, 1973, **47**, 1612.

[352] D. L. Hildenbrand, *Govt. Rep. Announce.* (*U.S.A.*), 1973, **73**, 52 (*Chem. Abs.*, 1973, **79**, 24 108).

[353] (a) J. Burgess, I. Haigh, and R. D. Peacock, *J. C. S. Dalton*, 1974, 1062; (b) G. Dittmer, A. Klopfer, D. S. Ross, and J. Schröder, *J.C.S. Chem. Comm.*, 1973, 846.

[354] J. Burgess, I. Haigh, R. D. Peacock, and P. Taylor, *J.C.S. Dalton*, 1974, 1064.

[355] F. Mathey and G. Muller, *Compt. rend.*, 1973, **277**, C, 45; F. Mathey and J. Bensoam, *ibid.*, 1973, **276**, C, 1569; *Tetrahedron Letters*, 1973, 2253.

[356] Y. Saeki and R. Matsuzaki, *J. Less-Common Metals*, 1973, **33**, 313.

[357] D. V. Drobot and E. A. Sapranova, *Zhur. neorg. Khim.*, 1974, **19**, 228.

[358] S. N. Shkol'nikov, M. I. Manenkov, and A. K. Yarmolovich, *Zhur. priklad Khim.*, 1973, **46**, 1918.

[359] J. C. Taylor and P. W. Wilson, *Acta Cryst.*, 1974, **30B**, 1216.

[360] Y. Virmani, D. S. Barnes, and H. A. Skinner, *J.C.S. Dalton*, 1974, 399.

Oxides. The nucleation and the growth of oxide crystals on a W(110) surface at 650—1100 K have been studied by LEED and Auger electron emission spectroscopy.[361] The mechanism of the reduction by H_2 or CO of MoO_3 to MoO_n ($n = 2.9$, 2.7, or 2) and WO_3 to WO_n ($n = 2.98$, 2.9, or 2) has been investigated[362a] and the enthalpies of formation have been determined.[362b] The Mo—O bond energy on the surface of MoO_3 has been estimated from a measurement of the vapour pressure of O_2 above this surface.[363] Mass spectrometric studies have shown that the evaporations of MoO_3 and WO_3 are accompanied by vapour-phase association which produces polymeric species.[364] Electron diffraction has been used to probe the structure of the new cubic oxide phases of WO_3 and W_2O_3. The former appears[365] to be isostructural with ReO_3 and the latter with Mo_2O_3. Synthetic procedures suitable for the preparation of single crystals of MoO_2 and WO_2 have been described.[366]

The phase relationships and physical properties of the systems $V_{1-x}M_xO_2$ (M = Mo or W; $x = 0$—1) have been reviewed.[51]

Chalcogenides. The use of CS_2 in the preparation of transition-metal sulphides has been described and a variety of such compounds obtained with this agent, including MoS_2 and WS_2.[367] The binding energies of core electrons in MoS_3 and MoS_2 have been reported.[368] A new calculation of the energy bands of MoS_2 has been performed which provides a new interpretation of the electronic spectrum of this compound.[369] The long-wavelength lattice vibrations of MoS_2 have been interpreted by a linear-chain model which incorporates both intra- and inter-layer interactions.[370] Thermal decomposition of MoS_2 to form Mo_2S_3 has been investigated,[371] and the structural and textural evolution of MoS_2 from MoS_3 studied using d.t.a., X-ray diffraction, and electron microscopic techniques.[372] The kinetics of the reaction of annealed tungsten with pure selenium vapour have been investigated by t.g.a.; WSe_2 was the only detectable product.[373] U.v. reflectance spectra have been recorded for WTe_2 and these data correlated with band-structure models.[374]

The low-valent chalcogenide halides $Mo_6X_{10}A$ (X = Br, A = S or Te; X = I, A = Se or Te) have been prepared by heating appropriate mixtures of MoX_2, Mo,

361 N. R. Avery, *Surface Sci.*, 1974, **40**, 533 (*Chem. Abs.*, 1974, **80**, 75039).
362 (*a*) O. Ya. Miroshinichenko and K. P. Smirnov, *Termodin. Kinet. Protsessov. Vosstanov. Metal., Mater. Konf.*, 1969, 113 (*Chem. Abs.*, 1973, **79**, 10339); (*b*) T. V. Charlu and O. J. Kleppa, *J. Chem. Thermodynamics*, 1973, **5**, 325 (*Chem. Abs.*, 1973, **79**, 10588).
363 Yu. A. Mischchenko. N. D. Gol'dshtein, and A. I. Gel'bshtein, *Zhur. fiz. Khim.*, 1973, **47**, 511.
364 E. K. Kazenas, D. M. Chizhikov, and Yu. V. Tsvetkov, *Termodin. Kinet. Protsessov Vosstanov. Metal., Mater. Konf.* 1969. Ed. D. M. Chizhikov. Nauka. Moscow, 1972, p. 14 (*Chem. Abs.*, 1973. **79**. 23676).
365 L. S. Palatnik, O. A. Obol'yaninova, M. N. Naboka, and N. T. Gladkikh, *Izvest. Akad. Nauk. S.S.S.R.. neorg. Materialy*, 1973, **9**, 801.
366 L. E. Conroy and L. Ben-Dor, *Inorg. Syn.*. 1973. **14**. 149.
367 N. V. Vekshina and V. M. Lantratov. *Zhur. priklad Khim.*. 1973. **46**. 917.
368 P. Ratnasamy, *Indian J. Chem.*, 1973, **11**, 695.
369 R. V. Kasowski, *Phys. Rev. Letters*, 1973, **30**, 1175.
370 T. J. Wieting, *Solid State Comm.*, 1973, **12**, 931.
371 R. A. Isakova, L. E. Ugryumova, K. S. Amosova. and N. A. Potanina, *Izvest. Akad. Nauk Kaz. S.S.R.. Ser. khim.*, 1973, **23**, 6.
372 P. Ratnasamy, L. Rodrigue, and A. J. Leonard, *J. Phys. Chem.*, 1973, **77**, 2242.
373 J. E. Dutrizac, *J. Less-Common Metals*, 1973, **33**, 341.
374 H. P. Hughes and W. Y. Liang, *J. Phys. (C)*, 1974, **7**, 1023.

and A at 800—1150 °C for 24 h in evacuated silica tubes and their X-ray powder diffraction patterns determined.[375] A crystal-structure determination of $Ni_{0.33}Mo_3Se_4$ has shown that the compound consists of nickel atoms occupying sites in the channels of the Mo_3Se_4 structure.[376a] This structure is also adopted by a wide range of such $M_xMo_3Se_4$ (M = transition metal, $0 < x \leqslant 1$) compounds,[376b] and also by $NiMo_3Si_4$.[376c] Intercalation of the lanthanides Eu, Yb, and Sm into the MoS_2 lattice has been studied at low temperatures using liquid ammonia solutions of these metals. The intercalated species inserts between every layer of the structure and NH_3 is also fully intercalated to give compound phases, *e.g.* $MoS_2(Eu)_{0.9-1.0}$-$(NH_3)_{1.0-1.5}$.[377]

Nitrides. Nitride formation on the surface of molybdenum has been revealed by high-temperature etching with ammonia, and the nitrides were characterized by X-ray diffraction spectrometry.[378] The diffusion layer formed on molybdenum consists of Mo_2N at >950 °C and MoN at $\leqslant 950$ °C. The interaction of ammonia with a tungsten surface can generate dense adlayers such as W_2N_3H.[379]

Carbides. A review of the preparation and properties of transition-metal carbides has been published.[62] Phase equilibria in the Mo—W—C system have been examined using X-ray diffraction techniques. At 1000 °C Mo_2C and WC are the stable phases.[380] The lattice constants of the f.c.c. phases formed during the thermal decomposition of $W(CO)_6$ at 400—800 °C have been determined. The composition of the phase obtained is related to the decomposition temperature, $WC_{0.59}$ being formed at 400 °C and $WC_{0.55}$ at 800 °C.[381] Cubic α-MoC_{1-x} (x ca. 0.16) has been obtained by carbonizing graphite–MoO_3 mixtures in a plasma arc and then quenching.[382] Thermodynamic calculations and experimental data have been presented for the Mo–C system which show that monocarbides are stabilized at high pressures.[383]

ζ-$(MoCr)_4C_{3-x}$ has been shown by X-ray powder diffractometry to be isomorphous with ζ-Hf_4N_3.[64] The standard free energy of formation of UWC_2 has been determined by e.m.f. measurements for the temperature range 981—1094 K.[384] The crystal structure of $PuWC_{1.15}$ has been reported.[385]

Silicides. The M–Si (M = Mo or W) phase diagrams in argon plasmas have been investigated using X-ray and other techniques to characterize the quenched products.

[375] M. Sergent and J. Prigent. *Compt. rend.*, 1973, **277**, C, 465.
[376] (a) O. Bars, J. Guillevic, and D. Grandjean. *J. Solid State Chem.*, 1973, **6**, 335; (b) M. Sergent and R. Chevrel, *J. Solid State Chem.*, 1973, **6**, 433; (c) J. Guillevic, O. Bars, and D. Grandjean. *J. Solid State Chem.*, 1973, **7**, 158.
[377] G. V. S. Rao, M. W. Schafer, and L. Tao, *Amer. Inst. Phys. Conf. Proc.*, 1972, 1173 (*Chem. Abs.*, 1973, **79**, 25008).
[378] Yu. M. Lakhtin, Ya. D. Kogan, and G. V. Lebedeva, *Metalloved. Term. Obrab. Metal.*, 1973, 6 (*Chem. Abs.*, 1973, **79**, 56495).
[379] Y. K. Peng and P. T. Dawson, *Canad. J. Chem.*, 1974, **52**, 1147.
[380] L. V. Gorshkova, V. S. Telegus, F. I. Shamrai, and Yu. B. Kuz'ma, *Porosh. Met.*, 1973, **13**, 74 (*Chem. Abs.*, 1973, **79**, 8826).
[381] V. V. Toryanik, V. D. Krylov, and Ya. S. Umanskii, *Doklady Akad. Nauk S.S.S.R.*, 1973, **212**, 86.
[382] S. Matsumoto, M. Saito, and S. Ohtsuki, *Chem. Letters*, 1973, 649.
[383] A. Ya. Shinyaev, V. B. Fedorov, L. V. Gorshkova, and D. B. Chernov, *Doklady Akad. Nauk. S.S.S.R.*, 1973, **212**, 1336.
[384] H. Tanaka, Y. Kishida, and J. Moriyama, *Nippon Kinzoku Gakkaishi*, 1973, **37**, 564 (*Chem. Abs.*, 1973, **79**, 24132b).
[385] M. Ugajin and J. Abe. *J. Nuclear Mater.*, 1973, **47**, 117.

The only compounds identified in this way were $MoSi_2$ and M_5Si_3.[386] The heats of formation of these compounds and that of Mo_3Si have been determined at temperatures from 1200 to 2200 K[387] and the X-ray diffraction characteristics of Mo_3Si, Mo_5Si_3, and W_5Si_3 determined.[388] W_5Si_3 and WSi_2 have been prepared by mixing SiH_4 and WF_6 vapours at temperatures between 600 and 800 °C.[389] The band energies of Mo_3Si, Mo_3Si_{13}, and $MoSi_2$ have been investigated by recording their ultrasoft X-ray emission spectra.[390]

Borides and Gallides. A series of borides WB_n ($n = 4$—14) has been obtained by heating mixtures of powdered tungsten and crystalline boron. The best conditions for the production of WB_4 were found and the lattice constants of this compound determined.[391] Boron fibres have been shown to react with tungsten at 1100 °C to form W_2B, WB, W_2B_5, and/or WB_4, the nature of the product depending upon the reactant stoicheiometry.[392] The compounds YMB_4 (M = Mo or W) have been identified in the corresponding Y-M-B system and characterized by X-ray powder diffractometry.[393]

The crystal structure of Mo_6Ga_{31} has been solved. The most important structural features of this phase are the $MoGa_{10}$ polyhedra which are arranged at the corners of distorted cubes. There are no Mo—Mo contacts. Not all the lattice sites have a full occupancy, thus suggesting that the '$MoGa_5$'-phase has a range of homogeneity between the limits Mo_6Ga_{31} and Mo_7Ga_{30}.[394]

Compounds containing Mo—Mo, W—W, or Related Bonds. Some tentative spectral evidence has been obtained for the formation of Mo_2 dimers when isolated molybdenum atoms are allowed to diffuse through an Ar matrix.[273] The kinetics of the formation and dissociation of W_2 and WRe adatom species have been studied by field ion microscopy and their dissociation energies estimated as 30 ± 5 and 14 ± 2 kJ mol^{-1}, respectively, at *ca.* 400 K.[395]

The di-μ-diethylphosphido-complexes $[M(CO)_4(PEt_2)]_2$ (M = Mo or W) are isomorphous and involve Mo—Mo and W—W bond lengths of 306 and 305 pm, respectively.[396] $[M(CO)_4(SCF)_3)]_2$ (M = Mo or W) have been prepared by the photolysis of the corresponding hexacarbonyl in the presence of CF_3SSCF_3 and

[386] L. Angelin and G. Cevales, *Rev. Internat. Hautes. Temp. Refract.*, 1973, **10**, 103 (*Chem. Abs.*, 1973, **79**, 58212); Yu. A. Kocherzhinskii, O. G. Kulik, E. A. Shishkin, and L. M. Yupko, *Doklady Akad. Nauk S.S.S.R.*, 1973, **212**, 642.

[387] V. P. Bondarenko, E. N. Fomichev, and A. A. Kalasknik, *Heat. Transfer-Sov. Res.*, 1973, **5**, 76 (*Chem. Abs.*, 1973, **79**, 58375).

[388] N. S. Poltavtsev, A. P. Patokin, and E. N. Pryanchikov, *Zasch. Vysokotemp. Polrytiya. Tr. Vses. Soveshch. Zahrostorkim. Pokrytiyam*, 5th., 1970, 343 (*Chem. Abs.*, 1973, **79**, 35975).

[389] J. Lo, *Diss. Abs. (B)*, 1973, **34**, 1179.

[390] A. S. Shulakov, T. M. Zimkina, V. A. Fomichev, and V. S. Neshpor, *Fiz. Tverd. Tela*, 1974, **16**, 401 (*Chem. Abs.*, 1974, **80**, 138986).

[391] L. G. Bodrova, M. S. Koval'chenko, and T. I. Serebryakova, *Porosh. Met.*, 1974, 1 (*Chem. Abs.*, 1974, **80**, 124048).

[392] Yu. V. Dzyadykevich, *Izvest. Akad. Nauk S.S.S.R., neorg. Materialy*, 1974, **10**, 44.

[393] Yu. B. Kuz'ma, S. J. Svarichevskaya, and A. S. Sobolev, *Izvest. Akad. Nauk S.S.S.R., neorg. Materialy*, 1973, **9**, 1697.

[394] K. Yvon, *Acta Cryst.*, 1974, **B30**, 853.

[395] D. W. Bassett and D. R. Tice, *Surface Sci.*, 1973, **40**, 499 (*Chem. Abs.*, 1974, **80**, 7531).

[396] M. H. Linck and L. R. Nassimbeni, *Inorg. Nuclear Chem. Letters*, 1973, **9**, 1105; M. H. Linck, *Cryst. Struct. Comm.*, 1973, **2**, 379.

these compounds appear to involve a metal–metal bond.[397] An X-ray crystallographic study of di-μ-bromobis(η⁴-tetraphenylcyclobutadiene)tetracarbonyldimolybdenum(I), $[(PhC)_4Mo(CO)_2Br]_2$, has been completed and the metal–metal separation determined as 295.4 pm.[398] The crystal structure of $[(\eta^5\text{-}C_5H_5)W(CO)_3]_2$ has been determined and that of the molybdenum analogue redetermined with an increased accuracy. The W—W distance (322.2 pm) is slightly shorter than the Mo—Mo distance (323.5 pm) in these compounds, so that the metal–metal separations in compounds of this type varies as Cr ≫ Mo > W. The fluxional behaviour of these molecules has been followed by variable-temperature ^1H n.m.r. spectroscopy and discussed with reference to the crystallographic data.[399] The full report of the study of intramolecular ligand-scrambling processes in the related molybdenum-methyl cyanide derivative, $[(\eta^5\text{-}C_5H_5)(OC)_3MoMo(CO)_2(CNMe)(\eta^5\text{-}C_5H_5)]$ has been published. This compound forms crystals exclusively of the *trans*-derivative with Mo—Mo 323.0 pm.[400] The metal–metal bond in $[(\eta^5\text{-}C_5H_5)Mo(CO)_3]_2$ can be cleaved by Yb in refluxing THF. Derivatives such as $[(\eta^5\text{-}C_5H_5)Mo(CO)_3X]$ (X = H or I) and $[(\eta^5\text{-}C_5H_5)Mo(CO)_2NO]$ are obtained rapidly and in good yields when this solution is treated with H_2O. I_2. or 'Diazalid', respectively.[401]

$[W_2(OAc)_4]$ has been obtained by treating $W(CO)_6$ with HOAc (1:2) in boiling benzene under CO_2 and with u.v. irradiation. The compound has been characterized by i.r., u.v., and mass spectral and magnetic measurements and it appears to have a polymeric structure which involves W—W bonds and bidentate acetato-groups.[402] The preparation and properties of $[CrMo(OAc)_4]$ are described on p. 90.[8] The syntheses of some new tetra-μ-carboxylatodimolybdenum(II) complexes have been described and a wide range of air stabilities was observed. Those compounds which contain structural features to impede binding co-axial to the metal–metal bonding are the most stable in this respect.[403a] Raman spectra have been presented for a large number of complexes which contain metal–metal bonds, including dimolybdenum(II) carboxylato- and halogeno-species, and the utility of Raman spectroscopy in structural studies of such compounds has been discussed.[403b]

$Li_4[Mo_2Me_8]$,4ether compounds have been prepared from LiMe and $[Mo_2(OAc)_4]$ in the appropriate ether. They are thermally stable at 25 °C but extremely pyrophoric. The crystal structure of the THF adduct has been determined and the anions shown to have virtual D_{4h} symmetry with Mo—Mo = 214.8 pm; none of the THF molecules is co-ordinated.[404]

The resonance Raman spectrum of $K_4[Mo_2Cl_8]$ has been reinvestigated using 488.0 and 514.5 nm excitation. An enormous enhancement of the intensity of the Mo—Mo stretching mode relative to the intensity of other fundamentals was observed and an overtone progression in v_1 to $5v_1$ identified. From these data the harmonic frequency ω_1, and anharmonicity constant X_{11} were calculated as 347.1 $\pm$ 0.5 cm^{-1}

[397] J. L. Davidson and D. W. A. Sharp, *J.C.S. Dalton*, 1973, 1957.

[398] M. Mathew and G. J. Palenik, *J. Organometallic Chem.*, 1973, **61**, 301.

[399] R. D. Adams, D. M. Collins, and F. A. Cotton, *Inorg. Chem.*, 1974, **13**, 1086.

[400] R. D. Adams, M. Brice, and F. A. Cotton, *J. Amer. Chem. Soc.*, 1973, **95**, 6594.

[401] A. E. Crease and P. Legzdins, *J.C.S. Chem. Comm.*, 1973, 775.

[402] G. Holsk, *Z. anorg. Chem.*, 1973, **398**, 249.

[403] (a) E. Hochberg, P. Walks, and E. H. Abbott, *Inorg. Chem.*, 1974, **13**, 1824; (b) J. San Filippo, jun., and H. J. Snaidoch, *ibid.*, 1973, **12**, 2326.

[404] F. A. Cotton, J. M. Troup, T. R. Webb, D. H. Williamson, and G. Wilkinson, *J. Amer. Chem. Soc.*, 1974, **96**, 3824.

and 0.50 ± 0.08 cm^{-1}, respectively. This study was extended to the purple solution formed when $K_4[Mo_2Cl_8]$ dissolves in 1M–HCl. The preservation of the Mo_2^{4+} entity was confirmed by the observation of a Raman progression, similar to that identified for $K_4[Mo_2Cl_8]$, with 514.5 nm radiation; $\omega_1 = 358.7 \pm 0.7$ cm^{-1} and $X_{11} = 1.5 \pm 0.2$ cm^{-1}. These results confirm that the lowest lying electronic transition of the Mo_2^{4+} unit is localized within the dimetal centre.[405] This conclusion is supported by low-temperature polarized electronic spectral data obtained for the isoelectronic Re_2^{6+} systems,[406] and the results of an SCF MO calculation performed[407] for $[Mo_2Cl_8]^{4-}$.

$K_4[Mo_2Cl_8]$ reacts with en at 100 °C to form the orange-brown compound $[Mo_2(en)_4]Cl_4$. Cyclic voltametric studies of this complex have identified an oxidation wave at 0.78 V *vs.* SCE, but the process appears to be irreversible. This suggests that a major structural reorganization follows rapidly upon the removal of an electron, possibly producing mononuclear Mo^{III} species. The electronic spectrum of $[Mo_2(en)_4]^{4+}$ has been recorded $[\lambda_{max} = 478\ (483),\ 360\ (36),\ \text{and}\ 255\ (966)\ \text{nm}]$ and that of Mo_2^{4+}aq further investigated $[\lambda_{max}\ 504\ (337)\ \text{and}\ 370\ (40)\ \text{nm}]$. $[Mo_2(bipy)_2Cl_4]$ was prepared by treating $K_4[Mo_2Cl_8]$ with bipy in MeOH solution[408] and the complexes $[Mo_2Cl_4L_4]$ (L = py, L_2 = bipy or phen) have been prepared by the reaction of $[Mo_2Cl_8]^{4-}$ with the appropriate ligand, and characterized by magnetic measurements and i.r. spectra.[409a] Other preparative routes to these py and bipy complexes have been described in a study which characterized a large number of other such tetrahalogeno-dimolybdenum(II) complexes, including new $[Mo_2X_4L_4]$ (X = Cl or Br; L = R_2S, RCN, or R_3P) derivatives.[409b] Mo_2-$(diphos)_3Cl_4,2C_6H_6$ has been obtained by reducing a THF solution of $[Mo_2(diphos)-Cl_6],2C_6H_6$ with Zn in the presence of diphos.[309]

Improved synthetic routes to the $[Mo_2X_9]^{3-}$ (X = Cl or Br) anions have been described and the preparation of $[Mo_2Cl_9]^{2-}$ reported for the first time. $[MoCl_6]^{2-}$ reacts with $[Mo(CO)_4Cl_3]^-$ in CH_2Cl_2 over 24 h with the complete evolution of CO to produce the dark green-brown ion $[Mo_2Cl_9]^{2-}$ in quantitative yield. This ion appears to have a structure analogous to that of $[W_2Cl_9]^{2-}$, it reacts with most donor solvents, and is rapidly and quantitatively reduced to $[Mo_2Cl_9]^{3-}$ by Sn–NH_4Cl in CH_2Cl_2.[410] Metal–metal interactions within the ions $[M_2Cl_9]^{3-}$ (M = Cr, Mo, or W) have been assessed using extended Hückel MO calculations[41] and, for $[Mo_2X_9]^{3-}$ (X = Cl or Br) salts, by magnetic susceptibility measurements between 80 and 295 K.[411] $[Mo_2(MeCN)_3Cl_6]$ has been prepared by refluxing $MoCl_5$ with MeCN under N_2 and the compound is suggested to contain three chloride atoms bridging the metal centres.[412] Preliminary evidence suggests that the neutral tris(*o*-tetrachlorobenzoquinone) complexes of molybdenum and tungsten, prepared by

405 R. J. H. Clarke and M. L. Franks, *J.C.S. Chem. Comm.*, 1974, 316.
406 C. D. Cowman and H. B. Gray, *J. Amer. Chem. Soc.*, 1973, **95**, 8177.
407 J. G. Norman. jun. and H. J. Kolari, *J.C.S. Chem. Comm.*, 1974, 303.
408 A. R. Bowen and H. Taube, *Inorg. Chem.*, 1974, **13**, 2245.
409 (*a*) J. V. Brencic, D. Dobcnik, and P. Segedin, *Monatsh.*, 1974, **105**, 142; (*b*) J. San Filippo, jun. H. J. Snaidoch, and R. L. Grayson, *Inorg. Chem.*, 1974, **13**, 2121.
410 (*a*) W. H. Delphin and R. A. D. Wentworth, *J. Amer. Chem. Soc.*, 1973, **95**, 7920; (*b*) W. H. Delphin and R. A. D. Wentworth, *Inorg. Chem.*, 1974, **13**, 2037; (*c*) W. H. Delphin, *Diss. Abs.* (B), 1974, **34**, 4268.
411 V. V. Zelentsov, H. C. Nguyen, A. T. Fal'kengof, N. A. Subbotina, and V. I. Spitsyn, *Zhur. neorg. Khim.*, 1973, **18**, 2790.
412 C. Miniscloux, G. Martino, and L. Sajus, *Bull. Soc. chim. France*, 1973, 2179.

refluxing the corresponding hexacarbonyl with *o*-tetrachlorobenzoquinone, contain direct metal–metal interactions.[124] The first molybdenum(III) dialkylamido- and dialkoxido-complexes have been prepared. The addition of $MoCl_3$ to an ice-cooled, stirred THF–hexane solution of $LiNMe_2$ (1:3) gives a dark brown solution from which $[Mo_2(NMe_2)_6]$ may be obtained. This compound is quantitatively converted into $[Mo_2(OBu^t)_6]$ by reaction with Bu^tOH.[283a] An *X*-ray crystallographic study of the former compound has shown that it consists of molecules which possess almost D_{3d} symmetry and which contain a triple bond of length 221.4 pm between the metal atoms. The Mo—N separations are 198 pm, but it is not clear what role Mo—N π-bonding plays in this compound.[283b] The auto-oxidations of molybdenum(III) and tungsten(III) alkyls, M_2R_6, have been shown to proceed through a series of fast, free radical displacements at the metal centre.[413]

The reduction of higher molybdenum or tungsten halides by aluminium in the appropriate sodium tetrahalogenoaluminate melt has been developed as a convenient and safe synthesis of the corresponding molybdenum(II) or tungsten(II) chloride or bromide in good yield.[414] *X*-Ray photoelectron spectroscopy has been used to distinguish between the bridging and terminal chlorine atoms in a variety of molybdenum(II) chloride adducts. The data for β-$MoCl_2$ suggest that it does not have a structure related to that of $[Mo_6Cl_8]Cl_4$.[415] The results of new extended Hückel MO calculations of the $[M_6X_8]^{4+}$ (M = Mo, X = Cl or Br; M = W, X = Cl) ions have been published.[416] The redox characteristics of $[Mo_6Cl_8]Cl_4$, including its reactions with halogens, have been studied[417] and oxidation of $[M_6Cl_8]Cl_4$ (M = Mo or W) with octachlorocyclopentane has been described. This latter reagent provides an alternative route to the preparation of $[Mo_6Cl_{12}]Cl_3$ and $[W_6Cl_{12}]Cl_6$ from their corresponding metal(II) chlorides.[418]

An *X*-ray crystallographic study of $Pb_{0.92}Mo_6S_{7.5}$, a non-stoicheiometric deviant of $PbMo_6S_8$, has shown that the compound contains octahedral Mo_6 clusters.[419]

Several compounds of Mo^V which contain μ-oxo- or μ-sulphido-groups and also involve significant metal–metal interactions are discussed on p.135.

Molybdenum(II) and Tungsten(II).—Apart from the molybdenum(II) and tungsten(II) species described above, studies of this oxidation state have largely been confined to carbonyl complexes.

$[Mo(diars)_2(CO)_2Cl]I_3,2CHCl_3$, $[W(CO)_4(diars)I]I_3$, and $[W(CO)_3(PMe_2Ph)_3I]$ BPh_4 all contain cations in which the metal ion is seven co-ordinate. The first two complexes involve a capped trigonal-prismatic environment for the cation, with the halogen atom in the capping position, whilst the third contains a cation whose geometry is intermediate between the capped trigonal prism and the capped octahedron, the halogen atom again occupying the unique position.[420] A crystal-structure determination of $[(\eta^5\text{-}C_5H_5)Mo(CO)_2(N\equiv CBu^t)_2]$ has shown that the (di-t-butyl-

[413] P. B. Brindley and J. C. Hodgson, *J. Organometallic Chem.*, 1974, **65**, 57.

[414] W. C. Dorman and R. E. McCarley, *Inorg. Chem.*, 1974, **13**, 491.

[415] A. D. Hamer and R. A. Walton, *Inorg. Chem.*, 1974, **13**, 1446.

[416] D. V. Korol'kov and V. N. Pak, *Zhur. strukt. Khim.*, 1973, **14**, 109.

[417] H. Schaefer, H. Plautz, and H. Baumann, *Z. anorg. Chem.*, 1973, **401**, 63.

[418] D. L. Kepert, R. E. Marshall, and D. Taylor, *J.C.S. Dalton*, 1974, 506.

[419] M. Marezio, P. D. Dernier, J. P. Remeika, E. Corenzurit, and B. T. Mathias, *Mater. Res. Bull.*, 1973, **8**, 657.

[420] M. G. B. Drew and J. D. Wilkins, *J.C.S. Dalton*, 1973, 2664; *J. Organometallic Chem.*, 1974, **69**, 271; *J.C.S. Dalton*, 1974, 1654.

methylene)amino-group is attached to the metal atom by a short bond of length 189.2 pm with $\angle$ MoNC of 172°. These data clearly suggest the existence of substantial Mo—N multiple bonding.[421] The reactions of the halogenocarbonyl complexes $[M(CO)_4X_2]$ (M = Mo or W; X = Cl or Br) with neat py, 2,6-lutidine, and PhCN have been shown to afford products which contain no carbonyl groups. However, in such cases the molybdenum compounds produced undergo facile disproportionations to give compounds of Mo^0 and Mo^{III}. Although similar disproportionations occur in the case of tungsten, they proceed less readily and the complexes $WX_2(NCPh)_2$ have been prepared and characterized. $Mo(phen)_2I_2$ has been prepared by treating MoI_3 with phen (1:3) at 180°C in PhCN.[422]

Molybdenum(III) and Tungsten(III).—The phase diagram of the $KCl-MoCl_3$ system has been investigated and compounds K_3MoCl_6, m.p. *ca.* 870°C, $K_3Mo_2Cl_9$, m.p. 745°C, and $KMoCl_4$ have been characterized by X-ray powder diffractometry.[423] D.t.a. studies of the system $NaCl-MoCl_3-TiCl_3$ have been completed but no ternary compounds were identified in this system.[424] E.s.r. evidence has been presented for Mo^{III} centres in Cs_2MCl_6 (M = Zr or Hf) lattices.[425]

$NaMoO_2$ has been identified as a product of the reduction of MoO_2 by sodium vapour.[426] Further studies have now clarified the confusion concerning the electronic spectrum of $[Mo(H_2O)_6]^{3+}$ and have improved the preparative route to this ion. The optimum conditions for the aquation of $[MoCl_6]^{3-}$ are an 0.0025M solution of the complex in 0.5M toluene-*p*-sulphonic acid (HPTS) over two days at room temperature, the products then being separated by cation exchange using $\leqslant 1$M-HPTS as the eluant. The electronic spectrum of $[Mo(H_2O)_6]^{3+}$ contains two *d–d* transitions at 308(28) and 370(16) nm.[408] Kinetic data have been obtained for the substitution reactions of this ion with Cl^- and NCS^- and these results shown to be consistent with an S_N2 mechanism.[427] The green molybdenum(III) species produced on the reduction of acidic solutions of molybdenum(VI) have been investigated. In aqueous HCl or H_2SO_4 a mixture of molybdenum(III) species are produced, but in aqueous HPTS only a single such species is formed which is blue-green $[\lambda_{max} = 360$ (306), 572 (39), and 624 (43) nm]. Cation exchange studies and the diamagnetism of the species have led to the structure (10) being suggested (p. 112). The presence of anions such as Cl^- or SO_4^{2-} results in some substitution of the water molecules in the $[(H_2O)_4MoO_2Mo(H_2O)_4]^{2+}$ units, thus explaining the mixtures obtained in aqueous solutions of the corresponding acids.[284]

X-Ray crystallographic data have shown that MMo_2S_4 (M = Co or Fe) are isostructural and contain MoS_6 octahedra.[428] $[Mo(S_2C_6H_4)_3]$ molecules, however, possess the expected trigonal-primatic MoS_6 unit and a comparison of the dimensions

[421] H. M. M. Shearer and J. D. Sowerby. *J.C.S. Dalton.* 1973, 2629.

[422] A. D. Westland and N. Muriithi, *Inorg. Chem.*, 1973, **12**, 2356.

[423] D. V. Drobot and E. A. Sapranova, *Zhur. neorg. Khim.*, 1973, **18**, 2008; E. N. Ryabov, R. A. Sandler. I. I. Kozhina, E. F. Klyuchnikova, and L. P. Dorfman, *Izvest. V.U.Z. Tsvet. Met.*, 1973, **16**, 125 (*Chem. Abs.*, 1973, **79**, 97 580).

[424] N. C. Nguyen, R. A. Sandler, I. V. Vasil'kova, and E. N. Ryabov, *Zhur. neorg. Khim.*, 1973, **18**, 1139.

[425] S. Maniv, W. Low, and A. Gabay, *J. Phys. Chem. Solids*, 1974, **35**, 373.

[426] H. Kessler, A. Hatterer, and C. Ringenbach, 'Proceedings of the International Conference on Liquid Alkali Metals', 1973, p. 21, ed. M. Monro, Brit. Nucl. Energy Soc. London; *Compt. rend.*, 1973, **277**, C, 763.

[427] Y. Sasaki and A. G. Sykes, *J.C.S. Chem. Comm.*, 1973, 767.

[428] J. Guillevic, J. Y. Le Marouille, and D. Grandjean, *Acta Cryst.*, 1974. **B30**, 111.

of these molecules with the isoelectronic anions of zirconium and niobium has been presented.[429]

Some developments of the ammine complex chemistry of molybdenum(III) have been reported. When K_3MoCl_6 is maintained at 25 °C for three weeks under liquid ammonia containing NH_4Cl, $[Mo(NH_3)_3Cl_3]$ is formed. However, this complex does not appear to be a good starting material for the preparation of other molybdenum(III) ammino-complexes. A better such material appears to be $[Mo(NH_3)_4-(H_2O)(HCO_2)]^{2+}$, which may be prepared by the addition of $K_3[Mo_2(OH)_3(CO)_6]$ to concentrated aqueous ammonia, followed by oxidation with silver tosylate, acidification with HSO_3CF_3, and purification by ion-exchange.[408] The crystal structure of *mer*-$[Mo(py)_3Cl_3]$ has been reported and the cell dimensions of the corresponding bromo-derivative were determined.[172] Various molybdenum(III) and tungsten(III) halogeno-complexes with unsaturated N-donor ligands have been synthesized and characterized. These include $MoCl_3(NCPh)_4,2CH_2Cl_2$, $MoBr_3(2,6-$lutidine)$_2$, $WX_3(py)_3$ (X = Cl or Br), $[MoL_2X_2]X$ (L = phen or bipy; X = Br or I), $Mo_2X_3(bipy)_3$ (X = Cl or Br), and the mixed oxidation state compounds $Mo_3Br_8-(NCPh)_6$ and $W_3I_8(NCMe)_6$.[422] The complex $[\{Mo(H_2O)_2L\}_2O]$, containing the terdentate Schiff base *N*-(salicylidene)-2-hydroxybenzamine (H_2L), has been prepared and characterized by i.r. and electronic spectroscopy.[430]

Molybdenum(IV) and Tungsten(IV).—A wide variety of molybdenum(IV) complexes dissolved in non-aqueous media have been shown to catalyse the reduction of acetylene by sodium borohydride. The optimum yield was 2.8 moles of C_2H_2 reduced per g. atom of Mo^{IV}, with both C_2H_4 and C_2H_6 being formed.[265]

The reactions of $[WF_6]^-$ with binary fluorides may be explained by the disproportionation of this ion to WF_6 and WF_4 followed, in some instances, by the complexing of the WF_4 with the fluoroanion.[431a] WCl_6 or WCl_5 react with an excess of a dialkyl sulphide, R_2S (R = Me or Et), *via* an S-dealkylation sequence to produce $(R_3S)_2[WCl_6]$.[431b] Synthetic procedures suitable for the preparation of WOX_2 (X = Cl or Br) have been described[432] and thermochemical data have been obtained for WSF_2.[352]

A stable aquo-ion of molybdenum(IV) has been prepared and characterized. When an equimolar mixture of 0.01M-Mo^{III} and Mo^V in 1M toluene-*p*-sulphonic acid is heated at 90 °C in an inert atmosphere for 1 h, the solution turns red. The species responsible for this colour has been purified and characterized using ion-exchange techniques and is suggested to have the structure (11) (p.112). This $[(H_2O)_4MoO_2Mo-(H_2O)_4]^{4+}$ ion is remarkably stable towards disproportionation and air oxidation.[285] The formation of complexes between molybdenum(IV) and tartaric acid (H_2L) in aqueous solution has been investigated by potentiometry, spectrophotometry, and polarography. In the presence of an excess of the acid a series of complexes is formed, most of which are dimeric with two L groups per molybdenum(IV). With H_2L:Mo *ca.* 1:1, another series of complexes exists which contains the $MoO(OH)^+$ moiety.[433]

[429] M. J. Bennett, M. Cowie, J. L. Martin, and J. Takats, *J. Amer. Chem. Soc.*, 1973, **95**, 7504.
[430] K. H. Chjo, *Daehan Hawak Hwoejee*, 1973, **17**, 169 (*Chem. Abs.*, 1973, **79**, 38 088).
[431] (*a*) S. Brownstein, *J. Inorg. Nuclear Chem.*, 1973, **35**, 3575; (*b*) P. M. Boorman, T. Chivers, and K. N. Mahadev, *J.C.S. Chem. Comm.*, 1974, 502.
[432] J. Tillack, *Inorg. Syn.*, 1973, **14**, 109.
[433] B. Viossat, *Rev. Chim. Minerale*, 1972, **9**, 737 (*Chem. Abs.*, 1974, **80**, 90 563).

[Mo(S$_2$CNEt$_2$)$_4$] has been synthesized by refluxing Mo(CO)$_6$ with tetraethyl-thiuram disulphide (1:2) in acetone under N$_2$. Voltametric measurements on this compound in CH$_2$Cl$_2$ have shown evidence for one-electron reduction and one- and two-electron oxidation. The corresponding tungsten(IV) derivative appears to undergo readily only the oxidation processes.[434] W(CO)$_6$ reacts with thiodibenzoyl-methane (HL) (1:2) in refluxing py to produce the corresponding WL$_4$ derivative.[435]

The reaction of [ReNCl$_2$(PMe$_2$Ph)$_3$] with [MoCl$_4$(MeCN)$_2$] in CH$_2$Cl$_2$ produces [(Me$_2$PhP)$_3$Cl$_2$ReNMoCl$_4$(MeCN)] with a μ-nitrido-group.[270] Various halogeno-complexes of molybdenum(IV) and tungsten(IV) with unsaturated N-donor ligands have been prepared and characterized. These include Mo(py)$_3$Cl$_4$, Mo(bipy)Cl$_4$,[412] and W(py)$_2$X$_4$ (X = Cl, Br, or I).[422] Electrochemical reductions of K$_2$[WCl$_6$], WCl$_4$(bipy), and WCl$_4$(py) in MeCN solution have been carried out to demonstrate the stabilization of low oxidation states by these π-acceptor, N-donor ligands.[436]

The crystal structure of *trans*-dichlorobis(*N*-methylsalicylaldiminato)molyb-denum(IV) has been determined and the compound shown to consist of discrete monomeric units which have a crystallographic centre of symmetry. The bonds involving the metal atom, Mo—O, Mo—N, and Mo—Cl, are 195, 214, and 239 pm, respectively.[437] The complex [Mo(SCN)$_2$(H$_2$O)L] [H$_2$L = *N*-(salicylidene)-2-hydroxybenzamine] has been prepared and its i.r. and electronic spectra recorded.[430]

The complexes MoL$_2$Cl$_4$ (L = Ph$_3$P or tricyclohexylphosphine) have been prepared by allowing the ligand L to react with an excess of Mo$_2$(MeCN)$_3$Cl$_6$ in a non-aqueous medium.[412] Metal–phosphorus coupling constants have been deter-mined for the complexes *cis-mer*-[MOCl$_2$(PMe$_2$Ph)$_3$] (M = Mo or W) and *cis-mer*-[WOCl$_2$(PMePh$_2$)$_3$] and these, and similar data for other metals, discussed with reference to the corresponding metal–phosphorus separations.[438]

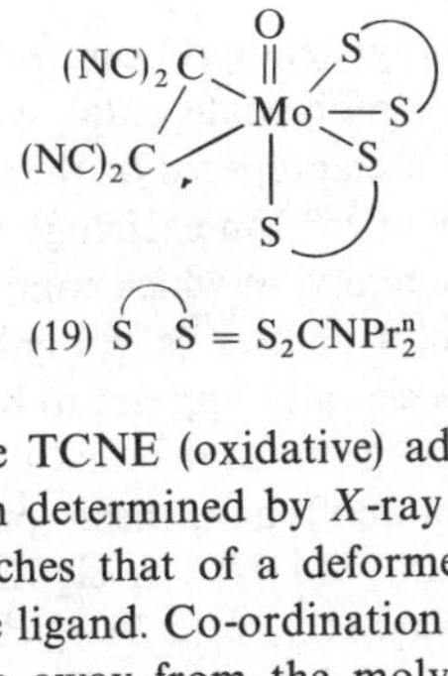

(19) S⌒S = S$_2$CNPr$^n_2$

The stereochemistry of the TCNE (oxidative) addition compound formed with [MoO(S$_2$CNPr$^n_2$)$_2$] has been determined by *X*-ray crystallography. The geometry of this complex (19) approaches that of a deformed pentagonal bipyramid, con-sidering TCNE as a bidentate ligand. Co-ordination of the olefin has resulted in the displacement of cyano-group away from the molybdenum atom; however, these groups retain their eclipsed configuration.[439] The crystal and molecular structure of [(η^5-C$_5$H$_5$)$_2$W(S$_2$C$_6$H$_4$)] has been determined. The co-ordination polyhedron about the metal atom, as defined by the C$_5$H$_5$ ring centroids and the sulphur atoms,

[434] Z. B. Varàdi and A. Nieuwport. *Inorg. Nuclear Chem. Letters.* 1974, **10**, 801.
[435] E. Uhlemann and U. Eckelmann. *Z. Chem.*, 1974, **14**, 66.
[436] H. F. Hagedorn, R. T. Iwamoto, and J. Kleinberg. *J. Electroanalyt. Chem., Interfacial Electrochem.*, 1973, **46**, 307 (*Chem. Abs.*, 1973, **79**, 142 281).
[437] J. E. Davies and B. M. Gatehouse, *J.C.S. Dalton*, 1974, 184.
[438] G. G. Mather, A. Pidcock, and G. J. N. Rapsey, *J.C.S. Dalton*, 1973, 2095.
[439] L. Ricard and R. Weiss, *Inorg. Nuclear Chem. Letters*, 1974, **10**, 217.

consists of a distorted tetrahedron of approximately C_{2v} symmetry.[440] The full account of the oxidation of amines by 'electron-rich' molybdenum and tungsten bis-$(\eta^5\text{-}C_5H_5)$ complexes, such as $[(\eta^5\text{-}C_5H_5)_2Mo(SMe_2)Br]^+$, has been published.[441] The preparation and characterization of $[(\eta^5\text{-}C_5H_5)_2W\{CH_2C_6H_nMe_{5-n}\}_2]$ ($n = 4$ or 3) are described on (p.120)[282] and the reactions of $[(\eta^5\text{-}C_5H_5)_2MH_2]$ (M = Mo or W) and related complexes on p.121.

Molybdenum(v) and Tungsten(v).—*Mononuclear Complexes.* MoF_3Br_2 has been prepared by the oxidation of MoF_3 by Br_2 at *ca.* 100 °C, the value of μ_{eff} was found to be 1.54 BM and the nature of this compound's thermal decomposition has been characterized.[442] The enthalpies of formation of $M^1M^2F_6$ (M^1 = K, Rb, or Cs; M^2 = Mo or W) have been determined by measuring their respective heats of alkaline hypochlorite hydrolysis at 298 K. The ΔH_f° values range from -2075 to -2224 kJ mol^{-1} and these data have allowed the fluorine affinities of MoF_5 and WF_5 to be estimated as -412 ± 6 and -491 ± 5 kJ mol^{-1}, respectively.[354] The characteristics of Mo^V centres produced in Cs_2MCl_6 (M = Zr or Hf) lattices have been reported.[425]

Synthetic procedures suitable for the preparation of WOX_3 (X = Cl or Br) by chemical transport techniques have been described.[432] A large number of salts containing the $[MoOF_5]^{2-}$ anion have been prepared and their magnetic moments obtained,[443] and a theoretical model has been presented for the interpretation of such data.[444] The e.s.r. spectra of the $[MOX_5]^{2-}$ (M = Mo or W; X = F, Cl, or Br) ions incorporated in a single crystal of a suitable host lattice have been determined and interpreted using an MO model.[227] The compounds $(Et_4N)_2[MOCl_4Br]$ (M = Mo or W) have been obtained by treating the corresponding $MOCl_4$ derivative with Et_4NBr. Spectroscopic data have been obtained for these ions which suggest that they possess C_{4v} symmetry.[445]

The e.s.r. characteristics of molybdenum(v) centres in several MO_2 (M = Ti, Sn, or Ge) lattices[446] and a variety of 'molybdate' catalysts[290,291] have been determined. Na_3MO_4 (M = Mo or W) have been prepared by the reduction of the corresponding Na_2MO_4 salt with sodium vapour.[426] Ion-exchange studies have shown that in all dilute acid solutions of molybdenum(v) in which complex formation does not occur. the predominant molybdenum(v) species is probably $[(H_2O)_3Mo(O)O_2Mo(O)(H_2O)_3]^{2+}$ (12) (p.112). This species also appears to be the major product of the air oxidation of solutions containing molybdenum-(II) or -(III).[286] Stable aqueous solutions containing molybdenum(v) have been prepared by the reduction of $[MoO_4]^{2-}$ with metallic Mo in 6—8M–HCl or H_2SO_4, and by the cathodic reduction of a solution containing Na_2MoO_4 at a Pt electrode.[447] The molybdenum(v) species formed in aqueous H_2SO_4 solution have been investigated and various ions

[440] T. Debaerdemaecker and A. Kutoglu, *Acta Cryst.*, 1973, **B29**, 2664.

[441] F. W. S. Benfield and M. L. H. Green, *J.C.S. Dalton*, 1974, 1244.

[442] K. A. Khaldoyanidi and A. A. Opalovskii, *Izvest. Sib. Otd. Akad. Nauk. S.S.S.R., Ser. khim. Nauk.*, 1973. 142 (*Chem. Abs.*, 1973, **79**, 12 985).

[443] M. C. Chakravorti and S. C. Pandit, *Indian J. Chem.*, 1973, **11**, 601; *J. Indian Chem. Soc.*, 1973, **50**, 618.

[444] I. De, V. P. Desai, and A. S. Chakravarty, *Phys. Rev.* (B), 1973, **8**, 3769.

[445] J. P. Brunette and M. J. F. Leroy, *J. Inorg. Nuclear Chem.*, 1974, **36**, 289.

[446] P. Meriaudeau, F. Lecomte, and P. Vergnon, *Compt. rend.*, 1973, **277**, C, 1061.

[447] A. A. Rozbianskaya, *Metody. Khim. Anal. Gorn. Porod. Miner*, 1973, 67, ed. I. D. Borneman-Starynkevich, 'Nauka'. Moscow. U.S.S.R.. (*Chem. Abs.*. 1974, **80**, 77 980).

identified, including $[MoO(H_2O)_{3-n}(SO_4)_n]^{(3-2n)+}$ ($n = 1, 2,$ or 3) which are formed at low pH in this medium.[448] A study of the Overhauser effect in ethylene glycol complexes of molybdenum(v) has been completed and the results obtained discussed in terms of the electron-density distributions.[228] A 2:1 complex of molybdenum(v) with poly(vinyl alcohol) has been prepared by treating $MoCl_5$ with this ligand; the electronic spectrum was recorded.[449] The reactions between $MoCl_5$ and Ph_2SO or 5-oxothianthrene have been investigated and 1:1 complexes apparently identified in each case.[450] The extraction of molybdenum(v) from acidic aqueous solutions using cyclohexanone solutions of Bu_3PO_4 has been investigated and the nature of the extracted complexes characterized.[451] E.s.r. spectroscopy has been used to monitor the formation of complexes between molybdenum(v) and diphenylphosphinic acid (HL) in organic solvents and the complexes $[MoOCl_nL_{3-n}]$ ($n = 0,$ 1, or 2) and $[MoOCl_3L]$ thus characterized. This system involves rapid solution equilibria which shift towards $[MoOCl_4]^-$ on addition of HCl.[452]

Thermochemical data for WSF_3 have been published.[352] $[MoOCl_3(SPPh_3)]$ has been prepared by treating $MoOCl_3$ with Ph_3PS (1:1—5) in CH_2Cl_2 and X-ray crystallographic studies have confirmed that the compound consists of discrete five-co-ordinate molecules which have an essentially square-pyramidal geometry. These results suggest that the compounds $MoOCl_3(SR_2)$ (R = Me, Et, or Pr^n), previously considered to be six-co-ordinate *via* halogeno-bridges, may have five-co-ordinate geometries related to that of $[MoOCl_3(SPPh_3)]$.[453] The e.s.r. characteristics of the complexes $[MOX_2\{S_2P(OEt)_2\}]$ (M = Mo or W; X = Cl or NCS),[233] $MoOCl_3L$, and $MoOCl_2L$ [HL = $Ph_2P(S)(OH)$][452b] have been reported. Similarly, e.s.r. spectral studies of the interactions between molybdenum(v) and *OO*-dialkyl-dithiophosphates[454] and diethyldithiocarbamate[455] ions have suggested the formation of particular complexes including $[MoOX\{S_2P(OEt)_2\}]$ (X = F, Cl, or Br), $[MoOX(S_2CNEt_2)_2]$ (X = Cl or Br), and $[MoO(S_2CNEt_2)_2]$. $[W(S_2CNEt_2)_4]Br$ has been shown by X-ray crystallography to involve an eight-co-ordinate trigonal-dodecahedral WS_8 moiety.[456] Related Mo^v complexes, $[Mo(S_2CNMe_2)_4]X$ (X = Br, Br_3, I, I_3, or I_7), have been prepared by oxidation of $[Mo(S_2CNMe_2)_4]$ with the appropriate halogen and characterized by magnetic, e.s.r., and i.r. spectral measurements.[457]

The i.r. spectra of the molybdenum(v) complexes formed by treating $MoCl_5$ or $MoCl_6$ with CCl_3CN indicate that a triple Mo—N bond is formed in these reactions.

[448] I. P. Kharlamov and Z. P. Korobova, *Ref. Zhur. Khim.*, 1973, 14V41 (*Chem. Abs.*, 1974, **80**, 64 396).

[449] I. Ya. Kalontarov, G. I. Konovalova, and K. M. Makhkamov, *Doklady Akad. Nauk Tadzh., S.S.R.*, 1973, **16**, 40 (*Chem. Abs.*, 1973, **79**, 79 428).

[450] N. Gunduz, *Commun. Fac. Sci. Univ. Ankara, Ser. B.*, 1971, **18**, 43 (*Chem. Abs.*, 1973, **79**, 26 639).

[451] A. M. Golub, B. A. Sternik, and N. V. Ul'ko, *Zhur. neorg. Khim.*, 1973, **18**, 2778.

[452] (a) G. M. Larin, A. A. Kuznetsova, L. F. Yankina, and Yu. A. Buslaev, *Zhur. neorg. Khim.*, 1973, **18**, 1819; (b) I. N. Marov, V. K. Belyaev, I. A. Zakharova, M. K. Grachev, A. A. Kuznetsova, and Yu. A. Buslaev, *Zhur. neorg. Khim.*, 1974, **19**, 413.

[453] P. M. Boorman, C. D. Garner, T. J. King, and F. E. Mabbs, *J.C.S. Chem. Comm.*, 1974, 663.

[454] G. M. Larin, E. V. Semenov, and P. M. Solozhenkin, *Doklady Akad. Nauk. S.S.S.R.*, 1974, **215**, 131; P. M. Solozhenkin, E. V. Semenov, and O. N. Grishina, *Doklady Akad. Nauk. Tadzh. S.S.S.R.*, 1973, **16**, 47 (*Chem. Abs.*, 1974, **80**, 59 098).

[455] G. M. Larin, P. M. Solozhenkin, and E. V. Semenov, *Doklady Akad. Nauk S.S.S.R.*, 1974, **214**, 1343.

[456] J. G. Wijnhoven, *Cryst. Struct. Comm.*, 1973, **2**, 637.

[457] A. Nieuwpoort, J. H. E. Moonen, and J. A. Cras, *Recl. Trav. chim.*, 1973, **92**, 1086 (*Chem. Abs.*, 1973, **79**, 152 379).

$MoOCl_3$ forms simple adducts with CCl_3CN but no complex could be isolated.[458] The nitrido-complex $[MoNCl_2(PPh_3)_2]$ has been obtained by treating $[MoCl_4-(THF)_2]$ or $[MoCl_3(THF)_3]$ with Me_3SiN_3 in CH_2Cl_2, followed by the addition of PPh_3. $[MoNCl_2(bipy)]$ has been prepared.[270] The reaction of the nitrido-complexes $[MoN(S_2CNR_2)_3]$ $[R_2 = Me_2, Et_2,$ or $(C_2H_2)_5]$ with sulphur to form the first thionitrosyl derivatives is described on p.118.[271]

The electrochemical reductions of $WCl_5(bipy)$ and $WCl_5(py)_2$ dissolved in MeCN have been carried out as a demonstration of the stabilization of low oxidation states by these N-donor ligands.[436] The complexes *trans*-$[MoOX_3L]$ (X = Cl or Br; L = phen or bipy) have been synthesized by the new method of refluxing the corresponding $LH_2[MoOX_5]$ salt with Hacac at >1 atm.[459] Hexamethylmelamine has been shown to form a 2:1 adduct with $MoCl_5$ and preliminary indications are that this is a seven-co-ordinate molybdenum(v) complex.[460] A number of salts containing the $[MO(NCS)_5]^{2-}$ (M = Mo or W) ions have been synthesized and their structures discussed with reference to their magnetic, i.r., and electronic spectral properties. All the available data indicate that the ions are mononuclear, with a C_{4v} structure involving N-bonded thiocyanato-groups.[461a] The phenothiazine derivatives chloro-promazine, methopromazine, or diethazine (L) react with molybdenum(v) and thiocyanate in aqueous acidic media to precipitate the corresponding $HL[MoO-(NCS)_4]$ salts.[461b] E.s.r., i.r., and electronic spectral data have been recorded for the complexes $[MoOCl_{4-n}(NCSe)_n]^-$ ($n = 1$—4) and $[MoO(NCSe)_5]^{2-}$ and these show that the potentially ambidentate ligands are N-bonded in all of these anions.[462] The salts $KMo[Fe(CN)_6],3H_2O$ and $MoO[Fe(CN)_5(H_2O)],9H_2O$ have been prepared by treating MoO^{3+} halogeno-complexes with $K_2[Fe(CN)_6]$ in 8M-HCl.[463] The complex $[MoOCl(H_2O)L]$ $[H_2L = N$-(salicylidene)-2-hydroxybenzamine] has been prepared and characterized by i.r. and electronic spectroscopy.[430] The reaction of octaethylporphyrin (H_2L) with $MoCl_2$, $K_3W_2Cl_9$, or H_2WO_4 in phenol has been shown to produce the corresponding $[MO(OPh)L]$ (M = Mo or W) complex; mass, i.r., and ^1H n.m.r. spectral characteristics were determined.[177]

A molybdenum(v) derivative of n-propyldiphenylphosphine imine $[MoCl_4-(NPPr^nPh_2)(OPPr^nPh_2)]$ has been formed by oxidation of the tertiary phosphine ligands of $[MoCl_4(PPr^nPh_2)_2]$ by tolyl-*p*-sulphonyl azide.[464] The crystal structure of $[MoCl_4(diars)_2]I_3$ has been determined[465] and the cation seen to have D_{2d} symmetry, with arsenic atoms on the A sites (Mo—As = 265.3 pm) and chlorine atoms on the B sites (Mo—Cl = 245.1 pm) in an arrangement found for the analogous titanium(IV) complex.

Cyclopentadiene has been shown to react with $MoCl_3(OR)_2$ (R = Me, Pr, Pr^i, Bu^n, Bu^i, or Am^i) in THF solution to produce the corresponding $[(\eta^5-C_5H_5)MoCl-(OR)_2]$ complexes, the i.r. spectra of which have been recorded.[466]

[458] G. W. A. Fowles, K. C. Moss, D. A. Rice, and M. Rolfe, *J.C.S. Dalton*, 1973, 1871.
[459] H. K. Saha, M. Bagchi, and S. Chakravorty, *J. Inorg. Nuclear Chem.*, 1974, **36**, 455.
[460] N. Gunduz, *Comm. Fac. Sci. Univ. Ankara, Ser. B*, 1972, **19**, 95 (*Chem. Abs.*, 1973, **79**, 70 456).
[461] (*a*) H. Sabat, M. F. Rudolf, and B. Jezowska-Trzebiatowska, *Inorg. Chim. Acta*, 1973, **7**, 365; (*b*) M. Tara-siwicj and H. Puzanowska-Tarasiewicj, *Chem. Anal.* (*Warsaw*), 1973, **18**, 599.
[462] A. M. Golub, V. A. Grechnikhina, V. V. Trachevskii, and N. V. Ul'ko, *Zhur. neorg. Khim.*, 1973, **18**, 2119.
[463] H. K. Saha and B. C. Saha, *J. Inorg. Nuclear Chem.*, 1974, **36**, 2115.
[464] D. Scott and A. G. Wedd, *J.C.S. Chem. Comm.*, 1974, 527.
[465] M. G. B. Drew, G. M. Egginton, and J. D. Wilkins, *Acta Cryst.*, 1974, **B30**, 1895.
[466] R. K. Multani, *J. Inorg. Nuclear Chem.*, 1974, **36**, 1411.

Dinuclear and Polynuclear Complexes. The electronic structure of the model complex (20) has been investigated using the all-valence-electron SCMO method. A significant feature of the interactions within this compound is the existence of a strong Mo—Mo π-bond.[467] Synthetic, spectroscopic, and theoretical studies have been described for a series of dimeric and tetrameric μ-oxomolybdenum(V) complexes. SCCC MO and angular overlap calculations have been presented for the $[Mo_2O_4Cl_4(H_2O)_2]^{2-}$ and $[Mo_2O_3(NCS)_8]^{4-}$ ions, and the origin of spin-pairing between the metal centres in these and other polymeric μ-oxomolybdenum(V) complexes discussed in some detail. The character of Mo—O bonding in these complexes has been discussed with reference to normal-co-ordinate analyses carried out for di-μ-oxo-bridged dimeric and μ-oxo-bridged tetrameric molybdenum(V) systems.[468]

$$
\begin{array}{c}
\text{HC} \overline{} \text{S}_{\diagdown} \overset{\text{O}}{\underset{\|}{}} \qquad \overset{\text{HO}}{\diagdown}\overset{\text{O}}{\underset{\|}{}} \\
\text{S} \overline{} \text{Mo} \overset{\diagup}{\diagdown} \text{O} \overline{} \text{Mo} \overset{\diagup}{\diagdown} \text{OH}_2 \\
\end{array}
$$

(20)

The kinetics of the equilibration of the $Mo_2O_4^{2+}$ aq and $[Mo_2O_4(edta)]^{2-}$ entities have been studied in aqueous $HClO_4$ solution in the presence of excess edta[469a] and the stability constant of $[W_2O_4(edta)]^{2-}$ has been determined (log $\beta = 29.1$).[469b] The preparation and characterization of the diamagnetic compounds $KMo_2O_4[Fe(CN)_5(H_2O)].H_2O^{463}$ and $[Mo_2O_4X_2L_2]$ (X = Cl or Br; L = bipy or phen),[459] and the complexes $[Mo_2O_4(C_2O_4)(H_2O)_2]^{2-}$, $[Mo_2O_3(C_2O_4)_2L_2']^{2-}$, $[Mo_2O_3L_4']$, and $[Mo_2O_4L_2']$ (HL' = *e.g.* N^1-*p*-chlorophenyl-N^5-isopropylbiguanidine)[470] have been described.

Further studies have been reported for the molybdenum-(VI) and -(V) 8-quinolinol-(Q) system suggested as a model for molybdenum–flavin interactions. ^{1}H N.m.r. spectra for $[Mo_2O_3Q_4]$ indicate that the ligands are co-ordinated with the N-atom *trans* to an oxo-group and thus the Mo—N interactions are weak and labile. The implications of such bonding for Mo–flavin interactions in the molybdoflavoprotein systems have been considered.[471] The reduction of flavins by $[Mo_2O_3(OH)_4(L-cysteinate)_2]^{4-}$ has been investigated in basic solution (pH 8—11), also with reference to molybdenum(V)–flavin electron-transfer reactions in biological systems. This reaction was found to proceed rapidly by initial one-electron transfer, followed by disproportionation of the flavosemiquinone. The rate of reaction increased with increasing pH and was catalysed by cysteine. The implications of these results for electron transfer processes in the various molybdenum flavoenzymes have been discussed.[472] Further study of dimeric molybdenum(V) NN-dialkyldithiocarbamato-

[467] B. Blues, D. H. Brown, P. G. Perkins, and J. T. P. Stewart, *Inorg. Chim. Acta.*, 1974, **8**, 67.
[468] B. Jezowska-Trzebiatowska, M. F. Rudolf, L. Natkaniec, and H. Sabat, *Inorg. Chem.*, 1974, **13**, 617.
[469] (a) Y. Sasaki and A. G. Sykes, *J.C.S. Dalton*, 1974, 1468; (b) J. Novak and J. Podlaha, *J. Inorg. Nuclear Chem.*, 1974, **36**, 1061.
[470] L. Antonescu, *An. Univ. Bucuresti Chim.*, 1972, **21**, 135 (*Chem. Abs.*, 1973, **79**, 99 954); C. Gheorghiu and L. Antonescu, *An. Univ. Bucuresti Chim.*, 1971, **20**, 113 (*Chem. Abs.*, 1973, **79**, 86 963).
[471] L. W. Amos and D. T. Sawyer, *Inorg. Chem.*, 1974, **13**, 78.
[472] P. Kroneck and J. T. Spence, *Biochemistry*, 1973, **12**, 5020.

complexes of the types $[Mo_2O_3(dtc)_4]$ and $[Mo_2O_4(dtc)_2]$ has resolved the ambiguities in earlier reports concerning the preparation and properties of $[Mo_2O_3-(S_2CNEt_2)_4]$. It is suggested that these two types of complex and their interrelationship are a relevant model of the active site of nitrogenase.[473] 5-Phenyl-1-pyrazolinedithiocarbamate (L) has been shown to form the complex $[Mo_2O_3L_2]^{474}$ and an e.s.r. study of $[Mo_2O_3(S_2CNEt_2)_4]$ in $CHCl_3$ solution at room temperature has suggested the presence of two isomeric paramagnetic molybdenum(v) complexes.[475] An improved preparation of xanthato-complexes of the type $[Mo_2O_3(S_2COR)_4]$ has been developed and the conversion of some of these, by reaction with H_2S, into the di-μ-sulphido-derivatives $[Mo_2O_2S_2(S_2COR)_2]$ has been described.[476] The di-μ-sulphido-ion $Mo_2S_2O_{2aq}^{2+}$ has been separated from $Na_2[Mo_2O_2S_2(cysteinate)_2]$ by reaction with 1M–HCl. This ion is diamagnetic, stable towards acids, and far more resistant to dissociation into paramagnetic monomeric species than its di-μ-oxo-bridged analogue.[287]

WCl_6 reacts with $PhCSCH_2COPh$ (HL) (1 : 6) in refluxing C_6H_6 over 6 h to produce the compound $Cl_2L_2WCl_2WL_2Cl_2$.[435] Di-μ-chloro-bis{chlorobis(*NN*-diethylselenocarbamato)molybdenum(v)} dichloride has been prepared by treating $MoCl_5$ in MeOH solution with $(Et_2NH_2)(Se_2CNEt_2)$. This compound appears to be paramagnetic and its e.s.r. characteristics have been determined.[477]

The e.s.r. spectrum of $[PMo_{12}O_{40}]^{4-}$ doped in $(Bu_4N)_4[SiW_{12}O_{40}]$ have been recorded for sample temperatures between 6 and 300 K. These spectra reveal a valence oscillation that may be considered as an intramolecular electron-hopping process and they indicate for the first time that at low temperatures the electron becomes trapped on a single metal atom[478].

E.s.r. spectra have also been presented for molybdenum(v) centres in other iso- and hetero-polymolybdate blues[479] and a theoretical description of these types of systems has been given.[480] Results of an investigation of the photochromism of amine molybdates are consistent with the formation of mixed isopolyanions containing molybdenum-(v) and -(vi) centres.[481] The lilac colouration formed during sulphide separations of molybdenum and tungsten appears to be due to the formation of a unit in which six tungsten(vi) atoms surround a molybdenum(v) centre.[482]

Molybdenum and Tungsten Bronzes. A single-crystal e.s.r. study of the recently discovered triclinic $Na_{0.33}WO_3$ bronze at 15 K has provided the first conclusive evidence for the existence of W^V centres in the tungsten bronzes. These centres are present in a concentration approximately equal to that of the alkali metal. Two different W^V resonances have been observed, suggesting two different W^V point

[473] W. E. Newton, J. L. Corbin, D. C. Bravard, J. E. Searles, and J. W. McDonald, *Inorg. Chem.*, 1974, **13**, 1100.
[474] V. M. Byr'ko, M. B. Polinskaya, and A. I. Busev, *Zhur. neorg. Khim.*, 1973, **18**, 2783.
[475] R. Kirmse, *Z. Chem.*, 1973, **13**, 187.
[476] W. E. Newton, J. L. Corbin, and J. W. McDonald, *J.C.S. Dalton*, 1974, 1044.
[477] R. Kirmse and E. Hoyer, *Z. anorg. Chem.*, 1973, **398**, 136.
[478] R. A. Prados, P. T. Meiklejohn, and M. T. Pope, *J. Amer. Chem. Soc.*, 1974, **96**, 1261.
[479] M. Otake, Y. Komiyama, and T. Otaki, *J. Phys. Chem.*, 1973, **77**, 2896; L. A. Furtune, N. A. Polotebnova, and A. A. Kozlenko. *Zhur. neorg. Khim.*, 1973, **18**, 2185; H.-C. Wang, Y. Yang, and Y.-W. Chen, *Sci. Sinica*, 1974, **17**, 17 (*Chem. Abs.*, 1974, **80**, 126494).
[480] H. So, *Dachan Hwahak Hwoejee*, 1973, **17**, 387 (*Chem. Abs.*, 1974, **80**, 20902).
[481] F. Arnaud-Neu and J. M. Schwing-Weill, *Bull. Soc. chim. France*, 1973, **12**, 3233; T. Yamase, T. Ikawa, H. Kokado, and E. Inoue, *Chem. Letters*, 1973, 615.
[482] Z. G. Golubtsova and L. I. Lebedeva, *Zhur. neorg. Khim.*, 1974, **19**, 393.

defects in the lattice. The results have confirmed that the triclinic $Na_{0.33}WO_3$ and $Li_{0.33}WO_3$ bronzes have properties remarkably different from those of the other semiconducting bronzes.[293]

An electrochemical method suitable for the preparation of hydrogen tungsten bronzes has been developed.[483] and the ΔH_f° values for $H_{0.35}WO_3(c)$ and $H_{0.18}WO_3$ (c) have been determined by solution calorimetry as -9.6 ± 0.8 and -4.8 ± 0.6 kJ mol^{-1}, respectively.[484] The crystal structure of $D_{0.53}WO_3$ has been determined by a room-temperature neutron diffraction study and shown to possess an atomic arrangement similar to that in $Sc(OH)_3$. It is considered that the compound is best considered as a non-stoicheiometric oxide hydroxide. The mobility of the hydrogen atoms in $H_{0.46}WO_3$. followed by pulse ^1H n.m.r. spectroscopy, was also investigated in this study.[485]

A review of the physical properties of the sodium tungsten bronzes has been published[486] and their reduction characteristics with H_2 have been described.[487] An inhibition effect has been observed during the cathodic reduction of H_2O_2 on Na_xWO_3 and the competing reaction is believed to be due to the formation of a hydrogen–sodium bronze phase.[488] The superconductivity of $Rb_{0.3}WO_3$ has been studied[489] and the formation of the hexagonal ammonium bronze $(NH_4)_xWO_3$ from $(NH_4)_{10}[W_{12}O_{41}].5H_2O$ monitored using i.r. and X-ray diffraction spectroscopy.[490]

The reaction of WO_3 and tungsten metal in the presence of dilute solutions of HF under hydrothermal conditions has been shown to produce the series of tetragonal oxyfluoride bronzes $WO_{3-x}F_x$ $(0.03 \leqslant x \leqslant 0.09)$.[491] The phases $Cs_xM_{2-y}W_yX_6$ (M = V. Cr. Fe. Ga. Ti. Nb. or W; X = F or F + O; $x = 0.35$—0.55; $0 \leqslant y \leqslant 1$) have been obtained by heating mixtures of the corresponding fluorides and oxides. These materials all crystallize with the hexagonal tungsten bronze structure and their lattice constants have been determined.[98]

New phases of the hexagonal bronze type have been identified in the Na_2O–$2CrO_3$–Cr_2O_3–WO_3 system and characterized by magnetic measurements and X-ray diffraction studies. $Na_{0.20}Cr_{0.14}(Cr_{0.25}W_{0.75})O_{3.145}$ contains tetrahedral $[CrO_4]^{2-}$ ions and Cr^{III} and W^{VI} atoms in octahedral sites.[492] The phases $M_xV_xMo_{1-x}O_3$ (M = K. Rb. or Cs; $x = 0.12$—0.14) contain octahedral (V. Mo)O_6 units linked by vertex-sharing to give a triple chain; these in turn share edges with three other such chains to produce wide tunnels which accommodate the additional metal atoms.[493] The series of bronzes Hg_xWO_3 $(0 \leqslant x \leqslant 0.30)$ have been prepared by treating WO_3. WO_2. and $HgWO_4$ at $620\,^\circ C$ under pressure and three distinct phases identified.[494]

[483] G. Siclet, J. Chevrier, J. Lenoir, and C. Eyraud, *Compt. rend.*, 1973, **277**, C, 227.
[484] P. G. Dickens. J. H. Moore. and D. J. Nield, *J. Solid State Chem.*, 1973, **7**, 241.
[485] P. J. Wiseman and P. G. Dickens, *J. Solid State Chem.*, 1973, **6**, 374; P. G. Dickens. D. J. Murphy. and T. K. Halstead, *ibid.*, p. 370.
[486] J. P. Randin, *J. Chem. Educ.*, 1974, **51**, 32.
[487] M. S. Wittingham and P. G. Dickens. Proceedings of the Seventh International Symposium on 'The Reactivity of Solids'. 1972. p. 640. ed. J. S. Anderson. Chapman and Hall. London (*Chem. Abs.*, 1974, **80**, 30 989).
[488] J. P. Randin, *Canad. J. Chem.*, 1974, **52**, 2542.
[489] D. R. Wanlass, *Diss. Abs. (B)*, 1973, **34**, 1238.
[490] A. B. Kiss and L. Chudik-Major, *Acta Chim. Acad. Sci. Hung.*, 1973, **78**, 237 (*Chem. Abs.*, 1973, **79**, 142 461).
[491] T. G. Reynolds and A. Wold, *J. Solid State Chem.*, 1973, **6**, 565.
[492] R. Salmon. G. Le Flem. and P. Hagenmuller, *Bull. Soc. chim. France*, 1973, 1889.
[493] B. Darriet and J. Galy, *J. Solid State Chem.*, 1973, **8**, 189.
[494] T. Plante, *Bull. Soc. chim. France*, 1973, 3301.

Molybdenum(VI) and Tungsten(VI).—*Halide, Oxyhalide, and Related Complexes.* The enthalpies of formation of $[WF_5Cl]$, $[WF_4Cl_2]$, $[WF_5(OMe)]$, *cis*-$[WF_4(OMe)_2]$ and *cis*-$[WF_2(OMe)_4]$ have been reported and redistribution and decomposition reactions in the fluoride–chloride or fluoride–methoxide series have been discussed in the light of these results.[495] Vapour density measurements on $MoOF_4$ have shown that this compound is monomeric in the gas phase. Similar measurements suggest that WOF_4 is also essentially monomeric as a gas, although some association may occur at low temperatures and/or high pressures. Gas-phase Raman spectra obtained for WOF_4 are invariant between 200 and 600°C and closely resemble those of the melt and solid phases. Thus solid WOF_4 is not principally an oxygen-bridged polymer as has been suggested previously. The details of the vibrational spectra obtained for the gaseous and matrix-isolated WOF_4 molecules are consistent with a C_{4v} structure. The spectra obtained for $MoOF_4$ closely resemble those for WOF_4 in all phases.[496] Thermodynamic properties for gaseous $MoOF_4$ and WOF_4 have been calculated from published spectroscopic data.[497] The enthalpies of formation of $MoOF_4(c)$ and $WOF_4(c)$ have been estimated as -1380 and $-1500 \, kJ \, mol^{-1}$, respectively, by measuring their enthalpies of alkaline hydrolysis.[353a] The preparation and properties of $NO_2[WOF_5]$ have been described[498] and a series of salts containing the $[MoO_2F_4]^{2-}$ and $[MoO_2F_3]^{-}$ anions with organic nitrogenous base cations have been isolated.[499] The extraction of molybdenum or tungsten from aqueous HF solutions by tri-n-octylamine or tri-n-octylbenzylamine (R_3N) has been investigated and the extracted species shown to be the $(R_3NH)_2[MO_2F_4]$ salts.[500] The crystal structure of $(NH_4)_2[MoO_3F_2]$ has been determined by X-ray crystallography and its i.r. and Raman spectra have been reported. Planar, *cis*-dioxodifluoro-groups are linked to give endless chains *via* non-linear μ-oxo-bridges.[501] Hydrogen-bonding in a series of ammonium oxyfluoro-molybdates and -tungstates has been investigated by i.r., ^1H n.m.r., and ^{19}F n.m.r. spectroscopy.[502] Stability constant data for the

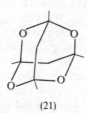

(21)

[495] J. Burgess, C. J. W. Fraser, R. D. Peacock, P. Taylor, A. Majid, and J. M. Winfield, *J. Fluorine Chem.*, 1973, **3**, 55.

[496] L. E. Alexander, I. R. Beattie, A. Bukovszky, P. J. Jones, C. J. Marsden, and G. J. Van Schalkwyk, *J.C.S. Dalton*, 1974, 81.

[497] E. G. Rakov, L. G. Koshechko, B. N. Sudarikov, and V. V. Mikulenok, *Trudy Moskov Khim.-Tekhnol. Inst.*, 1972, **71**, 25 (*Chem. Abs.*, 1974, **80**, 100 792).

[498] D. D. Gibler, *Nuclear Sci. Abst.*, 1973, **28**, 26892 (*Chem. Abs.*, 1974, **80**, 66 270).

[499] (a) M. C. Chakravorti and S. C. Pandit, *J. Inorg. Nuclear Chem.*, 1973, **35**, 3644; (b) R. G. Beiles, V. S. Samorodova, and Z. E. Rozmanova, *Zhur. neorg. Khim.*, 1973, **18**, 2978.

[500] Yu. S. Tseryuta, V. V. Bagreev, N. A. Agrinskaya, N. V. Gushchin, A. S. Basov, and Yu. A. Zolotov, *Zhur. analit. Khim.*, 1973, **28**, 946; Yu. S. Tseryuta, V. V. Bagreev, and E. S. Pal'shin, *Zhur. neorg. Khim.*, 1974, **19**, 200.

[501] R. Mattes and G. Müller, *Naturwissen*, 1973, **60**, 550.

[502] A. A. Opalovskii and Z. M. Kuznetsova, *Izvest. Sib. Otd. Akad. Nauk S.S.S.R., Ser. khim., Nauk*, 1973, 62 (*Chem. Abs.*, 1974, **80**, 7304).

$[MoO_2F_4]^{2-}$ and $[MoO_2F_3]^-$ ions in aqueous HF solution have been determined by an ion-exchange method.[503]

Procedures have been described which are suitable for the preparation of the oxyhalides WOX_2^1 (X^1 = Cl, Br, or I) and WOX_4^2 (X^2 = Cl or Br) using chemical transport techniques.[432] $MoOCl_4$ reacts with pentane-2,4-dione to yield 1,3,5,7-tetramethyl-2,4,6,8-tetraoxa-adamantane (21), a facile dimerization which contrasts with previous unsuccessful attempts. The corresponding reaction with $WOCl_4$ is described on p. 145.[504] The structure of MoO_2Cl_2,H_2O has been re-investigated and the appearance of the diffuse diffraction patterns explained in terms of two polytypes which contain the same basic structural element. This is considered to involve a distorted octahedron of two *trans* chlorine atoms, one terminal oxo-ligand *trans* to a co-ordinated water molecule, and two *trans* μ-oxo-groups.[505] The lattice constants for orthorhombic MoO_2Cl_2 have been determined and evidence has been obtained to suggest that the oxychlorides $Mo_2O_3Cl_6$ and $Mo_3O_5Cl_8$ do not exist.[506]

Thermodynamic data have been derived[507] for $WOBr_4(g)$ from vapour pressure measurements; $\Delta H_f^\circ = -412 \pm 9\,kJ\,mol^{-1}$. The value of this quantity for WO_2Br_2 (g) has been estimated as $-567 \pm 4\,kJ\,mol^{-1}$ from mass spectrometric and other data.[508] The extraction of molybdenum and tungsten from aqueous HBr solutions by trialkylamines in organic media has been studied and the extracted anions shown to be $[MoO_2Br_3]^-$ and $[WO_2Br_3]^-$.[509]

The chemical transport of molybdenum and tungsten as oxides or sulphides in the presence of iodine has been shown to proceed by the formation of the corresponding MO_2I_2 complex.[510]

O-*Donor Ligands*. A large number of these studies concern molybdates, tungstates, and mixed-oxide compounds, and the results of many of them are summarized in Tables 4, 5, and 6.

The vapour phases in equilibrium with M_2MoO_4 and M_2WO_4 (M = Li or Na) have been investigated by mass spectrometry. The sodium salts were reduced by the crucible material, but the lithium ones evaporated mainly as the molecular species.[511] An electron diffraction study of Ba_2WO_4 in the gas phase has shown that each WO_4^{2-} ion co-ordinates as a bidentate ligand to one Ba^{2+} ion.[512]

The kinetics of the solid-phase reactions of MoO_3 with M^1O, $M_2^1CO_3$, or $M^1(OH)_2$ (M^1 = Ca, Sr, or Ba) and WO_3 with $M_2^2CO_3$ (M^2 = Li, Na, or K) have been determined.[513] A new dense form of $BaWO_4$ has been prepared under pressure[514] and

503 N. A. Parpiev, I. A. Maslennikov, and Kh. S. Abdullaeva, *Uzbek. Khim. Zhur.*, 1973, **17**, 3 (*Chem. Abs.*, 1974. **80**. 137605).
504 M. G. B. Drew, G. W. A. Fowles, and D. A. Rice, *J.C.S. Chem. Comm.*, 1974, 614.
505 H. Schulz and F. A. Schröder, *Acta Cryst.*, 1973, **29A**, 322, 333.
506 F. A. Schröder and J. Scherle, *Z. Naturforsch.*, 1973, **28B**, 216.
507 A. Pebler, *Ind. and Eng. Chem. (Product Res. and Development)*, 1973, **12**, 301 (*Chem. Abs.*, 1973, **79**, 150152).
508 J. H. Dettingmeijer and B. Meinders, *Z. anorg. Chem.*, 1973, **400**, 10.
509 Yu. S. Tseryuta, V. V. Bagreev, Yu. A. Zolotov, and N. A. Agrinskaya, *Zhur. neorg. Khim.*, 1973, **18**, 1634.
510 H. Schaefer, T. Grofe, and M. Trenkel, *J. Solid State Chem.*, 1973, **8**, 14.
511 R. Yamdagni, C. Pupp, and R. F. Porter, *Govt. Rep. Announce.* (*U.S.A.*), 1973, **73**, 66 (*Chem. Abs.*, 1973, **79**, 108625).
512 A. A. Ivanov, V. P. Sprirdinov, E. V. Erokhin, and V. A. Levitskii, *Zhur. fiz. Khim.*, 1973, **47**, 3030.
513 V. M. Zhukovskii and S. F. Veksler, *Kinetika i Kataliz*, 1973, **14**, 248 (*Chem. Abs.*, 1973, **79**, 10323); V. F. Annopol'skii, E. K. Belyaev, and I. P. Knigavko, *Ukrain. Khim. Zhur.*, 1973, **39**, 883 (*Chem. Abs.*, 1973, **79**, 140057).
514 T. Fujita, S. Yamaoka, and O. Fukunaga. *Mater. Res. Bull.*, 1974, **9**, 141.

Table 4   *Molybdates, tungstates, and mixed-oxide compounds of molybdenum*(VI) *and tungsten*(VI) *with s- and p-block elements*

Compound	Comments and reported properties	Ref.
$K_6M_2O_9$ (M = Mo or W)	prepared by reduction of K_2MO_4 by K vapour	a
M_2MoO_4 (M = Na or Cs)	st.	b
$MgWO_4.nH_2O$ (n = 2 or 3)	i.r., t.d., X	c
$Cs_2Mg(MoO_4)_2.4H_2O$	st.	d
$CaMoO_4$	m.p. 1445 °C	
Ca_3MoO_6	m.p. 1370 °C. t.d., X	e
$Li_2Ba_5W_3O_{15}$	st; tetragonal bronze phase	f
M_2WO_5 (M = Sr or Ba)	X	g
M_3MoO_6		
M_3WO_6 (M = Ca, Sr. or Ba)	X	h
Ba_2MgWO_6	e	i
$Al_2(WO_4)_3$	st	j
$M_8Al_2W_{12}O_{43}$	X; cubic lattice	
$M_6Al_2W_{21}O_{69}$ (M = K. Rb. or Cs)	X; hexagonal lattice	k
$M(OH)WO_4$	⎫	
$M_2(OH)_4(WO_4)$	⎪ prepared in the	
$M_2(WO_4)_3$	⎬ $M(NO_3)_3-Na_2WO_4-H_2O$ system	l. m
$NaM(WO_4)_2$ (M = Al, Ga. or. In)	⎭	
$MIn(WO_4)_2$ (M = Li, Na, or K)	m.p.	n
$Tl_2Mo_3O_{10}$	X	o
$Tl_6Mo_7O_{24}$	prepared by heating Tl_2MoO_4 with	
$Tl_2Mo_4O_{13}$	MoO_3	p
$PbWO_4$	sol., st.	q
$B_6Pb_{15}Mo_2O_{30}$	m.p. 772 °C	r
$K_2P_2MO_9$ (M = Mo or W)	prepared in the corresponding KPO_3-MO_3 system; X, isomorphous	s
Bi_2MoO_6	st.	
$Bi_2Mo_2O_9$	X	
$Bi_2Mo_3O_{12}$	e, st.	t
$MBiMoO_4$		
$MBiWO_4$ (M = Li—Cs)	X	u
Bi_2TeWO_8	X, isotypic with $Bi_2Te_7O_8$	v
$Mo_2O_3(SO_4)_3.4H_2O$	t.d.	w

(a) Ref. 426; (b) K. Okada, H. Morikawa, F. Maruno, and S. Iurai, *Acta Cryst.*, 1974, **B30**, 1872; W. Gonschorek and Th. Hahn. *Z. Krist.*, 1973. **138**, 167; (c) T. I. Bel'skaya, V. E. Plyushchev, V. M. Kazan, S. V. Morozova and G. N. Boronskaya, *Izvest. V. U.Z. Khim. i Khim. Tekhnol.*, 1974, **17**, 14 (*Chem. Abs.*, 1974, **80**, 103285); A. N. Borsheh, Yu. G. Dorokhov, and A. M. Golub, *Ukrain Khim. Zhur.*, 1973, **39**, 724 (*Chem. Abs.*, 1973, **79**, 84458); (d) S. Peytavin and E. Philippot, *Cryst. Struct. Comm.*, 1973, **2**, 355; (e) T. M. Yanushkevich and V. M. Zhukovskii, *Zhur. neorg. Khim.*, 1973, **18**, 2234; T. M. Yanushkevich, N. N. Shevchenko, V. M. Zhukovskii, V. M. Ust'yantsev, and L. N. Lykova, *Zhur. neorg. Khim.*, 1973, **18**, 2931; (f) A. J. Jacobson, B. M. Collins, and B. E. F. Fender, *Acta Cryst.*, 1974, **B30**, 816; T. Negas, R. S. Roth, H. S. Parker, and W. S. Brower, *J. Solid State Chem.*, 1973, **8**, 1; (g) L. M. Kovba, L. N. Lykova, M. V. Paromova, and N. N. Shevchenko, *Zhur. neorg. Khim.*, 1973, **18**, 835; (h) N. N. Shevchenko, L. N. Lykova, and L. M. Kovba, *Zhur. neorg. Khim.*, 1974, **19**, 971; L. M. Kovba, L. N. Lykova, and N. N. Shevchenko, *ibid.*, 1973, **18**, 1991; (i) G. Blasse, and A. F. Corsmit, *J. Solid State Chem.*, 1973, **6**, 513; (j) J. J. De Boer, *Acta Cryst.*, 1974, **B30**, 1878; (k) M. V. Mokhosoev, S. A. Pavlova, E. I. Get'man, and N. G. Kisel, *Zhur. neorg. Khim.*, 1973, **18**, 2124; (l) K. G. Shcherbina, M. V. Mokhosoev, and A. I. Gruba, *Zhur. neorg. Khim.*, 1974, **19**, 396; *Izvest. V. U.Z. Khim. i khim. Tekhnol.*, 1973, **16**, 1442 (*Chem. Abs.*, 1973, **79**, 152369); (m) M. V. Mokhosoev, K. G. Shcherbina, A. I. Gruba, and V. I. Krivobok, *Zhur. neorg. Khim.*, 1974, **19**, 966; (n) V. N. Karpov, I. B. Korotkevich. M. M. Minkova, and O. V. Sorokina, *Zhur. neorg. Khim.*, 1973, **18**, 1314; (o) M. Toubol, P. Toledano, and G. Perez, *Compt. rend.*, 1974, **278**, C, 417; (p) M. R. Kini, M. R. Udupa, and G. Aravamudan, *Current Sci.*, 1973, **42**, 536 (*Chem. Abs.*, 1973, **79**, 86903); (q) A. A. Bulatova, V. B. Aleskovskii, and M. I. Bulatov, *Zhur. priklad Khim.*, 1973, **46**, 2761; (r) V. T. Mal'tsev, P. M. Chobanyan, and V. L. Volkov, *Zhur. neorg. Khim.*, 1973, **19**, 497; (s) L. V. Semenyakova and I. G. Kokarovtseva, *Zhur. neorg. Khim.*, 1973, **18**, 3068; (t) A. F. Van den Elzen

Table 5   *Molybdates, tungstates, and mixed-oxide compounds of molybdenum*(VI) *and tungsten*(VI) *with* d-*block elements*

Compound	Comments and reported properties	Ref.
$M(MoO_4)_2$ (M = Zr or Hf)	X	a
$M_2V_2Mo_3O_{15}$ (M = Li, Na, or K)		
$KVMo_7O_{24}$		
$K_3V_3Mo_2O_{15}$	prepared in the MVO_3–MoO_3 system	
$K_7V_7Mo_3O_{10}$	m.p., X	b
$Cs_4V_6Mo_2O_{23}$		
$Cs_4V_{10}Mo_2O_{33}$		
$K_2Nb_2W_2O_{12}$	X, pyrochlore structure	c
$AMWO_6$		
(A = H_3O or NH_4; M = Nb or Ta)	t.d., X	d
$M_2Cr_2(MoO_4)_3SO_4$ (M = Na, K, or Cs)	i.r., X	e
MnW_4O_{13}, $12.6 H_2O$		f
$Mn_5W_{12}O_{41}$, $35 H_2O$	i.r., t.d.	
$MFe(WO_4)_2$ (M = Li or Cs)	X	g
$CoWO_4$, $4H_2O$		
$Co_3W_7O_{24}$, $19H_2O$	obtained from aqueous solutions of	
CoW_4O_{13}, $8H_2O$	$Co(NO_3)_2$ and K_2WO_4; i.r., ^1H	h
$K_{10}Co_{10}W_{32}O_{101}$	n.m.r., X	
Rh_2WO_6	i.r., X	i
$NiMoO_4$	t.d., X	j
Sr_2NiWO_6	st; perovskite lattice	k
NiO,nWO_3,xH_2O	formed by the addition of $Ni(NO)_3$ aq. to	
(n = 1, 1.5, 2, 2.7 or 3)	K_2WO_4 at various pH values;	l
$K_6Ni_6W_{20}O_{27}$, $48H_2O$	i.r., ^1H n.m.r., t.d.	
Li_2CuWO_4	m.p. 765 °C, X	m
$CuMoO_4$	high pressure form. X isostructural with	
	$CuWO_4$	n
$ASbWO_6$ (A = Ag, NH_4, or H_3O)	X	o
$ZnWO_4$	m.p. 1230 °C	p
$Na_{12}Zn_4W_{10}O_{40}$	m.p. 645 °C, X	m
MWO_4 (M = Mn—Ni, Zn)		
Sr_2MWO_6 (M = Ni, Cu, or Zn)	i.r., R	q
Ba_2NiWO_6		

(a) N. V. Gul'ko, A. G. Karaulov, N. M. Taranukha, and A. M. Gavrish, *Izvest. Akad. Nauk. S.S.S.R., neorg. Materialy*, 1973, **9**, 1766; (b) B. V. Slobodin, M. V. Mokhosoev, and N. G. Kabanova, *Zhur. neorg. Khim.*, 1973, **18**, 2231; 1974, **19**, 388; (c) T. Ikeda and S. Shimizu, *Jap. J. Appl. Phys.*, 1974, **13**, 256 (*Chem. Abs.*, 1974, **80**, 113450); (d) D. Groult, C. Michel, and B. Raveau, *J. Inorg. Nuclear Chem.*, 1974, **36**, 61; (e) Ref. 112; (f) V. N. Pitsyuga, M. V. Mokhosoev, M. N. Zayats, and T. E. Frantsuzova, *Zhur. neorg. Khim.*, 1973, **18**, 1166; (g) E. M. Novikova, A. A. Maier, and T. V. Kulakova, *Izvest. Akad. Nauk. S.S.S.R., neorg. Materialy*, 1973, **9**, 997; P. V. Klevtsov, L. P. Kozeeva, R. F. Klevtsova, and N. A. Novgorodtseva, *Ref. Zhur. Khim.*, 1973, Abs. 2B535 (*Chem. Abs.*, 1973, **79**, 58510); (h) M. V. Mokhosoev, M. N. Zayats, and N. I. Loshkareva, *Zhur. neorg. Khim.*, 1973, **18**, 1829; M. V. Mokhosoev, N. A. Taranets, and M. N. Zayats, *ibid.*, p. 3329; (i) J. P. Badaud and J. Omaly, *Compt. rend.*, 1974, **278**, *C*, 521; (j) F. S. Pilipenko, A. L. Tsailingol'd, V. A. Levin, R. A. Buyanov, and M. M. Andrushkevich, *Kinetika i Kataliz*, 1973, **14**, 752 (*Chem. Abs.*, 1973, **79**, 129753); (k) P. Koehl, *Z. anorg. Chem.*, 1973, **401**, 121; (l) M. V. Mokhosoev and M. N. Zayats, *Zhur. priklad Khim.*, 1973, **46**, 2139; (m) V. A. Efremov and V. K. Trunov, *Zhur. neorg. Khim.*, 1974, **19**, 501; (n) A. W. Sleight, *Mater. Res. Bull.*, 1973, **8**, 863; (o) D. Groult, C. Michel, and B. Raveau, *J. Inorg. Nuclear Chem.*, 1973, **35**, 3095; (p) I. P. Kislyakov, I. N. Smirnova, and G. I. Boguslavskaya, *Izvest. V. U.Z. Khim. i khim. Tekhnol.*, 1973, **16**, 1440 (*Chem. Abs.*, 1974, **80**, 19979); (q) J. P. Lesne and P. Caillet, *Ann. Chim.* (*France*), 1974, **9**, 57 (*Chem. Abs.*, 1974, **80**, 89061); A. Lentz, *Z. anorg. Chem.*, 1973, **402**, 153.

and G. D. Rieck, *Acta Cryst.*, 1973, **B29**, 2436, 2433; T. Chen, *J. Cryst. Growth*, 1973, **20**, 29; (u) P. V. Klevtsov, V. A. Vinokurov, and R. F. Klevtsova, *Kristallografiya*, 1973, **18**, 1192; (v) B. Frit, *Compt. rend.*, 1973, **277**, *C*, 1227; (w) A. A. Palant, V. A. Reznichenko, and G. A. Meny ailova, *Doklady Akad. Nauk. S.S.S.R.*, 1973, **212**, 886.

Table 6   *Molybdates, tungstates, and mixed-oxide compounds of molybdenum*(VI) *and tungsten*(VI) *with* f-*block and related elements*

Compound	Comments and reported properties	Ref.
Y_2MoO_6		
$Y_2Mo_3O_{12}$	prepared in the Y_2O_3—MoO_3 system; X	a
Y_6MoO_{12}		
$MSc(WO_4)_2$		
$M_3Sc(WO_4)_2$ (M = Li—Cs)	i.r., m.p., t.d., X	b
La_6WO_{12}	ΔH_f°	c
α-LiLa(MoO$_4$)$_2$	X	d
$CsLaW_2O_8$	m.p. 1050 °C, t.d., X	e
$MLa_4W_7O_{28}$ (M = Ba or Pb)		
$PbLa_6W_{10}O_{40}$	t.d., X	f
$CeMo_2O_8$	X	g
$Ce_2W_2O_9$	m.p. 1395 °C, t.d., X, μ	h
$Ce_4W_9O_{33}$	m.p. 1026 °C, t.d., X, μ	h
$Nd_5Mo_3O_{16}$	st.	i
$KEuMo_2O_8$	st.	j
$MEuMo_2O_8$ (M = Na—Cs)		
$M_5EuMo_4O_{16}$ (M = K or Rb)	prepared in the M_2MoO_4–$Eu(MoO_4)_3$ system; m.p., t.d., X	k
$Na_8Eu_2Mo_7O_{28}$		
$LiGdMo_2O_8$	kinetics of solid state synthesis reported	l
$Gd_{x-y}Y_y(Ta_{3x}W_{1-3x})O_3$	phases characterized by X	m
α-MHo(WO$_4$)$_2$ (M = Rb or Cs)	i.r., X	n
$MLnW_2O_8, nH_2O$ (M = Na, K or Rb; Ln = Sc, Y, Nd or Er)	t.d., X	o
$KLn_3W_5O_{20}$ (Ln = Nd or Er)	m.p., t.d., X	
Ln_2MO_6		
$Ln_2(MO_4)_3, xH_2O$ (M = Mo or W; Ln = Sc, Y, La—Lu)	i.r., m.p., t.d., X	p
$M^1Ln(M^2O_4)_2, xH_2O$ (M^1 = Li—Cs, M^2 = Mo or W; Ln = Sc, Y, La—Lu)	^{1}H n.m.r., m.p., t.d., X	q
$Ln(OH)WO_4, xH_2O$ (Ln = Sc, Y, Sm—Lu)	e, t.d.	r
$LnWO_4Cl$ (Ln = Y, La, Sm, Gd)	X	s
$UMoO_6, xH_2O$		
$UMo_2O_9 \cdot xH_2O$		
$UMo_{10}O_{32}$	prepared in the UO_3–MoO_3–H_2O system, t.d., X	t
$U_2Mo_{21}O_{64}$		
$U_3Mo_{20}O_{64}$		

(a) N. V. Gul'ko, A. G. Karaulov, N. M. Taranukha, and A. M. Gavrish, *Izvest. Akad. Nauk S.S.S.R., neorg. Materialy*, 1973, **9**, 1766; (b) V. N. Karpov, O. V. Sorokina, V. I. Pakhomov, P. M. Fedorov, and T. Ya. Khandozhko, *Zhur. neorg. Khim.*, 1973, **18**, 1475; V. N. Karpov, and O. V. Sorokina, *ibid.*, p. 1663; (c) V. N. Chentsov, Yu. Ya. Skolis, V. A. Levitskii, and Yu. Khekimov, *Izvest. Akad. Nauk S.S.S.R., neorg. Materialy*, 1973, **9**, 1591; (d) P. V. Klevtsov, V. I. Protasova, L. Yu. Kharchenko, and R. F. Klevtsova, *Kristallografiya*, 1973, **18**, 833; (e) E. M. Novikova, A. A. Maier, and O. Ya. Tabunchenko, *Izvest. Akad. Nauk. S.S.S.R., neorg. Materialy*, 1973, **9**, 874; (f) A. A. Evdokimov and V. K. Trunov, *Zhur. neorg. Khim.*, 1973, **18**, 2834; 1974, **19**, 232; (g) L. H. Brixner, J. F. Whitney, and M. S. L. Kay, *J. Solid State Chem.*, 1973, **6**, 550; (h) M. Yoshimura, T. Sata, and T. Nakamura, *Nippon Kagaku Kaishi*, 1973, 2287 (*Chem. Abs.*, 1974, **80**, 77710); (i) P. H. Hubert, P. Michel, and A. Thozet, *Compt. rend.*, 1973, **276**, C, 1779; (j) R. F. Klevtsova, L. P. Kozeeva, and P. V. Klevtsov, *Kristallografiya*, 1974, **19**, 89; (k) M. V. Savel'eva, I. V. Shakhno, V. E. Plyushchev, and O. M. Kuperman, *Izvest. Akad. Nauk S.S.S.R., neorg. Materialy*, 1973, **9**, 632; (l) A. A. Maier, M. V. Provotorov, and E. T. Medvednikova, *Trudy Moskov Khim.-Tekhnol. Inst.*, 1973, **72**, 92 (*Chem. Abs.*, 1974, **80**, 90472); (m) B. Raveau and F. Studer, *Rev. Chim. Minerale*, 1973, **10**, 151; (n) N. S. Zhigubina, M. V. Mokhosoev, E. I. Get'man, and T. A. Ugnivenko, *Ukrain. Khim. Zhur.*, 1973, **39**, 1092 (*Chem. Abs.*, 1974, **80**, 43540); (o) A. M.

the synthesis of a crystalline basic zirconium tungstate reported which is isomorphous with $ZrMo_2O_7(OH)_2(H_2O)$.[515]

The crystallographic data available for molybdate and tungstate salts with heavy cations have been reviewed.[516] The crystal structure of white α-molybdic acid has been determined. This compound consists of isolated double chains of $\{MoO_5(H_2O)\}$ octahedra, each of which share *cis*-edges with two other such octahedra. The structure therefore can be considered to be derived from that for MoO_3, by co-ordination of one H_2O molecule to each metal centre.[517] Several papers have been published concerned with the structures of uranium molybdates. The structure of $UMo_2O_9,3H_2O$ is very similar[518a] to that described for α-molybdic acid above. $UMo_{10}O_{32}$, and α-, β-, and γ-$U_3Mo_{20}O_{64}$ all consist of endless Mo_2O_7 layers, which are two octahedra thick formed by MoO_6 octahedra linked at five vertices. Between these layers U, Mo, and O atoms link to form chains.[518b] U_2MoO_8 contains octahedral $\{Mo(O_2)O_4\}$ units.[518c]

The heats of formation of $Rb_2MoO_4(c)$ and $Cs_2MoO_4(c)$ have been determined as -1492 and $-1407\ kJ\ mol^{-1}$, respectively.[519] The structures and the electronic absorption and emission spectral characteristics of ordered perovskites of the type A_2BWO_6 have been investigated.[520] The preparation, structural data, and vibrational spectra of $MoO_4{}^{2-}$ salts of the cations $[M(NH_3)_4]^{2+}$ (M = Cu, Zn, or Cd) have been described and discussed.[245] The relative Raman intensities of the $v_1(A_1)$ line for several oxyanions, including $MoO_4{}^{2-}$ and $WO_4{}^{2-}$, have been measured and related

515 A. Clearfield and R. H. Blessing, *J. Inorg. Nuclear Chem.*, 1974, **36**, 1174.
516 T. M. Polyanskaya, S. V. Borisov, and N. V. Belov, *Kristallografiya*, 1973, **18**, 1141.
517 I. Böschen and B. Krebs, *Acta Cryst.*, 1974, **30B**, 1795.
518 (a) V. N. Serezhkin, V. F. Chuvaev, L. M. Kovba, and V. K. Trunov, *Doklady Akad. Nauk. S.S.S.R.*, 1973, **210**, 873; (b) V. N. Serezhkin, L. M. Kovba, and V. K. Trunov, *Zhur. strukt. Khim.*, 1973, **14**, 742; *Kristallografiya*, 1973, **18**, 961; *Doklady Akad. Nauk. S.S.S.R.*, 1973, **210**, 1106; (c) V. N. Serezhkin, L. M. Kovba, and V. K. Trunov, *Kristallografiya*, 1973, **18**, 514.
519 P. A. G. O'Hare and H. R. Hoekstra, *J. Chem. Thermodynamics*, 1973, **5**, 851; 1974, **6**, 117; D. W. Osborne, H. E. Flotow, and H. R. Hoekstra, *J. Chem. Thermodynamics*, 1974, **6**, 179.
520 P. Köhl. *Z. anorg. Chem.*, 1973, **401**, 121; G. Blasse and A. F. Corsmit. *J. Solid State Chem.*, 1973, **6**, 513.

Golub, V. I. Maksin, A. P. Perepelitsa, V. N. Solomakha, and K. Aganiyazov, *Ikh. Soedin. Splavov*, 6th, 1969, 42 (*Chem. Abs.*, 1973, **79**, 60982); E. S. Razgon, V. M. Amosov, and V. E. Plyushchev, *Izvest. V. U.Z. Khim. i khim. Tekhnol.*, 1973, **16**, 504 (*Chem. Abs.*, 1973, **79**, 24019); (p) L. H. Brixner, *Rev. Chim. Mineralé*, 1973, **10**, 47; J. H. G. Bode, H. R. Kuijt, M. A. J. Th. Lahey, and G. Blasse, *J. Solid State Chem.*, 1973, **8**, 114; L. H. Brixner, A. W. Sleight, and C. M. Foris, *J. Solid State Chem.*, 1973, **7**, 418; L. H. Brixner and A. W. Sleight, *Mater. Res. Bull.*, 1973, **8**, 1269; A. A. Evdokimov and V. K. Trunov, *Zhur. neorg. Khim.*, 1973, **18**, 3072; F. P. Alekseev, E. I. Get'man, G. G. Koshcheev, and M. V. Mokhosoev, *Ukrain. Khim. Zhur.*, 1973, **39**, 655 (*Chem. Abs.*, 1973, **79**, 86895); A. M. Golub, A. P. Perepelitsa, V. I. Maksin, A. A. Govorov, and A. M. Kalinichenko, *Ikh. Soedin. Splavov*, 6th, 1969, 35 (*Chem. Abs.*, 1973, **79**, 60961); P. V. Klevtsov, L. P. Koseeva, R. F. Klevtsova, and N. A. Novgorodtseva, *Ref. Zhur. Khim.*, 1973, Abs. 2B535 (*Chem. Abs.*, 1973, **79**, 58510); (q) N. N. Bushuev, V. K. Trunov, and A. R. Gizhinksii, *Zhur. neorg. Khim.*, 1973, **18**, 2865; V. K. Rybakov and V. K. Trunov, *ibid.*, p. 3152; 1974, **19**, 347; A. A. Evdokimov and V. K. Trunov, *ibid.*, p. 3157; V. E. Plyushchev and E. S. Razgon, *ibid.*, 1974, **19**, 85; V. K. Trunov, and A. A. Evdokimov, *Rev. Int. Hautes. Temp. Refract.*, 1973, **10**, 109 (*Chem. Abs.*, 1973, **79**, 58193); A. A. Maier, M. V. Provotorov, and V. A. Balashov, *Uspekhi Khim.*, 1973, **42**, 1788 (*Chem. Abs.*, 1974, **80**, 33336); E. M. Novikova and A. A. Maier, *Trudy Moskov Khim.-Tekhnol. Inst.*, 1973, **72**, 94, 102; M. V. Saval'eva, I. V. Shakno, and V. E. Plyushchev, *Ikh. Soedin. Splavov*, 6th, 1969, 32 (*Chem. Abs.*, 1973, **79**, 129734); A. P. Perepelitsa, V. I. Naksin, and A. M. Kalinichenko, *Ikh. Soedin. Splavov*, 6th, 1969, 38 (*Chem. Abs.*, 1973, **79**, 60979); (r) A. M. Golub, V. I. Maksin, V. N. Solomakha, and K. Aganiyazov, *Zhur. neorg. Khim.*, 1973, **18**, 3203; A. M. Golub, V. I. Maksin, A. P. Perepelitsa, and A. A. Govorov, *Ikh. Soedin. Splavov*, 6th, 1969, 40 (*Chem. Abs.*, 1973, **79**, 60951); (s) P. N. Yocom and R. T. Smith, *Mater. Res. Bull.*, 1973, **8**, 1293; (t) V. N. Serezhkin, L. M. Kovba, and V. K. Trunov, *Radiokhimiya*, 1973, **15**, 282 (*Chem. Abs.*, 1973, **79**, 10735); O. A. Ustinov, M. A. Andrianov, N. T. Chebotarev, and G. P. Novoselov, *At. Energy*, 1973, **34**, 155 (*Chem. Abs.*, 1973, **79**, 10448); V. N. Serezhkin, G. N. Ronami, L. M. Kovba, and V. K. Trunov, *Zhur. neorg. Khim.*, 1974, **19**, 1036.

to the extent of π-bonding within these species.[246] The force-field appropriate to these tetrahedral oxyanions has been discussed.[224b]

The synthesis and ion-exchange properties of ceric tungstate have been investigated.[521] The acidic properties of WO_3 in a KCl–NaCl melt have been studied,[522a] and the reduction characteristics of MoO_4^{2-} in fused LiCl–KCl reported.[522b]

The thermal decomposition of the peroxytungstates $K_2WO_n.xH_2O$ ($n = 5$—8) have been described.[523] A theoretical study by the all-valence SCMO method has been made of the $[Mo(O_2)_4]^{2-}$ and $[MoO(O_2)_2(C_2O_4)]^{2-}$ ions. In both cases the highest filled MO's are located mainly on the oxygen atoms and the lowest virtual orbitals on the central Mo^{VI} atom. The energies obtained from excited-state calculations were found to be in reasonable agreement with experimental values.[524] ^{19}F N.m.r. spectroscopy has been used to characterize the complexes $[WO(O_2)F_2-(R^1COCHCOR^2)]$ ($R^1 = R^2 = $ Me, Ph, or CF_3; $R^1 = CF_3$, $R^2 = $ Ph or C_4H_3S) and suggest the nature of the various isomeric species obtained.[525] The formation of $LiOBu^n$ when $LiBu^n$ reacts with peroxo-complexes has been studied. In the case of $[MoO(O_2)_2(HMPA)]$, isotopic labelling experiments have confirmed that the peroxo-oxygen atoms are incorporated into the butoxide product.[251] The fluoro-peroxo-tungstate anions $[WO(O_2)F_4]^{2-}$,[490b] $[WO(O_2)F_3L]$ (L = H_2O, N-oxide of py, picoline, p-methoxy- or p-nitropyridine, or 2,6-lutidine)[526] have been prepared and characterized.

The dissolution of WO_4^{2-} ions in aqueous oxalic acid has been studied and the ion $[WO_3(C_2O_4)]^{2-}$ identified at both high and low acidity.[527] Stability constant data have been reported for molybdenum(VI) or tungsten(VI) complexes with formate[528a] or succinate[528b] and for molybdenum(VI) with maleate.[528c] The molybdenum(VI) species formed in lactic or malic acid solutions over a wide pH range have been studied by dialysis and ion-exchange chromatography.[529] The oxides MO_3 (M = Mo or W) react with ethylene glycol or propanediols (H_2L) to form derivatives of the type ML_3, $MO_2(L)$, or $M(OH)_2L_2$.[530] Several investigations of the complexes formed between molybdenum and tungsten (M) with o-diphenols, such as pyrocatechol, pyrogallol, or 2,3-dihydroxynaphthalene (H_2L) have been reported and the corresponding cis-$[MO_2L_2]$ complexes characterized and their physical and chemical properties investigated.[531] The kinetics of the complex formation between WO_4^{2-}

[521] S. N. Tandon and J. S. Gill, *Talanta*, 1973, **20**, 585.
[522] (a) V. I. Shapoval, V. F. Grischenko, and L. I. Zarubitskaya, *Ukrain. Khim. Zhur.*, 1973, **39**, 867 (*Chem. Abs.*, 1973, **79**, 142 304); (b) P. C. Jain and S. P. Bannerjee, *J. Indian Chem. Soc.*, 1973, **50**, 357; B. N. Popov and H. A. Laitinen, *J. Electrochem. Soc.*, 1973, **120**, 1346.
[523] N. A. Korotchenko, K. K. Fornicheva, G. A. Bogdanov, and R. I. Tereshkina, *Zhur. neorg. Khim.*, 1973, **18**, 1224.
[524] D. H. Brown and P. G. Perkins, *Inorg. Chim. Acta*, 1974, **8**, 285.
[525] J. Y. Calves and J. E. Guerchais, *Rev. chim. Minerale*, 1973, **10**, 733.
[526] J. Y. Calves and J. E. Guerchais, *Z. anorg. Chem.*, 1973, **402**, 206; *Bull. Soc. chim. France*, 1973, 1220.
[527] Yu. M. Potashnikov, and O. I. Kovalevskii, *Zhur. obshchei Khim.*, 1973, **43**, 1869; Yu. M. Potashnikov, M. V. Mokhosoev, and O. I. Kovalevskii, *Zhur. neorg. Khim.*, 1973, **18**, 880.
[528] (a) D. A. Shishkov, *God. Vissh. Khimikotekhnol. Inst. Sofia*, 1970, **15**, 415 (*Chem. Abs.*, 1973, **79**, 10478); (b) D. A. Shishkov and H. I. Doichinova, *Doklady Bolg. Akad. Nauk.*, 1973, **26**, 927 (*Chem. Abs.*, 1973, **79**, 150048); (c) P. C. Jain and S. P. Banerjee, *Madhya Bharati*, Pt. 2, Sect. A, 1970, **18**, 105 (*Chem. Abs.*, 1974, **80**, 41 534).
[529] I. I. Somova, Yu. K. Tselinskii, M. V. Mokhosoev, and O. P. Karpov, *Zhur. neorg. Khim.*, 1974, **19**, 73.
[530] F. A. Schröder and J. Scherle, *Z. Naturforsch.*, 1973, **28B**, 46.
[531] J. Zelinka, M. Bartusek, and A. Okac, *Coll. Czech. Chem. Comm.*, 1973, **38**, 2898; M. Bartusek, *ibid.*, p. 2255; A. Bartecki, M. Cieslak, and M. Raczko, *Roczniki Chem.*, 1973, **47**, 693; B. Grebenova and M. Vrchlabsky, *Coll. Czech. Chem. Comm.*, 1973, **38**, 379.

and catechol and its derivatives have been studied and it appears that only the partially and fully protonated ligand species are reactive in this respect.[532] The reactions of $NH_2CH_2CH_2OH$ or $NH(CH_2CH_2OH)_2$ with $(NH_4)_2MoO_4$ under reflux in benzene over 8 h have been investigated and shown to form *cis*-[MoO_2-$(OCH_2CH_2NH_2)_2$] and *cis*-[$MoO_2\{(OCH_2CH_2)NH\}$], respectively. ^{1}H N.m.r. spectra of these and related complexes have been determined.[533]

The crystal structure of dioxobis-(1,3-diphenylpropanedionato)molybdenum(VI) has been determined by *X*-ray crystallography. The two *cis*-dioxo-groups involve Mo—O bond lengths of 169.6 pm, and the length of the Mo—O bonds in the chelate rings average 216.7 pm *trans* to the oxo-atoms, and 199.3 pm *cis* to these atoms.[534]

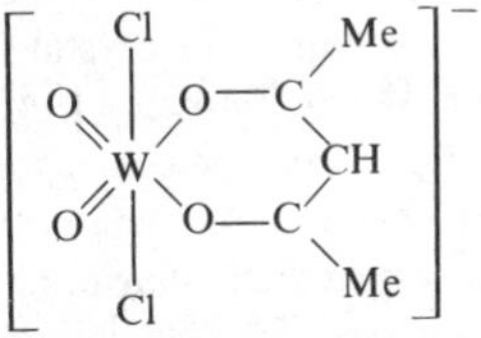

(22)

The series of *cis*-dioxo-*trans*-dichloro-[$MoO_2Cl_2(R^1COCHCOR^2)$] ($R^1 = R^2 =$ Me, CF_3, or Ph; $R^1 = CF_3$, $R^2 = $ Ph or C_4H_3S) complexes have been prepared and characterized.[534a] The configurations of the corresponding tungsten-difluoro-[534b] and tungsten-peroxo-difluoro-[525] complexes have been determined from ^{19}F n.m.r. studies. Substituted azobenzenes may be oxidized to azoxy-derivatives using Bu^tO_2H in the presence of catalytic quantities of [$MoO_2(dpm)_2$].[535] $WOCl_4$ reacts with azoxybenzene to give a stable covalent adduct in which the organic molecule is co-ordinated *via* its O-atom *trans* to W=O.[536] The reaction between $WOCl_4$ and pentane-2,4-dione contrasts with that of $MoOCl_4$ (see p. 139). The product has been characterized by *X*-ray crystallography and shown to be a salt comprising 2,4,6-trimethyl-3-acetylpyrilinium cations and the anions (22).[504] $MoOCl_4$ reacts with anisole (1:2) to produce [$MoOCl_2(C_6H_4OMe)_2$] the i.r. spectrum of which has been recorded.[537] I.r. spectra have also been used to characterize the complexes [MO_2Cl_{2-n} $\{OP(O)MeOH\}_n$] ($n = 1$ or 2) and [$MO_2\{O_2P(OMe)\}$] which are formed in the reactions between MO_2Cl_2(M = Mo or W) and methylphosphinic acid.[538] Di-butylhydroxy-methylphosphonate has been used to extract molybdenum from aqueous HCl solution; the extracted species was identified as $H_3O[MoO_2Cl_2\{(BuO)_2-POCH_2OH\}]$.[539] The formation of 1:1 complexes between WCl_6 or $WOCl_4$ and

[532] K. Kustin, and S.-T. Liu, *Inorg. Chem.*, 1973, **12**, 2362.
[533] M. G. Voronkov, A. Lapsina, V. A. Pestunovich, and J. Popelis, *Zhur. obshchei Khim.*, 1973, **43**, 2687.
[534] (*a*) B. Kojic-Prodic. Z. Ruzic-Toros. D. Grdenic. and L. Golic. *Acta Cryst.*, 1974, **30B**, 300; (*b*) R. Kergoat and J. E. Guerchais, *Rev. chim. Minerale*, 1973, **10**, 585; R. Kergoat, J. M. Mauguen, and J. E. Guerchais, *J. Inorg. Nuclear Chem.*, 1973, **35**, 3970; J. Y. Calves and J. E. Guerchais, *Bull. Soc. chim. France*, 1973, 1222.
[535] N. A. Johnson and E. S. Gould, *J. Org. Chem.*, 1974, **39**, 407.
[536] A. Greco, F. Pirinoli, and G. Dall'asta. *J. Organometallic Chem.*, 1974, **69**, 293.
[537] I. A. Glukhov, S. S. Eliseev, R. M. Narzikulova, and V. G. Kozlov, *Zhur. neorg. Khim.*, 1974, **19**, 963.
[538] Yu. G. Podzolko, A. A. Kuznetsova, L. F. Yankina, and Yu. A. Buslaev, *Zhur. neorg. Khim.*, 1973, **18**, 1255.
[539] S. D. Abaeva, N. N. Ivanova, A. K. Shokanov, and R. K. Zaripova, *Trudy Khim.-Met. Inst. Akad. Nauk. Kaz. S.S.R.*, 1972, 161 (*Chem. Abs.*, 1973, **79**, 117155).

F

$POCl_3$ in the gas phase has been confirmed by vapour pressure measurements[540a] and the solid complexes $[MoO_2(O_2PCl_2)(POCl_2)]$ and $[MoO_2Cl_2(POCl_3)]$ prepared and characterized by i.r. and Raman spectroscopy.[540b]

Iso- and Hetero-polyanions. Urea reacts with H_2MoO_4 in aqueous media to produce a mixture of $H_2Mo_2O_7,2urea$ and $H_3Mo_3O_{10},3urea$.[541] $Na_2[Mo_3O_{10}(H_2O)],H_2O$ has been characterized by a variety of spectral techniques and an equilibrium between this structure and $Na_2[Mo_3O_9(OH)_2],2H_2O$ was identified, the latter arrangement being favoured at higher temperatures.[542]

The crystal structure of the low-temperature form of lithium tetramolybdate, $L-Li_2Mo_4O_{13}$ has been determined and the compound shown to be a derivative of the V_6O_{13} structure with oxygen atoms in a f.c.c.-type array and Li^I and Mo^{VI} atoms in octahedral sites.[543] The full account of the crystal structure of $[H_3N_3P_3(NMe_2)_6]$ $[Mo_6O_{19}]$ has been published.[544] $Na_6[Mo_7O_{24}],14H_2O$ has been shown by X-ray crystallography to be isomorphous with the corresponding ammonium tetrahydrate salt.[545] $(NH_4)_6[Mo_8O_{27}],4H_2O$ contains octamolybdate ions which are polymerized to form infinite chains by vertex sharing[546a] and the mechanism of such polymerization has been discussed.[546b] The anions of $Na_2(NH_4)_8[H_2W_{12}O_{42}],$ $12H_2O$ and $Mg_5[H_2W_{12}O_{42}],38H_2O$ have been characterized by X-ray crystallography and shown to consist of four groups of three edge-linked octahedra arranged to give $2/m$ pseudo-symmetry.[547]

Polymerization processes which occur on acidification of Na_2MoO_4 solutions have been investigated by thermometric titration. Polymerization was found to proceed rapidly to $Mo_7O_{24}^{6-}$, with only $HMo_2O_8^{3-}$ identified as an intermediate species. Protonation of the heptamolybdate is then followed by formation of the octamolybdate ion. The enthalpies of association have been calculated for each of these condensation steps.[548] Similar data have been reported in separate studies.[549] Cation-exchange[550a] and precipitation studies[550b] of the molybdenum-containing species present in acidic solutions have also only obtained evidence for heptamolybdate species. The extraction of polymolybdates from aqueous solutions by tri-n-octylamine hydrochloride in $CHCl_3$ has been investigated and species of the type $[H_aMo_nO_b]^{z-}$ were extracted. The ratio $n:z$ was determined by chemical analysis, and vapour-pressure studies for the organic phase allowed a direct calculation of n and z. These studies identified continuous polymerization in the pH range 1—5, with ions of $n = 6$ or 10 persisting for the pH ranges 4—5 and 2—2.5, respectively. Two

[540] (a) V. I. Trusov and A. V. Suvorov, *Zhur. neorg. Khim.*, 1974, **19**, 549; (b) K. Dehnicke and A. F. Shihada, *Z. Naturforsch.*, 1973, **28B**, 148.

[541] Y. Umetsu and H. Mase, *Nippon Kagaku Kaishi*, 1973, 930 (*Chem. Abs.*, 1973, **79**, 38 123).

[542] J. Chojnacki and S. Hodorowicz, *Roczniki Chem.*, 1973, **47**, 2213 (*Chem. Abs.*, 1974, **80**, 115 534).

[543] B. M. Gatehouse and B. K. Miskin, *J. Solid State Chem.*, 1974, **9**, 247.

[544] H. R. Allcock, E. C. Bissell, and E. T. Shawl, *Inorg. Chem.*, 1973, **12**, 2963.

[545] K. Sjoborn and B. Hedman, *Acta Chem. Scand.*, 1973, **27**, 3673.

[546] (a) I. Boeschen, B. Buss, and B. Krebs, *Acta Cryst.*, 1974, **30B**, 48; (b) K. H. Tytko, *Z. Naturforsch*, 1973, **B28**, 272.

[547] H. D'Amour and R. Allmann, *Z. Krist.*, 1973, **138**, 5; Y. H. Tsay and J. V. Silverton, *ibid.*, 1973, **137**, 256.

[548] N. D. Jesperson, *J. Inorg. Nuclear Chem.*, 1973, **35**, 3873; N. Kiba and T. Takeuchi, *J. Inorg. Nuclear Chem.*, 1974, **36**, 847; *Bull. Chem. Soc. Japan*, 1973, **46**, 3086.

[549] S. Ahn and E. Park, *Daehan Hwahak Hwoejee*, 1973, **17**, 145 (*Chem. Abs.*, 1973, **79**, 97634); A. N. Zelikman and N. N. Rakova, *Ref. Zhur. Met.*, 1973, Abs. 5G270 (*Chem. Abs.*, 1974, **80**, 64394).

[550] (a) G. M. Vol'dman, A. N. Zelikman, and I. Sh. Khutoretskaya, *Zhur. neorg. Khim.*, 1973, **18**, 3046; (b) B. Grebenova and M. Vrchlabsky, *Coll. Czech. Chem. Commun.*, 1973, **38**, 394.

possible schemes for formation of the $n = 6$ ions and the possible conversion of these into $n = 7$, 8, and 10 ions were discussed.[551]

The heat of dehydration of borotungstic acid has been determined[552] and the preparation and i.r. spectrum of $LiH_2[Al(HMoO_4)_6],11H_2O$ have been described.[553] Crystals of β-$K_4SiW_{12}O_{40}.9H_2O$ have been shown by X-ray diffraction studies to contain $[SiW_{12}O_{40}]^{4-}$ anions which are geometrical isomers of the Keggin structure. This isomerism is obtained by a 60° rotation of one of the trigonal W_3O_{13} units of the Keggin model around its three-fold axis.[554] The condensation of H_2MoO_4 with $Si(OH)_4$ (12:1) in 5M–HCl has been studied potentiometrically and the formation of α-$[SiMo_{12}O_{40}]^{4-}$ observed, *via* the probable intermediates $[HMo_3O_{10}]^-$ and $[H_2Mo_6O_{20}]^{2-}$.[555] The electrochemical reduction characteristics of $H_4SiW_{12}O_{40}$ have been investigated and two primary reduced forms identified.[556] The 1:1 ion-association complex between Rhodamine B and $H_4SiMo_{12}O_{40}$ has been characterized,[557] and stability constants have been determined for the 1:1 complexes formed between $SiW_{11}O_{39}^{8-}$ and M^{II}_{aq} (M = Mn–Zn) species.[558] $[SiMo_{11}CrO_{39}$-$(H_2O)]^{5-}$ has been prepared by the addition of $Cr(NO_3)_3$ to a solution of α-molybdo-silicic acid and the complex characterized by polarographic and other studies. The water molecule appears to be bound to the Cr^{III} centre of this ion.[106] Magnetic moment measurements for $K_6[SiM_{11}FeO_{40}H],nH_2O$ (M = Mo or W) have been determined and the values obtained ($\mu_{eff} = 4.1$ and 1.9 BM, respectively) indicate that the Fe^{III} centre is high-spin in the molybdo-anion but low-spin in the tungsto-anion.[559] The polarographic characteristics have been reported for several members of the 9-tungstoheteropoly anion series.[560] I.r. and Raman spectral data have been recorded for the acids $H_n[XM_{12}O_{40}],nH_2O$ (X = P or Si; M = Mo or W)[561] and the role of water in these and the related germanium species has been discussed with reference to ^1H n.m.r. spectral data.[562]

X-Ray crystallographic studies have identified a novel structure for the ammonium phosphomolybdate $(NH_4)_5[(MoO_3)_5(PO_4)(HPO_4)],3H_2O$. The anion consists of a ring of five distorted MoO_6 octahedra, four of these links consist of shared edges and the fifth a shared apex of MoO_6 octahedra. The Mo_5O_{21} crown thus formed is capped on one face by a PO_4^{3-} ion and on the other by a HPO_4^{2-} ion. The cavity created in this ion appears to be too small to accommodate a heteroatom.[563] The crystal structures of $Na_6[P_2Mo_5O_{23}],13H_2O$,[564] $Na_4H_2[P_2Mo_5O_{23}],10H_2O$,[565] and

551 J. I. Bullock, R. Pathak, A. Rusheed, and J. E. Salmon, *J. Inorg. Nuclear Chem.*, 1974, **36**, 1881.
552 M. M. Sadykova, G. V. Kosmodem'yanskaya, and V. I. Spitsyn, *Zhur. neorg. Khim.*, 1974, **19**, 408.
553 Q. S. Holguin, R. Gruz, and C. A. Campero, *Rev. Inst. Mex. Petrol.*, 1973, **5**, 57 (*Chem. Abs.*, 1974, **80**, 43531).
554 K. Yamamura and Y. Kazuko, *J.C.S. Chem. Comm.*, 1973, 648.
555 E. Mars, *Roczniki Chem.*, 1973, **47**, 701 (*Chem. Abs.*, 1973, **79**, 100029).
556 A. M. Baticle, R. Rudelle, J. L. Sculfort, and P. Vennereau, *Compt. rend.*, 1973, **277**, C, 679.
557 A. Golkowska and L. Pszonicki, *Talanta*, 1973, **20**, 749.
558 A. Teze and P. Souchay, *Compt. rend.*, 1973, **276**, C, 1525.
559 P. Rigny, L. Pinsky, and J. M. Weulersse, *Compt. rend.*, 1973, **276**, C, 1223.
560 R. Contant, J. M. Fruchart, G. Herve, and A. Teze, *Compt. rend.*, 1974, **278**, C, 199.
561 R. Thouvenot, C. Rocchiccioli-Deltcheff, and P. Souchay, *Compt. rend.*, 1974, **278**, C, 455.
562 V. F. Chuvaev, E. V. Vanchikova, L. I. Lebedeva, and V. I. Spitsyn, *Doklady Akad. Nauk S.S.S.R.*, 1973, **210**, 370.
563 J. Fischer, L. Ricard, and P. Toledano, *J.C.S. Dalton*, 1974, 941.
564 R. Strandberg, *Acta Chem. Scand.*, 1973, **27**, 1004.
565 B. Herman, *Acta Chem. Scand.*, 1973, **27**, 3335.

$Na_3[H_3PMo_9O_{34}],10H_2O$[566] have also been determined. The kinetics of the formation of 12-molybdophosphoric acid in $HClO_4$ and H_2SO_4 solution have been studied and an initial reaction between PO_4^{3-} and Mo^{VI} entities suggested prior to polymerization.[567] A series of 12-molybdophosphate salts with biguanidium and related cations has been prepared.[568] The results of polarographic studies for 18-tungstomolybdates of the type $(NH_4)_6[P_2W_{18-n}Mo_nO_{62}]$ $(n = 0$—18) suggest that these species are further polymerized in solution.[569] The ion-exchange properties of zirconium molybdophosphates[570a] and molybdovanadates[570b] have been further investigated.

Preparative routes to heteropolytungstates containing X^{III} and X^V (X = P, As, Sb, or Bi) atoms have been further developed[571] and $H_3[SbMo_{12}O_{40}].48H_2O$ has been prepared by refluxing a dilute solution containing MoO_4^{2-} and $Sb_2O_7^{4-}$ ions.[572] New stability constant and polarographic data have been obtained for the 12- and 16-molybdoarsenic acids.[573] An X-ray crystallographic study of $(NH_4)_6[TeMo_6O_{24}]$, $Te(OH)_6,7H_2O$ has shown that the $[TeMo_6O_{24}]^{6-}$ anion consists of a central TeO_6 octahedron, surrounded by six MoO_6 octahedra which are condensed in a flat arrangement with $3m$ symmetry.[574] The role of tellurium(VI) in the formation of heteropoly-acids has been further described,[575] and the heats of solution of various Te–W heteropoly-compounds have been reported.[576]

The yellow heteropolyanion $[H_2W_{11}VO_{40}]^{7-}$ has been obtained by chlorine oxidation of the ion $[H_2W_{11}VO_{40}]^{8-}$. The former is an 11-tungstovanadate(v) ion which has a Keggin structure based on that of metatungstate. A summary of the interconversion and chemical reduction of tungstovanadates has also been presented.[577] The best conditions for the reduction of $H_3[VW_5O_{19}]$ by $TiCl_3$ or $SnCl_2$ have been determined[578] and the e.s.r. spectrum has been determined for the vanadium(IV) centre of $[VW_5O_{19}]^{4-}$ doped in a lattice of the corresponding penta-tungstovanadate(v) species.[579] The basicity and redox properties of the molybdo-phosphovanadic acids have been studied in non-aqueous media and both the basicity and the redox potential found to increase with increasing vanadium content.[580] ^{1}H N.m.r. spectra have been reported for the acids $H_4[PW_{11}VO_{40}].nH_2O$ and $H_5-[PW_{10}V_2O_{40}].nH_2O$ and for $n > 20$, all the protons appear to be present as H_3O^+

566 R. Strandberg, *Acta Chem. Scand.*, 1974, **28A**, 217; H. D'Amour, and R. Allmann, *Naturwissen.* 1974, **61**, 31.
567 P. M. Beckwith, *Diss. Abs.* (B), 1973, **33**, 5179.
568 A. Maitra, *J. Indian Chem. Soc.*, 1973, **50**, 361.
569 V. K. Yakushin, L. P. Maslov, and N. A. Tsvetkov, *Ref. Zhur. Khim.*, 1973, Abs. 4V72 (*Chem. Abs.*, 1973, **79**, 60956).
570 (a) I. K. Vinter, E. S. Boichinova, and N. E. Denisova, *Zhur. priklad Khim.*, 1973. **46**, 1471; I. K. Vinter, E. S. Boichinova, N. E. Denisova, and R. B. Chetverina, *Zhur. priklady Khim.*, 1973, **46**, 997, 1117; (b) R. G. Safina, N. E. Denisova, and E. S. Boichinova, *Zhur. priklad Khim.*, 1973, **46**, 2432.
571 C. Tourne, A. Revel, G. Tourne, and M. Vendrell, *Compt. rend.*, 1973, **277**, C, 643.
572 G. C. Bhattacharya and S. K. Roy, *J. Indian Chem. Soc.*, 1973, **50**, 359.
573 S. V. Lugovoi and Z. Z. Odud, *Zhur. neorg. Khim.*, 1973, **18**, 1076; *Izvest. V. U.Z. Khim. i khim. Tekhnol.*, 1973, **16**, 1826 (*Chem. Abs.*, 1974, **80**, 74809).
574 H. T. Evans, jun., *Acta Cryst.*, 1974, **30B**, 2095.
575 E. Sh. Ganelina, and L. A. Bubnova, *Zhur. neorg. Khim.*, 1973, **18**, 2180.
576 E. Sh. Ganelina, V. P. Kuz'micheva, L. A. Bubnova, and M. B. Krasnopol'skaya, *Zhur. neorg. Khim.*, 1973, **18**, 1893.
577 C. M. Flynn, jun., M. T. Pope, and S. O'Donnell, *Inorg. Chem.*, 1974, **13**, 831.
578 E. F. Tkach and N. A. Polotebnova, *Zhur. neorg. Khim.*, 1974, **19**, 155.
579 H. So, C. M. Flynn, jun., and M. T. Pope, *J. Inorg. Nuclear Chem.*, 1974, **36**, 329.
580 V. C. Ngyuen and N. A. Polotebnova, *Zhur. neorg. Khim.*, 1973, **18**, 2189.

ions.[581] The incorporation of Cr[III] atoms into paratungstate ions has been followed spectrophotometrically and lithium and sodium 6- and 12-tungstochromate(III) salts have been characterized by X-ray powder diffractometry.[107]

Cyclic voltammetric studies of $[CeMo_{12}O_{42}]^{8-}$ have defined a one-electron reduction peak at $+0.45$ V vs. SCE to give a species which, from e.s.r. studies, appears to contain Ce[III].[582] The reactions of $(NH_4)_6H_2[CeMo_{12}O_{42}],8H_2O$ and $(NH_4)_6H_2$-$[CeMo_{10}O_{36}],7H_2O$ with $K_8Ta_8O_{19},6H_2O$ have been followed spectrophotometrically and the salts $(NH_4)_{12}[CeTa_2Mo_{12}O_{49}],22.5H_2O$ and $(NH_4)_8[CeTa_2$-$Mo_{10}O_{41}],14H_2O$ isolated and characterized.[583] [1]H N.m.r. spectral data recorded for the acids $H_{12}[CeNbMo_{12}O_{46}]$ and $H_{12}[CeTa_2Mo_{12}O_{49}]$ have been reported and seen to conform with earlier predictions of their structure.[584] Further developments in the preparation of heteropolytungstatothorates have been presented[585] and the electronic absorption spectra of salts containing the $[MMo_{12}O_{42}]^{8-}$ (M = Ce, Th, or U) described.[586] [1]H N.m.r. spectral data obtained for $H_8[UMo_{12}O_{42}],18H_2O$ suggest that five of the eight protons are present as H_3O^+ ions and the other three are bonded as OH groups in the anion.[587]

S- *and* Se-*Donor Ligands.* The compounds WSF_4 and WS_2F_2 have been prepared by treating SF_6 with tungsten; some thermochemical data were obtained by mass spectrometric techniques.[352] [19]F N.m.r. spectroscopy has been used to characterize the products $[W_2S_2F_9]^-$ and $[WSF_{5-x}Cl_x]^-$ ($x = 0$—4) during the reaction of $WSCl_4$ with HF in MeCN at -30 to $0 °C$.[588]

The polythiomolybdate $3MoS_4.2NH_4OH$ has been prepared by heating an ammonical solution of $(NH_4)_2MoO_4$ with $(NH_4)_2S$ containing dissolved sulphur.[589] The syntheses of $CuNH_4(MoS_4)$[590] and $(LH)_2MS$ (L = hexamethyltetramine; M = Mo or W) have been described and the X-ray powder diffraction patterns of the latter two salts indicate that they are isomorphous.[591] X-Ray crystallographic data have been presented for compounds of the type A_2MX_4, A_2MOX_3, and $A_2MO_2X_2$ (A = K, Rb, Cs, or NH_4; M = Mo or W; X = S or Se). Molybdates and tungstates of the types A_2MX_4 and A_2MOX_3 crystallize with the β-K_2SO_4 structure, and the $(NH_4)_2MO_2X_2$ salts crystallize in the monoclinic space group C_2/c.[592] The absolute intensity of the $v_1(A_1)$ Raman line of the $[WS_4]^{2-}$ ion has been compared to that of the corresponding lines for several tetrahedral oxyanions and the former ion seen to involve considerable W—S π-bonding. The preparation, vibrational

[581] V. F. Chuvaev, G. M. Shinik, N. A. Polotebnova, and V. I. Spitsyn, *Doklady Akad. Nauk S.S.S.R.*, 1973, **211**, 614.

[582] L. McKean and M. T. Pope, *Inorg. Chem.*, 1974, **13**, 747.

[583] E. A. Torchenkova, D. Nguyen, L. P. Kazanskii, and V. I. Spitsyn, *Izvest. Akad. Nauk S.S.S.R., Ser. khim.*, 1973, 734.

[584] V. F. Chuvaev, D. Ngyuen, E. A. Torchenkova, and V. I. Spitsyn, *Doklady Akad. Nauk S.S.S.R.*, 1973, **209**, 635.

[585] A. V. Botar and T. J. R. Weakley, *Rev. Roumaine Chim.*, 1973, **18**, 1155 (*Chem. Abs.*, 1973, **79**, 99929).

[586] L. P. Kazanskii, E. A. Torchenkova, and V. I. Spitsyn, *Doklady Akad. Nauk S.S.S.R.*, 1973, **213**, 118.

[587] A. M. Golubev, L. P. Kazanskii, V. F. Chuvaev, E. A. Torchenkova, and V. I. Spitsyn, *Doklady Akad. Nauk S.S.S.R.*, 1973, **209**, 1340.

[588] Yu. A. Buslaev, Yu. V. Kokunov, and Yu. D. Chubar, *Doklady Akad. Nauk S.S.S.R.*, 1973, **213**, 1083.

[589] C. R. Kurtak and L. D. Hartzog, *U.S. P.*, 3764649 (*Chem. Abs.*, 1973, **80**, 85277).

[590] M. J. Redman, *Inorg. Syn.*, 1973, **14**, 95.

[591] M. R. Udupa and G. Aravamudan, *Current Sci.*, 1973, **42**, 676 (*Chem. Abs.*, 1973, **79**, 142398).

[592] A. Müller and W. Sievert, *Z. anorg. Chem.*, 1974, **403**, 251, 267; W. Gonschorek, Th. Hahn, and A. Müller, *Z. Krist.*, 1973, **138**, 380.

spectra, and X-ray diffraction data for $[Cr(NH_3)_6]^{3+}$ and alkali-metal salts of thio-anions of molybdenum(VI) and tungsten(VI) have been described.[593] A detailed interpretation of the electronic transition in the thioanions and oxythioanions of these elements has been presented[594a] and m.c.d. data obtained for the $[MS_4]^{2-}$ (M = Mo or W) ions.[594b] Condensation products of the $[MoO_2S_2]^{2-}$ and $[MoOS_3]^{2-}$ ions have been isolated following the acidification of the corresponding ammonium salt.[595] $[WO_2S_2]^{2-}$ has been shown to co-ordinate transition-metal ions, such as Ni^{II} or Co^{II}, as a bidentate S,S-donor ligand,[596a] and the preparation and characterization of $[Ni(en)_2(WS_4)]$ has been described.[596b]

$[WO_2(acac)_2]$ reacts with thiodebenzoylmethane (HL) (1:2) in refluxing xylene to produce the complex $[WO_2L_2]$.[435] The complex $[MoO_2L'_2]$ (HL' = 5-phenyl-1-pyrazoline-dithiocarbamate)[474] has been prepared and studies concerning the interactions of molybdenum(VI) with o-hydroxy-4-benzamidothiosemicarbazide[201] and thioglycolhydroxamic acid[597] have been described. A further report of the catalysis of phosphine oxidation by $[MoO_2(S_2CNR_2)_2]$ (R = Et, Pr, or Bu^i) complexes has been published,[598] and X-ray emission data for several dithiolato-molybdenum(VI) complexes have been reported.[599]

N-*Donor Ligands.* The full report of the reactions of WCl_6 or WCl_5 with CCl_3CN to give the nitrido-tungsten(VI) complex $[Cl_3CCCl_2NWCl_4,CCl_3CN]$ has been published.[458] Addition of trimethylsilylazide to a suspension of $[MoCl_4L_2]$ (L = MeCN or THF) in a suitable dry solvent gives a clear orange-red solution with evolution of N_2. From these solutions molybdenum(VI) nitrido-complexes such as $[MoNCl_3(bipy)]$, $[MoNCl_3(OPPh_3)_2]$, and $(Et_4N)_2[MoNCl_5]$ have been obtained by the appropriate addition of ligands. $[MoN(S_2CNEt_2)_3]$ is conveniently prepared by treating $[MoCl(S_2CNEt_2)_3]$ with Me_3SiN_3 or NaN_3 in MeCN solution.[270] WF_6 reacts with Me_3SiX (X = NEt_2 or C_6F_5O) to produce the compounds $WF_2(NEt_2)_4$, $[WF_4(NEt_2)_2]_n$, $[WF_5(OC_6F_5)]$, or $[WF_4(OC_6F_5)_2]$.[600a] $(Me_3Si)_2NH$ and Me_3SiN-$(Me)PF_2$ react with WF_6 in MeCN solution to produce $[WF_4(NMe)(NCMe)]$ (m.p. 132 °C). The co-ordinated MeCN in this latter compound is readily displaced by py to produce $[WF_4(NMe)(py)]$ (m.p. 117 °C).[600b]

The reasonably stable isothiocyanato-derivatives $(R_4N)_2[MO_2(NCS)_4]$ (R = Me or Et; M = Mo or W) may be precipitated by addition of an R_4N^+ salt to freshly acidified MO_4^{2-} solutions in the presence of a large excess of NCS^-.[601] The complexes

[593] A. Müller, I. Böschen, and E. J. Baran, *Monatsh.,* 1973, **104**, 821; A. Müller, N. Weinstock, N. Mohan, C. W. Schlaepfer, and K. Nakamoto, *Appl. Spectroscopy,* 1973, **27**, 257 (*Chem. Abs.,* 1973, **79**, 71831); A. Müller, H. Schulze, W. Sievert, and N. Weinstock, *Z. anorg. Chem.,* 1974, **403**, 310.

[594] (a) A. Bartecki and D. Dembicka, *Inorg. Chim. Acta,* 1973, **7**, 610; *Roczniki Chem.,* 1973, **47**, 477 (*Chem. Abs.,* 1973, **79**, 11562); (b) R. H. Petit, B. Briat, A. Müller, and E. Diemann, *Chem. Phys. Letters,* 1973, **20**, 540.

[595] Ya. D. Fridman, L. Ya. Mikhailyuk, G. I. Chalkov, and G. A. Tursunova, *Zhur. neorg. Khim.,* 1973, **18**, 1836.

[596] (a) A. Müller, H. H. Heinsen, and G. Vandrish, *Inorg. Chem.,* 1974, **13**, 1001; (b) M. C. Chakravorti and A. Müller, *Inorg. Nuclear Chem. Letters,* 1974, **10**, 63.

[597] R. F. Zagrebina, *Ref. Zhur. Khim.,* 1973, Abs., 2G14 (*Chem. Abs.,* 1973, **79**, 61150).

[598] R. Barral, C. Bocard, I. Serree de Roch, and L. Sajus, *Kinetika i Kataliz,* 1973, **14**, 164 (*Chem. Abs.,* 1973, **79**, 10281).

[599] J. Finster, N. Müsel, P. Müller, W. Dietzsch, A. Meisal, and E. Hoyer, *Z. Chem.,* 1973, **13**, 146.

[600] (a) A. Majid, D. W. A. Sharp, J. M. Winfield, and I. Hanby, *J.C.S. Dalton,* 1973, 1876; (b) M. Harman, D. W. A. Sharp, and J. M. Winfield, *Inorg. Nuclear Chem. Letters,* 1974, **10**, 183.

[601] B. J. Brisdon, and D. A. Edwards, *Inorg. Nuclear Chem. Letters,* 1974, **10**, 301.

formed between molybdenum(VI) or tungsten(IV) and *N*-phenyl-*N'*-2-4-phenyl-thiazolyl)guanidine[602a] and molybdenum(VI) and enanthohydroxamic acid[602b] have been investigated. The structures of the 8- and 2-methyl-8-quinolato-molybdenum(VI) complexes, $[MoO_2Q_2]$, have been investigated by ^1H n.m.r. spectroscopy. These ligands appear to co-ordinate with their N atom *trans* to an oxo-group and thus the Mo—N interactions are weak and labile. This effect is magnified by steric repulsions in the case of the 2-methyl complex.[471] I.r. spectral data have been presented for series of molybdenum(VI) and tungsten(VI) 1:1 and 1:2 8-quinolato-complexes and some structural information was obtained.[603] The complex $[MoO_2(H_2O)L]$ $[H_2L =$ *N*-(salicylidene)-2-hydroxybenzamine] has been described.[430]

Organometallic Compounds. The mechanism for the 'homogeneous' catalysis of olefin metathesis using co-ordination catalysts derived from WCl_6 has been re-investigated. The results obtained have shown that the active catalysts are insoluble in the reaction medium and therefore should be considered as heterogeneous catalysts. For example, activation of WCl_6 in benzene by alkylating agents such as $LiBu^n$ affords solids which contain virtually all the tungsten of the system.[604] Treatment of silica or γ-alumina with a solution of WMe_6, followed by activation at temperatures > 373 K, yields an alkene disproportionation catalyst with activities comparable to catalysts derived from $Mo(CO)_6$.[289] The stereochemistry of the oxidative addition compound (19) formed[439] between $[MoO S(S_2CNPr^n)_2]$ and tetracyanoethylene is described on p. 131, and the full account of the crystal structure of $[Me_4W\{ON(Me)-NO\}_2]$ has been published.[605] New derivatives of $[(\eta^5\text{-}C_5H_5)_2MoCl_2]$ (M = Mo or W) have been obtained by treating these compounds with alkali-metal salts of various pseudohalides or thiols. The compounds $[(\eta^5\text{-}C_5H_5)_2MoX_2]$ (M = Mo or W; X = CN, NCO, SCN, N_3, SMe, SEt, SPr^n, or SBu^n) have been prepared in this manner.[606]

3 Technetium and Rhenium

Introduction.—Reviews have been presented concerning the uses, exploitation, and resources of rhenium,[607] and the recent developments in the synthetic[608] and organometallic[609] chemistry of technetium and rhenium have also been described. The nature of rhenium–ligand bonding in a variety of compounds has been probed using ESCA[610a] and *X*-ray absorption[610b] spectroscopy.

Interesting new developments in the chemistry of these elements include the

602 (a) W. U. Malik, P. K. Srivastava, and S. C. Mehra, *J. Indian Chem. Soc.*, 1973, **50**, 739; (b) Zh. G. Gukasyan, N. M. Sinitsyn, K. I. Petrov, and I. P. Kislyakov, *Izvest. V. U.Z. Khim. i khim. Tekhnol.*, 1973, **16**, 998 (*Chem. Abs.*, 1973, **79**, 111280).

603 A. Doadrio and J. Martinez, *An. Quim.*, 1973, **69**, 879 (*Chem. Abs.*, 1974, **80**, 8606).

604 E. L. Muetterties and M. A. Busch, *J.C.S. Chem. Comm.*, 1974, 754.

605 S. R. Fletcher and A. C. Skapski, *J. Organometallic Chem.*, 1973, **59**, 299.

606 S. P. Anand, *J. Inorg. Nuclear Chem.*, 1974, **36**, 925.

607 R. U. King, *U.S. Geol. Surv. Prof. Pap.*, 1973, **820**, 557 (*Chem. Abs.*, 1973, **79**, 81515).

608 G. Davies, *Ann. Reports Inorg. Gen. Synthesis*, 1973, **1**, 100.

609 P. M. Treichel, *J. Organometallic Chem.*, 1973, **58**, 273.

610 (a) D. G. Tisley and R. A. Walton, *J. Mol. Structure*, 1973, **17**, 401; (b) A. V. Pendharkar and C. Mande, *Pramana*, 1973, **1**, 102 (*Chem. Abs.*, 1973, **79**, 130180).

preparation of the pentacarbonyl radicals[611] and $Me_4N[6-(CO)_3-6-ReB_9H_{13}]$, the anion of which is a member of a new class of metalloboranes.[612] The μ-dinitrogenyl complex $[(Me_2PhP)_4ClReN_2MoCl_4(OMe)]$ has been prepared and its structure (7) determined by X-ray crystallography (see p.110).[267] The preparation and characterization of $[\{ReCl(N_2)(PMe_2Ph)_4\}_2TiCl_4]$ and $[\{ReCl(N_2)(PMe_2Ph)_4\}TiCl_4(S)]$ (S = THF, Et_2O, or CH_2Cl_2) have been described. The first of these complexes appears to be trinuclear with two μ-dinitrogenyl ligands.[613]

Low-temperature studies of polarized single-crystal electronic spectra of $(Bu_4N)_2$-$[Re_2Cl_8]$ and $[Re_2Cl_6(PEt_3)_2]$ have shown that the low-energy band in these complexes corresponds to a $\delta \rightarrow \delta^*$ transition.[406] U.v. irradiation of a solution of $[Re_2Cl_8]^{2-}$ cleaves the metal-metal bond and provides a convenient synthetic route to monomeric chloro-complexes of rhenium(III).[614] This result suggests that the energy of the Re—Re bond in rhenium(II) dimers is significantly less than the value 1200—1600 kJ mol^{-1} previously suggested. A new class of rhenium(II) tertiary phosphine halogeno-complexes of the type $[Re_2Cl_4(PR_3)_4]$ has been isolated from the reaction of Re_3Cl_9 or $[Re_2Cl_8]^{2-}$ with PR_3. X-Ray crystallographic studies have confirmed that $[Re_2Cl_4(PEt_3)_4]$ has the expected eclipsed (D_{2d}) structure with a strong Re—Re bond of length 223.2(6) pm.[615] $[Re_2Cl_6(diphos)_2(MeCN)_2]$ has been prepared by treating $(Bu_4N)_2[Re_2Cl_8]$ with diphos and shown to have a centrosymmetric chlorine-bridged dimeric structure with each metal atom possessing a distorted octahedral environment. Surprisingly, there is no significant metal interaction within these units (Re $\cdots$ Re = 380.9 pm).[616]

$ReOMe_4$ has been prepared by the reaction of $[ReOCl_3(PPh_3)_2]$ or $ReOCl_4$ with LiMe in Et_2O at $-35°C$. Spectral data are consistent with this d^1 metal alkyl possessing a square-pyramidal C_{4v} structure.[617]

Carbonyl Complexes.—For a comprehensive discussion of this topic the reader should consult the Organometallic Specialist Periodical Report.

A species believed to be the $^{99}Tc(CO)_5$ radical has been produced in high yield by the β-decay of $^{99}Mo(CO)_6$. This species reacts with $Mn(CO)_5I$ to yield $Tc(CO)_5I$ and with photochemically produced $Mn(CO)_5$ to give $TcMn(CO)_{10}$.[611a] The co-condensation of rhenium atoms with pure CO at 15 K affords a product whose i.r. spectrum is consistent with a square-pyramidal $Re(CO)_5$ unit.[611b] A gas-phase electron-diffraction study of $Re_2(CO)_{10}$ has afforded the following intramolecular dimensions for a D_{4h} molecular structure, Re—Re = 304(1), Re—C = 201(1), C—O = 116(2) pm, and $\angle$ Re—Re—C = 88.0(0.5)°.[618] The i.r. spectra of $M_2(CO)_{10}$ (M = Mn, Tc, or Re) have been re-investigated in the ν(C—O) stretching region, with special emphasis on the weak isotopic bands and the equatorial and axial C—O stretching force constants found to be highest for $Tc_2(CO)_{10}$.[619] A report concerning the thermal

[611] (a) I. G. De Jong and D. R. Wiles, *Inorg. Chem.*, 1973, **12**, 2519; (b) H. Huber, E. P. Kundig, and G. A. Ozin, *J. Amer. Chem. Soc.*, 1974, **96**, 5585.

[612] J. W. Lott and D. F. Gaines, *Inorg. Chem.*, 1974, **13**, 2261.

[613] R. Robson, *Inorg. Chem.*, 1974, **13**, 475.

[614] G. L. Geoffroy, H. B. Gray, and G. S. Hammond, *J. Amer. Chem. Soc.*, 1974, **96**, 5565.

[615] F. A. Cotton, B. A. Frenz, J. R. Ebner, and R. A. Walton, *J.C.S. Chem. Comm.*, 1974, 4.

[616] J. A. Jaecker, W. R. Robinson, and R. A. Walton, *J.C.S. Chem. Comm.*, 1974, 306.

[617] K. Mertis, J. F. Gibson, and G. Wilkinson, *J.C.S. Chem. Comm.*, 1974, 93.

[618] N. I. Gapotchenko, Yu. T. Struchkov, N. V. Alekseev, and I. A. Ronova, *Zhur. strukt. Khim.*, 1973, **14**, 419.

[619] G. Bor and G. Sbrignadello, *J.C.S. Dalton*, 1974, 440.

decomposition of $Re_2(CO)_{10}$ has been published.[620a] The heat of formation of this compound has been estimated as $-1560 \pm 7\,kJ\,mol^{-1}$ from combustion enthalpy measurements[620b] and certain of its other thermodynamic characteristics have also been reported.[620c]

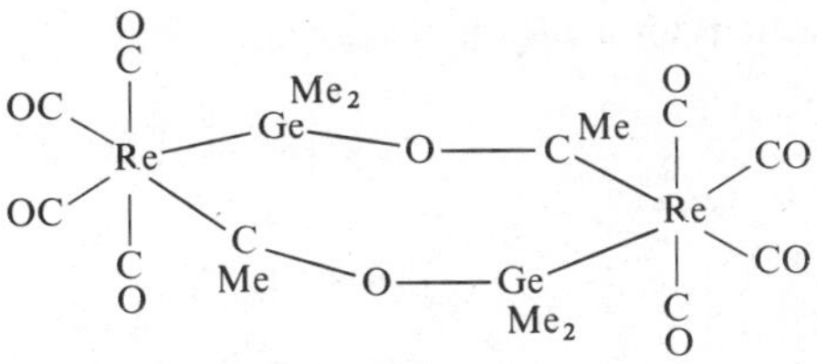

(23)

The preparation and characterization of the new mixed-metal carbonyl $[TcCo(CO)_9]$ have been described. I.r. spectral data for this compound and its manganese and rhenium analogues are consistent with free rotation about their metal–cobalt bonds.[621] X-Ray crystallographic studies have shown that $[Me_2GeRe(CO)_4\text{-}C(O)Me]_2$ possesses the structure (23) with a novel eight-membered central ring in which each metal carbene centre is planar.[622] The Raman spectra obtained for the molecules $X_3AM(CO)_5$ (X = Me or Cl, A = Ge or Sn, M = Mn or Re) have been investigated and the corresponding M—A bond orders assessed. These bond orders are in good agreement with the mean of the corresponding homonuclear values and do not indicate any significant π-bonding interactions.[623]

$Re(CO)_5Br$ reacts with KB_9H_{14} in refluxing THF to produce $[6\text{-}(CO)_3\text{-}6\text{-}ReB_9H_{13}]^-$ which has been isolated as its Me_4N^+ salt. An X-ray crystallographic study has shown that the manganese analogue of this compound is structurally similar to $B_{10}H_{14}$ and contains terdentate 'B$_9$-ligands' bonded to the $Mn(CO)_3$ group by two Mn—H—B bridge bonds and an Mn—B σ-bond.[612] Further studies of the $M(CO)_3$ (M = Tc or Re) complexes formed with porphyrins (H_2L) have been reported (see Volume 3, p. 170). ^1H N.m.r. studies of these $HLM(CO)_3$ complexes indicate that a ready intramolecular arrangement proceeds with the $M(CO)_3$ group changing its position of binding to three out of the four porphyrin nitrogen atoms in concert with the movement of the N—H group.[624]

$[Re(CO)_3(MeCN)_3]PF_6$ has been prepared and this compound appears to be a useful precursor for the preparation of cationic rhenium carbonyl compounds.[625] Carbonylation of tertiary phosphino-complexes under reducing conditions provides a convenient synthetic route to rhenium mono- and poly-carbonyl complexes.[626]

[620] (a) P. Lemoine and M. Gross, *J. Therm. Analysis*, 1974, **6**, 159; (b) V. J. Chernova, M. S. Sheiman, I. B. Rabinovich, and V. G. Syrkin, *Trudy Khim. i khim. Tekhnol.*, 1973, 43 (*Chem. Abs.*, 1974, **80**, 87924); (c) G. Burchalova, M. S. Sheiman, V. G. Syrkin, and I. B. Rabinovich, *ibid.*, p. 45 (*Chem. Abs.*, 1974, **80**, 149417).
[621] G. Sbrignadello, G. Tomat, L. Magon, and G. Bor, *Inorg. Nuclear Chem. Letters*, 1973, **9**, 1073; G. Bor, *J. Organometallic Chem.*, 1974, **65**, 81.
[622] M. J. Webb, M. J. Bennett, L. Y. Y. Chan, and W. A. G. Graham, *J. Amer. Chem. Soc.*, 1974, **96**, 5932.
[623] A. Terzis, T. C. Strekas, and T. G. Spiro. *Inorg. Chem.*, 1974, **13**, 1346.
[624] M. Tsutsui and C. P. Hrung, *Chem. Letters*, 1973, 941; *J. Amer. Chem. Soc.*, 1974, **96**, 2638; T. S. Srivstava, C. P. Hrung, and M. Tsutsui. *J.C.S. Chem. Comm.*, 1974, 447.
[625] R. H. Reimann and E. Singleton, *J. Organometallic Chem.*, 1973, **59**, C24.
[626] J. Chatt, J. R. Dilworth, H. P. Gunz, and G. J. Leigh, *J. Organometallic Chem.*, 1974, **64**, 245.

$[Re(CO)_5NO_3]$ reacts with Ph_4AsNO_3 to produce *cis*-$[Re(CO)_4(NO_3)_2]^-$.[627] The product of the reaction of $CF_3Re(CO)_5$ with CS_2 has been identified by *X*-ray crystallography as the trithiocarbonato-complex $[(OC)_4ReS_2CS\{Re(CO)_4\}_2SCS_2-Re(CO)_4]$.[628] The preparation and reactions of dithiocarbamato-complexes of rhenium have been examined using $Re(CO)_5Cl$ as the starting point. Scheme 3 summarizes these reactions for diethyldithiocarbamate (dtc).

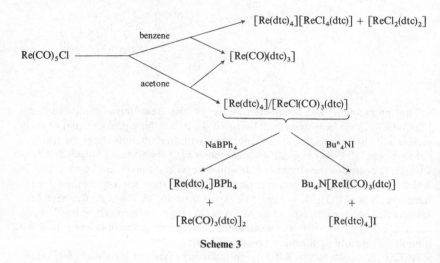

Scheme 3

Tl(dtc) reacts with $Re(CO)_5Cl$ to produce $[Re(CO)_3(dtc)]_2$, which is easily cleaved by a variety of ligands (L = CO, py, or Ph_3P) to produce the corresponding $[Re(CO)_3(dtc)L]$ complex.[629]

Dinitrogenyl Complexes.—The preparation and characterization of the μ-dinitrogenyl complex $[(PhMe_2P)ClReN_2MoCl_4(OMe)]$ are described on p. 110.[267] The red and blue products obtained from the reaction mixture of *trans*-$[ReCl(N_2)(PMe_2Ph)_4]$ with either $TiCl_3$ or $TiCl_3L_3$ (L = MeCN or THF) have been re-investigated. These products have been shown to involve titanium(IV); the composition of the blue form being $[\{ReCl(N_2)(PMe_2Ph)_4\}_2TiCl_4]$ and that of the red form $[\{ReCl(N_2)-(PMe_2Ph)_4\}TiCl_4(L)]$ (L = THF, Et_2O, or CH_2Cl_2). An excess of $TiCl_4$ reacts with $[ReCl(N_2)(PMe_2Ph)_4]$ in the presence of Et_2O to yield $[\{ReCl(N_2)(PMe_2Ph)_4\}-(Ti_2Cl_6O)(Et_2O)]$ which shows a $\nu(N—N)$ stretching mode at 1622 cm^{-1}. These studies have indicated that $[ReCl(N_2)(PMe_2Ph)_4]$ is a much more powerful Lewis acid towards $TiCl_4$ than is THF.[613] The first reported preparation of a dinitrogenyl complex of rhenium with an arsine ligand, $[ReCl(N_2)(arphos)_2]$ (arphos = 1-di-phenylphosphino-2-diphenylarsinoethane), has been described. The complexes $[ReCl(N_2)(LL)_2]$ (LL = diphos or arphos) react with $InCl_3$ in the absence of O_2 to produce the corresponding rhenium(II) $[ReCl(N_2)(LL)_2]Cl$ complex.[630]

[627] C. C. Addison, R. Davis, and N. Logan, *J.C.S. Dalton*, 1974, 1073.
[628] G. Thiele, G. Liehr, and E. Lindner, *J. Organometallic Chem.*, 1974, **70**, 427.
[629] J. F. Rowbottom and G. Wilkinson, *J.C.S. Dalton*, 1974, 684.
[630] D. J. Darensbourg, and D. Madrid, *Inorg. Chem.*, 1974, **13**, 1532.

Nitrosyl Complexes.—A general method for the synthesis of transition-metal nitrosyl complexes has been described.[314] The dinuclear compound [Re(CO)$_2$NO-(NO$_3$)$_2$]$_2$ has been characterized as a minor product of the reaction between Re$_2$(CO)$_{10}$ and N$_2$O$_4$; a structure containing two μ-nitrato-groups was proposed on the basis of spectral evidence.[631a] Syntheses of the new rhenium complexes [Re(CO)$_2$(NO)-(PPh$_3$)$_2$] and [Re(H)(NO)$_2$(PPh$_3$)$_2$] have been described. Some of the reactions of these complexes have been investigated and shown to lead to the preparation of the halogenonitrosyl complexes [Re(NO)$_2$(PPh$_3$)$_2$X], [Re(NO)$_2$(PPh$_3$)$_2$X$_2$]X$_3$, and [Re(NO)$_2$(PPh$_3$)$_2$X]$_2$ (X = Cl, Br, or I).[631b]

NO reacts with an aqueous suspension of ReO$_2$ at 70—80 °C to produce Re(NO)-(OH)$_3$, and an NO–HCl mixture reacts with an ethanolic or aqueous suspension of ReO$_2$ to yield [Re(NO)Cl$_5$]$^{2-}$. The passage of NO over (phenH$_2$)[ReX$_6$] (X = Cl or Br) at 220 °C produces the corresponding [Re(NO)(phen)X$_3$] complex.[632] ReCl$_5$ reacts with NO in benzene or CCl$_4$ solution to produce [ReCl$_4$(NO)$_2$].C$_6$H$_6$ or [ReCl$_5$NO], respectively. This latter compound is converted on treatment with a variety of ligands (L = MeCN, py, Ph$_3$P, or phen) into an ReCl$_4$(NO),L$_x$ (x = 1 or 2) complex. One such complex, [ReCl$_4$(NO)MeCN], reacts with O$_2$ in MeCN solution to produce [ReOCl$_3$(NO)MeCN].[633]

Binary Compounds and Related Systems.—*Oxides.* A mass spectrophotometric study of the thermodynamics of the evaporations of ReO$_2$ and ReO$_3$ has been described.[364] An alternative route to the preparation of ReO$_2$,2H$_2$O has been given[634] and X-ray crystallographic parameters have been determined for the orthorhombic and monoclinic forms of ReO$_2$ and for the cubic form of ReO$_3$ (prepared by oxidizing ReO$_2$,2H$_2$O in air at 260 °C).[635] The formation of monoclinic ReO$_2$ from Re$_2$O$_5$,[636a] and the relationship of the Re$_2$O$_5$ structure to that of V$_2$O$_5$,[636b] have been examined by electron microscopy. Oxidation of ReO$_2$,2H$_2$O to ReO$_3$ has been similarly studied and shown to proceed *via* the Re$_2$O$_5$ phase.[637] The structure of a defect ReO$_3$ species has been described[638] and molecular beam techniques have been employed to probe the interaction of atomic rhenium with molecular oxygen leading to formation of ReO$_3$.[639] Effusion-cell mass spectrophotometry has been used to study the nature of the vapour-phase species above the Re—O system, when the heats of formation of ReO$_3$(g) and Re$_2$O$_6$(g) were calculated as − 280 ± 17 and − 887 ± 83 kJ mol^{-1} at 298 K.[640]

N.q.r. studies of Re$_2$O$_7$ indicate a strong asymmetry in the electronic charge distribution at the metal centres, this result was discussed with reference to the mole-

[631] (a) R. Davis, *J. Organometallic Chem.*, 1973, **60**, C22; (b) G. La Monica, M. Freni, and S. Cenini, *ibid.*, 1974, **71**, 57.

[632] S. Rakshit, B. K. Sen, and P. Bandyopadhyay, *Z. anorg. Chem.*, 1973, **401**, 212.

[633] Yu. A. Buslaev, M. A. Glushkova, and N. A. Ovchinnikova, *Zhur. neorg. Khim.*, 1974, **19**, 743.

[634] V. S. Khain, I. N. Ushatinskii, and N. N. Antsygina, *Ref. Zhur. Khim.*, 1973, Abs. No. 15V11 (*Chem. Abs.*, 1974, **80**, 90437).

[635] D. Colaitis, C. Lecaille, and D. Lebas, *Rev. chim. Minerale*, 1972, **9**, 709.

[636] (a) D. Colaitis, D. Lebas, and C. Lecaille. *Mater. Res. Bull.*, 1973, **8**, 1153; (b) D. Colaitis, D. Lebas and C. Lecaille *ibid.*, 1973, **8**, 627.

[637] D. Colaitis, C. Lecaille, and D. Lebas, *Mater. Res. Bull.*, 1974, **9**, 211.

[638] D. Colaitis, C. Lecaille, and D. Lebas, *Mater. Res. Bull.*, 1974, **9**, 295.

[639] L. A. Kaye, *Diss. Abs.* (*B*), 1973, **34**, 199.

[640] H. B. Skinner and A. W. Searcy, *J. Phys. Chem.*, 1973, **77**, 1578.

cular structure.[641] The i.r. spectrum of Re_2O_7 molecules isolated in various matrices at *ca.* 20 K has been recorded and the matrices seen to exhibit significant perturbation effects.[642] The vapour pressure of Re_2O_7 has been recorded over the temperature range 200—325 °C[643] and an apparatus described suitable for the preparation of this compound by the thermal decomposition of $(NH_4)[ReO_4]$.[644]

Carbides, Silicides, and Borides. $UReC_{2-x}$ has been prepared and characterized by *X*-ray powder diffractometry.[645]

The heat of formation of $ReSi_2$ has been determined over the temperature range 1200—1900 K.[646]

The phases $(Re,Co)_3B$, $ReCoB$, and $(Re,Co)_7B$ have been characterized in the Re—Co—B system and their structures discussed with reference to *X*-ray diffraction studies.[647]

Compounds containing Tc—Tc or Re—Re Bonds.—Low-temperature studies of the polarized single-crystal spectra of $(Bu_4^nN)_2[Re_2Cl_8]$ and $[Re_2Cl_6(PEt_3)_2]$ have shown that the lowest energy electronic transition in these complexes corresponds to a $\delta \to \delta^*$ transition. The electronic origin of this transition for the former compound occurs at 13.597 kK and upon this is built a progression of the metal–metal stretching frequency with an average spacing of 255 cm^{-1}.[406] The Re—Re bond energy in $[Re_2Cl_8]^{2-}$ appears to be rather lower than the earlier estimates of 1200—1400 kJ mol^{-1} following a demonstration of its easy photolytic cleavage using u.v. light.[614] A MO assessment of the electronic structure of the $[Re_2Cl_8]^{2-}$ ion has been presented and some modifications to the description due to Cotton and Harris were proposed.[648] The uses of Raman[403b] and *X*-ray photoelectron[415] spectra in the investigation of the electronic and atomic structures of compounds containing metal–metal bonds have been discussed and such data presented for several rhenium(II) and/or rhenium(III) compounds.

X-Ray crystallographic studies have shown that $M_2[Re_2Cl_8],2H_2O$ (M = NH_4 or Rb) have the same structure as the corresponding potassium compound and that $Cs_2[Re_2Cl_8],H_2O$ contains both $[Re_2Cl_8]^{2-}$ and $[Re_2Cl_8(H_2O)_2]$ anions, with Re—Re distances of *ca.* 222 pm.[649] The mixed halogeno-complexes $[Re_2Cl_xBr_{8-x}]$ (x = 2, 3, or 6) have been prepared by the addition of stoicheiometric quantities of 48 % aq. HBr to a methanolic solution of $(Bu_4^nN)_2[Re_2Cl_8]$. The structures of these salts have been investigated by i.r. spectroscopy which has confirmed that they are genuine compounds and not mixtures containing $[Re_2Cl_8]^{2-}$ and $[Re_2Br_8]^{2-}$ ions. The lowering of the symmetry in these mixed halogeno-complexes allowed the $\nu(Re-Re)$ mode to be detected at *ca.* 274 cm^{-1} in the i.r. The structures of $[Re_2Cl_5(dth)_2]$, $[Re_2X_6(dth)_2]$ (X = Cl or Br), $[Re_2Br_4(dth)_2]$ (dth = 2,5-dithiahexane), and $[Re_2X_6(PPh_3)_2]$ (X = Cl or Br) have also been discussed with reference to vibrational

[641] P. W. Burkert and M. F. Eckel. *Z. Naturforsch.*, 1973. **28B**, 570.
[642] M. Spoliti, C. S. Nunziante, E. Coffari and G. De Maria, *Gazzetta*, 1973, **103**, 585.
[643] H. Martens and S. Ziegenbalg, *Z. anorg. Chem.*, 1973, **401**, 145.
[644] V. I. Bibikova, I. I. Vasilevskaya, A. G. Vasil'eva, and .L A. Nisel'son, *Zhur. priklad Khim.*, 1973, **46**, 1115.
[645] Z. M. Alekseeva. *J. Nuclear Mater.*, 1974, **49**, 333 (*Chem. Abs.*, 1974, **80**, 125679).
[646] V. P. Bondarenko, L. A. Dvorina, E. N. Fornichev, N. P. Slyussar, and A. D. Krivorotenko. *Zhur. fiz. Khim.*, 1973, **47**, 1044.
[647] M. V. Chepiga and Yu. B. Kuz'ma, *Izvest. Akad. Nauk S.S.S.R., neorg. Materialy*, 1973, **9**, 1688.
[648] R. A. Evarestov, *Zhur. strukt. Khim.*, 1973, **14**, 955; V. N. Pak and D. V. Korol'kov, *ibid.*, p. 956.
[649] P. A. Koz'min, G. N. Novitskaya, and V. G. Kuznetsov, *Zhur. strukt. Khim.*, 1973, **14**, 680.

spectroscopic data.[650a] Molten halogenoacetic acids XCH_2CO_2H (X = Cl or Br) react with $[Re_2Cl_8]^{2-}$ under an inert atmosphere to produce the corresponding $[Re_2(O_2CCH_2X)_4X_2]$ complex. In the case of the bromo-acid, the reaction can be stopped at an intermediate stage and $[Re_2Br_8]^{2-}$ isolated, thus indicating that the first step in this reaction is halide exchange between $[Re_2Cl_8]^{2-}$ and $BrCH_2CO_2H$.[650b] The crystal structure of $(NH_4)_2[Re_2Cl_6(O_2CH)_2]$ has been determined. Each anion has a Re—Re bond length of 226.0(5) pm and a *cis*-structure of C_{2v} symmetry with each metal atom being six-co-ordinate and a chloride ion occupying each of the axial positions.[651] A new class of tertiary phosphine complexes of the type $[Re_2X_4(PR_3)_4]$ (X = Cl or Br) have been isolated from the reaction of Re_3Cl_9 or $[Re_2X_8]^{2-}$ with PR_3. A single-crystal structure determination for $[Re_2Cl_4(PEt_3)_4]$ has shown that this rhenium(II) derivative has a strong Re—Re bond of length 223.2(6) pm and an eclipsed (D_{2d}) configuration. This short metal–metal separation clearly indicates that the two electrons which are added beyond those in the $\sigma^2\pi^4\delta^2$ set of the Re_2 unit, must occupy a non-bonding orbital. This is presumed to be the a_{2u} σ-non-bonding orbital. Preliminary studies of these new compounds have indicated that they are potentially useful synthetic intermediates. Thus $[Re_2X_4(PEt_3)_4]$ (X = Cl or Br) react with CCl_4 to produce the corresponding $(Et_3PCl)_2[Re_2Cl_4X_4]$ salt.[615] The magnetic moments of $M_3[Tc_2Cl_8],2H_2O$ (M = Cs or K) have been determined as 2.16 and 1.46 BM, respectively.[652]

$Re(CO)_5Br$ reacts with pentane-2,4-dione in the presence of KOH to form the complex $[Re_2(CO)_6(acac)_2(H_2O)_2]$, which appears to contain μ-aquo-ligands and a Re—Re bond.[653]

Halogen exchange and positive-ion fragmentation patterns have been determined for $[Re_3X_9]$ (X = Cl, Br, or I) by mass spectrometry. In the case of the mixed halide derivatives $[Re_3Cl_{9-n}X_n]$ (X = Br or I; n = 0—9) the observed relative abundance of the various fragments followed a statistical distribution with all the halogen positions being equivalent in this respect.[654a] Mass spectral and thermodynamic measurements for Re_3Cl_9 and Re_3Br_9 have led to estimates of the energies of the Re—X_t (300 and 240 kJ mol^{-1}), Re—X_b—Re (380 and 310 kJ mol^{-1}) and Re—Re (350 and 430 kJ mol^{-1}) bonds.[654b] The polarographic reduction of $[Re_3Cl_{12}]^{3-}$ has been studied and shown to give a single wave corresponding to a two-electron reversible process.[655] Re_3Cl_9 in dry CH_2Cl_2 under reflux reacts with 1,2,5-triphenyl-phosphole (TPP), its oxide (TPPO), selenide (TPPSe), or 5-phenyl-5H-dibenzo-phosphole (DBP) (L) to give the corresponding $Re_3Cl_9L_2,CH_2Cl_2$ derivatives.[656] Complex formation between Re_3Cl_9 and poly(vinyl alcohol) has been monitored by u.v. spectroscopy.[449]

[650] (a) C. Oldham and A. P. Ketteringham. *J.C.S. Dalton*. 1973, 2304; (b) C. Oldham and A. P. Ketteringham. *Inorg. Nuclear Chem. Letters*. 1974, **10**, 361.

[651] P. A. Koz'min, M. D. Surazhaskaya, and T. B. Larina, *Zhur. strukt. Khim.*, 1974, **15**, 64.

[652] V. V. Zelentsov, N. A. Subbotina, and V. I. Spitsyn. *Zhur. neorg. Khim.*, 1973, **18**, 1709.

[653] A. N. Nesmeyanov, A. A. Ioganson, N. E. Kolobova, and K. N. Anisimov, *Izvest. Akad. Nauk S.S.S.R.*, 1973, 2388.

[654] H. Schaefer, K. Rinke, and H. Rabeneck. *Z. anorg. Chem.*, 1973, **403**, 23; D. V. Korol'kov, Kh. Missner, and K. V. Ovchinnikov, *Zhur. strukt. Khim.*, 1973, **14**, 717.

[655] J. M. Schub. P. Lemoine, and M. Gross. *Electrochim. Acta*. 1973, **18**, 767.

[656] D. G. Holah, A. N. Hughes, and K. Wright, *Inorg. Nuclear Chem. Letters*. 1973, **9**, 1265.

Rhenium(III).—The majority of studies concerning this oxidation state are included in the previous section. Photolysis of the metal–metal bond of $[Re_2Cl_8]^{2-}$ by u.v. irradiation appears to provide a convenient route to monomeric chloro-complexes of rhenium(III).[614] The complexes $[Re_2Cl_3(diphos)_2]_2$ and $[ReCl_{2.5}(diphos)]_n$ have been formed by treating $(Bu_4N)_2[Re_2Cl_8]$ with diphos. An X-ray crystallographic study has confirmed that the former does not involve a metal–metal bond. The dimeric molecules of the compound are centrosymmetric and chlorine-bridged with Re $\cdots$ Re $= 380.9$ pm. This compound is oxidized by $Cl_3CX(X = Cl, NO_2, CN, or CO_2H)$ to produce the new rhenium(IV) solvates $[ReCl_4(diphos)]_xCCl_3X$ ($x = 0.75$ or 1). The nature of $[ReCl_{2.5}(dppe)]_n$ has not yet been established but it does not appear to have a strong metal–metal bond.[616]

The cathodic reduction of $[ReI_6]^{2-}$ has been studied in aqueous media and $[ReI_5(OH)]^{3-}$ identified as a reduction product.[657] A compound obtained in large yields from the reaction of pentane-2,4-dione with $[ReOCl_2(OEt)(PPh_3)_2]$ in benzene, has been shown by a single crystal X-ray crystallographic study to be *cis*-dichloro-pentane-2,4-dionato-*trans*-bis(triphenylphosphino)rhenium(III), $[ReCl_2(acac)-(PPh_3)_2]$. The ligands are arranged about this metal centre to give a distorted octahedral co-ordination geometry with average bond lengths Re—Cl $= 237$, Re—O $= 202$, and Re—P $= 247$ pm.[658] The preparation of the dithiocarbamato-complexes $[ReCO(S_2CNR_2)_3]$ (R $=$ Me, Et, or PhCH$_2$) has been achieved by treating Re(CO)$_5$Cl with the corresponding thiuram disulphides.[629] The crystal structure of this diethyldithiocarbamato-complex has been determined and shown to contain monomeric molecules in which the rhenium(III) centre is seven-co-ordinate with a distorted pentagonal-bipyramidal geometry, the carbonyl group occupying an axial position.[659] $[ReCl_2(NH_3)(N_2Ph)(PMe_2Ph)]$ reacts with the halogeno-acids HX (X $=$ Cl or Br) to produce the corresponding $[ReCl_2(NH_3)(N_2HPh)(PMe_2Ph)]X$ complex. This is the first known example of such protonation occurring at the unco-ordinated nitrogen atom of the phenylazo-group and this has been confirmed by an X-ray crystallographic study of the bromide salt.[660]

Technetium(IV) and Rhenium(IV).—The salts $M_2[TcBr_6](M = K, Cs, Rb, or NH_4)$ have been prepared by the reaction of MTcO$_4$ with HBr and MBr. X-Ray diffraction data indicate that the Rb and Cs compounds crystallize with cubic unit cells.[661] Preparative details have been published for the production of $K_2[ReCl_6]$ by the reduction of KReO$_4$ using tin(II) chloride.[662] The results of several extraction studies involving the $[ReX_6]^{2-}$ (X $=$ Cl or Br) ions have been described[663] and the thermal stability of $Ag_2[ReBr_6]$ has been examined.[664] The intensities of Raman-active fundamental vibrational modes of several octahedral hexahalogeno-anions, including

[657] R. Muenze, *Z. phys. Chem. (Leipzig)*, 1973, **253**, 283.
[658] I. D. Brown, C. J. L. Lock, and C. Wan, *Canad. J. Chem.*, 1974, **52**, 1704.
[659] S. R. Fletcher and A. C. Skapski, *J.C.S. Dalton*, 1974, 486.
[660] R. Mason, K. M. Thomas, J. A. Zubieta, A. R. Galbraith, and B. L. Shaw, *J. Amer. Chem. Soc.*, 1974, **96**, 260.
[661] L. L. Zaitseva, M. I. Konarev, V. S. Il'yashenko, I. V. Vinogradov, S. V. Shepel'kov, A. A. Kruglov, and N. T. Chebotarev, *Zhur. neorg. Khim.*, 1973, **18**, 2410.
[662] M. Pavlova, N. Iordanov, and D. Staikov, *Doklady Bolg. Akad. Nauk.*, 1974, **27**, 67 (*Chem. Abs.*, 1974, **80**, 152240).
[663] K. A. Bol'shakov, N. M. Sinitsyn, and T. M. Gorlova, *Doklady Akad. Nauk S.S.S.R.*, 1974, **214**, 1073; K. A. Bol'shakov, N. M. Sinitsyn, V. F. Travkin, and L. N. Antimonova, *Zhur. neorg. Khim.*, 1973, **18**, 2209; S. Tribalat and A. Jamard, *Ann. Chim. (France)*, 1973, **8**, 87.
[664] R. A. Dovlyatshina, *Ref. Zhur. Khim.*, 1973, Abs, No 5B774 (*Chem. Abs.*, 1973, **79**, 73102).

$[ReCl_6]^{2-}$, have been studied[665] and a force field appropriate to the hexahalogeno-anions of rhenium(IV) and technetium(IV) has been presented.[666] The optical spectra of technetium(IV) and rhenium(IV) centres contained in the $(EtNH_3)_2[SnX_6]$ (X = Cl or Br) lattices have been recorded at 77 and 298 K and discussed in terms of ligand-field theory.[667] The magnetic properties of the salts $(HL)_2[TcCl_6]$ (L = quinoline or py) have been investigated over the temperature range 80—290 K.[652]

The behaviour of ReO_2 in a variety of aqueous acids has been studied and the nature of the dissolved species discussed.[668] The mechanism of the controlled cathodic reduction of $[ReO_4]^-$ to produce rhenium(IV) has been investigated[669] and reduction of $KReO_4$ in concentrated HCl by iodides in an atmosphere of CO_2 has been shown to afford $K_4[Re_2OCl_{10}]$. The anion of this latter salt, and that of $Cs_4[Re_2OCl_8(OH)_2]$, have been shown by i.r. spectroscopy to involve a linear Re—O—Re arrangement.[670]

The compounds $[ReCl_2(S_2CNR_2)_2]$ (R = Et or PhCH$_2$) have been prepared by the interaction of $Re(CO)_5Cl$ with the corresponding thiuram disulphides (see Scheme 3). $[ReCl_4(PPh_3)_2]$ reacts with $Et_2NCS_2^-$ to yield $[Re_2Cl_2(S_2CNEt_2)_5]Cl_2$ and $[ReCl_2(S_2CNEt_2)_2]$ as two of the reaction products.[629]

The irradiation of perrhenate solutions, followed by the addition of thiocyanate has been shown to produce rhenium(IV) thiocyanato-complexes.[671]

Oxidation of $[ReCl_3(diphos)]_2$ is readily accomplished by molecules such as Cl_3CX (X = Cl, NO_2, CN, or CO_2H) to produce the corresponding $[ReCl_4(diphos)],xCCl_3X$ (x = 0.75 or 1) complex.[616] An X-ray crystallographic study of $[ReCl_4(diphos)],0.75$ CCl_4 has shown that the metal has the expected distorted *cis*-octahedral configuration with an occluded molecule of CCl_4 in the lattice.[672]

Rhenium(v).—The nature of the rhenium(v) species in aqueous HCl solutions has been further investigated.[673] The $[ReOCl_5]^{2-}$ ion undergoes acid hydrolysis with subsequent oxidation thus ruling out the disproportion mechanism proposed earlier. $Cs_2[Re_2O_2(H)Cl_{10}],H_2O$ has been prepared by oxidation of solutions of $[Re_2OCl_{10}]^{4-}$ with H_2O_2, followed by acidification of HCl and addition of CsCl. I.r. spectral data suggest that the anion of this dinuclear rhenium(v) salt involves μ-oxo- and μ-hydroxo-groups.[670] $Cd_2Re_2O_7$ has been prepared by the vapour-phase reaction between Re_2O_7 and cadmium and its X-ray diffraction pattern seen to be characteristic of a f.c.c. pyrochlore lattice.[674] The dithiocarbamato-complexes $[Re(S_2CNEt_2)_4]^+$ and $[Re_2O_2S\{S_2CN(PhCH_2)_2\}_4]$ have been characterized in an extensive study.[629] Complex formation between *o*-mercaptobenzoic acid and rhenium(v) has been investigated and a 2:1 complex characterized.[673]

The reaction of $[ReNCl_2(PMe_2Ph)_3]$ with $[MoCl_4(MeCN)_2]$ in CH_2Cl_2 yields the

665 Y. M. Bosworth and R. J. H. Clark, *J.C.S. Dalton*, 1974, 1749.
666 N. K. Sangal and L. Dixit, *Indian J. Pure Appl. Phys.*, 1973, **11**, 452 (*Chem. Abs.*, 1974, **80**, 53989).
667 H. J. Schenk and K. Schwochau, *Z. Naturforsch.*, 1973, **28a**, 89; L. Pross, K. Roessler, and H. J. Schenk, *J. Inorg. Nuclear Chem.*, 1974, **36**, 317.
668 Yu. Ya. Gukova and M. I. Ermolaev, *Zhur. neorg. Khim.*, 1973, **18**, 2296.
669 S. I. Sinyakova, A. M. Demkin, and L. V. Borisova, *Zhur. analit. Khim.*, 1973, **28**, 1974.
670 B. Jezowska-Trzebiatowska, L. Szternberg, and J. Mrozinski, *Pr. Nauk. Inst. Chem. Nieorg. Met. Pierwiastkow. Rzadkich. Politech. Wroclaw*, 1973, **16**, 147 (*Chem. Abs.*, 1974, **80**, 127634).
671 I. P. Kharlamov and Z. N. Korobova, *Ref. Zhur. Khim.*, 1973, Abs. No. 14G84 (*Chem. Abs.*, 1974, **80**, 77824).
672 J. A. Jaecker, W. R. Robinson, and R. A. Walton, *Inorg. Nuclear Chem. Letters*, 1974, **10**, 93.
673 M. Pavlova, *J. Inorg. Nuclear Chem.*, 1974, **36**, 1623.
674 J. M. Longo, P. C. Donohue, and D. A. Batson, *Inorg. Synthesis*, 1973, **14**, 146.

bridged nitrido-complex $[(Me_2Ph)_3Cl_2ReNMoCl_4(MeCN)]$, which is a purple-red, air-stable solid with $\mu_{eff} = 2.3$ BM.[270] The reaction of $[ReOCl_4(H_2O)]^-$ with thiocyanate salts has been shown to form $[ReOCl_4(NCS)]^{2-}$, and $ReCl_5$ reacts with $KX (X = NCS$ or $NCSe)$ in acetone to produce the corresponding $[ReX_4Cl],0.5Me_2CO$ complex.[675] Thiourea and NN'-ethylenethiourea (L) have been shown to react with $H_2[ReOCl_5]$ in aqueous 6M-HCl to produce the complexes $[ReOCl_{5-x}L_x],nH_2O$ $(x = 2$ or $4)$.[676] The complex $[ReO(OPh)L']$ $(H_2L' = $ octaethylporphyrin) has been

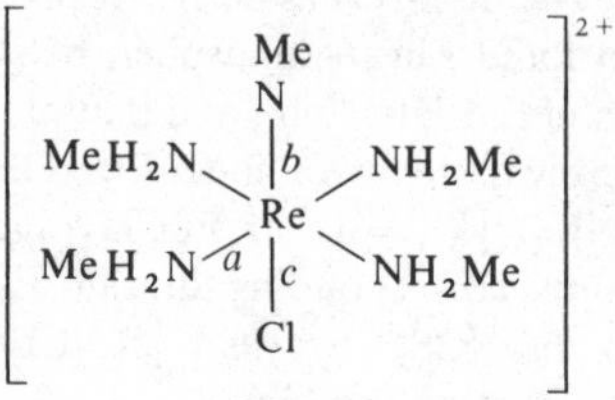

$$\left[\begin{array}{c} Me \\ N \\ MeH_2N \diagdown \ \big|b \diagup NH_2Me \\ Re \\ MeH_2N \diagup a \ \big|c \diagdown NH_2Me \\ Cl \end{array}\right]^{2+} \qquad \begin{array}{l} a = 218 \text{ pm} \\ b = 169 \text{ pm} \\ c = 240 \text{ pm} \end{array}$$

(24)

prepared by the reaction of H_2L' with Re_2O_7 in PhOH and characterized by mass, i.r., and ^1H n.m.r. spectral studies.[177] The two new complexes $[ReO_2(py)_3F]$ and $[ReO_2(py)_2F]$ have been obtained by heating $[ReO_2(py)_4]F$ at 70 and 100 °C, respectively. Digestion of an acetone solution of $[ReO_2(py)_4]F$ in the presence of dimethyl sulphate has been shown to precipitate $[ReO_2(py)_2(H_2O)F]$, and treatment of $[ReO_2(py)_4]F$ with 40% aq. HF to yield the salt $[ReO_2(py)_2(H_2O)_2]F$. This latter cationic species is new and several other of its salts have been prepared.[677] An X-ray crystallographic study has identified the structure (24) for $[Re(NH_2Me)_4(NMe)Cl]^{2+}$ in its perchlorate salt. The Cl—Re—N—C arrangement involving the methylimido-group is linear. The rates of chloride hydrolysis and exchange of this complex dissolved in dilute aqueous HCl have been determined.[678]

Technetium(VI) and Rhenium(VI).—Values for certain thermodynamic parameters of $TcOF_4$ and $ReOF_4$ have been computed.[237] The i.r. spectra have been recorded for $ReOF_4$ as a thin film and isolated in an Ar matrix and these data compared to Raman data obtained for this compound dissolved in HF solutions. All of this information is consistent with a C_{4v} molecular geometry for $ReOF_4$ monomers.[679]

The best conditions for the preparation of $[ReOCl_5]^-$ have been determined; the anion was isolated as its Ph_4P^+ salt.[680]

The behaviour of ReO_3 in a variety of aqueous, acidic solutions has been studied and the nature of the species formed discussed.[668] A low-temperature neutron-diffraction study of cubic Ba_2CoReO_6 has shown that the oxygen atoms are slightly displaced from the positions of an ideal perovskite towards the rhenium atoms, and

[675] A. M. Golub, B. O. Sternik, and N. V. Ul'ko, *Ref. Zhur. Khim.*, 1972, Abs. No. 8B402 (*Chem. Abs.*, 1974, **80**, 152281).

[676] K. V. Kotegov, N. V. Fadeeva, and Yu. N. Kukushkin, *Zhur. obshchei Khim.*, 1973, **43**, 1182.

[677] M. C. Chakravorti and M. K. Chaudhuri, *J. Inorg. Nuclear Chem.*, 1974, **36**, 757.

[678] R. S. Shandles, R. K. Murmann, and E. O. Schlemper, *Inorg. Chem.*, 1974, **13**, 1373.

[679] R. T. Paine, K. L. Treuil, and F. E. Stafford, *Spectrochim. Acta*, 1973, **29A**, 1891.

[680] L. V. Borisova, A. N. Ermakov, and O. D. Prasolova, *Doklady Akad. Nauk S.S.S.R.*, 1974, **215**, 589.

the magnetic properties of this compound have been discussed with respect to this structure.[681]

The interaction of $[ReOCl_3(PPh_3)_2]$ or $ReOCl_4$ in Et_2O at $-35\,°C$ with LiMe gives $ReOMe_4$, an air-sensitive, red-purple crystalline solid, m.p. $44\,°C$. E.s.r. spectral data obtained for this paramagnetic compound are consistent with a square-pyramidal, C_{4v}, molecular structure at $-175\,°C$; however, at temperatures above $-150\,°C$ the molecules are fluxional.[617]

Technetium(VII) and Rhenium(VII).—New i.r. Raman data for TcO_3F are consistent with a C_{3v} molecular structure. Comparison with vibrational data obtained for other MO_3X compounds indicates a necessity to revise some of the original assignments.[682] Re_2O_7 or $KReO_4$ dissolve in anhydrous HF to form ReO_3F and the Raman spectra of these solutions indicate that this latter molecule has C_{3v} molecular symmetry.[683] Formation of $[ReO_4F]^{2-}$ in aqueous HF solution containing $KReO_4$ has been indicated by conductometric and thermometric titration studies.[684] Normal-co-ordinate analyses have been performed to analyse the vibrational modes of the TcO_3Cl, ReO_3Cl, and ReO_3Br molecules.[239] The conditions suitable for the prepara-tion of ReO_2Cl_3[685] and ReO_3Br[686] have been described. Study of the chemical transport of rhenium as ReO_2, ReO_3, or ReS_2 in the presence of I_2 and/or H_2O has shown the important gas-phase species to be ReO_3I and ReO_3OH.[687]

$(Me_4N)[ReO_4]$ has been prepared and shown by X-ray powder diffractometry to be isomorphous with Me_4NClO_4.[688] Sublimation of $RbReO_4$ and $CsReO_4$ has been investigated by mass spectrometry and $MReO_4$ and $(MReO_4)_2$ species were identified in the gas phase.[689] The vibrational spectra of the $[ReO_4]^-$ ion in several lattices have been described and discussed[224b, 245] and the intensity of its $\nu_1(A_1)$ Raman line has been measured.[246] Complex formation between $[ReO_4]^-$ and several lanthanide ions (M^{3+}) has been studied spectrophotometrically and stability constants for $M(ReO_4)^{2+}$ and $M(ReO_4)_2^+$ entities were reported.[690] The distribution of $[ReO_4]^-$ ions between eutectic mixtures of $LiNO_3$–KNO_3 or $NaSCN$–$KSCN$ and a chloro-naphthalene solution of secondary or quaternary alkylammonium nitrates or thiocyanates has been studied between 190 and 140 °C. The distribution data obtained were interpreted in terms of anion exchange equilibria.[691]

The room-temperature m.c.d. spectrum of $[ReS_4]^-$ has been recorded and the data obtained indicate that the lowest energy band corresponds to a $t_1 \rightarrow 2e$ transition.[594b]

[681] C. P. Khattak, D. E. Cox, and F. F. Y. Wang, *Amer. Inst. Phys. Conf. Proc.*, 1972, 674 (*Chem. Abs.*, 1973, **79**, 36497).
[682] J. Binenboym, U. El-Gad, and H. Selig, *Inorg. Chem.*, 1974, **13**, 319.
[683] H. Selig and U. El-Gad, *J. Inorg. Nuclear Chem.*, 1973, **35**, 3517.
[684] M. C. Chakravorti and M. K. Chaudhuri, *Z. anorg. Chem.*, 1973, **398**, 221.
[685] I. A. Glukhov, N. A. El'manova, S. S. Eliseev, and M. T. Temurova, *Doklady Akad. Nauk Tadzh. S.S.S.R.*, 1973, **16**, 29 (*Chem. Abs.*, 1974, **80**, 66304).
[686] K. V. Ovchinnikov, L. B. Mel'nikov, and N. I. Kolbin, *Zhur. obshchei Khim.*, 1973, **43**, 699.
[687] H. Schäfer, M. Bode, and M. Trenkel, *Z. anorg. Chem.*, 1973, **400**, 253.
[688] S. Okransinski and G. Mitra, *J. Inorg. Nuclear Chem.*, 1974, **36**, 1908.
[689] K. Skudlarski and W. Lukas, *J. Less-Common Metals*, 1973, **31**, 329; K. Skudlarski, *Roczniki Chem.*, 1973, **47**, 1611 (*Chem. Abs.*, 1974, **80**, 64012).
[690] N. A. Orbin, *Ref. Zhur. Khim.*, 1973, Abs. No., 16V77 (*Chem. Abs.*, 1974, **80**, 43576).
[691] J. David-Auslaender, M. Zanger and A. S. Kertes, *J. Inorg. Nuclear Chem.*, 1974, **36**, 425.

2
Elements of the First Transitional Period

BY R. DAVIS, C. A. McAULIFFE, AND W. LEVASON

PART I: Manganese and Iron by R. Davis

1 Manganese

Carbonyl Compounds.—The sublimation of $Mn_2(CO)_{10}$ in the presence of traces of oxygen has been reinvestigated with the aid of e.s.r. spectroscopy. The results indicate the formation of the radical species $Mn(CO)_5O_2\cdot$, not $Mn(CO)_5\cdot$ as had been previously suggested.[1] The crystal and molecular structure of the trinuclear trihydride $H_3Mn_3(CO)_{12}$ has been reported.[2] The metal atoms form an equilateral triangle (Mn—Mn = 3.11 Å), each being bonded to four terminal carbonyl groups. The hydrogen atoms lie in the plane formed by the three metal atoms and symmetrically bridge manganese–manganese pairs (Mn—H = 1.72 Å, $\angle$ MnHMn = 131°, $\angle$ HMnH = 108°). Treatment of the trihydride with alcoholic bases does not yield the expected anion $[H_2Mn_3(CO)_{12}]^-$, but does give $[Mn_3(CO)_{14}]^-$ in low yield. This anion has a linear structure, analogous to the isoelectronic compound $Mn_2Fe(CO)_{14}$, in which the equatorial carbonyl groups of the terminal $Mn(CO)_5$ units are staggered with respect to those on the central manganese atoms and are bent away from the axial CO ligands. These findings are rationalized in terms of considerable interaction between the central metal atom and the CO groups on adjacent metal atoms.[3] The reactions of $[MnFe_2(CO)_{12}]^-$ and a variety of unidentate phosphorus donor ligands (L) have been investigated. Only the monosubstituted anions $[MnFe_2(CO)_{11}L]^-$ were isolated.[4]

The vibrational spectra of $[(CO)_5MnM(CO)_5]^-$ (M = Cr, Mo, or W) have been examined in detail. The spectra were interpreted in terms of C_{4v} symmetry for the anions and force constants for the Mn—M bonds calculated. These follow the order $K_{Mn-W} > K_{Mn-Mo} > K_{Mn-Cr}$ and comparison with the neutral molecules $(CO)_5MnM(CO)_5$ (M = Mn or Re) shows that for isoelectronic pairs, $K_{Mn-Re} > K_{Mn-W}$ and $K_{Mn-Mn} > K_{Mn-Cr}$. The authors have shown that simplified models are inadequate for estimation of such force constants, and have discussed the implications of these results in terms of Mn—M bonding. It is suggested that the extent of overlap between the $3d_{z^2}$ orbital of manganese and the nd_{z^2} orbital of M is largely responsible for the variation in bond strengths. In the case of the isoelectronic pairs, the degree of overlap should be approximately constant and thus the differences are due to the lower orbital energies in the case of Mn^0-Mn^0 and Mn^0-Re^0 compared with

[1] S. A. Fieldhouse, B. W. Fullam, G. W. Neilson, and M. C. R. Symons, *J. C. S. Dalton*, 1974, 567.
[2] S. W. Kirtley, J. P. Olsen, and R. Bau, *J. Amer. Chem. Soc.*, 1973, **95**, 4532.
[3] R. Bau, S. W. Kirtley, T. N. Sorrell, and S. Winarko, *J. Amer. Chem. Soc.*, 1974, **96**, 988.
[4] C. G. Cooke and M. J. Mays, *J. Organometallic Chem.*, 1974, **74**, 449.

those in Mn^0—Cr^{-1} and Mn^0—W^{-1}, respectively.[5] An examination of the i.r. spectra of $MMn(CO)_5$ (M = Li or Na) in ethereal solvents has led to the suggestion that ion-pairs of C_{3v} symmetry are present in which the alkali metal is bonded to a carbonyl oxygen atom. In THF, $NaMn(CO)_5$ exhibits equilibrium behaviour between solvent-separated and contact ion-pairs and treatment of $NaMn(CO)_5$ with Mg^{2+} ions in THF gives species whose i.r. spectra are also consistent with Mn—C—O—Mg interactions.[6] The structure of $[\{[MeO(CH_2)_2]_2O\}Cd\{Mn(CO)_5\}_2]$ has been determined. The two $Mn(CO)_5$ units exhibit very distorted octahedral co-ordination and the ether acts as a terdentate ligand, giving rise to distorted trigonal-bipyramidal co-ordination of the cadmium atom.[7] The same authors have reported the structures of the related compounds $[(bipy)Cd\{Mn(CO)_5\}_2]$ and $[(1,10\text{-phen})Cd\{Mn(CO)_5\}_2]$. These two compounds have very similar structures, in which the manganese atoms are also in distorted octahedral environments, the cadmium atoms being in distorted tetrahedral situations.[8]

There have been several publications reporting cationic manganese carbonyl complexes and a number of new synthetic routes to complexes of this type have been developed. The most unusual of these is:

$$[LMn(CO)_5]_2 \xrightarrow{\text{Na–Hg}} Na[LMn(CO)_5] \xrightarrow[\text{ii, BF}_3\text{ or HBF}_4]{\text{i, ClCO}_2\text{Et}} [LMn(CO)_5][BF_4] + EtOH + NaCl$$

The hexafluorophosphate salts of the same cations, $[Mn(CO)_5L][PF_6]$ can be obtained by treatment of $[Mn(CO)_5(MeCN)][PF_6]$ with $L[L = PPh_3, P(OPh)_3, (p\text{-MeC}_6\text{H}_4)_3P, PMe_2Ph, diphos, and py]$. Cotton–Kraihanzel C—O stretching force constants have been calculated for all the cations.[9] $Fac\text{-}[Mn(CO)_3(MeCN)_3]PF_6$ can be prepared by refluxing $Mn(CO)_5Br$ in acetonitrile, followed by addition of PF_6^-. Treatment of this salt with propylamine in refluxing chloroform gives $[Mn(CO)_3\text{-}(PrNH_2)_3]PF_6$; however, when pyridine or PMe_2Ph is employed as potential ligands, the complexes $fac\text{-}[Mn(CO)_3L_2(MeCN)]PF_6$ (L = py or PMe_2Ph) are formed and use of $P(OMe)_3$ or Et_2S yields $fac\text{-}[Mn(CO)_3L(MeCN)_2]PF_6$ [L = $P(OMe)_3$ or Et_2S]. If $fac\text{-}[Mn(CO)_3(MeCN)_3]PF_6$ is treated with stoicheiometric amounts of $P(OMe)_3$ in refluxing acetonitrile, $cis\text{-}[Mn(CO)_2L_2(MeCN)_2]PF_6$ and $mer\text{-}cis\text{-}[Mn(CO)_2L_3(MeCN)]PF_6$ are formed.[10] The reactions of other phosphine and phosphite donor ligands with the tris-acetonitrile complex have also been investigated and yield products of the type outlined above depending on the nature of the ligand and the conditions employed. Many of the cationic complexes can also be prepared by the reaction of $[Mn(CO)_{5-x}L_xBr]$ with $AgPF_6$ in acetonitrile. Cations of the type $cis\text{-}[Mn(CO)_2L_2(MeCN)_2]^+$ and $mer\text{-}cis\text{-}[Mn(CO)_2L_3(MeCN)]^+$ react further with $NOPF_6$ to give $trans\text{-}[Mn(CO)_2L_2(MeCN)_2]^{2+}$ and $mer\text{-}trans\text{-}[Mn(CO)_2L_3(MeCN)]^{2+}$, respectively. The latter undergoes reduction by hydrazine, yielding $mer\text{-}trans\text{-}[Mn(CO)_2L_3(MeCN)]^+$, which reacts further with L to yield $[Mn(CO)_2L_4]^+$.[11] The change in configuration when the mono-cations are further

[5] J. R. Johnson, R. J. Ziegler, and W. M. Risen, *Inorg. Chem.*, 1973, **12**, 2349.

[6] C. D. Pribula and T. L. Brown, *J. Organometallic Chem.*, 1974, **71**, 415.

[7] W. Clegg and P. J. Wheatley, *J. C. S. Dalton*, 1974, 424.

[8] W. Clegg and P. J. Wheatley, *J. C. S. Dalton*, 1974, 511.

[9] D. J. Darensbourg, M. Y. Darensbourg, D. Drew, and H. L. Conder, *J. Organometallic Chem.*, 1974, **74**, C33.

[10] R. H. Reimann and E. Singleton, *J. Organometallic Chem.*, 1973, **59**, C24.

[11] R. H. Reimann and E. Singleton, *J. C. S. Dalton*, 1974, 808.

oxidized is fascinating and undoubtedly merits a detailed investigation. Treatment of $Mn(CO)_5Br$ or cis-$[Mn(CO)_4(PMe_2Ph)Br]$ with $NOPF_6$ in acetonitrile yields $[Mn(CO)_5(MeCN)]^+$ and cis-$[Mn(CO)_4(PMe_2Ph)(MeCN)]^+$, respectively. Similar treatment of fac- and mer-isomers of $Mn(CO)_3L_2Br$ [$L = P(OMe)_3$, PMe_3, PMe_2Ph, and $Ph_2PCH_2PPh_2$] gives fac-$[Mn(CO)_3L_2Br](PF_6)$, whereas mer-cis-$[Mn(CO)_2L_2Br]$ gives mer-cis-$[Mn(CO)_2L_3Br](PF_6)$ ($L = PMe_3$) or mer-$trans$-$[Mn(CO)_2L_3Br](PF_6)$ [$L = P(OMe)_3$, $P(OEt)_3$, $P(OMe)_2Ph$ and PMe_2Ph]. These last named mer-$trans$-biscarbonyl cations undergo further reaction with L to yield cis-$[Mn(CO)_2L_4]^+$, which, in the case of $L = P(OMe)_2Ph$, undergoes further reaction with $NOPF_6$ to yield $trans$-$[Mn(CO)_2L_4]^{2+}$; again an interesting change of configuration on oxidation. Hydrazine reduction of the di-cation, however, yields $trans$-$[Mn(CO)_2L_4]^+$. Treatment of $Mn(CO)L_4Br$ with $NOPF_6$ produced the nitrosyl complex $[Mn(CO)(NO)L_4][PF_6]_2[L = P(OMe)_3]$.[12]

It has been found that NO_2 possesses similar oxidizing properties to $NOPF_6$ towards metal carbonyl derivatives.[13] For example, treatment of $[Mn(CO)_4(PPh_3)]_2$ with NO_2 in acetonitrile gives $[Mn(CO)_4(PPh_3)(MeCN)]^+$. This paper urges the need for great care to be taken in the use of nitric oxide as a nitrosylating agent and, in a re-investigation of several nitrosylation reactions, has shown that the presence of NO_2 in samples of nitric oxide can radically affect the products of such reactions. For example, when $Mn(CO)_3[P(OPh)_3]_2Br$ is treated with pure NO, $Mn(NO)_2[P(OPh)_3]_2Br$ is formed, but in the presence of small quantities of air, the product is fac-$\{Mn(CO)_3[P(OPh)_3]_2Br\}^+$. mer-cis-$Mn(CO)_2[P(OMe)_2Ph]_3Br$ undergoes a similar reaction with NO–air mixtures or pure NO_2 leading to mer-$trans$-$\{Mn(CO)_2[P(OMe)_2Ph]_3Br\}^+$. This cation reacts with excess phosphine to give cis-$\{Mn(CO)_2[P(OMe)_2Ph]_4\}^+$, which in turn, on further treatment with NO_2, yields $trans$-$\{Mn(CO)_2[P(OMe)_2Ph]_4\}^{2+}$.

Brunner has continued his studies on optically active manganese carbonyl complexes and has reported that treatment of $Mn(CO)_5Br$ with $ortho$-$Me_2NC_6H_4PPh_2$ (PN) yields two enantiomers of fac-$[Mn(CO)_3(PN)Br]$. Treatment of this complex with carbon monoxide in the presence of $AlCl_3$ produces the cation $[Mn(CO)_4(PN)]^+$, which was isolated as its hexafluorophosphated salt. Addition of menthoxide anions to the manganese carbonyl cation yields the diastereoisomers of $Mn(CO)_3(PN)$ $(CO_2C_{10}H_{19})$; however, these could not be separated due to their instability. Reaction of $Mn(CO)_5Br$ with the Schiff base NN* (1) leads to formation of two isomers of

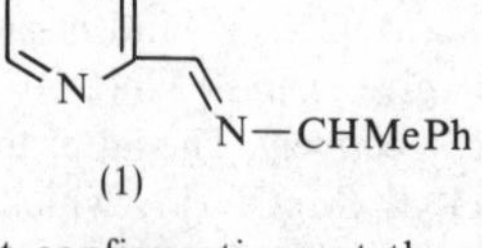

N—CHMePh

(1)

$Mn(CO)_3(NN^*)Br$, with different configurations at the metal atom. The diastereoisomers exhibit different n.m.r. spectra and can be obtained in optically active forms.[14] Direct reaction of $[(h^5$-$MeCp)Mn(CO)_2(NO)]PF_6$ with $Ph_2PNMeCHMePh$ (L) yields diastereoisomeric pair, $[(h^5$-$MeCp)Mn(CO)(NO)L]PF_6$, which are also

[12] R. H. Reimann and E. Singleton, *J. C. S. Dalton*, 1973, 2658.
[13] R. H. Reimann and E. Singleton, *J. Organometallic Chem.*, 1973, **57**, C75.
[14] H. Brunner and M. Lappus, *Z. Naturforsch.*, 1974, **29b**, 363.

separable.[15] Carboxamide complexes are formed when $[(h^5\text{-}Cp)Mn(NO)(CO)L]^+$ cations are treated with amines; the products can be isolated when L = CO. However, when L = PPh_3, the reversibility of the reactions leads to the isolation of starting materials only.[16]

The complexes $H_2SiCl[Mn(CO)_5]$, $H_2Si[Mn(CO)_5]_2$, and $Cl_2Si[Mn(CO)_5]_2$ have been prepared,[17] as have $(Me_3Si)_nMe_{3-n}SiM$ [M = $Mn(CO)_5$ or $Mn(CO)_4PPh$, $n = 1$–3].[18] Spectral data have been reported for the polysilane derivatives and the existence of Si–Mn π-bonding is inferred from the calculation of Cotton–Kraihanzel C—O stretching force constants. Low-frequency (below 600 cm^{-1}) Raman spectra of $X_3M'M(CO)_5$(X = Me or Cl; M' = Ge or Sn; M = Mn or Re) have been reported.[19] The M—M' symmetric stretching frequencies were assigned and force constants calculated. Bond orders were calculated from Raman band intensities and failed to reveal any substantial π-bonding. The intensities of the C—O stretching bands in the i.r. spectra of $X_3M'Mn(CO)_5$ complexes (X = Cl, Br, or I; M' = Si, Ge, or Sn) have also been measured[20] and the intensity variation of the A_1 mode was found to correlate with Mn—M' stretching force constants. This has been interpreted in terms of interaction between Mn—M' d-orbitals and the π* (radial) orbitals of the four equatorial carbonyl groups.

The reactions of HCl, HBr, Cl_2, I_2, ICl, and CF_3I with $Me_3MMn(CO)_5$, $Ph_3MMn(CO)_5$, and $Ph_n(C_6F_5)_{3-n}MMn(CO)_5$ (M = Si, Ge, or Sn, but not all possible reaction combinations) have been investigated. The reactions of HCl and Cl_2 with Mn—Sn complexes gave either partial or complete replacement of the organic groups; in all the other cases, cleavage of the Mn—M bond occurred.[21] Treatment of $Ph_{4-n}GeCl_n$ ($n = 1$ or 2) with $[Mn(CO)_5]^-$ in THF yields $Ph_{3-n}GeCl_n[Mn(CO)_5]$. The analogous silicon compounds were prepared by reaction of the appropriate phenylchloro- or phenylpentafluorophenyl-silanes with $Mn_2(CO)_{10}$ and the reactivities of the two series of complexes towards C_6F_5Li were investigated. The nature of the products was found to depend on M (M = Ge or Si) and the number of phenyl groups bonded to M, and the reactivity decreased in the order; Sn ∼ Ge > Si.[22] The vinyl-substituted germyl derivatives, $Me_2(C_2H_3)GeM$ [M = $Mn(CO)_5$ or $Mn(CO)_4PPh_3$] have been prepared. Photolysis of the former proceeds *via* rupture of the Ge—M bond yielding complex mixtures, from which only $Mn_2(CO)_{10}$ was isolable. It is probable that this product arises from formation and recombination of $[Mn(CO)_5]$ free radicals.[23] In contrast, photolysis of $Me_2Ge[Mn(CO)_5]_2$ gives $Me_2GeMn_2(CO)_9$ which, in solution, does not contain bridging CO groups: it has the structure $(CO)_4Mn \rightarrow Mn(CO)_5GeMe_2$. Photolysis of $Me_2GeCl[Mn(CO)_5]$ proceeds *via* formal loss of CO and Cl, the main product being $(CO)_4Mn(\mu\text{-}GeMe_2)_2Mn(CO)_4$.[24]

[15] H. Brunner and W. Rambold, *Angew. Chem. Internat. Edn.*, 1973, **12**, 1013.

[16] L. Busetto, A. Palazzi, D. Pietropaolo, and G. Dolcetti, *J. Organometallic Chem.*, 1974, **66**, 453.

[17] K. M. Abraham and G. Urry, *Inorg. Chem.*, 1973, **12**, 2850.

[18] B. K. Nicholson and J. Simpson, *J. Organometallic Chem.*, 1974, **72**, 211.

[19] A. Terzis, T. C. Strekas, and T. G. Spiro, *Inorg. Chem.*, 1974, **13**, 1346.

[20] S. Onaka, *J. Inorg. Nuclear Chem.*, 1974, **36**, 1721.

[21] R. E. J. Bichler, H. C. Clark, B. K. Hunter, and A. T. Rake, *J. Organometallic Chem.*, 1974, **69**, 367.

[22] H. C. Clark and A. T. Rake, *J. Organometallic Chem.*, 1974, **74**, 29.

[23] R. C. Job and M. D. Curtis, *Inorg. Chem.*, 1973, **12**, 2510.

[24] R. C. Job and M. D. Curtis, *Inorg. Chem.*, 1973, **12**, 2514.

Reaction of $(C_5H_5)_2Sn$ with $HMn(CO)_5$ in benzene at room temperature yields orange crystals of $H_2Sn_2[Mn(CO)_5]_4$. An X-ray study of this molecule shows it to have the structure $[(CO)_5Mn]_2(H)Sn-Sn(H)[Mn(CO)_5]_2$ (Sn-Mn = 2.67 and 2.73 Å, Sn-Sn = 2.89 Å).[25] $(CH_2CH)_2Sn(O_2CCF_3)_2$ reacts with $NaMn(CO)_5$ to give $(CH_2CH)_2Sn(O_2CCF_3)_n[Mn(CO)_5]_{2-n}$ ($n = 0$ or 1) depending on the reaction conditions employed. An X-ray structural study of the complex with $n = 0$ has been briefly reported and it was found that the tin atom has approximately tetrahedral geometry, with the two $Mn(CO)_5$ units adopting a pseudo-staggered geometry, thereby reducing the interactions between them.[26] Treatment of $NaMn(CO)_5$ with Me_3SnCH_2I in THF did not yield the expected product, but a mixture of $Me_3SnMn(CO)_5$ and $MeMn(CO)_5$.[27]

Indium or gallium metal reacts with either $Mn_2(CO)_{10}$, $Hg[Mn(CO)_5]_2$, or mercury and $Mn_2(CO)_{10}$ to form $M_2Mn_4(CO)_{18}$ (M = In or Ga). The indium compound can also be prepared by thermolysis of $In[Mn(CO)_5]_3$ and an X-ray study showed it to have structure (2) in which the $Mn(CO)_5$ units are *trans* with respect to

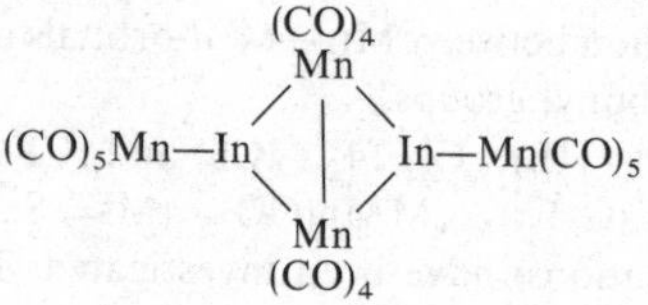

$$\begin{array}{c} (CO)_4 \\ Mn \\ (CO)_5Mn-In \quad \Big| \quad In-Mn(CO)_5 \\ Mn \\ (CO)_4 \end{array}$$

(2)

the planar In_2Mn_2 ring (Mn-Mn = 3.22, Mn-In = 2.59 Å). The gallium analogue is isomorphous with the indium compound, and reaction of either with pyridine or acetone leads to the isolation of $M_2Mn_4(CO)_{18},2D$ (M = In, D = py or Me_2CO; M = Ga, D = py), for which structure (3) is proposed. In pyridine solution, the

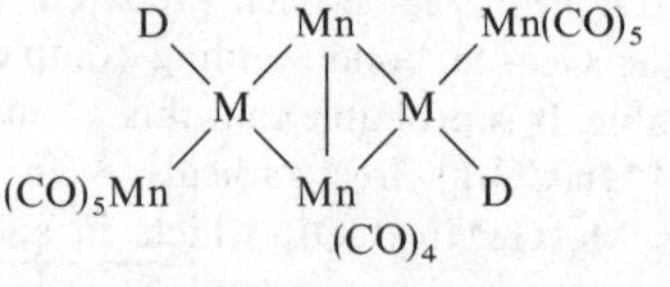

$$\begin{array}{c} (CO)_4 \\ D \quad Mn \quad Mn(CO)_5 \\ M \quad M \\ (CO)_5Mn \quad Mn \quad D \\ (CO)_4 \end{array}$$

(3)

$[Mn(CO)_5]^-$ ions dissociate from the cluster without cleavage of the Mn_2M_2 ring.[28] Treatment of $Mn(CO)_5Br$ with finely divided lanthanons in THF gives $[Mn(CO)_5]_x\text{-}LnBr_y$ (Ln = Y, Pr, Sm, Dy, Ho, Er, and Yb; x and y are integers, such that $x + y = 2$ or 3 depending on Ln). I.r. evidence indicates that the lanthanons are not bonded to the carbonyl oxygen atoms. $[Mn(CO)_5]_2Ho(acac),2Et_2O$ was also isolated. The

[25] K. D. Bos, E. J. Bulten, J. G. Noltes, and A. L. Spek, *J. Organometallic Chem.*, 1974, **71**, C52.
[26] C. D. Garner and B. Hughes, *J. C. S. Dalton*, 1974, 1306.
[27] R. B. King and K. C. Hodges, *J. Organometallic Chem.*, 1974, **65**, 77.
[28] H. J. Haupt and F. Neumann, *J. Organometallic Chem.*, 1974, **74**, 185.

lanthanide Grignard-type reagents react with Ph_3SnCl and MeI to yield Ph_3Sn-$[Mn(CO)_5]$ and $MeMn(CO)_5$, respectively. Reaction with donor ligands, L_2, gives $Mn_2(CO)_8L_2$ (L_2 = bipy, phen, or $2PPh_3$) and with *N*-methyl-*N*-nitrosotoluene-*p*-sulphonamide gives $Mn(CO)_4NO$. No reaction was observed with CO, H_2, cyclo-C_5H_6, or olefins. Similar Grignard-type solutions can also be prepared by reaction of $Mn(CO)_5Br$ or $Mn_2(CO)_{10}$ with metallic manganese or from $Mn_2(CO)_{10}$ or $NaMn(CO)_5$ with $MnCl_2$.[29]

The positional isomer *axial*-$[(CO)_5MnRe(CO)_4(PPh_3)]$ has been isolated and characterized and evidence obtained for the existence in solution of the less stable form *axial*-$[(Ph_3P)(CO)_4MnRe(CO)_5]$.[30] A preliminary account of the molecular structure of *axial*-$[Mn_2(CO)_9(PMe_2Ph)]$ has been published.[31] The photolysis of $(CO)_4Fe(\mu\text{-}AsMe_2)Mn(CO)_5$ yields (4). An *X*-ray structural study shows the metal atoms to be indistinguishable and yields values of Mn—Fe = 2.85 and M—As = 2.35 Å.[32]

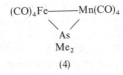

(4)

The complex $Mn(CO)_3(PH_3)_2Br$ has been reported.[32a] $Mn(CO)_5X$ (X = Br or I) reacts with $[Ph_2PCH_2CH_2PPhCH_2]_2$ (tetraphos) to give *fac*-$[Mn(CO)_3(tetraphos)$-Br] (two isomers), *cis*-$[Mn(CO)_2(tetraphos)X]$ and $[Mn(CO)_2Br]_2$ (tetraphos) as the main products. It is possible that the two isomers are produced as a consequence of isomerism in the ligand.[33] Similar treatment of $Mn(CO)_5X$ (X = Cl, Br, and I) with $PhP(CH_2CH_2PPh_2)_2$ (triphos) yields *fac*-$[Mn(CO)_3(triphos)X]$ complexes in which the phosphine ligand is only bonded in a bidentate manner. These complexes function as unidentate ligands towards a variety of metals and $[Br(CO)_3Mn(triphos)Cr(CO)_5]$ and $[I(CO)_3Mn(triphos)Mn(CO)_4I]$ have been isolated. The related complex $[Mn(CO)_2(triphos)\{P(OMe)_3\}Br]$ reacts with $Cr(CO)_5THF$ to form $Br(CO)_2$-$\{(MeO)_3P\}Mn(triphos)Cr(CO)_5$ and treatment of $Mn(CO)_3(triphos)X$ (X = Br or I) with either oxygen or ozone yields *fac*-$[Mn(CO)_3(triphos\text{—O})X]$.[34] The structures of both isomers of $Br(CO)_3Mn(triphos)Cr(CO)_5$ have now been reported in detail[34a] (see Vol. 2, p. 183). Co-ordination of triphos ligand to manganese occurs through two adjacent phosphorus atoms and results in formation of a five-membered chelate ring. Isomerism arises as a consequence of the orientation of the phenyl ring attached to the central phosphorus relative to the bromine atom. Thus, in one form, the bromine atom and the phenyl group on the same side of the five-membered chelate ring (mutually *cis*) and in the other, these groups are on opposite sides (mutually *trans*).[34a] Connor and Hudson have reported the preparation of the phosphine-

[29] A. E. Crease and P. Legzdins, *J.C.S. Chem. Comm.*, 1973, 775.

[30] J. P. Fawcett, A. J. Poe, and M. V. Twigg, *J. Organometallic Chem.*, 1973, **61**, 315.

[31] M. Laing, E. Singleton, and R. H. Reimann, *J. Organometallic Chem.*, 1973, **56**, C21.

[32] H. Vahrenkamp, *Chem. Ber.*, 1973, **106**, 2570.

[32a] R. A. Schunn, *Inorg. Chem.*, 1973, **12**, 1573.

[33] I. S. Butler and N. J. Colville, *J. Organometallic Chem.*, 1974, **66**, 111.

[34] N. J. Colville and I. S. Butler, *J. Organometallic Chem.*, 1973, **57**, 355.

[34a] P. H. Bird, N. J. Colville, I. S. Butler, and M. L. Schneider, *Inorg. Chem.*, 1973, **12**, 2902.

bridged complex $(CO)_5Mo(dmpe)Mn(CO)_4Br$ (dmpe = $Me_2PCH_2CH_2PMe_2$).[35] These authors have also found that reaction of $Et_4N[Mn(CO)_4X_2]$ (X = Cl or Br) with dmpe, dpe ($Ph_2PCH_2CH_2PPh_2$), or ape ($Ph_2PCH_2CH_2AsPh_2$) in the presence of either $AgBF_4$ or Et_3OBF_4 yields the seven-co-ordinate complexes, cis-$[Mn(CO)_4$-$(L_2)X]$ (L_2 = dmpa, dpe, or ape). At room temperature, these are rapidly converted into fac-$[Mn(CO)_3(L_2)X]$. This paper also reports the preparation of cis-$[Mn(CO)_4$-$(PMe_3)Br]$ and fac-$[Mn(CO)_3(PMe_3)_2Br]$.

The complexes $(h^5$-Cp)Mn(CO)_2(PPhX_2)$ (X = CN, NCO, NCS, or N_3) have been prepared by exchange reactions with the corresponding phenyldichlorophosphine complex. When X = NCO, the complex reacts with alcohols to form $(h^5$-Cp)Mn(CO)_2$ $[PPh(NHCO_2R)_2]$ (R = Me and Et) and reaction of phenyldiazidophosphine complex with CO and PPh_3 yields $(h^5$-Cp)Mn(CO)_2[PPh(NCO)_2]$ and $(h^5$-Cp)-$Mn(CO)_2[PPh(N_3)(NPPh_3)]$, respectively.[36] An X-ray structural study of $(h^5$-Cp)Mn(CO)_2PPh_3$ shows the structural parameters are almost identical with those of $(h^5$-Cp)Mn(CO)_3$ except that the M—C(O) bond lengths are reduced from 1.80 to 1.73 Å.[37] The structure of $(h^5$-Cp)Mn(CO)(PPh_3)_2,C_6H_6$ has also been reported. The relevant dimensions are Mn—C(Cp) = 2.16(1), M—C(O) = 1.748(9), C—O = 1.172(11), Mn—P = 2.237(3) Å, $\angle$ P—Mn—P = 104°. The carbonyl group is perpendicular to the plane formed by the metal atom and the two phosphorus atoms and the opening of the P—Mn—P angle by 14° is considered to be a consequence of electrostatic repulsion between the phosphorus atoms.[37a]

Reaction of cis-$Me_2AsC(CF_3)$=$C(CF_3)AsMe_2$ with $Mn_2(CO)_{10}$ yields (5), which contains a π-allyl-manganese bond. However, reaction with $Re_2(CO)_{10}$ in C_6H_6 gives a different product, for which the authors suggest (6) as the most likely structure.

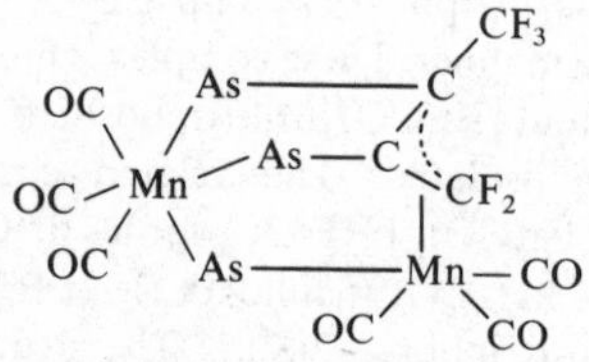

(5) Me groups
omitted for clarity

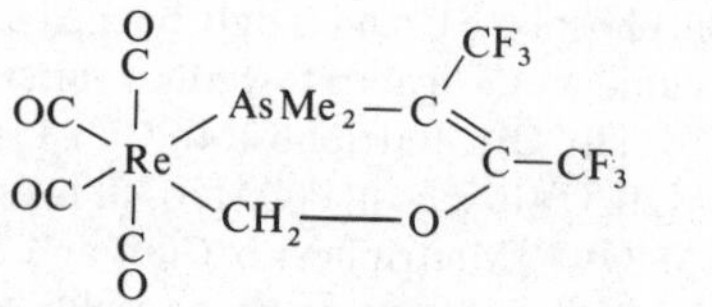

(6)

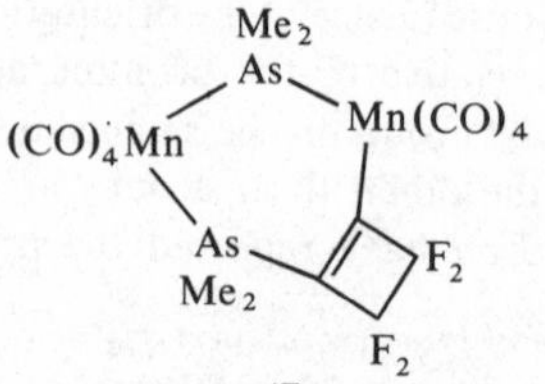

(7)

[35] J. A. Conner and G. A. Hudson, *J. Organometallic Chem.*, 1973, **73**, 351.
[36] M. Höfler and M. S. Schnitzler, *Chem. Ber.*, 1974, **107**, 194.
[37] C. Barbeau, K. S. Dichmann, and L. Ricard, *Canad. J. Chem.*, 1973, **51**, 3027.
[37a] C. Barbeau and R. J. Dubey, *Canad. J. Chem.*, 1974, **52**, 1140.

The origin of the methylene group in (6) is unknown, as reaction employing $[^2H_6]$-benzene as solvent does not lead to deuterium incorporation.[38] Thermal rearrangement of $(f_4fars)Mn_2(CO)_8$ yields (7), an *X*-ray study of which shows that ligand rearrangement and cleavage of the Mn—Mn bond has taken place.[39]

Treatment of $Mn(CO)_5Br$ with $PhN{=}NSiMe_3$ at $25\,°C$ for 6 days gives $[PhN{=}NMn(CO)_4]_2$. *X*-Ray crystallography shows that the PhNN ligands are in bridging positions and give no indication of direct Mn—Mn bonding. This dimer undergoes reaction with PPh_3 to yield $[(Ph_3P)_2Mn(CO)_2NNPh]$, the isolation of which is somewhat surprising as, firstly, carbonyl substitution is not usually observed in such reactions and secondly, the product is five-co-ordinate.[40] I.r. data have been reported for the pyrazolylborato-complexes, $\{[PzB(Pz)_3]Mn(CO)_2L\}$ [L = PCl_3, $P(OPh)_3$, PPh_3, PBu^n_3, or $P(cyclohexyl)_3$] and $\{[HB(CH_3PzCH_3)_3]Mn(CO)_2P\text{-}(OPh)_3\}$ (Pz = 1-pyrazolyl, CH_3PzCH_3 = 3,5-dimethyl-1-pyrazolyl). A method has been proposed for calculating inter-carbonyl angles from band intensities and the calculated angles for the above complexes indicate that the steric properties of the ligands influence their magnitude. N.m.r. data have also been reported for several of the complexes and it was found that $\{[PzB(Pz)_3]Mn(CO)_2L\}$ [L = CO, $P(OMe)_3$, and PMe_3] and $\{[HB(CH_3PzCH_3)_3]Mn(CO)_3\}$ show free rotation about the B—Mn axis at $5\,°C$, whereas $\{[HB(CH_3PzCH_3)_3]Mn(CO)_2L\}$ [L = $P(OMe)_3$, PMe_3, and $P(OPh)_3$] are all rigid.[41] The complexes $Mn(CO)_3(o\text{-}NCC_6H_4PPh_2)X$ (X = Cl, Br, or I) have been isolated and their i.r. spectra are consistent with a co-ordinated phosphorus atom and a π-bonded nitrile group. However, the analogous rhenium complexes are dimeric with the ligands acting in a bridging fashion, being co-ordinated by the phosphorus atoms and π-bonded C≡N groups. The authors suggest that the ligand $o\text{-}NCC_6H_4PPh_2$ is less efficient than $o\text{-}CH_2{=}CHC_6H_4PPh_2$, but more efficient than either $NC(CH_2)_nCN$ (n = 2 or 3), PPh_3, or NCC_6H_4CN·in competition for metal π-electron density.[42] $Mn(CO)_5Br$ reacts with $AgN(SO_2F)_2$ in dichloromethane at room temperature giving $\{Mn(CO)_5[N(SO_2F)_2]\}$ which dissociates in acetonitrile into $[Mn(CO)_5(MeCN)]^+$ and $[N(SO_2F)_2]^-$ ions.[43] Similar treatment of $Mn(CO)_5Br$ with $AgNSOF_2$ led to the isolation of $[Mn(CO)_4(NSOF_2)]_2$ for which structure (8) is suggested. The analogous reaction of $Re(CO)_5Br$ yielded the monomer $[Re(CO)_5(NSOF_2)]$ and it was presumed that, in the case of the manganese

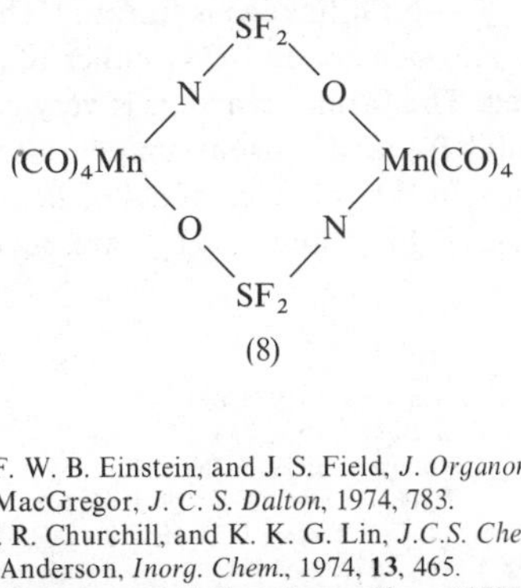

$$
\begin{array}{ccc}
 & SF_2 & \\
 & N \diagdown\diagup O & \\
(CO)_4Mn & & Mn(CO)_4 \\
 & O \diagup\diagdown N & \\
 & SF_2 & \\
\end{array}
$$

(8)

[38] W. R. Cullen, L. Mihichuk, F. W. B. Einstein, and J. S. Field, *J. Organometallic Chem.*, 1974, **73**, C53.

[39] F. W. B. Einstein and A. C. MacGregor, *J. C. S. Dalton*, 1974, 783.

[40] E. W. Abel, C. A. Burton, M. R. Churchill, and K. K. G. Lin, *J.C.S. Chem. Comm.* 1974, 268.

[41] A. R. Schoenberg and W. P. Anderson, *Inorg. Chem.*, 1974, **13**, 465.

[42] D. H. Payne, Z. A. Payne, R. Rohmer, and H. Frye, *Inorg. Chem.*, 1973, **12**, 2540.

[43] R. Mews and O. Glemser, *Z. Naturforsch.*, 1973, **28b**, 362.

reaction, the dimer was formed *via* such a monomeric species, although this could not be isolated.[44]

The physical properties of $Mn(CO)_5(TFA)$ (TFA = trifluoroacetate) have been reported and the derivatives $Mn(CO)_3L_2(TFA)$ [L = py or PPh_3, L_2 = bipy, phen, or $RSCH_2CH_2SR$ (R = Me or Et)] prepared.[45] $Mn(CO)_5Br$ reacts with the thiophosphinic acids $R_2P(S)(OH)$ (R = Et or Ph) to yield $[R_2PSOMn(CO)_3]_2$ for which structure (9) is suggested.[46] The pentacarbonylbromide also reacts with Me_2AsS_2Na

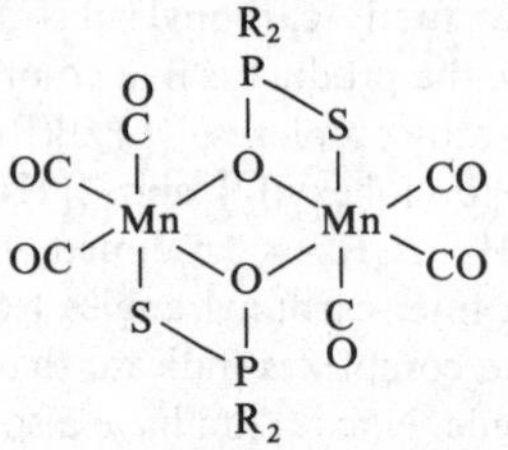

(9)

to form $[Me_2AsS_2Mn(CO)_4]$, which on heating forms the polymeric compound $[Me_2AsS_2Mn(CO)_3]_n$ and on treatment with pyridine gives $[(Me_2AsS_2)Mn(CO)_3py]$.[47] Photolytic reaction between disulphides and manganese carbonyl derivatives yields the unstable compounds $Mn(CO)_5SR$ (R = CF_3 or C_6F_5).[48] An i.r. spectral study of the complexes $[(h^5\text{-}Cp)Mn(CO)_2SR_2]$ (R = Me, Et, Pr^n or Bu^n) and $[(h^5\text{-}Cp)Mn(CO)_2THT]$ (THT = tetrahydrothiophen) has been reported and the results are consistent with the presence of conformational isomers (about the Mn—S bonds) in solution.[49] The structure of $[(h^5\text{-}Cp)Mn(CO)_2SO_2]$ has been determined and the Mn—S bond length is smaller than the sum of the covalent radii. It is suggested that the instability of the complex might, in part, be explained by crowding of the SO_2 and CO ligands.[50]

A structural study of $Mn(CO)_3(CNPh)_2Br$ has shown that the octahedral complex contains mutually *cis*-isocyanide groups and that the bromine atom is *trans* to a carbonyl ligand. The Mn—C—NPh linkage is linear.[51] The reaction of $(h^5\text{-}Cp)Mn\text{-}(CO)_2(C_8H_{14})$ (C_8H_{14} = cyclo-octene) and CS_2 either alone or in the presence of PPh_3 has been investigated. The former reaction is very complex, yielding $(h^5\text{-}Cp)\text{-}Mn(CO)_2(CS)$, $(h^5\text{-}Cp)Mn(CO)_3$, and various unidentified carbonyl- and/or thiocarbonyl-containing species. In the presence of a stoicheiometric quantity of PPh_3, only $(h^5\text{-}Cp)Mn(CO)_2(CS)$, $SPPh_3$, and C_8H_{14} are formed. Spectroscopic and

[44] R. Mews and O. Glemser, *J.C.S. Chem. Comm.*, 1973, 823.
[45] C. D. Garner and B. Hughes, *J. C. S. Dalton*, 1974, 735.
[46] E. Lindner and H. M. Ebinger, *Z. Naturforsch.*, 1973, **28b**, 113.
[47] E. Lindner and H. M. Ebinger, *J. Organometallic Chem.*, 1973, **66**, 103.
[48] J. L. Davidson and D. W. A. Sharp, *J. C. S. Dalton*, 1973, 1957.
[49] I. S. Butler and T. Sawai, *Inorg. Chem.*, 1973, **12**, 1994.
[50] C. Barbeau and R. J. Dubey, *Canad. J. Chem.*, 1973, **51**, 3684.
[51] D. Bright and O. S. Mills, *J. C. S. Dalton*, 1974, 219.

kinetic evidence suggests that the initial product of this reaction is $(h^5\text{-Cp})\text{Mn(CO)}_2$-$(\pi\text{-CS}_2)$. Similar treatment of $(h^5\text{-Cp})\text{Mn(CO)(CS)(C}_8\text{H}_{14})$ and $(h^5\text{-Cp})\text{Mn(CS)}_2$-$(C_8\text{H}_{14})$ with CS_2 and PPh_3 yields $(h^5\text{-Cp})\text{Mn(CO)(CS)}_2$ and $(h^5\text{-Cp})\text{Mn(CS)}_3$, respectively. $[(h^5\text{-Cp})\text{Mn(CO)(CS)PR}_3]$ (R = Ph or OCH_3), $[(h^5\text{-Cp})\text{Mn(CS)(diphos)}]$, and $\{[(h^5\text{-Cp})\text{Mn(CO)(CS)}]_2(\text{diphos})\}$ were also prepared.[52] Photochemical reactions of $(h^5\text{-Cp})\text{Mn(CO}_2)(\text{CS})$ with Group Va donor ligands yield $[(h^5\text{-Cp})\text{Mn(CO)(CS)L}]$ [L = PPh_3, $AsPh_3$, $SbPh_3$, P(cyclohexyl)$_3$, PMe_2Ph, or P(OR)$_3$ (R = Me, Et, CH_2CH_2Cl, or Ph)] and $(h^5\text{-Cp})\text{Mn(CS)L}_2$ [L = PPh_3 or P(OR)$_3$ (R = Me, Et, Ph, or CH_2CH_2Cl)]. This is in line with earlier predictions *vis-à-vis* M—CS *versus* M—CO bond strengths. Consequently, it appears that the present lack of thiocarbonyl complexes is due to the difficulty in preparing the CS ligand rather than to their inherent instability.[53]

Nitrosyl and Nitrogenyl Compounds.—Photolytically induced nitrosylation of $Mn_2(CO)_{10}$ in n-pentane initially yields $Mn(CO)_4NO$ which undergoes further reaction to give $Mn(CO)(NO)_3$. In THF, complete decarbonylation is observed yielding the $[Mn(NO)_3]$ group, which is probably solvated in solution. On treatment with unidentate ligands L, this moiety forms the complexes $Mn(NO)_3L$ [L = PPh_3, $AsPh_3$, or P(OMe)$_3$] and reaction with bidentate donors (L—L) gives $(ON)_3Mn$-$(L—L)Mn(NO)_3$ (L—L = 1,2-di-2'-pyridylethylene, 1,2-di-4'-pyridylethylene, $Ph_2PCH_2CH_2PPh_2$).[54] An X-ray crystallographic study has shown that $(h^5\text{-Cp})_3$-$Mn_3(NO)_4$ has the structure $(h^5\text{-Cp})_3Mn_3(\mu_3\text{-NO})(\mu_2\text{-NO})_3$ (10). The Mn—Mn bond

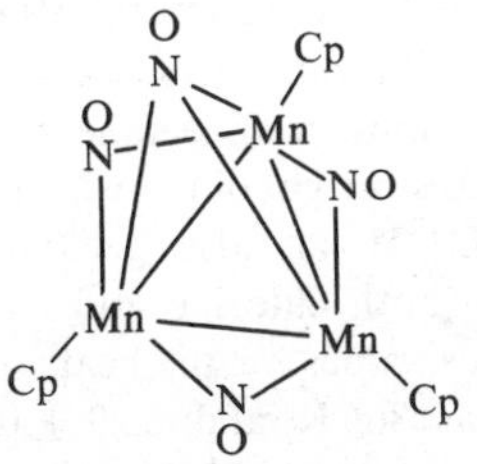

(10)

length in the equilateral triangular array of metal atoms is 2.506(3) Å. The N—O distance in the triply bridging nitrosyl group is 1.247(5) Å while those in the doubly bridging ligands are 1.212(6) Å. The Mn—N—O bond angles to both doubly and triply bridging ligands are all between 130 and 140°.[55]

$Mn(NO)_2[P(OMe)_2Ph]_2Cl$ forms crystals of two types. Those obtained from methylene dichloride–ethanol are triclinic with two molecules per asymmetric unit, while those from benzene are monoclinic with one molecule per asymmetric unit. Thus there are three independent molecules, but these have remarkably similar

[52] A. E. Fenster and I. S. Butler, *Inorg. Chem.*, 1974, **13**, 915.

[53] N. J. Colville and I. S. Butler, *J. Organometallic Chem.*, 1974, **64**, 101.

[54] M. Herberhold and A. Razavi, *J. Organometallic Chem.*, 1974, **67**, 81.

[55] R. C. Elder, *Inorg. Chem.*, 1974, **13**, 1037.

structures in which the $[MnP_2Cl]$ unit is coplanar and the apical NO groups complete trigonal-bipyramidal co-ordination ($\angle MnNO = 165 \pm 1°$).[56] The structure of $[(h^5\text{-Cp})Mn(CO)(NO)]_2$ has also been reported. It was not possible to distinguish crystallographically between the nitrosyl and carbonyl groups and thus the structure must be described as $(h^5\text{-Cp})(OX)Mn(\mu\text{-XO})_2Mn(XO)(h^5\text{-Cp})$ (X = C or N); however, there is strong chemical evidence that one CO group and one NO group bridge the metal atoms and hence one CO group and one NO group are terminally bonded [Mn—Mn = 2.571, Mn—X (bridging) = 1.906(average) and M—X (terminal) = 1.723 Å(average)]. The C_5H_5 ligands are in a *trans* configuration. The fluxional behaviour of this complex together with that of $[(h^5\text{-Cp})ML_2]_2$ (M = Cr, L = NO; M = Fe, L = CO) in solution were also studied. All three complexes exist as mixtures of *cis* and *trans* bridged isomers in solution and the *cis*:*trans* ratio follows the order Fe > Mn > Cr. The activation energies for *cis* $\rightleftharpoons$ *trans* and for bridged $\rightleftharpoons$ terminal ligand exchange processes follow the order Cr > Mn > Fe and, for the manganese and iron complexes, bridged $\rightleftharpoons$ terminal exchange is more rapid for the *trans*- than the *cis*-isomer.[57]

ESCA studies have been reported for the complexes $[(h^5\text{-Cp})Mn(CO)_2N_2]$, $[(h^5\text{-Cp})Mn(CO)_2]_2N_2H_2$, $[(h^5\text{-Cp})Mn(CO)_2N_2H_4]$, and $[(h^5\text{-Cp})Mn(CO)_2NH_3]$, which are possible analogues for the intermediate in nitrogen fixation. These show that in all cases the Mn—N bonds have about the same polarity, although the $1s$ binding energies of the nitrogen atoms bound directly to manganese do not show a regular progression on going to more negative formal oxidation numbers.[58]

Manganese(II).—*Halides.* An *X*-ray structural study of $CsMn_4Cl_9$ showed the caesium and chlorine atoms to be cubic close-packed and the manganese atoms to be co-ordinated to chlorine atoms in an octahedral fashion. Each $[MnCl_6]$ octahedron is linked to its six nearest neighbours by sharing five edges and one vertex.[59] The structure of $MnCl_2,14EtOH$ has also been reported. The metal atoms are octahedrally co-ordinated to both chlorine and oxygen atoms.[60] Calorimetry of $CsMnCl_3$ and $CsMnCl_4$ yields values for the heats of fusion of the two compounds of 15.70 ± 0.41 kcal mol^{-1} at 866 K and 6.29 ± 0.80 kcal mol^{-1} at 811 K, respectively; $\hat{C}_p$ values were also reported.[61] The electronic spectra of Mn^{2+} ions in $CsCdCl_3$ and Cs_2ZnCl_4 melts and NH_4SCN and $KSCN$ melts have been measured and compared with those obtained from Mn^{2+} in the glasses $CsCd(Mn)Cl_3$ and $Cs_2Zn(Mn)Cl_4$ at 25 °C and from crystalline $MnCl_2$. Ligand field parameters were determined in each case.[62] Reaction of MnI_2 with triethylaluminium in THF gives $MnHI(THF)_{1.5}$, and treatment of MnI_2 with iodine in THF yields '$MnI_{10}(THF)_n$'. The latter is presumably either $MnI_2,4I_2(THF)_n$ or $Mn(I_3)_2,2I_2(THF)_n$ and this reacts with trialkylaluminium compounds to give $MnAl_2R_4I_4(THF)_4$ (R = Me or Et).[63]

[56] M. Laing. R. H. Reimann, and E. Singleton, *Inorg. Nuclear Chem. Letters*, 1974, **10**, 557.

[57] R. M. Kirchner, T. J. Marks, J. S. Kristoff, and J. A. Ibers, *J. Amer. Chem. Soc.*, 1973, **95**, 6602.

[58] H. Binder and D. Sellmann, *Angew. Chem. Internat. Edn.*, 1973, **12**, 1016.

[59] J. Goodyear and D. J. Kennedy, *Acta Cryst.*, 1973, **B29**, 2677.

[60] P. L'Haridon and M. T. LeBihan, *Acta Cryst.*, 1973, **B29**, 2195.

[61] A. S. Kucharski and S. N. Flengas, *Canad. J. Chem.*, 1974, **52**, 946.

[62] S. V. Volkov and N. I. Buryak, *Russ. J. Inorg. Chem.*, 1973, **18**, 1260.

[63] A. Yamamoto, K. Kato, and S. Ikeda, *J. Organometallic Chem.*, 1973, **60**, 139.

Complexes. N-Donor ligands. Treatment of manganese(II) oxalate dihydrate with hydrazine in aqueous solution gives $Mn(ox)_xN_2H_4$ ($x = 1$ and 2). Spectral measurements and thermal decomposition studies on the two compounds were reported. It was suggested that addition of the first molecule of hydrazine, in place of water, does not alter the co-ordination of the metal, *i.e.* it remains in an octahedral environment with oxygen atoms of the oxalate groups in the equatorial plane, the oxalate acting as a bridging ligand, and the hydrazine molecule bridging between vertices of two octahedra. In the bis-compound, hydrazine replaces some co-ordinated oxygen atoms and is also hydrogen-bonded to the oxalate.[64] Reaction of MnF_2 with hydrazine hydrate and anhydrous hydrazine gives MnF_2,N_2H_4 and $MnF_2,2N_2H_4$, respectively.[65] Spectral studies on $Mn(N_2H_5)_2(SO_4)_2$ indicates the metal atom to be in a distorted octahedral environment similar to that reported previously for the zinc analogue. The hydrazinium cation appears to behave as a normal N-donor ligand.[66] The polymeric compounds $Mn(PhNHNH_2)_2X_2$ (X = Cl, Br, or I) have also been reported.[67]

The magnetic susceptibilities of $[Mn(bipy)_3],2THF$ and $Li[Mn(bipy)_3],4THF$ have been determined over the range 1.5 to 300 K. The susceptibility of the latter can be best explained on a radical-ion model in which three bipyridyl radical-ion ligands are co-ordinated to manganese(II): *i.e.* $[Mn^{2+}(bipy^-)_3]^-$ rather than $[Mn^-(bipy)_3]^-$. A similar model is also proposed for the other complex.[68] The octahedral polymers $Mn(4,4'-bipy)X_2$ (X = Cl or Br) have been prepared.[69]

The existence of pure colourless manganese(II) complexes containing four phen ligands per metal atom has been substantiated, and some new eight-co-ordinate complexes of this type, $Mn(phen)_4X_2,nH_2O$, have been isolated in which X is a weakly- or non-co-ordinating anion such as BPh_4^-, I^-, or 2-bromocamphor-π-sulphonate (n = 2,4, or 6 depending on X). A number of yellow compounds with two or three phen ligands per metal atom were also prepared for comparison with the tetrakis compounds. The compounds $M(phen)_4(ClO_4)_2,nH_2O$ (M = Sr, Ba, or Pb, n = 0, 4, or 5) were also prepared and the ligands *o*-phenylenediamine and 8-amino-quinoline (L) also yield eight-co-ordinate tetrakis-complexes with $Mn(ClO_4)_2$, as well as the octahedral complexes $[MnL_3](ClO_4)_2$ and MnL_2X_2 (X = Cl, or Br), However, bidentate N_2-donors with more flexible chelate rings yield only the tris- and bis-complexes.[70]

Previous magnetic studies on $[MnCl_2(pyrazole)_2]$ could only be interpreted if it was assumed that the compound was built up of rows containing manganese atoms bridged by chlorine atoms. *X*-Ray crystallography has been employed to confirm this structure in which rows of $[Mn(pyrazole)_2]$ units are bridged by chlorine atoms. The co-ordination around the metal atoms is *trans*-$[MnN_2Cl_4]$.[71] The complex

[64] T. M. Zhdenovskikh, E. I. Krylov, V. A. Sharov, and N. G. Fedotovskikh, *Russ. J Inorg. Chem.*, 1973, **18**, 651.
[65] P. Glavic, A. Bole and J. Slivnik, *J. Inorg. Nuclear Chem.*, 1973, **35**, 3979.
[66] A. Nieuwpoort and J. Reedijk, *Inorg. Chim. Acta*, 1973, **7**, 323.
[67] W. G. Glass and J. O. McBreen, *J. Inorg. Nuclear Chem.*, 1974, **36**, 747.
[68] K. Hara, M. Inoue, T. Horiba, and M. Kubo, *Chem. Letters*, 1974, 419.
[69] A. Anagostopoulos, *J. Inorg. Nuclear Chem.*, 1973, **35**, 3366.
[70] B. Chiswell and E. J, O.'Reilly, *Inorg. Chim. Acta*, 1973, **7**, 707.
[71] S. Gorter, A. D. van Ingen Schenau, and G. C. Verschoor, *Acta Cryst.*, 1974, **B30**, 1867.

[Mn(5-aminoindazole)Cl$_2$] has been reported; it is probably a tetrahedrally co-ordinate polymer.[72]

The Schiff base (11) can be prepared by heating together a 2:1 mole ratio of 2-acetylpyridine and 3,3'-iminobispropylamine in ethanol. If this is followed by the addition of Mn^{2+} ions, MnL(ClO$_4$)$_2$ [L = (11)] is formed. Attempts to achieve a

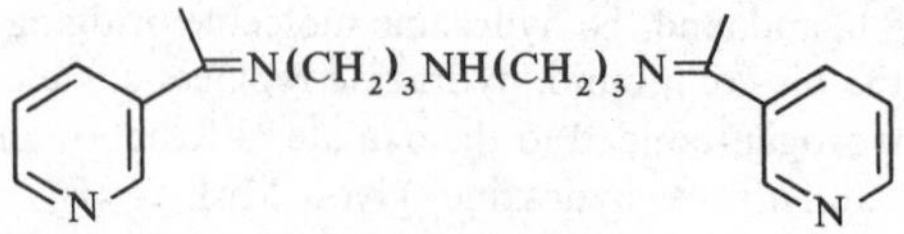

(11)

template synthesis of this complex gave only MnO$_2$, although similar template reactions were successful in the case of other metals.[73] Erythrocyte hemolysate prepared from blood withdrawn six days after injection of ^{54}MnCl$_2$ into rats contains significant amounts of labelled manganese. Gel chromatography indicates that a manganese(II) porphyrin is formed *in vivo*.[74] The magnetic susceptibility of mangan-ese(II) phthalocyanine has been measured between 13.8 and 280 K and deviation from the Curie–Weiss law was observed above 70 K. A qualitative description of exchange interactions in this complex and the cobalt(II) and copper(II) analogues has been presented on the basis of electronic states.[74a]

O-Donor ligands. The crystal structures of hydrated and vacuum-dehydrated manganese(II)-exchanged zeolites have been determined. Chemical analyses of the zeolites indicated 4.5 manganese(II) ions per unit cell after exchange. In the hydrated structure, Mn^{II} ions are found to be five-co-ordinate with trigonal-bipyramidal geometry, there being three oxygen atoms in the plane and two axial water mole-cules [Mn—OH$_2$ = 2.03(6) and 2.06(7) Å; Mn—O (in plane) = 2.28 Å]. In the dehydrated structure the manganese ions have slightly distorted trigonal-planar co-ordination [Mn—O = 2.11(1) Å].[75] The structure of Mn$_2$P$_2$O$_7$,2H$_2$O has also been reported. The manganese(II) ions are each co-ordinated to five oxygen atoms and one water molecule in a distorted octahedral manner [Mn—O = 2.192(1) Å (average)]. Chains of octahedra with shared edges run through the structure and the two ends of the pyrophosphate ion are twisted from the eclipsed conformation by 20°.[76] The structure of hureaulite, Mn$_5$(HOPO$_3$)$_2$(PO$_4$)$_2$(H$_2$O)$_4$, has also been determined.[77]

Preparative procedures for Mn(HSO$_4$)$_2$[78] and manganese(II) malonate dihydrate[79] have been published. Ethylene glycol (EG) forms the complexes Mn(EG)$_x$A$_n$

[72] S. A. A. Zaidi and A. S. Farooqi, *J. Inorg. Nuclear Chem.*, 1973, **35**, 4320.
[73] R. H. Prince and D. A. Stotter, *Inorg. Chim. Acta*, 1974, **10**, 89.
[74] R. G. V. Hancock and K. Fritze, *Bio-inorg. Chem.*, 1973, **3**, 77.
[74a] H. Miyoshi, *Bull. Chem. Soc. Japan*, 1974, **47**, 561.
[75] R. Y. Yanagida, T. B. Vance, and K. Seff, *Inorg. Chem.*, 1974, **13**, 723.
[76] S. Schneider and R. L. Collin, *Inorg. Chem.*, 1973, **12**, 2136.
[77] S. Menchetti and C. Sabelli, *Acta Cryst.*, 1973, **B29**, 2541.
[78] G. Palavit and S. Noel, *Rec. Trav. chim.*, 1974, **92**, 977.
[79] L. Walter-Levy, J. Perrotey, and J. W. Visser, *Bull. Soc. chim. France*, 1973, 2596.

(A $= NO_3$, $n = 2$; A $= SO_4$, $n = 1$; $x = 1$—4) and i.r. data indicates that the anions are co-ordinated in most of the complexes, the glycol ligands probably being unidentate.[80] A structural study on $MnCl_2(EG)_2$ shows the complex to have a *cis*-octahedral configuration, the glycol ligands being bidentate.[81] 1,3-Dioxan (L) forms the complexes $MnLX_2$ (X = Cl or Br) which are polymeric with octahedral co-ordination, the dioxan ligands forming the bridges. No mixed complexes were isolated from the MnX_2–L–H_2O system, although such products were obtained with 1,4-dioxan.[82]

E.s.r. data have been obtained for $[Mn(pyNO)_6](ClO_4)_2$ doped into the corresponding zinc and cadmium compounds (pyNO = pyridine *N*-oxide, 3-methyl-, 4-methyl-, 6-nitro-, and 4-cyano-pyridine *N*-oxide). Hyperfine coupling constants indicate that Mn^{II}—ONpy bonding is more covalent than Mn^{II}—OH_2.[83] The related compounds $Mn(pyNO)_3X_2(H_2O)$ (X = NCS or NCSe), $[Mn(qNO)_6]$ $[Mn(NCSe)_4]$, and two forms of $Mn(qNO)_2(NCS)_2$ (pyNO = pyridine *N*-oxide, qNO = quinoline *N*-oxide) have been prepared.[84] Pyrazine *N*-oxide (pz-NO) forms the complex $[Mn(pz-NO)Cl_2]_x$, in which the metal atom is five-co-ordinate, with bridging, NO-bonded ligands and both bridging and terminal chlorine atoms.[85]

The ligand $(EtO)_3PO$ reacts with manganese(II) halides at 100—250 °C yielding $\{Mn_2[O(O)P(OEt)_2]_2[O(O)P(OEt)\cdot O\cdot P(OEt)(O)O]\}$.[86] There has been further work in characterizing some phosphonate and phosphate ester complexes and it has been found that $Mn(DIMP)_4(ClO_4)_2$ (DIMP = di-isopropylmethylphosphonate) should be formulated as $[Mn(DIMP)_4OClO_3]ClO_4$. However, dimethylmethylphosphonates and triphenylphosphine oxide complexes of the same type do not show perchlorate ion co-ordination and for these complexes the presence of polynuclear cations, $[MnL_4]_n^{2n+}$, is favoured. It is proposed that these cations contain both bridging and terminal ligands and five-co-ordinate metal atoms.[87] Treatment of MnX_2(X = Cl, Br, or I) with Ph_3E (E = P or As) leads to formation of $[MnX_2$-$(OEPh_3)_2]$ and the related bidentate ligands, $(Ph_2P)_2CH_2$, $Ph_2P(CH_2)_2PPh_2$, *cis*-Ph_2PCH=$CHPPh_2$, and $Ph_2As(CH_2)_2AsPh_2$ (L) yield the dioxide complexes, $Mn(LO_2)X_2$. Reaction of MnX_2 (X = Cl or Br) with Ph_2PH gives bis-(diphenylphosphinato)-manganese(II) and this reaction involves a novel ligand deprotonation and oxidation.[88] The hexamethylphosphoramide and nonamethylimidodiphosphoramide (L) complexes $MnLnX_2$ ($n = 1$—4; X = BF_4, NO_3, Cl, Br, I, and NCS) have been prepared and it is suggested that the Mn—L bond is essentially ionic in character.[89]

Reaction of $MnCl_2$ with *N*-methylhydroxylamine in hot alcoholic solution gives $MnCl_2(MeHNOH)_2$, for which it is suggested that the ligand co-ordinates with the

[80] D. Knetsch and W. L. Groeneveld, *Rec. Trav. chim.*, 1973, **92**, 855.
[81] B. M. Antti, *Acta Chem. Scand.*, 1973, **27**, 3513.
[82] J. C. Barnes and C. S. Duncan, *Inorg. Chim. Acta*, 1973, **7**, 404.
[83] G. M. Wolterman and J. R. Wasson, *Inorg. Chem.*, 1973, **12**, 2366.
[84] G. B. Aitkin and G. P. McQuillan, *J. C. S. Dalton*, 1973, 2637.
[85] A. N. Speca, L. L. Pytlewski, and N. M. Karayannis, *J. Inorg. Nuclear Chem.*, 1973, **35**, 4029.
[86] C. M. Mikulski, L. L. Pytlewski, and N. M. Karayannis, *Z. anorg. Chem.*, 1974, **403**, 200.
[87] N. M. Karayannis, C. M. Mikulski, M. J. Strocko, L. L. Pytlewski, and M. M. Labes, *Inorg. Chim. Acta*, 1974, **8**, 91.
[88] S. Casey, W. Levason, and C. A. McAuliffe, *J. C. S. Dalton*, 1974, 886.
[89] M. W. G. de Bolster, B. Nieuwenhuijse, and J. Reedijk, *Z. Naturforsch.*, 1973, **28b**, 104.

zwitterion structure, $Mn \leftarrow \overset{-}{O}—\overset{+}{N}H_2Me$.[90] The manganese(II)–urea (U) complexes MnU_2X_2 (X = Cl or Br), MnU_4X_2 (X = Cl or Br), MnU_6X_2 (X = Br, I, or ClO_4), and $MnU_{10}X_2$ (X = Br or I) have been isolated. All are high spin, with octahedral co-ordination, and contain O-bonded urea ligands. The first has a chain structure with bridging X ligands, the second is a *trans*-octahedral monomer and the third and fourth should be formulated as $[MnU_6]X_2$ and $[MnU_6]X_2,4U$, respectively. In this last complex, the four extra urea molecules are hydrogen-bonded to the $[MnU_6]^{2+}$ cation.[91]

S-Donor ligands. As an extension of previous studies on related complexes of iron(II), cobalt(II), nickel(II), and zinc(II), the structure of $\{Mn[SPPh_2 \cdot N \cdot PPh_2S]_2\}$ has been determined by *X*-ray methods. The metal atom is co-ordinated in an approximately tetrahedral manner by the four sulphur atoms. The two MnS_2P_2N rings adopt the twisted boat conformation with S and P atoms at the apices. The single-crystal electronic spectrum has been measured and interpreted.[92]

Mixed donor ligands. $\Delta G°$, $\Delta H°$, and $\Delta S°$ values for the formation of complexes between asparaginate and Mn^{2+} as well as H^+, Fe^{2+}, Co^{2+}, Ni^{2+}, and Zn^{2+} in aqueous solution at 25 °C have been reported.[93] Cytostatic Hadacidin, *N*-formyl-*N*-hydroxyglycine (HAD), forms the complex Mn(HAD) for which spectroscopic and stability constant data have been reported.[94] Manganese(II) complexes of human serum albumin have also been studied.[95]

1,10-Phenanthroline *N*-oxide (phenNO) forms $[Mn(phenNO)_3](ClO_4)_2$ in which the ligand is chelating. Spectrochemical parameters suggest phenNO is a considerably weaker ligand than phen.[96] The complex $Mn_2[Mn(nta)_2], 2H_2O$ (nta = nitrilotriacetate) has been prepared.[97] Mn^{II}(salen) (salen = *NN'*-disalicylaldehydeethylenediamine) and its analogues react with oxygen in organic solvents to give three types of product which contain the structural units, $Mn^{III}—O_2—Mn^{III}$, $(Mn^{IV}—O—)_n$ and $Mn^{IV}=O$ respectively[98] [see also manganese(III)]. Salicylaldehyde salicylhydrazine [SSH;(12)] forms the complex Mn(SSH − H)$_2$ for which structure (13) is suggested. The complex has an anomalously high magnetic moment (7.7 BM) which is difficult to rationalize, as the presence of diamagnetic impurities or other oxidation states of manganese would presumably lower the value below that normally expected for

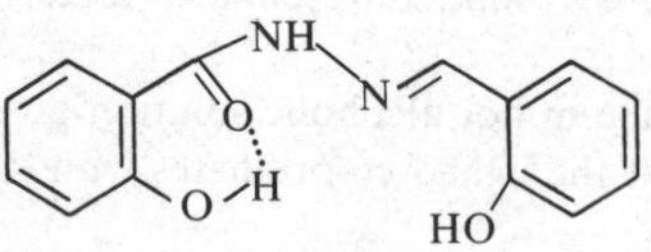

(12) 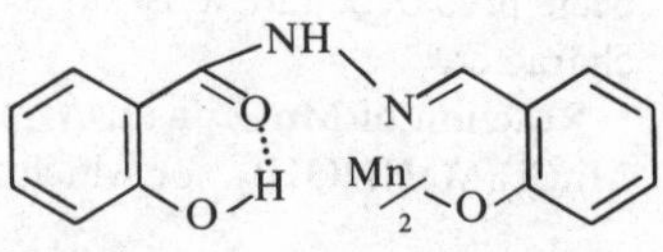   (13)

[90] M. A. Sarukhanov, S. S. Val'dman, and N. A. Parpiev, *Russ. J. Inorg. Chem.*, 1973, **18**, 439.

[91] J. P. Barbier and R. Hugel, *Inorg. Chim. Acta*, 1974, **10**, 93.

[92] O. Siiman and H. B. Gray, *Inorg. Chem.*, 1974, **13**, 1185.

[93] A. C. Baxter and D. R. Williams, *J. C. S. Dalton*, 1974, 1117.

[94] H. P. Fritz and O. von Stetten, *Z. Naturforsch.*, 1973, **28b**, 772.

[95] B. E. Chapman, T. E. MacDermott, and W. J. O.'Sullivan, *Bio-inorg. Chem.*, 1973, **3**, 27.

[96] A. N. Speca, L. L. Pytlewski, and N. M. Karayannis, *J. Inorg. Nuclear Chem.*, 1974, **36**, 1227.

[97] F. J. M. Rajabalee, *J. Inorg. Nuclear Chem.*, 1974, **36**, 557.

[98] T. Matsushita, T. Yarino, I. Masuda, T. Shono, and K. Shinra, *Bull. Chem. Soc. Japan*, 1973, **46**, 1712.

high-spin manganese(II). The authors also point out that the electronic spectrum of the complex does not unambiguously relate to either octahedral or tetrahedral co-ordination.[99] o-Hydroxy-4-benzamidothiosemicarbazide (HO·BTSC) forms the high-spin complexes, $[Mn(O·BTSC)_2],4H_2O$ and $[Mn(O·BTSC)(H_2O)_2(OH)]$, $3H_2O$, in both of which the metal ions are co-ordinated in an octahedral fashion.[100]

Manganese(III).—*Complexes.* Structural studies on $K_3[Mn(CN)_6]$ show it to be isomorphous with the analogous iron(III) complex. The octahedral anion has independent Mn—C bond lengths of 1.990(15), 1.940(15), and 2.003(15) Å.[101]

The enthalpies of solvation of $Mn(acac)_3$ by acetone, benzene, and chloroform have been reported.[102] Treatment of $Mn(acac)_3$ with either HCl, HBr, or HI in benzene–diethyl ether at room temperature yields the high-spin complex $Mn(acac)_2X$ (X = Cl, Br, or I). In nitromethane and dichloromethane, these molecules exist as covalent monomers, but in methanol dissociation occurs, giving $[Mn(acac)_2-(MeOH)_2]^+X^-$. Recrystallization from hot acetylacetone yields $[Mn(acac)_2-(H_2O)_2]X,2H_2O$ (X = Cl or Br).[103] The complexes $Mn(acac)_2X$ (X = N_3 or NCS) and $Mn(acac)_2(NCS)B$ (B = py or MeOH) have also been reported[104] and the structures of $Mn(acac)_2X$ (X = N_3 or NCS) determined.[104,105] Both contain bridging X groups (end-to-end in the case of the azide) in polymeric chains of six-co-ordinate manganese(III) polyhedra, and both compounds have a large tetragonal distortion. Treatment of $Mn(acac)_2N_3$ with H_2TPP (H_2TPP = tetraphenylporphyrin) in methanol at room temperature yields $[Mn(TPP)(N_3)(MeOH)],MeOH$ in high yield and purity. The structure of this complex and that of the new γ-form of $Mn(acac)_3$ have also been determined[105] and again both show tetragonal distortions.

The complex cations $[Mn^{III}_3O(O_2CMe)_6L_3]^+$ (L = py or β-picoline) have been isolated and their electrochemical behaviour studied. There is evidence for reduced species in which the oxo-centred metal triangle is maintained, but these could not be isolated.[106] A variable-temperature n.m.r. study of $Mn(S_2CNR_2)_3$ [R_2 = MePh or $(PhCH_2)_2$] has been carried out. Both are non-rigid, and kinetic parameters have been determined for the intramolecular metal-centred rearrangement which results in optical inversion. The trigonal-twist mechanism is assigned to the primary rearrangement pathway and a comparison with the In, Ga, V, and Cr analogues shows the rates of optical inversion to follow the order, In, Ga, V > Mn > Cr.[107]

An X-ray structural study on $[Mn(salen)OAc]$ shows that linear polymeric chains of Mn(salen) units are bridged by a single acetate group in an *anti–anti* configuration. Magnetic data are consistent with this structure.[108] The complexes $Mn(4-Bu^s-salen)X$ (X = F, Cl, Br, I, N_3, NCS, OCN, NO_2, and OAc) have been prepared and all have magnetic moments of about 4.9 BM in chloroform and DMSO solutions. In the solid state the magnetic moments vary from 4.2 to 5.0 BM depending on X, and

[99] K. K. Narang and A. Aggarwal, *Inorg. Chim. Acta*, 1974, **9**, 137.
[100] M. P. Swami, P. C. Jain and A. K. Srivastava, *Roczniki Chem.*, 1973, **47**, 2013.
[101] M. P. Gupta, H. J. Milledge, and A. E. McCarthy, *Acta Cryst.*, 1974, **B30**, 656.
[102] R. J. Irving and R. A. Schulz, *J. C. S. Dalton*, 1973, 2414.
[103] K. Isobe, K. Takeda, Y. Nakamura, and S. Kawaguchi, *Inorg. Nuclear Chem. Letters*, 1973, **9**, 1283.
[104] B. R. Stults, V. W. Day, E. L. Tasset, and R. S. Marianelli, *Inorg. Nuclear Chem. Letters*, 1973, **9**, 1259.
[105] V. W. Day, B. R. Stults, E. L. Tasset, R. O. Day, and R. S. Marianelli, *J. Amer. Chem. Soc.*, 1974, **96**, 2650.
[106] S. Uemura, A. Spencer, and G. Wilkinson, *J. C. S. Dalton*, 1973, 2565.
[107] L. Que and L. H. Pignolet, *Inorg. Chem.*, 1974, **13**, 351.
[108] J. E. Davies, B. M. Gatehouse, and K. S. Murray, *J. C. S. Dalton*, 1973, 2523.

G

it is suggested that the lower values of the magnetic moments arise as a consequence of dimer formation in the solid state.[109] The related Schiff-base complexes, [Mn(SB)Cl] have also been prepared in which SB is either a 4-substituted derivative of salen (substituents = H, Me, Bu^s, OMe, or Br) or 3,4-disubstituted derivative of hydroxyacetophenone ethylenedi-imine (substitutents = H or Me).[110] Boucher has also reported the new complexes [Mn(SB)X] (X = Cl or NO$_2$; SB = the optically active terdentate Schiff base derived from *o*-hydroxyacetophenone and (−)-1,2-propylenediamine) (14). Electronic and c.d. spectra were measured and assignments

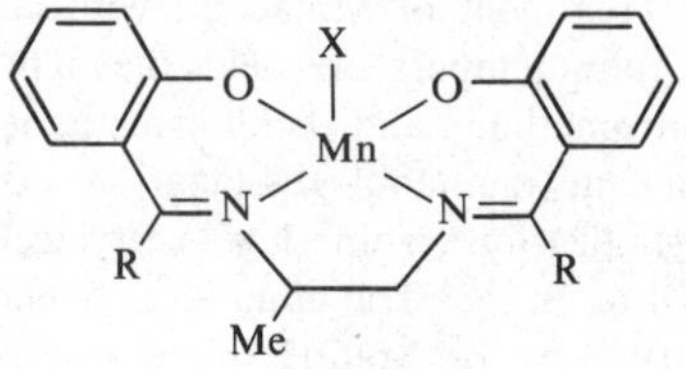

R = H or Me
X = Cl or NO$_2$

(14)

proposed. These are consistent with a preferred conformation of the central chelate ring of δ (axial methyl) and with an absolute configuration of Λ for the flattened tetrahedral array of donor atoms.[111] Manganese(III) complexes of bis(salicylidene)-*o*-phenylenediamine and its 5-bromo-, 5-chloro-, 4-methyl-, 4 hydroxy-, and 4-nitro-derivatives have also been prepared.[112] All obey the Curie–Weiss law and the temperature dependence of the magnetic moments is attributed to weak antiferromagnetic interactions. Oxygenation of Mn^{II}(salpn)H$_2$O [salpn = NN'-trimethylene-bis-(salicylaldiminate)] by O$_2$ in pyridine gives [(salpn)Mn^{III}(μ-OH)$_2$Mn^{III}(salpn)], 4py, the structure of which has been confirmed by X-ray methods. Thus oxidation of manganese(II) has taken place rather than direct oxygen addition; however, this does not preclude reaction *via* an initial dioxygen intermediate.[113]

Manganese(IV).—The single-crystal optical spectrum of K$_2$MnCl$_6$ doped into K$_2$SnCl$_6$ has been reported and the B and Dq parameters evaluated. The e.s.r. spectrum of polycrystalline manganese(IV) in K$_2$SnCl$_6$ has also been recorded. Variable-temperature studies show that Mn^{IV} is thermally unstable in K$_2$SnCl$_6$ yielding Mn^{II}, probably *via* the intermediacy of Mn^{III}.[114] MnCl$_2$ reacts with *o*-phenylene-bis(dimethylphosphine) (L) to give high-spin [MnCl$_2$L$_2$], whereas no reaction was observed with the corresponding diarsine. Treatment of the diphosphine complex with trityl hexafluorophosphate gives high-spin [MnCl$_2$L$_2$]PF$_6$, which

[109] L. J. Boucher and M. O. Farrell, *J. Inorg. Nuclear Chem.*, 1973, **35**, 3731.
[110] L. J. Boucher, *J. Inorg. Nuclear Chem.*, 1974, **36**, 531.
[111] L. J. Boucher and D. R. Herrington, *Inorg. Chem.*, 1974, **13**, 1105.
[112] V. V. Zelentsov and I. K. Somova, *Russ. J. Inorg. Chem.*, 1973, **18**, 1125.
[113] H. S. Maslen and T. N. Waters, *J.C.S. Chem. Comm.*, 1973, 760.
[114] P. J. McCarthy and R. D. Bereman, *Inorg. Chem.*, 1973, **12**, 1909.

reacts further with nitric acid, yielding the manganese(IV) cation $[MnCl_2L_2]^{2+}$ ($\mu_{eff} = 3.99$ BM).[115]

Higher Oxidation States of Manganese.—The reactions of various oxo-complexes of manganese in fused salts have been determined. In $NaNO_2$ and Na_2O_2–$NaNO_2$ melts, the most stable species appears to be $[MnO_4]^{3-}$, whereas in molten potassium nitrite and in nitrate melts in which K^+ ions predominate over Na^+ ions, $[MnO_4]^{2-}$ is the favoured species.[116] A kinetic study[117] of the disproportionation of manganate ions in acid solution is consistent with the following mechanism:

$$H^+ + MnO_4^{2-} \rightarrow HMnO_4^- \quad \text{(rapid, at equilibrium)}$$
$$HMnO_4^- \rightarrow MnO_3 + OH^- \quad \text{(first order)}$$
$$MnO_3 + HMnO_4^- \rightarrow MnO_4^- + Mn^V \quad \text{(second order)}$$

However, the third step may be:

$$MnO_3 + HMnO_4^- + H_2O \rightarrow HMnO_4 + H_2MnO_4^-$$

The decomposition of permanganate in a molten potassium–lithium nitrate eutectic has been studied at 150 and 200 °C. The rate of decomposition in a Pyrex vessel is increased in the presence of traces of NO_2^-, Br^-, and CN^- and in the presence of the manganate precipitate produced on decomposition. Various heavy-metal and transition-metal ions are also found to accelerate the decomposition, and Ni^{2+} and Co^{2+} ions are oxidized by permanganate in the melt to insoluble nickel(II) and cobalt(III) manganates.[118] The only solid products obtained on thermal decomposition of $KMnO_4$ are K_2MnO_4 and '$K_4Mn_7O_{16}$', thus refuting the earlier suggestion that $K_3(MnO_4)_2$ is an intermediate phase in the decomposition.[119]

Oxides and Sulphides.—Manganese(II) tungstates of different compositions have been reported.[120] Thus, $MnWO_4$ is formed when the pH of the sodium tungstate solution is between 8 and 9 and $5MnO,12WO_3,xH_2O$ (paratungstate) is precipitated at pH 3.3 and from the mother liquors in the pH range 4.75—3.60. $MnO,4WO_3,xH_2O$ (metatungstate) is obtained from the mother liquors in narrow pH range 4.9—5.0. Water is present in the tungstates both as water of hydration and as hydroxy-groups; its removal produces new phases. A structural study of $La_{32.66}Mn_{11}S_{60}$ shows there are 12 non-equivalent lanthanum sites, four of which are not completely filled, and the lanthanum ions are seven- or eight-co-ordinate. Four of the manganese sites are in octahedral environments of sulphur atoms and the other two sites have occupancies of 0.5.[121]

[115] L. F. Warren and M. A. Bennett, *J. Amer. Chem. Soc.*, 1974, **96**, 3340.
[116] R. B. Temple and G. W. Thickett, *Austral. J. Chem.*, 1973, **23**, 2051.
[117] J. H. Sutter, K. Colquitt, and J. R. Sutter, *Inorg. Chem.*, 1974, **13**, 1444.
[118] R. B. Temple and G. W. Thickett, *Austral. J. Chem.*, 1974, **27**, 943.
[119] F. H. Herbstein and A. Weissman, *J. C. S. Dalton*, 1973, 1701.
[120] V. N. Pitsyuga, M. V. Mokhosoev, M. N. Zayats, and T. E. Frantsuzova, *Russ. J. Inorg. Chem.*, 1973, **18**, 615.
[121] G. Collin and P. Laruelle, *Acta Cryst.*, 1974, **B30**, 1134.

2 Iron

Carbonyl Compounds.—Electrolysis of $Fe(acac)_3$ in pyridine in the presence of carbon monoxide yields $Fe(CO)_5$ as the principal product, together with a little $Fe_3(CO)_{12}$. However, when the reaction is carried out in DMSO, $[Fe_4(CO)_{13}]^{2-}$ and small amounts of $[Fe_2(CO)_8]^{2-}$ are formed.[122] The structure of $Fe_2(CO)_9$ has been redetermined and a greater accuracy achieved than was previously reported. The relevant dimensions are Fe—Fe = 2.523(1), Fe—C(terminal) = 1.838(3), Fe—C-(bridging) = 2.016(3), C—O(terminal) = 1.156(4), C—O(bridging) = 1.176(5) Å; $\angle$ C(terminal)FeC(terminal) = 96.1(1), $\angle$ FeC(bridging)Fe = 77.6(1) and $\angle$ FeCO-(terminal) = 177.1(3)°.[123] In THF solution, $Fe_2(CO)_9$ shows a pattern of reactivity different to that in other solvents and it proved possible to isolate the previously unknown compounds, $LFe(CO)_4$ (L = pyridine or pyrazine), by this method. The crystal structures of both compounds were determined and in both cases the N-donor ligand occupies an axial position in a trigonal-bipyramidal array. ^{13}C N.m.r. spectroscopy shows the pyridine complex to be fluxional down to $-90°$.[124] Reaction of $Fe_2(CO)_9$ with 2,2′-bipyridyl in THF yields $(OC)_2(bipy)Fe(\mu\text{-}CO)_2Fe(CO)_3$, the structure of which was also determined. The Fe—Fe bond is a little longer (2.611 Å) than that of $Fe_2(CO)_9$ and one bridging carbonyl is asymmetrically bonded to the metal atoms. The bond length to the chelated iron atom is 2.37 Å, whereas the other is 1.80 Å. The authors term this carbonyl ligand a 'semi-bridging CO (SBCO)' and suggest that this arrangement provides a mechanism for one metal atom, that would otherwise tend to be excessively negative, to transfer some of its electron density to a carbonyl group on a less negatively charged atom.[125] The same authors have also redetermined the structure of $Fe_3(CO)_{12}$, again with a higher refinement than previously achieved. The results confirm previous suggestions that the two bridging carbonyl groups are asymmetrically bonded. The Fe—C bond lengths associated with each of these ligands are 1.96 and 2.11 Å and 1.93 and 2.21 Å. The terminal carbonyl groups are essentially linear [$\angle$ FeCO = 173°(mean); Fe—C(terminal) = 1.82 Å]. On the basis of this structure, considerable insight has been gained into the dynamic properties of $Fe_3(CO)_{12}$ in solution. Firstly, the presence of two asymmetric bridging carbonyl groups (*i.e.*, intermediate between terminal and symmetric bridging) indicated a potential low-energy pathway to the unbridged structure and, secondly, the structure suggests a straightforward pathway for complete scrambling of all CO groups. This scrambling mechanism can be appreciated with the aid of figure (15), which shows one edge of the Fe_3 triangle and omits the $Fe(CO)_4$ group for clarity.[126]

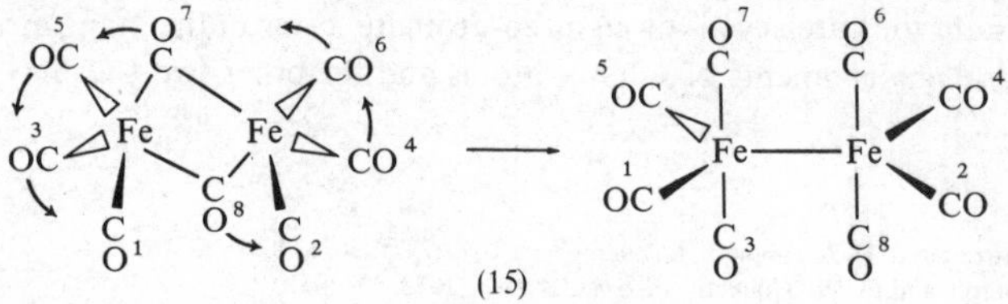

(15)

[122] J. Grobe, J. Kaufmann, and F. Kober, *Z. Naturforsch.*, 1973, **28b**, 691.

[123] F. A. Cotton and J. M. Troup, *J. C. S. Dalton*, 1974, 800.

[124] F. A. Cotton and J. M. Troup, *J. Amer. Chem. Soc.* 1974, **96**, 3438.

[125] F. A. Cotton and J. M. Troup, *J. Amer. Chem. Soc.*, 1974, **96**, 1233.

[126] F. A. Cotton and J. M. Troup, *J. Amer. Chem. Soc.*, 1974, **96**, 4155.

The i.r. spectrum of matrix-isolated $Fe(CO)_5$ has been recorded at 4 and 20 K. U.v. photolysis of the matrix induces the appearance of new bands that are assigned to $[Fe(CO)_4]$, which probably has C_{3v} symmetry. Prolonged u.v. photolysis leads to the formation of $[Fe(CO)_3]$, the i.r. spectrum of which has also been reported.[127] I.r. evidence has also been presented for the possible formation of $[Fe(CO)_4]^-$ by vacuum u.v. photolysis of matrix-isolated $Fe(CO)_5$.[128]

Crystallographic studies have been reported for $[Rh\{Fe(h^5\text{-}MeCp)(CO)_2(PPh_2)\}_2]$-$PF_6$ (16)[129] and $[(h^5\text{-}Cp)Fe(CO)(\mu\text{-}CO)_2Co(CO)_2(PMePh_2)]$.[130] The former adopts

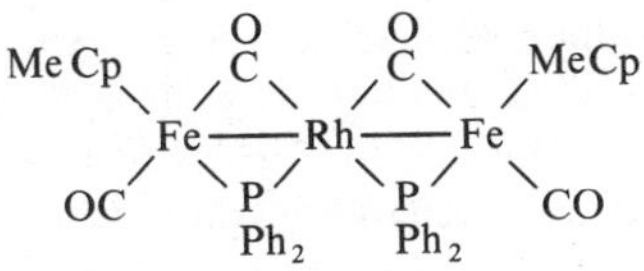

(16)

the non-closed form (Fe—Rh = 2.659 and 2.674 Å), whereas a closed triangulo-structure is predicted on the basis of the 18-electron rule and is observed for related molecules. It is suggested that the triangulo-structure is unstable with respect to the open structure as a result of the substitution of CO ligands by the highly basic phosphido-groups. The $Fe(CO)_2Co$ bridging system in the latter is non-planar with an angle of 25.4° between the two FeCCo planes. The Fe—C and Co—C bond lengths in the bridges are not the same and the authors suggest that this indicates a *trans*-influence of $PMePh_2$ ligand. However, there may be other explanations such as electron-density redistribution (see above).

The reaction of $Na[Cr(CO)_5]$ with $Fe_2(CO)_9$ in THF yields $Fe(CO)_5$ and the anion $[(CO)_5CrFe(CO)_4]^{2-}$, which was isolated as its sodium salt.[131] The formation of anions, especially $[HFe(CO)_4]^-$, from $Fe(CO)_5$ in methanol in the presence of tertiary amines, water, or hydrogen and carbon monoxide has been studied.[132] The structure of $[(Ph_3P)_2N][Fe(CO)_4CN]$ has been determined and the anion, which is isoelectronic with $Fe(CO)_5$, has a distorted trigonal-bipyramidal structure with the cyano-group in an axial position, as predicted from a simple π-bonding model. The difference between axial and equatorial Fe—C(O) bond lengths of 0.045(12) Å is in good agreement with the latest data for $Fe(CO)_5$.[132a]

$[Fe(CO)(h^5\text{-}Cp)]_4$ reacts with BCl_3 in liquid hydrogen chloride to yield $\{[Fe(CO)\text{-}(h^5\text{-}Cp)]_4H_2\}^{2+}$ and with PF_5 in the same medium to give $[(h^5\text{-}Cp)Fe(CO)]^+$. This paper also reports an improved preparative procedure for $[Fe(CO)(h^5\text{-}Cp)]_4$.[133] Treatment of $[(h^5\text{-}Cp)Fe(CO)_2]_2$ with AgX (X = ClO_4, PF_6, or SbF_6) in polar solvents (S) gives $[(h^5\text{-}Cp)Fe(CO)_2S]^+$ and silver metal. The complex with S = Me_2CO

[127] M. Poliakoff and J. J. Turner, *J. C. S. Dalton*, 1973, 1351; M. Poliakoff, *J. C. S. Dalton*, 1974, 210.
[128] J. K. Burdett, *J.C.S. Chem. Comm.*, 1973, 763.
[129] R. Mason and J. A. Zublieta, *J. Organometallic Chem.*, 1974, **66**, 279.
[130] G. Davey and F. S. Stephens, *J. C. S. Dalton*, 1974, 698.
[131] E. Lindner, H. Behrens, and D. Uhlig, *Z. Naturforsch.*, 1973, **28b**, 276.
[132] F. Wada and T. Matsuda, *J. Organometallic Chem.*, 1973, **61**, 365.
[132a] S. A. Goldfield and K. N. Raymond, *Inorg. Chem.*, 1974, **13**, 770.
[133] D. A. Symon and T. C. Waddington, *J. C. S. Dalton*, 1973, 1879.

was fully characterized. With other silver salts, AgR [NO_3, CF_3CO_2, SCN, NCO, $PhCO_2$, p-$MeC_6H_4SO_3$, or $OP(O)(OPh)_2$], the neutral compounds [(h^5-Cp)-$Fe(CO)_2R$] were obtained. The cations react with donor ligands, L, to yield [$Fe(h^5$-Cp)-$(CO)_2L$]X [L = CO, PPh_3, $AsPh_3$, $SbPh_3$, MeCN, $PhCH_2CN$, Ph_2CHCN, PhSMe, or (h^5-Cp)$Fe(CO)_2I$] and with the anions, Y^-, to form [$Fe(h^5$-Cp)$(CO)_2Y$] (Y = Cl or I).[134] Oxidation of the neutral dimer by oxygen in acetone and aqueous HBF_4, followed by the reaction with the neutral ligands, L, or the anions X^- yields [(h^5-Cp)-$Fe(CO)_2L$]$^+$ (L = PPh_3, py, N_2H_4, or SBu_2) and (h^5-Cp)$Fe(CO)_2X$ (X = Cl, Br, NCO, CN, NO_3, N_3, O_2CH, or O_2CCCl_3), respectively. Spectroscopic evidence indicates that the intermediate formed in these reactions is [(h^5-Cp)$Fe(CO)_2$-(H_2O)]$^+$.[135] The cation [(h^5-Cp)$Fe(CO)_3$]$^+$ has also been prepared by oxidation of [(h^5-Cp)$Fe(CO)_2$]$_2$ by NO_2 in chloroform.[13] The structure of [(h^5-Cp)$Fe(CO)_3$]PF_6 has been determined[135] and has shorter C—O and longer M—C bond lengths than the neutral species. The C—O distances are also shorter than those in the isoelectronic complex, (h^5-Cp)$Mn(CO)_3$.[136] (h^5-Cp)$Fe(CO)_2X$ (X = Cl, Br, or I) reacts with either $AgPF_6$ or H_2SO_4 followed by HPF_6 to form {[(h^5-Cp)$Fe(CO)_2$]$_2X$}PF_6[137] and treatment of [(h^5-Cp)$Fe(CO)_2$]$^+$ with (h^5-Cp)$Fe(CO)_2I$ and $AgBF_4$ yields {[(h^5-Cp)-$Fe(CO)_3$]$_2I$}BF_4.[138] The structure of this last compound has been determined and the two [(h^5-Cp)$Fe(CO)_2$] units are bridged by the iodine atom (Fe—I = 2.588, Fe⋯⋯Fe = 4.26 Å; ∠Fe—I—Fe = 110.8°). The large Fe—I—Fe bond angle presumably results from a compromise between the attempts of the iodine atom to maximize the p-character in its bonding orbitals and the necessity of minimizing the non-bonding contacts between the metal-containing units.

Radiochemical yields of (h^5-Cp)^{59}Fe$(CO)_2\cdot$ formed by neutron irradiation of [(h^5-Cp)$Fe(CO)_2$]$_2$ and (h^5-Cp)$Fe(CO)_2I$ have been reported. Other prominent products are [(h^5-Cp)^{59}Fe$(CO)_2$]$_2$, ^{59}Fe$(CO)_5$, and ^{59}Fe(h^5-Cp)$_2$ in both cases. Production of (h^5-Cp)^{59}Fe$(CO)_2\cdot$ and [(h^5-Cp)^{59}Fe$(CO)_2$]$_2$ is also observed when solutions of $Fe(CO)_5$ with Fe(h^5-Cp)$_2$ or cyclo-C_5H_6 are irradiated with neutrons.[139] The isomerization of [(h^5-Cp)$Fe(CO)_2$]$_2$ is discussed in detail under manganese carbonyl compounds.[57]

The interaction of the organolanthanons $(Cp)_3Ln$ ($Cp = C_5H_5$ or MeC_5H_4; Ln = Nd, Sm, Gd, Dy, Ho, Er, or Yb) with metal carbonyls and nitrosyls has been studied. From the reaction with [(h^5-Cp)$Fe(CO)_2$]$_2$ was isolated the adduct {[(MeC_5H_4)$_3$Sm]$_2$Fe$_2(CO)_4$(h^5-Cp)$_2$}, and reactions with $Co_2(CO)_8$, [(h^5-Cp)-$(CO)Fe(CO)_2Ni(h^5$-Cp)], [(h^5-Cp)$Ni(CO)$]$_2$, (h^5-Cp)$Mn(CO)_3$, and [(h^5-Cp)$Mn(CO)$-(NO)]$_2$ were also investigated. Adduct formation occurred with all complexes containing bridging CO ligands, but only in the case of {[($MeC_5H_4)_3$Sm]$_2Co_2(CO)_8$} was the adduct isolated. Interestingly, the interaction appears to occur not at bridging carbonyl groups, but at terminal nitrosyl groups.[140]

(h^5-allyl)$Fe(CO)_3I$ reacts with either Yb, Sm, Y, or Mn in THF to give mauve solutions which appear to contain an equilibrium mixture of (h^5-allyl)$Fe(CO)_3\cdot$ and

[134] W. E. Williams and F. J. Lalor, *J. C. S. Dalton*, 1973, 1329.
[135] B. D. Dombek and R. J. Angelici, *Inorg. Chim. Acta*, 1973, **7**, 345.
[136] M. E. Gress and R. A. Jacobson, *Inorg. Chem.*, 1973, **12**, 1746.
[137] D. A. Symon and T. C. Waddington, *J. C. S. Dalton*, 1974, 78.
[138] F. A. Cotton, B. A. Frenz, and A. J. White, *J. Organometallic Chem.*, 1973, **60**, 147.
[139] W. Kanellakopulos-Drossopulos and D. R. Wiles, *Canad. J. Chem.*, 1974, **52**, 894.
[140] A. E. Crease and P. Legzdins, *J. C. S. Dalton*, 1973, 1501.

$[(h^3\text{-allyl})\text{Fe(CO)}_3]_2$.[29] The reaction of $[(h^5\text{-Cp})\text{Fe(CO)}_2]_2$ with triethylaluminium in hydrocarbon solvents leads to the isolation of red air-sensitive crystals of $\{[\text{Fe}(h^5\text{-Cp})(\text{CO})_2]_2,2\text{AlEt}_3\}$. An *X*-ray crystallographic study of this adduct shows that the *cis*-$[(h^5\text{-Cp})\text{Fe(CO)}_2]_2$ unit is co-ordinated to the two aluminium atoms *via* the oxygen atoms of the bridging carbonyl groups (Al—O = 1.98 Å).[141] The adducts $\{[(h^5\text{-Cp})\text{Fe(CO)}_2]_2,\text{BX}_3\}$ (X = F or Br), $\{[(h^5\text{-Cp})\text{Fe(CO)}]_4,x\text{BX}_3\}$ (X = F, x = 1, 2, or 4; X = Cl, x = 1 or 2; X = Br, x = 1 or 2), $\{[(h^5\text{-Cp})\text{Fe(CO)}]_4,x\text{AlBr}_3\}$ (x = 1, 2, 3, or 4), $[\text{Fe}_2(\text{CO})_9,\text{AlBr}_3]$, and $[\text{Fe}_3(\text{CO})_{12},\text{AlBr}_3]$ have been studied. Again, the Group III atoms co-ordinate only to bridging CO groups; however, $[\text{Ru}_3(\text{CO})_{12},$ $\text{AlBr}_3]$ was also formed and this involves a terminal-to-bridging carbonyl shift. It is also suggested that $\text{Fe}_2(\text{CO})_9,\text{AlBr}_3$ is formed by a converse shift, the structure of the product being (17).[142]

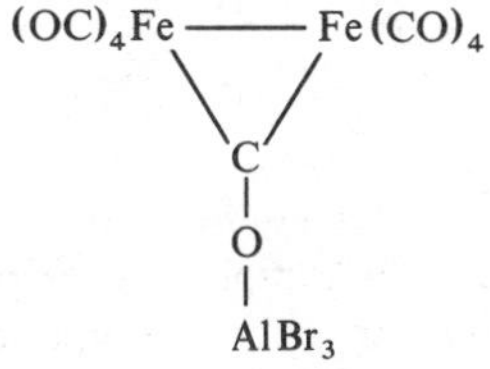

(17)

Photolysis of Fe(CO)_5 with $\text{Hg(SiMe}_3)_2$ yields a mixture of *cis*-$[\text{Fe(CO)}_4(\text{SiMe}_3)_2]$ and $[\textit{cis}\text{-Fe(CO)}_4\text{SiMe}_3]_2\text{Hg}$, which can be separated by fractional crystallization and sublimation. The spectroscopic properties of the new compounds have been reported and mechanisms suggested for their formation.[143] A similar reaction with $\text{Me}_2\text{SiCH}_2\text{CH}_2\text{SiMe}_2$ yields $(\text{OC})_4\overline{\text{FeSiMe}_2\text{CH}_2\text{CH}_2\text{SiMe}_2}$ which is less stable than the ruthenium and osmium analogues, but is stereochemically rigid at ambient temperature.[144] The reaction between $[\text{Fe}(h^5\text{-Cp})(\text{CO})_2]^-$ and $\text{Cl}\cdot\text{SiMe-}$ $\overline{\text{CH}_2\text{CH}_2\text{CH}_2}$ yields $[(h^5\text{-Cp})(\text{OC})_2\overline{\text{FeSiMeCH}_2\text{CH}_2\text{CH}_2}]$, and the reactions of this complex with KOH in methanol, chlorine in CCl_4, PPh_2Me, $[\text{C}_2\text{H}_4\text{PtCl}_2]_2$, and HCl in benzene have been studied.[145] The polysilane complexes $\{[(\text{Me}_3\text{Si})_n\text{Me}_{3-n}\text{Si}]\text{-}$ $\text{Fe(CO)}(h^5\text{-Cp})\}$ have been reported.[18] Treatment of Fe(CO)_5 with SiI_4 (5:1 mole ratio) in pentane under u.v. irradiation gives the volatile dark-red compound, $\text{SiI}_4\text{Fe}_2(\text{CO})_6$ for which structure (18) is suggested.[146]

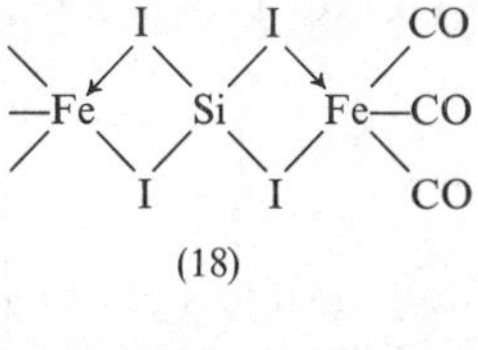

(18)

[141] N. E. Kim, N. J. Nelson, and D. F. Shriver, *Inorg. Chim. Acta*, 1973, **7**, 393.
[142] J. S. Kristoff and D. F. Schriver, *Inorg. Chem.*, 1974, **13**, 499.
[143] W. Jetz and W. A. G. Graham, *J. Organometallic Chem.*, 1974, **69**, 383.
[144] L. Vancea and W. A. G. Graham, *Inorg. Chem.*, 1974, **13**, 511.
[145] C. S. Cundy and M. F. Lappert, *J. Organometallic Chem.*, 1973, **57**, C72.
[146] G. Schmid and H.-P. Kempny, *Angew. Chem. Internat. Edn.*, 1973, **12**, 670.

The complex $[H_2SiCl \cdot Mn(CO)_5]$ reacts with $Co_2(CO)_8$ to yield $H(Cl)SiMn(CO)_5 \cdot Co(CO)_4$ and $ClSiMn(CO)_5 \cdot Co_2(CO)_7$ and the first of these undergoes reaction with $Fe_3(CO)_{12}$, giving $Fe(CO)_4[ClSiMn(CO)_5]_2Co_2(CO)_7$ which probably has structure (19).[17] The reactivities of $Ph_x(C_6F_5)_{3-x}M^1 \cdot M^2$ [M^1 = Si, Ge, or Sn; M^2 = $Fe(CO)_2$-(h^5-Cp) or $Fe(CO)(h^5$-Cp)(PR_3) (R = Ph, Et, OPh, or OEt)] towards HX (X = Cl or

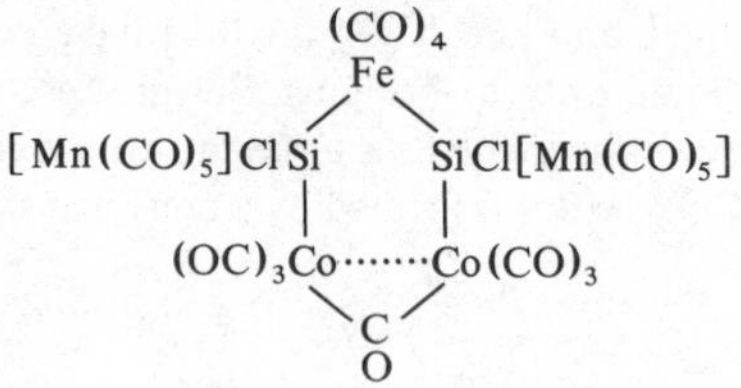

$$[Mn(CO)_5]Cl\,Si \overset{\displaystyle \overset{(CO)_4}{Fe}}{\underset{\underset{\displaystyle \overset{C}{O}}{(OC)_3Co \cdots\cdots Co(CO)_3}}{\big|\qquad\qquad\big|}} Si\,Cl[Mn(CO)_5]$$

(19)

Br), Cl_2, I_2, ICl, and CF_3I have been studied (not all possible combinations investigated). Reaction of HX with Sn—Fe complexes gives partial or complete replacement of the organic groups bonded to tin, but in all other cases, cleavage of the M^1—M^2 bond occurred.[21] Reaction of $Ph_{4-n}GeCl_n$ with $[(h^5$-Cp$)Fe(CO)_2]^-$ in THF yields $[Ph_{3-n}Cl_nGeFe(h^5$-Cp$)(CO)_2]$ (n = 1 or 2), and the corresponding silicon compounds were prepared by reaction of the appropriate phenylchloro-or phenylpentafluoro-phenyl-silane with $[(h^5$-Cp$)Fe(CO)_2]_2$. The reactivity of these complexes and their tin analogues towards C_6F_5Li was found to depend on the main-group metal (M) and the number of phenyl groups bonded to it. The reactions, leading to $[(C_6F_5)_n$-$Ph_{3-n}MFe(h^5$-Cp$)(CO)_2]$ occurred for all tin-containing species but only occurred with the trichloro-germyl and -silyl complexes.[22] The vinylgermyl complexes, $Me_2(C_2H_3)GeM$ [M = $Fe(CO)_3NO$ or $Fe(CO)_2(h^5$-Cp$)$] have been prepared, and photolysis of these compounds gives complex mixtures. The products are very unstable in the case of the nitrosyl complex and the primary product in the case of the cyclopentadienyl complex is (20), which exists in both *cis*- and *trans*-forms. The suggested mechanism for the formation (20) is given in Scheme 1.[23] Photolysis of $Me_2Ge[Fe(h^5$-Cp$)(CO)_2]_2$ also gives (20) and similar treatment of Me_2GeCl-

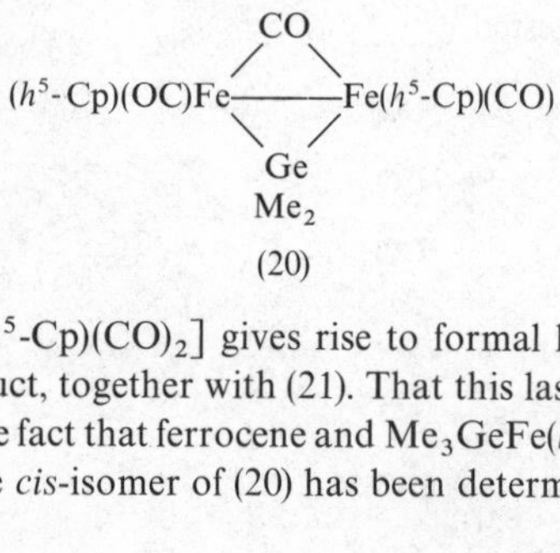

$$(h^5\text{-Cp})(OC)Fe\overset{\displaystyle \overset{CO}{\diagup \quad \diagdown}}{\underset{\displaystyle \underset{\underset{Me_2}{Ge}}{\diagdown \quad \diagup}}{\rule{2cm}{0.4pt}}}Fe(h^5\text{-Cp})(CO)$$

(20)

$$OC\overset{\displaystyle \overset{Me_2}{Ge}}{\underset{\displaystyle \underset{\underset{Me_2}{Ge}}{Fe\rule{1cm}{0.4pt}Fe}}{\diagup \quad \diagdown}}Cp$$

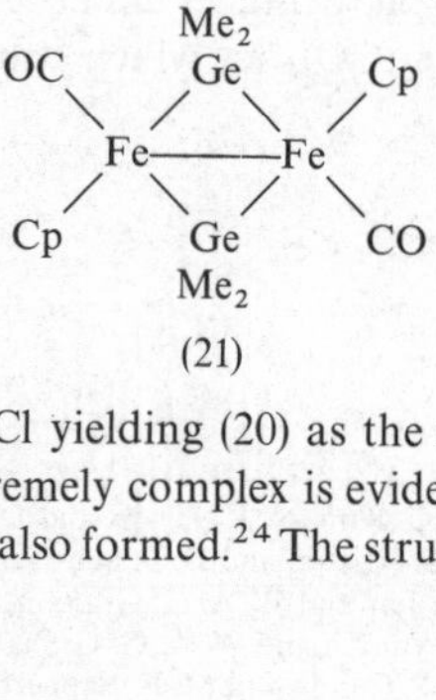

(21)

$[Fe(h^5$-Cp$)(CO)_2]$ gives rise to formal loss of CO and Cl yielding (20) as the main product, together with (21). That this last reaction is extremely complex is evidenced by the fact that ferrocene and $Me_3GeFe(h^5$-Cp$)(CO)_2$ are also formed.[24] The structure of the *cis*-isomer of (20) has been determined.[147]

[147] R. D. Adams, M. D. Brice, and F. A. Cotton, *Inorg. Chem.*, 1974, **13**, 1080.

$$Me_2(C_2H_3)GeFe(h^5—Cp)(CO)_2 \rightarrow Me_2GeFe(CO)_2(h^5—Cp)\cdot + C_2H_3\cdot$$
$$\longrightarrow Me_2Ge(C_2H_3)\cdot + Fe(CO)_2(h^5—Cp)\cdot$$
$$Me_2GeFe(CO)_2(h^5—Cp)\cdot + Fe(CO)_2(h^5—Cp)\cdot \rightarrow$$
$$Me_2Ge[Fe(CO)_2(h^5—Cp)]_2 \rightarrow (20)$$
$$Fe(CO)_2(h^5 - Cp)\cdot \rightarrow [Fe(CO)_2(h^5—Cp)]_2 \rightarrow [Fe(CO)(h^5=Cp)]_4$$

Scheme 1

One of the main products of the oxidation of (20) by air is $\{[(h^5\text{-Cp})Fe(CO)_2\text{-}GeMe_2]_2O\}$. In polar solvents, the n.m.r. spectrum of this compound suggests the presence of major and minor conformers which are interconverting slowly on the n.m.r. time-scale at 25 °C. An X-ray structural study shows the two $[(h^5\text{-Cp})Fe(CO)_2\text{-}GeMe_2]$ units to be linked by an oxygen bridge between the germanium atoms. Two rotational isomers are present in the unit cell in a disordered fashion (Fe—Ge = 2.372 (average), Ge—O = 1.785 Å (average); $\angle FeOGe = 134°$).[148]

The reaction of $(CH=CH)_2Sn(O_2CCF_3)_2$ with $Na_2Fe(CO)_4$ yields $(CH_2=CH)_2\text{-}Sn[Fe(CO)_4]_2$, which can also be prepared from $(CH=CH)_2Sn(O_2CCF_3)_2$ and $Fe(CO)_5$ in refluxing nonane or at room temperature in the presence of a trace of trifluoroacetic acid.[26] A by-product of the reaction involving $Na_2Fe(CO)_4$ is $(CH_2=CH)_2Sn_2Fe_4(CO)_{16}$, which has been characterised by spectral methods and for which structure (22) is suggested.[149] Reaction of Me_3SnCH_2I with $Na[Fe(h^5\text{-}Cp)(CO)_2]$ in THF did not yield the expected product but gave a mixture of $Me_3SnFe\text{-}(CO)_2(h^5\text{-Cp})$ and $MeSnFe(CO)_2(h^5\text{-Cp})$.[27]

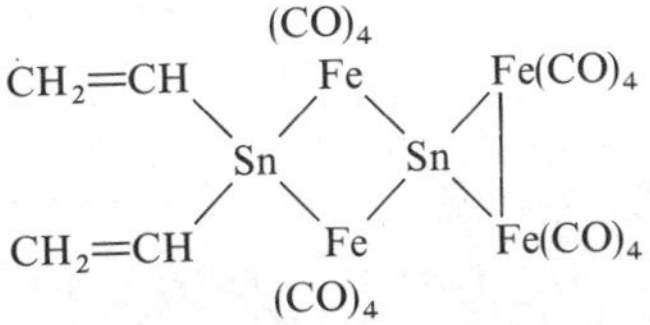

$$
\begin{array}{c}
\qquad\qquad (CO)_4 \\
CH_2=CH \qquad Fe \qquad\quad Fe(CO)_4 \\
\diagdown \quad\diagup \diagdown\quad\diagup \\
Sn \qquad Sn \\
\diagup\quad\diagdown\quad\diagup\diagdown \\
CH_2=CH \qquad Fe \qquad Fe(CO)_4 \\
\qquad\qquad (CO)_4
\end{array}
$$

(22)

A study of the ESCA and ^{119}Sn Mössbauer spectra of $Bu^t_2SnFe(CO)_4$ suggests a formal oxidation of IV for the tin atom.[150] The ^{57}Fe and ^{121}Sb Mössbauer spectra of $[X_nSb\{Fe(CO)_2(h^5\text{-Cp})\}_{4-n}]^+$ cations (X = Cl, Br, I, CF_3, Ph, or Bu^n; $n = 1—3$; but not all combinations) have also been studied and compared with those of the nominally isoelectronic complexes $[X_nSn\{Fe(CO)_2(h^5\text{-Cp})\}_{4-n}]$. ^{121}Sb isomer shifts fall between the ranges typical for Sb^{III} and Sb^V and thus formal oxidation states have little meaning. The data suggest that, in Fe—M (M = Sn or Sb), π-bonding is of greater importance in the antimony complexes than in the tin complexes, and the results also suggest that the positive charges on the antimony complexes are not extensively delocalized onto the ligands.[151] An X-ray structural study of $[Cl_2Sb$-

[148] R. D. Adams, F. A. Cotton, and B. A. Frenz, *J. Organometallic Chem.*, 1974, **73**, 93.
[149] C. D. Garner and R. G. Senior, *Inorg. Nuclear Chem. Letters*, 1974, **10**, 609.
[150] G. W. Grynewich, B. Y. K. Ho, T. J. Marks, D. L. Tomaja, and J. J. Zuckerman, *Inorg. Chem.*, 1973, **12**, 2522.
[151] W. R. Cullen, D. J. Patmore, J. R. Sams, and J. C. Scott, *Inorg. Chem.*, 1974, **13**, 649.

$\{Fe(CO)_2(h^5\text{-}Cp)\}](Sb_2Sl_7)$ shows the antimony atom of the cation to be in a distorted tetrahedral environment.[152] The structure of (23) has also been reported.[153] Reaction of $Fe(CO)_2(h^5\text{-}Cp)Cl$ with $(CH_2\!\!=\!\!CHCH_2)SbR^1R^2$ in THF, followed by treatment with $NaBPh_4$ in water yields $[R^1R^2Sb\{Fe(CO)_2(h^5\text{-}Cp)\}_2]^+$ ($R^1 = R^2 = Me$, Ph, or $CH_2\!\!=\!\!CHCH_2$; $R^1 = CH_2\!\!=\!\!CHCH_2$, $R^2 = Me$ or Ph), which is a new route to complexes of this type.[154]

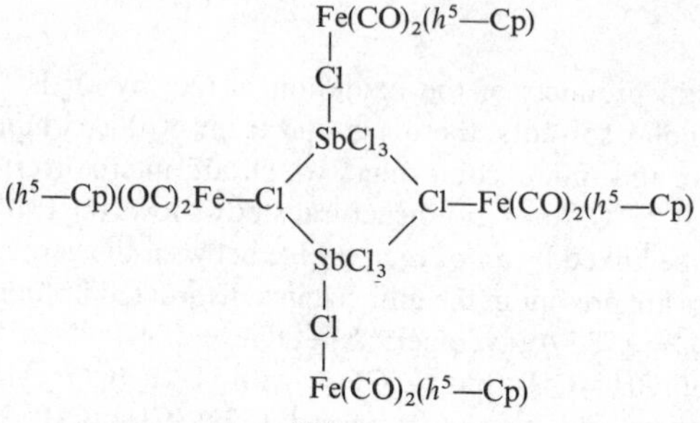

$$Fe(CO)_2(h^5\text{—Cp})$$
$$|$$
$$Cl$$
$$|$$
$$SbCl_3$$

$(h^5\text{—Cp})(OC)_2Fe\text{—Cl} \qquad Cl\text{—}Fe(CO)_2(h^5\text{—Cp})$

$$SbCl_3$$
$$|$$
$$Cl$$
$$|$$
$$Fe(CO)_2(h^5\text{—Cp})$$

(23)

The existence of isomers which differ in the orientation of bent bridging isocyanide ligands has been demonstrated for $(h^5\text{-}Cp)_2Fe_2(CO)(CNMe)_3$ and $(h^5\text{-}Cp)_2Fe_2(CO)_2\text{-}(CNMe)_2$ and the interconversion of such isomers studied by n.m.r. For the former, three isomers are possible, two *syn* and one *anti* (24). The *anti-* and one of the *syn*-forms

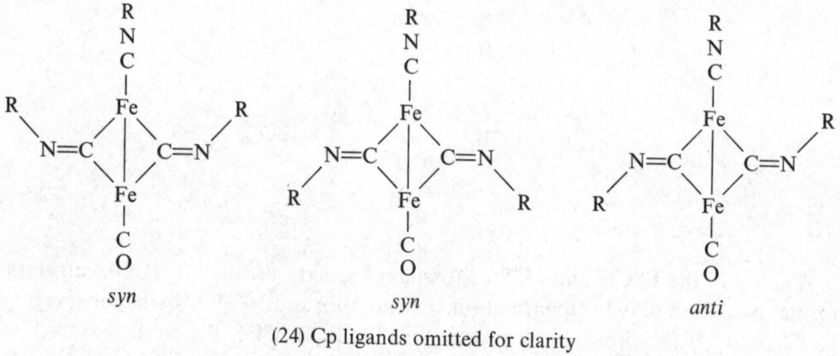

(24) Cp ligands omitted for clarity

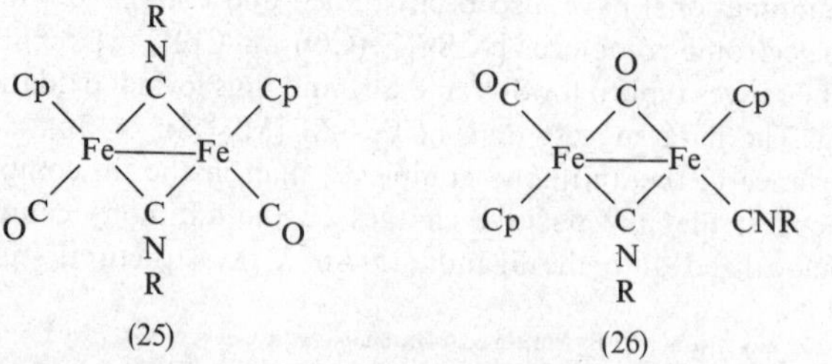

(25)   (26)

[152] F. W. B. Einstein and R. D. G. Jones, *Inorg. Chem.*, 1973, **12**, 1690.
[153] F. W. B. Einstein and A. C. MacGregor, *J. C. S. Dalton*, 1974, 778.
[154] Y. Matsumura, M. Harakawa, and R. Okawara, *J. Organometallic Chem.*, 1974, **71**, 403.

are said to predominate and interconvert rapidly on the n,m,r, time-scale at 25 °C. At − 100 °C, sharp signals due to the separate forms are observed. However, the authors do not appear to have considered the possibility of *cis-* and *trans*-isomers being present. In the case of the latter complex, the isomers (25) and (26) were observed.[155] One of two tautomers postulated as being present in solutions of $(h^5\text{-Cp})_2\text{Fe}_2(\text{CO})_2$-$(\text{CNMe})_2$ has been isolated and its structure determined. This is isomer (25) in which the cyclopentadienyl rings are *cis* and the bridging isocyanide groups are in the *anti* configuration. The detailed structure of the molecule is very similar to that of *cis*-$[(h^5\text{-Cp})_2\text{Fe}_2(\text{CO})_4]$, thus validating the argument that isocyanide complexes can be used to gain information on the dynamic properties of the unsubstituted analogues. On the basis of this structure, it is suggested that the bridged isomers of t-butyl isocyanide complexes would encounter severe steric strain, in agreement with the fact that no detectable amount of this isomer of $(h^5\text{-Cp})_2\text{Fe}_2(\text{CO})_3(\text{CNBu}^t)$ is observed in solution.[156] I.r. and n.m.r. spectra of this complex show it to exist primarily as a mixture of interconverting *cis* and *trans* forms in which the CNCMe_3 group passes from one iron atom to the other *via* a bridged intermediate. *Cis–trans* isomerization is rapid (ΔG of activation < 7.0 kcal mol^{-1}), whereas CNCMe_3 exchange is much slower (ΔG of activation $= 14.7 \pm 0.7$ kcal mol^{-1}). The solid complex was isolated and has structure (27).[157]

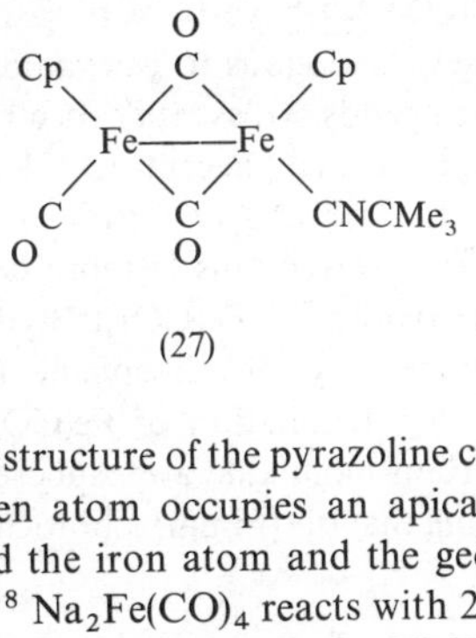

(27)

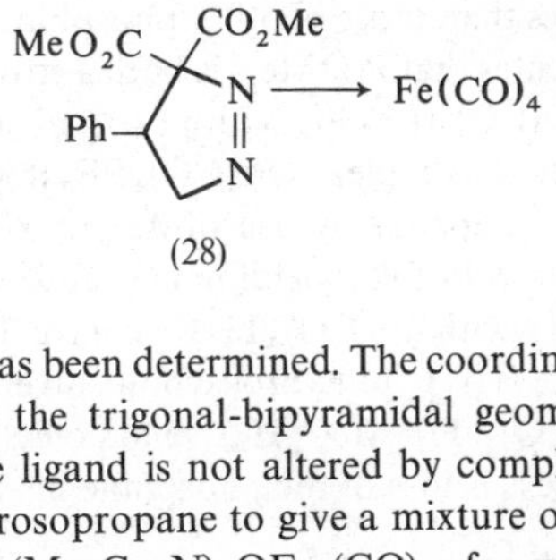

(28)

The structure of the pyrazoline complex (28) has been determined. The coordinated nitrogen atom occupies an apical position in the trigonal-bipyramidal geometry around the iron atom and the geometry of the ligand is not altered by complexation.[158] $\text{Na}_2\text{Fe}(\text{CO})_4$ reacts with 2-bromo-2-nitrosopropane to give a mixture of the volatile complexes $(\text{Me}_2\text{C}{=}\text{N})_2\text{Fe}_2(\text{CO})_6$ and $(\text{Me}_2\text{C}{=}\text{N})_2\text{OFe}_2(\text{CO})_6$, for which

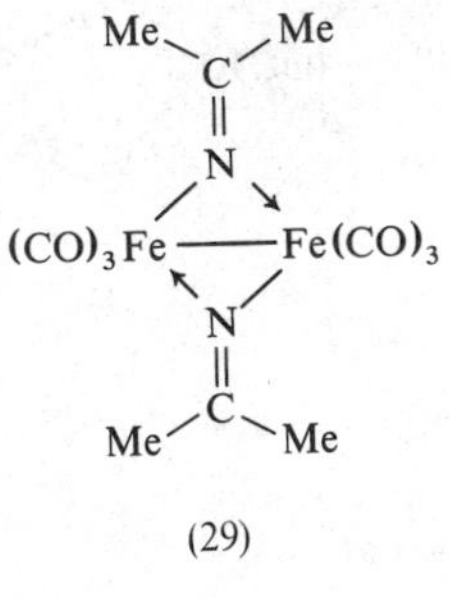

(29)

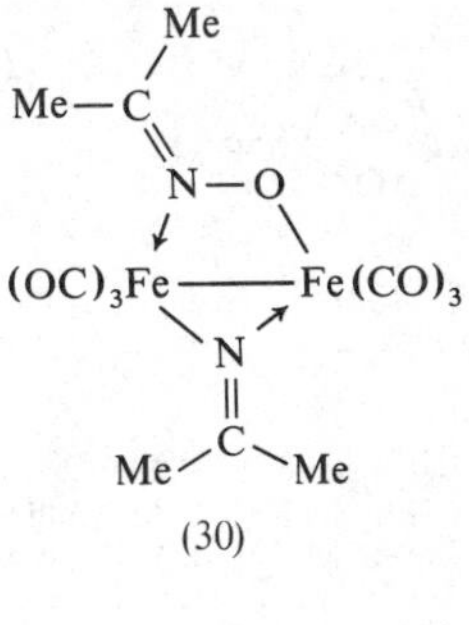

(30)

[155] R. D. Adams and F. A. Cotton, *Inorg. Chem.*, 1974, **13**, 249.
[156] F. A. Cotton and B. A. Frenz, *Inorg. Chem.*, 1974, **13**, 253.
[157] R. D. Adams, F. A. Cotton, and J. M. Troup, *Inorg. Chem.*, 1974, **13**, 257.
[158] C. Kruger, *Chem. Ber.*, 1973, **106**, 3230.

structures (29) and (30), respectively, are proposed.[159] Benzo[c]cinnolinebis(tricarbonyliron) [$N_2Fe_2(CO)_6$ (31)] forms both monosubstituted and disubstituted derivatives, $N_2Fe_2(CO)_5L$ [L = $P(OPh)_3$, $AsPh_3$, PBu_3^n, PPh_3, or $PPhEt_2$] and $N_2Fe(CO)_4L_2$ [L = PBu_3^n, $PHPh_2$, PPh_3, $P(OMe)_3$, or cyclo-$C_6H_{11}NC$]. Spectral data indicate the preferred site of substitution is *trans* to the metal–metal bond.[160]

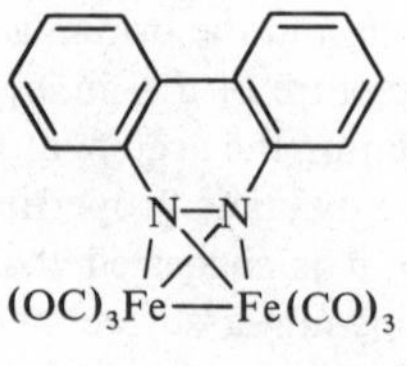

(31)

Group Va substituted complexes, $L_nFe(CO)_{5-n}$ ($n = 1$ or 2), can be prepared in high yield by use of $Fe(CO)_5$ as both solvent and substrate.[161] The complexes $Fe(CO)_3I_2P$, $Fe(CO)_2I_2P_2$, *trans*-[$Fe(CO)IP_4$]I, *trans*-[$Fe(CO)IP_4$]BPh_4, *mer*-[Fe(CO)_2IP_3]BPh_4, and *mer*-[$Fe(CO)BrP_4$]BPh_4, together with *cis*-[$Fe(SnCl_3)ClP_4$], [$Fe(SnCl_3)P_5$]BPh_4, and [FeP_6](BPh_4)_2 [P = $P(OMe)_3$] have been prepared. All are low-spin iron(II) complexes and are more closely analogous to isocyanide complexes than those of other phosphine and phosphite ligands. Mössbauer spectroscopy indicates that $P(OMe)_3$ is both a strong σ-donor and a strong π-acceptor.[162] Reaction of $Fe(CO)_4(PF_2Br)$ with potential sources of the oxide ion (Ag_2O, Cu_2O, *etc.*) gives the new complex, [$Fe(CO)_4PF_2$]O[$PF_2Fe(CO)_4$]; however, this is more conveniently prepared by use of $AgMnO_4$ instead of the oxides.[163] $Fe(CO)_4(PF_2Br)$ also reacts with (32), yielding (33) and not (34).[164] Triphenylcyclotriphosphane, (PhP)_3, reacts with $Fe(CO)_4THF$ to give $Fe(PhP)_3(CO)_4$.[165] Irradiation of $Fe(CO)_3[P(p-MeC_6H_4)_3]_2$ in hydrocarbon solvents under a carbon monoxide atmosphere yields $Fe(CO)_4[P(p-MeC_6H_4)_3]$ but no $Fe(CO)_5$, suggesting that the primary photochemical process is loss of the phosphine ligand.[166]

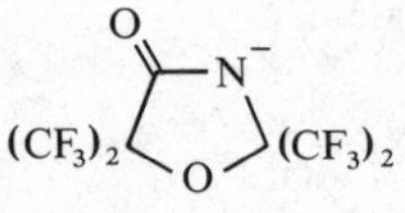

(32) 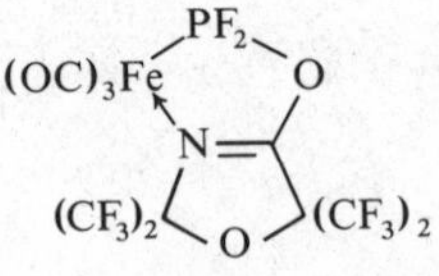   (33) 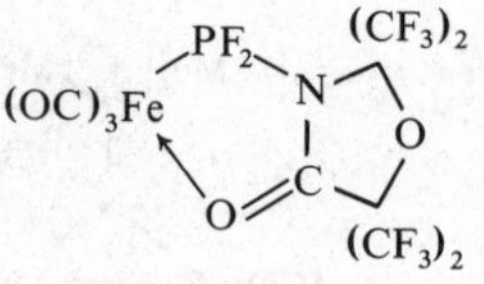   (34)

[159] R. B. King and W. M. Douglas, *Inorg. Chem.*, 1974, **13**, 1339.
[160] P. C. Ellgen and S. L. McMullin, *Inorg. Chem.*, 1973, **12**, 2004.
[161] H. L. Conder and M. Y. Darensbourg, *J. Organometallic Chem.*, 1974, **67**, 93.
[162] E. T. Libbey and G. M. Bancroft, *J. C. S. Dalton*, 1974, 87.
[163] W. M. Douglas, R. B. Johannesen, and J. K. Ruff, *Inorg. Chem.*, 1974, **13**, 371.
[164] D. P. Bauer, W. M. Douglas, and J. K. Ruff, *J. Organometallic Chem.*, 1973, **57**, C19.
[165] M. Baudler and M. Bock, *Angew. Chem. Internat. Edn.*, 1974, **13**, 147.
[166] J. D. Black, M. J. Boyland, P. S. Braterman, and W. J. Wallace, *J. Organometallic Chem.*, 1973, **63**, C21.

Reactions of $[FeCo_3(CO)_{12}]^-$ and $[MnFe_3(CO)_{12}]^-$ with unidentate phosphorus donors (L) leads only to the monosubstituted derivatives $[FeCo_3(CO)_{11}L]^-$ and $[MnFe_3(CO)_{11}L]^-$ but, with $Ph_2PCH_2CH_2PPh_2$ (LL), $[FeCo(CO)_{10}(LL)]^-$ was obtained. Protonation of the substituted iron–cobalt anions leads to the neutral hydrides. The kinetic isotope effect for $\{HFeCo_3(CO)_{11}[P(OPr)_3]\}$ is large, but not as large as for $HFeCo_3(CO)_{12}$. In the latter case this large effect was attributed to the proton having to 'tunnel' through a cluster face in order to reach the site of attachment and a possible explanation for the difference is that the phosphorus ligand opens up one face of the cluster, thereby lowering the barrier to 'tunnelling'. Substitution of $[FeCo_3(CO)_{12}]^-$, followed by protonation gives better yields of the hydrido-complexes than direct substitution of $HFeCo_3(CO)_{12}$.[4]

Shifts in the C—O stretching frequencies of $[(h^5\text{-Cp})Fe(CO)_2P(CF_3)_2]$ compared to those of $[(h^5\text{-Cp})Fe(CO)_2P(O)(CF_3)_2]$ have been interpreted in terms of reduced Fe—CO back π-bonding, and hence presumably an increase in Fe—P π-bonding, in the latter. The structures of both complexes have now been reported and the bond lengths agree with these postulates.[167] $[(CF_3)_2P]_2E$ reacts with $Fe(CO)_2(NO)_2$ and $Fe_2(CO)_9$ to yield $Fe(CO)(NO)_2[(CF_3)_2PEP(CF_3)_2]$ (E = O, S, or Se) and $Fe(CO)_4$-$[(CF_3)_2PEP(CF_3)_2]$ (E = O or S) respectively, all of which contain unidentate phosphorus co-ordinated ligands. In the case of $Fe_2(CO)_9$, the selenide also yields $Fe_2(CO)_6P_2(CF_3)_4Se$, which probably has structure (35).[168] $[Fe(h^5\text{-Cp})(CO)ER_n]_2$

$$(OC)_3Fe \diagdown \diagup Fe(CO)_3$$
$$\text{P} \diagup$$
$$\text{P—Se}$$

(35) CF_3 groups omitted for clarity

(E = P, R = Ph, n = 2; E = S, R = Me, Et, Pr^i, Bu^t, Bu^n, Ph, or $PhCH_2$, n = 1) complexes are readily oxidized by oxygen, iodine, or silver cations to $[Fe(h^5\text{-Cp})\text{-}(CO)(ER_n)]^+$, and treatment with excess bromine, Ag^+, or NO^+ yields the dications. E.s.r. and Mössbauer studies indicate that the electron removed on oxidation is of essentially metal-d character.[169] Photolysis of $Fe(CO)(h^5\text{-Cp})BrL$ [L = $P(OPh)_3$ or PPh_3] in pyridine or DMSO yields $[(h^5\text{-Cp})Fe(CO)_2]_2$ and $[(h^5\text{-Cp})Fe(CO)(PPh_3)]Br$, respectively.[170]

The dynamic properties of $(h^5\text{-Cp})_2Fe_2(CO)_3[P(OPh)_3]$ have been studied by n.m.r. Two processes occur: *cis–trans* isomerism and scrambling of bridging and terminal carbonyl groups. It was found that both processes have the same rate and activation parameters within experimental error and this is in complete agreement with the mechanism proposed by Adams and Cotton and discussed earlier in this

[167] M. J. Barrow, G. A. Sim, R. C. Dobbie, and P. R. Mason, *J. Organometallic Chem.*, 1974, **69**, C4.
[168] R. C. Dobbie and M. J. Hopkinson, *J. C. S. Dalton*, 1974, 1290.
[169] J. A. de Beer, R. J. Haines, R. Greatrex, and J. A. van Wyk, *J. C. S. Dalton*, 1973, 2341.
[170] D. M. Allen, A. Cox, T. J. Kemp, and L. H. Ali, *J. C. S. Dalton*, 1973, 1899.

section. Thus, the mechanism involves concerted opening and closing of pairs of ligand bridges and hindered rotations in the non-bridged tautomer. The Arrehenius activation energy for the process is 20.2(2) kcal mol^{-1} compared to 12 kcal mol^{-1} for $(h^5\text{-Cp})_2\text{Fe}_2(\text{CO})_4$. This is in agreement with the Adams and Cotton mechanism, as replacement of one CO group by the very bulky $P(\text{OPh})_3$ ligand should appreciably raise the barrier to rotation in the non-bridged intermediate.[171] The crystal structure of the *cis*-form of this molecule has also been determined[172] and the two molecules in the asymmetric unit cell have Fe—Fe bond lengths of 2.543(3) and 2.548(3) Å. Other features of the structure are very similar to the unsubstituted analogue.

Reaction of $\text{Fe}_2(\text{CO})_9$ with $\text{Ph}_2\text{PCH}_2\text{PPh}_2$ in THF at 23 °C for three hours, followed by chromatography gives the brown crystalline compound $[\text{Fe}_2(\text{CO})_7\text{-}(\text{Ph}_2\text{P})_2\text{CH}_2]$. An X-ray study shows that three terminal carbonyl ligands are bonded to each iron atom and that the Fe—Fe bond is long [2.709(2) Å]. The metal atoms are bridged by the diphosphine ligand and a symmetrically bonded carbonyl group, the co-ordination at each being approximately trigonal-bipyramidal with the phosphorus atom at the apex. ^{13}C N.m.r. in the carbonyl region shows that the carbonyl groups are rapidly scrambling over all sites, but the precise mechanism of this process has not yet been established.[173] The structure of the compound previously characterized as $\{\text{Fe}_2(\text{CO})_6[\mu\text{-P}(p\text{-MeC}_6\text{H}_4)_2]\text{H}\}$ has been determined and it is now reformulated as $\{\text{Fe}_2(\text{CO})_6(\mu\text{-OH})[\mu\text{-P}(p\text{-MeC}_6\text{H}_4)_2]\}$. The Fe—Fe bond length is 2.511(2) Å and the Fe—O bond lengths are 1.969(6) and 1.974(6) Å. The origin of the oxygen atom is still unknown.[174]

$\text{Fe}_2(\text{CO})_9$ reacts with $\text{Me}_2\text{AsNMe}_2$ to yield $(\text{OC})_4\text{FeAsMe}_2\text{NMe}_2$, which on treatment with HCl gives $(\text{OC})_4\text{FeAsMe}_2\text{Cl}$. This complex reacts with metal carbonyl anions, M, to give the bridged complexes, $(\text{OC})_4\text{FeAsMe}_2\text{M}[\text{M} = \text{Mo}(\text{CO})_3(h^5\text{-Cp})$, $\text{W}(\text{CO})_3(h^5\text{-Cp})$, $\text{Mn}(\text{CO})_5$, $\text{Fe}(\text{CO})_2(h^5\text{-Cp})$, or $\text{Co}(\text{CO})_4]$.[175] Photolysis of $(\text{OC})_4\text{-}$ $\text{FeAsMe}_2\text{Mn}(\text{CO})_5$ yields $(\text{OC})_4\text{FeAsMe}_2\text{Mn}(\text{CO})_4$, structural studies on which show the Fe—Mn bond length to be 2.85 Å.[32] Complexes of the type $L_n\text{M--AsR}_2$ are regarded as being capable of existence provided electronegative R groups can reduce the donor strength of the arsenic atom sufficiently to prevent association *via* ligand elimination yielding doubly bridged species. In support of this, the stable complex $(h^5\text{-Cp})\text{Fe}(\text{CO})_2\text{-}$ AsMe_2 has now been isolated by treating either $\text{Na}[(h^5\text{-Cp})\text{Fe}(\text{CO})_2]$ or $(h^5\text{-Cp})\text{Fe-}$ $(\text{CO})_2\text{SiMe}_3$ with Me_2AsCl.[176]

The kinetics and mechanism of substitution of $\text{Fe}_2(\text{CO})_6(\text{SR})_2$ by ER_3 $[(\text{SR})_2 = \text{S}_2\text{C}_6\text{H}_3\text{Me}, \text{ER}_3 = \text{PPh}_3, \text{P}(\text{OPh})_3, \text{PBu}_3^n, \text{P}(\text{OMe})_3, \text{PHPh}_2, \text{PEtPh}_2, \text{P}(\text{cyclo-}$ hexyl$)_3, \text{P}(o\text{-C}_6\text{H}_4\text{Me})_3$, or $\text{AsPh}_3; (\text{SR})_2 = \text{S}_2\text{C}_2\text{H}_4, \text{ER}_3 = \text{PBu}_3^n$ or $\text{PPh}_3; \text{R} = \text{Me}$, Et, CH_2Ph, Ph, or $p\text{-MeC}_6\text{H}_4, \text{ER}_3 = \text{PPh}_3]$ have been investigated. Complexes with rigid disulphide bridges show kinetics which are strictly first-order in both reagents. The complexes with simple sulphide bridges show *syn-anti* isomerism which proceeds at rates comparable with substitution; however, they still show the same rate law. An S_N2 or $S_{N}1_a$ mechanism is proposed with a seven-co-ordinate activated

[171] F. A. Cotton, L. Kruzynski, and A. J. White, *Inorg. Chem.*, 1974, **13**, 1402.

[172] F. A. Cotton, B. A. Frenz, and A. J. White, *Inorg. Chem.*, 1974, **13**, 1407.

[173] F. A. Cotton and J. M. Troup, *J. Amer. Chem. Soc.*, 1974, **96**, 4422.

[174] P. M. Treichel, W. K. Dean, and J. C. Calabrese, *Inorg. Chem.*, 1973, **12**, 2908.

[175] W. Ehrl and H. Vahrenkamp, *Chem. Ber.*, 1973, **106**, 2556.

[176] W. Malisch and M. Kuhn, *Angew. Chem. Internat. Edn.*, 1974, **13**, 84.

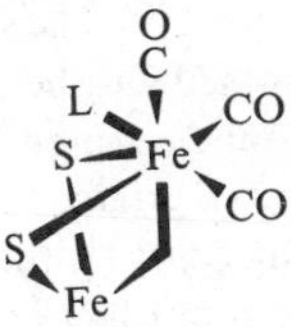

(36)

complex of the type (36).[177] Electrochemical oxidation of $[(h^5\text{-Cp})Fe(CO)(SR)]_2$ (R = alkyl or aryl) leads to the detection of + 1 and + 2 cations. The univalent cations can also be prepared by chemical oxidation with HPF_6, acetic acid, or HCl in the presence of BF_4^- ions.[178] $(h^5\text{-Cp})Fe(CO)_2SCF_3$ undergoes decarbonylation on photolysis yielding isomers of $[(h^5\text{-Cp})Fe(CO)SCF_3]_2$, which on treatment with NO gives $[Fe(NO)_2SCF_3]_2$.[48] Adamantanethioketone $[RC{=}S$ (37)] reacts with $Fe_2(CO)_9$ to yield (38) as the major product[179] and similar reactions with $RCS.NMe_2$ (R = NMe_2, Me, or Ph) gives (39), $RC(NMe_2)SFe(CO)_4$, and $S_2Fe_3(CO)_9$.[180] Treatment of $Fe(CO)_5$ with thiocyanogen yields the polymeric compound $[Fe(CO)_2(SCN)_2]$.[181]

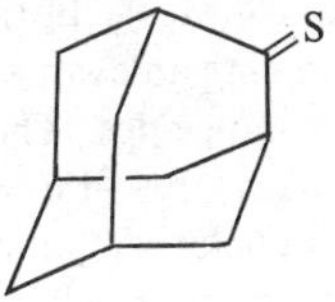

(37) 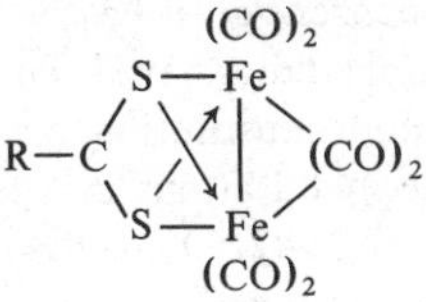   (38) 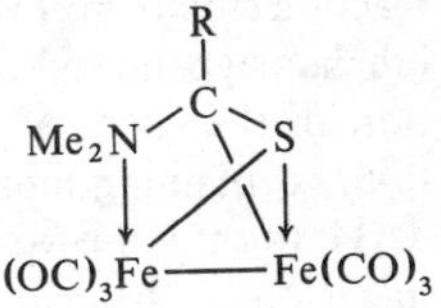   (39)

Structural studies on $\{[(h^5\text{-Cp})Fe(CO)_2]_2\text{-}(\mu\text{-}SO_2)\}$ show the SO_2 molecule to be symmetrically inserted between the two iron atoms (Fe—S = 2.2790 and 2.2814 Å)[182] and the structure of $cis\text{-}[(h^5\text{-Cp})_2Fe_2(CO)_3SO_2]$ shows the two $(h^5\text{-Cp})Fe(CO)$ units to be linked by a direct Fe—Fe bond and bridged by carbon monoxide and S-bonded SO_2.[183] The structure of $(h^5\text{-Cp})Fe(CO)_2(SO_2C_6F_5)$ has also been reported.[184]

Nitrosyl and Aryldiazo Complexes.—Addition of $NOPF_6$ to the cyclobutadiene complexes, $(R_4C_4)Fe(CO)_3$, yields $[(R_4C_4)Fe(CO)_2NO]PF_6$ (R = H, Me, or Ph). These react further with donor ligands, L, to form $[(R_4C_4)Fe(CO)(NO)L]PF_6$ (R = H, L = Ph_3P, Ph_3Sb; R = Ph, L = Ph_3P or Ph_3As). $\{(Ph_4C_4)Fe(NO)[P(OPh)_3]_2\}PF_6$ was also reported.[185] The cations $[Fe(CO)_2(NO)L_2]^+$ react with the appropriate sodium salts, NaX, to give $FeX(NO)(CO)L_2$ (L = PPh_3; X = N_3, SCN, NCO, CN, or

[177] P. C. Ellgen and J. N. Gerlach, *Inorg. Chem.*, 1973, **12**, 2526.

[178] P. D. Frisch, M. K. Lloyd, J. A. McCleverty, and D. Seddon, *J. C. S. Dalton*, 1973, 2268.

[179] H. Alper and A. S. K. Chan, *Inorg. Chem.*, 1974, **13**, 232.

[180] H. Alper and A. S. K. Chan, *Inorg. Chem.*, 1974, **13**, 225.

[181] J. Granifo and H. Müller, *J. C. S. Dalton*, 1973, 1891.

[182] M. R. Churchill, B. G. de Boer, and K. L. Kalra, *Inorg. Chem.*, 1973, **12**, 1646.

[183] M. R. Churchill and K. L. Kalra, *Inorg. Chem.*, 1973, **12**, 1650.

[184] A. D. Redhouse, *J. C. S. Dalton*, 1974, 1106.

[185] A. Efraty, R. Bystrek, J. A. Geaman, S. S. Sandhu, M. H. A. Huang, and R. H. Herber, *Inorg. Chem.*, 1974, **13**, 1269.

SeCN; L = $PMePh_2$, X = N_3 or SCN).[186] The structures of $Fe(NO)_2(CO)PPh_3$ and $Fe(NO)_2(PPh_3)_2$ have been determined; in the former the NO and CO groups are disordered. However, both compounds exhibit distorted tetrahedral co-ordination spheres and both contain linear nitrosyl groups.[187] Mössbauer studies on $Fe(NO)_2LL'$ {LL' = $(CO)_2$, $[P(OPh)_3]_2$, (PPh_3,CO), $(\frac{1}{2}(PhPC_2H_4)_2$, CO), $(PPh_3)_2$, $(CNPh)_2$, $(CNEt)_2$, $[(Me_2N)_3P]_2$, σ-phen, or bipy} have been reported and it was found that the quadrupole splitting increases as the N—O stretching force constant decreases. This can be explained qualitatively by strong π-bonding in the complexes, but in order to achieve a quantitative description it appears necessary to suppose that the NFeN angle depends on L and L' and that this is larger than *ca.* 58° in complexes of ligands with a smaller π-bonding ability than CO.[188] The reactions of $[Fe(NO)_2X]_2$ with thio-containing ligands have been reinvestigated and evidence presented that the products should be formulated as $Fe(NO)_2LX$ not $[Fe(NO)_2L]X$ (L = thiourea, thioacetamide, thiobenzamide, o-$H_2NC_6H_4SH$ or diphenylthiourea).[189]

Structural studies on $[(h^5\text{-Cp})Fe(\mu\text{-NO})_2Fe(h^5\text{-Cp})]$ shows the molecule to possess symmetrically bridging NO groups as well as an Fe—Fe double bond, the Fe—N—Fe—N ring being planar, [Fe—Fe = 2.326(4) Å].[190] Reaction of $[Fe(CO)_3\text{-}NO]_2Hg$ with sulphur in refluxing toluene yields $Fe_4(NO)_4(\mu_3\text{-S})_4$ and a similar reaction employing $(Me_3CN)_2S$ in benzene gives $Fe_4(NO)_4(\mu_3\text{-S})_2(\mu_3\text{-NCMe}_3)_2$. Both are diamagnetic and X-ray structural studies have been reported allowing an evaluation of the effect of bridging ligand substitution on cubane-like molecules. The $(\mu\text{-S})_4$-containing molecule has the metal atoms in a tetrahedral array [Fe—Fe = 2.634 (mean), Fe—S = 2.208(2) − 2.224(2), Fe—N = 1.661(5) − 1.666(5) Å] and linear N—O groups. The bis-nitrile complex has a similar basic structure in which the Fe—Fe bond lengths for iron atoms bound to sulphur compare well with those in the above compound, whereas those of iron atoms bound to $NCMe_3$ are shorter [2.562(1) Å]. The Fe—S distances are virtually identical in both compounds; however, the Fe—N bond lengths in the nitrile complex are extended to 1.914(3) and 1.908(3) Å. The paper contains a detailed discussion of the bonding in these molecules and also reports a one-electron oxidation step and a one-electron reduction step for the tetra-sulphur complex and a one-electron oxidation step and four one-electron reduction steps in the bis-nitrile case, as determined by cyclic voltammetry.[191]

The photochemical reaction of $Na_2[Fe(CN)_5NO]$ follows the route outlined in Scheme 2 and it was found that after long periods of exposure, secondary reactions

$$[Fe(CN)_5NO]^{2-} \xrightarrow{h\nu} [Fe(CN)_5NO^{2-}]^* \rightarrow Fe(CN)_5^{3-} + NO^+$$
$$Fe(CN)_5^{3-} + H_2O \rightarrow [Fe(CN)_5H_2O]^{3-}$$
$$NO^+ + H_2O \rightarrow HNO_2 + H^+$$

Scheme 2

[186] G. Dolcetti, L. Busetto, and A. Palazzi, *Inorg. Chem.*, 1974, **13**, 222.
[187] V. G. Albano, A. Araneo, P. L. Bellon, G. Ciani, and M. Manassero, *J. Organometallic Chem.*, 1974, **67**, 413.
[188] H. Mosbaek and K. G. Poulsen, *Acta Chem. Scand.* (*A*), 1974, **28**, 157.
[189] T. Birchall and K. M. Tun, *J. C. S. Dalton*, 1973, 2521.
[190] J. L. Calderon, S. Fontana, E. Frauendorfer, V. W. Day, and S. D. A. Iske, *J. Organometallic Chem.*, 1974, **64**, C16.
[191] R. S. Gall, C. T. W. Chu, and L. F. Dahl, *J. Amer. Chem. Soc.*, 1974, **96**, 4019.

to those given in Scheme 2 occurred depending on the pH of the solution and the wavelength used. It was found that $[Fe(CN)_5H_2O]^{3-}$ in aqueous solution exists in equilibrium with $[Fe(CN)_5OH]^{4-}$ and at pH > 7, irradiation effects the equilibrium. Furthermore, irradiation of alkaline solutions with light of wavelength greater than 300 nm accelerates the thermal reactions of $[Fe(CN)_5OH]^{4-}$, viz:

$$[Fe(CN)_5OH]^{4-} \xrightarrow[OH^-]{hv} [Fe(CN)_4(OH)_2]^{4-} + CN^-$$
$$[Fe(CN)_5OH]^{4-} + CN^- \rightleftharpoons Fe(CN)_6^{4-} + OH^-$$

In acidic solutions, thermal reactions of $[Fe(CN)_5H_2O]^{3-}$ take place, their rates increasing only after irradiation with longer wavelength light ($\lambda > 300$ nm).[192] Spectroscopic and magnetic neasurements on $M^{II}Fe(CN)_5NO,2H_2O$ (M = Mn, Fe, Co, Ni, or Cu) indicate the presence of polymeric structures with Fe—CN—M bridges, the M^{II} ion being high-spin, while the Fe^{II} ions are low-spin.[193] Reduction of $[Fe(CN)_5NO]^{2-}$ with sodium in liquid ammonia in the presence of Et_3N^+ cations yields $(Et_3N)_2[Fe^I(CN)_4NO]$, structural studies on which show the anion to have square-pyramidal co-ordination with a linear NO group at the apex.[194]

Solid FeCl(TPP) (TPP = tetraphenylporphinate) slowly takes up NO to form solid [FeCl(TPP)NO], in which the nitrosyl group is only weakly bonded; the reaction rapidly reverses on removal of the NO atmosphere. In solution, the same reaction is rapid, and purple crystals of [FeCl(TPP)NO] can be isolated. Spectral and magnetic data for the complex indicate that the iron atom may be considered as being high-spin octahedral Fe^{II}, although the authors could not exclude the possibility of a spin equilibrium. The relevance of these results to related systems in surface chemistry and biochemistry are discussed.[195] Other workers have also studied this reaction and found that the product had very similar spectra to those of Fe^{II}(myoglobin)CO and the NO complex of ferric peroxidase. Reaction of FeCl(TPP) with NO and methanol in toluene leads to $[Fe^{II}(TPP)NO]$, which forms a variety of six-co-ordinate adducts and whose e.s.r. spectrum suggests a bent Fe—N—O linkage.[196]

In the presence of base and alcohol, NO functions as a reducing agent and a nitrosylating agent towards $FeCl_2$, the reaction products being RNO_2 and an equilibrium mixture of $[Fe(NO)_2Cl]_2$ and $[Fe(NO)Cl_2]^-$. The anion reacts with PPh_3 to yield $Fe(NO)_2(PPh_3)Cl$ and $Fe(NO)_2(PPh_3)_2$. $CoCl_2$ shows similar behaviour and there is evidence for the reaction proceeding by nucleophilic attack of ROH or RO$^-$ on co-ordinated NO.[197]

$Fe(CO)_3(PPh_3)_2$ reacts with aryldiazonium salts to form $[ArN_2Fe(CO)_2(PPh_3)_2]$-BF_4 (Ar = Ph, p-F-, p-NO$_2$- and m-F-C$_6$H$_4$). Spectroscopic data have been reported for the complexes.[198] Unlike their nitrosyl analogues, the cations $[PhN_2Fe-(CO)_2L_2]^+$ (L = PPh$_3$, PMePh$_2$, or PMe$_2$Ph) are almost inert to substitution of a carbonyl ligand by excess L. Only when L = $[P(OCH_2)_3CMe]$ has formation of the

[192] A. Lodzinska and R. Gogolin, *Roczniki Chem.*, 1973, **47**, 888, 1111.
[193] H. Inoue, H. Iwase, and S. Yanagisawa, *Inorg. Chim. Acta*, 1973, **7**, 259.
[194] J. Schmidt, H. Kühr, W. L. Dorn, and J. Kopf, *Inorg. Nuclear Chem. Letters*, 1974, **10**, 55.
[195] L. Vaska and H. Nakai, *J. Amer. Chem. Soc.*, 1973, **95**, 5431.
[196] B. B. Wayland and L. W. Olson, *J.C.S. Chem. Comm.*, 1973, 897.
[197] D. Gwost and K. C. Caulton, *Inorg. Chem.*, 1973, **12**, 2095.
[198] W. E. Carroll and F. J. Lalor, *J. C. S. Dalton*, 1973, 1754.

trisubstituted cation been detected. Both $[(p\text{-MeC}_6\text{H}_4\text{N}_2)\text{Fe(CO)}_2(\text{PPh}_3)_2]^+$ and $(\text{ON})\text{Fe(CO)}_2(\text{PPh}_3)_2$ react smoothly with $[\text{Ph}_3\text{P}{=}\text{N}{=}\text{PPh}_3]\text{X}$ (X = Cl or Br) to yield $[(p\text{-MeC}_6\text{H}_4\text{N}_2)\text{Fe(CO)}(\text{PPh}_3)_2\text{X}]$ and $(\text{ON})\text{Fe(CO)}(\text{PPh}_3)_2\text{X}$, respectively, and the corresponding azide salt reacts with the diazo-cation to give $[(p\text{-MeC}_6\text{H}_4\text{N}_2)$-$\text{Fe(CO)}(\text{PPh}_3)_2\text{N}_3]$. However, the azide reacts with the nitrosyl complex to give both $(\text{ON})\text{Fe(CO)}(\text{PPh}_3)_2\text{N}_3$ and $(\text{ON})\text{Fe(CO)}(\text{PPh}_3)_2\text{NCO}$. The results are interpreted in favour of ArN_2^+ being a poorer π-acceptor than NO^+.[199] Reaction of aryldiazonium tetrafluoroborates with $\text{Fe(CO)}_3(\text{PR}_3)_2, \text{Fe(mnt)}_2^-, \text{Fe(cystH)}_2^{2+}$, and $\text{Fe(CO)}_2(\text{cyst})_2$ [mnt = $\text{S}_2\text{C}_2(\text{CN})_2$, cyst = cysteine] has also been found to result in ligand abstraction yielding arenediazophosphonium salts and S-(arenediazo)cysteine or the N_2-extrusion product $\text{ArS(CN)C}{=}\text{C(CN)SAr}$, these workers being unable to obtain arylazo-complexes under the conditions employed. The paper also raises the possibility of involvement of nucleophilic sulphur as a site for the activation and reduction of complexed dinitrogen in biological nitrogen fixation.[200]

Other Iron(0) Compounds.—Electrochemical reduction of Fe(bipy)_2 and Fe(bipy)_3 has been studied.[201]

Iron(II).—*Halides, Cyanides and Hydrides*. The structures of bis-3,5-diphenyl- and bis-3,5-dimethyl-1,2-dithioliumtetrachloroferrate(II) have been determined. In both compounds the FeCl_4^{2-} anion shows a marked deviation from T_d symmetry which is associated with Cl$\cdots$S interaction. The dimethyl-compound shows a broad absorption band at *ca*. 20000 cm^{-1} which arises from charge transfer associated with close contacts between each anion and two pairs of cations. The relevance of such interactions in electron-transfer reactions in the metallo-enzymes, cytochrome c, ferredoxin, and rubredoxin, is also discussed.[202] The compound (3,5-diphenyl-1,2-dithiolium)$_2$-$(\text{Fe}^{\text{III}}\text{Cl}_4)\text{Cl}$ has also been isolated from the reaction medium which initially yields the tetrachloroferrate(II). Structural studies indicate no short cation–Cl contacts, although four equivalent Cl$\cdots$S contacts of 3.18 Å were observed.[203] Electron-transfer between Fe^{II} and Fe^{III} in DMF is influenced by the presence of HClO_4, Cl$^-$, and H_2O in the reaction medium and this suggests the involvement of species such as partially hydrated iron, FeCl^{2+}, and FeCl_2^+.[204]

The anions $[(\text{NC})_5\text{FeNCCo}^{\text{III}}(\text{CN})_5]^{n-}$ (n = 6 and 7), containing Fe^{III} and Fe^{II} respectively, have been prepared by reaction of $[\text{Fe(CN)}_5\text{H}_2\text{O}]^{2-}$ and $[\text{Co(CN)}_6]^{3-}$. The related anion $[(\text{NC})_5\text{Ru}^{\text{II}}{-}\text{CN}{-}\text{Fe}^{\text{II}}(\text{CN})_5]^{6-}$ has also been reported by the similar reaction of $[\text{Fe(CN)}_5\text{H}_2\text{O}]^{2-}$ and $[\text{Ru(CN)}_6]^{4-}$.[205] Salts of $[(\text{NC})_5\text{Co}^{\text{II}}$-$\text{NCFe}^{\text{II}}(\text{CN})_5]^{7-}$ have been prepared by heating $[\text{Co(NH}_3)_6]_4[\text{Fe(CN)}_6]_3$ at H_2SO_4 and metal cations. ESCA data indicates that the binuclear cation contains low-spin iron(II).[206] Reaction of $[\text{Fe(CN)}_5\text{NH}_3]^{3-}$ with DMSO in cold water gives $[\text{Fe(CN)}_5$-

[199] W. E. Carroll, F. A. Deeney, and F. J. Lalor, *J. C. S. Dalton*, 1974, 1430.

[200] J. A. Carroll, D. R. Fisher, G. W. Rayner-Canham, and D. Sutton, *Canad. J. Chem.*, 1974, **52**, 1914.

[201] A. Misono, Y. Uchida, M. Hidai, T. Yamagishi, and H. Kageyama, *Bull. Chem. Soc. Japan*, 1973, **46**, 2769.

[202] H. C. Freeman, G. H. W. Milburn, C. E. Nockolds, R. Mason, G. B. Robertson, and G. A. Rusholme, *Acta Cryst.*, 1974, **B30**, 886; R. Mason, G. B. Robertson and G. A. Rusholme, *ibid.*, p. 894.

[203] R. Mason, G. B. Robertson, and G. A. Rusholme, *Acta Cryst.*, 1974, **B30**, 906.

[204] G. Wada, Y. Sahira, K. Ohsaki, and F. Shinoda, *Bull. Chem. Soc. Japan*, 1974, **47**, 851.

[205] G. Emschwiller and C. Friedrich, *Compt. rend.*, 1974, **278**, C, 1271, 1335.

[206] N. A. Verendyakina, G. B. Seifer, Y. Y. Kharitonov, and B. V. Borshagovskii, *Russ. J. Inorg. Chem.*, 1973, **18**, 395; N. A. Verendyakina, G. B. Seifer, Y. Y. Kharitonov, and V. I. Nefedov, *ibid.*, 549.

DMSO]$^{3-}$. I.r. and n.m.r. data indicate the DMSO ligand is S-co-ordinated and spectral and kinetic evidence suggests substantial π-interaction between iron and sulphur.[207] Addition of $K_4Fe(CN)_6$ to an SbX_3 melt (X = Cl or Br) or to SbF_3 in SO_2 solution yields $K_4[Fe(CNSbX_3)_6]$ (X = F or Cl) and $K_4[Fe(CNSbX_3)_4(CN)_2]$ (X = Cl or Br). Mössbauer data suggest that there is a small but significant increase in *s*-electron density in the anions, which is explained in terms of the increasing influence of the π-acceptor function of the ligands.[208]

I.r. and Raman data show there to be two isomers of $H_2Fe[PPh(OEt)_2]_4$ present in solution. The predominant isomer above 0 °C is the *cis*-octahedral form and the other isomer, which is present at low temperatures, is thought to have a structure based on a very slightly distorted FeP_4 tetrahedron with hydrogen atoms in either a face–face or face–edge arrangement.[209]

Complexes. N-*Donor ligands.* Spectroscopic data indicate that $Fe(N_2H_5)_2(SO_4)_2$ has distorted octahedral co-ordination and that the hydrazinium ligand behaves as a normal N-donor.[66] D.c. polarographic and cyclic voltammetry of iron(II) complexes of phen and its 4,7-dimethyl and 4,7-diphenyl derivatives in acetonitrile have been reported. All show a five-step reduction wave, in which the first three steps are thought to involve reduction at the metal and the others, reduction at the ligand.[210] Similar studies on $Fe(terpy)_2(ClO_4)_2$ show a four-step reduction wave, in which the first two steps are reductions at the metal, the remainder being at the ligand.[211] The structure of $[L-Fe(1,10-phen)_3][Sb(d-tartrate)]_2 8H_2O$ has been reported.[211a] Temperature dependence of the quadrupole splitting of $Fe(isoquinoline)_4X_2$ (X = I, Cl, or Br), $Fe(py)_4X_2$ (X = Br, OCN, or N_3), and $Fe(phen)_2X_2$ (X = Cl, Br, or SCN) have been studied in order to determine the splitting of the cubic $t_{2g}(π^*)$ orbitals. The relative energies of the $e_g(π^*)$ orbitals in FeL_4X_2 (L = py, isoquin) follows the order I ⩽ Br ⩽ Cl and analysis of the spectra of the 1,10-phen complexes suggests a trigonal distortion from octahedral symmetry.[212] The octahedral polymer $Fe(4,4'-bipy)Cl_2$ has been reported.[69]

5-Aminoimidazole (L) forms the complex $[FeLSO_4]$, for which no definite conclusions were reached regarding its structure.[72] The complexes $Fe(PI)_2X_2$ [PI = 2-(2'-pyridyl)imidazole; X = Cl, Br, NCS, N_3, or CN) have been prepared. The thiocyanate groups are N-bonded and both this and the cyanide complexes have *cis*-arrangements. The cyanide complex is low spin, the others all being high spin.[213] The related ligand, 2-(2'-pyridyl)benzimidazole (PBZ) forms the complexes, $[Fe(PBZ)_3]X_2, nH_2O$ (X = Cl, Br, I, NO_3, ClO_4, or $\frac{1}{2}SO_4$), $[Fe(PBZ)_2X_2], 2H_2O$ [X = NCS (*cis* and *trans* isomers) or CN] and $[Fe(PBZ)_2(H_2O)_2]X$ (X = SO_4, C_2O_4, or S_2O_3). All are high-spin.[214]

The 1,2,4-triazine derivatives (40) (L_1), and (41) (L_2) form the complexes $[Fe(L_1)_3]$-$(BF_4)_2, 2H_2O$ and $[Fe(L_2)_2]X_2, 4H_2O$ (X = BF_4 or I). Both ligands produce a strong

[207] H. E. Toma, J. M. Malin, and E. Giesbrecht, *Inorg. Chem.*, 1973, **12**, 2084.

[208] H. G. Nadler, J. Pebler, and K. Dehnicke, *Z. anorg. Chem.*, 1974, **404**, 230.

[209] A. Schweizer, D. D. Titus, and H. B. Gray, *J. Amer. Chem. Soc.*, 1973, **95**, 4522.

[210] S. Musumeci, E. Rizzarelli, I. Fragala, S. Sammartano, and R. P. Bonomo, *Inorg. Chim. Acta*, 1973, **7**, 660.

[211] S. Musumeci, E. Rizzarelli, S. Sammartano, and R. P. Bonomo, *J. Inorg. Nuclear Chem.*, 1974, **36**, 853.

[211a] A. Zalkin, D. H. Templeton, and Y. Ueki, *Inorg. Chem.*, 1973, **12**, 1641.

[212] P. B. Merrithew, J. J. Guerrera, and A. J. Modestino, *Inorg. Chem.*, 1974, **13**, 644.

[213] Y. Sasaki and T. Shigematsu, *Bull. Chem. Soc. Japan*, 1974, **47**, 109.

[214] S. P. Ghosh and L. K. Mishra, *Inorg. Chim. Acta*, 1973, **7**, 545.

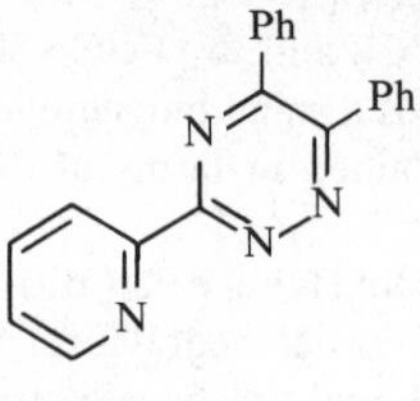

(40)

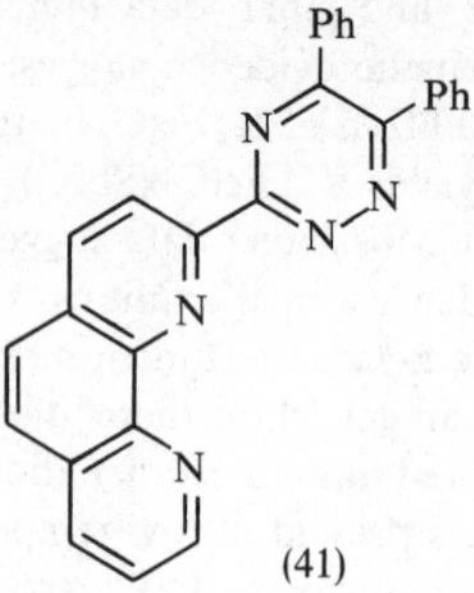

(41)

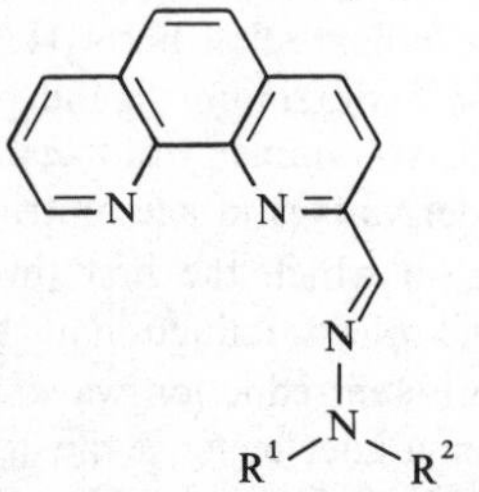

(42)

	R^1	R^2
	H	Ph
	H	Me
	Me	Me
	Me	Ph
	Ph	Ph
	H	2-pyridyl

ligand field, L$_1$ acting as a bidentate ligand, whereas L$_2$ is terdentate.[215] The ligands (42) form the complexes [Fe(42)$_2$](BF$_4$)$_2$nH$_2$O (n = 0, 0.5, or 1, depending on the ligand). The ligands show a gradation of field strengths and this is illustrated by the spin states of the complexes. When R^1 = H, R^2 = Me, the complex is low-spin over the temperature range 83–363 K, but some high-spin species are present at higher temperatures. In the same range, the complex with R^1 = R^2 = Me is essentially high spin, but undergoes significant spin-pairing and the R^1 = H, R^2 = Ph complex shows a complete temperature dependant $^5T_2 \rightarrow {}^1A_1$ spin transition, which is very sharp and gives rise to a pronounced change in colour and magnetism over a few degrees. The other complexes are high-spin over the entire range studied.[216] The 1:2 iron(II) complexes of 3,3',5,5'-tetramethyl- and 3,3',4,4',5,5'-hexamethyldipyrromethane are both high-spin species with distorted tetrahedral co-ordination.[217]

The new low-spin complexes [Fe(43)L$_1$L$_2$](PF$_6$)$_2$ (L$_1$ = L$_2$ = imidazole, MeCN; L$_1$ = MeCN, L$_2$ = CO) have been isolated. The bis-acetonitrile complexes undergo reversible substitution reactions with both CO and imidazole. In the case of imidazole substitution, one or more of the following are present in solution depending on the substrate concentration: [Fe(43)(MeCN)$_2$]$^{2+}$, [Fe(43)(MeCN)(imid)]$^{2+}$, and [Fe(43)-(imid)$_2$]$^{2+}$. For CO substitution, only [Fe(43)(MeCN)$_2$]$^{2+}$ and [Fe(43)(CO)-(MeCN)]$^{2+}$ are present in solution when the pressure of CO gas is less than or equal to one atmosphere.[218] The structure of high-spin [Fe^{II}Cl(44)]I has been determined

215 H. A. Goodwin and F. E. Smith, *Inorg. Chim. Acta*, 1973, **7**, 541.
216 H. A. Goodwin and D. W. Mather, *Austral. J. Chem.*, 1974, **27**, 965.
217 Y. Murakami, Y. Matsuda, K. Sakata, and K. Harada, *Bull. Chem. Soc. Japan*, 1974, **47**, 458.
218 D. A. Baldwin, R. M. Pfeiffer, D. W. Reichgott, and N. J. Rose, *J. Amer. Chem. Soc.*, 1973, **95**, 5152.

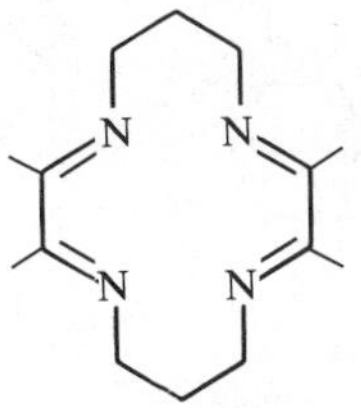

(43) 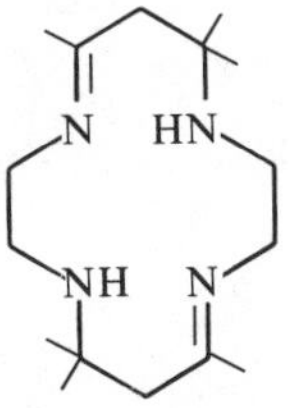   (44)

and studied as a possible model for the five-co-ordinate deoxy form of haemoglobin and myoglobin. The metal atom is bound to the four N atoms of the folded form of (44), the Cl ligand being axial. The Fe—N and Fe—Cl bond lengths are normal for high-spin iron(II) and the unusual co-ordination number arises because of steric constraints of the cyclic ligand, which do not permit the metal atom to fit into the N_4-plane and the fact that the ligand is not flexible enough to permit folding to accommodate a sixth ligand *cis* to the chlorine.[219] There has been disagreement over the magnetic moment of the tetraphenylporphinate-complex, [Fe(TPP)]. A new method of preparation, by treating H_2TPP with iron(II) propionate in propionic acid, followed by sublimation, has been reported and the magnetic moment determined to be 4.85 BM at 22 °C, in good agreement with spin-only values for high-spin iron(II).[220]

$FeCl_2,4H_2O$ reacts with dioximeH (dioximeH = dmgH or α-diphenylglyoxime) in alcohol in the presence of a slight excess of base A (A = 3- and 4-picoline, 3-chloro-, 3-bromo-, and 4-vinyl-pyridine) under an argon atmosphere to yield the low-spin complexes $[Fe(dioxime)_2A_2]$.[221] The di-imine complex $[Fe(gmi)_3]^{2+}$ (gmi = MeN= CHCH=NMe) is oxidized stoicheiometrically by Ce^{IV} in $10M\text{-}H_2SO_4$ to $[Fe(gmi)_3]^{3+}$. On dilution to lower pH, the iron(III) complex disproportionates to the starting material and the new cations $[Fe(gmi)_2GA]^{3+}$ and $[Fe(gmi)_2GH]^{2+}$ (GA = MeN=CHCONHMe or MeN=CHC(OH)=NMe, GH = MeN=CHCH= NCH_2OH). The same products are formed when $[Fe(gmi)_3]^{2+}$ reacts with Ce^{IV} in $1\text{--}4M\text{-}H_2SO_4$ indicating that reaction proceeds *via* formation and disproportionation of the $[Fe(gmi)_3]^{3+}$ cation. Eventually, $[Fe(gmi)_3]^{2+}$ consumes *ca.* 4.5 oxidation equivalents of Ce^{IV} yielding, in addition to $[Fe(gmi)_2GA]^{3+}$, further ligand-oxidized iron(III) products formed apparently from $[Fe(gmi)_2GH]^{2+}$ by a succession of oxidations to the iron(III) form and disproportionation of this. On the basis of a kinetic study, the participation of free-radical intermediates in the disproportionation of $[Fe(gmi)_3]^{3+}$ is proposed.[222]

The substituted quinuclidinones (45)—(48) (L) form the complexes $FeLX_2$ (X = Cl, Br, I, or ClO_4), which are high-spin and pseudo-tetrahedral. Only $Fe(46)Cl_2$ and $Fe(48)Cl_2$ were obtained in high purity, the results for the other preparations being unreproducible. The complexes contain an $[FeN_2Cl_2]$ core and it is suggested that the

[219] V. L. Goedken, J. Molin-Case, and G. G. Christoph, *Inorg. Chem.*, 1973, **12**, 2894.
[220] S. M. Husain and J. G. Jones, *Inorg. Nuclear Chem. Letters*, 1974, **10**, 105.
[221] A. V. Ablov and V. N. Zubarev, *Russ. J Inorg. Chem.*, 1973, **18**, 297.
[222] H. L. Chum and P. Krumholz, *Inorg. Chem.*, 1974, **13**, 514, 519.

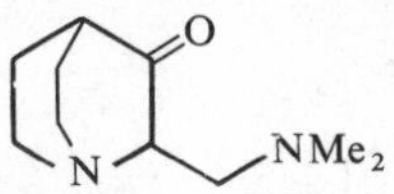

(45) 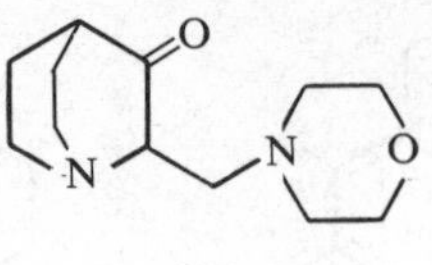   (46) 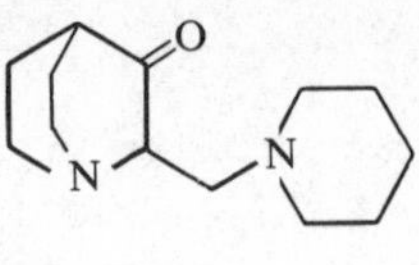   (47)

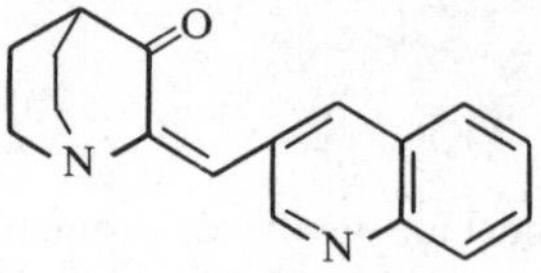

(48)

fact that complexes with 1:1 metal-to-ligand ratios only were isolated is a consequence of the bulk of the quinuclidinone group.[223] The polymeric complex $[FeCl_2(V-2P)_2]_{100}$ (V-2P = a monomer unit of poly-1-vinyl-2-pyrrolidinone) has also been reported.[224]

O-*Donor ligands.* Oxidation of $[Fe(H_2O)_6]^{2+}$ with $HClO_4$ yields a green compound containing an $Fe^{II}:Fe^{III}$ ratio of 1:3, for which a structure containing Fe^{II}—O—Fe^{III} linkages (49) has been proposed. The same compound is also obtained by aerial oxidation $[Fe(H_2O)_6]^{2+}$ in neutral or slightly acidic sulphate solutions. The results of this investigation have been used to estimate the iron composition of solid-green rust(II).[225]

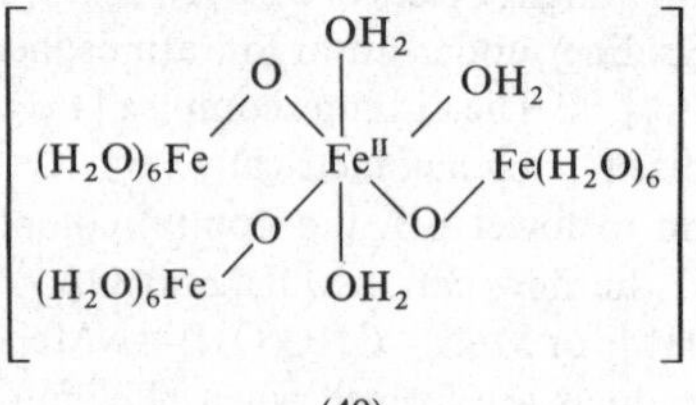

(49)

Treatment of iron(II) acetate with $Al(OR)_3$ (R = Bu^n, Pr^i, or Bu^i) in decalin at 500 °C yields $[Fe^{II}\{OAl(OR)_2\}_2]$. This compound rapidly takes up oxygen in heptane until the $Fe:O_2$ ratio is 4:1 and this is followed by a slower uptake until the ratio reaches 2:1. The product of oxygenation contains iron(III) and the process is reversible.[226] $(EtO)_3PO$ reacts with iron(II) halides at 100—250 °C yielding $Fe[OOP-(OEt)_2]_2$. The complex $Fe(DIMP)_4(ClO_4)_2$ (DIMP = di-isopropylmethylphosphon-ate) contains both unidentate and ionic perchlorate group, but the related dimethyl-methylphosphonate and Ph_3PO complexes do not show perchlorate co-ordination and are formulated as polynuclear cations, $[FeL_4]_n^{2+}$ involving both terminal and bridging ligands and five-co-ordinate iron(II).[87]

[223] G. J. Long and D. L. Coffin, *Inorg. Chem.*, 1974, **13**, 270; R. C. Dickinson and G. J. Long, *ibid.*, p. 262.
[224] H. G. Biedermann and W. Graf, *Z. Naturforsch.*, 1974, **29b**, 65.
[225] T. Misawa, K. Hashimoto, W. Suetaka, and S. Shimodaira, *J. Inorg. Nuclear Chem.*, 1973, **35**, 4159; T. Misawa, K. Hashimoto, and S. Shimodaira, *ibid.*, p. 4167.
[226] T. Ouhadi, A. J. Hubert, P. Teyssie, and E. G. Derouane, *J. Amer. Chem. Soc.*, 1973, **95**, 6481.

Pyrazine *N*-oxide (L) reacts with hydrated metal salts in a dehydrating solvent to form $[L_2(H_2O)(ClO_4)Fe(L)_2Fe(ClO_4)(H_2O)L_2](ClO_4)_2$, containing terminal O-bonded and bridging NO-bonding ligands. The complex $FeL_3(ClO_4)_2$ probably either has a dinuclear structure, $[L_2FeL_2FeL_2](ClO_4)_4$, or contains polynuclear cations $[FeL_4]_x(ClO_4)_{2x}$.[227] $[FeLCl_2]_x$ containing five-co-ordinate iron(II) and bridging NO-bonded L ligands, as well as both bridging and terminal chlorine atoms, has also been reported.[85]

S-donor and P-donor ligands. $Fe(S_2CNR_2)_2$ and the new complexes, $[Fe(S_2CNR_2)_3]^-$ (R = Me, Et, Pr^n, Bu^n, or Ph) have been prepared and studied. The small isomer shift observed in the Mössbauer spectrum of the bis-complex is attributed to considerable Fe–S covalency and the presence of five-co-ordinate iron(II). This study has demonstrated that the tris-dithiocarbamate complexes of iron-(III) and -(IV) can also be extended to iron(II) and, in general, a decrease in oxidation state lowers the degree of covalency. In accordance with this, the iron(IV) complexes are all low-spin, the iron(III) complexes show spin-equilibria and the iron(II) complexes are all high-spin. The isomer shifts for $[Fe(S_2CNR_2)_3]^-$ ions are smaller than in the case of $[Fe(acac)_3]^-$, which reflects increasing Fe–S covalency over that of Fe—O bonds. Because of this increased covalency, five-co-ordination is usually observed for bis-dithiocarbamate complexes; however, $Fe(S_2CNMe_2)_2$ is both five- and six-co-ordinate.[228] Other workers have also studied the air-sensitive compounds $Fe(S_2CNR_2)_2$. For R = Et, the structure consists of intramolecular antiferromagnetic dimers involving five-co-ordinate iron(II); however, the dimethyl analogue is an antiferromagnetic polymer which probably contains six-co-ordinate iron(II).[229] Pyridine-2-thiol (L) forms the complexes, FeL_2X_2 (X = Cl or Br).[230]

Treatment of $FeCl_2(dppe)$ with thallous cyclopentadienide in benzene yields $(Cp)FeCl(dppe)$ $(dppe = Ph_2PCH_2CH_2PPh_2)$. This reacts with $SnCl_2$ to give $(Cp)Fe(dppe)SnMe_3$. The chloride also reacts with BH_4^- and MeMgI, yielding $(Cp)FeX(dppe)$ (X = H and Me, respectively).[231] $(Cp)Fe(dppe)Br$ reacts with $BrCH_2CH_2Br$ and excess magnesium in dry THF to give $(Cp)(dppe)FeMgBr,3THF$ in which the $[(Cp)(dppe)Fe]$ unit is linked to the $MgBr(THF)_2$ moiety by an Fe—Mg bond [bond length = 2.593(7) Å]. The magnesium atom is in a distorted tetrahedral environment similar to that in other Grignard reagents.[232] $(Cp)Fe[P(OPh)_3]_2Sn_4Me_9$ reacts with X_2 (X = I or Br) in a 1:3 mole ratio to give $(Cp)Fe[P(OPh)_3]_2SnX_3$, which on treatment with excess X_2 yields $(Cp)Fe[P(OPh)_3]_2X$.[233]

$Ph_2P(CH_2)_2PPh(CH_2)_2PPh(CH_2)_2PPh_2$ (P_4) forms the square-pyramidal complexes $[Fe(P_4)X](BPh_4)$ (X = Cl, Br, or I) which exhibit spin-equilibrium.[234] Reaction of $FeCl_2$ with the phosphobenzenes (50), under conditions similar to those used for the preparation of ferrocene, gives (51) rather than (52).[235]

[227] A. N. Speca, N. M. Karayannis, and L. L. Pytlewski, *J. Inorg. Nuclear Chem.*, 1973, **35**, 3113.

[228] J. L. K. F. de Vries, J. M. Trooster, and E. de Boer, *Inorg. Chem.*, 1973, **12**, 2730.

[229] L. F. Larkworthy, B. W. Fitzsimmons, and R. R. Patel, *J.C.S. Chem. Comm.*, 1973, 902.

[230] I. P. Evans and G. Wilkinson, *J. C. S. Dalton*, 1974, 946.

[231] M. J. Mays and P. L. Sears, *J. C. S. Dalton*, 1973, 1873.

[232] H. Felkin, P. J. Knowles, B. Meunier, A. Mitschler, L. Ricard, and E. R. Weiss, *J.C.S. Chem. Comm.*, 1974, 44.

[233] W. Klaui and H. Werner, *J. Organometallic Chem.*, 1973, **60**, C19.

[234] M. Bacci, S. Midollini, P. Stoppioni, and L. Sacconi, *Inorg. Chem.*, 1973, **12**, 1801.

[235] G. Märkl and C. Martin, *Angew. Chem. Internat. Edn.*, 1974, **13**, 408.

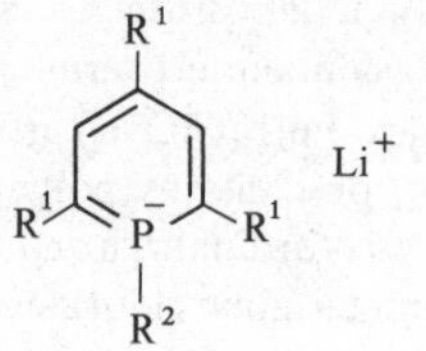

(50) $R^1 = R^2 = Ph$
$R^1 = Ph, R^2 = Me$

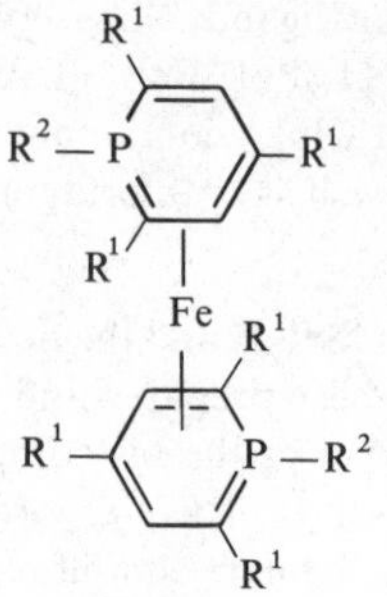

(51)

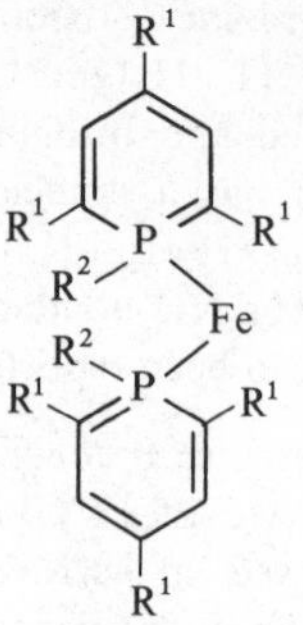

(52)

Mixed donor ligands. Thermodynamic parameters have been evaluated for the interaction of asparaginate with Fe^{2+} in aqueous solution at 25 °C. On the basis of the results it is calculated that quite a large proportion of asparaginate in blood plasma may be complexed to iron(II) as well as to the expected copper(II).[93] 1,10-Phenanthroline N-oxide (phenNO) forms the complex $[Fe(phenNO)_3](ClO_4)_2$ in which the ligands chelate by N- and O-donor atoms.[96]

Salicylaldehyde-salicylhydrazone (12) (SSH) forms the high-spin complex $Fe(SS)_2$ which is thought to have a structure similar to (13).[99] The planar pentadentate ligand, 2,6-diacetylpyridine-bis-semicarbazone (DAPSC) forms the pentagonal-bipyramidal complex $Fe(DAPSC)Cl_2,3H_2O$ in which chlorine and water ligands occupy the apical positions.[236] The complex $Fe(pa)Cl_2$ (pa = pyridine-4-carbaldehyde-oxime) has also been reported, which has an octahedral halogen bridged polymeric structure, in which the ligand is bonded by the pyridine N-atom only. Reaction of dipyridylketone (pyCOpy) with $FeCl_2$ gave high-spin $[Fe(pyCOpy)_2Cl_2]$, despite the fact that no precautions were taken to exclude ethanol or water and hence inhibit acetal formation. However attempts to prepare the iron(II) perchlorate analogue gave products containing both iron(II) and iron(III) and reaction with $Fe(ClO_4)_3$ gave $[Fe(pyCOpy)\{pyC(OEt)(OH)py\}](ClO_4)_3H_2O$. The low-spin complex $[Fe(pyCOpy)_2(NCS)_2]$ was isolated from an attempt to form an iron(III)-thiocyanate complex and reaction with $FeCl_3$ yields the high-spin polymeric complex $[Fe[pyC(OEt)_2py]Cl_3]_n$.[238] Dipole moment measurements on Fe^{II} complexes of $R^1C(SH)=CHC(O)R^2$ (R^1 = Me or Ph, R^2 = Ph; R^1 = Me, 2-thienyl, Ph, p-MeC_6H_4, or p-BrC_6H_4; $R^2 = CF_3$) did not clearly distinguish between *fac-* and *mer-*octahedral configurations.[239] The ligand $O=\overline{CNHCH_2CH_2SCH_2}$ (tm) forms the four-co-ordinate complexes $[Fe(tm)_2I_2]$, $[Fe(tm)_3I]I$, and $[Fe(tm)_4]I_2$.[240]

[236] D. Webster and G. J. Palenik, *J. Amer. Chem. Soc.*, 1973, **95**, 6505.
[237] P. S. Gomm, G. I. L. Jones, and A. E. Underhill, *J. Inorg. Nuclear Chem.*, 1973, **35**, 3745.
[238] V. Rattanaphani and W. R. McWhinnie, *Inorg. Chim. Acta*, 1974, **9**, 239.
[239] M. Das, S. E. Livingstone, S. W. Filipczuk, J. W. Hays, and D. V. Radford, *J. C. S. Dalton*, 1974, 1409.
[240] D. De Filippo, F. Devillanova, and G. Verani, *J. Inorg. Nuclear Chem.*, 1974, **36**, 1017.

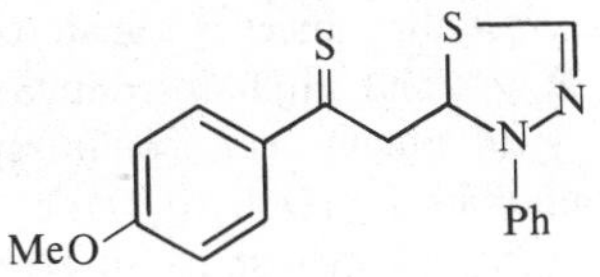

(53) 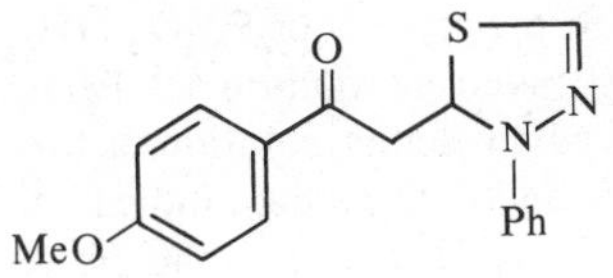　　　　　　　　　　　　　　　(54)

Ligands (53) and (54) (L) form the complexes FeL_2X_2 (X = Cl, Br, or I), for which spectral data suggest square-pyramidal structures.[241] 1,10-Phenanthroline-2-carbothioamide and its *N*-phenyl-derivative form both mono- and bis-ligand complexes with iron(II), which exhibit temperature-dependent spin-equilibria. In the mono-ligand complexes the metal atom is believed to be six-co-ordinate.[242]

Isocyanide complexes. $[Fe(p\text{-tolyl-NC})_5I]I_3$ has been isolated as a by-product of the preparation of *cis*-$[Fe(p\text{-tolyl-NC})_4I_2]$.[243] Reaction of $FeCl_3$ with MeNC at room temperature yields $[Fe^{II}(CNMe)_6][Fe^{III}Cl_4]_2$. Structural studies on this and $[Fe(NCH)_6][FeBr_4]_2$ have been reported.[244]

Iron(III).—*Halides and Cyanides.* The reaction $2FeF_3 + Fe \rightarrow 3FeF_2$ has been studied by calorimetry in the temperature range 25—650°C. It is an endothermic reaction that takes place at *ca.* 200°C. The $\Delta H°_{298}$ value for $FeCl_3$ was determined to be 19 ± 1 kcal mol^{-1} and specific heats of ferrous and ferric hydrofluorides were determined in the temperature range 200—400°C. Structural studies on $CsMgFeF_6$ and $CsZnFeF_6$ show these molecules to have $RbNiCrF_6$-type structures.[246] $BaLiFeF_6$ has also been reported.[247] $CsFeF_4$ is of a hitherto unknown structural type that is a variation on the $TlAlF_4$ structure. The caesium atoms are 12-co-ordinate [Cs—F (mean) = 3.25 Å] and the Fe—F bond lengths in the FeF_6 octahedra are characteristically shortened compared with those in normal FeF_4^- layers [Fe—F = 1.96(2) and 1.86(1) Å]. $RbFeF_4$ shows the same structural features.[248] $K_{0.6}FeF_3$ consists of a framework of FeF_6 octahedra joined at the corners.[249]

For many years, $FeCl_3$ has been believed to exist as a monomer in strong donor solvents and a dimer in weak donors such as benzene. However, on the basis of i.r. data, it has been reported that this is an oversimplification and that $FeCl_3$ is monomeric in benzene and is thought to be present as $(\pi\text{-}C_6H_6)FeCl_3$.[250] When very dilute aqueous solutions of $FeCl_3$ are treated with weak bases, deprotonation of $Fe(H_2O)_6^{3+}$ occurs initially giving mono- and di-nuclear hydroxy-aquo-complexes. The second stage of the reaction is condensation of four $[(H_2O)_4Fe(OH)_2Fe(OH_2)_4]^{3+}$ dimers around a chloride ion to give an eight-membered ring, which

[241] F. Y. Petillon, J. Y. Calves, J. E. Guerchais, and Y. M. Poirier, *J. Inorg. Nuclear Chem.*, 1973, **35**, 3751.

[242] H. A. Goodwin, D. W. Mather, and F. E. Smith, *Austral. J. Chem.*, 1973, **26**, 2623.

[243] J. W. Schnidler, J. R. Luoma, and J. P. Cusick, *Inorg. Chim. Acta*, 1973, **7**, 563.

[244] G. Constant, J. C. Daran, and Y. Jeannin, *J. Inorg. Nuclear Chem.*, 1973, **35**, 4083.

[245] Y. Macheteau and P. Barberi, *Bull. Soc. chim. France*, 1974, 34.

[246] R. Jesse and R. Hoppe, *Z. anorg. Chem.*, 1974, **403**, 143.

[247] W. Viebahn and D. Babel, *Z. anorg. Chem.*, 1974, **406**, 38.

[248] D. Babel, F. Wall, and G. Heger, *Z. Naturforsch.*, 1974, **29b**, 139.

[249] A. M. Hardy, A. Hardy, and F. Ferey, *Acta Cryst.*, 1973, **B29**, 1654.

[250] R. A. Work and R. L. McDonald, *Inorg. Chem.*, 1973, **12**, 1936.

quickly forms crystals of $Fe_4O_3(OH)_5Cl$. These crystalline micelles remain colloidably suspended in solution.[251] Ferric chloro-, aquo-, and sulphato-complexes are hydrolysed in acidic solution below 70°C to yield polynuclear iron(III) species. Magnetic and spectral data indicate that the compounds $Fe_2(OH)_2Cl_2O$, $Fe_2(OH)_3$-$(NO_3)_3$, $Fe_2(OH)_3(SO_4)_{3/2}$ and $Fe_3(OH)_2(SO_4)_{7/2}$ are all high-spin. All except the last of these are thought to contain edge-sharing octahedral dimeric units linked by OH- and oxo-bridges. At higher temperatures, $Fe_2(OH)_2Cl_2O$ is converted into the α- and β-forms of FeO(OH), whereas $Fe_2(OH)_3(SO_4)_{3/2}$ and $Fe_3(OH)_2(SO_4)_{7/2}$ yield $MFe_3(OH)_6(SO_4)_2$ (M = Na, K, or NH_4) and $Fe(OH)_3(NO_3)_3$ gives either α-FeO-(OH) or α-Fe_2O_3 depending on the temperature.[252]

X-Ray structural studies on $LaFe(CN)_6,3H_2O$ show that $[FeC_6]$-octahedra are linked by CN-bridges to nine-co-ordinate $[LaN_6(H_2O)_3]$ units.[253] Oxidation of $[Fe(CN)_6]^{3-}$ ions by hydroxylamine is catalysed by $[(edta)Fe]^{3+}$ and aquo-copper(II) species, but pentacyanoaquoferrates and other secondary reaction products are not significant catalysts.[254]

The structure of $Na_2[Fe(CN)_5NH_3],2H_2O$ has been determined. The octahedral anions contain almost linear cyanide linkages and the sodium ions are present in double chains with octahedral co-ordination from four nitrogen atoms and two water molecules. The chains are held together by the anions and there appears to be no hydrogen bonding between the anions and the water molecules.[255] Substitutions of $[Fe(CN)_5X]^{2-}$ ions (X = NH_3 or H_2O) by Y^- (Y = N_3, SCN, or OH) and $[Co(CN)_6]^{3-}$ ions are catalysed by $[Fe(CN)_5X]^{3-}$ ions. At iron(II) concentrations of $>1\%$ of the total iron present, the rate-determining step is substitution of $[Fe-(CN)_5X]^{3-}$ by Y^-.[256] The formula of $Na_2[Fe(CN)_5(enH)],6H_2O$, first described by Manchot, has been confirmed and detailed spectral and magnetic data have now been reported.[257] Two-electron oxidation of o-$C_6H_4(NH_2)_2$ yields o-$C_6H_4(NH)_2^{2+}$, which can be stabilized as $\{Fe(CN)_5[o$-$C_6H_4(NH)_2]\}^{2-}$. The structure of the anion has been determined and the bond lengths are consistent with α-di-imine character of the co-ordinated ligand.[258] Oxidation-reduction properties of $[(NC)_5Co(NC)Fe-(CN)_5]^{6-}$, the product of a redox addition reaction between $Co(CN)_5^{3-}$ and $Fe(CN)_6^{3-}$, have been studied. In the course of the redox reaction, the electron density is shifted entirely from cobalt to iron and the resulting dinuclear complex has high thermodynamic and kinetic stability.[259] Evidence for the formation of stable di- and tri-nuclear μ-cyano-species by $Fe(CN)_6^{3-}$ and bis-(histidinato)-cobalt(II) has been presented.[260] Previous work has shown that iron(III) and $Mo(CN)_8^{4-}$ forms soluble complexes in water of stoicheiometry $[Fe^{III}Mo(CN)_8]^-$. These have been re-examined in acidic solution in an attempt to prevent hydrolysis and to isolate

[251] W. Feitknecht, R. Giovanoli, W. Michaelis, and M. Müller, *Helv. Chim. Acta*, 1973, **56**, 2847.
[252] M. Kiyama and T. Takada, *Bull. Chem. Soc. Japan*, 1973, **46**, 1680.
[253] W. E. Bailey, R. J. Williams, and W. O. Milligan, *Acta Cryst.*, 1973, **B29**, 1365.
[254] G. J. Bridgart, W. A. Waters, and I. R. Wilson, *J. C. S. Dalton*, 1973, 1582.
[255] A. Tullberg and N. G. Vannerberg, *Acta Chem. Scand. (A)*, 1974, **28**, 340.
[256] A. D. James, R. S. Murray, and W. C. E. Higginson, *J.C.S. Dalton*, 1974, 1273.
[257] J. A. Olabe and P. J. Aymonino, *J. Inorg. Nuclear Chem.*, 1974, **36**, 1221.
[258] G. G. Christoph and V. L. Goedken, *J. Amer. Chem. Soc.*, 1974, **95**, 3869.
[259] J. Hanzlik and A. A. Vlcek, *Coll. Czech. Chem. Comm.*, 1973, **38**, 3019.
[260] S. Bagger and K. Gibson, *Acta Chem. Scand.*, 1973, **27**, 3227.

$[FeNCMo(CN)_7]^-$. It was found that the formation of this species was dependent on pH, and that the following equilibria could also be set up:

$$Fe^{3+}(aq) + H_2O \rightleftharpoons Fe(OH)^{2+}(aq) + H^+$$
$$Fe^{3+}(aq) + ClO_4^- \rightleftharpoons FeClO_4^{2+} \text{ (In } HClO_4 \text{ solution)}$$

the first of these being observed only at high pH's. The dinuclear anion could not be isolated, but the amorphous compound $Fe_4[Mo(CN)_8]_3,12H_2O$ was formed and this is thought to be polymeric with cyanide bridges.[261]

Complexes. N-Donor ligands. I.r. data have been reported for thiocyanato-iron(III) complexes, both in the solid state and in solution. There is some evidence that the octahedral anion $[Fe(CN)_6]^{3-}$ changes configuration to a tetrahedral species in low-polarity solvents.[262]

New intercalated compounds of FeOCl and pyridine have been synthesized by soaking FeOCl in pyridine vapour at temperatures between 30 and 120 °C. The species $FeOCl,py_x$ were obtained, $x = \frac{1}{3}$ or $\frac{1}{4}$ depending on the temperature. E.s.r. of $FeOCl,py_{1/3}$ suggests the existence of conduction electrons resulting from partial transfer of the nitrogen lone-pair into the FeOCl layer.[263] Measurement of the magnetically perturbed Mössbauer spectrum of $[Fe(terpy)_2](ClO_4)_3$ shows that the cation has a non-degenerate orbital ground state and this, together with other studies, shows that such a ground state is favoured by low-spin iron(III) complexes, thus avoiding Jahn–Teller distortions and allowing the maximum degree of π-bonding.[264] A kinetic study of fission of the first Fe—N bond in the acid aquation of $[Fe(bipy)_3]^{3+}$ indicates that there is a spin state change associated with the fission and that the only reactant is low-spin.[265] The octahedral polymer $[FeCl_3(4,4'-bipy)]$ has been reported.[69] Reaction of cyanide with $[L_2(H_2O)Fe—O—Fe(H_2O)L_2]^{4+}$ (L = phen or bipy) is both pH and temperature dependent. At 40—50 °C and pH 4.0, $[FeL_2-(CN)_2]CN$ is formed, whereas at room temperature and pH 1.2—1.5 the product is $[FeL_2(CN)_2]$. At room temperature and pH 3.8—4.2, addition of cyanide gives a green complex which rapidly adopts a yellow-green colour in solution and is thought to be $[L_2(NC)Fe—O—Fe(CN)L_2]^{2+}$. This cation ultimately decomposed to yellow $[FeL_2(CN)_2]^+$.[265a]

During the catalytic cycle of hydroperoxidases, two-electron oxidation of ferri-hemeprotein yields an enzymatically active species. Among the structures suggested for this is an iron(III) isoporphyrin. The model compound [*meso*-tetraphenylmethoxy-isoporphinatoiron(III)]hexafluorophosphate (55) has been prepared by controlled potential oxidation of tetraphenylporphinatoiron(III) chloride, Fe(TPP)Cl, in CH_2Cl_2–MeOH. The spectral properties of this compound are different from those of the enzyme and thus the latter cannot be an isoporphyrin.[266] The equilibria between Fe(TPP)Cl and imidazole in non-polar media has been investigated. Reaction proceeds *via* chlorine displacement by two imidazole units yielding the tightly bound

[261] G. F. Knight and G. P. Haight, *Inorg. Chem.*, 1973, **12**, 1934.
[262] A. Miezis, *Acta Chem. Scand. A*, 1974, **28**, 415.
[263] F. Kanamaru, S. Yamanaka, and M. Koizumi, *Chem. Letters*, 1974, 373.
[264] W. M. Reiff, *J. Amer. Chem. Soc.*, 1974, **96**, 3829.
[265] M. V. Twigg, *Inorg. Chim. Acta*, 1974, **10**, 17.
[265a] B. J. C. Lima, F. David, and P. G. David, *J.C.S. Dalton*, 1974, 680.
[266] J. A. Guzinski and R. H. Felton, *J.C.S. Chem. Comm.*, 1973, 715.

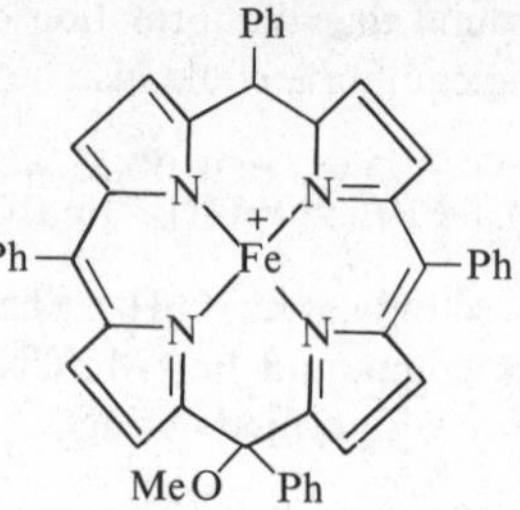

Ph

(55)

ion-pair, $[Fe(TPP)(imid)_2]Cl$. No evidence was found for the intermediates $[Fe-(TPP)(imid)Cl]$ or five-co-ordinate $[Fe(TPP)imid]Cl$. The bis-imidazole product is low spin, whereas the chloride is high-spin and the absence of intermediates is attributed to the stability of the low-spin product.[267] Reversible binding of NO and CO to chloro-iron(III)-protoporphyrin IX and -tetraphenylporphyrin, $[ClFe(por)]$, has been studied. In piperidine, low-spin $[(pip)_2Fe^{II}(por)]$ is formed which forms both $[(pip)Fe(por)XO]$ (X = C or N). Equilibrium constants have been determined in both cases.[268] The mononuclear complexes FeX(oep) (oep = octaethylporphyrin, X = OPh, Cl, OMe, or OAc) yield the μ-oxo complexes, $[(oep)Fe-O-Fe(oep)]$ on hydrolysis and condensation in the presence of alkali or during chromatography on alumina. The dinuclear products are solvolysed by PhOH, MeOH, or $MeCO_2H$ yielding FeX(oep), but are unaffected by water.[269]

3,3′,4-Trimethyldipyrromethane (dpm) yields the low-spin six-co-ordinate tris-complex with iron(III). Strong ligand-field bands in this complex are consistent with the presence of a trigonal distortion and highly covalent metal–ligand bonds.[217] The complex $[Fe(pbz)_3]Cl_3$ [pbz = 2-(2′-pyridyl)benzimidazole] has also been reported.[214]

O-*Donor ligands.* It has been established that the basic sulphate, $Fe_4(OH)_2(SO_4)_5$, $18H_2O$, or mixtures of this compound with $Fe_2(SO_4)_3,9H_2O$ are precipitated from the $Fe_2(SO_4)_3-Me_2CO-H_2O$ system, depending on the conditions employed.[270] Thermolysis of $(H_3O)[Fe_3(OH)_6(SO_4)_2]$ follows the course:

$$(H_3O)[Fe_3(OH)_6(SO_4)_2] \xrightarrow{\geq 230\,°C} Fe(OH)SO_4 + (FeO_2)SO_4 \xrightarrow{>500\,°C} Fe_2O(SO_4)_2 + \tfrac{1}{2}Fe_2O_3$$

followed by either:

$$3Fe_2O(SO_4)_2 \xrightarrow{560-620\,°C} 2Fe_2(SO_4)_3 + Fe_2O_3 \xrightarrow{>620\,°C} 3Fe_2O_3 + 6SO_3$$

or:

$$Fe_2O(SO_4)_2 \xrightarrow{\geq 620\,°C} Fe_2O_3 + 2SO_3$$

[267] C. L. Coyle, P. A. Rafson, and E. H. Abbot, *Inorg. Chem.*, 1973, **12**, 2006.
[268] D. V. Stynes, H. C. Stynes, B. R. James, and J. A. Ibers, *J. Amer. Chem. Soc.* 1973, **95**, 4087.
[269] J. W. Buchler and H. H. Schneehage, *Z. Naturforsch.*, 1973, **28b**, 433.
[270] E. V. Margulis, L. A. Savchenko, M. M. Shokarev, L. I. Beisekeeva, and G. A. Kapatsina, *Russ. J. Inorg. Chem.*, 1974, **18**, 1128.

depending on the nature of $Fe_2O(SO_4)_2$. Similar treatment of $Fe(OH)SO_4$ follows the course:

$$Fe(OH)SO_4 \xrightarrow{>500\,°C} 3Fe_2O(SO_4)_2 \rightarrow 2Fe_2(SO_4)_3 + Fe_2O_3 \rightarrow 3Fe_2O_3 + 6SO_3$$

however, $Fe_2O(SO_4)_3$ is also formed in small quantities.[271] The hydrolysis of iron(III) in 1M-NaNO$_3$ solution is an extremely slow process, reaching equilibrium after at least 4 weeks and giving a precipitate of $Fe(OH)_2NO_3$.[272] The kinetics of the reaction between iron(III) and either phenol or *o*-aminophenol indicate the former reacts in the protonated form with either Fe^{3+} or $Fe(OH)^{2+}$, whereas as the protonated form of the latter reacts with $Fe(OH)^{2+}$.[273]

Phosphosiderite and strengite, the naturally occurring forms of iron(III) ortho-phosphate, have been synthesized in crystalline form.[274] The compounds $FeAsO_4$, $2H_2O$ and $Fe_{1+x}H_{3-x}AsO_4,2H_2O$ $(0 \leqslant x \leqslant 0.07)$, which have the scordite structure, have also been prepared.[275] Crystallographic data for $Na_xFe^{III}H_{3-x}AsO_4,1—1.5H_2O$ and $NaFeAs_2O_7$ have been reported.[276] Thermolysis of $Fe(H_2AsO_4),nH_2O$ at 800°C yields $Fe_2As(AsO_4)_2$.[277]

The enthalpies of solvation of $Fe(acac)_3$ by acetone, benzene, and chloroform have been determined.[102] The structure of the 1:1 adduct between $AgClO_4$ and $Fe(acac)_3$, H_2O has been reported. The silver ion is located near two chelate rings but is apparently bonded to the active methylene group of only one of these [Ag—C = 2.29(2) Å]. It is also bonded to an oxygen atom of the perchlorate group [Ag—O = 2.50(s) Å] and to the water molecule [Ag—O = 2.25(2) Å]. Co-ordination around the iron atom remains octahedral. The adduct is prepared by mixing stoicheiometric quantities of $AgClO_4$ and $Fe(acac)_3$ in toluene; it forms dark-red air-stable crystals.[278] Treatment of $Fe(acac)_3$ with dry HX in dichloromethane at room temperature yields the high-spin monomeric complexes $Fe(acac)_2X$ (X = Cl or Br).[103] Magnetic and i.r. spectral data for $Fe(OMe)_2L$ (L = alkanoate) complexes has been interpreted in terms of tetrameric structures involving trigonal-prismatic, antiferromagnetically interacting iron atoms arranged in a planar cluster with equivalent bridging OMe groups (56).[279]

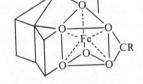

(56) Me groups omitted for clarity

[271] E. V. Margulis, L. A. Savchenko, M. M. Shokarev, L. I. Beisekeeva, and F. I. Vershinina, *Russ. J. Inorg. Chem.*, 1973, **18**, 666.
[272] P. R. Danesi, R. Choarizia, G. Scibona, and R. Riccardi, *Inorg. Chem.*, 1973, **12**, 2089.
[273] S. Gouger and J. Stuehr, *Inorg. Chem.*, 1974, **13**, 379.
[274] L. S. Eschchenko, L. N. Schegrov, V. V. Pechkovskii, and A. B. Ustimovich, *Russ. J. Inorg. Chem.*, 1974, **18**, 478.
[275] M. Ronis and F. d'Yvoire, *Bull. Soc. chim. France*, 1974, 78.
[276] F. d'Yvoire and M. Screpel, *Bull. Soc. chim. France*, 1974, 1211.
[277] F. d'Yvoire, M. Ronis, and H. Guerin, *Bull. Soc. chim. France*, 1974, 1215.
[278] L. R. Nassimbeni and M. M. Thackeray, *Acta Cryst.*, 1974, **B30**, 1072.
[279] E. Kokot, G. M. Mockler, and G. L. Sefton, *Austral. J. Chem.*, 1973, **26**, 2105; *Inorg. Chem.*, 1974, **13**, 1520.

The cations $[Fe_3O(O_2CMe)_6L_3]^+$ (L = py or β-picoline) have been prepared and their electrochemical behaviour has been studied. There is evidence for reduced species in which the oxo-centred triangular array is maintained, but these could not be isolated.[106]

$(EtO)_3PO$ reacts with iron(III) halides at 100—250°C to give $Fe[OOP(OEt)_2]_3$,[86] and further characterization of $Fe(OPPh_3)_4(ClO_4)_3$ has been reported.[87] The complexes $[Fe(nta)_2OR]_2$ (nta = naphthoyltrifluoroacetone; R = Me, Et, or Pr) have been prepared and, on the basis of magnetic and spectroscopic studies, a dimeric structure is proposed which agrees well with their antiferromagnetic behaviour.[280] The ligands o-$O_2NC_6H_4CO_2Me$ and o-$O_2NC_6H_4OMe$ (LH) form the complexes FeL_3, which have tentatively been assigned *trans*-structures.[281] The complexes $Fe(MHTP - H)_3$ and $Fe(MHTP - H)_3[MHTP = (57)]$ have also been prepared.[282]

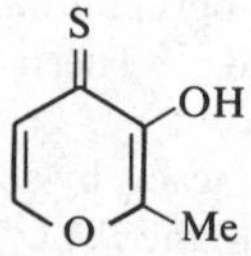

(57)

The solution chemistry of ferric fructose has been examined at various $OH^-:Fe^{3+}$ ratios and the reactivity towards transferrin, the serum iron-binding protein, studied. The implications of the results towards a mechanism by which sugars facilitate the absorption of dietary iron have been discussed.[283]

S-Donor ligands. The structure of $Fe(S_2CNEt_2)_3$ has been determined at 297 and 79 K in order to obtain a correlation with magnetic data ($\mu_{eff} = 4.3$ and 2.2 BM at 297 and 79 K, respectively). At room temperature the metal atoms are located on a pseudo two-fold axis which becomes a true two-fold axis at low temperature. There is a contraction of the Fe—S bond lengths of 0.05 Å and a small, but significant, decrease in the distortion from octahedral symmetry of the $[FeS_6]$ core at low temperature. When the mean Fe—S bond lengths of all iron(III) dithiocarbamate complexes are plotted against μ_{eff}, a smooth curve is obtained, indicating a strong correlation between geometry and magnetic behaviour. Analysis of the temperature parameters at room temperature indicates that the present data cannot, however, distinguish between a mixed-spin state and the existence of a mixture of two different spin states in the crystal lattice as an explanation of the variability of magnetic moments.[285] The dynamic stereochemistries of $M(dtc)_3$ (M = Fe, Co, or Rh), $[M(dtc)_3]^+$, and $[Fe(dtc)_2(o\text{-phen})]$ (dtc = dithiocarbamate) complexes has been studied. All the iron complexes are non-rigid and metal-centred rearrangements occur resulting in optical inversion *via* the trigonal twist mechanism. The cobalt complex is also non-rigid, but the mechanism of optical inversion could not be

[280] R. Grobelny, B. Jezowska-Trzebiatowska, B. Modraz, and Z. Olejnik, *Bull. Acad. polon. Sci., Ser. Sci. chim.*, 1973, **21**, 381.

[281] R. Astolfi, I. Collamti, and C. Ercolani, *J.C.S. Dalton*, 1973, 2238.

[282] E. Uhlemann, H. Motzny, and G. Wilke, *Z. Anorg. Chem.*, 1973, **401**, 255.

[283] G. Bates, J. Hegenauer, J. Renner, P. Saltmann, and T. Spiro, *Bio-inorg. Chem.*, 1973, **2**, 311.

[284] J. G. Leipoldt and P. Coppens, *Inorg. Chem.*, 1973, **12**, 2269.

[285] M. C. Palazzotto, D. J. Duffy, B. L. Edgar, L. Que, and L. H. Pignolet, *J. Amer. Chem. Soc.*, 1973, **95**, 4537.

determined; however, the trigonal twist mechanism is the most likely. The rhodium complex is stereochemically rigid up to 200 °C. The overall metal ion dependence on the rate of optical inversion *via* the trigonal twist mechanism is Fe^{II} ($S = 2$) > Fe^{III} ($S = \frac{1}{2} \rightleftharpoons \frac{5}{2}$) ~ Fe^{IV} ($S = 1$) > Co^{III} ($S = 0$) $\gg$ Rh^{III} ($S = 0$) and within the iron(III) group of complexes, those that are 'more high-spin' generally rearrange faster.[285] Reaction of $FeCl_3$ with $Me_2Sn(SSeCNR_2)$ (R = Me or Et) yields $Fe(SSeCNR_2)_2Cl$ not the tris-complex. Square-pyramidal structures are suggested for these molecules with electron distributions of $(d_{xy})^2$, $(d_{xz}, d_{yz})^2 (d_{z^2})^1 (d_{x^2-y^2})^0$.[286]

Previous studies have shown the mixed complexes, $[M(S_2CNR^1R^2)_2(S_2C_2R_2^3)]^z$ to be stereochemically non-rigid in solution and to undergo inversion of molecular configuration and C—N bond rotation. Singlet–triplet spin equilibria and the redox capacity of the neutral iron complexes with $R^1R^2 = (CH_2)_5$, R = CF_3 (tfd), and $R^1 = R^2 = Et$, $R^3 = CF_3$ or CN (mnt) have now been reported. In dichloromethane, $[Fe(S_2CNEt_2)_2mnt]$ has a temperature-independent magnetic moment of 2.47 BM which is interpreted in terms of a 75% population of the triplet state ($\mu_{trip} = 2.85$ BM). Solid-state susceptibilities for the three compounds approach zero at low temperatures and for the two diethyldithiocarbamate complexes approach maxima at 240 K (mnt complex) and 280 K (tfd complex), consistent with thermal distribution over a singlet ground state and a low-lying triplet state. $[Fe(S_2CNEt_2)_2$-mnt]$^-$ has a doublet ground state and shows Curie–Weiss behaviour below *ca.* 170 K. Polarography reveals a four-membered series with $z = -2, -1, 0$, and $+1$ for both iron and ruthenium complexes, the steps being reversible in the case of ruthenium. Mössbauer data have also been reported for the iron complexes and this, together with other spectroscopic evidence, suggests the oxidation $z = -1$ to $z = 0$ decreases the electron density at the metal atom, and that the $z = 0$ complex can be represented as the resonance hybrid:

$$[(R^1R^2NCS_2^-)_2Fe^{III}(S_2C_2R_2^{\cdot-})] \leftrightarrow [(R^1R^2NCS_2^-)_2Fe^{IV}(S_2C_2R_2^-)]$$

in which the iron(IV) contribution is significant.[287]

The kinetics of substitution of $\{Fe[S_2C_2(CN)_2]_2X\}^-$ (X = a phosphine or phosphite) have been reported and the X ligand is replaced by uni- and bi-dentate nucleophiles such as other phosphines by a process consisting of both associative and dissociative pathways, with the former predominating. A purely dissociative route is definitely excluded. However, replacement of (LL) in $\{Fe[S_2C_2(CN)_2]_2(LL)\}^-$ (LL = bipy, *o*-phen, $Ph_2PCH_2CH_2PPh_2$, *cis*-$Ph_2PCH=CHPPh_2$) by PBu_3^n to yield $\{Fe[S_2C_2(CN)_2]_2PBu_3^n\}^-$ occurs by a dissociative process.[288] Primary, secondary, tertiary, and heterocyclic uni- and bi-dentate amines form adducts with $[Fe(mnt)_2]_2^{2-}$ and $[Fe(tfd)_2]_2^{2-}$. With unidentate amines the equilibrium:

$$[Fe(mnt)_2]_2^{2-} + 2L \rightleftharpoons 2[Fe(mnt)_2L]^-$$

is set up and the five-co-ordinate adducts are high-spin. No bis-adducts, $[Fe(mnt)_2L_2]^-$ were formed in the case of iron; however, the analogous cobalt complex forms both $[Co(mnt)_2L_n]$ ($n = 1$ and 2).[289]

[286] S. Nakajima and T. Tanaka, *Bull. Chem. Soc. Japan*, 1974, **47**, 763.

[287] L. H. Pignolet, G. S. Patterson, J. F. Weiher, and R. H. Holm, *Inorg. Chem.*, 1974, **13**, 1263.

[288] D. A. Sweigart, *Inorg. Chim. Acta*, 1974, **8**, 317.

[289] I. G. Dance and T. R. Miller, *Inorg. Chem.*, 1974, **13**, 525.

Mixed donor ligands. The structure of μ-oxo-bis-[bis-(N-p-chlorophenylsalicyl-aldiminato)iron(III)] has been determined. The co-ordination of the metal atoms is intermediate between trigonal-bipyramidal (two apical N atoms) and square-pyramidal (bridging O atom apical) the Fe—O(oxo) bond lengths are 1.76 Å and ∠FeOFe = 175°. This angle appears to depend primarily on the configuration of the Schiff base ligands[290] and this is borne out by similar studies on {[Fe(salen)]$_2$O} which has almost identical Fe—O(oxo) bond lengths (1.78 Å) but ∠FeOFe = 145°.[291] The structures of [Fe(salpa)Cl] and [Fe(sane)$_2$Cl] [salpa = N-(3-hydroxy-propyl)salicylaldiminate, sane = N-(2-phenylethyl)salicylaldiminate] have also been determined.[292] The former is dimeric with five-co-ordinate iron, intermediate between trigonal-bipyramidal and square-pyramidal, and magnetic studies show the presence of antiferromagnetic coupling. The latter is monomeric, but also shows five-co-ordination of intermediate geometry. Above 196 K, this complex has a magnetic moment typical of high-spin iron(III), however, below 140 K, the magnetic moment is greatly reduced and this is possibly due to dimer formation at low temperature. [Fe(salpa)Cl]$_2$ reacts with Na$_2$O$_2$ to give the red antiferromagnetic complex, [Fe$_2$(salpa)$_2$(salpaH)$_2$] (μ_{eff} per iron = 4.45 and 2.34 BM at 298 and 77 K, respectively). A structural study of the toluene solvate of the complex shows the two six-co-ordinate iron atoms are linked by bridging propoxide groups into dinuclear units containing a planar $\overline{Fe—O—Fe—O}$ ring. Hydrogen bonding links adjacent units into infinite doubly-linked chains. The electronic spectra of a range of other five- and six-co-ordinate iron(III)-Schiff base complexes were also reported in this paper and, whereas the five-co-ordinate complexes show two bands in the visible region, those with six-co-ordination show four such bands.[293]

The potentially quinquedentate ligands (58) have been prepared and the complexes [Fe(L − 2H)Cl] obtained, in which the chloride ion is definitely co-ordinated. However, it was not possible to make a definite structural assignment to either five- or six-co-ordination.[294] The related ligand (59) also forms an [Fe(L − 2H)Cl]

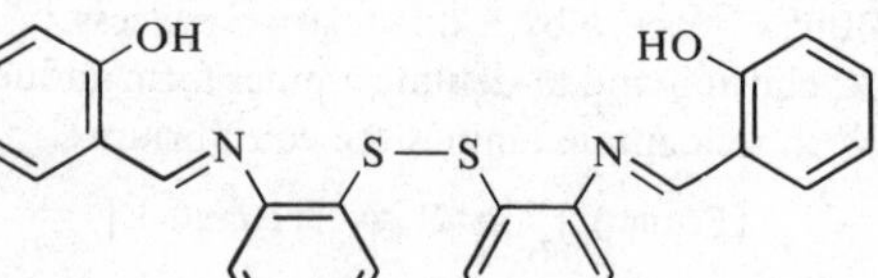

(58) n = 2, X = S or NH;
n = 3 X = NH

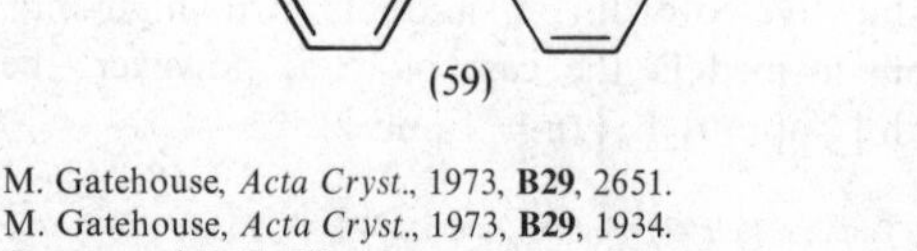

(59)

[290] J. E. Davies and B. M. Gatehouse, *Acta Cryst.*, 1973, **B29**, 2651.
[291] J. E. Davies and B. M. Gatehouse, *Acta Cryst.*, 1973, **B29**, 1934.
[292] J. A. Bertrand, J. L. Breece, and P. G. Eller, *Inorg. Chem.*, 1974, **13**, 125.
[293] J. A. Bertrand and P. G. Eller, *Inorg. Chem.*, 1974, **13**, 927.
[294] R. D. Patton and L. T. Taylor, *Inorg. Chim. Acta*, 1973, **7**, 191.

complex. In this case, structural studies showed the complex to be octahedral, the ligand functioning as an $[N_2O_2S]$donor.[295]

The mono-N-oxides, bipyNO and phenNO(L), form the complexes $[FeL_3](ClO_4)_3$ in which the ligands are bidentate. The ligand, bipyNO, exists in the *trans* form in the crystalline state, but assumes a configuration close to the *cis*-form on complex formation. Free phenNO is held rigidly in the *cis*-form and retains this configuration in complexes. *Dq* values show both ligands are weaker donors than their parent nitrogen bases, but are stronger than the bis-N-oxides, and both complexes exhibit temperature-dependent magnetic moments (3.02—3.67 BM for bipyNO complex and 1.98—2.52 BM for phenNO complex between 80 and 313 K) which are attributed to spin-free–spin-paired equilibria.[296] $[Fe(Hedta)]$ (Hedta = N-hydroxy-edta anion) reacts with H_2S and mercaptans to give unstable pink solutions. With H_2S a 2:1 complex Fe(Hedta): H_2S is formed, but in the cases of mercaptoethanol, ethylmercaptan, and L-cysteine, 1:1 complexes are obtained,[297] The photochemical decomposition of the Fe^{III}–edta complex has been studied.[298]

$PhC(CO_2H)=N \cdot NH \cdot C(Se) \cdot NH_2$ (H_2sesap) reacts with iron(III) salts in ammonical solution to give $NH_4[Fe(sesap)],3.5H_2O$. Similar complexes can be obtained by aerial oxidation of weakly alkaline iron(II)-H_2sesap solutions, however in a slightly acid medium, the iron(II) complex, $[Fe(Hsesap)_2],2H_2O$ is formed.[299] The anion $Fe(S_2C_2O_2)_3{}^{3-}$ has been prepared and contains S-bonded ligands. On treatment with $[(Ph_3P)_2M^1]^+$ (M^1 = Cu or Ag) cations, the complexes $\{Fe(S_2C_2O_2)_3-[M^1(PPh_3)_2]_3\}$ are obtained. These contain $Fe-O_2-C_2-S_2-M^1$ ligand bridging and the linkage isomerism has been confirmed by X-ray structural studies.[300] At low pH, iron(III) bonds to the ligands digly-NN-diacetic acid, trigly-NN-diacetic acid and tetragly-NN-diacetic acid by the diacetate functions and the complexes precipitate from solution. This is in contrast to Ni^{2+}, Cu^{2+}, and Co^{2+} complexes which remain in solution and are co-ordinated to the amide nitrogen atom of the peptide moieties.[301] Cytostatic hadacidin, N-formyl-N-hydroxylglycine (had), forms the complexes $Fe_2(had)_3, 3H_2O$ and $Fe(had - H)_2(acac),0.5H_2O^{94}$ and $O=CNHCH_2-CH_2SCH_2$ (tm) forms the four-co-ordinate complexes $[Fetm_2Cl_2]Cl$ and $[Fetm_3Br]-Br_2$.[240]

Iron(IV).—Tris-(diethyldithiocarbamato)iron(III) undergoes electrochemical oxidation in acetonitrile–benzene in the presence of Et_4NClO_4 to yield the $[Fe(S_2CNEt_2)_3]^+$ cation, which was isolated as its BF_4^- salt. This preparative method appears to be very satisfactory for such cations, as the product is free from impurities.[302] The structure of the related complex $[Fe(pyrol.dtc)_3](ClO_4)$ (pyrol.dtc = pyrolidyldithiocarbamate) has been reported. The $[FeS_6]$ core is intermediate between octahedral and trigonal-prismatic geometry (twist angle = 38°) and the Fe—S bond lengths are significantly shorter than those of related high-spin iron(III)

[295] J. A. Bertrand and J. L. Breece, *Inorg. Chim. Acta*, 1974, **8**, 267.
[296] A. N. Speca, N. M. Karayannis, and L. L. Pytlewski, *Inorg. Chim. Acta*, 1974, **9**, 87.
[297] C. V. Philips and D. W. Brooks, *Inorg. Chem.*, 1974, **13**, 384.
[298] J. H. Carey and C. H. Langford, *Canad. J. Chem.*, 1973, **51**, 3665.
[299] N. V. Gerbeleu and V. G. Bodyu, *Russ. J. Inorg. Chem.*, 1973, **18**, 1596.
[300] D. Coucouvanis and D. Piltingsried, *J. Amer. Chem. Soc.*, 1973, **95**, 5556.
[301] R. J. Motekaitis and A. E. Martell, *Inorg. Chem.*, 1974, **13**, 550.
[302] G. Cauquis and D. Lachenal, *Inorg. Nuclear Chem. Letters*, 1973, **9**, 1095.

H

complexes, which is consistent with depopulation of the antibonding e_g orbitals (*i.e.* $Fe^{III}t_{2g}{}^3e_g{}^2 \rightarrow Fe^{IV}t_2{}^4$) and the higher oxidation state. However, the former effect appears to dominate bond-length considerations, as there is little difference between those of iron(IV) and related low-spin iron(III) complexes.[303] $K_2(ded)$ [ded^{2-} = $(EtO_2C)_2CS_2{}^{2-}$] reacts with $Fe(ClO_4)_3,H_2O$ in water, followed by addition of the cation $[PhCH_2PPh_3]^+$ to yield $[PhCH_2PPh_3]_2[Fe(ded)_3]$ ($\mu_{eff} = 2.92$ BM). Structural studies again show the presence of distorted octahedral co-ordination (twist angle = 35.9°). Structural parameters of the ligands indicate that inter- or intra-ligand oxidation cannot adequately explain the formation of the anion, and comparison with $[Ni(ded)_2]^{2-}$ shows that charge delocalization in the ligand brought about by the metal ion is insensitive to formal oxidation state.[304] The co-ordination geometry of $(Ph_4As)_2[Fe^{IV}\{(S_2C_2(CN)_2\}_3]$ is described as trigonal-antiprismatic, closely approximating to octahedral. For any pair of *trans* sulphur atoms, $\angle$ SFeS is 171°.[305]

Iron(II) chloride reacts with $o\text{-}C_6H_4(PMe_2)_2$ (L) to form $[FeCl_2L_2]$, which on treatment with ferric chloride yields $[FeCl_2L_2]^+$ (low spin). This cation undergoes a reversible one-electron oxidation in acetonitrile, giving $[FeCl_2L_2]^{2+}$, isolated as its perrhenate salt but highly unstable ($\mu_{eff} = 3.65$ BM).[115]

Model Compounds for Iron in Biological Systems.—There have been a number of publications concerned mainly with the nature of iron–sulphur-containing proteins, nitrogen fixation, and oxygen transport. Most of these are concerned with studies on cluster compounds of the type $[Fe_4S_4(SR)_4]^{2-}$.

Various spectroscopic techniques have been employed in a study of the electronic states in $[Fe_4S_4(SR)_4]^{2-}$ anions and it has been concluded that they possess fully delocalized $[Fe_4S_4]$ cores. However, the results do not necessarily establish the presence of fractional valency states for the metal atoms and the possibility of the presence of four iron(III) ions and two delocalized ligand-based electrons could not be excluded.[306] The series of complexes $\{Fe_4S_4[S_2C_2(CN)_2]_4\}^{n-}$ (n = 0, 1, or 2) has also been studied. Structural investigations for the dianion shows it to contain a distorted $[Fe_4S_4]$ cube, and unperturbed Mössbauer spectra are consistent with four equivalent iron atoms. The overall electronic structure is very similar in all three cases, although the neutral complex and the dianion are diamagnetic while the monoanion is paramagnetic. Magnetically perturbed Mössbauer spectra suggest that in this last case the electron is highly delocalized over the dithiolene ligands, not primarily associated with the $[Fe_4S_4]$ core.[307] The spectral and redox properties of $[Fe_4S_4(SR)_4]^{2-}$ (R = alkyl, aryl, or Ac–Cys–NHMe) have also been examined in both non-aqueous and aqueous DMSO. $[Fe_4S_4\{S\text{-}Cys(Ac)NHMe\}_4]^{2-}$ and $[Fe_4S_4(SC_6H_4NMe_3)_4]^{2+}$ were generated in solution by ligand substitution reactions of $[Fe_4S_4(SBu^t)_4]^{2-}$. The alkyl-substituted tetramers possess two principal absorption bands which are similar to those in reduced 'high-potential' and oxidized ferredoxin (HP_{red} and Fd_{ox}, respectively). The maxima for the aryl-substituted tetramers occur

[303] R. L. Martin, N. M. Rohde, G. B. Robertson, and D. Taylor, *J. Amer. Chem. Soc.*, 1974, **96**, 3647.
[304] F. J. Hollander, R. Pedelty, and D. Coucouvanis, *J. Amer. Chem. Soc.*, 1974, **96**, 4032.
[305] A. Sequeira and I. Bernal, *J. Cryst. Mol. Structure*, 1973, **3**, 157.
[306] R. H. Holm, B. A. Averill, T. Herskovitz, R. B. Frankel, H. B. Gray, O. Siiman, and F. J. Grunthaner, *J. Amer. Chem. Soc.*, 1974, **96**, 2644.
[307] R. B. Frankel, W. M. Reiff, I. Bernal, and M. L. Good, *Inorg. Chem.*, 1974, **13**, 493.

at different positions. Polarography shows the presence of $[Fe_4S_4(SR)_4]^z$ ($z = 4-$, $3-$, $2-$, or $1-$), and the $2-$ to $3-$ process is reversible in nearly all cases. The spectral and polarographic results in aqueous DMSO, DMF, and non-aqueous DMSO are very similar to those for both HP_{red} and Fd_{ox}, suggesting that the protein structure and environment make a significant contribution to the redox potentials of the Fe—S clusters.[308]

The ligand substitution reactions:

$$[Fe_4S_4(SBu^t)_4]^{2-} + nRSH \rightleftharpoons [Fe_4S_4(SBu^t)_{4-n}(SR)_n]^{2-} + nBu^tSH$$

have also been studied (R = p-tolyl, N-acetyl-L-cysteinemethylamide, or $1,4-C_6H_4SH$) and, in the last case, the bridged anion $[(Bu^tS)_3Fe_4S_4—SC_6H_4S—Fe_4S_4(SBu^t)_3]^{4-}$ was tentatively identified. There appears to be no disruption of the $[Fe_4S_4]$ core on ligand substitution.[309] In a second paper, the authors reported detection of the ligand-substituted clusters $[Fe_4S_4(SBu^t)_{4-n}(SR)_n]^{2-}$ ($n = 1$—4) and drew up the following order of substitution:

MeCOSH $\sim$ p-$XC_6H_4SH \geqslant$ Ac-L-CysNHMe $\geqslant$ p-Y·$C_6H_4SH >$ PhCH$_2$SH $>$
  HOCH$_2$CH$_2$SH $>$ EtSH $\gg$ p-MeC$_6$H$_4$SH (X = H or NO$_2$; Y = NMe$_2$ or NMe$_3^+$)

The substitution tendencies roughly parallel aqueous acidities, at least up to pK_a ~ 6.5. This paper also includes the preparation of $[Fe_4S_4(SePh)_4]^{2-}$ by a similar ligand substitution procedure employing Ph_2Se in MeCN and reports the results of an X-ray structural study on $(Me_4N)_2[Fe_4S_4(SPh)_4]$. This anion contains an $[Fe_4S_4]$ core of effective D_{2d} symmetry analogous to that of the corresponding PhCH$_2$S tetramer. The principal difference between the two structures is in the detailed geometry of the Fe$_4$ portion, which in the present study more closely approaches T_d symmetry with Fe$\cdots$Fe distances of 2.736 Å (average). The Fe—S (core sulphur) bond lengths occur in sets of four at 2.267 Å and eight at 2.286 Å.[310] ^{1}H n.m.r. spectra of $[Fe_4S_4(SR)_4]^{2-}$ (R = Me, Et, Pr^n, Pr^i, $C_6H_{11}CH_2$, PhCH$_2$, m-C$_6$H$_4$(CH$_2$)$_2$, Bu^t, Ph, p-tolyl, p-NO$_2$C$_6$H$_4$) have been investigated in order to probe the electronic properties of the $[Fe_4S_4]$ core as manifested in such spectra. The magnitudes of CH$_2$—S chemical shifts are very similar to those of the HP_{red} and Fd_{ox} proteins and substantiate previous assignments to CH$_2$ protons of cysteinyl residues bound to iron. The results indicate that the shifts are dominated by contact interactions. The complex with R = m-C$_6$H$_4$(CH$_2$)$_2$ is thought to have the structure (60) in which the two phenyl rings are held approximately parallel to cluster faces, similar to the orientation of aromatic residues in 8-Fe ferredoxins. The ring-proton shifts in this case differ from those of the benzyl derivative, in which the rings have no fixed orientation.[311]

An important paper has reported that $[Fe_4S_4(SEt)_4]^{3-}$ reacts with $[(N_2)Mo-(Ph_2PCH_2CH_2PPh_2)_2(^{15}N_2)]$ in the presence of sodium fluoranthene to give 0.012 moles of $^{15}NH_3$ per mole of reagent. The authors suggest that this is a good

[308] B. V. DePamphilis, B. A. Averill, T. Herskovitz, L. Que, and R. H. Holm, *J. Amer. Chem. Soc.*, 1974, **96**, 4159.
[309] M. A. Bobrik, L. Que, and R. H. Holm, *J. Amer. Chem. Soc.*, 1974, **96**, 285.
[310] L. Que, M. A. Bobrik, J. A. Ibers, and R. H. Holm, *J. Amer. Chem. Soc.*, 1974, **96**, 4168.
[311] R. H. Holm, W. D. Phillips, B. A. Averill, J. J. Mayerle, and T. Herskovitz, *J. Amer. Chem. Soc.*, 1974, **96**, 2109.

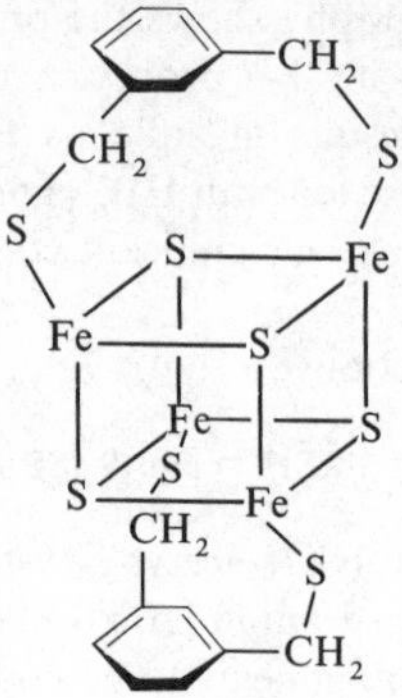

(60)

model for nitrogen fixation.[312] The possibility of nucleophilic sulphur as a site for the activation and reduction of complexed dinitrogen has also been raised by Sutton (see ref. 200). A verbal report by van Tamelen[313] that treatment of *trans*-$[Mo(N_2)_2$-$(Ph_2PCH_2CH_2PPh_2)_2]$ with $[Fe_4S_4(SEt)_4]^{2-}$, followed by acidification, also yields ammonia has been questioned by Chatt[314] who found no evidence for ammonia formation and reports the main products of the reaction as $[MoCl_2(Ph_2PCH_2CH_2$-$PPh_2)_2]^+$, H_2S, EtSH, H_2, iron chlorides, and N_2.[314]

$[(h^5$-Cp)Fe(SEt)$_2$S] has been studied by X-ray crystallography. It adopts structure (61) in which the Fe—S—S—Fe bridge is planar. On the basis of the structural evidence, the oxidation state of the disulphide bridge is between 0 and 2−. The authors suggest that such a structure might be a more acceptable model for plant ferredoxins than (62) and suggest (63) and (64) as possible structures. Structure (64) requires an extra cysteinyl residue, which is present in all plant ferredoxins except *Equisetum* ferredoxin.[315]

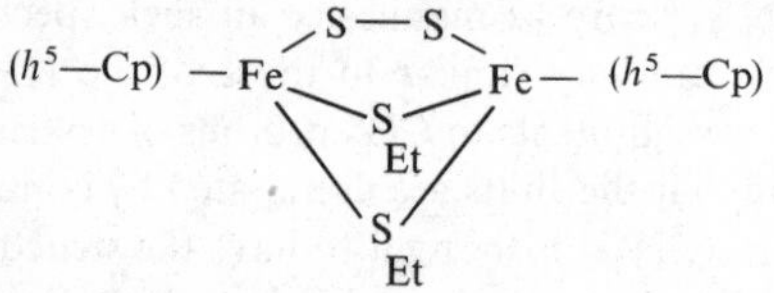

(61)

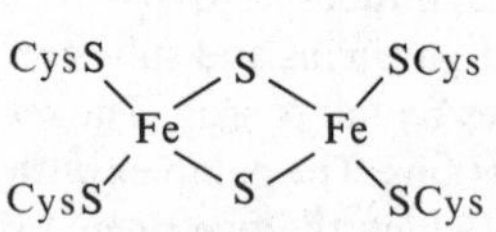

(62)

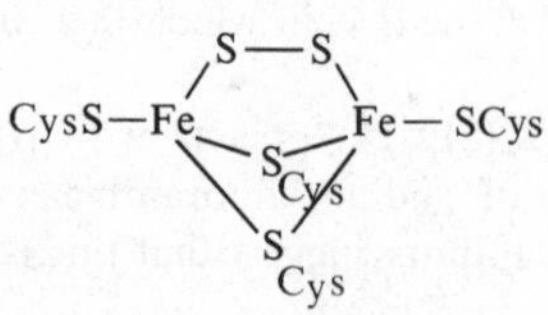

(63)

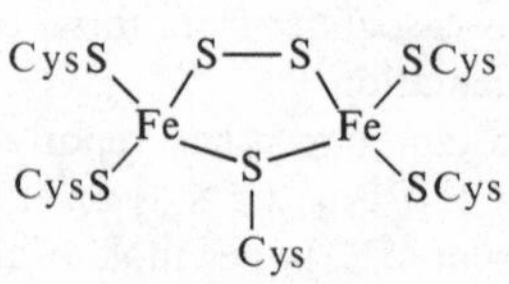

(64)

[312] E. E. van Tamelen, J. A. Gladysz, and C. R. Brulet, *J. Amer. Chem. Soc.*, 1974, **96**, 3020.
[313] E. E. van Tamelen, *Abstract of 24th Congress of IUPAC*, Hamburg, 1973, p. 152.
[314] J. Chatt, C. M. Elson, and R. L. Richards, *J.C.S. Chem. Comm.*, 1974, 189.
[315] A. Terzis and R. Rivest, *Inorg. Chem.*, 1973, **12**, 2132.

Following previous work on the Raman spectra of rubredoxin and adrenodoxin, Scovell and Spiro have reported a detailed study of the vibrational spectrum of $Fe_2S_2(CO)_6$ (*syn-* and *anti*-isomers), which they suggest may be used as a model compound.[316]

Two papers have discussed the nature of hydroxy-iron(III) polymers and their relevance as analogues of ferritin cores. In the first of these, the ageing of hydrolytic polymers of Fe^{III}, analogous to the cores of ferritin, has been monitored by the increase in light absorption, sedimentation velocity, and resistance to acid attack. Two concurrent, but distinct, processes are observed which can be associated with structural changes and growth *via* particle accretion.[317] The structure of the iron-containing polymer produced from bicarbonate hydrolysis of $Fe(NO_3)_3$ solutions (known as the Spiro–Saltmann Ball) has been in dispute, the suggested structures involving either tetrahedral $[FeO_2(OH)_2]$ units or octahedral iron(III). The preparation and structure of the related compound $\{[Fe(alanine)_2(H_2O)]_3O\}(ClO_4)_7$ have now been reported. It contains a trinuclear cation in which the oxygen atom lies at the centre of a plane of iron atoms. Each pair of iron atoms is bridged by two alanine molecules, and a water molecule completes six-co-ordination at each metal. The spectral parameters are very similar to those of both the Spiro–Saltmann Ball and ferritin, suggesting octahedral co-ordination in both these cases and the authors suggest that the $(Fe_3O)^{7+}$ unit must be considered as a strong candidate for the basic 'monomeric' unit in both compounds.[318]

Attempts to simulate oxygenation of haemoglobin have previously been thwarted by reaction (b) in the scheme:

$$Fe^{II} + O_2 \xrightleftharpoons{(a)} Fe^{II}\!-\!O_2 + Fe^{III} \xrightarrow{(b)} Fe^{II}\!-\!O\!-\!O\!-\!Fe^{II} \longrightarrow Fe^{III}$$

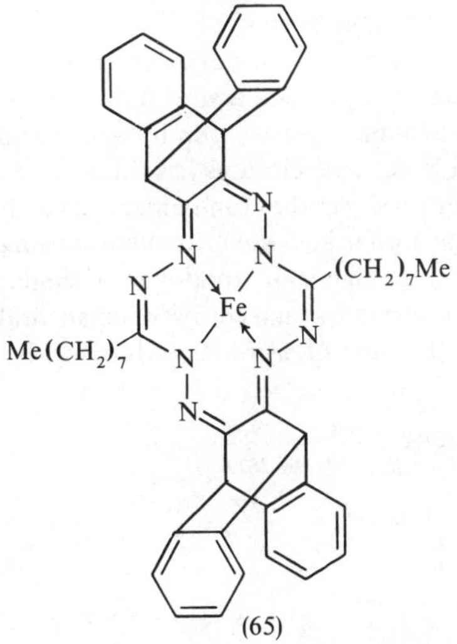

(65)

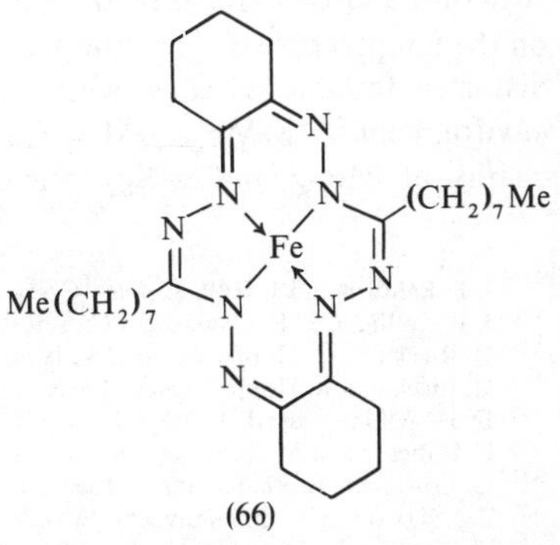

(66)

[316] W. M. Scovell and T. G. Spiro, *Inorg. Chem.*, 1974, **13**, 304.
[317] B. A. Sommer, D. W. Margerum, J. Renner, and P. Saltmann, *Bio-inorg. Chem.*, 1973, **2**, 295.
[318] E. M. Holt, S. L. Holt, W. F. Tucker, R. O. Asplund, and K. J. Watson, *J. Amer. Chem. Soc.*, 1974, **96**, 2621.

However, a complex (65) has now been synthesized which has a ligand capable of effectively blocking (b) and this indeed shows a reversible 1 : 1 uptake of oxygen. In the case of the less hindered complex (66), oxygen uptake is irreversible.[319] Similarly, the 'picket-fence' porphyrin complex $Fe(\alpha,\alpha,\alpha,\alpha\text{-TpivPP}),THF$ rapidly oxygenates in air to give a 90% yield of the paramagnetic complex $Fe(\alpha,\alpha,\alpha,\alpha\text{-TpivPP}),O_2$. The oxygenation was done by a gas–solid reaction and both the starting complex and the adduct have the same cell parameters.[320]

Oxides and Sulphides.—The first oxoferrate(II), $Na_3[FeO_3]$, has been prepared as a red powder from the reaction of Na_2O and FeO.[321] $K_6[Fe^{III}_2O_6]$ has been obtained from the potassium–iron–oxygen system by heating intimate mixtures of $KO_{0.56}$ and FeO to 600 °C for seven days. The compound contains discrete $[Fe_2O_6]^{6-}$ anions with a structure similar to that of Fe_2Cl_6.[322] The preparations and properties of $[Fe^{VI}O_4]^{2-}$ have been reviewed and a shortened preparative method reported. The paper also contains a study of the oxidation of polyhydridic alcohols by $[FeO_4]^{2-}$.[323] $BaTi_2Fe_4O_{11}$ (R-block) has been prepared from the R-layer of $BaFe_{12}O_{19}$ by substituting Ti^{4+} ions for Fe^{3+}. The ceramic reactions of the product were studied by d.t.a., thermal gravimetry, and X-ray diffraction.[324] The structures and electronic properties of solid solutions of $Zn(Fe,V)O_4$ with Fe_2O_3 and Zn_2VO_4 have been studied. The systems produced may be described by the general formulae, $Zn_{2-x}Fe^{3+}_x(Fe^{2+}_xFe^{3+}_{1.0}V^{3+}_{1-x})O_4$ and $Zn(Zn_yFe^{3+}_{1-y}V^{3+}_{1-y}V^{4+}_y)O_4$ in the ranges, $0 \leqslant x \leqslant 1$ and $0 \leqslant y < 0.5$.[325]

Iron metaniobates(IV) with structures of the natural minerals, ilmenite and pseudobrookite have been prepared by direct reaction of the oxides under vacuum at 1000—1100 °C. The compounds are stable in air up to 500 °C, but are oxidized at higher temperatures.[326] $\alpha\text{-}Sr_3FeO_{7-x}$ has been found to be isostructural with Sr_3TiO_7.[327] Synthesis and thermal decomposition of iron(III) normal selenite mono- or tri-hydrate, $Fe_2O_3,3SeO_2,xH_2O$ ($x = 1$ or 3) have been reported.[328]

Iron is incorporated into crystals of Ga_2S_3 by chemical transport with HCl. Two product phases are formed and then can be separated by careful temperature control. The material deposited at higher temperatures contains 5—10% iron by weight and has the $\gamma\text{-}Ga_2S_3$ structure and that obtained at lower temperatures has the $\alpha\text{-}Ga_2S_3$ structure and contains 0.1—0.5% iron. In both products the iron content depends on the temperature of deposition and most of the iron is high-spin iron(II) occupying distorted tetrahedral sites, while the remainder is high-spin iron(III) in a similar environment.[329] MMo_2S_4 (M = Co or Fe) are isostructural and consist of octahedral chains of MoS_6 and MS_6 units.[330] The structure of synthetic Mooihoekite,

[319] J. F. Baldwin and J. Huff, *J. Amer. Chem. Soc.*, 1973, **95**, 5757.
[320] J. P. Collman, R. R. Gagne, and C. A. Reed, *J. Amer. Chem. Soc.*, 1974, **96**, 2629.
[321] H. Rieck and R. Hoppe, *Naturwiss.*, 1974, **61**, 126.
[322] H. Rieck and R. Hoppe, *Angew. Chem. Internat. Edn.*, 1973, **12**, 673.
[323] D. H. Williams and J. T. Riley, *Inorg. Chim. Acta*, 1974, **8**, 177.
[324] F. Haberey and M. Velicescu, *Acta Cryst.*, 1974, **B30**, 1507.
[325] G. Colsmann, B. Reuter, and E. Riedel, *Z. anorg. Chem.*, 1973, **401**, 41.
[326] E. I. Krylov, G. C. Kasimov, and E. G. Vovkotrub, *Russ. J. Inorg. Chem.*, 1973, **18**, 287.
[327] E. Lucchini, D. Minichelli, and G. Sloccari, *Acta Cryst.*, 1973, **B29**, 2356.
[328] V. P. Volkova, G. F. Pinaev, and V. V. Pechkovskii, *Russ. J. Inorg. Chem.*, 1973, **18**, 309.
[329] J. V. Pivnichny and H. H. Brintzinger, *Inorg. Chem.*, 1973, **12**, 2839.
[330] J. Guillevic, J. Y. Le Marouille, and D. Grandjean, *Acta Cryst.*, 1974, **B30**, 111.

$Cu_9Fe_9S_{16}$, has been reported,[331] and $La_{32.66}Fe_{11}S_{60}$ has been found to contain 12 non-equivalent lanthanon sites, four of which are not completely filled, involving seven- or eight-co-ordination and four iron sites with octahedral co-ordination. Two other iron sites have occupancies of 0.5.[121] $Fe_2P_2X_6$ (X = S or Se) form layer lattices with double layers of X atoms. Iron atoms and pairs of phosphorus atoms occupy octahedral sites.[332]

Other Papers.—Various iron species prepared by the vacuum pyrolysis of acetyl-ferrocene–furfural resins at 400 °C have been studied by Mössbauer spectroscopy. These consist of an amorphous glass-like carbon matrix containing free iron atoms, Fe^+ ions, iron clusters, superparamagnetic iron, and ferromagnetic iron.[333] The effect of pressure of up to 50 kbar on the absorption spectra of five iron(III), two iron(II) and one mixed valence compound has been studied. In six of the compounds, but not in basic ferric acetate or soluble Prussian Blue, the observed pressure-induced bands were assigned to d–d transitions of converted iron(II) for the ferric compounds and to spin-forbidden d–d bands for the ferrous compounds. The charge-transfer band from iron(II) to iron(III) in soluble Prussian Blue showed a blue shift at pressures up to 7.2 kbar.[334]

3 Formation and Stability Constants

	Ref.
Manganese(II)	
Ethylenediamine and glycine	335
8-*N*-Arylhydroxamic acids	336
Glucose-6-phosphate	337
Uracil, thymine, or cytosine	338
2,4-Dihydroxyacetophenoneoxime	339
$[Co(NH_3)_5(ox)]^+$	340
Ethylidenetetrathioacetate	340*a*
Picolinic acid and styrene-vinylpicolinic acid copolymer	340*b*
Manganese(III)	
$HClO_4$ and H_2SO_4	340*c*
Iron(II)	
Thiocyanate	341
2-(2'-Pyridyl)-imidazole	342
N-(2-Acetylindane-1,3- dione)trialkylammonium iodide	343

[331] S. R. Hall and J. F. Rowland, *Acta Cryst.*, 1973, **B29**, 2365.

[332] W. Klingen, G. Eulenberger, and H. Hahn, *Z. anorg. Chem.*, 1973, **401**, 97.

[333] S. Yakima and M. Omori, *Chem. Letters*, 1974, 277.

[334] Y. Hara, I. Shirotani, N. Sakai, S. Nagakura, and S. Minomura, *Bull. Chem. Soc. Japan*. 1974, **47**, 434.

[335] K. T. Mui, W. A. E. McBryde, and E. Nieboer, *Canad. J. Chem.*, 1974, **52**, 1821.

[336] Y. K. Agrawal and S. G. Tandon, *J. Inorg. Nuclear Chem.*, 1974, **36**, 859.

[337] M. Asso and D. Benlian, *Compt. rend.*, 1974, **278**, C, 1373.

[338] M. M. Taqui Khan and C. R. Krishnamoorthy, *J. Inorg. Nuclear Chem.*, 1974, **36**, 711.

[339] V. Seshagiri and S. B. Rao, *J. Inorg. Nuclear Chem.*, 1974, **36**, 353.

[340] R. K. Nanda and A. C. Dash, *J. Inorg. Nuclear Chem.*, 1974, **36**, 1595.

[340*a*] P. Petras, J. Podlahova, and J. Podlaha, *Coll. Czech. Chem. Comm.*, 1973, **38**, 3221.

[340*b*] R. Paton, E. C. Watton, and L. R. Williams, *Austral. J. Chem.*, 1974, **27**, 1185.

[340*c*] V. P. Goncharik, L. P. Tikhonova, and K. B. Yatsimirskii, *Russ. J. Inorg. Chem.*, 1973, **18**, 658.

[341] S. Misumi, M. Aihara, and S. Kimoshita, *Bull. Chem. Soc. Japan*, 1974, **47**, 127.

[342] R. K. Boggess and R. B. Martin, *Inorg. Chem.*, 1974, **13**, 1525.

[343] M. K. Bachlaus, K. L. Menaria, and P. Nath, *Z. Naturforsch.*, 1973, **28b**, 317.

4-(2'-Pyridylazo)resorcinol	344
Histidine-containing peptides	345
Ethylidenetetrathioacetate	340a

Iron(III)
Diethylenetriaminepenta-acetate	346
Phosphate, fulvate, and carboxylates	347
Perchlorate and phosphate	348
Formate	349
Glycolate	350
Salicylate	351
3,4-(HO)$_2$C$_6$H$_3$CO$_2$H	352
Catechol-3,5-disulphonic acid and edta	353
Nitrilotriacetic acid and phenol derivatives	354
4-(2'-Pyridylazo)resorcinol	344

4 Bibliography

The following relevant reviews have been published:

Preparation and properties of high-valent first row transition-metal oxides and halides. (C. Rosenblum and S. L. Holt, *Transition Metal Chem.*, 1972, **7**, 87).

Higher oxidation state chemistry of iron, cobalt, and nickel. (W. Levason and C. A. McAuliffe, *Coord. Chem. Rev.*, 1974, **12**, 151).

The physical properties of transition-metal compounds of the ABX$_3$ type. (J. F. Ackerman, G. M. Cole, and S. L. Holt, *Inorg. Chim. Acta.*, 1974, **8**, 323).

High-pressure co-ordination chemistry. (E. Sinn, *Coord. Chem. Rev.*, 1974, **12**, 185).

Transition-metal atom inorganic synthesis in matrices. (G. A. Ozin and A. V. Voet, *Accounts Chem. Res.*, 1973, **6**, 313).

DMSO as a solvent and a ligand of metal complexes in inorganic chemistry. (A. Tenhunen, *Suomen Kemi*, 1973, **46**, 147).

Five-co-ordination in iron(II), cobalt(II), and nickel(II) complexes. (R. Morassi, I. Bertini, and L. Sacconi, *Coord. Chem. Rev.*, 1973, **11**, 343).

Cyanide phosphine complexes of transition metals. (P. Rigo and A. Turco, *Coord. Chem. Rev.*, 1974, **13**, 133).

Recent advances in the chemistry of isocyanide complexes. (F. Bonati and G. Minghetti, *Inorg. Chim. Acta*, 1974, **9**, 95).

The chemistry of co-ordinated azides. (Z. Dori and R. F. Zioli, *Chem. Rev.*, 1973, **73**, 247).

The synthesis and ion-binding of synthetic multidentate macrocyclic compounds. (J. J. Christensen, D. J. Eatough, and R. M. Izatt, *Chem. Rev.*, 1974, **74**, 351).

Macrocyclic ligands: complexing properties toward first and second row cations. (C. Kappenstein, *Bull. Soc. chim. France*, 1974, 89).

[344] D. Nonova and B. Evtimova, *J. Inorg. Nuclear Chem.*, 1973, **35**, 3581.
[345] A. Yokoyama, H. Aiba, and H. Tanaka, *Bull. Chem. Soc. Japan*, 1974, **47**, 112.
[346] N. A. Kostromina, N. V. Beloshitskii, and I. A. Sheka, *Russ. J. Inorg. Chem.*, 1973, **18**, 823.
[347] S. Ramamoorthy and P. G. Manning, *Inorg. Nuclear Chem. Letters*, 1974, **10**, 109.
[348] V. I. Sidorenko, E. F. Zhuravlev, V. I. Gordienko, and N. P. Grineva, *Russ. J. Inorg. Chem.*, 1973, **18**, 670.
[349] V. I. Paramonova, V. Y. Zamanskii, and V. B. Kolychev, *Russ. J. Inorg. Chem.*, 1973, **18**, 1132.
[350] S. Krzewska, *Roczniki Chem.*, 1974, **48**, 555.
[351] V. N. Vasil'eva, V. P. Vasil'ev, T. K. Korbut, and L. I. Bukoyazova, *Russ. J. Inorg. Chem.*, 1973, **18**, 974.
[352] P. K. Migal and V. A. Ivanov, *Russ. J. Inorg. Chem.*, 1973, **18**, 536.
[353] S. Koch and G. Ackermann, *Z. anorg. Chem.*, 1973, **400**, 29.
[354] S. Koch and G. Ackermann, *Z. anorg. Chem.*, 1974, **400**, 21.

Metal complexes of sulphur–nitrogen chelating ligands. (M. A. Ali and S. E. Livingstone, *Coord. Chem. Rev.*, 1974, **13**, 101).

Structural chemistry of transition-metal complexes of oximes. (A. Chakravorty, *Coord. Chem. Rev.*, 1974, **13**, 1).

Metal complexes of aromatic amine-*N*-oxides. (N. M. Karayannis, L. L. Pytlewski and C. M. Miκulski, *Coord. Chem. Rev.*, 1973, **11**, 93).

Thermochemistry of metal-polyamine complexes. (P. Paoletti, L. Fabrizzi, and R. Barbucci, *Inorg. Chim. Acta Rev.*, 1973, **7**, 43).

Reactions of co-ordinated pnictogen donor ligands. (C. S. Kraihanzel, *J. Organometallic Chem.*, 1974, **73**, 137).

Stereochemical studies of metal carbonyl–phosphorus trifluoride complexes. (R. J. Clark and M. A. Busch, *Accounts Chem. Res.*, 1973, **6**, 246).

Dinuclear oxo-bridged iron(III) complexes (K. S. Murray, *Coord. Chem. Rev.*, 1974, **12**, 1).

Organoiron complexes as potential reagents in organic synthesis. (M. Rosenblum, *Accounts Chem. Res.*, 1974, **7**, 122).

Nitrosonium salts as reagents in inorganic chemistry. (M. T. Mocella, M. S. Okamoto, and E. K. Barefield, *Syn. React. Inorg. Metal-org. Chem.*, 1974, **4**, 69).

Rapid intramolecular rearrangements in penta-co-ordinate transition metal compounds. (J. R. Shapley and J. A. Osborn, *Accounts Chem. Res.*, 1973, **6**, 305).

The application of *X*-ray photoelectron spectroscopy in inorganic chemistry. (W. L. Jolly, *Coord. Chem. Rev.*, 1974, **13**, 47).

Vacuum u.v. photoelectron spectra of inorganic molecules. (R. L. De Kock and D. R. Lloyd, *Adv. Inorg. Radiochem.*, 1974, **16**, 66).

Electronic transitions in transitional-metal compounds at high pressures. (H. G. Drickamer, *Angew. Chem. Internat. Edn.*, 1974, **13**, 39).

Magnetic anistropy. (S. Mitra, *Transition Metal Chem.*, 1972, **7**, 183).

Magnetic phase transitions at low temperatures. (J. E. River, *Transition Metal Chem.*, 1972, **7**, 1).

Iron–sulphur coordination compounds and proteins. (S. J. Lippard, *Accounts Chem. Res.*, 1973, **6**, 282).

Metalloprotein redox reactions. (L. E. Bennett, *Progr. Inorg. Chem.*, 1973, **18**, 1).

Novel metalloporphyrins-synthesis and implications. (D. Ostfield and M. Tsutsui, *Accounts Chem. Res.*, 1974, **7**, 52).

The reactivity of the porphyrin ligand. (J. H. Fuhrhop, *Angew. Chem. Internat. Edn.*, 1974, **13**, 321).

The chemistry of vitamin B_{12} and related inorganic model systems. (D. G. Brown, *Progr. Inorg. Chem.*, 1973, **18**, 177).

The activation of molecular oxygen. (G. Henrici-Olive and S. Olive, *Angew. Chem. Internat. Edn.*, 1974, **13**, 29).

The dioxygen ligand in mononuclear Group VIII transition-metal complexes. (J. S. Valentine, *Chem. Rev.*, 1973, **73**, 235).

Magnetic resonance studies in metal-enzymes. (P. J. Quilley and G. A. Webb, *Coord. Chem. Rev.*, 1974, **12**, 407).

Transition-metal complexes in cancer chemotherapy. (M. J. Cleare, *Coord. Chem. Rev.*, 1974, **12**, 349).

PART II: Cobalt, Nickel and Copper *by C. A. McAuliffe and W. Levason*

1 Cobalt

Carbonyls.—The kinetics of the formation of $Co_4(CO)_{12}$ from $Co_2(CO)_8$ in heptane solution have been reported.[1] The sublimation of $Co_2(CO)_8$ onto a cold finger in the presence of oxygen has been shown by e.s.r. studies to produce $\cdot O_2Co(CO)_4$ radicals in addition to the known $\cdot Co(CO)_4$.[2] The cluster carbonyl anion $[Co_4Ni_2(CO)_{14}]^{2-}$ has been X-rayed,[3] and the structure is composed of an octahedral cluster with statistical distribution of Co and Ni, six terminal carbonyls (one per metal) and eight carbonyls forming triple bridges over the faces of the octahedron. The crystal structure of π-benzene-enneacarbonyltetracobalt cluster (and the *o*- and *m*-xylene analogues) has been determined.[4] Mingos[5] has performed Wolfsberg–Helmholtz MO calculations on $[Co_6(CO)_{14}]^{4-}$, and suggests that the 86 valence electrons are accommodated in bonding and weakly antibonding MO's and consequently represent a stable closed-shell arrangement. The validity of analogies between octahedral boranes and metal clusters is discussed.[5]

Irradiation of $(\pi\text{-}C_5H_5)Co(CO)_2$ produces violet $(\pi\text{-}C_5H_5)(CO)Co(CO)Co(CO)\text{-}(\pi\text{-}C_5H_5)$, which can be converted thermally into the known $Co_3(\pi\text{-}C_5H_5)_3(CO)_3$ and a new cluster, $Co_4(CO)_2(\pi\text{-}C_5H_5)_4$ (1). Carbonylmethylidynetricobalt nonacar-

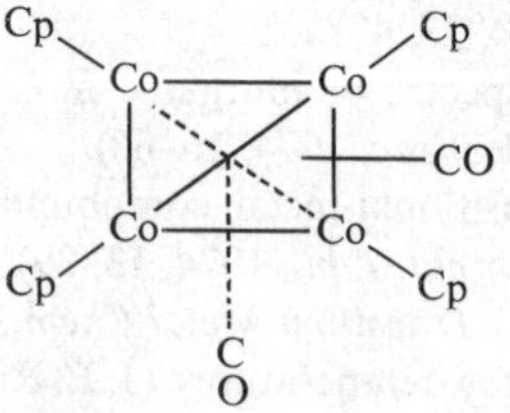

(1)

bonyl reacts with HPF_6 in propionic anhydride to form black crystals of composition $(CO)_9Co_3C_2O^+PF_6^-$, the proposed structure of which consists of a Co_3 triangle with a linear CCO group bonded to each cobalt and completing a tetrahedron. Reactions of this novel species with alcohols, amines and thiols are described.[7]

Mixed-metal carbonyls have attracted considerable effort. In THF under a CO atmosphere $CoTc(CO)_9$ has been synthesized from $Co(CO)_4^-$ and $BrTc(CO)_5$.[8] The crystal structures of $(\pi\text{-indenyl})Fe(CO)_2Co(CO)_4$,[9] $(\pi\text{-}C_5H_5)(CO)Fe(CO)_2Co\text{-}$

[1] F. Ungvary and L. Marko, *J. Organometallic Chem.*, 1974, **71**, 283.
[2] S. A. Fieldhouse, B. W. Fullam, G. W. Nielson, and M. C. R. Symons, *J.C.S. Dalton*, 1974, 567.
[3] V. G. Albano, G. Ciani, and P. Chini, *J.C.S. Dalton*, 1974, 432.
[4] P. H. Bird and A. R. Frazer, *J. Organmetallic Chem.*, 1974, **73**, 103.
[5] D. M. P. Mingos, *J.C.S. Dalton*, 1974, 133.
[6] K. P. C. Volhardt, J. E. Bercaw, and R. G. Bergman, *J. Amer. Chem. Soc.*, 1974, **96**, 4999.
[7] D. Seyferth, J. E. Hallgren, and C. S. Eschbach, *J. Amer. Chem. Soc.*, 1974, **96**, 1730.
[8] G. Sbrignadello, G. Tomati, L. Magon, and G. Bor, *Inorg. Nuclear Chem. Letters*, 1973, **9**, 1073.
[9] F. S. Stephens, *J.C.S. Dalton*, 1974, 13.

$(CO)_2(PPh_2Me)$,[10] and $(\pi\text{-}C_5H_5)Ni(CO)_2Co(CO)_2(PEt_3)$[11] all show the presence of non-planar $-M(CO)_2Co-$ bridges. For example, in the $(\pi\text{-indenyl})Fe(CO)_2Co(CO)_4$ compound the angle between the Fe(CO)Co planes is 148°, the Co environment being intermediate between trigonal bipyramidal and square pyramidal.

Thallium(I) tetracarbonylcobaltate can be obtained from $Tl[Co(CO)_4]_3$, $Co_2(CO)_8$, or $Hg[Co(CO)_4]_2$ and thallium, or from Tl^+ and $Co(CO)_4^-$ in aqueous solution. The i.r. spectrum in non-polar solvents indicates a $Tl-Co(CO)_4$ structure, but the X-ray crystal structure shows that the solid consists of discrete Tl^+ and $[Co(CO)_4]^-$ ions.[13] Some reactions are shown below. There is i.r. evidence for formation of an unstable $Tl[Co(CO)_3PPh_3]$ in solutions of $Na[Co(CO)_3PPh_3]$ and Tl(OAc) in THF–MeOH,[12] whilst treatment of $Tl[W(CO)_3(\pi\text{-}C_5H_5)]$ with $Co_2(CO)_8$ produces[12] $Tl[Co(CO)_4]_2[W(CO)_3(\pi\text{-}C_5H_5)]$.

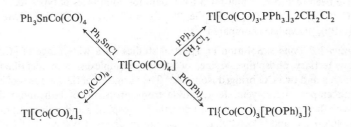

Group IVB Donors. $Co_2(CO)_8$ reacts with Me_3EH (E = Si, Ge, or Sn) or Et_3GeH to form $R_3ECo(CO)_4$.[14] The $Me_3SiCo(CO)_4$ reacts with Me_3EH (E = Ge or Sn) to afford $Me_3ECo(CO)_4$ and Me_3SiH. Other silyl complexes were obtained,[15] *e.g.*

$$SiH_2Cl_2 + NaCo(CO)_4 \rightarrow H_2SiClCo(CO)_4 + H_2Si[Co(CO)_4]_2$$

$$SiH_2Cl_2 + Co_2(CO)_8 \rightarrow HSiCl_2Co(CO)_4 + Cl_2SiCo_2(CO)_7$$

whilst heteropolymetallic silanes, *e.g.* $SiHClMn(CO)_5Co(CO)_4$ and $SiClMn(CO)_5\text{-}$ $Co_2(CO)_7$ can be isolated from reaction of $H_2SiClMn(CO)_5$ with $Co_2(CO)_8$. The disilane $HMe_2SiCH_2CH_2SiMe_2H$ and $Co_2(CO)_8$ afford unstable white $(OC)_4CoSi\text{-}$ $(Me_2)CH_2CH_2Si(Me_2)Co(CO)_4$, which rapidly decomposes to $Co_2(CO)_8$ in solution.[16]

Photolysis of $Me_2Ge[Co(CO)_4]_2$ proceeds with ring closure to give (2), whilst $Me_2GeClCo(CO)_4$ similarly forms (3); both are structurally related to $Co_2(CO)_8$ with Me_2Ge moieties replacing bridging carbonyls. The kinetics of the reversible ring-opening of $(OC)_3Co(GePh_2)(CO)Co(CO)_3$ with CO in decalin, and of the reactions

[10] G. Davey and F. S. Stephens, *J.C.S. Dalton*, 1974, 698.
[11] F. S. Stephens, *J.C.S. Dalton*, 1974, 1067.
[12] J. M. Burlitch and T. W. Theyson, *J.C.S. Dalton*, 1974, 828.
[13] D. P. Schussler, W. R. Robinson, and W. F. Edgell, *Inorg. Chem.*, 1974, **13**, 153.
[14] G. F. Bradley and S. R. Stobart, *J.C.S. Dalton*, 1974, 264.
[15] K. M. Abraham and G. Urry, *Inorg. Chem.*, 1973, **12**, 2851.
[16] L. Vancea and W. A. G. Graham, *Inorg. Chem.*, 1974, **13**, 511.
[17] R. C. Job and M. D. Curtis, *Inorg. Chem.*, 1973, **12**, 2514.

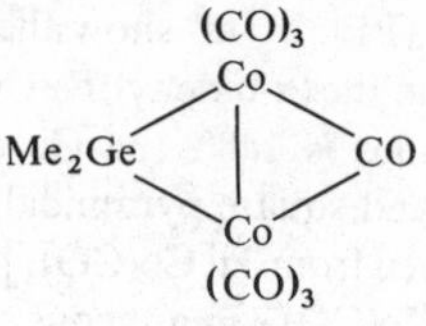

(2)

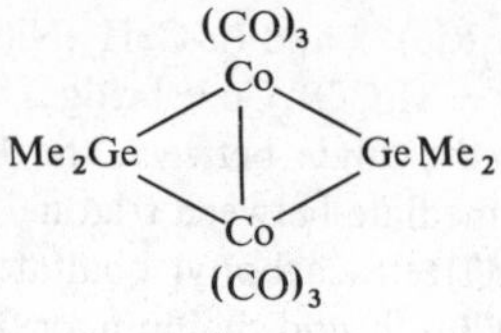

(3)

with PPh_3 or PBu_3^n to give mono- and di-substituted products have been examined.[18] The reaction of $(MeC_5H_4)_2Sn$ with $Co_2(CO)_8$ is complex and produces[19] $Sn\{Co(CO)_4\}_4$. Nesmeyanov *et al.*[20] have synthesized a range of complexes of type $Ph_3ECo(CO)_3MPh_3$ (E = Sn, Ge, or Pb; M = P, As, or Sb) and related Group IVB–cobalt carbonyl complexes. Mössbauer data for a series of Co–Sn carbonyls have been recorded,[21] and ^{59}Co n.q.r. data for complexes of types $XCo(CO)_3L$ and $XCo(CO)_2L_2$ {X = Ph_3Sn, Ph_3Ge, Ph_3Pb, or Cl_3Sn; L = PBu_3^n, $P(OMe)_3$, $P(OEt)_3$, $P(OPh)_3$} have been compared.[22]

Group VB Donors. Solution i.r. spectral studies on a wide range of $[Co(CO)_3L_3]_2$ (L = tertiary phosphine, arsine, or stibine) complexes have identified one non-bridged and two CO bridged isomers.[23] The variation of the amounts of each isomer with temperature, solvent, steric, and electronic properties of the ligands is discussed.[23] Dicobalt octacarbonyl reacts with ditertiary phosphines and arsines (L) to produce[24] $[Co(CO)_2L]_2$ and $[Co_2(CO)_6L]$ *via* the intermediates $[Co_2(CO)_4L_3][Co(CO)_4]$ and $[Co(CO)_3L][Co(CO)_4]$. The isomer obtained in the cases of the carbonyl-bridged $[Co_2(CO)_6L]$ [L = $Me_2As(CH_2)_nAsMe_2$] depends upon the length of the ligand backbone. The ligands $Ph_2ECH_2EPh_2$ (E = P, As, or Sb) and $R_2SbCH_2SbR_2$ (R = Et, Me, or p-MeC_6H_4) produce $[Co_2(CO)_6L]$ complexes which have structure (4) in solution.[25] The stibines have the same structure in the solid state, but the i.r. spectra of the diphosphine and diarsine complexes do not exclude the possibilities of the ligand chelating to one metal or bridging intermolecularly. Treatment of the

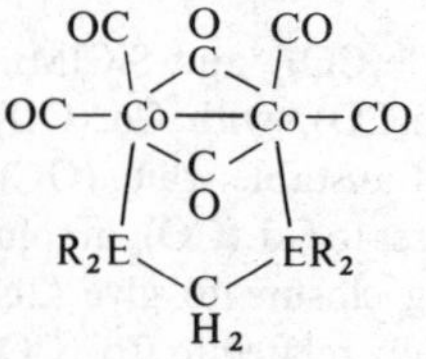

(4)

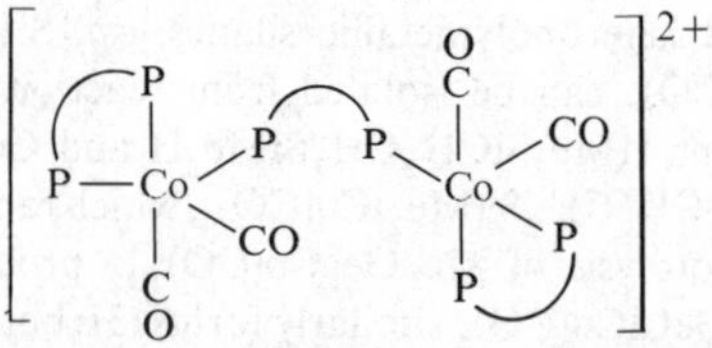

(5)

[18] M. Basato, J. P. Fawcett, and A. J. Poë, *J.C.S. Dalton*, 1974, 1350.

[19] A. B. Cornwell, P. G. Harrison and J. A. Richards, *J. Organometallic Chem.*, 1974, **76**, C26.

[20] A. N. Nesmeyanov, K. N. Anisimov, N. E. Kolobova, and V. N. Khandozhov, *Zhur. obshchei Khim.*, 1974, **44**, 1074, 1287. 1974, **44**, 1287.

[21] S. R. A. Bird, J. D. Donaldson, A. F. LeC. Holding, S. Cenini, and B. Ratcliff, *Inorg. Chim. Acta*, 1974, **8**, 149.

[22] T. E. Boyd and T. L. Brown, *Inorg. Chem.*, 1974, **13**, 427.

[23] D. J. Thornhill and A. R. Manning, *J.C.S. Dalton*, 1974, 6.

[24] D. J. Thornhill and A. R. Manning, *J.C.S. Dalton*, 1973, 2086.

[25] T. Fukumoto, Y. Matsumura, and R. Okawara, *J. Organometallic Chem.*, 1974, **69**, 437.

complexes with diarylacetylenes replaces the carbonyl bridges to give $[Co_2(CO)_4-(RC{\equiv}CR)L]$.[25] With $Co_2(CO)_8$ and cis-$Ph_2PCHCHPPh_2$, $Ph_2As(CH_2)_2AsPh_2$, $Ph_2P(CH_2)_2AsPh_2$, $PhP(CH_2CH_2PPh_2)_2$, $P(CH_2CH_2PPh_2)_3$, and $PhP(CH_2)_2-P(Ph)(CH_2)_2P(Ph)(CH_2)_2PPh_2$ were formed[26] complexes containing $[Co_2(CO)_4L_3]^{2+}$ ions of probable structure (5). The last two ligands also formed monocarbonyl cations $[Co(CO)L]^+$. The reaction of $Co_2(CO)_8$ with $(CF_3)_2PH$ produced $\{Co(CO)_3-[P(CF_3)_2]\}_2$ and a polymeric solid.[27] With $[(\pi-C_5H_5)Co(CO)_2]$ a complex reaction occurs[27] yielding $\{Co(CO)_3[P(CF_3)_2]\}_2$, $\{(\pi-C_5H_5)Co[P(CF_3)_2]\}_2$, and $\{(\pi-C_5H_5)Co(CO)[P(CF_3)_2]Co(CO)_3\}$, whilst the C_5H_5 ligand is also reduced to give C_5H_8 or $C_5H_7P(CF_3)_2$. An interesting comparison of the reaction of $Co_2(CO)_8$ with phosphites and phosphines has been reported.[28] Phosphites produce only mono- and di-substitution by direct reaction. It has been shown that $Co(CO)_4^-$ anions can nucleophilically attack $[Co(CO)_{4-n}(PR_3)_n]^+$ to form known dinuclear complexes.[28] However, when phosphite-substituted cations are used a Michaelis–Arbuzov rearrangement occurs to give an acyl–cobalt complex, and this reaction prevents the preparation of highly substituted phosphite complexes.[28]

The preparation of $[Co(CO)_2L_2I]$ and of $[Co(CO)_2L_3]^+$ (L = PR_3, R = Ph, p-MeC_6H_4, p-$MeOC_6H_4$, p-$NO_2C_6H_4$, o-MeC_6H_4, o-$Me_2C_6H_3$, or C_6H_{11}) has been achieved.[29] The structures are trigonal bipyramidal with equatorial CO groups, and the $[Co(CO)_2L_3]^+$ are non-rigid. The carbonylation of Co^{II} phosphine complexes in solution has been reported[30] to yield $Co(CO)_2(PR_3)_2X$ as well as $Co(CO)(PR_3)_2X_2$. Substitution of CO groups in $[FeCo_3(CO)_{12}]^-$ is more difficult than in neutral cluster carbonyls and only monosubstitution is achieved with $PMePh_2$, PPh_3, and $P(OPr^i)_3$, whilst $Ph_2P(CH_2)_2PPh_2$ gives $[FeCo_3(CO)_{10}L]^-$.[31]

Ehrl and Vahrenkamp[32–34] have produced several examples of arsenido-bridged mixed-metal carbonyls including $(CO)_4FeAsMe_2Co(CO)_4$, $(CO)_5CrAsMe_2Co(CO)_4$, and $(CO)_4\underline{FeAsMe_2Co}(CO)_3$.

Two rare examples of unsubstituted Group VB elements bonded to transition metals are $Co_2(CO)_6P_2$ and $Co_3(CO)_9PS$, which are produced from $NaCo(CO)_4$ and PX_3 and SPX_3, respectively.[35]

The reactions of CO with some higher valent cobalt compounds are discussed under the other ligands concerned.

Nitrosyls.—A cluster nitrosyl $Co_4(NO)_4(\mu_3\text{-}NCMe_3)_4$ has been prepared from $[Co(NO)(CO)_2(PPh_3)]$ and excess $(Me_3CN)_2S$.[36] An X-ray structural study shows it to consist of tetrahedral Co_4 grouping with triply bridging NR ligands over each face arranged so that the Co and the N occupy the corners of a distorted cube. Gilli

[26] R. L. Petersen and K. L. Walters, *Inorg. Chem.*, 1973, **12**, 3009.
[27] R. C. Dobbie and D. Whittaker, *J.C.S. Dalton*, 1973, 2427.
[28] M. S. Arabi, A. Maisonnat, S. Attali, and R. Poilblanc, *J. Organometallic Chem.*, 1974, **67**, 109.
[29] E. Bordignon, U. Croatto, U. Mazzi, and A. A. Orio, *Inorg. Chem.*, 1974, **13**, 935.
[30] G. Aklan, C. Foester, E. Hergovich, G. Speier, and L. Marko, *Verzpremi. Vegyip. Egypt. Kozlen*, 1973, **12**, 131 (*Chem. Abs.*, 1974, **80**, 77750).
[31] C. G. Cooke and M. J. Mays, *J. Organometallic Chem.*, 1974, **74**, 449.
[32] W. Ehrl and H. Vahrenkamp, *Chem. Ber.*, 1974, **106**, 2550.
[33] W. Ehrl and H Vahrenkamp, *Chem. Ber.*, 1974, **106**, 2556.
[34] W. Ehrl and H. Vahrenkamp, *Chem. Ber.*, 1974, **106**, 2563.
[35] A. Vizi-Orasz, G. Palyi, and L. Marko, *J. Organometallic Chem.*, 1973, **60**, C25.
[36] R. S. Gall, N. G. Connelly, and L. F. Dahl, *J. Amer. Chem. Soc.*, 1974, **96**, 4017.

et al.[37, 38] have X-rayed the tetrahedral $[Co(NO)(CO)_2(EPh_3)]$ (E = As or Sb); in both cases the CO and NO groups could not be distinguished from each other. The reaction[26] of $(CF_3)_2PH$ with $[Co(CO)_3(NO)]$ forms $\{Co(CO)_2(NO)[P(CF_3)_2H]\}$ which on pyrolysis produces $\{Co(CO)(NO)[P(CF_3)_2]\}_2$ and $\{Co(NO)_2[P(CF_3)_2]\}$. However, $(CF_3)_2PH$ does not react with $[Fe(CO)_2(NO)_2]$ at 80 °C, and a mixture of the iron complex and $[Co(CO)_3(NO)]$ formed only $[Co(CO)_2(NO)P(CF_3)_2H]$, but at higher temperatures a mixture of products, including $\{FeCo(CO)(NO)_3[P(CF_3)_2]_2\}$, is formed.

Black crystals produced on reflux of $[Co(NO)_2(Ph_2PCH_2CH_2PPh_2)]PF_6$ with Et_4NI in THF were identified as $[Co(NO)_2(Ph_2PCH_2CH_2PPh_2O)I]$ by X-ray methods.[39] The co-ordination about Co is distorted tetrahedral with the phosphorus ligand functioning as a unidentate P donor. The nitrosyl groups are disordered between two alternative positions and one of the M—N—O linkages is distinctly bent, the first reported example of this in a four-co-ordinate nitrosyl complex. The $[Co(NO)(PF_3)_3]$ complex reacts[40] with alkoxide or phenoxide ions with fluorine substitution to produce $\{Co(NO)[P(OR)]_3\}$ (R = Me, Et, Pr^n, Pr^i, or Bu^n), whilst hydroxide ions afford $[Co(NO)(PF_3)(PF_2O)_2]^{2-}$.

In dichloromethane solution $Co(sacsac)_2$ (sacsac = anion of dithioacetylacetone) reacts with nitric oxide at 0 °C to form monomeric $Co(sacsac)_2(NO)$, $\mu_{eff} = 0$ BM, formally a Co^{III} complex of NO$^-$. At room temperature a very soluble $Co(sacsac)_2$- $(NO)_2$ is formed. In solution $Co(sacsac)_2(NO)$ undergoes disproportionation into $Co(sacsac)(NO)_2$ and $Co(sacsac)_3$.[41]

Disproportionation of co-ordinated NO has been reported. A methanolic solution of $CoCl_2$ and $H_2NCH_2CH_2NH_2$ reacts with NO to form $Co(en)_2(NO)Cl_2$ which, on passage of more NO, forms $Co(en)_2(NO_2)Cl_2$ and nitrous oxide.[42] Similarly, $Co(dmg)_2(NO)$,MeOH reacts with NO in the presence of pyridine to form $Co(NO_2)$- $(dmg)_2(py)$.[42] Upon reaction of 2,2'-bipyridyl with a cobalt(II) salt and NO in the presence of a co-ordinating anion, species of type $[Co(bipy)_2X(NO)]^+$ form.[43] The magnetic and i.r. data are consistent with an NO$^-$ formulation. In solution the equilibrium Co^{III}–NO$^- \rightleftharpoons Co^{II}$ + NO is established.[43] The solid $[Co(bipy)_2$ (NO)- $(NO_2)]ClO_4$ is stable under nitrogen, but in air it undergoes irreversible partial oxidation to $[Co(bipy)_2(NO_2)_2]ClO_4$.[43] Substituted $[CoL(dmg)_2(NO)]$ (L = Bu^{t}py, CNpy, or PPh$_3$) reacts with oxygen to give $[CoL(dmg)_2NO_3]$, which contains O-bonded nitrate, and $[CoL(dmg)(NO_2)]$, which is the N-bonded (nitro) form.[44]

Cobalt(I).—Cobalt(I) carbonyl cations have already been discussed (p. 220). Trigonal-bipyramidal cations $[Co(CNR)_2L_3]^+$ and $[Co(CNR)_3L_2]^+$ [L = PPh(OEt)$_2$, R = C_6H_{11}, Ph, p-MeC$_6H_4$, p-MeOC$_6H_4$, p-NO$_2C_6H_4$, or o-MeC$_6H_4$] with equatorial CNR groups are formed from $[CoX\{(EtO)_2PPh\}_4]^+$ and RCN.[29] Five-co-ordinate trigonalbipyramidal complexes are also formed by reaction of $Co(PPh_3)_3X$

[37] G. Gilli, M. Sacerdoti, and G. Reichembach, *Acta Cryst.*, 1973, **B29**, 2306.
[38] G. Gilli, M. Sacerdoti, and G. Reichembach, *Acta Cryst.*, 1973, **B30**, 1485.
[39] J. S. Field, P. J. Wheatley, and S. Bhaduri, *J.C.S. Dalton*, 1974, 74.
[40] T. Kruck, J. Waldmann, M. Häfler, G. Birkenhäger, and C. Odenbrett, *Z. anorg. Chem.*, 1973, **402**, 16.
[41] A. R. Hendrickson, R. K. Y. Ho, and R. L. Martin, *Inorg. Chem.*, 1974, **13**, 1279.
[42] D. Gwost and K. G. Caulton, *Inorg. Chem.*, 1974, **13**, 414.
[43] A. Vlček and A. A. Vlček, *Inorg. Chim. Acta*, 1974, **9**, 165.
[44] W. C. Trogler and L. G. Marzilli, *Inorg. Chem.*, 1974, **13**, 1008.

or $Co(PPh_3)_2X_2$ (X = Cl or Br) with *p*-tolyl isocyanide, the products being $[Co(RNC)_4-PPh_3]X$ and $[Co(RNC)_3(PPh_3)_2]Y$ (Y = Cl, Br, I, ClO_4, or BPh_4).[45]

Gosser and Parshall[46] have investigated the reaction between $P(OEt)_3$ and cobalt(II) chloride. In the presence of NEt_3 this produces $\{Co[P(OEt)_3]_5\}Cl$, but interruption of the reaction before all the $CoCl_2$ is consumed results in the isolation of $[CoL_5]CoCl_3$ or $[CoL_5]_2[CoCl_4]$ [L = $P(OEt)_3$]. On heating $[CoL_5]Cl$ *in vacuo* the tetrahedral $[CoL_3Cl]$, $\mu_{eff} = 3.15$ BM, is produced. In ethanol $[CoL_3Cl]$ reacts with $NaBPh_4$ to yield $[CoL_5]BPh_4$ and 'CoL_2BPh_4'—the latter is probably a cobalt analogue of $Rh\{P(OEt)_3\}_2BPh_4$ which contains a phenyl ring π-bonded to the cobalt. It is suggested[46] that the previously reported CoL_4Cl and CoL_3Cl complexes are, in fact, $[CoL_5]Cl$ and $[CoL_5]_{n-2}[CoCl_n]_{2-n}$. In alcohol in the presence of acetyl chloride $[Co\{P(OMe)_3\}_5]^+$ can be prepared from $Co(C_8H_{12})(C_8H_{13})$ and $P(OMe)_3$.[47] A detailed n.m.r. study has shown that the pentakisphosphite complexes are stereochemically non-rigid at ambient temperatures, but become rigid at low temperatures.[48]

The $CoH(PF_3)_{4-n}(PPh_3)_n$ (n = 1—3) complexes have been prepared for n = 1 or 2 from $CoH(N_2)(PPh_3)_3$ and PF_3, but the third PF_3 group cannot be introduced by this route; $[CoH(PF_3)_3(PPh_3)]$ is formed by reaction of $Co(PPh_3)_3Cl$ with PF_3 in benzene.[49]

Holah *et al.*[50, 51] have reported that the products of the reaction of $CoCl_2$, $NaBH_4$, and tertiary phosphines in ethanol are markedly dependent upon the phosphines and the conditions. With triphenylphosphine, $Co(PPh_3)_3Cl$ and then $Co(BH_4)(PPh_3)_3$ ($\mu_{eff} = 2.71$ BM) are formed; both are tetrahedral d^8 systems. Recrystallization of $Co(BH_4)(PPh_3)_3$ from nitrogen-saturated benzene produced $Co(N_2)(BH_4)(PPh_3)_3$.

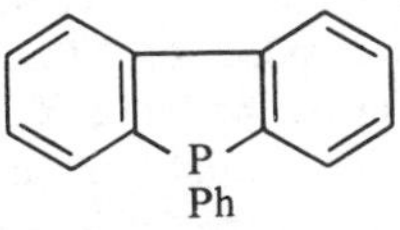

dbp

With the ligand dbp similar reactions yielded $CoCl(dbp)_3$ and $Co(dbp)_4$, but no $Co(BH_4)(dbp)_3$. Triphenyl-arsine and -stibine and $CoCl_2$ were reduced to black products, probably cobalt metal. From $NaBH_3CN$ is afforded a $Co(BH_3CN)(Ph_2-PCH_2CH_2PPh_2)$ complex, but this does not subsequently yield a dinitrogen complex.

Tris-(2-diphenylphosphinoethyl)amine, NP_3, forms a trigonal-bipyramidal hydridocobalt(I) complex, $[Co(NP_3)H]$.[52]

Cobalt(I) complexes $[Co(chel)]$, where chel = NN'-ethylene(acetylacetonedi-iminato), NN'-ethylenebis(NN'-dimethylsalicylideneiminato), NN'-ethylenebis(salicyli-

45 S. Otsuka and M. Rossi, *Bull. Chem. Soc. Japan*, 1973, **46**, 3411.
46 L. W. Gosser and G. W. Parshall, *Inorg. Chem.*, 1974, **13**, 1947.
47 P. Meakin and J. P. Jesson, *J. Amer. Chem. Soc.*, 1974, **96**, 5751.
48 J. P. Jesson and P. Meakin, *J. Amer. Chem. Soc.*, 1974, **96**, 5760.
49 M. A. Cairns and J. F. Nixon, *J. Organometallic Chem.*, 1974, **74**, 263.
50 D. G. Holah, A. N. Hughes, and B. C. Hui, *Inorg. Nuclear Chem. Letters*, 1974, **10**, 427.
51 D. G. Holah, A. N. Hughes, B. C. Hui, and K. Wright, *Canad. J. Chem.*, 1974, **52**, 2990.
52 C. A. Ghilardi and L. Sacconi, *Cryst. Structure Comm.*, 1973, **3**, 415.

deneiminato), or *NN-o*-phenylenebis(salicylideneiminato) react[53] with alkyl-ammonium or sulphonium salts to generate Co^{II} or Co^{III} species by reactions such as

$$[Co^{I}(salen)]^{-} + R^{1}NR_{3}^{2+} \rightarrow [Co^{II}(salen)] + NR_{3}^{2} + R^{1}$$

$$[Co^{I}(salen)]^{-} + R^{1}NR_{3}^{2+} \rightarrow [Co^{III}(salen)R^{1}] + NR_{3}^{2}$$

The cobalt(II) complex (6) undergoes disproportionation in the presence of CO under basic conditions to give $Co^{I}L(CO)$, where L is derived from the original ligand by introduction of a C=C bond into one of the six rings and subsequent loss of one H atom. An *X*-ray structure shows the complex to be square pyramidal with an axial CO group.[54]

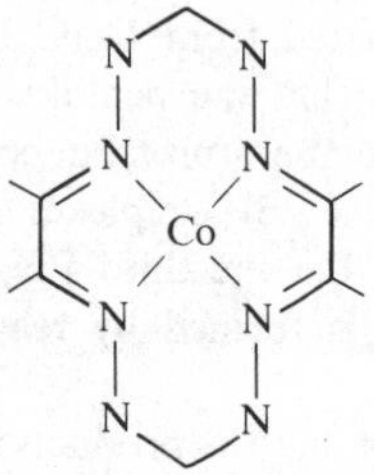

(6)

Cobalt(II) Complexes.—*Halides and Pseudohalides.* The electronic spectrum of $CoBr_{4}^{2-}$ in molten $Bu^{n}_{4}PBr$ has been studied in detail,[55] and photoelectron spectra of $[R_{4}N]_{2}CoX_{4}$ (X = Cl, Br, NCO, NCS, or NCSe) recorded.[56] The equilibria in cobalt(II) perchlorate solutions in the presence of LiCl were examined spectrophotometrically and potentiometrically and formation of $CoCl(ClO_{4})$, $CoCl_{2}$, $LiCoCl_{3}$, and $Li_{2}CoCl_{4}$ established.[57] Large cations such as ethylenedimorpholinium $(EDMH_{2}^{2+})$,[58] morpholium (MH^{+}),[58] and dabconium $\{H\overset{+}{N}(CH_{2}CH_{2})_{3}\overset{+}{N}H\}$ (DH_{2}^{2+})[59] form chlorocobaltate(II) species—$[EDMH_{2}][Co(H_{2}O)_{2}Cl_{4}]$ (which easily loses water to form the $[CoCl_{4}]^{2-}$ ion), $[MH]_{2}[CoCl_{4}]$, and $[DH_{2}][CoCl_{4}]$. The anilinium cation (LH^{+}) forms $(LH)_{2}CoX_{4}·6H_{2}O$ (X = Cl or Br).[60] Misumi *et al.* [61] have measured the formation constant of $[Co(NCS)_{4}]^{2-}$, $\log \beta_{2} = 15.5$. The electronic and i.r. spectra and magnetic moments of the novel $[CoX_{2}Y_{2}]^{2-}$ and $[CoX_{3}Y]^{2-}$ [X = NCS or NCO, Y = $N(CN)_{2}$] are consistent with a pseudo-tetrahedral structure.[62] Miezis[63] has examined the frequency and intensity of $\nu(CN)$

[53] G. Costa, A. Puxeddu, and E. P. Reisenhofer, *J.C.S. Dalton*, 1973, 2034.
[54] V. L. Goedken and S.-M. Peng, paper given at 16th International Coordination Chemistry Conference, Dublin, 1974.
[55] N. Islam, *Austral. J. Chem.*, 1973, **26**, 2371.
[56] J. Escard, G. Marvel, J. E. Guerchais, and R. Kergoat, *Inorg. Chem.*, 1974, **13**, 695.
[57] K. Sawada and M. Tanaka, *J. Inorg. Nuclear Chem.*, 1974, **36**, 1971.
[58] D. Venkappayya and G. Aravamudsen, *Current Sci.*, 1974, **43**, 277.
[59] G. Brun and G. Jourdan, *Compt. rend.*, 1974, **279** C, 129.
[60] G. Brun, *Compt. rend.*, 1974, **279** C, 105.
[61] S. Misumi, M. Aihara, and S. Kinoshita, *Bull. Soc. Chem. Japan*, 1974, **47**, 127.
[62] H. Köhler, T. P. Lischko, H. Hartung, and A. M. Golub, *Z. anorg. Chem.*, 1974, **403**, 35.
[63] A. Miezis, *Acta Chem. Scand.*, 1973, **27**, 3801.

of Co—NCS complexes in non-aqueous solvents. The anions are N-bonded and the first complex formed between Co^{II} and the anion is the octahedral $[Co(solvent)_5NCS]^+$.

Amine Complexes. Cobalt(II) iodide complexes of hexamethylenetetramine of stoicheiometry $CoLI_2$, CoL_2I_2, and CoL_4I_2 have been prepared.[64] Stability constants have been reported for the Co^{II} derivatives of NN'-bis-(2-aminoethyl)propane-1,3-diamine and the values compared with those of analogues with different ring sizes.[65] The cyclic diamines 1,5-diazocyclo-octane (daco) and 1,4-diazocycloheptane (dach) yield Co^{II} complexes of types $CoLX_2$, CoL_2X_2, CoL_2XY and $Co(daco)_2Y_2$ (X = Cl, Br, I, or NCS, Y = BF_4 or ClO_4).[66, 67] The $CoLX_2$ are non-conductors and have pseudo-tetrahedral structures, whilst the CoL_2X_2, formed from anhydrous cobalt(II) halide and excess ligand in ethanol, have a five-co-ordinate structure with an apical X group. The ionic halide is readily replaced by ClO_4^- or BF_4^- to form $[CoL_2X]Y$. Both the $[CoL_2X]X$ and $[CoL_2X]Y$ are unique examples of five-co-ordinate cobalt(II) which is not bonded to π-bonding ligands.[67] The $[Co(daco)_2](ClO_4)_2$ complex is either essentially square planar with weak tetragonal interaction with the ClO_4^- groups or is slightly tetrahedrally distorted from planarity. If $[Co(daco)_2](ClO_4)_2$ is indeed square planar it would seem to be the first example of this geometry for high-spin cobalt(II), $\mu_{eff} = 4.18$ BM.

Hydrazine forms[68] a 2:1 complex with cobalt(II) bromide $Co(N_2H_4)Br_2,2H_2O$ and phenylhydrazine forms[69] 2:1 complexes with all cobalt(II) halides. The latter seem to be octahedral in the solid state but dissolve in acetone to give blue solutions with characteristic pseudotetrahedral electronic spectra.

Pyridine and Related Ligands. Two forms (buff and orange) of $[Co(NCS)_2(4\text{-Bzpy})_4]$ (4-Bzpy = 4-benzoylpyridine) can be isolated, which may be the *cis-* and *trans-*isomers.[70] They have identical electronic spectra in solution, possibly due to formation of a bis(pyridine) complex, although this could not be isolated. The $CoBr_2$ adducts of 3-benzoylpyridine $[Co(3\text{-Bzpy})_2Br_2]_n$ and $[Co(3\text{-Bzpy})_2Br_2]$ represent an interesting example of a temperature-dependent change in cobalt co-ordination number by controlled heating of the former (polymeric octahedral) to give the tetrahedral monomer.[70] The octahedral $[Co(3\text{-Bzpy})_2Cl_2]_n$ loses ligand on heating to form the novel $[Co(3\text{-Bzpy})Cl_2]$, $\mu_{eff} = 5.7$ BM. Monomeric octahedral $[Co(4\text{-vpy})_4X_2]$ (X = Cl, Br, I, NCO, or NCS), polymeric octahedral $[Co(4\text{-vpy})_2X_2]_n$ (X = Cl or N_3), and tetrahedral $[Co(4\text{-vpy})_2X_2]$ (X = Cl, Br, or I) and $[Co(2\text{-vpy})_2X_2]$ (X = Cl, Br, I, or NCS) are formed from 2-vinylpyridine and 4-vinylpyridine.[71] Electronic and i.r. spectra and magnetic data were reported for these complexes and some structures were confirmed by X-ray powder patterns. The azido-complex $[Co(4\text{-vpy})_2(N_3)_2]$ exhibits antiferromagnetism.

[64] D. T. Altybaeva, D. B. Serukieva, B. I. Imanakunov, and P. T. Yun, *Izvest. Akad. Nauk Kirg. S.S.R.*, 1973, 51.
[65] S. K. Srivastava and H. B. Mathur, *Indian J. Chem.*, 1973, **11**, 1293.
[66] E. D. Steffen and E. D. Stephens, *Inorg. Nuclear Chem. Letters*, 1973, **9**, 1011.
[67] W. K. Musker and E. D. Steffen, *Inorg. Chem.*, 1974, **13**, 1951.
[68] K. P. Mamedov, R. Y. Aliev, Z. I. Suleimanov, A. D. Kuliev, and V. Z. Zeinalov, *Zhur. fiz. Khim.*, 1973, **47**, 2696.
[69] W. K. Glas and J. O. McBreen, *J. Inorg. Nuclear Chem.*, 1974, **36**, 747.
[70] J. de O. Cabral, M. T. P. Leite, and M. F. Cabral, paper given at 16th International Coordination Chemistry Conference, Dublin, 1974.
[71] N. H. Angew, R. J. Collin, and L. F. Larkworthy, *J.C.S. Dalton*, 1974, 272.

Formation of octahedral 2:1 adducts of piperidine, *N*-methylpiperazine, and morpholine with cobalt(II) acetate and propionate has been reported.[72] Piperazine itself gives 1:1 adducts. Adducts of cobalt(II) trifluoroacetate with pyridine, β- and γ-picoline, and β- and γ-ethylpyridine are of type $CoL_x(CF_3CO_2)_2$ ($x = 2$ or 4) where the trifluoroacetate group is either uni- or bi-dentate.[73] Thermal decomposition of the $CoL_4(CF_3CO_2)_2$ complexes produced the $CoL_2(CF_3CO_2)_2$ compounds, although the latter were very difficult to obtain pure by this method. ^1H and ^{19}F n.m.r. studies[74] in $CDCl_3$ solution indicated the presence of rapid *cis–trans* equilibria for both series. The exchange became slow at lower temperatures, whilst in $(CD_3)_2CO$ solution there was evidence for polymeric species.[74] An extended Hückel MO treatment has been used to explain the measured n.m.r. isotropic shifts.[75] Lincoln[76] has reported a ^{14}N n.m.r. study of pyridine exchange in *trans*-$[Co(acac)_2(py)_2]$.

The pink $[Co(4,4'-bipy)(CNS)_2]$ has a pseudo-octahedral structure ($\mu_{eff} = 5.15$ BM), the ligand being bridging bidentate and six-co-ordination being completed by —NCS— bridges.[77] The crystal structure of bis-(2,2':6':2''-terpyridyl)cobalt(II) bromide trihydrate contains two strained terdentate terpy ligands, the central Co—N distance being shorter than the outer (1.89, 2.10 Å); the bromide ions are disordered between lattice sites.[78] It is suggested that the potential barrier between these two sites is very small and may be a determining factor in the 'crossover' involving the temperature-dependent equilibria between spin states in the cation. Anomalous magnetic behaviour is also shown[79] by $[Co(tpa)_2](ClO_4)_2$ (tpa = tri-2-pyridylamine), attributed to the $^4T \rightleftharpoons {}^2E$ equilibrium. The corresponding hexafluorophosphate and the analogous 4-, 5-, and 6-methyl substituted tri-2-pyridylamine complexes are magnetically normal despite the fact that introduction of the substituents would be expected to favour the ^2E state.[79] The 4- and 5-methyl substituted ligand complexes can be oxidized to cobalt(III) derivatives in contrast to those of tpa itself.

2,2'-Dithiodipyridine (tdp) forms $Co(tdp)X_2$, which are tetrahedral monomers (X = Cl, Br, or I) or octahedral polymers (X = NCS); 4,4'-dithiodipyridine (fdp)

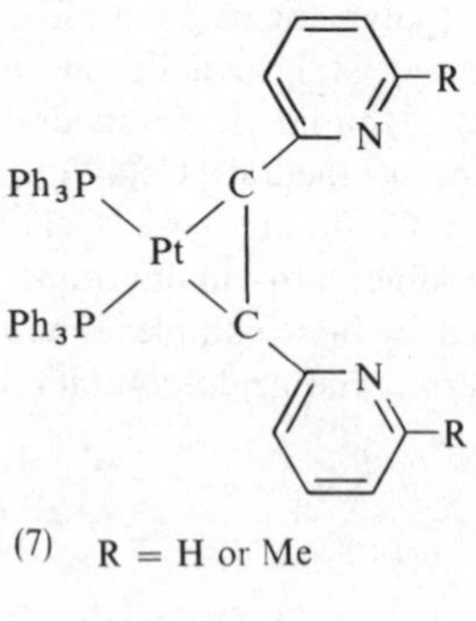

(7)   R = H or Me

[72] G. Marcotrigiano, G. C. Pellacani, and C. Preti, *Z. anorg. Chem.,* 1974, **408**, 313.
[73] C. A. Agambar, P. Anstey, and K. G. Orrell, *J.C.S. Dalton,* 1974, 864.
[74] P. Anstey and K. G. Orrell, *J.C.S. Dalton,* 1974, 870.
[75] P. Anstey and K. G. Orrell, *J.C.S. Dalton,* 1974, 1711.
[76] S. F. Lincoln, *J.C.S. Dalton,* 1973, 1896.
[77] I. S. Ahuja and R. Singh, *Indian J. Chem.,* 1974, **12**, 107.
[78] E. N. Maslen, C. L. Raston, and A. H. White, *J.C.S. Dalton,* 1974, 1803.
[79] P. F. B. Barnard, J. C. Lancaster, M. E. Fernandopulle, and W. R. McWhinnie, *J.C.S. Dalton,* 1973, 2172.

forms tetrahedral $Co(fdp)Cl_2$ and polymeric octahedral $Co(fdp)_2X_2$ (X = I or NCS).[81] I.r. spectra suggest that both ligands behave as N-donors only. The platinum complexes (7) can behave as ligands towards cobalt(II) forming $LCoCl_2$ species, where weak interaction occurs between the cobalt and the pyridyl nitrogens.[81]

The formation of hydrated cobalt(II) complexes of pyridine carboxylic acids and the subsequent thermal decomposition to lower hydrates has been documented.[82, 83] Cobalt(II) halides react with 6-methylpicolinic acid (6-mpaH), picolinic acid (paH), nicotinic acid (naH), and pyridine-2,6-dicarboxylic acid (2,6-py) to form $Co(6\text{-mpa})$ (6-mpaH)X (X = Cl, Br, or NCS), $Co(naH)_nX_2$ (n = 2, X = Cl, Br; n = 3, X = NCS), and $Co(pa)(paH)X,\frac{1}{2}EtOH$ (X = Cl, Br, or NCS) which are all probably octahedral.[83] 6-Methylpicolinic acid also formed 'Co(6-mpaH)$_4$X$_2$,2HX' (X = Cl or Br) which were formulated $[(6\text{-mpaH})_2H]_2[CoX_4]$, since the electronic spectra show absorptions characteristic of tetrahalogenocobaltate(II) ions.[83]

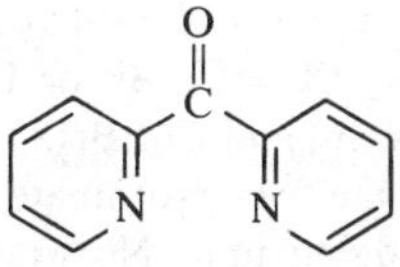

(8a) 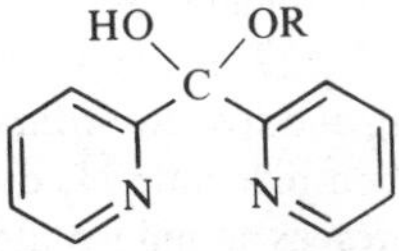   (8b)

Di-(2-pyridyl)ketone (8a) can form complexes containing the ligand in the ketone or hemiacetal (8b) form.[84, 85] Ortego *et al.*[85] obtained a series of complexes CoL_2X_2 (X = Cl, Br, I, NCS, ClO$_4$, or NO$_3$), all of which appear to contain the ligand in form (8a) behaving as an NN-donor. Pseudo-octahedral structures were proposed, except for X = I which appeared to be planar. The bromo- and thiocyanato-complexes exhibited a temperature-dependent structure in DMSO, octahedral at low and tetra-hedral at higher temperatures.[85] Complexes formulated as $CoL(CH_3CO_2)_2,H_2O$ and $3CoSO_4,2L(H_2O)_{8-10}$ were also reported.[85] In contrast, Rattanaphani and McWhinnie[84] prepared complexes containing the hemiacetal form (8b): $[CoL'Br_2]$, $[CoL'Cl_2]_n$, $[CoL_2L'](ClO_4)_2$, and $[CoLBr_2]$ [L = (8a), L' = (8b); R = Et or Me)]. The bisper-chlorate complex appears to be octahedral, whilst the dichloro-complex is polymeric octahedral, and the $[CoL'Br_2]$ is tetrahedral. Ethanol can easily be removed from $[CoL'Br_2]$ to form $[CoLBr_2]$.[84]

Di-(2-pyridyl)ketoxime [dpkx (9)] forms tetrahedral $Co(dpkx)Cl_2,H_2O$ and $Co(dpkx)_3Cl_2$, which may be octahedral but has $\mu_{eff} = 2.1$ BM, possibly due to Curie–Weiss dependence of the moment upon temperature.[86] The bonding in these complexes is probably *via* one pyridyl-N and the oxime-N. Pyridine-4-carbaldehyde

[80] M. E. Brisdon and W. R. Walker, *Austral. J. Chem.*, 1974, **27**, 87.
[81] G. R. Newkome and G. L. McClure, *J. Amer. Chem. Soc.*, 1974, **96**, 617.
[82] A. Anagnostopoulos, *Inorg. Nuclear Chem. Letters*, 1974, **10**, 525.
[83] V. M. Ellis, R. S. Vagg, and E. C. Walton, *J. Inorg. Nuclear Chem.*, 1974, **36**, 1031.
[84] V. Rattanaphani and W. R. McWhinnie, *Inorg. Chim. Acta*, 1974, **9**, 239.
[85] J. D. Ortego, D. D. Waters, and C. S. Steele, *J. Inorg. Nuclear Chem.*, 1974, **36**, 751.
[86] M. W. Blackmore, C. G. Sceney, D. R. O'Neill, and R. J. Magee, *J. Inorg. Nuclear Chem.*, 1974, **36**, 1170.

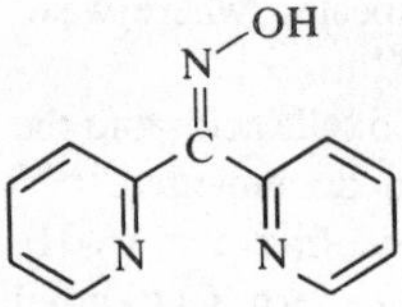

(9) 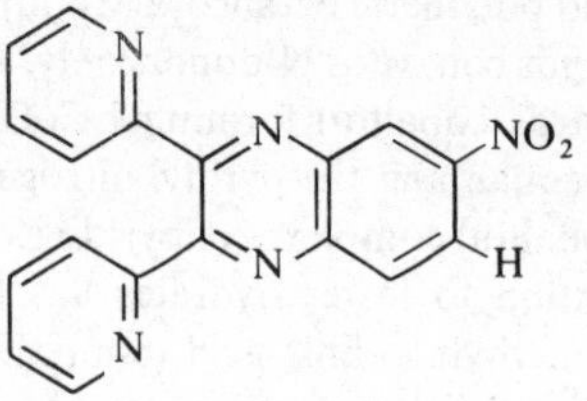   (10) 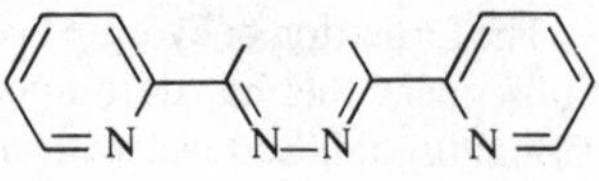   (11)

oxime forms octahedral halogen-bridged polymers CoL_2X_2 (X = Cl or Br), μ_{eff} = 4.8—5.2 BM, with co-ordination *via* the pyridine nitrogens only.[87]

The i.r. spectra of pyrazine complexes of cobalt(II) have been examined.[88] In the octahedral $[CoL_2X_2]_n$, L is unidentate, whilst in the $[CoLX_2]_n$ there are sheets of $[CoX_2]_n$ linked by bidentate pyrazine ligands. Complexes of 6-nitro-2,3-dipyridyl-quinoxaline (10)[89] are of types octahedral CoL_2X_2 (X = Cl, Br, or I), tetrahedral $CoLX_2$ (α-form), and five-co-ordinate $CoLCl_2$ (β-form) and $CoLBr_2$. The bis-ligand complexes contain bidentate N_2 donors, whilst in the five-co-ordinate moieties (10) probably behaves as a unidentate–bidentate bridging unit. No examples of N_4 donation were found, probably due to the reduced electron density produced by the nitro-group. Low frequency i.r. data were reported for these complexes, and for some other substituted quinoxaline complexes.

Boyd *et al.*[90] have determined the structure of $Co_2(pmk)_3ZnCl_4(ZnCl_3,H_2O),4H_2O$ [pmk = (11)]. The two cobalt atoms are linked by a helical arrangement of the three quadridentate ligands, so that there are three Co—N—N—Co bridges with a Co—Co separation of 3.81 Å. Each cobalt is in a trigonal-prismatic N_6 environment. Dinuclear complexes have been obtained containing ligands (12) and (13);[91] $Co_2L_2X_4$,-nH_2O [X = Cl or Br, L = (12); X = NO_3 or ClO_4, L = (13)] have reflectance spectra characteristic of octahedral cobalt(II) and μ_{eff} = 4.2—4.5 BM (300 K); weak antiferro-magnetic exchange was observed.[91]

Bidentate behaviour has also been observed for 3-(2-pyridyl)-5,6-diphenyl-1,2,4-triazine in $[CoL_3](BF_4)_2$ and $[CoL_2(H_2O)_2](BF_4)_2$, whilst 3-{2-(o-phen)}-5,6-diphenyl-1,2,4-triazine (L') is terdentate in $[CoL'_2]X_2$ (X = Br, BF_4, or NO_3). The spin state of the cobalt is temperature dependent for both types of complex.[92]

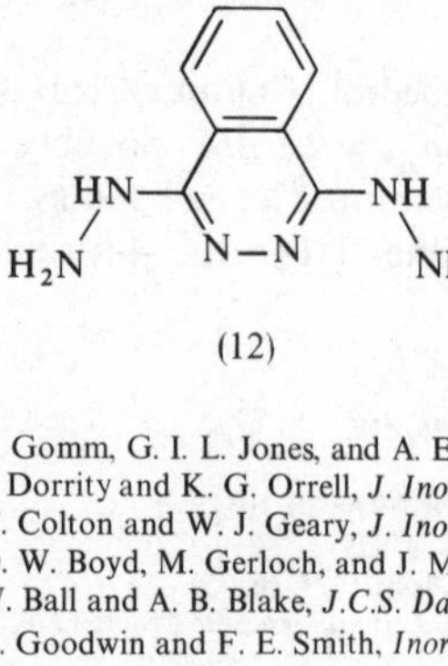

(12)

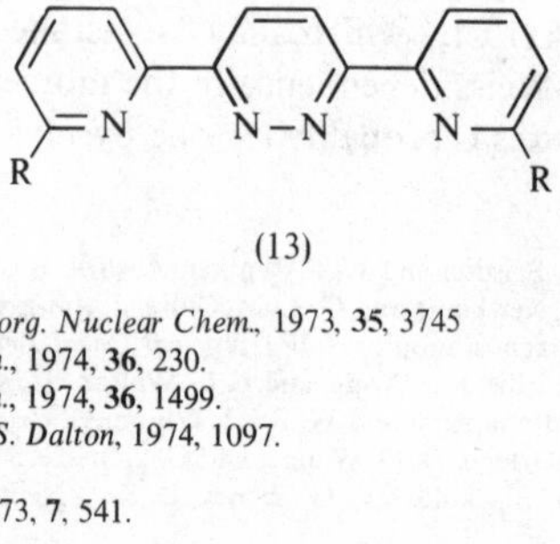

(13)

[87] P. S. Gomm, G. I. L. Jones, and A. E. Underhill, *J. Inorg. Nuclear Chem.*, 1973, **35**, 3745
[88] I. A. Dorrity and K. G. Orrell, *J. Inorg. Nuclear Chem.*, 1974, **36**, 230.
[89] D. F. Colton and W. J. Geary, *J. Inorg. Nuclear Chem.*, 1974, **36**, 1499.
[90] P. D. W. Boyd, M. Gerloch, and J. M. Sheldrick, *J.C.S. Dalton*, 1974, 1097.
[91] P. W. Ball and A. B. Blake, *J.C.S. Dalton*, 1974, 852.
[92] H. A. Goodwin and F. E. Smith, *Inorg. Chim. Acta*, 1973, **7**, 541.

The crystal structure[93] of $[Co(SCN)_2L]_2$ [L = 2(2'-pyridyl)-3-(N-2-picolylimino)-4-oxo-1,2,3,4-tetrahydroquinazoline] (14) shows each cobalt to be six-co-ordinate (N_5S donor set) bonded to two bridging —NCS— groups, one terminal —NCS group, and to ligand (14) *via* the three nitrogens (N*); $\mu_{eff} = 4.99$ BM, and visible spectra were also measured.[93]

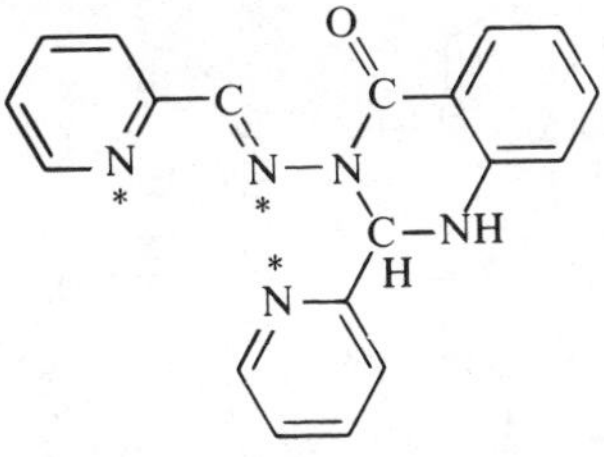

(14) 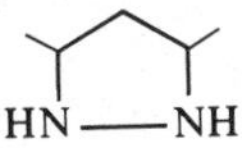   (15)

Imidazole and Related Ligands. The structures of the tetrahedral $[Co(imidazole)_2-(MeCO_2)_2]$[94] and the octahedral $[Co(imidazole)_6](CH_3CO_2)_2,H_2O$[95] have been reported. Octahedral complexes CoL_2X_2,H_2O (X = Cl or Br), CoL_2I_2, and $CoL_3X_2,-2H_2O$ are formed by 2-(2'-pyridyl)imidazole, and complexes of stoicheiometry $CoLX_2$ (X = Cl or Br) are trigonal-bipyramidal dimers with halogen bridges.[96] 3,5-Dimethyl-pyrazole, (15), forms CoL_2Cl_2 and $[Co(L - H)_2]_n$.[97] In the tetrahedral CoL_2Cl_2 complex of 5-aminoimidazole the ligand appears, on i.r. evidence, to be bonded through the pyridyl nitrogen.[98]

Macrocyclic N-Donors. Glick et al.[99] have proposed that the greater difference in Co—X axial bond length between the cobalt(II) and cobalt(III) complexes of (16) compared with the corresponding complexes of (17) accounts for the unusually slow self-exchange rate of the former. The electronic spectra of the five-co-ordinate cobalt(II) complexes of the macrocycles (18) and (19) have been reported.[100]

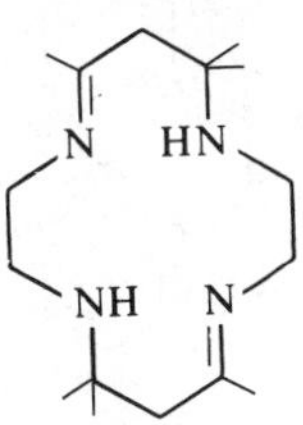

(16) 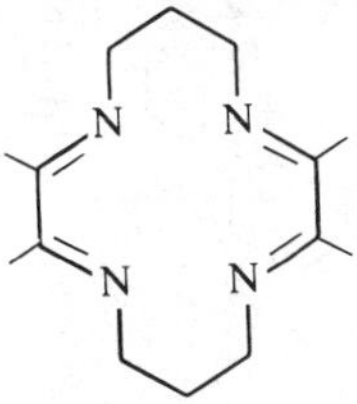   (17) 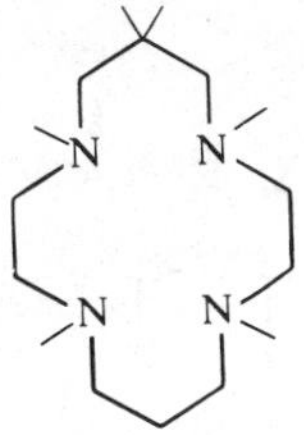   (18) 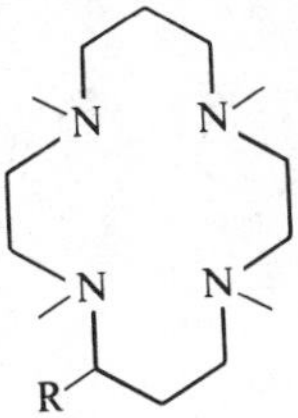   (19)

[93] A. Mangia, M. Nardelli, and G. Pellizzi, *Acta Cryst.*, 1974, **B30**, 487.
[94] A. Gadet, *Acta Cryst.*, 1974, **B30**, 349.
[95] A. Gadet and D. L. Soubeyran, *Acta Cryst.*, 1974, **B30**, 716.
[96] R. S. Dosser and A. E. Underhill, *J. Inorg. Nuclear Chem.*, 1974, **36**, 1231.
[97] C. B. Singh, S. Satpathy, and B. Sahoo, *J. Inorg. Nuclear Chem.*, 1973, **35**, 3947.
[98] S. A. A. Zaidi and A. S. Farooqi, *J. Inorg. Nuclear Chem.*, 1973, **35**, 4320.
[99] M. D. Glick, W. G. Schmonsees, and J. F. Endicott, *J. Amer. Chem. Soc.*, 1974, **96**, 5661.
[100] R. Buxtorf, W. Steinmann, and T. A. Kaden, *Chimia (Switz.)*, 1974, **28**, 15.

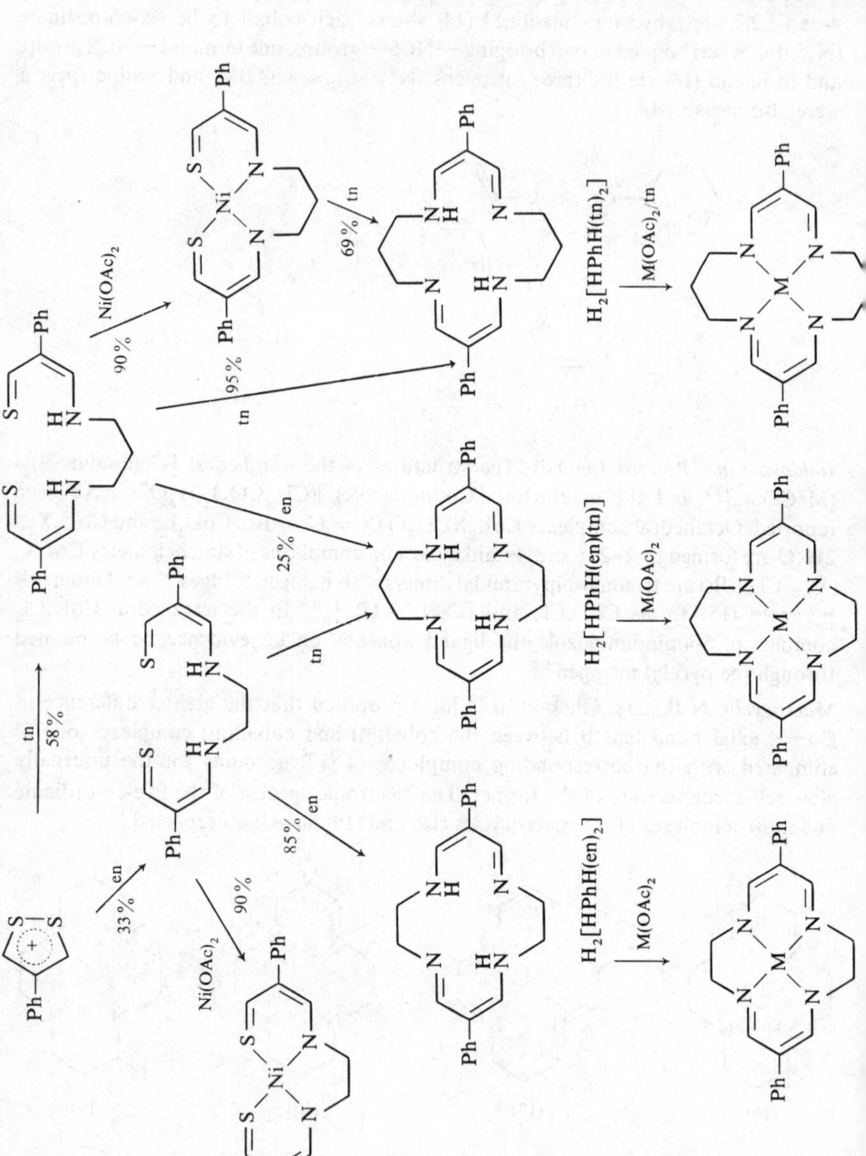
tn 58%
en 33%
Ni(OAc)$_2$ 90%
Ni(OAc)$_2$ 90%
85% en
25% en
tn
95% tn
69% tn
H$_2$[HPhH(en)$_2$]
M(OAc)$_2$
H$_2$[HPhH(en)(tn)]
M(OAc)$_2$
H$_2$[HPhH(tn)$_2$]
M(OAc)$_2$/tn

Holm and co-workers have reported a non-template synthesis of bis(β-iminoamine) macrocycles with 14-, 15-, and 16-membered rings (Scheme 1).[101] The complexes Co[HPhH(en)$_2$], Co[HPhHen,tn], and Co[HPhH(tn)$_2$] were obtained by reaction of the ligand with hydrated metal acetate in DMF under nitrogen.

Busch *et al.*[102] have reported the electrochemical reduction of complexes of tetra-aza[16]annulene (20), to the porphyrin-like dianion. The $Co^{II}(taab)^{2+}$ ion can be

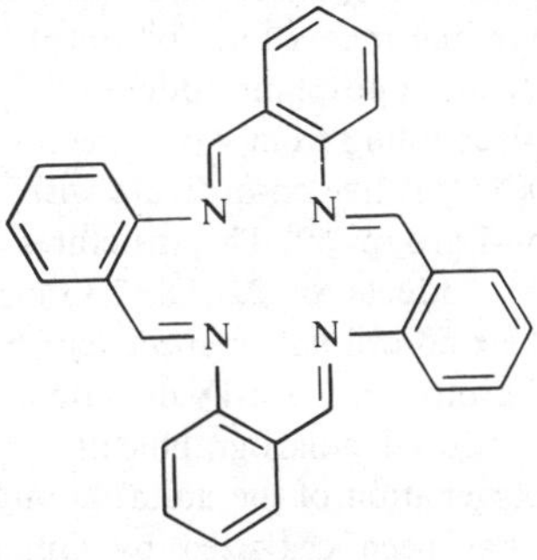

(20)   taab

electrochemically reduced in acetonitrile in three consecutive one-electron steps to a formally Co^{-I}, d^{10}, configuration. However, the electrochemical properties of the reduced species suggest that they are better viewed as complexes of the dianion (taab)$^{2-}$. $Co^{II}(taab)^{2+}$ undergoes an extensive series of redox reactions in acetonitrile (Scheme 2), and the complexes A, B, and C have been isolated. Mössbauer data have been reported for the α,β-polymorphs of cobalt phthalocyanine.[103]

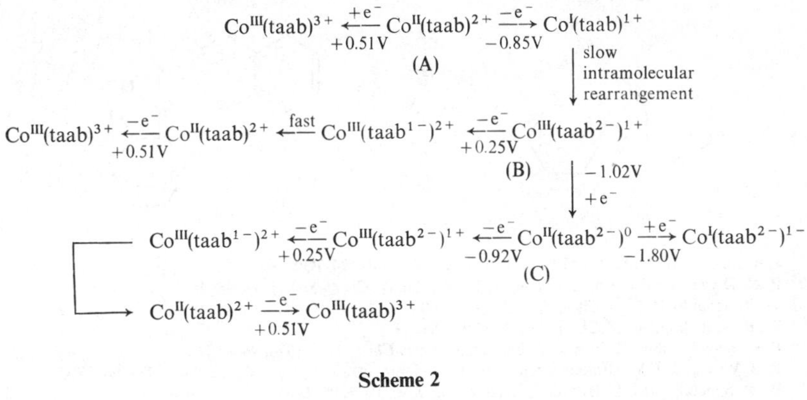

Scheme 2

[101] S. C. Tang, S. Kochi, G. N. Weinstein, R. W. Lane, and R. H. Holm, *Inorg. Chem.*, 1973, **12**, 2589.

[102] N. Takvoryan, K. Farmery, V. Katovic, F. V. Lovecchio, E. S. Gore, L. B. Anderson, and D. H. Busch, *J. Amer. Chem. Soc.*, 1974, **96**, 731.

[103] T. S. Srivastava, J. L. Przybylinski, and A. Nath, *Inorg. Chem.*, 1974, **13**, 1562.

Cobalt(II) porphyrins have been extensively studied and considerable structural data are available.[104] The structures of the five-co-ordinate square-pyramidal CoN$_5$ complexes (1,2-dimethylimidazole)-αβγδ-tetraphenylporphinatocobalt(II)[105] and (1-methylimidazole)-αβγδ-tetraphenylporphinatocobalt(II)[106] have been determined. Piperidine forms an octahedral complex bis(piperidine)-αβγδ-tetraphenylporphinatocobalt(II) which contains a CoN$_6$ grouping, the axial Co—N *ca.* 2.44 Å equatorial Co—N *ca.* 1.99 Å.[107] Cobalt(II) tetraphenylporphine (TPP) adducts with CO, NO, O$_2$, (MeO)$_3$P, MeCN, and py have been studied by e.p.r. and electronic spectroscopy[108, 109] and a bonding model for Co^{II} porphyrins has been discussed. The σ-donor interaction in tertiary phosphine adducts of Co(TPP) have been calculated using the ^{59}Co and ^{31}P coupling from e.p.r. spectra.[108] The structure of the nitric oxide adduct Co(TPP)(NO) is five-co-ordinate with a non-linear Co—N—O linkage and disordered nitrosyl groups.[110] The structures of 1-methylimidazole[111] and bis-(3-methylpyridine)[112] adducts of 2,3,7,8,12,13,17,18-octaethylporphinato-cobalt(II), which are five- and six-co-ordinate respectively, have been determined.

The cobalt(II) complexes of protoporphyrin-IX dimethylester and mesoporphyrin-IX dimethylester have been reduced polarographically.[113] The original study by Stynes and Ibers[114] of the oxygenation of the amine complexes of cobalt(II) proto-porphyrin-IX dimethylester has been criticized by Guidry and Drago,[115] who suggest that the treatment of the data was incorrect and that the system is ill-defined. Ibers *et al.*[116] have reconsidered their data and claim that this substantiates the earlier work and proves that the system is clearly defined.

Other N-*Donor Ligands.* Octahedral CoLX$_2$ (X = NO$_2$ or NO$_3$), μ_{eff} *ca.* 4.7 BM, are

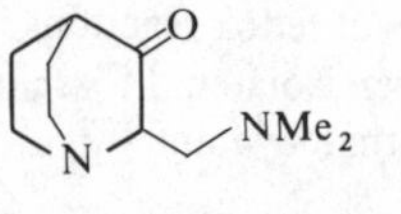

(21) dqn

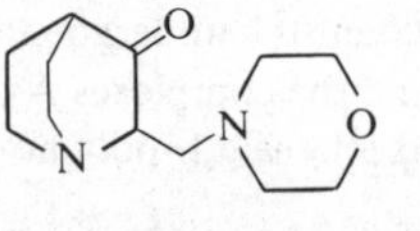

(22) mqn

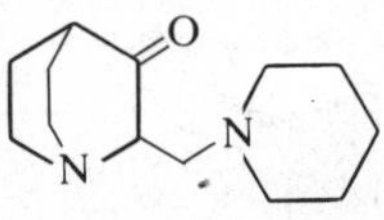

(23) pqn

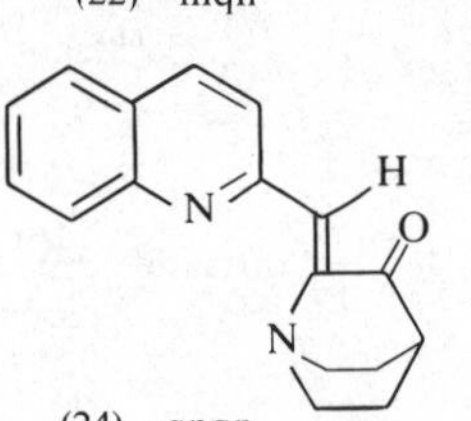

(24) qnqn

[104] J. A. Ibers, J. W. Lauher, and R. G. Little, *Acta Cryst.*, 1974, **B30**, 268.
[105] P. N. Dwyer, P. Madura, and W. R. Scheidt, *J. Amer. Chem. Soc.*, 1974, **96**, 4815.
[106] W. R. Scheidt, *J. Amer. Chem. Soc.*, 1974, **96**, 90.
[107] W. R. Scheidt, *J. Amer. Chem. Soc.*, 1974, **96**, 84.
[108] B. B. Wayland and M. E. Abd-Elmageed, *J. Amer. Chem. Soc.*, 1974, **96**, 4809.
[109] B. B. Wayland, J. V. Minkjewicz, and M. E. Abd-Elmageed, *J. Amer. Chem. Soc.*, 1974, **96**, 2795.
[110] W. R. Scheidt and J. L. Hoard, *J. Amer. Chem. Soc.*, 1973, **95**, 8281.
[111] R. G. Little and J. A. Ibers, *J. Amer. Chem. Soc.*, 1974, **96**, 4452.
[112] R. G. Little and J. A. Ibers, *J. Amer. Chem. Soc.*, 1974, **96**, 4440.
[113] T. Kakutani, S. Totsuka, and M. Seida, *Bull. Chem. Soc. Japan*, 1973, **46**, 3652.
[114] H. C. Stynes and J. A. Ibers, *J. Amer. Chem. Soc.*, 1972, **94**, 1559.
[115] R. M. Guidry and R. S. Drago, *J. Amer. Chem. Soc.*, 1973, **95**, 6645.
[116] J. A. Ibers, D. V. Ibers, H. C. Stynes, and B. R. James, *J. Amer. Chem. Soc.*, 1974, **96**, 1358.

formed by 1,2-dimorpholinoethane and 1,2-dipiperidinoethane.[117] The six-co-ordination is achieved *via* bidentate nitrite or nitrate ligands producing N_2O_4 donor sets. The quinuclidinone derivatives (21)—(24) form predominantly pseudotetrahedral complexes. The pseudotetrahedral Co(mqn)X_2 (X = Cl, Br, or I) are increasingly distorted in the order Cl > Br > I, whilst a comparison with Co(pqn)Cl_2 and Co(dqn)-Cl_2 shows that in the solid state the distortion increases mqn < pqn < dqn.[118] The [Co(mqn)$_2$](ClO$_4$)$_2$ complex also contains a tetrahedral cation. In contrast, magnetic, i.r. and electronic spectral data suggest[119] that Co(mqn)(NCS)$_2$,EtOH is octahedral, with structure (25). [Co(qnqn)Cl$_2$] is also pseudotetrahedral, μ_{eff} = 4.46

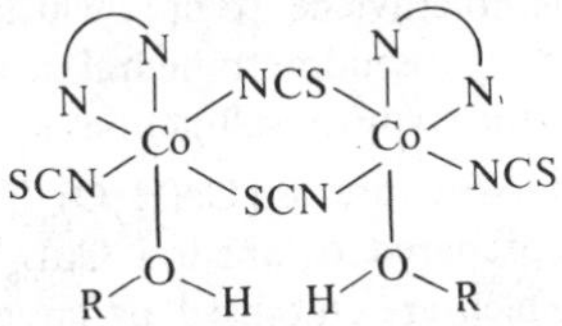

(25)

BM, with an N_2Cl_2 donor set.[120] The effect of pressure on the electronic spectrum of this complex has been examined.[121] Thiomorpholine-3-thione, $S=\overline{CNHCH_2CH_2SCH_2}$, and thiazolidine-2-selenone, $Se=\overline{CNHCH_2CH_2S}$, form tetrahedral Co$L_2X_2$ (X = Cl, Br, or I) complexes; spectral evidence suggests that co-ordination is *via* the N rather than the S or Se atoms.[122]

P-*Donor and* As-*Donor Ligands.* Trimethylphosphine forms both blue tetrahedral Co(PMe$_3$)$_2$Br$_2$, μ_{eff} = 4.4 BM, and green five-co-ordinate Co(PMe$_3$)$_3$Br$_2$, μ_{eff} = 2.20 BM, complexes.[123] The mixed complex [Co(PPh$_3$)(OPPh$_3$)I$_2$] has been formed from [Co(PPh$_3$)$_2$I$_2$] and [Co(OPPh$_3$)$_2$I$_2$] or OPPh$_3$.[124] The same complex also forms to some extent when CoI$_2$ and PPh$_2$ react in organic solvents in the presence of air, thus accounting for the variable properties of 'Co(PPh$_3$)$_2$I$_2$' reported in the literature. Similar complexes can be isolated using CoCl$_2$ and CoBr$_2$.[125] Difluorophenyl-phosphine, PF$_2$Ph, forms Co(PF$_2$Ph)$_3$X$_2$ (X = Br or I), but CoCl$_2$ and PF$_2$Ph undergo a redox disproportionation to yield Co^I(PF$_2$Ph)$_4$Cl and Co^{III}(PF$_2$Ph)$_3$Cl$_3$.[126] The formation of green [Co(pm)$_2$Cl$_2$] [pm = (26)], μ_{eff} = 1.97 BM, has been briefly reported.[127]

[117] A. L. Lott, *Inorg. Chem.*, 1974, **13**, 667.
[118] R. C. Dickinson and G. J. Long, *Inorg. Chem.*, 1974, **13**, 262.
[119] R. C. Dickinson and G. J. Long, *J. Inorg. Nuclear Chem.*, 1974, **13**, 1235.
[120] G. J. Long and D. L. Coffen, *Inorg. Chem.*, 1974, **13**, 270.
[121] G. J. Long and J. R. Ferraro, *Inorg. Nuclear Chem. Letters*, 1974, **10**, 393.
[122] C. Preti, G. Tosi, D. De Filippo, and G. Verani, *Canad. J. Chem.*, 1974, **52**, 2021.
[123] M. Zinoune, M. Dartiguenave, and Y. Dartiguenave. *Compt. rend.*, 1974, **278** C, 849.
[124] J. Rimbault and R. Hugel, *Rev. chim. Minerale*, 1973, **10**, 773.
[125] R. P. Hugel, J. C. Pierrand, and J. Rimbault, Proceedings 16th International Coordination Chemistry Conference, Dublin, 1974.
[126] O. Stelzer, *Chem. Ber.*, 1974, **107**, 2329.
[127] L. F. Warren and M. A. Bennett, *J. Amer. Chem. Soc.*, 1974, **96**, 3340.

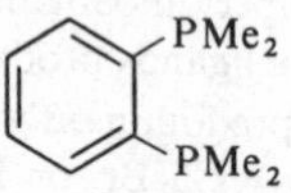

(26)

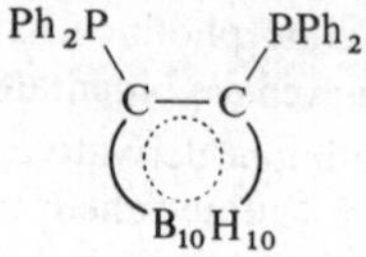

(27)

The diphosphinocarbaborane (27) forms tetrahedral $CoLX_2$ (X = Cl or NCS) and five-co-ordinate $[CoL_2X]X$ (X = Br or I).[128] The complexes are less stable than those of $Ph_2P(CH_2)_2PPh_2$ and rapidly decompose in hydroxylic solvents. The ligand *trans*-1,2-bis(diphenylphosphino)ethylene (tvpp), which is sterically incapable of chelation, forms $[Co(tvpp)X_2]_n$; pseudotetrahedral structures are indicated as are terminal Co—X bonds, and hence tvpp must function as a bidentate bridging ligand.[129]

O-*Donor Ligands.* The X-ray structure of $[Co(H_2O)_6][SiF_6]$ has been reported.[130] De Bolster *et al.*[131] have prepared octahedral CoL_6^{2+} moieties, containing O-bonded carbonyl ligands, which are stabilized by large non-co-ordinating anions: BF_4^-, $SbCl_6^-$, $InCl_4^-$. Formamide, 2-nitrobenzamide, and 2,6-dichlorobenzamide did not yield solid complexes. Ligand field parameters were determined and variations between different complexes were mainly attributed to steric effects.

1,3-Dioxan is bridging in the polymeric pseudo-octahedral $[CoLX_2]_n$ (X = Cl or Br).[132] Similar complexes are formed by 1,4-dioxan, as well as a 2:1 complex with $CoBr_2$,[132] but whereas the complexes of the 1,4-ligand readily hydrate, those of 1,3-dioxan do not. The structure of $Co(MeCONMe_2)_2Cl_2$ shows the cobalt to be in a pseudotetrahedral environment with a Cl_2O_2 donor set.[133] In $[Co(urea)_4](NO_3)_2$, however, the cobalt is co-ordinated to two bidentate (N,O) urea molecules in the equatorial plane and to two axial unidentate (O) ureas.[134] Terminal N-bonded selenocyanate groups have been found in $CoL_2(NCSe)_2$ (L = HMPA).[135]

The structure of a cobalt(II) complex of 2,2'-dithiobisbenzoate, $Co(H_2O)_6,(HO_2-CC_6H_4SSC_6H_4CO_2)_2,(MeOH)_4$, shows the Co to be bonded to six water molecules with two unco-ordinated hydrogen 2,2'-dithiobisbenzoate anions.[136] 5,5'-Thiodisalicylic acid (LH_4) forms[137] the pink octahedral $Na_2CoL,2H_2O$, $\mu_{eff} = 4.59$ BM. The nitrilotriacetates $NaCo(nta),H_2O$, $HCo(nta),H_2O$, and $Co_2[Co(nta)_2].3H_2O$ have been prepared.[138] The i.r. spectrum of the cobalt(II) complex of mandelic acid suggests structure (28).[139] The substituted malonic acids $RCH=C(CO_2H)_2$ form paramagnetic ($\mu_{eff} = 5.45$—6.15 BM) complexes $K_2Co\{RCH=C(CO_2)_2\},(H_2O)_n$

[128] W. E. Hill, W. Levason, and C. A. McAuliffe, *Inorg. Chem.*, 1974, **13**, 244.
[129] K. K. Chow, W. Levason, and C. A. McAuliffe, *Inorg. Chim. Acta*, 1973, **7**, 589.
[130] S. Ray, A. Zalkin, and D. H. Templeton, *Acta Cryst.*, 1973, **B29**, 2741.
[131] M. W. G. De Bolster, W. L. Driessen, W. L. Groeneveld, and C. J. Van Kerkwijk, *Inorg. Chim. Acta*, 1973, **7**, 439.
[132] J. C. Barnes and C. S. Duncan, *Inorg. Chim. Acta*, 1973, **7**, 404.
[133] E. Lindner, B. Pertikatsis, and A. Thasitis, *Z. anorg. Chem.*, 1973, **402**, 67.
[134] P. S. Gentile, J. White, and S. Haddad, *Inorg. Chim. Acta*, 1974, **8**, 97.
[135] A. S. Tryashin, V. V. Skopenko, and D. A. Stakhov, *Zhur. neorg. Khim.*, 1973, **18**, 2658.
[136] T. Offerson, L. G. Warner, and K. Seff, *Acta Cryst.*, 1974, **B30**, 1188.
[137] P. C. Srivastava, K. B. Pandeya, and H. L. Nigam, *J. Inorg. Nuclear Chem.*, 1973, **35**, 3613.
[138] F. J. M. Rajabalee, *J. Inorg. Nuclear Chem.*, 1974, **36**, 557.
[139] P. V. Khadikav, R. L. Ameria, M. G. Kekre, and S. D. Chauhan, *J. Inorg. Nuclear Chem.*, 1973, **35**, 4301.

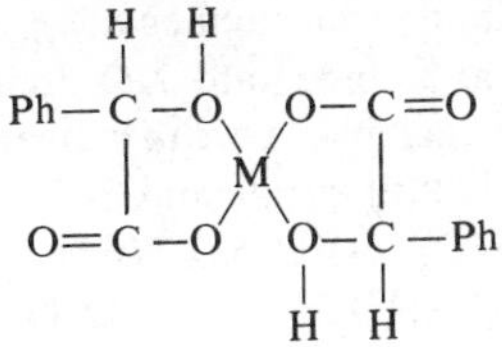

$$
\begin{array}{c}
\underset{H}{\overset{H}{|}}\ \ \ \ \underset{H}{\overset{H}{|}} \\
Ph-C-O\ \ \ \ O-C=O \\
|\ \ \ \ \ \ \diagdown M \diagup\ \ \ \ | \\
O=C-O\ \ \ \ O-C-Ph \\
\overset{|}{H}\ \ \ \ \ \ \overset{|}{H}
\end{array}
$$

(28)

(R = Ph, o-ClC$_6$H$_4$, p-ClC$_6$H$_4$, p-MeOC$_6$H$_4$, or p-NO$_2$C$_6$H$_4$).[140] The magnetic properties[141] of cobalt monoglycerolate, which is known to have a two-dimensional polymeric structure with the cobalt in a distorted trigonal-pyramidal (O$_5$) environment, have been studied over the range 1.6—50 K.

There has been considerable recent interest in heterocyclic amine N-oxide complexes. Detailed i.r. data have been reported for the octahedral Co(pyO)$_6^{2+}$ ion (pyO = pyridine N-oxide).[142] With cobalt(II) thiocyanate and selenocyanate pyO forms a series of complexes[143] including octahedral [Co(pyO)$_4$(NCS)$_2$], [Co(pyO)$_3$(NCS)$_2$-(H$_2$O)], [Co(pyO)$_3$(NCSe)$_2$(H$_2$O)], and [Co(pyO)$_2$(NCS)$_2$(EtOH)]. The last complex contains only terminal NCS groups (i.r. evidence) and hence either bridging EtOH or pyO ligands must be present. Removal of the ethanol produced [Co(pyO)$_2$-(NCS)$_2$], also octahedral but with —NCS— bridges; this loses pyO on further heating to yield a violet [Co(pyO)(NCS)$_2$], in which all three ligands are probably bridging. The thermal degradation reactions are outlined in Scheme 3. Quinoline N-oxide (qnO) forms [Co(qnO)$_3$(NCX)$_2$] (X = S or Se) which, on the basis of their electronic spectra are formulated as [Co(qnO)$_6$][Co(NCX)$_4$].[143] Cobalt(II) sulphate forms a hydrated 1:1 complex with 2,6-lutidine N-oxide (LNO) which contains ionic sulphate groups; the cation probably has the structure [(H$_2$O)$_4$Co(LNO)$_2$Co(H$_2$O)$_4$]$^{4+}$ with LNO bridges, and is assigned a six-co-ordinate structure based on electronic

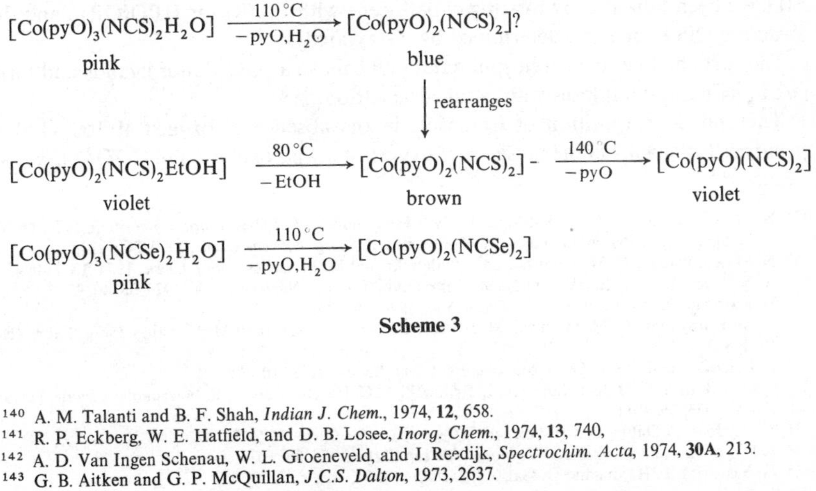

$$
\text{[Co(pyO)}_3\text{(NCS)}_2\text{H}_2\text{O]} \xrightarrow[-\text{pyO,H}_2\text{O}]{110\,^\circ\text{C}} \text{[Co(pyO)}_2\text{(NCS)}_2\text{]?}
$$

pink · blue

rearranges ↓

$$
\text{[Co(pyO)}_2\text{(NCS)}_2\text{EtOH]} \xrightarrow[-\text{EtOH}]{80\,^\circ\text{C}} \text{[Co(pyO)}_2\text{(NCS)}_2\text{]} \xrightarrow[-\text{pyO}]{140\,^\circ\text{C}} \text{[Co(pyO)(NCS)}_2\text{]}
$$

violet · brown · violet

$$
\text{[Co(pyO)}_3\text{(NCSe)}_2\text{H}_2\text{O]} \xrightarrow[-\text{pyO,H}_2\text{O}]{110\,^\circ\text{C}} \text{[Co(pyO)}_2\text{(NCSe)}_2\text{]}
$$

pink

Scheme 3

[140] A. M. Talanti and B. F. Shah, *Indian J. Chem.*, 1974, **12**, 658.

[141] R. P. Eckberg, W. E. Hatfield, and D. B. Losee, *Inorg. Chem.*, 1974, **13**, 740,

[142] A. D. Van Ingen Schenau, W. L. Groeneveld, and J. Reedijk, *Spectrochim. Acta*, 1974, **30A**, 213.

[143] G. B. Aitken and G. P. McQuillan, *J.C.S. Dalton*, 1973, 2637.

spectral and magnetic susceptibility measurements ($\mu_{eff} = 4.6$ BM).[144] o-Phenanthroline mono-N-oxide behaves as a bidentate NO-donor in $[Co(phenO)_3](ClO_4)_2$; the electronic spectra suggest that it is a weaker ligand than o-phen itself.[145] The complexes formed between cobalt(II) nitrate and 2-, 3-, and 4-picoline N-oxides have been formulated[146] $[Co(2-picO)_2(ONO_2)(O_2NO)]$, $[Co(3-picO)_2(O_2NO)_2]$, $[Co-(2-picO)_4(ONO_2)]NO_3$, $[CoL_6](NO_3)_2$ (L = 3-picO or 4-picO), and a complex of stoicheiometry $Co(NO_3)_2,(4-picO)$ which may be $[(ONO_2)_2Co(4-picO)_2Co-(O_2NO)_2](O_2NO$ and ONO_2 signify bidentate and unidentate nitrate ligands, respectively). Electronic spectral and magnetic data indicate a five-co-ordinate environment for the 2-picO complexes and a six-co-ordinate distorted octahedral geometry for the 1:1, 2:1, or 6:1 complexes of 3-picO or 4-picO, which demonstrates the effect of steric hindrance introduced by the 2-Me substituent in 2-picO. Pyrazine mono-N-oxides forms $[Co_2L_3Cl_4]_n$ ($n = 1$ or 2) which contains[147] five-co-ordinate cobalt, terminal N-bonded pyzO, bridging NO-bonded pyzO, and bridging and terminal chlorines.

The m.c.d. spectra of tetrahedral $Co(Ph_3PO)_4^{2+}$, $Co(Ph_3PO)_2X_2$ (X = Cl, Br, I, or NCS), $Co(Bu^n_3PO)_4^{2+}$, as well as of $Co(NCS)_4^{2-}$ and $Co(PPh_3)_2X_2$, have been examined.[148] Triphenylphosphine oxide does not form a complex with cobalt(II) sulphate,[146] but forms $[Co(PPh_3O)_4](ClO_4)_2$ with the perchlorate.[149] The dioxide $Ph_2As(O)CH_2CH_2As(O)Ph_2$ (daeO$_2$) forms 2:1 and 3:1 complexes with cobalt(II) perchlorate.[150] The former is probably five-co-ordinate, $\mu_{eff} = 4.24$ BM, with one co-ordinated ClO_4^- group, whilst the latter is octahedral, $\mu_{eff} = 5.06$ BM, with $[Co(daeO_2)_3]^{2+}$ cations containing seven-membered chelate rings.

The organophosphoryl ligands di-isopropylmethylphosphonate (dimp), dimethylmethylphosphonate (dmp), and trimethylphosphate (tmp) form the complexes $[Co(dimp)_4OClO_3]ClO_4$, $[Co(dmp)_4(H_2O)_2](ClO_4)_2$, and $[Co(tmp)_5](ClO_4)_2$, respectively.[149] However, on heating tetramethylethylenediphosphonate with $CoCl_2$ the ligand rearranges into diethylmethylenediphosphonate and $Co_2L(LH)$ is isolated; this is probably a square-pyramidal moiety, $\mu_{eff} = 5.12$ BM.[151]

Very stable paramagnetic complexes $(\pi$-$C_5H_5)_2Co_3\{PO(OR)_2\}_6$ (R = Me or Et) have been obtained by treating cobaltocene with $P(OR)_3$ or $HPO(OR)_2$, and the structure (29) of one was determined by X-ray analysis.[152]

The carbamylmethylenephosphonates (30) behave as OO-donor ligands and form $[CoL_3](ClO_4)_2$ complexes with octahedral cations.[153]

Thermal decomposition of $Co(acac)_3$ in the absence of oxygen at 100—130 °C produces $Co(acac)_2$; in the presence of oxygen stepwise oxidation to $Co(OAc)_2,xH_2O$ occurs.[154]

[144] N. M. Karayannis, C. M. Mikulski, L. L. Pytlewski, and M. M. Labes, *Inorg. Chim. Acta*, 1974, **10**, 97.
[145] A. N. Speca, L. L. Pytlewski and N. M. Karayannis, *J. Inorg. Nuclear Chem.*, 1974, **36**, 1227.
[146] N. M. Karayannis, C. M. Mikulski, L. L. Pytlewski, and M. M. Labes, *Inorg. Chem.*, 1974, **13**, 1146.
[147] A. N. Speca, L. L. Pytlewski, and N. M. Karayannis, *J. Inorg. Nuclear Chem.*, 1973, **35**, 4029.
[148] H. Kato and K. Akimoto, *J. Amer. Chem. Soc.*, 1974, **96**, 1351.
[149] N. M. Karayannis, C. M. Mikulski, M. J. Strocker, L. L. Pytlewski, and M. M. Labes, *Inorg. Chim. Acta*, 1974, **8**, 9.
[150] B. J. Brisdon and D. Cocker, *Inorg. Nuclear Chem. Letters*, 1974, **10**, 179.
[151] C. M. Mikulski, N. M. Karayannis, L. L. Pytlewski, R. O. Hutchins, and B. E. Maryanoff, *J. Inorg. Nuclear Chem.*, 1973, **35**, 4011.
[152] V. Hardner, E. Dubler and H. Werner, *J. Organometallic Chem.*, 1974, **71**, 427.
[153] B. D. Catsikis and M. L. Good, *J. Inorg. Nuclear Chem.*, 1974, **36**, 1039.
[154] G. Vasvari, I. P. Hajdu, and D. Gal, *J.C.S. Dalton*, 1974, 465.

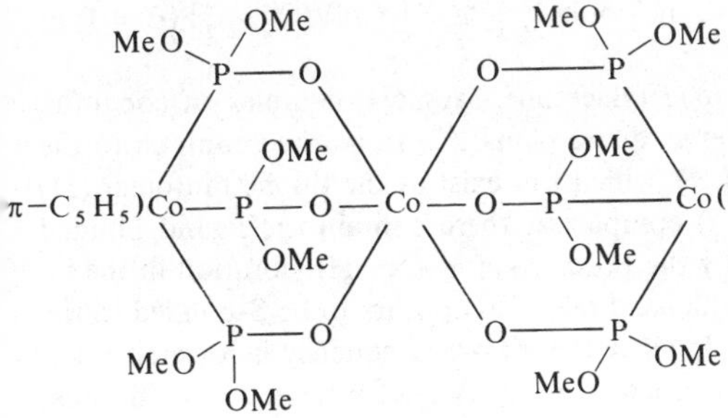

(29)

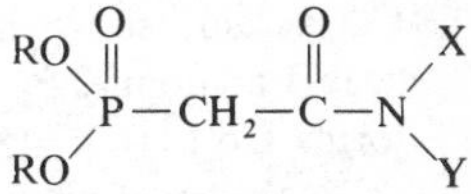

(30)

$R = X = Y = Me$

$R = Pr, X = H, Y = H$ or Bu^t

$R = Bu, X = Y = Et$

n-Hexyl-, n-octyl-, n-undecyl-, and n-hexadecyl-amine form bis-adducts with cobalt(II) trifluoroacetylacetonate, whilst hexamethylenediamine, 1,11-diaminoundecane, and 1,12-diaminododecane form mono-adducts with both bis(trifluoroacetylacetonato)cobalt(II) and $Co(acac)_2$. All these complexes appear to have polymeric structures, but substantial dissociation of the amines occurs in solution.[155] An X-ray structure of $Co(antipyrine)_2(NO_3)_2$ shows the presence of a distorted CoO_6 octahedron formed from two unidentate (O-donor) antipyrine ligands and two bidentate nitrato-groups.[156]

S-Donor Ligands. Dance[157,158] has examined the adducts formed between cobalt dithiolenes, $[CoL_2]_2^{x-}$ [$L = (CF_3)_2C_2S_2^{2-}$ and $(NC)_2C_2S_2^{2-}$(mnt)], and primary, secondary, tertiary, and heterocyclic unidentate and bidentate amines in both solution and the solid state. Both five- and six-co-ordinate adducts are formed. Novel structures result from $Co(mnt)_2^{2-}$ and $[Co(mnt)_2]_2^{2-}$ with 1,4-disubstituted pyridinium cations in which the anions are 'sandwiched' between two cations, and anion → cation charge-transfer spectra have been measured.[159]

Sodium cyclopentadienedithiocarboxylate, Et_4NBr, and $CoBr_2$ react in acetonitrile at $-70\,°C$ to produce $[Et_4N]_2Co(C_5H_4—CS_2)_2$, $\mu_{eff} = 2.55$ BM, which contains cobalt in a distorted square-planar environment.[160] Cobalt(II) diphenyl- and di(p-tolyl)-dithiophosphinate (L) form adducts with heterocyclic nitrogen bases of types $[CoL_2Q]$ ($Q = \alpha$-picoline, β-picoline, 2-aminopyridine, quinoline, or isoquinoline), which are five-co-ordinated CoS_4N chromophores, and $[CoL_2Q_2]$ ($Q =$ pyridine, γ-picoline, or isoquinoline), which are six-co-ordinate CoS_4N_2 moieties.[161] The reaction of $CoCl_2$ with potassium diphenylmonothioarsinate, dibenzylmonothioarsinate, and dibenzyldithioarsinate produces $Co\{R_2As(X)S\}$ ($X = O$ or S).[162] A spectrochemical series for these and related ligands has been established as

$$Bz_2As(O)S^- \sim Ph_2As(O)S^- < Bz_2AsS_2^- < Ph_2AsS_2^- \sim Ph_2PS_2^-.$$

[155] D. A. Fine, *J. Inorg. Nuclear Chem.*, 1973, **25**, 4023.

[156] C. Brassy, J. P. Mornon, and J. Delettre, *Acta Cryst.*, 1974, **B30**, 2243.

[157] I. G. Dance and T. R. Miller, *Inorg. Chem.*, 1974, **13**, 525.

[158] I. G. Dance, *Inorg. Chem.*, 1973, **12**, 2381.

[159] I. G. Dance and P. J. Solstad, *J. Amer. Chem. Soc.*, 1973, **95**, 7256.

[160] B. J. Kalbacher and R. D. Bereman, *Inorg. Chem.*, 1973, **12**, 2917.

[161] R. N. Mukherjee, M. S. Venkateshan, and M. D. Zingde, *J. Inorg. Nuclear Chem.*, 1974, **36**, 547.

[162] A. Müller, P. Werle, P. Christophliemk, and I. Tossides, *Chem. Ber.*, 1973, **106**, 3601.

The $WO_2S_2{}^{2-}$ ion behaves as an S-donor only in $[Ph_4E][Co(WO_2S_2)_2]$ (E = P or As).[163]

The crystal structure of the dithioacetylacetone (sacsacH) complex of cobalt(II), $Co(sacsac)_2$, shows it to be monomeric square planar,[164] in marked contrast to the acacH analogue. Pyridine 2-thiol (LH), which can exist as the thione tautomer (31) forms $Co(LH)_2X_2$ (X = Cl, Br, or I) complexes; these contain the ligand bonded through the sulphur, as evidenced by the presence of a $v(N—H)$ vibration in the i.r. spectra.[165] Morpholine 4-thiocarbonic acid (32) also appears to be S-bonded in the CoL_2X_2 complexes (X = Cl, Br, or I), since the $v(C=S)$ frequency is lowered upon co-ordination.[166] The electronic spectra of the CoL_2Cl_2 complexes (L = thiourea, N-phenylthiourea, and NN'-diphenylthiourea) have been recorded.[167]

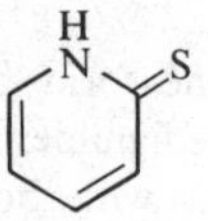

(31) 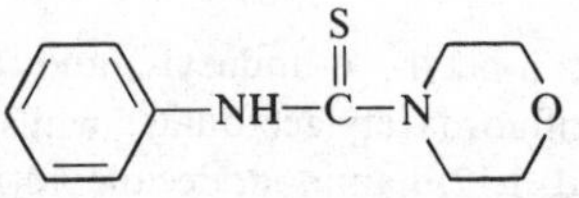   (32)

The open-chain quadridentate thioether, 1,2-bis(*o*-methylthiophenylthio)ethane (33), forms a *trans*-octahedral $[CoLI_2]$ complex, readily decomposed by donor solvents. Other cobalt(II) halides did not react with (33), and no cobalt(II) complexes formed with the 1,3-propane or 1,4-butane analogues.[168]

The $[(\pi\text{-}C_5H_5)Co(SR)]_2$ (R = Me, *p*-MeC_6H_4, or *p*-ClC_6H_4) complexes undergo voltammetric oxidation to the monocation derivatives, but, unlike the Ni or Fe analogues, they do not form dications.[169]

Formation of five-co-ordinate distorted square-pyramidal complexes CoL_2X_2 (X = Cl, Br, or I) with the ligand (34) has been reported, and the possible modes of co-ordination were discussed. I.r. spectra suggest that the ligand functions as a bidentate S_2-donor. The corresponding ligand containing oxygen instead of the

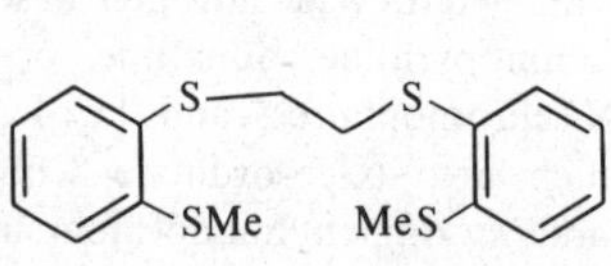

(33) 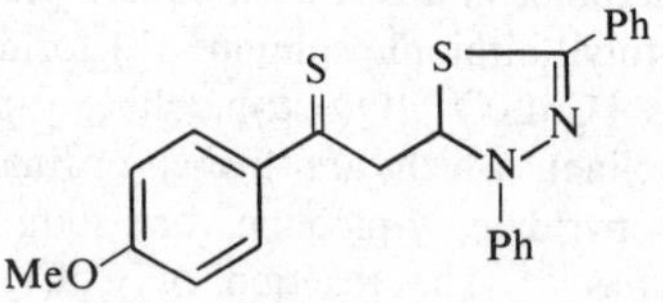   (34)

[163] A. Müller, H. N. Heinsen, and G. Vandrish, *Inorg. Chem.*, 1974, **13**, 1001.
[164] R. Beckett and B. F. Hoskins, *J.C.S. Dalton*, 1974, 622.
[165] I. P. Evans and G. Wilkinson, *J.C.S. Dalton*, 1974, 946.
[166] D. Venkappayya and D. H. Brown, *J. Inorg. Nuclear Chem.*, 1974, **36**, 1023.
[167] F. Pruchnik, *J. Mol. Structure*, 1973, **19**, 447.
[168] W. Levason, C. A. McAuliffe, and S. G. Murray, Proceedings 16th International Coordination Chemistry Conference, Dublin, 1974.
[169] P. Douglas, M. K. Loyd, J. A. McCleverty, and D. Seddon, *J.C.S. Dalton*, 1973, 2268.

thione sulphur also forms complexes.[170] Cobalt(II) complexes of 8-selenoquinoline have been prepared.[171]

Mixed Donor Ligands. A seven-co-ordinate cobalt(II) ion is completely enclosed in a macrocyclic polyether, $C_{10}H_{32}N_2O_5$, in $[CoL][Co(NCS)_4]$.[172] Tri-propan-2-olamine behaves as a terdentate NO_2-donor with one unco-ordinated OH group in $[CoL_2]SO_4$.[173] The cobalt(II) complexes of $NNN'N'$-tetrakis-(2-hydroxyethyl)-ethylenediamine, $CoLX_2$ (X = Cl, NCS, or NO_3) and Co(L − H)(NCS) have low magnetic moments, indicating a polymeric constitution,[174] probably with octahedrally co-ordinated cobalt(II). A complex Co_2LCl_4 may, however, be tetrahedral.[174] NN'-Ethylenediaminediacetic acid (H_2L) forms CoL, nH_2O (n = 3 or 4).[175] Motekaitis and Martell[176] reported the formation constants and structures of compounds formed as a function of pH in the cobalt(II) di-, tri-, and tetra-glycine-NN'-diacetic acid systems. At low pH co-ordination is *via* the acetate end-groups, but as the pH increases deprotonation at the amido nitrogen occurs.

Aroylhydrazines, $H_2NNHC(O)R$ (R = Ph, p-MeC_6H_4, o-HOC_6H_4, or p-$NO_2C_6H_4$), form CoL_3X_2 (L = benzoylhydrazine; X = Cl, Br, I, NO_3, or $\frac{1}{2}SO_4$) and CoL_2Cl_2 (L = o-hydroxy-, p-nitro-benzoylhydrazine), and all four ligands form bis-ligand thiocyanate complexes.[177] Magnetic and electronic spectral results suggest octahedral structures, and the ligands function as bidentate NO-donors through the carbonyl oxygen and the amino nitrogen. N-Acetyl-N'-benzoylhydrazine (abH_2) forms octahedral $Co(abH)_2,2H_2O$ (35), in which the ligand is terdentate.[178]

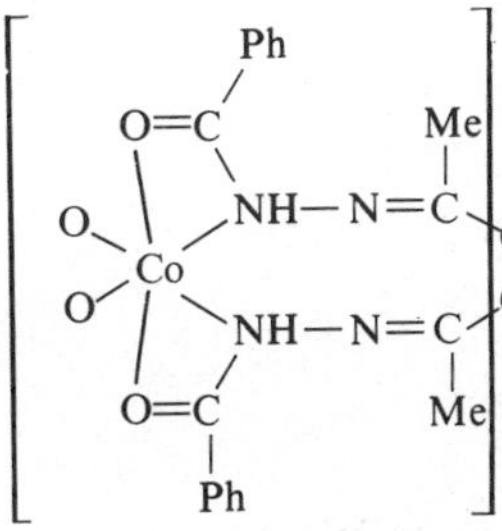

(35)

$R(CH_2)_nNHCOC_5H_4N$

(36)

R	n	
NH_2	2, 3	enpH, tnpH
NMe_2	2, 3	dmepH, dmppH
SEt	2	etepH
OMe	2	moepH

Salicylaldehyde salicylhydrazone (ssh), $HOC_6H_4CONHN{=}CHC_6H_4OH$, forms Co(ssh − H)$_2$ and Co(ssh − 2H), $2H_2O$. Both contain octahedral cobalt(II); in the former co-ordination is *via* OH and CN, whilst in the latter the OH, C=O, and CN

[170] F. Y. Petillon, J. Y. Calves, J. E. Guerchais, and Y. M. Poirer, *J. Inorg. Nuclear Chem.*, 1973, 35, 3751.
[171] Y. Mido, I. Fujiwara, and E. Sekido, *J. Inorg. Nuclear Chem.*, 1974, 36, 537, 1003.
[172] F. Mashieu and R. Weiss, *J.C.S. Chem. Comm.*, 1973, 816.
[173] S. C. Rastorgi and G. N. Rao, *J. Inorg. Nuclear Chem.*, 1974, 36, 1889.
[174] D. N. Zimmerman and H. H. Downs, *Inorg. Nuclear Chem. Letters*, 1973, 9, 1089.
[175] E. M. Urinovich, N. F. Shugal, Y. V. Oboznenko, and N. M. Dyaltlova, *Zhur, neorg. Khim.*, 1974, 19, 425.
[176] R. J. Motekaitis and A. E. Martell, *Inorg. Chem.*, 1974, 13, 550.
[177] M. F. Iskander, S. E. Zayan, M. A. Khalifa, and L. El-Sayed, *J. Inorg. Nuclear Chem.*, 1974, 36, 551.
[178] R. C. Aggarwal and K. K. Narang, *Inorg. Chim. Acta*, 1973, 7, 651.

are co-ordinated.[179] Complexes of salicylic acid hydrazide (LH), $Co(LH)_2(NCS)_2$ and $CoL_3,1.5H_2O$ have also been obtained.[180]

The picolinamide ligands (36) form octahedral $Co(moepH)_2Br_2,4H_2O$, Co-$(moepH)_2(NCS)_2$, and $Co(etepH)_2(NCS)_2$, $\mu_{eff} = 4.6$—4.85 BM containing ligands bonded through picoline N and the CO group.[181]

Orange $Co_3(pap)_2(SO_4)_2,12H_2O$ is formed on reaction of N-(2-pyridylmethylene-amino)-2-pyridine-carboxamide (papH) with $CoSO_4$ in aqueous ethanol in the presence of NaOH.[182] Spectral data suggest that octahedral cobalt(II) is present and that the bonding is as shown in (37).

Complexes of benzoylformic acid selenosemicarbazide, $Ph(CO_2H)C=NNHC$-$(Se)NH_2$ (H_2L), $[NH_4][CoL_2],H_2O$ and $[Co(HL)_2],1.5H_2O$, are formed by reaction of acetone selenosemicarbazone and $PhC(O)CO_2H$ with Co^{2+} in aqueous ammonia.[183] The preparation of bis-(5-ethyl-5-phenylbarbiturato)-bisimidazolecobalt(II), which is a tetrahedral complex containing a CoN_4 donor set, has been briefly reported.[184] This contrasts with the structure of the nickel(II) analogues, in which the ligand is an ON-donor. N-Salicylideneglycine is terdentate, forming complexes of types (38) and (39).[185] Mixed complexes of diphenylglycine (HL) and some sulphanil-

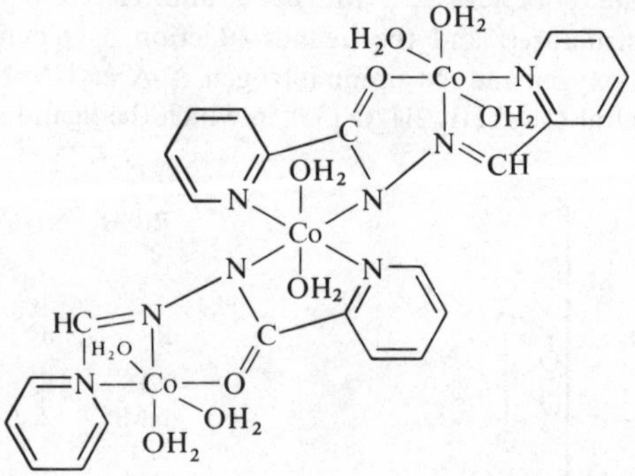

(37)

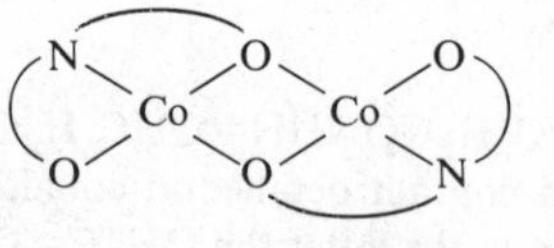

(38) 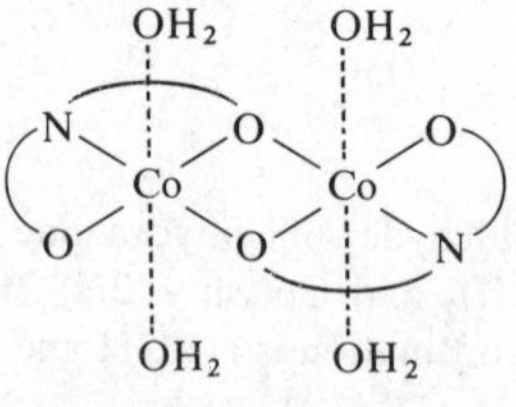   (39)

[179] K. K. Narang and A. Aggarwal, *Inorg. Chim. Acta*, 1974, **9**, 137.
[180] Ya. Ya. Kharitonov, R. I. Machkhoshvili, and N. B. Generalova, *Zhur. neorg. Khim.*, 1974, **19**, 270.
[181] M. Nonoyama and K. Yamasaki, *Inorg. Chim., Acta* 1973, **7**, 676.
[182] M. Nonoyama, *Inorg. Chim. Acta.* 1974, **10**, 133.
[183] N. V. Gerbelau and V. G. Bodyu, *Zhur. neorg. Khim.*, 1973, **18**, 3001.
[184] B. C. Wang, W. R. Walker, and N. C. Li, *J. Coord. Chem.*, 1973, **3**, 179.
[185] R. Pani and B. Behera, *Indian J. Chem.*, 1974, **12**, 215.

amides, (Q), *trans*-CoL_2Q_2 and $[CoL_2QQ']$ (Q = *N*-cyanosulphanilamide, Q' = sulphanilamide or sulphadimezine) have been prepared; co-ordination is *via* CN of Q and the amine groups of the benzene.[186]

Tetrahedral $[CoA_2]$ and octahedral $[CoB(H_2O)_2]$ and $[CoB(NH_3)_2]$ are formed by 8-amino-7-hydroxy-4-methylcoumarin (AH) and 3,3'-diamino-4,4'-dihydroxy-diphenylsulphone (BH_2).[187] The ligand (40) forms a tetrahedral 2:1 complex with cobalt(II).[188] I.r.[189] data indicate that in the CoL_2 (LH = 1-substituted tetrazoline-5-thione) and CoL'_2py_2 [L'H = 3-(4-pyridyl)triazoline-5-thione] NS-donor ligands are present, whilst thiazoline-2-thione (L'') is S-bonded in CoL''_2Cl_2, and thio-carbohydrazine (L''') is N-bonded in CoL'''_2X_2 (X = Cl, Br, I, NO_3, or NCS). The cobalt(II) complex, [CoL(OH)], $3H_2O$, of 1-phenyl-*o*-hydroxy-4-benzamidothiosemi-carbazone (LH) contains the ligand bonded through the thioketone S, amido N, and phenolic O.[190] A comparison[191] of 2-amino- and 2-acetylamino-thiazole shows that the former co-ordinates through the amine group and the latter *via* the CO in CoL_2X_2 complexes.

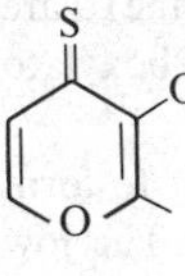

(40)

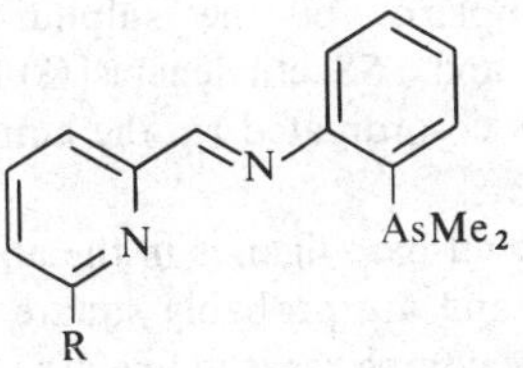

(41) R = H, NNAs
      R = Me, MeNNAs

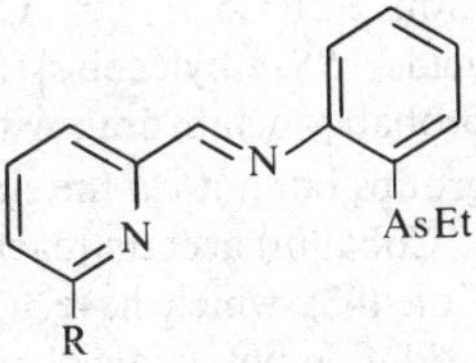

(42) R = H, NNAsEt
      R = Me, MeNNAsEt

The series $[Co(pc)_2]X_2$ [X = ClO_4, BF_4, NO_3, or $Co(NCS)_4$] and $Co(pc)X_2$ (X = Cl, or Br) are formed with 1,10-phenanthroline-2-carbothioamide (pc) in which the ligand is co-ordinated through the ring nitrogens and the sulphur. Both types of complex contain octahedral cobalt(II), the $Co(pc)X_2$ by virtue of bridging X groups, and the $[Co(pc)_2]^{2+}$ cations show anomalous temperature-induced spin trans-itions.[192] The *N*-phenyl-substituted analogue reacts with cobalt(II) with deprotonation at the NPhH group and hence the formation of cobalt(III).[192]

Chiswell and Lee[193] have synthesized cobalt(II) complexes of four terdentate ligands, (41) and (42), with an AsNN donor sequence. Complexes of formula $[CoL_2]X_2$ (X = ClO_4, I, or BPh_4) have been isolated for all four ligands, but for X = NO_3 only with NNAs and NNAsEt. The 6-methyl-substituted ligands produce $[CoL(NO_3)_2]$. Cobalt(II) iodide formed both 1:2 and 2:1 complexes, but the chloride, bromide, and thiocyanate produced only 1:1 complexes of either $CoLX_2$ or $[CoL_2][CoX_4]$ types.

[186] V. N. Shafranski and I. L. Fusu, *Izvest. Akad. Nauk Mold. S.S.R., Ser. biol. khim. Nauk,* 1973, 79.
[187] W. U. Malik, D. K. Rastorgi, and M. P. Teotia, *Indian J. Chem.,* 1973, **11**, 1303.
[188] E. Uhlemann, H. Motzny, and G. Wilke, *Z. anorg. Chem.,* 1973, **401**, 205.
[189] B. Singh and K. P. Thakur, *J. Inorg. Nuclear Chem.,* 1974, **36**, 1735.
[190] A. K. Srivastava. V. B. Rana, and M. Mohan, *J. Inorg. Nuclear Chem.,* 1974, **36**, 2118.
[191] P. P. Singh and A. K. Srivastava, *Austral. J. Chem.,* 1974, **27**, 504.
[192] H. A. Godwin, D. W. Mather, and F. E. Smith, *Austral. J. Chem.,* 1973, **26**, 2623.
[193] B. Chiswell and K. W. Lee, *Inorg. Chim. Acta,* 1973, **7**, 509.

I

The octahedral iodo or perchlorato compounds of MeNNAsEt have anomalous magnetic moments due to a high-spin ⇌ low-spin equilibrium. The low-spin $CoLI_2$ (L = NNAs or NNAsEt) have moments of 2.48—3.03 BM, and in conjunction with the electronic spectral data which are inconsistent with tetrahedral or five-co-ordinate geometry, this suggests that the compounds may be rare examples of planar cobalt(II).

Octahedral $[CoL_2(NCS)_2]$ and five-co-ordinate $[CoL_2I]I$ and $[CoL_2(OClO_3)]$-ClO_4 are formed from 2-diethylphosphinomethylpyridine. The structures of the 'CoLX$_2$' (X = Cl or Br) are less clear, although the data point to a $[CoL_2][CoX_4]$ formulation, possibly with cation–anion interaction, e.g. $[L_2Co–Cl–CoCl_3]$.[194] Under high pressure $[Co\{Et_2N(CH_2)_2NH(CH_2)_2PPh_2\}(NCS)_2]$ undergoes a high-spin→low-spin change.[195]

Octahedral $[CoL_3]X_2$ (X = ClO$_4$ or $\frac{1}{2}$CoBr$_4$) containing the zwitterionic amino-acid, DL-methylsulphonium-methioninate, $Me_2\overset{+}{S}CH_2CH_2CH(NH_2)COO^-$, contain bidentate aminocarboxylate co-ordination.[198] Facile deprotonation of the amino nitrogen produces the amido-acid, $Me_2\overset{+}{S}CH_2CH_2CH(NH^-)COO^-$, (L'), which forms $Li[CoL'_3]$.[196] Cobalt(II) complexes of the sulphur-containing amino-acids SS'-ethylenebis[(R)-cysteine] and SS'-ethylenebis[(S)-homocysteine] are probably octahedral, with the ligands co-ordinated *via* the amino and carboxylato groups but not *via* the sulphur.[197]

Cobalt(II) acetate reacts with the Schiff-base ligands in the absence of air to form CoL (43), which have $\mu_{eff} \simeq 2.0$ BM and are probably square planar.[198] The low oxidation potentials observed for these complexes is in line with their ability to form oxygen adducts.[199] NN'-Bis(o-aminobenzylidene)-(−)(R)-propane-1,2-diamino-cobalt(II), the (R, R)-cyclohexane-1,2-diamine analogue,[200] and NN'-bis-(2-pent-2-en-4-oxo)-(−)(R)-propanediaminocobalt(II) are planar complexes. A detailed electronic spectral study has established the d-orbital ordering to be $xy > z^2 > yz > x^2 - y^2 > xy$. The failure of the first two complexes to react with oxygen, compared with the highly oxygen-sensitive third complex, is attributed to stabilization of the $3d_{z^2}$ orbital by the stronger in-plane ligand field in the N_4 complexes.[201] The condensation of salicylaldehyde and the appropriate diamine and subsequent reaction

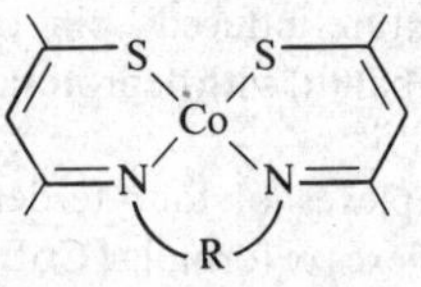

(43) R = —CH$_2$CH$_2$—

—CH$_2$CHMe—

cis-CH——CH—

(CH$_2$)$_4$

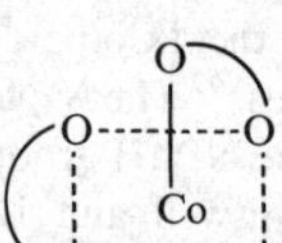

(44)

[194] M. Schäfer and E. Uhlig, *Z. anorg. Chem.*, 1974, **407**, 23.
[195] L. Sacconi and J. R. Ferraro, *Inorg. Chim. Acta*, 1974, **9**, 49.
[196] C. A. McAuliffe and W. D. Perry, *Inorg. Chim. Acta*, 1974, **10**, 215.
[197] R. J. Magee, W. Mazurek, M. J. O'Connor, and A. T. Phillip, *Austral. J. Chem.*, 1974, **27**, 1629.
[198] P. R. Blum, R. M. C. Wei, and S. C. Cummings, *Inorg. Chem.*, 1974, **13**, 450.
[199] W. R. Pangratz, F. L. Urbach, P. R. Blum, and S. C. Cummings, *Inorg. Nuclear Chem. Letters*, 1973, **9**, 1141.
[200] F. L. Urbach, R. D. Bereman, J. A. Topich, M. Harihan, and B. J. Kalbacher, *J. Amer. Chem. Soc.*, 1974, **96**, 5063.
[201] J. A. Fanikan, K. S. Patel, and J. C. Bailar, *J. Inorg. Nuclear Chem.*, 1974, **36**, 1547.

with cobalt(II) acetate under nitrogen leads to the isolation of $Co^{II}(Sal_2Bg)$ $(Sal_2Bg = o\text{-}HOC_6H_4C{=}N{-}L{-}N{=}CC_6H_4OH$, $-L-$ $= CH_2CHMe$, CH_2CMe_2, $o\text{-}C_6H_4$, and *trans*-1,2-cyclohexyl).[202] Complexes of the ligand NN'-bis-(salicylidene)-1, 1-(dimethyl)ethylenediamine with cobalt(II), nickel(II), and copper(II) have been synthesized and their i.r. spectra recorded.[203]

The Schiff bases $o\text{-}HOC_6H_4C(R^1){=}N(CH_2)_2NR_2^2$ (LH) react with $Co(acac)_2$ to form $Co(acac)L$ which appear to be indefinitely stable as solids but oxidize in solution.[203] When $R^2 = Et$ and $R^1 = H$ the product has structure (44) and is monomeric with $\mu_{eff} = 4.32$ BM. However, for $R^2 = Me$, $R^1 = Me$ or H, the products are octahedral polymers.

An unusual reaction occurs on mixing Cu(salen) and $Co(hfac)_2$, or Co(salen) and $Cu(hfac)_2$, in dichloromethane (hfac = hexafluoroacetylacetone).[204] In both cases $Cu(salen)Co(Hfac)_2$ is formed, in which the copper atom is in a square-planar environment and the cobalt is octahedral, bonded to the hfac and to the phenolic oxygens of the salen which bridge the two metals. The formation from Co(salen) and Cu(hfac) requires the two metals to interchange co-ordination spheres. A similar reaction occurs between $Cu(CF_3acac)_2$ and Co(salen), but the exchange is slower $[CF_3acacH = CF_3C(O)CH_2C(O)CH_3]$.[204]

The bonding of imidazole, pyrazole, and N-methylimidazole to Co(saloph) (saloph = NN'bis(salicylidene)-o-phenylenediamine) and $Co(acac)_2$ in dichloromethane has been studied.[205] Imidazole and pyrazole form two types of compound with Co(saloph), one bonded through the amine N, the other through the imine N, whilst $Co(acac)_2$ predominantly bonds to the imine N.

Tetrahedral $Co(R^1R^2C{=}N{-}NCSSMe)_2$ $(R^1, R^2 = Me; R^1, R^2 = C_5H_{10}; R^1 = Me, R^2 = Ph)$[206] and octahedral CoL_2 $[LH = MeC(O)CH_2C(Ph){=}N(CH_2)_2$-OH]^{207} have been reported. Prince and Stotter[208] have prepared $CoL(ClO_4)_2$ with the acetylpyridine-derived Schiff base $(2\text{-}C_5H_4N)C(Me){=}N(CH_2)_3N{=}C(Me)(2\text{-}C_5H_4N)$.

Other Compounds. Cobaltocene, Cp_2Co, has been electrolytically reduced in dimethylethane to form the very air-sensitive Cp_2Co^-, the first metallocene anion.[209] The diamagnetic modification of $(PhNC)_4CoI_2$ contains a strictly linear $Co{-}I{-}Co$ bridge, $[I{-}Co(PhNC)_4{-}I{-}Co(PhNC)_4{-}I]I$, through which spin-exchange occurs.[210] X-Ray crystal structures have been reported for $Co_3(PO_4)_2$,[211] $K_2Co(PO_3)_4$,[212] and one form of cobalt(II) arsenate, $Co_{6.95}As_{3.62}O_{16}[CoOAs_2O_5]$.[213]

Interactions with Dioxygen. Formation of a 1:1 diamagnetic adduct at low temperatures in HMPA solutions of [Co(pc)(4-Mepy)] (pc = phthalocyanine; 4-Mepy =

[202] R. M. McAllister and J. H. Weber, *J. Organometallic Chem.*, 1974, **77**, 91.
[203] R. H. Babundgi and A. Chakravorty, *Inorg. Nuclear Chem. Letters*, 1973, **9**, 1045.
[204] N. B. O'Bryan, T. O. Maier, I. C. Paul, and R. S. Drago, *J. Amer. Chem. Soc.*, 1973, **95**, 6640.
[205] B. S. Tovey and R. S. Drago, *J. Amer. Chem. Soc.*, 1974, **96**, 2743.
[206] L. El-Sayed, M. F. Iskander, and A. El-Touky, *J. Inorg. Nuclear Chem.*, 1974, **36**, 1739.
[207] S. L. Pania, K. N. Kuhl, and R. K. Mehta, *Current Sci.*, 1974, **13**, 43.
[208] R. H. Prince and D. A. Stotter, *Inorg. Chim., Acta*, 1974, **10**, 89.
[209] W. E. Geiger, *J. Amer. Chem. Soc.*, 1974, **96**, 2632.
[210] D. Baumann, H. Endress, H. J. Keller, and J. Weiss, *J.C.S. Chem. Comm.*, 1973, 853.
[211] A. G. Nord, *Acta Chem. Scand.*, 1974, **A28**, 150.
[212] M. Laüght, I. Tordjman, G. Bassi, and J. C. Guitel, *Acta Cryst.*, 1974, **B30**, 1100.
[213] N. Krishnamachari and C. Calvo, *Canad. J. Chem.*, 1974, **52**, 46.

4-methylpyridine) has been detected by e.p.r. spectroscopy.[214] Basolo *et al.*[215] have obtained thermodynamic data for reversible oxygen uptake by cobalt(II) complexes of the Schiff bases (45). A linear correlation between the equilibrium constant for oxygen-adduct formation and the ease of $Co^{II} \rightarrow Co^{III}$ oxidation, as measured by cyclic voltammetry, was found, which suggests that the electron density on cobalt (related to the redox potential) is an important factor in determining oxygen affinity in these compounds. An e.p.r. spectroscopic study of $[Co(bzacen)(py)(O_2)]$ {bzacen = *NN*'-ethylenebis(benzoylacetoniminate)}, using ^{17}O dioxygen suggests that the oxygen atoms are magnetically equivalent.[216]

X-Ray structures have been reported for {*NN*'-butylenebis(salicylideneiminato)} (pyridine)cobalt(II),[219] and the *meso* form {*NN*'-cyclohexylenebis(salicylideneiminato)} cobalt(II).[218] The former contains two types of cobalt environment, one essentially square-pyramidal and the other intermediate between square-pyramidal

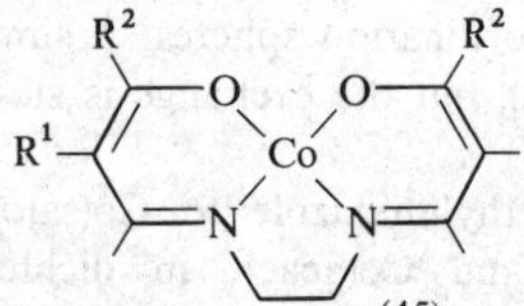

(45) 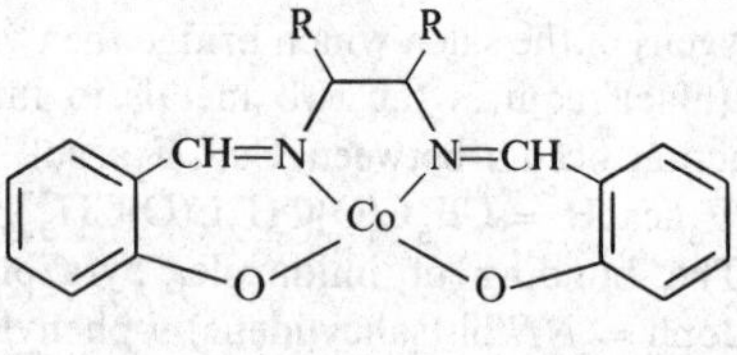   (46)

and trigonal-bipyramidal, whilst the latter compound consists of square-pyramidal molecules. The different behaviour of these molecules towards oxygenation was discussed in terms of steric effects.[217] The reversible addition of dioxygen to (46) shows that generally the adduct is more stable with Schiff bases containing *meso* diamines rather than racemic diamines.[219] The dioxygen complexes $[Co(salen)Z]_2,O_2$ (Z = imidazole, *etc.*) and $[Co(salen)Q_2]_2,O_2$ (Q = picoline or 2,6-lutidine) have been studied.[220] A new type of cobalt(II) complex, which exhibits reversible oxygen uptake, contains amino-acids and *N*-ethylimidazole or 1,2-dimethylimidazole ligands.[221] The aquoterpyridine-(1,10-phenanthroline)cobalt(II) cation readily takes up oxygen in solution at low pH to give an adduct which may rearrange or be readily oxidized to the superoxide.[222] Slightly acidic solutions of $[Co(pfp)_2(H_2O)_2]^{2-}$ (H_2pfp = perfluoropinacol) reversibly absorb oxygen to form $[Co(pfp)(Hpfp)(O_2)]^-$, isolated as R_4N^+ salts. Magnetic data and e.p.r. spectra suggest that they are best formulated as cobalt(II) rather than cobalt(III) species.[223] In contrast, the cobalt(II)–histidine

[214] C. Busetto, F. Cariati, D. Galizzioli, and F. Morazzoni, *Gazzetta*, 1974, **104**, 161.
[215] M. J. Carter, D. P. Rillema, and F. Basolo, *J. Amer. Chem. Soc.*, 1974, **96**, 392.
[216] E. Melamud, B. L. Silver, and Z. Dori, *J. Amer. Chem. Soc.*, 1974, **96**, 4689.
[217] N. Bresciani, M. Calligaris, G. Nardin, and L. Randaccio, *J.C.S. Dalton*, 1974, 498.
[218] N. Bresciani, M. Calligaris, G. Nardin, and L. Randaccio, *J.C.S. Dalton*, 1974, 1606.
[219] E. Cesarotti, M. Gullotti, S. Pasini, and R. Ugo, Proceedings 16th International Coordination Chemistry Conference, Dublin, 1974.
[220] A. F. Savitsku and I. N. Ivleva, *Zhur. obschei Khim.*, 1974, **44**, 248.
[221] B. Jezowska-Trzebiatowska and A. Vogt, Proceedings 16th International Coordination Chemistry Conference, Dublin, 1974.
[222] D. M. Huchital and A. E. Martell, *J.C.S. Chem. Comm.*, 1973, 868.
[223] D. F. Cristian and C. J. Willis, Proceedings 16th International Coordination Chemistry Conference, Dublin, 1974.

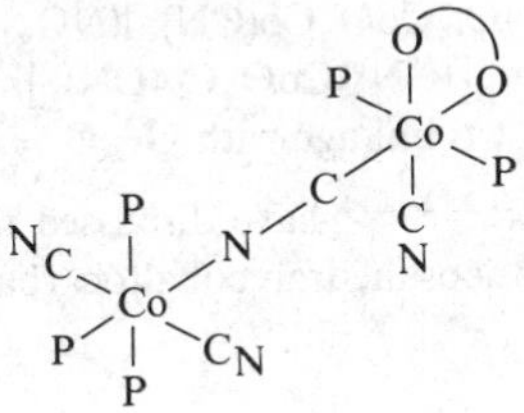

(47)

system takes up oxygen in strongly basic solution.[224] The $[Co(PMe_2Ph)_3(CN)_2]$ complex absorbs oxygen in benzene to form (47), which reacts with more PMe_2Ph to form $OPMe_2Ph$ and re-form the original complex.[225]

The reaction of oxygen with cobalt(II) chloride and triphenylphosphine (L) in allylamine (AA) proceeds[226] as shown in Scheme 4; [A] is probably a mixture of peroxidic species which dissolves on addition of BTH and precipitates an orange solid on irradiation. The same novel complex (48) is slowly deposited when solution [B] is allowed to stand in the dark; oxidation of the PPh_3 to $OPPh_3$ has occurred under mild conditions.

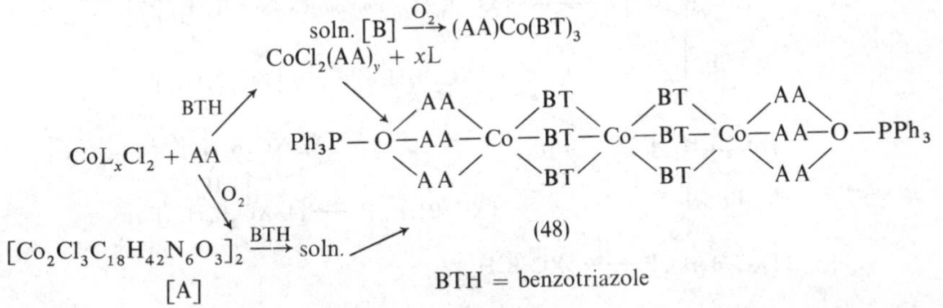

BTH = benzotriazole

Scheme 4

Co-ordination of a free superoxide ion by the cobalt(III) complex of aquocobalamin (vitamin B_{12a}) in DMF produces superoxocobalamin.[227] The i.r. spectra[228] of $[(NH_3)_5CoOCo(NH_3)_5]^+$ have been examined; laser-Raman spectra of the μ-superoxobispentamminecobalt(III) and μ-amido-μ-superoxo-bis[tetramminecobalt(III)] have intense absorptions at *ca.* 1100 cm^{-1}, attributed to O—O stretching.[229]

A redetermination[230] of the structure of $[(NH_3)_5CoO_2Co(NH_3)_5](NCS)_4$ has established the O—O bond length as 1·469(6) Å, considerably shorter than that

[224] K. L. Waters and R. G. Wilkins, *Inorg. Chem.*, 1974, **13**, 752.
[225] J. Halpern, B. L. Goodall, G. P. Gnare, H. S. Lim, and J. J. Pluth, Proceedings 16th International Coordination Chemistry Conference, Dublin, 1974.
[226] J. Drapier and A. J. Hubert, *J. Organometallic Chem.*, 1974, **64**, 385.
[227] J. Ellis, M. Green, and J. M. Pratt, *J.C.S. Chem. Comm.*, 1973, 781.
[228] S. Ikawa, K. Hasebe, and M. Kimura, *Spectrochim. Acta*, 1974, **30A**, 151
[229] T. Shibahara, *J.C.S. Chem. Comm.*, 1973, 864.
[230] R. F. Fronczek, W. P. Schaeffer, and R. E. Marsh, *Acta Cryst.*, 1974, **B30**, 117.

previously reported. In $K_8[(NC)_5CoO_2Co(CN)_5](NO_3)_2.4H_2O$, which is a double salt consisting of K^+, NO_3^-, and $[(CN)_5CoO_2Co(CN)_5]^{6-}$ ions and water molecules, there is a planar Co—O—O—Co linkage with O—O = 1.447(4) Å.[231]

Carbaboranes. Hawthorne *et al.*[232-234] have discussed the thermal rearrangements in both icosahedral and non-icosahedral cobaltocarbaboranes. Examples of the reactions are:

$$(\eta\text{-}C_5H_5)CoC_2B_8H_{10} \xrightarrow{\Delta} (\eta\text{-}C_5H_5)_2Co_2C_2B_8H_{10} \text{ (six isomers)}$$

$$[(C_5H_5)_2Co][Co(C_2B_8H_{12})_2] \xrightarrow{\Delta} (\eta\text{-}C_5H_5)_2Co_2C_2B_8H_{10} \text{ (five isomers)}$$

$$2,6,1,10\text{-}(\eta\text{-}C_5H_5)_2Co_2C_2B_6H_8 \xrightarrow{280\,^\circ C} 2,7,1,10\text{-}(\eta\text{-}C_5H_5)_2Co_2C_2B_6H_8$$

$$1,7,2,4\text{-}(\eta\text{-}C_5H_5)_2Co_2C_2B_7H_9 \xrightarrow{98\,^\circ C} 1,7,2,3\text{-}(\eta\text{-}C_5H_5)Co_2C_2H_7H_9$$

$$\downarrow 150\,^\circ C$$

$$1,10,2,3\text{-}(\eta\text{-}C_5H_5)_2Co_2C_2B_7H_7$$

The chemistry of the metalloborane polyhedra, including the reactions with Lewis bases, protonation, and elimination of hydrogen, have been described.[235] The reactions are illustrated in Scheme 5.

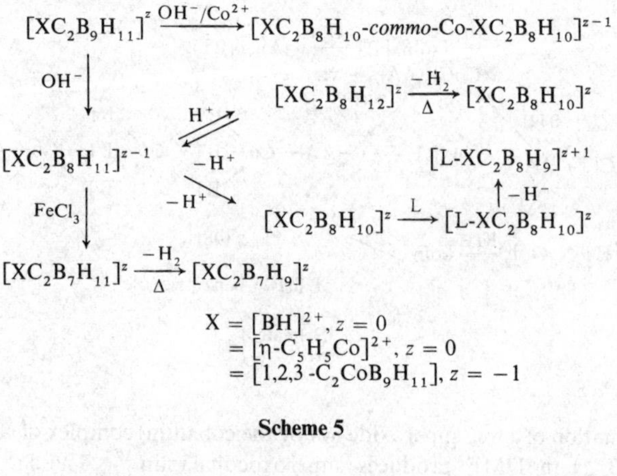

$$[XC_2B_9H_{11}]^z \xrightarrow{OH^-/Co^{2+}} [XC_2B_8H_{10}\text{-}commo\text{-}Co\text{-}XC_2B_8H_{10}]^{z-1}$$

$$X = [BH]^{2+}, z = 0$$
$$= [\eta\text{-}C_5H_5Co]^{2+}, z = 0$$
$$= [1,2,3\text{-}\dot{C}_2CoB_9H_{11}], z = -1$$

Scheme 5

Reduction[236] of *closo*-cobaltocarbaboranes with sodium–naphthalene in the presence of $CoCl_2$ and C_5H_5Na produces $(\eta\text{-}C_5H_5)_2Co_2B_nH_{n+2}$ and $(\eta\text{-}C_5H_5)_3Co_3\text{-}C_2B_nH_{n+2}$, whilst the reaction[237]

$$2\{4\text{-}(\eta\text{-}C_5H_5)\text{-}4\text{-}Co\text{-}1,8\text{-}C_2B_{10}H_{12}\} \xrightarrow[C_5H_6]{KOH/EtOH} \xrightarrow{CoCl_2} 2[4,5\text{-}(\eta\text{-}C_5H_5)_2\text{-}4,5\text{-}Co_2\text{-}1,8\text{-}C_2B_9H_{11}]$$

[231] F. R. Fronczek and W. P. Schaeffer, *Inorg. Chim. Acta*, 1974, **9**, 143.
[232] W. J. Evans, C. J. Jones, B. Štíbr, and M. F. Hawthorne, *J. Organometallic Chem.*, 1973, **60**, C27.
[233] W. J. Evans and M. F. Hawthorne, *J. Amer. Chem. Soc.*, 1974, **96**, 301.
[234] D. F. Dustin, W. J. Evans, C. J. Jones, R. J. Wiersema, H. Gong, S. Chan, and M. F. Hawthorne, *J. Amer. Chem. Soc.*, 1974, **96**, 3085.
[235] C. J. Jones, J. N. Francis, and M. F. Hawthorne, *J. Amer. Chem. Soc.*, 1973, **95**, 7633.
[236] W. J. Evans and M. F. Hawthorne, *Inorg. Chem.*, 1974, **13**, 869.

is an example of the conversion of a 13-vertex cobaltocarbaborane into a 13-vertex bicobaltocarbaborane which has one carbon atom fewer than the starting material. Similarly, $2,1,6\text{-}(C_5H_5)CoC_2B_7H_9$ is converted into $1,8,2,3\text{-}(C_5H_5)_2Co_2C_2B_7H_9$ and probably also into $1,2,8,3,6\text{-}(C_5H_5)_3Co_3C_2B_7H_9$.[238]

A direct synthesis of cobaltocarbaboranes by reaction of, for example, $1,5\text{-}C_2B_3H_5$ with $(\eta\text{-}C_5H_5)Co(CO)_2$ in the gas phase or in solution has been reported.[239] This produces the six-vertex $(\eta\text{-}C_5H_5)CoC_2B_3H_5$ and the seven-vertex $(\eta\text{-}C_5H_5)_2Co_2C_2\text{-}B_3H_5$. The novel $(\eta\text{-}C_5H_5)_3Co_3C_2B_5H_7$ (two isomers) can be obtained analogously.[239] Formation of $(\eta\text{-}C_5H_5)Co(\pi\text{-}2,3\text{-}C_2B_4H_6)$, $(\eta\text{-}C_5H_5)Co(\pi\text{-}2,3\text{-}C_2B_3H_7)$ and $(\eta\text{-}C_5H_5)_2Co_2(\pi\text{-}2,3\text{-}C_2B_3H_5)$ by reaction of $NaC_2B_4H_7$ with cobalt(II) chloride and NaC_5H_5 in the presence of air, water and acetone has been reported.[240]

^{11}B and ^{13}C n.m.r. isotropic shifts in $(\eta\text{-}C_5H_5)Co(C_2B_nH_{n+2})$ and $Co(C_2B_nH_{n+2})_2$ ($n = 6$—9) show that delocalization of electrons is ligand → metal, except for icosahedral cobalt(II) complexes when it is reversed.[241]

The structures of several cobaltocarbaboranes have been determined by X-ray analysis. That of $2,6\text{-}(\eta\text{-}C_5H_5)_2\text{-}2,6\text{-}Co_2\text{-}1,10\text{-}C_2B_6H_8$ at $-150\,^{\circ}C$ is a distorted bicapped square antiprism with carbons at the caps, and a Co—Co bond (the first in a bimetallocarbaborane) of 2.489 Å.[242] In $(\eta\text{-}C_5H_5)_2Co_2C_2B_8H_{10}$ there exists a distorted icosahedron comprising two Co, two C, and eight B, with the cobalt atoms occupying adjacent vertices [Co—Co = 2.387 Å].[243] The red $(\eta\text{-}C_5H_5)Co(B_{10}C_2H_{12})$ is a fluxional molecule which contains a triangulated 13-apex docosahedron in which the equatorial belt (CBCBBB) is bonded to the cobalt.[244]

The anion in $Cs[(\eta\text{-}C_5H_5)Co(CB_7H_8)]$ contains the cobalt(III) atom sandwiched between $C_5H_5^-$ and $CB_7H_8^{3-}$ moieties, the $CoCB_7$ skeleton approximating to a tricapped trigonal prism.[245] The cobalt(III) carbollide, $(\eta\text{-}C_5H_5)Co(\eta\text{-}7\text{-}B_{10}CH_{11})$ is formed[246] by the reaction

$$B_{10}CH_{13}^- \xrightarrow{\text{KOH–EtOH}} \xrightarrow[C_5H_6]{\text{CoCl}_2} \xrightarrow{\text{oxidize}} (\eta\text{-}C_5H_5)Co(\eta\text{-}7\text{-}B_{10}H_{11})$$

The benzodicarbollide compounds $[Me_4N][Co(49)_2]$ and $[Me_4N][Co^{III}(50)_2]$ have

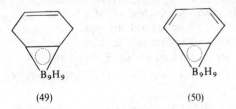

(49)   (50)

[237] D. F. Dustin and M. F. Hawthorne, *J. Amer. Chem. Soc.*, 1974, **96**, 3462.

[238] W. J. Evans and M. F. Hawthorne, *J.C.S. Chem. Comm.*, 1973, 706.

[239] V. R. Miller, L. G. Sneedon, D. C. Beer, and R. N. Grimes, *J. Amer. Chem. Soc.*, 1974, **96**, 3090.

[240] R. N. Grimes, D. C. Beer, L. G. Sneddon, V. R. Miller, and R. Weiss, *Inorg. Chem.*, 1974, **13**, 1138.

[241] R. J. Wiersema and M. F. Hawthorne, *J. Amer. Chem. Soc.*, 1974, **96**, 761.

[242] E. L. Hoel, C. E. Strousse, and M. F. Hawthorne, *Inorg. Chem.*, 1974, **13**, 1388.

[243] K. P. Callahan, C. E. Strousse, A. L. Sims, and M. F. Hawthorne, *Inorg. Chem.*, 1974, **13**, 1397.

[244] M. R. Churchill and B. G. DeBoer, *Inorg. Chem.*, 1974, **13**, 1411.

[245] K. P. Callahan, C. E. Strousse, A. L. Sims, and M. F. Hawthorne, *Inorg. Chem.*, 1974, **13**, 1393.

[246] R. R. Rietz, D. F. Dustin, and M. F. Hawthorne, *Inorg. Chem.*, 1974, **13**, 1580.

been isolated.[247] Mixed-metal bimetallic metallocarbaboranes have also been obtained, *e.g.* $(C_5H_5)Co^{III}(C_2B_7H_9)Fe^{III}(C_5H_5)$, $(C_5H_5)Co^{III}(C_2B_9H_{11})Fe^{III}(C_5H_5)$,[248] and the $(C_5H_5)Co^{III}(\pi\text{-}CB_7H_8)Ni^{IV}(C_5H_5)$ monocation.[249]

Oxides and Other Simple Anions. The non-stoicheiometry of CoO_{1+y} in the temperature range 500—1200 °C and at pressures of 152 to 10^{-3} mm O_2 was examined, and it was found that y varies between 0.0159 and 0.0015. A number of the properties of CoO_x were discussed in relation to this non-stoicheiometry.[250] On heating together Li_2O and CoO a blue Li_6CoO_4 ($\mu_{eff} = 5.11$ BM) with the β-Li_5AlO_4 structure is formed.[251] Blue-black $NaCoO_2$ is formed on heating a mixture of Na_2O_2 and Co_3O_4 ($Na:Co = 1:1$) in a bomb.[251] This has the α-$NaFeO_2$ structure and exhibits essentially temperature-independent paramagnetism (-0.7 BM) over the range 73—297 K. Oxides ACo_2O_4 ($A = Na$,[251] K, Rb, or Cs[253]) are obtained from A_xO and Co_3O_4 in a $1:2$ $A:Co$ ratio. Surprisingly, these compounds do not react with water. X-Ray studies have shown that they have different superstructures but the same substructure which consists (KCo_2O_4) of successive O^{2-} layers along [001] AABB ... with a distorted octahedral environment about the cobalt and a trigonal-prismatic arrangement about the K. A similar preparation using a $2:1$ $A:Co$ ratio ($A = K$, Rb, or Cs) produced[254] black oxocobaltates(IV), A_2CoO_3, which are rapidly decomposed by water. The crystal structure of Cs_2CoO_3 is similar to Rb_2TiO_3 and consists of infinite $[CoO_3]$ chains with the Co^{4+} tetrahedrally co-ordinated to O^{2-}. The Rb and Cs compounds exhibit quite complicated magnetic behaviour, μ_{eff} dropping from *ca.* 1.7 BM (297 K) to *ca.* 0.9 BM (77 K).

The X-ray structures of the skutterudites $CoAs_3$ and $CoSb_3$ have been reported.[255] A study of a series of $CoXY$ ($X = P$, As, or Sb; $Y = S$ or Se) has established the structures of $CoPS$ (tetragonal), $CoAsS$ (cubic), $CoSbS$ and $CoPSe$ (orthorhombic), and $CoAsSe$ and $CoSbSe$ (anomalous marcasites). The magnetic and electrical properties of these compounds are discussed.[256] The partially cobalt(II)-exchanged zeolite A, $Co_4Na_4Si_{12}Al_{12}O_{48}$, and its carbon monoxide adduct, $Co_4Na_4Si_{12}Al_{12}O_{48}.4CO$, have been X-rayed.[257]

Cobalt(III) Complexes.—*Halides.* The cubic elpasolithes Cs_2KCoF_6, Rb_2KCoF_6, Rb_2NaCoF_6, and the hexagonal Cs_2NaCoF_6 (Cs_2NaCrF_6 type) were prepared by fluorination of $[Co(NH_3)_6]Cl_3$ mixed with the appropriate amounts of alkali chloride or carbonate,[258] and these complexes have μ_{eff} corresponding to high-spin cobalt(III). The $CsACoF_6$ ($A = Mg$ or Zn) have $RbNiCrF_6$ structures.[259]

Ammine Complexes. The X-ray structure[260] of $[Co(NH_3)_5SO_3]Cl, H_2O$ shows the

[247] D. S. Matheson and R. E. Grunzinger, *Inorg. Chem.*, 1974, **13**, 671.
[248] D. F. Dustin, W. J. Evans, and M. F. Hawthorne, *J.C.S. Chem. Comm.*, 1973, 805.
[249] C. G. Salentine and M. F. Hawthorne, *J.C.S. Chem. Comm.*, 1973, 560.
[250] J. S. Choi and C. Y. Ho, *Inorg. Chem.*, 1974, **13**, 1720.
[251] H.-N. Migeon, M. Zanne, F. Jeannot, and C. Gleitzer, *Compt. rend.*, 1974, **278**, C, 4531.
[252] M. Jansen and R. Hoppe, *Z. anorg. Chem.*, 1974, **408**, 104.
[253] M. Jansen and R. Hoppe, *Z. anorg. Chem.*, 1974, **408**, 97.
[254] M. Jansen and R. Hoppe, *Z. anorg. Chem.*, 1974, **408**, 95.
[255] A. Kjekshus and T. Rakke, *Acta Chem. Scand.*, 1974, **A28**, 99.
[256] H. Nahigian, J. Steger, H. L. McKinzie, R. J. Arnott, and A. Wold, *Inorg. Chem.*, 1974, **13**, 1498.
[257] P. E. Riley and K. Seff, *Inorg. Chem.*, 1974, **13**, 1355.
[258] E. Alter and R. Hoppe, *Z. anorg. Chem.*, 1974, **407**, 313.
[259] R. Jesse and R. Hoppe, *Z. anorg. Chem.*, 1974, **403**, 143.
[260] R. C. Elder and M. Tikula, *J. Amer. Chem. Soc.*, 1974, **96**, 2635.

large *trans* effect of the S-bonded sulphito-group, Co—N (*trans* N, *cis* S) = 1.966(4) Å compared with Co—N (*trans* S) = 2.055(2) Å. Siebert and Wittke[261] have shown that the sulphito-group in this complex can be protonated to yield $[Co(NH_3)_5HSO_3]^{2+}$, whilst *cis*-$[NH_4][Co(SO_3)_2(NH_3)_4]$ reacts with HCl to form $[Co(HSO_3)SO_3$-$(NH_3)_4]$, $\frac{1}{2}H_2O$. Several $[Co(HSO_3)X(NH_3)_4]^+$ (X = Cl, Br, or CN) cations have also been prepared.[261] The i.r. spectra of these complexes are consistent with the bisulphito-group bonded through sulphur. Unstable *trans*-$Na[Co(NH_3)_4(SO_3)_2]$, $2H_2O$ is formed from *trans*-$[Co(NH_3)_4(H_2O)SO_3]^+$ in aqueous sulphite solution. The *cis*-anion has been reported, and it has been shown on the basis of u.v.–visible and i.r. spectra, that previous assignment of these isomers was incorrect and should be reversed.[262] The reaction of *cis*- and *trans*-$[Co(en)_2(NH_3)(NCS)]^{2+}$ (en = 1,2-ethylenediamine) with hydrogen peroxide oxidizes the thiocyanate group to produce $[Co(en)_2(NH_3)CN]^{2+}$ and $[Co(en)_2(NH_3)_2]^{3+}$ with retention of configuration.[263]

[1-Cyano-2,3,4-triammine(ethylenediamine)cobalt(III)]chloride has been prepared from $Na_3[Co(CN)_2(SO_3)(NH_3)_2]$,$6H_2O$ by treatment successively with aqueous ethylenediamine, hydrobromic acid, and liquid ammonia, and isolated chromatographically.[264] The electronic spectrum was compared with those of *cis*- and *trans*-$[Co(en)_2(NH_3)CN]Cl_2$ and $[Co(NH_3)_5CN]Cl_2$. *trans*-$[Co(NH_3)_4Cl(N_3)]ClO_4$ has been obtained from *trans*-$[Co(NH_3)_4Cl_2]Cl$ and sodium azide in aqueous solution, producing initially *trans*-$[Co(NH_3)_4Cl_2]N_3$. This complex was dissolved in boiling aqueous NH_4Cl solution and the required complex is precipitated by ether and then treatment with 72% $HClO_4$.[265] *trans*-$[Co(NH_3)_4Br(CN)]Br$ has been prepared by dissolving *trans*-$[Co(NH_3)_4CN(OH_2)]Cl_2$ in boiling 48% HBr.[265]

A series of complexes $[Co(NH_3)_5L]^{3+}$ (L = 1,2-, 1,3-, or 1,4-dicyanobenzene, 3- or 4-cyanobenzaldehyde, 4-acetylbenzonitrile, and 2-, 3-, or 4-cyanopyridine), which contain L bonded to the metal *via* a C≡N nitrogen, has been prepared.[266] These all undergo hydrolysis to the corresponding carboxamide, which is conclusive proof of the presence of a Co^{III}—N≡CR bond. The linkage isomer, $[Co(NH_3)_5L']$ (L' = 4-cyanopyridine), where L' is bonded through the pyridine nitrogen, has also been obtained.[266] There is a report[267] of $[Co(NH_3)_5(LH_2)]ClO_4$ [LH_3 = N-(COOH)$_3$]. The circular dichroism of $[Co(NH_3)_5(X–X)]^{n+}$ (X–X = *S*-lactate, n = 2; *S*-2-aminopropan-1-ol, n = 3),[268] of $[Co(NH_3)_5A]^{n+}$ (A = amino-acid ester, *S*-mandelate, or *S*-phenylacetate,[269] and of $[Co(NH_3)_5B]^{2+}$ (B = unidentate asymmetric amine, RNH_2)[270] has been discussed.

Diamine Complexes. The reaction of *trans*-$[Co(en)_2Cl_2]^+$ with unidentate alkyl or aryl amines produces *cis*-$[Co(en)_2(amine)Cl]^{2+}$, the stereochemistry of the product being indicated by resolution.[271] The spontaneous resolution of $[Co(C_2O_4)(en)_2]X$,

[261] H. Siebert and G. Wittke, *Z. anorg. Chem.*, 1974, **405**, 63.
[262] K. L. Scott, *J.C.S. Dalton*, 1974, 1486.
[263] A. R. Norris and J. W. L. Wilson, *Canad. J. Chem.*, 1973, **51**, 4152.
[264] A. R. Norris, *Canad. J. Chem.*, 1974, **52**, 477.
[265] C.-K. Poon and H.-W. Tong, *J.C.S. Dalton*, 1974, 1.
[266] R. J. Balahura, *Canad. J. Chem.*, 1974, **52**, 1762.
[267] R. D. Cannon and J. Gardiner, *Inorg. Chem.*, 1974, **13**, 390.
[268] C. J. Hawkins and G. A. Lawrence, *Inorg. Nuclear. Chem. Letters*, 1973, **9**, 1183.
[269] G. G. Dellesbaugh and B. E. Douglas, *Inorg. Nuclear Chem. Letters*, 1973, **9**, 1255.
[270] C. J. Hawkins and G. A. Lawrence, *Austral. J. Chem.*, 1973, **26**, 2401.
[271] I. J. Kindred and D. A. House, *Inorg. Nuclear Chem. Letters*, 1974, **10**, 25.

$[Co(CO_3)(en)_2]X$, cis-$[Co(NO_2)_2(en)_2]X$ (X = Cl or Br) and $MCo(edta)$, $2H_2O$ (M = NH_4 or K) has been reported.[272] The results of studies on the solubilities of the racemic and optically active forms show that spontaneous resolution is predictable from the relationship between the solubilities. Thermal ligand exchange has been observed in cis-$[Co(en)_2X_2]NCS$, when cis-$[Co(en)_2X(NCS)]X$ (X = Cl or Br) is formed. In cis-$[Co(en)_2ClBr]NCS$, the bromide is more likely to exchange than the chloride. Exchange was not apparent in the corresponding $trans$-isomers.[273] The preparation of cis- and $trans$-$[Co(en)_2(NO_2)_2]ClO_4$ by various routes (Scheme 6) has been reported.[274]

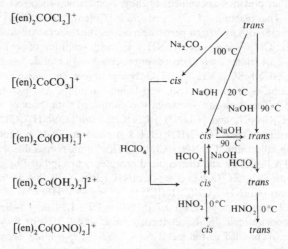

Scheme 6

$trans$-$[CoL_2(N_3)_2]^+$ and $trans$-$[CoL_2(CN)_2]^+$ (L = trimethylenediamine) have been prepared from $trans$-$[CoL_2Cl_2]^+$ in methanol. The cis-isomers are obtained similarly from cis-$[Co(en)_2(NO_3)_2]^+$ in aqueous solution.[275] The $[Co(en)_2X_2]X$ (X = Cl or Br) compounds form 1:1 and 1:3 adducts with thiourea and selenourea. X-Ray studies have shown that the thiourea is not bonded to the cobalt.[276] The complex thermal decompositions of $[Co(en)_3](NO_3)_3$ and $[Co(en)_3](HSO_4)_3$ have been investigated.[277] The former produces $Co(NO_3)_2$, NH_4NO_3, en, and organic products, and ultimately Co_3O_4, CO, N_2, and nitrogen oxides. The latter decomposes *via* $Co(en)(HSO_4)_2$ and $Co(HSO_4)_2$ to $CoSO_4$, H_2O, sulphur oxides, and organic products.

The $[Co(en)_3]^{3+}$ cation undergoes N-deprotonation to give $[Co(en - 2H)_3]^{3-}$ when treated with MeLi or Bu^nLi in diethyl ether or THF, from which acetic acid

[272] K. Yamanari, J. Hidaka, and Y. Shimura, *Bull. Chem. Soc. Japan*, 1973, **46**, 3724.
[273] R. Tsuchiya, T. Murakami and E. Kyuno, *Bull. Chem. Soc. Japan*, 1973, **46**, 3119.
[274] F. Seel and D. Meyer, *Z. anorg. Chem.*, 1974, **408**, 275.
[275] H. Kawaguchi and S. Kawaguchi, *Bull. Chem. Soc. Japan*, 1973, **46**, 3453.
[276] A. V. Ablov, N. N. Proshina, Y. A. Simono, and T. T. D. Chan, *Zhur. neorg. Khim.*, 1974, **19**, 95.
[277] L. W. Collins, W. W. Wendlandt, E. K. Gibson, and G. W. Moore, *Thermochim. Acta*, 1973, **7**, 209.

regenerates the starting material.[278] The $[Co(en)_2X(H_2NCH_2CN)]^{2+}$ (X = Cl or Br) ions undergo rapid base-catalysed ring-closure to form $[Co(en)\{H_2NCH_2C-(NH_2)=NCH_2CH_2NH_2\}X]^{2+}$ containing the terdentate amidine ligand.[279] The mercury(II) catalysed reactions of $[Co(en)_2X(H_2NCH_2CN)]^{2+}$ in acidic solution produced $[Co(en)_2(H_2NCH_2CONH_2)]^{3+}$ containing a chelated glycine amide ligand.[279] The tetramethylenediamine complex $[Co(tmd)_3]Br_3$ has been prepared by reaction of cobalt(II) nitrate with tmd in DMSO, followed by aerial oxidation and anion exchange. Resolution into optical isomers and c.d. spectra were reported.[280] The structure $(+)_{589}$-$[Co(tmd)_3]Br_3$ has been elucidated. The configuration of the seven-membered chelate rings is λ and the absolute configuration of the complex is designated $\Delta(\lambda\lambda\lambda)$.[281] The preparation and resolution of 2,2'-diaminobiphenyl-(ethylenediamine)cobalt(III) chloride, and of several related compounds of 2,2'-diaminobiphenyl with bidentate or quadridentate amine ligands has been achieved.[282] The structure of *trans*-$[Co(en)_2LCl]^+$ (L is the anion of the substituted purine, theophilline, $C_7H_7N_4O_2^-$) has been elucidated.[283] It has been noted that addition of alcohol to the aqueous $[Co(en)_3]^{3+}$–d-tartrate system causes the preferential precipitation of Δ-$[Co(en)_3]_2(d$-tart$)_3$ in 54–55% yield.[284]

The application of ^{59}Co n.m.r. spectroscopy to the study of cobalt(III) complexes is beginning to expand.[285, 286] An n.m.r. study of the stereoisomers of propane-1,2-diamine (pn), en, acac, CF_3acac, *etc.*, showed that the chemical shifts do not appear to be related to the absolute configuration, but that larger linewidths are observed with lower symmetry complexes.[286] The chemical shifts are larger for *ob-lel* than for *mer-fac* isomerism in $Co(pn)_3^+$.[285]

Further separation of some of the possible isomers of $[Co(\pm)pn_3]Cl_3$ and their characterization have been achieved.[287] Ion exchange chromatography separated the racemic lel_3, lel_2ob, ob_2lel, and ob_3, which were further separated into the Λ and Δ forms. The structures of $(-)_{589}$-$[Co(pn)_3]Cl_3,H_2O$, which has the Λ configuration,[288] and of $(-)_{546}$-tris-$(R,R$ – 2,4-diaminopentane)cobalt(III) chloride dihydrate, also Λ and in which the six-membered rings are in the twisted boat form with the λ configuration,[289] have been determined. One form of tris[(R)-propylenediamine-cobalt(III)][hexacyanocobaltate] has been X-rayed. The cation has the $(+)_{589}$-$[Co-(R)pn_3]^{3+}$ $\Lambda(ob)(fac)$ structure.[290] The solution and single-crystal d–d c.d. and ^{59}Co n.m.r. spectra of $Co(en)_3^{3+}$, $Co(d$–$pn)_3^{3+}$, and $Co(1,3$–$pn)_3^{3+}$ have been examined, and the results interpreted successfully on the basis of Liehr's and Piper's equations, indicating that these correctly describe the origin of optical activity in the visible region.[291] Four components have been separated by column chromatography

[278] W. K. Musker and M. J. Lukes, *Inorg. Nuclear Chem. Letters*, 1974, **10**, 405.
[279] K. B. Nolan and R. W. Hay, *J.C.S. Dalton*, 1974, 914.
[280] J. Fujita and H. Ogino, *Chem. Letters*, 1974, 57.
[281] S. Sato, Y. Saito, J. Fujita, and H. Ogiow, *Inorg. Nuclear Chem. Letters*, 1974, **10**, 669.
[282] T. Tanimura, H. Ito, J. Fujita, K. Saito. S. Hirai, and K. Yamasaka, *J. Coord. Chem.*, 1973, **3**, 161.
[283] L. G. Marzilli, T. J. Kistenmacher, and C.-H. Chang, *J. Amer. Chem. Soc.*, 1973, **95**, 7507.
[284] H. Yoneda, G. Takemoto, and U. Sakaguchi, *Bull. Soc. Chem. Japan*, 1974, **47**, 513.
[285] Y. Yoika, F. Yajima, A. Yamasaki, and S. Fujiwaru, *Chem. Letters*, 1974, 177.
[286] A. Johnson and G. W. Everett, *Inorg. Chem.*, 1973, **12**, 2801.
[287] S. E. Harming, S. Kallesøe, A. M. Sargeson, and C. E. Schäffer, *Acta Chem. Scand.*, 1974, **A28**, 385.
[288] R. Nagaro, F. Marumo, and Y. Sato, *Acta Cryst.*, 1973, **B29**, 2438.
[289] A. Kobayashi, F. Marumo, and Y. Sato, *Acta Cryst.*, 1973, **B29**, 2443.
[290] R. Kuroda and Y. Sato, *Acta Cryst.*, 1974, **B30**, 2126.
[291] R. R. Judkins and D. J. Royer, *Inorg. Chem.*, 1974, **13**, 945.

from $[\{(R)\text{-propylenediamine}\}_n(\text{en})_{3-n}\text{Co}]^{3+}$, and identified as $[\text{Co(en)}_2(R\text{-pn})]^{3+}$, $[\text{Co(en)}(R\text{-pn})_2]^{3+}$, and $[\text{Co}(R\text{-pn})_3]^{3+}$. The $[\text{Co(en)}_2(R\text{-pn})]^{3+}$ was separated into the Δ and Λ configurational isomers.[292] The correlation of absolute configuration by X-ray powder diffraction data of a series of Co^{III}, Cr^{III}, Rh^{III}, and Ir^{III} complexes of *trans*-cyclohexane-1,2-diamine, $[\text{ML}_3]\text{Cl}_3$, and of mixed-metal active racemates, $[\text{CoL}_3][\text{ML}_3]\text{Cl}_6$, $n\text{H}_2\text{O}$ (M = Cr, Rh, or Ir) has been reported.[293] The structure of $(-)_{581}[\text{Co}(+)(\text{*trans*-cyclohexane-1,2-diamine})_3]^{3+}$ is Λ and the ring configuration is δ. In $(+)_{510}$-[oxalatobis-(R,S-2,4-diaminopentane]cobalt(III) perchlorate monohydrate the cation has absolute configuration Λ, with the six-membered rings in chair conformations.[294] The complex $(+)_{589}$-tris(biguanide)cobalt(III) chloride monohydrate contains cobalt bonded to the terminal NH of the ligand, Co—N = 1.912 Å and the absolute configuration is Δ.[295] The tris$(-)_D$ and tris$(+)_D$ cobalt(III) complexes of 1-(2-pyridyl)ethylamine were prepared, and three different isomers and their enantiomers separated by column chromatography and identified using spectral data.[296]

Polyamine Donors. The distribution of the three isomers of $[\text{Co(dien)}_2]^{3+}$ (dien = diethylenetriamine) depends upon solvation, counter-ions present, and temperature.[297] The structure of *trans*-$[\text{Co(trien)(CN)}_2]\text{ClO}_4$ (trien = triethylenetetramine) contains a distorted octahedral cation with the trien equatorially co-ordinated and the axial cyano-groups C-bonded.[298] Searle *et al.*[299] have prepared the $[\text{Co(dmtrien)-}X_2]^{n+}$ (dmtrien = dimethyltriethylenetetramine = 4,7-dimethyl-1,4,7,10-tetra-azadecane; X = Cl, OH, NO_2, or $\frac{1}{2}\text{CO}_3$) and identified the interconvertible RR,SS-α and RR,SS-β geometric isomers from their ^{1}H n.m.r. spectra. From the tendency of the $\beta \rightarrow \alpha$ isomerization to occur, and by comparison with trien analogues, it was concluded that methyl substitution enhances the stability of the angular configuration of the central amine donor such that *trans* complexes may not be obtainable.[299]

Formation of complexes such as (51) and (52), which have a very high ionic charge and contain several metal centres, has been noted.[300]

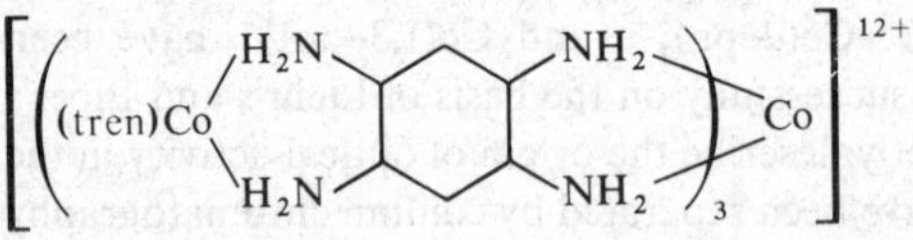

(51) 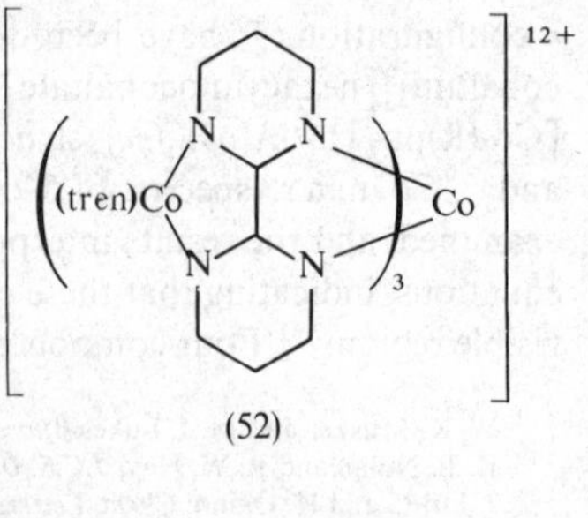   (52)

[292] M. D. Alexander and C. R. Spillert, *Synth. React. Inorg. Metal.-Org. Chem.,* 1974, **4**, 97.
[293] P. Anderson, F. Galsbøl, S. E. Harnung, and T. Laier, *Acta Chem., Scand.,* 1973, **27**, 3973.
[294] I. Oonishi, S. Sato, and Y. Sato, *Acta Cryst.,* 1974, **B30**, 2256.
[295] M. R. Snow, *Acta Cryst.,* 1974, **B30**, 1850.
[296] K. Michelson, *Acta Chem. Scand.,* 1974, **A28**, 428.
[297] F. R. Keene and G. H. Searle, *Inorg. Chem.,* 1974, **13**, 2173.
[298] R. K. Wismer and R. A. Jacobson, *Inorg. Chim. Acta,* 1973, **7**, 477.
[299] G. H. Searle, M. Petkovic, and F. R. Keene, *Inorg. Chem.,* 1974, **13**, 399.
[300] D. Baker and J. C. Bailar, Proceedings 16th International Conference on Chemistry of Coordination Compounds, Dublin, 1974.

Pentaethylenehexamine (1,14-diamino-3,6,9,12-tetra-azatetradecane) reacts with $[Co(NH_3)_5Br]Br_2$ in aqueous solution in the presence of active charcoal to form $[CoL]Cl_3, nH_2O$ (in the presence of NaCl). Geometrical and optical isomers were separated by column chromatography, and structures assigned on the basis of ^1H n.m.r., absorption, i.r., and c.d. spectra.[301] The structure of $(-)_{589}$-$[CoL][Co(CN)_6]$, $3H_2O$ [L = $NNN'N'$-tetrakis-(2'-aminoethyl)-1,2-diaminopropane] has been reported, and has the absolute configuration $\Delta\Lambda\Delta$.[302]

The isomers of $[\beta_1$-glycinatotriethylenetetra-aminecobalt(III)]$^{2+}$ have been separated by a cation-exchange resin,[303] and an X-ray investigation was performed on a crystal containing both the Δ-$(-)_{589}$-(RR) and -(RS) diastereoisomers.

Two studies[304, 305] have dealt with the hydrolysis of peptides and peptide esters by β-hydroxoaquotriethylenetetraaminecobalt(III) ion. The mechanism was elucidated and intermediates $[Co(trien)L]^{3+}$ (L = dipeptide or dipeptide ester) were isolated. The base hydrolysis of cis-$[Co(en)_2X(H_2NCH_2CH_2OCOMe)]^{2+}$ (X = Cl or Br) produces both cis-$[Co(en)_2OH(H_2NCH_2CH_2OCOMe)]^{2+}$ and $[Co(en)_2OH(H_2NCH_2CH_2OH)]^{2+}$. Acid hydrolysis in 6M-HCl produces cis-$[Co(en)_2Cl(H_2NCH_2CH_2OH)]^{2+}$. The hydrolysis of cis-$[Co(en)_2X(H_2NCH_2-COOR)]$ (X = halide) in the presence of Hg^{II} and ClO_4^- ions has also been examined.[307]

Macrocyclic N-*Donors.* The oxidation of Co(TPP)(NO) (TPP = αβγδ-tetraphenylporphinate) in the presence of 3,5-lutidine produces the cobalt(III) nitro-complex $[Co(TPP)(lut)(NO_2)]$. An X-ray determination has shown that the cobalt is octahedrally co-ordinated in an N_6 environment.[308] X-Ray analysis has established the identity of the product formed in small yield from a CHCl$_3$ solution of Co(TPP)(NO) on evaporation in the presence of piperidine as a cobalt(III) complex, $[Co(TPP)(pip)_2]X,2pip$, which contains an essentially octahedral CoN$_6$ group. The anion was not unequivocally established, either NO_3^- or HCO_3^- is possible.[309] A crystal structure analysis of meso-tetraphenylporphinatobis(imidazole)cobalt(III) acetate monohydrate revealed two $[Co(TPP)(imid)_2]^+$ cations bridged by acetate ions which hydrogen bond to the imidazole H; water is also H-bonded to the acetate groups. The cobalt(III) complexes of tetra-(4-N-methylpyridyl)porphine[311] has been briefly mentioned, as has $[CoL(PR_3)_2]Br$ (L = octaethylporphin).[312]

The configurations of trans-$[Co(cyclam)(CN)Br]^+$ and trans-$[Co(cyclam)(N_3)Cl]^+$ (cyclam = 1,4,8,11-tetra-azacyclotetradecane) have been assigned on the basis of i.r. and absorption spectral evidence. trans-$[Co(cyclam)X(OH_2)]^{2+}$ (X = CN, NCS, or

[301] Y. Yoshikawa and K. Yamasaki, *Bull. Chem. Soc. Japan*, 1973, **46**, 3448.
[302] A. Kobayashi, F. Marumo, and Y. Sato, *Acta Cryst.*, 1974, **B30**, 1495.
[303] D. A. Buckingham, P. J. Creswell, R. J. Delluca, M. Dwyer, G. J. Gainsford, L. G. Marzilli, I. E. Maxwell, W. T. Robinson, A. M. Sargeson, and K. R. Turnbull, *J. Amer. Chem. Soc.*, 1974, **96**, 1713.
[304] E. Kimura, *Inorg. Chem.*, 1974, **13**, 951.
[305] K. W. Bentley and E. H. Creaser, *Inorg. Chem.*, 1974, **13**, 1115.
[306] K. B. Nolan, B. R. Coles, and R. W. Hay, *J.C.S. Dalton*, 1973, 2503.
[307] K. Nomiya and H. Kobayashi, *Inorg. Chem.*, 1974, **13**, 409.
[308] J. A. Kaduk and W. R. Scheidt, *Inorg. Chem.*, 1974, **13**, 1875.
[309] W. R. Scheidt, J. A. Cunningham, and J. L. Hoard, *J. Amer. Chem. Soc.*, 1973, **95**, 8289.
[310] J. W. Lauher and J. A. Ibers, *J. Amer. Chem. Soc.*, 1974, **96**, 4447.
[311] R. F. Pasternak, E. G. Spiro, and M. Teach, *J. Inorg. Nuclear Chem.*, 1974, **36**, 599.
[312] E. Cetinkaya, A. W. Johnson, M. F. Lappert, G. M. McLaughlin, and K. W. Muir, *J.C.S. Dalton*, 1974, 1236.

NO_2) and *trans*-[Co(cyclam)OH(NH$_3$)]$^{2+}$ have also been prepared and the pK_a's of the aquo-complexes determined.[313]

Curtis[314] has prepared cobalt(III) complexes of 5,7,12,14-tetraethyl-7,14-dimethyl-1,4,8,11-tetra-azacyclotetradec-4,11-diene, L, by reaction of [H$_2$L](ClO$_4$)$_2$ either with Na$_3$Co(CO$_3$)$_3$.3H$_2$O and HX, or with Co(O$_2$CMe)$_2$ followed by H$_2$O$_2$ oxidation. The complexes isolated were *trans*-[CoLX$_2$]ClO$_4$ (X = Cl, Br, OAc, or NO$_2$) and *cis*-[CoL(acac)](ClO$_4$)$_2$, and ^{1}H n.m.r. spectra of the *trans* compounds indicate an N-rac configuration.

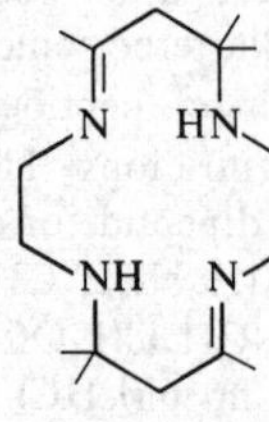

Me$_6$[14]dieneN$_4$ (53a)

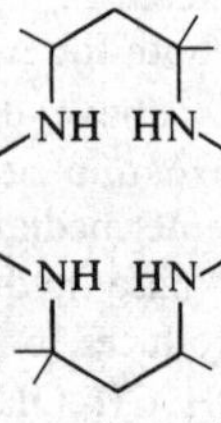

Me$_6$[14]aneN$_4$ (53b)

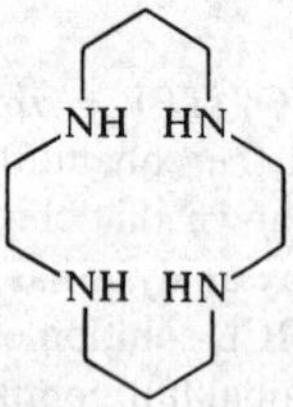

14[ane]N$_4$ (53c)

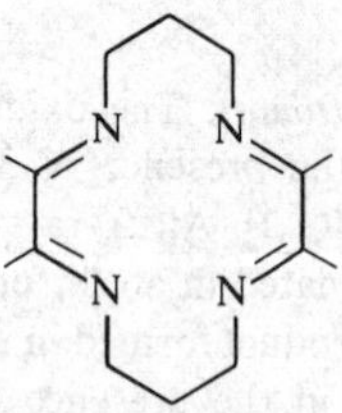

Me$_4$[14]tetraene (53d)

A comparison of the ligand-field splitting parameters, Dq, for a series of [CoLCl$_2$]$^+$ complexes, where L is a 12-, 13-, 14-, 15-, or 16-membered ring N$_4$ macrocycle showed that the 14-ring was the best fit for Co^{III} compared with the 15-ring for Ni^{II}.[315] It was concluded that rings smaller than the best fit exert anomalously high Dq values due to the distribution of strain over the whole complex and short M—N distances; oversize rings exert low ligand field strengths. Formation of [Co^{III}(taab^{2-})]$^+$ is discussed on p. 231. Alkyl-cobalt(III) bonds in complexes of macrocyclic N$_4$ ligands (53a—d) have been prepared;[316] these are the first such linkages in complexes with saturated ligands reported. The preparations involved photolysis of [Co(NH$_3$)$_5$-O$_2$CR]$^{2+}$ in acidic aqueous solution in the presence of [Co^{II}(N$_4$)]CoCl$_4$, the carboxylate functioning as a source of R· radicals, which produced [Co(N$_4$)RX](ClO$_4$)$_n$ (R = Me, Et; X = OH$_2$, CN, or NCS). The complexes are highly stable and not sensitive to light or air. Attempts to extend the synthetic method to produce analogues where the N$_4$ donor set is (NH$_3$)$_4$ or (en)$_2$ were unsuccessful.[316]

313 C.-K. Poon and H.-W. Tong, *J.C.S. Dalton*, 1974, 930.
314 N. F. Curtis, *Austral. J. Chem.*, 1974, **27**, 71.
315 L. Y. Martin, L. J. DeHayes, L. J. Zompa, and D. H. Busch, *J. Amer. Chem. Soc.*, 1974, **96**, 4047.
316 T. S. Roche and J. F. Endicott, *Inorg. Chem.*, 1974, **13**, 1575.

The optical and e.p.r. spectra of dipyridinecobalt(III) corrole, the cobalt(II) corrole anion, and the Co^{II}-1,19-diethoxycarbonyltetrahydrocorrin cation have been obtained in THF and pyridine solution. The anion and cation have different electronic ground states in spite of the similarity of the ligands.[317]

Other N-Donors. The structure of K$_3$Co(HNCONHCONH)$_3$,xH$_2$O, the potassium salt of the tris(biuretato)cobaltate(III) ion, consists of an almost regular octahedron of nitrogen atoms about cobalt (Co—N = 1·916 Å) with slightly puckered biuretato ligands.[318] Three four-membered rings are present in molecules of tris-(1,3-diphenyl-triazenato)cobalt(III),toluene, Co(PhNNNPh)$_3$,tol, the bonding being with the six terminal nitrogens.[319]

Contrary to previous indications, a recent study[320] of the products of various synthetic routes to [Co(phen)$_2$(NO$_2$)$_2$]$^+$ and [Co(bipy)$_2$(NO$_2$)$_2$]$^+$ has shown that only the *cis*-nitro-isomers can be isolated. Although different crystalline forms of several compounds containing these cations were recognized, a thorough search did not reveal any *trans*-isomers or nitrito co-ordination.

Stereoselective reduction of DL-[Co(en)$_2$(N—N)]$^{3+}$ (N—N = phen or bipy) by the bacteria *Pseudomonas stutzeri* to form D-[Co(en)$_2$(N—N)]$^{2+}$ has been found.[321] Racemic KCo(edta) is also reduced in the presence of 0.1 % w/v NH$_4$Cl, but a number of other cobalt(III) compounds were unaffected, *e.g.* [Co(en)$_3$]Cl$_3$ or *abd*-[Co(gly-cinate)$_3$]. The cobaltous complexes of 4- and 5-methyl substituted tri-2-pyridylamine (*q.v.*) are readily oxidized to cobaltic species, in contrast to the complexes of the parent ligand. The products [Co(2pyA)$_2$](ClO$_4$)$_3$, which contain octahedral CoN$_6$ chromo-phores and are 1:3 electrolytes in solution, undergo ready substitution with nucleo-philes such as nitrite ions to form [Co(2pyA)(NO$_2$)$_3$]. It is also claimed that, in addition, the [{Co(5-Me2pyA)}$_2$(NCS)$_3$][Co(NCS)$_4$]ClO$_4$,H$_2$O species is formed.

Oximato Compounds. The structure of [Co(dmg)$_2$(aniline)$_2$]Cl (dmgH = dimethyl-glyoxime) consists of a *trans* octahedral cation with Co—N(aniline) = 2.001 Å, longer than Co—N(oxime) (1.889, 1.885 Å).[322] The structure of [Co(dmg)$_2$(PBu$^n_3$)-(py)] reveals that the pyridine is co-ordinated to the cobalt by a Co—C bond from the 4-position of the ring, and not from the nitrogen.[323] The X-ray structures[324] of [Co(dmg-)$_2$(PPh$_3$)Cl]$^{2+}$ and [Co(dmg-)$_2$(NH$_3$)Cl]$^{2+}$ reveal the expected *trans* arrangement, and the evidence for *trans*-influence in this type of complex is discussed. The X-ray crystal data indicate that the PPh$_3$ exerts a greater *trans*-influence than the amine group, but much less than an alkyl group. The structure of aquomethylbis-(dimethylglyoximato)cobalt(III) contains a distorted octahedral arrangement about the cobalt, and a comparison of the Co—C(sp^3) axial bond length in this and in similar complexes showed no clear evidence of a *cis* influence.[325]

[317] N. S. Hush and I. S. Woolsey, *J.C.S. Dalton*, 1974, 24.
[318] P. J. M. W. L. Birker, J. M. M. Smits, J. J. Bour, and P. T. Beurskens, *Rec. Trav. chim.*, 1973, **92**, 1240.
[319] M. Corbett and B. F. Hoskins, *Austral J. Chem.*, 1974, **27**, 665.
[320] M. Davidson, T. W. Faulkner, M. A. Green, and E. D. McKenzie, *Inorg. Chim. Acta*, 1974, **9**, 231.
[321] L. S. Dollimore, R. D. Gillard, and I. H. Mather, *J.C.S. Dalton*, 1974, 518.
[322] L. P. Battaglia, A. B. Corradi, G. G. Palmieri, M. Nardelli, and M. E. V. Tani, *Acta Cryst.*, 1974, **B30**, 1114.
[323] W. W. Adams and P. G. Lenhert, *Acta Cryst.*, 1973, **B29**, 2412.
[324] S. Bruckner and L. Randaccio, *J.C.S. Dalton*, 1974, 1017.
[325] D. L. McFadden and A. T. McPhail, *J.C.S. Dalton*, 1974, 363.

Photoelectron spectra have been reported for a range of dimethylglyoxime derivatives, $H[Co(dmg)_2X^1X^2]$ ($X^1X^2 = Cl_2$, Br_2, I_2) and $[Co(dmg)_2X^3X^4]$ ($X^3X^4 = NH_3,NH_3$, NH_3,Cl, H_2O,Cl). LaRossa and Brown[327] have used ^{59}Co n.q.r. and n.m.r. to examine the bonding in $[XCo(dmg)_2L]$ compounds (X = Me, $CHCl_2$, Cl, or Br; L = Lewis base). Proton n.m.r. spectra of $[LCo(dmg)_2X]$ (X = NO_3, SeCN, Cl, Br, NCS, N_3, SCN, CN, NO_2, or CH_3; L = 4-Bu^{t}py or PBu$^n_3$) show that, except for the linear triatomics which form a bent bond to cobalt, the chemical shifts of the *trans* α-H resonance and the *cis*-dioxime resonance in $[(4-Bu^tpy)Co(dmg)_2X]$ are linearly related. The dependence of the chemical shifts on X, and the variations of the coupling constants in the PBu$^n_3$ series, suggest rehybridization of the Co—X bond; and specifically that the 4s character of the Co—X bond increases along the series. Differences between these results and those of a previous study were attributable to previously incorrectly formulated complexes.[328] It is pointed out that the tendency of $[L_2Co(dmg)_2][Co(dmg)_2X_2]$ to form during aerial oxidation of the L + dmgH + Co^{II} mixture is due to the presence of the equilibrium

$$2LCo(dmg)_2X \overset{Co^{II}}{\rightleftharpoons} [L_2Co(dmg)_2][Co(dmg)_2X_2]$$

which is markedly affected by the purity of the reagents, the nature of X and L, and the solvent used.[328] Dissolution of *trans*-$[Co(dmg)_2(NO_2)(OH_2)]$ in DMSO at 70 °C produces *trans*-$[Co(dmg)_2(NO_2)(DMSO)]$. Anation reactions of this complex are appreciably slower than those of *trans*-$[Co(dmg)_2(CH_3)(DMSO)]$, confirming the *trans* labilizing effect of CH_3^-.[329]

The base exchange rates[330-332] in RCo(chel)B {R = alkyl; chel = dmg, diacetylmono-oximinodiacetylmonoximato-1,3-propane, methylato-*NN'*-ethylenebis-(acetylacetoniminato), or methylato(cobyrinic acid heptamethyleter); B = Lewis base, $P(OMe)_3$, 1-(2-trifluoromethylphenylimidazole)} have been examined by n.m.r. and the relative importance of steric and electronic effects was discussed.

Formation of $H[Co(mpg)_2X_2]$ (mpgH = methylisopropylglyoxime; X = Cl, Br, or I) occurs on aerial oxidation of CoX_2 + mpgH in solution. Metathesis produces a range of compounds with complex cations, e.g. $[Co(en)_2Cl_2]^+$, $[Co(py)_4Cl_2]^+$.[333] On heating in aqueous solution $H[Co(DO)_2Cl_2]$ (DO = methyl-, ethylmethyl-, methylpropyl-, or methylbutyl-glyoximate) are converted into $[Co(DO)_2(H_2O)Cl]$.[334] The synthesis of cyano-bridged dicobaloximes, $BCo(dmg)_2$-CN-$Co(dmg)_2R$ (B = C_5H_5N, $C_5H_{11}N$, or NH_3; R = Me, Et, or Pr^n), results from $[Co(dmg)_2R(H_2O)]$ and $[Co(dmg)_2(CN)B]$ in chloroform.[335]

The ligand 2,2'(1,3-diaminopropane)bis-(2-methyl-3-butanone)dioxime (54) forms *trans*-$[CoLX_2]$ (X = NO_2, N_3, Cl, or OH) complexes. The *trans* ligands are labile

[326] K. Burger, E. Fluck, C. Varhelyi, H. Binder, and I. Speyer, *Z. anorg. Chem.*, 1974, **408**, 304.
[327] R. A. LaRossa and T. L. Brown, *J. Amer. Chem. Soc.*, 1974, **96**, 2072.
[328] W. C. Trogler, R. C. Stewart, L. A. Epps, and L. G. Marzilli, *Inorg. Chem.*, 1974, **13**, 1564.
[329] S. T. D. Lo and D. W. Watts, *Inorg. Chim. Acta*, 1974, **9**, 217, 221.
[330] R. J. Gusch and T. L. Brown, *Inorg. Chem.*, 1973, **12**, 2815.
[331] R. J. Gusch and T. L. Brown, *Inorg. Chem.*, 1974, **13**, 417.
[332] R. J. Gusch and T. L. Brown, *Inorg. Chem.*, 1974, **13**, 959.
[333] C. Varhelyi, Z. Finta, A. Benko, and A. Binder, *Monatsh.*, 1974, **105**, 490.
[334] G. P. Syrtsova, *Zhur. neorg. Khim.*, 1974, **19**, 430.
[335] A. L. Crumbliss and P. L. Gauss, *Inorg. Nuclear Chem. Letters*, 1974, **10**, 485.

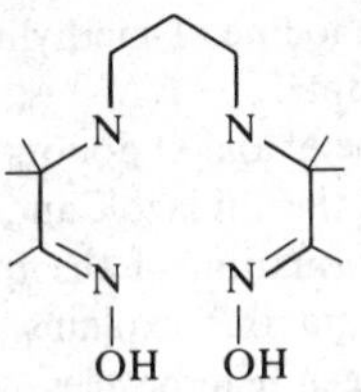

(54)

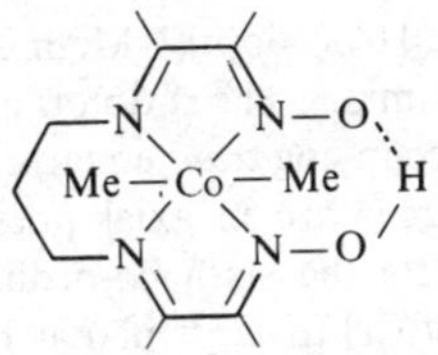

(55)

even in almost neutral aqueous solution.[336] An X-ray structure analysis of the complex $(X = NO_2)$ confirms the *trans* structure and the presence of N-bonded (nitro) NO_2 groups, the geometry about the cobalt being approximately octahedral.[337] The structure of dimethyl-{3,3'-(trimethylenedinitrilo)bis(butan-2-oneoximato)}cobalt(III) (55) shows the co-ordination about the cobalt, with axial methyl groups. The Co—Me distances are not particularly long, suggesting the absence of any strong *trans* influence.[338]

Schiff Bases. The reaction of sulphur (S_8) with Co(salen) in pyridine produces the dimeric S_4 bridged $[(py)_2Co(salen)]_2S_4$. Similar complexes are formed with ring-substituted salen ligands.[339] Mixed ligand complexes $[CoL(L-L)]_n.H_2O$ $(LH_2 =$ Schiff base; L–L = acac, glycinate, salicylaldehydato, or oxinate) and $[CoL(L'-L')]$ $(L'-L' = $ en or $HOCH_2CH_2NH_2)$ are octahedral with an unusual non-planar conformation of the Schiff base produced by the (necessarily) *cis* co-ordination of the bidentate ligand, whilst in the $[CoLL''_2]^+$ $(L'' = NH_3$, py, or $H_2O)$ normal (planar) co-ordination of the Schiff base is present.[340] Dinuclear $Co_2(bstn - H)_3.H_2O$ are formed with NN'-2-hydroxytrimethylenebis-salicylideneiminato (bstn).[340] The [Co-(salen)(acac)],H_2O complex is nitrated by copper(II) nitrate trihydrate in acetic anhydride to produce $[Co(salenNO_2)_2(acac)]$ and $[Co\{salen(NO_2)_2\}(acac.NO_2)]$.[341]

The ligands NN'-bis-(2-aminoacetophenone)ethylenediamine, NN'-bis-(2-amino-benzophenone)ethylenediamine, and NN'-bis-(2-aminobenzaldehyde)ethylene-diamine (LH_2) form mononuclear (CoLX) complexes and the more unusual bi- and tri-metallic derivatives (56) and (57).[342]

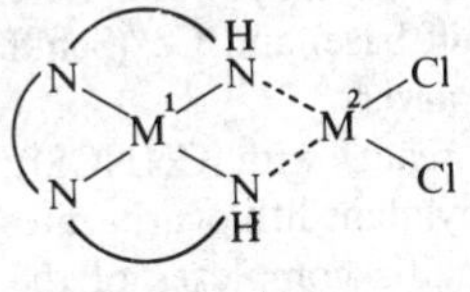

(56) $M^1 = $ Cu; $M^2 = $ Co

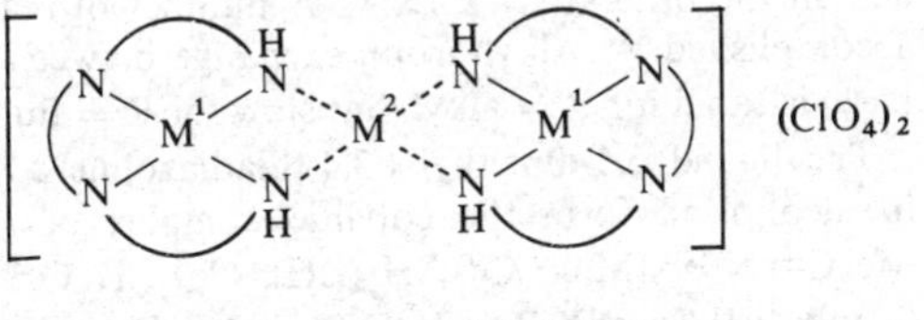

(57) $M^1 = $ Cu or Ni; $M^2 = $ Co

$(ClO_4)_2$

[336] H. Goff, S. Kidwell, J. Lauher, and R. K. Murmann, *Inorg. Chem.*, 1973, **12**, 2631.
[337] R. K. Murmann and E. O. Schlemper, *Inorg. Chem.*, 1973, **12**, 2625.
[338] M. Calligaris, *J.C.S. Dalton*, 1974, 1628.
[339] C. Floriani and G. Fachinetti, *Gazzetta*, 1973, **103**, 1317.
[340] K. Dey and R. L. De, *Z. anorg. Chem.*, 1973, **402**, 120.
[341] K. Dey and R. L. De, *J. Inorg. Nuclear Chem.*, 1974, 1182.
[342] E. Uhlmann and M. Plath, *Z. anorg. Chem.*, 1973, **402**, 279.

The reaction of [Co(saldpt)] successively with iodine, 1-methylimidazole, and LiBr yields [Co(saldpt)(1-Meimid)]Br,H_2O [saldptH_2 = NN'-bis(salicylidene)di-propylenetriamine). The cation is octahedral with the N and O donors of the salycili-dene rings occupying *trans* equatorial positions, and the imidazole and the secondary amine nitrogens are in axial positions. The conformation of the quinquedentate ligand restricts the sixth co-ordination position and this explains the selectivity of [Co(saldpt)Cl] towards planar ligands, *e.g.* nitrogen heterocycles, compared with the more basic but non-planar PR_3 ligands.[343]

Ligand (58) is synthesized *in situ* from Co(acac)$_2$ and 2-aminoethanol in the presence of acac and forms [CoL($H_2NCH_2CH_2OH$)$_2$]. In order to establish whether the 2-aminoethanol molecules are present as such or have further reacted, the complex

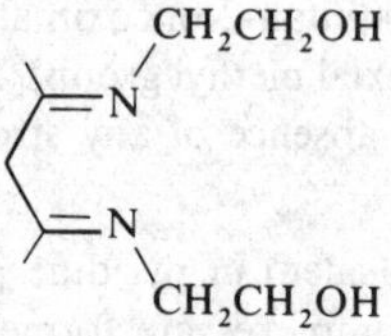

(58) LH_3

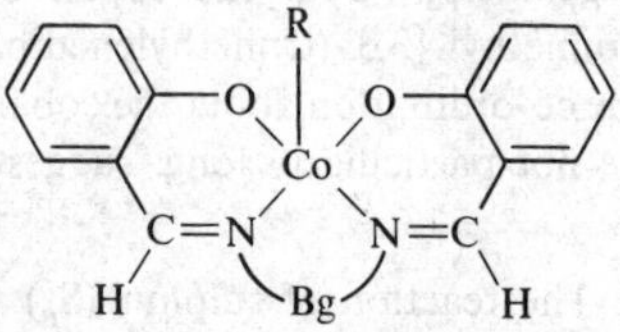

(59) Bg = CH_2CH_2 salen
CH$_2$CHMe sal$_2$-1,2-pn
CH$_2$CMe$_2$ sal$_2$-1,2-pn-2Me
o-C$_6$H$_4$ sal$_2$-oph
trans-1,2-cyclohexyl sal$_2$-cy

was X-rayed.[348] The complex contains the trianion of LH_3 functioning as a quadri-dentate, co-ordinated equatorially with two *trans* axial unidentate N-bonded 2-aminoethanol ligands. A series of RCo(sal$_2$Bg) (59) has been prepared by reduction of the cobaltous complex with NaBH$_4$ in alkaline H_2O–MeOH, followed by addition of RX in the presence of a trace of PdCl$_2$. Complexes were obtained for R = Me, Et, and Pr^n and sal$_2$Bg as in (59), and Pr^i/salen, but not for Schiff base complexes with —(CH$_2$)$_n$— (n = 3,4, or 5) bridges. The stability of the R—Co bond falls in the order 1° > 2° > 3°. A number of reactions of the Co–alkyl bond were accomplished.[345] Alkyl-group exchange between RCo^{III}(Schiff base) and Co^{II}(Schiff base) is rapid for R = alkyl, but slow for R = fluoroalkyl or aryl.[346]

The ligand 5,5-dimethyl-1,3,4-thiadiazoline-2-thione (60) reacts with Co(NCS)$_2$ in alcohol to form the cobaltic complex of the isopropylidenedithiocarbazate, Me$_2$C=N—N(Me)—C(S)SH (LH), CoL$_3$,H$_2$O.[347] The cobaltic complexes of the ligands (61) Co(pdX-R)$_3$,nH$_2$O (R = Me, Et, or Pr^n) tend to be formed from cobaltous complexes by air oxidation, whereas the Pr^i and Bu^t analogues under the same con-ditions do not oxidize.

[343] T. J. Kirstenmacher, L. G. Marzilli, and P. A. Marzilli, *Inorg. Chem.*, 1974, **13**, 2089.
[344] J. A. Bertrand, F. T. Helm, and L. J. Carpenter, *Inorg. Chim. Acta*, 1974, **9**, 69.
[345] A. van den Bergen and B. O. West, *J. Organometallic Chem.*, 1974, **64**, 125.
[346] R. M. McAllister and J. H. Weber, *J. Organometallic Chem.*, 1974, **77**, 91.
[347] M. A. Ali, S. E. Livingstone, and D. J. Philips, *Inorg. Chim. Acta*, 1973, **7**, 553.

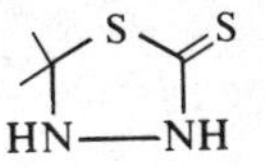

(60)

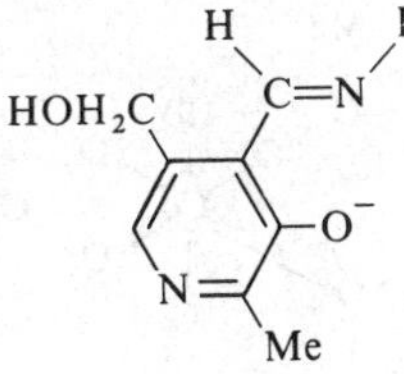

(61) pdX-R

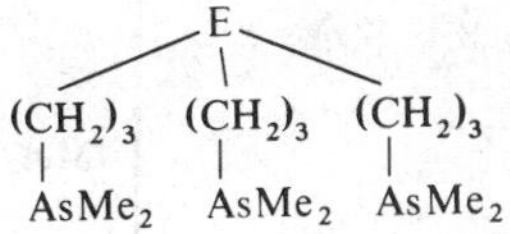

(62) E = P, tap
      E = As, taa
      E = Sb, tasb

P-*Donors*, As-*Donors*, *and* Sb-*Donors*. Formation of purple, diamagnetic $[Co(pmm)_2-Cl_2]^+$ [pmm = o-$C_6H_4(PMe_2)_2$] has been reported. Further oxidation to cobalt(IV) could not be achieved. Bosnich *et al.* have reported systematic syntheses for *cis*-$[Co(das)_2X_2]^{n+}$ (X = Cl, $\frac{1}{2}CO_3$, OH_2, NO_2, NO_3, or MeCN). The *cis*-$[Co(das)_2Cl_2]^+$ ion was resolved with sodium arsenyltartrate, and the optically active isomers could be converted into other *cis*-$[Co(das)_2X_2]^+$ with complete retention of optical activity. It is concluded that the *cis* → *trans* isomerization is catalysed by a cobalt(II) species, and this explains the failure to isolate pure *cis*-Co^{III} complexes with mildly reducing anions.[348]

The flexible quadridentate ligands (62) form octahedral $[CoLX_2]ClO_4$ (X = Cl, Br, I, or NCS) complexes, which of necessity have a *cis* structure.[349] The tasb complexes are the first examples of Co^{III}—Sb co-ordination. The racemic and *meso* forms of tetars, $Me_2AsCH_2CH_2As(Ph)CH_2CH_2As(Ph)CH_2CH_2AsMe_2$, have been separated *via* their cobalt(III) complexes.[350, 351] All five possible topological isomers of $[Co(tetars)Cl_2]^+$ (three racemic and two *meso* ligand) have been isolated. The *meso* ligand generally prefers a *trans* structure, but the racemic form imposes no strong preference on the complexes. Detailed ^{1}H n.m.r., absorption, and c.d. spectra are reported.

O-*Donors*. The oxidation of Co^{II} in glacial acetic acid solution with ozone produces 'cobalt(III) acetate', which is $Co_3O(O_2CMe)_6(MeCO_2H)_3$, a Co^{II}–Co^{III} species, the magnetic properties of which depend upon the drying technique used. The probable structure consists of a central oxygen atom co-ordinated to three cobalt atoms, each cobalt being surrounded by six oxygens.[352] Wilkinson and co-workers[353] found that the green product 'cobalt(III) acetate' reacted with pyridine or β-picoline and $NaClO_4$ in methanol to form $[Co_3O(O_2CMe)_6,L_3]ClO_4$ (L = py, or β-pic). Analogues with H_2O or $MeCO_2H$ replacing the amine could not be obtained. These diamagnetic amine adducts appear to have a structure (63) different from their Cr, Mn, Fe analogues. Cobalt(III) trifluoroacetate, $Co(OCCF_3)_3$, is formed from 'cobalt(III) acetate' and CF_3CO_2H–$(CF_3CO)_2O$. It is a powerful oxidant, *e.g.* it converts benzene into phenyl trifluoroacetate in >95% yield in *ca.* 30 mins.[354]

[348] B. Bosnich, W. G. J. Jackson, and J. W. McLaren, *Inorg. Chem.*, 1974, **13**, 1133.
[349] G. Kordosky, G. S. Benner, and D. W. Meek, *Inorg. Chim. Acta*, 1973, **7**, 605.
[350] B. Bosnich, W. G. J. Jackson, and S. B. Wild, *J. Amer. Chem. Soc.*, 1973, **95**, 8269.
[351] B. Bosnich, W. G. J. Jackson, and S. B. Wild, *Inorg. Chem.*, 1974, **13**, 1121.
[352] J. J. Ziolkowski, F. Pruchnik, and T. Szymansh-Buzar, *Inorg. Chim. Acta*, 1973, **7**, 473.
[353] S. Uemura, A. Spencer, and G. Wilkinson, *J.C.S. Dalton*, 1973, 2565.
[354] R. Tang and J. K. Kochi, *J. Inorg. Nuclear Chem.*, 1973, **35**, 3845.

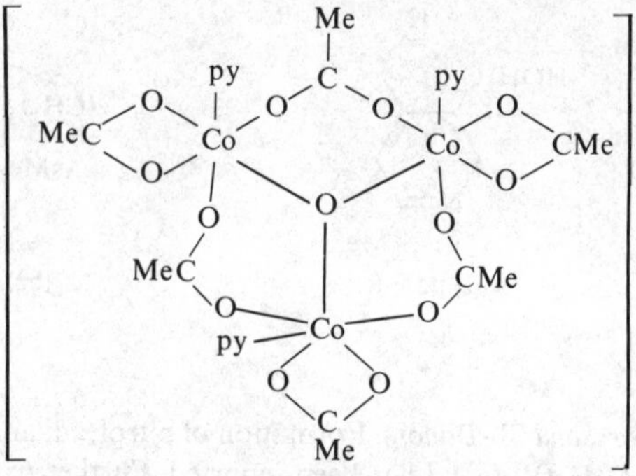

(63)

The structure of $Co(acac)_3$ is a distorted octahedron, $Co-O = 1.888(4)$ Å, $\angle OCoO = 96.5°$.[355] Thermal decomposition of $Co(acac)_3$ to $Co(acac)_2$ has been reported, and the mass spectra of $Co(acac)_2$,oxine and $Co(acac)(oxine)_2$ have been studied.[356] Acetic acid and $[Co(CO)_3L_2]^+$ (L = bipy or phen) in alcohol–water react to form $[Co(acac)L_2]^{2+}$ and $[Co(acac)_2L]^+$.[357] The *cis–trans* isomerization of $[Co(acac)_2(py)X]$ (X = N_3, NO_2, NCO, or CN) has been examined and the *trans* → *cis* rate is $N_3 > NCO > NO_2 > CN$.[358] Potassium cyanate reacts with $Co(acac)_2$ in aqueous solution in the presence of pyridine and hydrogen peroxide to form $[Co(acac)_2(NCO)(py)]$.[358] An extension of this work to KNCS produced *trans*-$[Co(acac)_2$-$(SCN)(py)]$. On standing this complex undergoes linkage isomerism to the isothiocyanato analogue, whilst under reflux *cis*-$[Co(acac)_2(NCS)(py)]$ forms.[359]

The $[Co(malonate)_3]^{3-}$ ion has been resolved by use of $(+)$-$[Co(en)_3]^{3+}$.[360] Formation of cobalt(III) complexes of 1-nitroso-2-naphthol-3,6-disulphonate by reaction of the ligand with cobalt(III) salts in the presence of air has been examined as a function of pH and concentration.[361, 362]

S-*Donors*. Formation of $[Ph_4E]_3[Co(CS_3)_3]$, which is a diamagnetic cobalt(III) species containing bidentate (S_2) CS_3^{2-} ligands, has been reported.[363] The structure of tris(ethylthioxanthato)cobalt(III), $Co(S_2CEt)_3$ shows $Co-S(mean) = 2.266(7)$ Å and $\angle SCoS(mean) = 76.2°$. The discrepancy of these results[364] and those previously reported[365] led Lippard and Li to suggest that the previous determination had, in fact, been performed on the Cr^{III} analogue. An examination of the redox behaviour of the tris-dithiocarbamates of Cr, Mn, Fe, and Co (ML_3) showed that the Cr^{III} and Co^{III} complexes were much more stable, suggesting that the $3d^n$ configuration

[355] G. J. Kruger and E. C. Reynhardt, *Acta Cryst.*, 1974, **B30**, 822.
[356] Y. Kidani, S. Naga, and H. Koike, *Chem. Letters.*, 1974, 781.
[357] K. Konya and B. E. Douglas, *Inorg. Nuclear Chem. Letters*, 1974, **10**, 491.
[358] D. R. Herrington and L. J. Boucher, *Inorg. Chem.*, 1973, **12**, 2378.
[359] L. J. Boucher, D. R. Herrington, and C. G. Coe, *Inorg. Chem.*, 1974, **13**, 2290.
[360] R. L. C. Russell and B. E. Douglas, *Inorg. Nuclear Chem. Letters*, 1973, **9**, 1251.
[361] G. C. Lalor and G. A. Taylor, *J. Inorg. Nuclear Chem.*, 1973, **35**, 4221.
[362] S. P. Bajui and G. C. Lalor, *J. Inorg. Nuclear Chem.*, 1973, **35**, 4231.
[363] A. Müller, P. Christophliemk, I Tossidis, and C. K. Jørgensen, *Z. anorg. Chem.*, 1973, **40**, 274.
[364] T. Li and S. J. Lippard, *Inorg. Chem.*, 1974, **13**, 1791.
[365] A. C. Filla, A. G. Manfredotti, C. Guastini, and M. Nardelli, *Acta Cryst.*, 1972, **B28**, 2231.

had a marked effect upon the redox potentials.[366] The temperature-dependent n.m.r. spectra of $Co(dtc)_3$ (dtc = NN-dibenzyldithiocarbamate) shows that the complex is stereochemically non-rigid, probably inverting by a trigonal twist mechanism.[367]

Optically active dithiocarbamates $Co(S_2CR_2)_3$ (R = Me, Et, Pr^i, Bu^n, Bu^i, Bz, Cy, or pip) can be prepared from the potassium dithiocarbamate salts and the optically active K[Co(edta)] or K[Co(pdta)] (pdta = 1,2-propanediaminetetra-acetate).[368] The optical inversion of $(+)_{546}$-[Co(pyr-dtc)$_3$] (pyr-dtc = NN-dipyrrolidyldithio-carbamate) in chloroform has been studied by loss of optical activity by polarimetry.[368]

In air $Co(sacsac)_2$ reacts with bipy or phen (L) to form octahedral [Co(sacsac)$_2$L$_2$]X (X = Br or NO$_3$), which are diamagnetic 1:1 electrolytes with a CoS_4N_2 chromophore.[369] A similar type of complex could not be isolated with pyridine ligands. The structures of [Co(en)$_2$(SCH$_2$CH$_2$NH$_2$)](NCS)$_2$ and [Co(en)$_2$(SCH$_2$CO$_2$)]Cl,H$_2$O have been determined, and the *trans* influence of the sulphur is shown by Co—N *trans* to S being longer than Co—N *cis* to S. The mercaptoacetate complex exhibits both $\Delta\lambda\delta$ and $\Lambda\delta\lambda$ configurations, and the β-mercaptoethylamine is $\Delta\lambda\lambda\lambda$ and $\Lambda\delta\delta\delta$.[370]

Amino-acid Complexes. X-Ray crystal structures have been reported for many cobalt(III) amino-acid complexes. Potassium dinitrobis(β-alaninato)cobaltate(III) has octahedral co-ordination about the cobalt, *trans* nitro-groups, and a *trans* arrangement of amino N- and carboxylato O-donors from the bidentate β-alaninates.[371] In Ca[Co(aspartate)$_2$] there are two isomeric ions: the *cis*(N)*trans*(O$_5$) (64) and *cis*(N)*trans*(O$_6$) (65),

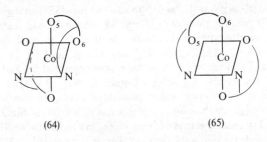

(64)　　　　　　　(65)

with distorted octahedral configurations.[372] In the silver salt of $(+)_{584}$-*cis*-(dinitro)-bis(D-alaninato)cobaltate(III) the octahedral cobalt is co-ordinated by two *cis* nitro-groups, two *trans* amino-groups, and two *cis* carboxylate groups.[373] The structure of *cis*-[Co(adeninate)(en)$_2$Cl]Br,H$_2$O consists of cobalt(II) octahedrally co-ordinated to two *cis* en ligands, one chloride ligand, and an N-bonded unidentate monoanion of adenine.[374]

[366] R. Chant, A. R. Hendrickson, R. L. Martin, and N. M. Rohde, *Austral. J. Chem.*, 1973, **26**, 2533.

[367] M. C. Palazzotto, D. J. Duffy, B. L. Edgar, L. Que, and L. H. Pignolet, *J. Amer. Chem. Soc.*, 1973, **95**, 4537.

[368] L. R. Graham, J. G. Hughes, and M. J. O'Connor, *J. Amer. Chem. Soc.*, 1974, **96**, 2271.

[369] R. K. Y. Ho and R. L. Martin, *Austral. J. Chem.*, 1973, **26**, 2299.

[370] R. C. Elder, L. R. Florian, R. E. Lake, and A. M. Yacynych, *Inorg. Chem.*, 1973, **12**, 2690.

[371] B. Prelesnik, M. B. Celap, and R. Herak, *Inorg. Chim. Acta*, 1973, **7**, 569.

[372] T. Oonishi, M. Shibata, F. Marumo, and Y. Sato, *Acta Cryst.*, 1973, **B29**, 2448.

[373] R. Herak, B. Prelesnik, Lj. Manojlović-Muir, and K. W. Muir, *Acta Cryst.*, 1974, **B30**, 229.

[374] T. J. Kistenmacher, *Acta Cryst.*, 1974, **B30**, 1610.

Bis(aminoacidato)cobalt(III) complexes of L-methionine and S-methyl-L-cysteine, $Co(S,N,O)_2^+$, have been prepared and separated into three geometrical isomers by ion-exchange chromatography, and configurations assigned on the basis of electronic absorption, c.d., and ^1H n.m.r. spectra.[375] Mixed complexes of L-proline and D- or L-aspartic acid $[Co(L-pro)_{3-n}\{L(D)asp\}_n]^{n-}$ ($n = 1$ or 2) have been prepared and separated into stereoisomeric forms by ion-exchange chromatography.[376] For $[Co(L-pro)_2(D-asp)]$ and $[Co(L-pro)(D-asp)]^{2-}$ all three possible *mer* isomers have been isolated in the Δ forms. For *mer*-$[Co(L-pro)(L-asp)_2]^{2-}$ only one isomer (Λ) appears to be produced, that with the *cis*(N)*trans*(O) configuration.[376] There is a marked difference in stereoselectivity between the L and D aspartato complexes, which were qualitatively explained in terms of interaction between the dangling β-carboxylate in the chelated aspartato group and the neighbouring ligand. Okamoto *et al.*[377] have prepared *cis*(O) and *trans*(O) triammine(sarcosinate-*N*-monopropionato)cobalt(III) complexes, and the *cis*(O)-isomer was resolved by use of $(+)_{346}$-$[Co(edta)]^-$. The absolute configurations were established by comparison of the c.d. spectra with those of the analogous complexes of L-alaninate- and L-prolinate-*N*-monocarboxylate ligands. Two diastereomeric isomers of *cis*(O)-(L-alaninate-*N*-monopropionato)-triamminecobalt(III) perchlorate are known.[377]

The cobalt(III) complexes of *N*-methyl-(*S*)- and -(*R*)-alaninate [NMe-(*S/R*)ala] and *NN'*-bis(β-aminoethyl)-1(*R*),2(*R*)-diaminocyclohexane(*R*-ba) have been characterized. They are the Λ-β_2(SSR) isomer of $[Co\{NMe-(S)ala\}R-ba]^{2+}$ and the two isomers of $[Co\{NMe-(R)ala\}R-ba]^{2+}$ with the Λ-β_2(SSR) and Λ-β_2(SSS) configurations.[378]

Four of the six possible isomers of $[Co(acac)(N-methyl-S-alaninato)_2]$ have been separated by chromatography, and identified as $\Delta trans$N, $\Lambda trans$N, ΛcisN, and Δcis-N—C$_1$, the ΔcisN—C$_2$ and ΛcisN—C$_1$ were not isolated.[379] In the $[Co(acac)-(S-aminoacidate)_2]$ systems, six isomers have been separated for *S*-alanine, and five for *S*-valine,[380] the ΛcisN—C$_2$ being obtained in very small yield for *S*-alanine and being absent in the *S*-valine system, probably due to steric effects.

NN'-Ethylenebis[(*S*)-methionine] (L) (66) reacts with KOH in MeOH, followed by $CoCl_2,6H_2O$ and air oxidation to yield $[CoL]ClO_4$, in which both the NH and carboxylate groups are bonded to the metal. Only one isomer was isolated, $(-)_{589}$-$[CoL]ClO_4$, and identified by X-ray analysis as having the Λ configuration.[197]

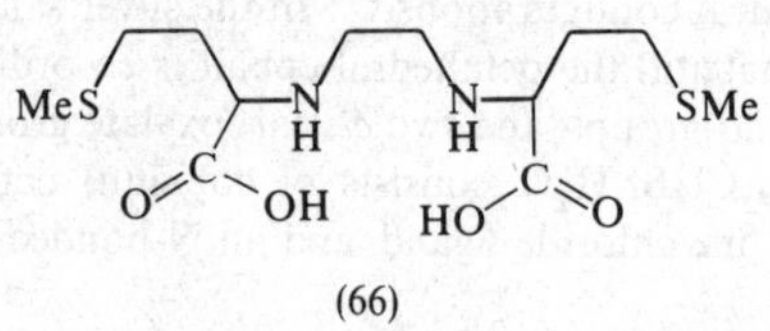

(66)

[375] J. Hidaka and Y. Shimura, Proceedings 16th International Coordination Chemistry Conference, Dublin, 1974.
[376] T. Matsuda and M. Shibata, *Bull. Chem. Soc. Japan*, 1973, **46**, 3104.
[377] K. Okamoto, J. Hidaka, and Y. Shimura, *Bull. Chem. Soc. Japan*, 1973, **46**, 3134.
[378] M. Saburi and S. Yoshikawa, *Bull. Chem. Soc. Japan*, 1974, **47**, 1184.
[379] L. A. Wingert, D. J. Seematter, J. G. Brushmiller, G. W. Everett, and K. S. Finney, *Inorg. Nuclear Chem. Letters*, 1974, **10**, 71.
[380] L. A. Wingert, D. J. Seematter, J. G. Brushmiller, G. W. Everett, and K. S. Finney, *Inorg. Nuclear Chem. Letters*, 1974, **10**, 75.

A novel light-reversible redox system has been discovered[381] with glycylglycine; it is summarized in Scheme 7. Preparations of mixed cobalt(III) complexes of macrocycles and amino-acids, *trans*-[Co{[14]aneN$_4$}(amino-acid)$_2$]$^{3+}$ and *trans*-[Co{Me$_4$-[14]tetraeneN$_4$}(amino-acid)$_2$]$^{3+}$, have been reported using glycine, *S*-alanine, *S*-phenylalanine, and *S*-leucine.[382]

$$\text{Co}^{\text{II}}\text{L}_2 \underset{\text{N}_2}{\overset{\text{O}_2}{\rightleftharpoons}} [\text{Co}^{\text{III}}\text{L}_2]_2 \text{O}_2^-(\text{OH})_x \overset{\text{O}_2}{\underset{}{\rightleftharpoons}} \text{Co}^{\text{III}}\text{L}$$

$$\underset{hv}{\underline{\hspace{6cm}}}$$

Scheme 7

Everett *et al.*[383] have examined the c.d. spectra of $[\text{Co}(\beta\text{-diket})_n(\text{en})_{3-n}]^{-(3-n)}$ and $[\text{Co}(\beta\text{-diket})_n\{(S)\alpha\text{-amino-acidate})\}_{3-n}]$ ($n = 0$—3) to ascertain whether the Cotton-effect configuration rule for assigning structure to low-symmetry mixed ligand complexes is valid. It was concluded that if reasonable assumptions about the energy separation of ligand field terms and the contribution of vicinal effects of the chiral ligands were made, then the rule may be applied successfully.[383] The interaction between tris(L-alaninato)cobalt(III) complexes and basic polyelectrolytes has been examined.[384] Gillard *et al.* have pointed out that care must be taken when relating solution and solid structures by comparison of the c.d. spectra.[385]

Other Mixed Donor Ligands. The α-*cis*- and β-*cis*-complexes of *NN*′-ethylenediaminediacetic acid (H$_2$edda), [Co(CO$_3$)(edda)]$^-$, [Co(OH$_2$)$_2$(edda]$^+$, [Co(NO$_2$)$_2$(edda)]$^-$, [CoCl$_2$(edda)]$^-$, [Co(mal)(edda)]$^-$, and [Co(OH$_2$)Cl(edda)] exist as Λ or Δ optical isomers, whilst in the case of unsymmetrical bidentate or non-identical unidentate ligands two further β-isomers are possible. Absolute configurations were assigned on the basis of o.r.d. measurements, and by an *X*-ray structure of β-*cis*-[Co(OH$_2$)$_2$-(edda)]ClO$_4$, which was identified as the Λ-β-*cis*-(*SR*), Δ-β-*cis*(*RS*) racemate.[386] Aquation of both α-*cis*- and β-*cis*-forms of [CoCl$_2$(edda)]$^-$ occurs with total retention,[387] whilst aquation of α-*cis*- or β-*cis*-[CoCO$_3$(edda)]$^-$ occurs through both water and acid attack.[388]

The resolution of uns-*cis* (or β-*cis*)-[CoL(edda)]$^-$ (L = CO$_3$, ox, mal) and their c.d. spectra have been compared with those of the *S-cis* (or α-*cis*) isomers, in an attempt to assess the contribution of vicinal and configurational effects and chelate ring size to rotatory strengths.[389] The changes observed in the c.d. spectra could be related to the arrangement of the chelate rings and to the configuration about the asymmetric

[381] I. Rosenthall, *Inorg. Nuclear Chem. Letters*, 1973, **9**, 1053.
[382] J. Cragel and B. E. Douglas, *Inorg. Chim. Acta*, 1974, **10**, 33.
[383] G. W. Everett, K. S. Finney, J. G. Brushmiller, D. J. Seematter, and L. A. Wingert, *Inorg. Chem.*, 1974, **13**, 536.
[384] B. M. Barteri, M. Branca, and B. Bispisa, *J.C.S. Dalton*, 1974, 543.
[385] R. D. Gillard, S. H. Laurie, D. C. Price, D. A. Phipps, and C. F. Weick, *J. C. S. Dalton*, 1974, 1385.
[386] P. J. Garnett and D. W. Watts, *Inorg. Chim. Acta*, 1974, **8**, 293.
[387] P. J. Garnett and D. W. Watts, *Inorg. Chim. Acta*, 1974, **8**, 307.
[388] P. J. Garnett and D. W. Watts, *Inorg. Chim. Acta*, 1974, **8**, 313.
[389] L. J. Halloran and J. I. Legg, *Inorg. Chem.*, 1974, **13**, 2193.

nitrogen. The reaction of uns-*cis*-[Co(CO$_3$)(edda)]$^-$ with en, (*R*)-1,2-diaminopropane (*R*-pn) and L-alanine (L) produced the uns-*cis*-[CoL(edda)].[390] The diastereoisomers of the *R*-pn and L-ala compounds were separated chromatographically, as was the en optical isomer with *d*-α-bromocamphor-π-sulphonate.[390] The preparation of two isomers (α and β) of [Co(edda)(H$_2$NCH$_2$CH$_2$CH$_2$NH$_2$)]ClO$_4$, and the β → α isomerization, have been briefly noted.[391] *trans*-(O)-[Co(edma)$_2$]$^+$ and *trans*(O)-[Co(ttda)]$^+$ (edma = ethylenediaminemonoacetate; ttda = triethylenetetramine-N^2N^3-diacetate) were resolved using silver-*d*-tartrate and *d*-tartaric acid. Both (+)$_{589}$-[Co(edma)$_2$]$^+$ and (+)$_{589}$-[Co(ttda)]$^+$ are tentatively assigned as ΛΛΔ, or a net Λ configuration.[392] The trimethylene-*NN'*-diacetate ligand forms uns-*cis*-[Co(tmdda)L]$^+$ [L = (NH$_3$)$_2$, en, 1,3-pn, bipy, or phen]. The effect of the chelate ring sizes in these complexes, and in the edda analogues, upon the shape of the c.d. spectra was discussed.[393] Two isomers of K[Co{(*R*)-(−)-Pd3A}NO$_2$] (Pd3A = 1,2-propanediaminetriacetate) have been isolated, and identified as the *cis* and *trans* equatorial isomers (67). The differences in the c.d. spectra of these two complexes suggest that the asymmetric nitrogen donor is the major source of the optical activity.[394]

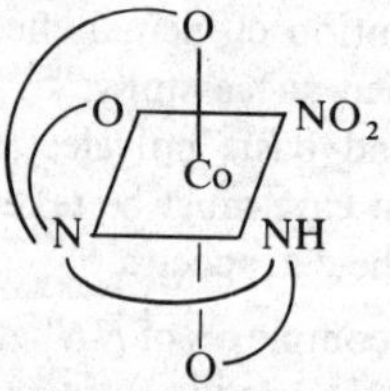

(67a) *cis*-equatorial 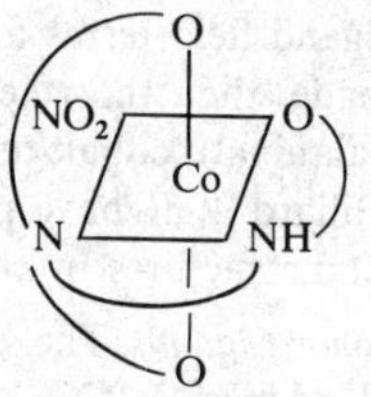   (67b) *trans*-equatorial

The *N*-hydroxyethyliminodiacetate ion, [(O$_2$CCH$_2$)$_2$NCH$_2$CH$_2$OH]$^{2-}$, usually functions as a terdentate O$_2$N donor with the alcohol group unco-ordinated, and forms complexes such as K[Co^{III}L$_2$], in which the —OH group can be acetylated without complex decomposition. A cation-exchange resin in the acid form converts the potassium salt into [Co(HL)L(H$_2$O)] in which one of the carboxylate groups has been protonated, and is now unco-ordinated, and replaced by the water. However, in [Co(en)L'] the ligand L' is believed to be deprotonated at the alcohol function and behaves as a quadridentate N$_2$O$_2$ donor.[395] A convenient method for the large-scale resolution of Co(edta)$^-$ using the L-histidinium cation has been reported.[396]

Celap *et al.* have prepared three of the five possible isomers of [Co(glycinate)$_2$-(NO$_2$)$_2$]$^-$ and have shown that these rearrange into the stable fourth isomer which

[390] L. J. Halloran and J. I. Legg, *Inorg. Chem.*, 1974, **13**, 2197.
[391] K. Kuroda, *Chem. Letters*, 1974, 17.
[392] K. Igi and B. E. Douglas, *Inorg. Nuclear Chem. Letters*, 1974, **10**, 587.
[393] K. Igi and B. E. Douglas, *Inorg. Chem.*, 1974, **13**, 425.
[394] C. W. Maricondi and C. Maricondi, *Inorg. Chem.*, 1974, **13**, 1110.
[395] P. Horrigan, R. A. Canelli, J. P. Kashman, C. A. Hoffman, R. Nauer, D. R. Boston, and J. C. Bailar, *Inorg. Chem.*, 1974, **13**, 1108.
[396] R. D. Gillard, P. R. Mitchell, and C. F. Weich, *J.C.S. Dalton*, 1974, 1635.

has *cis*-NO$_2$ groups and *trans*-NO glycinates.[397] The [Co(dea)(dapo)]X$_2$,nH$_2$O (X = Cl, Br, I, NCS, NO$_3$, or ClO$_4$) contain 1,3-diaminepropan-2-ol (dapoH) functioning as a terdentate chelate.[398] The cobalt(III) complex of acetylpyruvic acid [Co(en)$_2$(ap)]I is hydrolysed in the presence of KI to [Co(en)$_2$(ox)]I (ox = oxalate).[399] Condensation of [Co(NH$_3$)$_5$(pyruvilidinato)]$^{2+}$ with acetylacetone produces an imine chelate (68).[400]

The first example of absolute chiral recognition of a prochiral centre by a small molecule (*cf.* enzyme reactions) is the reaction of $\Lambda(-)_{436}$-α[(2S,9S)-2,9-diamino-4,7-diazadecanecobalt(III)]dichloride with α,α-aminomethylmalonate, which produces $\Lambda(-)_{436}$-β_2[(2S,9S)-2,9-diamino-4,7-diazadecanecobalt(III)-(R)-aminomethylmalonate]$^+$,[401] and a crystal structure determination of the product has been performed.[402]

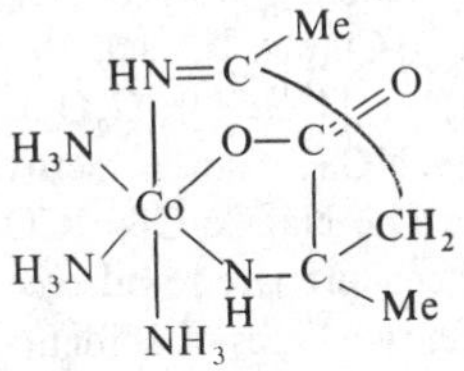

(68)　　　　　　　　　(69)

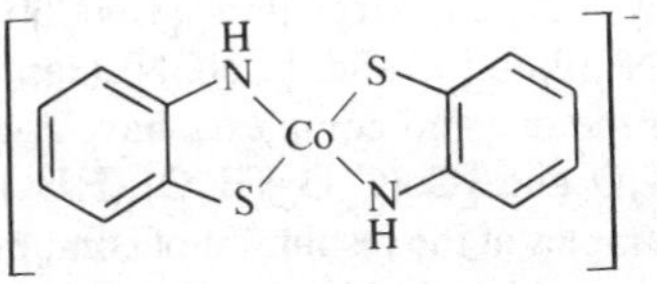

X-Ray crystal structure data on [Co(ete)(NO$_2$)Cl]Cl (ete = 1,9-diamino-3,7-dithianonane) indicates the complex to have the uns-*cis* structure with the nitro-group *trans* to the nitrogen and the chlorine *trans* to the sulphur, showing that in its formation from u-*cis*[Co(ete)(NO$_2$)$_2$]$^+$ the greater *trans* effect of sulphur allows the *trans*-nitro-group to be replaced.[403]

Dithio-oxamide and its *NN'*-dimethyl or -dicyclohexyl substituted derivatives (dto) form Co(dto)$_3$X$_3$ (X = Cl, Br, I, or ClO$_4$), which are octahedral CoN$_3$S$_3$ chromophores, as evidenced by the electronic spectra and by i.r. data. The complexes are diamagnetic and show varying degrees of anion–cation association in solution.[404] The dimeric Co$_2$(dto − 2H)$_3$$xH_2$O ($x$ = 0 or 4; dto − 2H = dianion of dithio-oxamide) are also octahedral with SN co-ordination, whilst the dicyclohexyl ligand affords Co(dto − 2H),EtOH, a polymeric cobaltous complex.[405] The reaction of aminothiophenol with potassium in EtOH followed by CoCl$_2$,6H$_2$O in presence of air produces the Co(abt)$_2^-$ anion (69), which can be isolated as Bu$_4^n$N[Co(abt)$_2$], μ_{eff} = 3.25 BM. The complex can be reduced polarographically.[406] Cobalt(III)

[397] M. B. Celap, J. K. Beattie, T. J. Janjic, and P. N. Radwojsa, *Inorg. Chim. Acta*, 1974, **10**, 21.
[398] V. N. Evreev, S. V. Marashkov, and V. E. Petrinkim, *Zhur. neorg. Khim.*, 1974, **19**, 103.
[399] J. G. Hughes and M. J. O'Connor, *Austral. J. Chem.*, 1974, **27**, 1161.
[400] B. T. Golding, J. M. Harrowfield, G. B. Robertson, A. M. Sargeson, and P. O. Whimp, *J. Amer. Chem. Soc.*, 1974, **96**, 3691.
[401] R. C. Job and T. C. Bruice, *J. Amer. Chem. Soc.*, 1974, **96**, 809.
[402] J. P. Glusker, H. L. Carrell, R. C. Job, and T. C. Bruice, *J. Amer. Chem. Soc.*, 1974, **96**, 5741.
[403] J. Murray-Rust and P. Murray-Rust, *Acta Cryst.*, 1973, **B29**, 2606.
[404] G. C. Pellacani and G. Peyronel, *Inorg. Chim. Acta*, 1974, **9**, 189.
[405] L. Menabue, G. C. Pellacani, and G. Peyronel, *Inorg. Nuclear Chem. Letters*, 1974, **10**, 187.
[406] P. J. M. W. L. Birker, E. A. De Boer, and J. J. Bour, *J. Coord. Chem.*, 1973, **3**, 175.

complexes of 1,10-phenanthroline carbothiamide and N-phenyl-1,10-phenanthroline-2-carbothiamide have been isolated and both ligands function as SN_2 terdentates.

Far-i.r. and Raman spectra have been reported for $K_3Co(CN)_6$ and $[Co(NH_3)_6]$-$[Co(CN)_6]$.[407] The absorption of hydrogen by aqueous solutions of $[Co(CN)_5]^{3-}$ produces $[HCo(CN)_5]^{3-}$ as the major product, and in the presence of excess CN^- some $[Co(CN)_6]^{3-}$ is also formed.[408] The reaction of $[Co(CN)_5]^{3-}$ with amines produces $[Co(CN)_m(am)_n]^{2-m}$ (am = en, bipy, or phen) and these have been investigated with respect to CN:Co ratio and activity in hydrogenation reactions, the active species probably being $[Co(CN)_3(am)_2]^-$.[409] Maki has examined the reduction of $[Co(CN)_5X]^{3-}$ (X = Cl, Br, I, NCS, NO_2, N_3, $S_2O_3^{2-}$, or SO_3^{2-}) at the dropping mercury electrode to form $[Co^I(CN)_5]^{4-}$. The final reduction product is $[(CN)_5-Co-H-Co(CN)_5]^{7-}$ rather than the earlier suggestion of $[Co^I(CN)_5H]^{3-}$.[410] The complex anion $[Co(CN)_2(CO_3)_2]^{3-}$ is formed by reaction of $[Co(CO_3)_3]^{3-}$ with KCN in aqueous solution and isolated as its $(+)_{589}$-$[Co(en)_3]^{3+}$ salt.[411] This anion may be converted into $[Co(CN)_2(CO_3)(C_2O_4)]^{3-}$, $[Co(CN)_2(C_2O_4)_2]^{3-}$, $[Co-(CN)_2(dtc)_2]^-$, and $[Co(CN)_2(xan)_2]^-$ (dtc = Me_2NCS_2, xan = methylxanthate). All the dicyano complexes have a *cis* structure.[411] The $Na[Co(CN)_2(CO_3)(NH_3)_2]$, $2H_2O$ and $[Co(C_2O_4)(H_2O)_2(NH_3)_2]NO_3$ have been prepared and resolved—rare examples of the resolution of compounds containing two types of unidentate and one unsymmetrical bidentate ligand.[412]

Polynuclear Bridged Complexes. Pentacyanocobalt(III)-μ-isocyano-pentamminecobalt(III) monohydrate has been X-rayed.[413] Potassium pentacyanocobaltate(II) reacts with p-benzoquinones to form $K_6[(CN)_5CoLCo(CN)_5]$ which are considered to be cobalt(III) complexes with p-hydroquinone dianion bridges.[414] Two forms of $K_6Co_2(CN)_{10}N_2O_2,nH_2O$, orange $n = 2$ and yellow $n = 4$, have been prepared from $[Co(NO)(NH_3)_5]Cl_2$ and KCN in aqueous solution. Both contain *trans*-hyponitrite groups, the differences being in the co-ordination (70a and b). On heating, varying amounts of NO and N_2O are evolved, whilst in aqueous KCN solution decomposition produces only N_2O.[415]

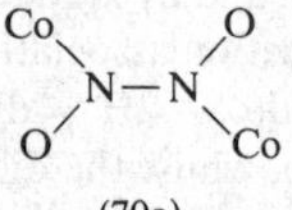

(70a) 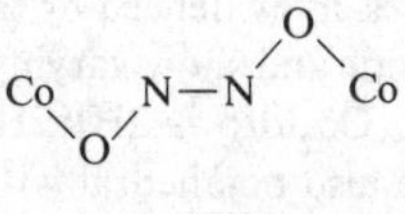   (70b)

Reaction of *trans*-$[Co(en)_2(H_2O)SO_3]^+$ with $[Fe(CN)_6]^{4-}$, $[Fe(CN)_6]^{3-}$, $[Co-(CN)_6]^{3-}$, or $[Fe(CN)_5(NO)]^{2-}$ produces μ-cyano-complexes, displacement of the water molecule in the cation occurring.[416]

[407] I. Nakagawa, *Bull. Chem. Soc. Japan*, 1973, **46**, 3690.
[408] H. J. Clase, A. J. Cleland, and M. J. Newland, *J.C.S. Dalton*, 1973, 2546.
[409] T. Funabiki, S. Kasaoka, M. Matsumoto, and K. Taruma, *J.C.S. Dalton*, 1974, 2043.
[410] A. H. Maki, *Inorg. Chem.*, 1974, **13**, 2180.
[411] S. Fujinami and M. Shibata, *Bull. Chem. Soc. Japan*, 1973, **46**, 3443.
[412] Y. Enomoto, T. Ito, and M. Shibata, *Chem. Letters*, 1974, 423.
[413] F. R. Fronczek and W. P. Schaefer, *Inorg. Chem.*, 1974, **13**, 727.
[414] J. Hanzlick and A. A. Vlcek, *Inorg. Chim. Acta*, 1974, **8**, 247.
[415] H. Okamura, E. Miki, K. Mizumachi, and T. Ishimori, *Chem. Letters*, 1974, 103.
[416] K. L. Scott, R. S. Murray, W. C. E. Higginson, and S. W. Foony, *J.C.S. Dalton*, 1973, 2335.

The first example of a sulphito bridge is present in $[Co_2(SO_3)(OH)_2(NH_3)_6]^{2+}$, formed from sulphur dioxide and $[Co_2(OH)_3(NH_3)_6]^{3+}$ and isolated with I_2, Br_2, $(NO_3)_2$, or $S_2O_5^{2-}$ anions. The SO_3^- bridge is oxygen bonded.[417] The reactions of μ-peroxodicobalt(III) complexes with SO_2 in the gas and liquid phases and in aqueous solution have been examined in detail, and those with NO briefly.[418]

The structure of di-μ-hydroxo-*trans*-diaquobis{triamminecobalt(III)}tetranitrate dihydrate, $[Co_2(OH)_2(H_2O)_2(NH_3)_6](NO_3)_4,2H_2O$, has been determined. The H_2O and NH_3 ligands were distinguished and the hydrogen bonds unambiguously assigned (71). The two H_2O ligands are *trans*.[419] Tri-μ-hydroxobis{triamminecobalt-(III) perchlorate reacts readily with carboxylic acids in acidic aqueous solution to produce carboxylate bridged species (72). A wide variety of RCO_2^- can be used, *e.g.* benzoate, pyridinecarboxylates, malonate. Dicarboxylic acids produce tetranuclear complexes in which four cobalt atoms are attached to two carboxylate groups, whilst with 1,3,5-benzenetricarboxylic acid (H_3L) a hexanuclear complex $\{[Co_2(OH)_2-(NH_3)_6]_3(L)\}^{9+}$ is formed.[420]

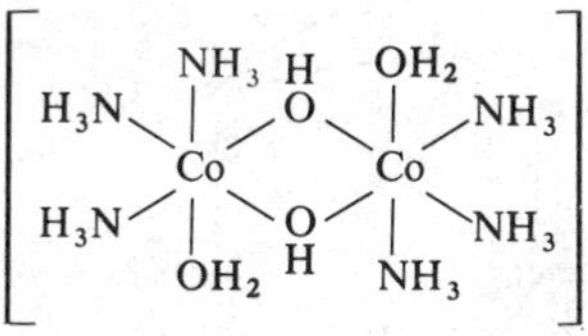

(71) 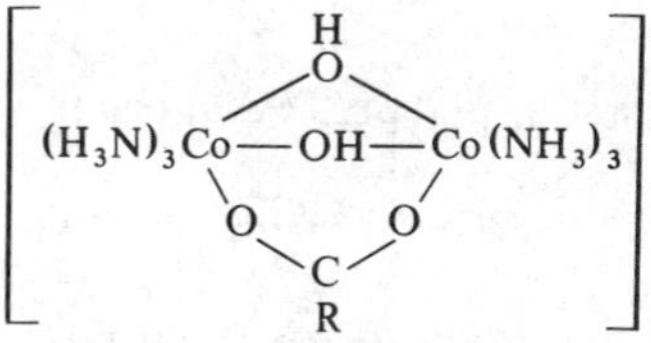   (72)

Cobalt(IV).—In addition to the oxocobaltates(IV) (*q.v.*), cobalt(IV) is present in the trisdithiocarbamates $[Co(S_2NR_2)_3]BF_4$ (R = Me, Et, Pr, or Cy) formed by oxidation of the appropriate cobalt(III) complex in toluene or benzene with BF_3. All the complexes are diamagnetic, which is thought to be because of spin-pairing (d^5, Co^{IV}) due to formation of associated species.[421]

2 Nickel

Carbonyls and Nitrosyls.—The preparation and i.r. spectra of $Ni(CO)_x$ ($x = 1$—4) in a matrix have been reported.[422] Nickel tetracarbonyl is reduced by sodium in THF or by potassium in methanol to $[Ni_3(CO)_3(\mu\text{-}CO)_3]_2^{2-}$, which may be identical with the previously reported $[Ni_4(CO)_9]^{2-}$.[423] This anion contains a trigonal antiprismatic Ni_6 group, six terminal carbonyls, and six bridging carbonyls, one along each edge of the two Ni_3 triangles. Surprisingly, the $[Pt_3(CO)_3(\mu\text{-}CO)_3]_2^{2-}$ ion is trigonal prismatic, and thus these two complexes constitute the first example of different structures between $3d$ and $5d$ elements in M—M bonded compounds.[423] In

[417] H. Siebert and G. Wittke, *Z. anorg. Chem.*, 1974, **406**, 282.
[418] C.-H. Yang, D. P. Keeton, and A. G. Sykes, *J.C.S. Dalton*, 1974, 1089.
[419] W. H. Bauer and K. Weighardt, *J.C.S. Dalton*, 1973, 2669.
[420] K. Weighardt, *J.C.S. Dalton*, 1973, 2548.
[421] L. W. Graham and M. J. O'Connor, *J.C.S. Chem. Comm.*, 1974, 68.
[422] E. P. Kundig, D. McIntosh, M. Moskovits, and G. A. Ozin, *J. Amer. Chem. Soc.*, 1973, **95**, 7234.
[423] J. C. Calebrese, L. F. Dahl, A. Cavalieri, P. Chini, G. Longoni, and S. Martinengo, *J. Amer. Chem., Soc.*, 1974, **96**, 2616.

dichloromethane $[NEt_4]GeCl_3$ reacts with $Ni(CO)_4$ to form $[Ni(CO)_3GeCl_3]^-$, and in THF to form $[Ni(CO)_2(GeCl_3)_2]^{2-}$.[424]

The reaction of $Ni(CO)_4$ with ER_3 (E = P, As, or Sb; R = Ph, $4-XC_6H_4$, or $3-XC_6H_4$; X = F or Cl) in diethyl ether under anaerobic conditions produces $Ni(CO)_3(ER_3)$. A study of the $\nu(CO)$ vibration frequencies suggests that the σ-donor capacity of the ligands with respect to substituents is Ph > $4-FC_6H_4$ > $4-ClC_6H_4$ > $3-FC_6H_4 \geqslant 3-ClC_6H_4$.[425] Pentane solutions of $Ni(CO)_4$ react with $(Me_3E)_3Sb$ (E = C, Si, Ge, or Sn) to form the monosubstituted product $Ni(CO)_3[(Me_3E)_3Sb]$.[426]

Nickel(II) phosphine complexes NiL_2Cl_2 (L = PPh_3, $MePh_2P$, $\frac{1}{2}Ph_2PCH_2CH_2$-PPh_2, or $\frac{1}{2}cis$-$Ph_2PCHCHPPh_2$) are reduced by carbon monoxide and NaSMe in ethanol to $NiL_2(CO)_2$, the MeS^- being converted into Me_2S_2.[427] The ligand $2-(CF_3)_2$-PB_5H_8 (L) forms $Ni(CO)_3L$ on reaction with $Ni(CO)_4$ without isomerization.[428] The structure of $Ni_2\{(CF_3)_2PSP(CF_3)_2\}_2(CO)_3$ is shown in (73), the Ni—Ni length is 2.572 Å.[429]

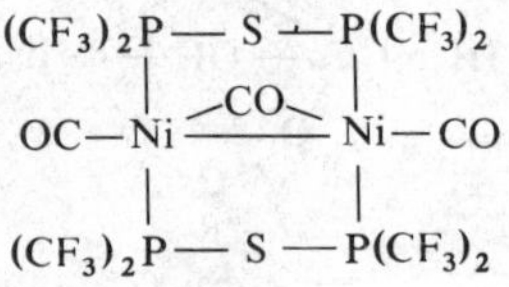

(73) 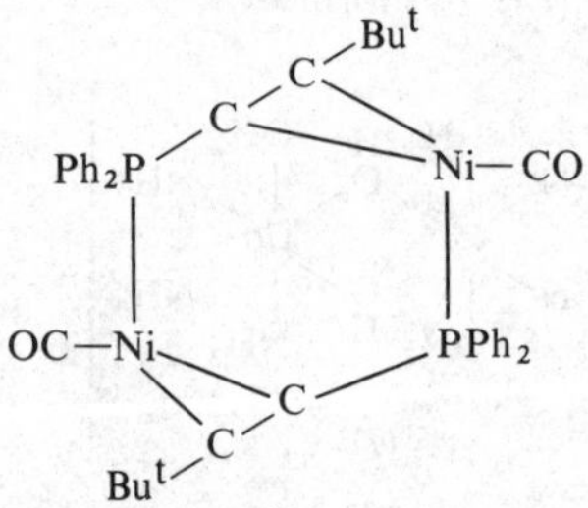   (74)

Dimethylchlorophosphine forms $Ni(CO)_3(PMe_2Cl)$ by direct reaction with $Ni(CO)_4$.[430] The chlorine is removed by reaction with $NaFe(CO)_2Cp$ to give the phosphido-bridged $(CO)_3Ni$-PMe_2-$Fe(CO)_2Cp$, but $NaMn(CO)_5$ or $NaRe(CO)_5$ form only secondary products:[430]

$$(CO)_3NiPMe_2Cl + NaM(CO)_5 \rightarrow NaCl + Ni(CO)_4 + \frac{1}{2}[(CO)_4MPMe_2]_2$$

M = Mn or Re

The first simple acetylene complex derived from direct reaction of $Ni(CO)_4$ and an acetylene is $Ni_2(CO)_2(Ph_2PC\equiv CBu^t)_2$, the structure (74) of which has been determined.[431] The ligands $Ph_2PC\equiv CPh$ and $Ph_2PC\equiv CBu^t$ react with $[(C_5H_5)Ni(CO)]_2$ or $(C_5H_5)_2Ni$ to form a number of complexes including $Ni(CO)_2(Ph_2PC\equiv CR)_2$, $[\{(\pi-C_5H_5)Ni\}_2(Ph_2PC\equiv CBu^t)]$ (75), $(OC)_3Ni(Ph_2PC\equiv CPh)\{Ni(C_5H_5)\}_2$ (76). Complex (75) contains an unco-ordinated phosphorus atom which can be readily converted into the phosphine oxide or quaternized to the phosphonium salts with

424 T. Kruck and W. Molls, *Z. Naturforsch.*, 1974, **29B**, 198.
425 F. T. Delbeke, G. P. Van der Kelen, and Z. Eeckhaut, *J. Organometallic Chem.*, 1974, **64**, 265.
426 H. Schumann and H. J. Breunig, *J. Organometallic Chem.*, 1974, **76**, 225.
427 K. Tanaka, Y. Kawata, and T. Tanaka, *Chem. Letters*, 1974, 831.
428 A. B. Burg, *Inorg. Chem.*, 1973, **12**, 3017.
429 H. Einspahro and J. Donahue, *Inorg. Chem.*, 1974, **13**, 1839.
430 W. Ehrl and H. Vahrenkamp, *J. Organometallic Chem.*, 1973, **63**, 289.
431 H. N. Paik, A. J. Carty, K. Dymock, and G. J. Palenik, *J. Organometallic Chem.*, 1974, **70**, C17.

RX (R = Me or Et; X = Br or I) without decomposition of the complex. The 'free' phosphine can also function as a donor towards another metal, *e.g. trans*-PdCl$_2$-(PhCN)$_2$ forms *trans*-PdCl$_2$L$_2$ where L = (75).[432]

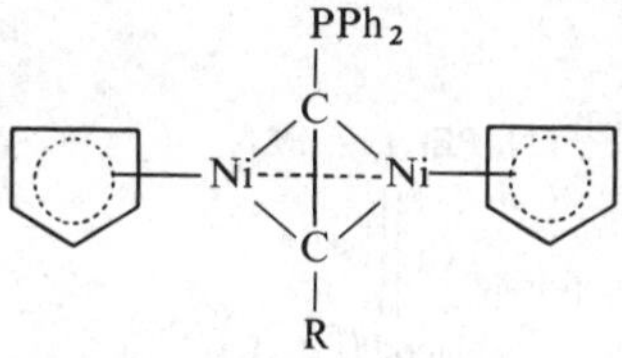

(75)

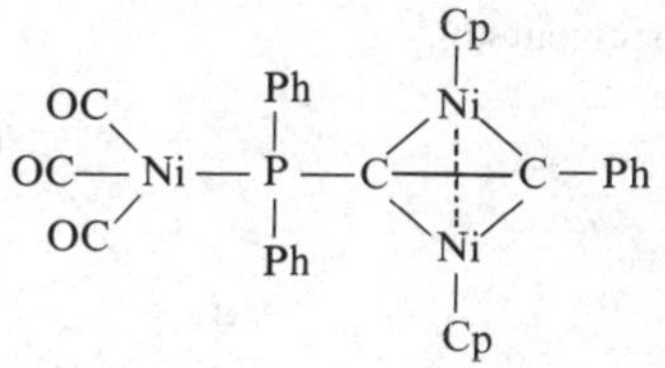

(76)

Nickel(0).—I.r. evidence for the production of a nickel thiocarbonyl moiety in argon matrices has been reported. In the mass spectrometer the species yields a parent ion consistent with the presence of Ni(CS)$_4$.[433] Benzilbisphenylimines (77) (L) form deeply coloured diamagnetic complexes obtained from NiCl$_2$ and the disodium salt of the ligand in THF or by reduction of Ni(acac)$_2$ and L with R$_2$AlOR. The (Ph$_3$P)$_2$NiL (R^1 = R^2 = H) is formed from (Ph$_3$P)Ni(C$_2$H$_4$) and L.[434] Tolman *et al.* have reported

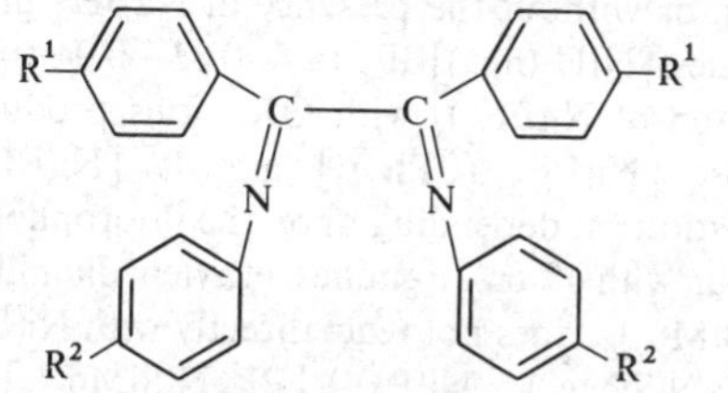

(77) R^1 = R^2 = H
      R^1 = H, R^2 = Me
      R^1 = MeO, R^2 = H

mol. wts., ^{1}H and ^{31}P n.m.r., and electronic spectra of a range of NiL$_4$ complexes (L = phosphine or phosphite). Whilst the rates of dissociation NiL$_4$ ⇌ NiL$_3$ + L are strongly affected by electronic factors, the extent of dissociation is largely attributable to steric effects. The new complexes Ni[P(O-*p*-C$_6$H$_4$Cl)$_3$]$_4$, Ni[P(O-*p*-Me-*o*-tolyl)$_3$]$_3$, Ni[P(O-*p*-Cl-*o*-tolyl)$_3$]$_3$, and Ni[P(OMe)Ph$_2$]$_4$ are reported.[435] Tetrakis-(triethylphosphine)nickel(0) has been isolated as a yellow pyrophoric solid by reaction of (cyclo-octa-1,5-diene)$_2$ Ni with PEt$_3$ under argon.[436, 437] In solution it dissociates

[432] A. J. Carty, H. N. Paik, and T. W. Ng, *J. Organometallic Chem.*, 1974, **74**, 279.
[433] L. W. Yarbrough, G. V. Calder, and J. G. Verkade, *J.C.S. Chem. Comm.*, 1973, 705.
[434] D. Walter, *Z. anorg. Chem.*, 1974, **405**, 8.
[435] C. A. Tolman, W. C. Seidel, and L. W. Gosser, *J. Amer. Chem. Soc.*, 1974, **96**, 53.
[436] C. A. Tolman, D. H. Gerlach, J. P. Jesson, and R. A. Schunn, *J. Organometallic Chem.*, 1974, **65**, C23.
[437] C. S. Cundy, *J. Organometallic Chem.*, 1974, **69**, 305.

reversibly into the purple $Ni(PEt_3)_3$ which appears to take up small amounts of nitrogen in solution to form $Ni(PEt_3)_3(N_2)$ and possibly $(Et_3P)_2Ni(N_2)Ni(PEt_3)_2$, although no compounds were isolated. Some other reactions are summarized in Scheme 8. Formation of $(Bu^n_3P)_2Ni(cod)$ and $Ni(Ph_2PCH_2CH_2PPh_2)_2$ is also documented.[437]

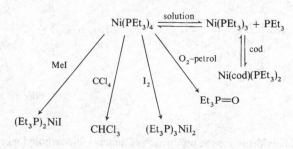

Scheme 8

Nickel(II) nitrate reacts with tris-(2-diphenylphosphinoethyl)amine (np_3) and $NaBH_4$ in acetone–ethanol to form $[Ni(np_3)]$. The absence of hydrido ligands was shown by chemical and i.r. spectral investigations. The X-ray structure shows the complex to be the first known example of trigonal-pyramidal co-ordination.[438] Similar reactions with or without the presence of $NaBH_4$ produce a range of non-stoicheiometric hydrides $[NiH_n(np_3)]BF_4$ ($n = 0.83$—0.04 with μ_{eff} in range 0.88—2.0 BM.[439] The reaction of $Ni(PF_3)_4$ with OR^- ions produces complexes such as $Ni[P(OMe)_3]_4$, whilst $[Ni(PF_3)_2(PPh_3)_2]$ affords $[Ni(PF_2OMe)_2(PPh_3)_2]$, the degree of fluorine substitution depending upon the fluorophosphine and the base.[440] Similar reactions occur with 1° or 2° amines, ethylenediamine, or hydrazine.[441, 442] The silylamine, $NH(SiMe_3)_2$, does not react directly with $Ni(PF_3)_4$, but on treatment of the latter with $NaN(SiMe_3)_2$, $Ni(PF_3)_3[PF_2N(SiMe_3)_2]$ and $Ni(PF_3)_2[PF_2N-(SiMe_3)_2]_2$ are formed.[443]

Nickel(I).—Ethanol solutions of nickel(II) halides mixed with PPh_3 and reduced with $NaBH_4$ under nitrogen produce $Ni(PPh_3)_3X$ (X = Cl, Br, or I), which are tetrahedral d^9 complexes. If the solutions of $Ni(PPh_3)_3X$ are treated with a further quantity of $NaBH_4$ the very air-sensitive $Ni(PPh_3)_3BH_4$ is formed. Surprisingly, this complex is diamagnetic, and thus may be dimeric.[50, 51] Under similar conditions 5-Ph-5-H-dibenzophosphole forms NiL_3Cl, NiL_2, and NiL_4, but not a borohydride complex. Nickel(II) chloride and $NaBH_3CN$ in the presence of PPh_3 produce only Ni^{II} complexes, but the Ni^I compound $Ni(PPh_3)_2BH_3CN$ can be isolated from $Ni(PPh_3)_3Cl$ and $NaBH_3CN$ in ethanol.[51]

[438] C. Mealli and L. Sacconi, *J.C.S. Chem. Comm.*, 1973, 886.
[439] L. Sacconi, A. Orlandini, and S. Midollini, Proceedings 16th International Coordination Chemistry Conference, Dublin, 1974.
[440] T. Kruck, M. Hofler, and H. Jung, *Chem. Ber.*, 1974, **107**, 2133.
[441] T. Kruck, H. Jung, M. Hofler, and H. Blume, *Chem. Ber.*, 1974, **107**, 2145.
[442] T. Kruck, H. Jung, M. Hofler, and H. Blume, *Chem. Ber.*, 1974, **107**, 2156.
[443] T. Kruck, G. Maueler, and G. Schmidgen, *Chem. Ber.*, 1974, **107**, 2421.

Nickel(II).—*Halides and Pseudohalides.* Hoppe has discussed the application of the MAPLE approach to hypothetical variants of the $RbNiCrF_6$ type and concludes that a statistical distribution of Ni^{II} and Cr^{II} ions shows practically the same value as does the real structure.[444] Spectroscopic studies of the NiF_2–BaF_2 system show the presence of Ba_2NiF_6 and $BaNiF_4$.[445] The structure of $CsNiCl_3$ consists of infinite linear arrays of $[NiCl_6]_n$ octahedra, sharing faces. The species $CsNiI_3$ is isomorphous with the chloro analogue and has the properties of an intrinsic semiconductor. The magnetic behaviour of $CsNiI_3$ deviates significantly from Curie–Weiss at low temperature, which is interpreted in terms of antiferromagnetic interaction.[446]

The anilinium cation, LH^+, produces octahedral $[LH]_2NiX_4,6H_2O$ (X = Cl or Br), which decompose on heating to form $NiX_4[LH]_2$ and $NiX_2(L)_2$, successively.[60] The dipositive dabconium ion $(L'H_2)^{2+}$ forms $[L'H_2]NiX_4$ (X = Cl or Br) which produce $[L'H]NiX_3$ and HX on heating.[59] The anions $\{Ni[N(CN)_2]_3\}^-$ and $\{Ni[N(CN)_2](NCS)_2\}^{2-}$ 'form polymeric octahedral complexes with bridging ligands.[447] Miezis has reported stability constant data for nickel(II) thiocyanate complexes formed in MeCN, propane-1,2-diolcarbonate, trimethylphosphate, NN-dimethylacetamide, and DMSO, and demonstrated that all contain N-bonded anions by i.r. measurements.[448] The i.r. spectrum of $K_2[Ni(CN)_2(NCS)_2]$ indicates that the thiocyanate groups are N-bonded.[449]

Nitrogen Donors. Pyridine and related ligands. The structure of *trans*-$Ni(py)_4I_2$ shows a NiN_4I_2 environment with long Ni—N bonds (2.127 Å).[450] 4-Vinylpyridine forms monomeric octahedral NiL_4X_2 (X = Cl, Br, I, NCO, or NCS). The 2:1 complexes NiL_2X_2 (X = Cl, Br, N_3, or NCS) are also octahedral *via* bridging X groups, the chloro- and bromo-compounds exhibiting weak ferromagnetic interaction, whilst the azide is antiferromagnetic.[71] Although 2-vinylpyridine readily forms tetrahedral complexes with cobalt(II), the corresponding nickel(II) complexes could not be isolated. Only NiI_2 complexes with 2-chloro- or 2-bromo-pyridine to form NiL_2I_2, which are diamagnetic, probably planar, species but which become paramagnetic in solution apparently owing to a change to tetrahedral geometry.[70]

The i.r. spectra of the NiL_2X_2 and $NiLX_2$ (L = pyrazine) are consistent with the structure assigned to the Co^{II} analogues (*q.v.*).[451] Pyridine, β- and γ-picoline, and β- and γ-4-ethylpyridine form octahedral $Ni(CF_3CO_2)_2L_x$ (x = 2 or 4) with the CF_3-CO_2^- groups being bi- and uni-dentate, respectively.[74] Detailed ^1H and ^{19}F n.m.r. studies show the presence of both *cis*- and *trans*-isomers, and that the *cis* ⇌ *trans* isomerization is slow < 60 °C, in contrast to the Co^{II} analogues.[75] ^{14}N n.m.r. spectroscopy has been used to study pyridine exchange in $Ni(acac)_2py_2$.[76] Formation of octahedral 2:1 adducts of nickel(II) carboxylates with piperidine, N-methylpiperazine, and morpholine has been reported. Piperazine forms only 1:1 adducts.[72]

Pyridine-3-carboxylic acid (LH) forms $Ni(LH)_2X_2$, which contain N-bonded

[444] R. Hoppe, *Z. anorg. Chem.*, 1973, **402**, 39.
[445] A. Lule and O. Schmitz-DuMont, *Monatsh.*, 1973, **104**, 1632.
[446] G. L. McPherson, J. E. Wall, and A. M. Hermann, *Inorg. Chem.*, 1974, **13**, 2230.
[447] H. Köhler, H. Hartung, and A. M. Gobub, *Z. anorg. Chem.*, 1974, **403**, 41.
[448] A. Miezis, *Acta Chem. Scand.*, 1974, **A28**, 407.
[449] A. N. Sergeyeva, L. I. Parlenko, L. N. Kohut, and J. I. Thachenko, *J. Mol. Structure*, 1973, **19**, 513.
[450] D. J. Hamm, J. Bordner, and A. F. Schreiner, *Inorg. Chim. Acta*, 1973, **7**, 637.
[451] J. Dorrity and K. G. Orrell, *J. Inorg. Nuclear Chem.*, 1974, **36**, 230.

ligands.[452] However, pyridine-2-carboxylic acid (L'H) and 6-methylpyridine-2-carboxylic acid (MeLH) deprotonate to form $Ni(L')(L'H)Cl,0.5H_2O$ and $Ni(MeL)$-$(MeLH)X,nH_2O$ (X = Cl or Br), which were assigned octahedral structures on the basis of their electronic spectra and magnetic moments (μ_{eff} = 2.85—3.17 BM).[452] The thermal decomposition of hydrated Ni^{II} complexes of several pyridine carboxylic acids has been examined.[82]

The nickel(II) thiocyanate complex of 4,4'-bipyridyl $NiL(NCS)_2$ is octahedral with both bridging ligand and bridging thiocyanate groups, μ_{eff} = 3.07 BM.[77] Far i.r. and Raman spectra of $NiLCl_2$ (L = 2,2'-bipy or o-phen) have been recorded.[453] The reaction of $[Ni(bipy)R_2]$ with the appropriate silane ($SiCl_3H$ or $SiMeCl_2H$) produces $[Ni(bipy)(SiX_3)_2]$ (X = Cl_3 or $MeCl_2$).[454] 6-Nitro-2,3-di(2-pyridyl)quinoxaline (10) forms $NiL(H_2O)_2Cl_2$ and NiL_2X_2 (X = Br or I) on reaction with nickel(II) halides in ethanol. All three are pseudo-octahedral, the quinoxaline ligand behaving as a bidentate chelate, and water is co-ordinated in the chloro-derivative. The far-i.r. spectra of these and some substituted quinoxaline complexes indicate a *trans* arrangement of the halides.[89]

A comparison of the electronic spectra of $[NiL_3](BF_4)_2$ [L = 3-(2-pyridyl)-5,6-diphenyl-1,2,4-triazine] and $[NiL'_2]$ [L' = 3-(2-o-phen)-5,6-diphenyl-1,2,4-triazine] showed that the ligand field strength of the bidentate L was greater than that of the terdentate L', probably due to the distortion of L' necessary to enable all three N-donors to co-ordinate to the nickel.[92]

Sacconi et al.[455] have obtained some unusual complexes with 1,8-naphthyridine (napy). With NiX_2 (X = Br or I) in n-butanol insoluble $Ni(napy)_2X_2$ compounds are obtained which have $\mu_{eff} \approx 3.2$ BM, and thus these are five- or six-co-ordinate complexes. In ethanol in the presence of $NaBPh_4$, nickel(II) chloride or bromide yield insoluble $[Ni(napy)_2X]BPh_4$ complexes which were assigned polymeric octahedral (halide bridged) structures. However, in boiling n-butanol black acetone-soluble complexes form, in addition to the $[Ni(napy)_2X]BPh_4$. Analytical data of the black compounds correspond to $[Ni_2(napy)_4X_2]BPh_4$, formally nickel in oxidation state 1.5. These complexes only form if $NaBPh_4$ is present, and reduction does not occur if $NaPF_6$ is used without added $NaBPh_4$. The species isolated were $[Ni_2$-$(napy)_4X_2]Y$ (X = Cl, Br, I, or NO_3, NCS; Y = BPh_4 or PF_6) and similar complexes are formed by 4-methyl-1,8-naphthyridine, μ_{eff} = 4.1—4.4 BM). The structure of $[Ni_2(napy)_4Br_2]BPh_4$ consists of two square-pyramidally co-ordinated nickel atoms with an N_4 set in the basal plane and apical bromides, the two NiN_4Br moieties being held together by four bridging napy ligands. The Ni—Ni distance is 2.41 Å, and both Ni atoms are in identical environments.[455]

The i.r. and visible spectra of $NiL_4(NO_3)_2$ (L = pyrazole) are consistent with a tetragonally distorted octahedral structure with *trans*-nitrato-groups.[456] In pseudo-tetrahedral NiL'_2Cl_2 (L' = 5-aminoindazole) the co-ordination is *via* the ring nitrogen.[98]

Amine donors. The reaction of nickel(II) nitrate, sodium nitrite, and ethylenediamine

[452] V. M. Ellis, R. S. Vagg, and E. C. Walton, *Austral. J. Chem.*, 1974, **27**, 1191.
[453] R. E. Wilde and T. K. K. Srinivasan, *J. Inorg. Nuclear Chem.*, 1974, **36**, 323.
[454] Y. Kiso, K. Tamao, and M. Kumada, *J. Organometallic Chem.*, 1974, **76**, 95.
[455] L. Sacconi, C. Mealli, and D. Gatteschi, *Inorg. Chem.*, 1974, **13**, 1985.
[456] J. Reedijk, H. T. Witteveen, and F. W. Klaaijsen, *J. Inorg. Nuclear Chem.*, 1973, **35**, 3439.

in aqueous solution produces $[Ni(en)_2NO_2]NO_3$, which contains bidentate nitrite, and *cis*-$[Ni(en)_2(NO_2)_2]$ with unidentate nitro-groups.[457] The conformation of the chelate rings in $[Ni(en)_3]^{2+}$ has been examined by n.m.r. spectroscopy and found to be both solvent and anion dependent.[458,459] Similar ^1H n.m.r. studies on a range of 1,3-diamine chelates found that, as expected, the chair conformation is most stable, although axial methyl groups in the chair conformer can exert a destabilizing effect and promote the formation of the twist conformer in the equilibrium distribution.[460]

A series of *trans*-octahedral complexes of N-benzylethylenediamine have been prepared: $Ni(mbe)_2X_2$ (X = Cl, Br, I, or NO_3), $Ni(mbe)_2I(NO_3)$, and $Ni(mbe)_2$-$(H_2O)SO_4$, whilst two isomers of $[Ni(mbe)_3](ClO_4)_2$ as blue and pink solids are formed, although structural differences are unclear.[461]

Nickel(II) complexes of *meso*-stilbenediamine (stien) analogues (substituted 1,2-diphenylethylenediamine) with a variety of substituents in the phenyl rings have been prepared by reaction between the ligands and the nickel(II) salt in hot alcoholic solution.[462] Often, like the stien analogues, both yellow and blue forms of the complexes NiL_2X_2 can be isolated, depending on the nature of X and the degree of hydration. The yellow forms are diamagnetic square-planar complexes, and the blue are paramagnetic and octahedral, and interconversion is sometimes possible. Similarities between ligand-field spectra of analogous complexes suggest that changes in stereochemistry with ligand must predominantly be influenced by steric factors, although increasing substituent bulk does not account for the observed trend.[462]

The $Ni(daco)_2^{2+}$ (daco = 1,5-diazacyclo-octane) ion can be N-deprotonated to give $Ni(daco - 2H)_2^{2-}$ by Bu^nLi or MeLi.[278] An X-ray structural investigation of the complex previously characterized as $[Ni(opda)_6]Cl_2$ (opda = *o*-phenylenediamine) has shown that the correct formulation is $[Ni(opda)_4]Cl_2,2opda$, with octahedral nickel co-ordinated to two bidentate and two *trans* unidentate opda ligands. The two free opda molecules are part of an extensive H-bonding network between the cations and the anions.[463]

Hexamethylenetetramine forms both 1:1 and 2:1 complexes with nickel(II) iodide.[64] Only the nitro-isomer of $[NiL(NO_2)]BPh_4,MeOH$ [L = $Et_2N(CH_2)_2$-$NH(CH_2)_2NEt_2$] has been isolated.[464] The structure of $[NiLCl]Cl$ [L = (78)] is square pyramidal with the four amine donors coplanar and the nickel slightly out of plane towards the apical chlorine. The ionic chloride is hydrogen-bonded to the cation

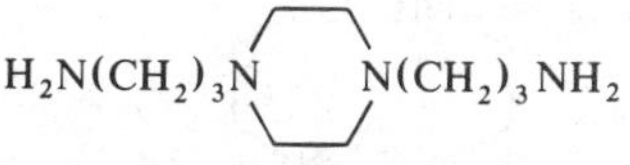

$$H_2N(CH_2)_3N\diagup\diagdown N(CH_2)_3NH_2$$

(78)

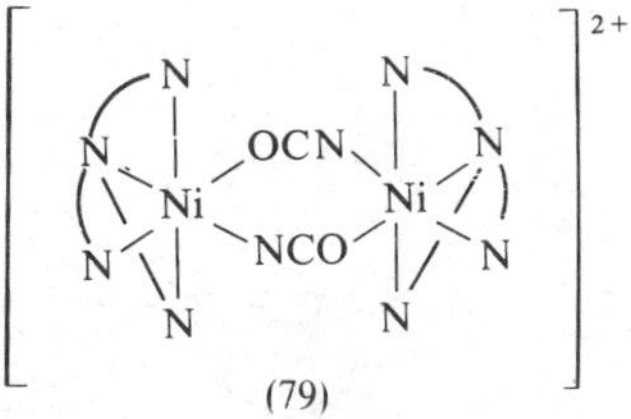

(79)

[457] A. E. Shvelashvili, L. P. Sarishvili, and R. M. Vashakidze, *Zhur. neorg. Khim.*, 1974, **19**, 568.
[458] R. E. Cramer and R. L. Harris, *Inorg. Chem.*, 1973, **12**, 2275.
[459] R. E. Cramer and R. L. Harris, *Inorg. Chem.*, 1974, **13**, 2208.
[460] J. E. Sarneski and C. N. Reilley, *Inorg. Chem.*, 1974, **13**, 977.
[461] K. C. Patel and D. E. Goldberg, *J. Inorg. Nuclear Chem.*, 1974, **36**, 565.
[462] W. A. Sadler and D. A. House, *J.C.S. Dalton*, 1973, 1937.
[463] R. C. Elder, D. Koran, and H. B. Mark, *Inorg. Chem.*, 1974, **13**, 1644.
[464] J. L. Burmeister, R. L. Hassel, K. A. Johnson, and J. C. Lim, *Inorg. Chim. Acta*, 1974, **9**, 23.

and is located near the sterically hindered 'sixth co-ordination site' of the nickel, $Ni—Cl = 3.395(3)$ Å.[465]

The bimetallic $[Ni_2(tren)_2(N_3)_2](BPh_4)_2$ [tren = 4-(2-aminoethyl)diethylenetri-amine] is prepared by reaction of $NiSO_4$, $NaBPh_4$, NaN_3, and the ligand in aqueous solution. The mode of azide bridging is thought to be $Ni\!\!\begin{array}{c} \diagup NNN \diagdown \\ \diagdown NNN \diagup \end{array}\!\!Ni$ and the magnetic moments is 4.2 BM at 283 K.[466] The complex $[Ni_2(tta)_2(NCO)_2](BPh_4)_2$ (tta = 2,2′,2″-triaminotriethylamine) contains essentially square-planar NCO bridges with the nickel atoms above and below the plane (79). This is the first authenticated example of O-bonded cyanate groups, and formation of the complex is thought to be due to favourable anion–cation packing.[467] The linear quadridentate 1,4,8,11-tetra-azaundecane forms complexes $[Ni_2L_3]X_2$ (X = ClO_4 or Cl,3H$_2$O), $[NiL(en)]X_2$ (X = ClO_4 or Cl,H$_2$O), $[NiLX_2]$,H$_2$O (X = NCS or N$_3$), $[NiL(OAc)_2]$,3Pr^{i}OH, $[NiLX]Y$ (X = NCS, N$_3$, OAc, or acac, Y = BF_4; X = N$_3$, Cl, NO$_3$, OAc, or acac, Y = ClO_4). On the basis of the splitting of the ν_1 band in the electronic spectrum the $NiLX_2$, $[NiLX]Y$, and $[NiL(OAc)_2]$ species were assigned *trans*-structures with 'planar' co-ordination of the quadridentate ligand. The other complexes were assigned structures with 'folded' ligands, and $[NiL]ZnCl_4$ contains a singlet ground state.[468]

The structure of 1,3-bis-(2-iminobenzylideneimino)propanenickel(II) consists of square planar nickel(II) with two markedly different Ni—N distances (1.923 and 1.860 Å).[469] Six-co-ordinate adducts are formed between $Ni(CF_3acac)_2$ and NR_3 (R = n-hexyl, n-octyl, n-undecyl, or n-hexadecyl) which, unlike the acac analogues, do not dissociate NR_3 in solution.[155]

Macrocyclic N-donors. The so-called 'macrocyclic effect' has been examined by equilibrium and calorimetric measurements of the formation of nickel(II) complexes of cyclam (80) and 2,3,2-tet (81). The cyclam complex is more stable owing to a more favourable ΔH change which overcomes a less favourable ΔS contribution. Solvation is less for cyclam since the H...N hydrogen bonding is less than in the open-chain quadridentate owing to steric effects.[470] A comparison of the ligand-field parameters for a series of $[NiLCl_2]$ complexes of 12-,13-,14-,15-, and 16-membered N_4 macro-cyclic ligands shows that the 15-ring is the best fit for nickel(II).[315]

1-Methylcyclam and 1,5-dimethylcyclam form planar nickel(II) complexes, but

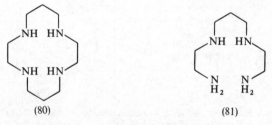

(80)   (81)

[465] J. G. Gibson and E. D. McKenzie, *J.C.S. Dalton*, 1974, 989.
[466] D. M. Duggan and D. N. Hendrickson, *Inorg. Chem.*, 1973, **12**, 2422.
[467] D. M. Duggan and D. N. Hendrickson, *Inorg. Chem.*, 1974, **13**, 2056.
[468] N. F. Curtis and N. B. Milestone, *Austral. J. Chem.*, 1974, **27**, 1167.
[469] N. A. Bailey, E. D. McKenzie, and T. M. Worthington, *J.C.S. Dalton*, 1974, 1363.
[470] F. P. Hinze and D. W. Margerum, *J. Amer. Chem. Soc.*, 1974, **96**, 4993.

1,4,8,11-tetramethylcyclam produces a complex which, on the basis of its electronic spectrum, contains five-co-ordinate nickel(II).[100,471] Barefield and Wagner prepared 1,4,8,11-tetramethyl-1,4,8,11-tetra-azocyclotetradecane by methylation of cyclam. It forms a red, diamagnetic planar $[NiL](ClO_4)_2$ species which exhibits a great tendency to become paramagnetic and five-co-ordinate in a variety of solvents by co-ordination of a solvent molecule. The square-pyramidal $[NiLX]ClO_4$ (X = Cl, Br, or NCS), $\mu_{eff} \sim 3.3$ BM, are readily prepared, as is the six-co-ordinate non-electrolyte $[NiL(NCS)_2]$.[472] The $[NiLX]^+$ species are thought to contain the ligand with all four N donors on one side of the plane and the halide on the other. Subsequently Ni(cyclam)$^{2+}$ has been alkylated (L') by deprotonation with KOH–DMSO followed by treatment with MeI. An $Ni_2L'_2(N_3)_3I$ complex has been subsequently X-rayed and contains two NiL' groupings bridged by an azide.[473]

Three configurational isomers of $[Ni(tet6)]^{2+}$ (tet6 = C-rac-5,5,7,12,12,14-hexa-methyl-1,4,8,11-tetra-azacyclotetradecane) have been prepared and X-rayed. All are planar and this result in considerable strain, which is most evident in the six-membered rings in the α-isomer, in the five-rings in the β-isomer, and in only one of the six-rings in the γ configuration.[474]

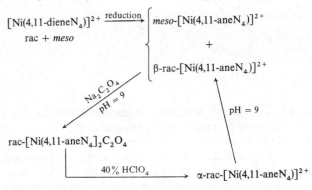

(82) 4,11-diene N$_4$

The catalytic hydrogenation of NiL [L = (82)] has been examined. When the reaction is performed under conditions where N-inversion is slow, the reduction is highly stereoselective (Scheme 9).[475]

$$[Ni(4,11\text{-diene}N_4)]^{2+} \xrightarrow{\text{reduction}} \begin{cases} meso\text{-}[Ni(4,11\text{-ane}N_4)]^{2+} \\ + \\ \beta\text{-}rac\text{-}[Ni(4,11\text{-ane}N_4)]^{2+} \end{cases}$$

rac + *meso*

Na$_2$C$_2$O$_4$, pH = 9

pH = 9

rac-[Ni(4,11-aneN$_4$)]$_2$C$_2$O$_4$

40% HClO$_4$ → α-rac-[Ni(4,11-aneN$_4$)]$^{2+}$

Scheme 9

[471] R. Buxtorf and T. A. Kaden, *Helv. Chim. Acta*, 1974, **57**, 1035.
[472] E. K. Barefield and F. Wagner, *Inorg. Chem.*, 1973, **12**, 2435.
[473] F. Wagner, M. T. Mocella, M. J. D'Amello, A. H. J. Wang, and E. K. Barefield, *J. Amer. Chem. Soc.*, 1974, **96**, 2625.
[474] N. F. Curtis, D. A. Swann, and T. N. Waters, *J.C.S. Dalton*, 1973, 1963.
[475] M. Wermeille, E. Sledziewska, and K. Bernauer, *Helv. Chim. Acta*, 1974, **57**, 180.

The ligand C-*meso*-(5,7,12,14-tetraethyl-7,14-dimethyl-1,4,8,11-tetra-azacyclo-tetra-4,11-diene) forms N-rac and N-*meso*[NiL](ClO$_4$)$_2$, N-*meso*-[NiL(NCS)]$_2$, N-rac-[{NiL}$_2$C$_2$O$_4$](ClO$_4$)$_2$. The ^{1}H n.m.r. spectra of [NiL′]$^{2+}$ (N-rac and N-*meso*) and of the isomeric [NiL″]$^{2+}$ [L′ = (83), L″ = (84)] have been reported and compared with those of the hexamethyl analogues.[314]

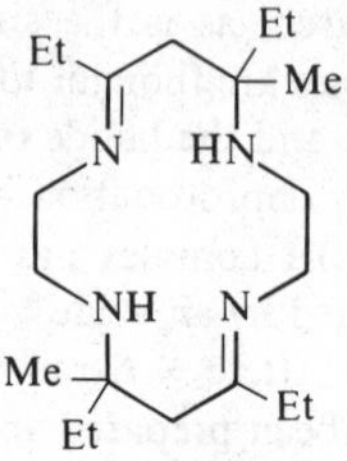

(83) 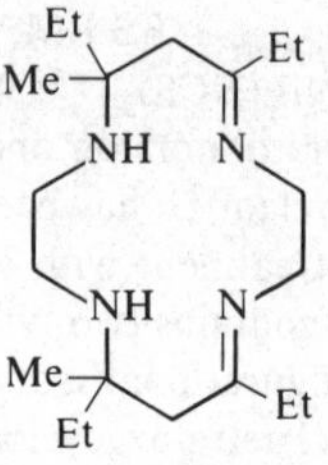   (84)

The stereochemistry of the 3,10-*meso*- and 3R,10R-isomers of 3,5,7,7,10,12,14,14-octamethyl-1,4,8,11-tetra-azacyclotetradeca-4,11-dienenickel(II) has been established by detailed study of their ^{1}H n.m.r. spectra.[476] The X-ray structures of 7,8,15,17,18,20-hexahydrodibenzo(*e,m*)pyrazino[2,3-*b*]-1,4,8,11-tetra-azacyclotetradecinatonickel(II) (85) and the related complex (86) have been determined. Both contain four-co-ordinate nickel(II), there being some slight distortion from planarity.[477]

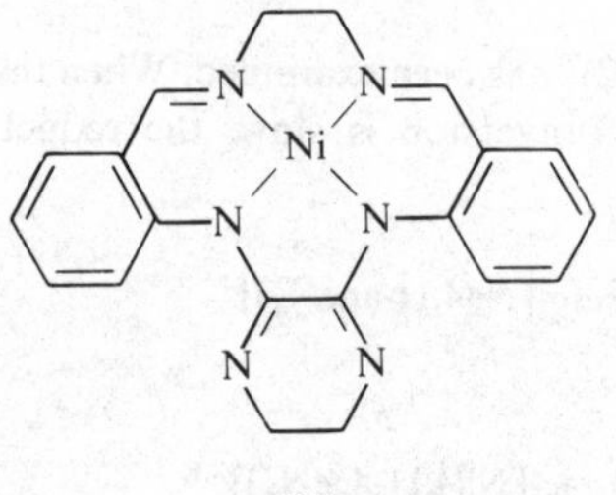

(85) 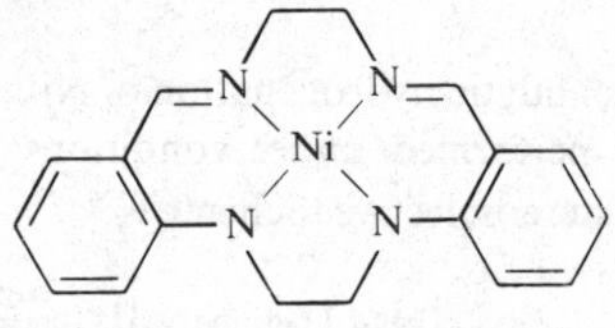   (86)

Busch and co-workers[478] have reported electrochemical data for a wide range of nickel(II) complexes of N$_4$ macrocycles (27 ligands in all). The one-electron reduction products could be formulated as Ni^I d^9 species or as Ni^{II}-stabilized anion radicals, depending upon the ligand; oxidation produces stable nickel(III) species. The ease of reduction or oxidation is markedly affected by the charge, stereochemistry, and degree of unsaturation of the ligand. Increase in ring size favours Ni^I and makes oxidation to Ni^{III} more difficult, which is probably due to the hole-size of the macrocycle. Ligand unsaturation makes Ni^{II} → Ni^{III} oxidation more difficult and favours the lower oxidation state.[478]

[476] T. Ito and D. H. Busch, *Inorg. Chem.*, 1974, **13**, 1770.
[477] E. N. Maslen, L. M. Englehardt, and A. H. White, *J.C.S. Dalton*, 1974, 1799.
[478] F. V. Lovecchio, E. S. Gore, and D. H. Busch, *J. Amer. Chem. Soc.*, 1974, **96**, 3109.

Electrochemical reduction of $Ni(taab)^{2+}$ [taab = (20), see p. 231] occurs in two one-electron steps, to complexes formulated as $[Ni^{III}(taab)]^+$ and $[Ni^{II}(taab)]^0$. The relationship between the annulene taab and the two-electron reduction product, the porphyrin-like $taab^{2-}$, is discussed.[102] The preparation of macrocycles of type (87) by a template synthesis requires a minimum ring size of $x = y = 3$ and depends upon the strong complexing of the metal ion at the pH of the reaction, otherwise the metal hydroxide precipitates. The $NiL(ClO_4)_2,nH_2O$ ($n = 0$, 1, or 2) species have been prepared for $x = 3$, $y = 4$.[479]

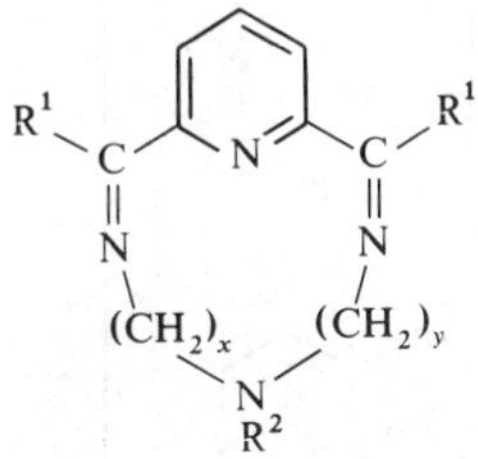

(87)

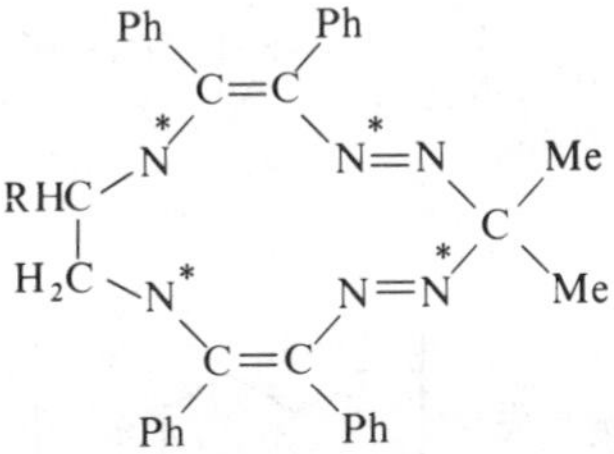

(88)   R = H or Me

The ligands $H_2[HPhHen_2]$, $H_2[HPhHentn]$, $H_2[HPhHtn_2]$ (see Scheme 1, p. 230) form diamagnetic planar nickel(II) complexes.[101] The ^1H n.m.r. spectra of the nickel(II) complexes of two macrocycles derived from (88) have been reported. The ligands co-ordinate in a planar manner *via* the four N* atoms.[480]

The bicyclic octadentate (89) forms Ni_2L in which the ligand has lost the four imino hydrogens.[481] A novel reaction occurs when an aqueous solution of 11,13-dimethyl-1,4,7,10-tetra-azacyclotrideca-10,12-dienenickel(II) iodide is allowed to stand at pH ~ 1. Black crystals separate and X-ray analysis shows them to contain a new octadentate bicyclic macrocycle formed by the linking of two monomeric macrocycles through the methine carbons. The nickel(II) is in a square-pyramidal N_4I environment.[482]

The air-stable reduced porphyrin $\alpha\gamma$-dimethyl-$\alpha\gamma$-dihydro-octaethylporphinato-nickel(II) contains a planar NiN_4 grouping with the porphyrin having a folded and ruffled core.[483] The triclinic form of 1,2,3,4,5,6,7,8-octaethylporphinatonickel(II) has very short Ni—N distances (1.929 Å) compared with the tetragonal form (1.957 Å).[485] The behaviour of tetra-(4-N-methylpyridyl)porphinatonickel(II) in aqueous acetone shows it to be four-co-ordinate and planar, but it adds pyridine or imidazole

[479] R. H. Prince, D. A. Stoffer, and P. R. Woolley, *Inorg. Chim. Acta*, 1974, **9**, 51.
[480] G. A. Melson, *Inorg. Chem.*, 1974, **13**, 994.
[481] E. B. Fleischer, A. Sklar, A. Kendall-Torry, P. A. Tasker, and F. B. Taylor, *Inorg. Nuclear Chem. Letters*, 1973, **9**, 1061.
[482] J. A. Cunningham and R. E. Sievers, *J. Amer. Chem. Soc.*, 1973, **95**, 7183.
[483] P. N. Dwyer, J. W. Buchler, and W. R. Scheidt, *J. Amer. Chem. Soc.*, 1974, **96**, 2789.
[484] D. L. Cullen and E. F. Meyer, *J. Amer. Chem. Soc.*, 1974, **96**, 2095.
[485] R. F. Pasternack, L. Francesconi, D. Raff, and E. Spiro, *Inorg. Chem.*, 1973, **12**, 2606.

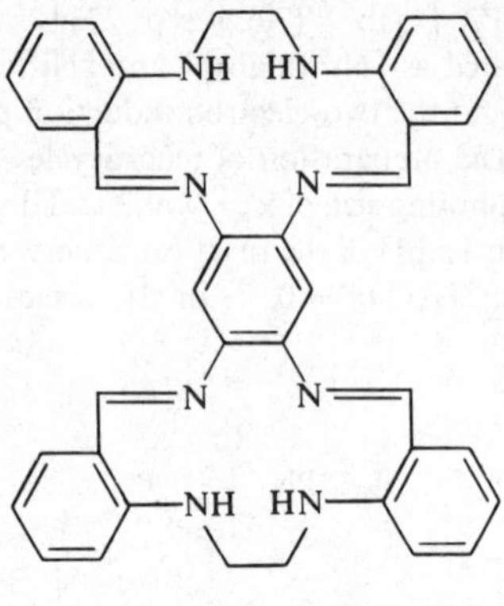

(89)

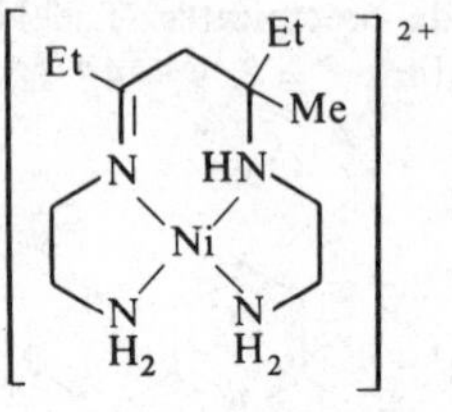

(90)

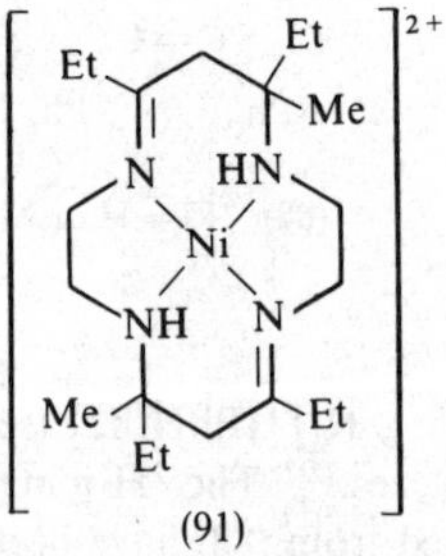

(91)

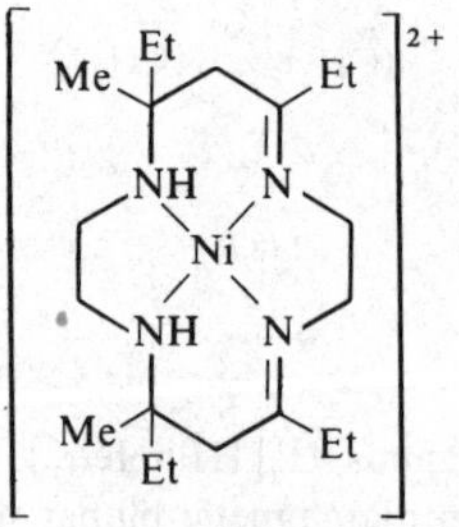

(92)

to become six-co-ordinate.[485] In dilute solution this complex does not aggregate, but tetracarboxyphenylporphinatonickel(II) tends to dimerize under the same conditions.[311] The polarographic reduction of the nickel(II) complexes of protoporphyrin-IX dimethyl ester[113] and 1,19-diethyloxycarbonyltetradehydrocorrin[486] have been reported, and the electronic spectrum of the corrole anion recorded.[487]

The reaction of butanone with $[Ni(en)_2](ClO_4)_2$ at room temperature forms the bisligand nickel(II) complex of the β-diketone $H_2NCH_2CH_2NHC(Me)(Et)CH_2COEt$ (Bk) in which the ligand behaves as a terdentate N_2O donor. The same reaction in the presence of co-ordinating anions (Cl, NO_3, or NCS) or solvents (py, MeCN, *etc.*) produces the complex of the quadridentate $H_2N(CH_2)_2N{=}C(Et)CH_2C(Me)(Et)$-$NH(CH_2)_2N{=}C(Me)(Et)$ which can be hydrolysed to yield (90). The Bk complex in pyridine forms the Ni^{II} complex of the N_4 macrocycle (91) by intramolecular imine formation. The diperchlorate salt of the ligand $H_2L(ClO_4)_2$ forms on reaction of $HenClO_4$ with butanone, and this reacts with nickel(II) acetate to form (92). Only very slow reaction of $[Ni(en)_3](ClO_4)_2$ with butanone occurs to form (92).[488]

Other N-*donors.* The preparations of $Ni(N_2H_4)_2X_2,nH_2O$ (X = F or Br) have been

[486] N. S. Hush and J. M. Dyke, *J. Inorg. Nuclear Chem.*, 1973, **35**, 4341.
[487] N. S. Hush, J. M. Dyke, M. L. Williams, and I. S. Woolsey, *J.C.S. Dalton*, 1974, 395.
[488] N. F. Curtis, *J.C.S. Dalton*, 1974, 347.

reported. Thermal decomposition of $[Ni(N_2H_4)_6]X_2$ proceeds *via* the tris-, bis-, and mono-hydrazine complexes for X = Cl or Br and *via* the tetrakis- and bis-complexes for X = I.[68] I.r. spectra show that the $[Ni(N_2H_4)_6]^{2+}$ contains unidentate hydrazine, the $[Ni(N_2H_4)_n]^{2+}$ (n = 3 or 2) bidentate bridging hydrazine, and the tetrakis-complex both unidentate and bridging bidentate ligands.[489] The phenylhydrazine complexes $[Ni(PhHNNH_2)_2X_2]$ (X = Cl, Br, or I) probably have a tetragonal octahedral structure (μ_{eff} = 2.9—3.4 BM), but it is not known whether this is achieved *via* X bridges or bidentate hydrazine.[69] The substituted phenylhydrazines $XC_6H_4NHNH_2$ (X = *o*- or *p*-NO$_2$) form both 2:1 and 4:1 nickel(II) chloride complexes in which the ligand bonds as a unidentate *via* the NH_2 group.[490] The X-ray structure of $[Ni(NH_2OH)_6]SO_4$ confirms that the hydroxylamine is N-bonded in the octahedral cation.[491]

Piperidine forms six-co-ordinate $[Ni(pip)_4(NCX)_2]$ (X = S or Se) and five-co-ordinate $[Ni(pip)_3(NCO)_2]$, but 2,6-dimethylpiperidine forms only planar $[Ni(2,6-Me_2pip)_2(NCX)_2]$ (X = O or S). In solution the electronic spectra show that the isothiocyanate and isocyanate complexes of piperidine exist as five-co-ordinate $[NiL_3X_2]$ species in the presence of excess piperidine, and as four-co-ordinate NiL_2X_2 in the absence of added ligand.[492] 1,2-Dimorpholinoethane (edm) and 1,2-dipiperidinoethane (edp) form $[NiLX_2]$ complexes (L = edm or edp, X = NO_2 or NO_3), which are non-electrolytes, $\mu_{eff} \simeq 3.24$ BM, with molecular weights corresponding to discrete octahedral complexes. I.r. spectra suggest bidentate anionic ligands producing an N_2O_4 donor set.[117]

Long *et al.*[118—120, 493, 494] have reported extensively on Ni^{II} complexes of quinuclidinone derivatives. Data on Ni(mqn)(NCS)$_2$,MeOH [mqn = (22)] indicate a dimeric octahedral structure analogous to the Co^{II} complex.[119] The ligand mqn forms two complexes $[Ni(mqn)Cl_2]$ with nickel(II) chloride, depending on the reaction conditions, a purple tetrahedral monomer and an off-white, probably chlorine-bridged, octahedral polymer.[118] The purple chloride and the $[Ni(mqn)X_2]$ (X = Br or I) have electronic and i.r. spectra and magnetic properties characteristic of tetrahedral Ni^{II} with the ligand bonding through both nitrogen atoms. Ni(mqn)$_2$(ClO$_4$)$_2$ has μ_{eff} = 4.02 BM, high for a Ni^{II} complex although not unexpected for a tetrahedral $[NiL_2]^{2+}$ cation with identical donors; the tetrahedral structure is also indicated by the electronic spectrum. Tetrahedral complexes[120] are also formed by 2-(2'-quinolyl)methylene-3-quinuclidinone (qnqn). The violet Ni(qnqn)X$_2$ (X = Cl or Br) are monomeric with N_2X_2 donor sets, with room temperature magnetic moments of 3.3—3.4 BM. A dimeric yellow $[Ni(qnqn)Cl_2]_2$ can be prepared under different conditions, and the yellow and purple isomers are interconvertible (Scheme 10).[120] The violet Ni(qnqn)Cl$_2$

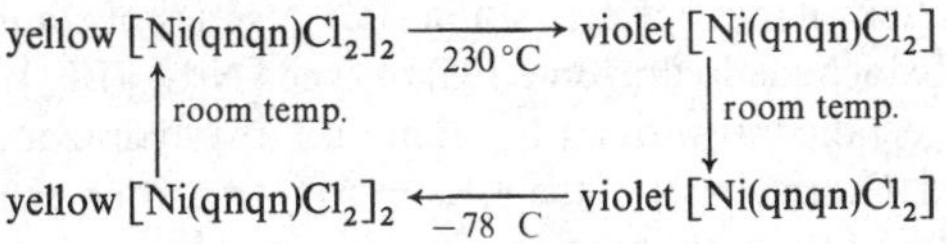

$$\text{yellow } [Ni(qnqn)Cl_2]_2 \xrightarrow[230\,°C]{} \text{violet } [Ni(qnqn)Cl_2]$$

room temp. ↑   ↓ room temp.

$$\text{yellow } [Ni(qnqn)Cl_2]_2 \xleftarrow[-78\,C]{} \text{violet } [Ni(qnqn)Cl_2]$$

Scheme 10

[489] R. Tsuchiya, M. Yonemura, A. Uehara, and E. Kyuno, *Bull. Chem. Soc. Japan*, 1974, **47**, 660.
[490] M. S. Novakovskii and L. K. Bessarabenko, *Zhur. obschei Khim.*, 1974, **44**, 764.
[491] L. M. Englehardt, P. W. G. Newman, C. L. Raston, and A. H. White, *Austral. J. Chem.*, 1974, **27**, 503.
[492] R. Eggli and W. Ludwig, *Inorg. Chim. Acta*, 1973, **7**, 697.

is converted into the yellow dimer by compression, the change being irreversible.[493] An X-ray structure determination of the yellow isomer confirms the dinuclear $[Ni(qnqn)Cl_2]_2$ constitution, each nickel is square pyramidally co-ordinated, with slightly asymmetric equatorial halide bridges;[494] Ni—Cl = 2.408(2) and 2.422(2) Å. The other equatorial positions are occupied by a terminal chlorine [Ni—Cl = 2.296(2) Å] and the quinoyl N [Ni—N = 2.047(3) Å], and the apical position by the quinuclidinone N[Ni—N = 2.067(2) Å].

2-Aminothiazole (93), which has three potential co-ordination sites, forms octahedral NiL_4X_2 (X_2 = Cl, Br, I, or ClO_4) on refluxing with NiX_2 in 1,2-dichloroethane. The ligand is bonded only *via* the NH_2 group. In acetone solution, the electronic spectrum suggests a change in stereochemistry to tetrahedral.[495] Pyridine-4-carbaldehyde oxime forms octahedral polymeric complexes NiL_2X_2 (X = Br or Cl) (94) with halide bridges, in which the ligand co-ordinates *via* the ring N, but not *via* the oxime group.[87] The $([NiL_2],2H_2O)_n$ (μ_{eff} = 3.20 BM) of 3,5-dimethylpyrazolate contains octahedrally co-ordinated nickel(II), the ligand probably functioning as a bridging NN donor.[97]

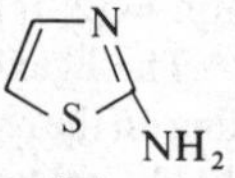

(93)

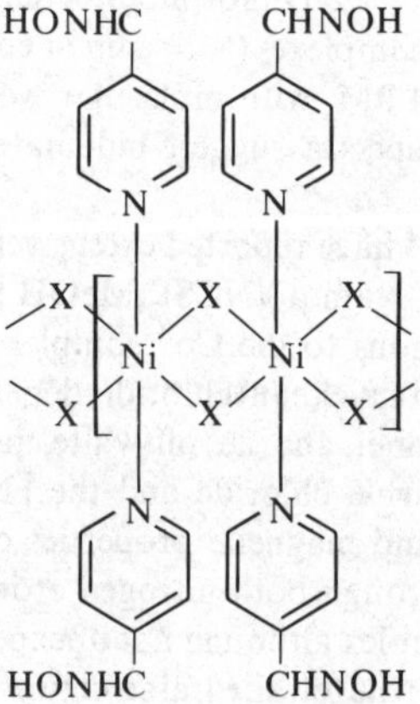

(94)

The structure of bis[dimethyl(1-pyrazolyl)borato]nickel(II) $[(Et_2B)(C_2N_2H_3)_2]_2Ni$ contains an essentially square-planar NiN_4 grouping. The structure shows that the methylene (Et) groups are in different environments, which differs from the conclusions derived from pervious n.m.r. spectral studies.[496] A series of complexes of 1,10-phenanthroline-2-carbaldehyde hydrazones (95) of type $[NiL_2](BF_4)_2$ has been prepared. All are high-spin octahedral with an N_6 donor set, the hydrazone group bonding *via* the azomethine N. On the basis of the $^3A_{2g} \rightarrow {}^3T_{2g}$ transition energy, the ligand field strength was deduced (b) > (a) > (d) > (f) > (c) > (e).[497]

[493] G. J. Long and J. R. Ferraro, *J.C.S. Chem. Comm.*, 1973, 719.
[494] G. J. Long and E. O. Schlemper, *Inorg. Chem.*, 1974, **13**, 279.
[495] P. P. Singh and A. K. Srivastava, *J. Inorg. Nuclear Chem.*, 1974, **36**, 928.
[496] H. F. Echols and D. Dennis, *Acta Cryst.*, 1974, **B30**, 2173.
[497] H. A. Goodwin and D. W. Mather, *Austral. J. Chem.*, 1974, **27**, 965.

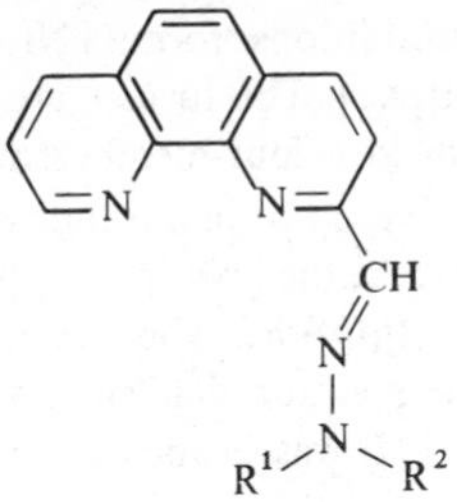

	R^1	R^2
(95) a;	H	Ph
b;	H	Me
c;	Me	Me
d;	Me	Ph
e;	Ph	Ph
f;	H	2-pyridyl

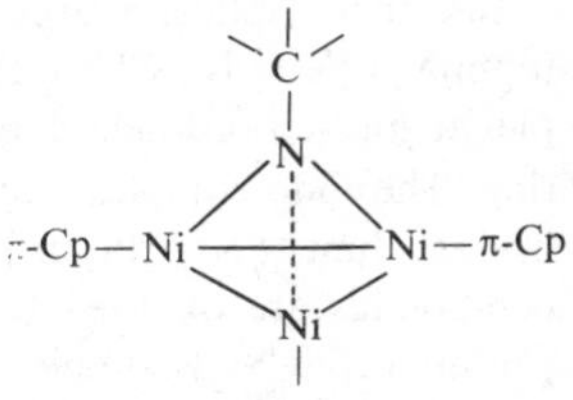

$$\pi\text{-Cp—Ni} \overset{\displaystyle \underset{|}{\overset{|}{C}}}{\underset{\displaystyle Ni}{\underset{|}{\bigtriangleup}}} \text{Ni—}\pi\text{-Cp}$$

(96)

Iminodiacetonitrile $HN(CH_2CN)_2$ forms $[NiL_2Cl_2]$, which from the similarity of its properties to those of $[CuL_2Cl_2]$ is assumed to have a similar structure (*q.v.*).[498] An *X*-ray structural investigation of $Bu^tN(\pi\text{-CpNi})_3$ (μ_{eff} = 1.68 BM) has revealed a trigonal-pyramidal structure (96).[499]

P-*Donors, As-Donors, and* Sb-*Donors.* The structure of $[Ni(PMe_3)_3Br_2]$ is *cis* trigonal-bipyramidal, with the two bromines equatorial, and axial phosphines. Similar structures were assigned to $[Ni(PMe_3)_3X_2]$ (X = Cl or I) in the solid state on the basis of the similar electronic spectra. In solution, dissociation into *trans*-$[Ni(PMe_3)_2-X_2]$ occurs, which can be suppressed by the presence of excess PMe_3. A large excess (*ca.* 20-fold) of PMe_3 in solutions of $Ni(PMe_3)_3Br_2$ produced spectral evidence for formation of $[Ni(PMe_3)_5]^{2+}$ at 77 K.[500] In contrast, $Ni(PMe_3)_3(CN)_2$ has a *trans* trigonalbipyramidal structure,[501] the differences between this and the halides being attributed to a second-order Jahn–Teller effect.[500] Ligand exchange in the five-co-ordinate $Ni(PR_3)_3(CN)_2$ proceeds by a dissociative mechanism.[502] Exchange studies have also been reported on the $NiH(PEt_3)_3X$ system.[503] $Ni(PPh_3)_2(BH_3CN)H$, $Ni(Ph_2PCH_2CH_2PPh_2)(BH_3CN)Cl$ and $Ni(dbp)_3H(BH_3CN)$ (dbp = 5-Ph-5-H-dibenzophosphole) are formed from $NaBH_3CN$ and $NiCl_2$ in ethanol in the presence of the appropriate phosphine.[51]

Difluorophenylphosphine forms $Ni(PF_2Ph)_3X_2$ (X = Br or I) which can be obtained either from anhydrous NiX_2 and PF_2Ph in benzene, or by halogenation of $[Ni(PF_2Ph)_4]$. They are diamagnetic and have a trigonal-bipyramidal structure.[126] The first complexes of a unidentate stibine with Ni^{III} were $Ni(SbMe_3)_3X_2$ (X = Br, Cl, I, or CN).[504] The complexes dissociate extensively in solution but can be obtained in the presence of excess $SbMe_3$. Electronic spectra are consistent with a *trans* trigonal-bipyramidal structure. The iodo-complex is diamagnetic but the bromo-complex has μ_{eff} = 1.60 BM, said to be due to singlet $\rightleftharpoons$ triplet state thermal equilibrium.[504]

498 G. W. Watt and P. H. Javora, *J. Inorg. Nuclear Chem.*, 1974, **36**, 1745.
499 N. Kamijgo and T. Watanabe, *Bull. Chem. Soc. Japan*, 1974, **47**, 373.
500 J. W. Dawson, T. J. McLennan, W. Robinson, A. Merle, M. Dartiguenave, Y. Dartiguenave, and H. B. Gray, *J. Amer. Chem. Soc.*, 1974, **96**, 4428.
501 A. Merle, M. Dartiguenave, Y. Dartiguenave, J. W. Dawson, and H. B. Gray, *J. Coord. Chem.*, 1974, **3**, 199.
502 C. G. Grimes and R. G. Pearson, *Inorg. Chem.*, 1974, **13**, 970.
503 P. Meakin, R. A. Schunn, and J. P. Jesson, *J. Amer. Chem. Soc.*, 1974, **96**, 277.
504 M. F. Ludmann-Obier, M. Dartiguenave, and Y. Dartiguenave, *Inorg. Nuclear Chem. Letters*, 1974, **10**, 147.

Bis(diphenylphosphino)methane (dpm), under controlled conditions, forms [Ni-(dpm)X$_2$] (X = Br, Cl, or I), as well as the known [Ni(dpm)$_2$X$_2$]. The former are planar four-co-ordinate complexes, containing chelating dpm in a four-membered ring. The iodo-complex can be reversibly converted into a five co-ordinate iodo-bridged dimer [Ni$_2$I$_4$(dpm)$_2$]. On heating above 220 °C in oxygen, the [Ni(dpm)X$_2$] compounds are oxidized to Ni^{II} compounds of Ph$_2$P(O)CH$_2$P(O)Ph$_2$. The *trans*-[Ni(dpm)$_2$(NCS)$_2$] complex in concentrated dichloromethane solution deposits an insoluble red [Ni(dpm)(CNS)$_2$]$_n$ polymer for which structure (97) was proposed.[505]

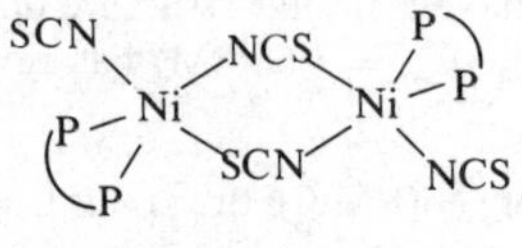

(97) 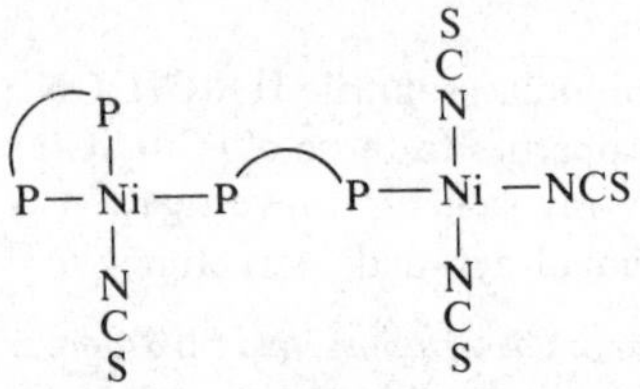   (98)

However, Chow and McAuliffe[506] suggest that this complex is more probably (98), a structure more consistent with the electronic spectral data. Five-co-ordinate [Ni(dpm)$_2$X]Y (X = Br, Cl, I, NCS, or NO$_2$, Y = BPh$_4$ or ClO$_4$) also contain chelating dpm, and have an essentially square-pyramidal structure, in both the solid state and solution, except for the iodo-complex, when substantial dissociation to a planar species seems to occur in solution. With Ni(CN)$_2$ a yellow complex [Ni-(dpm)$_2$CN]CN is formed, which probably has structure (99).[506] The ligand *o*-phenylenebis(dimethylphosphine) forms both plana [Ni(pmm)$_2$]$^{2+}$ and five-co-ordinate [Ni(pmm)$_2$X]$^+$ (X = Br or Cl).[127] The only five-co-ordinate Ni species isolable with bis(diphenylphosphino) *o*-carborane is [Ni(L$_2$)I]I, which appears to be trigonal bipyramidal, rather than square pyramidal as might have been expected.[128] The ligand *trans*-1,2-bis(diphenylphosphino)ethylene (tvpp) forms pseudotetrahedral complexes with nickel(II) halides NiX$_2$(tvpp). Since the ligand is incapable of chelation a structure similar to types (100a and b) is probably present. The corresponding Ni(tvpp)X$_2$ (X = NCS or CN) are diamagnetic and planar.[129]

The unusual ligand (101) forms planar NiLX$_2$ complexes, in which the diphosphine co-ordinates *trans* across the square plane.[507] The structure of the nickel(II) chloride complex of the optically active diphosphine (102) has been briefly reported.[508]

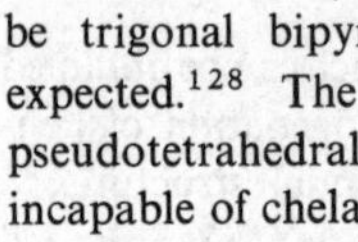

(99) 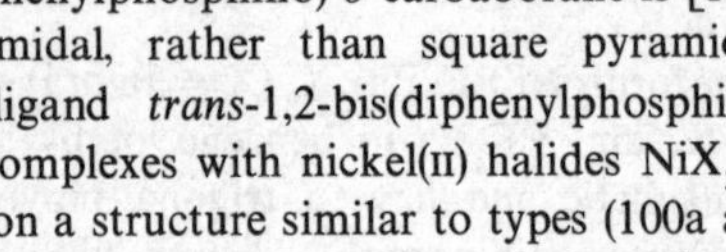   (100a) 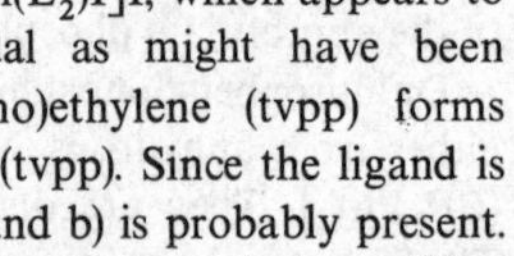   (100b)

[505] C. Ercolani, J. V. Quagliano, and L. M. Vallarino. *Inorg. Chim. Acta*, 1973, **7**, 413.
[506] K. K. Chow and C. A. McAuliffe, *Inorg. Chim. Acta*, 1974, **10**, 197.
[507] N. J. DeStefano, D. K. Johnson, and L. M. Venanzi, *Angew. Chem.*, 1974, **86**, 133.
[508] V. Gramlich and C. Saloman, *J. Organometallic Chem.*, 1974, **73**, C61.

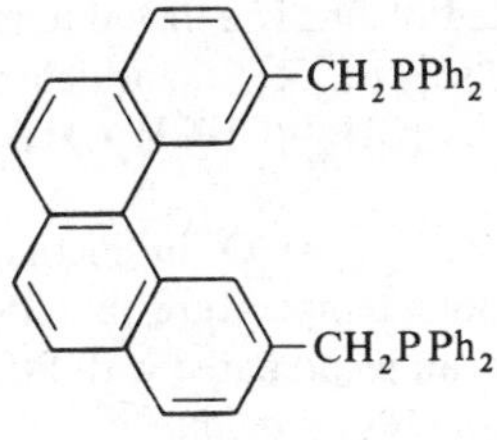

(101)

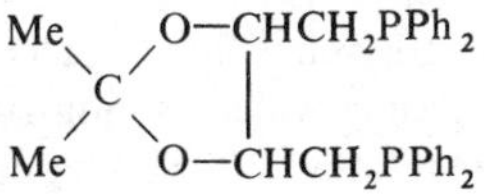

(102)

The ligand exchange between nickelocene and $Ni(dpe)X_2$ (X = Cl, Br, I, or CN) (dpe = $Ph_2PCH_2CH_2PPh_2$) in organic solvents produces $[\pi\text{-}C_5H_5NiX]_2(dpe)$ (X = CN or I) and $[\pi\text{-}C_5H_5Ni(dpe)]X$ (X = $\frac{1}{2}NiCl_4^{2-}$ or $NiBr_4^{2-}$), the former with a diphosphine bridge, the latter with a chelating diphosphine.[509] Novel metal–carbon bond formation occurs on treatment of $[Ni(PEt_3)_2Cl_2]$ or $[Ni(SEt_2)Cl_2]$ with $K(o\text{-}Ph_2PC_6H_4CH_2^-)$ when $Ni(o\text{-}Ph_2PC_6H_4CH_2^-)_2$ is produced.[510]

The reaction of nickel halide with $N(CH_2CH_2AsPh_2)_3$ (NAs_3) and $NaBPh_4$ produced the unexpected trigonal-bypyramidal product $[Ni(NAs_3)(o\text{-}Ph)]BPh_4$, with an NAs_3C set about the Ni, the σ-bonded phenyl group being in an axial position.[511]

The quadridentate $(o\text{-}C_6H_4AsMe_2)_3Sb$ (Sbta) forms diamagnetic trigonal-bipyramidal $[Ni(Sbta)X]Y$ (X = Br, Cl, I, CNS, or NO_3, Y = BPh_4) and square-pyramidal $[Ni(Sbta)_2]Y_2$ (Y = ClO_4 or BPh_4). The $SbAs_3$ donor set in Sbta exerts a stronger ligand field than the As_4 set in $(o\text{-}C_6H_4AsMe_2)_3As$, due to compression of the apical donor onto the metal, the compression naturally increasing with increasing size of the apical donor.[512] The $[Ni(Sbta)_2]Y_2$ complexes are thought to have structure (103) on the basis of ^1H n.m.r. spectra.[512]

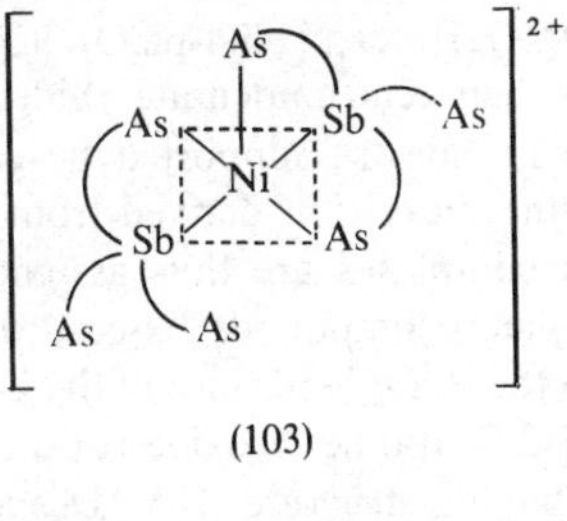

(103)

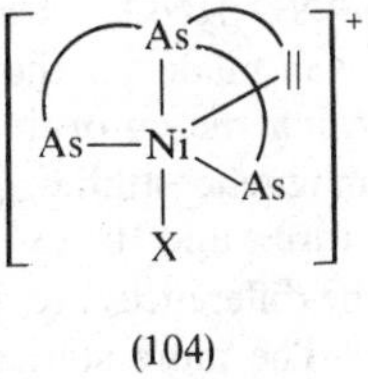

(104)

The triarsine olefin ligand $(CH_2=CHCH_2CH_2)As(CH_2CH_2CH_2AsMe_2)_2$ forms deep purple trigonal-bipyramidal complexes $[NiLX]Y$ (X = Br, Cl, or I; Y = Br, Cl, or ClO_4) which contain co-ordinated olefin (104). However, NiI_2 forms $[NiLI_2]$

[509] L. A. Kaempfe and K. W. Barnett, *Inorg. Chem.*, 1973, **12**, 2578.
[510] G. Longoni, P. Chini, F. Canziani, and P. Fantucci, *Gazzetta*, 1974, **104**, 249.
[511] P. Dapporto and L. Sacconi, *Inorg. Chim. Acta*, 1974, **9**, L2.
[512] L. Baracco, M. T. Halfpenny, and C. A. McAuliffe, *J.C.S. Dalton*, 1973, 1945.

which contains an As_3I_2 set and unco-ordinated olefin. The ligand $(CH_2=CHCH_2-CH_2)_2As(CH_2CH_2CH_2AsMe_2)$ (L') forms only $[NiL'_2X]^+$ in which no olefin co-ordination is present, whilst the triolefin arsine $(CH_2=CHCH_2CH_2)_3As$ does not complex with Ni^{II}.[513]

Excess trimethylphosphite reacts with $Ni(BF_4)_2, 6H_2O$ in methanol to form $[Ni(P[OMe]_3)_5]^{2+}$ which is non-rigid at room temperature, but becomes rigid at lower temperatures.[47, 48] Similar complexes can be prepared with $P(OCH_2)_3CMe$, $P(OCH_2)_3CEt$, and $P(OCHCH_2)_3$; all are trigonal bipyrimidal.[48]

O-*Donors.* The X-ray structures of $[Ni(H_2O)_6]SiF_6$[130] and $[Ni(H_2O)_6]L_2,4MeOH$ (L = monoanion of 2,2'-dithiobisbenzoic acid $[SC_6H_4CO_2H]_2)$[136] show that both contain hexaquonickel(II) cations. A series of octahedral Ni^{II} complexes of carbonyl ligands has been prepared as BF_4^-, $SbCl_6^-$, or $InCl_4^-$ salts and their reflectance and solution electronic spectra recorded.[131] Ligand-field parameters were calculated, and it was shown that no correlation exists between these parameters, and parameters describing electronic effects of the R^1 and R^2 substituents in the ligands R^1R^2CO. 1,3-Dioxan forms polymeric $[NiX_2L]_n$ (X = Cl or Br) which are pseudo-octahedral structures, and similar complexes are formed by 1,4-dioxan (L'), although $NiBr_2L'_2$ can also be isolated.[132] The complex $NiL_2(NCSe)_2$ is pseudo-octahedral with bridging NCSe groups.[135] A number of butylphthalate complexes of nickel(II) salts have been prepared for testing as therapeutic agents.[514]

Pyridine *N*-oxide forms $[Ni(pyO)_3(NCX)_2(H_2O)]$ (X = S or Se) and $[Ni(pyO)_2-(NCS)_2(EtOH)]$, both octahedral, the latter possibly *via* bridging EtOH.[143] Detailed i.r. spectra[142] have been reported for $[Ni(pyO)_6](BF_4)_2$ which contains approximately octahedrally co-ordinated nickel(II) with disordered BF_4^- groups.[515] The quinoline *N*-oxide compounds $[Ni(qnO)_2(NCX)_2]$ (X = S or Se) $(\mu_{eff} \sim 2.93$ BM) have electronic spectra consistent with octahedral nickel(II), and as the i.r. spectra indicate terminal N-bonded NCX groups, it was concluded that the (qnO) functions as a bridging ligand.[143]

2-, 3-, and 4-Picoline *N*-oxide react with nickel(II) nitrate in a 2:1 ratio to produce $[Ni(2-picO)_2(ONO_2)(O_2NO)]$, $[Ni(3-picO)_2(O_2NO)_2]$, and $[Ni(4-picO)_2(O_2NO)_2]$, whilst an 8:1 ratio produced $[Ni(2-picO)_4(ONO_2)]NO_3$, $[Ni(4-picO)_6](NO_3)_2$, and $[Ni(3-picO)_5(ONO_2)]NO_3$. The distinction between unidentate, bidentate, and ionic nitrate was made on the basis of the i.r. spectra supported by conclusions drawn concerning the co-ordination about the nickel ion derived from electronic spectra and magnetic studies. The (2-picO) complexes are thus assigned five-co-ordinate structures, and the (3-picO) and (4-picO) compounds six-co-ordinate configurations. The differences are due mainly to the steric hindrance of the 2-Me group in (2-picO).[146] The more sterically hindered 2,6-lutidine *N*-oxide (LO) forms $[Ni-(LO)_2SO_4]_2$ with nickel sulphate, probably having structure (105). Despite this dinuclear structure, the magnetic moment is normal for a five-co-ordinate d^8 metal ion (3.32 BM).[144] Ligand exchange in $[Ni(LO)_4](ClO_4)_2$ and $[Ni(LO)_5](ClO_4)_2$ has been studied by n.m.r. spectroscopy.[516]

[513] C. A. McAuliffe and D. G. Watson, *J. Organometallic Chem.*, 1974, **78**, C51.
[514] M. J. Cleare, J. Hoeschle, and C. A. McAuliffe, Proceedings 16th International Conference on Co-ordination Chemistry, Dublin, 1974.
[515] A. D. Van Ingen Schenau, G. C. Verschoor, and C. Romers, *Acta Cryst.*, 1974, **B30**, 1686.
[516] H. D. Gilman, *Inorg. Chem.*, 1974, **13**, 1921.

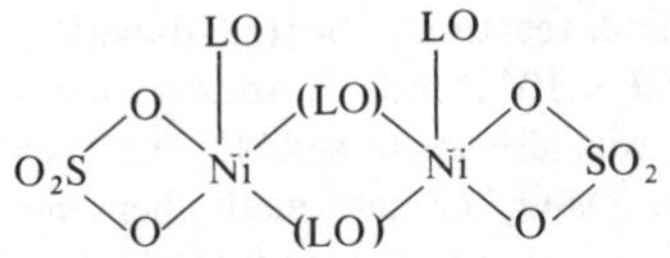

(105)

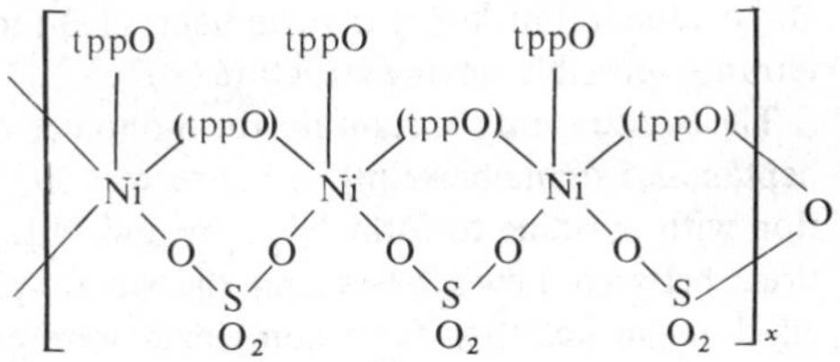

(106)

1,10-Phenanthroline N-oxide (phenNO) reacts with $Ni(ClO_4)_2,6H_2O$ in EtOH–triethylorthoformate to form $[Ni(phenNO)_3](ClO_4)_2,2H_2O$. Spectroscopic and magnetic data are consistent with an octahedral $NiN_3O_3^{2+}$ chromophore, ionic perchlorate groups, and lattice water.[145] The NiL_3Cl_2 complex of pyrazine N-oxide is probably a six-co-ordinate dimer containing both terminal N-bonded and bridging NO-bonded pyrazine oxide and terminal chlorines.[147]

Triphenylphosphine oxide forms $[Ni(tppO)_2SO_4]_n$ containing bridging SO_4 groups and probably both bridging and terminal tppO ligands, on the basis of the i.r. spectrum. A polymeric structure (106) is suggested on the basis of the magnetic moment value of $\mu_{eff} = 3.10$ BM and electronic spectrum characteristic of five-co-ordination.[144]

Formation of complexes of $Ph_2P(O)CH_2P(O)Ph_2$ occurs when the complexes $[Ni(Ph_2PCH_2PPh_2)X_2]$ are heated in air. The diphosphine dioxide ligand complexes directly with nickel(II) salts in ethanol forming $NiX_2,3dpmO_2,nH_2O$ and $2NiX_2,3dpm-O_2$ (X = Cl, Br, or I) depending upon the ratio of reactants used. The former are formulated as $[Ni(dpmO_2)_3]X_2$, and the latter as $[Ni(dpmO_2)_3]NiX_4$, the cations containing six-membered chelate rings producing an NiO_6^{2+} chromophore.

The diarsine dioxide ligands $Ph_2As(O)CH_2CH_2As(O)Ph_2$ ($daeO_2$) and $Ph_2As(O)-CH_2As(O)Ph_2$ ($damO_2$) form $Ni(daeO_2)_2(ClO_4)_2$ ($\mu_{eff} = 3.07$ BM) which is probably five-co-ordinate with one ionic and one unidentate perchlorate group, and $Ni-(damO_2)_3(ClO_4)_2$ ($\mu_{eff} = 3.38$ BM), octahedral with only ionic perchlorate.[150]

The nickel(II) complex $Ni_2L(LH)_2$ with diethylmethylenediphosphonato (L) ligands forms on heating nickel(II) chloride with tetraethylmethylenediphosphonate [*cf.* cobalt(II) salts].[151] Di-isopropylmethylphosphonate (dimp), dimethylmethylphosphonate (dmp), and trimethylphosphate (tmp) form nickel(II) perchlorate complexes $[Ni(dimp)_4,OClO_3]ClO_4$, $[Ni(dmp)_4]_n(ClO_4)_{2n}$, and $[Ni(tmp)_5](ClO_4)_2$, respectively. The i.r. spectra of the polymeric dimethylmethylphosphonate complex shows absorptions attributable to both bridging and terminal phosphonate groups.[516] Nickel(II) octylphosphinate $Ni[OP(C_8H_{17})_2O]_2$ is a yellow octahedral polymer with asymmetric phosphinate bridges which on heating changes to a purple tetrahedral structure.[516] The carbamylmethylenephosphonate (30; R = Pr; X, Y = H) forms the octahedral $[NiL_3](ClO_4)_2$ in which the ligand functions as a bidentate dioxygen donor.[150]

The trimeric $[Ni(acac)_2]_3$ (acacH = acetylacetone) has been converted into a different polymeric unit by dissolution in CS_2, and ether extraction of the resulting solid. There is electronic spectral and magnetic susceptibility evidence for both planar and octahedral nickel(II) environments (1:3 ratio), and the new unit is suggested as a tetramer possibly having structure (107).[517]

The equilibrium constant for the monomer ⇌ trimer reaction of bis[(2,6-dimethylheptane-3,5-dione)nickel(II), in benzene at 30 °C, is $2 \times 10^3 \, l^2 \, mol^{-2}$. Adduct formation with pyridine to form Ni_2L_4py and NiL_2py_2 was also examined.[518] The reactions between Lewis bases and the square-planar bis-(2,2,6,6-tetramethylheptane-3,5-dione)nickel(II) to form complexes were examined in solution and equilibrium constants and enthalpies determined.[519]

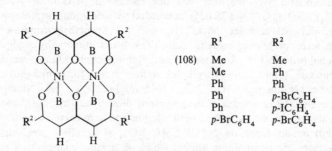

(107)

1,3,5-Triketones form bimetallic $[Ni_2L_2B_4]$ (B = H_2O or py) which have the structure (108) as shown by mass, i.r. and electronic spectral studies, and in the case of bis-(1,5-diphenylpentane-1,3,5-trionato)tetrapyridinedinickel(II) (B = py, $R^1 = R^2 =$ Ph) by an X-ray analysis.[520] Magnetic susceptibilities were measured over a range of temperatures for several of these complexes and for some 1,3-diketone analogues, and the magnetic exchange interactions discussed.[520] The monoclinic form of bis-(2-hydroxy-5-methylacetophenonato)nickel(II) contains a square-planar NiO_4 grouping.[521]

The structure and magnetic and spectroscopic properties of nickel(II) squarate dihydrate have been examined. The nickel is approximately octahedrally co-ordinated to four squarate oxygens and two water oxygens, the nickel lying on the edges of a cube with the squarate ligands on the cube faces.[522]

	R^1	R^2
(108)	Me	Me
	Me	Ph
	Ph	Ph
	Ph	$p\text{-BrC}_6H_4$
	Ph	$p\text{-IC}_6H_4$
	$p\text{-BrC}_6H_4$	$p\text{-BrC}_6H_4$

[517] J. A. Siegel and D. A. Rowley, *Inorg. Chim. Acta*, 1974, **9**, 19.
[518] D. P. Graddon and T. T. Nyein, *Austral. J. Chem.*, 1974, **27**, 407.
[519] M. J. Collins and H. F. Henneike, *Inorg. Chem.*, 1973, **12**, 2983.
[520] R. L. Lintvedt, L. L. Borer, D. P. Murtha, J. M. Kuszaj, and M. D. Glick, *Inorg. Chem.*, 1974, **13**, 18.
[521] R. D. Mounts and Q. Fernando, *Acta Cryst.*, 1974, **B30**, 542.
[522] M. Habenschuss and B. C. Gerstein, *J. Chem. Phys.*, 1974, **61**, 852.

3-Formylsalicylic acid forms dinuclear $Ni_2L_2,4H_2O$ which has structure (109).[523] The electronic solution spectra of a series of nickel(II) complexes of mandelic, malic, gluconic, and aspartic acids (all octahedral) have been reported and $10Dq$, B, and β calculated.[524] The i.r. spectrum of nickel(II) mandelate suggests[525] a structure similar to that of the cobalt(II) complex (q.v.). The probably octahedral complexes of several substituted malonic acids have been synthesized: $K_2Ni[RCH=C(CO_2)]_2,(H_2O)_n$ ($R = Ph$, $o\text{-}ClC_6H_4$, $p\text{-}ClC_6H_4$, $p\text{-}MeOC_6H_4$, or $p\text{-}NO_2C_6H_4$).[140]

Hexamethylphosphoramide forms a $NiL_2(NCSe)_2$ complex which, unlike its cobalt(II) analogue, contains both bridging and terminal NCSe ligands.[135] Nonayama has obtained some nickel(II) complexes of quinquedentate amides.[526]

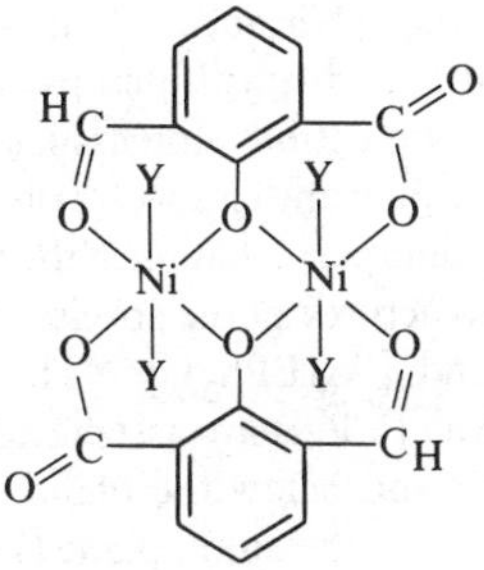

(109) 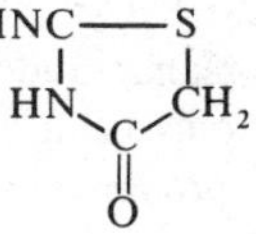   (110)

The ligand pseudothiohydantoin (110) forms $trans\text{-}NiL_4X_2$ (Cl, Br, I, ClO_4) and also a $cis\text{-}NiL_4Br_2$ which probably contains the ligand bonding *via* the carbonyl oxygen.[527] The structures of the octahedral perovskite $Sr_2[NiW]O_6$, and the monoclinic Sr_2-$[NiTe]O_6$ have been discussed.[528]

S-*Donors*. The nickel(II) chloride complex of 1,5-dithiocyclo-octane NiL_2Cl_2 contains the nickel ion in a distorted octahedral environment with four equatorial S-donors, and axial chlorines.[529] The linear tetrathioether chelates (111) form tetragonal $NiLX_2$ ($L = C_2$, $X = Br$, I; $L = C_3$, $X = I$) which are insoluble in all common solvents in which they do not decompose. The analogous ligand with the tetramethylene backbone does not form a stable nickel(II) complex.[168] The complex previously reported as $Ni(NH_3)_3CS_3$ has been shown to be $[Ni(NH_3)_6][Ni(CS_2)_2]$.[363]

The structure of cis-bis-(N-isopropyldithiocarbamato)nickel(II) is the first example of cis configuration in complexes of this kind, all previously reported examples have had a $trans$ structure. The adoption of a cis stereochemistry is due to the strong hydrogen bonding between the 'HN' hydrogen and sulphur atoms in neighbouring molecules.[530]

[523] M. Tanaka, H. Okawa, I. Hanaoka, and S. Kida, *Chem. Letters*, 1974, 71.
[524] P. V. Khadikar and M. G. Kekre, *Indian J. Chem.*, 1974, **12**, 213.
[525] P. V. Khadikar, R. L. Ameria, M. G. Kekre, and S. D. Chauhan, *J. Inorg. Nuclear Chem.*, 1973, **35**, 4301.
[526] M. Nonayama, *Inorg. Chim. Acta*, 1974, **10**, 59.
[527] P. P. Singh and U. P. Shukla, *Inorg. Chim. Acta*, 1973, **7**, 493.
[528] P. Kohl, *A. anorg. Chem.*, 1973, **401**, 121.
[529] N. L. Hill and H. Hope, *Inorg. Chem.*, 1974, **13**, 2079.
[530] C. L. Raston and A. H. White, *J.C.S. Dalton*, 1974, 1790.

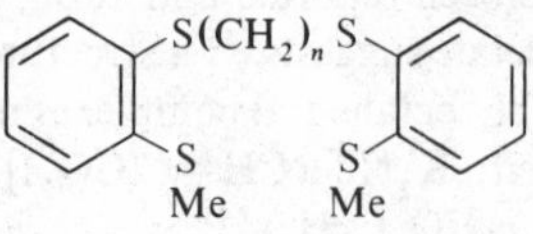

(111) $n = 2, C_2$
        $n = 3, C_3$

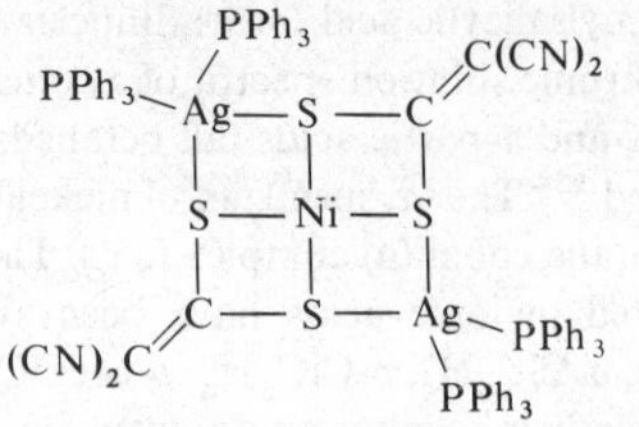

(112)

The $[Ni(tfd)_2]^{2-}$ [tfd = $(CF_3)_2C_2S_2^{2-}$] and $[Ni(mnt)_2]^{2-}$ ions form similar structures to the $[Co(mnt)_2]^{2-}$ analogue, with 1,4-disubstituted pyridinium cations, in which the anions are sandwiched between two cations. Ion association persists in solution as evidenced by anion → cation charge-transfer spectra; however, normal magnetic properties indicate a negligible electronic perturbation in the ground state.[159] Coucouvanis *et al.*[531] have determined the structures of the species formed between $[Ni(mnt)_2]^{2-}$ or $[Ni\{S_2C=C(CN)_2\}_2]^{2-}$ and $[Ag(PPh_3)_2]^+$. The NiS_4 units in both complexes are planar, and the $[Ag(PPh_3)_2]^+$ ions are associated with the NiS_4 moiety, the silver atoms being located above and below the plane with the P, Ag, and Ni in a plane approximately perpendicular to the anion plane (112).

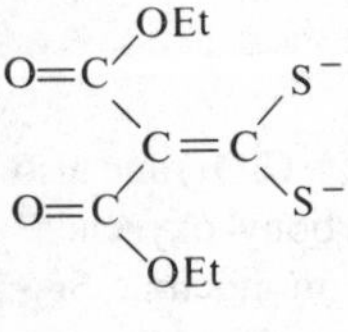

(113) 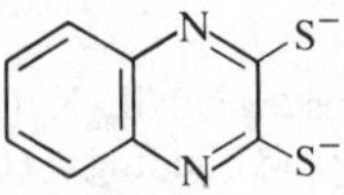   (114) 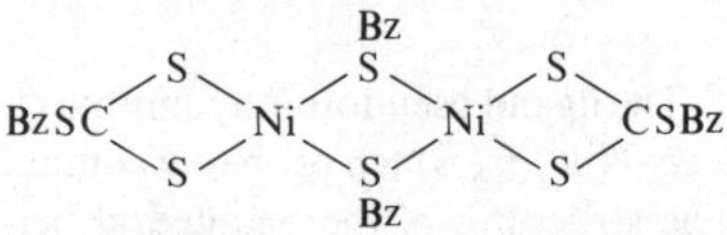   (115)

The nickel(II) complex of 1,1-diethoxycarbonyl-2,2-dithiolate (ded) (113) $[PhMe_3N]_2Ni(ded)_2$ contains a rigorously planar NiS_4 group with no axial interactions.[532] The bis-(2,3-quinoxalinedithiolato)nickel(II) anion in $[Et_4N]_2[NiL_2]$,-$2H_2O$ contains normal Ni—S bond lengths and the NiS_4 chromophore is planar.[533] The ligand has structure (114).

The mercaptide-bridged xanthate $[Ni_2(SCH_2Ph)_2(S_2CSCH_2Ph)_2]$ (115) has a central non-planar *syn-endo* Ni_2S_4 rhombus with a Ni $\cdots$ Ni separation of 2.795 Å.[534] Bis(ethylxanthato)nickel(II) forms 1:2 adducts with pyridine and 4-picoline in nitrobenzene solution.[535]

The *X*-ray structural analysis of $K_2Ni(S_2C_4O_2)_2$, where the ligand is the dithiosquarate anion, reveals an essentially planar NiS_4 grouping with the potassium ions interacting with the squarate oxygens of several ligands and with water molecules.[536]

[531]   D. Coucouvanis, N. C. Baenziger, and S. M. Johnson, *Inorg. Chem.*, 1974, **13**, 1191.
[532]   F. J. Hollander, M. L. Coffery, and D. Coucouvanis, *J. Amer. Chem. Soc.*, 1974, **96**, 4682.
[533]   A. Pignedoli, G. Peyronel and L. Antolini, *Acta Cryst.*, 1974, **B30**, 2181.
[534]   J. P. Fackler and W. J. Zegarski, *J. Amer. Chem. Soc.*, 1973, **95**, 8566.
[535]   L. Y. Yan, *Austral. J. Chem.*, 1974, **27**, 209.
[536]   D. Coucouvanis, F. J. Hollander, R. West, and D. Eggerding, *J. Amer. Chem. Soc.*, 1974, **96**, 3007.

The structure of Ni(sacsac)$_2$ (sacsacH = dithioacetylacetone) consists of discrete monomeric units with an essentially planar configuration.[164] *O*-Ethyl thioacetato-thioacetate (OEtsacsac) forms a diamagnetic monomeric tris ligand complex on reaction of the sodium salt with nickel(II) acetate in aqueous solution. N.m.r. spectral studies have established the presence of both *cis*- and *trans*-isomers but these have not been separated. The electronic spectrum compared with the other dithio chelates shows that the ligand field strength of OEtsacsac is about the same as for sacsac$^-$ or Et$_2$PS$_2^-$ and less than for ethylxanthate.[537]

The electronic spectra of a range of dithio- and perthio-carbonylatonickel(II) complexes and their pyridine adducts have been measured. The spectra show the presence of a variety of structures in solution, but complete interpretation of the spectra was prevented by the lack of a complete MO treatment of these complexes.[538]

Bis(diaryldithiophosphinato)nickel(II) (aryl = Ph or *p*-MeC$_6$H$_4$) from paramagnetic five-co-ordinate, 1:1 adducts with quinoline, α- and β-picoline, 2-aminopyridine, and 2,6-lutidine, with an S$_4$N donor set, and 2:1 adducts with pyridine, β- and γ-picoline, 4-phenylpyridine, isoquinoline, phenanthroline, and bipyridyl, having an S$_4$N$_2$ donor set and octahedral stereochemistry.[539] Two molecules of bipyridyl or phenanthroline per Ni(S$_2$PR$_2$)$_2$ also produce octahedral adducts, but here the donor set is N$_4$S$_2$ with unidentate dithiophosphinato groups. Ethylenediamine forms 1:1, 2:1, and 3:1 complexes with Ni(S$_2$PPh$_2$)$_2$ which contain N$_2$S$_4$, N$_4$S$_2$, and N$_6$ donor sets, respectively.[540] *NN'*-Dimethylformamide also forms 1:1 and 1:2 adducts with Ni(S$_2$PPh$_2$)$_2$ in benzene solution.[541] The polarized single-crystal electronic spectrum of Ni{S$_2$P(OMe)$_2$}$_2$L (L = 2,9-dimethylphenanthroline) has been been recorded and assigned.[542]

The reaction of bis(dimethylphosphorodithioato)nickel(II) Ni{S$_2$P(OMe)$_2$}$_2$ with Ph$_2$PCH$_2$CH$_2$PPh$_2$(dpe) and Ph$_2$PCH$_2$CH$_2$AsPh$_2$ (ape), does not result in simple adduct formation but in the formation of the complexes Ni[S$_2$PO(OMe)]L (L = dpe or ape) containing the novel S$_2$PO(OMe)$^{2-}$ anion. The structure of the ape complex has been determined, the nickel(II) ion being in a square-planar environment and co-ordinated to an S$_2$PAs donor set. The diarsine Ph$_2$AsCH$_2$CH$_2$AsPh$_2$, however, does not produce a similar complex, but instead yields the unstable Ni{S$_2$P(OMe)$_2$}L.[543]

Polymethylenebis(phenylthiourea)s PhNHCSNH(CH$_2$)$_n$NHCSNHPh (n = 2—10) behave as bidentate S$_2$ donors towards nickel(II). Various structural types are formed; tetrahedral monomeric [NiLX$_2$], tetrahedral polymeric [NiLX$_2$]$_n$, both with S$_2$X$_2$ donor sets, tetragonal monomer [NiL$_2$X$_2$] (S$_4$X$_2$ set), and planar polymeric [NiL$_2$X$_2$] (S$_4$ set), the type of complex depending upon the particular ligand and the reaction conditions. The ligand dependence is related to the number of methylene groups in the backbone: for $n > 5$ polymeric complexes result, whilst for $n = 3$, chelation occurs to form monomeric tetrahedral or tetragonal complexes. The ligand

[537] A. R. Hendrickson and R. L. Martin, *Inorg. Chem.*, 1973, **12**, 2582.
[538] C. Furlani, A. Flamini, A. Scamellotti, C. Bellitto, and O. Piovesana, *J.C.S. Dalton*, 1973, 2404.
[539] R. N. Mukherjee, M. S. Venkateshan, and M. D. Zingde, *J. Inorg. Nuclear Chem.*, 1974, **36**, 1043.
[540] R. N. Mukherjee, M. S. Venkateshan, and M. D. Zingde, *J. Inorg. Nuclear Chem.*, 1974, **36**, 1418.
[541] A. Furuhashi, T. Nomura, and M. Sugimoto, *J. Inorg. Nuclear Chem.*, 1974, **36**, 1415.
[542] A. A. G. Tomlinson and C. Furlani, *J.C.S. Dalton*, 1974, 1420.
[543] L. Gastaldi, P. Porta, and A. A. G. Tomlinson, *J.C.S. Dalton*, 1974, 1424.

PhNHCSNHCH$_2$CH(Me)NHCSNHPh and some selenium analogues were also prepared and complexed.[544]

Pyridine-2-thiol (LH) forms NiX$_2$(LH)$_2$ and NiX$_2$(LH)$_4$ (X = Cl, Br, or I) complexes. The 2:1 complexes are probably tetrahedral, although the data do not definitely rule out an octahedral polymeric constitution. The 4:1 complexes are diamagnetic and there is no evidence for halide co-ordination. Both types of complex exhibit a v(N—H) frequency in the i.r. spectra showing that co-ordination is *via* the sulphur with the ligand in the thione form.[165] The nickel(II) complexes of dibenzyldithioarsinic acid[162] and of the WO$_2$S$_2^{2-}$ ions are S-bonded like their cobalt(II) analogues, but as expected the nickel complex is planar.[163]

[(π-C$_5$H$_5$)Ni(CO)]$_2$ reacts, under irradiation, with CF$_3$SSCF$_3$ to give [(π-C$_5$H$_5$)-Ni(SCF$_3$)]$_2$, but the intermediate carbonyl containing monomer, although detected by i.r. spectroscopy, could not be isolated.[545] The [(π-C$_5$H$_5$)Ni(SR)]$_2$ (R = alkyl or aryl) can be oxidized voltammetrically to mono- and di-cationic species.

Carbaboranes. These are treated together irrespective of oxidation state. Formation of [NMe$_4$][(benzodicarbollyl)$_2$Ni^{III}] and (benzodicarbollylNi^{II})$_2$ has been reported, although the complexes were incompletely characterized.[247] The reaction of Ni-(dpe)Cl$_2$ with Na[C$_2$B$_4$H$_7$] in THF produced Ni(dpe)(C$_2$B$_4$H$_6$) probably with structure (116). Nickel(II) bromide, Na(C$_5$H$_5$), and Na(C$_2$B$_4$H$_7$) react to form (π-C$_2$B$_5$H$_7$)Ni$_2$(π-C$_5$H$_5$) and possibly traces of (C$_2$B$_7$H$_9$)Ni$_2$(C$_5$H$_5$). The former is thought to have an opened or distorted tricapped trigonal prismatic structure.[240] 1,2,4-(PPh$_3$)$_2$NiC$_2$B$_4$H$_6$ is formed from (π-C$_2$H$_4$)Ni(PPh$_3$)$_2$ and either 2,4-C$_2$B$_5$H$_7$ or 1,6-C$_2$B$_4$H$_6$ in THF.[239]

The addition of NaC$_5$H$_5$ and NiBr$_2$,2C$_2$H$_4$(OMe)$_2$ to sodium naphthalenide reduced [Me$_4$N][(π-C$_5$H$_5$)Co^{III}(π-C$_2$B$_7$H$_6$)] in THF produced (π-C$_5$H$_5$)Co^{III}(π-CB$_7$H$_8$)Ni^{IV}(π-C$_5$H$_5$) in several isomeric forms.[249] The (π-C$_5$H$_5$)Ni^{IV}(π-7-B$_{10}$CH$_{11}$) complex was prepared by reaction of (B$_{10}$CH$_{11}$)$^{3-}$ (prepared from B$_{10}$H$_{12}$CNMe$_2$ and Na–THF) with NaC$_5$H$_5$ and NiBr$_2$,2C$_2$H$_4$(OMe)$_2$, and air oxidation of the product.[246] Another route to the same compound is treatment of the 'triple-decker sandwich', [(π-C$_5$H$_5$)Ni^{II}(π-C$_5$H$_5$)Ni^{II}(π-C$_5$H$_5$)]BF$_4$ with B$_{10}$CH$_{11}$$^{3-}$ in THF, and air oxidation. On heating above 350 °C this complex isomerizes.[246]

$$(\pi\text{-C}_5\text{H}_5)\text{Ni}^{IV}(\pi\text{-7-B}_{10}\text{CH}_{11}) \xrightarrow{\Delta} (\pi\text{-C}_5\text{H}_5)\text{Ni}^{IV}(\pi\text{-2-B}_{10}\text{CH}_{11}) + (\pi\text{-C}_5\text{H}_5)\text{Ni}^{IV}(\pi\text{-1-B}_{10}\text{CH}_{11})$$

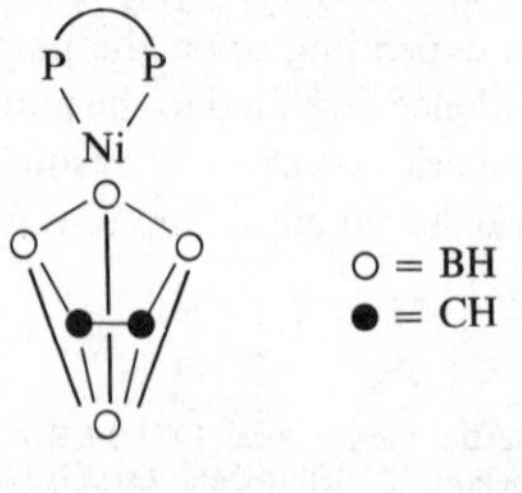

O = BH
● = CH

(116)

[544] T. Tarantelli, *J.C.S. Dalton*, 1974, 837.
[545] J. L. Davidson and D. W. A. Sharp, *J.C.S. Dalton*, 1973, 1957.

The ^{11}B and ^{13}C n.m.r. isotropic shifts for $Ni^{III}(C_5H_5)Ni(C_2B_nH_{n+2})$ and Ni-$(C_2B_nH_{n+2})_2$ ($n = 9, 8, 7,$ or 6) show L → Ni charge transfer.[241]

Mixed Donor Ligands. The possible structures of the complexes formed in solution between Ni^{2+} ions and the di-, tri- and tetra-glycine-NN-diacetic acids were established by potentiometric and i.r. solution studies; at high pH deprotonation of the amido-N occurs for the tri- and tetra-glycine compounds. Ethylenediamine NN'-diacetic acid (H_2L) forms NiL,nH_2O on reaction with aqueous nickel(II) nitrate solution.[175] Sodium nitrilotriacetate $Na_3(nta)$ forms $NaNi(nta),2H_2O$, $H_4Ni(nta)_2,2H_2O$ and $HNi(nta),3H_2O$ on reaction with nickel salts under the appropriate conditions.[547]

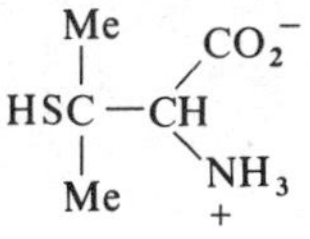

(117)

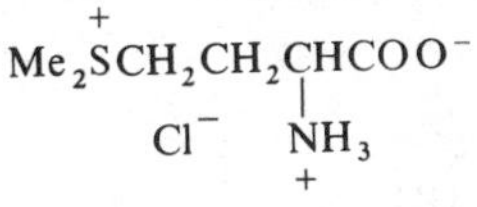

(118)

The thermal decomposition of $Ni(\alpha\text{-Ala})_2,4H_2O$ and $Ni(\beta\text{-Ala})_2,2H_2O$ and of some amine adducts has been examined.[548] DL-Penicillamine H_2pen (117) forms $Ni(pen)$, a polymeric diamagnetic planar complex, with the ligand bonded with O, N and bridging S-donors.[549] DL-Methylsulphonium methionine (Hmsm) (118) forms $[Ni(Hmsm)_2](ClO_4)_2$ ($\mu_{eff} = 3.12$ BM) containing octahedral nickel(II) and $[Ni(Hmsm)Cl_2]_n$, also octahedral probably *via* bridging chlorine and carboxylato-groups.[196] In ethanolic solution the lithium salt of the amino-acids reacts with nickel(II) chloride to form $[NiL_2]$, $Li[NiL_3]$, and $[NiLCl]$ under different conditions, all containing the deprotonated amino-acid, and all apparently octahedral.[196]

The three sulphur-containing amino-acids, SS'-ethylenebis[(R)-cysteine] (RR-ebc), SS'-ethylenebis[(S)-homocysteine] (SS-ebhc), and NN'-ethylenebis[(S)-methionine] SS-ebm), form NiL (L = RR-ebc, SS-ebhc) and $Ni_2(SS\text{-ebm})Cl_2$.[197] All three are slightly tetragonally distorted octahedral complexes, the latter with chlorine bridges. Angelici and co-workers[550—552] have examined the formation, stability constants, and stereochemistry of several Ni^{II} amino-acid complexes. N-(2-Pyridylmethyl)-L-aspartic acid and N-(6-methyl-2-pyridylmethyl)-L-aspartic acid complexes preferentially co-ordinate the L enantiomers of alanine, phenylalanine, threonine, leucine, and valine, in preference to the D enantiomers. The proton n.m.r. spectra of square-planar Ni^{II} complexes of the tripeptides of glycine and DL-α-alanine have been reported, and the NiL_2 complexes were prepared where L = triglycine, glycylglycyl-DL-α-alanine, glycine-DL-alanylglycine and DL-α-alanylglycylglycine.[553]

[546] R. J. Motekaitis and A. E. Martell, *Inorg. Chem.*, 1974, **13**, 550.
[547] F. J. M. Rajabalee, *J. Inorg. Nuclear Chem.*, 1974, **36**, 557.
[548] M. B. Bernard, N. Bois, and M. Daireaux, *Bull. Soc. Chim. France*, 1974, 31.
[549] S. T. Chow, C. A. McAuliffe, and B. J. Sayle, *J. Inorg. Nuclear Chem.*, 1973, **35**, 4349.
[550] R. Nakon, E. M. Beadle, and R. J. Angelici, *J. Amer. Chem. Soc.*, 1974, **96**, 719.
[551] R. V. Snyder and R. J. Angelici, *Inorg. Chem.*, 1974, **13**, 14.
[552] R. Nakon, P. R. Rechani, and R. J. Angelici, *Inorg. Chem.*, 1973, **12**, 2431.
[553] Y. Nakao, O. Oyama, and A. Nakahara, *J. Inorg. Nuclear Chem.*, 1974, **36**, 685.
[554] S. Guka, *Acta Cryst.*, 1973, **B29**, 2167.
[555] S. C. Rastorgi and G. N. Rao, *J. Inorg. Nuclear Chem.*, 1974, **36**, 1161.

The structure of bis(sarcosinato)nickel(II)dihydrate, $(CH_3NHCH_2COO)_2Ni,2H_2O$, is octahedral with axial water molecules, and NO bonded sarcosinate groups in the equatorial plane.[554]

Ethanolamine[555] and propan-2-olamine[555] form octahedral $[NiL_3]SO_4,nH_2O$ complexes which are NO co-ordinated, whilst tripropan-2-olamine[173] behaves as a terdentate NO_2 donor forming $[NiL_2]SO_4$, probably with structure (119). With diethanolamine (HL) several types of complex have been prepared.[556] $Ni(HL)_2X_2$ (X = Cl, Br, I, or NO_3) formed from NiX_2 + HL in a 1:2 ratio in ethanol in which the ligand behaves as a neutral bidentate ON donor, $[NiL,H_2O]Cl$ and $[NiLCl]$ formed from a 2:3 molar ratio under reflux in ethanol, and containing the monodeprotonated $HN(CH_2CH_2OH)CH_2CH_2O^-$. The $[Ni(HL)LY]$ (Y = NCS, Br, or ClO_4) and $Ni(LH)_2(NCS)_2$ were also isolated.[556]

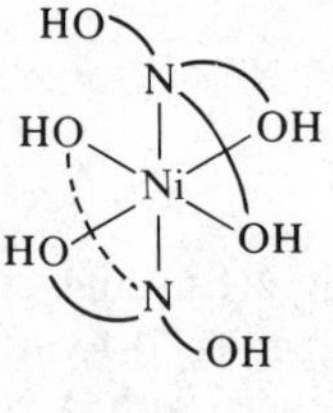

HO⟍
   N⟍
HO⟍ | ⟋OH
  '⟍Ni⟋
HO⟋ | ⟍OH
  ⟍'N
   ⟍OH

(119)

$R(CH_2)_nNHCOC_5H_4N$

LH (120) R = NH_2, NHMe, NMe_2, NEt_2, NHPh, OMe, SEt, $n = 2$
          R = NH_2, NHMe, NMe_2 $n = 3$

The aroylhydrazines $H_2NNHC(=O)R$ (R = Ph, p-$MeOC_6H_4$, o-HOC_6H_4, or p-$NO_2C_6H_4$) form $[NiL_3]X_2$ (X = Cl, Br, $\frac{1}{2}SO_4$) and $NiL_2(CNS)_2$, in which the hydrazines function as bidentate ligands bonding through the H_2N and the carbonyl oxygen.[177] The ligand N-acetyl-N'-benzoylhydrazine (L) forms $Ni(L - H)_2,2H_2O$ in which it bonds in the enol form as a terdentate uninegative ligand through the benzoyl oxygen and the neighbouring nitrogen to one metal, and through the acetyl oxygen to a second metal ion, each metal ion being co-ordinated to four ligands and having an N_2O_4 set.[178] The ligands (120) form[557] octahedral $Ni(LH)_2X_2,nH_2O$ (X = NCS or Cl, R = SEt; X = NCS, R = NHPh or OMe) co-ordinating *via* the amide oxygen and the ring nitrogen. However, the thiocyanate complexes $[NiL-(NCS)],xH_2O$ contain a terdentate, N_3 co-ordination (121) of the ligand, except for R = NHMe, $n = 3$, R = SEt or OMe, $n = 2$. The ligand $HMeN(CH_2)_3$-$NHCOC_5H_4N$ forms a paramagnetic tetragonal $[NiLNCS(H_2O)_2]$, but this can be converted on heating into $[NiLNCS]$, which is planar. The analogous ligands with a $(CH_2)_2$ linkage or with NH_2 terminal groups show no evidence of octahedral complex formation.[557]

N-(2-Pyridylmethyleneamino)-2-pyridinecarboxamide (papH) forms[182] $Ni_2(pap)$-$Cl_3,3H_2O$ and $Ni_3(pap)_2(SO_4)_2,14H_2O$. On the basis of i.r. and electronic spectra both were assigned octahedral structures (122) and (37).

Bis-(5-ethyl-5-phenylbarbiturato)bis(imidazole)nickel(II) is obtained from nickel(II) chloride, sodium phenobarbital, and imidazole, in aqueous solution.[184] The complex

[556] B. G. Sejekan, M. R. Udopa, and G. Aravamudan, *Indian J. Chem.*, 1974, 12, 533.
[557] M. Nonoyama and K. Yamasaki, *Inorg. Chim. Acta*, 1973, 7, 373.

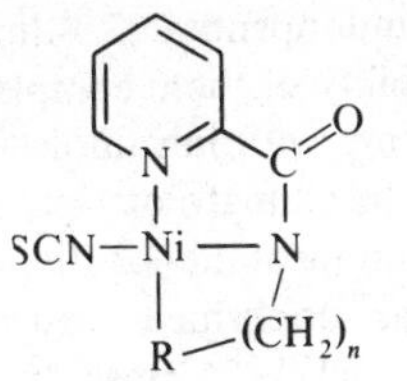

(121)

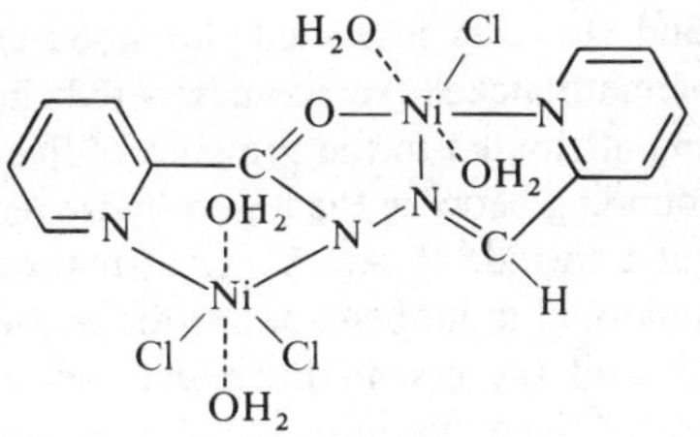

(122)

is square planar and this is in marked contrast to the complexes of the 5,5-diethyl analogue which gives two octahedral forms in addition to the square-planar isomer.

Semicarbazide, and the derived semicarbazones $H_2NNHCONR^1R^2$ ($R^1 = R^2 =$ Me, $R^1 =$ Me, $R^2 = CH_2CHMe_2$, $R^1 = R^2 =$ cycloheptane), form tetragonal NiL_2X_2 (X = Cl or Br) which bond *via* the CO and the hydrazine N. The complexes $[NiL_2(H_2O)_2]X_2$ (L = $H_2NNHCONMe_2$) contain co-ordinated water,[558] since the electronic spectra are insensitive to the change of anion, and far-i.r. spectra are identical.

Salicylaldehyde salicylhydrazone, $HOC_6H_4CONHN=CHC_6H_4OH$, forms Ni-(SS − 2H), containing a terdentate ligand bonding through OH, CO and C=N to give the polymeric octahedral complex, which has a subnormal magnetic moment.[179] Salicylaldehyde hydrazone (HL) forms $[NiL_2]_n$ in which the ligand co-ordinates *via* C=N and the phenolic oxygen.[559]

1-(2-Thiazolylazo)-2-naphthol (HL) (123) forms $[NiL_2]$, the X-ray structure of which reveals co-ordination *via* the phenolic O, the azo-N adjacent to the naphthol ring, and the thiazole N.[560]

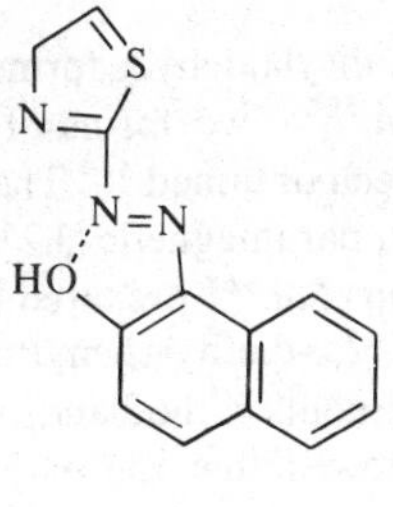

(123)

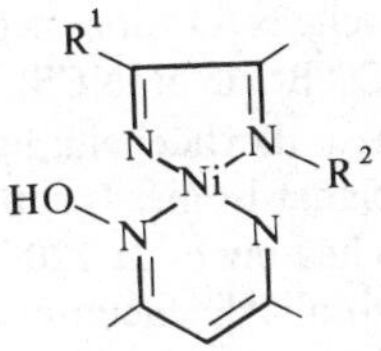

(124)　$R^1 =$ Ac　$R^2 =$ H or Me
　　　　$R^1 = R^2 =$ Me

Tetramminenickel(II) nitrite reacts with acetylacetone in refluxing ethanol to form (4-iminopentane-2,4-dione 3-oximato)(4-aminopent-3-en-2-onato)nickel(II). ^{1}H n.m.r. spectra of this and several related complexes confirmed the composition and structure (124), with a five-membered chelate ring of the 4R-iminopentane-2,4-dione 3-oximato

[558] B. Beecroft, M. J. M. Campbell, and R. Grzeskowiak, *J. Inorg. Nuclear Chem.*, 1974, **36**, 55.

[559] H. L. Ray, K. B. Pandeya, and R. P. Singh, *Indian J. Chem.*, 1974, **12**, 532.

[560] M. Kurahashi, *Chem. Letters*, 1974, 63.

ligand. However, in (3-methyliminobutan-2-one oximato)-(4-iminopentane-2,3-dione-3-oximato)nickel(II) a six-membered ring is present, the stability of these complexes being attributed to the presence of five- or six-membered rings with intramolecular H-bonding between the imino-hydrogen on one ligand and the oximato oxygen on a second ligand.[561] An X-ray structure of the (4-methyliminopentane-2,3-dione-3-oximato)-(4-iminopentane-2,3-dione-3-oximato)nickel(II) has confirmed structure (125) and the essentially planar co-ordination about the nickel.[562] Bose[563] has reported the preparation of (125) and its ethyl and n-propyl analogues by the reaction of the nickel(II) acetate, isonitrosoethylmethylketone, bis(isonitrosoacetylacetone-imino)nickel(II) and the appropriate RNH_2.

An X-ray examination of one of the minor products of nitrosation of 4-chlorophenol in the presence of a nickel(II) salt, has identified it as $KNiL_3,Me_2CO$ where L is 4-chloro-1-quinone-2-oximate, bonding as an NO donor, the nickel(II) being in a distorted octahedral environment.[564] Salicylic acid hydrazide (HL) forms $Ni(HL)_3$-Cl_2,H_2O containing HL in the amide form.[565]

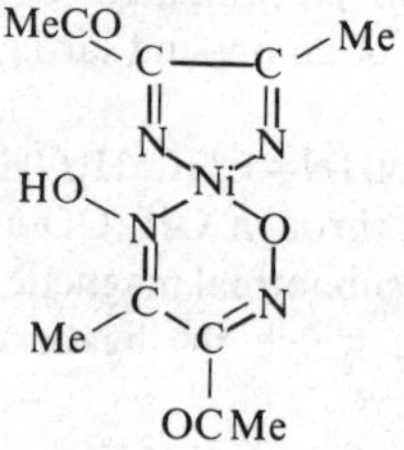

(125) 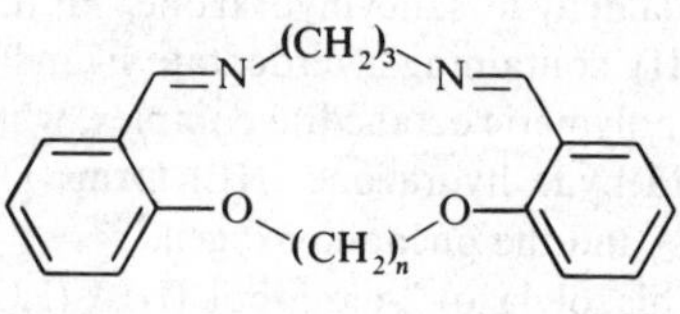   (126)

The macrocyclic N_2O_2 donors (126) derived from salicylaldehyde forms tetragonal $NiLX_2$ (X = Cl, Br, I, or NCS), μ_{eff} = 3.2—3.3 BM.[566] Two forms of bis-(N-iso-propyl-5,6-benzosalicylideneiminato)nickel(II) have been obtained.[567] The previously known green planar isomer is converted into a brown paramagnetic (3.22 BM) tetrahedral form on heating to ca. 220 °C for ca. 2 h. Higson et al.[568] prepared bis-[N-(2,6-dimethylphenyl)salicylideneiminato]nickel(II), bis-[N-(2,6-diethylphenyl)salicylidene-iminato]nickel(II) and some related species with substituents in the 3 and/or 5 position of the salicyl moiety. Electronic spectral studies showed that the majority of the complexes preferred to be planar, and only 3-nitro-substituted complexes showed the presence of significant amounts of the tetrahedral form in $CHCl_3$ solution,

[561] M. J. Lacey, J. S. Shannon, and C. G. MacDonald, *J.C.S. Dalton*, 1974, 1215.
[562] J. F. McConnell, M. J. Lacey, C. G. MacDonald, and J. S. Shannon, *Acta Cryst.*, 1973, **B29**, 2477.
[563] K. Bose, *Inorg. Chim. Acta*, 1973, **7**, 578.
[564] P. W. Correck, J. Charambous, M. J. Kensett, M. McPartlin, and R. Sims, *Inorg. Nuclear Chem. Letters*, 1974, **10**, 749.
[565] Yu. Ya. Kharitonov, R. I. Machkhoshvili, and N. B. Generalova, *Zhur. neorg. Khim.*, 1974, **19**, 270.
[566] L. G. Armstrong and L. F. Lindoy, *Inorg. Nuclear Chem. Letters*, 1974, **10**, 349.
[567] A. Takeuchi and S. Yamada, *Inorg. Chim. Acta*, 1974, **8**, 225.
[568] B. M. Higson, D. A. Lewton, and E. D. McKenzie, *J.C.S. Dalton*, 1974, 1690.

although even these are planar in the solid state. In donor solvents (A) (py, α- or γ-pic) two complexes are formed

$$NiL_2 \underset{A}{\rightleftharpoons} NiL_2A \underset{A}{\rightleftharpoons} NiL_2A_2$$

some of which can be isolated as solids.

Formation of paramagnetic five-co-ordinate α-picoline adducts of a range of salicylidenimine complexes (127) has been shown to be incomplete even in neat α-picoline solution.[569] A similar study showed that the nickel(II) complexes of *N*-n-butylsalicylaldimine, *N*-n-butyl-5-chlorosalicylaldimine, and *N*-n-butyl-5-chloro-2-hydroxybenzophenoneimine, add pyridine or γ-picoline to form five-co-ordinate mono-adducts and then six-co-ordinate bis-adducts in a stepwise manner, although only the latter can be isolated.[570] Stepwise stability constants and the corresponding $\Delta H°$ and $\Delta S°$ were calculated from a combination of magnetic and spectroscopic studies and thermometric titrations. Thermodynamic data for trimerization of bis-(*N*-phenyl-5-chlorosalicylideneiminato)nickel(II) and for formation of other pyridine adducts with nickel(II) complexes and salicylaldehyde derivatives have been reported.[571] Formation of imidazole and pyrazole adducts of Ni(saloph) and Ni(acac)$_2$ in solution has been investigated, and the results are similar to those obtained for the analogous Co^{II} system[205] (*q.v.*). Detailed i.r. spectra have been reported for Ni(N-Mesalen)$_2$[572] and for complexes of *NN'*-bis(salicylidene)-1,1-(dimethyl)ethylenediamine,[201] and the m.c.d. spectra of a range of nickel(II) Schiff base complexes in solution have been recorded.[573]

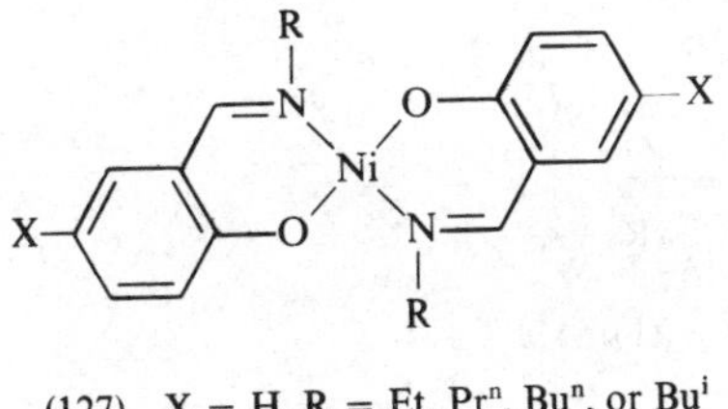

(127) X = H, R = Et, Pr^n, Bu^n, or Bu^i
      X = Cl, R = Pr^n or Bu^n

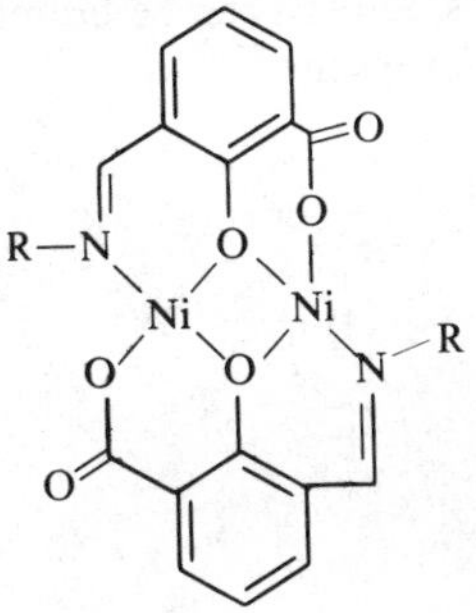

(128)

The octahedral nickel(II) NiL$_2$ complex derived from 4-benzoylacetone-ethanolamine(LH) adds pyridine to form NIL$_2$(py)$_2$, with the pyridine occupying the two positions previously occupied by the OH groups of the Schiff base.[207]

The Schiff bases H$_2$fsacR derived from 3-formylsalicylic acid and amines form dinuclear complexes with the essential structure (128). The complexes were isolated as hydrates and the electronic spectra indicate that the nickel is six-co-ordinate.[574]

[569] L. F. Lindoy and G. M. Mockler, *J. Coord. Chem.*, 1973, **3**, 169.
[570] D. R. Dakternieks, D. P. Graddon, L. F. Lindoy, and G. M. Mockler, *Inorg. Chim. Acta*, 1973, **7**, 467.
[571] D. R. Dakternieks and D. P. Graddon, *Austral. J. Chem.*, 1973, **26**, 2379.
[572] A. Bigotto and G. De Alti, *Spectrochim. Acta*, 1974, **30A**, 27.
[573] H. Kato and T. Sakamoto, *J. Amer. Chem. Soc.*, 1974, **96**, 4131.
[574] M. Tanaka, H. Okawa, T. Tamura, and S. Kida, *Bull. Chem. Soc. Japan*, 1974, **47**, 1669.

The condensation of *o*-aminobenzamide and pyridine-2-carbaldehyde produced the terdentate ligand (129) which partially isomerizes in methanol, in the presence of nickel(II) nitrate, to form the cyclic ligand (130). The nickel complex [Ni(129)(130)-(H$_2$O)](NO$_3$)$_2$,4H$_2$O has been X-rayed, and the results show that (129) bonds as a terdentate N$_2$O donor, and (130) as a bidentate N$_2$ ligand, the co-ordinated water molecule completing the irregular octahedron about the nickel.[575]

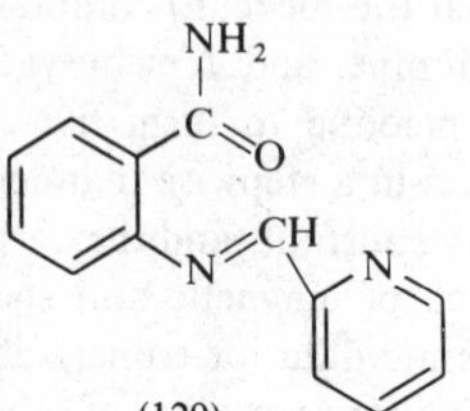

(129) 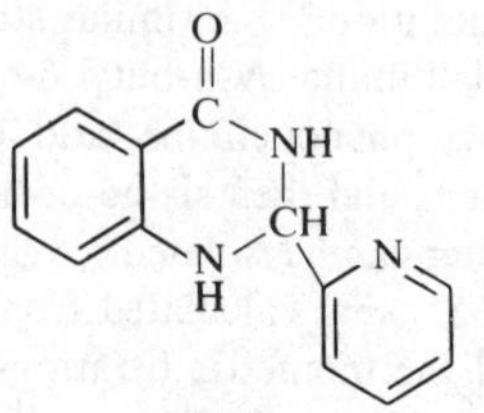   (130)

The bromination of complexes of the type (131) with *N*-bromosuccinimide in aceto-nitrile produces substances with two bromine atoms per molecule.[576] The brominated complexes are planar, diamagnetic non-electrolytes, and ^{1}H n.m.r. spectra are consistent with the substitution having occurred at the positions marked *. The free ligands, R = H, can be brominated and on reaction with nickel(II) acetate the same complex results.[576]

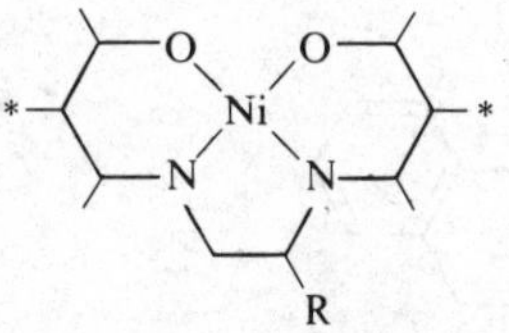

(131)   R = H or Me

The reaction of Ni(acac)$_2$ with a range of Schiff bases (HB) in refluxing toluene, followed by exposure to moist air produced [Ni(acac)(B)(OH$_2$)] complexes. The complexes have an octahedral structure, and on heating lose water being converted into square-pyramidal [Ni(acac)B]. In moist solvents the latter partially rehydrate.[577]

The isolation and stability of a range of Ni^{II} Schiff-base complexes of type [NiLX] (L = terdentate Schiff base) have been examined.[578] All are essentially planar, and are of the unusual type with one co-ordinated X group.

The nickel(II) complex of the O$_2$N$_2$ quadridentate (132) derived from benzil mono-hydrazone reacts with en or 1,2-pn at room temperatures, to produce complexes

[575] A. B. Corradi, C. G. Palmieri, M. Naldelli, and C. Pelizzi, *J.C.S. Dalton*, 1974, 150.
[576] L. F. Lindoy, H. C. Lip, and W. E. Moody, *J.C.S. Dalton*, 1974, 44.
[577] R. H. Balundge and A. Chakravorty, *Inorg. Chim. Acta*, 1974, 8, 261.
[578] G. O. Dudeck and E. P. Dudeck, *Inorg. Chim. Acta*, 1974, 8, 219.

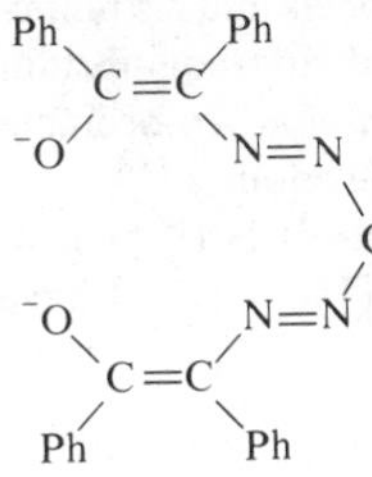

(132) 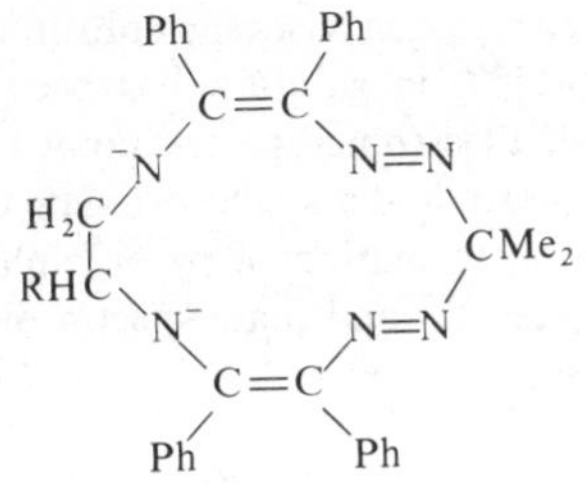   (133) 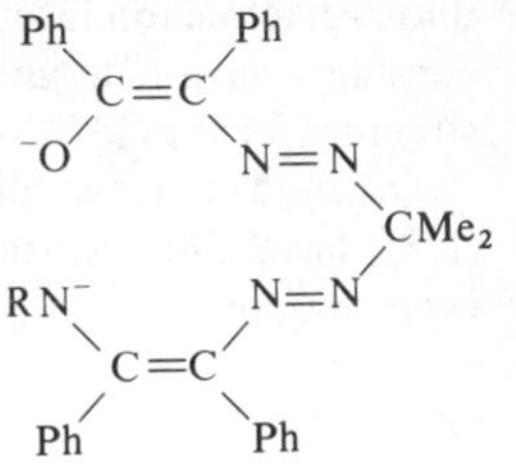   (134)   R = Et or CH$_2$CH$_2$CH$_2$NH$_2$

of the N$_4$ macrocycle (133). However, with ethylamine or 1,3-pn ring-closure does not occur, new ligands such as (134) being produced.[579]

The reaction of diethylenetriamine with acacH or CF$_3$acacH in the presence of NaI or NH$_4$PF$_6$ yielded the salts of the monoprotonated quadridentate Schiff base, which react with nickel(II) ions in strongly basic solution to yield square-planar (135) as shown by their i.r. and ^{1}H n.m.r. spectra. These complexes contain rare examples of unsymmetrical anionic singly condensed Schiff-base ligands.[580] The quinquedentate ketoamine (136) is formed from CF$_3$acacH and bis-(3,3'-diaminopropyl)amine in methanol, and this reacts with nickel(II) acetate in DMF to form the Ni(1tfacDPT) μ_{eff} = 3.05 BM, which appears to have a structure intermediate between trigonal-bipyramidal and square-pyramidal.[581] However, H$_2$tfacdien, which is similarly derived using diethylenetriamine, does not form a corresponding five-co-ordinate complex, but undergoes hydrolysis of one azomethine linkage, in the presence of base, to form the complex of the quadridentate Schiff base described above (135).[581]

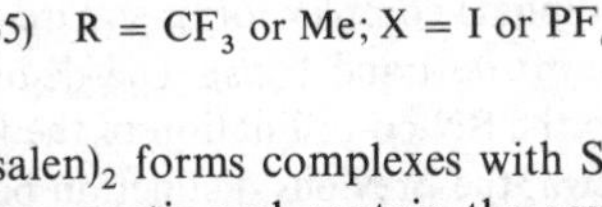

(135)   R = CF$_3$ or Me; X = I or PF$_6$ 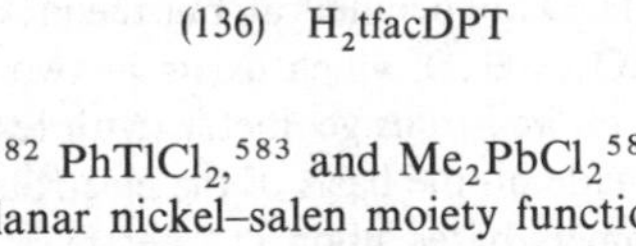   (136)   H$_2$tfacDPT

Ni(salen)$_2$ forms complexes with SnCl$_4$,[582] PhTlCl$_2$,[583] and Me$_2$PbCl$_2$[584] which are diamagnetic and contain the square-planar nickel–salen moiety functioning as a bidentate O$_2$ donor to the Group IIIB or IVB element. The nickel(II) complex of β-mercaptoethylamine behaves similarly as a S$_2$ donor.[583] Mössbauer data have been reported for a range of Ni(salen)$_2$-organotin chloride adducts.[584]

[579] C. M. Kerwin and G. A. Melson, *Inorg. Chem.*, 1973, **12**, 2410.
[580] W. N. Wallis and S. C. Cummings, *Inorg. Chem.*, 1974, **13**, 991.
[581] W. N. Wallis and S. C. Cummings, *Inorg. Chem.*, 1974, **13**, 988.
[582] M. Calligaris, L. Randaccio, R. Barbieri, and L. Pellerito, *J. Organometallic Chem.*, 1974, **76**, C56.
[583] L. Pellerito, R. Cefalù, and A. Gianguzza, *J. Organometallic Chem.*, 1974, **70**, C27.
[584] L. Pellerito, R. Cefalù, A. Gianguzza, and R. Barbieri, *J. Organometallic Chem.*, 1974, **70**, 303.

Adduct formation between nitrogen bases (py, substituted py, *o*-phen) and bis(mono-thio-β-diketonato)nickel(II) (137) in solution has been examined and equilibrium constants were determined. Electronic spectra point to a planar four-co-ordinate structure for the (137) complexes and a six-co-ordinate one for the adducts.[585]

Thermodynamic studies on formation of py or *o*-phen adducts of NiL_2 [HL = (138)] have also been reported,[586] and mass spectra of $[NiL_2]$ [HL = (139)] have been recorded.[587]

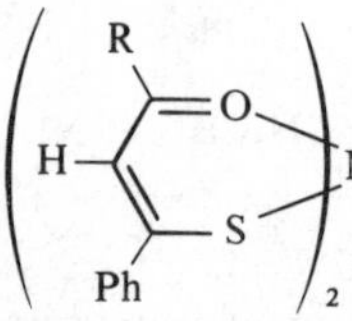

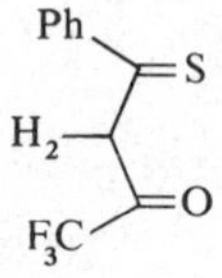

(138)   (139)   R = Ph or p-MeC$_6$H$_4$

(137)   R = OEt, SEt, Ph, Me, NMePh, NHPh,
          NH(p-ClC$_6$H$_4$), or NHEt

RC(SH)=CHCOCF$_3$

An *X*-ray study of $[Ni(monothioacac)_2]$ showed it to have an essentially planar *cis* structure, with the bond lengths suggesting that C⋯O has more double-bond character than C⋯S.[588] The structure of the nickel(II) complex of *cis*-1-mercapto-2-(*p*-bromobenzoyl)ethylene also reveals a *cis* configuration.[590]

The structure determination[591] of bis(di-n-butylmonothiocarbamato)nickel(II) $[(Bu^n_2NCSO)_2Ni]$ established the octahedral environment of the nickel, which is S_4O_2 co-ordinated, each sulphur bridging two nickel atoms, and producing a hexameric ring. The nickel phosphine complexes $Ni(PPh_3)_2Cl_2$ and $Ni(vpp)Cl_2$ [vpp = *cis*-1,2-bis(diphenylphosphino)ethylene] react with $[Et_2NH_2][SeC(O)NEt_2]$ in THF to form selenocarbamate complexes $L_2Ni[SeC(O)NEt_2]_2$ (L = PPh$_3$ or $\frac{1}{2}$vpp), which are planar with Se-bonded ligands in the solid state, but tend to dissociate PPh$_3$ with the selenocarbamate becoming bidentate (OSe). These complexes take up CO in solution to form $L_2Ni(CO)_2$.[592] The ligand [LH = (40)] forms square-planar $[NiL_2]$ which adds nitrogen bases to produce octahedral $[NiL_2(base)_2]$.[188]

Thiosemicarbazide $H_2NHNCSNH_2$ forms NiL_2X_2 (X = Cl or Br) which are planar with an S_2N_2 donor set.[558, 593] The $[NiL_2(H_2O)_2](NO_3)_2$ and $[NiL_2(H_2O)_2]-(ClO_4)_2$,$H_2O$ are octahedral, but the nickel(II) sulphate complex forms square-planar $[NiL_2]SO_4$, nH_2O which exists in two forms: α(*trans*) and β(*cis*). The distinction between *cis* and *trans* geometry (with respect to the SN co-ordination of the ligand) can be made on the basis of the electronic spectra; the previous distinction between these isomers based upon i.r. spectra is invalid owing to exchange which occurred in the preparation of the KBr discs.[593]

[585] M. Chikuma, A. Yokoyama, and H. Tanaka, *J. Inorg. Nuclear Chem.*, 1974, **36**, 1243.
[586] D. R. Dakternieks and D. P. Graddon, *Austral. J. Chem.*, 1974, **27**, 1351.
[587] M. Das and S. E. Livingstone, *Austral. J. Chem.*, 1974, **27**, 53.
[588] O. Siiman, D. D. Titus, C. D. Cowman, J. Freser, and H. B. Gray, *J. Amer. Chem. Soc.*, 1974, **96**, 2353.
[590] L. Kutschabsky, *Z. anorg. Chem.*, 1974, **404**, 239.
[591] B. F. Hoskins and C. D. Pannan, *Inorg. Nuclear Chem. Letters*, 1974, **10**, 229.
[592] K. Tanaka and T. Tanaka. *Inorg. Nuclear Chem. Letters*, 1974, **10**, 605.
[593] A. Sirota and T. Sramko, *Inorg. Chim. Acta*, 1974, **8**, 289.

An X-ray structure determination of $[NiL_3]Cl_2,H_2O$ shows that the thiosemicarbazide is NS-bonded, and that the cation has a meridional configuration.[594] Thiosemicarbazones $H_2NCSNHNR^1R^2$ ($R^1, R^2 = Me, R^1 = Me, R^2 = Bu^i, R^1R^2 =$ cycloheptane) form NiL_2X_2 (X = Cl or Br) which are trigonal-bipyramidal $[NiL_2X]X$ complexes with an N_2S_2X donor set.[558]

Dithiocarbazic acids $R^1R^2NNR^3CSSH$ form planar $Ni(R^1R^2NNR^3CSS)_2$ (R^1, $R^2,R^3 = H$, alkyl, or aryl). ^1H n.m.r. spectra of a variety of these in DMSO solution were generally consistent with NS-co-ordination for unsubstituted and 2-substituted acids and SS-co-ordination for the 3-substituted complexes. The complexes with an $[NiS_2N_2]$ chromophore produce evidence for *cis–trans*-isomers in solution.[595]

Pyridine-2-carbaldehyde-S-methylthiosemicarbazone (140) forms $Ni(pmts)X_2$ (X = Cl or Br) which are six-co-ordinate nickel with a N_2SX_3 donor set, two of the groups being bridging.[596] In the 2:1 complexes $[Ni(pmts)_2]Y_2$ [Y = ClO_4, BF_4, or (NO_3, H_2O)] the pmts are also terdentate N_2S donors, but the structure of $Ni(pmts)_2(CNS)_2$ is less clear, the available evidence could be interpreted as consistent with one or both —CNS— groups co-ordinated and hence one or both pmts functioning as bidentate NN ligands.[576] The hydrazine-S-methyldithiocarboxylate Schiff bases (141) form planar diamagnetic bis-ligand complexes with a NiS_2N_4 chromophore. The i.r. spectra exhibit no $\nu(N—H)$ vibration, and suggest that co-ordination is *via* the azomethine N and the thiol S. The ligand $R^1 = R^2 = Me$ forms a five-co-ordinate monopyridine adduct.[206]

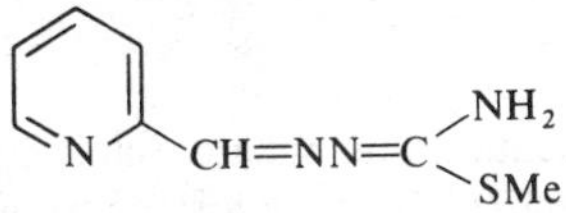

(140)   pmts

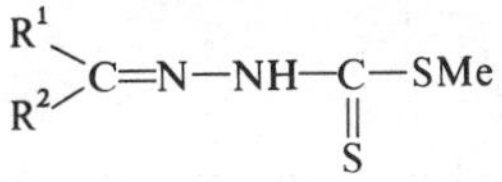

(141) $R^1 = R^2 = Me$
$R^1 = R^2 = C_5H_{10}$
$R^1 = Me, R^2 = Ph$

The 1-substituted tetrazoline-5-thiones (LH) (1-Ph or 1-MeClC$_6$H$_4$) are NS-bonded in the $[NiL_2]$ complexes, whilst thiazoline-2-thione is S-bonded in NiL_2I_2, and thiocarbohydrazide N-bonded in NiL_2Cl_2.[189] In the presence of nickel(II) salts 1,3,4-thiadiazoline-2-thiones rearrange to form the metal complexes of the tautomeric Schiff bases. 3,5,5-Trimethyl-1,3,4-thiadiazoline-2-thione forms the nickel complex $[NiL_2]$ of α-N-methyl-β-N-isopropylidenedithiocarbazate (LH).[347] Similarly the 5,5-dimethyl- and 5-(2-pyridyl) substituted ligands rearrange to produce the

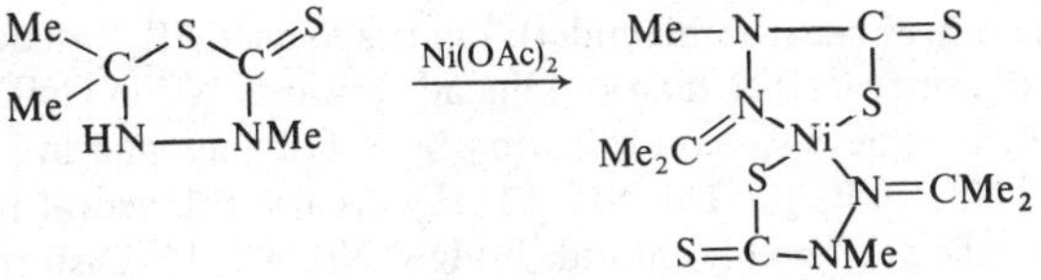

594 R. E. Ballard, D. B. Powell, and U. A. Jayasooriya, *Acta Cryst.*, 1974, **B30**, 1111.
595 D. Gattegno and A. M. Guiliani, *J. Inorg. Nuclear Chem.*, 1974, **36**, 1553.
596 M. A. Malik and D. J. Phillips, *Austral. J. Chem.*, 1974, **27**, 1133.

nickel complexes of isopropylidenedithiocarbazate and β-N-(2-pyridyl)methylene-dithiocarbazate.[347] The ligand 2-mercaptoquinazol-4-one reacts with nickel nitrate to form complex (142).[597]

Dithio-oxamide (LH_4) and dimethyldithio-oxamide (LMH_2) form complexes of types $Ni_4(LH_2)_5$ and $Ni_8(LM)_9$ which have anomalous magnetic moments.[405] The complex of NN-diethylphenylazothioformamide (143) $[NiL_2]^{2+}$, formed by reaction of nickel perchlorate with the ligand in acetone, is a member of a facile redox series

$$NiL_2^{2+} \overset{e^-}{\rightleftharpoons} NiL_2^+ \overset{e^-}{\rightleftharpoons} NiL_2^0 \overset{e^-}{\rightleftharpoons} NiL_2^- \overset{e^-}{\rightleftharpoons} NiL_2^{2-}$$

which differ from the well-known dithiolenes in that the parent complex is *not* planar, but has a pseudo-tetrahedral structure. Electrochemical, magnetic, e.s.r., and electronic spectral data were reported and a bonding model proposed.[598]

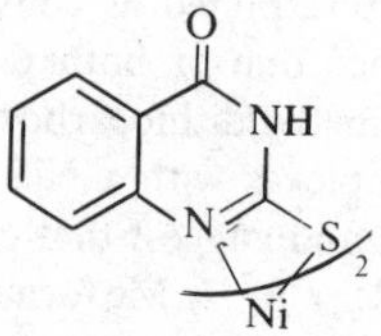

(142) 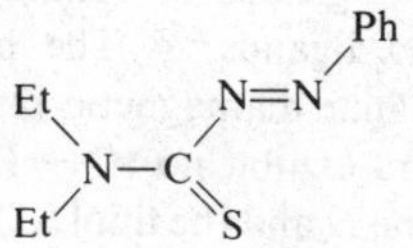   (143)

Square-planar complexes of three quadridentate thioiminato Schiff base ligands $[Ni(N_2S_2)_2]$ (43) which are monomeric, non-electrolytes are formed from $Ni(OAc)_2$ and the ligands in MeOH.[198, 199] Spectroscopic studies show that more charge delocalization is present in these complexes than in the oxygen analogues. Mixed (dithiolene)(α-di-imine)Ni^{II} complexes can be synthesized from Ni(dithiolene) + α-di-imine, Ni(α-di-imine)$_2$ + dithiolene or by mixing the Ni(α-di-imine)$_2$ and Ni(dithiolene)$_2$ in an appropriate solvent. These mixed (S_2N_2) donor species undergo electron-transfer reactions voltammetrically (changes $+1$, 0, -1, -2), but only the neutral species can be isolated.[599]

Benzoylformic acid selenosemicarbazide (H_2L) can function as either a mononegative $[Ni(LH)_2],4H_2O$ or a dinegative anion $NiLQ.xH_2O$ (Q = pyridine base).[183] The ligand 1-phenyl-o-hydroxy-4-benzamidothiosemicarbazone forms $[NiL_2]$ with the ligand behaving as a terdentate through the thioketo-S, amido-N, and the phenolic O.[190]

Both 1,10-phenanthrolinecarbothiamide (L′) and its N-phenyl (L″) analogue beaMaLj as terdentate N_2S donor (rather than N_3) ligands towards Ni^{II} in $[NiL'X_2]$ (X = Cl, Br, or NCS), which achieve six-co-ordination by X-bridging, and in $[Ni(L' - H)_2]$, $[Ni(L'' - H)_2]$, $[NiL''_2](BF_4)_2$. The $NiL''Cl_2,H_2O$ is also octahedral and loses water on heating to give the presumably chloride bridged $NiL''Cl_2$.[192] Distorted octahedral

[597] L. D. Dave and M. U. Cyriac, *J. Indian Chem. Soc.*, 1974, **51**, 383.
[598] K. Bechgaard, *Acta Chem. Scand.*, 1974, **A28**, 185.
[599] T. R. Millar and I. G. Dance, *J. Amer. Chem. Soc.*, 1973, **95**, 6970.

Ni^{II} is present in aquo(1-oxa-7,10-dithia-4,13-diazacyclopentadecane)nickel(II) di-nitrate, the co-ordination being *via* two *trans* N, two *cis* S, and one O of the ligand, and the O of the water molecule; the nitrate groups are ionic.[600]

The Schiff base α-*N*-methyl-*S*-methyl-β-*N*-(2-hydroxyphenyl)methylenedithiocar-bazate (144) loses a proton on co-ordination to nickel(II) to function as a terdentate NOS donor in [Ni(L − H)X] (X = Cl, Br, or NCS), all planar diamagnetic complexes. Similarly the Ni(L − H)$_2$ complex is octahedral μ_{eff} = 3.10 BM.[601]

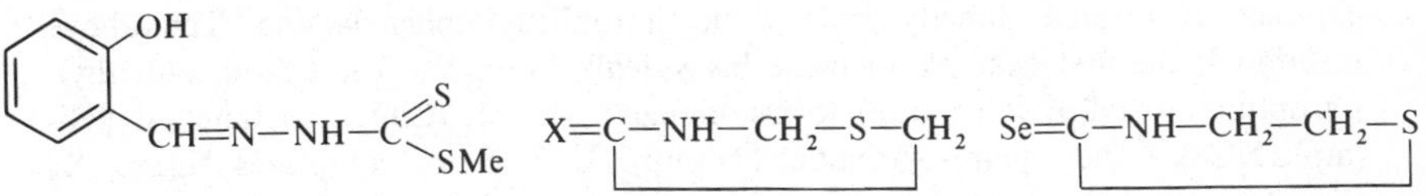

(144)　　　　　　　　　L′ and L″　　　　　　L‴

The three ligands thiomorpholine-3-one (L′) (X = O), thiomorpholin-3-thione (L″) (X = S), and thiazolidine-2-selenone (L‴) form nickel(II) complexes with different modes of ligand co-ordination. In [NiL′$_6$]X$_2$ (X = Cl, Br, or I) L′ is O-bonded and the cations are octahedral (μ_{eff} = 2.99–3.30 BM).[602] In NiL″$_2$X$_2$ (X = Cl or Br) the ligand is N-bonded and the X-groups bridge to achieve octahedral co-ordination whilst NiL″$_2$I$_2$ is tetrahedral and hence monomeric. The selenone L‴ produces NiL‴X$_2$ which are pseudotetrahedral again with X-bridges, and with the ligand probably bonded *via* the selenium to the nickel.[122]

Bis(monoethanol- and diethanol-dithiocarbamato)nickel(II) are planar complexes with exclusively Ni—S bonding; the oxygen atoms do not bond to Ni, but participate in considerable hydrogen-bonded interactions.[603] *NN′*-Dimethyldithiomalonamide, MeHNC(S)CH$_2$C(S)NMeH (LH), forms planar [NiL$_2$], which are S$_4$ co-ordinated, obtained from the ligand and nickel(II) salts in neutral solution.[604] In acidic solution the ligand does not deprotonate but bonds as a neutral NS-donor forming planar [Ni(HL)$_2$]X$_2$ (X = Cl, Br, I, or ClO$_4$), and octahedral [Ni(LH)$_3$]X$_2$ (X = Cl, ClO$_4$, or BF$_4$). The [Ni(HL)$_2$Cl$_2$] has a *trans*-octahedral structure, but the bromo- and iodo-derivatives are formulated [Ni(HL)$_2$][NiX$_4$] with planar cations.[604] In solution [Ni(LH)$_2$]I$_2$ and [Ni(HL)$_2$][NiI$_4$] have identical electronic spectra showing that the same species is present. 2,4-Dithiobiuret (HL) forms diamagnetic planar [NiL$_2$], Ni(HL)$_2$(OAc)$_2$, [Ni(HL)$_2$]Y$_2$ (Y = ClO$_4$, NO$_3$, Cl, Br, or $\frac{1}{2}$SO$_4$). The *X*-ray structure of the diperchlorate was also reported.[605] Ternary (1:1:1) complexes are formed between Ni^{II} salts, glycine and ethylenediamine, propylenediamine, *o*-phenylenediamine, or naphthalenediamine.[606]

The reaction of the bis(*o*-diphenylarsinophenylthio)alkanes (145) with NiX$_2$ (X = ClO$_4$, NO$_3$, Cl, or Br) in acetone solution at >30 °C proceeds *via* a deep blue five-co-ordinate intermediate to form green crystals of Ni(*o*-Ph$_2$AsC$_6$H$_4$S)$_2$. The same

[600] R. Louis, B. Metz, and R. Weiss, *Acta Cryst.*, 1974, **B30**, 774.

[601] M. A. Ali, S. E. Livingstone, and D. J. Phillips, *Inorg. Chim. Acta*, 1973, **7**, 531.

[602] D. de Filippo, F. Devillanova, G. Verani, and C. Preti, *J. Inorg. Nuclear Chem.*, 1974, **36**, 1017.

[603] B. Annuar, J. O. Hill, and R. J. Magee, *J. Inorg. Nuclear Chem.*, 1974, **36**, 1253.

[604] G. C. Pellacani, G. Peyronel, and W. Malavasi, *Inorg. Chim. Acta*, 1974, **8**, 49.

[605] A. Pignedoli, G. Peyronel, and G. Giovetti, *Gazzetta*, 1973, **103**, 1237.

[606] V. Kumari, R. C. Sharma, and G. K. Chaturvedi, *Monatsh.*, 1974, **105**, 629.

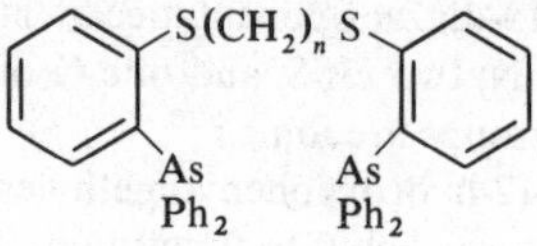

(145) $n = 2$—4   (146)

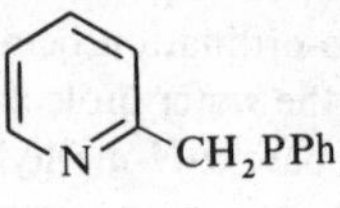

complex is formed directly from *o*-mercaptophenyldiphenylarsine. The above reaction is the first example of facile bis-S-dealkylation.[607] The ligand 2-(diethyl-phosphinomethyl)pyridine (mp) forms planar $[Ni(mp)_2](ClO_4)_2$, octahedral $[Ni(mp)_2(NCS)_2]$ and square-pyramidal $[Ni(mp)_2I]I$. The two complexes $Ni(mp)_2X_2$ (X = Cl or Br) have anomalous magnetic properties. and are interallogones probably containing both high-spin and low-spin five-co-ordinate Ni^{II}. Also isolated were $[Ni(mp)_2][NiX_4]$ (X = Cl or Br) complexes.[194] The structure of $[NiL_2I]I$ [L = (146)] is *cis* square-pyramidal with *cis* N_2P_2 set in the base plane and an apical iodine; the Ni—I bond is unusually long [3.047(3) Å].[608]

The AsN_2 donor terdentates (41) and (42) form high-spin octahedral $[NiL_2](Y)_2$ (L = NNAs, MeNNAs, NNAsEt, or MeNNAsEt, Y = BPh_4, ClO_4, or I; L = NNAs or NNAsEt, Y = NO_3). The 1:1 complexes $NiLX_2$ (X = Cl or Br) are five-co-ordinate high-spin ($\mu_{eff} = 3.1$—3.4 BM) in the solid state and in solution in non-co-ordinating solvents, but in MeOH they rearrange into $[NiL_2]^{2+}$ ions. The iodo-complexes of MeNNAs and MeNNAsEt are similar, but $Ni(NNAs)I_2$, whilst five-co-ordinate in the solid, becomes planar in nitromethane and $Ni(NNAsEt)I_2$ exists solely in the diamagnetic planar form. The thiocyano $NiL(NCS)_2$ complexes all appear to be polymeric octahedral entities, as are $NiL(NO_3)_2, nH_2O$ (L = MeNNAs or MeNNAsEt).[609]

Other Compounds. Phenyl isocyanide (L) forms complexes of stoicheiometry Ni_3L_{11}-$(ClO_4)_6$ and $Ni_4L_4(ClO_4)_8$ with nickel(II) perchlorate.[610] The complex $[Cr(en)_3]$$[Ni(CN)_5],1.5H_2O$ which normally consists of both square-pyramidal and trigonal-bipyramidal anions is converted into exclusively the former under moderate pressure, the effect reversing when the pressure is released.[611] The structures of the two clath-rates $Mn(NH_3)_2Ni(CN)_4,2C_6H_6$ and $Ni(NH_3)_2Ni(CN)_4,2C_{12}H_{10}$ have been reported.

The electrical conductance of $Ni(dpg)_2$ (dpg = diphenylglyoxime anion) becomes much greater on reaction with halogens, forming $Ni(dpg)_2X$ (X = Cl or Br).[614] Similar effects were observed in the $Ni(bdq)_2I_x$ (x = 0.6—0.7) (bdq = 1,2-benzo-quinone dioxime) system.[615] The structure of $Ni_3(BO_3)_2$ has been determined.[616]

607 C. A. McAuliffe, *Inorg. Chem.*, 1973, **12**, 2477.
608 W. Haase, *Z. anorg. Chem.*, 1974, **404**, 273.
609 B. Chiswell, *Inorg. Chim. Acta*, 1973, **7**, 517.
610 T. J. Weaver and C. A. L. Becker, *J. Inorg. Nuclear Chem.*, 1973, **35**, 3739.
611 L. J. Basile, J. R. Ferraro, M. Chosa, and K. Nakamoto, *Inorg. Chem.*, 1974, **13**, 496.
612 R. Kuroda and Y. Sasaki, *Acta Cryst.*, 1974, **B30**, 687.
613 T. Iwamoto, T. Miyoshi, and Y. Sasaki, *Acta Cryst.*, 1974, **B30**, 292.
614 A. E. Underhill, D. M. Watkins, and R. Pethig, *Inorg. Nuclear Chem Letters*, 1973, **9**, 1269.
615 H. Endes, H. J. Keller, M. Mègnamisi-Bélombè, W. Moroni, and D. Nötke, *Inorg. Nuclear Chem. Letters*, 1974, **10**, 467.
616 J. Pardo, M. Martinez-Ripoll, and S. Garcia-Blanco, *Acta Cryst.*, 1974, **B30**, 37.

Nickel(III).—Direct fluorination of nickel(II) chloride or sulphate produced $NiF_{2.09\pm0.01}$, whilst solvolysis of K_3NiF_6 or K_2NiF_6 in liquid HF yielded substances of composition variable from $NiF_{2.3}$ to $NiF_{2.5}$. All were unstable, decomposing rapidly even under dry nitrogen. Reaction of K_2NiF_6 with AsF_5 or BF_3 in anhydrous HF produced black solids, $\mu_{eff} \sim 2.4$ BM, analysing as $NiF_3,2KL$ (L = AsF_6^- or BF_4^-), which decomposed when attempts to wash out the KL with HF were made. The 'nature' of K_3NiF_6 was discussed and it was concluded that spectral and magnetic data are more consistent with a $K_3Ni^{III}F_6$ structure than with the ($K_2Ni^{IV}F_6$, $K_2Ni^{II}F_4$ mixture) previously proposed.[617]

The new compounds Cs_2KNiF_6, Rb_2NaNiF_6, and K_2NaNiF_6, all cubic elpasolithes, and Cs_2NaNiF_6, which is hexagonal with the Cs_2NaCrF_6 structure, are obtained by fluorination of $[Ni(NH_3)_6]Cl_2$ and the appropriate alkali chlorides or carbonates. Variable-temperature magnetic measurement on Cs_2KNiF_6 and K_3NiF_6 indicated the existence of a low-spin $\rightleftharpoons$ high-spin equilibrium.[618] The preparation of $CsCuNi^{III}F_6$, $CsZnNi^{III}F_6$ with the $RbNiCrF_6$ structure has been reported.[259] Reiner *et al.* examined in detail the e.s.r. and ligand-field (*d–d*) spectra of Cs_2KNiF_6, Rb_2KNiF_6, K_3NiF_6, and Na_3NiF_6 in the range 298—4.2 K. All contain Ni^{3+} in a $t_{2g}^6e_g^1$ (low-spin) configuration, which is stabilized with respect to the $t_{2g}^5e_g^2$ (high-spin) state by Jahn–Teller splitting of the ^2E_g state by *ca.* 7000 cm^{-1}. The NiF_6 octahedra are tetragonally elongated, dynamic distortion being present at 298 K. In the hexagonal Cs_2NaNiF_6, half the Ni^{3+} ions are in octahedral sites sharing corners, whilst the others share faces of the octahedra, with high- and low-spin Ni^{3+} ions side by side.[619]

Ferric chloride oxidizes $[Ni(pmm)_2Cl]^+$ [pmm = *o*-$C_6H_4(PMe_2)_2$] to green $Ni(pmm)_2Cl_2]^+$, $\mu_{eff} = 1.90$ BM.[127] Konya and Nakamoto, using the metal-isotope technique, assigned $\nu(Ni—Cl) = 240$ cm^{-1} in $[Ni(das)_2Cl_2]^+$ and $\nu(Ni—Br) = 183.5$ cm^{-1} in $[Ni(das)_2Br_2]^+$.[620]

The one-electron reduction of $Ni^{II}(taab)^{2+}$ produced $Ni^{III}(taab^{2-})^+$ [taab = (20)], this formulation being favoured over other possibilities in view of the e.s.r. spectral evidence, and since the product contains the porphyrin-like (18 π-electron) dianion, which is expected to be very stable.[102] Busch and co-workers[478] have reported the electrochemical oxidation of a large series of Ni^{II} complexes of N_4 macrocycles to either octahedral (from neutral ligands) or planar (from dianionic ligands) Ni^{III} compounds. The variation of the ease of oxidation with ring size of the macrocycle and with the degree of unsaturation of the ligand was discussed. Increase in ring size, ligand unsaturation, and also the presence of axial methyl substituents all increase the $Ni^{II} \xrightarrow{e} Ni^{III}$ potential (see also p. 276).

Nickel(IV).—The production of single crystals of $BaNiO_3$ by reaction of $Ba_2Ni_2O_5$ and oxygen at 600 °C and 2000 bar has been reported. The product is diamagnetic with a hexagonal perovskite structure.[621] Solvolysis of K_2NiF_6 in anhydrous HF has

[617] T. L. Court and M. F. A. Dove, *J.C.S. Dalton*, 1973, 1995.

[618] E. Alter and R. Hoppe, *Z. anorg. Chem.*, 1974, **405**, 167.

[619] D. Reinen, C. Friebel, and V. Propach, *Z. anorg. Chem.*, 1974, **408**, 187.

[620] K. Konya and K. Nakamoto, *Spectrochim. Acta*, 1973, **29B**, 1965.

[621] Y. Takeda, M. Shimada, F. Kanamaru, M. Koizumi, and N. Yamamoto, *Chem. Letters*, 1974, 107.

been noted(p. 303).[617] The structure of K_2NiF_6 has been redetermined by neutron diffraction and the Ni—F distance established more accurately [1.776(8) Å].[622]

Deep purple $[Ni(pmm)_2Cl_2]^{2+}$ is formed by electrochemical or concentrated nitric acid oxidation of the Ni^{III} complex. The electrochemical oxidation is reversible in acetonitrile.[127] The $\nu(Ni—Cl)$ in $[Ni(das)_2Cl_2](ClO_4)_2$ is at 421·5 cm^{-1}, markedly higher than the frequency in the Ni^{III} analogue. The sharp increase in frequency from $Ni^{III} \rightarrow Ni^{IV}$ is interpreted as being due to the removal of an antibonding electron from the d_{z^2} orbital (which lies along the Cl—Ni—Cl axis).[620]

The structure of bis-(2,6-diacetylpyridine dioximato)nickel(IV) shows the co-ordination about the nickel to be distorted octahedral (N_6). The average Ni—N distance is 1.93 Å (*cf.* 2.10 Å in similar Ni^{II} complexes).[623]

ESCA Measurements.—Several recent papers have described photoelectron spectroscopic studies on nickel compounds in a variety of formal oxidation states.[624—627] The care needed to obtain reproducible results has been stressed; oxidation or decomposition of the sample can often cause difficulties, whilst care must be taken in calibration of the energy. The expected trend of increasing energy of the Ni $2p_{\frac{3}{2}}$ binding energies as the formal oxidation state changes, $Ni^0 \rightarrow Ni^{II} \rightarrow Ni^{IV}$, is often observed, but the energies span a range of values within a particular oxidation state and there is sometimes considerable overlap in the ranges *e.g.* between Ni^0 and Ni^{II}.[624] The results on about 70 compounds were summarized[626] by Matienzo *et al.* (i) Binding energy alone is insufficient to determine the oxidation state of the metal ion. At a given oxidation state the metal–ion binding energy is proportional to its positive charge and inversely proportional to covalency. (ii) For the same ligand in different geometries the $2p$ binding energies of the nickel ion increase in the order planar < tetrahedral < octahedral. (iii) Shake-up satellites are found only for paramagnetic compounds. (iv) The separation of shake-up satellites from the main line is proportional to the degree of covalency, essentially following the nephelauxetic series.

The application of ESCA measurements to nickel dithiolenes were originally interpreted in the results as due to Ni^0, although more recent work[627] shows that the values of the Ni $2p_{\frac{3}{2}}$ bonding energies are not inconsistent with Ni^{II} and that the values are essentially constant along the series $[Ni(S_2C_2R_2)_2]^n$ ($n = 0, \rightarrow -1, \rightarrow -2$). This shows that the electrons gained in the reduction reside primarily on the ligand, a result confirmed by the absence of shake-up satellites in the paramagnetic compounds. ESCA studies on some Ni^{II} halide and pseudohalide complexes have also been reported.[56]

3 Copper

Copper(I).—*Halides and Pseudohalides.* The enthalpies of mixing CuCl and CuBr with the corresponding caesium halides have been measured at 810 and 663 °C (CuCl). There are analogies between these systems and the LiX–CsX systems, and the

[622] J. C. Taylor and P. W. Wilson, *J. Inorg. Nuclear Chem.*, 1974, **36**, 1561.
[623] G. Sproul and G. D. Stucky, *Inorg. Chem.*, 1973, **12**, 2898.
[624] C. A. Tolman, W. M. Riggs, W. J. Linn, C. M. King, and R. C. Wendt, *Inorg. Chem.*, 1973, **12**, 2770.
[625] L. O. Pont, A. R. Siedle, M. S. Lazarus, and W. C. Jolly, *Inorg. Chem.*, 1974, **13**, 483.
[626] L. J. Matienzo, L. I. Yin, S. O. Grim, and W. E. Swartz, *Inorg. Chem.*, 1973, **12**, 2762.
[627] S. O. Grim, L. J. Matienzo, and W. E. Swartz, *Inorg. Chem.*, 1974, **13**, 447.

results could be interpreted on the basis of the Reiss, Katz, and Kleppa theory.[628]
The mixed valence compound $[Cu^{II}(bipy)_2][Cu^{I}Cl_2]_2$ was obtained by reaction of
2,2-bipyridyl, $CuCl_2,2H_2O$ and KCl in aqueous ethanolic solution with CuCl and
KCl in aqueous acid. The black crystals consist of isolated linear $CuCl_2^-$ ions, trigonal-
bipyramidal $Cu^{II}(bipy)_2Cl$ groups, and infinite chains of alternating $Cu^{I}Cl_2$ and
$Cu^{I}Cl_4^-$ groups. In the isolated $Cu^{I}Cl_2$ groups the Cu—Cl = 2·091 Å. The chains
consist of linear $Cu^{I}Cl_2$ groups, Cu—Cl = 2.140 Å, and $Cu^{I}Cl_4^-$ tetrahedra, Cu—Cl =
2.393, 2.339, 2.384 Å.[629] [2-(Diphenylphosphino)ethyl]diethylamine reacts with
copper(II) chloride in THF or EtOH to yield either [2-(diphenylphosphinoyl)ethyl]-
diethylammonium dichlorocuprate(I), $[Ph_2P(O)(CH_2)_2NEt_2H]CuCl_2$, or [2-(diphenylphosphino)ethyl]diethylammonium dichlorocuprate(I), $[Ph_2P(CH_2)_2NEt_2H]$-
$CuCl_2$. Both are diamagnetic, consistent with Cu^{I}, and the i.r. spectra show v(Cu—Cl)
at 408 cm^{-1}. The former contains discrete cations and $Cu^{I}Cl_2$ anions, (the P=O
group is not co-ordinated), and the linear [ClCuCl] linkage has Cu—Cl = 2.090
Å.[630] The structure of $[Cu(NH_3)_6]_4Cu_5Cl_{17}$ consists of $Cu_5Cl_{16}^{11-}$ and isolated
Cl^- ions, the former consisting of $CuCl_4$ tetrahedra linked by the fifth Cu atom.[631]

The i.r. and Raman spectra of several iodocuprate(I) compounds have been examined
in an attempt to identify the presence of discrete anions.[632] $CsCu_2I_3$ and $[EPh_3Me]$-
Cu_2I_3 do not appear to contain discrete $Cu_2I_3^-$ ions, and are probably polymeric.
However, the yellow $[EPh_3Me]_2Cu_4I_6$ (E = P or As) obtained from CuI and EPh_3-
MeI in nitromethane, may well consist of discrete $Cu_4I_6^{2-}$ groups with a tetrahedral
Cu_4 arrangement and iodine bridges along the edges.[632]

The acidification of $K_3Cu(CN)_4$ in aqueous solution results in the successive
formation of $Cu(CN)_3^{2-}$, $Cu(CN)_2^-$, and Cu^+ with decreasing pH.[633] The formation
constant (β_2) of $Cu(CN)_2^-$ is log $\beta_2 = 16.26$. Copper(I) selenocyanate can be prepared
by thiosulphate reduction of Cu^{II} in the presence of KCNSe.[634]

N-*Donors*. A series of Cu^{I} amine complexes $CuL_n(ClO_4)$, (L = py, γ-pic, or
quinoline, n = 4; L = α-pic, 2-Etpy, 2,5-lutidine, or 2-Pr^{i}py, n = 3, L = 2,6-lutidine,
2,4-lutidine, 2-methylquinoline, 8-methylquinoline, or 2-benzylpyridine, n = 2) have
been prepared by copper reduction of a mixture of $Cu(ClO_4)_2$ and the appropriate
amine.[635] None of the complexes contain evidence of perchlorate co-ordination, but
several exhibit a small magnetic moment <0.9 BM, probably due to Cu^{II} impurity.
I.r. spectra suggest that the tris complexes are three-co-ordinate, but the bis
complexes may well have a higher co-ordination (than 2) in the solid state, although
the factors promoting higher co-ordination numbers are unclear.[635] Pyrazine,
5-ethylpyrimidine and 2,2'-bipyrimidine form 1:2 complexes, $(CuCl)_2L$, with CuCl,
but 5-ethyl-4-methoxypyrimidine forms only a 1:1 complex.[636] The structure of the
1:1 adduct of copper(I) chloride and 2,3-diazabicyclo[2,2,1]hept-2-ene (147) consists

[628] P. Dantzer and O. J. Kleppa, *Inorg. Chem.*, 1973, **12**, 2699.
[629] J. Kaizer, G. Brauer, F. A. Schröder, I. F. Taylor, and S. E. Rasmussen, *J.C.S. Dalton*, 1974, 1490.
[630] G. Newton, H. D. Caughman, and R. C. Taylor, *J.C.S. Dalton*, 1974, 258.
[631] P. Murray-Rust, *Acta Cryst.*, 1973, **B29**, 2559.
[632] G. A. Bowmaker, L. D. Brockless, C. D. Earp, and R. Whiting, *Austral. J. Chem.*, 1973, **26**, 2593.
[633] C. Kappenstein and R. Hugel, *J. Inorg. Nuclear Chem.*, 1974, **36**, 1821.
[634] E. Söderbach, *Acta Chem. Scand.*, 1974, **A28**, 116.
[635] A. H. Lewin, I. A. Cohen, and R. J. Michl, *J. Inorg. Nuclear Chem.*, 1974, **36**, 1951.
[636] A. Teuerstein, B. A. Feit, and G. Navon, *J. Inorg. Nuclear Chem.*, 1974, **36**, 1055.

L

of a one-dimensional polymer with a linear array of copper atoms bridged by ligand and chloride, the geometry about the copper being distorted tetrahedral.[637]

An X-ray study of [Cu(phen)(thiourea)I], probably the complex previously identified as Cu(phen)(thiourea)$_2$I, has shown the copper to be essentially tetrahedrally co-ordinated with a N$_2$SI donor set.[638]

(147) 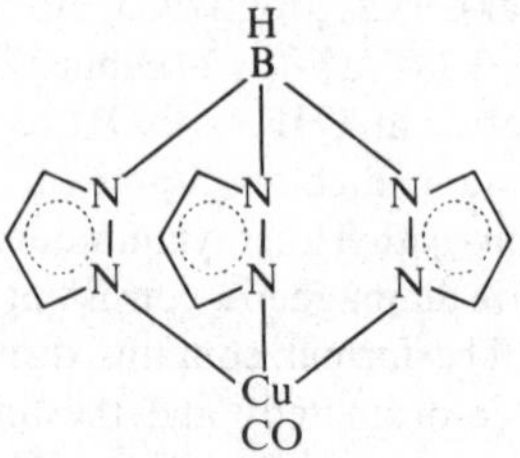   (148)

Copper(I) chloride and potassium hydrotripyrazol-1-ylborate in the presence of carbon monoxide in THF or acetone produced [Cu{HB(pz)$_3$}CO] (148), ν(CO) = 2083 cm^{-1}. On heating it decarbonylates to [Cu$_2${HB(pz)$_3$}$_2$] and then disproportionates to Cu and Cu[HB(pz)$_3$]$_2$. The CO can be displaced by a variety of ligands to give {HB(pz)$_3$CuL} [L = PPh$_3$, PMePh$_2$, P(OMe)$_3$, P(OPh)$_3$, $\frac{1}{2}$Ph$_2$PCH$_2$- CH$_2$PPh$_2$, AsPh$_3$, SbPh$_3$, or Bu^{t}NC).[639]

P-*Donors and* As-*Donors.* The structure of the tetrameric [PPh$_3$CuCl]$_4$ consists of the copper and chlorine atoms at the corners of a slightly distorted cube with Cu $\cdots$ Cu distances of 3.118, 3.337, 3.417, and 3.430 Å, suggesting that no appreciable Cu $\cdots$ Cu interaction is present.[640] The analogue [PPh$_3$CuBr]$_4$,2CHCl$_3$ is not based upon a cube, however, but has a 'step' structure (149), again with no Cu $\cdots$ Cu interaction; Cu $\cdots$ Cu 2.991 and 3.448 Å.[641] Both [PEt$_3$CuI]$_4$ and [AsEt$_3$CuI]$_4$ have cubane-type structures, and whilst the Cu $\cdots$ Cu separations are different, 2.927 and 2.783 Å, respectively, the I $\cdots$ I separations are very similar, 4.380 and 4.442 Å, respectively.[642]

The copper atom in PPh$_3$CuBF$_4$ is trigonally distorted tetrahedrally co-ordinated with the tetrafluoroborate group weakly bonded *via* a terminal F—Cu linkage (2.31 Å). The i.r. and Raman spectra of the complex were recorded, and the vibrations of the co-ordinated BF$_4$ group assigned.[643]

Formation of white Cu(pmm)$_2^+$ [pmm = o-C$_6$H$_4$(PMe$_2$)$_2$] has been mentioned.[127] The structure of (CuCl$_2$(dpm),xC$_2$H$_4$Cl$_2$ (dpm = Ph$_2$PCH$_2$PPh$_2$) consists of tetrametallic (CuCl)$_4$(dpm)$_2$ units and separate dichloroethane molecules. Two of the copper atoms are trigonally co-ordinated by two chlorines and one phosphorus, and

[637] G. S. Chandler, C. L. Raston, G. W. Walker, and A. H. White, *J.C.S. Dalton*, 1974, 1797.
[638] A. K. Jain and A. J. Smith, *Inorg. Nuclear Chem. Letters*, 1974, **10**, 707.
[639] M. I. Bruce and A. P. P. Ostazewski, *J.C.S. Dalton*, 1973, 2433.
[640] M. R. Churchill and K. L. Kalra, *Inorg. Chem.*, 1974, **13**, 1065.
[641] M. R. Churchill and K. L. Kalra, *Inorg. Chem.*, 1974, **13**, 1427.
[642] M. R. Churchill and K. L. Kalra, *Inorg. Chem.*, 1974, **13**, 1899.
[643] A. P. Gaughan, Z. Dori, and J. A. Ibers, *Inorg. Chem.*, 1974, **13**, 1657.

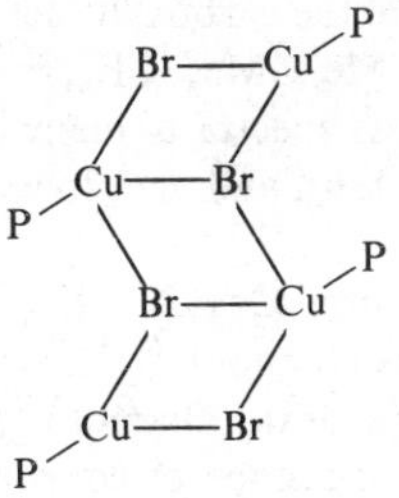

(149) 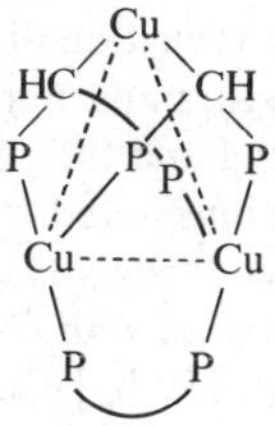   (150)

two distorted tetrahedrally co-ordinated to one P and three Cl, the overall geometry being of the step rather than the cubane type.[644]

Copper diphenylphosphinite complexes, $\{Cu[P(OR)Ph_2]_4\}BPh_4$ (R = Me or Et), were prepared from $P(OMe)Ph_2$ and CuCl under reflux in methanol, followed by addition of $NaBPh_4$.[645]

The reaction of aryl copper compounds with dpm produced a yellow diamagnetic solid formulated as $Cu(dpm), \frac{2}{3}PhMe$, which readily forms chloro-complexes on reaction with halogenated organic solvents. X-Ray examination showed the structure (150) to consist of an isosceles arrangement of copper atoms, the basal pair being bridged by three 'dpm' molecules, and the apical copper interacting with the methylene backbone of two of the ligands.[646] Two possible formulations are $[Cu^I(Ph_2P\dot{-}\dot{-}CH\dot{-}\dot{-}PPh_2)]_3$ and $[Cu^{II}(Ph_2P\dot{-}\dot{-}CH\dot{-}\dot{-}PPh_2)_2.Cu_2^{\frac{1}{2}}(Ph_2P\dot{-}\dot{-}CH\dot{-}\dot{-}PPh_2)]$, the latter being favoured by the non-equivalence of the phosphine ligands.[646]

Copper(II) or copper(I) salts in ethanolic solution react with $NaBH_4$ or $NaBH_3CN$ in the presence of the appropriate phosphine to form $CuBH_4(PR_3)_2$ [PR_3 = PPh_3 or 5-phenyl-5H-dibenzophosphole (dpb)] and $Cu(BH_3CN)(PR_3)_3$ (PR_3 = PPh_3, dpb, or $\frac{1}{2}Ph_2PCH_2CH_2PPh_2$). The $Cu(BH_3CN)(PPh_3)_3$ complex is different from the complex of this empirical formula obtained previously, and thought to contain N-bonded BH_3CN. The complex produced in the more recent study may well contain $Cu-HBH_2CN$ linkages.[51]

Copper(II) salts react with arsine, AsH_3, to produce products identified as $Cu_{5-x}As_2$, $Cu_{3-x}As_2$, and Cu_3As.[647]

O-Donors and S-Donors. Copper(I) acetate consists of infinite chains of dinuclear dimeric $Cu_2(OAc)_2$ units with Cu and O bridges, each copper being bonded to three oxygens and a second copper atom in a distorted planar arrangement. The Cu—Cu distance of 2.556(2) Å is significantly shorter than any reported copper(II) carboxylates.[648] New syntheses for a range of copper(I) carboxylates have been devised[649] using the reduction of the copper(II) compounds in non-aqueous solvents by copper,

[644] G. Nardin and L. Randaccio, *Acta Cryst.*, 1974, **B30**, 1377.
[645] D. A. Couch and S. D. Robinson, *Inorg. Chem.*, 1974, **13**, 456.
[646] A. Camus, N. Marisch, G. Nardin, and L. Randaccio, *J. Organometallic Chem.*, 1973, **60**, C39.
[647] P. Priest, *Mater. Res. Bull.*, 1974, **9**, 337.
[648] R. D. Mounts, T. Ogura, and Q. Fernando, *Inorg. Chem.*, 1974, **13**, 802.
[649] D. A. Edwards and R. Richards, *J.C.S. Dalton*, 1973, 2463.

hydrazine hydrate *etc.*, and from copper(I) oxide and the carboxylic acid in xylene or toluene. The copper reduction of $Cu(O_2CR)_2$ (R = Me, CMe_3, CF_3, or CH=CHPh) was also used by Ogura and Fernando.[650] The mass spectra of many of these complexes were examined and the presence of dimeric ions, and very low abundance of tri- and tetra-meric ions noted.[649, 650]

The tetrathioether ligands (111; $n = 2$, $n = 3$) form Cu_2LX_2 (X = Cl, Br, or I) complexes upon reaction with CuX–HX in aqueous ethanol.[168]

The structure of $[PPh_4]_4Cu_8L_6$ (L = dithiosquarato dianion), formed from $(Ph_4P)_2(C_4S_2O_2)$ and $Cu(NCMe)_4ClO_4$, consists of a cube of copper atoms, with ligands bridging the six faces of the cube, each copper being co-ordinated to three sulphurs (of different ligands) and each sulphur to two copper atoms. The structure is very similar to that of $Cu_8(i\text{-mnt})_6^{2-}$ (i-mnt = 1,1-dicyanoethylenedithiolate).[651] ^{63}Cu and ^{65}Cu n.q.r. spectral data on some dithiocarbamate complexes of Cu^I have been reported.[652]

The complexes of thiomorpholin-3-one, (CuLI) and $[Cu_2L_4Cl_2]$ are diamagnetic compounds probably containing Cu–S co-ordination.[602] Chlorobis(2-thiouracil)-copper(I),DMF, formed from copper(II) chloride and thiouracil, contains almost trigonal-planar Cu^I with S_2Cl co-ordination.[653]

Diaryl tellurides react with copper(I) halides (CuX) in ethanol–HX or KX solution to form the following complexes, irrespective of the ratio of the reactants: CuL_nX (L = Ph_2Te, $n = 3$, X = Cl; L = Ph_2Te, $n = 2$, X = Br; L = Ph_2Te or (p-EtOC$_6$-H$_4$)$_2$Te, $n = 1$, X = I; L = (p-tolyl)$_2$Te, $n = 2$, X = Cl, Br, or I; L = (p-EtOC$_6$H$_4$)$_2$-Te, $n = 2$, X = Cl or Br). The available evidence suggests that $(Ph_2Te)_3CuCl$ may be tetrahedral, the $(R_2Te)_2CuX$ complexes halogen-bridged dimers, and $\{(p\text{-tolyl})_2Te\}_2$-CuI three-co-ordinate.[654]

Mixed Donors and Other Components. An *O*-ethylboranocarbonate ($H_3BCO_2Et^-$) complex of Cu^I, $(PPh_3)_2Cu(H_3BCO_2Et)$ is formed from KH_3BCO_2Et and $(PPh_3)_3$-CuCl in $CHCl_3$–EtOH. The i.r. and ^{11}B n.m.r. spectra suggest the bonding is *via* the terminal B—H linkages of the anion (151).[655]

The structure of $(PPh_3)_4Cu_2(BH_3CN)_2$ contains H_3BCN^- ligands bridging two copper atoms to give a ten-membered nonplanar ring with a Cu···Cu separation of 5.63(2) Å. Each copper atom is tetrahedrally co-ordinated to two PPh_3 groups, one H and one N from different H_3BCN groups.[656]

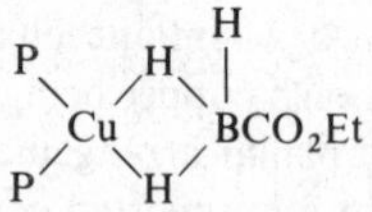

(151) 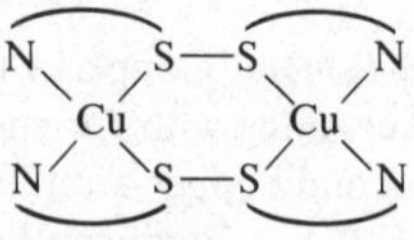   (152)

[650] T. Ogura and Q. Fernando, *Inorg. Chem.*, 1973, **12**, 2611.
[651] F. J. Hollander and D. Coucouvanis, *J. Amer. Chem. Soc.*, 1974, **96**, 5646.
[652] T. J. Barstow and H. J. Whitfield, *J. Inorg. Nuclear Chem.*, 1974, **36**, 97
[653] G. W. Hunt and E. L. Amma, *J.C.S. Chem. Comm.*, 1973, 869.
[654] W. R. McWhinnie and V. Rattanaphani, *Inorg. Chim. Acta*, 1974, **9**, 153.
[655] J. C. Bommer and K. W. Morse, *J. Amer. Chem. Soc.*, 1974, **96**, 6222.
[656] K. M. Melmed, T. Li, J. J. Mayerle, and S. J. Lippard, *J. Amer. Chem. Soc.*, 1974, **96**, 69.

The reaction of CuI with KNH_2 or $CsNH_2$ in liquid ammonia in the presence of hydrogen produced thermally unstable hydrides $K_2Cu_3H_5,xNH_3$ and $CsCuH_2$, xNH_3.[657] Potassium tri-butylborohydride (KBu^s_3BH) reduced CuI to a hydrido species, probably $KCuH_2$, which is a powerful reductant towards organic halogen compounds.[658]

A novel Cu_4Ir_2 cluster is present in $Cu_4Ir_2(PPh_3)_2(C{\equiv}CPh)_8$ formed from *trans*-$Ir(CO)(PPh_3)_2Cl$ and $(PhC{\equiv}CCu)_n$. The metal atoms are in a slightly distorted octahedral arrangement with mutually *trans* Ir atoms ($Ir-Cu_{av} = 2.870$ Å, $Cu-Cu_{av} = 2.739$ Å). The phosphines are bonded to the Ir atoms, which are also each σ-co-ordinated to $C{\equiv}CPh$ groups. The $C{\equiv}CPh$ groups undergo π-interaction with the Cu atoms, two acetylenic groups bonding to each copper. The formal oxidation states of the metals were deduced to be Cu^0 and Ir^{IV}.[659]

Copper(I) monothiocarbonates, $Cu^I(SOCOR)$, (R = Me, Et, or Bu^n), are formed by reaction of the potassium monothiocarbonates with copper(II) salts.[660]

The structure of cyclo-di-μ-{bis-[2-(NN-dimethylamino)ethyl]disulphide}dicopper(I) tetrafluoroborate, $Cu_2(Me_2NCH_2CH_2SSCH_2CH_2NMe_2)(BF_4)_2$, contains a central CuSSCuSS ring (152), the Cu^I being in a very badly distorted tetrahedral environment. The complex is prepared by reaction of the ligand with either Cu^I or Cu^{II} ions; in the latter case, the ligand reduces the metal to the required oxidation state.[661]

Copper(II).—*Halides and Pseudohalides.* Formation of complexes between Cu^{II} and F^- ions in aqueous solution has been examined by use of a fluoride-specific aqueous electrode, and the production of both CuF^+ and CuF_2 demonstrated.[662] A range of $M^ICu^{II}M^{III}F_6$ complexes (M^I = K, Rb, or Cs; M^{III} = Sc, Al, Ga, In, Tl, Fe, Co, Mn, or Rh), some belonging to the $RbNiCrF_6$ type and others non-cubic, has been synthesized; many of them show complicated magnetic behaviour.[259, 663] The tetragonally distorted M_2CuF_4 (M = Na, K, or Rb) and Ba_2CuF_2 have been studied by e.s.r. and electronic spectroscopy, and the Jahn–Teller distortion compared with that found in other cupric systems.[664]

The species present in aqueous hydrochloric acid solutions of copper(II) depends upon the copper concentration.[665]

A considerable number of chlorocuprates(II) have been examined by spectroscopic and magnetic techniques. The dipositive dabconium ion LH_2 forms LH_2CuX_4 (X = Cl or Br).[59] Octahedral $CuCl_6$ arrangements are contained in $CsCuCl_3$: they are slightly tetragonally distorted and not isostructural with $CsNiCl_3$ or $CsMgCl_3$.[666] Several tetrachlorocuprates(II) have been prepared and X-rayed and the structures of the dianions are found to be markedly dependent upon the cation present. In [thiamineH_2]$CuCl_4$ the tetrahedral anion is flattened with four of the ∠ ClCuCl

[657] K. A. Strom and W. L. Jolly, *J. Inorg. Nuclear Chem.*, 1973, **35**, 3445.
[658] T. Yoshida and E. Negishi, *J.C.S. Chem. Comm.*, 1974, 762.
[659] M. R. Churchill and S. Bezman, *Inorg. Chem.*, 1974, **13**, 1418.
[660] M. A. Khivaja and R. J. Magee, *Inorg. Nuclear Chem. Letters*, 1974, **10**, 87.
[661] T. Ottersen, L. G. Warner, and K. Seff, *J.C.S. Chem. Comm.*, 1973, 876: *Inorg. Chem.*, 1974, **13**, 1904.
[662] S. Gifford, W. Cherry, J. Jecmen, and M. Readnour, *Inorg. Chem.*, 1974, **13**, 1434.
[663] R. Hoppe and R. Jesse, *Z. anorg. Chem.*, 1973, **402**, 29.
[664] C. Friebel and D. Reinen, *Z. anorg. Chem.*, 1974, **407**, 193.
[665] J. L. Tyvoll and D. L. Wertz, *J. Inorg. Nuclear Chem.*, 1974, **36**, 1319.
[666] C. J. Kroese, W. J. A. Maaskant, and G. C. Verschoor, *Acta Cryst.*, 1974, **B30**, 1053.

$\sim 99°$ and one each of 136 and 131°.[667] In the complex with the 1,5-dihydrodeca-methylpentaphosphonitrilium cation, $[(NPMe_2)_5H_2]CuCl_4,H_2O$, the anion has a similar distorted tetrahedral structure.[668] However, in the green low-temperature form of $[Pr^iNH_3]_2CuCl_4$ the discrete $CuCl_4^{2-}$ ions are packed together to form infinite ribbons, one-third of the groups being planar and the other two-thirds planar with a slight distortion towards tetrahedral.[669] Planar anions are also present in the anilinium salt $[PhNH_3]_2CuCl_4$, with more distant association with two other chlorines to give the familiar $4 + 2$ co-ordination.[670] A particularly interesting case is that of $[PhCH_2CH_2NH(CH_3)H]_2CuCl_4$, which exists as a green form at 25 °C with essentially planar anions, and as a yellow form at ~ 90 °C with distorted flattened tetrahedral anions.[671] The $Cu_2Cl_6^{2-}$ ion, as its Ph_4P^+ [672] or Ph_4As^+ [673] salts, is non-planar, the co-ordination of the copper being halfway between tetrahedral and planar; the difference between this structure and the planar $Cu_2Cl_6^{2-}$ type found with smaller cations may be due to the lack of interaction between the anions, in the former as a result of the larger cations. Aqueous solutions of $Cu(NO_3)_2,3H_2O$ and KNCSe produce either $Cu(SeCN)_2$ or $KCu(SeCN)_3$ depending upon the conditions.[634]

N-*Donors.* $NNN'N'$-Tetramethylethylenediamine and a range of β-diketones (*e.g.* acacH, hfacH, dipivaloylmethane) form $[Cu(tetren)(\beta\text{-diket})]X,nH_2O$ (X = ClO_4 or NO_3), which have been examined spectroscopically to establish the effect of the various substituents upon the observed spectra.[674]

A series of five-co-ordinate $[CuLL'](NO_3)_2$ [L = 2,2'-bipyridyl or 2-aminoethyl-pyridine; L' = $H_2N(CH_2)_nNR(CH_2)_mNH_2$, R = H or Me, n, m = 2 or 3] which are 2:1 electrolytes with magnetic moments ~ 1.83—1.86 BM have been prepared.[675] A particularly interesting complex is $[Cu(H_3BCN)_2(Me_5dien)]$ (Me_5dien = 1,1,4,7,7-pentamethyldiethylenetriamine) obtained from $CuBr_2$, Me_5dien, and $NaBH_3CN$ in ethanol. The copper is five-co-ordinate distorted square-pyramidal with one apical (Cu—N = 2.153 Å) and one basal (Cu—N = 1.980 Å) $NCBH_3$ group, the large variation in the Cu—N distances explaining why two ν(CN) vibrations are observed.[676] The $[Cu_2(tren)_2(CN)_2](BPh_4)_2$ complex contains two trigonal-bipyramidally co-ordinated copper atoms with axial CN groups; the dimeric cation being held together in a most unusual way by H-bonding between the CN group of one Cu atom and the HN of the second copper. The magnetic interaction observed is notable in view of the nature of the bridge.[676]

N-Benzylethylenediamine forms distorted octahedral CuL_2X_2 (X = Cl, Br, I, NO_3, or ClO_4), the extent of distortion depending upon the anion; $[CuLX_2]$ (X = Cl or Br) are also distorted polymeric octahedral complexes.[677] The copper(II) nitrite and nitrate complexes of 1,2-dimorpholinoethane (edm) and 1,2-dipiperidinoethane,

[667] M. R. Cania, G. V. Fazakerley, P. W. Linder, and L. R. Nassimbeni, *Acta Cryst.*, 1974, **B30**, 1660.
[668] M. P. Calhoun and J. Trotter, *J.C.S. Dalton*, 1974, 382.
[669] D. N. Anderson and R. D. Willet, *Inorg. Chim. Acta*, 1974, **8**, 167.
[670] K. P. Larsen, *Acta Chem. Scand.*, 1974, **A28**, 194.
[671] R. L. Harlow, W. J. Wells, G. W. Watt, and S. H. Simonsen, *Inorg. Chem.*, 1974, **13**, 2106.
[672] M. Textor, E. Dubler, and H. R. Oswald, *Inorg. Chem.*, 1974, **13**, 1361.
[673] R. D. Willett and C. Chow, *Acta Cryst.*, 1974, **B30**, 207.
[674] Y. Fukuda, A. Shimura, M. Mukaida, E. Fujita, and K. Sone, *J. Inorg. Nuclear Chem.*, 1974, **36**, 1265.
[675] H. Ojima and K. Nonayama, *Z. anorg. Chem.*, 1973, **401**, 207.
[676] D. M. Duggan and D. N. Hendrickson, *Inorg. Chem.*, 1974, **13**, 1911.
[677] K. C. Patel and D. E. Goldberg, *J. Inorg. Nuclear Chem.*, 1973, **35**, 4041.

[CuLX$_2$], are all six-co-ordinate with bidentate anions, and this is confirmed by an X-ray study of Cu(edm)(NO$_2$)$_2$ which clearly shows 4 + 2 (N$_2$O$_2$ + O$_2$) co-ordination[117] is present.

The copper(II) chloride complex of 2-(2'-quinolyl)methylene-3-quinuclidinone has i.r. and electronic spectra consistent with the presence of a very distorted tetrahedral environment about the copper.[120] Copper(II) complexes of the 2,3-dipyridylquinoxaline (10),[89] 5-aminopyrazoles,[98] pyrazine and various substituted pyrimidines,[636] and several substituted 3-methylpyrazoles[678] have been reported. Two different forms of Cu(py)$_2$(CNS)$_2$ can be obtained depending upon the method of preparation.[679]

Although simple copper(II) cyanides cannot be prepared, stable complexes can be obtained with *o*-phenanthroline: [Cu(phen)$_2$CN]Y,nH$_2$O {Y = Cl, Br, I, NO$_3$, ClO$_4$, or [Cu(phen)CN$_2$]$^-$}. The i.r. and electronic spectra of these compounds are consistent with trigonal-bipyramidal copper with an equatorial CN group. A square-planar [Cu(phen)(CN)$_2$] was also prepared.[680]

Nitromethane solutions of [Cu^I(Rphen)$_2$](ClO$_4$) (Rphen = *o*-phen, 5-methyl-*o*-phen, 5,6-dimethyl-*o*-phen, or 5-chloro-*o*-phen) absorb oxygen to form copper(II) complexes containing the nitromethanide ion, CH$_2$NO$_2^-$, which can also be obtained directly using sodium nitromethane. The products [Cu^{II}(Rphen)$_2$CH$_2$NO$_2$](ClO$_4$) seem to be six-co-ordinate with the nitromethanide ion functioning as a bidentate O$_2$ donor.[681] The complex Cu(dpkc)Cl$_2$ (dpkc = di-2-pyridylketoxime) probably contains square-planar copper, with the ligand bonded as a bidentate N$_2$ donor.[86] The [CuL$_2$X$_2$]$_n$ (X = Cl or Br; L = pyridine-4-carbaldehyde oxime) probably have an analogous halide-bridged structure to the nickel and cobalt complexes.[87]

Pyridine carboxylic acids, 6-methylpyridine-2-carboxylic acid (6MPH), pyridine-2-carboxylic (P2CH), pyridine-3-carboxylic (P3CH), form Cu(6MP)(6MPH)X (X = Cl or Br), Cu(P2C)(P2CH)X, and Cu(P3CH)$_2$X$_2$, which are probably five-co-ordinate.[83] Pyridine-2,6-dicarboxylic acid (DPCH$_2$) forms Cu(DPC),2H$_2$O, which forms a range of nitrogenous base adducts, Cu(DPC),L,xH$_2$O (L = terpy, bipy, py, or α-pic), whilst complexes containing the monoanionic form Cu(DPCH)(terpy)X (X = ClO$_4$ or PF$_6$) can be obtained from Cu(terpy)(OAc)X and DPCH$_2$.[682]

Macrocycles. Tetramethylcyclam (1,4,8,11-tetramethyl-1,4,8,11-tetrazacyclotetradecane) forms four-co-ordinate CuL(ClO$_4$)$_2$ and five-co-ordinate [CuLBr]ClO$_4$ upon reaction with the appropriate copper salts.[472] The ligands H$_2$[HPhH(en)$_2$], H$_2$[HPhHen.tn], and H$_2$[HPhH(tn)$_2$] (Scheme 1, p. 230) form planar CuL complexes.[101] The [Cu^{II}(taab)]$^{2+}$ [taab = (20)] in acetonitrile undergoes two one-electron reduction steps to form formally Cu^I and Cu^0 complexes.[102] However, by similar arguments to those used in the cobalt case (*q.v.*) Busch and co-workers[102] have proposed that the ultimate reduction is better formulated as a Cu^{III} entity [Cu^{III}-(taab^{2-})]$^+$ containing the annulene-like taab^{2-}. A number of other copper(II) complexes of nitrogen macrocycles have been prepared.[468,471,488,683]

[678] Ya. A. Shuster, V. A. Kozlova, L. E. Semikinn, B. P. Zhantalai, N. Y. Rostchakovskaya, and V. I. Seralya, *Zhur. obschei Khim.*, 1974, **44**, 1379.

[679] L. Macaskova, M. Kabesova, J. Garaj, and J. Gazo, *Monatsh.*, 1973, **104**, 1473.

[680] M. Wicholas and T. Wolford, *Inorg. Chem.*, 1974, **13**, 316.

[681] J. Zagal, E. Spodine, and W. Zamudio, *J.C.S. Dalton*, 1974, 85.

[682] D. P. Murtha and R. A. Walton, *Inorg. Chim. Acta*, 1974, **8**, 279.

[683] E. P. Fleischer, *Inorg. Nuclear Chem. Letters*, 1973, **9**, 1061.

O-Donors and S-Donors. The spinel $CuAl_2O_4$ is formed by heating the constituent oxides together;[684] the effect of varying the $CuO-Al_2O_3$ ratio was examined. $CuCO_3$ has been prepared by reaction of CuO and CO_2 under high pressure.[685]

1,3-Dioxan forms $Cu(1,3-D)X_2$ (X = Cl or Br) which are polymeric pseudo-octahedral complexes, with terminal dioxan ligands. On standing at ambient temperatures, they lose dioxan to form $(CuX_2)_3,2(1,3-D)$.[132] Copper(II) nitrate reacts with heterocyclic *N*-oxides in a 1:2 ratio to form $[Cu(2-picO)_2(ONO_2)_2]$ and $[CuL-(O_2NO)_2]$ (L = 3-picO or 4-picO), and in a 1:3 ratio to produce $[Cu(2-picO)_4]-(NO_3)_2$ and $[Cu(4-picO)_4ONO_2]NO_3$; again the steric hindrance of the 2-methyl group in 2-picO promotes a lower co-ordination number than is found in the 3-picO or 4-picO complexes.[146] $Cu(NO_3)_2,4(3-picO)$ (μ_{eff} = 1.95 BM) is of less certain structure; the nitrate groups are clearly ionic, but the cation may be either distorted four-co-ordinate or dinuclear *N*-oxide-bridged, $[(3-picO)_3Cu(3-picO)_2Cu(3-picO)_3]^{4+}$. The copper(II) sulphate complex of 2,6-lutidine *N*-oxide, $[Cu(LNO)_2SO_4]_2$, is most probably sulphate-bridged.[144] The pyridine *N*-oxide complex $Cu(pyO)_2(CNS)_2$ is six-co-ordinated thiocyanate-bridged.[143] $[Cu(phenNO)_2](ClO_4)_2$ (phenNO = 1,10-phenanthroline mono-*N*-oxide) appears to contain four-co-ordinate copper,[145] but pyrazine mono-*N*-oxide forms six-co-ordinate $[CuL_2Cl_2]_n$, which probably contains both N-bonded terminal and NO-bonded bridging pyrazine *N*-oxide ligands and terminal and bridging Cl.[147] The complexes of dimethylmethylphosphonate, trimethylphosphate, and di-isopropylmethylphosphonate have the same stoicheiometry in the nickel analogues, *viz.* $[Cu(DMP)_4]_n(ClO_4)_{2n}$, $[Cu(tmp)_5](ClO_4)_2$ and $[Cu(dimp)_4OClO_3]ClO_4$, respectively.[149] Copper bisoctylphosphinate, $[Cu(OP-(C_8H_{17})_2O)_2]_n$, is trimeric in CCl_4 solution, but differences in the i.r. spectra suggest a different geometry occurs in the solid. The diarsine dioxide ligands, $daeO_2$ and $damO_2$, produce $Cu(daeO_2)_2(ClO_4)_2$ and $Cu(damO_2)_3(ClO_4)_2$, respectively.[150] Trichloro-{[2-(diphenylphosphinyl)ethyl]dimethylammonium}copper(II) is formed from [(2-diphenylphosphinyl)ethyl]dimethylamine and $CuCl_2$ in ethanolic hydrochloric acid solution. The *X*-ray structure reveals the presence of the zwitterionic ligand which is bonded to the copper *via* the phosphinyl oxygen, three chloride ligands completing a very distorted tetrahedron.

A series of hydroxyquinoid compounds Cu(L − 2H) with polymeric structures, *e.g.* (153) has been obtained.[687] 3-Formylsalicylic acid $[LH_2]$ forms dinuclear $[CuL]_2$, which is oxygen-bridged, and has a reduced magnetic moment of 0.32 BM.[523]

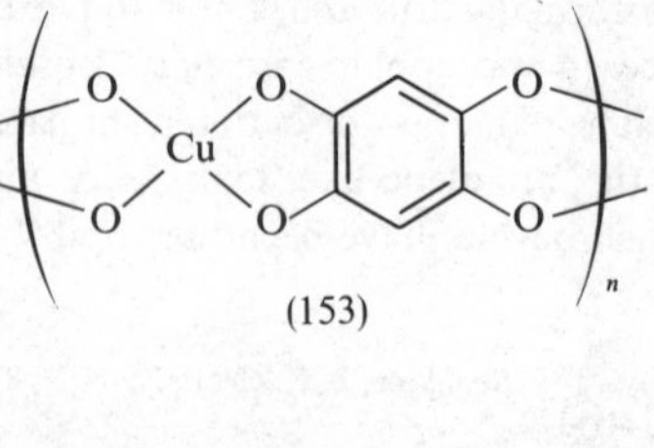

(153)

[684] H. Paulson and E. Rosén, *Z. anorg. Chem.*, 1973, **401**, 172.
[685] H. Ehrhardt, W. Johannes, and H. Seidel, *Z. Naturforsch.*, 1973, **B28**, 682.
[686] M. G. Newton, H. D. Caughman, and R. C. Taylor, *J.C.S. Dalton*, 1974, 1031.
[687] H. D. Cable and H. F. Holtzclaw, *J. Inorg. Nuclear Chem.*, 1974, **36**, 1049.

The benzeneseleninic acids (as their Na salts), $NaRSeO_2$ (R = Ph, *m*- or *p*-ClC_6H_4, *m*- or *p*-BrC_6H_4, *p*-MeC_6H_4, or *p*-$NO_2C_6H_4$) form pseudotetrahedral $Cu(RSeO_2)_2$ which have i.r. spectra suggesting OO-bonded anions.[688] The sulphinate complex $Cu(PPh_3)_2O_2SPh$ is oxidized by air, in THF or $CHCl_3$ solution, to the corresponding sulphonate, $Cu(PPh_3)_2O_3SPh$.[689]

α-Nitro-ketones form $[CuL_2]$, $[CuL_2Q_2]$, and $[CuL_2L']$ (L = nitroacetone, 3,3-dimethyl-1-nitrobutan-2-one, or 3-nitrocamphor; Q = water or acetone; L' = bipy or phen) complexes.[690] $[CuL_2]$ are essentially planar and $[CuL_2Q_2]$ tetragonal octahedral with axial Q. Spectroscopic examination of the $[CuL_2L']$ complexes shows the N_2 ligand to be co-ordinated in the equatorial plane, with two O of the ketone equatorial, and two further O's in axial positions.[690, 691] $[Cu(benzoylacac)_2-(quin)]$ is five-co-ordinate ($\mu_{eff} = 1.85$ BM).[692]

The linear quadridentate S_4 donors (33) form[168] dimeric, probably tetrahedral, $Cu_2(S_4)X_4$ (X = Cl or Br) complexes.

Unlike thiazolidine-2-thione and its simpler derivatives, the 3-alkyl-5-hydroxy-5-(1,2,3,4-tetrahydroxy-*n*-butyl)thiazolidine-2-thiones (which are specific spectrophotometric reagents for Cu^{II}) undergo rearrangement in the presence of the Cu^{II} ions as shown to form complexes of dithiocarbamates.[693] The analytical specificity is explained by the inability of most other metal ions to effect this rearrangement.

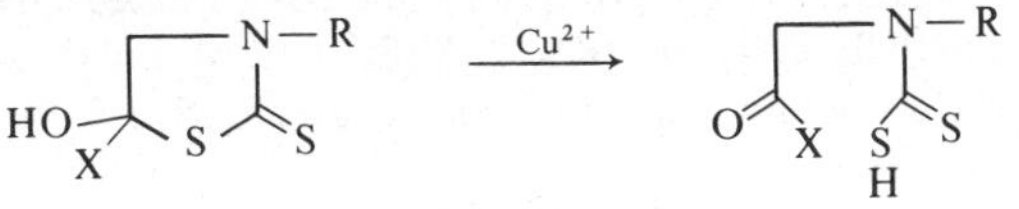

Thiomorpholin-3-one forms $Cu(tm)_2X_2$ (X = Cl, Br, or I), $Cu_2(tm)_2(SO_4)_2$ and $[Cu(tm)_4](NO_3)_2$ which are probably S-bonded in contrast to the O-bonding found in the nickel(II) complexes.[602] Diethanol dithiocarbamic acid forms $Cu[(HOC_2H_4)_2-NCS_2]_2$ which is a monomeric planar CuS_4 chromophore in solution, but $Cu\cdots S$ interaction probably occurs between neighbouring molecules in the solid state.[603]

Mixed Donors. Ethanolamine and propan-2-olamine function[555] as neutral ON-donors in $[CuL_3]SO_4$, but propan-1-olamine (HL) behaves[694] as either a neutral ON-ligand or as a uninegative ligand by deprotonation of the alcohol group in complexes such as $[CuLX]_2$ (X = Cl, Br, SCN, OH, or OAc), $[Cu(HL)L]Y$ (Y = NO_3 or OAc), $[Cu(HL)LX]$ (X = Cl or Br), or $[Cu(HL)X_2]$. Tripropan-2-olamine[173] behaves as a terdentate NO_2 donor in $[CuL_2]SO_4$. The ligands $H_2N(CH_2)_m$CHRNH-$(CH_2)_n$OH (LH) (m = 1 or 2, n = 2 or 3; R = H or Me) form $[Cu(LH)Cl_2]$ and Cu-$(LH)_2X_2$ which contain LH as a bidentate N_2 donor, and $CuL(ClO_4)$ which contain

[688] C. Preti, G. Tosi, D. De Filippo, and G. Verani, *Inorg. Nuclear Chem. Letters*, 1974, **10**, 541.
[689] J. Bailey and M. J. Mays, *J. Organometallic Chem.*, 1973, **63**, C24.
[690] D. Altanasio, I. Collamati, and C. Ercolani, *J.C.S. Dalton*, 1974, 1319.
[691] D. Altanasio, I. Collamati, C. Ercolani, and G. Rotillo, *J.C.S. Dalton*, 1974, 2242.
[692] B. K. Mohapatra, *J. Inorg. Nuclear Chem.*, 1973, **35**, 3930.
[693] W. D. Basson and A. L. du Preez, *J.C.S. Dalton*, 1974, 1708.
[694] M. Brezeanu and I. Jilani, *Rev. Roumaine Chim.*, 1973, **18**, 2039.

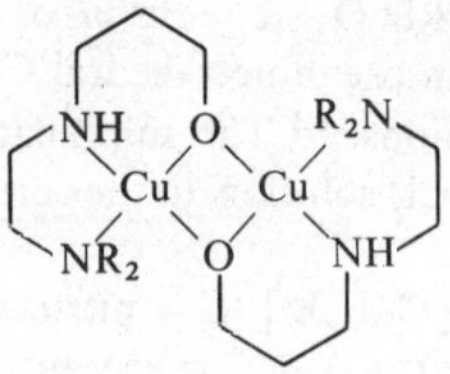

(154)   R = H, Me, or Et 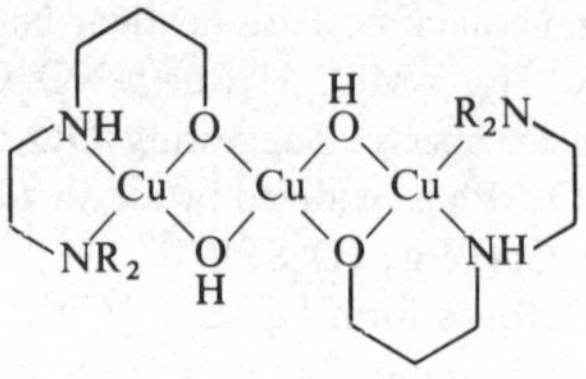   (155)   R = Me or Et

dinuclear units with alkoxide bridges (154);[695] when L = $H_2N(CH_2)_3NH(CH_2)_2O^-$, the complex is tetrameric.

Using $R_2N(CH_2)_2NH(CH_2)_3OH$ with $Cu(ClO_4)_2$ and NaOH in aqueous solution, two types of complex can be isolated, green crystals of type (154), and blue crystals containing trinuclear complexes (155).[696]

The amine alcohols derived from ethylenediamine, $HO(CH_2)_2NH(CH_2)_2NH$-$(CH_2)_2OH$ (L'), $HO(CH_2)_2NH(CH_2)_2NH_2$ (L''), and $HO(CH_2)_3NH(CH_2)_2NH_2$ (L'''), form distorted octahedral CuL_2Y_2 (L = L', L'', or L'''; Y = ClO_4 or $\frac{1}{2}SO_4$) which contain a planar CuN_4 group, with axial O_2 co-ordination from the alcohol groups. In the CuL_2Cl_2 (L = L'' or L'''), the axial ligands are chlorine, whilst CuL''-$(NO_3)_2$ contains one axial ONO_2 group and one co-ordinated OH group.[697] The amine oxides $R_2N(O)CH_2CH_2OH$ (R = Me, Et, Pr^n, or Bu^n) form Cu(L − H)X (X = Cl or Br) which contain the deprotonated uninegative $R_2N(O)CH_2CH_2O^-$, possibly dimeric species with alcoxide bridges.[698] The ligand 2,6-bis-(N-2'-amino-ethylaminomethyl-p-cresol forms dinuclear planar Cu_2LXY compounds (156), five-co-ordinate $Cu_2L[X_mY_n]$ [$m + n = 3$; XY = $(CNS)_2ClO_4$, $(CNS)_3$, Br_2ClO_4, or Cl_2ClO_4] and a mononuclear $CuL(ClO_4)$ which may be distorted trigonal-bipyramidal.[699]

The aroylhydrazines $H_2NNHC(O)R$ (R = Ph, p-MeOC$_6$H$_4$, or p-NO$_2$C$_6$H$_4$) form $CuL_2(Y)_2$ (Y = NO_3 or $\frac{1}{2}SO_4$), and both mono- and bis-chelate complexes with

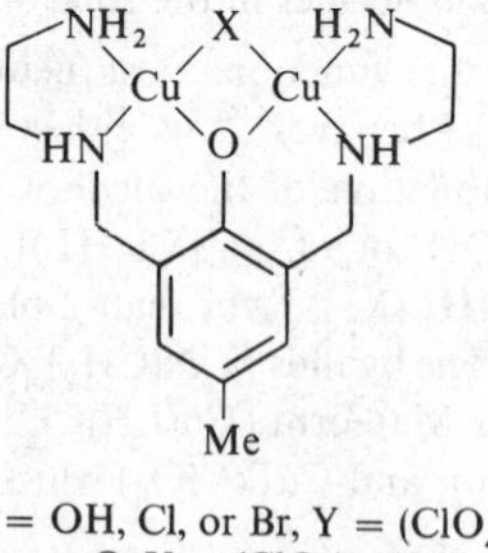

(156)   X = OH, Cl, or Br, Y = $(ClO_4)_2$
        X = pyO, Y = $(ClO_4)_3$, H_2O

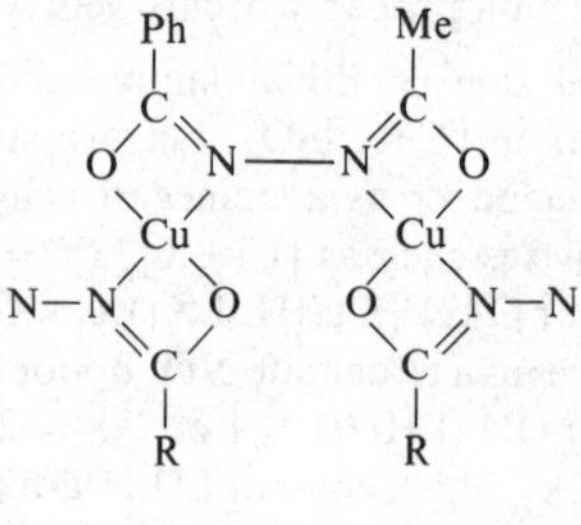

(157)

[695] Y. Ishimura, Y. Nonaka, Y. Nishida, and S. Kida, *Bull. Chem. Soc. Japan*, 1973, **46**, 3728.
[696] Y. Nishida and S. Kida, *Chem. Letters*, 1974, 339.
[697] D. N. Zimmerman and J. L. Hall, *Inorg. Chem.*, 1973, **12**, 2616.
[698] M. Okamura and S. Kida, *J. Inorg. Nuclear Chem.*, 1974, **36**, 1413.
[699] I. E. Dickson and R. Robson, *Inorg. Chem.*, 1974, **13**, 1301.

$CuCl_2$ at low temperatures, but at higher temperatures a redox reaction occurs, and mixed oxidation-state complexes $Cu_2L_2Cl_3$ and $Cu_4L_2Cl_5$ are formed.[177]

N-Acetyl-*N'*-benzoylhydrazine forms Cu(L − 2H) (157) which are planar, but differ in structure from the nickel and cobalt analogues.[178]

The copper(II) complexes of a range of substituted picolinamides $R(CH_2)_n$-$NHCOC_5H_4N$ (36) have been synthesized.[181] They are of two main types, CuX_2L (X = Cl, Br, ClO_4, or NCS) where co-ordination is *via* the amide oxygen and the ring nitrogen, and CuX(L − H) where the ligand has deprotonated at the amide nitrogen and is co-ordinated through this and the ring nitrogen. *N*-(2-Pyridylmethyleneamino)-2-pyridinecarboxamide (papH) forms two isomers of Cu_2Cl_3(pap) which may differ in the way the basic units (158) are linked together, $Cu_3(NO_3)_4(pap)_2,2H_2O$ and Cu_2-$(OH)(pap)SO_4,4H_2O$. All the complexes have subnormal magnetic moments, and may contain the pap functioning as a quinquedentate N_4O donor to the two different metal atoms.[182]

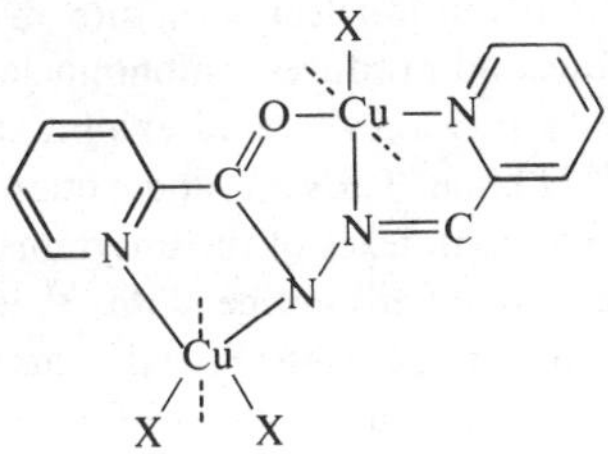

(158) 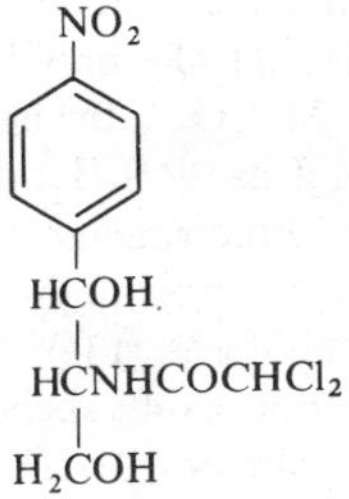   (159)

Chloroamphenicol (159) forms a 2:1 complex with Cu^{II} from methanol solution but in water the amide function hydrolyses and a 2:1 complex of p-$NO_2C_6H_4$-$CH(OH)CH(NH_2)CH_2OH$ is formed.[700]

Dithio-oxamide (LH_2) and its dimethyl and dicyclohexyl derivatives form polymeric $[CuL]_n$ probably with S_2N_2 co-ordinated copper.[405] The copper complexes of 1,10-phenanthroline-2-carbothiamide, $CuLX_2$ (X = Cl, Br, or NCS), are six-co-ordinate with X-bridges. The *N*-phenyl-1,10-phenanthroline-2-carbothiamide complexes of copper are similar to those of Co^{III}, in that only species containing the deprotonated ligand can be obtained, Cu(L' − H)X (X = Cl, Br, or NO_3).

Alcoholic copper(II) acetate brings about the rearrangement of 3,3,5-trimethyl-1,3,4-thiadiazolidine-2-thione into the (complexed) Schiff base α-*N*-methyl-β-*N*-isopropylidenedithiocarbazate.[601] The condensation of the appropriate salicylaldehyde with 2-*N*-methyl-*S*-methyldithiocarbazate or *SS'*-dimethyldithiocarbazate produces Schiff bases (160) and (161) which can lose a proton to form uninegative ONS donor ligands; copper complexes are of several types.[347] The $[CuL_2]$ [L = $R^1R^2C=N-NH-C(S)SMe$, $R^1 = R^2 = Me$, $R^1 = Me$, $R^2 = Ph$; $R^1R^2 = C_5H_{10}$) complexes are possibly five-co-ordinate dimers with planar CuS_2N_2 chromophores

[700] G. Fazakerley, P. W. Linder, and L. R. Nassimbeni, *Inorg. Nuclear Chem. Letters*, 1973, **9**, 1069.

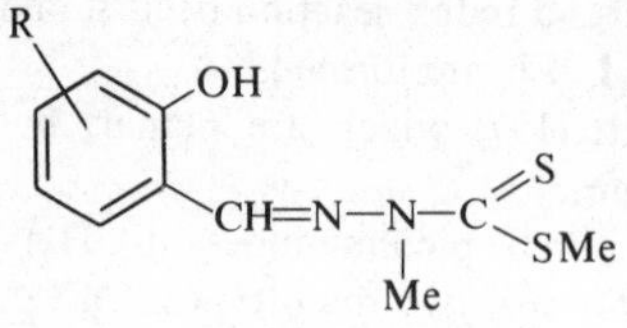

(160) 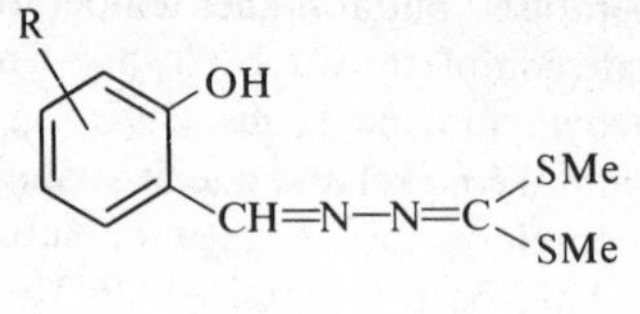   (161) 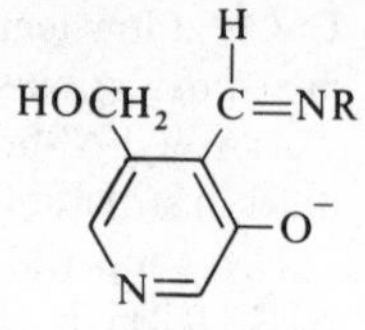   (162)

and intermolecular Cu—S interaction.[206] Pyridine-2-carbaldehyde-*S*-methylthio-semicarbazone forms $[CuLX_2]$ (X = Cl, Br, or NO_3), which are probably X-bridged six-co-ordinate complexes, the ligand functioning as a N_2S donor.[596] Square-planar copper complexes are formed by thioiminato Schiff bases with S_2N_2 donor sets.[198, 199]

The structures of the complexes formed by the Schiff bases (162) depend upon the nature of R; when R is a simple alkyl or aryl group the complexes are of the type CuL_2 with an essentially planar structure,[701] but when R contains a potential donor, $CH_2CH_2O^-$ or $CH_2CH_2CH_2O^-$, a multinuclear structure results, whilst when R = $MeCO_2^-$, the ligand is terdentate, and produces a mononuclear complex.

The Schiff bases (LH_2) derived from 3-formylsalicylic acid exhibit strong intra-molecular antiferromagnetic coupling.[574] Planar CuN_3O and square-pyramidal CuN_3O_2 groupings are present in the copper complexes of the unsymmetric quadri-dentate Schiff bases (135)[580] and quinquedentate keto-amine (136).[581] Bromination of 5,8-diazadodeca-4,8-diene-2,11-dione (baen) produces the 3,10-dibromo-derivative which co-ordinates to copper(II) to form the planar $Cu(baenBr_2)$, also obtained directly by bromination of Cu(baen).[576]

Trimeric $Cu_3(dmg)_2X_4,nH_2O$ (dmgH = dimethylglyoxime) (163), contain co-ordinated halide when X = Cl, but co-ordinated water when X = Br.

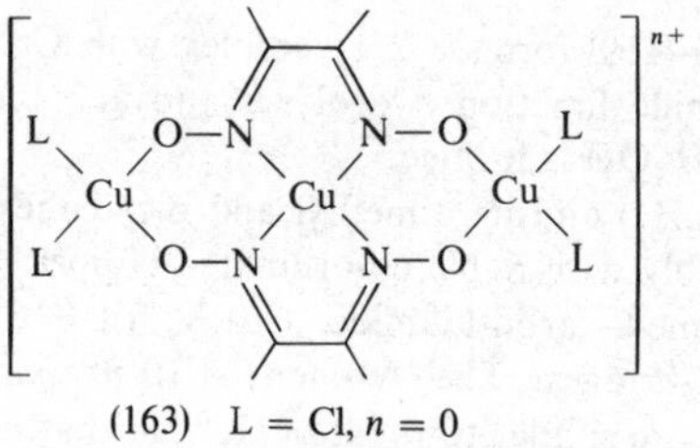

(163)   L = Cl, n = 0
         L = H_2O, n = 4

$Cu_3(dmg)_4(NO_3)_2$ probably contains a linear Cu_3 system, the central Cu being in an O_4 environment and the other two being N_4 co-ordinated, the nitrate groups being unco-ordinated.[702]

The complexes of *NN'*-bis-(2-pyridylmethyl)oxamide (L') and *NN'*-bis-(2-pyridyl-ethyl)oxamide (L'') with Cu^{II}, [Cu(L − 2H)], can function as ligands towards further copper atoms, *e.g.* they react with $Cu(bipy)_2(NO_3)_2$ (or copper malonate) to form

[701] S. Yamada, Y. Kuge, and T. Yamayoshi, *Inorg. Chim. Acta*, 1974, 9, 29.
[702] C. B. Singh and B. Sahoo, *J. Inorg. Nuclear Chem.*, 1974, 36, 1259.

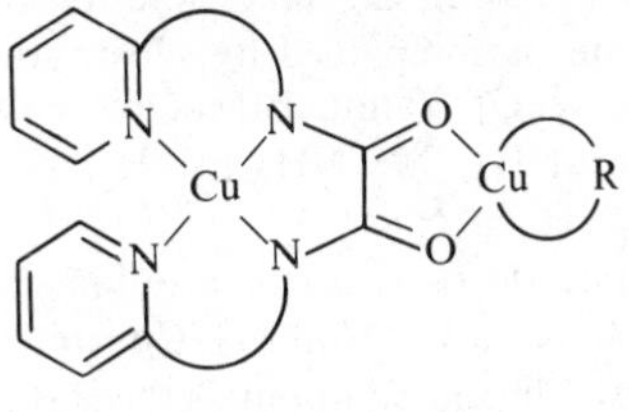

(164)   R = bipy or CH$_2$(CO$_2$)$_2^{2-}$ 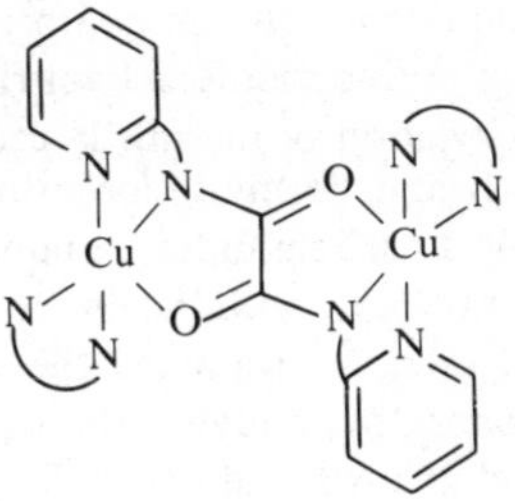   (165)

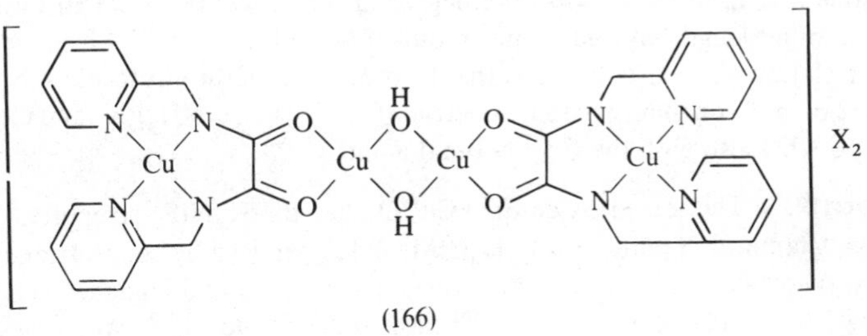

(166)

(164) and (165). Under the appropriate conditions tetranuclear (166) is produced. There is considerable Cu—Cu interaction as evidenced by the sub-normal magnetic moments.[703]

SS′-ethylenebis[(R)-cysteine] and SS′-ethylenebis[(S)-homocysteine] (L) function as ON-donors in Cu(L − 2H),nH$_2$O (n = 0 or 1).[197] The CuL$_2$Cl$_2$ complex of *dl*-methylsulphonium-methioninate is probably octahedral, whilst the CuL′Cl of the derived amide acid may be tetrahedral.[196] Tetrahedral CuLCl$_2$ complexes of the alkaloid (−)sparteine and its (−)α- and (−)β-isosparteine diastereoisomers have been prepared.[704]

The complexes of 3,3′-diamino-4,4′-dihydroxydiphenylsulphone are planar with structure (167).[705]

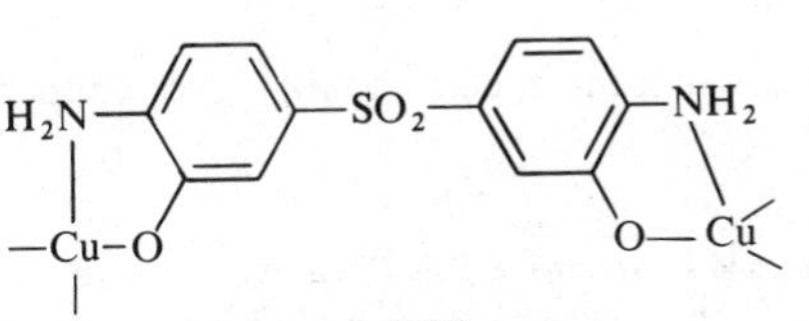

(167)

[703] H. Ojima and K. Nonayama, *Z. anorg. Chem.*, 1973, **401**, 195.
[704] E. Boschmann, L. M. Weinstock, and M. Carmack, *Inorg. Chem.*, 1974, **13**, 1297.
[705] W. U. Malik, M. P. Teotia, and D. K. Rastogi, *J. Inorg. Nuclear Chem.*, 1973. **35**. 4047.

Iminodiaceto- and iminodipropio-nitrile both form CuL_2X_2 complexes,[498] whilst reaction of (2-pyridinecarbonitrile)$_2$Cu with amines in methanol solution results[706] in alkoxylation of the nitrile group, sometimes accompanied by substitution of an amine ligand, forming, for example, $CuCl_2$(MeOpy)(amine) (MeOpy = o-methyl-pyridine-2-carbonimidate, amine = NH_3, $MeNH_2$, Me_2NH, $EtNH_2$, or pip) or $CuCl_2$(MeOpy)$_2$ with the amines Me_3N, Et_2NH, Et_3N.

Sigel et al.[707] examined the stability of chelate rings versus size for a range of NO-donors by studying the equilibria $CuA + CuB \rightleftharpoons CuAB + Cu$ and $CuA_2 + CuB_2 \rightleftharpoons 2CuAB$, and concluded that for mixed ligand compounds the stability with respect to ring size is $5 + 6$ ring $> 6 + 6 > 5 + 5$.

Hathaway and Hodgson[708] have discussed the results of structural studies on a wide range of N- or O-donor complexes in terms of the various distorted geometries possible and have shown that the degree of tetragonality (equal to mean in-plane Cu—L bond length divided by mean out-of-plane Cu—L) varies from ca. 0.6 to 0.9. A correlation also exists between the band maxima of the electronic spectra and the square of the frequency of the asymmetric v(Cu—N) in $Cu(NH_3)_4X_2$ and $Cu(NH_3)_5X_2$, and possibly also with the Cu—N bond lengths.[709]

Copper(III).—The mixed oxide $SrLaCu^{III}O_4$ has the K_2NiF_4 structure.[856] The Cu^I cation, Cu(pmm)$^+$ [pmm = o-$C_6H_4(PMe_2)_2$] is oxidized by concentrated nitric acid to the yellow $[Cu(pmm)_2]^{3+}$, which is square-planar and diamagnetic (d^8). This will add Cl^- to form $[Cu(pmm)_2Cl]^{2+}$. The less stable $[Cu(das)_2]^{3+}$ and $[Cu(das)_2Cl]^{2+}$ [das = o-$C_6H_4(AsMe_2)_2$] have been similarly prepared. These are the first reported Cu^{III} complexes with heavy Group VB donor ligands.[127]

The $K_2Cu^{II}(ded)_2$ complex (ded = 1,1-diethoxycarbonyl-2,2-dithiolate dianion) is oxidized by excess Cu^{II} or H_2O_2 to $KCu(ded)_2$, which has a very similar structure to its Ni^{II} d^8 analogue, the Cu^{III} being in a S_4 environment. The justification of this formulation rather than as a Cu^I disulphide has been discussed.[532]

Unusual mixed valence complexes of copper with dithiocarbamate ligands have been prepared. Bromine oxidation of Cu(Bu$_2$dtc)$_2$ mixed with MBr$_2$ (M = Zn, Cd, or Hg) in $CHCl_3$–EtOH produced $[Cu_3(Bu_2dtc)_6][MBr_3]_2$. The structure of the cadmium compound consists of a cation composed of three nearly planar parallel Cu(Bu$_2$dtc)$_2$ units and $Cd_2Br_6^{2-}$ ions. The formal oxidation state of the copper is $2\frac{2}{3}$, and the cation is believed to contain a central copper(II) layer and two outer copper(III) layers.[857] Both Cu^I and Cu^{III} are present in complexes of type $Cu_7Br_7(Bu_2dtc)_2$, $Cu_5Br_5(Pr_2dtc)_2$, $Cu_5Br_5(Et_2dtc)_2$, and $Cu_3Br_3(Me_2dtc)_2$ (R$_2$dtc = dialkyldithio-carbamate), the Cu^{III} being present as $Cu(dtc)_2^+$ ions, and the Cu^I as halocuprate anions.

Copper (III) is formally present in $[Cu^{III}(taab)]^-$ (see copper(II) section).[102]

[706] S. Suzuki, M. Nakahara, and K. Watanabe, *Bull. Chem. Soc. Japan*, 1974, **47**, 645.

[707] H. Sigel, R. Caraco, and B. Prijis, *Inorg. Chem.*, 1974, **13**, 462.

[708] B. J. Hathaway and P. G. Hodgson, *J. Inorg. Nuclear Chem.*, 1973, **35**, 4071.

[709] R. J. Dudley, B. J. Hathaway, P. G. Hodgson, J. K. Mulcahy, and A. A. G. Tomlinson, *J. Inorg. Nuclear Chem.*, 1974, **36**, 1947.

[856] J. B. Goodenough, G. Demazeau, M. Pouchard, and P. Hagenmuller, *J. Solid State State Chem.*, 1973, **8**, 325.

[857] J. A. Cras, *Proceedings 16th International Conference on Coordination Chemistry, Dublin*, 1974.

4 X-Ray Data (Copper), Physical Data (Copper), Formation and Stability Constants (Cobalt, Nickel, and Copper)

X-Ray Data.—Numerous *X*-ray crystallographic studies of copper(II) complexes have been reported; several are mentioned above, the rest are listed here.

Compound	Comments	Ref.
[Cu{H$_2$N(CH$_2$)$_3$NH$_2$}NCS]ClO$_4$	distorted trigonal-bipyramidal with axial NCS	710
[CuHN{(CH$_2$)$_2$NH$_2$}$_2$(NCS)$_2$]	tetragonal pyramid, one axial NCS	711
[CuHN{(CH$_2$)$_2$NH$_2$}$_2$NCS]$_2$(ClO$_4$)$_2$	dinuclear NCS-bridged, distorted octahedral, four N-donors in plane, one S axial, weak axial ClO$_4^-$ co-ord	712
[CuHN{(CH$_2$)$_3$NH$_2$}$_2$(NCS)$_2$]	tetragonal pyramid, one NCS basal, one NCS axial	713
[Cu{HN(CH$_2$)$_3$NH$_2$}$_2$(NCS)]$_2$(ClO$_4$)$_2$	dinuclear, NCS-bridged, weak axial ClO$_4^-$ co-ordination	714
[Cu(Et$_2$NCH$_2$CH$_2$NEt$_2$)OH]$_2$(ClO$_4$)$_2$	almost planar, OH$^-$ bridged, Cu $\cdots$ Cu 2.978 Å	715
α-[CuL(OH)]$_2$(ClO$_4$)$_2$ L = 2-(2-dimethylaminoethyl)pyridine	distorted octahedron, bridging OH$^-$, bidentate bridging axial ClO$_4^-$	716
β-[CuL(OH)]$_2$(ClO$_4$)$_2$ L = 2-(2-dimethylaminoethyl)pyridine	distorted tetragonal pyramid, OH-bridged, unidentate axial ClO$_4^-$	717
CuL$_2$(ClO$_4$)$_2$ L = 2-(2-dimethylaminoethyl)pyridine	tetragonal octahedron, weak axial unidentate ClO$_4^-$ interaction	718
CuL(NO$_3$)$_2$ L = 2-(2-dimethylaminoethyl)pyridine	distorted octahedron, unsymm. bidentate NO$_3^-$ groups	719
[Cu(bipy)$_2$Cl]Cl,6H$_2$O	trigonal-bipyramidal, N$_2$Cl *eq*, N$_2$ *ax*	720
[CuL(NCBH$_3$)$_2$] L = 1.1.4,7,7-pentamethyldiethylene-triamine	distorted square pyramid, apical NCBH$_3$ group	721
[CuL$_2$](ClO$_4$)$_2$ L = *N* (2-hydroxyethyl)ethylenediamine	distorted octahedral, N$_4$ set *eq* N$_2$ set *ax*, terdentate ligand in *fac* configuration	722

[710] M. Cannas, G. Carta, and G. Marongiu, *J.C.S. Dalton*, 1974, 550.
[711] M. Cannas, G. Carta, and G. Marongiu, *J.C.S. Dalton*, 1974, 553.
[712] M. Cannas, G. Carta, and G. Marongiu, *J.C.S. Dalton*, 1974, 556.
[713] M. Cannas, G. Carta, A. Cristini, and G. Marongiu, *J.C.S. Dalton*, 1974, 1278.
[714] M. Cannas, G. Carta, and G. Marongiu, *Gazzetta*, 1974, **104**, 581.
[715] E. D. Estes, W. E. Hatfield, and D. J. Hodgson, *Inorg. Chem.*, 1974, **13**, 1654.
[716] D. L. Lewis, W. E. Hatfield, and D. J. Hodgson, *Inorg. Chem.*, 1974, **13**, 147.
[717] D. L. Lewis, K. T. McGregor, W. E. Hatfield, and D. J. Hodgson, *Inorg. Chem.*, 1974, **13**, 1013.
[718] D. L. Lewis and D. J. Hodgson, *Inorg. Chem.*, 1974, **13**, 143.
[719] D. L. Lewis and D. J. Hodgson, *Inorg. Chem.*, 1973, **12**, 2935.
[720] F. S. Stephens and P. A. Tucker, *J.C.S. Dalton*, 1973, 2293.
[721] B. G. Siegal and S. J. Lippard, *Inorg. Chem.*, 1974, **13**, 822.
[722] R. V. Chastain and T. L. Dominick, *Inorg. Chem.*, 1973, **12**, 2621.

Compound	Comments	Ref.
$Cu(py)_2L_2$	square planar	723
L = 5,5'-diethylbarbiturato anion		
$CuL_5(ClO_4)_2$	five-co-ordinate Cu bonded to ethylene-thiourea molecules *via* O, one ClO_4^- weakly bonded	724
L = 2-imidazolidinone		
$Na[CuN(CH_2CO_2)_3],H_2O$	distorted octahedral NO_3 set from one ligand, O_2 from neighbouring ligand	725
$[CuL_2Cl_2(H_2O)_2],3H_2O$	distorted octahedral, *trans*-bonded	
L = 9-methyl-6-oxypurine	purines, *trans*-H_2O's, weak *trans*-axial Cl's	726
$Cu(en)_2(acac)_2,2H_2O$	four-co-ordinate N_4 set, acac bonds axial *via* one O, second O H-bonded to amine	727
$[Cu(O_2CMe)_2]_2,H_2O$	neutron diffraction	728
$[Cu(O_2CMe)_2]_2,H_2O$	planar acetate bridges, Cu—Cu = 2.616 Å	729
$Na_2Cu(CO_3)_2,3H_2O$	square-pyramidal Cu, axial H_2O, one chelating CO_3^{2-}, one bridging CO_3 producing infinite chain	730
$K_2Cu(PO_4)_3$		731
$Bi_2Cu_3S_4Cl$		732
$(Ph_4As)_2Cu_2Cl_6$	isolated non-planar anions, Cu intermediate between planar and tetrahedral	733
$3Cu(IO_3)_2,2H_2O$ (bellingerite)	tetragonal distorted CuO_6 groups and $Cu_2O_8(H_2O)$ octahedral dimers	734
$[PhCH_2CH_2NH_2(Me)]_2CuCl_4$	green form planar; yellow form flattened tetrahedron	735
$[thiamineH_2]CuCl_4$	flattened distorted tetrahedron	736
$[Ph_4P]_2Cu_2Cl_6$	isolated anions, non-planar	737
$[(NPMe_2)_5H_2]CuCl_4,H_2O$	distorted tetrahedral anions	738
$[Ph_4As]_2Cu(NO_3)_4,CH_2Cl_2$	planar anion; either four-co-ordinate with unidentate NO_3^-, or eight-co-ordinate with bidentate NO_3^-	739
$Cu[Me_2NH(CH_2CH_2P(O)Ph_2)]Cl_3$	distorted tetrahedral Cl_3O donor set	740
$Cu(5,5\text{-diethylbarbiturate})_2(py)_2$	$N_4 + O_2$ co-ordination	741
$Cu(5,5'\text{-diethylbarbiturate})_2(pic)_2$	planar N_4 set	742

[723] M. R. Caira, G. V. Fazakerley, P. W. Linder, and L. R. Nassimbeni, *Inorg. Nuclear Chem. Letters*, 1973, **9**, 1101.

[724] R. J. Majeste and L. M. Trefonas, *Inorg. Chem.*, 1974, **13**, 1062.

[725] S. H. Whitlow, *Inorg. Chem.*, 1973, **12**, 2286.

[726] E. Sletten, *Acta Cryst.*, 1974, **B30**, 1961.

[727] M. Matsui, T. Kurauchi, S. Ooi, Y. Nakaraura, S. Kawayuchi, and H. Kuroya, *Inorg. Nuclear Chem. Letters*, 1973, **9**, 1041.

[728] G. M. Brown and R. Chidamboram, *Acta Cryst.*, 1973, **B29**, 2393.

[729] P. de Meester, S. R. Fletcher, and A. C. Skapski, *J.C.S. Dalton*, 1973, 2575.

[730] P. D. Brotherton and A. H. White, *J.C.S. Dalton*, 1973, 2338.

[731] M. Laught, I. Tordjman, G. Bassi, and J. C. Guitel, *Acta Cryst.*, 1974, **B30**, 1100.

[732] J. Lewis, jun. and V. Kupcik, *Acta Cryst.*, 1974, **B30**, 848.

[733] R. D. Willett and C. Chow, *Acta Cryst.*, 1974, **B30**, 207.

[734] S. Ghose and C. Wan, *Acta Cryst.*, 1974, **B30**, 965.

[735] R. L. Harlow, W. J. Wells, G. W. Watt, and S. H. Simonsen, *Inorg. Chem.*, 1974, **13**, 2106.

[736] M. R. Caira, G. V. Fazakerley, P. W. Linder, and L. R. Nassimbeni, *Acta Cryst.*, 1974, **B30**, 1660.

[737] M. Textor, E. Dubler, and H. R. Oswald, *Inorg. Chem.*, 1974, **13**, 1361.

[738] H. P. Calhoun and J. Trotter, *J.C.S. Dalton*, 1974, 382.

[739] T. J. King and A. Morris, *Inorg. Nuclear Chem. Letters*, 1974, **10**, 237.

[740] M. G. Newton, H. D. Caughman, and R. C. Taylor, *J.C.S. Dalton*, 1974, 1031.

[741] G. V. Fazakerley, P. W. Linder, L. R. Nassimbeni, and A. L. Rogers, *Inorg. Chim. Acta*, 1974, **9**, 193.

[742] M. R. Caira, G. V. Fazakerley, P. W. Linder and L. R. Nassimbeni, *Acta Cryst.*, 1973, **B29**, 2898.

Compound	Comments	Ref.
$Cu(iminodiacetamide)_2(ClO_4)_2$	elongated octahedron O_2 ax, O_2N_2 eq	743
CuL_2	planar; also discussed structural data for	
L = N-hexyl-7-methylsalicylaldimine	other salicylaldimato compounds	744
$[Cu(salen)CuCl_2]_2$	tetranuclear $Cu(O)_2Cu(Cl)_2Cu(O)_2Cu$	
	bridges	745
$[Cu(salen)]_2,NaClO_4,p$-xylene		746
CuL_2 (δ-form)	planar N_2O_2 set	747
L = 2-hydroxy-N-methyl-1-naphthylmethyleneiminato		
$Cu(NN$-dimethylglycinato$)_2(H_2O)$	five-co-ordinate, *trans*-glycinates, axial	
	H_2O	748
$[Cu(4$-methyl-1-quinone-2-oximato$)_2,py$	distorted square pyramid	749
$[CuL(H_2O)ClO_4]_2$	~ planar $Cu_2N_2O_2$ ring, weak axial ClO_4,	
L = 2(2-hydroxyethyl)imino-3-	H_2O co-ordinated	750
oximobutane		
$Cu(N$-carboxymethyl-L-valinate$)_2$	distorted square pyramid, NO_4 set	751
$CsCuCl_3$	trigonally distorted octahedral Cu	752
$[Pr^iNH_3]_2CuCl_4$	$\frac{1}{3}$ of $CuCl_4^{2-}$ planar; $\frac{2}{3}$ tetrahedrally	
	distorted	753
$[PhNH_3]_2CuCl_4$	almost planar anions (4 + 2)	754
$[Cu(H_2O)_6][SiF_6]$	$\frac{1}{4}$ regular octahedral; $\frac{3}{4}$ distorted octa-	
	hedral	755
$KCu_2(OH)(SO_4)_2,H_2O$	distorted octahedron	756
$Li_2Cu_2P_6O_{18}$	P_6O_{18} rings	757
$Cu(H_2O)(imidazole)_3SO_4$	distorted octahedral N_3O_3 set; bidentate	
	bridging SO_4^{2-}	758
$LiCu(succinimidato)_4,H_2O$	flattened tetrahedral anion	759
$KCu(succinimidato)_4,6H_2O$	planar anion	759
$[CuL,H_2O]_2,2H_2O$	distorted tetrahedral	760
L = N-(picolinoyl)-3-amino-1-propoxido		
$[Cu_2(tren)_2(CN_2)](BPh_4)_2$	trigonal bipyramidal Cu, N_4C donor set.	
tren = 2,2',2''-triaminoethylamine	dimeric cation due to H-bonds between	
	CN and amines	761
$[Cu(Schiff base)X]_2,[Cu_2(Schiff base)X_2]$	O_2 bridged, magnetic properties are	
	function of structure	762
$[CuL]_2,9H_2O$	ligand bridged, N_2O_2 set about each Cu,	
L = D-penicillamine disulphide	S of ligand weakly co-ordinated	763

[743] M. Sekizaki, *Bull. Chem. Soc. Japan*, 1974, **47**, 1447.
[744] P. C. Jain and U. K. Syal, *J.C.S. Dalton*, 1973, 1908.
[745] C. A. Bear, J. M. Waters, and T. N. Waters, *J.C.S. Dalton*, 1974, 1059.
[746] H. Milburn, M. R. Truter, and B. L. Vickery, *J.C.S. Dalton*, 1974, 841.
[747] D. W. Martin and T. N. Waters, *J.C.S. Dalton*, 1973, 2440.
[748] T. S. Cameron, K. Prout, F. J. C. Rossotti, and D. Steele, *J.C.S. Dalton*, 1973, 2626.
[749] M. McPartlin, *Inorg. Nuclear Chem. Letters*, 1973, **9**, 1207.
[750] J. A. Bertrand, J. H. Smith, and P. G. Eller, *Inorg. Chem.*, 1974, **13**, 1649.
[751] S. K. Porter, R. J. Angelici, and J. Clardy, *Inorg. Nuclear Chem. Letters*, 1974, **10**, 21.
[752] C. J. Kroese, W. J. A. Maazkant, and G. C. Verschoor, *Acta Cryst.*, 1974, **B30**, 1053.
[753] D. N. Anderson and R. D. Willett, *Inorg. Chim. Acta*, 1974, **8**, 167.
[754] K. P. Larsen, *Acta Chem. Scand.*, 1974, **A28**, 194.
[755] S. Ray, A. Zalkin, and D. H. Templeton, *Acta Cryst.*, 1973, **B29**, 2748.
[756] M. Tardy and J. M. Brégeault, *Bull. Soc. chim. France*, 1974, 1866.
[757] M. Laugt and A. Durif, *Acta Cryst.*, 1974, **B30**, 2118.
[758] G. Franson and B. K. S. Lundberg, *Acta Chem. Scand.*, 1974, **A28**, 578.
[759] T. Tsukihara, Y. Kasube, K. Fujimori, K. Kawashima, and Y. Kan-Nan, *Bull. Soc. chim. France*, 1974, **47**, 1582.
[760] J. A. Bertrand, E. Fujita, and P. G. Eller, *Inorg. Chem.*, 1974, **13**, 2067.
[761] D. M. Duggan and D. N. Hendrickson, *Inorg. Chem.*, 1974, **13**, 1911.
[762] R. M. Countryman, W. T. Robinson, and E. Sinn, *Inorg. Chem.*, 1974, **13**, 2013.
[763] J. A. Thick, D. Mastropaolo, J. Potenza, and H. J. Schugar, *J. Amer. Chem. Soc.*, 1974, **96**, 726.

Compound	Comments	Ref.
Cu[2-(2'-pyridyl)-3-(N-2-picolylimino)-4-oxo-1,2,3,4-tetrahydroquinazoline]Cl$_2$	five-co-ordinate, N$_3$Cl$_2$ set	764
Cu(hfac)$_2$(pyrazine)	octahedral pyrazine-bridged polymer	765
{Cu(hfac)$_2$}$_2$(pyrazine)	square pyramidal, pyrazine bridge	765
Cu$_2$LCl$_3$	L = NN'-bis-{2-(2-pyridyl)ethyl}-2,3-pyrazinedicarboxamide	766
[(CuL)$_2$Cu(H$_2$O)$_2$](ClO$_4$)$_2$,H$_2$O, [(CuL)$_2$Cu(H$_2$O)](ClO$_4$)$_2$	L = terdentate Schiff base, co-ordinated to central Cu via bridging O's	767
Cu(N-Me-salicylaldiminato)$_2$.1,3,5-C$_6$H$_3$(NO$_2$)$_3$		768
[Cu(bipy)(N$_3$)$_2$]	planar N$_4$ set, two further N's at long distance	769
[Cu-αβγδ-tetra-n-propylporphine]	planar N$_4$ set	770
[Cu(Me$_2$CNO$_2$)$_2$]	distorted octahedral O$_6$ set	771
[Cu(asparginate)$_2$]	dinuclear Cu$_2$L$_4$ units associated via Cu—O(acetyl)bonds	772
[Cu(2-diethylaminoethanolate)X]	X = Cl, tetrameric cubane Cu$_4$O$_4$	773
	X = Br dimeric O-bridged	773
[Cu$_3$L$_3$(H$_2$O)$_3$(OH)$_{1/2}$](ClO$_4$)$_{3/2}$. (H$_2$O)$_4$ L = 2-propylamino-2-methyl-3-butaneoximate	Cu$_3$ triangle about central O, each Cu co-ordinate to N$_2$O$_2$ set.	774
[Cu$_2$L$_2$(CN)]ClO$_4$	two trigonal bipyramidal Cu atoms with CN bridges	775
L = 5,7,7,12,14,14-hexamethyl-1,4,8,11-tetra-azacyclotetradeca-4,11-diene		
CuL(ClO$_4$)$_2$ L = 1,8-bis-(2'-pyridyl)-3,6-diazaoctane	CuN$_4$ planar set, weak ClO$_4^-$ co-ordination	777
[Cu(antipyrine)$_2$(NO$_3$)$_2$]	distorted octahedral CuO$_6$ set.	776
[Cu$_2$Cl$_2$L]$_2$,2HO L = 2,6-diacetylpyridinebis(picolinylhydrazonate)	two non-equivalent five-co-ordinate Cu, ligand is octadentate	778
Cu$_3$Cl$_6$(H$_2$O)$_2$(tetramethylenesulphone)$_2$	planar Cu$_3$Cl$_6$ trimer, Cl-bridged	779
Cu$_3$MS$_x$Se$_{4-x}$ (M = Nb or Ta)	pseudo-sulvanite type	780
[Cu$_2$(CF$_3$CO$_2$)$_3$(OH)(quinoline)$_2$]$_2$	distorted square pyramidal CuO$_4$N	781
Cu(L − H$_2$) L = H$_2$C[C(Me)=NCH$_2$CH$_2$OH]$_2$	planar	782

764 A. Mangia, M. Nardelli, C. Pelizzi, and G. Pelizzi, *Acta Cryst.*, 1974, **B30**, 17.
765 R. C. E. Belford, D. E. Fenton, and M. R. Truter, *J.C.S. Dalton*, 1974, 17.
766 E. B. Fleischer, D. Y. Jeter, and R. Florian, *Inorg. Chem.*, 1974, **13**, 1042.
767 J. M. Epstein, B. N. Figgis, A. H. White, and A. C. Willis, *J.C.S. Dalton*, 1974, 1954.
768 A. W. King, D. A. Swann, and T. N. Waters, *J.C.S. Dalton*, 1973, 1819.
769 G. W. Bushnell and M. A. Khan, *Canad. J. Chem.*, 1974, **52**, 3125.
770 I. Moustakali and A. Tulinsky, *J. Amer. Chem. Soc.*, 1973, **95**, 6811.
771 O. Simonsen, *Acta Cryst.*, 1973, **B29**, 2600.
772 Lj. Manojlović-Muir, *Acta Cryst.*, 1973, **B29**, 2033.
773 W. Haase, *Chem. Ber.*, 1974, **106**, 3132.
774 P. F. Ross, R. K. Murmann, and E. O. Schemper, *Acta Cryst.*, 1974, **B30**, 1120.
775 D. M. Duggan, R. G. Jungst, K. R. Mann, G. D. Stucky, and D. N. Hendrickson, *J. Amer. Chem. Soc.*, 1974, **96**, 3443.
776 C. Brassy, A. Renard, J. Delettre, and J. P. Mornon, *Acta Cryst.*, 1974, **B30**, 2246.
777 D. A. Wright and J. D. Quinn, *Acta Cryst.*, 1974, **B30**, 2132.
778 A. Mangia, C. Pelizzi, and G. Pelizzi, *Acta Cryst.*, 1974, **B30**, 2146.
779 D. D. Swank and R. D Willett, *Inorg. Chim. Acta*, 1974, **8**, 143.
780 A. Müller and W. Sievert, *Z. anorg. Chem.*, 1974, **406**, 80.
781 R. G. Little, J. A. Moreland, D. B. W. Yawnay, and R. J. Doedens, *J. Amer. Chem. Soc.*, 1974, **96**, 3834.
782 J. A. Bertrand and F. T. Helm, *J. Amer. Chem. Soc.*, 1973, **95**, 8184.

Copper(II) Compounds. Physical Data.—Detailed discussion of spectral and magnetic data on copper compounds is covered in the Specialist Periodical Report 'Electronic Structure and Magnetism of Inorganic Compounds'. References tabulated here are not specifically discussed above.

Complex	Physicochemical Method	Ref.
Cu^{II}–amino-acid complexes	e.p.r.	783
Cu(diamine)X$_2$	charge-transfer spectra	784
Cu^{II}–polyamines	e.p.r.	785
[Cu(guaninium)Cl$_3$]$_2$,2H$_2$O	magnetism	786
[Cu(bipy)$_2$Cl]Cl, Cu(py)$_3$(NO$_3$)$_2$	single-crystal spectra	787
Cu(py)$_3$(NO$_3$)$_2$	e.p.r.	788
[CuL$_3$](ClO$_4$)$_2$ (L = bipy, phen, *etc.*)	electronic spectra	789
Cu$_4$(py)$_4$OCl$_6$	magnetism	790
Cu(bipy)Cl$_2$, Cu(phen)Cl$_2$	Far-i.r. and raman	791
Cu$_2$(OAc)$_4$(pyrazine)	e.p.r.	792
Cu(acac)$_2$(quinolone)	single-crystal electronic spectra	793
Cu(R-pyO)$_2$(NO$_3$)$_2$ (R = H, Me, or OMe)	magnetism and e.p.r.	794
Cu–acetonitrile compounds	^{14}N n.m.r.	795
[Cu(py)(OAc)$_2$]$_2$, [Cu(quinoline)(OAc)$_2$]$_2$	e.p.r.	796
Cu(hfac)$_2$ + bipy, py	e.p.r.	797
Cu(R-acetoacetato)$_2$ (R = Bu^t or Et)	e.p.r.	798
Cu–N-donor ligands	e.p.r. (*N*-hyperfine structure)	799
Cu–β-diketones (Zn host)	e.p.r.	800
Cu(cyclobutane-1,1-dicarboxylate)	e.p.r.	801
Cu(O$_2$CSiPh$_3$)$_2$,H$_2$O; Cu(O$_2$CGePh$_3$)$_2$,H$_2$O	magnetism	802
Cu(pyO)$_6{}^{2+}$	i.r.	803
Cu$_3${MeC(NO)C(NEt)Me}$_3$ClO$_4$	magnetism	804
Cu(dmg)$_2$	e.p.r.	805
Cu complexes + galactose oxidase	e.p.r.	806
Cu–amino-acids	e.p.r., electronic spectra, magnetism	807
Cu(1,1-dicyanoethylene-2,2-diselenate)$_2{}^{2-}$	e.p.r.	808

[783] H. Yokoi, *Bull. Chem. Soc. Japan*, 1974, **47**, 639.

[784] B. P. Kennedy and A. B. P. Lever, *J. Amer. Chem. Soc.*, 1973, **95**, 6907.

[785] R. Barbucci, P. Paoletti, and M. J. M. Cambell, *Inorg. Chim. Acta*, 1974, **10**, 69.

[786] R. F. Drake, V. H. Crawford, N. W. Laney, and W. E. Hatfield, *Inorg. Chem.*, 1974, **13**, 1246.

[787] R. J. Dudley, B. J. Hathaway, P. G. Hodgson, P. C. Power, and D. J. Loose, *J.C.S. Dalton*, 1974, 1005.

[788] G. L. McPherson and C. P. Anderson, *Inorg. Chem.*, 1974, **13**, 677.

[789] B. J. Hathaway, P. G. Hodgson, and P. C. Power, *Inorg. Chem.*, 1974, **13**, 2009.

[790] R. F. Drake, V. H. Crawford, and W. E. Hatfield, *J. Chem. Phys.*, 1974, **61**, 4524.

[791] R. E. Wilde and T. K. K. Srinivason, *J. Inorg. Nuclear Chem.*, 1974, **36**, 323.

[792] J. S. Valentine, A. J. Silverstein, and Z. G. Soos, *J. Amer. Chem. Soc.*, 1974, **96**, 97.

[793] M. A. Hitchman, *Inorg. Chem.*, 1974, **13**, 2218.

[794] K. T. McGregor, J. A. Barnes, and W. E. Hatfield, *J. Amer. Chem. Soc.*, 1973, **95**, 7993.

[795] R. J. West and S. F. Lincoln, *J.C.S. Dalton*, 1974, 281.

[796] F. E. Mabbs, J. K. Porter, and W. R Smail, *J. Inorg. Nuclear Chem.*, 1974, **36**, 819.

[797] J. Pradilla-Sorzano and J. P. Fackler, *Inorg. Chem.*, 1974, **13**, 38.

[798] H. Yokoi, *Bull. Chem. Soc. Japan*, 1974, **47**, 497.

[799] Y. Nonaka, T. Tokai, and S. Kida, *Bull. Chem. Soc. Japan*, 1974, **47**, 312.

[800] S. Misumi, T. Isobe, and K. Sugiyama, *Bull. Chem. Soc. Japan*, 1974, **47**, 1.

[801] A. D. Toy, J. D. Smith, and J. R. Pilbrow, *J.C.S. Dalton*, 1973, 2498.

[802] D. W. Steward and S. R. Piskor, *J.C.S. Chem. Comm.*, 1973, 702.

[803] A. D. V. Ingen-Schenan, W. L. Groenveld, and J. Reedijk, *Spectrochim. Acta*, 1974, **30A**, 213.

[804] J. G. Mohanty, S. Baral, R. P. Singh, and A. Chakravorty, *Inorg. Nuclear Chem. Letters*, 1974, **10**, 655.

[805] A. D. Toy, J. D. Smith, and R. Pilbrow, *Austral. J. Chem.*, 1973, **26**, 2347.

[806] R. S. Giordano, R. D. Bereman, D. J. Kosman, and M. J. Ettinger, *J. Amer. Chem. Soc.* 1974, **96**, 1023.

[807] J. F. Villa, H. C. Nelson, and R. T. Curran, *Inorg. Chem.*, 1974, **13**, 1255.

[808] R. Kirmse, W. Dietzsch, and E. Hoyer, *J. Inorg. Nuclear Chem.*, 1974, **36**, 1959.

Complex	Physicochemical Method	Ref.
$Cu(semicarbazone)_2Cl_2$	e.p.r.	809
Cu^{II}–porphin	matrix electronic spectra	810
Cu^{II}–tetradentate Schiff base	magnetism	811
$[Cu(salen)]_2$	magnetism	812
$Cu-N_2S_2, N_2O_2$ Schiff bases	e.p.r.	813
$Cu(N\text{-}Mesalen)_2$	i.r.	814
Cu^{II}–Schiff bases	i.r.	815
$Cu(\alpha\text{-}NO_2\text{ketonates})_2$ + heterocycles	electronic spectra, e.p.r.	816
Cu^{II}–dinuclear, multidentate Schiff bases	variable temperature magnetism	817
$Cu(NH_3)_4X_2$	theory of d–d spectra	818
Cu–dipyrromethane chelate	e.p.r. and electronic spectra	819
Cu–Schiff bases + $1,3,5\text{-}C_6H_3(NO_2)_3$	^1H n.m.r.	820
Dinuclear, substituted Schiff base compounds	magnetism, i.r., electronic, e.p.r.	821
Cu^{II}–terdentate amines	e.p.r. and magnetism	822
$K_2CuCl_4,2H_2O$	e.p.r.	823
Cu^{II}–N_4 macrocycles	electronic spectra	824
$Cu(mandelate)_2$	i.r.	825
Cu^{II}–corrins	polarography	826
$Cu\{NN\text{-ethylenebis(acetylacetoiminato)}\}_2$	ESCA	827
$[Cu(OAc)_2(4\text{-}RpyO)]_2$	magnetism, i.r. and electronic spectra	828
$[Cu_4OX_6L_4], [Cu_4OX_{10}]^{4-}$	electronic and i.r. spectra	829
$[Cu_2(tren)_2(N_3)_2](BPh_4)_2$	magnetism	830
$[R_4N]_2CuX_4$ (X = Cl or Br)	ESCA	831
$[Cu(Et_2NCH_2CH_2NEt_2)(OH)]_2(ClO_4)_2$	magnetism	832
$[Co(en)_3][Cu_2Cl_8]Cl_2,2H_2O$	single-crystal magnetism	833
$[Et_2NH_2]_2CuCl_4$	e.p.r. and electronic spectra	834
$K_2CuCl_4,2H_2O$	i.r. spectra of H_2O co-ordination	835
$[(H_3NCH_2CH_2)_2NH_2][CuCl_4]Cl$	magnetism	836

[809] M. J. M. Cambell and R. Grzeskowiak, *Inorg. Nuclear Chem. Letters*, 1974, **10**, 473.
[810] J. J. Leonard and F. R. Longo, *J. Amer. Chem. Soc.*, 1973, **95**, 8506.
[811] E. F. Hasty, T. J. Colburn, and D. N. Hendrickson, *Inorg. Chem.*, 1974, **13**, 2414.
[812] G. O. Carlisle, G. D. Simpson, and W. E. Hatfield, *Inorg. Nuclear Chem. Letters*, 1973, **9**, 1247.
[813] R. S. Giordamo and R. D. Bereman, *Inorg. Nuclear Chem. Letters*, 1974, **10**, 203.
[814] A. Bigotto and G. De Alti, *Spectrochim. Acta*, 1974, **30A**, 27.
[815] J. A. Fanikan, K. S. Patel, and J. C. Bailar, *J. Inorg. Nuclear Chem.*, 1974, **36**, 1547.
[816] D. Altanasio, I. Collamati, and C. Ercolani, *J.C.S. Dalton*, 1974, 1319.
[817] J. P. Fishwick, R. W. Jotham, S. F. A. Kettle, and J. A. Marks, *J.C.S. Dalton*, 1974, 125.
[818] D. W. Smith, *J.C.S. Dalton*, 1973, 1853.
[819] Y. Murakami, Y. Matsuda, K. Sakata, and A. E. Martell, *J.C.S. Dalton*, 1973, 1729.
[820] H. A. O. Hill, A. D. J. Horsler, and P. J. Sadler, *J.C.S. Dalton*, 1973, 1805.
[821] P. Gluvchinsley, G. M. Mockler, P. C. Healy, and E. Sinn, *J.C.S. Dalton*, 1974, 1156.
[822] D. L. Lewis, K. T. McGregor, W. E. Hatfield, and D. J. Hodgson, *Inorg. Chem.*, 1974, **13**, 1013.
[823] L. Mauriello, R. Scaringe, and G. Kokasza, *J. Inorg. Nuclear Chem.*, 1974, **36**, 1565.
[824] R. Buxtorf, W. Steinmann, and T. A. Koden, *Chimia (Switz.)*, 1974, **28**, 15.
[825] P. V. Khadikar, R. L. Ameria, M. G. Kekre, and S. D. Chauhan, *J. Inorg. Nuclear Chem.*, 1973, **35**, 4301.
[826] N. S. Hush and J. M. Dyke, *J. Inorg. Nuclear Chem.*, 1973, **35**, 4341.
[827] G. Condorelli, I. Fragala, G. Centineo, and E. Tondello, *Inorg. Chim. Acta*, 1973, **7**, 725.
[828] D. Hibdon and J. H. Nelson, *Inorg. Chim. Acta*, 1973, **7**, 629.
[829] H. Dieck, *Inorg. Chim. Acta*, 1973, **7**, 397.
[830] D. M. Duggan and D. N. Hendrickson, *Inorg. Chem.*, 1973, **12**, 2422.
[831] J. Escard, G. Mavel, J. E. Guerchais, and R. Kergoat, *Inorg. Chem.*, 1974, **13**, 695.
[832] E. D. Estes, W. E. Hatfield, and D. J. Hodgson, *Inorg. Chem.*, 1974, **13**, 1654.
[833] K. T. McGregor, D. B. Lossie, D. J. Hodgson, and W. E. Hatfield, *Inorg. Chem.*, 1974, **13**, 756.
[834] D. R. Hill and D. W. Smith, *J. Inorg. Nuclear Chem.*, 1974, **36**, 466.
[835] G. H. Thomas, M. Falk, and O. Knop, *Canad. J. Chem.*, 1974, **52**, 1029.
[836] W. E. Hatfield and D. B Lossee, Proceedings 16th International Conference on Coordination Compounds, Dublin, 1974; *J. Amer. Chem. Soc.*, 1974, **95**, 8169.

Complex	Physicochemical Method	Ref.
Cu^{II}-corroles	electronic spectra	837
[Cu(oxime)(H$_2$O)(ClO$_4$)]$_2$	magnetism	838
Cu^{II}-porphyrins	polarography	839
Cu{N-(hydroxyalkyl)pyrrole-2-carboxaldimine}	magnetism	840
[Cu(Schiff base)X]$_2$	magnetism as function of structure	841
Cu^{II}-polypyridylethylenes	magnetism	842
Cu(subs. carboxylate)$_2$,nH$_2$O	magnetism and electronic spectra	843
Cu^{II}-arylacetato-compounds	magnetism	844
Cu^{II}-H$^+$-adenine, guanine	^{1}H n.m.r. spectra in water	845
Cu{NN'-ethylenebis(trifluoroacetoniminato)}	e.p.r.	846
Cu^{II}-galactose oxidase	e.p.r.	847
Cu^{II}-β-diketonates + heterocycles	thermodynamic data	848
xCu(OH)$_2$,yMCl$_2$ (M = Mn, Co, Ni, Cu, or Zn)	thermal analysis	849
A$_2$Cu(OH)$_6$,Ba$_2$CuF$_6$ (A = Sr or Ba)	electronic and e.p.r. spectra	850
Cu(NH$_3$)-(C$_2$O$_4$) complexes	i.r., e.p.r., and magnetism	851
Cu(anilines)(NCO)$_2$	e.p.r.	852
Cu(mnt)$_n^{n-}$ + pyridinium cations	charge-transfer spectra	853
Cu(Se$_2$CNEt$_2$)$_2$	e.p.r.	854
Aminopolycarboxylic acid compounds	polarography	855

Formation and Stability Constants (Cobalt, Nickel, and Copper)

Metal	Ligand	Ref.
Co^{II}	pyridine-2-carboxylic acid	858
Co^{II}, Ni^{II}, Cu^{II}	resacetophenoneoxime	859
Co^{II}, Ni^{II}	azelaic acid	860
Co^{II}, Ni^{II}, Cu^{II}	cyclopentamine, cyclohexamine	861
Co^{II}, Ni^{II}	glutaminate and asparaginate	862
Co^{II}, Ni^{II}, Cu^{II}	Schiff bases derived from benzoylacetone and β-alanine	863
Co^{II}, Ni^{II}, Cu^{II}	N-methylhistamine, 4-(2-methylaminoethyl)imidazole	864

[837] N. S. Hush, J. M. Dyke, M. L. Williams, and I. S. Woolsey, *J.C.S. Dalton*, 1974, 395.
[838] J. A. Bertrand, J. H. Smith, and P. G. Eller, *Inorg. Chem.*, 1974, **13**, 1649.
[839] T. Kakutani, S. Totsuka, and M. Seida, *Bull. Chem. Soc. Japan*, 1973, **46**, 2652.
[840] C. R. Pauley and L. J. Theriot, *Inorg. Chem.*, 1974, **13**, 2033.
[841] R. M. Countryman, W. T. Robinson, and E. Sinn, *Inorg. Chem.*, 1974, **13**, 2013.
[842] H. G. Biedermann, W. Graf, E. Griessl, and K. Wichmann, *Z. Naturforsch.*, 1974, **B29**, 267.
[843] J. Sokolik, M. Zemlicha, and J. Kratsmor-Smogrovic, *Z. Naturforsch.*, 1974, **B29**, 448
[844] J. Kratsmor-Smogrovic, V. Seressova, and O. Svajlenova, *Z. Naturforsch.*, 1974, **B29**, 449.
[845] W. R. Walker, J. M. Guo, and N. C. Li, *Austral. J. Chem.*, 1973, **26**, 2391.
[846] R. S. Giordano and R. D. Bereman, *J. Amer. Chem. Soc.*, 1974, **96**, 1019.
[847] R. S. Giordano, R. D. Bereman, D. J. Kosman, and M. J. Ettinger, *J. Amer. Chem. Soc.*, 1974, **96**, 1023.
[848] D. P. Graddon and W. K. Ong, *Austral. J. Chem.*, 1974, **27**, 741.
[849] A. D. Leu, P. Ramamurthy and E. A. Secco, *Can. J. Chem.*, 1973, **51**, 3882.
[850] C. Friebel, *Z. Naturforsch.*, 1974, **B29**, 295.
[851] M. Melnik, H. Longfelderova, J. Guraj, and J. Gazo, *Inorg. Chim. Acta*, 1973, **7**, 669.
[852] J. Kohout, M. Quastlerova-Hvastilijova, and J. Gazo, *Inorg. Chim. Acta*, 1974, **8**, 241.
[853] I. G. Dance and P. J. Solstad, *J. Amer. Chem. Soc.*, 1973, **95**, 7256.
[854] R. Kirmse and E. Hoyer, *Z. anorg. Chem.*, 1973, **401**, 295.
[855] R. Krannich and E. Uhlig, *Z. anorg. Chem.*, 1973, **402**, 285.:
[858] F. Glaizer, P. Buxbaum, E. Papp-Molnar, and K. Burger, *J. Inorg. Nuclear Chem.*, 1974, **36**, 859.
[859] V. Seshagiri and S. B. Rao, *J. Inorg. Nuclear Chem.*, 1974, **36**, 353.
[860] S. C. Tripathi and S. Paul, *Indian J. Chem.*, 1973, **11**, 1042.
[861] G. K. R. Maker and D. R. Williams, *J.C.S. Dalton*, 1974, 1121.
[862] A. C. Baxter and D. R. Williams, *J.C.S. Dalton*, 1974, 1117.
[863] R. K. Mehta, S. L. Pania, and R. K. Gupta, *Indian J. Chem.*, 1973, **11**, 1073.
[864] A. Braibanti, F. Dallavalle, E. Leporati, and G. Mori, *J.C.S. Dalton*, 1973, 2539.

Metal	Ligand	Ref.
Co^{II}	pyrimidines	865
Co^{II}	4-hydroxy-3-oximinomethylazobenzene	866
Co^{II}, Ni^{II}, Cu^{II}	6-aminopurine	867
Co^{II}	3-[(carboxymethyl)thiol]-L-alanine	868
Co^{II}, Ni^{II}, Cu^{II}	$(C_5H_4N)CH_2NH(CH_2)_nNHCH_2(NC_5H_4)$	869
Co^{II}, Ni^{II}, Cu^{II}	4-Me-cyclam	870
Co^{II}	di-, tri-, and tetra-glycine, NN-diacetic acids	167
Co^{III}, Cu^{II}	hydroxyhydroquinone	871
Co^{III}	oximes	872
Co^{III}	dioximates and iodide	873
Ni^{II}	mixed ligand complexes bipy/phen, en/pn	874
Co^{II}, Ni^{II}, Cu^{II}	N-phenyl-2-furohydroxamic acid	875
Ni^{II}	N-methylsubstituted ethylenediamines	876
Ni^{II}	glycolic acid	877
Ni^{II}, Cu^{II}	$[Ru(NH_3)_5N\bigcirc N]^{2+}$	878
Ni^{II}	salicylhydrazone	879
Ni^{II}	asparaginate	880
Ni^{II}, Cu^{II}	hydroxaminic acids	881
Ni^{II}, Cu^{II}	adrenaline	882
Ni^{II}	azide ions in various solvents	883
Ni^{II}/Cu^{II}	mixed-metal compounds of triethylenetetraminehexa-acetic acid	884
Co^{II}, Ni^{II}, Cu^{II}	NN'-bis-(2-aminoethyl)propane-1,3-diamine	885
Ni^{II}	thioglycolanilide	886
Cu^{II}	glycyl-L-histidylglycine, glycylglycyl-L-histidine	887
Cu^{II}	histidine	888
Ni^{II}, Cu^{II}	histidine, edta, polyphenols	889
Cu^{II}	$Cu(acac)_2$ + heterocyclic N-donors	890
Cu^{II}	dipeptide-amides	891

[865] M. M. Taqui Khan and C. R. Krishnamoorthy, *J. Inorg. Nuclear Chem.*, 1974, **36**, 711.
[866] P. N. Mohandas, O. P. Sunar, and C. P. Trivedi, *J. Inorg. Nuclear Chem.*, 1974, **36**, 937.
[867] G. K. R. Maker and D. R. Williams, *J. Inorg. Nuclear Chem.*, 1974, **36**, 1675.
[868] R. Nakon, E. M. Beadle, and R. J. Angelici, *J. Amer. Chem. Soc.*, 1974, **96**, 719.
[869] G. Anderegg, N. G. Podder, P. Blauenstein, M. Hangortner, and H. Stunzi, Proceedings 16th International Conference on Coordination Compounds, Dublin, 1974.
[870] L. Hertli and T. A. Kaden, *Helv. Chim. Acta*, 1974, **57**, 1328.
[871] M. N. Rusina and N. M. Dyatlova, *Zh. neorg. Khim.*, 1973, **18**, 2672.
[872] K. Burger, F. Glaizer, E. Papp-Molnar, and T. T. Bink, *J. Inorg. Nuclear Chem.*, 1974, **36**, 863.
[873] F. Glaizer, T. T. Bink, and K. Burger, *J. Inorg. Nuclear Chem.*, 1974, **36**, 1601.
[874] P. G. Parikh and P. K. Bhattacharya, *Indian J. Chem.*, 1974, **12**, 402.
[875] J. P. Shukla, Y. K. Agrawal, and D. R. Agrawal, *Indian J. Chem.*, 1974, **12**, 534.
[876] T. S. Turan, *Inorg. Chem.*, 1974, **13**, 1584.
[877] S. Harady, Y. Okuue, H. Kan, and T. Yasunaga, *Bull. Chem. Soc. Japan*, 1974, **47**, 769.
[878] M. S. Pereira and J. M. Malin, *Inorg. Chem.*, 1974, **13**, 386.
[879] R. Cetina, J. Gomez-Lara, and R. Contreras, *J. Inorg. Nuclear Chem.*, 1973, **35**, 4217.
[880] R. D. Graham and D. R. Williams, *J.C.S. Dalton*, 1974, 1123.
[881] Y. K. Agrawal and S. G. Tandon, *J. Inorg. Nuclear Chem.*, 1974, **36**, 869.
[882] B. Grgas-Kuzner, V. Simeon, and O. A. Weber, *J. Inorg. Nuclear Chem.*, 1974, **36**, 2151.
[883] R. Abu-Eittah and S. Elmakabaty, *Bull. Chem., Soc. Japan*, 1973, **46**, 3427.
[884] M. S. Haque and M. Kopanica, *Bull. Chem. Soc. Japan*, 1973, **46**, 3072.
[885] S. K. Srivastava and H. B. Mathur, *Indian J. Chem.*, 1973, **11**, 1243.
[886] S. N. Rakkar and P. V. Khadikar, *Indian J. Chem.*, 1973, **11**, 1189.
[887] H. Aiba, A. Yokayama, and H. Tanaka, *Bull. Chem. Soc. Japan*, 1974, **47**, 136, 1437.
[888] T. P. A. Kruk and B. Sarkar, *Canad. J. Chem.*, 1973, **51**, 3549.
[889] I. P. Mavani, C. R. Jejurkar, and P. K. Bhattacharya, *Bull. Chem. Soc. Japan*, 1974, **47**, 1280.
[890] T. Kogame, H. Yukawa, and R. Hirota, *Chem. Letters*, 1974, 477.
[891] O. Yamauchi, Y. Nakao, and A. Nakahara, *Bull. Chem. Soc. Japan*, 1973, **46**, 3749.

Metal	Ligand	Ref.
Cu^{II}	terdentate amines	892
Cu^{II}	halogenocarboxylates	893
Cu^{II}	chromotropic acid, chromotrope-2B	894
Cu^{II}	glycine, alanine, α-aminobutyric acid	895
Cu^{II}	Cu(theonyl-CF$_3$acac)$_2$; heterocyclic N, NO ligands	896
Cu^{II}	Schiff bases	897
Cu^{II}	nitrilotriacetate	898
Cu^{II}	dipeptides	899
Cu^{II}	bipy-adenosine or -inosine	900
Cu^{II}	aspartic, glutamic acids, glycine	901
Cu^{II}	4,4,9,9-tetramethyl-5,8-diazododecane-2,11-diamine	902
Ni^{II}	methylene-, ethylene-, and trimethylene-dithiodiacetic acids	903
Ni^{II}, Cu^{II}	mixed-ligand amino-acid compounds	904
Co^{II}, Ni^{II}, Cu^{II}	3-[(carboxymethyl)thiol]-L-alanine	905
Ni^{II}, Cu^{II}	3,7-diazanonadioic acid diamide and diethylamide	906
Cu^{II}	asparagine and flutamine	907
Cu^{II}	chloro-N-methyltetraphenylporphine	908
Cu^{II}	4,9-diazadodecane-1,12-diamine (spermine)	909
Cu^{II}	3-azaheptane-1,7-diamine and homologues	910
Co^{II}, Ni^{II}, Cu^{II}	4,7-diazadecane-1,10-diamine	911
Cu^{II}	ternary complexes: diamines + oxalate, glycinate	912
Cu^{II}	edta and diethylenetriaminepenta-acetic acid	913
Cu^{II}	tyrosine and tryptophan derivatives	914
Cu^{II}	imidazole and OH$^-$ ion	915

5 Bibliography

A number of review articles covering aspects of cobalt, nickel, or copper chemistry have appeared. These include metal cyanide complexes of phosphines,[916] higher

[892] O. Yamauchi, H. Benno, and A. Nakahara, *Bull. Chem. Soc. Japan*, 1973, **46**, 3458.

[893] R. S. Vaidya and S. N. Banerji, *Indian J. Chem.*, 1973, **11**, 1200.

[894] C. B. Riolo, T. F. Soldi, G. Gallotti, and M. Pesavento, *Gazzetta*, 1974, **104**, 193.

[895] A. Gergely and I. Sovayo, *J. Inorg. Nuclear Chem.*, 1973, **35**, 4355.

[896] N. S. Al-Niaimi and H. A. A. Rasoul, *J. Inorg. Nuclear Chem.*, 1974, **36**, 2051.

[897] R. P. Singh and A. A. Khan, *J. Inorg. Nuclear Chem.*, 1973, **35**, 3865.

[898] W. A. E. McBryde, J. L. McCourt, and V. Cheam, *J. Inorg. Nuclear Chem.*, 1973, **35**, 4193.

[899] R. Nakon and R. J. Angelici, *J. Amer. Chem. Soc.*, 1974, **96**, 4178.

[900] C. F. Nauman and H. Sigel, *J. Amer. Chem. Soc.*, 1974, **96**, 2750.

[901] I. Nagypal, A. Gergely, and E. Farkas, *J. Inorg. Nuclear Chem.*, 1974, **36**, 699.

[902] G. R. Hedvig and H. K. J. Powell, *J.C.S. Dalton*, 1973, 1942.

[903] M. Aplincourt and R. Hugel, *J. Inorg. Nuclear Chem.*, 1974, **36**, 345.

[904] R. Nakon, P. R. Rechani, and R. J. Angelici, *Inorg. Chem.*, 1973, **12**, 2431.

[905] R. Nakon, E. M. Beadle, and R. J. Angelici, *J. Amer. Chem. Soc.*, 1974, **96**, 719.

[906] T. A. Kaden and A. D. Zuberbühler, *Helv. Chim. Acta*, 1974, **57**, 286.

[907] A. Gergely, I. Nagypal, and E. Farkas, Proceedings 16th International Conference on Coordination Chemistry, Dublin, 1974.

[908] D. K. Lavallec and A. E. Gebala, *Inorg. Chem.*, 1974, **13**, 2004.

[909] B. N. Palmer and H. K. J. Powell, *J.C.S. Dalton*, 1974, 2086.

[910] B. N. Palmer and H. K. J. Powell, *J.C.S. Dalton*, 1974, 2089.

[911] R. Barbucci, L. Fabbrizzi, P. Paoletti, and A. Vacca, *J.C.S. Dalton*, 1973, 1763.

[912] H. Sigel, R. Caraco, and B. Prijo, *Inorg. Chem.*, 1974, **13**, 462.

[913] E. W. Baumann, *J. Inorg. Nuclear Chem.*, 1974, **36**, 1827.

[914] O. A. Weber, *J. Inorg. Nuclear Chem.*, 1974, **36**, 1341.

[915] S. Sjoberg, *Acta Chem. Scand.*, 1973, **27**, 3721.

[916] P. Rigo and A. Turco, *Coord. Chem. Rev.*, 1974, **13**, 133.

oxidation state chemistry of Co and Ni,[917] five-co-ordination in Co[II] and Ni[II],[918] complexes of SN chelates,[919] nitrosyl complexes,[920] aromatic amine-*N*-oxide complexes,[921] models for copper–protein interaction,[922] oxidation states and reversible redox reactions of metalloporphyrins,[923] the organic chemistry of copper,[924] and a comprehensive review of P-, As-, and Sb-donor complexes.[925]

[917] W. Levason and C. A. McAuliffe, *Coord. Chem. Rev.*, 1974, **12**, 151.
[918] R. Morassi, I. Bertini, and L. Sacconi, *Coord. Chem. Rev.*, 1973, **11**, 343.
[919] M. A. Ali and S. E. Livingstone, *Coord. Chem. Rev.*, 1974, **13**, 101.
[920] J. H. Enemark and R. D. Feltham, *Coord. Chem. Rev.*, 1974, **13**, 339.
[921] N. M. Karayannis, L. L. Pytlewski, and C. M. Mikulski, *Coord. Chem. Rev.*, 1973, **11**, 93.
[922] R. Österberg, *Coord. Chem. Rev.*, 1974, **13**, 309.
[923] J. H. Fuhrhop, *Structure and Bonding*, 1974, **18**, 1.
[924] A. E. Jukes, *Adv. Organometallic Chem.*, 1974, **12**, 215.
[925] 'Transition–metal complexes of Phosphorus, Arsenic, and Antimony Donor Ligands', ed. C. A. McAuliffe, McMillan, London, 1973.

3
The Noble Metals

BY L. A. P. KANE-MAGUIRE AND D. W. CLACK

PART I: Ruthenium, Osmium, Rhodium, and Iridium *by L. A. P. Kane-Maguire*

1 Ruthenium

Cluster Compounds.—While pyrolysis of solid $Ru_3(CO)_{12}$ is known[1] to lead to carbide complexes such as $Ru_6C(CO)_{17}$, the presence of small amounts of water has been shown to yield instead the polynuclear hydrides α-$H_4Ru_4(CO)_{12}$ and α-$H_2Ru_4(CO)_{13}$.[2] Use of deuterium oxide gave the corresponding deuteriated species. An alternative route to carbide compounds involves heating α-$H_4Ru_4(CO)_{12}$ at 130 °C under 10 atm of ethylene.[3] Chromatography identified at least eight products, including $Ru_6C(CO)_{17}$ (30%) and the new $Ru_5C(CO)_{15}$ (*ca.* 1%). The ethylene apparently plays no role in this reaction. An interesting report indicates that photolysis may lead to unusual reactions in ruthenium clusters.[4] Room temperature irradiation of n-heptane solutions of $Ru_3(CO)_{12}$ under carbon monoxide (10 p.s.i.) with Pyrex-filtered sunlight rapidly produces a quantitative yield of $Ru(CO)_5$. This contrasts with previous thermal reactions of donor ligands on $Ru_3(CO)_{12}$, which involve carbonyl substitution rather than Ru—Ru bond fission.

The adduct $Ru_3(CO)_{12}$,$AlBr_3$ is formed *via* treatment of $Ru_3(CO)_{12}$ with $AlBr_3$ in toluene.[5] This diamagnetic red solid reverts to the parent on exposure to air or acetone. It displays a low-frequency carbonyl stretching band at 1535 cm^{-1} characteristic of Lewis-acid-co-ordinated bridging carbonyls. Reaction of $Ru_3(CO)_{12}$ with 1,2-bis(dimethylsilyl)ethane has been shown[6] to yield the stable white chelate complex (1; M = Ru); ^{13}C n.m.r. studies indicate that this species is stereochemically rigid.

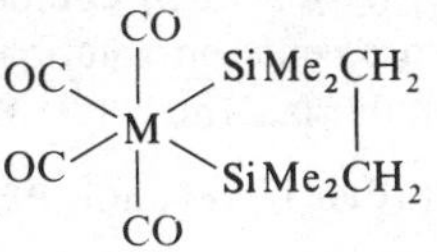

$$OC \underset{OC}{\overset{CO}{\underset{|}{\overset{|}{M}}}} \overset{SiMe_2CH_2}{\underset{SiMe_2CH_2}{\diagdown}}$$

CO

(1)

Ruthenium-(− II), -(0), and -(I).—A single-crystal *X*-ray analysis of the complex $[Ru(NO)_2(PPh_3)_2]$ has revealed a distorted tetrahedral co-ordination about the

[1] C. R. Eady, B. F. G. Johnson, and J. Lewis, *J. Organometallic Chem.,* 1972, **37**, C39.
[2] C. R. Eady, B. F. G. Johnson, and J. Lewis, *J. Organometallic Chem.,* 1973, **57**, C84.
[3] C. R. Eady, B. F. G. Johnson, J. Lewis, and T. Matheson, *J. Organometallic Chem.,* 1973, **57**, C82.
[4] B. F. G. Johnson, J. Lewis, and M. V. Twigg, *J. Organometallic Chem.,* 1974, **67**, C75.

ruthenium atom, and linear Ru—N—O bonds.[7] This result suggests a ruthenium $(-II)$ species with the nitrosyl ligand co-ordinated as NO^+. A novel nitrosyl-transfer reaction of this complex with $[RuCl_2(PPh_3)_3]$ was also reported, giving the known Ru^0 complex $[RuCl(NO)(PPh_3)_2]$ (equation 1). Pyrolysis of $Ru_3(CO)_{12}$ in the presence

$$[Ru(NO)_2(PPh_3)_2] + [RuCl_2(PPh_3)_3] \xrightarrow[\text{dust}]{\text{Zn}} 2[RuCl(NO)(PPh_3)_2] + PPh_3 \quad (1)$$

of donor ligands has been shown to provide convenient routes to a variety of Ru^0

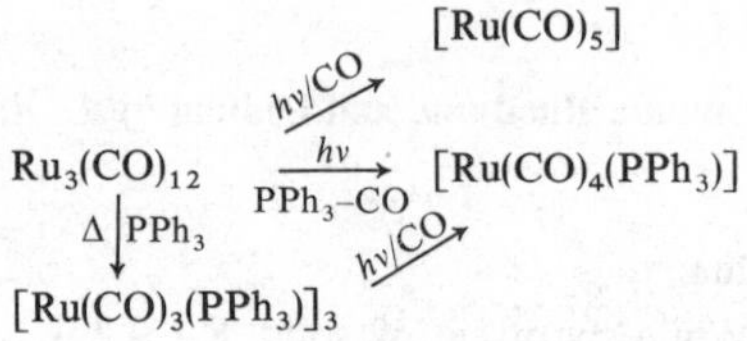

$$Ru_3(CO)_{12} \begin{array}{c} \xrightarrow{hv/CO} [Ru(CO)_5] \\ \xrightarrow[PPh_3-CO]{hv} [Ru(CO)_4(PPh_3)] \\ \xrightarrow{hv/CO} \end{array}$$
$$\Delta \downarrow PPh_3$$
$$[Ru(CO)_3(PPh_3)]_3$$

Scheme 1

complexes (Scheme 1).[4] The step-wise reduction of $RuCl_3,3H_2O$ by molecular hydrogen has been observed in NN-dimethylacetamide solution, leading finally to a Ru^I species.[8] A Ru^{II} hydride intermediate was proposed in the ultimate $Ru^{II} \rightarrow Ru^I$ step, which does not occur in aqueous solution.

Ruthenium(II).—*Group VII Donors. Hydrido-phosphine complexes.* Reaction of the ruthenium(II) species $[Ru_5Cl_{12}]^{2-}$ with an ethanolic solution of triphenylphosphine and $NaBPh_4$ has produced the complex $[RuH(BPh_4)(PPh_3)_3]$.[9] A high-field resonance at 19τ was observed for the metal–hydride, while i.r. data indicated co-ordination of the tetrahydroborato ligand *via* a double hydrogen-bridged structure.

Halogeno-carbonyl and -phosphine complexes. A convenient general synthesis for phosphine complexes of the type $[RuX_2(PR_3)_n]$ (X = halide; n = 3 or 4) has been described involving refluxing $[RuX_2(PPh_3)_4]$ in hexane with excess PR_3 (*e.g.* PMe_2Ph, $PMePh_2$, PEt_2Ph, $PEtPh_2$, or $PClPh_2$).[10] The ethylphosphine species are exclusively $n = 3$, while methylphosphines give six-co-ordinate products. Solvent polarity is critical in the above preparations, polar solvents yielding instead the dinuclear cation $[Ru_2Cl_3(PR_3)_6]^+$. Two related studies report the dissociative behaviour of $[RuCl_2(PPh_3)_n]$ (n = 3 or 4) complexes in solution.[11, 12] Variable-temperature [31]P n.m.r. measurements on solutions of $[RuCl_2(PPh_3)_4]$ in chloroform are consistent with the equilibria (2) and (3).[11] While dissociation (2) is essentially

$$[RuCl_2(PPh_3)_4] \rightleftharpoons [RuCl_2(PPh_3)_3] + PPh_3 \quad (2)$$

$$[RuCl_2(PPh_3)_3] \rightleftharpoons \text{`}RuCl_2(PPh_3)_2\text{'} + PPh_3 \quad (3)$$

[5] J. S. Kristoff and D. F. Schriver, *Inorg. Chem.*, 1974, **13**, 499.
[6] L. Vancea and W. A. G. Graham, *Inorg. Chem.*, 1974, **13**, 511.
[7] A. P. Gaughan, B. J. Corden, R. Eisenberg, and J. A. Ibers, *Inorg. Chem.*, 1974, **13**, 786.
[8] B. C. Hui and B. R. James, *Canad. J. Chem.*, 1974, **52**, 348.
[9] D. G. Holah, A. N. Hughes, B. C. Hui, and K. Wright, *Inorg. Nuclear Chem. Letters*, 1973, **9**, 835.
[10] P. W. Armit and T. A. Stephenson, *J. Organometallic Chem.*, 1973, **57**, C80.
[11] K. G. Caulton, *J. Amer. Chem. Soc.*, 1974, **96**, 3005.
[12] B. R. James and L. D. Markham, *Inorg. Chem.*, 1974, **13**, 97.

complete, only small amounts of 'RuCl$_2$(PPh$_3$)$_2$' are formed. The ^{31}P n.m.r. spectrum of the latter species indicates the chloride-bridged dimeric structure (2), where ruthenium has a square-pyramidal geometry. Equilibrium (3) has also been con-

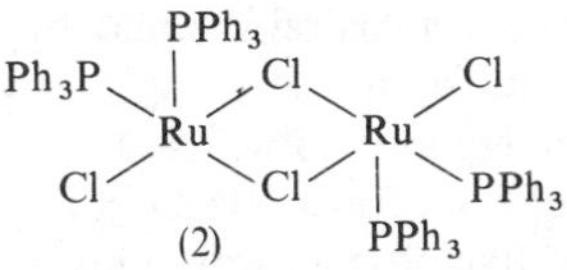

(2)

firmed in benzene and NN'-dimethylacetamide, with extensive dissociation.[12] Loss of a chloride ion from [RuCl$_2$(PPh$_3$)$_2$] was also observed in NN'-dimethylacetamide, and the equilibrium constant measured using u.v.–visible spectrophotometry.

Reaction of [RuCl$_2$(PPh$_3$)$_2$] and [RuCl$_2$(AsPh$_3$)$_2$]$_2$ with neat amine ligands has led to partial or complete replacement of the PPh$_3$ or AsPh$_3$ groups.[13] Mixed complexes obtained include dark violet [RuCl$_2$(EPh$_3$)(am)$_3$] (E = P or As; am = PhNH$_2$ or o-toluidine) and [RuCl$_2$(AsPh$_3$)$_2$(am)$_2$] (am = MeCN, Et$_2$NH, or Et$_3$N). In contrast, quinoline and substituted pyridines gave the fully substituted [RuCl$_2$(am)$_4$]. New thiocarbonyl complexes [RuCl$_2$(CS)$_2$(EPh$_3$)$_2$] were also obtained *via* direct reaction with carbon disulphide. Similar substitutions on the methanol complexes [RuX$_3$(EPh$_3$)$_2$(MeOH)] (E = P or As; X = Cl or Br) have been reported, using a variety of N,S,O, and C donors.[14] Species isolated include [RuX$_3$(EPh$_3$)$_2$L] (L = RCN, Me$_2$SO, CS$_2$, or CS), [RuX$_3$(EPh$_3$)$_2$L$_2$] (L$_2$ = Me$_2$S, bipy, or py), and [Ru(S$_2$PPh$_2$)$_3$]. Concomitant reduction to Ru^{II} was also observed under certain conditions, yielding [RuX$_2$(EPh$_3$)$_2$L$_2$] (L = CO or RCN) and [RuX$_2$L$_4$] (L = py or Me$_2$SO), and the cationic species [RuCl(PPh$_3$)(N–N)$_2$]Cl (N–N = 2,2'-bipyridine or 1,10-phenanthroline).

In the solvents NN'-dimethylformamide, NN'-dimethylacetamide, and dimethyl sulphoxide, the complexes [RuX$_2$(PPh$_3$)$_3$] and [HRuX(PPh$_3$)$_3$] (X = Cl or Br) have been shown to react with CO or CO–H$_2$ mixtures to give [RuX$_2$(CO)$_2$(PPh$_3$)$_2$], [HRuX(CO)$_2$(PPh$_3$)$_2$], and [RuX$_2$(CO)(PPh$_3$)$_2$(solvent)].[15] The solvent molecules are co-ordinated *via* O donors, and could be removed by recrystallization from CH$_2$Cl$_2$–MeOH producing the new five-co-ordinate complex [RuCl$_2$(CO)(PPh$_3$)$_2$]. This species proved to be an effective olefin-isomerization catalyst. Reaction of [RuCl$_2$(CS)(PPh$_3$)$_2$]$_2$ with HCl in acetone leads to chloride-bridge cleavage and isolation of Ph$_4$As[RuCl$_3$(CS)(PPh$_3$)$_2$].[16] In the absence of HCl, however, the dimeric species [(PPh$_3$)$_2$ClRuCl$_3$Ru(CS)(PPh$_3$)$_2$] is obtained, being also prepared *via* the action of CS$_2$ on [RuCl$_2$(PPh$_3$)$_3$]. On the basis of ^{31}P n.m.r. and X-ray studies (Table 1) it was assigned the tri-μ-chloro structure (3).

The new five-co-ordinate complex [RuCl$_2$(CO)(PCy$_3$)$_2$] (PCy$_3$ = tricyclohexylphosphine) has been prepared (equation 4).[17] Its i.r. spectrum suggested the structure

$$RuCl_3 + CO + PCy_3 \xrightarrow[\text{2-methoxyethanol}]{\Delta} [RuCl_2(CO)(PCy_3)_2] \qquad (4)$$

[13] R. K. Poddar and U. Agarwala, *J. Inorg. Nuclear Chem*., 1973, **35**, 3769.
[14] L. Ruiz-Ramirez, T. A. Stephenson, and E. S. Switkes, *J.C.S. Dalton*, 1973, 1770.
[15] B. R. James, L. D. Markham, B. C. Hui, and G. L. Rempel, *J.C.S. Dalton*, 1973, 2247.
[16] T. A. Stephenson, E. S. Switkes, and P. W. Armit, *J.C.S. Dalton*, 1974, 1134.
[17] F. G. Moers, R. W. M. Ten Hoedt, and J. P. Langhout, *J. Organometallic Chem*., 1974, **65**, 93.

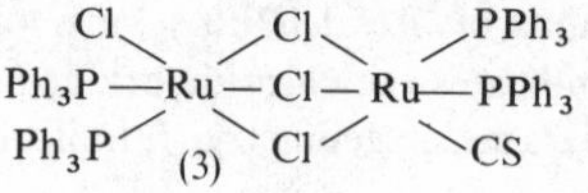

(3)

(4) in which the unoccupied octahedral site may be blocked by a bulky cyclohexyl group. However, it can pick up a CO group to give *trans*-$[RuCl_2(CO)_2(PCy_3)_2]$. A similar five-co-ordinate complex $[RuCl_2L_3]$ (L = PMe_2naphthyl) has been isolated from the reaction of L on $RuCl_3,3H_2O$.[18] Failure to obtain more than small amounts of $[Ru_2Cl_3L_6]^+$ (the product with other dialkylarylphosphines) was again ascribed to steric factors. This complex also adds a further ligand (L′) to yield octahedral compounds $[RuCl_2L_3L']$ (L′ = CO, py, or MeCN).

Carbonylation of several transition-metal phosphine complexes with carbon dioxide under mild conditions has been described, including reaction with $[RuCl_2(PPh_3)_3]$ to give $[Ru(CO)Cl_2(PPh_3)_3]$.[19] The first example of an *o*-metallated complex of Ru^{II}–phosphine systems has been reported.[20] Complex (5) was isolated from the stoicheiometric hydrogenation of olefins in vacuo using $[HRuCl(PPh_3)_3]$. Reaction of (5) with H_2–PPh_3 regenerates the hydrido species, while HCl produces a mixture of $[RuCl_2(PPh_3)_2]$ and $[RuCl_2(PPh_3)_3]$.

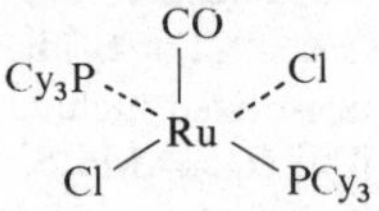

(4)

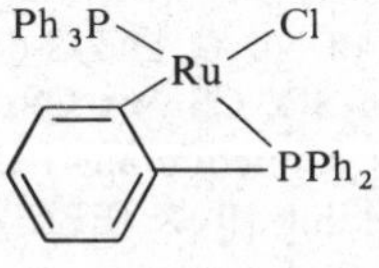

(5)

The red solution obtained from the treatment of ethanolic $RuCl_3,xH_2O$ with CO has been shown from i.r. studies to contain a mixture of $[RuCl_2(CO)_2]$ and *fac*-$[RuCl_3(CO)_3]^-$.[21] In addition, a carbonyl band at 2044 cm^{-1} was tentatively assigned to either $[RuCl_3(CO)]$ or $[RuCl_5(CO)]^{2-}$. The species $[RuClH(PPh_3)_3]$ has been proposed as the active catalyst in the hydrosilylation of ketones and aldehydes by Et_3SiH catalysed by $[RuCl_2(PPh_3)_3]$ (equation 5).[22]

$$[RuCl_2(PPh_3)_3] + Et_3SiH \rightarrow [RuClH(PPh_3)_3] \qquad (5)$$

Group VI Donors. O-Donor Ligands. Reaction of carboxylic acids with hydrido and triphenylphosphine complexes has provided a general route to a variety of carboxylato-complexes including $[RuH(OCOR)(PPh_3)_3]$, $[RuX(OCOR)(CO)(PPh_3)_2]$ (X = H or Cl), $[Ru(OCOR)_2(CO)(PPh_3)_2]$, and $[Ru(OCOR)_2(CO)_2(PPh_3)_2]$ (R = alkyl or aryl).[23] Molecular weight measurements indicate that all are monomeric in solution, thus excluding the presence of bridging carboxylate ligands. Both uni- and bi-dentate carboxylate groups were established by i.r. spectroscopy. Tris-

[18] P. G. Douglas and B. L. Shaw, *J.C.S. Dalton*, 1973, 2078.
[19] P. Svoboda, T. S. Belopotapova, and J. Hetflejs, *J. Organometallic Chem.*, 1974, **65**, C37.
[20] B. R. James, L. D. Markham, and D. K. W. Wang, *J.C.S. Chem. Comm.*, 1974, 439
[21] M. L. Berch and A. Davison, *J. Inorg. Nuclear Chem.*, 1973, **35**, 3763.
[22] C. Eaborn, K. Odell, and A. Pidcock, *J. Organometallic Chem.*, 1973, **63**, 93.
[23] S. D. Robinson and M. F. Uttley, *J.C.S. Dalton*, 1973, 1912.

{(+)-3-acetylcamphorato}ruthenium(III) (6) is known to exist in four diastereo-isomeric forms. The electrolytic oxidation and reduction of each of these isomers have now been studied, producing the corresponding Ru^{IV} and Ru^{II} species with complete retention of configuration.[24] Absolute configurations were assigned to

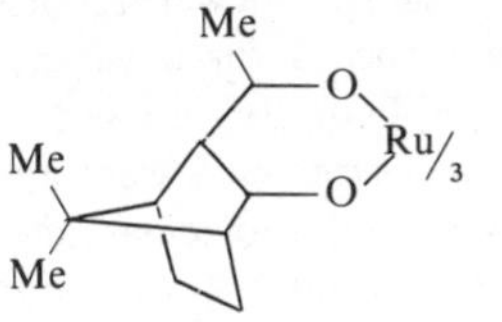

(6)

these Ru^{II} and Ru^{IV} diastereoisomers from their circular dichroism, employing the known configurations of the Ru^{III} precursors.

S-*donor ligands.* The stable green dinuclear species $[(NH_3)_5Ru\text{—}SS\text{—}Ru(NH_3)_5]^{4+}$ has been prepared by a number of routes including reaction of $[Ru(NH_3)_5(H_2O)]^{2+}$ with COS or Zn–Hg reduction of $[Ru(NH_3)_5(SO_2)]^{2+}$.[25] This complex represents a novel bonding mode for sulphur, involving either Ru^{II} atoms bridged by singlet disulphur, or Ru^{III} centres bridged by S_2^{2-} ($\mu_{eff} = 0.45$ BM at 25 °C, $v(S\text{–}S) = 519$ cm^{-1}). It reacts stoicheiometrically with cyanide ions according to equation (6).

$$[(NH_3)_5Ru\text{-}SS\text{-}Ru(NH_3)_5]^{4+} + 2SCN^- \rightarrow 2[Ru(NH_3)_5(SCN)]^+ \qquad (6)$$

E.s.r. spectra have been recorded for a large range of tris(bidentate sulphur) metal complexes, including $[Ru(S\text{–}S)_3]^{n-}$ (S–S = sacsac, *cis*-1,2- dicyanoethylene-1,2-dithio-late, dimethyldithiocarbamate, diphenyldithiophosphinate, or dithio-oxalate; $n = 0$ or 3).[26] Some of the complexes were prepared for the first time. The near isotropic g-values obtained indicate large low-symmetry distortion, both geometric and electronic in origin.

The synthesis of a series of compounds of general formula *cis*-$[Ru(S\text{–}S)_2L_2]$-[S–S$^-$ = $^-S_2PR_2$ (R = Me, Et, or Ph) or $^-S_2CNMe_2$; L = phosphines or phosphites] has been reported *via* treatment of suitable Ru^{II} or Ru^{III} substrates with Na(S–S).[27] For S–S$^-$ = $^-S_2PR_2$, ready carbonylation occurs to give $[Ru(S_2PR_2)_2\text{-}L(CO)]$. All the above compounds undergo ligand exchange with more basic ligands L', yielding $[Ru(S\text{–}S)_2LL']$ and/or $[Ru(S\text{–}S)_2L'_2]$. Most of the complexes also showed temperature-dependent ^1H n.m.r. spectra. Lineshape analysis indicated that while this is due to facile interconversion of optical enantiomorphs for the $^-S_2PR_2$ complexes, restricted rotation about C—N bonds was responsible in the case of the dithiocarbamate ($^-S_2CNMe_2$) species.[28]

Group V Donors. Molecular nitrogen complexes. The known complex $[Ru(en)_2(N_2)\text{-}(H_2O)]^{2+}$ has been prepared by treating $[Ru(en)_3]^{2+}$ with alkaline NO.[29] This

[24] G. W. Everett and R. R. Horn, *J. Amer. Chem. Soc.*, 1974, **96**, 2087.
[25] C. R. Brulet, S. S. Isied, and H. Taube, *J. Amer. Chem. Soc.*, 1973, **95**, 4758.
[26] R. E. De Simone, *J. Amer. Chem. Soc.*, 1973, **95**, 6238.
[27] D. J. Cole-Hamilton and T. A. Stephenson, *J.C.S. Dalton*, 1974, 739.
[28] D. J. Cole-Hamilton and T. A. Stephenson, *J.C.S. Dalton*, 1974, 754.
[29] S. Pell and J. N. Armor, *J.C.S. Chem. Comm.*, 1974, 259.

interesting reaction, similar to the recent report[30] in which $[Ru(NH_3)_5(N_2)]^{2+}$ was prepared from $[Ru(NH_3)_6]^{2+}$, apparently involves oxidation of ethylenediamine by NO, and C—N bond cleavage. Complete details have appeared for the preparation of pure samples of the ^{15}N-labelled complexes $[Ru(NH_3)_5(N_2)]$ Br_2 (where N_2 is $^{14}N_2$, $^{14}N^{15}N$, or $^{15}N_2$), and i.r. spectra were reported.[31] Low-frequency Raman and i.r. spectra have also been recorded for a variety of $[M(NH_3)_5Y]X_2$ complexes (M = Ru or Os; X = Cl, Br, I, or BF_4; Y = N_2, or CO).[32] Deuteriation, ^{15}N-substitution, and intensity measurements were employed to assist in the assignment of M—Y vibrational modes. Molecular nitrogen has been shown from i.r. and ^1H n.m.r. evidence to have a much greater affinity than pent-2-ene for $[RuH_2(PPh_3)_3]$.[33] This finding is in accord with a recent report of the retarding influence of nitrogen on the double bond isomerization and hydrogenation of pentenes catalysed by $[RuH_2(PPh_3)_3]$ or $[RuH_4(PPh_3)_4]$,[34] and suggests that more attention should be given to the possible poisoning effects of molecular nitrogen.

Nitrosyl complexes. An interesting communication suggests that there is a correlation between the behaviour of co-ordinated NO^+ (including Ru^{II} nitrosyl complexes) as an electrophile and the position of the NO stretching frequency.[35] Only complexes with $\nu(NO) > 1886\ cm^{-1}$ are attacked at the nitrosyl group by nucleophiles such as OH^-, OMe^-, or N_3^-. The first report has appeared of the interconversion of trigonal-bipyramidal and square-pyramidal structures for a nitrosyl complex, $[Ru(NO)(Ph_2PC_2H_4PPh_2)_2]PF_6$.[36] A proton-decoupled ^{31}P n.m.r. study indicated a turnstile rotation, with the NO group axial in the square-pyramid transition state.

Other N-*donor ligands.* Nitrile complexes such as $[RuX_2(RCN)_4]$ (R = Me, Et, Pr^n, Pr^i, Ph, or $PhCH_2$; X = Cl or Br) have been prepared from $RuCl_3,3H_2O$

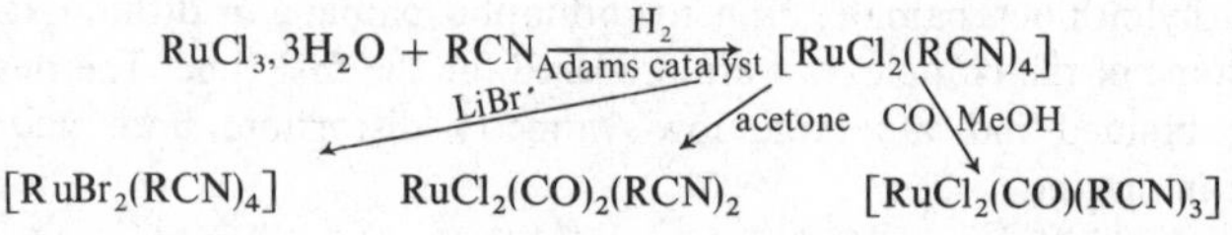

$$[RuBr_2(RCN)_4] \xleftarrow{\ LiBr\ } RuCl_3,3H_2O + RCN \xrightarrow[\text{Adams catalyst}]{H_2} [RuCl_2(RCN)_4]$$

$$RuCl_2(CO)_2(RCN)_2 \qquad\qquad [RuCl_2(CO)(RCN)_3]$$

acetone   CO MeOH

Scheme 2

(Scheme 2).[37] These complexes exhibit $\nu(CN)$ near $2250\ cm^{-1}$ indicating co-ordination to Ru *via* the N atom. Rupture of the chloride bridges in polymeric $[RuCl_2(diene)]_n$ by 15 different amines has been employed as a general route to complexes of the type $[RuCl_2(amine)_2]$.[38] With dimethylamine, piperidine, pyridine, or cyclohexylamine, a side-reaction leading to $[RuHCl(amine)_2]$ was also observed. A series of hexakis(amine) complexes of the type $[RuL_6]X_2$ (L = $MeNH_2$, $EtNH_2$,

[30] S. Pell and J. N. Armor, *J. Amer. Chem. Soc.*, 1972, **94**, 686.
[31] S. Pell, R. H. Mann, H. Taube, and J. N. Armor, *Inorg. Chem.*, 1974, **13**, 479.
[32] M. W. Bee, S. F. A. Kettle, and D. B. Powell, *Spectrochim. Acta*, 1974, **30A**, 585
[33] F. Pennella, *J. Organometallic Chem.*, 1974, **65**, C17.
[34] F. Pennella, R. L. Banks, and M. R. Rycheck, Proceedings of the Fourteenth International Conference Coordination Chemistry, Toronto, 1972, p. 78.
[35] F. Bottomley, W. V. F. Brooks, S. G. Clarkson, and S. B. Tong, *J.C.S. Chem. Comm.*, 1973, 919.
[36] P. R. Hoffman, J. S. Miller, C. B. Ungermann, and K. G. Caulton, *J. Amer. Chem. Soc.*, 1973, **95**, 7902.
[37] W. E. Newtown and J. E. Searles, *Inorg. Chim. Acta*, 1973, **7**, 348.
[38] C. Potvin and G. Pannetier, *Bull. Soc. chim. France*, 1974, 783.

Pr^nNH_2, Bu^nNH_2, or Bu^iNH_2) has been prepared *via* route (7),[39] which is similar to the method previously employed[39a] to synthesize the analogous hexa-ammine.

$$RuCl_2 \cdot xH_2O + RCH_2NH_2 + Zn \xrightarrow[H_2O, KX]{Ar} [Ru(RCH_2NH_2)_6]X_2 \qquad (7)$$

Subsequent reaction of these complexes with benzonitrile yielded the useful products $[RuL_5(PhCN)]X_2$.

Two papers report the isolation of arylazoruthenium complexes.[40,41] Reaction of $ArNH_2$ ($Ar = p\text{-}MeC_6H_4$ or $p\text{-}MeOC_6H_4$) with the co-ordinated nitrosyl ligand in $[RuCl(NO)(bipy)_2]^{2+}$ yielded complexes (7).[40] The position of the N–N

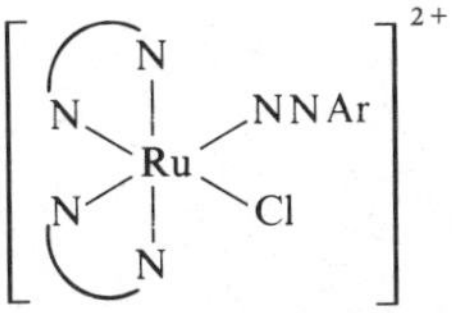

(7)

stretch in their i.r. spectra (*ca.* 2090 cm^{-1}) indicated that ArN_2 is co-ordinated as the cationic diazonium ligand $Ar—N{\equiv}N^+$, rather than as $ArN{=}N^-$ [ν (NN) generally *ca.* 1444—1642 cm^{-1}]. Aqueous solutions of (7) evolved nitrogen gas on warming, while treatment with KI produced $[RuI_2(bipy)_2]$, *p*-iodoanisole, and nitrogen. Direct reaction between $ArN_2^+Br^-$ and $[RuX_2(PPh_3)_3]$ also provided a route to the arylazo-complexes $[RuX_3(N_2Ar)(PPh_3)_2]$.[41] I.r. studies again suggest a linear $Ru—N—NAr$ linkage, *i.e.* $Ru^{II}—N^+{\equiv}NAr$. Alternatively, treatment of $RuCl_3$ with an alcoholic mixture of 1,3-diaryltriazene and PPh_3 may be employed. However, 1,3-diaryltriazenes have been shown to react generally with hydrido or low oxidation state phosphine complexes of platinum metals to afford a wide variety of complexes containing the 1,3-diaryltriazenido ligand $(ArNNNAr^-)$.[42] Products obtained in this manner include the monomeric, air-stable $[Ru(ArNNNAr)_2(PPh_3)_2]$ and $[RuX(ArNNNAr)(CO)(PPh_3)_2]$ ($X = H$ or Cl). I.r. and H n.m.r. data indicate bidentate $ArNNNAr^-$ ligands, and structures (8) and (9), respectively ($M = Ru$).

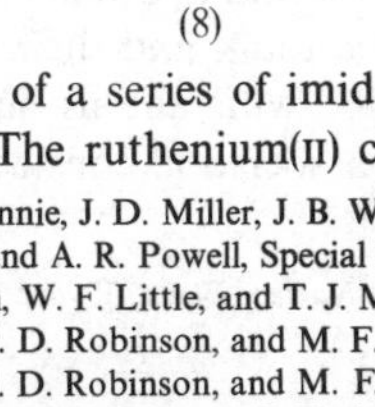

(8)

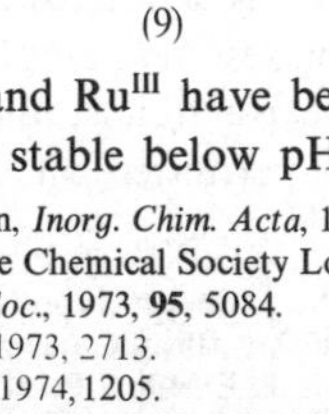

(9)

Complexes of a series of imidazoles with Ru^{II} and Ru^{III} have been prepared *via* Scheme 3.[43] The ruthenium(II) complexes are not stable below pH 2 and undergo

[39] W. R. McWhinnie, J. D. Miller, J. B. Watts, and D. Y. Waddan, *Inorg. Chim. Acta*, 1973, 7, 461.
[39a] F. M. Lever and A. R. Powell, Special Publication No. 12, The Chemical Society London 1959, p. 135.
[40] W. L. Bowden, W. F. Little, and T. J. Meyer, *J. Amer. Chem. Soc.*, 1973, 95, 5084.
[41] K. R. Laing, S. D. Robinson, and M. F. Uttley, *J.C.S. Dalton*, 1973, 2713.
[42] K. R. Laing, S. D. Robinson, and M. F. Uttley, *J.C.S. Dalton* 1974, 1205.
[43] R. J. Sundberg, R. F. Bryan, I. F. Taylor, and H. Taube, *J. Amer. Chem. Soc.*, 1974, 96, 381.

aquation to give $[Ru(NH_3)_5(H_2O)]^{2+}$ and a low conversion into derivatives of Ru^{II}–$(NH_3)_4$ in which an imidazole ligand is bonded to Ru *via* C-2. Two anionic isothiocyanato-complexes have been isolated from the reaction of NH_4CNS with

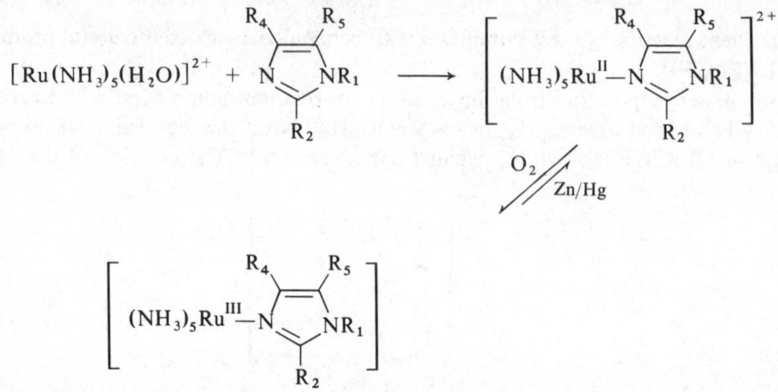

Scheme 3

$RuCl_3,H_2O$ in the presence of carbon monoxide.[44] The chloroform-soluble fraction yielded purple $(Et_4N)_2[Ru(CO)_2(NCS)_4]$, and conductivity data indicated the polymeric composition $(Et_4N)_2[Ru_2(CO)_4(CNS)_6]$ for the green residue, for which i r studies showed both N-bonded CNS and Ru—SCN—Ru bridging.

Synthesis of the new complexes $[Ru(en)_2(phen)]I_2$ and $[Ru(phen)_2(diim)]I_2$ (where diim is NH=CHCH=NH) has been reported (equations 8 and 9).[45] Reaction (8) is analogous to the previously reported oxidation of $[Ru(en)_3]^{2+}$ to give

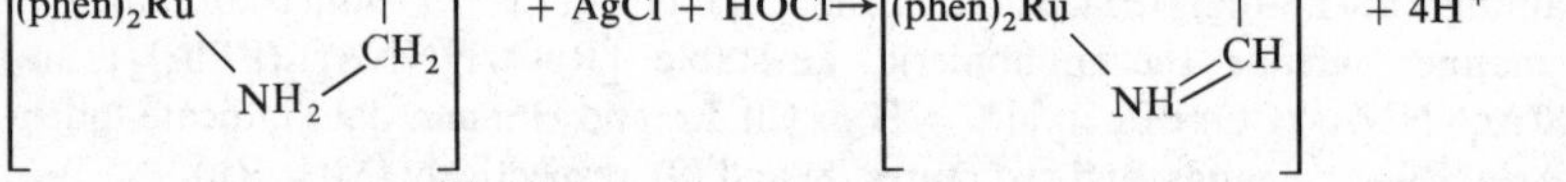

$$cis\text{-}[Ru(en)_2Cl_2]Cl + Zn/Hg \xrightarrow{phen} [Ru(en)_2(phen)]I_2 \qquad (9)$$

$[Ru(en)_2(diim)]^{2+}$, for which an improved synthesis was also given. An extensive study has been made of the reactivity of derivatives of 1,10-phenanthroline, both free and as tris-complexes with a variety of metals including ruthenium(II).[46] Reduction of $[Ru(5\text{-}NO_2phen)_3]^{2+}$ with $NaBH_4$ or Sn–HCl giving $[Ru(5\text{-}NH_2phen)_3]^{2+}$ was studied, as well as an interesting reversible reaction with hydroxide ion. Optically active Ru^{II} complexes were employed as mechanistic tools, and their activities discussed. Solvent extraction studies of $[Ru(phen)_3]^{2+}$ with anions such as ClO_4^-, I^-, or NCS^- has permitted calculation of ion-pair association constants in various organic solvents.[47]

[44] A. R. M. Abbadi, M. A. Shehadeh, and J. V. Kingston, *J. Inorg. Nuclear Chem.*, 1974, **36**, 1173.
[45] D. F. Mahoney and J. K. Beattie, *Inorg. Chem.*, 1973, **12**, 2561.
[46] R. D. Gillard and R. E. E. Hill, *J.C.S. Dalton*, 1974, 1217.
[47] T. Takamatsu, *Bull. Chem. Soc. Japan*, 1974, **47**, 118.

Interest continues in pyrazine-bridged polymeric ruthenium complexes. Following an earlier communication,[48] further details have been given of a general synthetic route to pyrazine-bridged Ru^{II} bipyridyl complexes (equations 10 and 11).[49] Employing the reaction of $Ru-NO_2$ complexes to yield $Ru-NO$, in combination with the

$$[RuCl(NO)(bipy)_2]^{2+} + N_3^- + \text{solvent} \rightarrow [RuCl(\text{solvent})(bipy)_2]^+ + N_2 + H_2O \qquad (10)$$

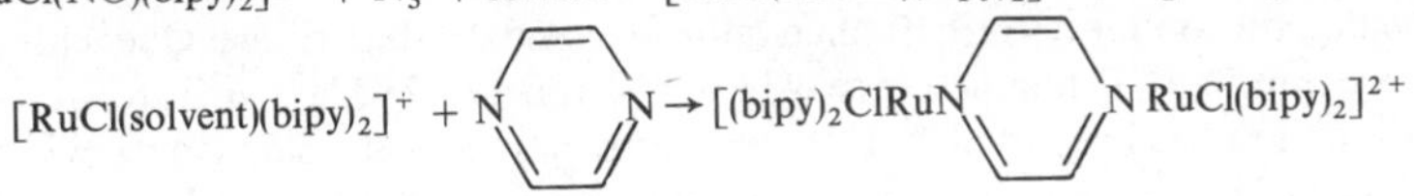

$$[RuCl(\text{solvent})(bipy)_2]^+ + N\!\!\diagdown\!\!\diagup\!\!N \rightarrow [(bipy)_2ClRuN\!\!\diagdown\!\!\diagup\!\!N\,RuCl(bipy)_2]^{2+} \qquad (11)$$

above reactions, has yielded more highly linked compounds (Scheme 4). ESCA spectra have been reported for a variety of such complexes including $[(NH_3)_5Ru(pyz)Ru-$

$$[(bipy)_2ClRu(pyz)Ru(NO)(bipy)_2]^{2+} + H^+ \rightarrow [(bipy)_2ClRu(pyz)Ru(NO)(bipy)_2]^{4+}$$

$$\big\downarrow \text{solvent} \atop N_3^-$$

$$[(bipy)_2ClRu(pyz)Ru(\text{solvent})(bipy)_2]^{3+}$$

$$\big\downarrow [RuX(pyz)(bipy)_2]^+$$

$$[(bipy)_2ClRu(pyz)Ru(bipy)_2(pyz)RuX(bipy)_2]^{4+}$$
$$(X = Cl \text{ or } NO_2)$$

Scheme 4

$(NH_3)_5]^{n+}$ ($n = 4$, 5, or 6) and $[(bipy)_2ClRu(pyz)RuCl(bipy)_2]^{n+}$ ($n = 2$, 3, or 4).[50] These studies reveal equivalent metal sites even in the mixed-valency $Ru^{II}-Ru^{III}$ species, despite the speed with which ESCA monitors electron distribution (10^{-17} s). Another investigation shows that the complex $[Ru(NH_3)_5(pyz)]^{2+}$ can associate with several aqueous first-row transition-metal ions to form pyrazine-bridged dinuclear complexes.[51] Equilibrium formation constants with Ni^{II}, Cu^{II}, and Zn^{II} were measured spectroscopically.

The redox behaviour of Ru^{II} porphyrins has been studied using e.s.r., i.r., and u.v.–visible spectroscopy.[52] Whereas $[Ru(CO)(porp)]$ containing an axially co-ordinated CO undergoes porphyrin ring oxidation as the first step, the four-co-ordinate $[Ru(porp)]$ is more readily oxidized giving the corresponding Ru^{III} species $[Ru(porp)]^+$.

Quantum yields have been measured for the photoaquation of a large range of substituted pyridine complexes of the type $[Ru(NH_3)_5(pyX)]^{2+}$.[53] The marked dependence of quantum yields on the nature of X indicates that metal-to-ligand-charge-transfer (MLCT) excited states are not involved in the photosubstitution. Presumably, a ligand-field excited state is responsible. Evidence has been reported for a simple outer-sphere reduction of cytochrome *c* by $[Ru(NH_3)_6]^{2+}$.[54] Such a

[48] S. A. Adeyemi, J. N. Braddock, G. M. Brown, J. A. Ferguson, F. J. Miller, and T. J. Meyer, *J. Amer. Chem. Soc.*, 1972, **94**, 2599.

[49] S. A. Adeyemi, E. C. Johnson, F. J. Miller, and T. J. Meyer, *Inorg. Chem.*, 1973, **12**, 2371.

[50] P. H. Citrin, *J. Amer. Chem. Soc.*, 1973, **95**, 6472.

[51] M. S. Pereira and J. M. Malin, *Inorg. Chem.*, 1974, **13**, 386.

[52] G. M. Brown, F. R. Hopf, J. A. Ferguson, T. J. Meyer, and D. G. Whitton, *J. Amer. Chem. Soc.*, 1973, **95**, 5939.

[53] G. Malouf and P. C. Ford, *J. Amer. Chem. Soc.*, 1974, **96**, 601.

[54] R. X. Ewall and L. E. Bennett, *J. Amer. Chem. Soc.*, 1974, **96**, 940.

mechanism is expected in view of the rapidity of the reaction, the nature of $[Ru(NH_3)_6]^{2+}$ and the positive charge on both reactants. In contrast an attempt to oxidize $[Ru(NH_3)_5(H_2O)]^{2+}$ or $[Ru(NH_3)_6]^{2+}$ with VO^{2+} has resulted in the isolation of the crystalline, dark green dinuclear species $[(NH_3)_5RuOV(H_2O)_n]^{4+}$.[55] This complex contains a bridging oxygen ligand and is very stable in solution.

Several papers have appeared concerning the interesting luminescence properties of ruthenium(II) complexes of 1,10-phenanthroline and 2,2'-bipyridine. Quenching of the charge transfer (CT) luminescence of $[Ru(CN)_2(phen)_2]$ by Cu^{2+} in water at room temperature has been reported.[56] This study provides the first unequivocal evidence for both static and dynamic quenching of a metal complex exhibiting CT luminescence. The unusual static quenching appears to occur *via* co-ordination of Cu^{2+} to the Ru^{II} complex *via* a CN-bridge. Oxygen quenching of the CT excited states of $[Ru(CN)_2(phen)_2]$ and $[Ru(bipy)_3]^{2+}$ in water and methanol has also been described by the same research group.[57] Singlet oxygen is produced. A related study has investigated the quenching of the phosphorescence of $[Ru(bipy)_3]^{2+}$ by a series of Cr^{III} complexes.[58] With anionic $[Cr(CN)_6]^{3-}$, quenching varied with the ionic strength as expected for diffusion dependence. The quenching ability of the cations $(CrXY(en)_2)^+$ (X = Br, NCS, Cl, or F) varied in the order $F < Cl < NCS < Br$.

Polarization of the spin-forbidden luminescence of $[Ru(bipy)_3]^{2+}$ has been achieved in EPA glass at 77 K.[59] The results indicate that the absorption and emission oscillators are in the same plane. The electrochemistry and electrogenerated chemiluminescence (ECL) of the complexes $[RuL_3]^{2+}$ (L = 1,10-phenanthroline, or 2,2'-bipyridine) and $[RuL'_2]^{n+}$ (L' = 2,2',2''-terpyridine, n = 2; L' = 2,4,6-tripyridyltriazine) have also been investigated.[60] All formed one-electron oxidation and reduction stages, with the bipy, phen, and terpy complexes producing ECL *via* redox reaction of oxidized and reduced forms. A triplet state was proposed as the emitting species. Another interesting report describes the use of $[Ru(bipy)_3]^{2+}$ as a sensitizer for the photoreactions of oxalato-complexes of Cr^{III}, Fe^{III}, and Co^{III}.[61] Redox decomposition of $[Co(ox)_3]^{3-}$ occurs with $[Ru(biby)_3]^{2+}$ acting as a one-electron reducing agent, whereas $[Cr(ox)_3]^{3-}$ racemizes *via* excitation energy transfer. In contrast, with the labile $[Fe(ox)_3]^{3-}$ the net reaction is one of quenching only.

P- and As-donor ligands. A series of related papers has reported the first systematic study of the co-ordinating ability of the ligands $P(OR)_2Ph$ (L) and $P(OR)Ph_2$ (L') (R = Me or Et) towards platinum metals.[62-64] The general synthetic route to cationic complexes involved addition of free ligand to a solution or suspension of a labile transition-metal precursor (usually a triphenylphosphine or olefinic complex) in the corresponding alcohol ROH, followed by addition of $NaBPh_4$.[62, 63] Among the new complexes thus isolated were $[RuHL_5]^+$, $[RuL_6]^{2+}$, and the dinuclear species $[B_3RuX_3RuB_3]$ (B = L or L'; X = Cl or Br). On the other hand, the white neutral

[55] H. De Smedt, A. Persoons, and L. De Maeyer, *Inorg. Chem.*, 1974, **13**, 90.
[56] J. N. Demas and J. W. Addington, *J. Amer. Chem. Soc.*, 1974, **96**, 3663.
[57] J. N. Demas, D. Diemente, and E. W. Harris, *J. Amer. Chem. Soc.*, 1973, **95**, 6864.
[58] F. Bolletta, M. Maestri, L. Moggi, and V. Balzani, *J. Amer. Chem. Soc.*, 1973, **95**, 7864.
[59] I. Fujita and H. Kobayashi, *Inorg. Chem.*, 1973, **12**, 2759.
[60] N. E. Tokel-Takvoryan, R. E. Hemingway, and A. J. Bard, *J. Amer. Chem. Soc.*, 1973, **95**, 6582.
[61] J. N. Demas and A. W. Adamson, *J. Amer. Chem. Soc.*, 1973, **95**, 5159.
[62] D. A. Couch and S. D. Robinson, *Inorg. Nuclear Chem. Letters*, 1973, **9**, 1079.
[63] D. A. Couch and S. D. Robinson, *Inorg. Chem.*, 1974, **13**, 456.

compounds $[RuCl_2(CO)_2B_2]$ and $[RuCl_2(CO)B_3]$ (B = L or L') were obtained *via* reaction of $P(OR)_2Ph$ or $P(OR)Ph_2$ with $[RuCl_2(CO)_2]_n$ in ethanolic solvents.[64] The new species were characterized by conductivity and ^1H and ^{31}P n.m.r. data.

In a similar study, ligand exchange reactions performed in polar solvents have afforded a convenient general route to an extensive series of cationic trimethyl- and triethyl-phosphite complexes, including $[Ru\{P(OR)_3\}_6]^{2+}$ and $[RuX\{P(OR)_3\}_5]^+$

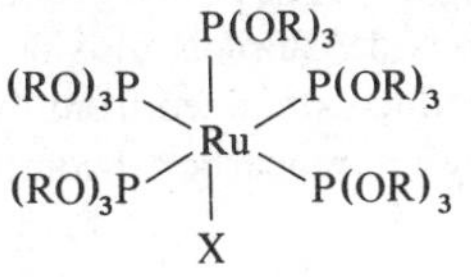

(10)

(X = H, Cl, or Br).[62, 65] The former complex provided the first example of ruthenium co-ordinated to six equivalent unidentate P-donor ligands. ^1H N.m.r. evidence indicates that $[RuX\{P(OR)_3\}_5]^+$ has structure (10). These complexes are somewhat less stable to air than the corresponding dialkylphosphonite and alkyldiphenyl phosphonite derivatives, stability decreasing down the series $P(OR)Ph_2 > P(OR)_2Ph > P(OR)_3$. Synthesis of the compounds $[(PhAsO_3)Ru(CO)_2L_2]$ and $[(PhAsO_{2.5})Ru(CO)_2L']$ [L = PPh_3, $AsPh_3$, $P(o\text{-tolyl})_3$, $As(o\text{-tolyl})_3$; L' = $Ph_2PC_2H_4PPh_2$, $Ph_2AsC_2H_4$ $AsPh_2$, or $Ph_2AsC_3H_6AsPh_2$] has been reported.[66] The phenylarsonate group is believed to be respectively bidentate and unidentate in these complexes. Two i.r. bands were observed in each case for the AsO_3 group between 690 and 750 cm^{-1}.

Group IV Donors. C-donor ligands. Treatment of $[Ru(CN)_6]^{4-}$ with dimethylsulphate for 6 h at 95 °C has produced the white isocyanide complex $[Ru(CNMe)_6]^{2+}$.[67] Electrochemical studies showed that the corresponding Ru^{III} species can only exist under strongly oxidizing conditions. Furthermore, addition of simple amines to $[Ru(CNMe)_6]^{2+}$ produces either (11) or (12) (containing either uni- or bi-dentate

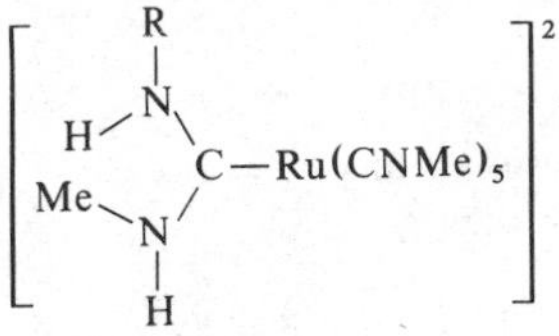

(11)

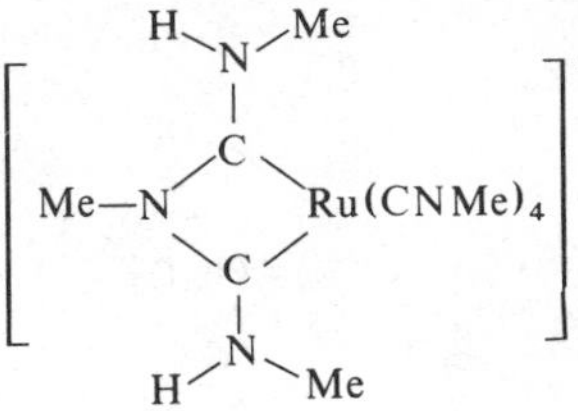

(12)

carbene ligands), depending on steric factors. The electrical conductivities and i.r. spectra of the compounds $Fe_4[M(CN)_6]_3$ (M = Fe^{III}, Ru^{II}, or Os^{II}) have been measured at different temperatures.[68] In an interesting experiment, oxidation of the

[64] D. A. Couch, S. D. Robinson, and J. N. Wingfield, *J. C. S. Dalton*, 1974, 1309.
[65] D. A. Couch and S. D. Robinson, *Inorg. Chim. Acta*, 1974, 9. 39.
[66] S. S. Sandu and A. K. Mehta, *Inorg. Nuclear Chem. Letters*, 1973, 9, 1197.
[67] D. J. Doonan and A. L. Balch, *Inorg. Chem.*, 1974, 13, 921.
[68] H. Inoue and S. Yanagisawa, *J. Inorg. Nuclear Chem.*, 1974, 36, 1409.

anions $[M(CN)_6]^{4-}$ (M = Fe, Ru, or Os) has been observed with OH radicals generated by radiolysis of the water solvent.[69] The mechanism for Ru and Os involves
single-electron transfer with formation of the intensely coloured $[M(CN)_6]^{3-}$
species. The gaseous molecule RuC_2 has been found in a high-temperature mass
spectrometer, and its atomization energy determined.[70]

Mixed Ruthenium-(II) and -(III) Complexes.—Interactions of the oxo-centred
Ru^{III} acetate $[Ru_3O(CO_2Me)_6(MeOH)_3]^+$ and the reduced species $[Ru_3O(CO_2Me)_6$
$(MeOH)_3]$ with carbon monoxide, sulphur dioxide, and NO have been studied.[71]
Scheme 5 summarizes the products obtained from the latter mixed-valency complex
following carbonylation. Only a monocarbonyl complex is seen to be formed, and the

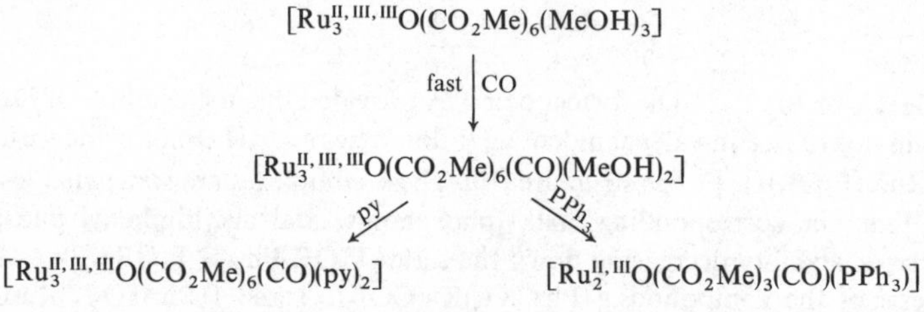

$$[Ru_3^{II,III,III}O(CO_2Me)_6(MeOH)_3]$$

$$fast \Big\downarrow CO$$

$$[Ru_3^{II,III,III}O(CO_2Me)_6(CO)(MeOH)_2]$$

py ↙   ↘ PPh_3

$$[Ru_3^{II,III,III}O(CO_2Me)_6(CO)(py)_2] \qquad [Ru_2^{II,III}O(CO_2Me)_3(CO)(PPh_3)]$$

Scheme 5

six bridging carboxylate groups of the original oxo-centred species rearrange to give
two bridging and four bidentate carboxylates. Electrochemical oxidation and reduction of each of the substrates and products was also investigated. A further new
$Ru^{II,III}$ dimer, $[(PPh_3)_2Cl_2RuCl_3Ru(CS)(PPh_3)_2],2Me_2CO$, has been prepared *via*
reaction (12).[16] The species is paramagnetic with $\mu = 2.0$ BM per dimer, while *X*-ray
studies reveal the tri-μ-chloro structure (13) (Table 1).

$$[(PPh_3)_2ClRuCl_3Ru(CS)(PPh_3)_2] + HCl \rightarrow [(PPh_3)Cl_2RuCl_3Ru(CS)(PPh_3)_2],2Me_2CO \quad (12)$$

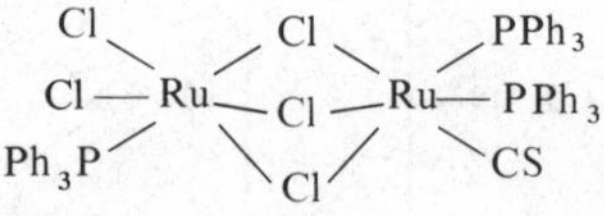

(13)

ESCA spectra have been recorded for a series of pyrazine-bridged ruthenium dimers
including $[(NH_3)_5Ru(pyz)Ru(NH_3)_5]^{5+}$ and $[(bipy)_2ClRu(pyz)RuCl(bipy)_2]^{3+}$.[50]
The data show that both Ru^{II} and Ru^{III} are present in those complexes although the
metal sites are equivalent. The speed with which ESCA monitors electron distribution
(*ca.* 10^{-17} s) would seem to make it an excellent spectrosocpic tool for such valence-
state assignments.

[69] W. L. Waltz, S. S. Akhtan, and R. L. Eager, *Canad. J. Chem.*, 1973, **51**, 2525.
[70] K. A. Gingerich, *J.C.S. Chem. Comm.*, 1974, 199.
[71] A. Spencer and G. Wilkinson, *J.C.S. Dalton*, 1974, 786.

Ruthenium(III).—*Group VII Donors.* The magnetic circular dichroism and absorption spectra of $[RuX_6]^{3-}$ (X = Cl or Br) have been measured in the charge-transfer region.[72] Band splittings were interpreted in terms of low-symmetry distortion and spin–orbit coupling. A related study has also reported the diffuse reflectance absorption spectrum of the fluoro-analogue $[RuF_6]^{3-}$.[73] Crystal field bands were observed between 4 and 50 kK and assigned to either spin-forbidden or spin-allowed $t_{2g}^5 \rightarrow t_{2g}^4 e_g$) transitions. Stepwise reduction of $RuCl_3,3H_2O$ (which is a mixture of Ru^{III} and Ru^{IV}) by molecular hydrogen has been observed in NN-dimethylacetamide solution, leading ultimately to a Ru^I species.[8] The initial steps involve activation of hydrogen by a Ru^{III} catalyst, and probably proceed *via* a Ru^{III}–hydride intermediate.

Group VI Donors. Stability constants have been measured for the formation of a range of metal complexes with *o*-coumaric acid, including a 2:1 (L:M) species with ruthenium(III).[74] pH titration techniques were employed. A synthetic study has been made of the interactions of the oxo-centred Ru^{III} acetate $[Ru_3O(CO_2Me)_6(MeOH)_3]$ CO_2Me with π-acid ligands such as CO, SO_2, NO, and MeNC.[71]

The electronic spectra of $[Ru(SCN)_6]^{3-}$ and several related thiocyanato-complexes have been recorded in both aqueous 2M-KSCN and molten KSCN.[75] The latter molten solvent displays interesting reducing properties, converting blue $[Ru(SCN)_6]^{3-}$ rapidly at 165 °C into its yellow Ru^{II} analogue. Conductivity and molecular weight measurements indicate that the known complex $RuCl_3L_{1.5}$ (L = $C_3H_7SeC_2H_4SeC_3H_7$) has the dimeric formula $[Ru_2Cl_6L_3]$.[76] Its ^1H n.m.r. spectrum in acetone reveals only one sharp singlet for the methylene protons at $\tau = 7.27$. In order to explain the 12 equivalent CH_2 protons, structure (14) was proposed containing bridging organoselenium ligands in which the central methylene groups have the eclipsed conformation. This is the first time such a *cis*-conformation has been observed for a 1,2-disubstituted ethane-type ligand.

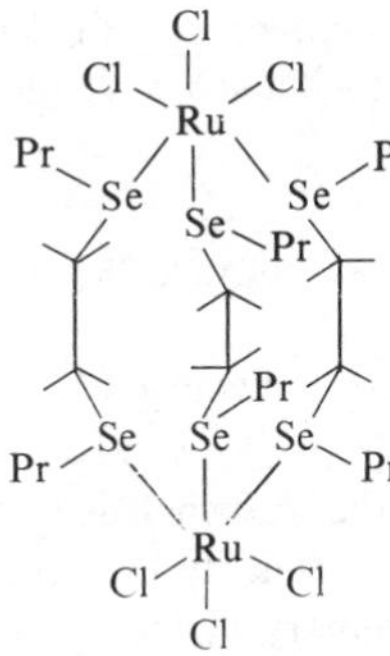

(14)

Group V Donors. N-Donor ligands. The photochemistry of the complexes $[RuX_2-(NH_3)_4]^+$ (X = Cl, Br, or I) has been studied in aqueous solution.[77] Whereas the

[72] A. J. McCaffery, M. D. Rowe, and D. A. Rice, *J.C.S. Dalton*, 1973, 1605.
[73] G. L. Allen, G. A. M. El-Sharkawy, and K. D. Warren, *Inorg. Chem.*, 1973, **12**, 2231.
[74] S. C. Tripathi and S. Paul, *J. Inorg. Nuclear Chem.*, 1973, **35**, 2465.
[75] K. S. De Haas, *J. Inorg. Nuclear Chem.*, 1973, **35**, 3231.
[76] G. Hunter and R. C. Massey, *Inorg. Nuclear. Chem. Letters*, 1973, **9**, 727.
[77] A. Ohyoshi, N. Takebayashi, Y. Hiroshima, K. Yoshikuni, and K. Tsuji, *Bull Chem. Soc. Japan*, 1974, **47**, 1414.

dichloro-species is photochemically inactive in the visible region, the aquation process (13) occurs for cis-$[RuX_2(NH_3)_4]^+$ (X = Br or I) at pH < 2.5. Quantum yields depended on the wavelength of irradiation, but were independent of the con-

$$cis\text{-}[RuX_2(NH_3)_4]^+ + H_2O \xrightarrow{hv} cis\text{-}[RuX(H_2O)(NH_3)_4]^{2+} + X^- \qquad (13)$$

centrations of X^- and acid. Thermal decompositions of the above compounds, as well as $[RuX(NH_3)_5]X_2$ (X = Cl, Br, or I), have also been investigated employing thermogravimetry, differential thermal analysis, and evolved gas analysis.[78] Thermal stabilities decreased in the order Cl > Br > I. For all the chloro- and bromo-complexes, the species $[RuX_3(NH_3)_3]$ were observed as intermediates on the path to ruthenium metal. In contrast, the simpler decomposition reactions (14) and (15) were observed for the iodo-compounds.

$$[RuI(NH_3)_5]I_2 \rightarrow RuI_3 + 5NH_3 \rightarrow Ru + \tfrac{3}{2}I_2 + 5NH_3 \qquad (14)$$

$$[RuI_2(NH_3)_4]I \rightarrow RuI_3 + 4NH_3 \rightarrow Ru + \tfrac{3}{2}I_2 + 4NH_3 \qquad (15)$$

Application of the simplified MO theory of Zwickel and Creutz[79] has permitted the prediction of both the band positions and intensities in the charge-transfer spectra of $[MX(NH_3)_5]^{2+}$ and cis- and trans-$[MX_2(NH_3)_4]^+$ (M = Ru or Os; X = Cl, Br, or I).[80] The bands were assigned as transitions from a halogen ligand π-orbital to a metal d-orbital. Interestingly, the Ru—X bonds were shown to be slightly more covalent than corresponding Os—X bonds, while covalency also increased from Cl to Br to I as expected. The m.c.d. and absorption spectra have also been recorded for $[Ru(NH_3)_6]Cl_3$ in the charge-transfer region, the first m.c.d. band being assigned as a transition to a $^2T_{1u}(\delta)$ level.[72] ESCA spectra have been recorded for a series of pyrazine-bridged ruthenium dimers including $[(NH_3)_5Ru(pyz)Ru(NH_3)_5]^{6+}$ and $[(bipy)ClRu(pyz)\text{-}RuCl(bipy)_2]^{4+}$, confirming the presence of only Ru^{III}.[50]

As-*Donor ligands.* The new complexes $[RuCl_3(AsPh_3)_3]$ and $[RuCl_3(AsPh_3)\text{-}(OAsPh_3)_2]$ have been prepared (Scheme 6).[81] In methanol solution the former loses a

$$RuCl_3,xH_2O + AsPh_3 \xrightarrow{\text{MeOH–HCl}} [RuCl_3(AsPh_3)_3] \xrightarrow[O_2]{C_6H_6} [RuCl_3(AsPh_3)(OAsPh_3)_2]$$

Scheme 6

triphenylarsine ligand to form the known $RuCl_3(AsPh_3)_2,MeOH$. From e.s.r. and u.v.–visible spectra, and dipole moment studies other workers have shown that while the latter complex has C_{2v} symmetry in the solid state, it is trigonal-bipyramidal (D_{3h}) in benzene solution.[82] On the other hand, $(Et_4N)[RuCl_4(AsPh_3)_2]$ is D_{4h} both as a solid and in solution. One electron levels of these two complexes were obtained in terms of the crystal-field parameters Dq, Ds, and Dt.

Ruthenium(IV).—*Group VII Donors.* Electronic spectra have been recorded for a

[78] A. Ohyoshi, S. Hiraki, and H. Kawasaki, *Bull Chem. Soc. Japan*, 1974, **47**, 841.
[79] A. M. Zwickel and C. Creutz, *Inorg. Chem.*, 1971, **10**, 2395.
[80] E. Verdonck and L. G. Vanquickenborne, *Inorg. Chem.*, 1974, **13**, 762.
[81] R. K. Poddar, I. P. Khullar, and U. Agarwala, *Inorg. Nuclear Chem. Letters*, 1974, **10**, 221.

series of hexafluoro-metallate anions including $[RuF_6]^{4-}$.[73] Crystal-field bands were assigned for this complex between 4 and 40 kK, while moderately strong bands below 4 kK were attributed to transitions between spin–orbit components of the ground state. Charge-transfer bands were also identified. A comparison of the asymmetric stretching frequency $v_3(M—X)$ and the unit cell length for a series of compounds $A_2[MX_6]$ (A = K, Rb, Cs, or NH_4; M = Ru, Os, Sn, or Te; X = Cl or Br) has been made.[83] The trends observed for the Te and Sn compounds are thought to be related to steric factors, while the opposite result observed for the Ru and Os analogues (v_3 increases with decrease in cation size) was explained in terms of $M \rightarrow X(d-p)\pi$-bonding.

Aquation of the oxo-bridged dimer $K_4[Ru_2OCl_{10}]$ has been re-investigated in aqueous HCl.[84] Various Ru^{IV} species were observed at different times and acid concentrations (0.002—12 M), although no definite assignments were made.

Group VI Donors. The electrolytic oxidation and reduction of each of the four known diastereoisomeric forms of tris-{(+)-3-acetylcamphorato}ruthenium(III) (6) have been reported, producing the corresponding Ru^{IV} and Ru^{II} species with complete retention of configuration.[22] Absolute configurations were assigned to these Ru^{IV} and Ru^{II} diastereoisomers from their circular dichroism, employing the known configurations of the Ru^{III} precursors.

Anhydrous $SrRuO_3$ has been prepared by fusing together metallic ruthenium and $SrCO_3$ in air at 1000—1050 °C, a refinement of a previous procedure.[85] As well as X-ray powder diffraction studies (Table 1), its stability in various mineral acids was investigated. Interestingly, Ru^{IV} has been suggested to exist as the hydroxy-species $[Ru_4(OH)_{12}]^{4+}$ in methanesulphonic and nitric acids of pH 1.5 to 4.[86] The hydroxide $Ru(OH)_4$ precipitated on addition of alkali. Whereas $[Ni(PF_3)_4]$ can be prepared by the direct reaction of PF_3 (4000 atm) on NiO at 300 °C, analogous conditions with RuO_2 have led only to reduction of the metal oxide.[87]

Oxidation of the corresponding tris(dithiocarbamato)M^{III} complexes with BF_3 in dry toluene has given complexes of type (15; M = Ru or Rh).[88] Both compounds show the characteristic shift of $v(C≡N)$ on oxidation to higher frequency near

$$\left[\begin{array}{c} Et \\ N—C \\ Et \end{array} \underset{S}{\overset{S}{\diagup\diagdown}} M/_3 \right]^+ BF_4^-$$

(15)

1500 cm^{-1} previously observed in the corresponding Fe^{IV} complex. Their unexpected diamagnetism in the solid state may be due to pairing of electron spins *via* association, which is supported by molecular weight determinations in chloroform.

Group V Donors. Following the recent report[89] of the preparation of $[Os(NPPh_3)$

[82] P. T. Manoharan, P. K. Mehrotra, M. M. Taquikhan, and R. K. Andal, *Inorg. Chem.*, 1973, **12**, 2753.

[83] J. E. Fergusson and P. F. Heveldt, *Austral. J. Chem.*, 1974, **27**, 661.

[84] I. P. Alimarin, V. I. Shlenskaya, and Z. A. Kuratashvili, *Russ. J. Inorg. Chem.*, 1973, **18**, 250.

[85] I. M. Kuleshov, N. F. Ahakhova, and V. K. Chubarova, *Russ. J. Inorg. Chem.*, 1973, **18**, 487.

[86] C. Bremard, G. Nowogrocki, and G. Tridot, *Bull. Soc. chim. France*, 1974, 392.

[87] A. P. Hagen and E. A. Elphingstone, *J. Inorg. Nuclear Chem.*, 1973, **35**, 3719.

[88] L. R. Gahan and M. J. O'Connor, *J.C.S. Chem. Comm.*, 1974, 68.

[89] D. Pawson and W. P. Griffith, *Inorg. Nuclear Chem. Letters*, 1974, **10**, 253.

$Cl_3(PPh_3)_2]$, containing a $Ph_3P=N^-$ ligand, this work has been extended to produce a range of related Ru and Os complexes *via* Scheme 7 (M = Ru or Os; $PR_3 =$ $PPhEt_2$, PPh_2Et, PPh_2Me or PEt_3).[89] The ruthenium species have magnetic moments ($\mu = 2.9$ BM) typical of Ru^{IV}, and exhibit strong broad i.r. bands in the region

$$Bu^nN[MNCl_4] + xsAsPh_3 \rightarrow [MNCl_3(AsPh_3)_2] \xrightarrow[\text{acetone}]{xs\ PR_3} [M(NPR_3)Cl_3(PR_3)_2]$$

Scheme 7

1050—1200 cm^{-1} assigned to the P$=$N stretch of coordinated $R_3P=N^-$. A preliminary X-ray structural study shows a linear Ru—N—P linkage.

Ruthenium(v) and Higher Oxidation States.—A series of dioxygenyl salts of complex fluorometallates, including the new $O_2^+[RuF_6]^-$, have been simply prepared by static heating of the metal at 300—500 °C with 5 atm of a 3:1 mixture of fluorine and oxygen.[90] This Ru species was characterized largely by correlating X-ray diffraction, Raman, and mass spectral data with analogous known systems. The diffuse reflectance electronic spectrum of $[RuF_6]^-$ has also been reported.[73] Crystal-field bands were observed and assigned between 9 and 32 kK, while an intense band at 40 kK was attributed to a Laporte-allowed $\pi \rightarrow t_{2g}$ charge-transfer transition.

Oxidation of ruthenate ion (RuO_4^{2-}) with periodate in alkaline solution has been shown to produce $[RuO_4]^-$.[91] This contrasts with reported oxidations of low oxidation state Ru by IO_3^- in acidic or neutral solution, which yield $[RuO_4]$. Two related papers have reported the photoelectron[92] and electronic[93] spectra of $[RuO_4]$. The photoelectron spectrum was poorly resolved due to instrumental corrosion by the sample, but permitted assignment of the five highest energy filled MO's. Electronic spectra were obtained both in the gas phase and at 20 K in an SF_6 matrix. The first virtual two-electron MO of $[RuO_4]$ was estimated to be of lower energy than in the corresponding $[OsO_4]$, leading to the lower energy d–d bands observed in the former species.

2 Osmium

Cluster Compounds.—Pyrolysis of $Os_6(CO)_{18}$ at 225 °C under vacuum has been shown to yield the carbide species $Os_5C(CO)_{15}$ in good yield.[3] Alternatively, this carbide can be obtained by direct pyrolysis of $Os_3(CO)_{12}$ but in only 10% yield, and the latter reaction obviously proceeds *via* intermediate formation of $Os_6(CO)_{18}$. These observations provide further evidence that the source of the carbido C is the reduction of co-ordinated CO. In contrast, pyrolysis of $Os_3(CO)_{12}$ in the presence of small traces of water leads to formation of a variety of polynuclear hydrides including the novel $H_2Os_5(CO)_{16}$, $H_2Os_5(CO)_{15}$, $H_2Os_6(CO)_{18}$, and $H_2Os_7C(CO)_{19}$.[2] Careful exclusion of water in studies of the high-temperature chemistry of $Os_3(CO)_{12}$ is therefore suggested.

A high-yield general route to organotin complexes of osmium has been found

[90] A. J. Edwards, W. E. Falconer, J. E. Griffiths, W. A. Sunder, and M. J. Vasile, *J.C.S. Dalton*, 1974, 1129.
[91] G. I. Rozovskii, Z. A. Poshkute, A. Y. Prokopchik, and P. K. Norkus, *Russ. J. Inorg. Chem.*, 1973, **18**, 1432.
[92] S. Foster, S. Felps, C. W. Johnson, D. B. Larson, and S. P. McGlynn, *J. Amer. Chem. Soc.*, 1973, **95**, 6578.
[93] S. Foster, S. Felps, L. C. Cusachs, and S. P. McGlynn, *J. Amer. Chem. Soc.*, 1973, **95**, 5521.

Table 1 X-*Ray data for ruthenium compounds*

Compound	R	Comments	Ref.
$[XeF_5][RuF_6]$	0.042	discrete XeF_5^+ and $[RuF_6]^-$ units. Octahedral Ru geometry	a
XeF_7RuF_6	0.07	discrete $XeRuF_7$ units 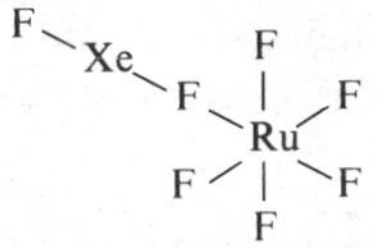	a

Compound	R	Comments	Ref.
$[RuH(COOMe)(PPh_3)_3]$	0.089	highly distorted octahedral Ru co-ordination. Acetate ligand is bidentate, weakly held (Ru—O *ca.* 2.2 Å)	b
$Na_{2.1}Ru_4O_9$	—	powder only. $[Ru_4O_9]_n^{2.1n-}$ network formed by single, double, and triple chains of RuO_6 octahedra sharing edges	c
$SrRuO_3$	—	powder only. Perovskite $CaTiO_3$ type lattice	d
$[Ru(HCS_2)_2(PPh_3)_2]$	0.059	octahedral co-ordination with dithio-formate ligands bidentate *via* S atoms	e
$[(CS)(PPh_3)_2RuCl_3Ru(PPh_3)_2Cl]$	0.115	both Ru atoms have distorted octahedral coordination. Ru—Ru = 3.35 Å. Disorder prevented satisfactory delineation of thiocarbonyl ligand	f
$[RuCl_3(NO)(PMePh_2)_2]$	0.044	octahedral Ru. Linear Ru—N—O ($\angle$ = 176.4°) indicating R^{II}—NO^+	g
$[Ru(NO)_2(PPh_3)_2], C_6H_6$	0.043	distorted tetrahedral Ru. Ru—N—O angles are 177.7° and 170.6°, indicating a Ru^{-II} complex of NO^+	h
$[Ru(NPEt_2Ph)Cl_3(PEt_2Ph)_2]$	0.035	contains an $Et_2PhP=N^-$ ligand with linear Ru—N—P	i
$[Ru_2(NH_3)_8(NH_2)_2]Cl_4$	0.036	dimer contains distorted octahedral Ru atoms. Ru—Ru = 2.625 Å. Bridging amido ligands (*via* N)	j
$[RuCl_3(PPh_3)_2(p-N_2C_6H_4Me)]$ Me_2CO	0.045	octahedral with *trans* PPh_3 ligands. Linear Ru—N—N suggests ArN_2 ligand is a three-electron donor	k
$[Ru(CO)(py)(tetraphyenyl-porphinato)]$	0.065	Ru atom is 0.079 Å out of porphyrin plane towards CO	l
$[Ru(\mu-PPh_2)(NO)(PMePh_2)]$	0.054	dimeric with bridging Ph_2P groups Distorted tetrahedral co-ordination about Ru atoms. Linear Ru—N—O	m
$[RuH(PPh_3)_2(\pi-Ph-PPh_2)]BF_4$	0.054	two PPh_3 ligands bonded to Ru *via* P, while other is π-bonded *via* one phenyl group	n

(a) N. Bartlett, M. Gennis, D. D. Gibler, B. K. Morell, and A. Zalkin, *Inorg. Chem.*, 1973, **12**, 1717. (b) A. C. Skapski and F. A. Stephens, *J.C.S. Dalton*, 1974, 390. (c) J. Darriet, *Acta Cryst.*, 1974, **B30**, 1459. (d) V. I. Spitsyn, I. M. Kuleshov, N. A. Shishakov, and N. F. Shakhova, *Russ. J. Inorg. Chem.*, 1974, **18**, 592. (e) R. O. Harris, L. S. Sadavoy, S. C. Nyburg, and F. H. Pickard, *J.C.S. Dalton*, 1973, 2646. (f) A. J. F. Fraser and R. O. Gould, *ibid.*, 1974, 1139. (g) A. J. Schultz, R. L. Henry, J. Reed, and R. Eisenberg, *Inorg. Chem.*, 1974, **13**, 732. (h) Ref. 7. (i) F. L. Phillips and A. C. Skapski, *J.C.S. Chem. Comm.*, 1974, 49. (j) M. T. Flood, R. F. Ziola, J. E. Earley, and H. B. Gray, *Inorg. Chem.*, 1973, **12**, 2153. (k) J. V. McArdle, A. J. Schultz, B. J. Corden, and E. Eisenberg *ibid.*, p. 1676. (l) R. G. Little and J. A. Ibers, *J. Amer. Chem. Soc.*, 1973, **95**, 8583. (m) J. Reed, A. J. Schultz, C. G. Pierpont, and R. Eisenberg, *Inorg. Chem.*, 1973, **12**, 2949. (n) J. C. McConway, A. C. Skapski, L. Phillips, R. J. Young, and G. Wilkinson, *J.C.S. Chem. Comm.*, 1974, 327.

using $Os_3(CO)_{12}$ as precursor (equation 16; $R_3 = Ph_3$, Bu_3^n, or $PhBu_2^n$).[94] With

$$Os_3(CO)_{12} + Na-NH_3 \rightarrow Na_2[Os(CO)_4] \xrightarrow{R_3SnX} trans\text{-}[Os(SnR_3)_2(CO)_4] \quad (16)$$

bifunctional organotins, *e.g.* R_2SnCl_2, the four-membered di-μ-tin rings (16) were produced. The related reaction of $Os_3(CO)_{12}$ with 1,2-bis(dimethylsilyl)ethane has been shown to give the stable white chelate complex (1; $M = Os$).[6] [13]C N.m.r. studies indicate that this species is stereochemically rigid.

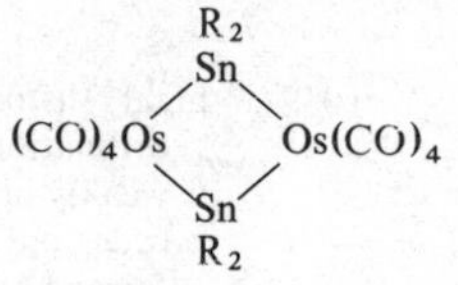

$$
\begin{array}{c}
R_2 \\
Sn \\
(CO)_4Os \diagup \quad \diagdown Os(CO)_4 \\
\diagdown \quad \diagup \\
Sn \\
R_2
\end{array}
$$

(16)

Osmium(II).—*Group VII Donors. Hydrido- and halogeno-carbonyl complexes.*
Carbonylation of $[OsHX(CO)(PCy_3)_2]$ has produced the novel dicarbonyl hydride $[OsHX(CO)_2(PCy_3)_2]$ ($X = Cl$ or Br; $Cy =$ tricyclohexyl).[17] I.r. and [1]H n.m.r. data indicate the geometry (17). Addition of other ligands to the five-co-ordinate

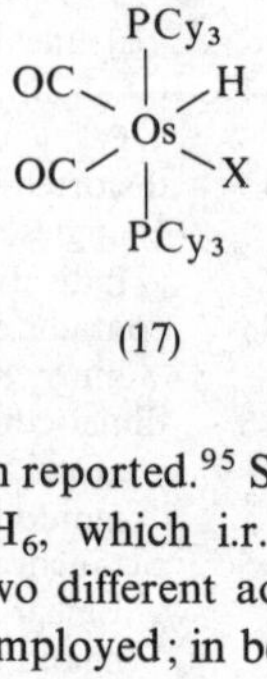

$$
\begin{array}{c}
PCy_3 \\
OC \diagdown \quad | \quad \diagup H \\
Os \\
OC \diagup \quad | \quad \diagdown X \\
PCy_3
\end{array}
$$

(17)

$[OsHCl(CO)(PCy_3)_2]$ has also been reported.[95] Sulphur dioxide gives yellow crystals of $[OsHCl(CO)(PCy_3)_2(SO_2)],C_6H_6$, which i.r. data show to contain S-bonded SO_2 and a weak Os—H bond. Two different addition products are obtained with CS_2 depending on the conditions employed; in benzene, the product β-$[OsCl(HCS_2)(CO)(PCy_3)_2]$ is indicated on the basis of i.r. data [ν(C—S) bands at 917 and 790 cm^{-1}].

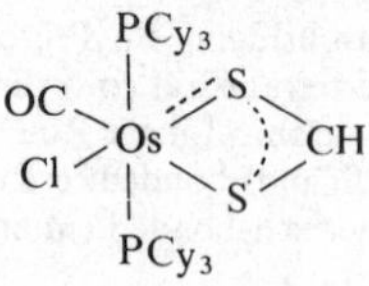

$$
\begin{array}{cc}
\begin{array}{c}
PCy_3 \\
OC \diagdown \quad | \quad \diagup S \diagdown \\
Os \quad \quad \diagdown CH \\
Cl \diagup \quad | \quad \diagdown S \diagup \\
PCy_3
\end{array}
&
\begin{array}{c}
PCy_3 \quad S \\
OC \diagdown \quad | \quad \quad \| \\
Os - SCH \\
Cl \diagup \quad | \\
PCy_3
\end{array}
\\
(18a) & (18b)
\end{array}
$$

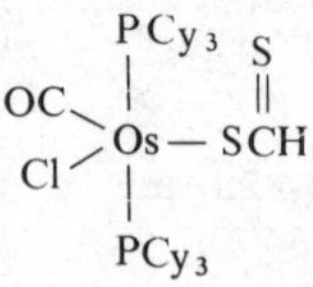

Structures (18a) or (18b) were suggested for this species. On the other hand, reaction in the solid state yielded both (18) and another product $[OsHCl(CS_2)(CO)(PCy_3)_2]$, containing terminally S-bonded CS_2 [ν(C—S) = 1510 cm^{-1}].

[94] J. P. Collman, D. W. Murphy, E. B. Fleischer, and D. Swift, *Inorg. Chem.*, 1974, **13**, 1.
[95] F. G. Moers, R. W. M. Ten Hoedt, and J. P. Langhout, *Inorg. Chem.*, 1973, **12**, 2196.

The related dihydrides $[OsH_2(CO)(NO)L_2]PF_6$ (L = PPh_3, or PCy_3) have been synthesized *via* reaction (17).[96] Their ^1H n.m.r. spectra show them to be stereo-

$$[OsCl(CO)(NO)L_2] + AgPF_6 + H_2 \xrightarrow{CH_2Cl_2} [OsH_2(CO)(NO)L_2]PF_6 \qquad (17)$$

chemically non-rigid, a rare occurence for six-co-ordinate metal complexes. Carbon monoxide displaces molecular hydrogen from both complexes to give the salts $[Os(CO)_2(NO)L_2]PF_6$. However, while PPh_3 reacts similarly producing $[Os(CO)(NO)(PPh_3)_3]PF_6$, the addition of PCy_3 to $[OsH_2(CO)(NO)(PCy_3)_2]^+$ results in H^+ abstraction and the isolation of neutral $[OsH(CO)(NO)]$.

Group VI Donors. Reaction of carboxylic acids with hydrido and triphenylphosphine substrates has provided a general route to a large range of carboxylato-complexes including yellow $[OsH(OCOR)(PPh_3)_3]$.[23] Molecular weight measurements indicate that all are monomeric in solution, thus excluding the presence of bridging carboxylate ligands. I.r. spectroscopy and isomorphism with the corresponding known ruthenium compounds indicated structure (19).

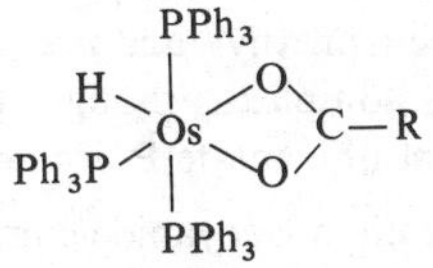

(19)

Group V Donors. N-Donor ligands. Low-frequency Raman and i.r. spectra (650—40 cm^{-1}) have been reported for the complexes $[M(NH_3)_5Y]X_2$ (M = Ru or Os; Y = N_2 or CO; X = Cl, Br, I or BF_4).[32] The Os—Y solid state vibrational modes were assigned using intensity measurements, deuteration, and ^{15}N substitution. The salt $[(C_8H_{17})_3NH]_2[OsCl_5(NO)]$ has been isolated during the extraction of $[OsCl_5(NO)]^-$ from HCl solution with tri-n-octylamine.[97] Thermographic studies were also reported. More interestingly, a correlation has been suggested between the reactivity of a large range of M—NO$^+$ complexes, including Os^{II} species, and the frequency of their ν(NO) stretch.[35] Only complexes with ν(NO) greater than 1886 cm^{-1} are observed to react with nucleophiles such as OH$^-$, OMe$^-$, N$_3^-$, *etc.*

Direct reaction between $ArN_2^+BF_4^-$ and $[OsBr_2(PPh_3)_3]$ has provided a route to arylazo-complexes of the type $[OsBr_3(N_2Ar)(PPh_3)_2]$.[41] I.r. studies suggest a linear Os—N—NAr linkage. On the other hand, similar reactions with $[OsHCl(CO)(PPh_3)_3]$ and $[OsH_2(CO)(PPh_3)_3]$ afford the new aryldi-imine derivatives $[OsCl(NH=NAr)(CO)(PPh_3)_2]BF_4$ and $[OsH(NH=NAr)(CO)(PPh_3)_3]BF_4$. In a related study, a wide variety of complexes containing the 1,3-diaryl-triazenido ligand (ArNNNAr$^-$) have been produced *via* the reaction of 1,3-diaryltriazene with hydrido or low oxidation state complexes of platinum metals.[42] Products obtained by this novel and convenient route include $[Os(ArNNNAr)_2(PPh_3)_2]$ and $[OsH(ArNNNAr)(CO)(PPh_3)_2]$. I.r. and ^1H n.m.r. data suggest structures (8) and (9) respectively (M = Os).

[96] B. F. G. Johnson and J. A. Segal, *J.C.S. Dalton*, 1974, 981.
[97] K. A. Bol'shakov, N. M. Sinitsyn, V. F. Travkin, and G. L. Plotinskii, *Russ. J. Inorg. Chem.*, 1973, **18**, 755.

New octaethylporphinato (oep) Os^{II} complexes of the type $[Os(oep)(CO)L]$ (L = MeOH, EtOH, THF, py, PPh_3, or $AsPh_3$) have been formed using Scheme 8.[98]

$$[OsO_4] + H_2(oep) \xrightarrow[\substack{\text{monomethyl ether} \\ 200\,°C}]{\text{diethyleneglycol}} \text{soln} \xrightarrow{\text{L}} [Os(oep)(CO)L]$$

Scheme 8

I.r. data suggest that back-bonding onto the CO ligand increases in such porphinato complexes in the order Fe < Ru < Os. Treatment of the new oep-complexes with trimethylphosphite yielded $[Os(oep)\{P(OMe)_3\}_2]$.

P-donor ligands. A general route to neutral complexes of the little-studied ligands $P(OR)_2Ph$ and $P(OR)Ph_2$ (L; R = Me or Et) has been described.[64] Among the numerous noble metal derivatives prepared were species of the type $[OsCl_2(CO)_2L_2]$, isolated from the reaction of L with $[OsCl_2(CO)_3]_2$ in ethanol. The new species were characterized by conductivity, ^1H and ^{31}P n.m.r. data. In a similar study, ligand exchange reactions in polar solvents have afforded convenient syntheses for an extensive series of cationic trimethyl- and triethyl-phosphite complexes including $[Os\{P(OR)_3\}_6]^{2+}$.[62,65] This complex is thought to be the first example of osmium co-ordinated to six equivalent unidentate P-donor ligands.

Group IV Donors. Various osmium isocyanide complexes have been prepared including the Os^{II} species $[Os(MeNC)_6]^{2+}$, $[OsCl_2(CO)(PhNC)(PEt_3)_2]$, $[OsCl_2(R^1NC)_2(PR_3^2)_2]$, and $[OsCl(CO)(R^1NC)(PR_3^2)_3]^+$ (R^2 = mixed alkyl, aryl).[99] Of these complexes, only the dicationic species contains a co-ordinated isocyanide ligand which is susceptible to nucleophilic attack (*e.g.* equation 18). This property appears

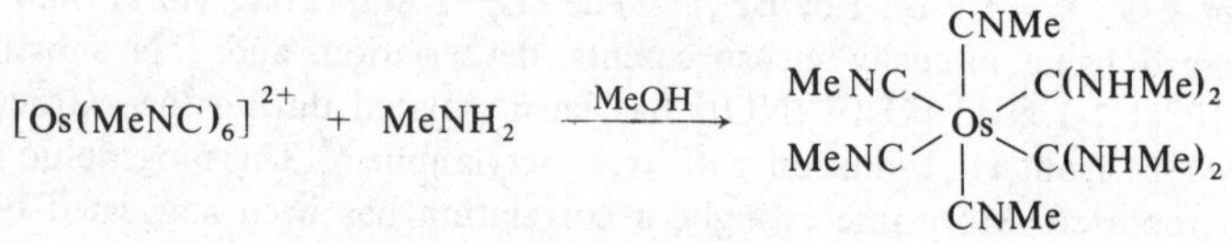

$$[Os(MeNC)_6]^{2+} + MeNH_2 \xrightarrow{MeOH} \begin{array}{c} \text{CNMe} \\ | \\ \text{Me NC} \diagdown \diagup \text{C(NHMe)}_2 \\ \text{Os} \\ \text{Me NC} \diagup \diagdown \text{C(NHMe)}_2 \\ | \\ \text{CNMe} \end{array}$$

$$(18)$$

to be related to the increase in frequency of the $\nu(N{\equiv}C)$ stretch of co-ordinated compared to free cyanide. The electrical conductivities and i.r. spectra of the compounds $Fe_4[M(CN)_6]_3$ (M = Fe^{II}, Ru^{II}, or Os^{II}) have been measured at different temperatures.[68] In an interesting experiment, oxidation of the anions $[M(CN)_6]^{4-}$ (M = Fe, Ru, or Os) has been observed with OH radicals generated by radiolysis of the water solvent.[69] The mechanism for osmium involves single-electron transfer with the formation of the intensely coloured $[Os(CN)_6]^{3-}$ species.

Treatment of $[OsH_2(CO)_4]$ with tin oxides, alkoxides, or amines has yielded organotin complexes of the type *trans*-$[Os(SnR_3)_2(CO)_4]$ (R_3 = Ph_3, Bu_3^n, or

$$[OsH_2(CO)_4] + R_3Sn{-}B \rightarrow trans\text{-}[Os(SnR_3)_2(CO)_4] + 2BH \qquad (19)$$

$PhBu_2^n$) (equation 19).[94] Phenyl groups on tin co-ordinated to osmium were selec-

[98] J. W. Buchler and K. Rohbock, *J. Organometallic Chem.*, 1974, **65**, 223.

[99] J. Chatt, R. L. Richards, and G. H. D. Royston, *J.C.S. Dalton*, 1973, 1433.

tively cleaved by electrophilic reagents to give a number of halogen-functionalized Sn—Os—Sn complexes, as in equation (20) (X = Cl or Br). An interesting extension

$$trans\text{-}[Os(SnPh_3)_2(CO)_4] + HX \rightarrow trans\text{-}[Os(SnCl_2Ph)_2(CO)_4] \qquad (20)$$

of this work was the treatment of $trans$-[Os(SnClBu$_2$)$_2$(CO)$_4$] with [Re(CO)$_5$]$^-$ which produced [{(CO)$_5$ReSnBu$_2$}$_2$Os(CO)$_4$], a rare example of a pentametallic chain involving transition metals.

Osmium(III).—Whereas treatment of the bulky phosphine (L = PMe$_2$naphthyl) with RuCl$_3$ yielded five-co-ordinate [RuCl$_2$L$_3$], similar reaction of OsO$_4$ with L in HCl solution gave octahedral [OsCl$_3$L$_3$].[18] In addition, the internally metallated complex [OsCl$_2$(P–C)L] (P–C = PMe$_2$C$_{10}$H$_6$) was obtained as a minor product. The e.s.r. spectra have been recorded for a large range of tris(bidentate sulphur) metal compounds including [Os(sacsac)] and the novel [(Ph$_4$P)$_3$[Os(mnt)$_3$] (mnt = cis-1,2-dicyanoethylene-1,2-dithiolate).[26] The near isotropic g-values obtained indicate large low-symmetry distortion both geometric and electronic in origin. Electronic spectra of [Os(SCN)$_6$]$^{3-}$ and several related thiocyanato-complexes have been reported both in aqueous 2M-KSCN and in molten KSCN as solvent.[75]

A relatively inexpensive route to [Os(NH$_3$)$_6$]$^{3+}$ has been found using the commercially available (NH$_4$)$_2$[OsCl$_6$] as starting material (equation 21).[99a] It is interesting that a similar reaction with ruthenium chloride is known[100] to yield instead the

$$(NH_4)_2[OsCl_6] + NH_3\text{--}Zn \xrightarrow{5\,h} \text{soln} \xrightarrow{KI} [Os(NH_3)_6]I_3 \qquad (21)$$

Ru^{II} species [Ru(NH$_3$)$_6$]$^{2+}$. The [Os(NH$_3$)$_6$]X$_3$ (X = Cl, Br, I or ClO$_4$) salts are stable in dry air but decompose when moist or in aqueous solution, even under argon. Application of the simplified MO theory of Zwickel and Creutz[79] has permitted the prediction of both the band positions and intensities in the charge-transfer spectra of [MX(NH$_3$)$_5$]$^{2+}$ and cis- and $trans$-[MX$_2$(NH$_3$)$_4$]$^+$ (M = Ru or Os; X = Cl, Br or I).[80] The bands were assigned as transitions from a halogen ligand π-orbital to a metal d-orbital. Interestingly, the Os—X bonds were shown to be slightly less covalent than corresponding Ru—X bonds.

Intensely coloured air-stable isocyanide complexes of the type [OsCl$_2$(R^{1}NC)(PR$_3^2$)$_3$]ClO$_4$ (R^2 = mixed alkyl, aryl) have been prepared (equation 22).[99] From their single ν(Os—Cl) stretches and comparison with corresponding Os^{III} complexes they were assigned structure (20). Related cations [OsCl$_2$(PhNC)$_2$(PEt$_3$)$_2$]ClO$_4$ and [OsCl$_2$(PhNC)(PEt$_3$)$_2$]ClO$_4$ were also synthesized by oxidation of their Os^{II}

$$mer\text{-}[OsCl_3(PR_3^2)_3] + R^1NC + AgClO_4 \xrightarrow[20\,°C]{acetone} [OsCl_2(R^1NC)(PR_3^2)_3]ClO_4 + AgCl \qquad (22)$$

analogues with a HNO$_3$–HCl mixture.

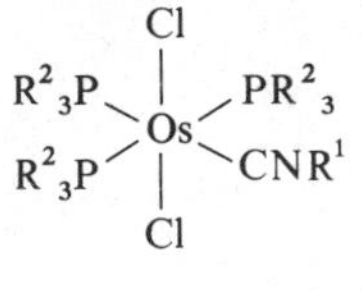

(20)

[99a] F. Bottomley and S. B. Tong, *Inorg. Chem.*, 1974, **13**, 243.
[100] F. M. Lever and A. R. Powell, *J.C.S. Dalton*, 1969, 1477.

Osmium(IV).—Treatment of $[OsH_4(PPh_3)_3]$ with 1,3-diaryltriazenes in boiling 2-methoxyethanol has produced $[OsH_3(ArNNNAr)(PPh_3)_2]$.[42] These species were formulated as Os^{IV} trihydrides rather than the Os^{II} monohydrides $[OsH(ArNNNAr)(PPh_3)_2]$ on the basis of i.r. and ^1H n.m.r. evidence. The apparent equivalence of the three hydrides in these complexes ($\tau \sim 19$) suggested stereochemically non-rigid seven-co-ordination.

The complexes $[OsNX_3(APh_3)_2]$ (X = Cl or Br, A = As; X = Cl, A = Sb) have been prepared *via* reaction (23).[89, 101] These novel compounds were confirmed to be

$$Bu_4^n N[OsNX_4] + APh_3 \xrightarrow[\text{MeOH}]{70\,^\circ C} [OsNX_3(APh_3)_2] \qquad (23)$$

Os^{VI} nitrido species from their diamagnetism and i.r. spectra [$\nu(Os\equiv N)$ near 1060 cm^{-1}]. In contrast, the corresponding reaction with PPh_3 yielded products which from their paramagnetism ($\mu = 1.8$ BM) and i.r. spectra were formulated as the triphenylphosphine-imine derivatives $[OsX_3(NPPh_3)(PPh_3)_2]$. It was suggested that the $^-NPPh_3$ ligand was formed by nucleophilic attack of PPh_3 on the co-ordinated nitrogen, behaving here as an electrophilic nitrene. Oxidation of these $[OsX_3(NPPh_3)(PPh_3)_2]$ species with chlorine generated diamagnetic Os^{VI} nitrido complexes (equation 24).

$$[OsCl_3(NPPh_3)(PPh_3)_2] + Cl_2 \xrightarrow{\text{MeOH}} [OsNCl_3(PPh_3)_2] \qquad (24)_{\prime}$$

Far-i.r. spectra have been studied for a large range of hexachlorometallates of the type $K_2[MCl_6]$ (M = Os, Ir, Pd or Pt) at pressures of up to 20 kbar using a new design cell.[102] The $\nu_4\{\sigma(ClMCl)\}$ mode was observed to be much more pressure-sensitive than the corresponding $\nu_3(M\text{—}Cl)$ stretch. In a related study other workers have compared the stretching frequency $\nu_3(M\text{—}X)$ and the unit cell length for a series of compounds $A_2[MX_6]$ (A = K, Rb, Cs or NH_4; M = Ru, Os, Sn or Te; X = Cl or Br).[83] Whereas the trends observed for the Te and Sn compounds are thought to be related to steric factors, the opposite result found with the Ru and Os analogues (ν_3 increases with decrease in cation size) are explained in terms of $M \rightarrow X$ $(d–p)$ π-bonding.

Sharp line luminescence has also been observed for $[OsBr_6]^{2-}$ doped in crystals of $Cs[ZrBr_6]$ at 20 K.[103] The magnetic susceptibility calculated from optical data agrees well with available susceptibility data for $Cs_2[OsBr_6]$.

Osmium(VI).—$[OsO_4]$ has been shown to react with monoalkenes (R) to give five-co-ordinate mono- and di-esters of the saturated diolato ligand $(O_2R)^{2-}$, namely $[Os_2O_4(O_2R)_2]$ and $[OsO(O_2R)_2]$.[104] These complexes were assigned structures (21) and (22), respectively, on the basis of i.r. and molecular weight measurements.

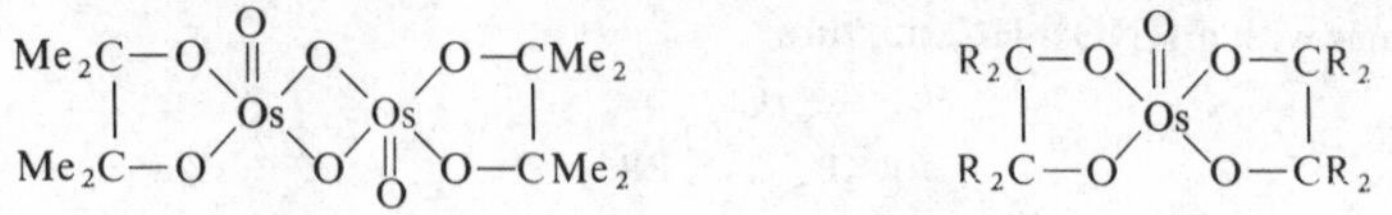

(21)   (22)

[101] W. P. Griffith and D. Pawson, *J.C.S. Chem. Comm.*, 1973, 418.
[102] D. M. Adams and S. J. Payne, *J.C.S. Dalton*, 1974, 407.
[103] J. L. Nims, H. H. Patterson, S. M. Khan, and C. M. Valencia, *Inorg. Chem.*, 1973, **12**, 1602.
[104] R. J. Collin, J. Jones, and W. P. Griffith, *J.C.S. Dalton*, 1974, 1094.

In contrast, treatment of $[OsO_4]$ with alkenes (R) or acetylenes in the presence of tertiary nitrogen bases (L = pyridine or isoquinoline) yielded the octahedral 'osmyl' species (23) and (24). From measurements of the ionic products of the sparingly

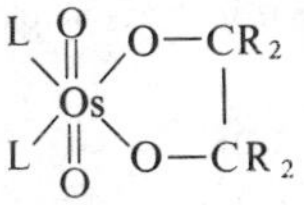

(23) 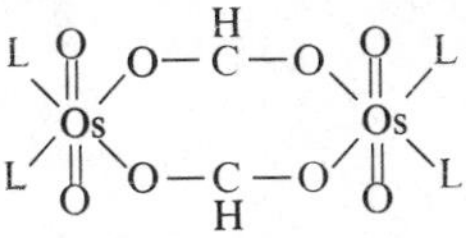   (24)

soluble diethyl-, butyl-, and benzyl-dithiocarbamato Os^{VI} complexes $[OsO_2L_2]$ it has been concluded that these species are of high stability.[105]

Oxidation of the octaethylporphinato (oep) complex $[Os(oep)(CO)(py)]$ by hydrogen peroxide in CH_2Cl_2 has been reported yielding the unusual high oxidation state metalloporphyrin $[Os(oep)O_2]$.[106] U.v.–visible spectra confirm that the porphyrin ligand has remained intact, while $v(OsO_2)$ appears at 825 cm^{-1} in the i.r. The black crystalline complex is reduced by $SnCl_2$ in methanol to an Os^{IV} species, and further reduction with dithionite gives the Os^{II} derivative. New mixed-ligand Os^{VI} nitrido complexes have been synthesized using Scheme 9 (X = Cl or Br; L–L = 1,10-phen or 2,2'-bipy).[107] All exhibit $v(Os\equiv N)$ bands near 1100 cm^{-1}.

$$K[OsNX_4(H_2O)],H_2O$$

oxine / EtOH   EtOH \ L–L

$[OsNX(oxine)_2]$ $[OsNBr_3(L-L)]$

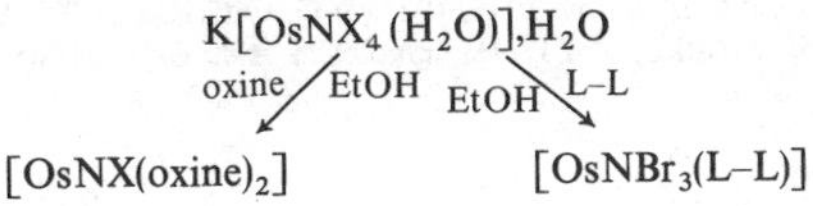

Scheme 9

Osmium(VIII).—Two papers have reported the photoelectron[92] and electronic[93] spectra of $[OsO_4]$. The photoelectron spectrum allowed the assignment of the five highest energy MO's. Electronic spectra were obtained both in the gas phase and at 20 K in an SF_6 matrix. The first virtual two-electron MO of $[OsO_4]$ was estimated to be of higher energy than in the corresponding $[RuO_4]$, leading to the higher energy d–d bands observed in the former species. Stability constants have been measured for complex formation between o-coumaric acid and a variety of noble metal ions, including a 1:1 complex with Os^{VIII}.[74]

The mixed-ligand nitrido-complex $(NOsO(bigH)_2](OH)_3$ (bigH = $C_2N_5H_7$) has been prepared (equation 25).[107] Its unusually low frequency $v(Os\equiv N)$ stretch was

$$KOsNO_3 + \text{biguanidine sulphate} \xrightarrow{pH\ 10} [NOsO(bigH)_2](OH)_3 \quad (25)$$

attributed to very strong donation by the biguanide ligand, and the trans-effect of Os=O where the valency of osmium is eight. Whereas reaction of PF_3 (400 atm) with NiO at 300 °C is known to give the trifluorophosphine complex $[Ni(PF_3)_4]$, a similar reaction with $[OsO_4]$ resulted only in reduction of the metal oxide.[87]

105. L. F. Shuydka, Y. I. Usatsenko, and F. M. Tulyupa, *Russ. J. Inorg. Chem.*, 1973, **18**, 396.
106. J. W. Buchler and P. D. Smith, *Angew. Chem. Internat. Edn.*, 1974, **13**, 341.
107. N. Ta, P. Pramanik, and D. Sen, *J. Indian Chem. Soc.*, 1974, **51**, 374.

Table 2 X-*Ray data for osmium compounds*

Compound	R	Comments	Ref.
$Ph_4As[OsNCl_4]$	0.036	anion has a square-base pyramidal shape with Os atom 0.58 Å above plane 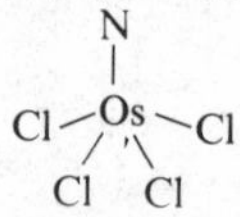	a
$[Os(SnPh_3)_2(CO)_4]$	0.063	six-co-ordinate Os, with a linear Sn—Os—Sn geometry. The Os—Sn distance of 2.712 Å indicates little double bond character	b
$[Os_3(CO)_7(C_6Ph_6)]$	0.078	triangular cluster of Os atoms, with only terminal CO ligands	c
$[OsO_4]$	—	refinement of earlier ^{e} crystal data. Showed that the 0.03 Å difference between Os—O bond lengths reported for gaseous and crystalline $[OsO_4]$ is not due to neglect of anomalous dispersion corrections	d

For the $Ph_4As[OsNCl_4]$ structure:

$$Cl \overset{\displaystyle N}{\underset{\displaystyle Cl \quad Cl}{\overset{|}{\diagup Os \diagdown}}} Cl$$

(a) S. R. Fletcher, W. P. Griffith, D. Pawson, F. L. Phillips, and A. C. Skapski, *Inorg. Nuclear Chem. Letters*, 1973, **9**, 1117. (b) J. P. Collman, D. W. Murphy, E. B. Fleischer, and D. Swift, *Inorg. Chem.*, 1974, **13**, 1. (c) G. Ferraris and G. Gervasio, *J.C.S. Dalton*, 1973, 1933.(d) G. Gilli and D. W. J. Cruickschank, *Acta Cryst.*, 1973, **B29**, 1983. (e) T. Ueki, A. Zalkin, and D. H. Templeton, *Acta Cryst.*, 1965, **19**, 157.

3 Rhodium

Cluster Compounds.—Following the recent[108] preparation of $[Rh_6C(CO)_5]^{2-}$, the first carbido-cluster of the cobalt subgroup, two related species have now been obtained (Scheme 10).[109] Preliminary single-crystal X-ray studies were carried out

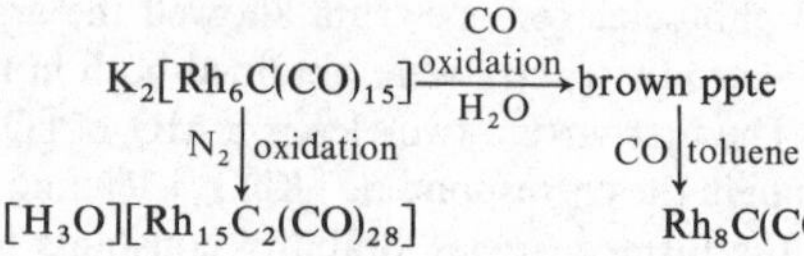

$$K_2[Rh_6C(CO)_{15}] \xrightarrow[H_2O]{\overset{CO}{oxidation}} \text{brown ppte}$$

$$N_2 \Big| oxidation \qquad\qquad CO \Big| toluene$$

$$[H_3O][Rh_{15}C_2(CO)_{28}] \qquad\qquad Rh_8C(CO)_{19}$$

Scheme 10

on both these black, diamagnetic products (Table 3). $Rh_8C(CO)_{19}$ disproportionates in acetonitrile to regenerate $[Rh_6C(CO)_{15}]^{2-}$ and $[Rh(CO)_2(MeCN)_2]^+$. In addition, a mixed metal-carbido complex $[Co_2Rh_4C(CO)_{15}]^{2-}$ was reported from the interaction of $[Rh(CO)_4]^-$ and $Co_3ClC(CO)_9$.

^{13}C N.m.r. has been used by several groups as a structural tool in rhodium clusters. The spectrum of $[Rh_{12}(CO)_{30}]^{2-}$ is essentially the same at -70 and $52\,°C$, and separate resonances due to terminal, doubly bridging, and triply-bridging carbonyl ligands could be distinguished.[110] Their relative intensities (20:2:8 respectively) agreed with the results from a previous X-ray analysis.[110a] On the other hand, while

[108] V. G. Albano, M. Sansoni, P. Chini, and S. Martinengo, *J.C.S. Dalton*, 1973, 651.
[109] V. G. Albano, P. Chini, S. Martinengo, M. Sansoni, and D. Strumolo, *J.C.S. Chem. Comm.*, 1974, 299.
[110] P. Chini, S. Martinengo, D. J. A. McCaffrey, and B. T. Heaton, *J.C.S. Chem. Comm.*, 1974, 310.
[110a] V. G. Albano and P. L. Belon, *J. Organometallic Chem.* 1969, **19**, 405.

the ^{13}C n.m.r. of $RhCo_3(CO)_{12}$ at $-85\,°C$ confirmed structure (25), at higher temperatures the resonance lines changed suggesting two types of CO site exchange *via* a concerted carbonyl-terminal exchange process.[111] A low-temperature ^{13}C n.m.r. study of the related $Rh_4(CO)_{12}$ in CD_2Cl_2 solution revealed four resonances of equal

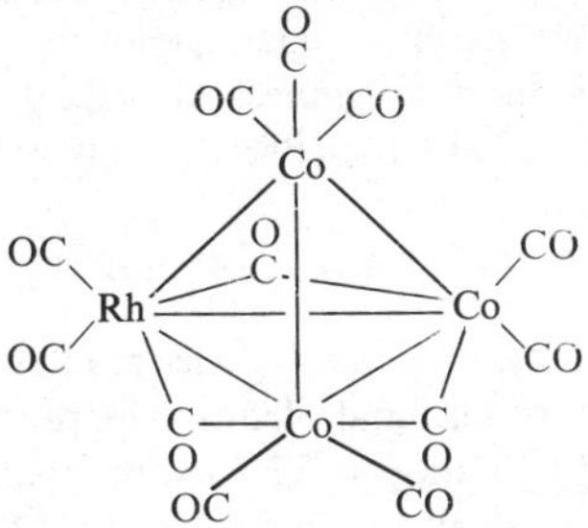

(25)

intensity, confirming that this cluster has the same structure in solution as in the crystalline form.

The observed oxidation of many low-valent transition metal complexes with Karl Fischer reagent, including that of $Rh_4(CO)_{12}$ to Rh^{III}, has potential as a method for the oxidimetric estimation of these compounds in solution.[113]

Rhodium(0).—Low-temperature (10 K) matrix isolation techniques have produced the novel dinitrogen complexes $[Rh(N_2)_x]$ $(x = 1—4)$.[114] The $[Rh(N_2)_4]$ species was shown from i.r. studies to have a distorted tetrahedral geometry in solid nitrogen, but an approximately regular tetrahedral symmetry in solid argon.

Synthesis of the pale yellow dimer $[Rh_2\{P(OMe)_3\}_8]$ has been reported *via* ligand displacement from $[CpRh(C_2H_4)_2]$ (Cp = cyclopentadienyl) using excess $P(OMe)_3$ at $60\,°C$.[115] Its ^{31}P n.m.r. spectrum at room temperature indicates the non-fluxional bicapped trigonal antiprismatic structure (26). However, at $100\,°C$ the spectrum is

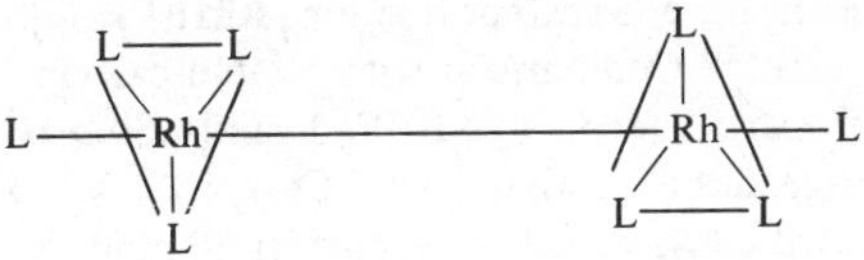

(26)

consistent with a specific equatorial intermolecular ligand exchange (rather than an intramolecular exchange between equatorial and axial sites). A further new example of a carbon dioxide complex, $[Rh_2(PPh_3)_3(CO)_2(CO_2)_2],C_6H_6$, has been synthesized *via* direct reaction of CO_2 on a low-valent substrate (equation 26).[116] Three medium

$$[Rh(PPh_3)(CO)_2]_2,C_6H_6 + PPh_3 + CO_2 \xrightarrow[\text{temp.}]{\text{room}} [Rh_2(PPh_3)_3(CO)_2(CO_2)_2],C_6H_6 \quad (26)$$

[111] B. F. G. Johnson, J. Lewis, and T. W. Matheson, *J.C.S. Chem. Comm.*, 1974, 441.
[112] J. Evans, B. F. G. Johnson, J. Lewis, J. R. Norton, and F. A. Cotton, *J.C.S. Chem. Comm.*, 1973, 807.
[113] I. Bosnyák-Ilcsik, S. Papp, L. Bencze, and G. Pályi, *J. Organometallic Chem.*, 1974, **66**, 149.
[114] G. A. Ozin and A. V. Voet, *Canad. J. Chem.*, 1973, **51**, 3332.
[115] R. Mathieu and J. F. Nixou, *J.Z.S. Chem. Comm.*, 1974, 147.
[116] I. S. Kolomnikov, T. S. Belopotapova, T. V. Lysyak, and M. E. Vol'pin, *J. Organometallic Chem.*, 1974, **67**, C25.

or intense i.r. bands in the product at 1600, 1355, and 825 cm^{-1} were assigned to modes of co-ordinated CO_2 from ^{13}C substitution studies. Although stable in air, the yellow compound liberates 1 mol of CO_2 per Rh atom on heating or addition of sulphuric acid.

Rhodium(I).—*Group VII Donors. Hydrido-phosphine complexes.* The trifluorophosphine complex $[RhH(PF_3)(PPh_3)_3]$, an analogue of the well-known homogeneous catalyst $[RhH(CO)(PPh_3)_3]$, has been synthesized by the displacement process (27).[117] Similarly, white crystalline $[RhH(PF_3)_2(PPh_3)_2]$ was obtained using a two-molar

$$[RhH(PPh_3)_4] + PF_3(equimolar) \xrightarrow[C_6H_6]{\substack{room \\ temp.}} [RhH(PF_3)(PPh_3)_3] + PPh_3 \qquad (27)$$

excess of PF_3. Not surprisingly, these new species are also good catalysts for hydrogenation and isomerization of terminal olefins. The preparation of the interesting chiral hydride $[RhH(diop)_2]$ [diop = 2,3-*O*-isopropylidene-2,3-dihydroxy-1,4-bis (diphenylphosphino)butane] has also been described.[118] It is an efficient catalyst for homogeneous asymmetric hydrogenation of terminal olefins and unsaturated carboxylic acids. Another new homogeneous hydrogenation catalyst reported is the bright orange $[RhH(dbp)_4]$ [dbp = (27); equation 28].[119] Molecular weight measurements

$$[RhCl_3(dbp)_3] + NaBH_4 \xrightarrow[\Delta]{EtOH} [RhH(dbp)_4] \qquad (28)$$

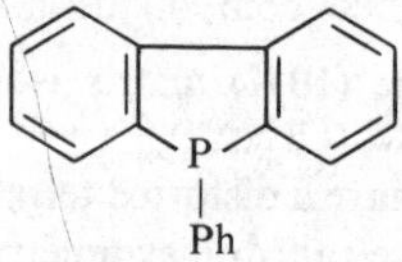

P
|
Ph
(27)

indicate dissociation in solution to $[RhH(dbp)_3]$, which can be isolated from benzene solution by addition of ethanol. The presence of a hydride ligand was confirmed from i.r. [ν(Rh—H) at 2070 cm^{-1}] and ^1H n.m.r. data.

Photoelectron spectra have been reported for $[RhH(PF_3)_4]$ and some related trifluorophosphine species.[120] Comparison with related carbonyl complexes suggests similar π-acceptor properties for CO and PF_3 ligands. Studies have shown that the activity of the hydrogenation catalyst $[RhH(CO)_2(PPh_3)_3]$, which decreases with time, can be regenerated using weak u.v.–visible irradiation.[121]

Halogeno-carbonyl and -phosphine complexes. Ligand exchange reactions in polar solvents have afforded a convenient general route to a wide range of cationic trimethyl- and triethyl-phosphite noble metal complexes including $[Rh\{P(OR)_3\}_5]^+$.[19] Following a preliminary communication,[122a] details have appeared for a general simple synthesis of compounds of the type $[RhCl(CO)L]_2$ [L = n-C$_3$H$_7$OPF$_2$, (POMe)$_3$, PPh$_3$, P(NMe$_2$)$_3$, PMe$_2$Ph, PMe$_3$, or PEt$_3$].[122b] I.r. and ^1H n.m.r. data suggest structures involving a double square plane in either a bent or planar configuration,

[117] J. F. Nixon and J. R. Swain, *J. Organometallic Chem.*, 1974, **72**, C15.
[118] W. R. Cullen, A. E. Fenster, and B. R. James, *Inorg. Nuclear Chem. Letters*, 1974, **10**, 167.
[119] D. E. Budd, D. G. Holah, A. N. Hughes, and B. C. Hui, *Canad. J. Chem.*, 1974, **52**, 775.
[120] J. F. Nixon, *J.C.S. Dalton*, 1973, 2226.
[121] W. Strohmeier and G. Csontos, *J. Organometallic Chem.*, 1974, **67**, C27.
[122] (a) A. Maisonnat, P. Kalck, and R. Poilblanc, *Compt. rend.*, 1973, **276**, 1263; (b) A. Maisonnat, P. Kalck and R. Poilblanc, *Inorg. Chem.*, 1974, **13**, 661.

with bridging chloride groups. Chatt and co-workers have reported the similar preparation of a variety of water-soluble tertiary phosphine complexes, for example equation (29) [L = $PMe(CH_2OCOMe)_2$, $PEt(CH_2OCOMe)_2$, $PPh_2(CH_2OCOMe)$,

$$[RhCl(CO)_2]_2 + 4L \rightarrow trans\text{-}[RhCl(CO)L_2] \qquad (29)$$

$PBu^i_2(CH_2OCOMe)$, and $P(CH_2OH)_3$].[123] N.m.r. measurements in benzene solution indicated that some of these substituted tertiary phosphine ligands undergo exchange at room temperature.

Reaction of excess PCy_3 (Cy = cyclohexyl) with $[RhCl(C_8H_{14})_2]_2$ in benzene has yielded the solvated species $[RhCl(PCy_3)_2(S)]$.[124] The solvent molecule is displaced by other ligands to give $trans\text{-}[RhCl(PCy_3)_2L]$ (L = O_2, N_2, H_2, CO or C_2H_4). Particularly interesting is the spontaneous formation of the dinitrogen complex (1 atm of N_2 for 5 days), which was ascribed to a combination of the steric requirements and electronic properties of PCy_3. Similarly, the yellow crystalline cis-$[RhCl(CO)_2(OPCy_3)]$ has been obtained in high yield from the bridge-splitting reaction of $OPCy_3$ on $[RhCl(CO)_2]_2$.[125] An X-ray analysis was also reported (Table 3).

The composition of $[RhClL_3]$ (L = triarylphosphine) species in solution has aroused considerable recent interest. Further information has now been obtained from a study combining molecular weight, ^{31}P, ^{13}C, ^1H n.m.r., i.r., and u.v.–visible measurements on the equilibria present in solutions containing mixtures of Rh^I chloride, PPh_3, or $P(p\text{-tolyl})_3$, and molecular hydrogen.[126] Results indicated that the $[RhClL_3]$ complexes do not dissociate to $[RhClL_2]$ to a spectroscopically significant extent, but are in equilibrium with a chloride-bridged dimer $[RhClL_2]_2$. This dimer reacts with H_2 to give $H_2[RhClL_2]_2$, and is readily cleaved by L regenerating $[RhClL_3]$.

Intermolecular redistribution of CO, halide, and phosphine ligands has been observed in various $[MX(CO)(PR_3)_2]$ (M = Rh or Ir, X = Cl or Br) systems using i.r. and n.m.r. spectroscopy.[127] The authors suggested that CO and halide exchange occurs *via* the doubly five-co-ordinate, bridged intermediates (28a) and (28b) (M =

$$
\begin{array}{ccc}
R_3P & X & PR_3 \\
 \diagdown\diagup\diagdown\diagup & \\
OC\!-\!M & & M\!-\!CO \\
 \diagup & \diagdown\diagup & \diagdown \\
R_3P & X & PR_3 \\
\end{array}
\qquad
\begin{array}{ccc}
 & O & \\
R_3P & C & PR_3 \\
 \diagdown\diagup\diagdown\diagup & \\
X\!-\!M & & M\!-\!X \\
 \diagup & \diagdown\diagup & \diagdown \\
R_3P & C & PR_3 \\
 & O & \\
\end{array}
$$

| (28a) | (28b) |

Rh). Co-ordinative unsaturation may be necessary for redistribution of ^{13}CO since no exchange was observed with the related $[RhCl_3(CO)(PPh_3)_2]$.

Rapid exchange of ligands between $[RhCl(CO)_2]_2$ and $[RhCl(PF_3)_2]_2$ has been confirmed, yielding mainly $[RhCl(CO)(PF_3)]_2$, which from i.r. data was believed to be a mixture of the isomers (29a), (29b), and (30).[128] Addition of excess PF_3 to

[123] J. Chatt, G. J. Leigh, and R. M. Slade, *J.C.S. Dalton*, 1973, 2021.
[124] H. L. M. Van Gaal, F. G. Moers, and J. J. Steggerda, *J. Organometallic Chem.*, 1974, **65**, C43.
[125] G. Bandoli, D. A. Clemente, G. Daganello, G. Carturan, P. Uguagliati, and U. Belluco, *J. Organometallic Chem.*, 1974, **71**, 125.
[126] C. A. Tolman, P. Z. Meakin, D. L. Lindner and J. P. Jesson, *J. Amer. Chem. Soc.*, 1974, **96**, 2762.
[127] P. E. Garrou and G. E. Hartwell, *J.C.S. Chem. Comm.*, 1974, 381.
[128] M. A. Bennett and T. W. Turney, *Austral. J. Chem.*, 1973, **26**, 2321.

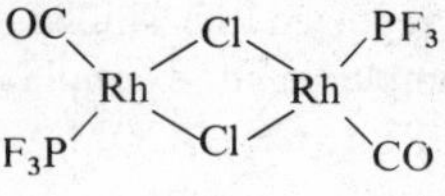

OC、   、Cl、   ／PF₃   OC、   ／Cl、   ／CO   OC、   ／Cl、   ／PF₃
   Rh   Rh   Rh   Rh   Rh   Rh
F₃P／   ＼Cl／   ＼CO   F₃P／   ＼Cl／   ＼PF₃   OC／   ＼Cl／   ＼PF₃

      (29a) 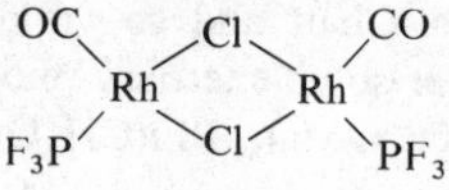   (29b) 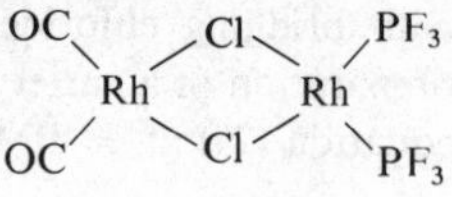   (30)

$[RhX(PF_3)_2]_2$ afforded the apparently five-co-ordinate $[RhX(PF_3)_4]$ (X = Cl, Br, or I). On the other hand, while $P(OPh)_3$ and $Ph_2PC_2H_4PPh_2$ displace all PF_3 ligands from $[RhCl(PF_3)_2]_2$, PPh_3 and $AsPh_3$ lead instead to $[RhCl(PF_3)L_2]$. An i.r. study of the exchange reaction [30; $X \neq Y$; X, Y = halide, $RCOO^-$, or RO^-

$$[Rh_2(CO)_4X_2] + 2MY \rightarrow [Rh_2(CO)_4Y_2] + MX \qquad (30)$$

(R = alkyl, aryl)], commonly used for the synthesis of bis (μ-anion)dirhodium(I) tetracarbonyls, has revealed the presence of $[Rh_2(CO)_4XY]$ type intermediates containing two different bridging ligands.[129]

The first examples of the carbonylation of transition-metal complexes by carbon dioxide under mild conditions have been reported.[19] For example, $[RhCl(PPh_3)_3]$ reacts with CO_2 at room temperature in the presence of silicon hydrides to give the known $[RhCl(CO)(PPh_3)_2]$. A polymeric complex has been produced from the interaction of $[RhCl(CO)(PPh_3)_2]$ with a resin containing grouping (31), as part of a

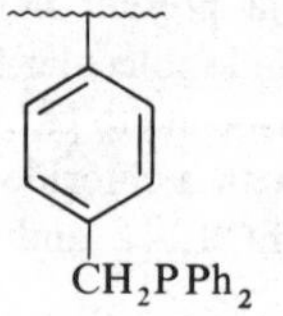

CH₂PPh₂
(31)

systematic study of the catalytic properties of resin-bound transition-metal compounds.[130] Efficient catalysis of hydroformylation and isomerization processes was observed.

Whereas SCl_2 behaves as an oxidative addition reagent towards Ir^I substrates, treatment of $[RhCl(CO)(PPh_3)_2]$ with SCl_2 has been observed to cause chlorination only, producing $[RhCl_3(CO)(PPh_3)_2]$.[131] The dark purple product obtained from the reaction of octaethylporphyrin (H_2oep) with $[RhCl(CO)_2]_2$ in hot chloroform has been shown from i.r. and X-ray studies (Table 3) to be $[H_4oep][RhCl_2(CO)_2]$.[132] Since the corresponding reaction in benzene is known[132a] to give the metalloporphyrin $H[Rh_2(oep)(CO)_4Cl]$, the presence of HCl in the former reaction was suggested, deriving from the chloroform solvent.

Group VI Donors. O-Donor ligands. A general route to noble metal carboxylate derivatives has been described involving reaction of low oxidation state complexes

[129] G. Pályi, A. Vizi-Orosz, L. Markó, F. Marcati, and G. Bor, *J. Organometallic Chem.*, 1974, **66**, 295.
[130] G. O. Evans, C. U. Pittman, R. McMillan, R. T. Beach, and R. Jones, *J. Organometallic Chem.*, 1974, **67**, 295.
[131] W. H. Baddley and D. S. Hamilton, *Inorg. Nuclear Chem. Letters*, 1974, **10**, 143.
[132] E. Cetinkaya, A. W. Johnson, M. F. Lappert, G. M. McLaughlin, and K. W. Muir, *J.C.S. Dalton*, 1974, 1236.
[132a] Z. Oshida, H. Ogoshi, T. Omura, E. Watanabe, and T. Kurosaki, *Tetrahedron Letters*, 1972, 1077.

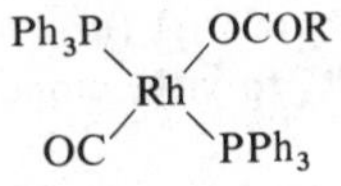

(32)

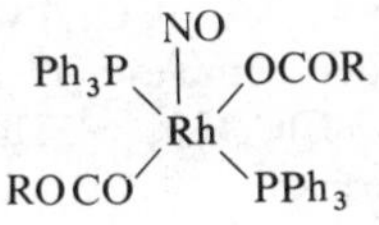

(33)

with carboxylic acids.[23] Among the numerous compounds isolated were [Rh(OCOR)(CO)(PPh$_3$)$_2$] and [Rh(OCOR)$_2$(NO)(PPh$_3$)$_2$], which were assigned structures (32) and (33), respectively, on the basis of i.r. and n.m.r. data. A more novel route to carboxylate species is reaction (31) in which CO_2 inserts into the Rh—C bond form-

$$[Rh(Ph)(PPh_3)_3] + CO_2 \xrightarrow[C_6H_6]{room\ temp.} [Rh(OCOPh)(PPh_3)_3] \qquad (31)$$

ing a red crystalline benzoate complex.[135] X-Ray studies confirmed the product to be monomeric and to contain a unidentate benzoate ligand (Table 3). Other workers have employed a similar CO_2 insertion process to prepare bicarbonato compounds such as [M(OCO$_2$H)(CO)(PPh$_3$)$_2$] (M = Rh or Ir, equation 32).[134] The diamagnetic

$$[M(OH)(CO)(PPh_3)_2] + CO_2 \xrightarrow{EtOH} [M(OCO_2H)(CO)(PPh_3)_2] \qquad (32)$$

complexes were tentatively assigned unidentate bicarbonate co-ordination, although neither apparently showed an O—H stretch in the i.r.

Dinuclear complexes of general formula [L(CO)RhQRh(CO)L] [L = PPh$_3$ or CO; Q = (34a, b, c) *etc.*], containing bridging quinone-type ligands, have been synthesized *via* equation (33).[135] I.r. studies support O-co-ordination. Related dimers

$$2[Rh(CO)(acac)L] + QH_2 \rightarrow [L(CO)RhQRh(CO)L] \qquad (33)$$

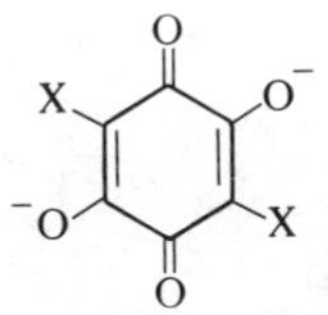

(34a) X = Cl or Br

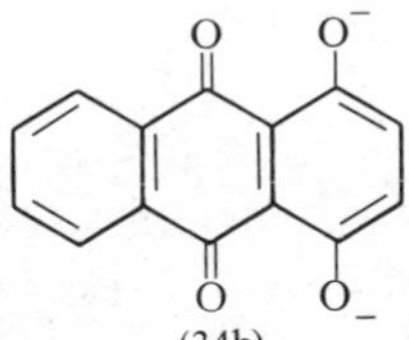

(34b)

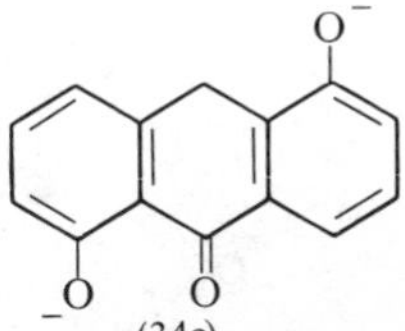

(34c)

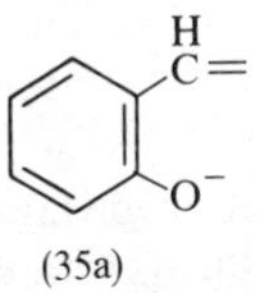

of the type [(CO)$_2$RhLRh(CO)$_2$] [L = HsalNNsalH, HsalN(CH$_2$)$_6$NsalH, HsalylNHNsalH, or HAcpHNNsalH; sal = (35a), salyl = (35b), AcpH = (35c)]

(35a)

(35b)

(35c)

were also prepared by similar routes, as well as the orange monomer [Rh(PPh$_3$)$_2$(CO)(salylNHNsalH)].[136] The latter complex contains a free chelate centre.

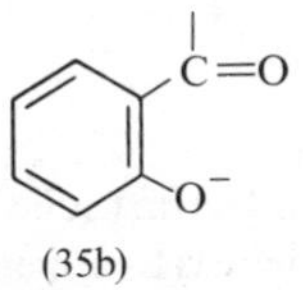

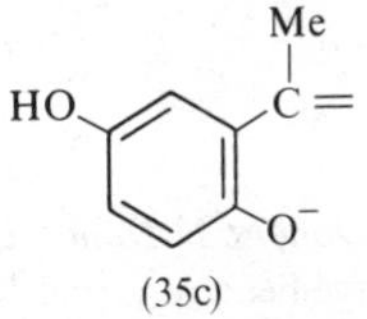

[133] I. S. Kolomnikov, A. O. Gusev. T. S. Belopotapova, M. K. Grigoryan, T. V. Lysyak, Y. T. Struchkov, and M. E. Vol'pin, *J. Organometallic Chem.*, 1974, **69**, C10.
[134] B. R. Flynn and L. Vaska, *J. Amer. Chem. Soc.*, 1973, **95**, 5081.
[135] L. V. Revenko, Y. A. Shvetsov, and M. L. Khidekel, *Bull. Acad. Sci. USSR*, 1973, **22**, 2572.
[136] B. G. Rogachev, A. S. Astakhova, and M. L. Khidekel, *Bull Acad. Sci. USSR*, 1973, **22**, 2544.

Treatment of $[Rh_2Cl_2(CO)_4]$ with NaOR in hexane has led to a series of rather unstable alkoxide complexes of formula $[Rh_2(OR)_2(CO)_4]$ (R = Me, Et, Pr, iso-pentyl, Ph, or p-ClC$_6$H$_4$).[137] These react with PR$_3$ to yield monomeric $[Rh(OR)(CO)(PR_3)_2]$ species.

Measurement of the polarized i.r. spectrum of single crystals of $[Rh(acac)(CO)_2]$ has enabled the direct and unambiguous assignment for the first time of the in-plane and out-of-plane modes.[138] Chlorine oxidation of this complex has yielded the Rh^{III} dimer $[RhCl_3(CO)_2]_2$, while similar reactions with $[Rh(acac)(PPh_3)(CO)]$ gave $[RhX_3(PPh_3)(CO)]$ (X = Cl or Br).[139] Karl Fischer reagent (which contains py, SO$_2$, and I$_2$) has also been shown to oxidize Rh^I species such as $[Rh(OCOCH_2Me)(CO)(PPh_3)_2]$ and $[Rh_2(OCOCH_2Me)_2(CO)_3(PPh_3)]$ to Rh^{III}.[113] This latter reaction may be useful for the oxidimetric estimation of such complexes in solution.

S-Donor ligands. New complexes of the ligand $^-SC_6Cl_5$ have been isolated using Scheme 11 (L = CO; L$_2$ = cyclo-octa-1,5-diene, norbornadiene, Cp$_2$).[140] Related,

$$[RhClL_2]_2 + HSC_6Cl_5 \rightarrow [Rh(SC_6Cl_5)L_2]_2 \xrightarrow{PPh_3} [Rh(SC_6Cl_5)(CO)(PPh_3)_2]$$

Scheme 11

presumably dimeric, compounds of type $[RhCl\{P(C_6Cl_5)Ph_2\}_2]$ were also obtained from the action of $P(C_6Cl_5)Ph_2$ on RhCl$_3$. Reaction sequence (34) has yielded the diamagnetic monothiocarbamates $[Rh_2(SCONR_2)_2(CO)_4]$.[141] Their i.r. spectra

$$RhCl_3 + CO \xrightarrow{EtOH} soln \xrightarrow[R_2NH]{COS} [Rh_2(SCONR_2)_2(CO)_4] \qquad (34)$$

suggested the bridged structure (36).

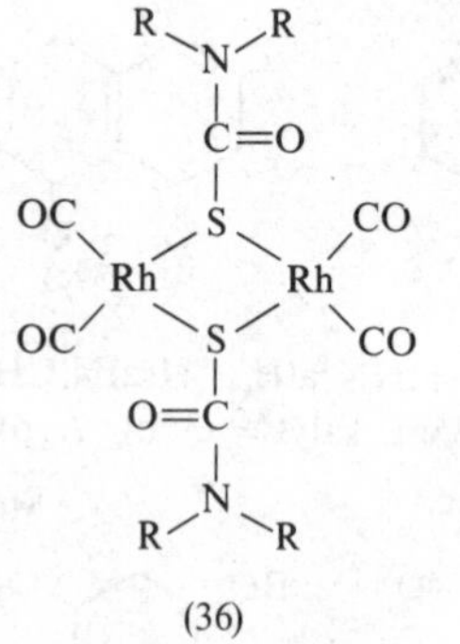

(36) 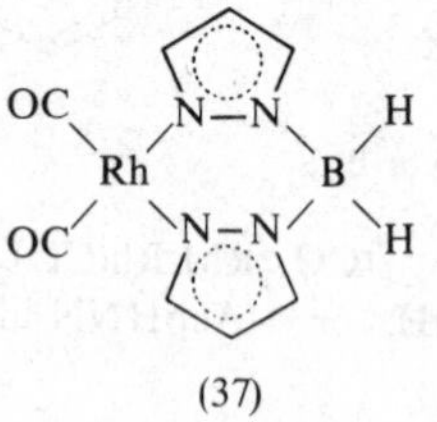   (37)

Group V Donors. N-Donor ligands. The first solid linkage isomers involving the cyanate group have been reported.[142] Whereas treatment of $[RhCl(PPh_3)_3]$ with Ph$_4$AsNCO in acetonitrile yields yellow $[Rh(NCO)(PPh_3)_3]$, the addition of ethanol to the reac-

[137] A. Vizi-Orosz, G. Pályi, and L. Markó, *J. Organometallic Chem.*, 1973, **57**, 379.
[138] D. M. Adams and W. R. Trumble, *J.C.S. Dalton*, 1974, 690.
[139] Y. S. Varshavsky, T. G. Cherkasova, and N. A. Buzina, *J. Organometallic Chem.*, 1973, **56**, 375.
[140] R. D. W. Kemmitt and G. D. Rinmer, *J. Inorg. Nuclear Chem.*, 1973, **35**, 3155.
[141] E. M. Krankovits, R. J. Magee, and M. J. O'Connor, *Inorg. Chim. Acta*, 1973, **7**, 528.
[142] S. J. Anderson, A. N. Norbury, and J. Songstad, *J.C.S. Chem. Comm.*, 1974, 37.

tion mixture causes the orange $[Rh(OCN)(PPh_3)_3]$ to precipitate. These isomers exhibited $\nu(CN)$ stretches at $2230\,cm^{-1}$ (broad) and $2215\,cm^{-1}$ (sharp), respectively. The related isocyanato-complex *trans*-$[Rh(NCO)(PPh_3)_2(CO)]$ has been isolated from the reaction of $MeNO_2$ on $[RhCl(PPh_3)_2(CO)]$, *via* a fulminato intermediate.[143]

Several research groups have reported the preparation of pyrazolylborate complexes.[144-146] Treatment of $[RhCl(CO)_2]_2$ with potassium bis(pyrazolyl)borate $\{K^+[H_2BPz_2]^-(Pz = C_3H_3N_2)\}$ gave a quantitative yield of yellow $[Rh(H_2BPz_2)(CO)_2]$, which was assigned structure (37) from i.r. and ^1H n.m.r. studies.[144] Similar reactions with the tripyrazolylborate anion $[HBPz_3]^-$ have provided an alternative route to the known species $[Rh_2(HBPz_3)_2(CO)_3]$ and $[Rh(HBPz_3)_2]PF_6$ (equations 35a and 35b).[145] Synthesis of the tetrapyrazolylborato complex $[Rh_2(BPz_4)_2(CO)_3]$

$$[RhCl(CO)_2]_2 + KHBPz_3 \rightarrow [Rh_2(HBPz_3)_2(CO)_3] \qquad (35a)$$

$$RhCl_3, xH_2O + KHBPz_3 \rightarrow [Rh(HBPz_3)_2]PF_6 \qquad (35b)$$

was also described employing $K^+BPz_4^-$ as reagent. Isolation of the simple pyrazole derivative $[Rh(Pz)(CO)_2]_2$ (using PzH in the presence of Et_3N) completed this series of related complexes. The latter complex was shown to be fluxional from ^1H n.m.r. studies, suggesting intramolecular exchange 'shuttling' of the H and Rh between the two N atoms (equation 36). Other workers have prepared novel $[Rh_2(RBPz_3)_2$-

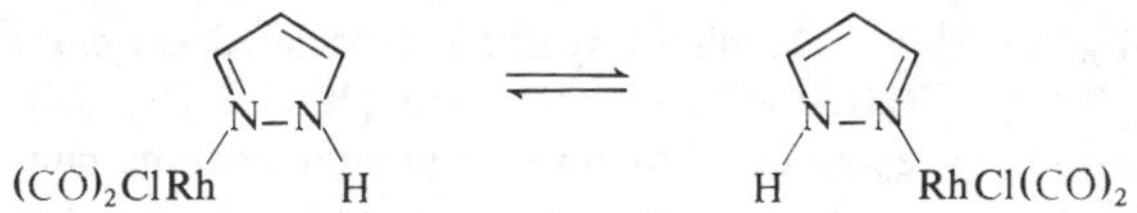

$$(36)$$

$(CO)_3]$ (R = H or Pz) complexes *via* carbonylation of $[Rh(RBPz_3)(L-L)]$ (L–L = C_2H_4; cyclo-octa-1,5-diene) substrates.[146] Oxidation of this product with iodine in dichloromethane yielded the deep red Rh^{III} derivative $[Rh(RBPz_3)I_2(CO)]$, which ^1H n.m.r. evidence showed to be an h_3-polypyrazolylborate complex.

Reaction of toluene-*p*-sulphonyl azide (RN_3) with $[M(NO)(CO)(PPh_3)_2]$ (M = Rh or Ir) in benzene has afforded urea derivatives of the type $[M(NO)(PPh_3)_2(RNCONR)]$ (38).[147] The related compounds $[RhCl(PPh_3)_2(RNCONR)]$ were

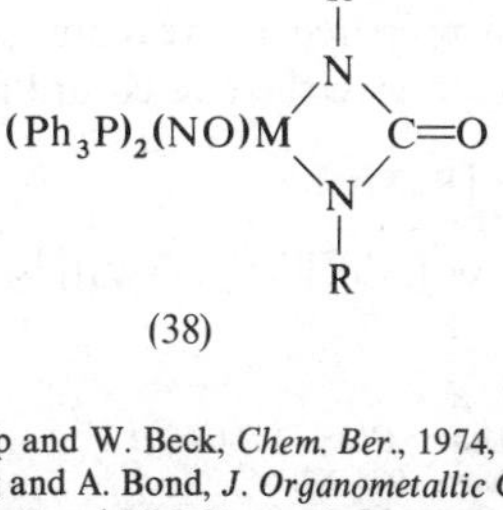

(38)

(39)

[143] K. Schorpp and W. Beck, *Chem. Ber.*, 1974, **107**, 1371.
[144] R. B. King and A. Bond, *J. Organometallic Chem.*, 1974, **73**, 115.
[145] N. F. Borkett and M. I. Bruce, *J. Organometallic Chem.*, 1974, **65**, C51.
[146] D. J. O'Sullivan and F. J. Lalor, *J. Organometallic Chem.*, 1974, **65**, C47.
[147] W. Beck, W. Rieber, S. Cenini, F. Porta, and G. La Monica, *J.C.S. Dalton*, 1974, 298.

similarly isolated from the reaction of $[RhCl(PPh_3)_3]$ with either RNCO or RNHCONHR. In contrast, treatment of the hydride $[RhH(Ph_2PC_2H_4PPh_2)_2]$ with RN_3 gave a dimeric complex (39) which was believed to contain a di-imide bridge. A novel and convenient route to a wide range of noble metal-triazenido derivatives has been reported, involving reaction of 1,3-diaryltriazenes (ArN=NNAr) with hydrido and low oxidation state phosphine complexes.[42] While the reported $[Rh(\underline{N}O)(ArNNNAr)_2]$ (Ar = *p*-tolyl) is probably five-co-ordinate (^1H n.m.r.), i.r. and ^{13}C n.m.r. studies could not distinguish between a five-co-ordinate (bidentate ArNNNAr) and a fluxional four-co-ordinate structure for $[Rh(ArNNNAr)(CO)(PPh_3)_2]$ species.

Preparation of new complexes of the types $[Rh(L-L)(CO)_3]^+$ and $[Rh(L-L)(CO)X]$ (L–L = 1,10-phen or 2,2'-bipy; X = Cl, Br or I) has been described using Scheme 12.[148] The $[Rh(L-L)(CO)X]$ products undergo oxidative addition with MeI

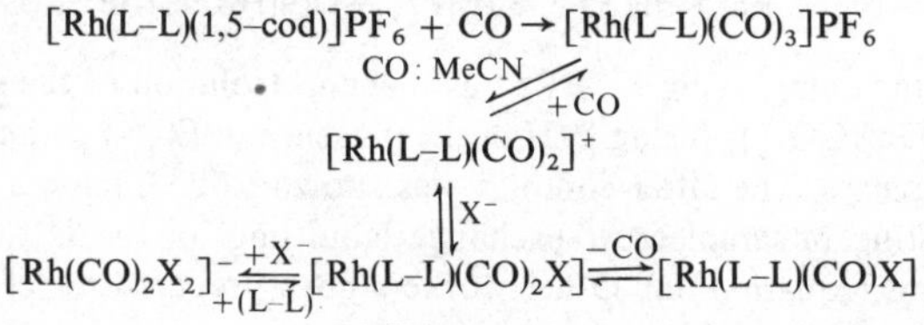

$$[Rh(L-L)(1,5-cod)]PF_6 + CO \rightarrow [Rh(L-L)(CO)_3]PF_6$$

$$\begin{array}{c} CO:MeCN \\ \underset{+CO}{\nearrow} \\ [Rh(L-L)(CO)_2]^+ \end{array}$$

$$[Rh(CO)_2X_2]^- \underset{+(L-L)}{\overset{+X^-}{\rightleftarrows}} [Rh(L-L)(CO)_2X] \underset{}{\overset{-CO}{\rightleftarrows}} [Rh(L-L)(CO)X]$$

Scheme 12

giving $[Rh(L-L)(CO)(Me)I]$. Details of synthesis of related cations of the types $[Rh(L-L)(CO)_2]ClO_4$, $[Rh(L-L)(CO)L_2']ClO_4$ and $[Rh(CO)_2L_3']ClO_4$ (L' = PR_3 or AsR_3) have also been reported,[149] following a preliminary communication.[149a]

An interesting paper with obvious environmental implications has indicated that reaction (37), which converts NO and CO into less noxious products, is catalysed by

$$2NO + CO \xrightarrow{M(NO)_2} N_2O + CO_2 \tag{37}$$

a variety of bis(nitrosyl) complexes including $[M(NO)_2Br(PPh_3)_2]$ and $[M(NO)_2(PPh_3)_2]PF_6$ (M = Rh or Ir).[150] These results suggest that such four- and five-co-ordinate bis(nitrosyl) complexes may be better formulated as 18-electron N–N bonded, *cis*-dinitrogendioxide species with significant N–N interaction, e.g. $[M(NO)_2Br(PPh_3)_2]$ becomes $[M(N_2O_2)Br(PPh_3)_2]$.

P- and *As-Donor ligands*. A series of dimeric $PF_2NR_2(L; R = Me \text{ or } Et)$ complexes of the type $[RhXL_2]_2$ (X = Cl or I) have been prepared *via* reaction (38).[151] The PF_2NEt_2 derivatives reversibly add a further L to give the five-co-ordinate species

$$[RhCl(C_2H_4)_2]_2 + L \rightarrow [RhXL_2]_2 \tag{38}$$

$[RhXL_4]$. In addition, thermal decomposition of $[RhCl(PF_2NMe_2)_2]_2$ yielded the

[148] C. Mestroni, A. Camus, and G. Zassinovich, *J. Organometallic Chem.*, 1974, **65**, 119.
[149] G. K. N. Reddy and B. R. Ramesh, *J. Organometallic Chem.*, 1974, **67**, 443.
[149a] G. K. N. Reddy and C. H. Susheelamma, *Chem. Comm.*, 1970, 54.
[150] B. L. Haymore and J. A. Ibers, *J. Amer. Chem. Soc.*, 1974, **96**, 3325.
[151] M. A. Bennett and T. W. Tuney, *Austral. J. Chem.*, 1973, **26**, 2335.

novel tris-complex $[RhCl(PF_2NMe_2)_3]$, which has no counterpart in PF_3 or CO chemistry. Spectroscopic studies suggested that these alkylaminofluorophosphine ligands are poorer π-acceptors and/or better σ-donors than PF_3. In contrast, the related ligand (40) (PN) has been shown to react with $[RhCl(C_2H_4)_2]_2$, to give the orange complex $[Rh(PN)_2]^+$ in which the ligand is bidentate, co-ordinating *via* both N and P atoms.[152, 153] However, carbon monoxide adds reversibly to this product

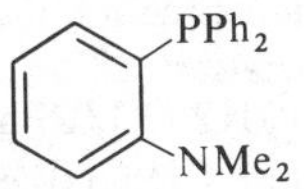

(40)

to give $[Rh(PN)_2(CO)_2]PF_6$ containing only unidentate PN (P-bonded). A variety of oxidative additions to $[Rh(PN)_2]^+$ were also studied, yielding Rh^{III} derivatives of the type $[RhXY(PN)_2]^+$ (XY = Cl_2, MeI, O_2, *etc*).

The stereorigidity of some five-co-ordinate $[Rh\{P(OR)_3\}_5]BPh_4$ complexes has been investigated by variable-temperature ^{31}P n.m.r.[154] Line-broadening observed below $0\,^\circ C$ was assigned to intramolecular exchange in view of the presence of $^{31}P-^{103}Rh$ coupling. Disappearance of such coupling above $0\,^\circ C$ suggested intermolecular exchange. Another study report the absorption and emission spectra of the square-planar compounds $[Rh(Ph_2PC_2H_4PPh_2)_2]Cl$ and $[Rh(Ph_2PC_2H_2PPh_2)_2]Cl$ both as solids and in frozen solution at $77\,K$.[155] Red shifts in the emission bands of these complexes indicated distortion towards tetrahedral geometry.

Re-investigation of the reactions between SbR_3 (R = Ph, o-MeC_6H_4, p-MeC_6H_4) and $[RhCl(CO)_2]_2$ and $[Rh(acac)(CO)_2]$ has shown the products to be $[RhCl(CO)-(SbR_3)_n]$ ($n = 2$ and/or 3, depending on solvent and size of R) and $[Rh(acac)(CO)-(SbR_3)_2]$, respectively.[156] Although five-co-ordination is favoured in the solid state there is considerable dissociation of SbR_3 in solution.

Group IV Donors. The fulminato complex *trans*-$[Rh(CNO)(PPh_3)_2(CO)]$ has been obtained as an intermediate in the formation of the corresponding isocyanato species from *trans*-$[RhCl(PPh_3)_2(CO)]$ and $MeNO_2$.[143] An alternative safe route to this and other Pd^{II} and Pt^{II} fulminato species is *via* attack of $(AsPh_4)CNO$ on appropriate substrates.[157]

A large variety of isocyanide complexes of general formula $[Rh(CNR)_4]PF_6$ have been prepared *via* equation (39) (M = Rh; R = Me, Pr^i, Cy, Bu^t or p-MeOPh).[158]

$$[MCl(CO)_2]_2 + CNR \xrightarrow[NH_4PF_6]{MeOH} [M(CNR)_4]PF_6 \qquad (39)$$

These products undergo numerous oxidative additions to yield $[Rh(CNR)_4XY]^+$

[152] T. B. Rauchfuss and D. M. Roundhill, *J. Organometallic Chem.*, 1973, **59**, C30.

[153] T. B. Rauchfuss and D. M. Roundhill, *J. Amer. Chem. Soc.*, 1974, **96**, 3098.

[154] P. Meakin and J. P. Jesson, *J. Amer. Chem. Soc.*, 1973, **95**, 7272.

[155] G. L. Geoffrey, M. S. Wrighton, G. S. Hammond, and H. B. Gray, *J. Amer. Chem. Soc.*, 1974, **96**, 3105.

[156] P. E. Garrou and G. E. Hartwell, *J. Organometallic Chem.*, 1974, **69**, 445.

[157] W. Beck, K. Schorpp, and C. Oetker, *Chem. Ber.*, 1974, **107**, 1380.

[158] J. W. Dart, M. K. Lloyd, R. Mason, and J. A. McCleverty. *J.C.S. Dalton*, 1973, 2039.

species, although no addition occurs with molecular hydrogen. The same authors reported synthetic methods for a wide range of mixed isocyanide-phosphine derivatives including $[Rh(CNR^1)_2(PR_3^2)_2]PF_6$, $[Rh(CNR)(PPh_2Me)_3]PF_6$, $[Rh(CNR^1)_3(PR_3^2)_2]PF_6$, and $[Rh(CNBu^t)_4(PPh_3)]PF_6$.[159] Protonation of $[Rh(CNR)_3(PPh_3)_2]^+$ with HPF_6 yielded the hydride $[RhH(CNR)_3(PPh_3)_2](PF_6)_2$, and several oxidative additions were described. A variable-temperature ^{13}C n.m.r. study of the related cation $[Rh(CNCMe_3)_4]^+$ revealed lineshape changes which were interpreted in terms of an intrinsically long relaxation time for the bound isocyanide C nucleus, rather than a facile Rh—C bond dissociation.[160]

Reaction of $NaCH(CN)_2$ on $[MCl(CO)(PPh_3)_2]$ (M = Rh or Ir) has afforded the dicyanomethanide complexes $[M\{CH(CN)_2\}(CO)(PPh_3)_2]$.[161] The presence of a strong i.r. band between 2120 and 2150 cm^{-1} indicated M—N bonding *i.e.*

$$M\text{---}N\text{=}C\text{=}C\overset{\displaystyle CN}{\underset{\displaystyle H}{\Big<}}$$

, in contrast to similar Pd^{II} and Pt^{II} compounds which were

C-bonded. Synthesis of the related complexes $[M\{C(CN)_3\}(CO)(PPh_3)_2]$ (M = Rh or Ir) has been described, and the mode of co-ordination of the methanide ligand discussed in the light of i.r., Raman, and ^{14}N n.m.r. data.[162]

Group III Donors. **B-*Donor ligands.*** Following two preliminary reports[163a,b] further studies have appeared on the preparation of complexes of hexaborane, B_6H_{10}, in which the metal is regarded as having inserted into the unique basal B—B bond to form a three-centre, two-electron bond.[164] Compounds described include $[Rh(B_6H_{10})_2Cl]_2$ and $[Rh(B_6H_{10})_2(acac)]$, the latter of which must be stored at $-78\,^\circ C$. The complexes were characterized by i.r. and ^1H and ^{11}B n.m.r. spectroscopy.

Rhodium(II).—*Group VI Donors.* Treatment of the Rh^{II} acetate $[Rh_2(MeCO_2)_4(H_2O)_2]$ with salicylic acid has yielded the salicylato species $[Rh_2(MeCO_2)_2(C_6H_4OHCO_2)_2(H_2O)_2]$.[165] The co-ordinated water molecules are readily displaced by neutral ligands to give a variety of derivatives of the type $[Rh_2(MeCOO)_2(C_6H_4OHCOO)_2L_2]$ (L = NH_3, py, $MeCSNH_2$). Similarities in properties and chemical behaviour of these new compounds to those of Rh^{II} acetate suggest the same dimeric structure (41). Similarly, reaction of Rh^{II}-formato and -acetato substrates or $H_3[RhCl_6]$ with thiosalicylic acid ($HSC_6H_4CO_2H$) led to dimers of the type $[Rh_2(thiosal)_4L_x]$ (x = 1 or 2; L = $HSC_6H_4CO_2H$, H_2O or HCl).[166] The ligands L could be readily displaced by a variety of other groups L' to give further new derivatives $[Rh_2(thiosal)_4L'_x]$ [x = 1 or 2; L' = NH_3, en, $CS(NH_2)_2$ or HBr].[167] All of

159 J. W. Dart, M. K. Lloyd, R. Mason, and J. A. McCleverty, *J.C.S. Dalton*, 1973, 2046.
160 E. L. Muetterties, *Inorg. Chem.*, 1974, **13**, 495.
161 W. H. Baddley and P. Chondhury, *J. Organometallic Chem.*, 1973, **60**, C74
162 W. Beck, K. Schorpp, C. Oetker, R. Schlodden, and H. S. Smedal, *Chem. Ber.*, 1973, **106**, 2144.
163 (a) A. Davison, D. D. Traficante, and S. S. Wreford, *J.C.S. Chem. Comm.*, 1972, 1155; (b) J. P. Brennan, R. Schaeffer, A. Davison, and S. S. Wreford, *J.C.S. Chem. Comm.*, 1973, 354.
164 A. Davison, D. D. Traficante, and S. S. Wreford, *J. Amer. Chem. Soc.*, 1974, **96**, 2802.
165 L. A. Nazarova and A. G. Maiorova, *Russ. J. Inorg. Chem.*, 1973, **18**, 904.
166 A. G. Maiorova, L. A. Nazarova, and G. N. Emil'yanova, *Russ. J. Inorg. Chem.*, 1973, **18**, 986.
167 A. G. Maiorova, L. A. Nazarova, and G. N. Emil'yanova, *Russ. J. Inorg. Chem.*, 1973, **18**, 989.

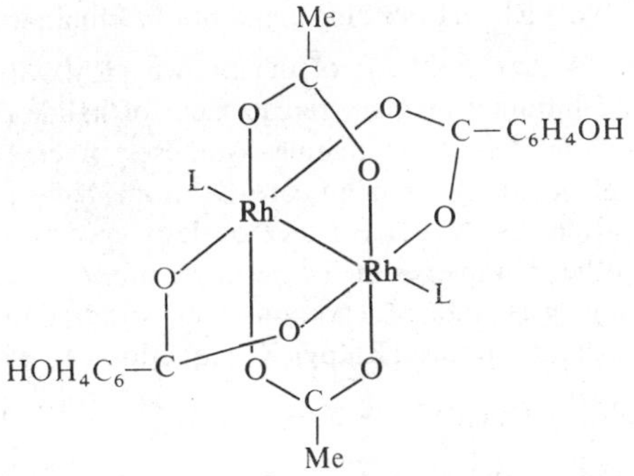

(41)

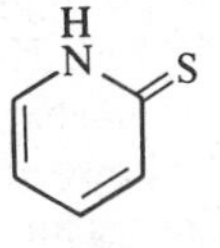

(42)

these new complexes are diamagnetic, their chemical behaviour suggesting Rh^{II} structures of the type (41).

In contrast, treatment of $[Rh_2(MeCO_2)_4]$ with pyridine-2-thiol, which exists as the tautomer LH (42), does not cause displacement of co-ordinated acetate but produces $[Rh_2(MeCO_2)_4(LH)_2]$;[168] its i.r. spectrum confirms S-co-ordination. A Raman spectral study has also been reported for rhodium(II) acetate and other dinuclear species, giving information on the Rh—Rh bond strength.[169]

Other Donor Ligands. The synthesis of a range of ethylene-1,2-bisdiphenylarsine complexes has been described, including $[RhCl_2L(H_2O)]_2$.[170] This complex is a non-electrolyte, and physical data suggest a dimeric octahedral structure with bidentate L. Its diamagnetism suggests the possible presence of a Rh^{II}—Rh^{II} bond. In the course of a systematic preparative study of Rh^I isocyanide derivatives, the presumably dimeric Rh^{II} species $[Rh(CNBu^t)_3(PPh_3)I]_2(PF_6)_2$ and $[Rh\{CN-(p-ClC_6H_4)\}_2(PPh_2Me)_2I]_2(PF_6)_2$ have been described.[159]

Rhodium(III).—*Group VII Donors.* Various bromo- and mixed bromoaquo-complexes such as $K_3[RhBr_6]$, $K_2[RhBr_5(H_2O)]$, *fac*-$K[RhBr_4(H_2O)_2]$, *mer*-$[RhBr_3(H_2O)_3]$, and the dimeric $K_3[Rh_2Br_9]$ and $K_4[Rh_2Br_{10}]$ have been obtained by crystallizing HBr–aqueous Rh mixtures, and separated by chromatographic techniques.[171] The anion $[RhBr_6]^{3-}$ has also been shown to be the absorbing species in a far-i.r. and laser Raman investigation of $RhCl_3$ (200 p.p.m.) doped in silver bromide crystals.[172] Apparently the Rh^{3+} ion enters the AgBr lattice substitutionally and covalently bonds to its six nearest-neighbour bromide ions.

The interesting cyanotrihydroborato complex $[RhH_2(BH_3CN)(PPh_3)_3]$ has been synthesized (equation 40).[9] I.r. data revealed an increase in the $\nu(C\equiv N)$ stretch-

$$[RhCl(PPh_3)_3] + NaBH_3CN \xrightarrow{EtOH-C_6H_6} [RhH_2(BH_3CN)(PPh_3)_3] \qquad (40)$$

ing frequency on co-ordination, suggesting N-bonding of the BH_3CN ligand,

[168] I. P. Evans and G. Wilkinson, *J.C.S. Dalton*, 1974, 946.
[169] J. S. Filippo and H. J. Sniadoch, *Inorg. Chem.*, 1973, **12**, 2326.
[170] R. K. Poddar and U. Agarwala, *J. Inorg. Nuclear Chem.*, 1974, **36**, 575.
[171] W. Robb and P. van Z. Bekker, *Inorg. Chim. Acta*, 1973, **7**, 626.
[172] T. S. Kuan, *Inorg. Chem.*, 1974, **13**, 1256.

although alternative structures involving Rh—H bonding could not be eliminated.

Group VI Donors. O-Donor ligands. A new synthesis of the known $[RhL_3X(ox)]$ compounds $[L = py, 4\text{-Mepy or } 3,5\text{-lutidine})$ involves replacement of halide (X^-) from *trans*-$[RhL_4X_2]X$ by oxalate.[173] The corresponding syntheses of *cis*-$[Rh\text{-}(en)_2L']^+$ $(L' = ox^{2-}$, malonate, glycinato-O or alaninato-O) from *trans*-$[Rh\text{-}(en)_2Cl_2]^+$, however, required catalysis by $NaBH_4$. These oxalato species were employed as useful precursors for other complexes including novel mixed halogen compounds. Particularly interesting was the observation that irradiation of $[Rh(py)_3(ox)X]$ led to the four-co-ordinate species $[Rh(py)_3X]$ (equation 41), which

$$[Rh(py)_3(ox)X] \xrightarrow{hv} [Rh(py)_3X] + 2CO_2 \qquad (41)$$

is readily oxidized by reagents such as Cl_2, HCl, H_2 or O_2, and appears to be a potential nitrogen analogue of Wilkinson's catalyst. The previously unknown carbonato complex $[Rh(CO_3)(NH_3)_5]ClO_4$ has also been isolated by treating $[Rh(NH_3)_5(H_2O)](ClO_4)_3$ with aqueous Li_2CO_3 at pH 8.5.[174]

The preparation and thermal decomposition of the novel $[RhCl(dmg)_2L]$ and $[RhX(dmg)_2(thiourea)]$ $(dmgH = dimethylglyoxime; L = PMe_3$ or $AsMe_3; X = Cl$ or Br) have been reported.[175] Decomposition begins at 110—140 °C and is complete with the formation of the intermediate $[RhX(dmg)_2]$, which is apparently dimeric with halide bridges. Various related Schiff base complexes of general structure (43: B = 1,2-phenylene, 4-Me-1,2-phenylene, 4,5-diMe-1,2-phenylene, C_2H_4,

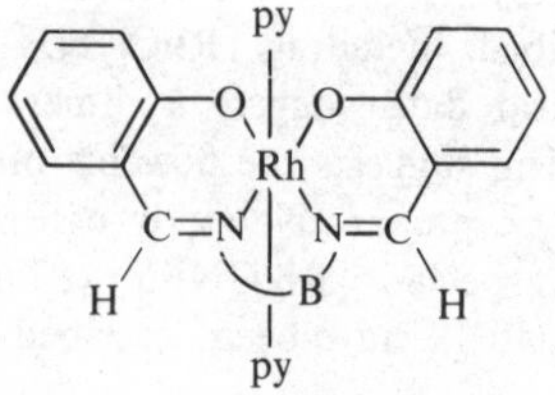

(43)

C_3H_6 or C_4H_8) have been synthesized by boiling *trans*-$[Rh(py)_4Cl_2]Cl$ with the Schiff bases in pyridine containing zinc catalyst.[176]

Ozonolysis of Rh^{II} acetate has been shown to yield the oxo-centred complexes $[Rh_3O(CO_2Me)_6L_3]ClO_4$ $(L = py, \beta\text{-pic, } H_2O$ or PPh_3).[177] The four diastereoisomers of $[Rh\{(+)\text{-atc}\}_3]$ (atc = 3-acetylcamphorato) have been isolated from the reaction of the β-diketone on $RhCl_3.xH_2O$ in the presence of KOH, and separated by t.l.c.[178] Their configurations were established from ^1H n.m.r. and c.d. measurements. In an interesting reaction, addition of MeI to $[RhI_2(CO)_2]^-$ has yielded an acetyl

[173] A. W. Addison, R. D. Gillard, P. S. Sheridan, and L. R. H. Tipping, *J.C.S. Dalton.* 1974, 709.
[174] D. A. Palmer and G. M. Harris, *Inorg. Chem.*, 1974, **13**, 965.
[175] I. S. Shaplygin, *Russ. J. Inorg. Chem.*, 1973, **18**, 400.
[176] C. A. Rogers and B. O. West, *J. Organometallic Chem.*, 1974, **70**, 445.
[177] S. Uemura, A. Spencer, and G. Wilkinson, *J.C.S. Dalton*, 1973, 2565.
[178] G. W. Everett and A. Johnson. *Inorg. Chem.* 1974, **13**, 489.

complex $[Me_3PhN]_2$ $[Rh_2I_6(OCMe)_2(CO)_2]$, whose di-μ-iodo structure was confirmed by X-ray analysis (Table 3).[179]

Stability constants have been measured by pH titration for the formation of a range of noble metal complexes with *o*-coumaric acid, including a 1:1 species with Rh^{III}.[74] Thermal decomposition and X-ray diffraction studies have shown that Rh_2O_3 reacts with PbO at 580—600 °C to yield blue-black crystals of the semiconducting $PbRh_2O_4$.[180]

S-Donor ligands. The S-bonded thio-oxalato complex $[Ph_4As][Rh(S_2C_2O_2)_3]$ has been obtained *via* direct treatment of $RhCl_3$ with $K_2S_2C_2O_2$.[181] This species can function as a ligand towards co-ordinatively unsaturated complexes such as $[Cu(PPh_3)_2]^+$, yielding interesting derivatives of the type (44). Two independent groups have reported the synthesis of $[Rh(LH)_3]Cl_3$ [LH = thiosemicarbazide or $H_2NNHC(S)NH_2$] from reaction between free LH and aqueous $RhCl_3$.[182,183] Deprotonation of this cation with aqueous NH_3 gave $[RhL_3]$.[182] Other species isolated were orange $[Rh(LH)_2Cl(H_2O)]Cl_2$ and dark-brown $[Rh(LH)Cl_3(H_2O)_2]$, the latter of which was shown by i.r. studies to contain unidentate S-bonded thiosemicarbazide. Complexes of the related ligand (45) have been described with a variety of noble metals, including a product formulated as $[Rh_2L_2Cl_2(OH)_2-(H_2O)]$.[184] I.r. data indicated N,S-bonding and the presence of a chloride bridge.

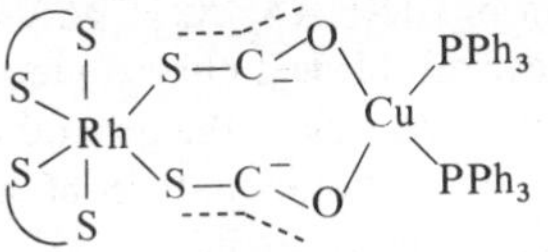

(44) 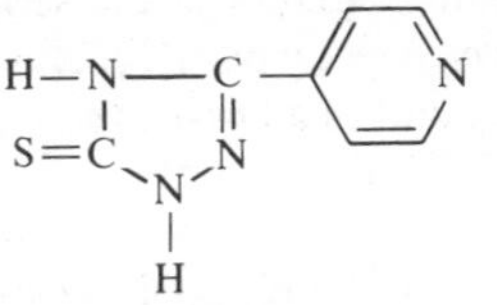   (45)

The OO'-dialkyldithiophosphate complexes $[Rh\{(OR)_2PS_2\}_3]$, and some similar Cr^{III}, Zn^{II}, Ni^{II}, Pd^{II}, and Pt^{II} compounds have been prepared, and separated by gas chromatography.[185] Mass spectra have been reported for Co^{III} and Rh^{III} chelates of the fluorinated monothio-β-diketones $RC(SH)=CHCOCF_3$ ($R = Ph$, *p*-MeC_6H_4, *p*-BrC_6H_4, or *p*-FC_6H_4).[186] Whereas cobalt undergoes valency changes to both Co^{II} and Co^I fragment ions, rhodium changes valency only to the Rh^{II} state.

Cation and anion exchange processes have established that the compound previously formulated as $RhCl_3(SEt_2)$ has the double complex structure $[RhCl_2(SEt_2)_4]$ $[RhCl_4(SEt_2)_2]$, in which both ions have a *trans* geometry.[187]

The electronic spectra of $[Rh(SCN)_6]^{3-}$ and several related thiocyanato-complexes have been recorded in both aqueous 2M-KSCN and molten KSCN.[75] The latter

[179] G. W. Adamson, J. J. Daly, and D. Forster, *J. Organometallic Chem.*, 1974, **71**, C17.
[180] I. S. Shaplygin and V. B. Lazarev, *Russ. J. Inorg. Chem.*, 1973, **18**, 275.
[181] D. Coucouvanis and D. Pittingsrud, *J. Amer. Chem. Soc.*, 1973, **95**, 5556.
[182] A. V. Ablov, D. A. Bologa, and N. M. Samus, *Russ. J. Inorg. Chem.*, 1973, **18**, 80.
[183] I. N. Kiseleva, N. A. Ezerskaya, and T. P. Solovykh, *Russ. J. Inorg. Chem.*, 1973, **18**, 408.
[184] B. Singh, R. Singh, K. P. Thakur, and M. Chandra, *Indian J. Chem.*, 1974, **12**, 631.
[185] T. J. Cardwell and P. S. McDonough, *Inorg. Nuclear Chem. Letters*, 1974, **10**, 283.
[186] M. Das and S. E. Livingstone, *Austral. J. Chem.*, 1974, **27**, 1177.
[187] A. P. Kochetkova, L. B. Sveshnikova, V. I. Sokol, and A. A. Gribenyuk, *Russ. J. Inorg. Chem.*, 1973, **18**, 1290.

solvent shows some interesting reducing properties. An i.r. study (4000–125 cm^{-1}) of $[M\{Se(S)P(OPh)_2\}_3]$ (M = Rh or Ir) has identified both the $\nu(M—S)$ and $\nu(M—Se)$ bands, showing the OO'-diphenylphosphoroselenothionate ligands to be bidentate.[188] Variable-temperature ^1H n.m.r. spectral studies of tris(NN-disubstituted dithiocarbamato)metal(III) complexes (M = Fe, Co, or Rh) have shown that the Rh complex is stereochemically rigid up to 200 °C.[189] The rate of optical inversion *via* a trigonal twist mechanism decreases in the order $Fe^{II} > Fe^{III} > Co^{III} > Rh^{III}$.

Group V Donors. N-Donor ligands. Several papers have appeared by Gillard and co-workers on Rh-amine complexes. The isolation of a series of hydrocarbon solvates of the type mer-$[RhL_3X_2Y]S_n$ (L = py or γ-picoline; X,Y = Br or Cl; S = CH_2Cl_2, $CHCl_3$, or $CDCl_3$) is described, i.r. and Raman data indicating that these solvent molecules are held by H-bonds.[190] Solvent loss occurs on heating, the change being monitored by X-ray powder studies. Also reported is the preparation of the complexes trans-$[M(py)_4X_2]^+$ (M = Rh or Ir; X = Cl or Br) with unusual anions such as Cl_3^-, Br_2Cl^-, Br_3^-, and I_3^-.[191] For example, bromination of trans-$[RhCl_2$-$(py)_4]Cl$ yielded trans-$[RhCl_2(py)_4]Br_2Cl$ in which the presence of Br_2Cl^- was confirmed from i.r. and Raman data. A further paper describes the separation of the cis-α-, cis-β-, and trans-isomers of $[RhCl_2(trien)]^+$ (trien = triethylenetetramine), which were characterized by i.r. and ^1H n.m.r. spectroscopy.[192] Base hydrolysis of the cis-isomers was observed to be much faster than that of the trans-form.

Photochemical studies of the series of amine cations $[Rh(AA)_2X_2]^+$ (AA = en, bipy, or phen; A = py or NH_3; X = Cl, Br, or I) have established photoaquation of the halide ligands in all cases, the yield of halide ion decreasing in the general order Cl < Br < I.[193] Only in the pyridine complexes was photoaquation of amine ligands observed. A re-investigation of the electrochemical reduction of trans-$[Rh(en)_2Cl_2]^+$ at the Hg pool suggests that the product is not a Rh^{II} dimer as earlier proposed, but a Hg^{II}-bridged complex of Rh^I, namely $\{[(en)_2Rh]_2Hg\}^{4+}$.[194]

$$[Rh(TPP)Cl(CO)] + NaOEt \xrightarrow{\text{EtOH}} [Rh(TPP)(CO_2Et)] \tag{42}$$

The complex $[Rh(TPP)(CO_2Et)]$ (reaction 42) is diamagnetic and monomeric in dichloromethane.[195] The presence of an ethoxycarbonyl ligand was supported by the strong carbonyl stretch at 1700 cm^{-1}, and the regeneration of $[Rh(TPP)Cl(CO)]$ and ethanol *via* treatment with HCl. Related metalloporphyrins of the type $[RhL$-$(Me_2NH)_2]$ (L = TPP or etioporphyrin) have been synthesized, and their structures

[188] I. M. Cheremisina, L. A. Il'ina, and S. V. Larionov, *Russ. J. Inorg. Chem.*, 1973, **18**, 675.
[189] M. C. Palazzotto, D. T. Duffy, B. L. Edgar, L. Que, and L. H. Pignolet, *J. Amer. Chem. Soc.*, 1973, **95**, 4537.
[190] A. W. Addison and R. D. Gillard, *J.C.S. Dalton*, 1973, 2002.
[191] A. W. Addison and R. D. Gillard, *J.C.S. Dalton*, 1973, 2009.
[192] P. M. Gidney, R. D. Gillard, B. T. Heaton, P. S. Sheridan, and D. H. Vaughan, *J.C.S. Dalton* 1973, 1462.
[193] M. M. Muir and W. L. Huang, *Inorg. Chem.*, 1973, **12**, 1831.
[194] J. Gulens and F. C. Anson, *Inorg. Chem.*, 1973, **12**, 2569.
[195] I. A. Cohen and B. C. Chow, *Inorg. Chem.*, 1974, **13**, 488.

determined by *X*-ray analysis (Table 3).[196, 197] Since the $[Rh(TPP)(Me_2NH)_2]$ species was isolated from the direct reaction of TPF with $[RhCl(CO)_2]_2$ in DMF, it must result from an amine impurity in the solvent.

A novel and convenient route to a wide range of noble metal-triazenido derivatives has been described, employing the reaction of 1,3-diaryltriazenes (ArN=NNAr) on hydrido or low oxidation state phosphine substrates.[42] Complexes isolated include $[RhH_2(ArNNNAr)(PPh_3)_2]$ (46a), $[RhCl(ArNNNAr)_2(PPh_3)]$ (46b), and [Rh-(ArNNNAr)_3] (46c), the last of which is the first known example of a tris(triazenido)

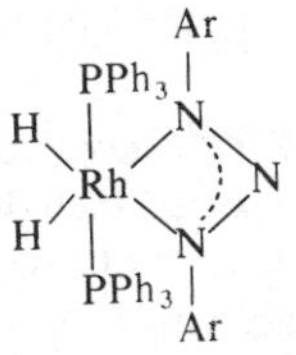

(46a) 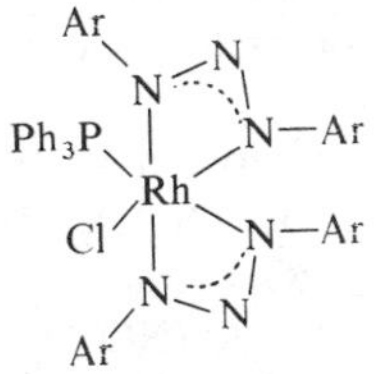   (46b) 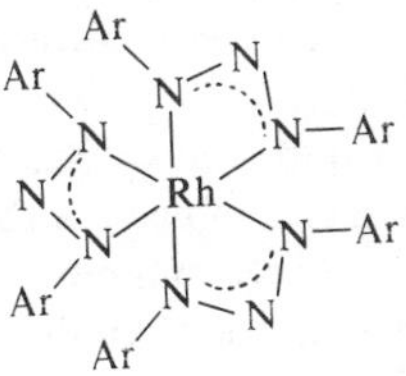   (46c)

noble metal species. Their stereochemistries and bidentate co-ordination were established by i.r. and ^{1}H n.m.r. Treatment of $[RhX(PPh_3)_3]$ (X = Cl or Br) with $ArN_2^+BF_4^-$, followed by addition of LiX, has yielded the related arylazo-complexes $[RhX_2(N_2Ar)(PPh_3)_2]$.[41] The presence of two i.r. bands for the N—N stretch suggested co-ordination of a bent $^-$N—NAr ligand to Rh^{II}, rather than a linear Rh$^I \rightarrow\ ^+$NNAr linkage. Such a bent arrangement for Rh—N_2Ar complexes has been recently confirmed by *X*-ray analysis for $[RhCl(N_2Ph)\{PPh(C_3H_6PPh_2)_2\}]$ (Table 3).[198] In contrast, addition of $ArN_2^+BF_4^-$ to $[RhHCl_2(PPh_3)_3]$ has been shown[41] to give $[RhCl_3(NH=NAr)(PPh_3)_2]$, rather than the solvated species previously suggested. Di-imine complexes could also be obtained by treating the corresponding arylazo-derivatives with HCl (equation 43).

$$[RhCl_2(N_2Ar)(PPh_3)_2] \underset{}{\overset{HCl}{\rightleftharpoons}} [RhCl_3(NH=NAr)(PPh_3)_2] \qquad (43)$$

A preliminary account of the reaction of the ligand 1-benzoyliminopyridinium betaine (L = $C_5H_5NNCOPh$) with transition metals includes the isolation of a compound formulated as $[Rh(L-H)_2(H_2O)_2]Cl$.[199] ^{1}H N.m.r. and i.r. evidence indicates o-metallation of the phenyl ring, causing the betaine to function as a bidentate ligand *via* an o-carbon of the phenyl ring and the imino-N atom.

Following a brief communication,[200] a detailed study has appeared of the photo-chemical reactions of $[Rh(N_3)(NH_3)_5]^{2+}$.[201] Photolysis in 1.0M-HCl at 350 nm was observed to follow equation (44), for which a coordinated nitrene intermediate

[196] E. B. Fleischer, F. L. Dixon, and R. Florian, *Inorg. Nuclear Chem Letters*, 1973, **9**, 1303.

[197] L. K. Hanson, M. Gouterman, and J. C. Hanson, *J. Amer. Chem. Soc.*, 1973, **95**, 4822.

[198] A. P. Gaughan, B. L. Haymore, J. A. Ibers, W. H. Meyers, T. E. Nappier, and D. W. Meek, *J. Amer. Chem. Soc.*, 1973, **95**, 6859.

[199] S. A. Dias, A. W. Downs, and M. R. McWhinnie, *Inorg. Nuclear Chem. Letters*, 1974, **10**, 233.

[200] J. L. Reed, F. Wang, and F. Basolo, *J. Amer. Chem. Soc.*, 1972, **94**, 7172.

[201] J. L. Reed, H. D. Gafney, and F. Basolo, *J. Amer. Chem. Soc.*, 1974, **96**, 1363.

$$[Rh(N_3)(NH_3)_5]^{2+} + 2H^+ + Cl^- \xrightarrow{h\nu} [Rh(NH_2Cl)(NH_3)_5] + N_2 \qquad (44)$$

was proposed. In contrast, irradiation in the charge-transfer region (*e.g.* 250 nm) caused formation of $[Rh(H_2O)(NH_3)_5]^{3+}$ to be more important, as a result of redox interactions. Comparison with previous data for other $[M(N_3)(NH_3)_5]^{2+}$ (M = Co or Ir) complexes reveals the following trend in photochemical behaviour: Co is predominently redox, Rh exhibits both nitrene and redox pathways, while Ir involves only a nitrene path. Other workers have also proposed the presence of the nitrene intermediates $[Rh(N)(NH_3)_5]^{2+}$ and $[Rh(NH)(NH_3)_5]^{3+}$ during the flash photolysis of $[Rh(N_3)(NH_3)_5]^{2+}$ in the presence of acid.[202] In addition, these nitrenes were believed to dimerize giving the coloured species $[Rh(N)(NH_3)_5]_2H^{5+}$ and $[Rh(N)-(NH_3)_5]_2^{4+}$.

Controlled studies of the reaction between $Na_3[RhCl_6]$ and KNO_2 have led to the isolation of each of the mixed $[RhCl_{6-x}(NO_2)_x]^{3-}$ (x = 2—6) species, including the previously unknown $K_3[RhCl_4(NO_2)_2]$ and $K_3[RhCl_2(NO_2)_4]$.[203] These intermediates are readily converted into $[Rh(NO_2)_6]^{3-}$ on heating with excess NO_2^-, indicating a *trans*-effect for the NO_2^- ligand somewhat greater than chloride. Metal–ligand bands have been assigned in the far-i.r. spectra of $[M(NH_3)_5(H_2O)]^{3+}$ (M = Co, Rh, or Ir) cations, permitting the evaluation of the sequence Co < Rh < Ir for their bond dissociation energies.[204] Raman data were also reported for the Rh^{III} complex.

P- *and* **As-***donor ligands.* The preparation of a series of water-soluble tertiary phosphine complexes such as *fac*-$[RhCl_3\{P(CH_2OH)_3\}_3]$ and *mer*-$[RhCl_3L_3]$ [L = $P(CH_2OCOMe)_3$, $PMe(CH_2OCOMe)_2$, $PEt(CH_2OCOMe)_2$, $PPh_2(CH_2OCOMe)$, $PBu^t_2(CH_2OCOMe)$, or $P(CH_2OH)_3]$ has been reported.[123] Another synthetic study shows that dimethylphosphite, $P(OMe)_2OH$ adds oxidatively to $[MCl(octene)_2]_2$ (M = Rh or Ir) complexes in the presence of PMe_2Ph to give $[MHCl(PMe_2Ph)_3\{P(O)(OMe)_2\}]$.[205]

I.r. and Raman spectra have been recorded for a series of octahedral trimethylphosphine complexes including *trans*-$[RhX_2(PMe_3)_4]^+$, and *mer*- and *fac*-$[RhX_3(PMe_3)_3]$ (X = Cl or Br), and vibrational assignments proposed.[206] ^{1}HN.m.r. spectra were also reported and discussed, together with data from ^{1}H–^{31}P INDOR measurements.[207]

Rhodium and iridium halides have been shown to react with n-propyldiphenylarsine (L = AsPr^{n}Ph$_2$) in hot alcoholic media containing the appropriate HX to yield $[MX_3L_3]$ (M = Rh or Ir; X = Cl or Br).[208] These compounds undergo a variety of reactions which are summarized in Scheme 13. Structures were assigned using i.r. and ^{1}H n.m.r. spectroscopy.

[202] G. Ferraudi and J. F. Endicott, *Inorg. Chem.*, 1973, **12**, 2389.
[203] L. A. Nazarova, I. I. Chernyaev, and A. S. Morozova, *Russ. J. Inorg. Chem.*, 1973, **18**, 983.
[204] D. Gattegno and F. Monacelli, *Inorg. Chim. Acta*, 1973, **7**, 370.
[205] M. A. Bennett and T. R. B. Mitchell, *J. Organometallic Chem.*, 1974, **70**, C30.
[206] P. L. Goggin and J. R. Knight, *J. C. S. Dalton*, 1973, 1489.
[207] P. L. Goggin, R. J. Goodfellow, J. R. Knight, and M. G. Norton, *J.C.S. Dalton*, 1973, 2220.
[208] G. K. N. Reddy and N. M. N. Gowda, *Indian J. Chem.*, 1974, **12**, 185.

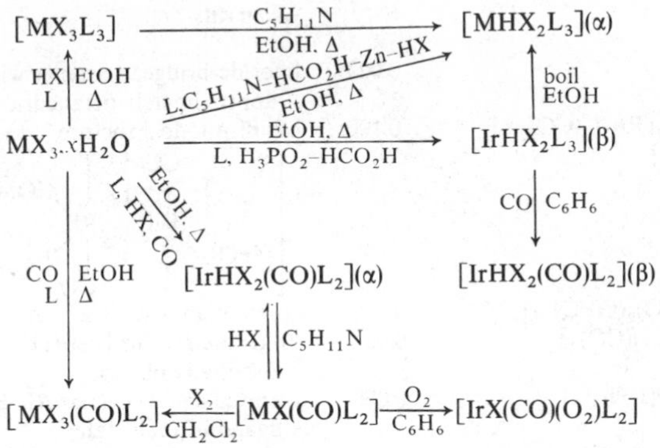

$$[MX_3L_3] \xrightarrow[\text{EtOH. }\Delta]{C_5H_{11}N} [MHX_2L_3](\alpha)$$

Scheme 13

Group IV Donors. Acidopentacyano-complexes of the type $[M(CN)_5X]^{n-}$ (M = Rh or Ir; X = Cl^-, Br^-, I^-, OH^-, or NCMe) have been readily prepared using the sequence (45).[209] An analysis of their u.v.–visible spectra in aqueous solution estab-

$$[M(CN)_6]^{3-} \xrightarrow[\text{H}_2\text{O}]{h\nu(254\,\text{nm})} [M(CN)_5(H_2O)]^{2-} + CN^-$$
$$X \downarrow \Delta \qquad\qquad (45)$$
$$[M(CN)_5X]^{n-} + H_2O$$

lished the following order of decreasing ligand field strength for X: NCMe > OH^- > H_2O > Cl^- > Br^- > I^-.

Single-crystal polarized Raman spectra and i.r. data have been obtained for the hexacyanides $Cs_2Li[M(CN)_6]$ (M = Cr, Fe, Co, or Ir), together with i.r. measurements for the corresponding rhodium complex.[210] Combination of these results allowed the unambiguous assignment of their optically active intra- and intermolecular vibrations. Detailed i.r., u.v.–visible and X-ray powder studies have also been made on related co-ordination polymers of the type $M_3[Rh(CN)_6]_2,xH_2O$ (M = Mn^{II}, Fe^{II}, Co^{II}, Ni^{II}, Cu^{II}, Zn^{II}; x = 7–15) (Table 3).[211] Their i.r. spectra revealed bridging cyanide ligands, while magnetic susceptibility measurements showed the presence of high-spin M^{II} and low-spin Rh^{III}.

Rhodium-(IV) and -(V).—The diffuse reflectance electronic spectrum of $[RhF_6]^{2-}$ has been measured between 4 and 50 kK as part of a systematic study of hexafluorometallate anions.[73] Its d–d bands were assigned using the strong field O_h matrices of Tanabe and Sugano. Oxidation of the tris(dithiocarbamato) M^{III} complexes with BF_3 in dry toluene has given corresponding M^{IV} compounds of type (15; M = Ru or Rh).[88] Both complexes show a shift of $v(C{=\!=}N)$ to higher frequency near 1 500 cm^{-1} on oxidation. Their unexpected diamagnetism in the solid state may be due to pairing of electron spins *via* association, which is supported by molecular weight determinations in chloroform.

[209] G. L. Geoffroy, M. S. Wrighton, G. S. Hammond, and H. B. Gray, *Inorg. Chem.*, 1974, **13**, 430.
[210] B. I. Swanson and L. H. Jones, *Inorg. Chem.*, 1974, **13**, 313.
[211] H. Inoue, Y. Morioka, and S. Yanagisawa, *J. Inorg. Nuclear Chem.*, 1973, **35**, 3455.

Table 3 X-*Ray data for rhodium compounds*

Compound	R	Comments	Ref.
[RhF$_5$]	0.029	fluoride-bridged tetramer with approximately octahedral Rh	a
[Me$_3$PhN]$_2$[Rh$_2$I$_6$(CO)$_2$ (COMe)$_2$]	0.100	has di-μ-iodo structure	b

$$\left[\begin{array}{c} \text{CO} \quad \text{I} \\ \text{I} \diagdown \mid \diagup \text{I} \diagdown \text{COMe} \\ \text{Rh} \quad \text{Rh} \\ \text{MeOC} \diagup \mid \diagup \text{I} \diagup \mid \diagdown \text{I} \\ \text{I} \quad \text{CO} \end{array} \right]^{2-}$$

Compound	R	Comments	Ref.
cis-[RhCl(CO)$_2$(OPCy$_3$)]	0.065	square-planar Rh. Carbonyls are cis	c
[oepH][RhCl$_2$(CO)$_2$]$_2$	0.089	cis-geometry. The [oepH]$^{2+}$ cation is essentially planar	d
[Rh(PhCO$_2$)(PPh$_3$)$_3$]	0.068	distorted square-planar Rh. Benzoate ligand is unidentate	e
[Rh{(+)-chxn}$_3$]Cl$_3$,H$_2$O	—	powder only. Isomorphous with Cr, Co, and Ir analogues	f
[RhCl(N$_2$Ph){PPh(C$_3$H$_6$PPh$_2$)$_2$}] PF$_6$	0.072	square-pyramidal Rh with PhN$_2$-ligand at apex. Doubly bent phenyldiazo-group	g

$$\text{Ph} \diagdown \text{N} = \text{N} \diagup ^{\text{Rh}}$$

Compound	R	Comments	Ref.
[Rh(Me$_2$NH)$_2$(TPP)]Cl	0.072	octahedral Rh	h
[(oep){Rh(CO)$_2$}$_2$]	0.062	two Rh ions co-ordinated to the octaethylporphinato group, above and below plane. Each Rh is surrounded by square plane of 2 CO and 2 N	i
[Rh(Me$_2$NH)$_2$(EP)]Cl,2H$_2$O	0.068	octahedral Rh, positioned in plane of of etioporphinato (EP) ring	j
[RhCl$_3$(PEt$_2$Ph)$_3$]	0.067	distorted octahedron	k
[Rh$_8$(CO)$_{19}$C]		preliminary. Metal cluster derived from a prismatic unit with 2 extra Rh atoms attached. Carbido carbon is in centre of	l

prism. 11 carbonyls linear, 6 edge-bridging, 2 face-bridging.

Compound	R	Comments	Ref.
[H$_3$O][Rh$_{15}$(CO)$_{28}$C$_2$]		preliminary. Rh$_{15}$ cluster is a centred tetracapped pentagonal prism. Carbide C's in centre of two octahedra. 14 terminal CO's, and remainder edge-bridging.	l
[Rh{P(OMe)$_3$}$_2$(BPh$_4$)]	0.051	a redetermination. Square-planar Rh, bonding to 2 P atoms and 2 double bonds of one arene ring.	m
M$_3^{II}$[Rh(CN)$_6$]$_2$,xH$_2$O		powders only. M = Mn, Fe, Co, Ni, Cu, Zn^{II}	n

(a) B. K. Morrell, A. Zalkin, A. Tressand, and N. Bartlett, *Inorg. Chem.*, 1973, **12**, 2640. (b) Ref. 179. (c) Ref. 125. (d) Ref. 132. (e) Ref. 133. (f) P. Andersen, F. Galsbol, S. E. Harnung, and T. Laier, *Acta Chem. Scand.*, 1973, **27**, 3973, chxn = trans-1,2-cyclohexanediamine. (g) Ref. 198. (h) Ref. 196, TPP = αβγδ-tetraphenyl-porphinato. (i) A. Takenaka and Y. Sasada, *J.C.S. Chem. Comm.*, 1973, 792. (j) Ref. 197. (k) A. C. Skapski and F. A. Stephens, *J.C.S. Dalton*, 1973, 1789. (l) Ref. 109. (m) M. L. Nolte and G. Gafner, *Acta Cryst.*, 1974, **B30**, 738. (n) Ref. 211.

A series of dioxygenyl salts of complex fluorometallates, including the new O_2^+-$[RhF_6]^-$, has been simply prepared by static heating of the metal at 300—350 °C with 5 atm of a $3:1$ mixture of fluorine and oxygen.[90] The deep red Rh species was characterized by correlating X-ray diffraction, Raman, and mass spectral data with analogous known systems.

4 Iridium

Cluster Compounds.—Air-stable clusters of general formula $Ir_4(CO)_8L_4$ [L = $PMePh_2$, PMe_2Ph, $PEtPh_2$, $AsMePh_2$, $P(OPh)_3$, or $\frac{1}{2}Ph_2PC_2H_4PPh_2$] have been prepared *via* direct reaction between $Ir_4(CO)_{12}$ and L.[212] ^1H and ^{13}C n.m.r. studies showed the compounds to be stereochemically non-rigid above room temperature, implicating an intramolecular carbonyl scrambling process. Addition of acid produced mainly diprotonated species which were also stereochemically non-rigid, probably involving both CO and proton intramolecular scrambling.

Iridium(I).—*Hydrido- and Halogeno-carbonyl or -phosphine Complexes.* Treatment of $[IrH(CO)(PPh_3)_3]$ with SiF_4 in benzene at room temperature has afforded a yellow product, which X-ray analysis showed to be $[IrH_2(CO)(PPh_3)_3]^+$ $[SiF_5]$,$3C_6H_6$ (Table 4).[213] Alternatively, the same compound could be rapidly and quantitatively obtained by adding concentrated aqueous HF to silica in a benzene–methanol solution of $[IrH(CO)(PPh_3)_3]$. Since boron fluoride was found to form an analogous product, this study suggests caution when interpreting the results of reactions of covalent fluorides with basic metal complexes when carried out in glass apparatus.

Interesting large-ring chelate complexes of the ligands $Bu_2^tP(CH_2)_nPBu_2^t$ ($n = 9$ or 10) have been isolated, including the monomers (47).[214] Contrary to previous opinion, these results suggest that complexes with large-ring bidentate ligands may be more stable than those with ligands of intermediate ring size.

A facile synthesis of the anionic complexes $AsPh_4[IrX_2(CO)_2]$ (X = Cl, Br, or I)

$$[IrCl(CO)_3]_n + AsPh_4X \xrightarrow{EtOH} AsPh_4[IrX_2(CO)_2] \tag{46}$$

has been reported (equation 46).[215] These anions undergo a wide range of ligand substitution processes some of which are collected in Scheme 14.

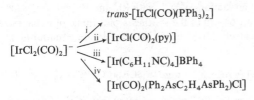

Reagents: i, PPh_3; ii, py; iii, $C_6H_{11}NC,CH_2Cl_2,NaBPh_4$; iv, $Ph_2AsC_2H_4AsPh_2$

Scheme 14

[212] P. E. Cattermole, K. G. Orrell, and A. G. Osborne, *J.C.S. Dalton,* 1974, 328.
[213] P. Bird, J. F. Harrod, and K. A. Than, *J. Amer. Chem. Soc.,* 1974, **96**, 1222.
[214] A. J. Pryde, B. L. Shaw, and B. Weeks, *J.C.S. Chem. Comm.,* 1973, 947.
[215] P. Piraino, F. Faraone, and R. Pietropaolo, *Inorg. Nuclear Chem. Letters,* 1973, **9**, 1237.

Also described are the new compounds *trans*-[IrCl(CO)L$_2$] [L = PMe$_2$(*o*-MeOC$_6$H$_4$) or PMe$_2$(*p*-MeOC$_6$H$_4$)].[216] Of particular interest were oxidative addition studies with reagents such as MeI, MeCOCl, PhCOCl, or CH$_2$=CHCH$_2$Cl, yielding octahedral derivatives of the type (48). The faster rate of these reactions compared

(47)   (48)

with those for corresponding PMe$_2$Ph substrates was interpreted in terms of a direct electronic interaction between the methoxy oxygen of L and the Ir. A related paper describes mechanistic studies on the oxidative addition of optically active α-bromo-esters, including (*RS,SR*)-PhCHFCHBrCO$_2$Et, to [IrCl(CO)(PMe$_3$)$_2$].[217] The reactions were shown from ^{1}H, ^{19}F, and ^{13}C data to involve loss of stereochemistry at carbon and to be retarded by electron scavengers, thus indicating a free radical mechanism. Also, in contrast to previous findings,[218] addition of MeCHBrCO$_2$Et on [IrCl(CO)L$_2$] (L = PMe$_3$ or PMePh$_2$) was shown to result in loss of optical activity, and to be retarded by electron scavengers.

Weak u.v. irradiation of [IrCl(CO)(PPh$_3$)$_2$] has been observed to enhance its activity as a hydrogenation catalyst by a factor of 40, *via* the formation of reactive intermediates.[219]

Group VI Donors. Stable bicarbonato complexes of the type [M(OCO$_2$H)(CO)-(PPh$_3$)$_2$] (M = Rh or Ir) have been isolated using the CO$_2$-fixation reaction (32).[134] These diamagnetic species were tentatively assigned a unidentate bicarbonate co-ordination, although neither exhibited an O—H stretch in the i.r.

The first reported complexes containing the chlorosulphido ligand, namely [IrX(CO)(PPh$_3$)$_2$(SCl$_2$)] (X = Cl, Br, or NCS) and [Ir(Ph$_2$PC$_2$H$_4$PPh$_2$)$_2$(SCl$_2$)] PF$_6$, have been prepared *via* room-temperature addition of SCl$_2$ on [IrX(CO)(PPh$_3$)$_2$] and [Ir(Ph$_2$PC$_2$H$_4$PPh$_2$)$_2$]$^+$, respectively.[131] I.r. data support structure (49) containing a bent Ir—S—Cl linkage for these adducts. In contrast to this oxidative addition behaviour towards Ir, similar reactions of SCl$_2$ with Rh^I and Pt^0 substrates led to chlorination only. Another report describes the syntheses of a range of noble metal SCF$_3$ complexes including that of *trans*-[Ir(SCF$_3$)(CO)(PPh$_3$)$_2$] using equation (47).[220]

$$trans\text{-}[IrCl(CO)(PPh_3)_2] + AgSCF_3 \rightarrow trans\text{-}[Ir(SCF_3)(CO)(PPh_3)_2] \qquad (47)$$

The high frequency of its ν(CO) band at 2007 cm^{-1} indicated that SCF$_3$ is a very effective M → L π-donor, while ^{19}F n.m.r. data on related Pt complexes reveal the *trans*-influence series H > PEt$_3$ > PPh$_3$ > SCF$_3$ > Cl.

[216] E. M. Miller and B. L. Shaw, *J.C.S. Dalton*, 1974, 480.
[217] J. A. Labinger, A. V. Kramer, and J. A. Osborn, *J. Amer. Chem. Soc.*, 1973, **95**, 7908.
[218] R. G. Pearson and W. R. Muir, *J. Amer. Chem. Soc.*, 1970, **92**, 5519.
[219] W. Strohmeier and G. Csontos, *J. Organometallic Chem.*, 1974, **72**, 277.
[220] K. R. Dixon, K. C. Moss, and M. A. R. Smith, *J.C.S. Dalton*, 1973, 1529.

An n.m.r. investigation of the cations $[Ir(Ph_2PC_2H_4PPh_2)_2X_2]^+$ ($X_2 = O_2$, S_2, or Se_2) and $[Ir(Ph_2PC_2H_4PPh_2)_2C_3S_2]^+$ has assigned triplets in the region τ 3.5—4.3 to certain *o*-phenyl hydrogens.[221] Since these are shifted upfield from the main phenyl multiplet, interaction with the non-phosphorus ligands was suggested, indicating *cis*-octahedral geometry.

Group V Donors. N-Donor ligands. Reaction of toluene-*p*-sulphonyl azide (RN_3) with $[M(NO)(CO)(PPh_3)_2]$ (M = Rh or Ir) in benzene has yielded urea derivatives of the type (38; M = Rh or Ir).[147] Alternatively, treatment of $[M(NO)(PPh_3)_3]$ with RNCO may be employed. The usual orange complex (50) has been isolated from the reaction between $[Ir(\pi\text{-}2MeC_3H_4)(CO)(PPh_3)_2]$ and trifluoroacetonitrile, its structure being established by X-ray analysis (Table 4).[222] On the other hand, the corresponding

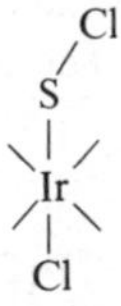

(49) 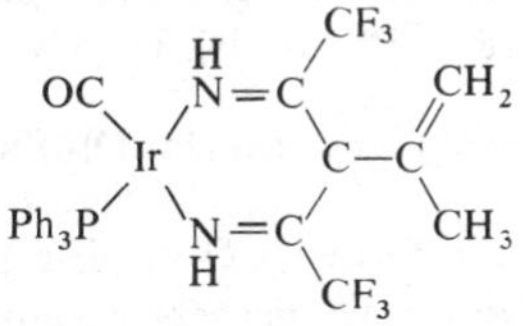   (50) 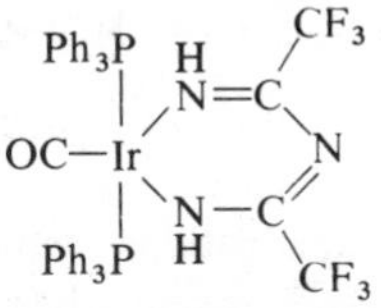   (51)

reaction with $[Ir(\pi\text{-}C_3H_5)(CO)(PPh_3)_2]$ produced an analogue of (50), together with a low yield of a yellow species tentatively formulated as (51).

Planar chelated complexes of the type $[Ir(CO_2(L\text{-}L)]$ [L–LH = (52a), (52b), and derivatives] have been described.[223] Also reported are the syntheses of a variety of cyanamide metal complexes, including $[Ir(NCNH)(CO)(PPh_3)_2]$ and $[(PPh_3)_2\text{-}(CO)IrNCNIr(CO)(PPh_3)_2]$.[224] I.r. and ^{14}N n.m.r. spectra were discussed.

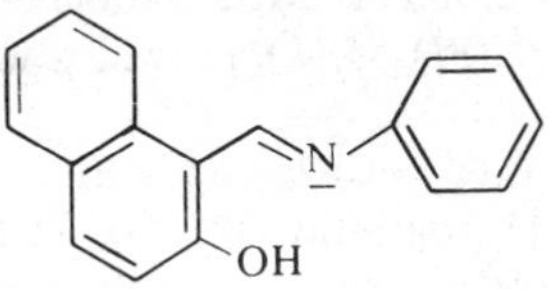

(52a) 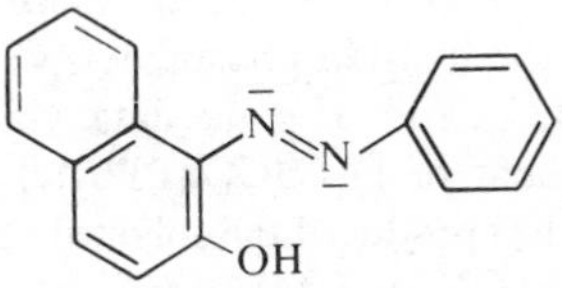   (52b)

An interesting communication suggests that there is a correlation between the behaviour of co-ordinated NO^+ (including Ir^I nitrosyl complexes) as an electrophile and the position of the NO stretching frequency.[35] Only complexes with $v(NO) >$ 1886 cm^{-1} are attacked at the nitrosyl group by nucleophiles such as OH^-, MeO^-, or N_3^-. Inspection of the N 1s portion of the X-ray photoelectron spectrum of *trans*-$[IrCl(N_2)(PPh_3)_2]$ reveals non-equivalent N atoms, confirming linear Ir—N—N co-ordination.[225] In addition, at least one of the nitrogens bears a negative charge,

[221] A. P. Ginsberg and W. E. Lindsell, *Inorg. Chem.*, 1973, **12**, 1983.
[222] M. Green, S. H. Taylor, J. J. Daly, and F. Sanz, *J.C.S. Chem. Comm.*, 1974, 361.
[223] R. Aderjan and H. J. Keller, *Z. Naturforsch.*, 1973, **28b**, 500.
[224] W. Beck, H. Bock, and R. Schlodder, *Z. Naturforsch.*, 1974, **29**, 75.
[225] V. I. Nefedov, V. S. Lenenko, V. B. Shur, M. E. Vol'pin, J. E. Salyn, and M. A. Porai-Koshits, *Inorg. Chim. Acta*, 1973, **7**, 499.

in contrast to organic diazonium species. Similar studies of the related complexes *trans*-[IrCl(L)(PPh$_3$)$_2$][(L = CO, CS, PPh$_3$ PhC≡CPh, C$_2$(CN)$_4$)] indicate that molecular nitrogen and CO are better acceptors than PPh$_3$, CS, or PhC≡CPh, but worse than tetracyanoethylene.

P- and As- ligands. In the first systematic study of the co-ordinating ability of the ligands P(OR)$_2$Ph and P(OR)Ph$_2$ (L; R = Me or Et) towards noble metals, a large variety of complexes have been obtained by treatment of labile phosphine or olefinic precursors with free ligand.[62,63] Among the derivatives described were the cations [Ir{P-(OMe)$_2$Ph}$_5$]$^-$ and [IrL$_4$]$^+$. Similar preparative methods also led to a range of trialkylphosphite complexes including [Ir{P(OR)$_3$}$_5$]$^+$ (R = Me or Et).[62,65] Qualitative observations led to the following relative stability order: P(OR)Ph$_2$ > P(OR)$_2$Ph > P(OR)$_3$.

Reaction (48) has provided a general route to tertiary phosphine complexes of the type *trans*-[Ir(CO)$_2$(PR$_3$)$_2$]BF$_4$ (R = Ph, *o*-tolyl, Pr^i, Cy, or Ph$_2$Bu^t).[226] These red,

$$[IrCl(CO)_3]_n + PR_3 \xrightarrow[\text{EtOH}]{\text{AgBF}_4} trans\text{-}[Ir(CO)_2(PR_3)_2]BF_4 \qquad (48)$$

air-stable cations readily pick up a further CO to yield five-co-ordinate [Ir(CO)$_3$-(PR$_3$)$_2$]$^+$, except when R = Ph or *o*-tolyl (perhaps because these are the least basic phosphines). Molecular hydrogen adds to give the octahedral dihydrides [IrH$_2$(CO)$_2$-(PR$_3$)$_2$]$^+$, although no reaction is observed with O$_2$ or SO$_2$. A similar study described the isolation of the chelated phosphine complexes [Ir(CO)(L–L)$_2$]Cl and [Ir(L–L)$_2$] Cl (L–L = Ph$_2$PC$_2$H$_4$PPh$_2$, Ph$_2$PCH=CHPPh$_2$ or Ph$_2$PCH=CHAsPh$_2$) from the reaction of L–L on [IrCl(CO)(PPh$_3$)$_2$].[227] The four-co-ordinate cations [Ir(L–L)$_2$]$^+$ absorb molecular oxygen irreversibly in benzene solution to produce the stable oxygen adducts [Ir(L–L)$_2$(O$_2$)]Cl. Other authors have also reported an interesting temperature-dependent reaction of PH$_3$ on [Ir(Ph$_2$PC$_2$H$_4$PPh$_2$)$_2$]$^+$.[228] Whereas attack at − 70 °C gave [Ir(PH$_3$)(L–L)$_2$]$^+$, reversible oxidative addition occurred at room temperature yielding the hydrido-phosphide species *cis*-[IrH(PH$_2$)(L–L)$_2$]$^+$, which was characterized by i.r. and ^{1}H n.m.r. data.

Treatment of [IrCl(CO)(PPh$_3$)$_2$] with AgY (Y = ClO$_4$$^-$ or BF$_4$$^-$) in various solvents has produced the solvated complexes [Ir(solvent)(CO)(PPh$_3$)$_2$]Y (solvent = MeCN, PhCN, or DMSO).[229] The solvent molecules could be easily substituted by CO, PR$_3$, isocyanides, and a range of anions to give four-co-ordinate *trans*-[IrX(CO)-(PPh$_3$)$_2$]$^{n+}$ n = 0 or 1).[229,230] In addition, all the neutral products were observed to form oxygen adducts of the type [IrX(O$_2$)(CO)(PPh$_3$)$_2$], sometimes reversibly. The toluene-*p*-sulphonate derivative [Ir(SO$_2$R)(CO)(PPh$_3$)$_2$] was particularly interesting, being shown from i.r. data to be O-bonded (*i.e.* Ir—OSOR) but changing to S-bonded on further co-ordination of molecular oxygen.

^{193}Ir Mössbauer and ESCA spectra have been recorded for the related species [IrX(CO)(PPh$_3$)$_2$] (X = Cl, Br, I, N$_3$, SCN, or NO).[231] Little change in Ir 4*f* core binding energies was observed with variation in X, indicating little modification of

[226] F. Faraone, P. Piraino, and R. Pietropaolo, *J.C.S. Dalton*, 1973, 1625.
[227] K. K. Chow, C. A. McAuliffe, and S. G. Murray, *Inorg. Chem.*, 1973, **12**, 1701.
[228] R. A. Schunn, *Inorg. Chem.*, 1973, **12**, 1573.
[229] C. A. Reed and W. R. Roper, *J.C.S. Dalton*, 1973, 1365.
[230] C. A. Reed and W. R. Roper, *J.C.S. Dalton*, 1973, 1370.
[231] F. Holsboer, W. Beck, and H. D. Bartunik, *J.C.S. Dalton*, 1973, 1828.

electronic charge on the Ir centre. Bonding of the nitrosyl ligand to Ir was also discussed in the light of N 1s-binding energy data. Investigation of the absorption and emission spectra of $[Ir(L-L)_2]Cl$ (L–L = $Ph_2PC_2H_4PPh_2$ or $Ph_2PC_2H_2PPh_2$) in both the solid state and frozen solution at 77K suggests that emission occurs from a square planar d^7 a_{2u} charge-transfer state of $E_u(^3A_{2u})$ symmetry.[155]

A range of complexes with the novel quadridentate and tridentate ligands (53; L = P or As, or mixed P, As) and (54; L = P or As) have been prepared and characterized by ^1H n.m.r. and u.v.–visible spectroscopy,[232, 233] Trigonal-bipyramidal or octahedral compounds were generaly obtained, for example, the species $[IrX\{(o-$

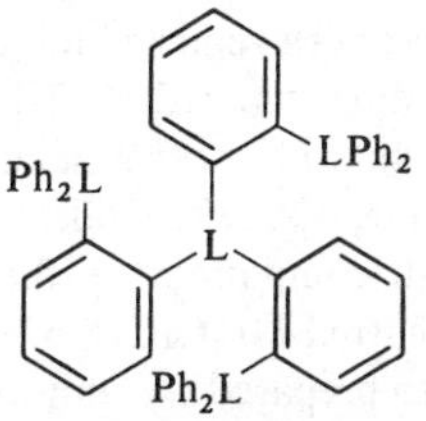

(53) 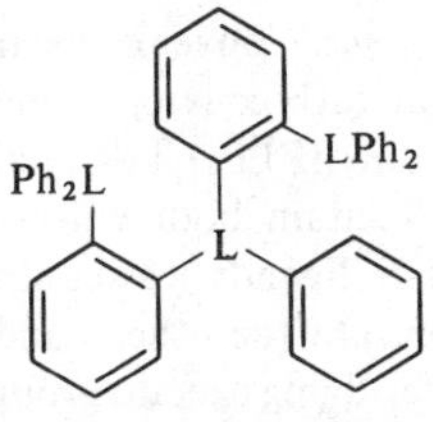   (54) 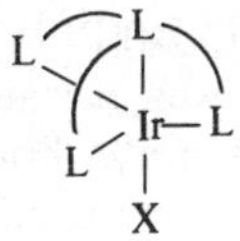   (55)

$Ph_2PC_6H_4)_3P\}]^{n+}$ and $[IrX\{(o-Ph_2AsC_6H_4)_3As\}]^{n+}$ (X = CO or Cl; n = 1 or 0) were shown to have the configuration (55).

The previously unknown $[IrF(AsPh_3)_3]$ and trans-$[IrF(CO)(AsPh_3)_2]$ have been prepared,[234] as has the related $[IrCl(CO)(Ph_2AsCH=CHAsPh_2)]$.[170] Magnetic and conductivity studies support a square-planar formulation for the last-mentioned complex.

Group IV Donors. A large variety of isocyanide complexes of general formula $[Ir(CNR)_4]PF_6$ have been prepared via equation (39) (M = Ir; R = Me, Pr^i, Cy, Bu^t, or p-MeOPh), and their oxidative additions yielding $[Ir(CNR)_4XY]^+$ species investigated.[158] The same authors reported synthetic methods for a wide range of mixed isocyanide–phosphine derivatives, including $[Ir(CNR^1)_3(PR^2_3)_2]Cl$ and $[Ir(CO)(CNPr^i)_2(PPh_3)_2]Cl$.[159] Protonation of the former species with HPF_6 yielded the hydride $[IrH(CNR)_3(PPh_3)_2](PF_6)_2$. Several oxidative additions were also described, these being more facile than with the corresponding Rh^I substrates. An alternative but similar route to $[Ir(CNR)_4]I$ (R = p-tolyl or p-MeOPh) has also appeared, involving treatment of $[IrCl(cyclo-octa-1,5-diene)]_2$ with RNC in dichloromethane.[235]

Reaction of $NaCH(CN)_2$ on $[MCl(CO)(PPh_3)_2]$ (M = Rh or Ir) has afforded the dicyanomethanide complexes $[M\{CH(CN)_2\}(CO)(PPh_3)_2]$ (see p. 362).[161] Synthesis of the related complexes $[M\{C(CN_3)\}(CO)(PPh_3)_2]$ (M = Rh or Ir) has been described and the mode of co-ordination of the methanide ligand discussed in the light of i.r., Raman, and ^{14}N n.m.r. data.[162]

232. D. G. E. Kerfoot, R. J. Mawby, A. Sgamellotti, and L. M. Venanzi, *Inorg. Chim. Acta.*, 1974, **8**, 195.

233 R. J. Mynott, E. F. Trogu, and L. M. Venanzi, *Inorg. Chim. Acta*, 1974, **8**, 201

234 C. A. McAuliffe and R. Pollock, *J. Organometallic Chem.*, 1974, **59**, C13.

235 K. Kawakami, M. Haga, and T. Tanaka, *J. Organometallic Chem.*, 1973, **60**, 363.

Group III Donors. B-Donor ligands. Following two preliminary reports[163a−b] further studies have been made on the preparation and characterization of complexes of hexaborane, B_6H_{10}, in which the metal is regarded as having inserted into the unique basal B—B bond to form a three-centre, two-electron bond.[164] Compounds described include $[Ir(B_6H_{10})_2Cl]_2$, obtained by treating $[IrCl(cyclo\text{-}octa\text{-}1,5\text{-}diene)]_2$ with free B_6H_{10}.

Iridium(III).—*Group VII Donors.* Pulse radiolysis of $[IrCl_6]^{3-}$ in aqueous solution has been shown to give $[IrCl_6]^{2-}$ *via* oxidation with the hydroxyl radicals produced.[236] On the other hand, when propan-2-ol was added to the solution, pulse radiolysis led to reduction by aquated electrons, hydrogen radicals, and R· to yield $[IrCl_6]^{4-}$.

Group VI Donors. O-Donor ligands. Convenient syntheses have been reported for an extensive series of noble-metal carboxylato-complexes including octahedral $[IrH_2(OCOR)(PPh_3)_3]$ and $[Ir(OCOR)_3(PPh_3)_2]$ (R = alkyl or aryl).[23] I.r. measurements showed the latter species to contain both uni- and bi-dentate OCOR, while the absence of bridging carboxylate ligands was confirmed in all compounds by their monomeric nature in solution. On the other hand the O-centred $[Ir_3O(CO_2Me)_6(H_2O)_3](CO_2Me)$, containing bridging acetato groups, has been prepared *via* addition of $AgCO_2Me$ to $IrCl_3,xH_2O$.[177] This green species could be reversibly oxidized with ozone to yield the blue $Ir^{III, III, IV}$ form $[Ir_3O(CO_2Me)_6(H_2O)_2(MeCO_2H)](CO_2Me)_2$.

The previously unknown carbonato-complex $[Ir(CO_3)(NH_3)_5]ClO_4$ has been isolated by treating $[Ir(NH_3)_5(H_2O)](ClO_4)_3$ with aqueous Li_2CO_3 at pH 8.5, and the kinetics of its formation and aquation were studied.[174] Stability constants have been measured using pH titration for a variety of *o*-coumaric acid complexes, including a 1 : 3 species with Ir^{III}.[74]

S- and Se-Donor ligands. Synthesis of the dimers $[IrX(SAr)(SAr')(CO)(PPh_3)]_2$ (X = Cl, Br, or I; Ar = *p*-substituted benzene; Ar' = 2,4-$(NO_2)_2C_6H_4$ has been described (equation 49), spectroscopic studies suggesting the bridged structure (56).[237]

$$2[IrX(CO)(PPh_3)_2] + 2ArSSAr' \rightarrow [IrX(SAr)(SAr')(CO)(PPh_3)]_2 \quad (49)$$

Thiosalicylamide (L) has been shown to be a useful reagent for the gravimetric or spectrophotomeric determination of Ir^{III}, reacting with such salts in acetate buffer (pH 4.6—5.8) to give a quantitative precipitate of the orange-yellow $[IrL_3]$.[238] The

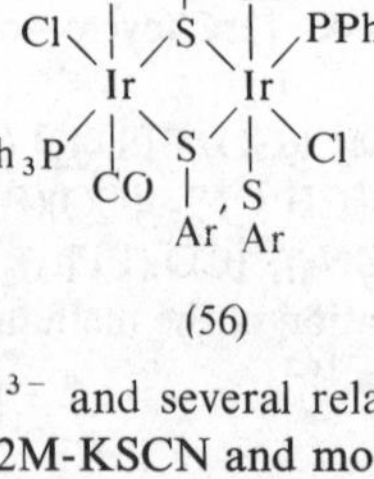

(56)

electronic spectra of $[Ir(SCN)_6]^{3-}$ and several related thiocyanato-complexes have been recorded in both aqueous 2M-KSCN and molten KSCN.[75]

[236] R. K. Broszkiewicz, *J.C.S. Dalton*, 1973, 1799.
[237] C. T. Lam and C. V. Senoff, *J. Organometallic Chem.*, 1973, **57**, 207.
[238] M. P. Ray, K. Sur, and S. C. Shome, *J. Indian Chem. Soc.*, 1974, **51**, 517.

Reaction of excess CSe_2 and areneselenols (equimolar) with *trans*-$[IrCl(CO)(PPh_3)_2]$ has yielded the oxidative addition products $[IrCl(CO)(PPh_3)_2(CSe_2)]$ and $[IrHCl(CO)(PPh_3)_2$ $(SeC_6H_4Me-p)_2]$, respectively.[239] I.r. and ^1H n.m.r. data indicate the structures (57a) and (57b), where the CSe_2 species has a configuration similar to the known CS_2 adduct.

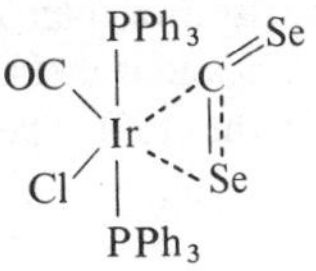

(57a) 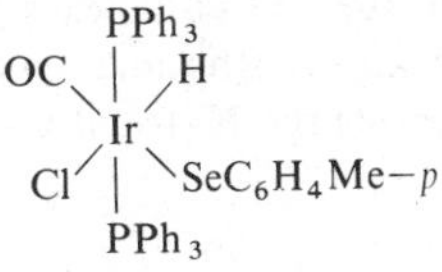   (57b)

Group V Donors. N- Donor ligands. Two interesting papers with obvious environmental implications have indicated that the dinitrosyl cation $[Ir(NO)_2(PPh_3)_2]^+$ reacts with CO to give $[Ir(CO)_3(PPh_3)_2]^+$ plus CO_2 and N_2O.[240, 150] A mechanism was suggested involving formation of $[Ir(NO)_2(CO)(PPh_3)_2]^+$, followed by oxygen transfer to CO leaving a metal–nitrene intermediate.[240] The reaction is reversed by addition of NO at 80 °C in acetone, providing a potentially useful continuous oxidation–reduction process. In addition, reaction (37) which converts NO and CO into less noxious products, was found to be catalysed by a variety of dinitrosyl complexes including $[M(NO)_2Br(PPh_3)_2]$ and $[M(NO)_2(PPh_3)_2]PF_6$ (M = Rh or Ir).[150] These latter results were suggested to indicate that such four- and five-co-ordinate dinitrosyl complexes may be better formulated as 18-electron N—N bonded, *cis*-dinitrogen-dioxide species, *e.g.* $[M(NO)_2Br(PPh_3)_2]$ becomes $[M(N_2O_2)Br(PPh_3)_2]$.

The cations *trans*-$[IrX_2(py)_4]^+$ (X = Cl or Br) have been isolated with the unusual anions Cl_3^-, Br_2Cl^-, and ClO_4^-, $MeNO_2$.[191] For example, chlorination of *trans*-$[Ir(py)_4Cl_2]Cl$ yielded $[Ir(py)_4Cl_2]Cl_3$ in which the presence of Cl_3^- was confirmed from i.r. and Raman data.

Quantum yields have been measured for the photoaquation of *trans*-$[IrCl_2(py)_4]^+$, *trans*-$[IrCl_2(en)_2]^+$, *cis*-$[IrCl_2(phen)_2]^+$, and *cis*$[IrX_2(bipy)_2]^+$ (X = Cl or Br) at both 350 and 254 nm.[241] Only *trans*-$[IrCl_2(py)_4]^+$ released a significant amount of amine, halide aquation predominating in all cases. Yields decreased in the order $Br^- > Cl^-$, as was previously observed with analogous Rh^{III} compounds. The action of chlorine and bromine on $K[IrX_4(en)]$ (X = Cl or Br) has been shown to give products of the same analytical composition but different properties.[242] While their diamagnetism precludes oxidation of the Ir^{III} centres, their i.r. spectra suggest that the co-ordinated ethylenediamine has been converted to the di-imine chelate HN=CH—CH=NH in a manner similar to that previously reported[243] for the oxidation of $[Ru(en)_3]^{n+}$ (n = 2 or 3) to $[Ru(en)_2(HN=C_2H_2=NH)]^{2+}$.

The novel pale-yellow $[Ir(bipy)_3](NO_3)_3$ has been synthesized and its structure confirmed by ^1H and ^{13}C n.m.r.[244] Its luminescence was also recorded at 77 K, the

[239] K. Kawakami, Y. Ozaki, and T. Tanaka, *J. Organometallic Chem.*, 1974, **69**, 151.
[240] B. F. G. Johnson and S. Bhaduri, *J.C.S. Chem. Comm.*, 1973, 650.
[241] M. M. Muir and W. L. Huang, *Inorg. Chem.*, 1973, **12**, 1930.
[242] I. B. Baranovskii, R. E. Sevast'yanova, and G. Y. Mazo, *Russ. J. Inorg. Chem.*, 1973, **18**, 759.
[243] B. C. Lane, J. E. Lester, and F. Basolo, *J.C.S. Chem. Comm.*, 1971, 1618.
[244] C. M. Flynn and J. N. Demas, *J. Amer. Chem. Soc.*, 1974, **96**, 1959.

mean lifetime of *ca.* 80 μs establishing the emission as a phosphorescence similar to that known for the corresponding Rh^{III} complex. Other authors have reported the luminescence of the related species (phenH)[IrCl$_4$(phen)].[245] Whereas frozen solutions revealed a green emission characteristic of both phenH$^+$ and co-ordinated phenanthroline, fluid solutions gave a blue fluorescence due only to phenH$^+$.

Treatment of Na$_3$[IrX$_6$] (X = Cl or Br) with ligand (HL = 58) in 2-methoxyethanol has afforded the new complexes [IrXL]$_2$.[246] Comparison of their i.r. and ^{1}H n.m.r. spectra with known Rh^{III} analogues indicated that the ligand is metallated by Ir^{III}, and co-ordinated *via* the N–1 and C–10 atoms as in structure (59). The chloro-bridge is

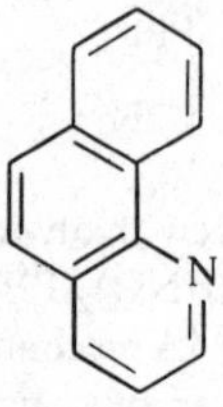

(58) 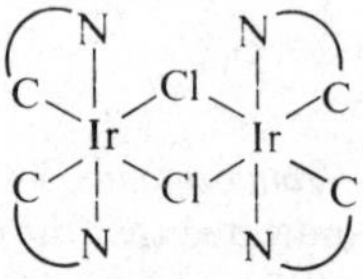   (59)

ruptured by reagents (L′) such as PBu$^n_3$ and SEt$_3$ yielding [IrL$_2$ClL′]. A novel and convenient route to a wide range of noble-metal triazenido-derivatives has been reported, involving reaction of 1,3-diaryltriazenes (ArN=NNAr) with hydrido and low oxidation state phosphine complexes.[42] The iridium compounds described, namely [IrH$_2$(ArNNNAr)(PPh$_3$)$_2$] and [IrHCl(ArNNNAr)(PPh$_3$)$_2$], were shown from i.r. data to contain bidentate ArNNNAr ligands.

The brown-green acetonitrile complex K$_2$[IrCl$_5$(MeCN)],H$_2$O has been prepared by a new route involving reaction of MeCN with K$_2$[IrCl$_5$(H$_2$O)] in boiling water.[247] Its i.r. spectrum showed the usual increase in ν(C=N) on co-ordination, and this together with ^{1}H n.m.r. data indicated an increased flow of σ-electrons from MeCN → M compared with the Rh^{III} analogue.

P- *and* As-*Donor ligands.* Two related papers have reported the first systematic study of the complexing ability of the ligands P(OR)$_2$Ph and P(OR)Ph$_2$ (R = Me or Et) towards noble metals.[62, 64] The neutral [IrHCl$_2$L] were prepared *via* direct action of the ligands on [IrHCl$_2$(C$_8$H$_{12}$)]$_2$,[64] while similar procedures gave the cations [IrH{P(OR)$_2$Ph}$_5$]$^{2+}$ and [IrCl{P(OR)$_2$Ph}$_4$]$^+$.[62] These complexes were somewhat more air-stable than the corresponding trialkylphosphite species. On the other hand, dimethylphosphite P(OMe)$_2$OH has been shown to add oxidatively to [IrCl-(C$_8$H$_{14}$)$_2$]$_2$ to yield the polymers [IrHCl{P(O)(OMe)$_2$}{P(OH)(OMe)$_2$}]$_n$(n = 2 or 3).[205]. The presence of PMe$_2$Ph or AsPh$_3$ in the reaction mixture led instead to [IrHCl(PMe$_2$Ph)$_3${P(O)(OMe)$_2$}] and [IrHCl(AsPh$_3$)$_2${P(O)(OMe)$_2$}{P(OH)(OMe)$_2$}].

Reaction of NaBH$_3$CN on [IrHCl$_2$(PPh$_3$)$_3$] has afforded the white cyanotri-hydroborato-complex [IrH(BH$_3$CN)$_2$(PPh$_3$)$_3$].[9] I.r. data revealed an increase in the ν(C=N) stretching frequency on coordination N-bonding of the BH$_3$CN

[245] J. E. Frey and W. E. Ohnesorge, *J. Inorg. Nuclear Chem.*, 1973, **35**, 4307.
[246] M. Nonoyama, *Bull. Chem. Soc. Japan*, 1974, **47**, 767.
[247] B. D. Catsikis and M. L. Good, *Inorg. Nuclear Chem. Letters*, 1973, **9**, 1129.

ligand, although alternative structures involving Ir—H bonding could not be eliminated. ^{1}H N.m.r. spectra have been reported and discussed for the compounds *trans*-$[IrX_2(PMe_3)_4]^+$, and *mer*- and *fac*-$[IrX_3(PMe_3)_3]$ (X = Cl or Br), together with $^1H\{^{31}P\}$ INDOR measurements.[207]

Iridium halides have been shown to react with n-propyldiphenylarsine (L = AsPr^{n}Ph$_2$) in hot alcoholic media containing the appropriate HX to afford $[IrX_3L_3]$ (X = Cl or Br).[208] These compounds undergo a variety of reactions which are summarized in Scheme 13 (M = Ir). Also described were oxidative addition reactions of X_2 and O_2 on the α-isomers of $[IrHX_2(CO)L_2]$ (L = AsMePh$_2$, AsEtPh$_2$, or AsPr^{n}Ph$_2$), yielding $[IrX_3(CO)L_2]$ and $[IrX(O_2)(CO)L_2]$, respectively.[248]

Dimeric $[Ir_2L_3Cl_6]$, containing bidentate Ph$_2$AsCH=CHAsPh$_2$, has been obtained *via* reaction of the bis(arsine) on IrCl$_3$,xH$_2$O.[170] Its diamagnetism and non-electrolytic character supported the given formulation.

Group IV Donors. Acidopentacyano-complexes of the type $[Ir(CN)_5X]^{n-}$ (X = Cl$^-$, Br$^-$, I$^-$, OH$^-$, or NCMe) have been readily synthesized using the sequence (45; M = Ir).[209] An analysis of their u.v.–visible spectra in aqueous solution established the following order of decreasing ligand field strength for X: NCMe > OH$^-$ > H$_2$O > Cl$^-$ > Br$^-$ > I$^-$.

Addition of SnCl$_4$ or MeSnCl$_3$ to *trans*-$[IrCl(CO)(PPh_3)_2]$ has yielded the monomeric complexes $[IrCl(CO)(PPh_3)_2(SnCl_4)]$ and $[IrCl(CO)(PPh_3)_2(MeSnCl_3)]$, respectively.[249] I.r. and ^{1}H n.m.r. studies of both products indicated the presence of two rapidly interconverting isomers both in solution and in the solid state. An intramolecular oxidation–reduction mechanism was proposed for the isomerization.

Iridium(IV).—*Group VII Donors.* Pulse radiolysis of $[IrCl_6]^{2-}$ in aqueous-alcoholic solvents has resulted in reduction (by aquated electrons, H radicals, and R$^\cdot$) to the corresponding Ir^{III} species. No aquation was observed in the early stages of reaction. Magnetic circular dichroism and absorption spectra have also been measured for $[IrCl_6]^{2-}$ in the charge-transfer region, and band splittings were interpreted in terms of low-symmetry distortion and spin–orbit coupling.[72] The extraction of $[IrCl_6]^{2-}$ from the HCl solutions has also been investigated using quaternary phosphonium salts.[250] In the case of tetradecylphosphonium bromide, the complex $[PDec_4][IrCl_6]$ was isolated and characterized.

Group VI Donors. O-Donor ligands. The hitherto unknown M$_2$$[Ir(NO_3)_6]$ (M = K, Rb, or Cs) have been isolated from the reaction of N$_2$O$_5$ with M$_2$$[IrBr_6]$.[251] Their i.r. spectra indicated bidentate co-ordination in all cases, while X-ray powder studies of the potassium salt show it to be isomorphous with K$_2$$[Pt(NO_3)_6]$ (Table 4). Of particular interest are e.s.r. measurements at 77 K which revealed extensive π-bonding between iridium and the nitrate ligands (equivalent to net back-bonding of 1.9 electrons into the π* orbitals on the six nitrates, almost three times that found in K$_2$$[IrCl_6]$).

In aqueous acidic sulphate or perchlorate media Ce^{IV} has been shown to rapidly and quantitatively oxidize $[Ir(ox)_3]^{3-}$ to $[Ir(ox)_3]^{2-}$.[252] The oxidation could also be

[248] G. K. N. Reddy and E. G. Leelamani, *J. Inorg. Nuclear Chem.*, 1974, **36**, 295.
[249] C. B. Dammann, J. L. Hughey, D. C. Jicha, T. J. Meyer, P. E. Rakita, and T. R. Weaver, *Inorg. Chem.*, 1973, **12**, 2206.
[250] N. M. Sinitsyn, T. M. Gorlova, Yu. S. Shorikov, and E. M. Efimova, *Russ. J. Inorg. Chem.*, 1973, **18**, 1498.
[251] B. Harrison, N. Logan, and H. B. Raynor, *J.C.S. Chem. Comm.*, 1974, 202.
[252] H. G. Kruszyna, I. Bodek, L. K. Libby, and R. M. Milburn, *Inorg. Chem.*, 1974, **13**, 434.

Table 4 X-*Ray data for iridium compounds*

Compound	R	Comments	Ref.
$[Ir_2(NO)_4(PPh_3)_2]$	0.033	contains Ir—Ir bond (2.717 Å), and tetrahedral geometry about each metal. IrNO angle = 167°	a
$[IrH_2(CO)(PPh_3)_3]SiF_5$. $3\frac{1}{2}C_6H_6$	0.058	irregular octahedron about Ir. First X-ray structure for SiF_5; shows it to be trigonal bipyramid	b
$[Ir(NCS)(NH_3)_5]Cl_2$	0.032	N-bonded CNS ligand. The isocyanate ligands are disordered	c
$[Ir(PPh_2)_2(CO)_2(PPh_3)_2]$	0.032	contains bridging diphenylphosphido ligands. The Ir—Ir distance (2.551 Å) suggests a strong bond	d
$[Ir(CO)(PPh_3)\{(HNCCF_3)_2C_4H_5\}]$	0.045	square-planar Ir, with chelated ligand (structure 50)	e
$K_2[Ir(NO_3)_6]$	—	powder only. Isomorphous with $K_2[Pt(NO_3)_6]$	f
$IrB_{1.35}$	0.082	puckered Ir double layers containing B atoms	g
$[Ir\{(+)-chxn\}_3]Cl_3,H_2O$	—	powder only. Isomorphous with Cr, Co, and Rh analogues	h

(a) M. Angoletta, G. Ciani, M. Manassero, and M. Sansoni, *J.C.S. Chem. Comm.*, 1973, 789. (b) Ref. 213. (c) P. H. Flack, *Acta Cryst.*, 1973, **B29**, 2610. (d) P. L. Bellon, C. Benedicenti, G. Caglio, and M. Manassero, *J.C.S. Chem. Comm.*, 1973, 946. (e) Ref. 222. (f) Ref. 251. (g) T. Lundstrom and L. E. Tergenius, *Acta Chem. Scand.*, 1973, **27**, 3705. (h) P. Andersen, F. Galsbol, S. E. Harnung, and T. Laier, *Acta Chem. Scand.*, 1973, **27**, 3973.

Table 5 *I.r. structural studies of iridium complexes*

Complex	Other studies	Results and conclusions	Ref.
$K_2[IrCl_6]$	$Os^{(IV)}$, $Pd^{(IV)}$, $Pt^{(IV)}$	far i.r. of $K_2[MCl_6]$ complexes at pressures up to 20 kbar. The $v_4[\sigma(ClMCl)]$ mode is more pressure-dependent than corresponding $v_3[v(M—Cl)]$	a
$[Ir\{(PhO)P(S)Se\}_3]$	$Rh^{(III)}$, $Pd^{(II)}$, $Pt^{(II)}$	region 4000—125 cm^{-1}. Identified $v(M—S)$ and $v(M—Se)$ bands, showing ligand to be bidentate	b
$[Ir(NH_3)_5(H_2O)]^{3+}$	$Co^{(III)}$, $Rh^{(III)}$	far i.r. only. The M–ligand frequencies and bond dissociation energies decrease in the order Ir > Rh > Co	c
$[IrCl_{6-n}(pyrazine)_n]X_{3-n}$ $X = Na, K, or Cl; n = 1—4$	Raman	region 4000—200 cm^{-1}. Assigned bands and discussed differences between free and co-ordinated pyrazine	d
$trans-[IrX_4(PMe_3)_2]^-$, $mer-$ and $fac-[IrX_3(PMe_3)_3]$, $trans-[IrX_2(PMe_3)_4]^+$ $X = Cl$ or Br	Raman $Rh^{(III)}$, $Ir^{(IV)}$ ^1H n.m.r.^{f}	the Ir—Cl stretch for *trans*-chlorides increases along the series anionic, that positive charge on metal increases with positive charge on complex	e

Table 5—*contd.*

Complex	Other studies	Results and conclusions	Ref.
$Cs_2Li[Rh(CN)_6]$	polarized Raman	comparison with similar spectra of related $Cs_2Li[M(CN)_6]$ species (M = Cr, Fe, Co, or Rh) allowed assignment of optically active intra- and inter-molecular vibrations established geometric structures.	g
$[IrXY(SiR_3)(CO)(PPh_3)_2]$ (X = H or D; Y = Cl or Br; R = F, Cl, OEt, Cl_2Me, or $ClMe_2$)		established geometric structures. Showed *trans*-influences to increase in order $SiF_3 \sim SiCl_3 < SiCl_2Me < Si(OEt)_3 \ll SiClMe_2$	h

(a) Ref. 102. (b) Ref. 188. (c) Ref. 204. (d) J. Zarembowitch and L. Sebagh, *J. Inorg. Nuclear Chem.*, 1973, **35**, 4113. (e) Ref. 206. (f) Ref. 207. (g) Ref. 210. (h) R. N. Haszeldine, R. V. Parish, and J. H. Setchfield, *J. Organometallic Chem.*, 1973, **57**, 279.

effected anodically or using aqueous Co^{3+}. Formulation of the product as an Ir^{IV} species rather than $[Ir^{III}(ox)_2(C_2O_4^{\cdot-})]$ was supported by its slow reaction with excess Ce^{IV}, since free or co-ordinated oxalate radicals $C_2O_4^{\cdot-}$ generally react very rapidly with Ce^{IV}.

Treatment of $Na_2[IrCl_6]$ with KOH under strictly defined conditions has yielded red $K_2[Ir(OH)_6]$, this novel species being characterized from thermogravimetric analyses. volt–ampere curves, and absorption spectra.[253]

PART II: Palladium, Platinum, Silver, and Gold *by D. W. Clack*

1 Palladium

Palladium(0).—Co-condensation of the metals Ni, Pd, and Pt with gaseous oxygen or an oxygen–argon mixture at 4.2—10 K gives binary dioxygen complexes of these metals.[1] Diffusion-controlled warm-up and other studies have established the existence of the two types of complex $M(O_2)$ and $(O_2)M(O_2)$, both of which possess side-on co-ordinated dioxygen. However, dinitrogen complexes of type $[M(N_2)(PEt_3)_3]$ (M = Pd or Pt) are not formed by reaction of $M(PEt_3)_4$ with N_2 although the analogous nickel complex does exist.[2] This may be due to the reduced ability of Pd and Pt to back-bond compared with Ni. I.r. evidence has shown that the species $[M(CO)_3(PPh_3)]$ (M = Pd or Pt) are reversibly formed from $[M_3(CO)_3(PPh_3)_4]$ and carbon monoxide.[3] No tetracarbonyls are produced.

A series of substituted phosphine complexes of palladium have been synthesized by displacement of the allyl ligand from (2-methylallyl)PdCl$_2$ with excess phosphine, and ^{13}C n.m.r. studies have shown that the equilibrium (1) lies well to the left;

$$PdL_4 \rightleftharpoons PdL_3 + L \ (L = PMePh_2) \tag{1}$$

253 M. B. Bardin and P. M. Ketrush, *Russ. J. Inorg. Chem.*, 1973, **18**, 693.

1 H. Huber, W. Klotzbücher, G. A. Ozin, and A. Vander Voet, *Canad. J. Chem.*, 1973, **51**, 2722.
2 C. A. Tolman, D. H. Gerlach, J. P. Jesson, and R. A. Schunn, *J. Organometallic Chem.*, 1974, **65**, C23.
3 R. Whyman, *J. Organometallic Chem.*, 1973, **63**, 467.

however, when L = PEt$_3$ or PBu$^n{}_3$ the complexes are extensively dissociated at room temperature.[4] The tendency to form low co-ordination numbers clearly depends on steric factors, the following order of ability being observed: PMe$_3$ ~ PMe$_2$Ph ~ PMePh$_2$ < PPh$_3$ ~ PEt$_3$ ~ PBu$^n{}_3$ < P(benzyl)$_3$ < PPr$^i{}_3$ < P(cyclohexyl)$_3$ ~ PBu$^t{}_2$Ph. The preparation of the two-co-ordinate phosphine complexes [Pd(PPhBu$^t{}_2$)$_2$] and [Pd(PBu$^t{}_3$)$_2$] *via* reaction (2) has substantiated the above sequence.[5] The former reacts with oxygen to yield the dioxygen complex

$$Pd(C_5H_5)(C_3H_5) + 2L \xrightarrow{\text{hexane}} Pd^0(L)_2 \qquad (2)$$

[Pd(PPhBu$^t{}_2$)$_2$(O$_2$)] as stable green crystals. Both two-co-ordinate complexes undergo oxidative addition with CF$_3$CO$_2$H to produce the hydride *trans*-[Pd(H)-(O$_2$CCF$_3$)L$_2$]. The X-ray structure of [Pd(PBu$^t{}_3$)$_2$] has shown that the palladium has almost linear (176.8°) co-ordination (Table 1, p. 399).

I.r. and n.m.r. have been used to study a series of maleic anhydride–phosphine complexes [PtL$_2$(CHCO)$_2$O] [L = (p-MeOC$_6$H$_4$)$_3$P, MePh$_2$P, (p-ClC$_6$H$_4$)$_3$P, (PhO)$_3$P, MeC(CH$_2$O)$_3$P, Ph$_3$P].[6] In all complexes ν(CO) is shifted to lower wavenumbers on co-ordination while the maleic anhydride protons are shifted to higher field, both changes being linearly related to the basicities of the phosphorus ligands. It was postulated that σ-donation from the double bond is a more important factor in the high-field shift than π-back-donation.

There have been several papers reporting new oxidative addition reactions of [Pd(PPh$_3$)$_4$]. Alkoxalyl compounds (1), formed by addition of ClCOCO$_2$R (R = Me or Et), decarbonylate in chloroform or benzene at room temperature to yield *trans*-[PdCl(CO$_2$R)(PPh$_3$)$_2$].[7] The X-ray structure (Table 1, p. 399) of the methyl derivative

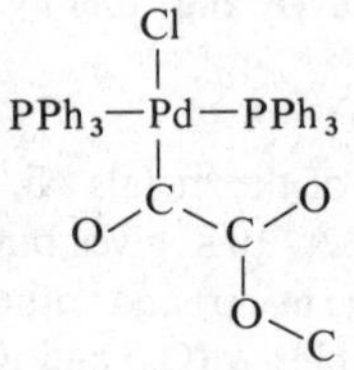

(1)

shows that the methoxalyl ligand has an *s-trans* planar conformation (1). Nitromethane reacts with [Pd(PPh$_3$)$_4$] to yield an isocyanato-complex [Pd(NCO)$_2$-(PPh$_3$)$_2$].[8]

The most common type of oxidative addition to [Pd(PPh$_3$)$_4$] has been shown to follow equation (3) (X = Cl, *e.g.* CCl$_4$, ICl, PhCOCl, SnCl$_2$, SnCl$_4$, Cl$_2$CS; X = Br,

$$[Pd(PPh_3)_4] + 2XY \rightarrow [PdX_2(PPh_3)_2] + Y_2 \qquad (3)$$

[4] A. Musco, W. Kuran, A. Silvani, and M. W. Anker, *J.C.S. Chem. Comm.*, 1973, 938.
[5] M. Matsumoto, H. Yashioka, K. Nakatsu, Y. Yoshida, and S. Otsuka, *J. Amer. Chem. Soc.*, 1974, **96**, 3322.
[6] H. Minematsu, Y. Nonaka, S. Takahashi, and N. Hagihara, *J. Organometallic Chem.*, 1973, **59**, 395.
[7] J. Fayos, E. Dobizynski, R. J. Angelici, and J. Clardy, *J. Organometallic Chem.*, 1973, **59**, C33.
[8] K. Schorpp and W. Beck, *Chem. Ber.*, 1974, **107**, 1371.

e.g. IBr, PhCOBr);[9] insertion reactions of type (4) are much less common, the only one

$$[Pd(PPh_3)_4] + XY \rightarrow [PdXY(PPh_3)_2] + 2PPh_3 \qquad (4)$$

observed being with 3-benzylideneacetylacetone.

Novel bimolecular palladium complexes of composition $[Pd_2(dBA)_3(solvent)]$ (solvent = $CHCl_3$, C_6H_6, or C_6H_5Me) have been obtained by recrystallizing bis-(dibenzylideneacetone)palladium from the appropriate organic solvent.[10] The X-ray structure of $[Pd_2(dba)_3(CHCl_3)]$ has shown that the two palladium atoms are bridged by three dba molecules, co-ordinating through the olefinic bond with trigonal geometry around the palladium (Table 1, p. 399).

Palladium(I).—Unusual carbonylhalide complexes of palladium(I), $[PdX(CO)]$ (X = Cl or Br), have been isolated by reduction of the dihalide with CO (equation 5).[11]

$$2PdX_2 + 2CO + H_2O \rightarrow 2[PdX(CO)] + CO_2 + 2HX \qquad (5)$$

X-Ray powder photographs showed that the chloride and bromide complexes were isomorphous. The i.r. spectrum has been interpreted in terms of both chloro bridges (2)[11] and mixed carbonyl-chloro bridges (3).[12] Both $[PdCl(CO)]$ and $[PdBr(CO)]$

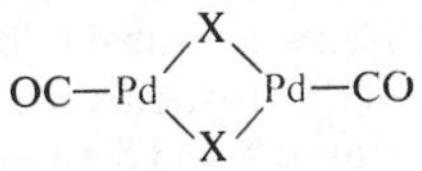

(2) 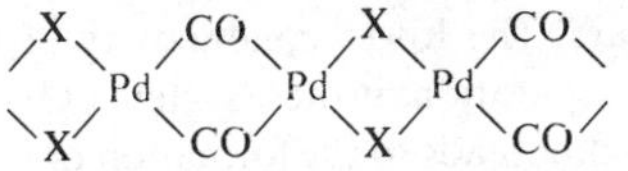   (3)

exhibit an e.s.r. spectrum, the latter showing a hyperfine pattern of seven lines centred at $g = 2.04$, and temperature variation studies indicate that the magnetism is not due to a thermally accessible triplet state. Bis(diphenyl-arso)methane reacts with $[PdX(CO)]$ to yield $[PdX(dam)]_2$ (X = Cl or Br) with the proposed structure (4) containing a Pd—Pd bond.

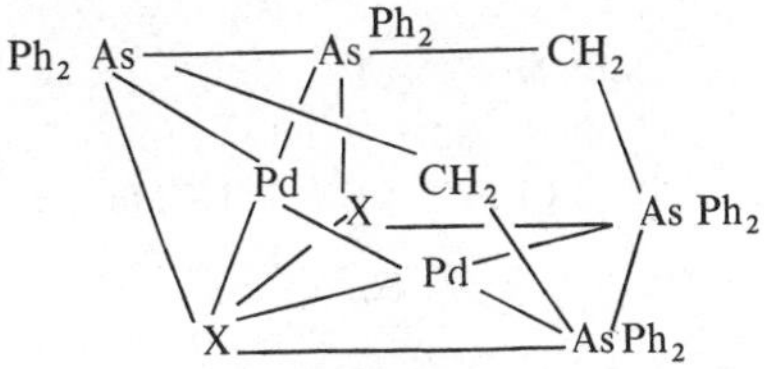

(4)

Complex halogeno-anions of type $[PdX_2(CO)]^-$ (X = Cl or Br) have also been obtained by treatment with MX (M = Cs, Pr^n_4N, Bu^n_4N, or Pn^n_4N; X = Cl or Br) or by carbonylation of the $[PdX_4]^{2-}$ salts in hot aqueous HX. Structures (5) and (6) have been independently proposed from i.r. data, and evidence is given that the anions are planar in the solid $Pr^n_4N^+$ salts, but in solution and as salts of the larger cations they appear to be bent about the carbonyl bridge.[12] A Pd—Pd bond is implied by their diamagnetism.

[9] H. A. Tayim and N. S. Akl, *J. Inorg. Nuclear Chem.*, 1974, **36**, 944.
[10] T. Ukai, H. Kawazura, Y. Ishii, J. J. Bonnet, and J. A. Ibers, *J. Organometallic Chem.*, 1974, **65**, 253.
[11] R. Colton, R. H. Farthing, and M. J. McCormick, *Aust. J. Chem.*, 1973, **26**, 2607.
[12] P. L. Goggin and J. Mink, *J.C.S. Dalton*, 1974, 534.

$$\text{Cl} \diagdown \underset{\text{Pd}}{} \diagup \text{Cl} \diagdown \underset{\text{Pd}}{} \diagup \text{CO}$$
$$\text{CO} \diagup \diagdown \text{Cl} \diagdown \text{Cl}$$

$$\text{Cl} \diagdown \underset{\text{Pd}}{} \diagup \text{CO} \diagdown \underset{\text{Pd}}{} \diagup \text{Cl}$$
$$\text{Cl} \diagup \diagdown \text{CO} \diagdown \text{Cl}$$

(5)   (6)

Palladium(II).—*Group VII and Hydride Donors.* SCF–Xα calculations have been made for the ions $[MCl_4]^{2-}$ (M = Pd or Pt), and the computed optical transitions were found to be in good agreement with experimental spectra.[13] The calculated d-level ordering for the complexes was $d_{x^2-y^2} > d_{xy} > d_{xz}, d_{yz} > d_{z^2}$ with the mixing between ligand and metal d-orbitals found to be considerably stronger than given by previous calculations. The u.v.–visible spectra of $[Pd_2X_6]^{2-}$ (X = Cl or Br) have been measured;[14] the band assignments include ligand $\rightarrow$ metal charge-transfer and $d \rightarrow p$ bands in addition to the low-intensity $d \rightarrow d$ bands.

Two papers have reported the electrical and photo conductivity of some one-dimensional mixed-valence complexes $[M(A)X_3]$ [M = Pd or Pt; A = en, $(NH_3)_2$; X = Cl, Br, or I].[15,16] The results suggest that the conventional band-type conduction model is not appropriate for discussing electron transport in these materials. The conductivity is pressure dependent and in some cases increases by a factor 10^9 in 140 kbars. The lower conductivity of the en complexes is probably due to long M^{II}—M^{IV} separations in these systems. Oxidation of bis-(1,2-benzoquinonedioximato)-Pd with iodine leads to the formation of PdL_2I_x ($x = 0.6$—0.7) and d.c. measurements indicate a conductivity of $10^{-5} \, \Omega^{-1} \, cm^{-1}$ which is 10^3 higher than that of PdL_2.[17] Phenylhydrazine complexes of type $[PdX_2(PhNHNH_2)_2]$ (X = Cl or Br) have been prepared by addition of phenylhydrazine to the palladium dihalide in aqueous hydrochloric acid solution.[18] The complexes are presumably square planar with phenylhydrazine acting as a unidentate ligand. On the other hand, electronic and Raman spectra and conductivity measurements on solutions of $PdCl_2$ in MeCN have established the presence of the dinuclear species $Pd_2Cl_4(MeCN)_2$ (7).[19] Addition of

$$\text{MeCN} \diagdown \underset{\text{Pd}}{} \diagup \text{Cl} \diagdown \underset{\text{Pd}}{} \diagup \text{Cl}$$
$$\text{Cl} \diagup \diagdown \text{Cl} \diagdown \text{NCMe}$$

(7)

more chloride to the solution produces $Pd_2Cl_6^{2-}$ and ultimately breaks up the dinuclear species. Other halogeno-bridged palladium complexes have been prepared from the dihalides and thiourea to give $PdX_2\{SC(NH_2)_2\}_2$ (X = Cl or Br).[20] I.r. studies indicate a bridge structure with thiourea bonding through S. The dimeric

[13] R. P. Messmer, L. V. Interrante, and K. H. Johnson, *J. Amer. Chem. Soc.*, 1974, **96**, 3847.
[14] J. L. H. Batiste and R. Rumfeldt, *Canad. J. Chem.*, 1974, **52**, 174.
[15] L. V. Interrante and K. W. Browall, *Inorg. Chem.*, 1974, **13**, 1162.
[16] L. V. Interrante, K. W. Browall, and F. P. Bundy, *Inorg. Chem.*, 1974, **13**, 1158.
[17] H. Endres, H. J. Keller, M. Mégnamisi-Bélombe, W. Morini, and D. Nöthe, *Inorg. Nuclear Chem. Letters*, 1974, **10**, 467.
[18] W. K. Glass and J. O. McBreen, *J. Inorg. Nuclear Chem.*, 1974, **36**, 747.
[19] I. I. Volchenskova and K. B. Yatsimirskii, *Russ. J. Inorg. Chem.*, 1973, **18**, 990.
[20] G. Marcotrigiano, R. Battistuzzi, and G. Peyronel, *J. Inorg. Nuclear Chem.*, 1973, **35**, 2265.

structure is broken up by electron donors such as *p*-toluidine, pyridine, or ethylene-diamine (equation 6).

$$[Pd_2X_4\{SC(NH_2)_2\}_2] + 2L \rightarrow [PdX_2L_2] + [PdX_2\{SC(NH_2)_2\}_2] \qquad (6)$$

In contrast, addition of ethanolic solutions of thioamides to aqueous $PdCl_2$ yields monomeric square-planar complexes of type $[PdCl_2\{SCR^1(NHR^2)\}]$ ($R^1 =$ Me, Ph,

or furan, $R^2 = $) and $[PdCl_2L_2]$ (L = range of saturated and

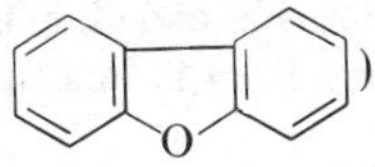

unsaturated thioamides).[21,22] I.r. and visible–u.v. spectra suggest the two square-planar structures (8) and (9) with co-ordination through both N and S of the thio-

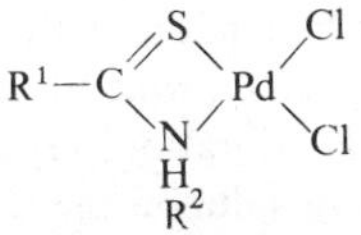

(8) 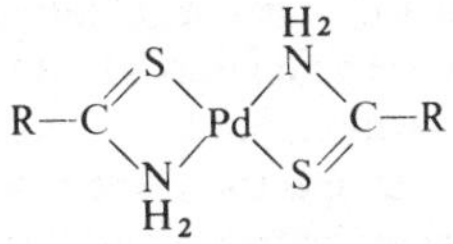   (9)

amide. A polynuclear chloropalladium complex has been synthesized according to reaction (7).[23] *X*-Ray structure determinations of this and similar platinum complexes

$$K_2PdCl_4 + 3HgCl_2 + 2Et_4NCl \rightarrow (Et_4N)_2[Hg_3PdCl_{10}] + 2KCl \qquad (7)$$

(Table 1, p. 400) show that they all possess the trinuclear anionic unit $[Hg_2MCl_8]$ (M = Pd or Pt) (10).

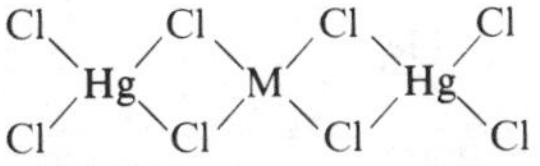

(10)

A tetrachloroaluminate has been isolated for the first time by reaction of $PdCl_2$ with excess $AlCl_3$ at $200\,^\circ C$.[24] An *X*-ray powder pattern was run and the low magnetic moment indicates a square-planar $PdCl_4$ co-ordination.

Carcinostatic metal complexes of Ni and Pd with asparaginate and chloride ligands have been studied using glass and chloride electrodes.[25] The equilibria in the Pd^{II}–asparaginate–chloride–hydroxide systems involve the complexes $[Pd(Asn)H^{2+}]$, $[Pd(Asn)OH]$, $[Pd(Asn)ClOH]^-$, and $[Pd(Asn)ClH]^+$. β-values are useful to indicate the strengths of the bonds involved and to suggest possible structures of these anti-cancer metal complexes. Resin-bound $[PdCl_4]^{2-}$ was found to act as a catalyst in the hydrogenation of allyl alcohol to yield propanol and propanal, and this system was compared with the usual palladium catalysts.[26]

[21] D. Negoiu, V. Muresan, and S. Floren, *Rev. Roumaine Chim.*, 1973, **18**, 1875.

[22] D. Negoiu, V. Muresan, and C. Fulea, *Rev. Roumaine Chim.*, 1973, **18**, 1749.

[23] R. M. Barr, M. Goldstein, N. D. Hairs, M. McPartlin, and A. J. Markwell, *J.C.S. Chem., Comm.*, 1974, 221.

[24] G. N. Papatheodorou, *Inorg. Nuclear Chem. Letters*, 1974, **10**, 115.

[25] R. D. Graham and D. R. Williams, *J.C.S. Dalton*, 1974, 1123.

[26] R. Linarte-Lazcano, M. Paez-Pedrosa, J. Sabadie, and J. E. Germain, *Bull. Soc. chim. France*, 1974, 1129.

The heats of the displacement reactions of chloropalladium complexes (equation 8)

$$[PdCl_2(A)] + 2B \rightarrow [PdCl_2B_2] + A$$
$$[A = (PhCN)_2 \text{ or cod}; B = py, Ph_3P, \text{ or } (PhO)_3P] \tag{8}$$

have given the relative displacement energies as $Ph_3P > (PhO)_3P > py \gg cod$.[27]

Stable dihydride complexes of Pd and Pt have been isolated from the reaction of tricyclohexylphosphine with the metal acetylacetonates and aluminium alkyls.[28] The pale yellow $[PdH_2(PCy_3)_2]$ has a *trans* structure from i.r. and n.m.r. studies, with $\nu(Pd—H)$ occurring at $1740\ cm^{-1}$.

Solvolysis of $[Pd(Bu^nOPPh_2)_4]$ in aqueous ethanol yields $[Pd_2(OPPh_2)_4-(HOPPh_2)_2]$, unlike its Pt analogue which gives rise to a hydrido-complex.[129]

Group VI. O-Donor ligands. The complex $[Pd(CH_2CO_2)(PPh_3)_2]$ has been shown to react with chelating ligands L (acac or oxine) to produce the yellow C-bonded complexes of acetic acid $[Pd(CH_2CO_2H)L(PPh_3)]$.[29] In addition the linkage isomer, O-bonded, has been prepared by treatment of $[Pd(acac)_2]$ with PPh_3 and $MeCO_2H$ in benzene. The i.r. spectra of the solids show $\nu(OH)$ vibrations in the range 2500—2700 cm^{-1}, thus indicating molecular association. In solution the C-bonded isomer has $\nu(CO)$ at $1720\ cm^{-1}$ and appears to be monomeric. On the other hand, $[Pd(acac)_2]$ reacts with nitrogen bases (L = py, Et_2NH, or PhMeNH) to produce complexes (11) where one acac is C-bonded and unidentate.[30] The analogous phosphine complex $[Pd(acac)_2(PPh_3)]$ yields $[Pd(PPh_3)_4]$ with excess PPh_3.

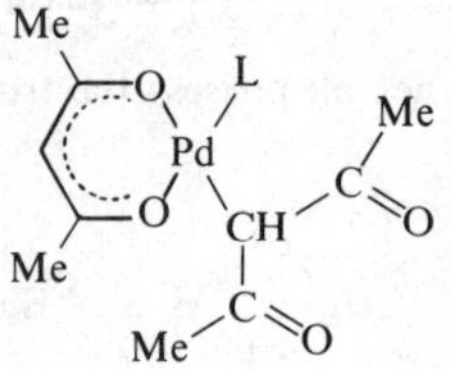

(11)

The formation and stability constants of complexes between Pd (Pt, Rh, Ir, Os, Ru) and *o*-coumaric acid have been determined by pH titration.[31] The results indicate that a 1:2 complex is formed with Pd.

A study of the olefin oxidation catalyst system, palladium acetate–MOAc (M = Li or Na), has shown that in the absence of acetate ion, Pd acetate–acetic acid exists as the trimeric species $[Pd_3(OAc)_6]$.[32] Reaction with MOAc is not instantaneous, and u.v.–visible spectra indicate an initial equilibrium involving trimer → dimer (9). When M = Na conversion into dimer is complete at 0.2M–NaOAc. Further addition of

$$2[Pd_3(OAc)_6] + 6MOAc \rightleftharpoons 3[M_2Pd_2(OAc)_6] \tag{9}$$

[27] W. Partenheimer and E. F. Hoy, *Inorg. Chem.*, 1973, **12**, 2805.
[28] K. Kudo, M. Hidai and Y. Uchida, *J. Organometallic Chem.*, 1973, **56**, 413.
[29] S. Baba and S. Kawaguchi, *Inorg. Nuclear Chem. Letters*, 1973, **9**, 1287.
[30] S. Baba, T. Ogura, and S. Kawaguchi, *Bull. Chem. Soc. Japan*, 1974, **47**, 665.
[31] S. C. Tripathi and S. Paul, *J. Inorg. Nuclear Chem.*, 1973, **35**, 2465.
[32] R. N. Pandy and P. M. Henry, *Canad. J. Chem.*, 1974, **52**, 1241.

NaOAc results in monomer formation (10).

$$Na_2[Pd_2(OAc)_6] + 2NaOAc \rightleftharpoons 2Na_2[Pd(OAc)_4] \qquad (10)$$

Treatment of bis(salicylaldehydo)Pd with salicylaldehydehydrazone in EtOH has given $[Pd(Sal=N-N=SalH)_2]$ (Sal =

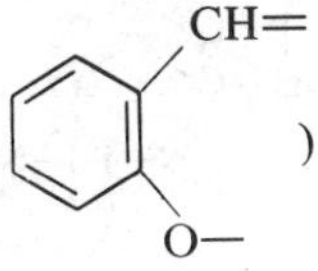

) as an orange precipitate.[33]

This complex has two free chelate centres and reacts with $[Rh(acac)(CO)_2]$ in C_6H_6 at 70 °C to yield the novel trinuclear complex $[Pd\{Sal=N-N=SalRh(CO)_2\}_2]$. The synthesis of the ternary oxide K_2PdO_2 has been effected by reaction between K_2O and PdO.[34] X-Ray studies (Table 1, p. 400). suggest a chain structure which corresponds to that of K_2PdS_2.

Mixed O- *and* N-*donor ligands.* Studies of the reactivities of co-ordinated ligands have shown that bis-salicylaldoxime chelates of Ni, Pd, and Cu react with $AlPh_3$ in benzene to yield grey diamagnetic complexes (12).[35] Dimethylglyoxime complexes undergo similar reactions (see below).

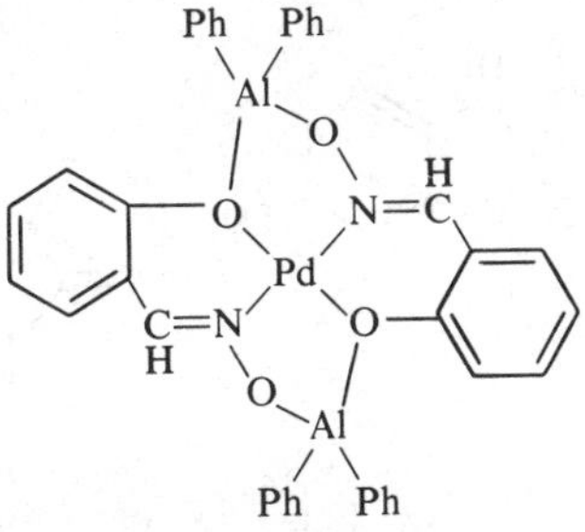

(12)

The synthesis of complexes of some cyclic tertiary amino-acids with Pd(Pt) has been reported (equation 11).[36] The complexes (L = 1-pyrrolidine acetate, 1-piperidine

$$K_2MCl_4 + 2L,HCl \xrightarrow[H_2O]{pH\ 7.4} ML_2 \qquad (11)$$

acetate, or hexahydro-1-azepine acetate) are chelated *via* N and O and have high stability constants, $K = 10^8 - 10^{11}$. In addition, i.r. and normal co-ordinate analyses of Pd and other metal DL-tyrosine complexes have been made.[37] The complex has the *trans* structure (13). The complexes of Pd and Pt with 3,3'-diamino-4,4'-dihydroxy-diphenylsulphone have been obtained as insoluble, probably polymeric, materials.[38] Analytical data indicate that 1:1 complexes are formed, while i.r. spectra show that

[33] B. G. Rogachev, A. S. Astakhova. and M. L. Khidekel, *Bull. Acad. Sci. U.S.S.R.*, 1973, 22, 2544.
[34] H. Sabrowsky, W. Bronger, and D. Schmitz, *Z. Naturforsch.*, 1974, 29, 10.
[35] N. Voiculescu, I. Popesen, and N. Luchian, *Rev. Roumaine Chem.*, 1973, 18, 1601.
[36] F. Hogue and H. Frye, *Inorg. Nuclear Chem. Letters*, 1974, 10, 505.
[37] Y. Inomata, T. Inomata, and T. Moriwaki, *Bull. Chem. Soc. Japan*, 1974, 47, 818.
[38] W. V. Malik, M. P. Teotia, and D. K. Rastogi, *J. Inorg. Nuclear Chem.*, 1973, 35, 4047.

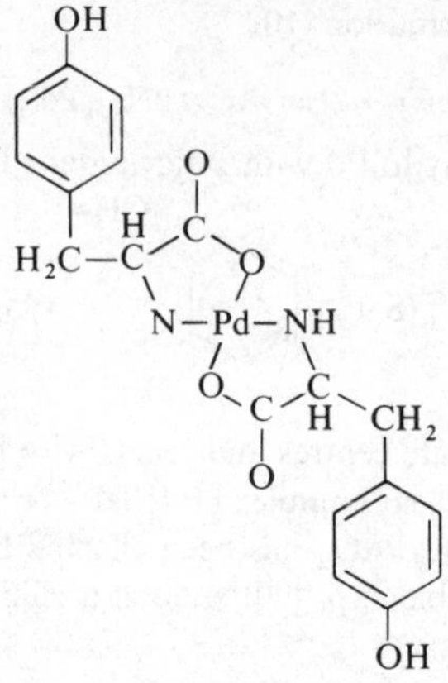

(13)

the S is not involved in co-ordination. The d–d bands were assigned for square-planar geometry around the metal (14).

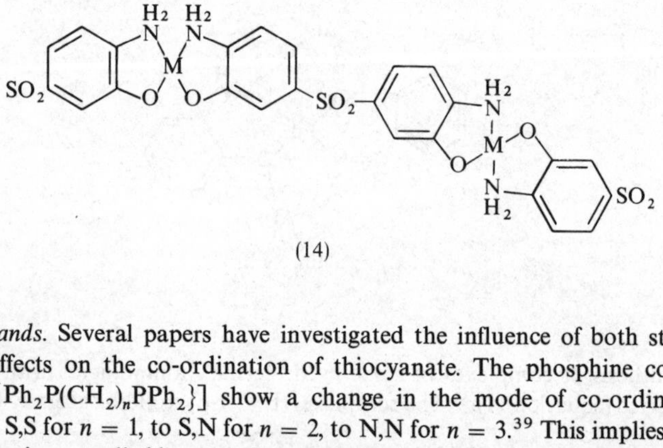

(14)

S-*Donor ligands.* Several papers have investigated the influence of both steric and electronic effects on the co-ordination of thiocyanate. The phosphine complexes $[Pd(SCN)_2\{Ph_2P(CH_2)_nPPh_2\}]$ show a change in the mode of co-ordination of SCN^- from S,S for $n = 1$, to S,N for $n = 2$, to N,N for $n = 3$.[39] This implies that the co-ordination is controlled by steric factors alone since π-bonding capabilities remain constant in this series. Preliminary X-ray data (Table 1, p. 400) show that an increase in n is accompanied by an increase in the PPdP angle, and a large increase in steric interaction between the phenyl groups and the co-ordinated thiocyanate. I.r. and u.v.-visible spectral studies on the complexes $[MXL]BPh_4$ (M = Pd or Pt; L = Et_4 dien or $MeEt_4$dien; X = CNS) reveal that for M = Pd the complexes undergo S → N bonded linkage isomerism in solution, with N → S reisomerizations in the solid state.[40] On the other hand, both $[Pd(NCS)(tripy)]BPh_4$ and $[Pd(NO_2)(Et_4dien)]BPh_4$ were isolated as N-bonded species. Further work on complexes of type $[Pd(CNR)_2$-$(SNC)_2]$ (R = Ph or cyclohexyl) and $[Pd(CNR)(SCN)_2(PPh_3)]$ (R = Ph, cyclohexyl,

[39] G. J. Palenik, W. L. Steffen, M. Mathew, M. Li, and D. W. Meek, *Inorg. Nuclear Chem. Letters*, 1974, **10**, 125.
[40] J. L. Burmeister, R. L. Hassel, K. A. Johnson, and J. C. Lim, *Inorg. Chim. Acta*, 1974, **9**, 23.

or *p*-nitrophenyl) has shown that the former are solely S-bonded while the latter may be either S- or N-bonded.[41] On the other hand, raising the temperature of the reaction between $[Pd(SCN)_4]^{2-}$ and PPh_3 or $AsPh_3$ in solution leads to products which contain increasingly larger proportions of the *trans*-N-bonded isomers which are unaffected by heating in the solid state.[132]

Dinuclear complexes $[M(SCN)\{(Ph_2PO)_2H\}]_2$ (M = Pd or Pt) have been prepared by metathetical reaction from the chloro-bridged dimers (see below).[42] The *X*-ray structure (15) (see Table 1, p. 400) shows square-planar co-ordination around the Pd (N, S, 2P) with an unusual symmetrically H-bonded anion $[(Ph_2PO)_2H]^-$. In view of the possibility of S- or N-bonding, molten thiocyanate is of some interest

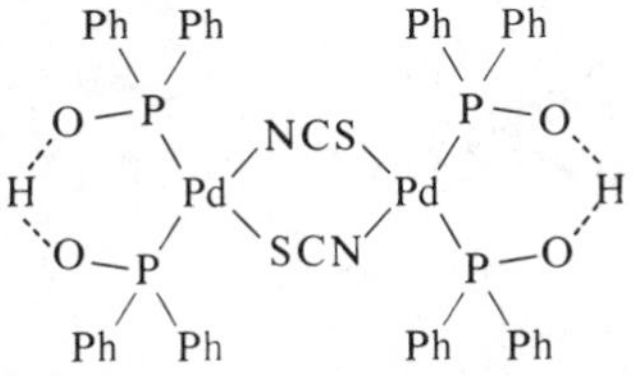

(15)

as a non-aqueous solvent. Electronic spectra of Pd^{II} in both aqueous and molten thiocyanate indicate that the species present is tetragonal $[Pd(SCN)_4]^{2-}$ ion.[43]

Treatment of $PdCl_2$ with naphthylthiourea (Ntu) in MeOH has yielded the five-co-ordinate complex $[Pd(Ntu)_4Cl]Cl$, which reacts further with N-heterocyclic amines to produce $[Pd(Ntu)_2LCl]Cl$ [L = 2,2'-bipy, *o*-phen] and $[Pd(Ntu)_2L'_2Cl]Cl$ [L' = py, γ-, β-, and γ-pic, (2,4 and 2,6 lutidines), quinoline, isoquinoline, or acridine].[44] The complexes are 1:1 electrolytes in MeOH, and i.r. spectra reveal co-ordination *via* the S atom of Ntu and N of the heterocyclics. Complexes of pyridine-2-thiol (LH) are not common and stimulate interest as to possible modes of co-ordination. Complexes of Pd (Rh, Ag) have been prepared.[45] I.r. spectra reveal an N—H stretch, indicating unidentate co-ordination through S for $[Pd(LH)_4Cl_2]$,EtOH, which is a 2:1 electrolyte in MeOH. The red diamagnetic complex $[PdL_2LH]$ appears to be five-co-ordinate with the anion L^- acting as a bidentate ligand.

Synthesis of both mono- and di-nuclear complexes of Pd and Pt with $(SCF_3)^-$ has been reported.[46, 47] Mononuclear species have been obtained by oxidative addition of $(SCF_3)_2$ to the $[M(PPh_3)_4]$ (M = Pd or Pt) complexes, and by metathetical reactions (equation 12). ^{19}F N.m.r. was used to establish the *trans*-influence series

$$[MCl_2(PPh_3)_2] + 2AgSCF_3 \rightarrow cis\text{-}[M(SCF_3)_2(PPh_3)_2] \qquad (12)$$

H > PEt_3 ~ PPh_3 > SCF_3 > Cl in analogous Pt complexes.[46] Dinuclear complexes of formula $[M_2(SCF_3)_2(PR_3)_4]X_2$ (M = Pd or Pt; R = Et or Ph; X = BF_4 or ClO_4)

[41] R. R. Cooke and J. L. Burmeister, *J. Organometallic Chem.*, 1973, **63**, 471.
[42] D. V. Naik, G. J. Palenik, S. Jacobson, and A. J. Canty, *J. Amer. Chem. Soc.*, 1974, **96**, 2286.
[43] K. S. De Haas, *J. Inorg. Nuclear Chem.*, 1973, **35**, 3231.
[44] M. M. Khan, *J. Inorg. Nuclear Chem.*, 1974, **36**, 299.
[45] I. P. Evans and G. Wilkinson, *J.C.S. Dalton*, 1974, 946.
[46] K. R. Dixon, K. C. Moss and M. A. R. Smith, *J.C.S. Dalton*, 1973, 1528.

were prepared according to equations (13) and (14).[47] ^{19}F N.m.r. allowed assignments

$$2[MCl(SCF_3)(PPh_3)_2] + 2AgX \rightarrow [M_2(SCF_3)_2(PPh_3)_4] X_2 + 2AgCl \quad (13)$$

$$[M_2Cl_2(PPh_3)_4]X_2 + 2AgSCF_3 \rightarrow [M_2(SCF_3)_2(PPh_3)_4] X_2 + 2AgCl \quad (14)$$

of the structures (16) and (16a), where the existence of *syn*- and *anti*-isomers in square-planar metal complexes is demonstrated for the first time. The equilibrium between these isomers does not appear to be labile, but the ease of bridge cleavage by donor ligands (PR$_3$, py, *p*-toluidine, or Cl$^-$) shows that SCF$_3$ bridges are rather weak, in sharp contrast with the strong bridging abilities of alkyl and aryl sulphides.

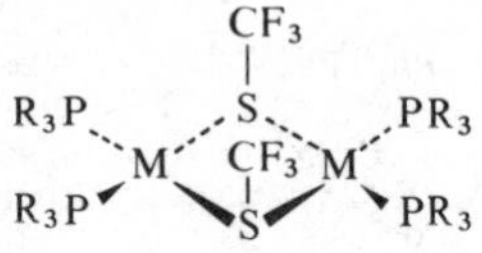

(16) *syn* 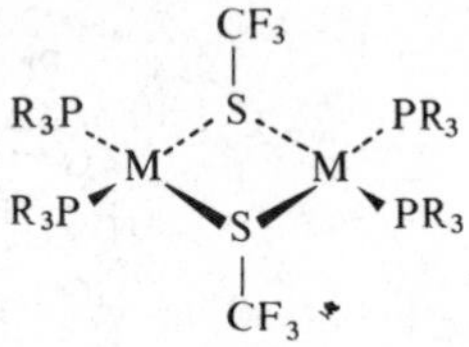   (16a) *anti*

Orange-red diamagnetic complexes of Pd and Pt with mercaptoacetic and 3-mercaptopropionic acid have been reported.[48] The complexes are non-electrolytes in DMSO; their general insolubility and i.r. measurements are indicative of polymeric species involving S-bridging. Treatment of an aqueous solution of PdCl$_2$ with a methanolic solution of appropriate sodium dithiocarbamates has yielded a large variety of Pd dithiocarbamates as yellow precipitates.[49] Their u.v.–visible spectra are reported and many of the i.r. bands assigned. I.r. studies of the monothiocarbamate complexes [Pd(OSCNR$_2$)$_2$(PMe$_2$Ph)$_2$] (R = Me, Et, or Bu) have shown that these are S-bonded.[50] Simple bis(monothiocarbamate) complexes for R = pyrrolidyl are not formed, rather a tetrakis species (M = Pd or Pt) which shows bonding through both O and S, although no definite structure has been proposed. Extraction of metallic Au, Ag, Pd, or Pt from aqueous solution by a large number of alkyl and alkene disulphides and sulphides in toluene depends on the length and degree of branching of the alkyl group.[51, 52] The following order of activity of extraction has been obtained (EtSCH$_2$)$_2$ > (EtS)$_2$CH$_2$ > (C$_4$H$_9$SCH$_2$)$_2$ > (EtSCH=)$_2$ > (i-C$_4$H$_9$SCH$_2$)$_2$ > (i-C$_4$H$_9$SCH=)$_2$ > (EtS)$_2$ > (MeS)$_2$ > (C$_3$H$_7$S)$_2$ > (C$_4$H$_9$S)$_2$ > (i-C$_4$H$_9$S)$_2$ > (i-C$_3$H$_7$S)$_2$ > (PhS)$_2$. In addition, the metal ions Pd^{II}, Pt^{II}, and Rh^{III} are precipitated quantitatively from aqueous chloride solution by NaSEt, and may be separated from Ir^{IV} in this way.[53]

Cleavage of disulphides (R$_2$S$_2$) by [Pd(PPh$_3$)$_4$] has yielded the red dimeric compound (17), which produces the di-iodo complex when treated with MeI.[54] Mono-

[47] K. R. Dixon, K. C. Moss and M. A. R. Smith, *J.C.S. Dalton*, 1974, 971.
[48] M. Chandrasekharan, M. R. Udupa, and G. Aravamudan, *J. Inorg. Nuclear Chem.*, 1974, **36**, 1153.
[49] C. G. Scenery and R. J. Magee, *Inorg. Nuclear Chem. Letters*, 1974, **10**, 323.
[50] E. M. Krankovits, R. J. Magee, and M. J. O'Connor, *Austral. J. Chem.*, 1973, **26**, 1645.
[51] V. A. Pronin, M. V. Usol'tseva, Z. N. Shastina, N. K. Gusanova, E. P. Vyalykh, S. V. Amosova, and B. A. Trofimov, *Russ. J. Inorg. Chem.*, 1973, **18**, 1016.
[52] V. A. Pronin, M. V. Usol'tseva, Z. N. Shastina, B. A. Trofimov, and E. P. Vyalykh, *Russ. J. Inorg. Chem.*, 1973, **18**, 1615.
[53] Y. N. Kukushkin, S. A. Simanova, V. K. Krylov, and L. G. Mel'gunova, *Russ. J. Inorg. Chem.*, 1973, **18**, 1364.
[54] R. Zanella, R. Ros, and M. Graziani, *Inorg. Chem.*, 1973, **12**, 2737.

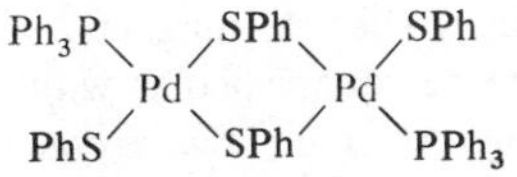

$$Ph_3P \diagdown \underset{PhS}{\overset{}{Pd}} \diagup SPh \diagdown \underset{SPh}{\overset{}{Pd}} \diagup SPh \diagup PPh_3$$

(17)

meric species are, however, formed with electronegatively substituted phenyl di-sulphides such as *o*- or *m*-nitrophenyl disulphide. The analogous Pt complex reacts with $[(MeCN)_2PdCl_2]$ to give the interesting hetero dinuclear complex $[(PPh_3)_2Pt-(SPh)_2PdCl_2]$ with bridging SPh groups. The four-co-ordinate complexes $[M(S_2PR^1_2)_2(PR^2_3)]$ and $[M(S_2PR^1_3)(PR^2_3)_2](S_2PR_2)$ (M = Pd or Pt; R = Me or Et) have been synthesized by treatment of the appropriate dimethylphosphino-dithiato-complex with phosphine.[55] Variable temperature ^{1}H n.m.r. indicates that rapid unidentate–bidentate exchange occurs at room temperature. A general method of synthesizing the complexes $[M(S_2PR^1_2)(PR^2_3)_2]X$ (M = Pd or Pt; R = Me or Ph; X = BPh$_4$ or PF$_6$) is given (equation 13), and an empirical method for

$$[M(S_2PR^1_2)_2] + PR^2_3 + NaX \rightarrow [M(S_2PR^1_2)(PR^2_3)_2]X \qquad (13)$$

distinguishing between uni- and bi-dentate and ionic modes of co-ordination of the S_2PMe_2 group is discussed.

The mass spectra of Pd (Pt) chelates of $RC(SH)=CHCOCF_3$ (R = Ph, *p*-MeC$_6$H$_4$, *p*-BrC$_6$H$_4$, *p*-NO$_2$C$_6$H$_4$, or 2-thienyl) and the Pd chelate of $MeC(SH)=CHCF_3$ have been recorded.[56, 57] Comparison with data for the chelates with Ni and Zn reveals that the fragmentation pattern is determined by the metal. An investigation into the internal metallation of ligand co-ordination has shown that the complexes formed from $[MCl_4]^{2-}$ (M = Pd or Pt) by treatment with n-butyl α-naphthyl sulphide have

the formula $\left[\left(\bigotimes^{SC_4H_9} \right)_2 MCl_2 \right]$.[58] ^{1}H N.m.r. has shown that internal

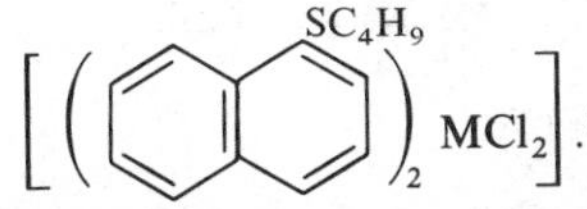

metallation does not occur, unlike the complexes with similar N-containing ligands.

A new series of complexes of Pd (Ni, Cu, Co) with quadridentate Schiff-base ligands containing two S and two N as donor ligands have been prepared with the general structure (18; R = C_2H_4, C_3H_6, or C_6H_{10}).[59] The complexes were characterized by

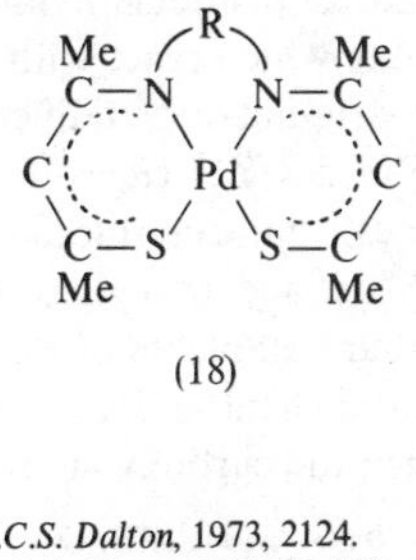

$$\begin{array}{c} Me \quad \overset{R}{\diagup \diagdown} \quad Me \\ C-N \quad N-C \\ / \quad \diagdown Pd \diagup \quad \backslash \\ C \quad \quad \quad C \\ \backslash \quad \diagup \quad \diagdown \quad / \\ C-S \quad \quad S-C \\ Me \quad \quad \quad Me \end{array}$$

(18)

[55] D. F. Steel and T. A. Stephenson, *J.C.S. Dalton*, 1973, 2124.
[56] M. Das and S. E. Livingston, *Austral. J. Chem.*, 1974, **27**, 749.
[57] M. Das and S. E. Livingston, *Austral. J. Chem.*, 1974, **27**, 53.
[58] V. I. Sokolov, *Bull. Acad. Sci. U.S.S.R.*, 1974, **22**, 1600.
[59] P. R. Blum, R. M. C. Wei, and S. C. Cummings, *Inorg. Chem.*, 1974, **13**, 450.

i.r., u.v.–visible, and ^{1}H n.m.r. spectra, magnetic susceptibility, and conductivity measurements, and shown to be square planar with more charge delocalization in the sulphur-containing rings than in the oxygen analogues. The complexes [M(OEtsacsac)$_2$] (M = Pd, Pt, or Ni) have been reported and although ^{1}H n.m.r. confirms the existence of both *cis*- and *trans*-isomers they have not been separated for M = Pd.[60] Spectroscopic studies suggest that substitution of Me by OEt in thioacetothioacetate complexes concentrates electronic charge in the C—C—C backbone.

Interest in potential anti-cancer activity has led to the preparation of a Pd–*S*-methyl-L-cysteine complex which analysed as [Pd$_3$(SMCH)$_2$Cl$_6$].[61] Its i.r. and visible spectra and conductivity are consistent with a chloro-bridged cation (19). On the other hand, the ethyl cysteinate complex of Pd, prepared according to equation

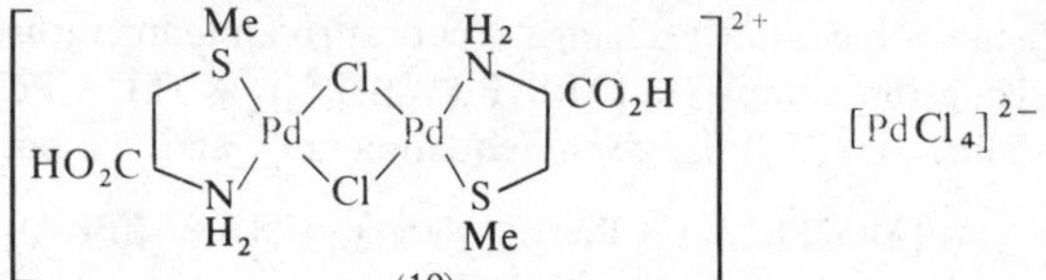

(19)

(14) [L – H = HSCH$_2$CH(NH$_2$)CO$_2$Et], is a non-electrolyte in H$_2$O; structure

$$4L,HCl + PdCl_2 \xrightarrow{H_2O} Pd(L - H)_2 \qquad (14)$$

(20) was proposed.[62]

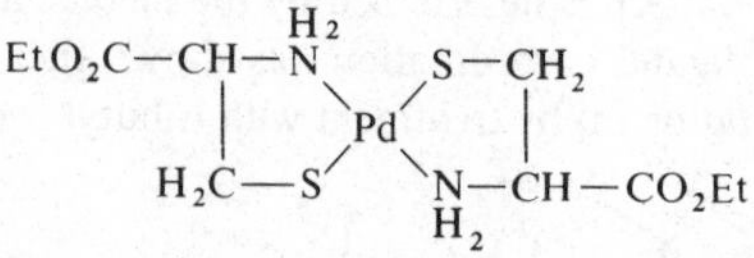

(20)

Both mono- and bis-chelated Pd complexes of *o*-methylthiophenyldiphosphine (P—SMe) and bis(*o*-methylthiophenyl)phenylphosphine (MeS—P—SMe) and their arsenic analogues have been synthesized.[63] The mono-chelated complexes of both ligands undergo S-demethylation in hot DMF to give red or red-brown products which are probably dimeric with thiol-bridged structures. In the presence of excess free ligand the dealkylation affords bis-demethylated products, [Pd(X—S)$_2$] and [Pd(MeS—X—S)$_2$] (X = P or As), which react with a variety of alkylating reagents RCH$_2$Y to give a complex of the original or of a different thioether ligand of the form [Pd(X—SR)Y$_2$] or [Pd(MeS—X—SR)Y$_2$] (R = Et, Bu or *p*-NO$_2$C$_6$H$_4$CH$_2$; Y = halogen). ^{1}H N.m.r. results suggest that one thiomethyl residue is not co-ordinated in all complexes of MeS—X—SMe, and that these complexes are four-co-ordinate square planar. Other square-planar complexes of Pd with S co-ordination have been prepared from α-mercaptopropionic acid and aqueous PdCl$_2$, and i.r. spectra show that the ligand is chelated through the carboxylate residue.[64]

[60] A. R. Hendrickson and R. L. Martin, *Inorg. Chem.*, 1973, **12**, 2583.
[61] C. A. McAuliffe, *Inorg. Chem.*, 1973, **12**, 1699.
[62] M. Chandrasekharan, M. R. Udupa and G. Aravamudan, *J. Inorg. Nuclear Chem.*, 1974, **36**, 1417.
[63] T. N. Lockyer, *Austral. J. Chem.*, 1974, **27**, 259.
[64] S. K. Srivastara, K. B. Pandeya, and H. L. Nigam, *Indian J. Chem.*, 1974, **12**, 530.

Binding energies have been measured for a series of dithiene complexes (Et_4N)-$[MS_4C_4R_4]$ (R = CN or Ph), $(Et_4N)_2[MS_4C_4(CN)_4]$, $[MS_4C_4Ph_4]$, and $(N_2H_5)_2$-$[MS_4C_4Ph_4]$ (M = Pd, Pt or Ni) (see p. 408).

The first report of gas chromatography of metal chelates involving ligands containing only S donors has appeared.[65] The chelates $[M\{(OR)_2PS_2\}_2]$ (M = Pd or Pt) were found to be thermally stable up to 170 °C and elution and separation on a gas chromatogram was possible.

Se- and Te-donor ligands. Variable-temperature n.m.r. studies for the complexes $[M(Et_2X)_2hal_2]$ (M = Pd or Pt; X = Se or Te; hal = Cl, Br or I) indicate that the Et_2X ligands do not dissociate but that rapid inversion occurs at X.[66] The barrier to inversion at X (S, Se, Te) is always higher for Pt than for Pd, and less for S than Se and Te. Series of selenocarbamate complexes of Pd and Pt of general formula $[M(L_2)(PPh_3)_2][M = Pd$ or Pt; L = SeC(O)NR_2; R = Me, Et, Pr^n or Bu^n$)$ and $[M\{SeC(O)NMe_2\}_2(PMePh_2)_2]$ (M = Pd or Pt) have been synthesized.[67] I.r. and molecular weight determinations indicate that in the solid state the selenocarbamate ligand is bonded through the selenium with a free CO group. In dilute solutions PPh_3 dissociates to yield the bidentate selenocarbamates (21).

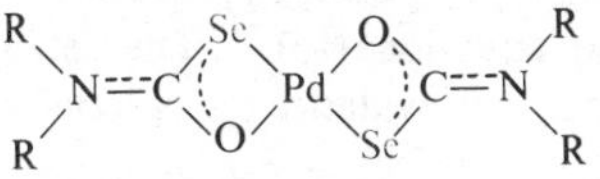

(21)

Group V. N-Donor ligands. Photoelectron spectra of $N N'$-ethylenebis(acetylacetoneiminato) complexes of Pd (Ni, Cu) have been measured and assignments of the valence bands have been made in conjunction with MO calculations.[68] Mass spectral studies of mixtures of homogeneous dioxime complexes have revealed for the first time evidence of the presence of mixed complexes.[69] Interest has been generated over possible high-temperature superconducting properties of complexes of type $[PdI(dpg)_2]$ (dpg = diphenylglyoximato) and the d.c. conductivity of this compound is a factor of 10^5 higher than that of $[Pd(dpg)_2]$.[70] This high conductivity was believed to be due either to chains of metal atoms in a non-integral oxidation state or to chains of halogens. Co-ordinated dimethylglyoximate with Pd (Ni, Cu) has been shown to undergo substitution of the H-bonded hydrogen with phenylaluminium.[71] Chemical analysis, magnetic measurements, and i.r. spectra suggest structure (22).

A very convenient general synthesis for an extensive range of metal triazenido-complexes has been described (see p. 412).

Complexes of the potentially bidentate ligands 3-(4-pyridyl)triazoline-5-thione, with Pd (Pt, Rh) have been prepared and were shown to be bonded through both N and S.[72] The thermal stability of the compounds $[PdX_2L_2]$ (X = Cl, I or NO_3;

[65] T. J. Cardwell and P. S. McDonough, *Inorg. Nuclear Chem. Letters*, 1974, **10**, 283.
[66] R. J. Cross, T. H. Green, and R. Keat, *J.C.S. Chem. Comm.*, 1974, 207.
[67] K. Tanaka and T. Tanaka, *Bull. Chem. Soc. Japan*, 1974, **47**, 847.
[68] G. Condorelli, I. Fragela, G. Centineo, and E. Tondello, *Inorg. Chim. Acta*, 1973, **7**, 725.
[69] A. V. Ablov, Z. Y. Vaisbein, and K. S. Khariton, *Russ. J. Inorg. Chem.*, 1973, **18**, 909.
[70] A. E. Underhill, D. M. Watkins, and R. Pethig, *Inorg. Nuclear Chem. Letters*, 1973, **9**, 1269.
[71] N. Voiculescu, I. Popescu, and L. Roman, *Rev. Roum. Chem.*, 1973, **18**, 1595.
[72] B. Singh, R. Singh, K. P. Thakur, and M. Chandra, *Indian J. Chem.*, 1974, **12**, 631.

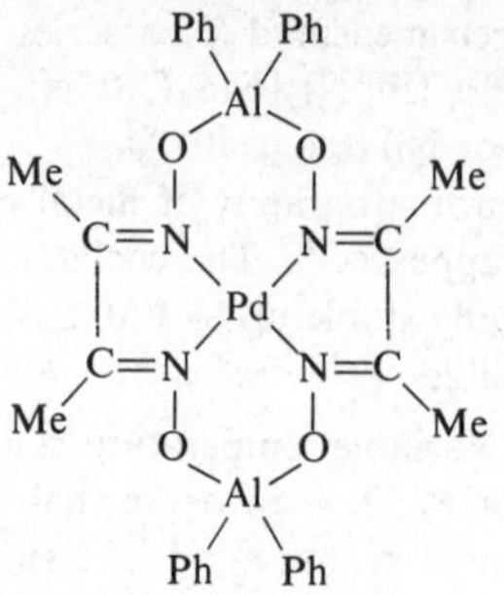

(22)

L = benzotriazole, benzoxazole, or benzothiazole) has also been studied.[73] Cyan-amide complexes $(AsPh_4)_2 [Pd(NCNR)_4]$ (R = H or Ph) have been prepared and their i.r. and ^{14}N n.m.r. reported.[74]

Carbodi-imide ($RN\!=\!C\!=\!NR$) may potentially co-ordinate *via* the π-bond or the N lone pair. Palladium complexes of this ligand have been obtained as diamagnetic, monomeric complexes of type $[PdX_2L_2]$ [X = Cl or Br: L = $(Bu^t N)C$ or $MeNCNBu^t$] according to equation (15).[75] The retention of the antisymmetric $N\!=\!C\!=\!N$ stretch at 2130 cm^{-1} is evidence for N lone-pair co-ordination. Although

$$R^1N\!=\!C\!=\!NR^2 + [PdX_2(PhCN)_2] \xrightarrow{CH_2Cl_2} [PdX_2(R^1N\!=\!C\!=\!NR^2)_2] \quad (15)$$

carbodi-imide is normally unreactive to alcohols, if reaction (15) is carried out in MeOH, 1,2-addition of MeOH occurs to yield the bis(isourea) complexes $[PdX_2\{R^1NHC(OMe)\!=\!NR^2\}_2]$. Urea derivatives of formula $[Pd(PPh_3)_2\{RN(CO)NR\}]$ have also been synthesized by reaction of toluene-*p*-sulphonyl azide with $[Pd(PPh_3)_3(CO)]$ in benzene or by treatment of $[M(PPh_3)_4]$ (M = Pd or Pt) with toluene-*p*-sulphonyl isocyanate.[75a]

Treatment of *trans*-$[PdCl_2(PhCN)_2]$ with biacetyl-bis(*N*-methyl, *N*-phenyl)osazone (L) in CH_2Cl_2 yields the square-planar osazone complex $[PdCl_2L]$, which under suitable conditions loses HCl to yield the internally metallated complex (23), the structure of which has previously been determined.[76, 77] Studies of the condensation

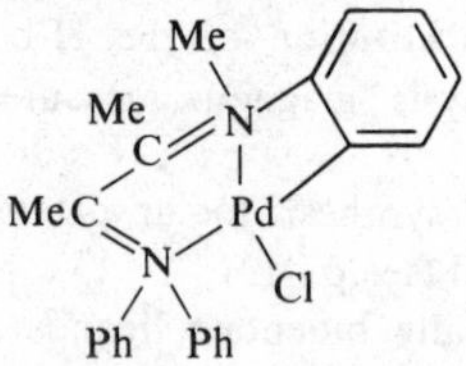

(23)

[73] J. E. House and P. S. Lau, *J. Inorg. Nuclear Chem.*, 1974, **36**, 223.
[74] W. Beck, H. Bock, and R. Schlodder, *Z. Naturforsch*, 1974, **29**, 75.
[75] B. M. Bycroft and J. D. Cotton, *J.C.S. Dalton*, 1973, 1867.
[75a] W. Beck, W. Rieber, S. Cenini, F. Porta, and G. La Monica, *J.C.S. Dalton*, 1974, 298.
[76] L. Caglioti, L. Cattalini, F. Gasparrini, M. Chedini, G. Paolucci, and P. A. Vigato, *Inorg. Chim. Acta*, 1973, **7**, 538.
[77] G. Bombieri, L. Caglioti, L. Cattalini, E. Forsellini, F. Gasparrini, R. Graziani, and P. A. Vigato, *J.C.S. Chem. Comm.*, 1971, 1415.

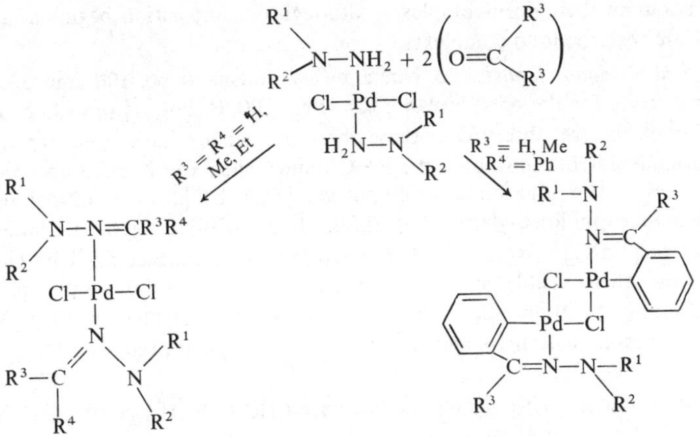

Scheme 1

between the NH_2 group of L, where L = NN-substituted hydrazine co-ordinated to Pd in the complex $[PdCl_2L_2]$, and a carbonyl function R^1R^2CO have shown that two types of reaction can occur which lead to either a hydrazone complex or to a cyclopalladated complex with a direct Pd—C σ-bond (Scheme 1).[78]

Three new molecular nitrogen complexes of Pd have been prepared from hydrazine sulphate.[79] A yellow precipitate, formulated as the hydrazine-bridged dimer

$$[(OH)N_2Pd \underset{N_2H_4}{\overset{N_2H_4}{\diagup\diagdown}} Pd(N_2)OH]SO_4 \text{ was obtained from } K_2Pd(NO_2)_4, \text{ which}$$

decomposed in aqueous NaOH to yield nitrogen and palladium metal. Reaction occurred with *trans*-$[Pd(NO_2)_2(NH_3)_2]$ to yield the green $[Pd(NO_2)(NH_3)(H_2O)$-

$$(H_2O)(N_2)] \text{ complex and the brown } [(H_2O)(N_2)Pd \underset{N_2H_4}{\overset{OH}{\diagup\diagdown}} Pd(NO_2)(NH_3)]SO_4.$$

All three species showed an i.r. band at 2100 cm^{-1} due to co-ordinated N_2.

The reactions of the complexes $[MX_2L]$ (X = Cl, Br, or I) and $[ML_2X_2]$ (M = Pd or Pt; X = Cl or I; L = $H_2NCH_2CH_2NHCH_2CH_2OH$) in liquid NH_3 at $-75\,°C$ have been studied in order to examine the inductive effect of the amino-group on the reactivity of the ethanolic oxygen and the lability of the hydrogen.[80] Various types of behaviour were observed ranging from no reaction to deprotonation and/or ligand substitution. The thermal decomposition of *trans*-$[PdI_2L_2]$ (L = py, 2-Mepy, 3-Mepy or 4-Mepy; 2,4- 3,4- 3,5- and 2,6-dimethylpy) has been studied using a differential scanning calorimeter.[81] Loss of the two substituted pyridines occurs analogous to

[78] P. Braunstein, J. Dehand, and M. Pfeffer, *Inorg. Nuclear Chem. Letters*, 1974, **10**, 521.

[79] A. I. Korosteleva, V. M. Volkov, and S. S. Chernikov, *Russ. J. Inorg. Chem.*, 1973, **18**, 1139.

[80] G. W. Watt and J. S. Thompson, *J. Inorg. Nuclear Chem.*, 1974, **36**, 1075.

[81] P. S. Lau and J. E. House, *J. Inorg. Nuclear Chem.*, 1974, **36**, 207.

the behaviour of the chloro-complexes, although decomposition begins at a lower temperature with the iodo complexes.

P-. As-, *and* Sb-*donor ligands*. [1]H N.m.r. measurements at 60, 100, and 220 MHz have been made for the complexes $[PdCl_2\{Ph_{3-n}P(CH_2Ph)_n\}_2]$ (n = 1, 2, or 3).[82] The different spectra observed at these three frequencies have been explained in terms of hindered rotation about the P—C bonds when n = 2 and 3 and probably also for n = 1. The square-planar complexes $[PdX_2L]$ [L = *cis*-1,2-difluoro-1,2-bis(diphenylphosphino)ethylene; X = Cl, Br, I, or SCN] have been obtained by treatment of $[PdX_4]^{2-}$ with L.[83] The complexes were characterized by chemical analysis and their visible spectra recorded. In addition, difluorophosphonato-complexes $[M(PF_2O)_4]^{2-}$ (M = Pd or Pt) have been isolated as their *N*-allyl-pyridinium salts (equation 16) and their n.m.r. and i.r. spectra reported.[84]

$$[PtCl_2py_2] + 4CH_2=CHCH_2OPF_2 \xrightarrow{C_6H_6} (CH_2=CHCH_2NC_5H_5)_2[Pt(PF_2O)_4] \quad (16)$$

Hydrolysis of the phosphine ligand in acetylenic phosphine complexes $[MCl_2-(Ph_2PC\equiv CCF_3)_2]$ (M = Pd or Pt) leads to cleavage of the alkynyl groups with formation of chloro-bridged dimers $[MCl(Ph_2PO)_2H]_2$. X-Ray determination of the structure of the CNS complex $[M(CNS)(Ph_2PO)_2H]_2$ (15), prepared by meta-thetical reaction from the chloro-dimer, has shown the presence of an unusual symmetrically H-bonded anion $[(Ph_2PO)_2H]^-$.[42] Complexes of type $[M(CNS)_2L_2]$ (M = Pd or Pt; L = Ph_3AsS, Ph_3PSe or Ph_3PS) have been prepared from $[M(SCN)_4]^{2-}$ and L (see p. 406).

Reported for the first time is direct evidence, ^{1}H and [31]P{[1]H} Fourier mode n.m.r., for the rates of ligand association in a series of $[MHL_3]^+$ species (M = Ni, Pd, or Pt; L = PEt_3) relative to the rate of intramolecular rearrangement within the five-co-ordinate transition state $[MHL_4]^+$.[85] Changes in the lineshapes suggest an associative mechanism for the exchange process and also that intramolecular rearrangement of $[MHL_4]^+$ is faster than loss of L by dissociation. N.m.r. results also suggest that exchange of phosphite ligand in the trigonal-bipyramidal complexes $[M\{P(OMe)_3\}_5](BPh_4)_2$ (M = Pd or Pt) is of the Berry type.[86] Moreover, a comparison with results for similar complexes of Ni and Rh showed that little change in the barrier height occurred for changes of metals from first row through to third row. The synthesis of new water-soluble tertiary phosphine complexes *cis*- and *trans*-$[MX_2(PR_{3-n}Q_n)_2]$ (M = Pd or Pt; R = Me, Et, Bu^t, or Ph; Q = CH_2OCOMe or CH_2OH; n = 0, 1 or 2) by treatment of the appropriate phosphine with $[MCl_4]^{2-}$ in aqueous ethanol has been reported.[87] The methylene protons of the acetoxymethyl and hydroxymethyl groups show no [31]P–[1]H coupling, which is attributed to $|^2J(PH) + ^4J(PH)|$ being accidentally nearly zero. [1]H N.m.r. spectra of $[MX(PMe_3)_3]NO_3$ (M = Pd or Pt) have been obtained and discussed together with [1]H{[31]P} INDOR

[82] J. H. Nelson and D. A. Redfield, *Inorg. Nuclear Chem. Letters*, 1973, **9**, 807.

[83] K. K. Chow and C. A. McAuliffe, *Inorg. Nuclear Chem. Letters*, 1973, **9**, 1189.

[84] J. Grosse and R. Schmutzler, *Z. Naturforsch*, 1973, **28b**, 515.

[85] P. Meakin, R. A. Schunn, and J. P. Jesson, *J. Amer. Chem. Soc.*, 1974, **96**, 277.

[86] J. P. Jesson and P. Meakin, *Inorg. Nuclear Chem. Letters*, 1973, **9**, 1221.

[87] J. Chatt, G. J. Leigh, and R. M. Slade, *J.C.S. Dalton*, 1973, 2021.

measurements.[88] In addition, the signs of some of the couplings have been determined by heteronuclear double resonance.

Interest in the internal metallation ability of the bulky dimethyl-(1-naphthyl)phosphine has stimulated an extension of previous work[89] with the preparation of the complexes $[MX_2\{EMe_2(1\text{-naphthyl})\}_2]$ (M = Pd or Pt; X = Cl, I or Me; E = P or As).[90] Although under certain conditions the Pt complexes could be metallated in the *peri*-position, the analogous Pd complexes decomposed without internal metallation. Square-planar complexes of the unusual ligand 2,11-bis(diphenylphosphinomethyl)-benzo[*c*]phenanthrene) [(44), see p. 414], which is capable of spanning *trans*-positions, have been obtained by reaction with suitable compounds of Pd (Pt, Ni).[91] These complexes are monomeric and non-electrolytes in CH_2Cl_2 and the *trans*-configuration of P atoms was established by [1]H n.m.r. Structural assignment of Pd complexes with methylenephosphine ligands by use of the pattern of the benzyl methylene protons in [1]H n.m.r. has been briefly reviewed and shown to be useful.[92]

Four similar papers reporting new cationic and neutral complexes of Pt group metals with phosphite, phosphonite, and phosphinite ligands have appeared.[93—96] The general preparative route involved addition of free ligand L = $P(OR)Ph_2$, $P(OR)_2Ph$, or $P(OR)_3$ (R = Me or Et) to a solution of a labile transition-metal precursor followed by addition of $NaBPh_4$. Complexes of type $[PdL_5]^{2+}$, $[PdL_4]^{2+}$, $[PdClL_3]^+$, $cis[PdCl_2L_2]$ were characterized by chemical analysis, [1]H n.m.r., and conductivity measurements.

A preliminary examination of the complexing ability of the Sb analogue of the much studied *o*-phenylenebis(dimethylarsine) ligand shows that pale yellow-orange crystalline $[MX_2L]$ (M = Pd or Pt; X = Cl, Br, I or SCN) are formed which are con-electrolytes in $PhNO_2$.[97] Mono- and di-nuclear complexes of type $[PdX_2L_2]$ and $[Pd_2X_4L_2]$ [X = Cl, Br, I, NO_2, CN or SCN; L = Sb(*o*-tolyl)$_3$, Sb(*p*-tolyl)$_3$, or Sb(*m*-tolyl)$_3$] have been obtained as diamagnetic lightly coloured compounds by treatment of the appropriate Pd salt with L.[98]

Group IV Donors. A safe synthesis of a number of metal fulminato-complexes, including $[Pd(R_3E)_2(CNO)_2]$ (E = P, As, or Sb), by reaction of $AsPh_4CNO$ with the appropriate Pd complex has been reported.[99] The complex $[Pd(CH_2CO_2)(PPh_3)_2]$ has produced the yellow C-bonded complexes of acetic acid $[Pd(CH_2CO_2H)L(PPh_3)]$ by treatment with chelating ligands L (acac, oxine).[29] I.r. spectra of the solids show $v(OH)$ vibration in the range 2500–2700 cm^{-1}, thus indicating molecular association. In solution the complexes appear to be monomeric. In addition $[Pd(acac)_2]$ reacts with the nitrogen bases (L = py, Et_2NH, or PhMeNH) to produce complexes (11)

[88] P. L. Goggin, R. J. Goodfellow, J. R. Knight, M. G. Norton, and B. F. Taylor, *J.C.S. Dalton*, 1973, 2220.
[89] J. M. Duff and B. L. Shaw, *J.C.S. Dalton*, 1972, 2219.
[90] J. M. Duff, B. E. Mann, B. L. Shaw, and B. L. Turtle, *J.C.S. Dalton*, 1974, 139.
[91] N. J. DeStefano, D. K. Johnson, and L. M. Venanzi, *Angew. Chem. Internat. Edn.*, 1974, 13, 133.
[92] B. E. Mann, *Inorg. Nuclear Chem. Letters*, 1974, 10, 273.
[93] D. A. Couch and S. D. Robinson, *Inorg. Nuclear Chem. Letters*, 1973, 9, 1079.
[94] D. A. Couch and S. D. Robinson, *Inorg. Chem.*, 1974, 13, 456.
[95] D. A. Couch, S. D. Robinson, and J. N. Wingfield, *J.C.S. Dalton*, 1974, 1309.
[96] D. A. Couch and S. D. Robinson, *Inorg. Chim. Acta*, 1974, 9. 39.
[97] E. Shewchuk and S. B. Wild, *J. Organometallic Chem.*, 1974, 71, Cl.
[98] D. Negoiu and L. Paruta, *Rev. Roumaine Chim.*, 1973, 18, 2059.
[99] W. Beck, K. Schorpp, and C. Oetker, *Chem. Ber.*, 1974, 107, 1380.

where one of the acac is C-bonded and unidentate.[30] Bifunctional amines YNH_2 (*e.g.* *o*-phenylenediamine, ethanolamine, 2-aminopyridine, allylamine) have been shown to react with complexes of type *cis*-[Pd(RNC)Cl$_2$L] (R = Ph, *p*-MeC$_6$H$_4$, *p*-MeOC$_6$H$_4$, or *p*-O$_2$NC$_6$H$_4$; L = PhNC, *p*-MeC$_6$H$_4$NC, or PPh$_3$) to yield neutral carbene derivatives *cis*-[Pd{C(NHR)NHY}Cl$_2$L] in which the functional group Y is not co-ordinated to the metal.[100] Co-ordination of Y can be achieved by treatment with NaClO$_4$ and/or AgBF$_4$.

Dicyanomethanide complexes of Pd (Pt, Ir, Rh) have been prepared, *e.g.* reaction (17).[101] Observation of ν(CN) at 2200 cm^{-1} indicates that the Pd and Pt complexes are C-bonded (24), unlike the Rh and Ir analogues (25) which show a strong M—N

$$[PdCl_2(NCPh)_2] + NaCH(CN)_2 \xrightarrow[Ph_4AsCl]{MeOH} (Ph_4As)_2[Pd(C_3HN_2)_4] \qquad (17)$$

band between 2120 and 2150 cm^{-1}. In addition, tricyanomethanide complexes of Pd and Pt have been shown to be produced from K[C(CN)$_3$] and the appropriate

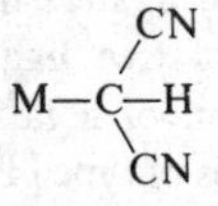

(24) 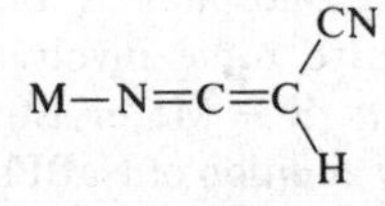   (25)

halogeno-metal complex.[102] The co-ordination of the tricyanomethanide ligand in the complexes [MXC(CN)$_3$L$_2$] (M = Pd, or Pt; X = Cl, Br, or C(CN)$_3$; L = PPh$_3$, or AsPh$_3$) was discussed using i.r., ^{14}N n.m.r., and Raman data. Treatment of salts of Pd (Pt, Rh, Ir) with 1-benzoyliminopyridinium (L) gives the complexes [PdCl-(L − H)L].[103] ^{1}H N.m.r. and i.r. measurements have indicated that L is metallated in the *ortho*-position and acts as a bidentate ligand through this carbon and the imino-N atom.

N-(Diethylboryl)- and *N*-(trimethylsilyl)-ketimine afford the yellow *N*-boryl- and *N*-silyl-ketimine complexes [PdCl$_2${Ph$_2$CNBEt$_2$}$_2$] and [PdCl$_2${Ph$_2$CNSiMe$_3$}$_2$] when treated with [PdCl$_2$(PhCN)$_2$].[104] ^{11}B N.m.r. and i.r. spectroscopy show that the ketimine molecules are co-ordinated to the Pd by the C≡N group in both complexes. Formation of sulphur ylide complexes of Pd (Pt) of type [PdX$_2$(Sy)$_2$] [X = Cl, Br, or I; Sy = Me(Ph)SCHC(O)C$_6$H$_4$Cl-*p*] has been reported (see p. 416).[105]

Studies of the electrode reactions of aqueous [Pd(CN)$_4$]$^{2-}$ by d.c. polarography have shown the presence of an irreversible two-electron step, while the oscillopolarogram showed three anodic waves.[106] It was concluded that cyano-complexes of Pd^0 and Pd^I were formed, the former of which decomposed rapidly into Pd metal (amalgam) and free cyanide.

[100] R. Zauella, T. Boschi, B. Crociani, and U. Belluco, *J. Organomentallic Chem.*, 1974, **71**, 135.
[101] W. H. Baddley and P. Choudhurg, *J. Organometallic Chem.*, 1973, **60**, C74.
[102] W. Beck, K. Schorpp, C. Oetker, R. Schlodder, and H. S. Smedal, *Chem. Ber.*, 1973, **106**, 2144.
[103] S. A. Dias, A. W. Downs, and W. R. McWhinnie, *Inorg. Nuclear Chem. Letters*, 1974, **10**, 233.
[104] G. Schmid and L. Weber, *Chem. Ber.*, 1974, **107**, 547.
[105] H. Koeguka, G. E. Matsubayashi, and T. Tanake, *Inorg. Chem.*, 1974, **13**, 443.
[106] M. Hirota and S. Fujiwara, *J. Inorg. Nuclear Chem.*, 1973, **35**, 3883.

Table 1 X-*Ray data for palladium compounds*

Compound	R	Comments	Ref.
$[PdCl_2(PPh_3)_2NEt]$	0.109	Ligand bonded through two P	a
$[Pd_2(SCMe_3)_2(S_2CSMe_3)_2]$	0.078	distorted PdS_4 planar trigonally bound bridging S with non-planar Pd_2S_2 rhombus; Pd—Pd = 3.162 Å	b
$[Pd_3(SEt)_3(S_2CSEt)_3]$	0.058	distorted PdS_4 planar with trigonally bound bridging S. Substantial distortion from trigonal 'chair' form of Pd_3S_3 ring. Three Pd—Pd distances of 3.655, 3.307, and 3.303 Å	b
$[Pd(HL)_2](ClO_4)_4.2H_2O$ L = *cis*-3,5-diaminopiperidine	0.046	Pd co-ordinated to the four primary amino-N in square plane; Pd—N 2.07 Å. Two ring N at larger distance of 3.358 Å. Pd—N (ring) direction is 40° to normal to square plane	c
$[PdL_2](ClO_4)_2$ L = *cis*-3,5-diaminopiperidine	0.052	square-planar co-ordination by four primary amino-N atoms. Ring N—Pd at 3.084 and 3.167 Å	c
$[Pd(L)(HL)](NO_3)_3$ L = *cis*-3,5-diaminopiperidine	0.10	square planar as above. Pd—ring N direction at 29° to normal to square plane	c
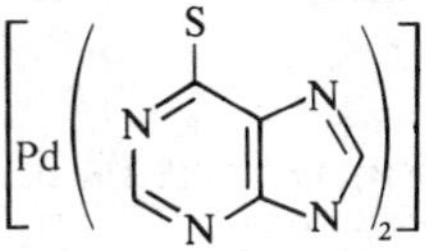 $[Pd(\text{purine})_2](MeCONMe_2)$	0.069	both purine ligands are co-ordinated *via* S and N-7 to form a five-membered chelate ring with slightly distorted square around Pd	d
$[Pd(NH_3)_4]C_4N_2H_2(CO_2)_2$	0.038	multiple H-bonds are formed between the square-planar $Pd(NH_3)_4^{2+}$ cations and the anions	e
$[PdCl_2(4,7,13,16\text{-tetraoxa-1,10-dithiacyclo-octadecane})]$	0.036	*cis* square-planar co-ordination of Cl and S atoms	f
$[PdCl_2(1,10\text{-diaza-4,7,13,16-tetraoxa-21,24-dithiabicyclo-8,8,8-hexacosane})]$	0.040	Pd is located outside the cavity of the macrobicycle with *cis*-square-planar co-ordination of Cl and S	g
$[Pd(acac)_2(PPh_3)]$	0.078	square-planar Pd with co-ordination *via* P, two O (acac bidentate), and γ-C (acac unidentate)	h
$[Pd_2Cl_2\{P(OMe)_3\}_2(CSNMe)_2]$	0.059	Pd atoms linked together through two bridging thiocarbamoyl groups	i
$[Pd_2(PhHC{=}CHCOCH{=}CHPh)_3 (CHCl_3)]$	0.067	Pd linked *via* three dibenzylidene-acetone groups. Co-ordination around Pd is trigonal through olefinic portion of ligand	10
$[Pd(PhHC{=}CHCOCH{=}CHPh)_3]$	0.084	Pd is trigonal with each pentadienone co-ordinating through one olefin group.	j
$[Pd(PBu^t_3)_2]$	0.062	almost linear co-ordination about Pd(PPdP = 176.8°). Molecule has approximately C_2 symmetry	5
$[PdCl(COCO_2Me)(PPh_3)_2]$	0.091	Methoxylalyl ligand has *s-trans* planar conformation; phosphine groups *trans*, Pd—C = 1.97(2) Å	7

Table 1—*contd.*

Compound	R	Comments	Ref.
K_2PdO_2	0.130	chain structure analogous to K_2PtS_2. Planar oxygen co-ordination around Pd	34
$[Pd(SCN)_2(PPh_2CH_2PPh_2)]$	0.049	see text. Steric interaction between phenyl groups and SCN	39
$[Pd(NCS)_2(PPh_2C_3H_6PPh_2)]$	0.037	leads to N-bonded thiocyanate	39
$[Pd_2(SCN)_2\{(Ph_2PO)_2H\}_2]$	0.053	see text. Square-planar co-ordination using N, S, and 2P. Bridging NCS groups. Phosphinato groups linked *via* a symmetrical O---H---O bond	42
$K_2[Pd(CN)_6]$	0.047	reinvestigation of structure octahedral co-ordination	k
$K_2[PdCl_6]$	—	powder. Cell constant 9.7066 Å	l
$(NH_4)_2[PdCl_6]$	—	cell constant 9.8222 Å	l
$(XeF_5)_2[PdF_6]$	0.0256	octahedral Pd, with two crystallographic distinct XeF_5^+ ions	m
$[Pd(AlCl_4)_2]$		powder	24
$(Et_4N)_2[Hg_3PdCl_{10}]$	0.130	contains trinuclear anionic unit Hg_2PdCl_3 with bridged chlorines	23
$[MX_3(A)]$ $[M = Pd$ or $Pt; X = Cl, Br,$ or $I;$ $A = (NH_3)_2 (EtNH_2)_4$ or en]		powder at 60 kbar pressure	16

(a) J. A. A. Mokudu, D. S. Payne, and J. C. Speakman, *J.C.S. Dalton*, 1973, 1443; (b) J. P. Fackler and W. J. Zegarski, *J. Amer. Chem. Soc.*, 1973, **95**, 8566; (c) H. Manchar and D. Schwarzenbach, *Helv. Chim. Acta*, 1974, **57**, 519; (d) H. I. Heitner and S. J. Lippard, *Inorg. Chem.*, 1974, **13**, 815; (e) R. L. Harlow and S. H. Simonsen, *Acta Cryst.*, 1974, **B30**, 1370; (f) B. Metz, D. Moras, and R. Weiss, *J. Inorg. Nuclear Chem.*, 1974, **36**, 785; (g) R. Louis, J. C. Thierry, and R. Weiss, *Acta Cryst.*, 1974, **B30**, 753; (h) M. Horike, Y. Kai, N. Yasuoka, and N. Kasai, *J. Organometallic Chem.*, 1974, **72**, 441; (i) S. K. Porter, H. White, C. R. Green, R. J. Angelici, and J. Clardy, *J.C.S. Chem. Comm.*, 1973, 493; (j) M. C. Mazza and C. G. Pierport, *Inorg. Chem.*, 1973, **12**, 2955; (k) J. Weiss, *Z. Naturforsch.*, 1974, **29**, 119; (l) J. M. Adams, *Acta Cryst.*, 1974, **B30**, 555; (m) K. Leary, D. H. Templeton, A. Zalkin, and N. Bartlett, *Inorg. Chem.*, 1973, **12**, 1726.

Table 2 *I.r. structural studies of palladium complexes*

Complex	Other studies	Results and conclusions	Ref.
$K_2[PdCl_6]$	Pt, Os, Ir	far i.r. at pressures up to 20 kbar. $v_4,\delta(ClMCl)$ mode generally more pressure sensitive than the $v_3, v(MCl)$	a
$[Pd(A_2NC_2H_4S)_2]$ (A = H or D)	Ni, Pt Raman	identified $v(Pt—N)$ and $v(Pd—S)$ bands. Complexes have *trans* configuration. Isomeric compound obtained from alkaline solution has a *cis*-structure	b
$[PdX_2(SbR_3)_2]$ (R = *o*-tolyl, *p*-tolyl, or *m*-tolyl; X = Cl, Br, I. or NO_2)	u.v.–vis	assigned bands in i.r. and u.v.–vis	c
$[PdX_2L_2]$ *cis* and *trans* (X = Cl, Br, or I; L = Ph_2MeP, Ph_2EtP, $PhMe_2P$, $PhEt_2P$, Pr^n_3P, Pr^i_3P)	Raman	assigned metal–ligand stretching frequencies	d

Table 2—*contd.*

Complex	Other studies	Results and conclusions	Ref.
[Pd{(PhO)₂PSSe}₂]	Rh, Ir, Pt	identified ν(M—S) and ν(M—Se)	e
[PdX₂L₂]	—	S-bonding capabilities are	
(X = SCN or NCS; L = PPh₃,		enhanced by electron withdrawing	f
AsPh₃, thiourea, bipy, or phen)		substituents on L	
[Pd(MeCOCHCSMe)₂]	Ni, Zn, Cd	Normal-co-ordinate analysis	g
		using C_{2v} symmetry. Bands at	
		224 and 173 cm⁻¹ assigned to	
		Pd—S stretching	
M₂[Pd(NO₂)₄]	Pt	In solution the complexes have	h
		D_{4h} symmetry. Spectra indicate	
(M = Na, K, Rb, Cs, or Ag)		that metal-N π-bonding is more	
		pronounced with Pt	

(a) D. M. Adams and S. J. Payne, *J.C.S. Dalton*, 1974, 407; (b) V. A. Jayasooriya and D. B. Powell, *Spectrochim. Acta*, 1974, **30A**, 553; (c) D. Negoiu and L. Paruta, *Rev. Roumaine Chim.*, 1974, **19**, 57; (d) E. A. Allen and W. Wilkinson, *Spectrochim. Acta*, 1974, **30A**, 1219; (e) I. M. Cheremisina, L. A. Il'ina, and S. V. Larionov, *Russ. J. Inorg. Chem.*, 1973, **18**, 675; (f) A. Miezis, *Acta Chem. Scand.*, 1973, **27**, 3746; (g) O. Suman, D. D. Titus, C. D. Cowman, J. Fresco, and H. B. Gray, *J. Amer. Chem. Soc.*, 1974, **96**, 2353; (h) D. W. James and M. J. Nolan, *Austral. J. Chem.*, 1973, **26**, 1433.

2 Platinum

Cluster Compounds.—An interesting new series of platinum carbonyl dianions of formula $[Pt_3(CO)_3(\mu_2\text{-}CO)_3]_2^{2-}$ ($n = 2$—5) have been synthesized by reduction of $[PtCl_6]^{2-}$ with CO and methanolic NaOH.[107] Species with decreasing n were isolated with increasing concentration of reducing agent (equation 18). The X-ray

$$[PtCl_6]^{2-} \rightarrow [PtCOCl_3]^- \rightarrow [Pt_{15}(CO)_{30}]^{2-} \rightarrow [Pt_{12}(CO)_{24}]^{2-} \rightarrow [Pt_9(CO)_{18}]^{2-}$$
$$\rightarrow [Pt_6(CO)_{12}]^{2-} \quad (18)$$

structures (Table 3, p. 420) for $n = 2$, 3, and 5 show that all the anions are based on the polymerization of a common $Pt_3(CO)_6$ component by further Pt—Pt bonding (26).

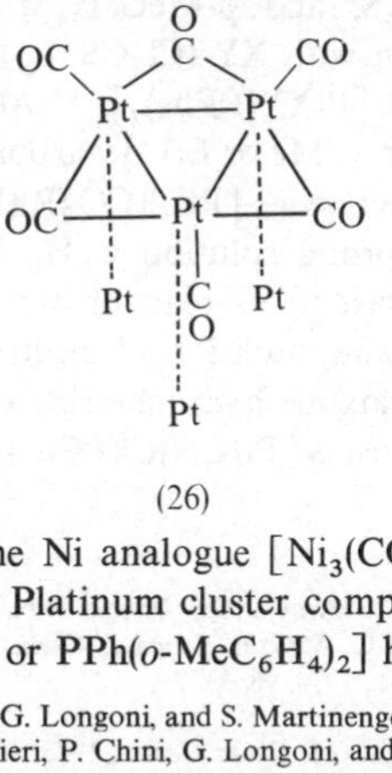

(26)

It is interesting to note that the Ni analogue $[Ni_3(CO)_3(\mu_2\text{-}CO)_3]_2$ shows a quite different metal architecture.[108] Platinum cluster compounds of type $[Pt_4(CO)_5L_4]$ $[L = PPh_3, PPh_2(o\text{-}MeC_6H_4)$, or $PPh(o\text{-}MeC_6H_4)_2]$ have been obtained as purple

[107] J. C. Calabrese, L. F. Dahl, P. Chini, G. Longoni, and S. Martinengo, *J. Amer. Chem. Soc.*, 1974, **96**, 2614.
[108] J. C. Calabrese, L. F. Dahl, A. Cavalieri, P. Chini, G. Longoni, and S. Martinengo, *J. Amer. Chem. Soc.*, 1974, **96**, 2618.

o

solids by reaction of $Li_2[PtCl_4]$ with CO in DMF–EtOH followed by addition of L and hydrazine hydrate.[109] The compounds were characterized by chemical and i.r. spectral analyses.

Platinum(0) and Platinum(I).—Co-condensation of the metals Ni, Pd, and Pt with gaseous O_2 or O_2–Ar mixture at 4.2—10 K gives binary dioxygen complexes.[1] The two types of complex, $M(O_2)$ and $(O_2)M(O_2)$, both possess side-on co-ordinated dioxygen. The bis-oxygen complex has a unique D_{2d} spiro-type structure. Matrix i.r. and Raman studies have also shown that condensation of Pt and N_2 at 4.2—10 K produces the species $[Pt(N_2)_n]$ ($n = 1$—3) where side-on co-ordinated N_2 is proposed for $n = 2$ with end-on N_2 for $n = 1$ and 3.[110] The relationship between dinitrogen complexes of Ni, Pd, and Pt and N_2 chemisorbed on these metals is also discussed. However, dinitrogen complexes of type $[M(N_2)(PEt_3)_3]$ (M = Pd and Pt) are not formed by reaction of $[M(PEt_3)_4]$ with N_2, although the analogous Ni complex does exist.[2] This may be due to the reduced ability of Pd and Pt to back-bond compared with Ni. Also reported for the first time are SCF-Xα scattered wave calculations for a dioxygen metal complex $[Pt(PH_3)_2(O_2)]$.[111] The calculations confirm previous pictures[112] of a shift in O–O bonding electrons towards the metal through mixing of oxygen $1\pi_u$ and $3\sigma_g$ orbitals with Pt orbitals, and a build-up of extra antibonding electron density in the $^1\pi_g$ orbitals. I.r. evidence has shown that the species $[M(CO)_3$- $(PPh_3)]$ (M = Pd or Pt) are reversibly formed from $[M_3(CO)_3(PPh_3)_4]$ and CO while no tetracarbonyls are produced.[3]

Several papers reporting new oxidative additions to $[Pt(PPh_3)_n]$ have appeared. E.s.r. evidence has been presented which shows that addition of alkyl halide proceeds *via* two one-electron steps with a Pt^I intermediate.[113] The rate-determining step (19) involves addition to $[Pt(PPh_3)_2]$ with generation of alkyl radicals (R = Me, CD_3, Et, $PhCH_2$, or Ph_3C) which have been trapped as the nitroxide $Bu^t(R)NO$. The

$$[Pt(PPh_3)_2] + RX \xrightarrow{\text{slow}} [PtX(PPh_3)_2] + R^{\bullet} \qquad (19)$$

second fast step involves addition of $R^{\bullet}$ to yield *trans*-$[PtX(R)(PPh_3)_2]$. Oxidative addition of HX (X = C_6H_5O, *o*- and *p*-BrC_6H_4O, *o*- and *p*-$NO_2C_6H_4O$, *o*- and *m*-$NH_2C_6H_4O$, *p*-$NH_2C_6H_4S$, and *p*-MeC_6H_4S) has led to products of type $[PtX_2(PPh_3)_2]$ while reaction with XY (Cl_2CS, I_2, HBr, and PhCOCl) resulted in insertion products of formula $[PtXY(PPh_3)_2]$.[114] Alkoxalyl compounds (1), formed by addition of $ClCOCO_2R$ (R = Me or Et), decarbonylate in chloroform or benzene at room temperature to yield *trans*-$[PtCl(CO_2R)(PPh_3)_2]$.[7] Nitromethane reacts with $[Pt(PPh_3)_4]$ in polar protic solution C_6H_6–EtOH–H_2O to produce *trans*- $[Pt(CNO)_2(PPh_3)_2]$ in contrast to its reaction with $[Pd(PPh_3)_4]$ which yields an isocyanate.[8] On the other hand, addition of 2-nitropropane in benzene gives *cis*- $[Pt(NO_2)_2(PPh_3)_2]$. Formaldoxime hydrochloride oxidatively adds to $[Pt(PPh_3)_4]$ to afford the cyano-complex *trans*-$[Pt(CN)Cl(PPh_3)_2]$.[8]

[109] C. J. Wilson, M. Green, and R. J. Mawby, *J.C.S. Dalton*, 1974, 421.
[110] E. P. Kundig, M. Moskovits, and G. A. Ozin, *Canad. J. Chem.*, 1973, **51**, 2710.
[111] J. G. Norman, *J. Amer. Chem. Soc.*, 1974, **96**, 3327.
[112] R. Mason, *Nature*, 1968, **217**, 543.
[113] M. F. Lappert and P. W. Ledner, *J.C.S. Chem. Comm.*, 1973, 948.
[114] H. A. Tayim, and N. S. Akl, *J. Inorg. Nuclear Chem.*, 1974, **36**, 1071.

Loss of PPh_3 from $[Pt(PPh_3)_4]$ provides polymeric compounds $[PtP_2Ph_4C_6H_4]_n$ ($n = 2$, 3, and 4) as orange-red solids.[115] The compounds, which have been characterized by mass spectroscopy and thermogravimetric analyses, are formed through o-metallation and were assumed to have bridging PPh_2 groups. Reactions of $[PtL_n]$ ($L = PPh_3$, PMe_2Ph, or PEt_3; $n = 3$, or 4) with secondary alkyl halides RX (2-bromobutane, 2-bromoethylpropionate) have yielded products of type $[PtX_2L_2]$.[116] However, addition of 2-iodobutane or 1-bromoethylbenzene to $[Pt(PMe_2Ph)_4]$ gives the yellow-orange $[PtX_2L_3]$ which are rare examples of five-co-ordinate Pt^{II}. Equation (20) was given as a possible mechanism.

$$[PtL_3] + RX \xrightarrow[\substack{\text{oxidative} \\ \text{addition}}]{-L} [PtXRL_2] \xrightarrow{\text{olefin}} [PtHXL_2]$$
$$[PtHXL_2] \rightarrow [PtX_2L_2] + ?$$
$$[PtX_2L_2] + L \rightleftharpoons [PtX_2L_3] \rightleftharpoons [PtXL_3]X \qquad (20)$$

The anions $[Pt_2X_4(CO)_2]^{2-}$ ($X = Cl$ or Br) have been isolated as their $[Pr_4N]^+$ salts in two crystallographic forms by reaction of K_2PtX_4 with CO in concentrated HX solution.[117] I.r. and Raman spectra are indicative of structure (27) having a Pt—Pt bond with 45° rotation of the PtX_2CO planes giving rise to *cisoid* and *transoid*

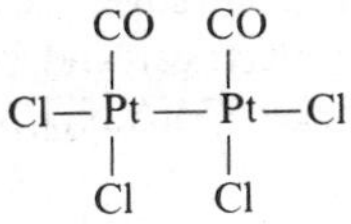

$$\begin{array}{ccc} CO & CO \\ | & | \\ Cl-Pt & -Pt-Cl \\ | & | \\ Cl & Cl \end{array}$$

(27)

structures. Reactions of these compounds with Cl_2, H_2, MeHgCl, and HCl lead to cleavage of the Pt—Pt bond.

Platinum (II)—*Group VII and Hydride Donors.* SCF—$X\alpha$ calculations have been made for the ions $[MCl_4]^{2-}$ (M = Pd or Pt) (see p. 384).[13] In addition the extended Hückel MO method has been employed to calculate activation energies of the halide-

$$[Pt^{II}L_4] \rightleftharpoons [Pt^{IV}XZL_4] \qquad (21)$$

bridged electrochemical interconversions (21), where $X = Cl$, Br, or I and $Z = X$ or H_2O.[118] The procedure appears to allow identification of the electronic states primarily responsible for electron transfer between typical electrodes and reactants and is therefore useful in interpreting the courses of electrode reaction. Another theoretical paper examines the influence of ligands on the acidic properties of an inner-sphere proton-containing molecule, from the viewpoint of rehybridization of the atomic orbitals of Pt.[119] The cases for ligands with σ-donor, π-donor, and π-acceptor properties were examined.

[115] F. Glockling, T. McBride, and R. J. I. Pollock, *J.C.S. Chem. Comm.*, 1973, 650.
[116] R. G. Pearson, W. J. Louw, and J. Rajaram, *Inorg. Chim. Acta*, 1974, **9**, 251.
[117] P. L. Goggin and R. J. Goodfellow, *J.C.S. Dalton*, 1973, 2355.
[118] C. N. Lai and A. T. Hubbard, *Inorg. Chem.*, 1974, **13**, 1199.
[119] Y. N. Kukushkin, *Russ. J. Inorg. Chem.*, 1973, **18**, 289.

The u.v.–visible spectra of $[Pt_2X_6]^{2-}$ (X = Cl or Br) have been measured;[14] the band assignments include ligand $\rightarrow$ metal charge-transfer and $d \rightarrow p$ bands, in addition to the low-intensity $d \rightarrow d$ bands. The purple vapour complex formed from $[Pt_6Cl_{12}]$ solid cluster and gaseous $AlCl_3$ has been studied spectrophotometrically.[120]

$$\begin{array}{c} \text{Cl} \\ \diagdown \\ \text{Al} \\ \diagup \\ \text{Cl} \end{array} \begin{array}{c} \text{Cl} \\ \\ \\ \\ \text{Cl} \end{array} \begin{array}{c} \text{Cl} \\ \diagdown \\ \text{Pt} \\ \diagup \\ \text{Cl} \end{array} \begin{array}{c} \text{Cl} \\ \\ \\ \\ \text{Cl} \end{array} \begin{array}{c} \text{Cl} \\ \diagdown \\ \text{Al} \\ \diagup \\ \text{Cl} \end{array}$$

(28)

Structure (28) was proposed with $\Delta H = 7.8$ kcal mol^{-1} and $\Delta S = 6.4$ e.u. for the reaction (22). Similar chloro-bridged species have also been obtained by treatment of $K_2[PtCl_4]$ with $HgCl_2$ and Et_4NCl.[23] Two compounds $(Et_4N)_2[HgPtCl_8]$ and $(Et_4N)_2[Hg_3PtCl_{10}]$ were formed, depending on the concentration used, both of which possess the trinuclear anionic unit (10).

$$\tfrac{1}{6}[Pt_6Cl_{12}] + 2AlCl_3 \rightarrow [PtAl_2Cl_8] \tag{22}$$

cis-$[Pt_2Cl_4(PPh_3)_2]$, characterized by i.r., mass, and visible spectroscopy, has been obtained as a pale yellow compound which contrasts with the orange trans-isomer.[121] The bridge-cleavage reaction with p-toluidine to form monomeric species was much faster than with the trans-isomer and moreover yielded the cis-$[PtCl_2(PPh_3)(p\text{-tol})]$.

Two papers have reported the electrical and photo conductivity of some one-dimensional mixed-valence complexes $[M(A)X_3]$ M = Pd or Pt; A = en, $(NH_3)_2$; X = Cl, Br, or I (see p. 384).[15, 16]

The enthalpies for the reactions of cis- and trans-$[Pt(NH_3)_2X_2]$ (X = Br or I) with 9.4% aqueous NH_3 to yield $[Pt(NH_3)_4X_2]$ have been measured calorimetrically.[122, 123] The results have been used to calculate the standard enthalpies of formation of the solid cis- and trans-isomers; the enthalpy of cis $\rightarrow$ trans isomerization has also been determined. The enthalpies of dissolution of the complexes $K_2[PtBr_4]$, $Rb_2[PtBr_4]$, and $K[PtBr_3(NH_3)]$ have been measured; the enthalpy of formation of $Rb_2[PtBr_4]$ was calculated to be -224.4 kcal mol^{-1}.[124]

Observations that the phosphines do not mix when a phosphine is used to catalyse the cis $\rightarrow$ trans isomerism of $[PtCl_2(PEt_3)_2]$, and also that $[PtCl(PEt_3)_3]^+$ does *not* react with Cl^- to produce cis- or trans-$[PtCl_2(PEt_3)_2]$ contradict the consecutive displacement mechanism (23) and instead suggest fluxional rotation.[125]

$$cis\text{-}[MX_2L^1_2] \underset{X}{\overset{L^2}{\rightleftharpoons}} [MXL^1_2L^2] \underset{L^1}{\overset{X}{\rightleftharpoons}} trans\text{-}[MX_2L^1L^2] \tag{23}$$

^{129}I Mössbauer spectroscopy has shown that neutral ligands L exert significant cis-influences opposite to their trans-influences in the series of complexes trans-$[PtL_2I_2]$ (L = PEt_3, Me_2S, Et_2S, β-picoline, py, or NH_3).[126] This observation may be explained by Syrkin's theory of the trans-effect in terms of d–s hybridization.

[120] G. N. Papatheodorou, *Inorg. Chem.*, 1973, **12**, 1899.
[121] F. R. Hartley and G. W. Searle, *Inorg. Chem.*, 1973, **12**, 1949.
[122] V. A. Palkin, T. A. Kuzina, and N. N. Kuz'mina, *Russ. J. Inorg. Chem.*, 1973, **18**, 406.
[123] V. A. Palkin and T. A. Kuzina, *Russ. J. Inorg. Chem.*, 1973, **18**, 246.
[124] T. A. Kuzina, V. A. Palkin, and N. N. Kuz'mina, *Russ. J. Inorg. Chem.*, 1973, **18**, 244.
[125] W. J. Louw, *J.C.S. Chem. Comm.*, 1974, 353.
[126] B. W. Dale, R. J. Dickinson, and R. V. D. Parish, *J.C.S. Chem. Comm.*, 1974, 35.

Studies of the oxidative addition of SCl_2 to complexes of Rh, Ir, and Pt have revealed that whereas Rh and Ir produce complexes of the chlorosulphide ligand (SCl), addition to Pt complexes yields only chloro-complexes (equation 24).[127]

$$[Pt(CO)_2(PPh_3)_2] + SCl_2 \rightarrow [PtCl_2(CO)(PPh_3)] \qquad (24)$$

Mass spectral data for the complexes $[PtCl_2(CO)(PR_3)]$ (R = Ph, *o*-, *m*-, *p*-tolyl) indicate formation of a five-co-ordinate species through addition of atomic oxygen.[128] Fragmentation patterns and i.r. spectra for these complexes are also given.

New hydrido-complexes have been obtained by solvolysis of alkyldiphosphinite complexes $[Pt(ROPPh_2)_4]$ (R = Bu^n, Pr^i, or Me) in aqueous ethanol.[129] I.r. and n.m.r. studies show that these complexes have the structure (29). The low $\nu(Pt—H)$

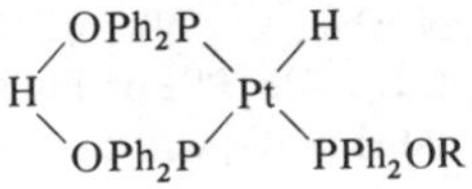

(29)

frequency of 2000 cm^{-1} is indicative of the known large *trans*-influence of diphenyl-phosphinato ligand. The complexes react with HCl to yield $[PtCl(OPPh_2)(HOPPh_2)_2]$ and similar reactions have also been observed for other complexes possessing hydride *trans* to high *trans*-effect ligands [equation (25), X = Cl, Br, or CF_3CO_2]. It has

$$[PtH(CN)(PPh_3)_2] + HX \rightarrow [PtXCN(PPh_3)_2] \qquad (25)$$

been shown that the compound previously[130] formulated as $[Pt(PPh_3)_2SiF_4]$ is in fact $[PtH(PPh_3)_3]SiF_5$, which may also be made by addition of concentrated HF to a solution of $[Pt(PPh_3)_4]$ containing silica.[131] It was also found that BF_3 forms analogous products, thus care must be taken when interpreting results of reactions of covalent fluorides with basic metal complexes when reactions are carried out in glass apparatus.

Stable dihydride complexes of Pd and Pt have been isolated from reaction (26) (R^1 = Cy$_2$Et, Cy$_2$Pr^i, or Cy$_3$; Cy = cyclohexyl).[28] A *trans*-structure is indicated

$$[Pt(acac)_2] + 2PR_3^1 \xrightarrow{Et_2O} [PdH_2(PR_3^1)_2] + [Pd(PR_3^1)_2] \qquad (26)$$

from ^1H n.m.r. and a single $\nu(Pt—H)$ of 1910 cm^{-1}. The complexes react with CCl_4 to give equimolar amounts of *trans*-$[PtHCl(PR_3)_2]$ and $CHCl_3$.

Group VI Donors. The synthesis of complexes of some cyclic tertiary amino-acids with Pd or Pt has been reported (equation 11).[36] The complexes are chelated *via* N and O and have high stability constants, $K = 10^8—10^{11}$. Complexes of Pt (Pd, Cu) with 3,3'-diamino-4,4'-dihydroxydiphenylsulphone have been obtained (see p. 387).[38]

Thiocyanato-complexes have attracted considerable attention during the past year. I.r. and u.v.–visible studies have shown that the complexes $[PtL(SCN)](BPh_4)$ (L = Et$_4$ dien or MeEt$_4$dien) are stable S-bonded species both in solution and in the

[127] W. H. Baddley and D. S. Hamilton, *Inorg. Nuclear Chem. Letters*, 1974, **10**, 143.
[128] K. L. Klassen and N. V. Duffy, *J. Inorg. Nuclear Chem.*, 1973, **35**, 2602.
[129] P. C. Kong and D. M. Roundhill, *J.C.S. Dalton*, 1974, 187.
[130] T. R. Durkin and E. P. Schram, *Inorg. Chem.*, 1972, **11**, 1048.
[131] P. Bird, J. F. Harrod, and K. A. Than, *J. Amer. Chem. Soc.*, 1974, **96**, 1222.

solid state, in contrast to analogous Pd complexes which undergo S → N linkage isomerism in solution.[40] However, the complexes cis-[Pt(SCN)₂L₂] (L = PPh₃ or AsPh₃) undergo exclusively cis S-bonded → cis N-bonded isomerization in the solid state on heating.[132] In addition, the results of $v(CN)$ integrated absorption intensity measurements indicate that the complexes cis-[Pt(SCN)ClL₂] (L = phosphine) and cis-[Pt(SCN)(CO)(PMe₂Ph)₂]X (X = BF₄ or PF₆), previously[133] formulated as containing S-bonded thiocyanate, actually involve N-bonding in the solid phase.

Dinuclear complexes of type [MX{(Ph₂PO)₂H}]₂ (M = Pd or Pt; X = Cl, SCN, or Br) have been obtained by hydrolysis of the phosphine ligand in acetylenic phosphine complexes [MCl₂(Ph₂PC≡CCF₃)₂].[42] Thiocyanato-complexes of type [M(CNS)₂L₂] (M = Pd or Pt) have been prepared from [M(SCN)₄]²⁻ and L (Ph₃AsS, Ph₃PSe, or Ph₃PS) by extraction into CHCl₃.[134] I.r. spectra show that only mononuclear complexes are formed when L = Ph₃AsS, but the $v(CN)$ = 2150 cm⁻¹ for the compounds obtained with L = Ph₃PSe or Ph₃PS is indicative of bridged structures which have been formulated as di-, tri-, and tetra-meric species (30).

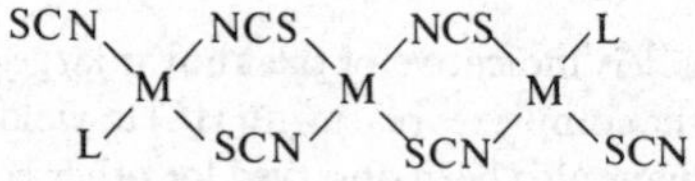

$$\begin{array}{ccccccc} SCN & & NCS & & NCS & & L \\ & \diagdown \ / & & \diagdown \ / & & \diagdown \ / & \\ & M & & M & & M & \\ & \diagup \ \diagdown & & \diagup \ \diagdown & & \diagup \ \diagdown & \\ L & & SCN & & SCN & & SCN \end{array}$$

(30)

Synthesis of series of both mono- and dinuclear complexes of Pd and Pt with (SCF₃)⁻ has been reported.[46, 47, 135] Mononuclear species were prepared by oxidative addition of (SCF₃)₂ to the [M(PPh₃)₄] complexes or by metathetical reactions (equation 12). ¹⁹F N.m.r. was used to establish the trans-influence series H > PEt₃ ~ PPh₃ > SCF₃ > Cl;[46] in addition ¹H n.m.r. coupling constant experiments for the series cis- and trans-[PtX(SCF₃)(PEt₃)₂] (X = CN, NO₂, SCF, N₃, I, NCS, NCO, Br, Cl, or NO₃) showed that ³J(Pt–F) can be used to establish scales of cis- and trans-influence for the X ligands where an excellent inverse correlation between the two scales was found.[135] However, this correlation could not be extended to neutral X ligands. Dinuclear complexes of formula [M₂(SCF₃)₂(PR₃)₄]X₂ (M = Pd or Pt; R = Et or Ph; X = BF₄ or ClO₄) have been prepared (see p. 390).[47]

Orange-red diamagnetic complexes of Pd and Pt with mercaptoacetic acid and 3-mercaptopropionic acid have been reported (see p. 390).[48] Simple bis-complexes of N-pyrrolidyl monothiocarbamate with Pd and Pt are not formed,[50] but i.r. studies indicate that the tetrakis-complexes are bonded through both O and S, although no definite structure has been proposed. Extraction of the metals Pd, Pt, Ag, and Au from aqueous solution by a large number of alkyl and alkene disulphides and sulphides has been reported (see p. 390).[51, 52]

S-Bridged compounds, with the structure (31) indicated from i.r. measurements, have been synthesized by the action of H₂S on K[Pt(am)Cl₃](am = NH₃ or

¹³² R. L. Hassel and J. L. Burmeister, *Inorg. Chim. Acta*, 1974, **8**, 155.
¹³³ W. J. Cherwinski and H. C. Clark, *Inorg. Chem.*, 1971, **10**, 2263.
¹³⁴ M. G. King, G. P. Mcquillan, and R. Milne, *J. Inorg. Nuclear Chem.*, 1973, **35**, 3039.
¹³⁵ K. R. Dixon, K. C. Moss, and M. A. R. Smith, *Inorg. Nuclear Chem. Letters*, 1974, **10**, 373.

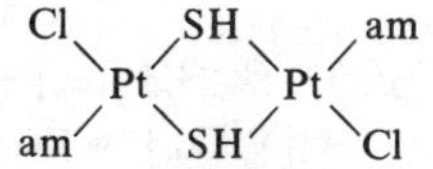

$$\begin{array}{ccccc} Cl & & SH & & am \\ & \diagdown & \diagup & \diagdown & \diagup \\ & Pt & & Pt & \\ & \diagup & \diagdown & \diagup & \diagdown \\ am & & SH & & Cl \end{array}$$

(31)

β-pic).[136] However, analogous compounds with am = py contain only one SH bridge and also a chloro-bridge. Compounds containing a straight S-bridge could not be prepared by addition of Na_2S to $K[Pt(am)Cl_3]$.

Cleavage of the S—S bond in disulphides (R_2S_2) by $[Pt(PPh_3)_4]$ yields monomeric species $[Pt(SPh)_2(PPh_3)_2]$, unlike the Pd analogues which give rise to SR-bridged dimers.[54] Reaction of the Pt complex with transition-metal complexes containing labile groups yields the interesting heterodinuclear complexes (32) and (33) with bridging SPh groups.

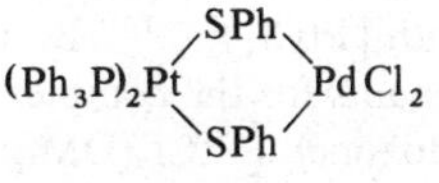

$$\begin{array}{ccc} & SPh & \\ & \diagup \diagdown & \\ (Ph_3P)_2Pt & & PdCl_2 \\ & \diagdown \diagup & \\ & SPh & \end{array} \qquad \begin{array}{ccc} PPh_3 & SPh & \\ \diagdown & \diagup \diagdown & \\ & Pt & Mo(CO)_4 \\ \diagup & \diagdown \diagup & \\ PPh_3 & SPh & \end{array}$$

(32) 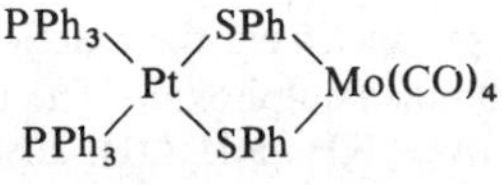   (33)

The S-bridged anions $[Pt_2X_6(SMe_2)]^{2-}$ (X = Cl or Br) and $[Pt_2X_5(SMe_2)]^-$ have been isolated as $(Pr_4N)^+$ salts.[137] I.r. and n.m.r. data indicate the former to have only SMe_2 as bridging ligand (34) while the latter has both SMe_2 and chloro-bridges (35). The vibrational spectra of $[Pt_2X_4(SMe_2)_2]$ (X = Cl or Br) are assigned

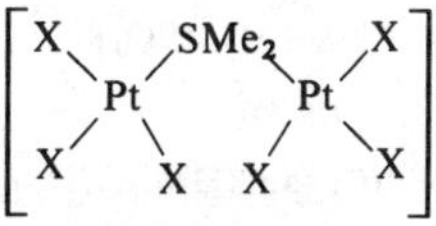

$$\begin{bmatrix} X & SMe_2 & X \\ \diagdown & \diagup \diagdown & \diagup \\ & Pt & Pt \\ \diagup & \diagdown \diagup & \diagdown \\ X & X \quad X & X \end{bmatrix} \qquad \begin{bmatrix} X & SMe_2 & X \\ \diagdown & \diagup \diagdown & \diagup \\ & Pt & Pt \\ \diagup & \diagdown \diagup & \diagdown \\ X & X & X \end{bmatrix}$$

(34) 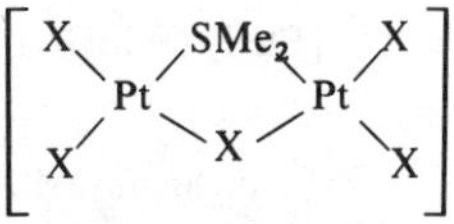   (35)

on the basis of a S-bridged structure, but it is proposed that the corresponding iodo-complex is halide-bridged as also are $[Pt_2X_4(SPh_2)_2]$.

The four-co-ordinate complexes $[M(S_2PR^1_2)_2(PR^2_3)]$ and $[M(S_2PR^1_3)(PR^2_3)_2]$ (S_2PR_2) (M = Pd or Pt; R = Me or Et) have been synthesized (see p. 391).[55] Variable temperature ^1H n.m.r. indicates rapid unidentate–bidentate interchange at room temperature. Full lineshape analyses for the complexes $[Pt(S-S)_2(ER^2_3)][S-S = S_2CNEt_2, S_2P(OEt)_2,$ or S_2PMe_2; E = P or As] suggest a concerted mechanism with bond-breaking and bond-making steps important.

Mass spectra for the chelates of $RC(SH)=CHCOCF_3$ with Pd and Pt (R = Ph, p-MeC$_6$H$_4$, p-BrC$_6$H$_4$, p-NO$_2$C$_6$H$_4$, or 2-thienyl) have been recorded.[56, 57] Comparison with data for the chelates with Ni and Zn reveals that the fragmentation pattern is determined by the metal.

Complexes formed from $[MCl_4]^{2-}$ (M = Pd or Pt) by treatment with n-butyl

[136] Y. N. Kukushkin, V. V. Sibirskaya, S. D. Bapzargashieva, and T. K. Mikhal'chenko, *Russ. J. Inorg. Chem.*, 1973, **18**, 1385.
[137] P. L. Goggin, R. J. Goodfellow, and F. J. S. Reed, *J.C.S. Dalton*, 1974, 576.

α-naphthyl sulphide have been reported (see p. 391),[58] as have complexes [M(OEtsac-sac)$_2$] (M = Pd, Pt, or Ni) (see p. 392).[60] The ethyl cysteinate complex [Pt(ec)$_2$] has been prepared by reduction of H$_2$[PtCl$_6$] with HSCH$_2$CH(NH$_2$)CO$_2$Et.[62] I.r. evidence suggests the structure (20).

The first reported thiocarbonate complexes of Pt have been synthesized by the reaction of K[PtLCl$_3$] (L = NH$_3$ or py) with Na$_2$CS$_3$.[138] The products of formula [Pt(CS$_3$)L] are dark red and probably polymeric. Monomeric complexes of type [Pt(CS$_3$)L$_3$] have also been obtained as orange products by treatment of [PtL$_4$]$^{2+}$ (L = NH$_3$, or py) with Na$_2$CS$_3$, which decompose on standing to yield [Pt(CS$_3$)L].

Experiments with ^{35}S-labelled DESO have shown that DESO does not undergo exchange in the complex [PtCl(DESO)(en)]$^+$.[139] The difference between this inertness and the rapid exchange observed for [PtCl$_3$(DESO)]$^-$ was attributed to the weak *trans*-effect of NH$_2$ and to the positive charge on the complex ion. In addition, i.r. data have established that the product of the reaction of [PtCl$_4$]$^{2-}$ with Pr$_2$SO is *trans*[PtCl$_2$(Pr$_2$SO)$_2$] which *cis*-isomerizes on heating, whereas *trans*-[PtCl$_2$(Pr$^i_2$SO)$_2$] is the only product from the reaction of Pr$^i_2$SO with [PtCl$_4$]$^{2-}$.[140] Both complexes contain S-bonded sulphoxide. The instability constants for the complexes [PtCl$_2$A-(DMSO)] (A = NH$_3$,NH$_2$OH, acetoxime, or aldoxime), [PtCl$_3$(DMSO)]$^-$, *trans*-[PtCl$_2$(H$_2$O)(DMSO)], and [PtCl(H$_2$O)$_2$ (DMSO)]$^+$ have been measured using a AgCl electrode.[140a, b]

A further series of sulphoxide–amine complexes of type *cis*[PtCl(am)$_2$(DMSO)]Cl, *cis*-[PtCl$_2$(am)(DMSO)] and [PtCl(am')(am)(DMSO)]Cl [am = alicyclic primary amine C$_n$H$_{2n-1}$NH$_2$ (n = 3—8); am' = cyclopropylamine] have been synthesized (Scheme 2).[141] I.r. again shows DMSO to be S-bonded.

$$[\text{PtCl}_2(\text{DMSO})_2] + \text{am} \rightarrow \textit{cis}\text{-}[\text{PtCl(am)}_2(\text{DMSO})]$$
$$\downarrow \text{HClO}_4$$
$$[\text{PtCl(am}')(\text{am})(\text{DMSO})] \xleftarrow{\text{am}'} \textit{cis}\text{-}[\text{PtCl}_2(\text{am})(\text{DMSO})]$$

Scheme 2

The first report of gas chromatography of metal chelates involving ligands containing only S donors has appeared.[65] The chelates [M{(OR)$_2$PS$_2$}$_2$] (M = Pd, or Pt) were found to be thermally stable up to 170 °C and elution and separation on a gas chromatogram were possible.

Binding energies have been measured for a series of dithiene complexes (Et$_4$N)-[MS$_4$C$_4$(CN)$_4$], (Et$_4$N)$_2$[MS$_4$C$_4$(CN)$_4$], [MS$_4$C$_4$Ph$_4$], (Et$_4$N)[MS$_4$C$_4$Ph$_4$] and (N$_2$H$_5$)$_2$[MS$_4$C$_4$Ph$_4$] (M = Ni, Pd, or Pt).[142] The results indicate that the charge on the particular metal remains essentially constant in the series of neutral, anionic,

[138] Y. N. Kukushkin, V. V. Sibirskaya, S. D. Bapzargashieva, and O. I. Arkhangel'skaya, *Russ. J. Inorg. Chem.*, 1973, **18**, 847.

[139] Y. N. Kukushkin, V. V. Golosov, *Russ. J. Inorg. Chem.*, 1973, **18**, 752.

[140] Y. N. Kukushkin, V. V. Spevak, *Russ. J. Inorg. Chem.*, 1973, **18**, 240.

[140a] Y. N. Kukushkin, A. I. Stetsenko, K. M. Trusova, and V. G. Dul'Banova, *Russ. J. Inorg. Chem.*, 1973, **18**, 1283.

[140a] Y. N. Kukushkin, A. I. Stetsenko, K. M. Trusova, and V. G. Dul'Banova, *Russ. J. Inorg. Chem.*, 1973,

[141] P. D. Braddock, R. Romeo, and M. L. Tobe, *Inorg. Chem.*, 1974, **13**, 1170.

[142] S. O. Grim, L. J. Matienzo, and W. E. Swartz, *Inorg. Chem.*, 1974, **13**, 447.

and dianionic complexes, and that the additional electronic charge in this sequence resides principally on the S atoms. Two of the compounds $(Et_4N)[MS_4C_4Ph_4]$ (M = Pd or Pt) are reported for the first time and both have paramagnetism expected for odd-electron species. In addition $Pt(4f_{7/2})$, $N(1s)$ and $Cl(2p_{1/2,\,3/2})$ binding energies are given for the series $[PtCl_{4-n}(NH_3)_n]^{(n-2)+}$ (n = 0—4).

New thioamide complexes $[PtCl_2L_2]$(L = phenylthioacetamide, o-tolylthioacet-amide, 2,5-dimethylphenylthioacetamide, cynamthioamide and p-methoxycynamthio-amide) and $[PtCl_2L]$ (L = diphenylthioacetamide) have been reported.[143] I.r. and u.v.–visible spectroscopy suggest the structures (36) and (37) with N- and S-bonded thioamide.

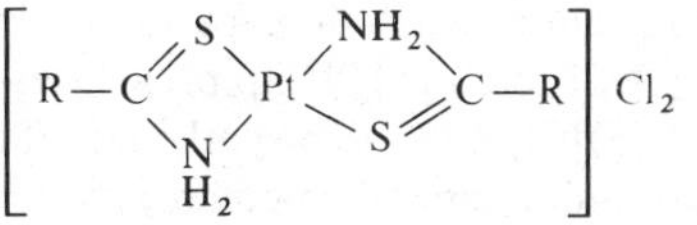

(36) 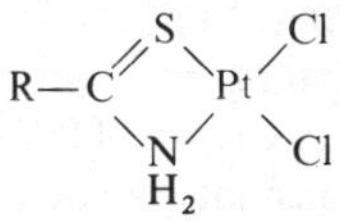 (37)

The complexes $[M(hal)_2(Et_2X)_2]$ (M = Pd or Pt; X = Se or Te; hal = Cl, Br, or I) have been studied by variable-temperature n.m.r. (see p. 393).[66] Reactions of excess CSe_2 and arene selenols (equimolar) with $[PtL_4]$ (L = PPh_3 or PPh_2Me) have yielded the oxidative addition products $[Pt(CSe_2)L_2]$ and $[PtH(SeR)L_2]$ (R = Ph or p-MeC_6H_4).[144] However, reactions involving excess arene selenols lead to formation of $[Pt(SeR)_2L_2]$. I.r. and ^1H n.m.r. data indicate the structures (38) and (39) with configurations similar to the corresponding CS_2 adducts.

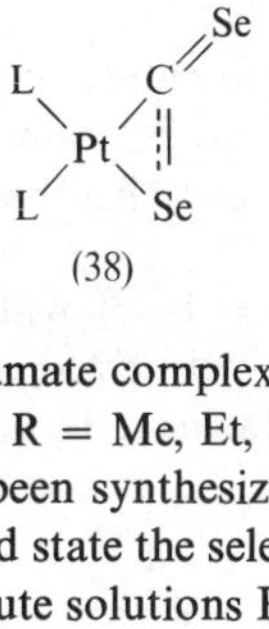

(38) 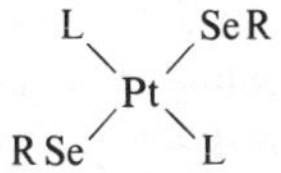 (39)

Series of selenocarbamate complexes of general formula $[M(L_2)(PPh_3)_2]$(M = Pd or Pt; L = $SeCONR_2$; R = Me, Et, Pr^n, or Bu^n) and $[M(SeCONMe_2)_2(PMePh_2)_2]$ (M = Pd or Pt) have been synthesized.[67] I.r. and molecular weight determinations indicate that in the solid state the selenocarbamate ligand is bonded through Se with a free CO group. In dilute solutions PPh_3 dissociates to yield bidentate selenocarba-mates (21) and (40).

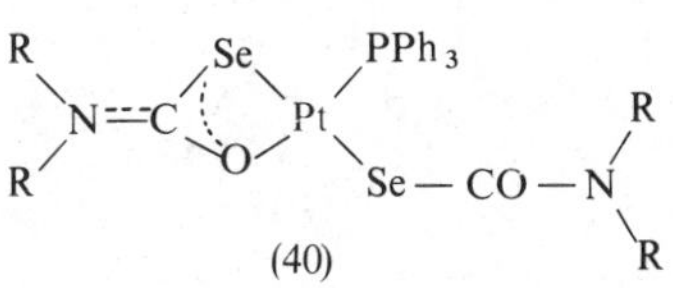

(40)

Group V. N-Donor ligands. The ylide complex $[PtCl_2(DMSO)\{Me_2SNC(O)R\}]$ (R = Ph or p-MeC_6H_4), formed by reaction of aqueous $K_2[PtCl_4]$ with MeSNC(O)R,

[143] D. Negoiu and V. Muresan, *Rev. Roumaine Chim.*, 1974, **19**, 1031.
[144] K. Kawakami, Y. Ozaki, and T. Tanaka, *J. Organometallic Chem.*, 1974, **69**, 151.

has been characterized by i.r. and ^{1}H n.m.r. and shown to have a *cis*-configuration co-ordinated by N of $Me_2SNC(O)R$ and S of $DMSO$.[145] The Pt chelates $[Pt(NH_3)_2L]$ Cl_2 (L = *meso*-1,2-dimethyl-1,2-ethanediamine, *l*-1,2-dimethyl-1,2-ethanediamine, or 1,1-dimethyl-1,2-ethanediamine) have been studied by ^{13}C n.m.r. to elucidate their conformational behaviour in aqueous solution.[146] The five-membered chelate rings may adopt either σ- or λ-forms which are interconvertible on ring inversion. The substituted Me groups may be axial or equatorial with respect to the ring. A useful general route for the preparation of tetramine complexes $[PtA_4X_2]$ (A = NH_3, py, γ-pic, $MeNH_2$, or $EtNH_2$, or $\frac{1}{2}$ en; X = Cl, Br, or I) has been developed, involving reaction of excess of A on the sulphoxide complexes *cis*-$[PtX_2(R_2SO)_2]$ (R = Me or Et). In addition the bridged complex $[PtI_2(DMSO)]_2$ was obtained for the first time by treatment of $[PtCl_2(DMSO)_2]$ with KI in aqueous solution. A series of new piperidine complexes including *cis*- and *trans*-$[PtCl_2(pip)_2]$, $[PtCl(pip)_3]Cl$, *cis*- and *trans*-$[Pt(pip)_2(NH_3)_2]Cl_2$, and $[Pt(NH_3)_3(pip)]Cl_2$ have been synthesized (Scheme 3).[148] The stability constants of these mixed amine–piperidine complexes were measured

$$K_2[PtCl_4] + pip \rightarrow \textit{cis-}[PtCl_2(pip)_2] \xrightarrow[\text{pip}]{\text{excess}} [PtCl(pip)_3]Cl$$

$$\downarrow HCl$$

$$\textit{trans-}[PtCl_2(pip)_2]$$

$$[PtCl_2(NH_3)_2] + pip \rightarrow \text{isomers } [Pt(pip)_x(NH_3)_y]Cl_2$$
$$(x = 1 \text{ or } 2: y = 3 \text{ or } 2)$$

Scheme 3

potentiometrically and it was concluded that piperidine was intermediate between pyridine and morpholine in its ability to stability Pt^{II} complexes. The new yellow complex *cis*-$[PtBr(NH_2CH_2CO_2H)(NH_2CH_2CO_2)]$ has been synthesized as part of a study of ring-closure of glycine complexes.[149]

The reaction of *cis*-$[PtCl_2(PMe_xPh_{3-x})]$ (x = 1—3) with either dibenzoyl- or diacetyl-hydrazine in EtOH has yielded new hydrazido-complexes;[150] ^{1}H n.m.r. and also unpublished X-ray data reveal the hydrazide to be in the enolized form (41).

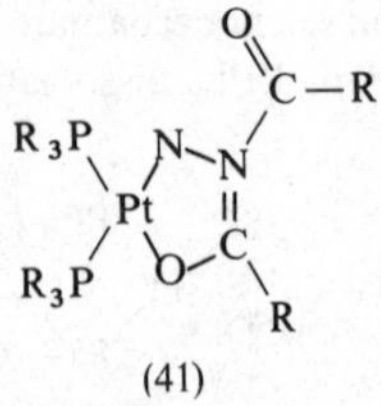

(41)

[145] G. Matsubayashi, M. Toriuchi, and T. Tanaka, *Bull. Chem. Soc. Japan*, 1974, **47**, 765.
[146] S. Bagger, *Acta Chem. Scand.*, 1974, **Á28**, 467.
[147] Y. N. Kukushkin, V. N. Spevak, and V. P. Kotel'nikov, *Russ. J. Inorg. Chem.*, 1973, **18**, 595.
[148] Y. N. Kukushkin and V. A. Yurinov, *Russ. J. Inorg. Chem.*, 1973, **18**, 91.
[149] L. M. Volshtein, L. F. Krylova, and A. V. Belyaev, *Russ. J. Inorg. Chem.*, 1973, **18**, 562.
[150] J. R. Dilworth, A. S. Kasenally, and F. M. Hussein, *J. Organometallic Chem.*, 1973, **60**, 203.

The complex reacts with HCl or HOAc to give dibenzoyl hydrazine (equation 27),

$$[Pt\{C(O)PhNNCOPh\}(PR_3)_2] + 2HCl \rightarrow [PtCl_2(PR_3)_2] + (PhCONH)_2 \quad (27)$$

but produces dibenzoyldi-imine with bromine. Complexes of the potentially bidentate ligand 3-(4-pyridyl)triazoline-5-thione with Pt (Pd, Rh) have been synthesized and were shown to be bonded through both N and S.[72] X-Ray structural measurements for the complex cis-$[PtCl(PEt_3)_2(phen)]BF_4$ have shown for the first time the presence of unidentate phenanthroline.[150a] In the solid there are two Pt—N distances (2.137 and 2.843 Å) with overall square-planar co-ordination round Pt rather than five-co-ordination. On the other hand, conductance and ^1H n.m.r. results for solutions of this complex have been interpreted in terms of a novel fluxional behaviour in which phenanthroline very rapidly exchanges its point of attachment to Pt.

Changes in the u.v.–visible and ^1H n.m.r. spectra of aqueous $[Pt(bipy)_2]^{2+}$ with pH have been attributed to the equilibrium (28), where nucleophilic attack has occurred on the ligand.[151]

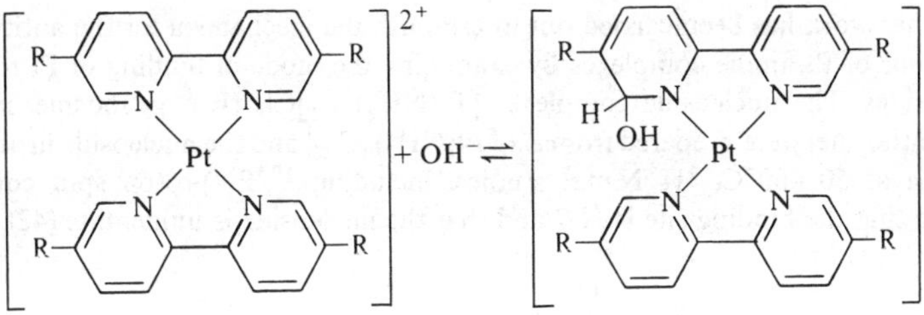

The reactions of the complexes $[MX_2L]$ (X = Cl, Br, or I) and $[ML_2X_2]$ (M = Pd or Pt; X = Cl or I; L = $H_2NCH_2CH_2NHCH_2CH_2OH$) in liquid NH_3 at $-75\,°C$ have been studied (see p. 395).[80]

The nitrosylation of Pt complexes with a gaseous mixture of NO and N_2O_4 in HCl has been used to prepare the novel nitroso-compounds containing a Cl—Pt—NO axis, cis-$[PtCl(NO)A_2]$ (A = NH_3, $MeNH_2$, Me_2NH, or $\frac{1}{2}$en) and $[PtCl_5NO]^{2-}$ and $[PtCl_4(NO)(NH_3)]^-$.[152] In addition the complexes $(C_9H_7NH)[PtCl_4NO]$ and $[PtCl_3(NO)A]$ (A = NH_3, $MeNH_2$, or py) have been isolated from sulphuric acid (equation 29).

$$K[PtCl_3(A)] + NO + NO_2 + H_2SO_4 \rightarrow [PtCl_3(NO)A] \quad (29)$$

Some new cationic aryldiazo-complexes of formula trans-$[Pt(N_2Ar)(PEt_3)_2L]^+$ ($N_2Ar = m$- or o-$N_2C_6H_4F$; L = NH_3, py, PEt_3, or EtNC) have been prepared in order to examine the effect of changing L on the nature of the Pt–diazo linkage.[153]

[150a] G. W. Bushnell, K. R. Dixon, and M. A. Khan, *Canad. J. Chem.*, 1974, **52**, 1367.
[151] R. D. Gillard and J. R. Lyons, *J.C.S. Chem. Comm.*, 1973, 585.
[152] A. I. Stetsenko, N. A. Sukhanova, and L. S. Tikhonova, *Russ. J. Inorg. Chem.*, 1973, **18**, 85.
[153] A. W. B. Garner and M. J. Mays, *J. Organometallic Chem.*, 1974, **67**, 153.

The salts were isolated as purple-red crystals with PF_6^- as anion. I.r. and ^1H and ^{19}F n.m.r. spectra indicate a *trans*-square-planar geometry thereby implying $(ArN_2)^-$ rather than $(ArN_2)^+$ co-ordination. Protonation of these complexes yields aryldi-imine complexes which undergo reduction with molecular hydrogen to give aryl hydrazine complexes, *trans*-$[Pt(NH_2NHAr)(PEt_3)_2L]^+$.

A very convenient general synthesis for an extensive range of metal triazenido-complexes has been described.[154] The complexes, which contain unidentate or chelated 1,3-diaryltriazene, were isolated as highly coloured, air-stable crystalline products. The new complexes *cis*- and *trans*-$[M(ArNNNAr)_2(PPh_3)_2]$ (M = Pd or Pt), $[Pt(ArNNAr)_2\{Ph_2P(CH_2)_nPPh_2\}]$ (n = 2—4), $[PtCl(ArNNNAr)(PR_3)_2]$ (R = Et or Ph) were characterized from i.r. and n.m.r. spectra.

Although ureylene complexes of Pd may be prepared by reaction of azide with $[Pd(PPh_3)_3(CO)]$, similar reactions with $[Pt(PPh_3)_n(CO)_{4-n}]$ (n = 2 or 3) yield a more complex compound $[Pt(PPh_3)_2(R_2N_4CO)]$ of uncertain structure.[75a] However, the ureylene complexes have been obtained by oxidative addition of NN'-ditoluene-*p*-sulphonyl urea to $[Pt(PPh_3)_4]$ or by treatment of $[M(PPh_3)_4]$ (M = Pd or Pt) with isocyanate.

Further work has been carried out to establish the mechanism for the antitumour behaviour of Pt amine complexes by examining the mode of binding of Pt to some nucleosides. The nucleoside complexes $[Pt(NH_3)_2L_2]Cl_2$ (L = guanosine, inosine, and xanthosine) were prepared from *cis*-$[Pt(NH_3)_2Cl_2]$ and the nucleoside in aqueous solution at 50—60 °C. ^1H N.m.r. studies, including ^{195}Pt–proton spin coupling, showed that the binding site is N-7 and that the nucleoside is unidentate (42).[155]

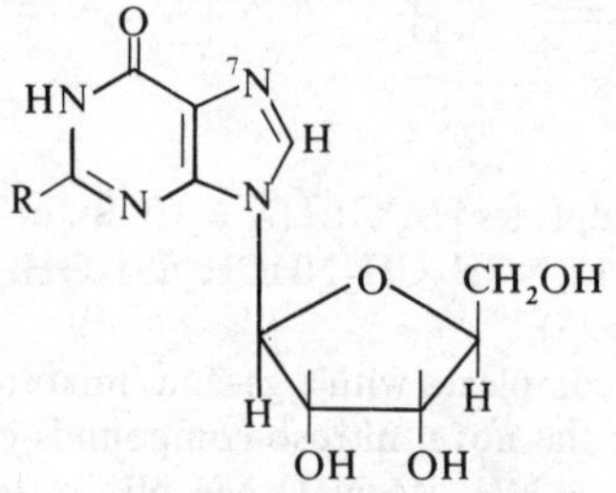

(42) R = NH_2, H, or OH

Bis-biguanide complexes $(C_2N_5H_7)$ of Pt have been obtained by treating the ligand with $PtCl_2$ in alkaline medium.[156] A series of mono- and bis-chelated proline complexes $[Pt(L-pro)Cl(NH_3)]$, $K[Pt(L-pro)Cl_2]$, $[Pt(L-pro)(NH_3)_3]Cl$, $[Pt(L-pro-H)Cl_2(NH_3)]$, and $K[Pt(L-pro - H)Cl_3]$ (L-pro = $C_4H_7NHCO_2^-$) and similar complexes of L-proline and glycine have been examined using c.d., o.r.d., and electronic spectroscopy.[157, 158] It was shown that unidentate co-ordination through N is not stereospecific in contrast to co-ordination as a bidentate ligand. Opening of the chelate ring at the carboxylate group causes racemization of the asymmetric donor

[154] K. R. Laing, S. D. Robinson, and N. F. Uttley, *J.C.S. Dalton*, 1974, 1205.
[155] P. C. Kong and T. Theophonides, *Inorg. Chem.*, 1974, **13**, 1167.
[156] D. Sen, *J. Indian Chem. Soc.*, 1974, **51**, 183.
[157] O. P. Slyudkin, O. N. Adrianova, P. A. Chel'tsov, and L. M. Volshtein, *Russ. J. Inorg. Chem.*, 1973, **18**, 1398.
[158] O. P. Slyudkin, O. N. Adrianova, and L. M. Volshtein, *Russ. J. Inorg. Chem.*, 1973, **18**, 1610.

N ligand. Treatment of *trans*-$[PtCl_2(LH_2)_2]$ ($L^{2-} = NH_2CH_2CO\overline{N}CH_2CO_2{}^-$) with base has given $[Pt(LH)_2]$ which is dibasic.[159] In addition $[PtL_2]^{2-}$ anion has been isolated as a potassium salt. U.v.–visible and i.r. spectra indicate that the glycylglycine is acting as a bidentate ligand, whereas in similar complexes of Ca^{II}, Ni^{II}, and Zn^{II} it behaves as a terdentate ligand. Chelation occurs through N of NH_2 and N of the peptide link. Amongst other amino-acid complexes studied has been that of norleucine (ZH), prepared according to equation (30).[160] In addition $[HPtZ_3]$ was also isolated and shown to have the structure (43). The ions $[H_2PtZ_3]^+$

$$K_2PtCl_4 + ZH \rightarrow PtZ_2 \text{ (cis and trans)} \tag{30}$$

and $[PtZ_3]^-$ were formed from $[HPtZ_3]$ in acid and base solutions, respectively.

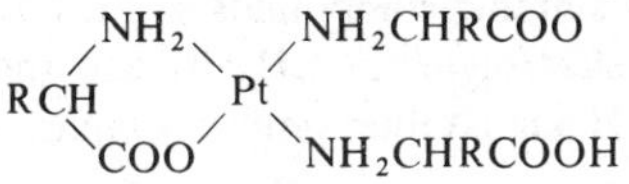

$$RCH \underset{COO}{\overset{NH_2}{<}} Pt \underset{NH_2CHRCOOH}{\overset{NH_2CHRCOO}{<}}$$

(43) R = Bu

$[H_2PtZ_4]$ was also obtained by reaction of *cis*-$[PtZ_2]$ with ZH and behaved as a dibasic acid.

The *cis*- and *trans*-isomers of $[Pt\{MeNH(OH)\}_2(NH_3)_2]$ have been synthesized and potentiometric titrations show that both are dibasic.[161] The difference between consecutive dissociation constants is much greater for the *cis*-isomer, which is attributed to *cis*-interaction of the ligands. In addition $[PtCl_2(NHMeOH)_4]$ was shown to be tetrabasic, the titration curve behaving as though two dibasic acids were present.

P-*Donor ligands.* A rare example of a five-co-ordinate Pt^{II} complex has been prepared [equation (31), RX = 2-iodobutane or 1-bromoethylbenzene]. A possible mechanism

$$[Pt(PMe_2Ph)_4] + RX \rightarrow [PtX_2(PMe_2Ph)_3] \tag{31}$$

(equation 20) was also given.[116] Phosphite ligand exchange in the trigonal-bipyramidal complexes $[M\{P(OMe)_3\}_5](BPh_4)_2$ (M = Pd or Pt) has been studied (see p. 396).[86] Difluorophosphonato-complexes $[M(PF_2O)_4]^{2-}$ (M = Pd or Pt) have been isolated as their n-allyl pyridinium salts (equation 16).[84]

Direct evidence for the rates of ligand association in a series of $[MHL_3]^+$ species (M = Ni, Pd or Pt; L = PEt_3) has been reported (see p. 396).[85] New water-soluble tertiary phosphine complexes *cis*- and *trans*-$[MX_2(PR_{3-n}Q_n)_2]$ (M = Pd or Pt; R = Me, Et, Bu, or Ph; Q = CH_2OCOMe or CH_2OH; n = 0, 1 or 2) have been synthesized (see p. 396).[87]

Interest in the internal metallation ability of the bulky dimethyl-(1-naphthyl) phosphine has stimulated an extension of previous work[89] with the preparation of the complexes $[MX_2\{EMe_2(1\text{-naphthyl})\}_2]$ (M = Pd or Pt; X = Cl, I, or Me; E = P or As)[90]. Although the Pt halide complexes showed no tendency to undergo internal metallation even on prolonged heating, the methyl analogues readily evolved CH_4

[159] M. F. Mogilevkina, V. I. Bessonov, and I. H. Cheremisina, *Russ. J. Inorg. Chem.*, 1973, **18**, 1396.
[160] L. M. Volshtein and A. E. Prosenko, *Russ. J. Inorg. Chem.*, 1973, **18**, 1393.
[161] A. I. Stetsento and S. G. Strelin, *Russ. J. Inorg. Chem.*, 1973, **18**, 1404.

on pyrolysis to give complexes in which the naphthyl group was metallated in the 8-(peri) position. However, in the presence of NaOAc the halide complexes were also readily metallated in this position.

Complexes of type $[PtX_2L_2]$ $[X = Cl$ or I; $L = PPh_2(2\text{-}MeOC_6H_4)$, $PBu^tMe\text{-}(2\text{-}MeO_6H_4)$, $PBu^t_2(2\text{-}MeOC_6H_4)$, or $PPh_2(2\text{-}EtOC_6H_4)]$ have been synthesized which on heating are converted into O-metallated chelates (equation 32).[162] C-metallated derivatives, characterized by ^1H and ^{31}P n.m.r., were also prepared by

$$trans\text{-}[PtCl_2\{PPh_2(2\text{-}MeOC_6H_4)\}_2] \rightarrow cis\text{-}[Pt(Ph_2PC_6H_4O)_2] \qquad (32)$$

heating L or $[PtCl_2L_2]$ with $[PtCl_2(RCN)_2]$ in xylene.

Square-planar complexes of the unusual ligand 2,11-bis(diphenylphosphinoethyl)-benzo[c]phenanthrene) (44), which is capable of spanning *trans* positions, have been obtained by reaction with suitable compounds of Pt, Pd, and Ni.[91] The complexes are monomeric and non-electrolytes in CH_2Cl_2 and the *trans*-configuration of P atoms was established by ^1H n.m.r. Other similar complexes containing a diphosphine

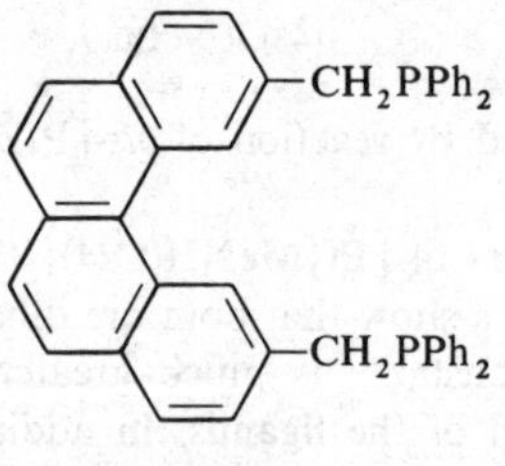

(44)

spanning *trans*-positions have been synthesized (equation 33).[163] The yellow mono-

$$[PtCl_2(PhCN)_2] + Bu^t_2P(CH_2)_{10}PBu^t_2 \rightarrow isomers [PtCl_2\{Bu^t_2P(CH_2)_{10}PBu^t_2\}]_x \quad (33)$$

meric isomer sublimes at 180—195 °C and was characterized by i.r. and mass spectroscopy. The other less soluble isomer was shown to have the structure (45) by i.r. and ^1H n.m.r. spectroscopy.

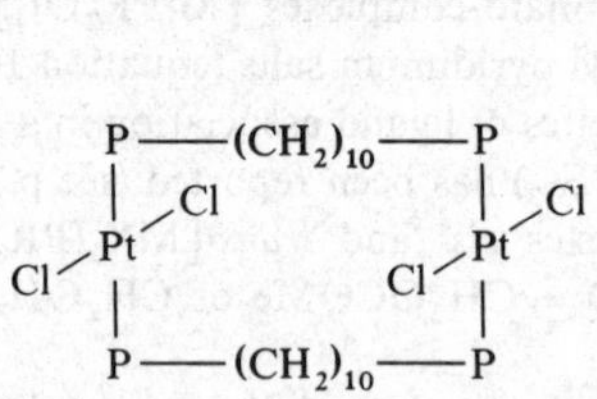

(45)

ESCA measurements for a wide range of phosphine complexes of type $[PtXYL_2]$ $[L = PPh_3; X, Y = H, Cl; H, C(CN)_3; Cl, Cl; SCN, SCN; N_3,N_3; NCO, NCO; CN, Cl; CN, CN; CNO, CNO]$ have yielded Pt $4f_{5/2}$ and $4f_{7/2}$ binding energies.[164]

[162] C. E. Jones, B. L. Shaw, and B. L. Turtle, *J.C.S. Dalton*, 1974, 992.
[163] A. J. Pryde, B. L. Shaw, and B. Weeks, *J.C.S. Chem. Comm.*, 1973, 947.
[164] W. Beck and F. Holsboer, *Z. Naturforsch*, 1973, **28b**, 511.

Stereospecific synthetic methods for the preparation of complexes *cis-* and *trans-*[PtClL(PEt$_3$)$_2$]$^+$ and *cis-*[PtClX(PEt$_3$)$_2$] (L = py, PPh$_3$, PClPh$_2$, or P(OPh)$_3$; X = SCN, NO$_2$, NO$_3$, N$_3$, NCO or Br) have been developed.[165] Three types of reaction were employed, (34)—(36), and it was shown by i.r., far-i.r. [ν(Pt–Cl)], and

$$[Pt_2Cl_2(PEt_3)_4](ClO_4)_2 + 2L \rightarrow 2\ cis\text{-}[PtClL(PEt_3)_2]ClO_4 \tag{34}$$

$$trans\text{-}[PtCl_2(PEt_3)_2] + L + NaClO_4 \rightarrow trans\text{-}[PtClL(PEt_3)_2]ClO_4 + NaCl \tag{35}$$

$$[Pt_2Cl_2(PEt_3)_4](ClO_4)_2 + 2NaX \rightarrow 2\ cis\text{-}[PtClX(PEt_3)_2] + NaClO_4 \tag{36}$$

^{1}H n.m.r. that bridge cleavage reactions in square-planar Pt complexes occur with retention of stereochemistry. This stereospecificity contrasts with the reported reaction of *cis-*[PtCl$_2$(PEt$_3$)$_2$] with L to give a mainly *trans* product,[166] and suggests that the rate of isomerization of the product is greater than the rate of substitution by L.

Four similar papers reporting new cationic and neutral complexes of Pt group metals with phosphite, phosphonite and phosphinite ligands have appeared (see p. 397).[93—96]

$$[PtX_2(PhCN)_2] + PPh_2H \xrightarrow{C_6H_6} \tag{37}$$

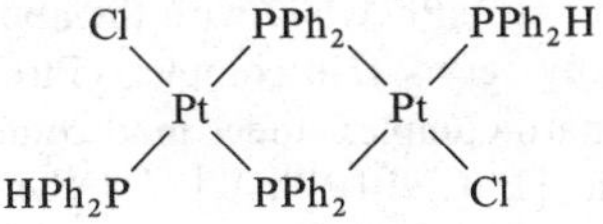

(X = Cl, Br, or I)

Secondary phosphine complexes of Pt, which have potential use as anticancer drugs, have been synthesized (equation 37). It was found that deprotonation occurs more easily in these Pt complexes than in the Pd analogues.[167]

Sb-*Donor ligands*. Stibine complexes of type [PtX$_2$(osb)$_2$] and [PtX$_2$(nsb)$_2$] (X = Cl, Br or I; osb = *o*-methoxyphenyldimethylstibine, nsb = *o*-dimethylaminophenyl-dimethylstibine) have been synthesized by treatment of K$_2$PtX$_4$ with the free ligand in ethanol.[168] These complexes dissociate in MeOH to produce halide ion which can be titrated with AgNO$_3$. In addition the cationic [PtCl(nsb)$_2$]$^+$ has been isolated as its tetraphenylborate salt. I.r., electronic, and ^{1}H n.m.r. spectra suggest the square-planar structure (46).

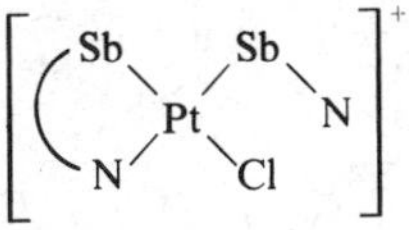

(46)

A preliminary examination of the complexing ability of the Sb analogue of the much studied *o*-phenylenebis(dimethylarsine) ligand shows that pale yellow-orange

^{165} K. R. Dixon, K. C. Moss, and M. A. R. Smith, *Canad. J. Chem.*, 1974, **52**, 692.
^{166} M. J. Church and M. J. Mays, *J. Chem. Soc. (A)*, 1968, 3074.
^{167} W. Levason, C. A. McAuliffe, and B. Riley, *Inorg. Nuclear Chem. Letters*, 1973, **9**, 1201.
^{168} B. Zarli, L. Volponi, and G. DePaoli, *Inorg. Nuclear Chem. Letters*, 1973, **9**, 997.

crystalline solids $[MX_2L]$ (M = Pd or Pt; X = Cl, Br, I, SCN) are formed which are non-electrolytes in $PhNO_2$.[97]

Group IV. C-Donor ligands. Treatment of salts of Pt (Pd, Rh, Ir) with 1-benzoyl-iminopyridinium (L) gives the complexes $[PtCl(L - H)L]$ (see p. 398).[103] Dicyano-methanide complexes of Pt (Pd, Rh, Ir) have been prepared, *e.g.* equation (38) (see p. 398).[101] Tricyanomethanide complexes of Pt have been synthesized by oxidative

$$[PtClX(PPh_3)_2] + NaCH(CN)_2 \xrightarrow{MeOH} [PtX(C_3HN_2)(PPh_3)_2] \qquad (38)$$

addition of dicyanoketen or tetracyanomethane to $[Pt(PPh_3)_4]$, and complexes of type $[MX\{C(CN)_3\}L_2]$ (M = Pd or Pt; X = Cl, Br or $C(CN)_3$; L = PPh_3 or $AsPh_3$) by treatment of $K[C(CN)_3]$ with the corresponding halogeno-complex have also been obtained.[102] The co-ordination of the $C(CN)_3$ ligand was discussed using i.r., ^{14}N n.m.r. and Raman data.

Formation of either *cis*- or *trans*-isomers of $[PtCl(CH_2CN)(PPh_3)_2]$ occurred when $[Pt(PPh_3)_4]$ was treated with $ClCH_2CN$ in acetone or benzene.[169] Both phosphine ligands are displaced by reaction with bidentate ligands, *e.g.* = $Ph_2PC_2H_4PPh_2$, $Ph_2PC_2H_4AsPh_2$, *cis*-$Ph_2PCH=CHPPh_2$.

A safe synthesis of a number of metal fulminato-complexes, including $[Pt(R_3E)_2(CNO)_2]$ (E = P, As or Sb) and $[PtX(CNO)(PPh_3)_2]$, (X = H, H, Me, CN or NCO) by reaction of $AsPh_4CNO$ with the appropriate Pt substrate has been reported.[99] The thermally very stable complex $[Pt(CNO)_2(PPh_3)_2]$ was shown to isomerize to the isocyanato-complex under mild conditions, and to be reduced by phosphines to the cyanide $[Pt(CN)_2(PPh_3)_2]$.[99] Other fulminato-complexes have been synthesized (equation 39) (R = Ph, Me or Et), and again i.r. evidence also shows isomerization

$$[PtCl_2(PPh_3)_2] \xrightarrow[C_6H_6]{CH(NO_2)COR} [PtCl(CNO)(PPh_3)_2] \qquad (39)$$

to the N-bonded isocyanato-complex.[170] The first step in the reaction was thought to be probably a cleavage of the C—C bond followed by co-ordination of the fragments and production of the corresponding carboxylic acid.

Formation of sulphur ylide complexes of Pt(Pd) of type $[PtX_2(R_2S)Sy]$ (X = Cl, Br or I; R = Me, Et; Sy = $Me(Ph)SCHC(O)C_6H_4Cl$-p) has been reported.[105] The $\nu(CO)$ band of the metal complexes occurs at higher frequencies than in the free ylide, and in addition spin–spin coupling between the ylide methine proton and the ^{195}Pt nucleus is observed, which is indicative of the Pt—C bond (47).

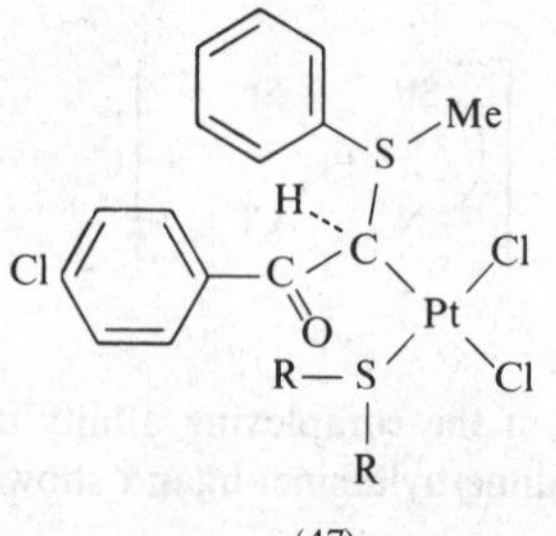

(47)

[169] K. Suzuki, H. Yamamoto, and S. Kanie, *J. Organometallic Chem.*, 1974, 73, 131.
[170] S. Alessandrini, I. Collamati, and C. Ercolano, *J.C.S. Dalton*, 1973, 2409.

Three double complex salts $[Pt(CNR)_4][Pt(CN)_4](R = Me, Et$ or $Bu^t)$ exhibit low-energy electronic absorption bands in the solids which are absent in solution spectra of the anion or cation complexes with simple counter-ions.[171] These bands, which are responsible for the intense colour of the solids, are ascribed to $M \rightarrow L$ charge transfer which has been red-shifted by Davydov interaction between anion and cation. Cation–anion association was also observed in MeCN solution and association constants were measured.

The gaseous molecule PtC_2 has been observed in a high-temperature mass spectrometer and its atomization energy determined.[172]

Si-Donor ligands. Labile Pt complexes $[Pt(PPh_3)_2(C_2H_4)]$ or $[PtMe_2(PMe_2Ph)_2]$ have been shown[173, 174] to react with organosilicon hydrides of type R_3SiH ($R_3 = Ph_3$, Ph_2Me, Ph_2H, $PhMe(CH_2=CH)$, Et_3, $(EtO)_3$, $(Me_3SiO)_2Me$ or $(p-FC_6H_4)_3$ to produce hydridosilyl complexes $[PtH(SiR_3)(PPh_3)_2]$ or $[PtH(SiR_3)(PMe_2Ph)_2]$. On the other hand, bis-(silyl) complexes $[Pt(SiMe_2Cl)_2(PPh_3)_2]$ and $[Pt(SiMePh_2)_2$-$(PMe_2Ph)_2]$ are formed from Me_2ClSiH or excess $MePh_2SiH$. The silicon dihydride Ph_2SiH_2, however, reacts more readily to give $[Pt(SiPh_2H)_2(PMe_2Ph)_2]$. I.r. spectra show that all complexes have the *cis*-configuration. Also reported are new Pt complexes containing chelating bis-(silyl) ligands.[175] These complexes with the ligands C_6H_4-*o*-$(SiHMe_2)_2$ and C_6H_4-*o*-$(SiHMe_2)(CH_2SiHMe_2)$ contain five- or six-membered rings, and further reaction with 1,2-bis(diphenylphosphino)ethane can occur to provide novel doubly-chelated complexes (48). In addition four-membered ring chelates

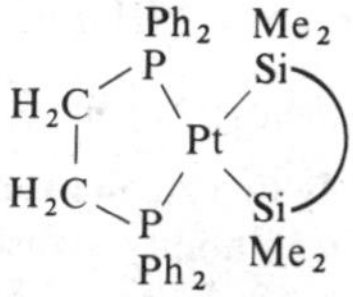

(48)

have been obtained by treatment of $[Pt(PPh_3)_2(CH_2=CH_2)]$ with the disiloxane $(HPh_2Si)_2O$ (equation 40). Seven-membered rings are not, however, formed by reaction with C_6H_4-*o*-$(HMe_2SiCH_2)_2$ but instead the hydrosilyl complex *cis*-[PtH-

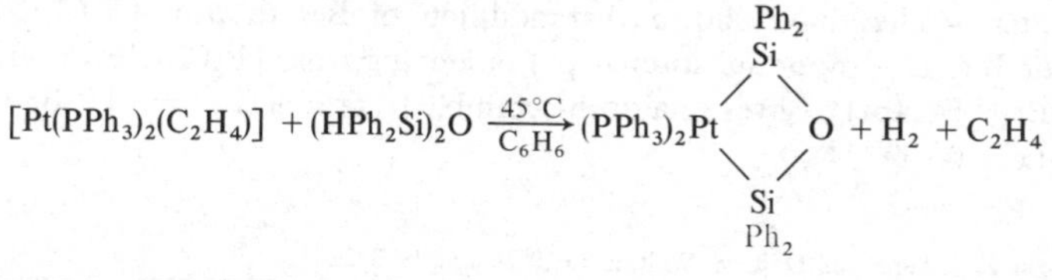

$$[Pt(PPh_3)_2(C_2H_4)] + (HPh_2Si)_2O \xrightarrow[C_6H_6]{45\,^\circ C} (PPh_3)_2Pt \underset{\underset{Ph_2}{Si}}{\overset{\overset{Ph_2}{Si}}{\diagdown}}O + H_2 + C_2H_4 \qquad (40)$$

$\{SiMe_2CH_2C_6H_4$-*o*-$(CH_2)SiMe_2H\}(PPh_3)_2]$ is produced. The stereochemistry of the formation and cleavage of Pt—Si bonds has also been examined.[176] Two

[171] H. Isci and W. R. Mason, *Inorg. Chem.*, 1974, **13**, 1175.
[172] K. A. Gingerich, *J.C.S. Chem. Comm.*, 1974, 199.
[173] C. Eaborn, B. Ratcliff, and A. Pidcock, *J. Organometallic Chem.*, 1974, **65**, 181.
[174] C. Eaborn, A. Pidcock, and B. Ratcliff, *J. Organometallic Chem.*, 1974, **66**, 23.
[175] C. Eaborn, T. M. Methan, and A. Pidcock, *J. Organometallic Chem.*, 1973, **63**, 107.

complexes have been formed (equations 41 and 42) with almost complete retention

$$cis\text{-}[PtCl_2(PMe_2Ph)_2] + (+)\text{-}R_3Si*H \xrightarrow[C_6H_6]{Et_3N} (+)\text{-}trans\text{-}[PtCl(Si*R_3)(PMe_2Ph)_2] \quad (41)$$

$$[Pt(PPh_3)_2C_2H_4] + (+)\text{-}R_3Si*H \rightarrow (-)\text{-}cis\text{-}[Pt(Si*R_3)H(PPh_3)_2] \quad (42)$$

$$R_3Si* = Me(1\text{-}C_{10}H_7)PhSi$$

of configuration around Si. In addition $(+)\text{-}R_3Si*H$ is regenerated from both complexes on treatment with $LiAlH_4$ and it is suggested that cleavages of the Pt—Si bond in these complexes, especially those involved in reaction or formation of R_3Si*H, occur *via* oxidative addition reductive elimination sequences with almost complete retention of configuration at Si. Similar reactions involving $Ge*R_3H$ also indicate that these complexes are formed with retention of configuration at Ge.

Group III. Boron donor ligands. The preparation of a range of metal complexes of B_6H_{10} has been reported (equation 42a) following preliminary publications.[177—179]

$$K[PtCl_3(C_2H_4)] + 2B_6H_{10} \rightarrow trans\text{-}[PtCl_2(B_6H_{10})_2] + C_2H_4 + KCl \quad (42a)$$

The complex possesses a novel three-centre two-electron bond, characterized by i.r., ^1H n.m.r., and ^{11}B n.m.r. measurements. The X-ray structure has already been reported.[179]

Platinum(IV). *Group VII Donors.* The equilibrium constants of equation (43) have been studied potentiometrically for am = NH_2EtOH, $MeNH_2$, $EtNH_2$, NH_3 or

$$[PtCl_2(am)_4]^{2+} + Br^- \rightleftharpoons [PtClBr(am)_4]^{2+} + Br^- \rightleftharpoons [PtBr_2(am)_4]^{2+} \quad (43)$$

$\frac{1}{2}$en,[180] and were found to depend only to a small extent on the nature of am. Similar features were also observed for the stability constants in aqueous solution of $[PtCl_2(am)_4]^{2+}$ and $[PtCl(H_2O)(am)_4]^{3+}$ (am = NH_3, $MeNH_2$, $EtNH_2$, py or $\frac{1}{2}$en).[181]

The thermal stabilities of the complexes $K_2[PtX_6]$ (X = F, Cl, Br or I) have been studied and it was found that the stabilities decrease from X = F to X = I.[182] An investigation into the photolysis of frozen $K_2[PtI_6]$ solution in LiCl (10M) glasses has shown that Pt^{II} and I_2^- are the principal photolysis products which are formed by electron phototransfer from ligand to metal.[183]

^{31}P N.m.r. studies have shown that addition of Br_2 to $trans\text{-}[PtCl_2(PR_3)L]$ (R = Et or Bu; L = py or substituted py) or mixing $trans\text{-}[PtCl_4(PEt_3)(py)]$ with $trans\text{-}[PtBr_4(PEt_3)(py)]$ gives halogen-scrambled products $trans\text{-}[PtBr_xCl_{4-x}(PEt_3)L]$ (x = 0—4).[184]

[176] C. Eaborn, D. J. Tune, and D. R. M. Walton, *J.C.S. Dalton*, 1973, 2255.
[177] A. Davison, D. D. Traficante, and S. S. Wreford, *J. Amer. Chem. Soc.*, 1974, **96**, 2802.
[178] A. Davison, D. D. Traficante, and S. S. Wreford, *J.C.S. Chem. Comm.*, 1972, 1155.
[179] J. P. Brennan, R. Schaeffer, A. Davison, and S. S. Wreford, *J.C.S. Chem. Comm.*, 1973, 354.
[180] Yu. N. Kukushkin, E. F. Strizhev, and B. A. Krasnov, *Russ. J. Inorg. Chem.*, 1973, **18**, 1607.
[181] Yu. N. Kukushkin, E. F. Strizhev, B. A. Krasnov and N. V. Kiseleva, *Russ. J. Inorg. Chem.*, 1973, **18**, 1144.
[182] L. K. Shubochkina, V. I. Gushchin, M. B. Varfolomeev, and E. F. Shubochkina, *Russ. J. Inorg. Chem.*, 1973, **18**, 850.
[183] G. A. Shagisultanova and A. A. Karaban, *Russ. J. Inorg. Chem.*, 1973, **18**, 905.
[184] B. T. Heaton and K. J. Timmons, *J.C.S. Chem. Comm.*, 1973, 931.

Oxidative addition of halogen (Cl_2, Br_2 or I_2) to $[PtX_3L]^-$ or *cis*-$[PtX_2L_2]$ (L = PMe_3, $AsMe_3$ or SMe_2; X = halogen) has led to the corresponding Pt^{IV} complexes.[185] The complexes were characterized by i.r. spectroscopy, and ^1H n.m.r. measurements were also made for these and other complexes of Pt and Pd. Values of $J(PtP)$ and δ_P have been determined by heteronuclear INDOR, and approximate values of $J(PtN)$ were obtained from $^1H(^{195}Pt)$ INDOR in three cases. It was concluded that the variations in the spectra were consistent with a significant π-contribution to the Pt—P bonding at least in Pt^{II} complexes.

Oxygen Donors. The formation and stability constants of complexes between Pt (Pd, Rh, Ir, Os, Ru) and *o*-coumaric acid have been determined by pH titration.[31] The results indicate that a 1:2 complex is formed with Pt. Acid–base properties of aquocomplexes formed from $[Pt(X)_2(OH)_2(NH_3)(MeNH_2)]$ (X = Cl, Br or NO_2) in aqueous solutions have been examined using potentiometric titration experiments.[186] The K_a of co-ordinated water was lower for $(X)_2 = (H_2O)_2$ than for $(X)_2 = (H_2O)$ (OH^-).

It has been shown by i.r. and ^1H n.m.r. that acacH and trifluoracetylacetone act as reducing agents on complexes of type $[PtX_6]^{2-}$ (X = Cl, Br or I) to produce the yellow Pt^{II} anions $[Pt(acac)_3]^-$ (X = I) and $[PtX(acac)_2]^-$ (X = Cl or Br).[187] In contrast, reaction with $PtCl_4$ (equation 44) leads to condensation of acacH to produce

$$PtCl_4 + acacH \xrightarrow{\Delta} (C_{10}H_{13}O_2)_2[PtCl_6] \qquad (44)$$

the unusual substituted pyrillium cation (49).

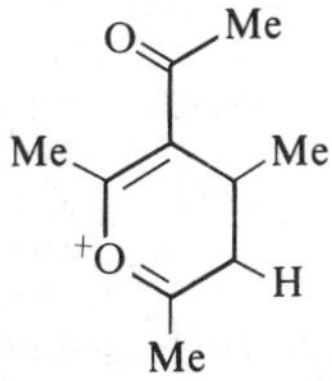

(49)

Nitrogen Donors. The reactions of the complexes $[PtBr_3(L - H)]$ and $[PtX_4L]$ (X = Cl, Br or I; L = $H_2NCH_2CH_2NHCH_2CH_2OH$) in liquid NH_3 at $-75\,°C$ have been studied in order to examine the inductive effect of the amino-group on the reactivity of the ethanolic oxygen and the lability of the hydrogen.[80] Various types of behaviour were observed, ranging from no reaction to deprotonation and/or ligand substitution.

Bis-biguanide complexes of Pt have been obtained by treating the ligand with $PtCl_4$ in alkaline medium.[156] Treatment of H_2PtCl_6 with biguanide hydrogen sulphate has yielded the tris-(biguanidinium) complex $[Pt(C_2N_5H_7)_3]^{4+}$.

Following the recent observation that en in the edge structure triamine [(en)-

[185] P. L. Goggin, R. J. Goodfellow, S. R. Haddock, J. R. Knight, F. J. S. Reed, and B. F. Taylor, *J.C.S. Dalton*, 1974, 523.
[186] N. N. Zheligooskaya, G. P. Zimtseva, and V. I. Spitsyn, *Bull. Acad. Sci. U.S.S.R.*, 1974, **22**, 2333.
[187] C. Oldham and A. P. Ketteringham, *Inorg. Chim. Acta*, 1974, **9**, 127.

Table 3 X-*Ray data for platinum complexes*

Complex	R	Comments	Ref.
$K_2[PtCl_6]$	0.067	octahedral Pt—Cl = 2.323 Å	a
$[FePt_2(CO)_5\{P(OPh)_3\}_3]$	0.062	triangular FePt$_2$ cluster. Four CO bonded to Fe, one CO and P(OPh)$_3$ to Pt with distorted square-planar structure. Pt—Pt = 2.633 Å	b
$(PPh_4)_2[Pt_3(CO)_3(\mu_2{-}CO)_3]_2$	0.059	all anions are based on polymerization of a triplatinum	107
$(PPh_4)_2[Pt_3(CO)_3(\mu_2{-}CO)_3]_3$	0.068	hexacarbonyl component by further Pt—Pt bonding (See	107
$(AsPh_4)_2[Pt_3(CO)_3(\mu_2{-}CO)_3]_5$	0.063	text)	107
$[Pt(PPh_3)_2(PhCONNCOPh)]EtOH$	0.037	bidentate (O,N) dibenzoylhydrazide with five-membered ring	c
$[Pt(PPh_3)_3(SO_2)], C_6H_6$	0.067	trigonal-pyramidal structure with Pt—S = 2.399 Å. Oxygen atoms of SO$_2$ are disordered about the three-fold axis	d
$[PtMe_3(rac\text{-}diars)I]$	0.060	pseudo-octahedral 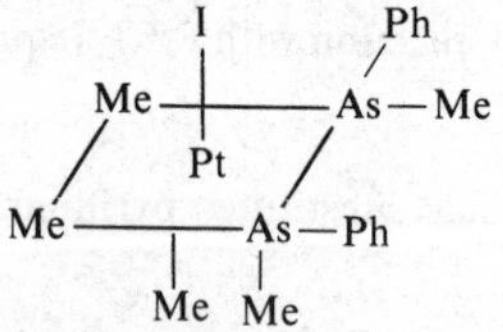	e

Complex	R	Comments	Ref.
$[Pt(PPh_3)_2(1\text{-}ethynylcyclohexanol)_2]$	0.052	*trans*-square-planar Pt—C(ethynyl) = 2.004 Å	f
$[Pt\{SC(NH_2)_2\}_4]Cl_2$	0.054	almost planar PtS$_4$, Pt—S average = 2.313 Å N—H·····Cl hydrogen bond	g
$[Pt(PPh_3)_2(C_4H_8)]$	0.074	distorted square-planar with co-ordination through terminal C of tetramethylene ring. Pt—C = 2.12 and 2.05 Å	h
$[Pt(Me)(MeC{\equiv}CMe)(PPhMe_2)_2]PF_6$	0.036	approximately square-planar if acetylene considered unidentate, phosphines *trans*-configuration	i
$[PtI(MeC{=}NC_6H_4Cl)(PEt_3)_2]$	0.038	structure shows long Pt—N of 3.037 Å, *i.e.* not a pseudo five-co-ordinate complex. Pt—C = 2.029 Å 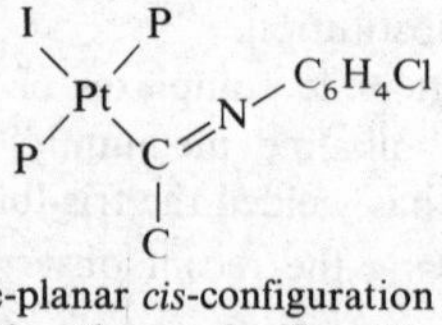	j

Complex	R	Comments	Ref.
$[PtF\{CH(CF_3)_2\}(PPh_3)_2]$	0.063	square-planar *cis*-configuration Pt—P lengths are significantly different. Pt—P (*trans* to F) = 2.218 Å; Pt—P [*trans* to CH(CF$_3$)$_2$] = 2.310 Å	k

Table 3—*contd.*

Complex	R	Comments	Ref.
$(Me_4N)_3[Pt(GeCl_3)_5]$	0.054	anion is distorted trigonal bipyramid. Equatorial angles are considerably different from 120°	*l*
cis-$[Pt(NCS)(SCN)(Ph_2PC\equiv CBu^t)_2]$	0.057	square planar 2P, S, N	*m*
cis-$[PtCl(PEt_3)_2(phen)]BF_4$	0.051	results indicate square-planar rather than five-co-ordinate. Pt—N(1) = 2.137 Å, Pt—N(2) = 2.843 Å. First example of unidentate phen. observed in the solid state. In solution there is a possible fluxional behaviour: see text	150a
$[PtCl_2(bipy)]$ (red form)	0.024	layers of monomeric molecules lying parallel and separated by 3.40 Å. Pt atoms are almost superimposed	*n*
$(Et_4N)_2[Pt_2(NO)_2Cl_6]$	0.07	two Pt bridged by Cl and NO One Pt is square planar while the other is octahedral. Terminal NO is *cis*- to bridging NO and bent towards it (∠ PtNO = 122°) 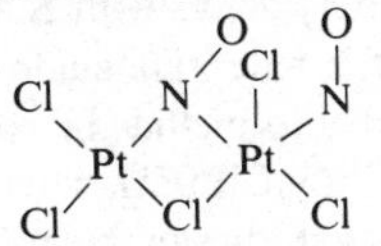	*o*

$$
\begin{array}{c}
\qquad\quad O \quad O \\
\qquad\quad \diagdown\diagup \\
\qquad\quad N \quad Cl \\
Cl \quad \diagup \quad | \quad N \\
\diagdown \diagup \diagdown \diagup \quad \diagdown \\
Pt \qquad Pt \\
\diagup \quad \diagdown \quad \diagdown \\
Cl \qquad Cl \qquad Cl
\end{array}
$$

Complex	R	Comments	Ref.
$[Pt(PF_2O)(Et_2PhP)_2]$		powder. PEt_2Ph groups *trans*-, PF_2O bonded *via* P	*p*
cis-$[PtCl_2(CO)(PPh_3)]$	0.067	Slightly distorted square planar. ∠ CPtCl = 172.4°	*q*
$CaPt_2O_4$		powder (*X*-ray and neutron diffraction). Involves the first example of a two-dimensional analogue of the general Pt bronze structure, with non-intersecting Pt chains in the *x* and *y* directions.	*r*
$Cd_{0.3}Pt_3O_4$		powder (*X*-ray and neutron diffraction). Has the $NiPt_3O_4$ structure with Pt chains in *x*, *y*, and *z* directions Pt—Pt = 2.80 Å indicates metallic bonding. Planar triangular co-ordination of O around Pt	*r*
$[PtX_4en][PtX_2en]$ (X = Cl, Br, or I)		powder. Unit cell data and indexed powder patterns given	*s*
$[PtFe_2(CO)_9(PPh_3)]$	0.072	each Fe has four terminal CO groups with Pt having CO and 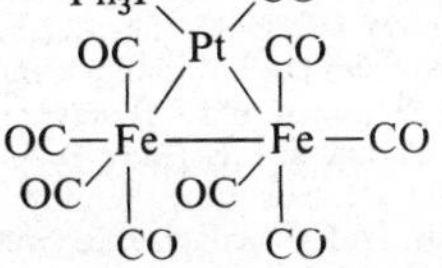 PPh₃ groups	*t*

$$
\begin{array}{c}
Ph_3P \qquad CO \\
\diagdown \qquad \diagup \\
OC \quad Pt \quad CO \\
| \quad \diagup\diagdown \quad | \\
OC-Fe \text{———} Fe-CO \\
OC \diagup \; | \quad OC \diagup \; | \\
\quad CO \qquad CO
\end{array}
$$

Table 3—*contd.*

Complex	R	Comments	Ref.
$[Pt(Me)\{MeC(OMe)\}(PMe_2Ph)_2]PF_6$	0.047	Me, Pt and carbene ligand all lie in the mirror plane. Carbene is in *trans*-configuration and is disordered in the mirror plane. $Pt—CH_3 = 2.13$ Å	*u*

(*a*) R. J. Williams, D. R. Dillin, and W. O. Milligan, *Acta Cryst.*, 1973, **B29**, 1369; (*b*) V. G. Albano and G. Ciano, *J. Organometallic Chem.*, 1974, **66**, 311; (*c*) S. D. Ittel and J. A. Ibers, *Inorg. Chem.*, 1973, **12**, 2290; (*d*) J. P. Linsky and C. G. Pierpont, *Inorg. Chem.*, 1974, **12**, 2959; (*e*) G. Casalone and R. Mason, *Inorg. Chim. Acta*, 1973, **7**, 429; (*f*) R. A. Maiezcurrena and S. E. Rasmussen, *Acta Chem. Scand.*, 1973, **27**, 2678; (*g*) R. L. Girling, K. K. Chattergee, and E. L. Amma, *Inorg. Chim. Acta*, 1973, **7**, 557; (*h*) C. G. Barefield, H. A. Eick, and R. H. Grubbs, *Inorg. Chem.*, 1973, **12**, 2166; (*i*) B. W. Davies and N. C. Payne, *Canad. J. Chem.*, 1973, **51**, 3477; (*j*) K. P. Wagner, P. M. Treichel, and J. C. Calabrasse, *J. Organometallic Chem.*, 1973, **56**, C33; (*k*) J. Howard and P. Woodward, *J.C.S. Dalton*, 1973, 1840; (*l*) E. D. Estes and D. J. Hodgson, *Inorg. Chem.*, 1973, **12**, 2932; (*m*) Y. S. Wong, S. Jacobson, P. C. Chieh, and A. J. Carty, *Inorg. Chem.*, 1974, **13**, 284; (*n*) R. S. Osborn and D. Rogers, *J.C.S. Dalton*, 1974, 1002; (*o*) J. M. Epstein, A. H. White, S. B. Wild, and A. C. Willis, *J.C.S. Dalton*, 1974, 436; (*p*) J. Grosse, R. Schmutzler, and W. S. Sheldrick, *Acta Cryst.*, 1974, **B30**, 1623; (*q*) L. Manojlović-Muir, K. W. Muir, and R. Walker, *J. Organometallic Chem.*, 1974, **66**, C21; (*r*) D. Cahen, J. A. Ibers, and M. H. Mueller, *Inorg. Chem.*, 1974, **13**, 110; (*s*) K. W. Browall, J. S. Kasper, and L. V. Interrante, *Acta Cryst.*, 1974, **B30**, 1649; (*t*) R. Mason and J. A. Zubieta, *J. Organometallic Chem.*, 1974, **66**, 289; (*u*) R. F. Stepaniak and N. C. Payne, *J. Organometallic Chem.*, 1973, **57**, 213.

(py)Cl(NO$_2$)ClPtCl] reacts with KNO$_2$ to produce [(NOen − H)(py)Cl(NO$_2$)ClPt], which contains the bidentate acido-amino ligand NONC$_2$H$_4$NH$_2^-$,[188] these studies have been extended to include facial triamine complexes [(en)(NO$_2$)Cl(NH$_3$)ClPtCl] and [(en)(NO$_2$)Cl(py)ClPtCl], where again inner-sphere nitrosation of en was found to occur.[189] Other studies of this nitrosation for the complexes *trans*-[PtCl$_2$(en)$_2$]Cl$_2$ and *cis*-[Pt(en)$_3$]Cl$_4$ in acidic media, pH 5—7, have indicated that co-ordinated en undergoes nitrosylation in a deprotonated form.[190]

The hitherto unknown hexanitrato-complexes K$_2$[M(NO$_3$)$_6$] (M = Pt or Ir) have been synthesized by reaction of K$_2$[PtBr$_6$] with N$_2$O$_5$.[191] I.r. data suggest that the nitrate groups are bidentate, and *X*-ray powder measurements show both complexes to be isomorphous.

Boiling of K$_2$[Pt(X,Y)$_3$] (X,Y = NO$_2$, NO$_3$; I,I; Br, Br) in concentrated HNO$_3$ has produced the brick-red complex K$_2$[Pt(NO$_3$OH)$_3$]; however, no reaction occurred when X,Y = F,F or Cl,Cl.[192] I.r. and *X*-ray photoelectron spectra indicate the presence of both co-ordinated NO$_3$ and OH groups. The complex hydrolyses (equation 45) in boiling water to give a brown precipitate which is probably polymeric.

$$4K_2[Pt(NO_3OH)_3] + 6H_2O \rightarrow H_2[Pt_4(OH)_{18}] + HNO_3 \tag{45}$$

Platinum(v).—Twelve dioxygenyl salts of complex fluorometallate anions, [MF$_6$]$^-$ (M = As, Sb, Bi, Ru, Rh, Au, or Pt) have been prepared by static heating of the metal at 300—350 °C with 5 atm of 3:1 fluorine–oxygen mixture for upwards of 10 h.[193] The salts were characterized by powder *X*-ray, Raman, and mass spectroscopy.

[188] I. Z. Chernyaev, O. N. Adrianova, and N. S. Gladkaya, *Zhur. neorg. Khim.*, 1967, **12**, 1877.

[189] O. N. Adrianova, I. F. Golovaneva, and A. I. Kravchenko, *Russ. J. Inorg. Chem.*, 1973, **18**, 753.

[190] N. S. Gladkaya and O. N. Adrianova, *Bull. Acad. Sci. U.S.S.R.*, 1973, **22**, 2607.

[191] B. Harrison, N. Logan, and J. B. Raynor, *J.C.S. Chem. Comm.*, 1974, 202.

[192] L. K. Shubochkina, V. I. Nefedov, E. F. Shubochkina, and M. A. Golubnichaya, *Russ. J. Inorg. Chem.*, 1973, **18**, 995.

[193] A. J. Edwards, W. E. Falconer, J. E. Griffiths, W. A. Sunder, and M. J. Vasile, *J.C.S. Dalton*, 1974, 1129.

Table 4 *I.r. structural studies of platinum complexes*

Complex	Other studies	Results and conclusions	Ref.
$M_2[Pt(CN)_5X]$ (M = alkali metal) X = Cl, Br, I, or CN)	Raman, far i.r.	assigned many of the fundamental frequencies. Apical Pt—C and C—N frequencies shift to lower energy for the series Cl > Br > I	a
$[Pt(NO_2)_{6-n}Cl_n]$ (n = 0—5)	Raman far i.r.	assigned bands	b
$K_2[PtX_6]$ (X = Br or I) $M_2[PtCl_6]$ (M = K, Rb, Cs, Tl, or NH$_4$)	far-i.r. Pd, Os, Ir	studied at pressures up to 20 kbar: ν_4, δ(ClMCl) mode generally more pressure sensitive than the ν_3, ν(MCl)	c
$[Pt(A_2NC_2H_4S)_2]$ (A = H or D)	Ni, Pd Raman	identified (Pt—N), (Pt—S) bands. Complexes have *trans*-configuration	d
$[Pt\{(PhO)_2PSSe\}_2]$	Rh, Ir, Pd	identified ν(M—S) and ν(M—Se)	e
$[PtCl(NH_3)_4(^{15}NO)]Cl_2$ $K_2[PtCl(NO_2)(^{15}NO)]$ $M_2[PtX_5(^{15}NO)]$ (M = K or Cs; X = Cl or Br)	^{14}NO and ^{15}NO	weak band at 290 cm^{-1} assigned to Pt—(NO) stretch, band ca. 530—570 cm^{-1} to Pt—(NO) band. Suggested that isotropic shift for NO stretch may be useful for determining whether NO$^+$ or NO$^-$	f
$M_2[Pt(NO_2)_4]$ (M = Na, K, Rb, Cs, or Ag)	Pd	in solution the complexes have D_{4h} symmetry. Spectra indicate that metal—N π-bonding is more pronounced for Pt than for Pd	g
cis-$[PtCl_2(PPh_3)_2]$ *cis*-$[Pt(PPh_3)_2(O_2)]$ *trans*-$[PtHCl(PPh_3)_2]$ *trans*-$[PtCl(COPh)(PPh_3)_2]$ $[PtCl(COOMe)(PPh_3)_2]$	far i.r.	found no correlation between number of bands in region 400—450 cm^{-1} and geometry and previous assignment of these bands is probably incorrect. Bands in region 200—160 cm^{-1} depend on geometry. *cis*-complexes have two bands while *trans* have only one	h
$[PtCl(NH_3)_2L]NO_3$ (L = PPh$_3$ or SbPh$_3$)	stability and acid dissociation constants	data indicate both σ-donating and π-accepting capacities diminish with PPh$_3$ > SbPh$_3$	i
$K[PtCl_3NH_3]H_2O$ $K[PtCl_3(ND_3)]D_2O$	Raman, single crystal powder	assigned vibrations	j
$[Pt(NH_3)_4][PtCl_4]$	Raman	A_{2u} lattice mode of cation–anion chain is at 81 cm^{-1} not at 201 cm^{-1} as previously assigned. Strong Pt—Pt interaction	k

(a) M. N. Memering, L. H. Jones, and J. C. Bailar, *Inorg. Chem.*, 1973, **12**, 2793; (b) M. J. Nolan and D. W. Jones, *Austral. J. Chem.*, 1973, **26**, 1413; (c) D. M. Adams, and S. J. Payne, *J.C.S. Dalton*, 1974, 407; (d) V. A. Jayasooriya and D. B. Powell, *Spectrochim. Acta*, 1974, **30A**, 553; (e) I. M. Cheremisina, L. A. Il'ina, and S. V.

Larinov, *Russ. J. Inorg. Chem.*, 1973, **18**, 675; (*f*) E. Miki, K. Mizamachi, T. Ishimori, and H. Okano, *Bull. Chem. Soc. Japan*, 1974, **47**, 656; (*g*) D. W. James and M. J. Nolan, *Austral. J .Chem.*, 1973, **26**, 1433; (*h*) S. H. Mastin, *Inorg. Chem.*, 1974, **13**, 1003; (*i*) I. V. Gavrilova, M. I. Gel'fman, N. V. Ivannikova, N. V. Kiseleva, and V. V. Razumovskii, *Russ. J. Inorg. Chem.*, 1973, **18**, 98; (*j*) D. M. Adams and R. E. Christopher, *J.C.S. Dalton*, 1973, 2298; (*k*) D. M. Adams and J. R. Hall, *J.C.S. Dalton*, 1973, 1450.

Table 5   *N.m.r. studies of platinum complexes*

Complex	Probe	Results and conclusions	Ref.
trans-[PtH(NCS)(PEt$_3$)$_2$]	^{1}H	the broadening of one of the sets of hydride ^{1}H n.m.r. patterns shown to be due to coupling with ^{14}N nucleus	a
[PtH(PEt$_3$)$_3$]$^+$ *trans*-[PtH(PEt$_3$)$_2$(PPh$_3$)] [PtH(PPh$_3$)$_3$]$^+$	^{1}H, ^{31}P	hydridic hydrogen resonances show different centre-band and ^{31}P side-band multiplet structures	b
[(Me$_2$S)$_2$PtX$_2$] (X = halide) [(MePhS)$_2$PtX$_2$] [(R$_2$Se)$_2$PtX$_2$] (R = Me or Et) [L$_2$PtX$_2$][L = Et$_3$P, Et$_2$PhP, Bz$_2$PhP, P(OMe)$_3$]	^{1}H	^{195}Pt shifts related to variations in electronic excitation energies when size of ligands remains approximately constant, but bulky ligand atoms produce large shifts to high field which dominate	c
[PtCl$_{4-n}$(R$_3$P)$_n$]$^{(n-2)+}$ [PtCl$_{6-n}$(R$_3$P)$_n$]$^{(n-2)+}$ (n = 1—3)	^{1}H	^1J(Pt—P) coupling constant obtained, interpretation in terms of s-orbital bond orders is supported by the correlation of bond lengths with ^{1}J for Pt^{II} complexes	d
[MX(PMe$_3$)$_3$]NO$_3$ (M = Pd or Pt) *mer*-[PtX$_3$(PMe$_3$)$_3$]NO$_3$ (X = Cl or Br) [PtI(PMe$_3$)$_3$]NO$_3$ [Pt(PMe$_3$)$_4$(BF$_4$)$_2$]	^{1}H	also ^{1}H$\{^{31}$P$\}$INDOR measurements made. Signs of some of couplings determined by heteronuclear double resonance	88
[Pt(PEt$_3$)(L)$_3$] (L = halide, hydride, silyl, germyl)	^{1}H	^{31}P and ^{195}Pt shifts and signs of coupling constants determined	e

(*a*) B. E. Mann, B. L. Shaw, and K. A. J. Stringer, *J. Organometallic Chem.*, 1974, **73**, 129; (*b*) T. W. Dingle, K. R. Dixon, *Inorg. Chem.*, 1974, **13**, 846; (*c*) W. McFarlane, *J.C.S. Dalton*, 1974, 324; (*d*) G. G. Mather, A. Pidcock, and G. J. N. Rapsey, *J.C.S. Dalton*, 1973, 2095; (*e*) D. W. W. Anderson, E. A. V. Ebsworth, and D. W. H. Rankin, *J.C.S. Dalton*, 1973, 2370.

3 Silver

Silver(I).—*Group VII Donors.* A study of the phase equilibria of the AgI–NaI–H$_2$O system has previously led to the isolation of the compound AgI,NaI,3H$_2$O;[194] u.v.–visible spectra and *X*-ray powder studies indicate the formulation Na[AgI$_2$],3H$_2$O.[195] The phase transition for the system AgI–I$^-$–(KNa)NO$_3$ at 280 °C has been studied at high iodide concentrations.[196] The results indicate the presence of [AgI$_4$]$^{3-}$ and polynuclear species. Thermodynamic calculations on the distribution of AgI between solid or liquid (K$_2$Ag)I and a nitrate melt have been reported.[197] It was concluded

[194] V. Krym, *J. Russ. Phys. Chem. Soc.*, 1909, **41**, 382.
[195] N. Nishimura and T. Higashiyama, *Bull. Chem. Soc. Japan*, 1974, **47**, 593.
[196] B. Holmberg, *Acta Chem. Scand.*, 1973, **27**, 3550.
[197] B. Holmberg, *Acta Chem. Scand.*, 1973, **27**, 3657.

that the species Ag^+, AgI, AgI_2^-, AgI_3^-, and $Ag_2I_6^{4-}$ were present, and stability constants for these complexes were computed. The solubilities of AgCl and AgBr in molten $Ca(NO_3)_2$–KNO_3 have been measured as a function of halide concentration using ^{110}Ag as tracer.[198] The solubility products and consecutive stability constants for the species AgCl, $AgCl_2^-$, AgBr, and $AgBr_2^-$ were also determined.

The formation of the transient species Ag^0 by reduction of Ag^+ from hydrated electrons using a double flash photolysis technique has been observed.[199] This species may be photodissociated at 315 nm, probably *via* charge-transfer to solvent, to produce Ag^+ and solvated electrons.

Group VI. O-Donor ligands. ^1H N.m.r. shifts have been measured for aqueous $AgNO_3$ and $AgBF_4$ between 15 and 100 °C.[200] The very small influence of the salts on the chemical shift of H_2O is attributed to almost exact cancellation of upfield anionic and downfield cationic contributions. The cationic hydration number appears to be less than 1 owing to the lability of the complex.

Conductivity measurements on the binary salt system Li_2CrO_4–Ag_2CrO_4 indicate the formation of the compound $2Li_2CrO_4,Ag_2CrO_4$, which was confirmed by an X-ray powder photograph.[201] The equilibrium diagram was constructed for this system. The liquidus diagram for the Ag_3PO_4–$AgPO_3$ system has been studied by differential thermal analysis.[202] The results showed that only one congruently melting compound of composition $Ag_4P_2O_7$ (m.p. 643 °C) was formed.

Studies of the electrical properties of polycrystalline $AgNO_3$ above and below room temperature have revealed the existence of a dielectric transition at -35 °C with a maximum in the dielectric constant occurring at 110 °C.[203]

Interest in the adsorbed oxygen species on Ag catalysts has stimulated the measurement of ESCA and Auger spectra of Ag, Ag_2O, and AgO.[204] The $3d$ binding energies were found to decrease in the order $AgO < Ag_2O < Ag$, which is in the opposite sense to that generally observed for metals and their oxides.

New sulphinato-complexes of Ag(Cu, Au) have been prepared by the metathetical exchange reaction (46).[205] The Ag and Cu complexes were shown by i.r. studies to

$$[Ph_3PMCl] + AgO_2SPh \xrightarrow[THF]{PPh_3} [(Ph_3P)_2MO_2SPh] \qquad (46)$$

be O-bonded.

S-Donor ligands. Direct reaction of pyridine-2-thiol with $AgClO_4$ or $AgBF_4$ in EtOH has yielded pale yellow complexes $[Ag(LH)_4]BF_4$ and $[Ag(LH)_2]ClO_4$.[45] I.r. evidence shows that LH is unidentate through sulphur. The mixed complex $[Co(SMC)_2Ag]NO_3,H_2O$ (SMC = S-methyl-L-cysteine) has been obtained by treatment of $[Co(SMC)_2]$ with $AgNO_3$ in water.[61] U.v.–visible and i.r. spectra suggest that Ag is co-ordinated through sulphur while the Co has N,O co-ordination.

Extraction of the metals Ag, Au, Pd, Pt from aqueous solution by a large number of alkyl and alkene disulphides and sulphides has been reported (see p. 390).[51,52]

[198] I. J. Gal, G. Djuric, and L. Melovski, *J.C.S. Dalton*, 1973, 2066.
[199] N. Basco, S. K. Vidyarthi, and D. C. Walker, *Canad. J. Chem.*, 1973, **51**, 2497.
[200] J. W. Akitt, *J.C.S. Dalton*, 1974, 175.
[201] M. P. Burmistrova and D. M. Shakirova, *Russ. J. Inorg. Chem.*, 1973, **18**, 1326.
[202] R. K. Osterheld and T. J. Mozer, *J. Inorg. Nuclear Chem.*, 1973, **35**, 3463.
[203] J. H. Fermor and A. Kjekshus, *Acta Chem. Scand.*, 1973, **27**, 3712.
[204] G. Schön, *Acta Chem. Scand.*, 1973, **27**, 2623.
[205] J. Bailey and M. J. Mays, *J. Organometallic Chem.*, 1973, **63**, C24.

Mixed cyanide–thiocyanate complexes of type $K[Ag(CN)(SCN)]$ and $K_2[Ag(CN)(SCN)_2]$ have been synthesized and characterized by conductivity and i.r. measurements.[206] The complexes decompose at *ca.* 250 °C and give Ag_2S at 360 °C.

New unusual polynuclear complexes have been obtained by replacing the inert cations in the dithio-oxalate compounds $(Ph_4X)[M(S_2C_2O_2)_3]$ ($X = P$ or As; $M = Fe$, Al or Cr), and in similar complexes of Ni and Zn, by co-ordinatively unsaturated species such as $[Ag(PPh_3)_2]^+$.[207] X-Ray structural studies to be published have shown that the $[Fe(S_2C_2O_2)_3]^{3-}$ ion undergoes a change of co-ordination from S-bonded to O-bonded upon reaction with $[Ag(PPh_3)_2]^+$, thereby leaving the S free to co-ordinate to Ag(50). I.r. and u.v.–visible spectra indicate that the complex $M = Cr$ linkage isomerizes on heating (equation 47). Further spectral studies for the Ni complex have led to the proposed structure (51) where the $[Ag(PPh_3)_2]$ moiety co-ordinates across two dithio-oxalate groups.

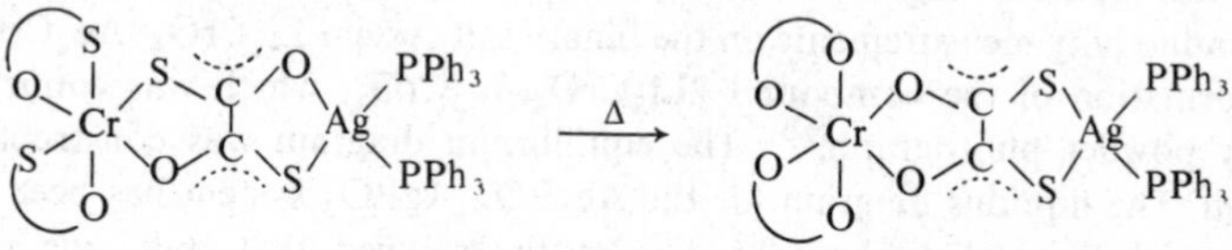

(47)

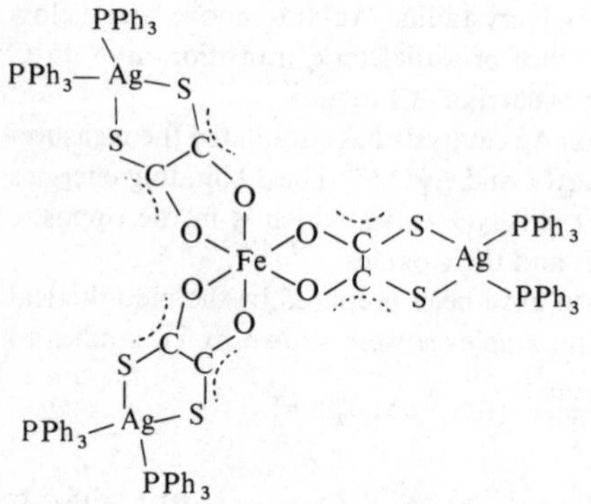

(50)

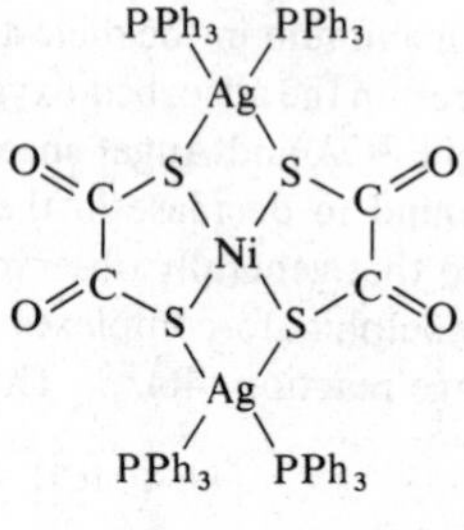

(51)

Group V. N-donor ligands. Complexes of pyrazine and 2,2'-bipyrimidine with $AgNO_3$ have been isolated as white solids, which are believed to be polymeric with the Ag having a co-ordination number of two.[203] Following previous work on complexes $[AgX_2NO_3]$ (X = quinoline or isoquinoline), where it was shown that the nitrate group is co-ordinated to Ag,[209] new complexes of type $[AgX_2]_3L\{L = [Cr(NCS)_6]^{3-}$, $[Cr(NCS)_4(aniline)_2]^-$, $[Cr(NCS)_4(o\text{-toluidine})_2]^-$, and $[Cr(NCS)_4(p\text{-toluidine})_2]^-\}$ have been synthesized which contain only two-co-ordinate Ag.[210] These complexes

206 A. N. Sergeeva, D. I. Semenishin, and A. V. Mazepa, *Russ. J. Inorg. Chem.*, 1973, **18**, 1572.
207 D. Coucouvanis and D. Piltingsrud, *J. Amer. Chem. Soc.*, 1973, **95**, 5556.
208 A. Teuerstein, B. A. Feit, and G. Navon, *J. Inorg. Nuclear Chem.*, 1974, **36**, 1055
209 R. N. Patel and D. W. R. Rao, *Current Sci.*, 1966, **35**, 618.
210 P. K. Mathur, *J. Inorg. Nuclear Chem.*, 1974, **36**, 943.

have been characterized by conductivity, analytical, and i.r. measurements, which indicate that all Cr cations are N-bonded.

Bright yellow complexes of halogeno derivatives of diazoaminobenzene and Ag have been prepared; the structure (52) has been proposed on the basis of i.r. evidence.[211] I.r. studies of mixed complexes between $AgNO_3$ and thiazole thiones (53) and

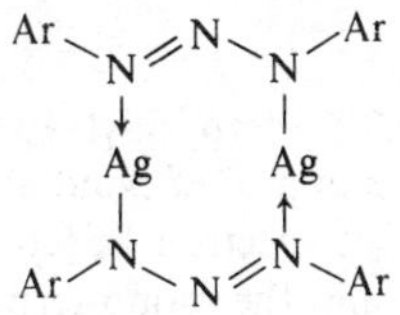

(52)

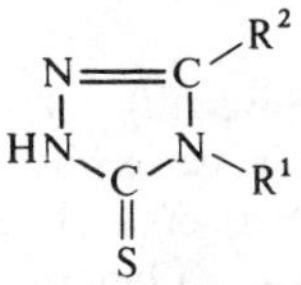

(53) R^1, R^2 = Ph, H; Et, Me; Me, Et;
Me, H; Ph, Me

thiazolidine-2-thione have shown the presence of both nitrate and nitrato complexes.[212] The former contain discrete NO_3^- ions whereas the latter are bonded to silver through O. Thermogravimetric analysis and differential scanning calorimetric measurements have shown that the thermal decomposition of $[Ag(NH_3)_2]_2SO_4$ occurs *via* a one-step process to give Ag_2SO_4 and NH_3.[213]

Complex formation of Ag with 2-(aminomethyl)py and 6-methyl-2-(methylaminomethyl)py has been studied potentiometrically in 0.5M–KNO_3 at 25 °C using glass and Ag_2S electrodes.[214] Both ligands form complexes of type $L_xAg_yH_z(x,y,z = 1,1,1;$ 1,1,0; 1,2,0; 2,2,0; 2,1,0).

P-*Donor ligands.* Four similar papers reporting new cationic and neutral complexes of Pt group metals with phosphite, phosphonite, and phosphinite ligands have appeared.[93—96] The general preparative route involved addition of the free ligand $L = P(OR)Ph_2$, $P(OR)_2Ph$, or $P(OR)_3$, (R = Me or Et) to a solution of a labile transition-metal precursor followed by addition of $NaBPh_4$. Complexes of type $[AgL_4]^+$ were characterized by chemical analyses, ^1H n.m.r., and conductivity measurements.

Diphosphine and diarsine complexes $[(AgNO_3)_2L]$ [L = $CH_2(PPh_2)_2$, {CH_2-(PPh_2)}$_2$, CH_2{$CH_2(PPh_2)$}$_2$, *cis*-PPh_2CH=$CHPPh_2$, *o*-$C_6H_4(AsMe_2)_2$] have been synthesized.[215] I.r. data indicate unidentate nitrato-groups and the hitherto unknown dimeric structure with a single ligand bridge was proposed (54).

$$O_2NO—Ag \leftarrow L \sim L \rightarrow AgONO_2$$

(54)

Group IV. C-donor ligands. The *highly explosive* yellow-orange disilver diazomethane has been obtained in high yield (equation 48).[216] The i.r. spectrum suggests

$$2AgOAc + 3CH_2N_2 + py \xrightarrow{Et_2O} [Ag_2CN_2]py + 2AcOMe + 2N_2 \quad (48)$$

[211] B. E. Zaitsev, V. A. Zaitseva, B. N. Ivanov-Emin, E. S. Lisityna, and A. I. Ezhoo, *Russ. J. Inorg. Chem.*, 1973, **18**, 30.
[212] A. Parentich, L. H. Little, and R. H. Ottewill, *J. Inorg. Nuclear Chem.*, 1973, **35**, 2271.
[213] S. M. Caulder, K. H. Stern, and F. L. Carter, *J. Inorg. Nuclear Chem.*, 1973, **36**, 234.
[214] A. M. Goeminne and Z. Eekhaut, *J. Inorg. Nuclear Chem.*, 1974, **36**, 357.
[215] W. Levason and C. A. McAuliffe, *Inorg. Chim. Acta*, 1974, **8**, 25.
[216] E. T. Blues, D. Bryce-Smith, J. G. Irwin, and I. W. Lawstron, *J.C.S., Chem. Comm.*, 1974, 466.

a C-bonded structure, with resonance forms (55).

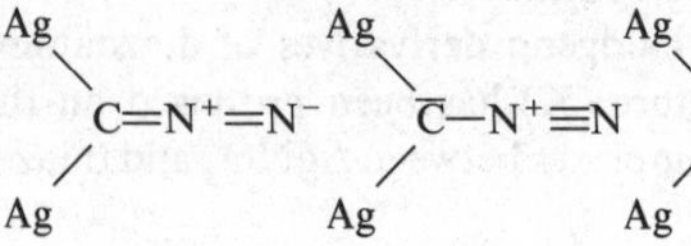

$$\underset{Ag}{\overset{Ag}{\diagdown}}C=N^+=N^- \qquad \underset{Ag}{\overset{Ag}{\diagdown}}\bar{C}-N^+\equiv N \qquad \underset{Ag}{\overset{Ag}{\diagdown}}\bar{C}-N=N^+$$

(55)

Silver(II) and Silver(III).—The cationic complex bis-(2,2',2''-terpyridyl)Ag^{2+} has been isolated as its peroxydisulphate salt.[217] The presence of a $d\rightarrow d$ band at 15 600 cm^{-1} is indicative of a six-co-ordinate ion. X-ray photoelectron spectra have provided Ag($3d_{5/2, 3/2}$) binding energies for this complex and the mono-terpy complex. The influence of pyridine and 2-, 3-, and 4-picoline on bis(diethyl dithiocarbamato)Ag^{II} has been examined by e.s.r. spectroscopy.[218] The results indicate formation of mono-addition products and the spin hamiltonian parameters for these complexes have been determined.

Silver(II) complexes of the macrocyclic tetra-aza ligands (56) and (57) have been prepared by disproportionation of Ag^I in the presence of the ligand.[219] Electrochemical or chemical oxidation with [NO]$^+$ salts has yielded Ag^{III} complexes of

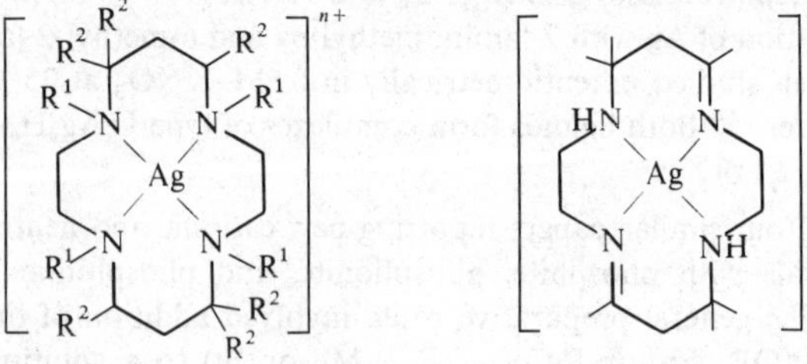

(56) R^1 = H, R^2 = Me, n = 2 or 3
R^1 = H, R^2 = H, n = 2 or 3
R^1 = Me, R^2 = Me, n = 2

(57)

structure (56). Planar structures are implied from e.s.r. measurements except for the complex (56) R^1 = Me, R^2 = Me, n = 2, where $g_{\parallel} < g_{\perp}$ and the structure is possibly folded with co-ordinated ClO$_4^-$ (58).

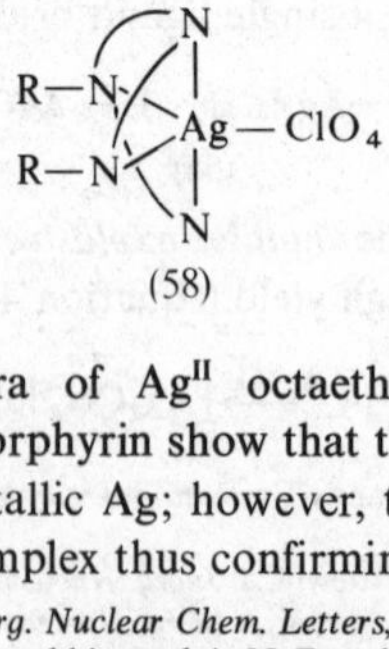

$$\underset{R-N}{\overset{R-N}{\diagup}}\!\!\underset{\diagdown}{\overset{\diagup}{>}}Ag-ClO_4$$

(58)

X-Ray photoelectron spectra of Ag^{II} octaethylporphyrin, Ag^{II} tetraphenylporphyrin, and Ag^{III} octaethylporphyrin show that the $3d_{5/2, 3/2}$ binding energies are only slightly higher than in metallic Ag; however, these binding energies lie 2.7 eV higher in energy for the Ag^{III} complex thus confirming the Ag^{III} oxidation state.[220]

[217] D. P. Murtha and R. A. Walton, *Inorg. Nuclear Chem. Letters*, 1973, **9**, 819.
[218] V. V. Zhukov, I. N. Marov, O. M. Petrukhin, and A. N. Ermakov, *Russ. J. Inorg. Chem.*, 1973, **18**, 1522.
[219] E. K. Barefield and M. T. Mocella, *Inorg. Chem.*, 1973, **12**, 2829.
[220] D. Karweik, N. Winograd, D. G. Davis, and K. M. Kadish, *J. Amer. Chem. Soc.*, 1974, **96**, 591.

Table 6 X-*Ray data for silver complexes*

Complex	R	Comments	Ref.
[AgC(CN)$_2$NO]	—	N and O atoms bond to Ag atom to give approximately tetrahedral co-ordination	a
[AgC(CN)$_2$NO$_2$]	0.035	four crystallographically independent [C(CN)$_2$NO$_2$]$^-$ ions, all close to planar. Co-ordination round four Ag is different and irregular. NO$_2$ group acts as bidentate ligand Ag—O 2.28—3.00 Å. Ag—N distance is short 2.12—2.34 Å. Ag----N≡C is approximately linear for seven out of eight CN groups	b
[Me$_4$N][Ag(NCO)$_2$]	0.048	Approximately linear [OCN—Ag—NCO]$^-$ ion. Me$_4$N ions subject to disorder or extreme thermal motion	c
[$\frac{1}{2}$AgNO$_2$(C$_{11}$H$_8$N$_4$O$_2$),2H$_2$O] [$\frac{1}{2}$Ag(NO$_2$)$_{0.55}$(NO$_3$)$_{0.45}$] [(C$_{11}$H$_8$N$_4$O$_2$),2H$_2$O]	0.062 0.069	Complexes are isomorphous. Flavin bound to Ag through N and O. Ag lies on two-fold axis with co-ordination midway between square-plane and tetrahedral	d
[C$_{24}$H$_{48}$N$_4$O$_6$.3AgNO$_3$]	0.038	Macrocyclic ligand encloses two Ag, with each Ag bound to heteroatoms plus an O atom from (NO$_3$)$^-$ ion	e

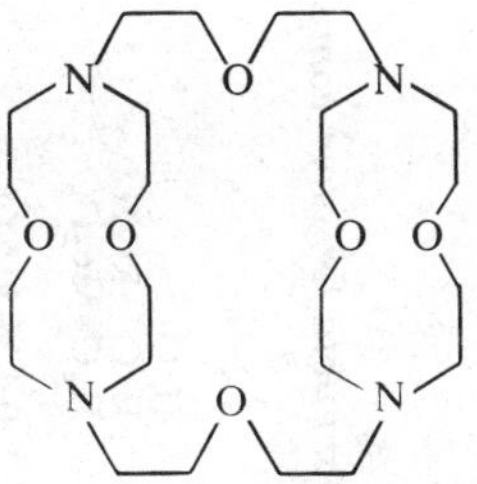

(a) Y. M. Chow and D. Britton, *Acta Cryst.*, 1974, **B30**, 1117; (b) Y. M. Chow and D. Britton, *Acta Cryst.*, 1974, **B30**, 147; (c) K. Aarflot and K. Ase, *Acta Chem. Scand.*, 1974, **A28**, 137; (d) R. H. Benno and C. J. Fritchie, *Acta Cryst.*, 1973, **B29**, 2493; (e) R. Wiest and R. Weiss, *J.C.S. Chem. Comm.*, 1973, 678.

Table 7 *Stability constant data for silver compounds*

Ligand	Solvent	Complex	Method	Comments	Ref.
4-Methylpyridine 2-Aminopyridine 4-Aminopyridine	H_2O H_2O-ROH H_2O-Me_2CO	AgL_2^+	Potentiometric	stability of 4-Mepy complex decreases with organic solvent added to H_2O whereas stabilities of both 2-NH_2py and 4-NH_2-py increase	a
$CH_2(OH)CH_2NH_2$ $(C_2H_4OH)_2NH$ $(C_2H_4OH)_3N$	0.4 M-$LiNO_3$ H_2O-Me_2CO	AgL_n^+ $n = 1, 2,$ or 3	Potentiometric	measured K_1, K_2, and K_3. Stabilities increase with added acetone	b
2,2′-bipy	$ROH-H_2O$ Me_2CO-H_2O	$Ag(bipy)_2^+$	Potentiometric	stabilities decrease with added organic solvent to 50% mixture, then increase with further increase in organic solvent	c
3,3′-Diaminodipropylamine	1.0 M-KNO_3 (aq)	$Ag_2L_2^{2+}$	Potentiometric Polarographic	measured formation constants between 25—55°C. Shown to be dimeric	d
$NH_2CH_2CH_2OH$ $S_2O_3^{2-}$	$MeOH-H_2O$ H_2O	AgL^+, AgL_2^+, AgL_3^+ $Ag(S_2O_3)_{3/2}^{3-}$	Potentiometric	measured between 25—45°C	e f

(a) G. V. Bundu and L. V. Nazarova, *Russ. J. Inorg. Chem.*, 1973, **18**, 807; (b) P. K. Migal and K. I. Ploae, *Russ. J. Inorg. Chem.*, 1973, **18**, 169; (c) G. V. Bundu and L. V. Nazarova, *Russ. J. Inorg. Chem.*, 1973, **18**, 1574; (d) E. Dazzi and M. T. Falqui, *Gazzetta*, 1974, **104**, 589; (e) V. V. Udovento and G. B. Pomerants, *Russ. J. Inorg. Chem.*, 1973, **18**, 937; (f) J. C. Ghosh, *J. Indian Chem. Soc.*, 1974, **51**, 361.

4 Gold

Gold(I).—*Group VII Donors.* I.r. and Raman spectra of the complexes $[\{(4\text{-}FC_6H_4)_3P\}\text{-}AuX]$ (X = Cl or Br) have been measured and assignments of $\nu(Au\text{---}X)$ and $\nu(Au\text{---}P)$ made.[221] Complexes of type $(R_4N)[AuX_2]$ (R = Et or Bu^n; X = Cl, Br or I) have been isolated (equations 49 and 50) by reduction of $[AuCl_4]$ with acetone or phenyl-hydrazinium chloride.[222] The compounds are stable in air but sensitive to water.

$$2R[AuCl_4] + PhNHNH_2HCl \rightarrow R[AuCl_2] + R[PhAuCl_3] + N_2 + HCl \quad (49)$$

$$R[AuBr_4] + Me_2CO \rightarrow R[AuBr_2] + CH_2BrCOMe + HBr \quad (50)$$

I.r. and Raman spectra were interpreted in terms of linear anions. In addition, far-i.r. and Raman spectra of the solid $[AuXL]$ (L = PPh$_3$ or AsPh$_3$; X = Cl or Br) and Raman polarization data in solution have enabled identification of the Au—P and Au—As stretching modes.[223]

Formation constants K_{AuCl} and $K_{AuCl_2^-}$ of the chloro-gold complexes formed in DMSO have been determined potentiometrically.[224] The reduction of the complexes was also studied electrochemically and possible oxidation and reduction sequences were given.

Group VI Donors. Sulphinato-complexes of Au, Ag, and Cu have been synthesized (equation 46).[205] I.r. studies indicate that the complex $[(Ph_3P)Au(SO_2Ph)]$ is S-bonded, unlike the Ag and Cu complexes which are O-bonded. Moreover, the Cu and Au complexes are readily oxidized by molecular oxygen (Scheme 4) to sulphonate complexes, and it was postulated that an equilibrium between S-bonded and O-

Scheme 4

bonded isomers was present for the Au complex.

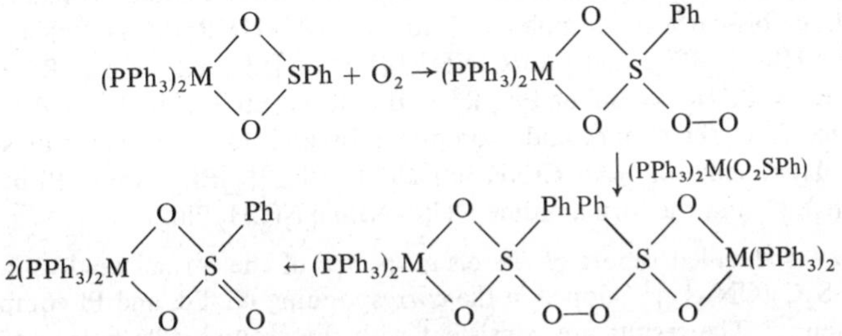

Group V. N-Donor ligands. Reaction of $[AuCl(Me_2S)]$ with 1-unsubstituted pyrazoles in methanolic KOH has yielded *N*-pyrazolyl gold(I) derivatives.[225] The white compounds are stable to air, light, and heat, and molecular weight measurements show they are trimeric in both vapour and CHCl$_3$ solution, and on this basis, and i.r. evidence, structure (59) was proposed.

I.r. spectra for a series of thiocyanato complexes $[LAu(SCN)]$ [L = P(OPh)$_3$, PMe$_3$, PPh$_3$, AsPh$_3$, *etc.*] have shown that the ratio of N-bonded NCS to S-bonded SCN increases as the *trans*-effect of the ligand L increases.[226]

[221] A. G. Jones and D. B. Powell, *Spectrochim. Acta*, 1974, **30A**, 1001.
[222] P. Braunstein and R. J. H. Clark, *J.C.S. Dalton*, 1973, 1845.
[223] A. G. Jones and D. B. Powell, *Spectrochim. Acta*, 1974, **30A**, 563.
[224] T. E. Suarez, R. T. Iwamoto, and J. Kleinberg, *Inorg. Chim. Acta*, 1973, 7, 458.
[225] F. Bonati, G. Minghetti, and G. Banditelli, *J.C.S. Chem. Comm.*, 1974, 88.
[226] J. L. Burmeister and J. B. Melpolder, *J.C.S. Chem. Comm.*, 1973, 613.

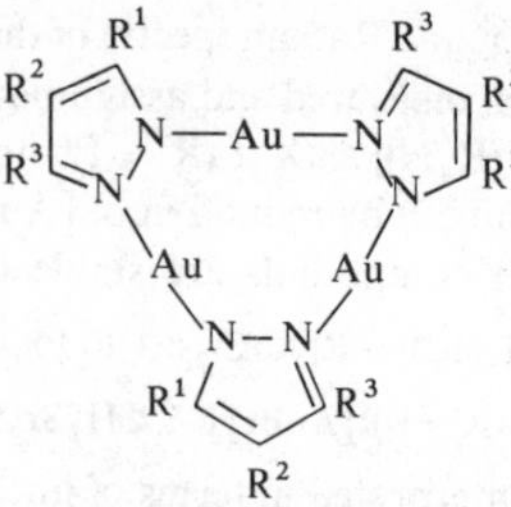

(59) R^1 = H or Me; R^2 = Et or H; R^3 = H, Me or Ph

P-Donor ligands. Four similar papers reporting new cationic complexes of Pt group metals with phosphite, phosphonite, and phosphinite ligands have appeared.[93—96] The general preparative route involved addition of free ligand L = $P(OR)Ph_2$, $P(OR)_2Ph$, or $P(OR)_3$ (R = Me or Et) to a solution of a labile transition-metal precursor followed by addition of $NaBPh_4$. Complexes of type $[AuL_4]^+$ were characterized by chemical analyses, ^1H n.m.r., and conductivity measurements.

Group IV. C-Donor ligands. The ability of Cu and Au to catalyse the reaction between isocyanides and amines to form formamidines has prompted the preparation of gold isocyanide complexes.[227] Complexes of type $[Au(CNR)Cl]$ and $[Au(CNR)_2]PF_6$ (R = Bu^t or Pr^i) were obtained from $[AuCl_4]^-$ and CNR. I.r. and ^1H n.m.r. evidence shows that bis-carbene complexes $[Au\{C(NHR^1)(NHR^2)\}(NH_2R^2)Cl]$, $[Au\{C-(NHR^1)(NHR^2)\}_2]PF_6$, and $[Au\{C(NHR^1)(NR_2^3)\}_2]PF_6$. ($R^1$ = Bu^t, R^2 = Bu^t, Pr^i or Bu^n; R^1 = Pr^i, R^2 = Bu^t or Pr^i; R^1 = Bu^t, R^3 = $EtCH_2Ph$; R^1 = Pr^i, R^3 = Et) are formed from these isocyanide complexes by addition of primary or secondary amines. Treatment of $[Au\{C(NHBu^t)[N(CH_2Ph_2)]\}_2]PF_6$ with PPh_3 yielded $[Au(PPh_3)_4]^+$ and the formamidine $CH(=NBu^t)[N(CH_2Ph)_2]$.

Gold(II).—A detailed report of the e.s.r. spectra of the formal gold(II) complexes $[Au\{1,2-S_2C_2(CN)_2\}_2]^{2-}$ doped in the corresponding Ni, Pd, and Pt complexes has been given.[228] The results are consistent with the ground-state hole configuration $(b_{1g})^2(a_g)^1$, where a_g is primarily a ligand-based orbital with *ca.* 15% 6s and small 5d admixtures.

Gold(III).—*Group VII Donors.* The intermediate in the reduction of $[AuCl_4]^-$ by NN'-diphenylthiourea has been isolated as $[AuCl_2(DPT)_2]Cl$.[229] The self-reduction of this complex was followed kinetically (equation 51). A new improved preparation of $AuCl_3$ has also been reported.[230] $HAuCl_4$ has been shown to oxidize (S)-methionine

$$[AuCl_2(PhNHCSNHPh)_2] \rightarrow [AuCl_2]^- + \{Ph_2NH(N:)CS\}_2 \qquad (51)$$

stereospecifically to methionine sulphoxide, and the oxidation also occurs at S when methionine is part of a peptide chain.[231]

The electrical conductance of the mixed-valence triple salts $(NH_4)_6[Ag_2AuX_{17}]$ (X = Cl or Br) and $Cs_4[MAuX_{12}]$ (M = Cu or Pd; X = Cl or Br) as both single

[227] J. A. McCleverty and M. M. M. da Mota, *J.C.S. Dalton*, 1973, 2571.
[228] R. L. Schupp and A. H. Maki, *Inorg. Chem.*, 1974, **13**, 44.
[229] V. M. Shulman, Z. A. Savel'era, and R. I. Novoselov, *Russ. J. Inorg. Chem.*, 1973, **18**, 376.
[230] T. Mundorf and K. Dehnicke, *Z. Naturforsch.*, 1973, **28b**, 506.
[231] E. Bordignon, L. Cattalini, G. Natile, and A. Scatturin, *J.C.S. Chem. Comm.*, 1973, 878.

crystals and compressed pellets has been measured.[232] It was found that single crystals behave as semiconductors both parallel and perpendicular to the needle axis, and that changing from Cl to Br has a much larger effect on the conductivity than changing M from Ag to Cu.

Group VI. S-Donor ligands. Extraction of the metals Au, Ag, Pd, Pt from aqueous solution by a large number of alkyl and alkene sulphides and disulphides has been reported (see p. 390).[51, 52]

Addition of a solution of dithio-oxamide (DH_4) or substituted dithio-oxamides to Au^{III} compounds has yielded dithio-oxamide complexes $[Au(DH_4)(DH_3)]Cl_2$, $[AuCl_2(Me_2DH)]$, $[AuBr_2(Me_2DH_2)]Br$, $[AuBr_2(DChDH_2)]Br$, $[AuCl_2(DChDH)]$ (DCh = dicyclohexyl), and $[AuX_2L][AuX_4]$ (X = Cl or Br; L = Me_4D or Et_4D).[233] The complexes are all diamagnetic, and i.r. spectra show that while the tetra-substituted dithio-oxamide complexes are S,S-bonded the remaining complexes are S,N-co-ordinated.

Group V. N-Donor ligands. Azo-compounds react with AuX_3 (X = Cl or Br) to provide N-co-ordinated complexes (60).[234] The thermal decomposition of these

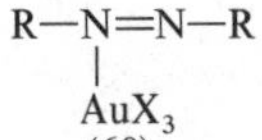

$$R—N{=}N—R$$
$$|$$
$$AuX_3$$
$$(60)$$

Table 8 X-*Ray structural data for gold compounds*

Compound	R	Comments	Ref.
$SmAu_6$	0.128	refinement. Hexagonal antiprisms of Au atoms centred by Sm atom	a
$(Xe_2F_{11})^+[AuF_6]^-$	0.052	anion essentially octahedral Au—F = 1.86 Å. Cation consists of two XeF_5 groups linked by F bridge	b
$[AuCl(PPh_3)_2].\frac{1}{2}C_6H_6$	0.064	first X-ray evidence for trigonal–planar co-ordination around Au. Au—Cl = 2.50 Å Au—P = 2.331 Å ∠ PAuP = 132.1°	c
$AuBr_3$ $AuCl$	—	powder	d
$Na_3[Au(S_2O_3)_2].2H_2O$	0.085	linear co-ordination. $S_2O_3^{2-}$ bonded through S. Au—Au distance is short (3.3 Å)	e
$[\{(p\text{-tolyl})_3P\}_6Au_6](BF_4)_2$	0.052	Au atoms form a distorted octahedron. Au—Au = 3.019 Å	f

(a) H. D. Flack, J. M. Moreau and E. Parthe, *Acta Cryst.* 1974, **B30**, 820; (b) K. Leary, A. Zalkin, and N. Bartlett, *Inorg. Chem.*, 1974, **13**, 775; (c) N. C. Baenziger, K. M. Dittemore, and J. R. Doyle, *Inorg. Chem.*, 1974, **13**, 805; (d) J. Strahle and K. P. Lorcher, *Z. Naturforsch.*, 1974, **29**, 266; (e) R. F. Baggio and S. Baggio, *J. Inorg. Nuclear Chem.*, 1973, **35**, 3191; (f) P. Bellon, M. Manassero, and M. Sansoni, *J.C.S. Dalton*, 1973, 2423.

[232] P. S. Gomm and A. E. Underhill, *Inorg. Nuclear Chem. Letters*, 1974, **10**, 309.
[233] A. C. Fabretti, G. C. Pellacani, G. Peyronel, and B. Scapivelli, *J. Inorg. Nuclear Chem.*, 1974, **36**, 1067.
[234] R. Huttel and A. Konietzny, *Chem. Ber.*, 1973, **106**, 2098.

compounds was shown to occur in two steps, the first of which produces 2-chloro-azobenzene and AuCl. This is followed by attack of AuCl at both *ortho-* and *para*-positions to produce dichloroazobenzene.

Gold(v).—Twelve dioxygenyl salts of complex fluorometallate anions $[MF_6]^-$ (M = As, Sb, Bi, Ru, Rh, Pt or Au) have been prepared by static heating of the metal at 300—350 °C with 5 atm. of 3:1 fluorine–oxygen mixture for upwards of 10 h.[193] The salts were characterized by powder X-ray, Raman, and mass spectroscopy.

4
Zinc, Cadmium, and Mercury

BY J. HOWELL AND M. HUGHES

1 Introduction

This chapter reviews the inorganic co-ordination chemistry of the Group IIB elements zinc, cadmium, and mercury from September 1973 to September 1974. As this is the first review of this particular group in this series of Reports, several references to earlier work have been included, mainly for the sake of completeness where a series of papers has been published. Where a report is concerned with the comparative chemistry of all three elements, it is discussed in the section devoted to zinc.

A separate section, compiled by Martin Hughes of the University of Cambridge, reviews advances in the bio-inorganic chemistry of the Group IIB elements, and reflects the rapid growth of interest in this area in recent years.

2 Zinc

The photoelectron spectra of gaseous zinc, cadmium and mercury halides have been reported.[1—3] Assuming a linear molecule, a simple MO scheme may be assigned, and the valence structure appears to be $\sigma_g^2 \sigma_u^2 \pi_u^2 \pi_g^4$. Spin—orbit splitting of the π_g and π_u orbitals is observed, with the π_g splittings generally being greater than those of the π_u orbitals, and the iodide splittings exceeding those of the bromides. It is notable that the π_u molecular orbital has metal contribution, whereas the π_g does not. There is a striking discontinuity in the energies of most of the molecular orbitals on going from cadmium to mercury (Zn > Cd < Hg), in common with the trends observed for several other properties of compounds of Group IIB. An order of electronegativity may be deduced which is in agreement with the Aldred—Rochow scale (Zn > Hg) but at variance with the Pauling scale.

The photoelectron spectra of the complexes $[NR_4]_2[ZnX_4]$ (X = Cl, Br, NCO, NCS, or NCSe) have been reported.[4] A decrease in binding energy is noted on increasing the R chain length in the tetra-alkylammonium ion, while the trend in binding energies associated with X follows the spectrochemical series. The crystal structure of Cs_2ZnCl_4 has been determined.[5] The anion shows only slight distortion from tetrahedral symmetry (Zn—Cl = 2.28—2.31 Å), in contrast to the same anion in the complex $[Co(NH_3)_6][ZnCl_4][Cl]$ where gross distortions are observed.[6] Calculations

[1] G. W. Boggess, J. D. Allen, and G. K. Schweitzer, *J. Electron Spectroscopy Related Phenomena*, 1973, **2**, 467.
[2] L. C. Cusacks, F. A. Grimm, and G. K. Schweitzer, *J. Electron Spectroscopy Related Phenomena*, 1974, **3**, 229.
[3] B. G. Cocksey, J. H. D. Eland, and C. J. Danby, *J.C.S. Faraday II*, 1973, **69**, 1558.
[4] J. Escard, G. Mavel, J. E. Guerchais, and R. Kergoat, *Inorg. Chem.*, 1974, **13**, 695.
[5] J. A. McGinnety, *Inorg. Chem.*, 1974, **13**, 1057.
[6] D. W. Meek and J. A. Ibers, *Inorg. Chem.*, 1960, **9**, 465.

indicate that in the latter complex, the major cause of bond length distortion is anisotropy in the applied electrostatic crystal forces. For the former complex, charge distribution values were obtained from crystal force, n.q.r. and MO calculations; the results were in overall agreement. The far-i.r. spectra of the $[ZnCl_4]^{2-}$ anion in salts with the 1,10-phenanthrolinium and 2,2'-bipyridinium cations have been reported.[7] The anion in the former salt appears to be tetrahedral, while that in the latter is believed to be halogen-bridged. The first crystallographic determination of the $[ZnI_4]^{2-}$ anion has been reported in its salt with the 2,4-dimethyl-1*H*-1,5-benzo-diazepinium cation.[8] It possesses rigorous C_2 symmetry (Zn—I = 2.60 Å) with deviations from the highest tetrahedral symmetry being due to packing forces.

Two ^{67}Zn (natural abundance = 4.12%; $I = \frac{5}{2}$) n.m.r. studies have been reported.[9, 10] The chemical shift of ^{67}Zn (4.81 MHz at 1.807 Tesla) in aqueous zinc chloride, bromide, and iodide solutions was found to be strongly concentration dependent, while no such dependence was noted in solutions of the perchlorate, nitrate, or sulphate. This behaviour resembles that found for analogous cadmium systems, and is attributed to the formation of mono- and poly-halogeno- complexes even at low salt concentrations. In addition, the zinc halide solutions show an anomalous shift to higher frequencies for their solutions in D_2O, compared with those in H_2O. The perchlorate, nitrate and sulphate show no solvent isotope effect.

^{35}Cl N.m.r. has been used as a probe of chelated zinc(II) environments.[11] Molar relaxivity is used as a parameter suitable for characterizing the zinc environment in terms of the quadrupolar relaxation of ^{35}Cl nuclei that it can produce in 0.5 M-NaCl. Bidentate chelation (glycinate, glutamate, succinate) is shown to increase its effectiveness in producing ^{35}Cl relaxation. Terdentate chelation (iminodiacetate, aspartate) can either increase or decrease the amount of relaxation caused by zinc. The least relaxation is produced when the ligand atoms have a formal negative charge and a large acidity constant. In some instances, zinc which is chelated by four ligand atoms (nitrilotriacetate) is effective at ^{35}Cl relaxation, and therefore, does not represent a co-ordinatively saturated environment. Molar relaxivities for 1:1 and 2:1 chelates usually differ and it is possible to derive formation constants which are consistent with literature values. In a number of systems, hydrolysis-type reactions are readily identified.

The reactivity of a variety of non-aqueous zinc, cadmium, and mercury systems has been described. A series of papers[12—16] reports several aspects of the inorganic chemistry of fused zinc chloride. The stoicheiometry of the reactions of 15 oxyanions with fused zinc chloride has been established.[12] With hydroxide, carbonate, nitrate, peroxide, and sulphite, oxide ion was the non-volatile product; bicarbonate was also decomposed to oxide, possibly *via* an intermediate basic carbonate chloride. Nitrite gave oxide as the final product, with partial intermediate formation of nitrate. Sulphate did not react, but its solution, and those of metaphosphate and pyrophosphate, could be quenched to glasses. Thiosulphate gave sulphide and sulphate as non-volatile products, while pyrosulphate, persulphate, and bisulphate gave sulphate. Metabisulphite produced sulphate, sulphide, and oxide, with the proportions varying with

[7] S. N. Ghosh, *J. Inorg. Nuclear Chem.*, 1973, **35**, 2329.
[8] P. L. Orioli and H. C. Lip, *Cryst. Struct. Comm.*, 1974, **3**, 477.
[9] B. W. Epperlein, H. Krüger, O. Lutz, and A. Schwenk, *Z. Naturforsch*, 1974, **29a**, 660.
[10] B. W. Epperlein, H. Krüger, O. Lutz, and A. Schwenk, *Phys. Letters* 1973, **45A**, 255.
[11] J. A. Happe, *J. Amer. Chem. Soc.*, 1973, **95**, 6232.
[12] D. H. Kerridge, and I. A. Sturton, *Inorg. Chim. Acta*, 1973, **7**, 701.

concentration. The stoicheiometries of the reactions of eight halogen and pseudo-halide anions with fused zinc chloride have been established.[13] Perchlorate and chlorate decompose to oxide, oxygen, and chlorine, the latter with partial intermediate formation of perchlorate. Both bromate and iodate gave oxide, oxygen, and free halogen, accompanied by extensive oxidation of the melt, while periodate gave oxygen and iodate, and at higher temperatures, oxide, accompanied by melt oxidation. Of the pseudohalides, cyanide did not react; cyanate gave cyanide as the main product, while thiocyanide formed sulphide, cyanide, and sulphur.

The reactions of four transition-metal ions and oxyanions with fused zinc chloride have been elucidated.[14] Potassium chromate dissolves without reaction at 350 °C ($\lambda_{max} = 26000$ cm^{-1}) but at higher temperatures reacts to give zinc chromite, with slight melt oxidation. Potassium dichromate also gave zinc chromite through a chromate intermediate, but with more extensive melt oxidation. Silver oxide reacted to give silver chloride with no decomposition to silver metal, while sodium metavanadate reacted probably to give a zinc metavanadate species. The solubilities of cobalt(II) chloride, nickel(II) chloride, chromium(III) chloride, and zinc oxide and sulphate in fused zinc chloride have also been measured.[15] The spectra of nickel(II) and cobalt(II) dissolved in a zinc chloride–zinc sulphate melt showed a tetrahedral–octahedral co-ordination equilibrium, the octahedral species being favoured by an increase in temperature and zinc sulphate concentration.[16]

E.m.f. measurements on the $ZnCl_2$–ACl (A = K, Rb, or Cs) systems have yielded partial molar free energies, enthalpies, and entropies, and the results have been interpreted in terms of an acid–base reaction.[17] The enthalpy of mixing consists of a contribution from the reaction to form the tetrahedral complex A_2ZnCl_4, and a contribution from the mixing of the products with remaining reactants to form a solution. The results indicate that as the polarizing power of the alkali-metal cation decreases, the 'reaction' part of the enthalpy becomes more pronounced, reflecting the increased basicity of the alkali-metal salt.

A dissolution mechanism for zinc, cadmium, and mercury in their molten halides has been proposed on the basis of new experimental and literature data.[18] Dissolution occurs at the metal–salt phase boundary. Adsorbed M^{2+} cations are reduced to M^+ ions which then migrate into the salt phase where M_2^{2+} dimers form. The stability of the M_2^{2+} ions was found to increase in the order Zn < Cd ≪ Hg (Hg^+ cannot be detected in solution).

The stability constants in melts of NH_4NO_3, nH_2O of ZnX^+, ZnX_2 ($n = 1$—3; X = Cl or Br), CdX^+, CdX_2 ($n = 1.5$–3; X = Cl or Br) and HgX^+, HgX_2 ($n = 2.5$; X = Cl or Br) have been determined.[19, 20] The behaviour of zinc is peculiar if the K_1 and K_2 values are compared with those of cadmium and mercury. The stability constants increase with temperature and the bromide is more stable than the chloride, trends which are opposite to those normally observed for the halide complexes of

[13] D. H. Kerridge, and I. A. Sturton, *Inorg. Chim. Acta*, 1973, **8**, 31.
[14] D. H. Kerridge, and I. A. Sturton, *Inorg. Chim. Acta*, 1973, **8**, 37.
[15] D. H. Kerridge, and I. A. Sturton, *Inorg. Chim. Acta*, 1973, **8**, 27.
[16] D. H. Kerridge, and I. A. Sturton, *Inorg. Chim. Acta*, 1974, **10**, 13.
[17] N. J. Robertson, and A. S. Kucharski, *Canad. J. Chem.*, 1973, **51**, 3114.
[18] Z. Gregorczyk, *Chem. Abs.*, 1974, **80**, 74848.
[19] R. M. Nicolic, and I. J. Gal, *J.C.S. Dalton* 1972, 162.
[20] R. M. Nicolic, and I. J. Gal, *J.C.S. Dalton*, 1974, 985.

most metals in anhydrous or aqueous melts. The data also show that hydration of M^{2+} cations competes more significantly with ion association in the case of zinc, which is consistent with the much smaller Pauling radius of Zn^{2+} as compared with Cd^{2+} and Hg^{2+}.

The Raman spectrum[21] of co-ordinated zinc(II) in KSCN at 200 °C shows the presence of the tetrahedral $[Zn(NCS)_4]^{2-}$ species, somewhat distorted by polymerization, and bands due to Zn—NCS—Zn bridges are also observed. Linkage isomerism is also present, as evidenced by bands due to both Zn—NCS and Zn—SCN bonding modes. A Raman study of zinc(II) co-ordination in aqueous thiocyanate solution has also been reported.[22] No bridging thiocyanate is observed, and bonding to the zinc is totally in the Zn—NCS mode. The results are consistent with the existence of all four species $Zn(NCS)_{1-4}$. There is strong evidence that the two lower species are octahedrally co-ordinated, while the higher species are tetrahedral. The Raman and i.r. spectra of the $[Zn(CNO)_4]^{2-}$ anion have also been reported.[23] The results are consistent with high tetrahedral symmetry, *i.e.*, linear metal—CNO groups.

Raman spectral studies of the species $MX_n^{(n-2)-}$ ($n = 2$—4; M = Zn, Cd, or Hg; X = Cl, Br, or I) in anhydrous tributyl phosphate have been reported.[24] For the MX_2 molecules, sufficient metal dihalide–solvent interaction exists to suggest bent X—M—X species with C_{2v} rather than $D_{\infty h}$ symmetry. The effect appears most marked for zinc(II) and least marked for mercury(II), which is in accord with the Lewis acidity sequence $ZnX_2 > CdX_2 > HgX_2$. A similar analysis of the anionic MX_3^- complexes formed from a 1:1 mixture of LiX and MX_2 again demonstrates solvent interaction, and a tetrahedral C_{3v} species is indicated, rather than the planar structure found in the solid state. Studies involving the halogeno-complexes of zinc, cadmium, and mercury in DMSO and DMF have also been reported.[25, 26]

A series of complex zinc hydrides of composition $M_n Zn_m H_{2m+n}$ (M = Li, Na, or K) has been synthesized by reaction of the appropriate $M_n Zn_m R_{2m+n}$ alkyl with either $LiAlH_4$, $NaAlH_4$, or AlH_3.[27] Thus, the reaction of Li_2ZnMe_4 with $LiAlH_4$ produces Li_2ZnH_4, which is also the product of the reactions of $LiZnR_2H$ (R = methyl, s-butyl) and $LiZn_2Me_4H$ with $LiAlH_4$. The reaction of $LiAlH_4$ with $LiZnMe_3$ produces $LiZnH_3$. In contrast to the reaction of $Zn(s-butyl)_2$ with KH, which yields K_2ZnH_4 directly, KH and $ZnMe_2$ react in a 1:1 ratio to produce $KZnMe_2H$. Reaction of this with a further mole of $ZnMe_2$ gives the complex KZn_2Me_4H, which, like its lithium and sodium analogues, decomposes when attempts are made to isolate it. The new complexes MZn_2H_5 may be prepared by the reduction of either $MZnMe_2H$ (M = K or Na) or MZn_2Me_4H (M = K) with AlH_3, while reaction of $MZnMe_2H$ (M = K or Na) with $MAlH_4$ (M = K or Na) gives the $MZnH_3$ complexes. Structural data are difficult to obtain because of the insolubility of the hydrides.

Tetrahedral and octahedral complexes of zinc, cadmium, and mercury with a wide variety of nitrogen and oxygen donor ligands have been reported. Metal complexes

[21] N. P. Evtushenko, V. A. Sushko, and J. V. Volkov, *Chem. Abs.*, 1974, **80**, 32 090.
[22] D. P. Strommen, and R. A. Plane, *J. Chem. Phys.*, 1974, **60**, 2643.
[23] W. Beck, C. J. Oetker, and P. Swoboda, *Z. Naturforsch.*, 1973, **28b**, 229.
[24] D. N. Waters, E. L. Short, M. Tharwat, and D.F.C. Morrison, *J. Cryst. Mol. Structure*, 1973, **17**, 389.
[25] E. Y. Gorenbein, A. K. Trotimchuk, M. N. Vainshtein, and E. P. Skorobogat'ko, *Zhur. obshchei. Khim.*, 1973, **43**, 1440.
[26] V. M. Samoilenko and V. I. Lyashenko, *Zhur. neorg. Khim.*, 1975, **18**, 2968.
[27] E. C. Ashby and J. J. Watkins, *Inorg. Chem.*, 1973, **12**, 2493.

of several zinc, cadmium, and mercury salts with 2-, 3-, and 4-cyanopyridine have been reported.[28] In none of the complexes was cyanide co-ordination observed. Zinc halides react with 3- and 4-cyanopyridine, but not with 2-cyanopyridine, to give 1:2 complexes which are assigned a monomeric tetrahedral structure on the basis of i.r. evidence. The cadmium halides also form 1:2 complexes with all the cyanopyridines, except cadmium chloride, which reacts with 2-cyanopyridine to give a 1:1 complex. The former contain octahedrally co-ordinated cadmium with halogen bridges, while the latter contains a dimeric, tetrahedrally co-ordinated cadmium. Mercuric chloride forms 1:2 complexes with 2- and 4-cyanopyridine and 1:1 complexes with 3- and 4-cyanopyridine, while mercuric bromide gives 1:1 complexes with 2- and 3-cyanopyridine, but a 1:2 complex with 4-cyanopyridine. These complexes have structures analogous to the cadmium compounds. Zinc thiocyanate yields a monomeric tetrahedral 1:2 complex with 4-cyanopyridine, but does not react with the 2- and 3-derivatives. Cadmium thiocyanate gives 1:2 complexes with 2- and 3-cyanopyridine which are assigned a polymeric octahedral structure containing M—SCN—M bridges. Mercuric thiocyanate reacts with 3-cyanopyridine to give a dimeric tetrahedral 1:1 complex. Only mercuric cyanide reacts with 3-cyanopyridine to give a monomeric tetrahedral 1:2 adduct. Zinc sulphate reacts with 4-cyanopyridine to give a 1:2 complex considered to have octahedrally co-ordinated zinc in a polymeric structure involving chelating sulphate.

The 2,2'-bipyridylamine complexes of some zinc, cadmium, and mercury salts have been reported.[29] All the metal halide complexes isolated are of the stoicheiometry MLX_2 (M = Zn, Cd, or Hg; X = Cl, Br, or I), and are assigned a monomeric, tetrahedral structure on the basis of i.r. evidence. The complexes $ML(SCN)_2$ (M = Zn, Cd, or Hg) are also monomeric. Only mercuric cyanide reacts to give the monomeric $HgL(CN)_2$, whereas zinc sulphate forms 1:1 and 1:1.5 complexes which possess highly polymeric structures. Zinc and cadmium acetates form 1:1 complexes having pseudo-octahedral structures with bidentate, symmetrically chelating acetates. Cadmium also forms a 1:3 complex which is best formulated as $[CdL_3][OAc]_2$.

Complexes with pyridine N-oxide of zinc thiocyanate and selenocyanate of the stoicheiometry ZnL_3X_2 (X = SCN, SeCN) have been reported.[30] I.r. evidence shows that they are best formulated as $[ZnL_6]^{2+}[ZnX_4]^{2-}$. A thermochemical study of the series $ML_2(XCN)_2$ (M = Zn or Cd; X = O, S, or Se; L = pyridine) has also been reported.[31] Zinc perchlorate complexes of monopyrazine N-oxide of the stoicheiometry $[Zn(ONC_4H_4N)_n(NC_4H_4NO)_{4-n}(OH_2)(OClO_3)]ClO_4$ (n = integer < 4) have been reported.[32] The most probable structure involves a monomeric octahedral co-ordination by water, perchlorate, and both O- and N-bonded unidentate pyrazine N-oxide ligands.

The preparation and spectral characterization of the complex $Zn[(NSiMe_3)_2]_2L$ (L = pyridine) have been reported.[33]

Reaction of the zinc halides with morpholine, thiomorpholine, and thioxan gives

[28] I. S. Ahuja, and R. Singh, *J. Inorg. Nuclear Chem.*, 1974, **36**, 1505.
[29] I. S. Ahuja and R. Singh, *Inorg. Chim. Acta*, 1973, **7**, 465.
[30] G. B. Aitken, and G. P. McQuillan, *J.C.S. Dalton*, 1973, 2637.
[31] I. Porubstky, A. Nemeth, and G. Liptoy, *Chem. Abs.*, 1974, **80**, 140 736.
[32] A. N. Speca, N. M. Karayannis, and L. L. Pytlewski, *J. Inorg. Nuclear Chem.*, 1973, **35**, 3113.
[33] K. J. Fisher, *Inorg. Nuclear Chem.Letters*, 1973, **9**, 921.

complexes of stoicheiometry ZnX_2L_2.[34] The results suggest that for the ligands under consideration, the order of co-ordinative ability for the donor atoms is $N > S > O$. Complexes of stoicheiometry $ZnLX_2$ ($X = NO_2$ or NO_3) have been isolated from the reaction of zinc nitrite or nitrate with 1,2-dimorpholinoethane and 1,2-dipiperidinoethane.[35] The complexes contain zinc in a distorted octahedron of two ligand nitrogens and two bidentate O-bonded nitrite or nitrate anions. The nitrite complexes show greater stability than the nitrates. The vibrational spectra of the ethyleneimine complexes ZnL_2Cl_2 and CdL_4Cl_2 have been reported.[36]

Ethylenediamine is known to react with zinc, cadmium, and mercury halides to give complexes possessing a polymeric chain structure with the ethylenediamine acting as a bidentate bridging ligand, and it has now been reported[37] to react with $Hg(SCN)_2$ and $M(CN)_2$ ($M = Zn, Cd, $ or Hg) to give 1:1 and 1:2 complexes containing monomeric structures with the ethylenediamine acting as a bidentate chelating ligand. Reaction of the more basic NN-diethylethylenediamine not only with mercuric halides and pseudo-halides, but also with zinc and cadmium thiocyanates gives monomeric, tetrahedral 1:1 complexes.[38] Complexes of zinc chloride and perchlorate with the carcinogenic NN'-β-chloroethylethylenediamine have been reported.[39] The solvolysis rates of the free ligand and of the zinc complex have been studied by n.m.r. spectroscopy.

The crystal structures of dichloro-($NNN'N'$-tetramethylethylenediamine)zinc and dibromo-($NNN'N'$-tetramethylethylenediamine)cadmium have been reported.[40, 41] In the former complex, the geometry about the zinc is distorted tetrahedral with average $Zn-Cl$ and $Zn-N$ distances of 2.21 and 2.08 Å, respectively. In the latter complex, each cadmium is octahedrally co-ordinated by two pairs of bromine atoms (average $Cd-Br$ distance $= 2.75$ and 2.84 Å) and a pair of nitrogen atoms in a *cis*-configuration (average $Cd-N$ distance $= 2.46$ Å). Bromine atoms provide bridges, resulting in a chain-like structure. The crystal structure of (ethylenediamine)zinc benzohydroxamate hydrate has been reported.[42] The two benzohydroxamate ions and an ethylenediamine form a distorted octahedron about the zinc. The asymmetric unit also contains a water molecule and a benzohydroxamic acid molecule.

The new mixed ethylenediamine complexes[43] $[Zn(en)_2NO_2]X(X = NO_2$ or $Br)$, containing bidentate nitrite, have been reported, as well as the complex $[Zn(en)_2NO_2]$ ClO_4 which is polymeric and contains nitrite bridging through both nitrogen and oxygen. The complexes $[Zn(en)_2C_2O_4]X_2$ ($X = NCS, NO_3$, or ClO_4) contain oxalate bridging two $Zn(en)_2$ moieties. The new hydrazine complexes $Zn(C_2O_4)(N_2H_4)$, H_2O and $Zn(C_2O_4)(N_2H_4)_2$ have been prepared.[44] The complexes $Zn(N_2H_4)_2Br_2$, H_2O and $[Zn(N_2H_4)_2(H_2O)_2]X$ ($X = SO_4, SO_3$, and CO_3) have also been prepared.[45]

The crystal structure of dichloro-bis(NN-dimethylacetamide)zinc shows[46] zinc

[34] E. A. Allen, and W. Wilkinson, *J. Inorg. Nuclear Chem.*, 1973, **35**, 3135.
[35] A. L. Lott, *Inorg. Chem.*, 1974, **13**, 667.
[36] E. P. Razumova, V. T. Aleksanyan, G. V. Borokin, and B. S. Nametkin, *Zhur. neorg. Khim.*, 1974, **19**, 329.
[37] I. S. Ahuja, and R. Singh, *Inorg. Nuclear Chem. Letters*, 1973, **9**, 289.
[38] I. S. Ahuja, and R. Singh, *Inorg. Nuclear Chem. Letters.*, 1974, **10**, 421.
[39] H. P. Fritz, and F. Tiedt, *Z. Naturforsch.*, 1973, **28b**, 176.
[40] S. Htoon and M. F. C. Ladd, *J. Cryst. Mol. Structure*, 1973, **3**, 95.
[41] S. Htoon and M. F. C. Ladd, *J. Cryst. Mol. Structure*, 1974, **4**, 97.
[42] S. Gothlicher and P. Ochsenreiter, *Chem. Ber.*, 1974, **107**, 391.
[43] A. E. Shvelashvili, *Zhur. neorg. Khim.*, 1974, **19**, 568.
[44] E. A. Nikonenko, E. I. Krylov, and V. A. Sharov, *Zhur. neorg. Khim.*, 1973, **18**, 2370.
[45] R. Y. Aliev and M. N. Guseinov, *Chem. Abs.*, 1974, **81**, 32671.
[46] M. Herceg and J. Fischer, *Acta Cryst.*, 1974, **30B**, 1289.

tetrahedrally co-ordinated by the two ligand oxygens (Zn–O = 1.96, 1.98 Å) and two chlorines (Zn—Cl = 2.21, 2.22 Å). Similarly, the structure determination[47] of bis-(5,5'-diethylbarbiturato)bis(picoline)zinc shows zinc tetrahedrally bonded to the deprotonated nitrogen atoms of the barbital anions (Zn—N = 1.99 2.01 Å) and to the nitrogen atoms of the picoline ligands (Zn—N = 2.07, 2.10 Å). The molecules are linked by N—H $\cdots$ O hydrogen bonds. The structure determination of di-μ-(*NN*-diethylnicotinamide)tetraisothiocyanatodizinc also shows zinc tetrahedrally co-ordinated, with *NN*-diethylnicotinamide molecules bridging through both nitrogen and oxygen.[48] A unique zinc complex containing a co-ordinated biradical has been postulated as an intermediate in the decomposition of the tetramethyl-2-tetrazene complex of zinc chloride (Scheme 1).[49]

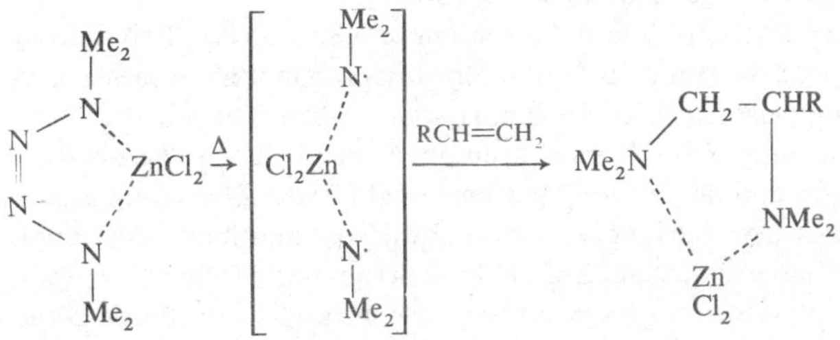

Scheme 1

The thermodynamics of complex formation of several linear aliphatic tetramines with zinc have been reported as part of a study to determine the effect of ring size on these thermodynamic functions. Previous results[50] on the $[Zn(2,2,2\text{-tet})]^{2+}$ complex assign it a tetrahedral configuration on the basis of a very large entropy of complexation. However, zinc complexes with 2,3,2-tet and 3,2,3-tet exhibit a higher heat of formation but a lower entropy, and favour either octahedral or square-pyramidal geometry.[51] The complex $[Zn(3,2,3\text{-tet})]^{2+}$ may be transformed into the species $[Zn(OH)(3,2,3\text{-tet})]^+$, and the thermodynamic functions associated with this reaction are in accord with simple deprotonation of a co-ordinated water molecule in square-pyramidal or octahedral geometry.

Zinc complexes of several polymeric N-donor ligands have been reported. Poly-(1-vinyl-2-pyrrolidinone) of various molecular weights forms the complex[52] $[ZnCl_2(C_6H_9NO)]_{100}$. Complexes with the polyvinylpyridines poly-(2-pyridylethylene) and poly-(4-pyridylethylene) have also been prepared.[53]

Several reports have appeared on complexes of zinc and cadmium with Schiff bases. The electronic absorption spectra of various salicylaldimine complexes of zinc have been determined[54] to clarify assignments made previously for similar copper(II)

[47] L. Nassimbeni and A. Rodgers, *Acta Cryst.*, 1974, **30**, 1953.
[48] F. Bigoli, A. Braibanti, M. A. Pellinghelli, and A. Tiripicchio, *Acta Cryst.*, 1973, **29B**, 2708.
[49] C. J. Michejda and D. H. Cambell, *J. Amer. Chem. Soc.*, 1974, **96**, 929.
[50] L. Sacconi, P. Paoletti, and M. Ciampolini, *J. Chem. Soc.*, 1961, 5115.
[51] R. Barbucci, L. Fabrizzi, P. Paoletti, and A. Vacca, *J.C.S. Dalton* 1973, 1763.
[52] H. G. Biedermann and W. Graf, *Z. Naturforsch.*, 1974, **29b**, 65.
[53] H. G. Biedermann and E. Griessl, *Z. Naturforsch.*, 1974, **29b**, 132.
[54] A. C. Braithwaite and T. N. Waters, *J. Inorg. Nuclear Chem.*, 1973, **35**, 3223.

Q

complexes. A band at *ca.* 41 000 cm^{-1} has been reassigned as an $n-\pi^*$ transition of the oxygen lone pair, rather than the $\sigma-3d$ transition proposed earlier. The i.r. spectra of ^{15}N-labelled complexes of *N*-*p*-tolysalicylaldimines with zinc, copper, and cobalt have yielded assignments of the metal–ligand stretching frequency and certain ligand vibrations.[55] The v(M—N) values are metal-ion dependent in the order Co $<$ Cu $>$ Zn as expected from crystal-field theory. Substituent-induced shifts are related to the residual polar effects of salicylaldimine substitution and to the inductive effects of *N*-aryl substituents.

Zinc and cadmium complexes have been obtained with the quadridentate thioiminato Schiff base *NN'*-ethylenebis(monothioacetonimine).[56] U.v. and visible spectra indicate square-planar geometry, rather than the tetrahedral geometry observed for the salicylaldimine complexes.

The complexes of zinc and cadmium with Schiff bases derived from salicylaldehyde and propan-2-olamine or 2-aminomethylpropanol have been reported,[57] as well as their complexes with ethanolamine and propan-2-olamine.[58, 59] Zinc complexes of stoicheiometry ZnL_2X_2 have been obtained where L = benzoyl- and salicyl-hydrazones of vanilline, furfural, and cinnamaldehyde. Zinc complexes of several substituted *o*-hydroxybutyrophenones and their oximes have been reported.[60—62] Complexes of stoicheiometery ZnL_2L' have been reported where L = the α-nitroketonatoanion obtained from nitroacetone, 3,3-dimethyl-1-nitrobutan-2-one or 3-nitrocamphor and L' = bipyridyl or 1,10-phenanthroline.[63]

Complexes of several S-donor ligands have also been reported. The reaction of dithio-oxamide and its tetramethyl and tetraethyl derivatives with zinc, cadmium, and mercury halides leads to complexes of stoicheiometry MLX_2 (M = Zn, Cd, or Hg; X = Cl, Br, or I).[64, 65] M—S bonding is involved; i.r. spectra show that the zinc and mercury complexes are four-co-ordinate, while the cadmium complexes are octahedral with halogen bridges. The complexes $[ML_3][ClO_4]_2$ have also been prepared and do not contain co-ordinated perchlorate. The complexes ZnL (H_2L = dithio-oxamide, *NN*-dimethyl- and *NN*-dicyclohexyl-dithio-oxamide) have been isolated,[66] and consist of linear chains containing four-co-ordinate zinc.

The crystal structure of the first monomeric thiocarbamate complex, bis(cyclopentamethylenethiocarbamato)bis(piperidine)zinc, has been reported.[67] The co-ordination geometry about the zinc is approximately tetrahedral, consisting of sulphurs of the two thiocarbamate ligands (Zn—S = 2.29, 2.31 Å) and nitrogens of two piperidine ligands. Hydrogen bonding of the unco-ordinated thiocarbamate oxygen with the proton of the co-ordinated piperidine occurs.

[55] G. C. Percy and D. A. Thornton, *J. Inorg. Nuclear Chem.*, 1973, **35**, 2319.
[56] P. R. Blum, R. M. C. Wei, and S. C. Cummings, *Inorg. Chem.*, 1974, **13**, 450.
[57] G. N. Rao and S. C. Rustagi, *Indian J. Chem.*, 1973, **11**, 1181.
[58] S. C. Rustagi and G. N. Rao, *J. Inorg. Nuclear Chem.*, 1974, **36**, 1161.
[59] J. R. Shah and R. P. Patel, *J. prakt. Chem.*, 1973, **315**, 843.
[60] I. M. Issa, R. M. Issa, Y. M. Temerk, and M. M. Ghoneim, *Monatsh.*, 1973, **104**, 963.
[61] J. R. Shah and R. P. Patel, *J. Inorg. Nuclear Chem.*, 1973, **35**, 2589, 3023.
[62] J. R. Shah and R. P. Patel, *Indian J. Chem.* 1973, **11**, 606, 607.
[63] D. Attanasio, I. Collamati, and C. Ercolani, *J.C.S. Dalton*, 1973, 2242.
[64] G. C. Pellacani, A. C. Fabretti, and G. Peyronel, *Inorg. Nuclear Chem. Letters*, 1973, **9**, 897
[65] A. C. Fabretti, G. C. Pellacani, and G. Peyronel, *J. Inorg. Nuclear Chem.*, 1974, **36**, 1751.
[66] L. Menabue, G. C. Pellacani, and G. Peyronel, *Inorg. Nuclear Chem. Letters*, 1974, **10**, 187.
[67] D. L. Greeve, B. J. McCormick, and C. G. Pierpont, *Inorg. Chem.*, 1973, **12**, 2148.

^{1}H n.m.r. spectra of some dithiocarbazo-complexes of zinc, ZnL_2 (HL = R^1R^2 NNHCSSH), have been determined,[68] and indicate pseudotetrahedral geometry with chelation by sulphur only. The trithiocarbamato-complex anions $[M(CS_3)_2]^{2-}$ (M = Zn or Cd) have been isolated as their tetraphenyl-phosphonium and -arsonium salts.[69] The trithiocarbamate is bidentate and forms four-membered chelate rings. Dithiolato-complexes of zinc and cadmium, $[NEt_4]_2MX_2$, where X is the dianion of cyclopentadienedithiocarboxylic acid, have been reported.[70, 71]

The chemistry of five-co-ordinate zinc has been the subject of interest. The crystal structure of μ-oxalato-bis[di-(3-aminopropyl)amine]zinc bis(perchlorate) has been determined.[72] The oxalate group is bridging and bichelate, bound in such a way as to complete a five-membered, rather than a four-membered, ring. Co-ordination of the three amine nitrogens completes a distorted trigonal-bipyramidal co-ordination about the zinc. Apical sites are occupied by one oxygen of the oxalate and by the tertiary nitrogen of the amine ligand. The crystal structure of bis-[(2-thiobenzaldimino)2,6-diacetylpyridine]zinc has been determined as a five-co-ordinate complex with the ligand donor atoms describing an approximate trigonal bipyramid.[73] The overall configuration of the ligand is decidedly helical, however, resulting from steric interactions between the methyl groups and the protons of the aromatic rings.

Complexes $ZnL(ClO_4)_2$ and $ZnLCl(ClO_4)$ (L = tetramethylcyclam) have been isolated.[74] The latter contains the five-co-ordinate $ZnLCl^+$ cation. N.m.r. results indicate it to possess a square-pyramidal configuration with the chlorine in the apical position and all four methyl groups on the same side of the plane.

The complexes ZnX_2L_2 [X = Cl, Br, or I; L = {3,5-diphenyl-Δ4-(1,3,4-thiadiazo-line)-2-ylidene}-*p*-methoxythioacetophenone or -*p*-methoxyacetophenone have been isolated.[75] On the basis of i.r. evidence the metal is believed to be trigonal bipyramidally co-ordinated with the ligand acting in a bidentate manner. The complexes are thus best formulated as $[ZnL_2X]X$.

A theoretical study of five-co-ordination in complexes of stoicheiometry bis(unidentate ligand)(terdentate ligand)metal has been reported.[76] Minimization of the repulsive energy terms shows that neither trigonal-bipyramidal nor square-pyramidal is the expected stereochemistry; this arises simply from the relative rigidity of the chelate rings and is quite distinct from any other steric interaction in the molecule.

Zinc in a trigonal-prismatic configuration has been found in the [fluoroborotris-(2-aldoximino-6-pyridyl)phosphine]zinc complex cation.[77] As in the analogous d^7Co^{II} and d^8Ni^{II} complexes, the clathrochelate zinc cation possesses approximate C_{3v} symmetry. The immediate environment of zinc consists of six nitrogen atoms which define a slightly tapered trigonal prism. Distortion from C_{3v} symmetry is probably caused by the encapsulated metal ion being larger than the cavity of the free unperturbed anion.

[68] D. Gattegnio and A. M. Guiliani, *J. Inorg. Nuclear Chem.*, 1974, **36**, 1553.
[69] A. Müller, P. Christophliemk, I. Tossidis, and C. K. Jørgensen, *Z. anorg. Chem.*, 1973, **401**, 274.
[70] P. C. Savino and R. D. Bereman, *Inorg. Chem.* 1973, **12**, 173.
[71] B. J. Kalbacher and R. D. Bereman, *Inorg. Chem.* 1973, **12**, 2997.
[72] N. F. Curtis, I. R. N. McCormick, and T. N. Waters, *J.C.S. Dalton*, 1973, 1537.
[73] V. L. Goedken and G. G. Christoph, *Inorg. Chem.*, 1973, **12**, 2136.
[74] E. K. Barefield and F. Wagner, *Inorg. Chem.*, 1973, **12**, 2435.
[75] F. Y. Petillon, J. Y. Calves, J. E. Guerchais, and Y. M. Pourier, *J. Inorg. Nuclear Chem.*, 1973, **35**, 3751.
[76] D. L. Kepert, *J.C.S. Dalton*, 1974, 612.
[77] M. R. Churchill and A. H. Reis, *Inorg. Chem.*, 1973, **12**, 2280.

The first pentagonal-bipyramidal complex of zinc, $ZnLCl_2,3H_2O$, has been reported, where L is the planar five-co-ordinate 2,6-diacetylpyridinebis(semicarbazone).[78] The crystal structure determination shows the complex to contain the $LZnCl(H_2O)$ cation, with three N- and two O-donor atoms forming a slightly distorted planar pentagon around the metal ion and the chlorine and water in axial positions.

The crystal structures of several carboxylato-complexes have been reported. The structure of zinc *o*-ethoxybenzoate monohydrate is polymeric[79] with distorted tetrahedral co-ordination of each zinc by three carboxylato-oxygens and one water molecule. The structure of (pyrazine-2,3-dicarboxylato)zinc dihydrate is also polymeric,[80] but each zinc is co-ordinated in a distorted octahedral manner by three carboxylato-oxygens, one nitrogen, and two water molecules. The preparation and crystallographic data for the malonato-complexes $Zn[CH_2(COO)_2],2H_2O$ and $Zn[HOOCCH_2COO]_2 2H_2O$ have also been reported.[81]

A review has been published on the X-ray structural determination of zinc, cadmium, and mercury complexes.[82] In addition there have been reports regarding the various systems detailed in Table 1.

Table 1

System		Data^a	Ref.
M = Zn, Cd, or Hg			
MH_2L	H_4L = phthalyltetrathioacetic acid	s	83
MX_2L_2	L = morpholine-4-thiocarbonic acid anilide	p, i.r., c	84
$MX_2(Hbn)$	X = Cl, Br, or I	p, i.r.,	85
$[MX(bn)]_2$	Hbn = benzoin	c	
M = Zn or Cd			
ML	H_2L = thiosalicylic acid	p	86
	H_2L = N-(acetylacetonyl)-anthranilic acid	p	87
ML_2	HL = 8-selenoquinoline	p, i.r.	88
	HL = dibenzodithioarsenic acid	p, i.r.	89
	HL = pyrazolo-[5-azo-1-(2-hydroxynaphthalene]-4-carboxylic acid	p, u.v.	90
	HL = 1-hydroxyphenazine	p, s	91
MAL	H_2L = salicylic, thiosalicylic acid A = dipyridyl	s	92
MLX	H_2L = N-(2-hydroxy-1-naphthalidene)anthranilic acid or β-alanine; X = Cl or Br	p	93
MX_2L_2	L = aniline, *o*-toluidine, *m*-xylylidine, pyridine X = Cl, Br, or I	p, u.v., X, t	94
ML_n^{2+}	n = 1 or 2; L = N-methyl-N-phenyl-2-mercaptoacetamide	s	95
	n = 2; L = formylferrocenethiosemicarbazone	p, i.r.	96

[78] D. Webster and G. J. Palenik, *J. Amer. Chem. Soc.*, 1973, **95**, 6505.
[79] S. Natarajan, D. S. Sake Gowda, and L. Cartz, *Acta Cryst.*, 1974, **30B**, 401.
[80] P. Richard, D. Tran Qui, and E. F. Bertaut, *Acta Cryst.*, 1974, **30B**, 628.
[81] L. Walter-Levy, J. Perrotey, and J. W. Visser, *Bull. Soc. chim. France*, 1973, 2596.
[82] M. B. Hursthouse, in 'Molecular Structure by Diffraction Methods', ed G. A. Sim and L. E. Sutton (Specialist Periodical Reports). The Chemical Society, London, 1973, Vol. 1, p. 716.

$ML_n^{(2-n)+}$	HL = 4-hydroxy-3-oximino-methylazobenzene; $n = 1$ or 2	s	97
M = Zn			
ZnL	L = cyclic trimeta-, tetrameta-, hexameta-, and octameta-phosphate anions	s	98
ZnLL'	H_2L = dibasic acid; L' = piperidine	p, i.r.	99
ZnX_2L_2	X = Cl; L = quinoline, piperidine	p	100
	X = NCO; L = substituted pyridines, quinoline, isoquinoline	p, i.r.	101
	X = Cl; L = quinoxaline	p, i.r. X	102
$Zn(OAc)_2L_n$	$n = 1$ or 2; L = urea, thiourea	p	103
$Zn(NCS)_2L_2,nH_2O$	L = various acid hydrazides	p, i.r.	104
ZnL_2SO_4	L = ethyleneglycol	p	105
$ZnL_3(NO_3)_2$			
$ZnL_2Cl_2,2H_2O$	L = anisic acid hydrazide	p, i.r.	106
ZnL_n^{2+}	$n = 1$ or 2; L = p-(mercaptoaceta-mido)chlorobenzene	s	107
$ZnL_n^{(2-n)+}$	$n = 1$, HL = 1-phenazinemethiol	p, i.r.	108
	HL = o-benzarsenous oxide	s	109
	HL = 9-(2-pyridylazo)-10-phenanthrol	s	110
	HL = various amino-acids	s	111
	$n = 1$ or 2, HL = 4-isonitroso-1-phenyl-3-methylpyrazolin-5-one	s	112
	HL = N-phenyl-o-methoxy-benzohydroxamic acid	s	113
	HL = N-methyl-, NN-dimethyl-histidine	s	114
	HL = m-mercaptoacetamidophenol	s	115
	HL = various hydroxycarboxylic acids	s	116
	$n = 2$, HL = 3,3',4,4'-tetrachloro-5,5'-diethoxycarbonyldipyrromethene	p, i.r., X	117
	HL = salicylic acid hydrazide	p, i.r.	118
ZnL^{4+}	L = pentammine(pyrazine)ruthenium(II)	s	119
$ZnL_n^{(2n-2)-}$	H_2L = 3-nitrosalicylic acid; $n = 1$ or 2	s	120
$[Zn(CN)_n(NH_3)_{4-n}]^{(n-2)-}$	$n = 0$—4	s	121
$Zn(NCS)_nI_{2-n}L_2$	$n = 0$—2; L = pyridine	s	122
$[ZnL]^-$	H_3L = nitrilotriacetic acid	s	123

(a) p = preparation; s = stability constants; c = conductivity; X = X-ray; t = thermochemistry.

[83] M. M. Jones, A. J. Banks, and C. H. Brown, *J. Inorg. Nuclear Chem.*, 1974, **36**, 1833.

[84] D. Venkappayya and D. H. Brown, *J. Inorg. Nuclear Chem.*, 1974, **36**, 1023.

[85] K. C. Malhotra and S. C. Chandry, *Austral. J. Chem.*, 1974, **27**, 79.

[86] N. S. Al Niaimi and B. M. Al-Saadi, *J. Inorg. Nuclear Chem.*, 1974, **36**, 1617.

[87] R. K. Mehta, R. K. Gupta, and V. C. Singhi, *Z. phys. Chem.*, 1973, **253**, 49.

[88] Y. Mido, I. Fujiwara, and E. Sekido, *J. Inorg. Nuclear Chem.*, 1974, **36**, 537, 1003.

[89] A. Müller, P. Werle, P. Christophliemk, and I. Tossidis, *Chem. Ber.*, 1973, **106**, 3601.

[90] B. Janik and T. Gancarczyk, *Chem. Abs.*, 1973, **79**, 86928.

[91] Y. Kidani, K. Inagaki, and H. Koike, *Yakugaku Zasshi*, 1973, **93**, 1089.

[92] J. D. Joshi, I. P. Mavani, and P. K. Bhattacharya, *Indian J. Chem.*, 1973, **11**, 820.

[93] R. K. Mehta, V. C. Singhi, R. K. Gupta, and S. L. Pania, *J. Indian Chem. Soc.*, 1973, **50**, 721.

3 Cadmium

Several studies on the high-resolution n.m.r. of ^{111}Cd (natural bundance $= 12.8\%$, $I = \frac{1}{2}$) and ^{113}Cd (natural abundance $= 12.3\%$, $I = \frac{1}{2}$) have been reported.

Studies of the ^{113}Cd (13.31 MHz at 14 kG) n.m.r. spectra of aqueous solutions of cadmium(II) salts in the presence and absence of diverse uni- and bi-dentate complexing agents have been reported.[124, 125] For the free halide salts, a large non-linear variation in chemical shift with concentration in the 0.1—5 mol l^{-1} range indicates formation of both mono- and poly-halogeno-complexes. Cadmium sulphate, perchlorate, and nitrate show linear relationships depending much less on concentration, suggesting little association, although the nitrate results are in contrast with previous Raman results which showed considerable co-ordinated nitrate. In the presence of a wide variety of organic and inorganic ligands, the shifts produced are essentially linear, and several generalizations can be made: (i) ligands binding through oxygen (sulphate, nitrate, nitrite, acetate, formate) cause increased shielding of the cadmium nucleus; (ii) those binding through nitrogen (ammonia, pyridine, azide, ethylenediamine) produce a marked deshielding; (iii) ligands in which binding *via* sulphur seems probable (thiourea, thiocyanate) cause very large deshielding. Many of the nitrogen ligands also produce line-broadening, which apparently stems from a slower dissociation rate of these complexes. Either their chemical shifts or the broadening may be used as a basis for a determination of binding constants using a titration method.

[94] P. B. Issopoulos, *Compt. rend.*, 1974, **278**, C, 263.
[95] C. S. Bhandari and N. C. Sogani, *J. Inst. Chem., Calcutta*, 1973, **45**, (Pt 4), 138.
[96] W. D. Fleischman and H. P. Fritz, *Z. Naturforsch.*, 1973, **28B**, 383.
[97] P. N. Mohandas, O. P. Sunar, and C. P. Trivedi, *J. Inorg. Nuclear Chem.*, 1974, **36**, 937.
[98] G. Kura and S. Ohashi, *J. Inorg. Nuclear Chem.*, 1974, **36**, 1605.
[99] R. I. Kharitoonva, M. G. Pivovarova, and T. I. Kirienko, *Zhur. obschei Khim.*, 1973, **43**, 2043.
[100] M. Barvinok, A. V. Panin, and L. A. Obozova, *Zhur. neorg. chim.*, 1973, **18**, 2442.
[101] B. K. Mohapatra, *Current Sci.*, 1973, **42**, 565.
[102] A. Tenhunnen, *Finn. Chem. Letters*, 1974, **1**, 28.
[103] Z. Ergeshbaev, B. Murzubraimov, and K. Sulaimankulov, *Chem. Abs.*, 1974, **80**, 59 404.
[104] Y. Y. Kharitonov, R. I. Machkhoshvili, N. B. Generalova ,and R. N. Shchelekov, *Zhur. neorg. Khim.*, 1974, **19**, 1124.
[105] D. Knetsch and W. K. Groenveld, *Rec. Trav. chim.*, 1973, **92**, 855.
[106] Y. Y. Kharitonov, R. I. Machkhoshvili, and N. B. Generalova, *Zhur. neorg. Khim.*, 1973, **18**, 2000.
[107] S. N. Kakkar and P. V. Khadikar, *Indian J. Chem.*, 1973, **11**, 1325.
[108] Y. Kidani, K. Anakawa, and H. Koike, *Yakugaku Zasshi*, 1973, **93**, 928.
[109] B. S. Sekhon, S. S. Parmar, S. K. Push-Karna, and S. L. Chopra, *Indian J. Chem.*, 1973, **11**, 835.
[110] K. Masayoshi and A. Kawase, *Bunseki Kagaku*, 1973, **22**, 860.
[111] S. K. Das, J. R. Shah, and R. P. Patel, *J. Indian Chem. Soc.*, 1973, **50**, 228.
[112] G. Reinhard, R. Dreyer, and R. Muenze, *Z. phys. Chem.*, 1973, **254**, 226.
[113] Y. K. Agrawal and I. P. Shukla, *Talanta*, 1973, **20**, 1353.
[114] A. Braibanti, F. Dallavalle, E. Leporati and G. Mori, *J.C.S. Dalton*, 1973, 2539.
[115] S. N. Kakkar, N. S. Poonia, and P. V. Khadikar, *Indian J. Chem.*, 1973, **11**, 709.
[116] I. Filipovic, I. Piljac. B. Bach-Dragutinovic, I. Kruhak, and B. Grabaric, *Croat. Chem. Acta*, 1973, **45**, 447.
[117] Y. Murakami, Y. Matsuda, K. Sakata, and A. E. Martell, *J.C.S. Dalton*, 1973, 1729.
[118] Y. Y. Kharitonov, R. I. Machkhoshvili, and N. B. Generalova, *Zhur. neorg. Khim.*, 1974, **19**, 270.
[119] M. S. Pereira and J. M. Malin, *Inorg. Chem.*, 1974, **13**, 386.
[120] P. V. Khadikar and R. L. Ameria, *J. Indian Chem. Soc.*, 1973, **50**, 503.
[121] V. A. Konev and A. O. Brandt, *Zhur. neorg. Khim.*, 1974, **19**, 319.
[122] Z. Mikulec and M. Valentova, *Coll. Czech. Chem. Comm.*, 1973, **38**, 2268.
[123] F. J. M. Rajabalee, *J. Inorg. Nuclear Chem.*, 1974, **36**, 557.
[124] G. E. Maciel and M. Borzo, *J.C.S. Chem. Comm.* 1973, 394.
[125] R. J. Kostelnik and A. A. Bothner-By, *J. Magn. Resonance* 1974, **14**, 141.

Similar conclusions have been reached from ^{111}Cd (16.31MHz at 1.807 Tesla) n.m.r. studies.[126] Large non-linear chemical shifts with concentration of cadmium chloride solutions were observed, while the nitrate, sulphate, and perchlorate show much smaller linear shifts, consistent with the formation of stable chlorocadmium complexes even at low concentration.

The relatively new technique of quantitative Raman spectroscopy has been applied[127] to the determination of the equilibrium constants for the reactions

$$[CdI_{3-n}Br_{n-1}]^{2-} + I^- \rightleftharpoons [CdBr_nI_{4-n}]^{2-} + Br^- \quad n = 0\text{---}3$$

The advantages of the technique include the direct proportionality of the signal obtained for a single complex species in a complicated equilibrium to the concentration of the species, which is particularly advantageous in cases of mixed complex formation, where most methods given only indirect evidence for the existence of mixed species, and where very complicated relationships exist between measurable quantities and the total concentration of reactants. The results obtained (K_0—K_3 = 5.3, 8.5, 5, 0.45) are in substantial agreement with previous values obtained from polarographic and potentiometric studies.

The preparation of some new ethylenediamine complexes of cadmium(II) has been reported.[128—130] The reaction of a mixture of cadmium oxalate and CdX$_2$ (X = Cl, Br, I, or SCN) with ethylenediamine gives the complexes Cd(en)X(C$_2$O$_4$)$_{0.5}$. The cadmium is octahedrally co-ordinated by ethylenediamine, X, and three oxygen atoms (two from bidentate oxalate, one from bridging oxalate). The *cis*-octahedral complexes Cd(en)$_2$(NCS)Cl and Cd(en)$_2$BrI have also been reported, as well as the complexes Cd(en)$_2$XX' (X = NO$_3$ or ClO$_4$; X' = Br or $\frac{1}{2}$C$_2$O$_4$), which contain bridging bromide and oxalate groups. The nitrite complexes Cd(en)$_2$NO$_2$(ClO$_4$) and CdenINO$_2$ have also been prepared, as well as the tris-complex Cd(en)$_3$X(ClO$_4$) (X = I or NO$_3$).

The thiourea complexes (CdL$_6$)(ClO$_4$)$_2$ have been prepared[131] [L = thiourea or RNHC(S)NHPh, R = Me, Ph, *m*- and *p*-MeC$_6$H$_4$, or α-naphthyl]. Their i.r. spectra show them to be S-bonded.

The crystal structures of several Lewis base adducts of Cd[Mn(CO)$_5$]$_2$ have been determined. In the cases where the ligand is the terdentate terpyridyl[132] or diglyme [di-(2-methoxyethyl)ether],[133] the co-ordination is best described as very distorted trigonal bipyramidal, with the Mn(CO)$_5$ groups occupying equatorial positions. In both complexes, deviations of the equatorial angles from 120° are attributed to the large size of the Mn(CO)$_5$ groups. In the former complex, considerable displacement of the nitrogen atoms from ideal axial geometry is due to restrictions imposed by the terpyridyl ligand, while in the latter complex, the OCdO angle of 126° is due

[126] H. Krüger, O. Lutz, A. Schwenk, and G. Stricker, *Z. Phys.*, 1974, **266**, 233.
[127] N. Yellin and Y. Marcus, *J. Inorg. Nuclear Chem.*, 1974, **36**, 1331.
[128] A. E. Shvelashvili and L. P. Sarishvili, *Zhur. neorg. Khim.*, 1973, **18**, 3133.
[129] A. E. Shvelashvili and L. P. Sarishvili, *Zhur. neorg. Khim.*, 1974, **19**, 1015.
[130] A. E. Shvelashvili and L. P. Sarishvili, *Chem. Abs.*, 1974, **80**, 55448.
[131] V. A. Fedorov, A. V. Fedorova, G. G. Nitanteva, L. G. Sobolova, and L. I. Gruber, *Zhur. neorg. Khim.*, 1973, **18**, 3341.
[132] W. Clegg and P. J. Wheatley, *J.C.S. Dalton*, 1973, 90.
[133] W. Clegg and P. J. Wheatley, *J.C.S. Dalton*, 1973, 424.

mainly to puckering of the chelate rings. The cadmium–manganese bond lengths (2.78 and 2.71 Å) are consistent with the view that weakening of the metal–metal bond occurs on adduct formation.

The structures of the bidentate bipyridyl and 1,10-phenanthroline adducts have also been reported.[134] The cadmium is tetrahedrally co-ordinated, distorted by considerable reduction of the NCdN angle due to the geometrical limitations of the bidentate ligand, and expansion of the MnCdMn angle due to the bulky nature of the $Mn(CO)_5$ groups. The manganese–cadmium bond length (2.68 Å) is in keeping with the observation that the covalent radius increases with co-ordination number.

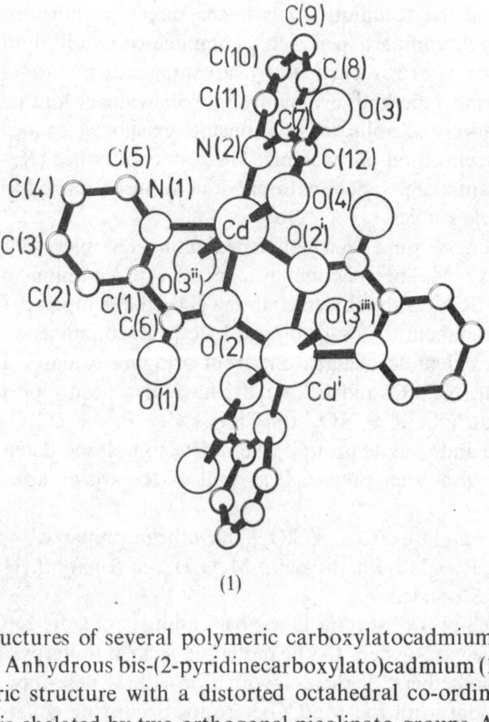

(1)

The crystal structures of several polymeric carboxylatocadmium complexes have been determined. Anhydrous bis-(2-pyridinecarboxylato)cadmium (1)[135] has a centro-symmetric dimeric structure with a distorted octahedral co-ordination about each cadmium, which is chelated by two orthogonal picolinato-groups. A sixth co-ordination position is occupied by the non-chelating oxygen atom of the non-bridging picolinato group of a neighbouring dimer, thus conferring a polymeric $[Cd(C_5H_4NCOO)_2]_{2n}$ structure on the complex. The structure determination of cadmium cyanoacetate (2)[136] shows the cadmium to be co-ordinated to one nitrogen and five oxygen atoms in a distorted octahedral geometry. Each of the two independent cyanoacetate ligands co-ordinates to three symmetry related cadmium atoms to give a three-dimensional polymeric lattice. One ligand co-ordinates through two oxygen

[134] W. Clegg and P. J. Wheatley, *J.C.S. Dalton*, 1973, 511.
[135] J. P. Deloume and H. Loiseleur, *Acta Cryst.*, 1974, **30B**, 607.
[136] M. L. Post and J. Trotter, *J.C.S. Dalton*, 1974, 285.

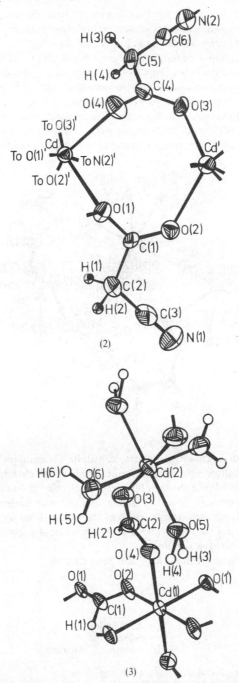

(2)

(3)

atoms, one of them bridging, while the other co-ordinates through two oxygen atoms and the nitrogen atom. In common with many other acetates, the metal atom is not chelated, with the ligands acting as bridging molecules.

The crystal structure of cadmium(II) formate dihydrate (3) has been determined[137] and consists of a three-dimensional polymer with each cadmium octahedrally co-ordinated. In a two-dimensional plane, the formates bridge in an *anti, anti*-configuration, while in the perpendicular plane, the bridging is in a *syn,anti*-manner. Water molecules occupy the remaining co-ordination positions, and the polymer is further strengthened by hydrogen bonds between water and formate. The compound is isomorphous with magnesium, zinc, manganese, and copper formate dihydrates.

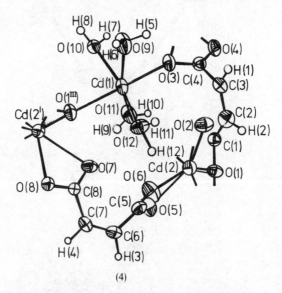

(4)

The structure of cadmium(II) maleate dihydrate (4) contains[138] two cadmium atoms and two maleate ligands, both pairs of which are chemically distinct. One cadmium is six-co-ordinate through four water molecules and two bridging maleate ligands, while the other is eight-co-ordinate, in a distorted dodecahedral geometry, through four chelated carboxy-groups from the two maleate ligands. The maleate ligands link the cadmium atoms into a three-dimensional polymer which is further strengthened by hydrogen bonding.

The systems shown in Table 2 have also been studied.

Table 2

System		Data[a]	Ref.
CdL^{2+}	L = 2,2′-bipyridyl	s	139
CdBr$_2$,L,H$_2$O	L = *p*-nitroaniline	p, i.r.	140
CdL$_2$	HL = benzo-2,1,3-thiadiazole	p	141

[137] M. L. Post and J. Trotter, *Acta Cryst.*, 1974, **30B**, 1880.
[138] M. L. Post and J. Trotter, *J.C.S. Dalton*, 1973, 674.

CdL_n^{2+}	$n = 1$ or 2; $L = $ 3-picoline, 3,5-lutidine	s	142
	$n = 1$—3; $L = $ N-substituted ethylenediamines	s	143
$CdLL_n'$	$H_2L = $ dibasic acid		
	$L' = $ piperidine, $n = 1$ or 2	p	144
$[Cd(C_2O_4)L_n]^{n-}$	$L = $ formate, acetate; $n = 1$ or 2	s	145
CdL_2SO_4	$L = $ monoethanolamine, propan-2-olamine	p, i.r. t	58, 59
CdL_2SO_4,H_2O	$L = $ anisic acid hydrazide	p, i.r.	106
$[CdL_{3-n}(en)_n]^{2+}$	$L = $ ethanolamine; $n = 1$—3	s	146
$(CdL_n)^{(2n-2)-}$	$H_2L = $ isophthalic acid; $n = 1$—3	s	147
$[Cd(NH_3)X_n]^{(2-n)+}$	$n = 1$—3		
$[Cd(NH_3)_2X_n]^{(2-n)+}$	or $\rbrace$ $X = $ Cl, Br, I, or SCN	s	148
$[Cd(NH_3)_3X]^+$	$n = 1$ or 2		
$[CdX_n]^{(n-2)-}$	$X = $ I or Br; $n = 1$—4	s	149
$[Cd(SCN)_4]^{2-}$		s	150
CdL_6SO_4		p, i.r.	105
$CdL(NO_3)_2$	$L = $ ethylene glycol		
$Cd(L)_{4,3}X_2$	$X = $ Cl or Br; $L = $ ethylene glycol.	p, i.r.	151
$CdL_5(H_2O)(ClO_4)_2$; $CdL_6(NO_3)_2$	$L = $ imidazole	p, s	152

(a) p = preparation; s = stability constant; t = thermochemistry.

4 Mercury

The structure of mercuric halides dissolved in aromatic solvents continues to be the subject of controversy. Careful vapour pressure osmometric studies[153, 154] now show conclusively that there is considerable self-association of mercuric halides in their solutions in benzene, toluene, xylene, and mesitylene. The association was found to be dimeric for chloride and bromide, and dimeric and possibly trimeric for iodide.

The solution to the problem of whether complex formation takes place is still not clear. Previously, the observation of a non-zero dipole moment, the observation of charge-transfer spectra in the u.v., and the results of far-i.r. studies indicated the

[139] G. C. Budu and L. V. Nazorova, *Zhur. neorg. Khim.*, 1973, **18**, 2960.
[140] A. V. Ablov, V. Y. Ivanova, and G. E. Volodina, *Zhur. neorg. Khim.*, 1973, **18**, 1740.
[141] S. A. D'yachenko, V. S. Tsveniashvili, S. O. Gerasimova, I. R. Belen'kaya, and N. S. Khavtasi, *Zhur. obschei Khim.*, 1974, **44**, 386.
[142] E. M'Foundou and G. Berthon, *Compt. rend.*, 1973, **277**, C, 965.
[143] H. S. Creyf and L. C. Van Ponche, *J. Inorg. Nuclear Chem.*, 1973, **35**, 3837.
[144] R. I. Kharitonova and M. G. Pivovarova, *Chem. Abs.*, 1974, **80**, 99 889.
[145] D. M. Czakis-Sulikowska, *Chem. Abs.*, 1973, **79**, 108 729.
[146] P. K. Migal and E. P. Koptenko, *Zhur. neorg. Khim.*, 1974, **19**, 322.
[147] A. Kumar and J. N. Guar, *J. Electroanalyt. Interfacial Electrochem.*, 1974, **49**, 317.
[148] Y. D. Fridman, D. Sarbaev, and T. V. Danilova, *Zhur. neorg. Khim.*, 1974, **19**, 867.
[149] V. A. Fedorov, L. I. Kiprin, N. S. Schekina, N. P. Samsonova, M. Y. Kutuzova, and V. E. Mironov, *Zhur. neorg. Khim.*, 1974, **19**, 872.
[150] S. Misumi, M. Aihara, and S. Kinoshuta, *Bull. Chem. Soc. Japan*, 1974, **47**, 127.
[151] D. Knetsch, and W. L. Groenveld, *Inorg. Chim. Acta*, 1974, **7**, 81.
[152] J. B. Jensen, *Acta Chem. Scand.*, 1973, **27**, 3563.
[153] I. Eliezer and G. Algavish, *Inorg. Chim. Acta*, 1974, **9**, 257.
[154] C. L. Cheng, R. K. Pierens, D. V. Radford, and G. L. D. Ritchie, *J. Chem. Phys.*, 1973, **59**, 5209.

presence of solute–solvent interactions. Recently, however, charge-transfer spectra have been observed in the spectra of the gaseous mercuric halides,[155] while more recent Raman and far-i.r. data have indicated linear solute species and the absence of significant intermolecular charge transfer.[156] Dielectric relaxation studies have now been reported which show conclusively that mercuric halides are non-polar in benzene,[154] and the authors have attributed the previous observations of a non-zero dipole moment to uncertainties relating to atomic polarization. Recently, however, solid adducts of the mercuric halides with aromatic molecules have been reported: $2HgBr_2,L$ (L = toluene or ethylbenzene), $HgBr_2,$acenaphthene, $HgI_2,$pyrene, $HgI_2,$2-anthracene.[157] No mercuric chloride complexes could be prepared. Solution association constants were calculated, but as the charge-transfer bands previously quoted were used as an indication of solvent–solute interaction, the results may be open to question.

The dipole moments of a large number of alkylmercuric halides in various organic solvents (hexane, benzene, dioxan, pyridine, ethylacetate) have been reported.[158] A new, more accurate electron diffraction study of gaseous mercuric chloride has been reported.[159] The interatomic distances (Hg—Cl = 2.25 Å, Cl—Cl = 4.48 Å) are shorter than previously reported values by 0.02 to 0.09 Å. A complete normal-co-ordinate analysis of bis(methylthio)mercury has also been reported.[160] The Raman spectra of gaseous mercuric chloride, bromide, and iodide have been reported.[161] The bond polarizability derivatives calculated from the data increase in the order Cl < Br < I, suggesting an increased degree of covalence in the mercury–halogen bond with increasing size of the halogen atom.

The results of calorimetric studies on the reactions of the mercuric halides with Lewis bases in benzene have been reported.[162] The bases used include those with oxygen, nitrogen, phosphorus, arsenic, and sulphur. In most cases, the reaction product in dilute solution is a three-co-ordinate 1:1 adduct with an enthalpy of formation of ca. $-70\,kJ\,mol^{-1}$, but considerably smaller with tributylamine and tetrahydrofuran. Dimerization of the initial three-co-ordinate adduct was observed in the $HgX_2–Bu_3P$ (X = Cl, Br, or I) and $HgI_2–AsPh_3$ systems. With phosphorus donors, unusually large free energies of formation are observed for the 1:1 adducts, and 2:1 adducts are also formed. The structure of the 1:1 adduct of mercuric bromide with bipyridine has been shown[163] to consist of discrete bromine-bridged dimers. Co-ordination about the mercury atom is distorted pyramidal, with the non-bridging bromine atom at the apex.

Several new polymeric mercury cations have been characterized.[164, 165] The reac-

[155] P. Templet, I. R. McDonald, S. P. McGlynn, C. H. Kendrow, I. L. Roebber, and K. Weiss, *J. Chem. Phys.*, 1972, **56**, 5746.
[156] T. B. Brill, *J. Chem. Phys.* 1972, **57**, 1534.
[157] I. M. Vezzosi, G. Peyronel, and A. F. Zamoli, *Inorg. Chim. Acta*, 1974, **8**, 229.
[158] T. Y. Mel'nikova, *Chem. Abs.*, 1973, **79**, 77825.
[159] K. Kashiwabara, S. Konaka, and M. Limura, *Bull. Chem. Soc. Japan*, 1973, **46**, 410.
[160] N. Iwasaki, J. Tomooka, and K. Toyoda, *Bull. Chem. Soc. Japan*, 1974, **47**, 1323.
[161] R. J. H. Clark and D. M. Rippon, *J.C.S. Faraday*, II, 1973, **69**, 1496.
[162] Y. Farhangi and D. P. Graddon, *Austral. J. Chem.*, 1973, **26**, 983.
[163] D. C. Craig, Y. Farhangi, D. P. Graddon, and N. C. Stephenson, *Cryst. Struct. Comm.*, 1973, **3**, 155.
[164] B. D. Cutforth, C. G. Davies, P. A. W. Dean, R. J. Gillespie, and P. K. Ummat, *Inorg. Chem.*, 1973, **6**, 1343.
[165] B. D. Cutforth, R. J. Gillespie, and P. R. Ireland, *J.C.S. Chem. Comm.*, 1973, 723.

tion of AsF_5 with mercury in sulphur dioxide yields initially a compound of formula $Hg_{2.85}AsF_6$, which has a novel structure containing infinite polymeric mercury cation chains. Full details of the structure have not yet been published. Further reaction with AsF_5 yields two compounds $Hg_4(AsF_6)_2$ and $Hg_3(AsF_6)_2$ according to the stoicheiometry

$$n Hg + 3AsF_5 \rightarrow Hg_n(AsF_6)_2 + AsF_3$$

The crystal structure of $Hg_4(AsF_6)_2$ shows the Hg_4^{2+} to be nearly linear, with an HgHgHg angle of $176°$, and with a *trans*-configuration. The terminal Hg—Hg bond length is 2.57 Å, while the central Hg—Hg bond length is 2.70 Å. The average bond length of 2.61 Å follows from the observation that Hg—Hg bond lengths increase with decreasing charge in the series Hg_2^{2+}, Hg_3^{2+}, Hg_4^{2+}. Two stretching frequencies are observed in the Raman spectrum at 95 and 79 cm^{-1}.

The structure of $Hg_3(AsF_6)_2$ may be described as consisting of discrete linear Hg_3^{2+} cations and octahedral AsF_6^- anions, although the Hg_3^{2+} cation in $Hg_3(AlCl_4)_2$ has been found to be non-linear (HgHgHg angle = $176°$).[166] The Hg—Hg bond length is 2.55 Å, slightly longer than that reported for mercurous halides. The Hg—F distances indicate that the terminal mercury atom forms bonds of more or less covalent character to the fluorine of the AsF_6^- anion. An analogous reaction occurs with SbF_5

$$3Hg + 5SbF_5 \rightarrow Hg_3(Sb_2F_{11})_2 + SbF_3$$

Further reaction of $Hg_n(AsF_6)_2(n = 3 \text{ or } 4)$ with AsF_5 yields $Hg_2(AsF_6)_2$. The stretching frequency in $Hg_2(AsF_6)_2$ occurs at $182\,cm^{-1}$, while that of the Hg_3^{2+} cation in Hg_3X_2 ($X = AsF_6^-$ or $Sb_2F_{11}^-$) occurs at $113\,cm^{-1}$. Oxidation of mercury by fluorosulphonic acid was also shown to produce Hg_3^{2+}, eventually yielding Hg_2^{2+}. All the above cations have been shown to exist in equilibrium in sulphur dioxide solution.

$Hg_2(AsF_6)_2$ has been found to react with arenes in sulphur dioxide according to the equation[167]

$$m Hg_2(AsF_6)_2 + n \text{ arene} \rightarrow m Hg_2(AsF_6)_2, n \text{ arene}$$

$m = n = 1$; arene = benzene, biphenyl, naphthalene, or *m*-dinitrobenzene

$m = 1, n = 2 \text{ or } 3$; arene = 9,10-benzophenanthrene

Interaction causes a lowering of the Raman stretching frequency from 166 to 134 cm^{-1}. ^{13}C N.m.r. spectra readily show the presence in solution of the complexes, and are consistent with the presence of two labile equilibria

$$Hg_2^{2+} + \text{arene} \rightleftharpoons Hg_2(\text{arene})^{2+} + \text{arene} \rightleftharpoons Hg_2(\text{arene})_2^{2+}$$

Bonding geometries are postulated to resemble those found in bis(benzene)dipalladium chloroaluminates.

[166] R. D. Ellison, H. A. Levy, and K. W. Fung, *Inorg. Chem.*, 1972, **11**, 833.
[167] P. A. W. Dean, D. G. Ibbott, and J. B. Stothers, *J. C. S. Chem. Comm.*, 1973, 626.

Polymeric anions containing mercury–platinum and mercury–palladium bonds have been characterized.[168] Reaction of K_2PtCl_4 with an equimolar amount of $(Et_4N)_2HgCl_4$ affords either $[Et_4N]_2Hg_2PtCl_8$ or $[Et_4N]_2Hg_3PtCl_{10}$ depending on the concentration of the solution. The common feature of both complexes is the trinuclear anionic unit $[Hg_2PtCl_8]^{2-}$: in the first, the anions are linked by bonds from terminal chlorine atoms to form polymeric zig-zag chain anions $[Hg_2PtCl_8]_n^{2n-}$, while in the second the anionic units are linked sideways by bonds from their four bridging chlorine atoms to linear $HgCl_2$ groups, giving the mercury atoms a characteristic distorted octahedral co-ordination. Reaction of H_2PdCl_4 with a 2:1 mixture of Et_4NCl and $HgCl_2$ gives the complex $[Et_4N]_2Hg_3PdCl_{10}$, whose structure is analogous to the platinum complex. The neutral compound $(Me_3P)_2PtCl_2(HgCl_2)_2$ has been reported,[169] and its i.r. spectrum suggests a structure which is closely related to the $[Hg_3MCl_{10}]^{2-}$ anion.

Stable mercurinium ions, often postulated as intermediates in solvomercuration reactions, have been observed for the first time.[170] They may be prepared in media of low nucleophilicity either by direct mercuration or by ionization of a β-substituted organomercurial. For example, the ethylene mercurinium ion may be prepared from 2-methoxyethylmercuric chloride in an ionizing medium such as HSO_3F–SbF_5–SO_2

$$MeOCH_2CH_2HgCl \xrightarrow[SO_2]{HSO_3F-SbF_5} \underset{Hg^{2+}}{CH_2\!-\!CH_2} + MeOH_2^+$$

while the propylenemethylmercurinium ion may be obtained by metallation of propylene with methylmercuric fluorosulphate in sulphur dioxide.

$$CH_3CH\!=\!CH_2 + CH_3HgOSO_2F \xrightarrow{SO_2} \left[\underset{CH_3-CH-CH_2}{\overset{\overset{\textstyle CH_3}{\underset{\textstyle Hg^+}{|}}}{}} \right] FSO_3^-$$

Other ions studied include the cyclohexene- and norbornylene-mercurinium ions. N.m.r. studies show deshielding of both protons and carbons, indicating charge transfer from mercury to the carbon skeleton; both mercury–proton and mercury–carbon spin–spin couplings are observed. Some of the spectra are best interpreted in terms of an equilibrium between the mercurinium ion and the free olefin. Quenching experiments are consistent with the presence of mercurinium ions.

Organomercury radicals have been obtained from the ^{60}Co γ-ray irradiation of alkylmercuric halides and dimethylmercury at 77 K.[171] For example, irradiation of

[168] R. M. Barr, M. Goldstein, T. N. D. Hairs, M. McPartlin, and A. J. Markwell, *J. C. S. Chem. Comm.*, 1974, 221.
[169] P. R. Brookes and B. L. Shaw, *J.C.S. Dalton*, 1973, 783.
[170] G. A. Olah and P. R. Clifford, *J. Amer. Chem. Soc.*, 1973, **95**, 6067.
[171] B. W. Fullam and M. C. R. Symons, *J.C.S. Dalton*, 1974, 1086.

ethylmercuric chloride generates the radical $CH_3\dot{C}HHgCl$, which shows little delocalization of the unpaired electron, as well as the $\dot{H}gEt$ and $\dot{H}gCl$ radicals, both of which show large hyperfine coupling constants. Irradiation of dimethylmercury produces the radical $\dot{C}H_2HgCH_3$, as well as a species thought to be $(Me_2Hg)\cdot$.

^{31}P N.m.r. studies of compounds of the type L_2HgX_2 and $L_2Hg_2X_4$ (L = tributyl-, triethyl-, dibutylphenyl-, butyldiphenyl-, and diethylphenyl-phosphine, triethylphosphite; X = Cl, Br, I, SCN, or CN) have been reported.[172, 173] Mercury–phosphorus coupling constants of 4000—7500 and 7800—12 000 Hz were obtained for the phosphine and phosphite complexes, respectively. Several trends are observed. For a particular phosphine, $J(Hg–P)$ increases in the order of electronegativity of the halogen or pseudo-halogen ligand (Cl > Br > SCN > I $\approx$ CN), consistent with the greater spin interaction resulting from an increase in mercury–phosphorus σ-bonding. Secondly, the $L_2Hg_2X_4$ dimers show larger coupling constants than the L_2HgX_2 monomers, again attributed to greater mercury–phosphorus σ-interaction. There is also a proportional relationship between the magnitude of the downfield chemical shift and the size of the coupling constant. Thirdly, as butyl groups are replaced by phenyl in the series $(Bu_{3-n}Ph_nP)_2HgX_2$ (n = 0—3), the coupling constant decreases. This corresponds to the order of basicity of the phosphines, and is the same relationship as has been found for cadmium and tin halide complexes, but opposite to that found for complexes where metal–phosphorus π-bonding may be important. Several of the complexes prepared showed no mercury–phosphorus coupling, which is explained in terms of a ligand dissociation which is rapid on the n.m.r. time-scale. Results on the analogous cadmium complexes L_2CdI_2 (L = tributyl- or trioctyl-phosphine) show $J(P–^{111}Cd) \approx J(P–^{113}Cd) \approx 1300$ Hz.[173]

The reaction of trimethylphosphine with MeHgCl results in chloride displacement to give $[MeHgPMe_3]Cl$.[174] In contrast, mercuric chloride, bromide, iodide, acetate, thiocyanide, thiocyanate and nitrate can add up to four phosphine molecules. The 1:1 halide complexes exist as halogen-bridged dimers, whereas the pseudohalides oligomerize through the cyanide and thiocyanide groups, respectively. The 1:2 complexes contain linear $[Me_3P]_2Hg^{2+}$ cations, while the 1:3 and 1:4 complexes consist of $[Me_3P]_3Hg^{2+}$ and $[Me_3P]_4Hg^{2+}$ cations, respectively. The species have been investigated by conductivity, i.r., Raman, and ^1H and ^{31}P n.m.r. techniques. In all cases, ligand exchange processes occur in solution.

The ^{13}C spectra of some β-methoxyalkyl mercuric chlorides have been reported.[175] The results indicate that $^1J(Hg–C)$ ($\sim$ 1600—1800 Hz) varies in direct proportion to the increase in electron density at the mercury–carbon bond, and that an increase in mercury–carbon σ-bonding is found with increasing alkyl substitution, an effect which is in direct contrast to that observed for platinum π-olefin complexes.

The class of chalconium compounds $[E(HgMe)_3]X$ (E = O or S) has been extended by the preparation of the new selenonium complex $[Se(HgMe)_3]NO_3$ from the reaction of H_2Se with $MeHgNO_3$.[176] The bis(methylmercuric)chalcogenides $E(HgMe)_2$ (E = Se or Te) were also obtained by the reaction of H_2E with methylmercuric bromide.

[172] S. O. Grim, P. J. Lui, and R. L. Keiter, *Inorg. Chem.*, 1974, **13**, 342.
[173] A. Yamasaki and E. Fluck, *Z. anorg. Chem.*, 1973, **396**, 297.
[174] A. Schmidbauer and K. H. Raethlein, *Chem. Ber.*, 1973, **106**, 2491.
[175] I. Ibusuki and Y. Saito, *Chem. Letters*, 1973, 1255.
[176] D. Breitinger and W. Morrell, *Inorg. Nuclear Chem. Letters*, 1974, **10**, 409.

I.r. and Raman data show C_{3v} symmetry for the $Se(HgMe)_3^+$ cation and C_{2v} symmetry for the $E(HgMe)_2$ compounds. A comparison of force constant values clearly indicates a weakening of the mercury–selenium bond on formation of the selenonium cation.

Three-co-ordinate mercury(II) has been reported in the compound $Hg_3O_2Cl_2$ (5).[177] The structure, which differs from one previously reported, contain two crystallographically distinct mercury atoms and can be described in terms of HgO_3 pyramids with the mercury atom at the apex. Each oxygen is co-ordinated to three mercury (2) atoms; thus each polyhedron shares its basal corners with three polyhedra and its basal edges with two polyhedra, building up an infinite layer. Each oxygen is also bonded to a mercury (1) atom, completing the distorted tetrahedral co-ordination of oxygen, and forming a three-dimensional network of formula $[Hg_3O_2]_n^{2+}$. Mercury–chlorine distances indicate ionic chloride; thus mercury (2) is three–co-ordinate, while mercury (1) is linearly co-ordinated.

The preparation and crystal structure determination of mercury(I) complexes of composition $Hg_2L_2X_2$, $Hg_2L_4X_2$, and $Hg_2L_6X_2$ have been described,[178—180] where L = 3-chloropyridine, pyridine N-oxide, and triphenylphosphine oxide, respectively.

The structure of the first complex consists of $[Hg_2(C_5H_4NCl)_2]^{2+}$ cations. The mercury is co-ordinated by the basic nitrogen of the pyridine to give effectively digonal co-ordination (Hg—Hg = 2.49 Å; ∠ HgHgN = 167°). Derivations from linearity and deviations of the ligands from planarity are attributed to repulsion between the chloro-substituents on neighbouring cations.[178]

The structure of the second complex (6) again shows the presence of Hg_2 dimers (Hg—Hg = 2.52 Å), but with different co-ordination numbers for each mercury. Mercury (1) is co-ordinated in a very irregular tetrahedron composed of mercury (2), bridging oxygens O–1^{II} and O–3^{II} and the terminal oxygen O–4. Mercury (2) is irregularly five-co-ordinated by mercury (1), bridging oxygens O–1, and O–3 to mercury (1^{II}) and bridging oxygens O–2 and O–2^{II} to mercury (2^{III}).[179]

The third complex consists of discrete $Hg_2(Ph_3PO)_6^{2+}$ cations (Hg—Hg = 2·52 Å), with each mercury in a distorted tetrahedral environment.[180]

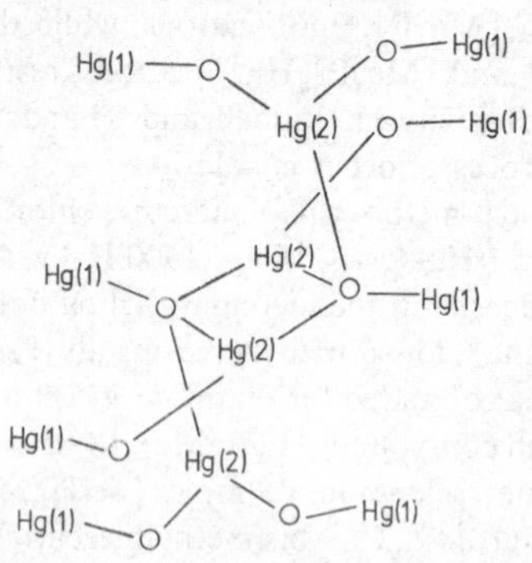

(5)

[177] K. Aurivillius and C. Stalhandske, *Acta Cryst.*, 1974, **30B**, 1907.
[178] D. L. Kepert, D. Taylor, and A. H. White, *J.C.S. Dalton*, 1973, 893.
[179] D. L. Kepert, D. Taylor, and A. H. White, *J.C.S. Dalton*, 1973, 392.
[180] D. L. Kepert, D. Taylor, and A. H. White, *J.S.C. Dalton*, 1973, 1658.

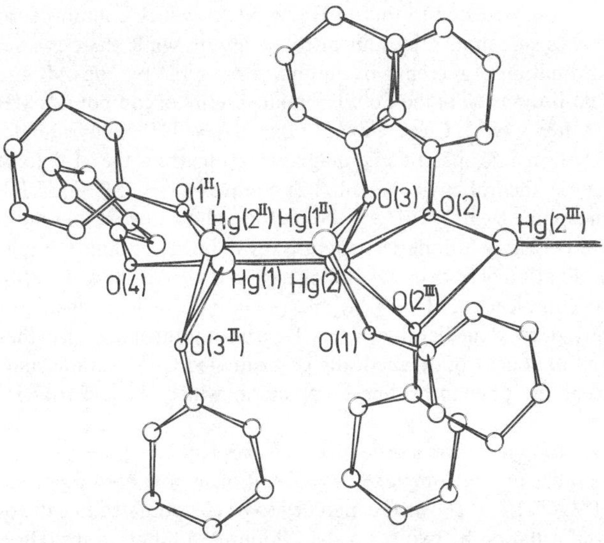

(6)

The preparation of a wide variety of complexes $Hg_2L_2X_2$ has been described, where L is a heterocyclic N-donor ligand.[181] Ligands of low basicity formed stable complexes, the critical base strength being between that of 4-benzoylpyridine ($pK_a = 3.35$) and pyridine ($pK_a = 5.21$). More basic ligands caused disproportionation, undoubtedly owing to the high affinity of mercury(II) for N-donor ligands.

The crystal structure of dinitrato(triphenylphosphine)mercury has been reported.[182] The compound consists of infinite chains rather than the simple dimers previously suggested. The co-ordination about mercury is best described as a very distorted tetrahedral arrangement consisting of triphenylphosphine, an unshared unidentate NO_3 group, and two bridging NO_3 groups. The geometry of the unshared NO_3 groups adds an additional weak mercury–oxygen interaction, in which case the geometry becomes distorted square-pyramidal.

Tetrahedral co-ordination is also found in the complex $Hg(SCN)_2(PPh_3)_2$[183] which contains discrete monomeric units. The linear SCN groups are bound *via* the sulphur atom (mean Hg—S = 2.57 Å) and define HgSC angles of *ca.* 100°.

The reaction of mercuric iodide with the chelating diphosphine bis(diphenylphosphinoethyl)sulphide gives the complex $[(Ph_2PC_2H_4)_2S]HgI_2$.[184] Structural analysis shows the mercury to be in a distorted tetrahedral co-ordination, with formation of an eight-membered puckered chelate ring.

A new series of octahedral Co^{II} complexes is formed when CoX_2 (X = Cl, Br, I, or SCN) reacts with mercuric thiocyanide in the presence of L to give the complexes $CoHg(SCN)_4.2L$ (L = THF, dioxan, pyridine, or aniline).[185] The structure of these

[181] D. L. Kepert and D. Taylor, *Austral. J. Chem.*, 1974, **27**, 1199.
[182] S. H. Whitlow, *Canad. J. Chem.*, 1974, **52**, 198.
[183] R. C. Makhija, A. L. Beauchamp, and R. Rivest, *J.C.S. Dalton*, 1973, 2447.
[184] K. Aurivillius and L. Falth, *Chem. Abs.*, 1974, **80**, 53286.
[185] R. C. Makhija, L. Pazdernik, and R. Rivest, *Canad. J. Chem.*, 1973, **51**, 438.

complexes is closely related to that of $HgCo(NCS)_4$ which contains cobalt in almost tetrahedral sites surrounded by four nitrogen atoms, while there is essentially tetrahedral co-ordination of mercury by sulphur atoms; the two ligands L complete the octahedral co-ordination of the Co^{II}. The related series of compounds $MHg(SCN)_4,2L$ ($M = Mn^{2+}, Fe^{2+}, Ni^{2+}, Cu^{2+}, Cd^{2+},$ or Zn^{2+}; L = THF or pyridine) has also been reported.[186] Electronic spectral and magnetic results show the M^{2+} to be a high-spin species in an octahedral environment. The complex $CoHg(SCN)_4,2THF$ reacts with triphenylphosphine to give $CoHg(SCN)_4,2PPh_3$, shown by i.r. and molecular weight data to have probably a dimeric structure involving thiocyanide bridges and tetrahedral co-ordination of both cobalt and mercury. It also reacts with various N-donors to give the complexes $CoHg(SCN)_4,nL$ ($n = 2$, L = triethylenetetramine; $n = 3$, L = bipyridyl or ethylenediamine; $n = 4$, L = phenanthroline). All of these complexes are proposed to consist of octahedrally co-ordinated Co^{2+} cations and $Hg(SCN)_4^{2-}$ anions, except the phenanthroline complex, in which the mercury is also six-co-ordinate.

Thiourea is known to form a series of complexes $HgCl_2L_n$ ($n = 1-4$, L = thiourea). Structural studies on the complexes $n = 2$ or 4 have now been reported. The crystal structure of $HgCl_2L_2$[187] shows the mercury to be co-ordinated in a distorted trigonal coplanar conformation by two equivalent thiourea sulphur atoms (Hg—S = 2·42 Å) and a chlorine atom (Hg—Cl = 2.57 Å). The other ionic chlorine atom lies normal to this plane, between and equidistant from a pair of mercury atoms (Hg—Cl = 3.22 Å).

The crystal structures of an α-[188, 189] and a β-form[190] of HgL_4Cl_2 have been reported. The molecular structures are essentially identical, consisting of discrete HgL_4^{2+} cations, with the mercury co-ordination being distorted tetrahedral. The ligands are planar, but not coplanar with the mercury, and are tilted towards two chlorine atoms to form four N—H $\cdots$ Cl hydrogen bonds. The mercury–sulphur distances (2.51—2.62 Å) are significantly longer than those observed in HgL_2Cl_2, presumably as a result of the increase in co-ordination number.

The preparation and structural characterization of the new complex $HgCl_2,\frac{2}{3}L$ (L = thiourea) have been described.[191] The complex is best described as $[(HgClLCl)]_n$, $HgCl_2$, and consists of infinite HgLCl chains bridged by further chlorine atoms, with the mercury adopting an irregular four-co-ordination. The chains are interlaced with discrete linear $HgCl_2$ molecules (Hg—Cl = 2.32 Å) constrained within the lattice by N—H $\cdots$ Cl hydrogen bonding. The mercury–sulphur distance (2.40 Å) is comparable with that observed for the HgL_2Cl_2 derivative.

The series of complexes HgL_2A_2 (A = ClO_4, BF_4, CF_3COO, or $HCOO$) has been prepared.[192, 193] Although they behave as 2:1 electrolytes, i.r. studies show the presence of co-ordinated perchlorate, tetrafluoroborate, and trifluoroacetate. They are thus best formulated as $[HgL_2A]A$, presumably containing mercury in a highly deformed trigonal symmetry. The series HgL_2XA (A = ClO_4, BF_4, or CF_3COO; X = Cl, Br, or I) has also been prepared.[193] Conductivity studies showed them to be 1:1

[186] R. C. Makhija, L. Pazdernik, and R. Rivest, *Canad. J. Chem.*, 1973, **51**, 2987.
[187] P. D. Brotherton, P. C. Healy, C. L. Raston, and A. H. White, *J.C.S. Dalton*, 1973, 334.
[188] A. Korczynski, M. Nardelli, and M. A. Pellinghelli, *Roczniki Chem.*, 1973, **47**, 905.
[189] A. Korczynski, M. Nardelli, and M. A. Pellinghelli, *Cryst. Structure Comm.*, 1972, **1**, 327.
[190] P. D. Brotherton and A. H. White, *J.C.S. Dalton*, 1973, 2696.
[191] P. D. Brotherton and A. H. White, *J.C.S. Dalton*, 1973, 2698.
[192] G. Peyronel, G. O. Marcotrigiano, and R. Battistuzzi, *J. Inorg. Nuclear Chem.*, 1973, **35**, 1117.

electrolytes, but i.r. studies failed to distinguish a tetrahedral dimeric halogen-bridged $[Hg_2L_4X_2]A_2$ structure and a trigonal $[L_2HgX]A$ structure.

The complex double salts $[HgL_4][HgX_4]$ (L = thiourea, allylthiourea; X = Cl, Br, or I) have been prepared by addition of thiourea to a basic HgX_2 solution.[194] The crystal structure of a related complex dichloro-bis(o-ethylthiocarbamate)mercury has been reported.[195] In contrast, the mercury is co-ordinated in a distorted tetrahedral manner by two chlorines and the sulphur atoms of two virtually planar thiocarbamate ligands (mean Hg—S = 2.44 Å). The packing is again characterized by extensive N—H $\cdots$ Cl hydrogen bonding.

A series of dimeric halogen-bridged mercury(II) complexes $[HgX_2L]_2$ (X = Cl or Br, L = R_2S, Et_2Se, or Et_2Te; X = I, L = Et_2Se) has been prepared.[196] I.r. spectra are similar to those reported for similar dimeric compounds.

The far-i.r. and Raman spectra of the complexes HgX_2py_2 (X = Cl, Br, or I) have been studied.[197] I.r. bands previously attributed to the chloride have been shown to arise from $HgCl_2py$ as an impurity. The correct spectral data show that the true description of the structure of $HgCl_2py_2$ is one involving infinite planar chains of $[HgCl_2py_2]$ containing octahedrally co-ordinated mercury, rather than the previously described alternative of $HgCl_2$ molecules with loosely held pyridine of crystallization. The complexes HgX_2py_2 (X = Br or I) were shown to consist of monomeric tetrahedrally co-ordinated units, rather than the $HgCl_2py_2$ structure previously proposed.

A series of five-co-ordinate cadmium and mercury complexes with the tripod-like tetramines tris-(2-aminoethyl)amine (tren) and tris-(2-dimethylaminoethyl)amine (Me_6tren) have been prepared.[198] A complete X-ray structural analysis showed that the compound $Cd_2(Me_6tren)I_4$ consists of the five-co-ordinate cation $[CdIMe_6tren]^+$ and dinuclear $Cd_2I_6^{2-}$. The cation is a slightly distorted trigonal bipyramid, whereas the anion is formed by edge-sharing between pairs of CdI_4 tetrahedra. This represents only the second isolation and characterization of these unusual dinuclear anions in the solid state.[199] The molecular structure of the $HgBr_6^{4-}$ anion in Tl_4HgBr_6 has recently been reported.[200] I.r. and conductivity data show that the complexes M_2trenI_4 (M = Cd or Hg), $Hg_2Me_6trenI_4$, and $Hg_2trenBr_4$ possess the same structure, as do the related complexes $[CdClL]BPh_4$ (L = tren or Me_6tren), $[HgClMe_6tren]ClO_4$, and $[HgSCNMe_6tren]BPh_4$.

The compounds $Hg_3(Me_6tren)_2X_6$ (X = Cl or Br) have been formulated on the basis of i.r. and conductivity data as $[HgXMe_6tren]_2HgX_4$. The complexes $Cd(Me_6tren)X_2$ (X = Cl or Br) show the presence of unco-ordinated NMe_2 groups in the i.r. and are assigned a five-co-ordinate structure where the tetramine acts as a terdentate ligand. In contrast, $CdBr_2$tren is best formulated as $[CdBrtren]Br$, with the tetramine acting as a quadridentate ligand. The complex $Hg(SCN)_2$tren also has this structure. The complexes $Cd(ClO_4)L,nH_2O$ (L = tren or Me_6tren) have also been prepared. Where $n = 1$, the complexes are best formulated as $[Cd(H_2O)L](ClO_4)_2$, containing bound

[193] G. O. Marcotrigiano, G. Peyronel, and R. Battistuzzi, *Inorg. Chim. Acta*, 1974, **9**, 5.

[194] M. Genchev and V. St. Krustev, *Chem. Abs.*, 1974, **80**, 66258.

[195] G. Bandoli, D. A. Clemente, L. Sindellari, and E. Tondello, *Cryst. Structure Comm.*, 1974, **3**, 289.

[196] J. E. Fergusson and C. S. Loh, *Austral. J. Chem.*, 1973, **26**, 2615.

[197] R. M. Barr, M. Goldstein, and W. D. Unsworth, *J. Cryst. Mol. Structure*, 1974, **4**, 165.

[198] M, Ciampolini, A. Cristini, A. Diaz, and G. Ponticelli, *Inorg. Chim. Acta*, 1973, **7**, 549.

[199] P. T. Beurskens, W. P. J. H. Bosman, and J. A. Cras, *J. Cryst. Mol. Structure*, 1972, **2**, 183.

[200] K. Brodersen, G. Thiele, and G. Gorz, *Z. anorg. Chem.*, 1973, **401**, 217.

water in the five-co-ordinate cation. Where $n = 0$, the i.r. shows the presence of both co-ordinated and ionic perchlorate (which is easily displaced in solution) and the complex is then formulated as $[CdLClO_4]ClO_4$. The related nitrate complex $CdL(NO_3)_2$ also contains both ionic and co-ordinated nitrate.

Ammonium trichloromercurate(II) has been shown to exist in two crystalline forms, known as α and β. The crystal structure of the α-form, obtained from a 1:1 NH_4Cl–$HgCl_2$ melt, was postulated to consist essentially of $HgCl_2$ molecules surrounded by four chloride ions, and it is now reported[201] that this structure is retained in the monohydrate NH_4HgCl_3,H_2O, obtained by crystallization from aqueous solution. However, Raman and far-i.r. studies now reported indicate that considerable covalent character exists in the long mercury–chlorine bonds, and the structure is best considered as infinite sheets of edge-shared $[HgCl_6]$ octahedra, with disordered ammonium ions dispersed between the sheets. $RbHgCl_3$ has been shown to be isostructural with this α-form.[202]

The β-form, obtained from 1:1 NH_4Cl–$HgCl_2$ in aqueous solution is isostructural with the $MCdX_3$ (M = NH_4, K, or Rb; X = Cl, Br, or I) compounds, possessing a double rutile chain structure.[203] Far-i.r. and Raman data for the β-form, and for the $MCdX_3$ compounds are reported. Raman and far-i.r. data for the complex K_2HgCl_4, H_2O, containing *trans*-edge-shared $[HgCl_6]$ octahedra, have been reported, as well as data for the compounds $KHgCl_3$, NH_4HgBr_3, and $NaHgCl_3,2H_2O$ (originally reported as anhydrous $NaHgCl_3$). The structures of these last compounds are not obvious from the data.

The crystal structures of several carboxylato-mercury complexes have been reported. That of mercury(I) trifluoroacetate[204] shows discrete $(CF_3COO)_2Hg_2$ molecules with C_2 symmetry. The Hg—Hg bond length (2.51 Å) agrees well with other values for mercury(I) oxyacid salts. The carboxy-group is bonded by one oxygen (Hg—O = 2.14 Å) with $\angle$ HgHgO = 166°. The molecules are connected by intermolecular Hg $\cdots$ O bonds (2.64 Å) to form puckered ribbons along the c direction. The crystal structure of mercury(I) orthoarsenate, $(Hg_2)_3(AsO_4)_2$, has been reported.[205] Effectively tetrahedral co-ordination about the mercury atom is established by one mercury atom from the same Hg_2^{2+}, one oxygen at 2.16—2.23 Å, and two further oxygens at 2.42—2.71 Å. The Hg—Hg—O bonds deviate from linearity by 23—34°.

The crystal structure of mercury(II) acetate has been reported,[206] and consists of discrete $(MeCO_2)_2Hg$ molecules (Hg—O = 2.07 Å; $\angle$ OHgO = 176°). Chains are formed in the c direction by two weak Hg $\cdots$ O interactions (2.73 Å). By the packing of these chains, the mercury gets a fifth oxygen neighbour (2.75 Å), yielding a tetragonal pyramid as the mercury co-ordination polyhedron. The structure of the related compound dipivaloylmethanemercuric acetate has also been reported.[207] The dipivaloylmethane is bound to the central carbon in the diketone form (Hg—C = 2.11 Å), while the acetate is bound unsymmetrically at one of the oxygens (Hg—O = 2.10 Å). Crystal packing again gives the mercury five-co-ordination.

[201] K. Sagisawa, K. Kitahama, H. Kiriyama, and R. Kiriyama, *Acta Cryst.*, 1974, 30B, 1603.
[202] R. M. Barr and M. Goldstein. *J.C.S. Dalton*, 1974, 1180.
[203] R. M. Barr and M. Goldstein. *Inorg. Nuclear Chem. Letters*, 1974, 10, 33.
[204] M. Sikirica and D. Grdenic, *Acta Cryst.*, 1974, 30B, 144.
[205] B. Kamenar and B. Kaitner, *Acta Cryst.*, 1973, 29B, 1666.
[206] R. Allmann, *Chem. Abs.*, 1974, 80, 75256.
[207] R. Allmann and H. Musso, *Chem. Ber.*, 1973, 106, 3001.

Two crystalline modifications of mercury(II) NN-diethyldithiocarbamate have been reported.[208] The α-form consists of dimeric $Hg_2(S_2CNEt_2)_4$ units, in which the metal atom is five-co-ordinate, in a manner similar to analogous zinc and cadmium complexes. The β-form is composed of essentially monomeric $Hg(S_2CNEt_2)_2$ units, with the two ligand molecules co-ordinated in a plane through the S atoms (Hg—S = 2.40 Å).

Other mercury systems are listed in Table 3.

Table 3

	System	Data^a	Ref.
ML_2	HL = dithiobenzoic acid; M = Cd or Hg	p, i.r. R	209
Hg^IL	HL = diphenylselenothiophosphinic acid	p, i.r.	210
HgL_n^{2+}	L = salicylamide; n = 1 or 2	s	211
HgX_2L	L = 2,2'-bipyridylNN'-dioxide; X = Cl, Br, or SCN	p, i.r.	212
$HgCl_2L$	L = various substituted sulphonamides	p, i.r.	213
$HgL_n^{(2-n)+}$	L = various amino-acids; n = 1 or 2	s	214
$HgXL$	X = Cl or Ph; L = 5-methyldithiazone	p	215
HgX_2L_2	X = OAc or Cl; L = 5-methyl-2-aminobenzothiazole	p	216
$[HgL_2]^{2+}$	L = monothiol	s	217
$[HgL_n]^{(n-2)-}$	HL = ethanethiosulphonic acid; n = 2—4	s	218
$[HgX_nI_{4-n}]^{2-}$	X = Cl or Br; n = 0—4	s	219
$[HgL_3]^-$	L = ethylxanthate	s	220
HgL_2	HL = methionine	^{1}H n.m.r.	221
$[HgL_n]^{(2n-2)-}$	H_2L = 4,6-dihydroxypyrimidine	p	222
	H_2L = thiosalicylic zcid	p	86, 223, 224
$M_n[HgL_2]_m$ ⎤	n = 2, m = 1, M = Na, NH$_4$, or K		
$Hg(HL)Cl$ ⎟	n = 1, m = 1, M = Ca or Sr	p, i.r.	123
$Hg(HL)_2$ ⎬	n = 2, m = 3, M = La, Pr, or Nd		
$[HgL]^-$ ⎟	H_3L = nitrilotriacetic acid		
$HgHL$ ⎦			
$Hg(OH)I$		s	225

(a) p = preparation; R = Raman spectrum; s = stability constant;

[208] H. Iwasaki, *Acta Cryst.*, 1973, 29B, 2115.
[209] R. Mattes, W. Stork, and I. Pernoll, *Z. anorg. Chem.*, 1974, 404, 97.
[210] A. Müller, V. V. K. Rao, and P. Christophliemk, *J. Inorg. Nuclear Chem.*, 1974, 36, 472.
[211] B. S. Pannu, B. S. Sekhon, and S. L. Chopra, *J. Indian Chem. Soc.*, 1973, 50, 629.
[212] I. S. Ahuja and R. Singh, *Indian J. Chem.*, 1973, 11, 1070.
[213] J. S. Shukla and P. Bhatia, *J. Inorg. Nuclear Chem.*, 1974, 36, 1422.
[214] W. E. Van der Linden and C. Beers, *Analyt. Chim. Acta*, 1974, 68, 143.
[215] H. M. N.-H. Irving, A. H. Nabilsi and S. S. Sahotra, *Analyt. Chim. Acta*, 1973, 67, 135.
[216] M. Chaurasia and P. K. Sharma, *Current Sci.*, 1974, 43, 211.
[217] D. P. Ryabushko, A. T. Pilipenko, and L. A. Krivokhizhiva, *Chem. Abs.*, 1974, 80, 74839.
[218] A. A. Gundorina and A. N. Sergeeva, *Zhur. neorg. Khim.*, 1974, 29, 334.
[219] V. I. Relevantiev, B. I. Peshchevitskii, and H. Bhadmaeva, *Zhur. neorg. Khim.*, 1973, 18, 2054.
[220] I. Kovac and B. Tokes, *Chem. Abst.*, 1973, 79, 12143.
[221] B. Birgesson, T. Drakenberg, and G. A. Veville, *Acta Chem. Scand.*, 1973, 27, 3953.
[222] F. Capitan, F. Salinas, and G. J. Rodriguez, *Chem. Abs.*, 1973, 79, 152412.
[223] M. K. Khoul and K. P. Dubey, *J. Inorg. Nuclear Chem.*, 1973, 35, 2567.
[224] M. K. Khoul and K. P. Dubey. *J. Inorg. Nuclear Chem.*, 1973, 35, 2571.
[225] I. Ahlberg, *Acta Chem. Scand.*, 1973, 27, 3003.

5 The Bio-inorganic Chemistry of Zinc, Cadmium, and Mercury

This group of metals contains one member essential to life and two others which are highly toxic. It thus attracts interest both from workers using biological materials and from others using simulated or model systems to investigate the function of zinc in those enzymes in which it occurs, including an examination of (a) the mode of inter-action of all three metals with proteins, dinucleotides, and amino-acids in the hope of elucidating the mechanisms of transport of such metals either in blood, in plants, or across cell membranes, and (b) the possible mechanisms of methylmercury toxicity in mammals.

This section is therefore divided into three parts: enzymic reactions involving zinc; metal–protein complex formation; and studies involving reaction mechanisms.

Enzymic Reactions involving Zinc.—The isolation of several new zinc metalloenzymes has been announced during the past year. In a remarkable piece of work involving a novel method of analysis, Vallee and co-workers[226] reported the characterization as a zinc metalloenzyme of the RNA-dependent DNA polymerase (*i.e.* reverse transcriptase) which exists in the oncogenic avian myeloblastosis virus associated with leukaemia in chicks. This demonstrates a postulate by Vallee of 25 years ago that disturbance in a zinc-dependent enzyme system is critical in the pathophysiology of myelogenous and lymphatic leukaemia. The analytical method incorporates microwave emission spectroscopy and gel exclusion chromatography, and is capable of determining metal content down to 10^{-14} gram atoms.

An alkaline phosphatase has been isolated from *Bacillus subtilis*, a gram-positive bacterium.[227] The enzyme consists of dimers having an approximate molecular weight of 100 000, with 2.48 ± 0.5 gram atoms of Zn^{2+} per mole. The specific activity is more than twice that of the more familiar phosphatase isolable from the gram-negative *E. coli*, and is reduced to zero on removal of the zinc ion. The activity is regenerated on replacement of the metal ion by either cobalt or zinc, but is different from the alkaline phosphatase from *E. coli* in that the cobalt-substituted enzyme has twice the activity of the zinc-regenerated enzyme. In the *E. coli* enzyme, the cobalt alkaline phosphatase has only 20% of the activity of the zinc enzyme. The u.v. spectrum of the cobalt-substituted enzyme is virtually identical with that of the cobalt-substituted *E. coli* alkaline phosphatase.

The regeneration of the activity of the enzyme by cobalt differs from the behaviour of an alkaline phosphatase isolated from human placenta.[228] This enzyme is reported to be re-activated by replacement of the native zinc ion by either zinc, magnesium, or mercury. No other metal is active.

Some insight into the function of the zinc ion in alkaline phosphatase from *E. coli* has resulted from ^{35}Cl n.m.r. studies.[229] The uncertainty which has surrounded the number of zinc ions required for activity of the enzyme has been resolved somewhat by the observation that alkaline phosphatase prepared in the absence of edta requires only 2 moles of Zn^{2+} per mole of enzyme for full activity. ^{35}Cl line-broadening by n.m.r. shows that on addition of two moles of zinc per mole enzyme, no broadening

[226] D. S. Auld, H. Kawaguchi, D. M. Livingston, and B. L. Vallee, *Proc. Nat. Acad. Sci. U.S.A.*, 1974, **71**, 2091.

[227] F. K. Yoshizumi and J. E. Coleman, *Arch. Biochem. Biophys.*, 1974, **160**, 255.

[228] F. R. Hindriks, A. Groen, and A. M. Kroon, *Biochem. Biophys. Acta*, 1973, **315**, 94.

[229] J.-E. Norne, H. Csopak, and B. Lindman, *Arch. Biochem. Biophys.*, 1974, **162**, 552.

occurs. At metal:enzyme ratios > 2, there is metal–chloride interaction. In the presence of inorganic phosphate, the apoenzyme shows no effect on the ^{35}Cl linewidth, while the native enzyme shows a decrease in linewidth, demonstrating that phosphate is displacing Cl^- from the metal. Thus direct phosphate–metal interaction is indicated.

The first neutral zinc-containing endopeptidase has been reported.[230] Isolated from the brush-border of rabbit kidney, it has a molecular weight of *ca.* 93 000 and 1 atom of Zn^{2+} per mole. It is inhibited by edta and reactivated by zinc and other bivalent metals, less zinc being required for full reactivation.

Garner and Behal have reported the isolation of a zinc-activated aminopeptidase from human liver.[231] The enzyme, which is capable of forming a cobalt–enzyme complex, has zinc located at or near the substrate binding site.

The reaction mechanism of carbonic anhydrase (CA) is still a matter of controversy. Pocker and Guilbert[232] have reported that the hydrolysis of diesters by CA, specifically the hydrolysis of methyl 4-nitrophenylcarbonate, is similar to that which occurs in CO_2 hydrolysis: the rate varies as though dependent on the ionization of a group in the enzyme with a pK near 7. Only the basic form is active.

Which group is ionizing is matter of uncertainty. Some workers, reassessing the available data,[233] have concluded that the traditional picture of a zinc-bound water molecule is still the most 'economical' mechanism, while others,[234] using data obtained from pH-titration studies of Zn^{II} imidazole systems, have concluded that the ionization is not that of a zinc-bound water, but rather of a co-ordinated histidine, as shown.

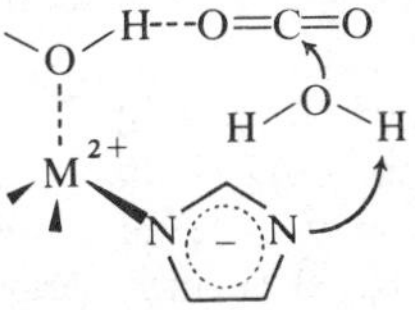

Studies on the various zinc-activated dehydrogenases continue apace. The reduction of *trans*-4-*NN*-dimethylaminocinnamaldehyde (A) by liver alcohol dehydrogenase (LADH) is reported to involve the zinc at the active site of the enzyme acting as a Lewis acid and co-ordinating the substrate *via* the aldehyde oxygen.[235] The kinetics of the reaction show that (A) + LADH + NADH form a stable intermediate at pH 9, the overall reaction sequence being:

$$E(NADH)_2 + 2(A) \overset{k_1}{\rightleftharpoons} E'(A)_2 \rightarrow products$$

The reaction is pH independent, k_1 is fast, and there is no deuterium isotope effect.

Further evidence for the binding of substrate aldehyde to the zinc ion comes from comparative kinetics of the reduction of *p*-substituted benzaldehydes by borohydride and LADH.[236] For borohydride, the ratio of rates for *p*-Cl to *p*-MeO is 100, while for

[230] M. A. Kerr and A. J. Kenny, *Biochem. J.*, 1974, **137**, 489.
[231] C. W. Garner and F. J. Behal, *Biochemistry*, 1974, **13**, 3221.
[232] Y. Pocker and L. J. Guilbert, *Biochemistry*, 1974, **13**, 70.
[233] R. H. Prince and P. R. Woolley, *Bio-org. Chem.*, 1973, **2**, 337.
[234] D. W. Appelton and B. Sarkar, *Proc. Nat. Acad. Sci. U.S.A.*, 1974, **71**, 1680.
[235] M. F. Dunn and J. S. Hutchinson, *Biochemistry*, 1973, **12**, 4882.

LADH the ratio is 2. This is interpreted as zinc(II) polarization of the benzaldehyde carbonyl group resulting in Lewis acid catalysis.

The mechanism of the overall reaction between LADH, NADH, and an aldehyde has been reported to proceed *via* formation of a LADH–NADH complex followed by reaction with substrate, according to studies on the 'unproductive' LADH–substrate complex formed between enzyme and 4-(2-imidazolylazo)benzaldehyde.[237] Substrate- and coenzyme-competitive inhibitors displace the azo-aldehyde, showing that the site of binding overlaps those of normal substrate and coenzyme. There is no reaction with the azo-aldehyde unless prior binding of coenzyme occurs.

Increased activity of LADH has been reported to result from changing the nature of the amino-groups at the active site.[238] An increase in size of the positively charged substituents on the amino-groups increases the Michaelis and inhibition constants for substrates and also increases the turnover numbers. Changing the *net* charge of the amino-groups results in a decrease in the rates of binding of NAD^+ and NADH. Increased activity has also been reported to follow methylation of 20 of the 60 lysine residues of LADH,[239] together with conformational changes and a decrease in substrate inhibition by ethanol.

A similarity between the location of zinc(II) in yeast alcohol dehydrogenase (YADH) and that of the metal in LADH has been established.[240] NADH and substrate proton relaxation rates using the paramagnetic iodoacetamide analogue complex of YADH show the coenzyme nicotinamide moiety to be situated *ca.* 7 Å from the zinc ion. This is virtually identical with the position of the metal in LADH determined by X-ray crystallography.[241] The substrate resides between the NADH and the metal ion, *i.e.* it is co-ordinated to the zinc.

The metal–NADH distance determined by fluorescence studies on the cobalt-substituted enzyme, using binary NADH, thio-NADH, and Rose Bengal/enzyme complexes, differs from the above.[242] The minimum metal–NADH distance determined by Forster calculations based on the energy transfer from bound ligand to cobalt(II) is reported as 19 Å. It is thought possible that the discrepancy between this value and that determined by X-ray diffraction may be due to conformational changes in solution.

The binding of zinc to metal-free beef-heart Rhodanese (which catalyses the reaction between cyanide and thiosulphate to produce thiocyanate) has been studied using microcalorimetry.[243] The zinc is reported to bind with a large endothermic enthalpy change similar to that observed with zinc model compounds, and with a large positive entropy change similar to that which occurs on binding zinc ion to apocarbonic anhydrase. Whether the zinc ion which exists in native Rhodanese functions as a catalytically active binding site for thiosulphate remains unresolved.

[236] J. W. Jacobs, J. T. McFarland, I. Wainer, D. Jeanmaier, C. Ham, K. Hamm, M. Wnuk, and M. Lam, *Biochemistry*, 1974, **13**, 60.
[237] J. T. McFarland, Y. H. Chu, and J. W. Jacobs, *Biochemisrry*, 1974, **13**, 65.
[238] M. Zoltobrocki, J. C. Kim, and B. V. Plapp, *Biochemistry*, 1974, **13**, 899.
[239] C. S. Tsui, Y-H. Tsui, G. Lauzon, and S. T. Cheng, *Biochemistry*, 440, 1974, **13**, 440.
[240] D. L. Sloan and A. S. Mildvan. *Biochemistry*, 1974, **13**, 1711.
[241] C. I. Branden, H. Ekland, B. Nordstrom, T. Boiwe, G. Soderlund, E. Zeppezauer, I. Ohlsonn, and A. Åkeson, *Proc. Nat. Acad. Sci. U.S.A.*, 1973, **70**, 2439.
[242] M. Takahushi and R. A. Harvey, *Proc. Nat. Acad. Sci. U.S.A.*, 1973, **70**, 4743.
[243] D. W. Bolen and S. Rajender, *Arch. Biochem. Biophys.*, 1974, **161**, 435.

Metal–Protein Complex Formation.—The suggestion that the increased uptake of metal and phosphate nutrients by plants which results from the addition of nitrilotriacetate (NTA) to soil occurs because of formation of mixed-ligand metal complexes has been investigated by Ramamoorthy and Manning.[244] A pH-titration study of the equilibria and distribution of species in the two-ligand systems involving fulvate, phosphate, and NTA showed that in weak acid or neutral systems, the predominant species were $M^{II}NTA$, $FulvM^{II}NTA$, $PhosM^{II}fulv$, $PhosM^{II}NTA$ (M = Zn or Cd). Lead complexes were insoluble.

The same workers have reported[245] that in three-ligand systems of zinc(II) or lead(II) and orthophosphate, cysteine, and a carboxylate, mixed-ligand complexes predominate at physiological pH. Such complexes solubilize lead(II).

A study of the mixed complexes formed by metal interactions with asparagine and glutamine has shown that quite a large proportion of asparagine in blood may be co-ordinated to zinc(II) and iron(II) as well as copper(II).[246]

The value of ligand therapy in reducing the concentration of cadmium in the blood following cadmium poisoning may be increased by a study of the formation constants of the complexes formed between cadmium and the biological and therapeutical ligands used.[248] Glass and solid-state cadmium-electrode potentiometric investigation of the cadmium complexes of three of the ligands most commonly used in ligand therapy (2,3-dimercaptopropan-1-ol, NTA, and H_4edta) shows that the ligands are non-specific for cadmium. Criteria for specific ligand–cadmium complexation are discussed.

The stability constants of zinc(II) complexes of uracil, thymine, and cytosine have been reported.[249] At 45 °C in 0.1M-KNO_3, 1:1 complexes are formed. The 2:1 ligand: metal complexes formed between thiosalicylic acid and zinc(II), mercury(II), cadmium-(II) and lead(II) have been isolated, and formation of the 1:1 complexes in solution has been characterized by pH-titration.[250] With mercury, the 2:1 complex has been assigned the structure (7), while the other metals form complexes of general structure (8). This is thought to be a consequence of the order S—M^{II} bond strength being Hg > Zn > Cd > Pb.

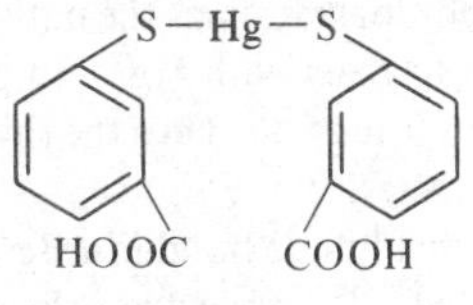

(7) 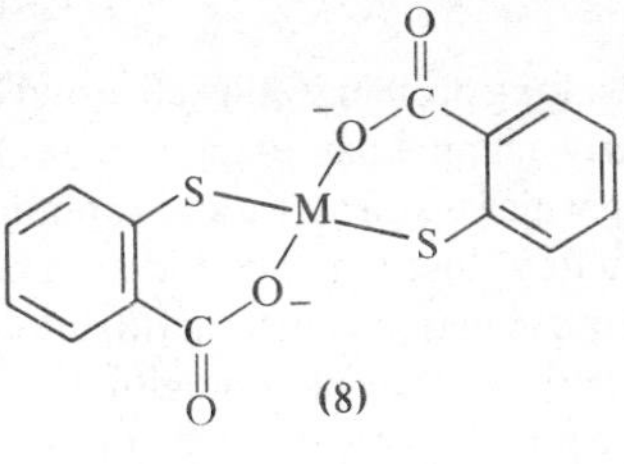   (8)

^{13}C N.m.r. has been used to determine the binding sites of zinc(II), cadmium(II), mercury(II) and lead(II) to glutathione.[251] All the metals bind to the potential co-ordi-

[244] S. Ramamoorthy and P. E. Manning, *J. Inorg. Nuclear Chem.*, 1974, **36**, 695.
[245] S. Ramamoorthy and P. E. Manning, *J. Inorg. Nuclear Chem.*, 1974, **36**, 1671.
[246] A. C. Baxter and D. R. Williams, *J.C.S. Dalton*, 1974, 1117.
[247] K. Nag and P. Banerjee, *J. Inorg. Nuclear Chem.*, 1974, **36**, 2145.
[248] M. D. Walker and D. R. Williams, *J.C.S. Dalton*, 1974, 1186.
[249] M. M. Taqui Kahn and C. R. Krishnamoorthy, *J. Inorg. Nuclear Chem.*, 1974, **36**, 711.
[250] N. S. Al-Niaimi and B. M. Al-Saadi, *J. Inorg. Nuclear Chem.*, 1974, **36**, 1671.
[251] B. J. Fuhr and D. L. Rabenstein, *J. Amer. Chem. Soc.*, 1973, **95**, 6944.

nation sites with a high degree of specificity, with the actual site being dependent on the metal ion and the solution pD. Mercury binds only to the thiol group at metal:glutathione ratios up to 0.5:1. Cadmium(II) and zinc(II) bind to both sulphate and aminogroups. In a similar piece of work,[252] the binding of mercury(II) and $MeHg^+$ to glutathione was characterized by ^{13}C n.m.r. with virtually identical results: the mercury binds the thiol group.

The novel technique of difference Raman spectroscopy has been used to determine the position of binding of $MeHg^+$ to uridine and cytidine.[253] The binding at pH 7 is such that the uridine interaction is stronger than the cytidine; the most likely coordination site is N_3.

The structure of the methylmercury–pencillamine complex has been determined by X-ray crystallography.[254] The mercury is bound to the sulphur as a linear Me—Hg—S system.

The 2:1 DL-penicillamine:zinc(II) complex has been isolated and its structure designated as (9) using i.r. spectra.[255]

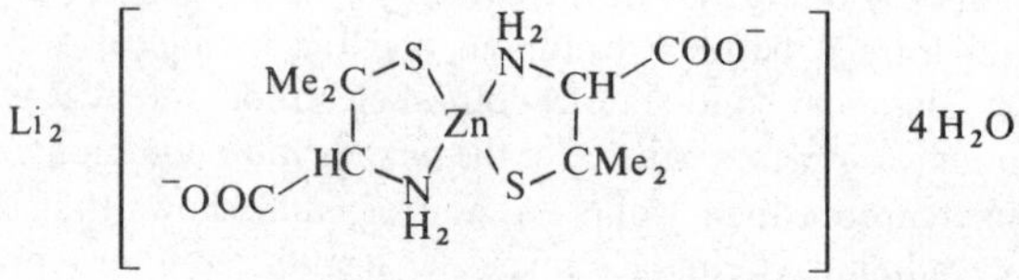

$$Li_2 \left[\begin{array}{c} Me_2C \overset{S}{\diagdown} \quad \overset{H_2}{\underset{}{N}} \diagdown \quad CH \diagup COO^- \\ | \qquad Zn \qquad | \\ {}^-OOC \diagup HC \diagdown N \diagup \quad \diagdown S \diagdown CMe_2 \\ \qquad \quad H_2 \end{array} \right] 4H_2O$$

(9)

The early suggestion that mercury–sulphur interactions might be used to produce heavy-atom derivatives of biological systems suitable for X-ray crystallographic studies has been investigated by several workers. Spenling and Steinberg[256] have reported that reaction of Hg(OAc) with insulin results in the selective reduction and mercuration of the A6–A11 bridge, according to the reaction sequence

$$Hg_2^+ + RSSR \rightarrow 2(RSHg^+) \rightarrow RSHgSR + Hg^{2+}$$

The resulting protein is still active and has the same conformation as the native insulin. Similarly, mammalian α-amylase has been reported to react with Hg^{2+} to produce a protein which is still active and contains an S—Hg—S bridge.[257] Thus the thiol groups are situated close together and are not required for activity.

The introduction of mercury(II) into an interchain disulphide bond in a Bence-Jones dimer derived from immunoglobin has been reported.[258] An electron density map of the λ-protein using the mercury(II) derivative has been obtained.[259] It is possible that this may show where the antigen activity of immunoglobin resides.

[252] G. A. Neville and T. Drakenberg, *Acta Chem. Scand.(B)*, 1974, **28**, 473.
[253] S. Mansey, T. E. Wood, J. C. Sprowles, and R. S. Tobias, *J. Amer. Chem. Soc.*, 1974, **96**, 1762.
[254] Y. S. Wong, P. C. Chieh, and A. J. Carty, *J.C.S. Chem. Comm.*, 1973, 741.
[255] S. T. Chow, C. A. McAuliffe, and B. J. Sayle, *J. Inorg. Nuclear Chem.*, 1973, **35**, 4349.
[256] R. Spenling and I. Z. Steinberg, *Biochemistry*, 1974, **13**, 2007.
[257] M. L. Steer, N. Tal, and A. Levitzki, *Biochem. Biophys. Acta*, 1974, **334**, 389.
[258] K. R. Ely, R. L. Girling, M. Schiffer, D. E. Cunningham, and A. B. Edmundson, *Biochemistry*, 1973, **12**, 4233.
[259] M. Schiffer, R. L. Girling, K. R. Ely, and A. B. Edmundson, *Biochemistry*, 1973, **12**, 4179.

An interesting result, which may have some bearing on the effect of replacing native metals in enzymes with other metals, has been reported by Nakon *et al.*[260] They found that the co-ordination sites of 3-[(ethoxycarbonyl ethyl)thio]-L-alanine (SCMC), and 3-[(2-aminoethyl)thio]-L-alanine in binding to bivalent zinc, copper, cobalt, and nickel are dependent on both pH and metal. The copper complex of SCMC has the structure (10), while the zinc complex has structure (11). Similar effects are seen on changing pH.

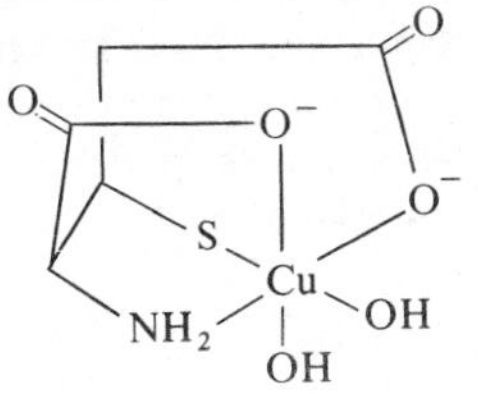

(10) 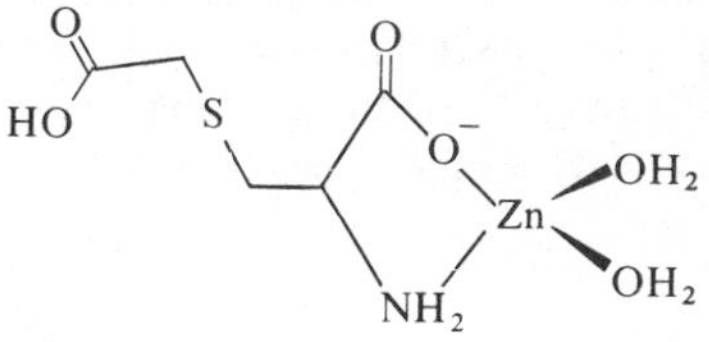   (11)

Nakon and Angelici have also determined the mode of binding of zinc and copper to the optically active dipeptides AlaAla, AlaPhe, LeuLeu, and LeuTyr.[261] The metals bind the dipeptides in the β-conformation in solution, and the value of K is greater for the 'mixed' (DL) dipeptides than for the 'pure' (LL).

A difference in the mode of metal binding by dopa from that of dopamine, adrenaline, noradrenaline, and tryamine has been reported;[262] dopa co-ordinates *via* amine and carboxylate groups, with the others *via* phenolic oxygen.

The effect of zinc on the oxygen affinity of red blood cells is thought to be a result of zinc–haeme interaction.[263] The amount of zinc required to increase the oxygen affinity has been determined as that which yields a metal:haeme ratio of 0.4:1. The workers conclude that of all the possible binding sites on haeme, the amino-acid residue with the highest binding capacity for zinc is also that residue involved in the oxygen affinity effect of zinc.

An investigation of the role played by zinc in sickle cell anaemia has shown[264] that zinc decreases the amount of haemoglobin associated with red-cell membranes and inhibits the effect of calcium in causing haemoglobin retention by membranes.

The crystal structures of the cadmium complexes bis(L-methionato) cadmium(II), bis(L-asparaginato) cadmium(II), triaqua-bis-(L-glutamato)dicadmium(II) hydrate, dichlorobis(glycylglycine)cadmium(II), aquachloro(glycylglycinato)cadmium(II) and aqua(L-glutamato)cadmium(II) hydrate have been reported.[265] Several distinct binding interactions are in evidence.

It has been established by polarography that cadmium(II) and mercury(II) form 1:1 and 2:1 complexes with vitamin B_6.[266] Activation parameters are reported.

[260] R. Nakon, E. M. Beadle, and R. J. Angelici, *J. Amer. Chem. Soc.*, 1974, **96**, 719.
[261] R. Nakon and R. J. Angelici, *J. Amer. Chem. Soc.* 1974, **96**, 4179.
[262] B. Grgas-Kuznar, V. Simeon, and O. A. Weber, *J. Inorg. Nuclear Chem.*, 1974, **36**, 2151.
[263] F. J. Oelshlegel, G. J. Brewer, C. Knutsen, A. S. Prasad, and E. B. Schoomaker, *Arch. Biochem. Biophys.*, 1974, **163**, 732.
[264] S. Dash, G. J. Brewer, and F. J. Oelshlegel, *Nature*, 1974, **250**, 251.
[265] R. J. Flook, H. C. Freeman, C. J. Moore, and M. L. Scudder, *J.C.S. Chem. Comm.*, 1973, 753.
[266] D. N. Chaturvedi and C. M. Gupta, *J. Inorg. Nuclear Chem.*, 1974, **36**, 2155.

Cadmium(II) will bind into the Ca^{2+} binding sites on trypsin.[267] It is thought that binding involves carboxylate and possibly tryptophan.

Bivalent zinc, cadmium, nickel, and copper have been found to form ternary mixed-ligand complexes with histidine or edta and polyphenols.[268] The formation constants for the ternary complexes are less than those for the binary systems.

Zinc and other metal ions have been found to promote pyrrole hydrogen ionization in 2-(2'-pyridyl)imidazole.[269] Complexation studies[114] on the systems N-methyl-histamine and NN-dimethylhistamine with bivalent zinc, copper, cobalt, and nickel have shown that the stabilities of the complexes follow the Irving–Williams series. With respect to the variation of a ligand with the same metal ion, the stability decreases in the series: histamine, N-methylhistamine, NN-dimethylhistamine, possibly as a result of steric hindrance. The complexes are assigned the structure (12).

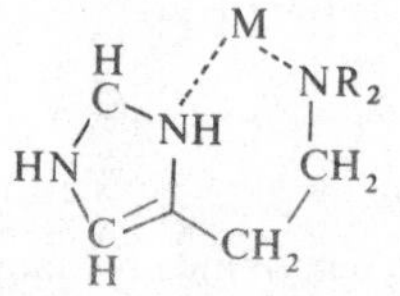

(12)

Studies involving Reaction Mechanisms.—It has been reported[270] that the zinc-catalysed splitting of dinucleotides occurs by a process involving preliminary forma-tion of a reactive metal-bridged 1:1 complex, followed by cleavage. The relative rates of reaction for a series of dinucleotides is the reverse of the order of the affinity of the base moiety towards the metal ion.

Zinc catalysis of transammination reactions involving Schiff-base formation between pyridoxamine phosphate and pyruvate has been reported[271] to be a result of the metal complexation preserving the integrity of the pyridinium group to a higher pH. The metal also promotes formation of the Schiff base and thereby increases the concen-tration of the reactive species, in addition to promoting base-catalysed pathways.

It has been established that zinc-catalysed decarboxylation of β-keto-acids involves a preliminary metal-promoted keto-enol tautomerism, as shown.[272]

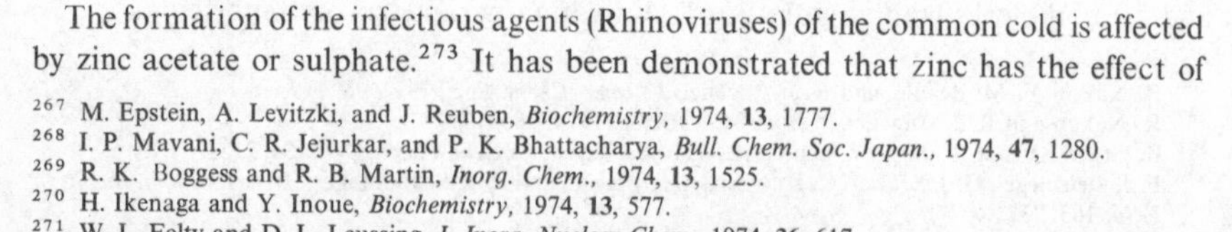

The formation of the infectious agents (Rhinoviruses) of the common cold is affected by zinc acetate or sulphate.[273] It has been demonstrated that zinc has the effect of

[267] M. Epstein, A. Levitzki, and J. Reuben, *Biochemistry*, 1974, **13**, 1777.
[268] I. P. Mavani, C. R. Jejurkar, and P. K. Bhattacharya, *Bull. Chem. Soc. Japan.*, 1974, **47**, 1280.
[269] R. K. Boggess and R. B. Martin, *Inorg. Chem.*, 1974, **13**, 1525.
[270] H. Ikenaga and Y. Inoue, *Biochemistry*, 1974, **13**, 577.
[271] W. L. Felty and D. L. Leussing, *J. Inorg. Nuclear Chem.*, 1974, **36**, 617.
[272] W. D. Covey and D. L. Leussing, *J. Amer. Chem. Soc.*, 1974, **96**, 3860.
[273] B. D. Korant, J. C. Kauer, and B. E. Butterworth, *Nature*, 1974, **248**, 588.

inhibiting cleavage in the RNA-directed synthesis of viral polypeptides, and thus acts as an anti-viral agent at less than millimolar concentrations.

It has been reported that zinc ion inhibits the uptake of Fe^{2+} by apo-ferritin as a result of zinc(II) binding (i) on the surface of the protein at those sites thought usually to be occupied by Fe^{2+}, and where catalytic hydrolysis of Fe^{2+} to $Fe^{III}OOH$ occurs, and (ii) within the protein, where it prevents the microcrystalline growth of $Fe^{III}OOH$.[274]

A method involving zinc in the preparation of pure crystalline samples of glutamine synthetase from *E. coli* has been reported.[265] Between nine and ten equivalents of zinc are bound per mole of enzyme.

Several reports concerning zinc metalloporphyrins have appeared. Spaulding *et al.*[276] have determined the crystal structure of the radical perchloratotetraphenyl-porphinatozinc(II). The zinc is five-co-ordinate, with a very strongly bound perchlorate forming the apex of a pyramid, the square base of which is formed by the four nitrogens of the porphyrin ring.

The fact that the zinc ion does not sit in the plane of the ring in tetracarboxyphenyl-porphinatozinc(II) is thought to be the reason why this particular metal porphyrin does not dimerize.[277]

The incorporation of metals into metal-free porphyrins has been reported to proceed by a dissociative mechanism.[278] The porphyrin and metal form an outer-sphere complex, the porphyrin nucleus deforms to provide a suitable configuration to complex with the metal ion, and then dissociative exchange occurs as shown:

$$H_2P + ML_4 \rightarrow (H_2P\text{-}\text{-}\text{-}LML_3) \rightarrow (H_2P\text{---}ML_3) + L$$

A study of the hyperfine splitting of the e.s.r. spectrum of the cation radical of oxidized zinc(II) bacteriochlorin, together with selective deuteriation, has allowed assignment of the electron density to the atoms of the system.[279] There is high spin density on the protons of the saturated rings of the bacteriochlorin with or without the zinc. A similar situation may pertain in metal-free bacteriochlorophyll.

E.s.r. studies on the cation radicals of several zinc(II) porphyrins have established that the ground state is $^2A_{2u}$ as opposed to $^2A_{1u}$.[280] It has been proposed that the ground state of such radicals in some enzymes may reflect their protein environment, and that differences in the ground states may control the enzyme's reaction with substrates. The interesting suggestion is made that electron transfer to and from haeme iron may proceed *via* radical transients with the π-electron configuration described.

Intense investigation into the toxicological properties of mercury continues. It has been reported that, in the interaction of mercury with chromatin in mice, the mercury is firmly bound to the protein, not directly to the DNA.[281]

The stereochemical course of the cleavage of cobalt–carbon bonds in alkyl

[274] I. G. Macara, T. G. Hoy, and P. M. Harrison, *Biochem. J.*, 1973, **135**, 785.
[275] R. E. Miller, E. Shelton, and E. R. Stadtman, *Arch. Biochem. Biophys.*, 1974, **163**, 155.
[276] L. D. Spaulding, P. G. Eller, J. A. Bertrand, and R. H. Felton, *J. Amer. Chem. Soc.*, 1974, **96**, 982.
[277] R. F. Pasternak, L. Francesconi, D. Raff, and E. Spiro, *Inorg. Chem.*, 1974, **12**, 2606.
[278] P. Hambright, and P. B. Chock, *J. Amer. Chem. Soc.*, 1974, **96**, 3123.
[279] J. Fajer, D. C. Borg, A. Forman, R. H. Felton, D. Dolphin, and L. Vegh, *Proc. Nat. Acad. Sci. U.S.A.*, 1974, **71**, 955.
[280] J. Fajer, D. C. Borg, A. Forman, A. D. Adler, and V. Varadi, *J. Amer. Chem. Soc.*, 1974, **96**, 1239.
[281] S. E. Bryan, A. L. Guy, and K. J. Hardy, *Biochemistry*, 1974, **13**, 313.

cobaloximes by mercuric ion has been determined.[282] The reaction, which is described as S_E2, proceeds with inversion at the α-carbon atom. The workers envisage an 'open' transition state:

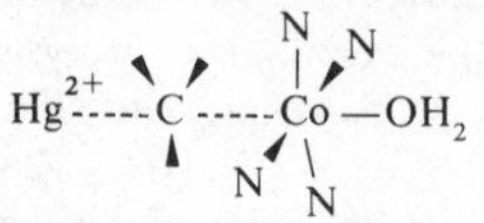

$$Hg^{2+}\text{-----}C\text{-----}Co\text{---}OH_2$$

The enzymic methylation of inorganic mercury by methanogenic bacteria in soil has been reported.[283] Heat-sterilization of soil prevents the methylation.

The observation that victims of the Japanese Minamata disaster had suffered considerable damage to plasma membranes has led to investigation of possible reaction mechanisms whereby such damage may result from $MeHg^+$ poisoning. It has been demonstrated that methylmercury is the most potent inhibitor of the enzyme adenyl cyclase yet reported.[284] The enzyme occurs in liver plasma membranes and plays a part in the metabolism of mammalian cells.

Segall and Wood[285] have suggested that $MeHg^+$ may display its toxic effect by reacting catalytically and directly with a group of phospholipids which are important in membrane structures for cells of the central nervous system. With results obtained from a 220 MHz n.m.r. study, the workers show that $MeHg^+$ reacts catalytically in promoting hydration and hydrolytic cleavage of the vinyl ether linkages in plasmogens to give a mixture of long-chain saturated aldehydes, as shown.

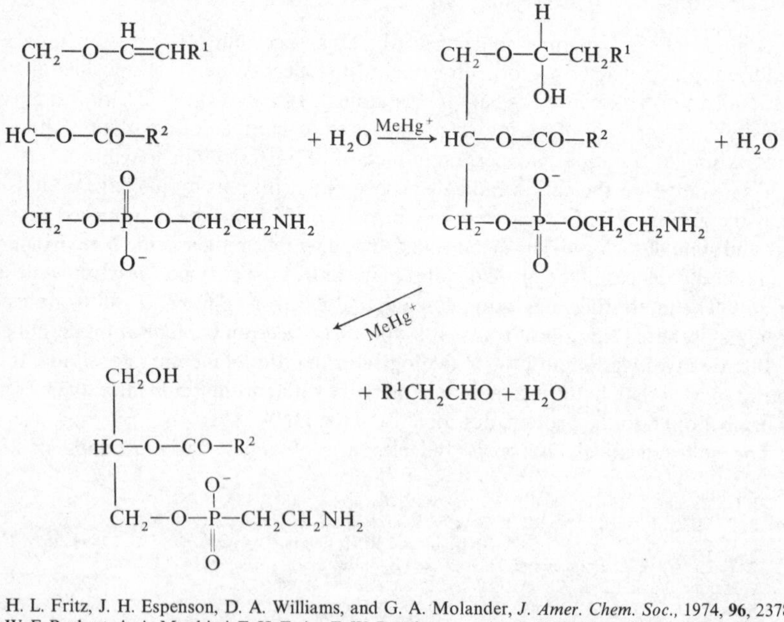

[282] H. L. Fritz, J. H. Espenson, D. A. Williams, and G. A. Molander, *J. Amer. Chem. Soc.*, 1974, **96**, 2378.
[283] W. F. Beckert, A. A. Moghissi, F. H. F. Au, E. W. Bretthauer, and J. C. McFarlane, *Nature*, 1974, **249**, 674.
[284] D. R. Strom and R. P. Gunsulus, *Nature*, 1974, **250**, 778.
[285] H. G. Segall and J. M. Wood, *Nature*, 1974, **248**, 456.

5
Scandium, Yttrium, the Lanthanides, and the Actinides

BY J. A. McCLEVERTY

1 Scandium and Yttrium

Structural Studies.—In $[Sc(\eta^5\text{-}C_5H_5)_2Cl]_2$, the Sc atoms are bridged[1] by chlorine atoms and have an 'oyster-like' co-ordination similar to that in $[Ti(\eta^5\text{-}C_5H_5)_2Cl_2]$ (bond lengths, co-ordination numbers, and other structural comments are given in Table 1, p. 475). The co-ordination environment in $Sc(O_2C_7H_5)_3$ ($HO_2C_7H_5$ = tropolone) lies[2] between trigonal antiprismatic and trigonal prismatic (the projected twist angle is 33°). It is proposed that the fluxional compound undergoes either a series of rapid intramolecular rearrangements due to the anticipated low-energy barrier to rotations about the C_3 axis, or a series of concerted rotations of the ligands about a C_2 axis.[3] The complex formulated[4] as $HSc(O_2C_7H_5)_4$ exists as a hydrogen-bonded dimer,[5] each Sc atom being eight-co-ordinate with respect to the O atoms of the tropolonate ligands. The co-ordination environment is close to D_{2d} symmetry, the O atoms being located at the vertices of a polyhedron best described as an irregular bicapped trigonal prism distorted towards a dodecahedron. The four ligands attached to each metal are of three types: two are symmetrically co-ordinated whereas the other two are asymmetrically co-ordinated and differ by virtue of varying degrees of involvement with hydrogen-bonding. The two quasi-dodecahedra are linked together about a centre of symmetry by two almost linear hydrogen bonds. The average ionic radius of the scandium(III) ion in this complex is 0.82 Å. $Sc(THF)_3Cl_3$ has a *mer*-octahedral structure,[6] and in $[YL_6]I_3$ (L = antipyrine), the metal atom is co-ordinated[7] by six O atoms of the carbonyl group of each antipyrine ligand.

Chemical Studies.—The dinuclear porphyrin complex [LScOScL] (L = octaethylporphinate) was synthesized[8] by reaction of $Sc(acac)_3$ with H_2L in molten imidazole.

The hydroxo-species $Sc(OH)_3,nH_2O$ and $M(OH)_{x\sim1}O_{y\sim1},nH_2O$ (M = Sc, Y, and lanthanide elements) have been prepared[9] by reaction of ammonia or alkaline solutions with the appropriate metal chlorides, nitrates, or sulphates. By heating Y_2O_3

[1] J. L. Atwood and K. D. Smith, *J.C.S. Dalton*, 1973, 2487.
[2] T. J. Anderson, M. A. Neuman, and G. A. Melson, *Inorg. Chem.*, 1974, **13**, 158
[3] E. L. Muetterties and L. J. Guggenberger, *J. Amer. Chem. Soc.*, 1972, **94**, 8046; S. S. Eaton, G. R. Eaton, R. H. Holm, and E. L. Muetterties, *J. Amer. Chem. Soc.*, 1973, **95**, 1116.
[4] E. L. Muetterties and C. M. Wright, *J. Amer. Chem. Soc.*, 1965, **87**, 4706.
[5] D. J. Olszanski, T. J. Anderson, M. A. Neuman, and G. A. Melson, *Inorg. Nuclear Chem. Letters*, 1974, **10**, 137; A. R. Davis and F. W. B. Einstein, *Inorg. Chem.*, 1974, **13**, 1880; T. J. Anderson, M. A. Neuman, and G. A. Melson, *ibid.*, p. 1884.
[6] J. L. Atwood and K. D. Smith, *J.C.S. Dalton*, 1974, 921.
[7] R. W. Baker and J. W. Jeffery, *J.C.S. Dalton*, 1974, 229.
[8] J. W. Buchler and H. H. Schneehage, *Z. Naturforsch.*, 1973, **28b**, 433.
[9] V. V. Sakharov, V. A. Musorin, I. M. Gavrilova, L. M. Zaitsev, and I. A. Apraksin, *Zhur. neorg. Khim.*, 1973, **18**, 3189; V. A. Musorin, V. V. Sakharov, and L. M. Zaitsev, *ibid.*, 1974, **19**, 1476.

with an excess of $NaVO_3$ or KVO_3 at 900—1000 °C, YVO_4 was formed.[10] Reaction of $Sc(OH)_3$ with H_2SeO_3 afforded[11] $Sc_2O(SeO_3)_3,3H_2O$, which could be dehydrated to give $Sc_2O(SeO_3)_2$ and finally Sc_2O_3 and SeO_2. With $NaH_3(SeO_3)_2$, $ScH(SeO_3)_2$,-H_2O was formed. An X-ray photoelectron spectral study has been made[12] of a series of Sc complexes of O-donor atom ligands. These include $Sc(acac)_3$, $Sc(O_2C_7H_5)_3$, $[HSc(O_2C_7H_5)_4]_2$, $Cs[Sc(O_2C_7H_5)_4]$, $Sc(dipic)(dipicH),7H_2O$, $Na[Sc(dipic)_2],4H_2O$ ($dipicH_2$ = pyridine-2,6-dicarboxylic acid), $Sc_2(O_2C_2O_2)_3,6H_2O$, $Na[Sc(O_2C_2O_2)_2]$,-$4H_2O$, $Sc_2(O_2C_2O_2)_3(o\text{-phen})_2$, and Sc_2O_3. The Sc $2p$ electron binding energies are relatively insensitive to the environment about the central metal atom.

The antipyrine (L) and diantipyrrylmethane (Q) complexes $[ScL_6][NCSe]_3$, $MQ_3(NCSe)_3$ (M = Sc or Y), $ScQ_4(NCSe)_3$, and $YQ_2(NCSe)_3$ have been synthesized.[13] Treatment of scandium(III) salts with 8-hydroxyquinoline (C_9H_6NOH) and its chloro-, methyl-, and nitro-derivatives, in aqueous media, gave[14] $Sc(C_9H_6NO)_3,H_2O$ and not $Sc(C_9H_6NO)_3,(C_9H_6NOH)$. The latter was obtained by reaction of the former with molten C_9H_6NOH. The thioxan oxide and tetramethylenesulphoxide (L) complexes $ScL_6(ClO_4)_3$, in which the ligands are co-ordinated *via* the sulphoxide O atom, have been reported,[15] as have yttrium complexes of glycinebis(methylphosphonic acid), $[\{(HO)_2POCH_2\}_2NCH_2CO_2H = H_5L]$,[16] and hydroxyethylidenediphosphonic acid, $[(HO)_2P(O)C(OH)MeP(O)(OH)]_2 = H_5L]$,[17] *viz.* YH_2L,nH_2O (n = 4 or 5).

Reaction of YCl_3 with $(RO)_3P{=}S$ gave[18] $Y\{(RO)_2PS\}_3(OH_2)_x$ (R = Et or Bu; x = 1—3); similar compounds could be obtained from lanthanide chlorides although when R = Et, species with fewer than two ethyl groups per P atom were obtained. On the basis of solubility and i.r. spectral measurements it was suggested that the complexes were polymeric, with both bidentate (O- and S-bonded) and unidentate (O-bonded) phosphorus ligands [perhaps as $(RO)_2POS^-$].

Reaction[19] of Y_2S_3 and its lanthanide analogues (Dy, Pr, Nd, Sm, and Gd) with GeS and GeS_2 at high temperatures afforded M_2GeS_4 and $M_{12}Ge_5S_{28}$, respectively. Treatment of $Sc(NO_3)_3$ with KF and KOH in water gave[20] $Sc(OH)_2F$ and $Sc(OH)$-$F_2,0.5H_2O$; dehydration of these afforded $ScOF$. No evidence for ScF_4^- could be detected[21] during the thermal decomposition of $[NH_4]_3[ScF_6]$, although ScF_3 was the product; conditions for making pure $[NH_4][ScF_4]$ were described, however. At 60—80 °C, M_2O_3 (M = Sc or Y) reacted[22] with NH_4HF_2 giving $[NH_4]_3[ScF_6]$ and $NH_4Y_2F_7,NH_4F$. Treatment of ScF_3 with B_2O_3 afforded[23] $Sc_2(B_2O_5)F_2$.

[10] A. A. Fotiev and M. Ya. Khodos, *Chem. Abs.*, 1974, **81**, 20263.
[11] L. N. Komissarova and A. S. Znamenskaya, *Zhur. neorg. Khim.*, 1974, **19**, 295.
[12] A. D. Hamer, D. G. Tisley, and R. A. Walton, *J. Inorg. Nuclear Chem.*, 1974, **36**, 1771.
[13] A. M. Golub and M. V. Kopa, *Chem. Abs.*, 1973, **79**, 61052.
[14] A. Corsini, F. Toneguzzo, and M. Thompson, *Canad. J. Chem.*, 1973, **51**, 1248.
[15] G. Vicentini and W. N. de Lima, *Anales Acad. Brasil, Cienc.*, 1973, **45**, 219.
[16] G. S. Tereshin and L. K. Karitonova, *Zhur. neorg. Khim.*, 1974, **19**, 1264.
[17] G. S. Tereshin, L. K. Karitonova, L. V. Krinitskaya, and T. P. Korabel'nikova, *Zhur. neorg. Khim.*, 1974, **19**, 1131.
[18] C. M. Mikulski, L. L. Pytlewski, and N. M. Karayannis, *J. Less-Common Metals*, 1973, **33**, 377.
[19] M. P. Stepanets, R. A. Beskrovnaya, and V. V. Serebrennikov, *Izvest. Akad. Nauk S.S.S.R., neorg. Materialy*, 1974, **10**, 372; V. P. Gus'kova and V. V. Serebrennikov, *Chem. Abs.*, 1974, **80**, 55403.
[20] L. N. Komissarova, V. M. Shatskii, L. A. Nesterova, and S. Ya. Semochkin, *Zhur. neorg. Khim.*, 1974, **19**, 31; L. M. Komissarova, V. M. Shatskii, and L. A. Nesterova, *ibid.*, p. 291.
[21] P. Bukovec and J. Sifta, *Monatsh.*, 1974, **105**, 510.
[22] M. A. Mikhailov, D. G. Epov, V. I. Sergienko, E. G. Rakov, and G. P. Shchetinina, *Zhur. neorg. Khim.*, 1973, **18**, 1508.
[23] L. R. Batsanova, L. A. Novosel'tseva, and A. I. Madaras, *Izvest. Akad. Nauk S.S.S.R., neorg. Materialy*, 1974, **10**, 621.

2 The Lanthanides

Structural Studies.—In $Nd(C_5H_4Me)_3$, each Nd atom is η^5-bonded[24] by three of the cyclopentadienyl rings and η^1-bonded by another ring of an $Nd(C_5H_4Me)_3$ group. This sharing of a C atom between two metals is repeated throughout, thereby rendering the complex tetrameric. The bonding between the rings and Nd is regarded as essentially ionic.

In $KLa_2(NH_2)_7$, obtained by reaction of K and La with NH_3 under pressure and at 350 °C, the lanthanum atom is eight-co-ordinate.[25] The co-ordination poly-hedron may be described as a deformed trigonal prism which is bicapped on two four-fold faces.

In $La_2O_2\{N(SiMe_3)_2\}_4(OPPh_3)_2$, obtained by reaction of $La\{N(SiMe_3)_2\}_3$ with an excess of Ph_3PO or with $(Ph_3PO)_2H_2O_2$, there is a bridging peroxo-group (1) and each metal atom is five-co-ordinate.[26] Similar species were produced containing Pr, Sm, and Eu, but with a 1:1 molar stoicheiometry, $La\{N(SiMe_3)_2\}_3(OPPh_3)$ could be obtained.

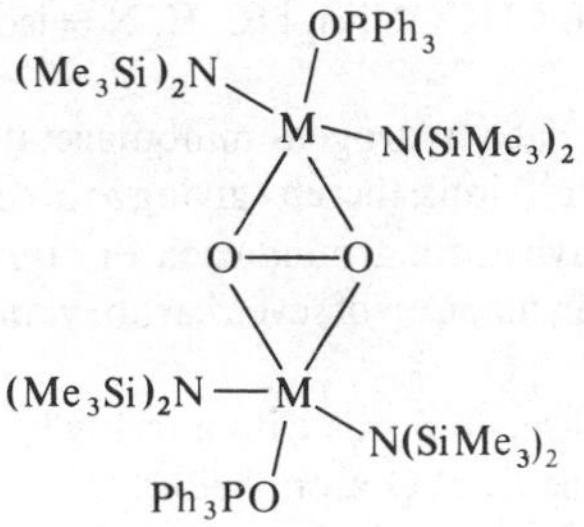

In the complex $GdL(NO_3)_3$, [L = (2)], the metal atom is ten-co-ordinate,[27] the co-ordination polyhedron being described as a distorted pentagonal bipyramid. Four of the five equatorial positions are occupied by the N atoms of L, and the remaining three positions by the three bidentate nitrate groups.

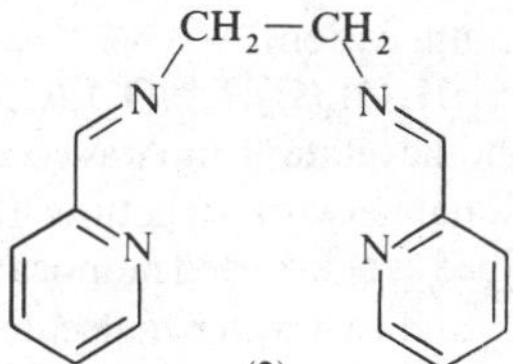

The anion in $[Ph_3EtP][Ce(NO_3)_5]$ has C_2 crystallographic symmetry,[28] the metal atom being ten-co-ordinate (bidentate nitrates). If each nitrate group is regarded as a single ligand site, the co-ordinate polyhedron approximates to a trigonal bipyra-mid. In YbP_3O_9, the metal has a distorted octahedral arrangement[29] of O-donor atoms; in NdP_3O_9 and NdP_5O_{14}, the metal is eight-co-ordinate with respect to oxygen, having a dodecahedral environment. Nine-co-ordinate samarium occurs[30] in $[NH_4]$ $[Sm(SO_4)_2,4H_2O]$, where each sulphate is terdentate with respect to the metal, the co-ordination sphere being completed by three water molecules. The co-ordination polyhedron may be described either as a tricapped trigonal prism or as a mono-

[24] J. H. Burns, W. H. Baldwin, and F. H. Fink, *Inorg. Chem.*, 1974, **13**, 1916.
[25] C. Hadenfeldt, B. Gieger, and H. Jacobs, *Z. anorg. Chem.*, 1974, **408**, 27.
[26] D. C. Bradley, J. S. Ghotra, F. A. Hart, M. B. Hursthouse, and P. R. Raithby, *J.C.S. Chem. Comm.*, 1974, 40.
[27] G. D. Smith, C. N. Caughlan, Mazhar-ul-Haque, and F. A. Hart, *Inorg. Chem.*, 1973, **12**, 2654.
[28] A. R. Al-Kharaghouli and J. S. Wood, *J.C.S. Dalton*, 1973, 2318.
[29] H. Y.-P. Hong, *Acta. Cryst.*, 1974, **B30**, 468, 1857.
[30] B. Eriksson, L. O. Larsson, L. Niinisto, and U. Skoglund, *Inorg. Chem.* 1974, **13**, 290.

R

capped square antiprism. A square antiprismatic arrangement of O atoms around Ce has been found[31] in $Na_6[CeW_{10}O_{36}H_2],30H_2O$.

The structures adopted by mono-adducts of lanthanide tris-β-diketonates have been rationalized[32] by a consideration of ligand–ligand repulsion energies. These energies were calculated using a 'normalized bite' for the diketonate ligands, the 'bite' being defined as the distance between two donor atoms of the same bidentate ligand divided by the metal–donor atom distance. Three minima of closely similar energies appear on the potential energy surface. The first corresponds to a capped octahedron (C_{3v}) and is found in $Ho(PhCOCHCOPh)_3(H_2O)$ and $Yb(PhCOCHCOMe)_3(H_2O)$. The second corresponds to an irregular polyhedron (C_1) and may be used to describe the geometry of $Yb(acac)_3\{MeCOCHC(NH_2)Me\}$, $Dy(BuCOCHCOBu)_3(H_2O)$, $Lu(Bu-COCHCOBu)_3(MeC_5H_4N)$, and $Lu(C_3F_7COCHCOBu)_3(H_2O)$; a related distorted capped trigonal prismatic structure is adopted by $Yb(MeCOCHCOMe)_3(H_2O)$. The third potential energy minimum corresponds to a geometry intermediate between a pentagonal bipyramid and a capped trigonal prism; no lanthanide β-diketonate complex has yet been found with this structure. A distorted dodecahedral arrangement of the O atoms occurs[33] in $[C_9H_8N][Ce(CF_3COCHCOCF_3)_4]$ $(C_9H_8N = $ iso-quinolinium cation).

In $[Pr_2(H_2O)_4(C_4H_6NO_4)_2(C_4H_5NO_4)]Cl_2,3H_2O$, each hydrogen-iminodiacetate and iminodiacetate group was co-ordinated[34] to four Pr^{3+} ions, thereby giving a three-dimensional network structure. The co-ordination polyhedron around each Pr atom is described as a distorted monocapped square antiprism made up of seven carboxylato O atoms and two water molecules.

In La_2O_2S, the lanthanum atom is located[35] on a three-fold axis with a triangle of sulphide ions above and a triangle of O atoms, and one axial O atom below.

In $KHoBeF_6$, BeF_4 tetrahedra connect[36] chains of HoF_8 antiprisms.

Chemical Studies.—Reaction of hydrogen at 300 °C with Eu or Yb metal afforded[37] MH_2, whereas preheated Yb reacted at slightly reduced hydrogen pressures and at room temperature to give $YbH_{2.6}$. These compounds appear to contain hydride ion, and LiH reacted with EuH_2 to afford $LiEuH_3$.

By arc melting the elements together under argon, $YbCo_3B_2$ and MCo_2B_2 (M = Pr or Ho) were prepared.[38] The cobaltoborides MCo_3B_2 (M = Sc, Y, Ce, Tb → Tm) have $CaCu_5$-type structures. Studies of the thermal stability of lanthanide tetraborides in the range 1500—2500 °C (*in vacuo*) have revealed[39] the existence of two classes of compounds: (i) MB_4, M = Y, Gd → Er, which sublime without decomposition, and (ii) MB_4, M = La, Ce → Sm, which dissociate into MB_6 and boron. The hexaborides exist in three classes: (i) MB_6, M = La, Ce → Sm, which sublime or melt without decomposition; (ii) EuB_6 and YB_6, which dissociate by preferential evaporation of the metal to give rhombohedral boron and YbB_{12}, respectively; and (iii)

[31] J. Iball, J. N. Low, and T. J. R. Weakley, *J.C.S. Dalton*, 1974, 2021.
[32] D. L. Kepert, *J.C.S. Dalton*, 1974, 617.
[33] A. T. McPhail and P.-S. W. Tschang, *J.C.S. Dalton*, 1974, 1165.
[34] J. Albertsson and A. Oskarsson, *Acta Chem. Scand.*, 1974, (*A*)28, 347.
[35] B. Morosin and D. J. Newman, *Acta Cryst.*, 1973, **B29**, 2647.
[36] Y. Le Fur, I. Tordjman, S. Aléonard, G. Bassi, and M. T. Roux, *Acta Cryst.*, 1974, **B30**, 2049.
[37] V. I. Mikheeva, M. E. Kosta, and A. I. Konstantinova, *Chem. Abs.*, 1974, **80**, 9898.
[38] P. Rogl, *Monatsh.* 1973, **104**, 1623.
[39] P. Hagenmuller, J. E. Tourneau, J. Mercurio, and R. N. Naslain, *Chem. Abs.*, 1974, **80**, 152305.

Table 1 *Structures of scandium, yttrium, and lanthanide complexes*

Compound	Ref.	Bond lengths/Å	Remarks
$[Sc(\eta\text{-}C_5H_5)_2Cl]_2$	1	Sc—Cl 2.58 Sc—C 2.46 (av.)	'oyster-like' with bridging Cl
$Sc(O_2C_7H_5)_3$	2		D_3 imposed molecular symmetry trigonal anti-prismatic, twist angle 60°: six-co-ordinate
$HSc(O_2C_7H_5)_4$	5	Sc—O 2.16—2.31	eight-co-ordinate, close to D_{2d} (dodecahedron)
$Sc(THF)_3Cl_3$	6	Sc—O 2.18 (av.) Sc—Cl 2.41 (av.)	*mer*-octahedral
$Y(antipyrine)_6I_3$	7		octahedral (O-donor atoms)
$[Nd(C_5H_4Me)_3]_4$	24		Each Nd η^5-bonded by three C_5H_4Me rings and η^1-bonded by another
$KLa_2(NH_2)_7$	25	La—N 2.55—2.92	eight-co-ordinate: bicapped trigonal prism (four-fold faces occupied)
$La_2O_2\{N(SiMe_3)_2\}_4(OPPh_3)_2$	26	La—O (peroxide) 2.33—2.35 La—O (P) 2.44 La—N 2.39—2.40 O—O 1.70	five-co-ordinate La
$GdL(NO_3)_3$	27	Gd—N 2.50—2.61 Gd—O 2.45—2.55	L = (2); nitrates bidentate; ten-co-ordinate
$[Ph_3EtP][Ce(NO_3)_5]$	28	Ce—O 2.57	nitrates bidentate; ten-co-ordinate
NdP_3O_9	29		eight-co-ordinate; dodecahedron
NdP_5O_{14}	29		eight-co-ordinate; dodecahedron
YbP_3O_9	29		six-co-ordinate; octahedral
$[NH_4][Sm(SO_4)_2,4H_2O]$	30	Sm—O 2.38—2.56 (H_2O) 2.44—2.51 (SO_4)	nine-co-ordinate; tri-capped trigonal prism or monocapped square antiprism
$Na_6CeW_{10}O_{36}H_2,30H_2O$	31		eight-co-ordinate; square antiprism
$[C_9H_8N][Ce\{CF_3COCHC\text{-}(O)(C_4SH_3)\}_4]$	33	Ce—O 2.47	eight-co-ordinate; distorted dodecahedral
$[Pr_2(C_4H_6NO_4)_2(C_4H_5NO_4)]Cl_2,3H_2O$	34	Pr—O 2.34—2.75	nine-co-ordinate; distorted monocapped square antiprismatic
La_2O_2S	35	La—O 2.42 La—S 3.04	nine-co-ordinate; mono-capped (three-fold face) trigonal prismatic
$KHoBeF_6$	36	Ho—F 2.25—2.39	eight-co-ordinate; square antiprismatic

MB_6, M = Y, Gd, Tb, and Dy, which dissociate into MB_4 and boron. Reaction of $H_2B_{12}H_{12}$,$6H_2O$ with lanthanide, scandium or yttrium hydroxides or carbonates afforded[40] $M_2(B_{12}H_{12})_3$,xH_2O (x = 15, 18, 20, or 21 depending on M); the compounds are ionic.

Codeposition of lanthanide atoms with CO in argon matrices has afforded[41] a series of metal carbonyls, $M(CO)_x$ (x = 1 — 6) (M = Pr, Nd, Eu, Gd, Ho, and Yb). The final product upon controlled annealing of the matrices was $M(CO)_6$. It was found that v_{CO} varies with co-ordination number in much the same way as in other compounds, but that it is relatively independent of the nature of M.

Treatment of $Ce(\eta^5-C_5H_5)_3Cl$ with NaH and with $NaNH_2$ in THF gave[42] $Ce(\eta^5-C_5H_5)_3X$ (X = H or NH_2); $Ce(\eta^5-C_9H_7)_2X_2$ was prepared similarly from the dichloride. Reaction of $M(\eta^5-C_5H_5)_2Cl$ with $LiC{\equiv}CPh$ gave[43] the paramagnetic acetylides $M(\eta^5-C_5H_5)_2C{\equiv}CPh$ (M = Gd, Er, and Yb). The mercaptides $Ce(\eta^5-C_5H_5)_3SR$ and $Ce(\eta^5-C_9H_7)_2(SR)_2$ (R = alkyl) are probably polymeric.[44]

When mixtures of stoicheiometric oxides and nitrides were heated *in vacuo*, a range of compounds of variable composition, MN_xO_{1-x} (M = La, Ce, Pr, or Tb), was obtained. [45] Reaction of La, Gd, or Sc with $NaNH_2$ in liquid NH_3 at high temperature and under pressure gave[46] $Na_3[M(NH_2)_6]$ (M = La or Gd) and Na-$[Gd(NH_2)_4]$; no Sc compounds could be isolated. In the presence of NH_4I, these reactions formed $La(NH_2)_3$, GdN, and ScH_2. The hydrazine complexes Sm-$(N_2H_4)_3X_3$,nH_2O (X = Cl, I, NO_3, or OAc) and $[Sm(N_2H_4)_6]_n[X]_3$,nH_2O (X = SO_4, n = 2; X = ClO_4, n = 1) have been reported.[47] NN'-Bis(1-acetonylethylidene)-ethylenediamine (aeH_2) reacted[48] with the lighter lanthanide perchlorates giving $M(aeH_2)_4(ClO_4)_3$ (M = La, Pr, Nd, or Sm) whereas the heavier lanthanides afforded $M(aeH_2)_2(ClO_4)_2(OH)$,H_2O (M = Gd, Dy, or Er); $Gd(aeH_2)_4(ClO_4)_3$ could be obtained from anhydrous $Gd(ClO_4)_3$ and aeH_2 in methanol. Treatment of $[NH_4]_2$-$[Ce(NO_3)_6]$ with Ph_3PO and o-phenanthroline in acetone afforded[49] $[Ce(o\text{-phen})_4]$-$[NO_3]_3$. The magnetic moments of one Yb and two Gd phthalocyanine (PcH_2) complexes were typical[50] for M^{III} and showed little or no evidence for coupling between the metal ions. The Yb complex was formulated as $(ClPc)YbCl$,$2H_2O$ in which a ligand aromatic ring had been chlorinated. Green and blue isomers of a Gd complex were attributed to the equilibrium $Gd(Pc)(PcH) \rightleftharpoons H^+ + [Gd(Pc)_2]^-$.

The complexes $[R_4N]_3[M(NCO)_6]$ (R = Et or Bu, M = Sc, Y, Eu, Gd, Dy, Ho, Er, and Yb) contained[51] six-co-ordinate M and N-bonded isocyanate. Reaction of $[Et_4N][NCO]$ with MX_3 (M = Y, La → Gd, Dy, Er, and Yb, X = Cl,[52] NCS,[52]

[40] O. A. Kanaeva and N. T. Kuznetsov, *Chem. Abs.*, 1974, **80**, 33 351.

[41] J. L. Slater, T. C. de Vore, and V. Calder, *Inorg. Chem.*, 1973, **12**, 1918; 1974, **13**, 1808.

[42] S. Kapur, B. L. Kalsotra, and R. K. Multani, *J. Inorg. Nuclear Chem.*, 1974, **36**, 932.

[43] M. Tsutsui and N. Ely, *J. Amer. Chem. Soc.*, 1974, **96**, 4042.

[44] S. Kapur, B. L. Kalsotra, and R. K. Multani, *J. Inorg. Nuclear Chem.*, 1973, **35**, 3966.

[45] R. C. Brown and N. J. Clark, *J. Inorg. Nuclear Chem.*, 1974, **36**, 1777, 2287.

[46] G. Linde and R. Juza, *Z. anorg. Chem.*, 1974, **409**, 191.

[47] R. Ya. Aliev, *Zhur. neorg. Khim.*, 1974, **19**, 274.

[48] N. Yoshida, A. Matsumoto, and J. Shiokawa, *Bull. Chem. Soc. Japan*, 1974, **47**, 648.

[49] F. Brezina, *Z. Chem.*, 1973, **13**, 383.

[50] A. G. MacKay, J. F. Boas, and G. J. Troup, *Austral. J. Chem.*, 1974, **27**, 955.

[51] R. L. Dieck and T. Moeller, *J. Inorg. Nuclear. Chem.*, 1973, **35**, 3781; M. B. Harris and L. Thompson, *ibid.*, 1974, **36**, 212.

[52] R. L. Dieck and T. Moeller, *J.Less-Common Metals*, 1973, **33**, 355.

or NO_3^{53}) gave $[Et_4N]_3[MX_3(NCO)_3]$; the nitrates were nine-co-ordinate (bidentate nitrate) and again isocyanate was N-bonded. The complexes $M(o\text{-phen})_3(NCS)_3$ and $ML_6(NCS)_3$ (L = antipyrine; M = Y, La, Sm, and Yb) have also been described,[54] together with $M(o\text{-phen})_3Br_3, nH_2O$ (n = 2 or 3) and $Y_2(o\text{-phen})_2(SO_4)_3, 2H_2O$.

When lanthanide metals (M = La → Lu) were heated directly with phosphorus, or M_2O_3 with PH_3 in air, the binary phosphides MP were produced.[55] These decomposed in water and, in air at 700 °C, were oxidized to phosphates; on heating PrP at 2850 °C, a refractory material, deficient in P but containing free Pr and adsorbed oxygen, was formed. EuP_2, obtained by reaction of Eu_2O_3 with P_n, contains europium(II).[56]

Table 2 *Redox potentials of lanthanide ion couples* M^{3+}/M^{2+} *in aqueous solution*

Element	E°/V	Element	E°/V
La	−5.8	Gd	−4.9
Ce	−4.2	Tb	−3.5
Pr	−3.0	Dy	−2.6
Nd	−2.8	Ho	−2.9
Pm	−2.5	Er	−3.0
Sm	−1.5 ± 0.2[a]	Tm	−2.1
Eu	−0.35 ± 0.03[b]	Yb	−1.10 ± 0.01[c]

[a] *J. Amer. Chem. Soc.*, 1948, **70**, 1347;   [b] *J. Chem. Thermodynamics*, 1973, **5**, 513;
[c] *J. Amer. Chem. Soc.*, 1942, **64**, 1133.

The values for the redox potential for the couple M^{3+}/M^{2+} have been estimated[57] using a simple ionic model and available thermodynamic data. The results (Table 2) correlate closely with the ionization potentials for the M^{2+} ions, and are in good agreement with both chemical observations and other estimates obtained by spectroscopic correlations. Irreversible oxidation of terbium(III) to terbium(IV) in aqueous K_2CO_3–KOH solutions has been observed electrochemically;[58] the discovery of an intermediate of mixed oxidation state explains partly the reduction behaviour of terbium(IV) deposits. Praseodymium(IV) and terbium(IV) have also been detected in nitrate solutions.

Thermal decomposition of $Ce_2(O_2C_2O_2)_3, 10H_2O$ afforded[59] CeO_2, and a series of lanthanide oxysulphides (La → Er) was prepared[60] by heating oxysulphites in H_2S–CO atmospheres. By reaction of gaseous sulphur with M_2O_3, oxalates, or oxycarbonates at high temperatures, the oxysulphides M_2O_2S (M = Y, Gd, or La) and $M_{2-x}^1 M_x^2 O_2 S$ (M^1 = Y or La; M^2 = Eu or Tb) were formed.[61] These decomposed thermally to give M_2S_3 (M = Y or La), and ThOS was produced from ThO_2 and sulphur at 1100 °C.

[53] R. L. Dieck and T. Moeller, *J. Inorg. Nuclear Chem.*, 1974, **36**, 2283.
[54] Yu. G. Eremin and G. I. Bondarenko, *Zhur. neorg. Khim.*, 1973, **18**, 1715.
[55] K. E. Mironov and I. G. Vasil'eva, *Chem. Abs.*, 1973, **79**, 60949.
[56] K. E. Mironov and G. P. Brygalina, *Izvest. Akad. Nauk S.S.S.R., neorg. Materialy*, 1974, **10**, 920.
[57] D. A. Johnson, *J.C.S. Dalton*, 1974, 1671.
[58] R. C. Propst, *J. Inorg. Nuclear Chem.*, 1974, **36**, 1085; N. S. Vagina, *Chem. Abs.*, 1973, **79**, 99923.
[59] E. Ishii and Y. Miyake, *Chem. Abs.*, 1974, **80**, 22194.
[60] M. Koskenlinna and L. Niinisto, *Suomen Kem. (B)*, 1973, **46**, 326.
[61] H. Rudolf and J. Loriers, *Bull. Soc. chim. France*, 1974, 377.

Treatment of MCl_3,$(Pr^iOH)_3$ (M = La → Sm, Gd, Ho, Er, or Yb) with $K[Ga(OPr^i)_4]$ in Pr^iOH, or of $M(OPr^i)_3$ with $Ga(OPr^i)_3$, gave[62] $M[Ga(OPr^i)_4]$. Calcium oxide reacted[63] with M_2O_3 giving CaM_4O_7, CaM_2O_4, $Ca_2M_2O_5$, and $Ca_3M_2O_6$ (M = Sm → Lu, Sc, or Y), and coprecipitation of lanthanide and aluminium hydroxides from ammonia or $(NH_4)_2CO_3$ solutions, followed by dehydration, afforded $MAlO_3$.[64]

Reaction of M_2O_3 with CO_2 saturated with water vapour gave[65] $M_2(CO_3)_3,nH_2O$ (M = Dy, Ho, or Er). Europium salts reacted[66] with K_2CO_3 in solution affording[66] $EuCO_3$, which has a KNO_3-type structure. The carbonate decomposed thermally to give $Eu_2O(CO_3)_2$, then $Eu_2O_2(CO_3)$, and finally Eu_2O_3. Treatment of MCl_3 with aqueous Na_2CO_3 provided[67] $Na[M(CO_3)_2],6H_2O$ (M = La → Sm, Gd, Dy, and Y). Alkali or ammonium carbonates reacted[68] with $Eu(NO_3)_3$ giving hydroxycarbonate, and $Eu_2(CO_3)_3,4H_2O$, $Q[Eu(CO_3)_2],6H_2O$ (Q = Na, NH_4, or Cs), and $Q_3[Eu(CO_3)_3]$. Lanthanide oxides reacted[69] with GeO_2 in aqueous ammonia affording M_2GeO_5, $M_2Ge_2O_7$, and $2M_2O_3,3GeO_2$ (M = La, Tb, Lu, or Y), as well as Y_4GeO_8.

The association between Er^{3+} and NO_3^- in methanol–water mixtures has been investigated[70] ultrasonically; outer-sphere complexes apparently predominate in more aqueous solutions. The nitrates $M(NO_3)_3,nH_2O$ (n = 0 — 6) decomposed[71] to Ln_2O_3 via $Ln(NO_3)_{3-x}(NO_2)_x$, $LnO(NO_3)$, and $LnO(NO_3),nLn_2O_3$. In pyridine–water mixtures, cerium(III) nitrate formed[72] $[pyrH]_5[Ce(NO_3)_8]$. The 4,4'-bipyridyl (bipy) complexes $M(bipy)_2(NO_3)_3,2H_2O$ contain[73] ten-co-ordinate metal (M = La, Nd, Sm, Tb, or Er) in which the NO_3 groups are bidentate, the bipy ligands are unidentate and the two water molecules are co-ordinated. The species $M(bipy)Cl_3$,-xH_2O (x = 1 — 4; M = La, Nd, Sm, or Er) are polymeric. The hydrothermal preparation of a series of lanthanide hydroxide nitrates $M_2(OH)_{6-x}(NO_3)_x$ (x = 2, M = La → Dy, Y; x = 0.85 ± 0.05, M = La → Nd; x = 0.6 ± 0.1, M = Sm → Yb) has been described.[74] These decompose thermally, giving $M_2O_2(OH)(NO_3)$ (M = La → Nd), $MO(NO_3)$ (M = La → Gd), $M_3O_4(NO_3)$ (M = La → Gd), and $M_4O_5(NO_3)_2$ (M = Dy → Yb); $M_2(OH)_5(NO_3),2H_2O$ (M = Y or Yb) was also reported. $GdONO_3$ exists as a $PbFCl$-type structure, and the essential features of the hydroxide nitrate phase equilibria could be explained in terms of alternating sheets of $Ln(OH)_2^+$ and NO_3^- ions.

Several series of phosphite and phosphate compounds, $M(H_2PO_3)_3$, (M = Y,

[62] R. C. Mehrotra, M. M. Agarwal, and A. Mehrotra, *Syn. Inorg. Metal-Org. Chem.*, 1973, **3**, 407.
[63] G. I. Gerasimyuk, Z. A. Zaitseva, L. M. Lopato, and S. G. Tresvyatskii, *Izvest. Akad. Nauk S.S.S.R., neorg. Materialy*, 1973, **9**, 1759.
[64] V. S. Krylov, I. L. Belova, R. Magunov, V. D. Kozlov, A. V. Kalinichenko, and N. P. Krot'ko, *Izvest. Akad. Nauk S.S.S.R., neorg. Materialy*, 1973, **9**, 1388.
[65] J. D. Coutures and P. Caro, *Compt. rend.* 1974, **278**, *C*, 861.
[66] Yu. S. Sklyarenko, N. Stroganova, and I. P. Galkina, *Chem. Abs.*, 1973, **79**, 60958.
[67] A. Mochizuki, N. Nagashima, and H. Wakita, *Bull. Chem. Soc. Japan*, 1974, **47**, 755.
[68] N. V. Mzarenlishvili, V. P. Natidze, and E. N. Zedelashvili, *Chem. Abs.*, 1974, **80**, 71700.
[69] L. P. Benderskaya, M. D. Kravchenko, and A. N. Tananaev, *Chem. Abs.*, 1974, **80**, 9908; V. D. Kozlov, R. L. Magunov, and V. S. Krylov, *ibid.*, 1973, **79**, 60981.
[70] J. Reidler and H. B. Silber, *J. Inorg. Nuclear Chem.*, 1974, **36**, 175.
[71] K. E. Mironov, A. P. Popov, E. V. Karaseva, E. D. Sinitsyna, and L. A. Khripin, *Chem. Abs.*, 1973, **79**, 60948.
[72] N. E. Mininkov and E. F. Zhuravlev, *Zhur. neorg. Khim.*, 1974, **19**, 1656.
[73] A. Anagnostopoulos, *J. Inorg. Nuclear Chem.*, 1973, **35**, 3611.
[74] J. M. Haschke, *Inorg. Chem.*, 1974, **13**, 1812.

La→Lu), $M_2(HPO_3)_2,nH_2O$ ($n = 2$ or 4, $M = La→Gd$, Dy, Er, or Yb), $M_2(PhPO_3)_3,$-nH_2O ($n = 0—2$; $M = La → Sm$, Dy, Er, or Yb), MPO_4,nH_2O ($n = 0$, $M = Tb$ → Yb; $n = 3$, $M = Eu$), $2EuPO_4$, Q_3PO_4,nH_2O, and $Q_3Eu_2(PO_4)_3,xH_2O$ ($Q = Na$, K, or NH_4) have been reported.[75, 76] The species $CeM(H_2PO_2)_6$ ($M = Er → Lu$) have been described,[77] and reaction of $Eu(NO_3)_3$ with $Q_4P_2O_7$ gave[76] $Eu_4(P_2O_7)_3$, $QEuP_2O_7$, and $Q_5[Eu(P_2O_7)_2]$. Enthalpy and entropy changes for the formation of 2:1 complexes of M^{3+} (La → Tm, Y) with the tripolyphosphate ions $[P_3O_9H]^{4-}$ and $[P_3O_9]^{5-}$ have been calculated.[78] M_2O_3 reacted[79] with As_2O_3 giving $MAsO_3$, and treatment of $MAsO_4$ with H_2S at $400\,°C$ gave $MAsOS_2$ ($M = La → Sm$) and $MAsO_2S_2$ ($M = Eu → Lu$, Y). The stibites and bismuthites $M^1M^2O_3$ ($M^1 = La → Lu$, except Ce; $M^2 = Sb$ or Bi) and $2M_2O_3,xSb_2O_3$ ($x = 3.0—3.8$; $M = La$, Pr, or Nd) have been described.[80]

Entropy titrations have been used[81] in a reassessment of the thermodynamics of the reaction between lanthanide ions and sulphate ions, and i.r. spectral studies have revealed[82] that in $Ce_2(SO_4)_3,nH_2O$ ($n = 2$, 4, 4.5, 6, 7, 7.5, or 8) there is hydrogen-bonding between the sulphate and water. The isostructural $M_2(SO_4)_3,Rb_2SO_4$ ($M = La$ or Ce) and $M_2(SO_4)_3,Rb_2SO_4,8H_2O$ ($M = Pr → Tm$),[83] and the sulphamates $M(SO_3NH_2)_3,xH_2O$ ($M = Y$, La → Sm, Gd, Dy, Ho, Er, or Yb; $x = 1.5—2.5$),[84] have been described. Series of double selenites [85] $Q[M(SeO_3)_3],nH_2O$ ($M = Nd$, Sm, Gd, Dy, or Er; $Q = NH_4$, K, or Na; $x = 2$ or 2.5) and selenates[86] $Q[M(SeO_4)_2],nH_2O$ ($M = La$ or Ce; $Q = Li$, Na, K, Rb, Cs, or NH_4; $n = 0—9$) have been prepared; in the latter both H_2O and selenate are bound to the lanthanide. The iodates $MH_2IO_6,3H_2O$ ($M = Nd → Lu$) have been dehydrated to give MIO_5.[87] Reaction of lanthanide oxides with M^2O_2 ($M^2 = Ti$, Zr, or Hf) gave[88] $M^1_2O_3,2M^2O_2$ ($M^1 = La → Lu$). Reduction of MVO_3 with carbon at $1400°C$ afforded[89] M_2O_3 and metallic vanadium; the europium oxide afforded Eu_2VO_4 and $Eu_3V_2O_7$, which contains europium(II), both of which subsequently gave EuO. The existence of MVO_4,nH_2O ($M = La$, Ce, Pr, Gd, Er, Yb, or Y; $n = 0$ or 2), $M_4(V_2O_7)_3$, and $M(VO_3)_3$ has been demonstrated;[90] dehydration of $MVO_4,2H_2O$ afforded MVO_3. The stannates $M_2(SnO_3)_3$ and $M_2Sn_2O_7$ (the latter obtained from the former on heating) were

[75] A. T. Malinina, V. N. Biryulina, and V. V. Serebrennikov, *Chem. Abs.*, 1973, **79**, 60950; K. Kalieva, Z. V. Stretkova, N. A. Babynina, L. N. Gordienko, M. Votoyarov, I. E. Sakavov, and A. K. Mustaev, *ibid.*, 1974, **80**, 55392.

[76] I. V. Tananaev and M. V. Landia, *Chem. Abs.*, 1973, **79**, 142408.

[77] T. G. Shvalova, G. E. Seryakova, and A. T. Malinina, *Chem. Abs.*, 1974, **81**, 32652.

[78] M. M. Taqui Khan and P. R. Reddy, *J. Inorg. Nuclear Chem.*, 1974, **36**, 607.

[79] L. E. Angapova and V. V. Serebrennikov, *Chem. Abs.*, 1974, **80**, 55399.

[80] G. Adachi, M. Ishihara, and J. Shiokawa, *J. Less-Common Metals*, 1973, **32**, 179; S. N. Nasonova, V. V. Serebrennikov, and G. A. Narnov, *Chem. Abs.*, 1974, **80**, 140623.

[81] H. K. J. Powell, *J.C.S. Dalton*, 1974, 1108.

[82] T. P. Spacibenko, *Zhur. neorg. Khim.*, 1974, **19**, 899.

[83] L. D. Iskhakova, and V. E. Plyushchev, *Zhur. neorg. Khim.*, 1973, **18**, 1500.

[84] M. L. Zimmerman and E. Giesbrecht, *Anales Acad. Brasil, Cienc.*, 1973, **45**, 99.

[85] O. Erametsa, T. Pakkanen, and L. Niinisto, *Suomen Kem. (B)*, 1973, **46**, 330.

[86] L. A. G. Madrazo, J. G. R. Bernat, and F. Z. A. Garcia, *Ion (Madrid)*, 1973, **33**, 242.

[87] A. H. J. Lokio and J. R. Kyrki, *Suomen Kem. (B)*, 1973, **46**, 206.

[88] N. I. Timofeeva, Z. I. Krainova, and V. N. Sakovich, *Izvest. Akad. Nauk S.S.S.R., neorg. Materialy*, 1973, **9**, 1756.

[89] T. Shinike, G. Adachi, and J. Shiokawa, *Chem. Abs.*, 1974, **80**, 115591.

[90] R. S. Saxena and M. C. Jain, *J. Indian Chem. Soc.*, 1973, **50**, 77; A. M. Golum and S. A. Nedil'ko, *Zhur. neorg. Khim.*, 1973, **18**, 1414; G. G. Mel'chenko and V. V. Serebrennikov, *ibid.*, p. 1172; A. A. Fotiev and M. Ya. Khodos, *Chem. Abs.*, 1974, **81**, 20263.

also described. Treatment of aqueous acidic solutions of $Na_5HV_{10}O_{28}$ with $M(NO_3)_3$ (M = La or Na) gave $M_2V_{10}O_{28},nH_2O$.[91] By heating CeO_2 or Yb_2O_3 with NbO_2, $MNbO_4$ (containing M^{III}) was formed;[92] the Nd analogue was obtained[93] by dehydration of $Nd[Nb(OH)_6]_3$. Calcination at 900 °C of the coprecipitated hydroxides afforded[94] $MTiNbO_6$ (M = Gd, Tb, or Dy) which has a euxenite-type structure. Similar treatment with La and Cr hydroxides formed[95] $LaCrO_4$ at 300 °C. As the temperature was raised to 750 °C, $La_2(CrO_4)_3$ and $LaCrO_3$ were produced. $La_2(CrO_4)_3$ reacted[96] with alkali chromates affording $M_2La_2(CrO_4)_4,nH_2O$ (M = Li, Na, or NH_4); $MgLa_2(CrO_4)_4,6H_2O$ was also prepared. From the system $M(NO_3)_3-Cs_2CrO_4-H_2O, CsM(CrO_4)_2,H_2O$ was isolated.[97] A series of molybdates $M_2(MoO_4)_3,-nH_2O$ (M = Sc, Y, Tb, Er, or Yb) was obtained[98] from the reaction of $M(NO_3)_3$ with Na_2MoO_4 under neutral conditions, but in the presence of NaOH, $M_2(OH)-(MoO_4),H_2O$ (M = Sc, Y, or Er) was obtained. Treatment of $M(NO_3)_3$ with Q_2MoO_4 in water gave $QM(MoO_4)_2$ (Q = K or Rb; M = Sc, Y, Tb, or Er). Spectroscopic studies of $[ZMo_{12}O_{42}]^{8-}$ (Z = Ce, Th, or U) revealed[99] that they have the same structure in the solid state and solution. Reaction of the cerium(IV) species with bivalent ions(M^{2+}) gave $[CeM_2Mo_{12}O_{42}]^{4-}$, and with Th^{4+} $[Th(CeThMo_{12}O_{42})_3]^{8-}$ was formed.[100] Reduction of the 12-molybdocerate(IV) voltammetrically gave, reversibly in acidic solution, $[CeMo_{12}O_{42}]^{9-}$, containing cerium(III).[101] In $[CeTa_2-Mo_{10}O_{41}]^{8-}$, obtained[102] by reaction of $[CeMo_{12}O_{42}]^{8-}$ with $K_8Ta_6O_{19},16H_2O$, MoO_6 and TaO_6 are arranged about the CeO_{12} icosahedra. Basic lanthanide tungstates $M(OH)(WO_4),nH_2O$ (M = Sm → Lu, Sc, or Y) and normal tungstates $M_2(WO_3)_3,nH_2O$ (M = Sc, Y, Er, Yb, or Tb; n = 8 or 10) have been described,[103] and reaction of the latter with Q_2WO_4 (Q = alkali metal) gave[104] $QM(WO_4)_2,nH_2O$ (n = 1—5). Two series of oxides of the perovskite type have been identified:[105] MWO_6 (M = Y, Pr → Yb) or $MCrWO_6$ (M = Y, Sm → Er) have a $CaTa_2O_6$ site in which the transition-metal ions occupy octahedral sites and M^{3+} lies in tunnels created by distortion of the perovskite structure; and $Ca_{1+x}M_{1-x}CrO_4$ or $Ca_{2-x}M_x$-

[91] L. K. Tolstov, A. A. Ivakin, B. V. Slobodin, and R. N. Pletnev, *Chem. Abs.*, 1974, **81**, 20262.
[92] G. G. Kasimov, E. G. Vovkotrub, I. G. Rozanov, and S. V. Smirnov, *Zhur. neorg. Khim.*, 1973, **18**, 1997; G. V. Bazuev and G. P. Shveikin, *ibid.*, p. 1930.
[93] A. M. Sych, L. A. Eremenko, L. A. Zastavker, and M. M. Nekrasov, *Izvest. Akad. Nauk S.S.S.R., neorg. Materialy*, 1973, **10**, 496.
[94] A. M. Sych and V. G. Klenus, *Izvest. Akad. Nauk S.S.S.R., neorg. Materialy*, 1974, **10**, 634.
[95] Ya. S. Rubinchik, T. P. Veremei, M. M. Pavlyuchenko, and I. A. Mochal'nik, *Doklady Akad. Nauk Belorussk. S.S.R.*, 1973, **17**, 830.
[96] I. J. Martin. *Ion (Madrid)*, 1973, **33**, 135.
[97] T. I. Kuzina, I. V. Shakhno, A. N. Krachak, and V. E. Plyushchev, *Zhur. neorg. Khim.*, 1973, **18**, 2727.
[98] A. M. Golub, A. P. Perepelitsa, V. I. Maksin, and A. M. Kalinichenko, *Chem. Abs.*, 1973, **79**, 60979.
[99] A. M. Golub, A. P. Perepelitsa, V. I. Maksin, A. A. Govorov, and A. M. Kalinichenko, *Chem. Abs.*, 1973, **79**, 60961.
[100] V. I. Spitsyn, E. A. Torchenkova, L. P. Kazanskii, and P. Bajdala, *Z. Chem.*, 1974, **14**, 1.
[101] L. McKean and M. T. Pope, *Inorg. Chem.*, 1974, **13**, 747.
[102] E. A. Torchenkova, Nguyen Dieu, L. P. Kazanskii, and V. I. Spitsyn, *Izvest. Akad. Nauk S.S.S.R., Ser. khim.*, 1973, 734.
[103] A. M. Golub, V. I. Maskin, V. N. Solomakha, and K. Aganiyazov, *Zhur. neorg. Khim.*, 1973, **18**, 3203; A. M. Golub, V. I. Maksin, A. P. Perepelitsa, and A. A. Govorov, *Chem. Abs.*, 1973, **79**, 60951.
[104] E. S. Razgon and V. E. Plyushchev, *Chem. Abs.*, 1973, **79**, 60975; V. E. Plyushchev and E. S. Razgon, *Zhur. neorg. Khim.*, 1974, **19**, 85; A. M. Golub, V. I. Maksin, A. P. Perepelitsa, V. N. Solomakha, and K. Aganiyazov, *Chem. Abs.*, 1973, **79**, 60982.
[105] J. Fava, A. Daoudi, R. Salmon, H. Le Flem, and P. Hagenmuller, *Chem. Abs.*, 1974, **81**, 32645.
[106] J. N. Patil and D. N. Sen, *J. Indian Chem. Soc.*, 1973, **50**, 413.

MnO_4 (M = Pr → Gd) have the K_2NiF_4-type structure. Included in the second group is $EuMAlO_4$ (M = La → Gd), in which the Eu^{2+} and M^{3+} ions coexist in identical sites.

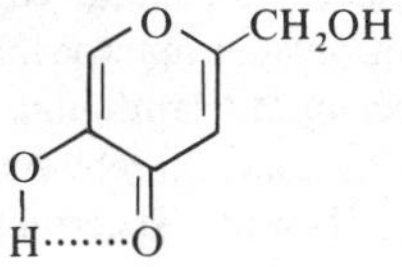

(3)

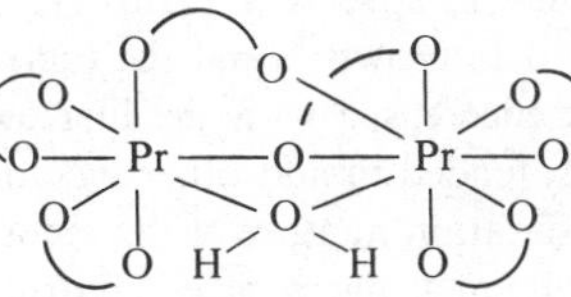

(4) O O ≡ β-diketone

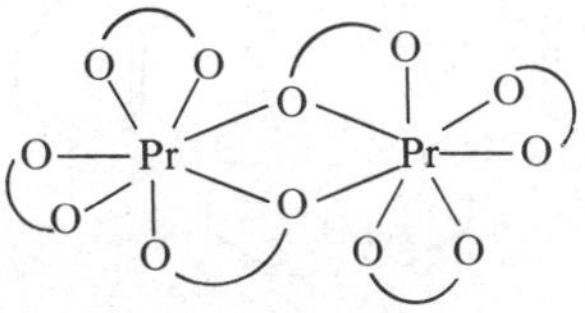

(5)

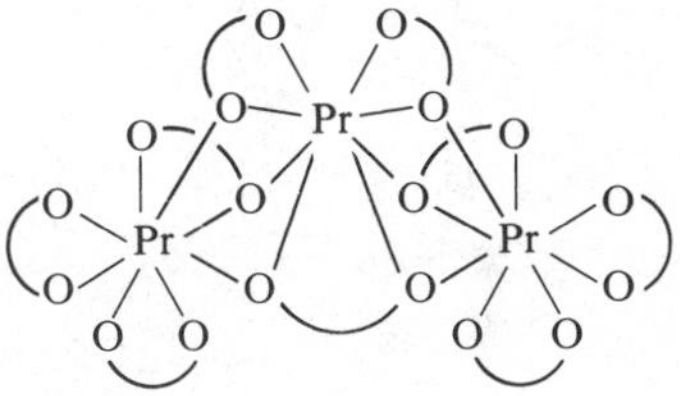

(6)

2-Hydroxypropane-1,3-diamine-$NNN'N'$-tetra-acetic acid or edta displaced[106] 2-hydroxy-1-naphthaldehyde from its lanthanide (La → Sm) complexes. In the 3,4-dihydroxybenzaldehyde (H_2L) complexes of Pr, Nd, Sm, and Gd [$M(HL)_3$], the ligand is apparently attached[107] to the metal *via* an O atom of the CHO group and a deprotonated *meta*-hydroxy-group. Kojic acid (3) appeared[108] to be bidentate towards lanthanide ions in $M(C_6H_5O_4)_3(H_2O)_2$ (M = La → Sm, Gd, Dy, Ho, or Y). The tris-fod [fod = $^-OC(C_3F_7)CHC(O)Bu^t$] complex of Pr^{III} undergoes[109] self-association in non-polar solvents. Thus, in the presence of traces of water, the equilibrium $Pr(fod)_3 + Pr(fod)_3(OH_2) \rightleftharpoons Pr_2(fod)_6(OH_2)$ (4) appears to be to the right, while under anhydrous conditions, dimers and trimers are produced, *viz.* $2Pr(fod)_3 \rightleftharpoons Pr_2(fod)_6(5) \rightleftharpoons Pr_3(fod)_9$ (6). In (4) and (6) the metal atoms are eight-co-ordinate, whereas in (5) they are seven-co-ordinate. The factors influencing the degree of self-association are: (i) the state of hydration which inhibits trimerization; (ii) restriction of association in polar solvents because these tend to form adducts themselves; (iii) greater association with the larger lanthanide ions than with the smaller ones under identical conditions; and (iv) a greater degree of association between fod chelates in non-polar solvents than between dpm chelates [dpm = $^-OC(Bu^t)CHC(O)Bu^t$]. $Eu(fod)_3$ formed adducts with $Me_2NC(O)OMe$ (L), *viz.* $Eu(fod)_3(OH_2)L$ and $Eu(fod)_3L_2$. Mössbauer spectral studies of [$Eu(dbm)_4$]$^-$ and $Eu(dbm)_3(o\text{-phen})$ [dbm = $^-OC(Ph)CHC(O)Ph$] provided[110] no evidence for the

[107] S. F. M. Ali, M. P. Gawande, and V. R. Rao, *Current Sci.*, 1973, **42**, 817.
[108] R. C. Agarwal, S. P. Gupta, and D. K. Rastogi, *J. Inorg. Nuclear Chem.*, 1974, **36**, 208; R. C. Agarwal and S. P. Gupta, *Current Sci.*, 1974, **43**, 263.
[109] A. H. Bruder, S. R. Tanny, H. A. Rockefeller, and C. S. Springer, jun., *Inorg. Chem.*, 1974, **13**, 880.
[110] M. F. Taragin and J. C. Eisenstein, *J. Inorg. Nuclear Chem.*, 1973, **35**, 3815.

participation of $4f$ electrons in the bonding with the donor atoms. Benzo-15-crown-5 polyether (7) forms[111] the complexes $M(NO_3)_3, C_{14}H_{20}O_5$ (M = La $\rightarrow$ Sm) and $M(NO_3)_3, C_{14}H_{20}O_5, 3H_2O$,(acetone) (M = Sm $\rightarrow$ Lu), while with dibenzo-18-crown-6 polyether (8) only $M(NO_3)_3, C_{20}H_{24}O_6$ (M = La $\rightarrow$ Nd) could be obtained as a stoicheiometric species. Dibenzo-18-crown-6 ether formed less stable complexes than benzo-15-crown-5, and the thermal stabilities of unsolvated macrocyclic ether complexes decreased with the increasing atomic number of the lanthanide. These complexes tended to add other ligands such as water or acetone, thereby increasing the co-ordination number of the metal ion. The isotropic ^1H n.m.r. spectral shifts of the co-ordinated macrocycles and the water protons arising from the metal ion in

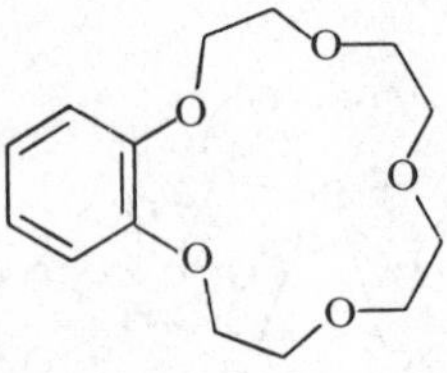

(7) 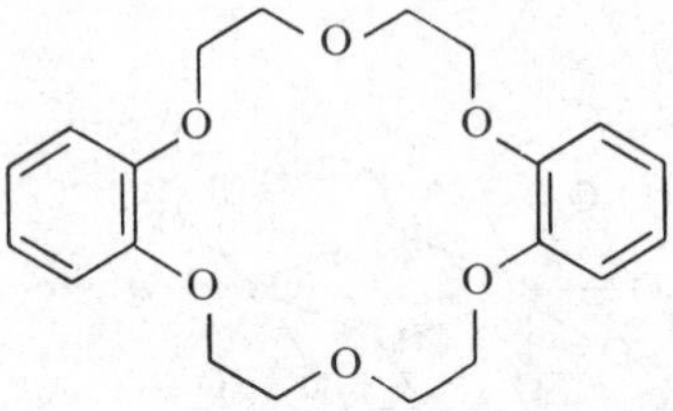   (8)

$M(NO_3)_3, C_{14}H_{20}O_5, 3H_2O$,(acetone) (M = Sm, Eu, Tm, or Yb) correlated well with previous observations made using $M(dpm)_3$ shift reagents. Partial separation of mixtures of Er and Pr nitrates was achieved using a chromatography column packed with (8). In the 2,5-piperazinedione (L) complexes, $ML_4(ClO_4)_3$ (M = La $\rightarrow$ Er, Y) the ligand is bonded[112] *via* O atoms, and ClO_4^- is also co-ordinated. Acetamide[113] formed a 4:1 adduct with $GdCl_3, 6H_2O$, and similar complexes were obtained[114] with NN-dimethylacetoacetamide and dipropionamide, *viz.* $[ML_4]$-$[ClO_4]_3$ (M = La $\rightarrow$ Lu. Y). In the last two complexes the metal atom is probably eight-co-ordinate with respect to the O-donor atoms of L. The $NNN'N'$-tetramethylmalonamide (tmma) complexes $[M(tmma)_2Cl]Cl_2$ (M = La $\rightarrow$ Lu, Y, except Pm and Tm), $[M(tmma)_3(NCS)_2][NCS]$ (M = La $\rightarrow$ Yb, Y, except Pm), and Lu-$(tmma)_2(NCS)_3$ have been reported,[115] as have the urea (L)[116] $Gd_2L_8(SO_4)_3$, Dy_2L_7-$(SO_4)_3$, and $Yb_2L_6(SO_4)_3$ and acetylurea (au)[117] species $M(au)_4(ClO_4)_3$ (M = La $\rightarrow$ Yb, Y).

Picolinate (pic$^-$) and quinolinate (quin$^-$) complexes $M(pic)_3(H_2O)$, $M(pic)_3$, and $M(quin)_3$ (M = La, Pr $\rightarrow$ Eu, Dy, or Ho) have been described;[118] analogous isoni-

[111] R. B. King and P. R. Heckley, *J. Amer. Chem. Soc.*, 1974, **96**, 3118.
[112] J. C. Prado and G. Vicentini, *Inorg. Nuclear Chem. Letters*, 1973, **9**, 693.
[113] A. A. Dilebaeva, M. K. Kydynov, and K. S. Sulaimankulov, *Chem. Abs.*, 1974, **80**, 9953.
[114] M. Perrier and G. Vicentini, *J. Inorg. Nuclear Chem.*, 1974, **36**, 1187; O. L. Alves, Y. Gushikem, and C. Airoldi, *J. Inorg. Nuclear Chem.*, 1974, **36**, 1079.
[115] G. Vicentini, M. Perrier, L. B. Zinner, and M. I. Amin, *J. Inorg. Nuclear Chem.*, 1974, **36**, 771.
[116] A. A. Sopueva, K. S. Sulaimankulov, and L. A. Tokmergenova, *Zhur. neorg. Khim.*, 1974, **19**, 1095.
[117] C. Airoldi and Y. Gushikem, *J. Inorg. Nuclear Chem.*, 1974, **36**, 1892.
[118] M. D. Zhuraleva, R. A. Chupakhina, and V. V. Serebrennikov, *Zhur. obshchei Khim.*, 1974, **44**, 621; N. B. Kelk, N. J. Hornung, and W. G. Bos, *J. Inorg. Nuclear Chem.*, 1974, **36**, 1521.

cotinates have also been reported.[119] Tryptophan (L) is bound to the complexes ML_3Cl_3,nH_2O (M = La, Sm, Gd, Dy, or Er) *via* the deprotonated carboxylate group and the NH_2 function of the amino-acid.[120] In the *o*-hydroxybenzoylhydrazide (HL) complexes $ML_3,3H_2O$, the ligand is co-ordinated[121] *via* both O atoms and the primary N atom. Lanthanide tris-salicylaldehydato (L^-) complexes formed[122] adducts with *o*-hydroxybenzoylhydrazide (Q), *viz*. $LaL_3Q_3,4H_2O$, LaL_3Q_6, and $YL_3Q_5,2H_2O$ (Q is bonded *via* the carbonyl O atom). Several ε-caprolactam (cap) complexes containing polybromide ions, $La(cap)_6Br_3(Br_2)_n$ (n = 1, 3, or 5) and $Sm(cap)_{12}Br_3$-$(Br_2)_n,xH_2O$ (n = 3, 6, or 9, x = 6 or 12) have been described.[123] Compounds containing α-benzoin oxime (H_2L), *viz*. $M_2L_3,nROH$ (M = La → Sm, R = Me, n = 1, 5, or 6; M = Sm, Eu, Tb, Dy, or Er, R = Et, n = 5 or 6) and Y_2L_3 have been reported,[124] and DMF and other amine *N*-oxide adducts (*e.g.* derived from quinoline, pyridine, or Et_3N), $[ML_n][Cr(NCS)_6]$ (n = 6, 7, or 8) have been characterized.[125] Pyrazine *N*-oxide (L), $[ML_8][ClO_4]_3$ (M = Pr, Eu, Er, or Y) and $[LaL_7(OH_2)_2]$ $[ClO_4]$[126] and *o*-hydroxyquinoline *N*-oxide (HL) complexes[127] $ML_2(NO_3)_3$,-$2H_2O$ (M = La, Nd, Gd, Dy, Er, or Y) have also been reported.

The lanthanide tris-5-chlorosalicylate complexes (Ce → Tb) formed[128] 1:2 adducts with Ph_3EO (E = P or As). *NN*-Dimethyldiphenylphosphinamide (L) formed[129] adducts with lanthanide thiocyanates, $ML_n(NCS)_4$ (n = 4, M = La → Gd; n = 3, M = Tb → Lu, Y), in which the phosphorus ligand is O-bonded, and ethylenedi-aminebis(isopropylphosphonic) acid (H_4L) reacted[130] with $M(NO_3)_3$, giving $[ML]^-$ (M = La → Sm).

The stability constants of lanthanide (Pr → Dy) complexes of 2-nitroso-1-naphthol-4,6- and -4,7-disulphonic acids have been measured[131] and seven-co-ordinate tetramethylene sulphoxide (TMSO) complexes $M(TMSO)_4(NCS)_3$ (M = Sm → Lu, Y) have been reported.[132] Various thioxan oxide (9) (TSO) complexes have been described:[133] $MX_3,nTSO$; M = La → Nd, n = 4 (X = NO_3); n = 6 (X = Cl), n = 9 (X = ClO_4); M = Sm → Lu, Y, n = 3 (X = NO_3), n = 3.5—4.5 (X = Cl), n = 7 or 8 (X = ClO_4). The nitrates are non-electrolytes in $MeNO_2$, the chlorides are 1:1 electrolytes in MeOH, and the perchlorates behave as 1:2 electrolytes; it was

(9)

[119] R. A. Chupakhina, G. Ya. Chuchelina, M. D. Zhuravleva, Z. Ya. Grankina, E. K. Kolyago, and V. V. Serebrennikov, *Chem. Abs.*, 1974, **80**, 90445.
[120] B. S. Manhas and V. K. Bhatia, *Indian J. Chem.*, 1973, **11**, 1068.
[121] J. Mach, *Monatsh.*, 1973, **104**, 564.
[122] J. Mach, *Monatsh.*, 1973, **104**, 1539.
[123] Yu. G. Eremin, T. I. Martyshova, T. A. Andreeva, and V. V. Smirnov, *Zhur. neorg. Khim.*, 1974, **19**, 1692; T. A. Andreeva and T. I. Martyshova, *Chem. Abs.*, 1974, **81**, 32688.
[124] R. Pastorek, *Chem. Abs.*, 1974, **81**, 44790.
[125] E. S. Pavlenko, V. N. Kumok, and V. V. Serebrennikov, *Chem. Abs.*, 1974, **80**, 55440.
[126] G. Vicentini and L. B. Zinner, *Inorg. Nuclear Chem. Letters*, 1974, **10**, 629.
[127] T. M. Shevchenko, E. S. Pavlenko, and V. V. Serebrennikov, *Chem. Abs.*, 1974, **81**, 32689.
[128] S. Plostinaru and P. Spacu, *Rev. Roumaine Chim.*, 1973, **18**, 2051; 1974, **19**, 567.
[129] G. Vicentini, L. B. Zinner, and B. L. Rothschild, *Inorg. Chim. Acta*, 1974, **9**, 213.
[130] V. P. Khramov and A. B. Ivanov, *Chem. Abs.*, 1973, **79**, 61054.
[131] H. Saarinen, *Acta Chem. Scand.* (B), 1974, **28**, 589.
[132] L. B. Zinner and G. Vicentini, *Anales Acad. Brasil, Cienc.*, 1973, **45**, 223.
[133] G. Vicentini and L. C. Garla, *J. Inorg. Nuclear Chem.*, 1973, **35**, 3973; G. Vicentini and M. Perrier, *ibid.*, 1974, **36**, 77.

suggested that when $n = 3.5$—4.5, the TSO ligands are shared between two metal atoms. Lanthanide complexes of o-(2-thiazolylazo)phenols $[ML]^{2+}$ (M = La → Yb, except Tb, Tm) have been described.[134]

A series of carboxylates $M(O_2CR)_3,nH_2O$ (M = Pr, Nd, or Sm, R = alkyl or chloromethyl; M = La → Sm, R = Ph, $n = 2$) has been prepared,[135] and the benzoates formed a 1:1 adduct with $(BuO)_3PO$. A conformational study has been made[136] of indol-3-ylacetate complexes of the lanthanide. Those complexes of Pr, Nd, Sm, Eu, and Gd^{III} are axially symmetric, with bidentate carboxylate, but in the Tm^{III} species, each ligand is unidentate with respect to the metal. In a thermodynamic study of lanthanide complexes of $[OCOCH_2XCH_2CH_2XCH_2CO_2]^{2-}$ (X = S, O, or NH) (A^{2-}) it was established[137] that when X = S, the main species formed are $[MA]^+$, $[MA_2]^-$, and $[MHA]^{2+}$. When X = O, mainly $[MA]^+$ and $[MA_2]^-$ are obtained, with small amounts of $[MHA]^{2+}$ at low pH, while when A^{2-} concentrations are high, $[MA_3]^{3-}$ is formed. When X = NH, $[MA]^+$, $[MA_2]^-$, and $[MH_2A]^{3+}$ are produced, the last being significant only at ca. pH 3.

NN'-Ethylenedianthranilic acid (H_2L) forms[138] bimetallic complexes M_2L_3,xH_2O (M = La → Yb, except Tm and Lu; $x = 1 - 2$). The ligands are quadridentate, one of them functioning as a bridge between the two metals. Paludrine hydrochloride (palHCl) formed complexes of the type $M(pal)_nCl_3$ (M = Nd → Yb, $n = 1, 2,$ or 3), $(MCl_3)_3(pal)$ (M = Dy or Ho), $Sm(pal)_4Q_3$, and $Gd(pal)_5Cl_3$; similar thiocyanates were also prepared. Iminodiacetic acid (H_2ida) formed[139] 1:1, 2:1, and 3:1 complexes with Sm^{3+}, and $QM(ida)_2,nH_2O$ (Q = alkali metals, M = Nd or Yb, $n = 4 → 5$) has been isolated. The acid has been used in conjunction with ion-exchange resins to separate the lanthanides from the transplutonium elements. The nitrilotriacetate complexes $K_3M(nta)_2,6H_2O$ (M = La → Lu) and $M(nta),3H_2O$ (M = Pr → Eu) have been reported.[140] In $Nd(nta),3H_2O$, the polyacid is bonded via its N atoms and three O atoms from each of the carboxylato-groups. Since the metal is co-ordinatively unsaturated, bonds are also formed between two neighbouring nta groups and two water molecules, and so a layered structure is built up. Neodymium oxide reacted with hexamethylenediaminetetra-acetic acid (H_4L), giving[141] $Nd_2(H_2L)_3,5H_2O$. $NdCl_3$ reacted with K_4L affording $KNdL,nH_2O$, and $NdHL,nH_2O$, $Nd_2L(OH)_3,5.5-H_2O$, and K_5NdL_2,nH_2O were also prepared. The oxalates $Gd_2(O_2C_2O_2),10H_2O$ and $Q[Gd(O_2C_2O_2)_2],nH_2O$ (Q = Rb or Cs) have been described[142] and stepwise formation constants for gadolinium(III) and dysprosium(III) complexes of aspartic, glutaric, NN'-ethylenediamine-disuccinic and -bis(α-glutaric) acids have been measured.[143] In the presence of amino-acids, lanthanide(III) ions form[144] polynuclear complexes at ca. pH 6.

[134] F. Kai and Y. Sadakane, *J. Inorg. Nuclear Chem.*, 1974, **36**, 1404.
[135] R. C. Paul, G. Singh, and J. S. Ghotra, *Indian J. Chem.*, 1973, **11**, 294; T. Chen, F. L. Chiu, L. Chen., L. P. Wang, and S. N. Li, *Chem. Abs.*, 1974, **80**, 43 532.
[136] B. A. Levine, J. M. Thornton, and R. J. P. Williams, *J.C.S. Chem. Comm.*, 1974, 669.
[137] I. Grenthe and G. Gardhammer, *Acta Chem. Scand.* (A), 1974, **28**, 125.
[138] P. Spacu and E. Ivan, *Rev. Roumaine Chim.*, 1973, **18**, 589; I. Albescu, *ibid.*, p. 599.
[139] A. D. Site and F. A. Kappelmann, *Inorg. Nuclear Chem. Letters*, 1974, **10**, 81; R. I. Badalova, N. D. Mitrofanova, and L. I. Martynenko, *Chem. Abs.*, 1974, **81**, 71 990.
[140] K. L. Belyaeva, M. F. Porai-Koshits, and T. I. Malinovskii, *Chem. Abs.*, 1974, **80**, 103 338.
[141] N. P. Kuz'mina, L. I. Martynenko, and V. I. Spitsyn, *Izvest. Akad. Nauk S.S.S.R.*, *Ser. khim.*, 1974, 523.
[142] E. G. Davitashvili, M. E. Modebadze, and N. G. Sheliya, *Chem. Abs.*, 1974, **80**, 103 355.
[143] O. P. Sunar, S. Tak, and C. P. Trivedi, *J. Inorg. Nuclear Chem.*, 1974, **36**, 1163.
[144] R. Prados, L. G. Stadtherr, H. Donato, and R. B. Martin, *J. Inorg. Nuclear Chem.*, 1974, **76**, 689.

The heats of atomization, dissociation energies, and heats of sublimation of EuS, MSe (M = La → Eu), SmTe, and EuTe have been reported.[145] While the dissociation energies and heats of atomization changed regularly with atomic number, the heats of sublimation were insensitive to the number of f-electrons. Sm and Yb trichlorides reacted with H_2Se giving [146] M_2Se_3, which were reduced by gaseous M to give MSe. Treatment of $M_2(MoO_4)_3$ (M = La → Sm, Y) with H_2S at 350—700 °C afforded[147] MoS_2 and M_2S_3. Furfuryl mercaptan (RSH) formed[148] $M(SR)_3$ with cerium(III) and lanthanum(III), and the thiourea (tu) complexes $4M(OAc)_3,5tu,xH_2O$ and $M(OAc)_3,$-$(tu),xH_2O$ (M = Tb → La, Y, x = 1, 2, 4, 6, or 8) have been reported.[149] $La_4Ge_3S_{12}$ contains[150] LaS_6 trigonal prisms which are connected together by GeS_4 tetrahedra. M_2S_3 reacted[151] with SnS_2 giving M_2SnS_5 (M = La, Pr, or Nd). MCl_3,nH_2O afforded,[152] with an excess of $NaS_2P(OEt)_2$, $[M\{S_2(OEt)_2\}_4]^-$ (M = La → Lu, except Y), but in a stoicheiometric reaction, pure $M\{S_2P(OEt)_2\}_3$ could not be isolated; when treated with Ph_3PO, $M\{S_2P(OEt)_2\}_3(Ph_3PO)_x$ (x = 2 or 3) was extracted. In $NdYbS_3$, the Yb has[153] a non-regular octahedral environment while the Nd atom has a bicapped (triangular faces) prismatic geometry. $PrCl_3$ and $GdCl_3$ reacted with Bi_2S_3 at 650—800 °C giving $MBiS_3$.[154]

Cerium and terbium trifluorides reacted[155] with XeF_2 giving MF_4, but the other lanthanide fluorides did not behave similarly. Addition of aqueous HF to $Nd(NO_3)_3$ afforded[156] NdF_3,xH_2O (x = 0.5 and 9). Reaction of La_2O_3 with Na_3AlF_6 gave[157] $NaLaF_4$ at 1050 °C while treatment of MF_3 (M = Tb → Lu, Y) with RbF (1:3 mole ratio) at 550—600 °C under argon produced Rb_3MF_6.[158] From the RbF–YbF_3 and TlF–YbF_3 systems, Q_3YbF_6 (Q = Rb or Tl), QYb_2F_7, and QYb_3F_{10} were obtained,[159] and Rb_2YbF_5 decomposed in the solid state into α-Rb_3YbF_6 and α-$RbYb_2F_7$. In the presence of QCl (Q = Na, K, Rb, or Cs), Tb_2O_7 reacted with fluorine giving[160] $QTbF_5$, $QTbF_6$, Q_3TbF_7, QTb_2F_9, $Cs_3Tb_2F_{11}$, $Na_5Tb_2F_{13}$, $Rb_2Tb_3F_{17}$, QTb_6F_{25}, and $Q_7Tb_6F_{31}$. Carbonyl chloride can effect the conversion[161] of MOCl into MCl_3 (M = Ce → Gd). The hydrothermal synthesis of $M(OH)_2Cl$ (M = La → Ho) from M_2O_3, NH_4Cl, and water has been reported.[162] Pure lanthanide tri-iodides can be prepared[163] by dehydration of MI_3,nH_2O, by conversion of halides or oxides

145 S.-I. Nagai, M. Shinmei, and T. Yokokawa, *J. Inorg. Nuclear Chem.*, 1974, **36**, 1904.
146 T. Petzel, *Inorg. Nuclear Chem. Letters*, 1974, **10**, 119.
147 V. U. Yampol'skaya and V. V. Serebrennikov, *Chem. Abs.*, 1974, **80**, 140751.
148 R. S. Saxena and S. S. Sheelwant, *J. Inorg. Nuclear Chem.*, 1973, **35**, 3963.
149 N. N. Sakharova, Yu. G. Sakharova, and G. M. Borisova, *Zhur. neorg. Khim.*, 1973, **18**, 1212.
150 A. Mazurier and J. Etienne, *Acta Cryst.*, 1974, **B30**, 759.
151 G. G. Mel'chenko and V. V. Serebrennikov, *Chem. Abs.*, 1974, **80**, 55405.
152 A. A. Pinkerton, *Inorg. Nuclear Chem. Letters*, 1974, **10**, 495.
153 D. Carré and P. Laruelle, *Acta Cryst.*, 1974, **B30**, 952.
154 P. G. Rustamov, V. B. Cherstvova, and G. G. Guseinov, *Chem. Abs.*, 1974, **81**, 48771.
155 V. I. Spitsyn, Yu. M. Kiselev, and L. I. Martynenko, *Zhur. neorg. Khim.*, 1974, **19**, 1152.
156 Yu. M. Kiselev, V. I. Spitsyn, and L. I. Martynenko, *Izvest. Akad. Nauk S.S.S.R., Ser. khim.*, 1973, 729.
157 S. F. Belov, K. I. Petrov, A. F. Gladneva, and Yu. B. Kirillov, *Izvest. Akad. Nauk S.S.S.R., neorg. Materialy*, 1974, **10**, 276.
158 I. B. Shaimuradov, L. P. Reshetnikova, V. A. Efremov, A. V. Novoselova, and A. I. Grigor'ev, *Zhur. neorg. Khim.*, 1974, **19**, 366.
159 A. Vedrine, R. Boutonnet, and J. C. Cousseins, *Compt. rend.*, 1973, **177**, C, 1129.
160 D. Avignant and J. C. Cousseins, *Compt. rend.*, 1973, **278**, C, 613.
161 A. N. Ketov and E. V. Burmistrova, *Chem. Abs.*, 1974, **80**, 152180.
162 L. N. Dem'yanets, V. I. Bukin, E. N. Emel'yanova, and V. I. Ivanov, *Kristallografiya*, 1973, **18**, 1283.
163 F. J. Arnaiz, J. G. Ribas, and L. A. Gomez, *Quim. Anal.*, 1974, **28**, 86; N. Haeberle, *Chem. Abs.*, 1974, **81**, 71961.

into iodides, or by direct synthesis from the elements. The best method appeared to involve direct elemental reaction. Reaction of $LaI_3, 9H_2O$ with NH_4I under nitrogen gave the anhydrous iodide, but dehydration at 373 °C afforded LaOI. While Eu reacted with I_2 to give only EuI_2, Sm and Yb gave both di- and tri-iodides; all other metals gave MI_3.

The elements Y, Pr, Sm, Dy, Ho, Er, and Yb reacted[164] with $Mn(CO)_5Br$ in THF giving red solids $[Mn(CO)_5]_xMBr_y$ which are sensitive to air and moisture; x and y are integers such that $x + y = 2$ or 3 depending on the lanthanide. Similar red solids were obtained from Yb, $Mn_2(CO)_{10}$, and $(CH_2Br)_2$. Solutions containing these species are conducting and there was no evidence for M–C–O–lanthanide interaction. $Mn(CO)_5Br$, Ho, and Na(acac) reacted in ether to give $[Mn(CO)_5]_2Ho(acac),2OEt_2$.

Lanthanide Shift Reagents.—The solvation numbers and kinetics of substrate exchange in lanthanide shift-reagent systems have been studied.[165] At low temperatures, substrate exchange was slow on the ^1H n.m.r. time-scale, thereby enabling solvation numbers, which were often solvent dependent, to be determined.

Methods have been devised[166] for the separation of pseudo-contact and contact contributions to the shifts of ^{31}P caused by $[M(edta)]^-$ (M = Pr → Y) and other complexes bonded to adenosine-5′-monophosphate and cytidine-5′-monophosphate at various pH values.

From studies of the ^1H n.m.r. spectra of $Eu(dpm)_3(py)_2$ and its 3-picolinyl analogue, it was deduced[167] for this type of complex that the spectral shifts cannot be explained on the basis of a single-term equation which assumes that the principal magnetic axis is coincident with a Eu–donor atom bond. A more complete interpretation was offered using a two-term equation. The barrier to rotation about the Eu—N bond in the 3-picolinyl complex is at least 7 kcal mol^{-1}, indicating[168] that there is no free rotation in this system and that all conformers are equally populated. However, using as a model a seven- or eight-co-ordinate lanthanide shift reagent complex which, by definition, is fluxional in solution, apparent axial symmetry of the adduct may be assumed.[69] Tests of this hypothesis were made with $M(dpm)_3Q_2$ (Q = 4-picoline, isoquinoline, or cyclohexanol; M = Pr → Yb except Gd), and it is known[170] that the 4-picoline adducts have 'non-axial structures' in the solid state. There was reasonable agreement between theory and measurement, within certain broad limits. Similar assumptions were made when γ-picoline N-oxide, pyridine N-oxide, aniline, p-toluidine, and tropone– and methylenecycloheptadienol–tricarbonyl iron complexes were investigated using $M(fod)_3$, $M(dpm)_3$ (M = Pr, Eu, Er, or Yb), and $Eu\{CF_3-COCHC(O)Bu^t)\}_3$.

From studies of the ^1H n.m.r. spectra of polyol–lanthanide(III) (La, Pr, Nd, Eu, Tb, Yb) complexes it has been established[172] that the contact interaction could be de-

[164] A. E. Craese and P. Legzdins, *J.C.S. Chem. Comm.*, 1973, 775.
[165] D. F. Evans and M. Wyatt, *J.C.S. Dalton*, 1974, 765.
[166] C. M. Dobson, R. J. P. Williams, and A. V. Xavier, *J.C.S. Dalton*, 1973, 2662; 1974, 1762; C. D. Barry, C. M. Dobson, R. J. P. Williams, and A. V. Xavier, *ibid.*, 1974, 1765.
[167] R. E. Cramer, R. Dubois, and K. Seff, *J. Amer. Chem. Soc.*, 1974, **96**, 4125.
[168] R. E. Cramer and R. Dubois, *J.C.S. Chem. Comm.*, 1973, 936.
[169] W. de W. Horrocks, *J. Amer. Chem. Soc.*, 1974, **96**, 3023.
[170] W. de W. Horrocks and J. P. Sipe, *Science*, 1972, **177**, 994.
[171] B. F. G. Johnson, J. Lewis, and P. McArdle, *J.C.S. Dalton*, 1974, 1253.
[172] S. J. Angyal, D. Greeves and V. A. Pickles, *J.C.S. Chem. Comm.*, 1974, 589.

tected over five intervening bonds, provided that these bonds formed a planar zig-zag arrangement.

Thiazoles form[173] 1:1 adducts with $Eu(dpm)_3$ and are N-bonded. Steric hindrance is low when the 2- or 4-positions are substituted, while the other position adjacent to the N atom is not.

Chiral complexes have been prepared[174] and tested for their utility in inducing chemical shifts between corresponding resonances of enantiomeric species. The complex (10) was the most effective for resolution of enantiotopic resonances, but while (11) and (12) were less effective, they were more easily synthesized.

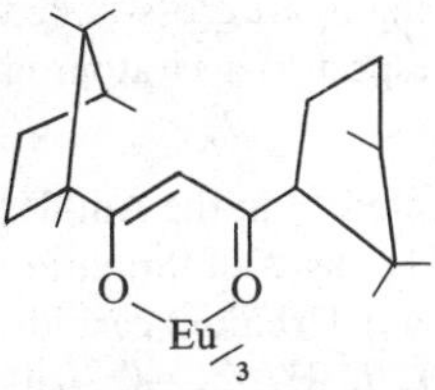

(10) 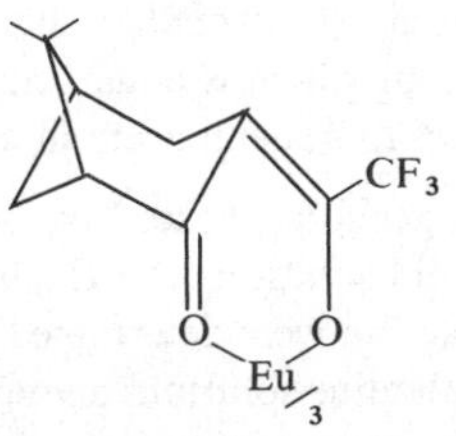   (11) 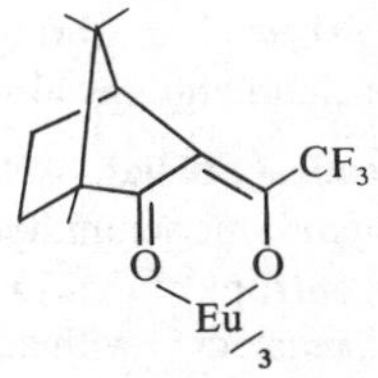   (12)

3 The Actinides, Uranyl, and Related Species

Structural Studies.—The deuterium atoms have been located[175] in $D_2U_3O_{10}$. This oxide contains two crystallographically independent U atoms, one octahedrally co-ordinated by oxide ions, and the other existing in a pentagonal-bipyramidal environment.

Preliminary X-ray diffraction studies of β-$Th(NO_3)_4$,3DMSO have established[176] that the metal atom is eleven-co-ordinate, the nitrates being bidentate and the sulphoxides O-bonded. The co-ordination polyhedron is close to that of a singly capped pentagonal antiprism (D_{5d}). Other DMSO complexes, *viz.* $M(NO_3)_4$,xDMSO ($x = 6$, M = Th, Np, and Pu; $x = 3$, M = U, Np, and Pu) were also reported. The metal atom in $Th(Ph_3PO)_2(NO_3)_4$ is ten-co-ordinate, being described[176a] as an irregular octahedron with apical phosphite ligands, the bidentate nitrates being regarded as occupying a single ligand site (see Table 3, p. 498).

In $[NH_4]_7[Th_2F_{15}]$,H_2O each Th atom shares a small triangular ring of three F atoms and is located in a puckered equatorial ring of six F atoms.[177] Each of the six-membered fluorine atom rings is capped by H_2O or NH_4^+, which are located on the three-fold axis. The structure of β-$[NH_4][UF_5]$ consists[177] of infinite sheets of UF_9 polyhedra. Three edges of the polyhedron are shared with three adjacent polyhedra, and two corners are shared with two additional polyhedra. The remaining corner is unshared, and the F atom at this position forms the cap of a rather square irregular

[173] M. Yu. Kornilov and A. V. Turov, *Ukrain. khim. Zhur.*, 1974, **40**, 214.
[174] M. D. McCreasy, D. W. Lewis, D. L. Wernick, and G. M. Whitesides, *J. Amer. Chem. Soc.*, 1974, **96**, 1038.
[175] J. C. Taylor and P. W. Wilson, *Acta Cryst.*, 1974, **B30**, 151.
[176] P. J. Alvey, K. W. Bagnall, and D. Brown, *J.C.S. Dalton*, 1973, 2326.
[176a] K. M. A. Malik and J. W. Jeffery, *Acta Cryst.*, 1973, **B29**, 2687.
[177] R. A. Penneman, R. R. Ryan, and E. Rosenzweig, *Acta Cryst.*, 1974, **B30**, 1966.
[177a] G. Bruntov, *Acta Cryst.*, 1973, **B29**, 2976.

pentagonal bipyramid of F atoms. This structure is substantially different from that of α-[NH$_4$][UF$_5$]. In Na$_3$BeTh$_{10}$F$_{45}$ there are three independent Th atoms, each surrounded[177a] by nine F atoms in a tricapped trigonal prismatic arrangement. A framework structure is built up by edge- and corner-sharing of the F atoms of the ThF and BeF polyhedra (see Table 3, p. 498).

The structure of UCl$_6$ has been refined[178] by neutron and X-ray powder diffraction, and has confirmed the octahedral geometry.[179] While the actinide tetrahalides usually form square antiprismatic (UF$_4$, ThI$_4$) or dodecahedral (UCl$_4$, ThCl$_4$, PaBr$_4$) co-ordination polyhedra, UBr$_4$ is the only compound of this class so far discovered to have a co-ordination number less than eight.[180] The configuration about each U atom is pentagonal bipyramidal, each metal being bound to seven Br atoms, the pentagons being edge (equatorially) fused to form endless chains. The chains are cross-linked into sheets, *via* a bridging Br atom which is equatorial with respect to a U atom in one chain and apical with respect to a metal atom in another.

Chemical Studies.—Three homogeneous phases have been detected[181] in the Am–H$_2$ system: americium hydrogen solid solutions, AmH$_2$, and AmH$_3$ (heats of formation and entropy of the Am + H$_2$ reaction were determined for AmH$_2$). Uranium carbide, UC, reacted[182] with acidic and alkaline solutions giving H$_2$, CH$_4$, C$_2$H$_6$, C$_2$—C$_4$ paraffins and olefinic hydrocarbons. Recent progress in the organometallic chemistry of uranium(IV) has been reviewed,[183] and the synthesis of Pu(C$_5$H$_5$)$_3$, from PuCl$_3$ or Cs$_2$[PuCl$_6$] and NaC$_5$H$_5$ or Mg(C$_5$H$_5$)$_2$, has been reported.[184] The bis-cyclo-octatetraenyl complex Pa(η-C$_8$H$_8$)$_2$ is isostructural with its Th and U analogues.[185]

Two separate linear relationships for lattice parameters in the composition range UN$_{1.40}$ $\rightarrow$ UN$_{1.76}$ (*i.e.* α-U$_2$N$_3$) as a function of N content have been calculated.[186] One is characteristic of U$_2$N$_3$ and the other of UN$_2$. ThCl$_4$ reacted[187] with NH(SiMe$_3$)$_2$ giving monomeric Th{N(SiMe$_3$)$_2$}$_3$Cl. Treatment of ThCl$_4$(NEt$_3$)$_2$ with phthalocyanine (H$_2$Pc) afforded Th(Pc)$_2$.[188] Aquated thiocyanate complexes, [M(NCS),aq]$^+$ (M = Am, Cm, Bk, or Eu), are predominantly of the outer-sphere type whereas M(NCS)$_3$ are mainly of the inner-sphere type.[189] Reaction of the metal filings with red phosphorus gave[190] UP and U$_3$P$_4$; the latter decomposed to the former on heating.

The monoxides MO (M = Th or U) are unstable with respect to disproportionation into M and MO$_2$.[191] The molar volumes of these and higher actinide monoxides, lie between those of the respective metals and their dioxides, and calculated cationic radii are significantly larger than those in MO$_2$ or MS$_2$. The sesquioxides Cm$_2$O$_3$,

[178] J. C. Taylor and P. W. Wilson, *Acta Cryst.*, 1974, **B30**, 1481.
[179] W. H. Zachariasen, *Acta Cryst.*, 1948, **1**, 285.
[180] J. C. Taylor and P. W. Wilson, *J.C.S. Chem. Comm.*, 1974, 598.
[181] J. W. Roddy, *J. Inorg. Nuclear Chem.*, 1973, **35**, 4141.
[182] M. I. Ermolaev and G. V. Tishchenko, *Chem. Abs.*, 1974, **80**, 55472.
[183] E. Cernia and A. Mazzei, *Inorg. Chim. Acta*, 1974, **10**, 239.
[184] L. R. Crisler and W. G. Eggerman, *J. Inorg. Nuclear Chem.*, 1974, **36**, 1424.
[185] J. Goffart, J. Fuger, D. Brown, and G. Duyckaerts, *Inorg. Nuclear Chem. Letters*, 1974, **10**, 413; D. F. Starks, T. C. Parsons, A. Streitwieser, jun., and N. Edelstein, *Inorg. Chem.* 1974, **13**, 1307.
[186] H. Tagawa and N. Masaki, *J. Inorg. Nuclear Chem.*, 1974, **36**, 1099.
[187] D. C. Bradley, J. S. Ghotra, and F. A. Hart, *Inorg. Nuclear Chem. Letters*, 1974, **10**, 209.
[188] I. S. Kirin and A. B. Kolyadin, *Zhur. neorg. Khim.*, 1973, **18**, 3140.
[189] W. F. Kinard and G. R. Choppin, *J. Inorg. Nuclear Chem.*, 1974, **36**, 1131.
[190] M. Takac and Z. Ban, *Croat. Chem. Acta*, 1973, **45**, 579.
[191] R. J. Ackermann and E. G. Rauh, *J. Inorg. Nuclear Chem.*, 1973, **35**, 3787.

Bk_2O, and Cf_2O_3 have been prepared, and their melting points determined.[192] Addition of water to the system $LiF-BeF_2-ThF_4-UF_4-PaF_5$ caused precipitation[193] of Pa_2O_5, and the point at which UO_2 coprecipitated was also determined. Reduction of $\gamma-U_3O_7$ with hydrogen afforded[194] phases corresponding to $\beta-U_3O_7$, U_4O_9, UO_{2+x} and traces of U_8O_{21+x}, and similar treatment of UO_3 gave U_3O_8 and UO_2.[195] X-Ray photoelectron spectroscopic studies[196] of UO_2, $UO_{2.1}$, $UO_{2.2}$, U_4O_9, U_3O_7, U_3O_8, and $\gamma-UO_3$ (when compared with data obtained from the metal[197]) revealed that the formal oxidation state varied from four to six. The significant chemical shifts observed for U $4f$ and O $1s$ peaks were explained in terms of structural and oxidation-state changes.

Ammonium diuranate decomposed[198] thermally to give $UO_3,2H_2O$, then UO_3 and U_3O_8. Treatment of the technical aqueous waste solutions containing NH_4F and UO_2^{2+} afforded[199] $UO_4,2NH_3,2HF$, which is isostructural with $UO_4,4H_2O$.[200] By heating Tl_2CO_3 with amorphous UO_3, Tl_4UO_5 was formed,[201] and reaction of PbO with U_3O_8 gave[202] $Pb_2U_2O_7$ and $Pb_3U_{11}O_{36}$. Bi_2UO_6 has a distorted cubic structure,[203] and in reactions of Li_2O with UO_3, the species Li_6UO_6, Li_4UO_5, Li_2UO_4, $LiU_{0.83}O_3$, $Li_2U_3O_{10}$, and $Li_2U_6O_{19}$ were formed.[204] Those lithium uranates having a high Li_2O content contained octahedrally co-ordinated uranium(VI).

Reaction of solid ThX_4,nH_2O (X = Cl, $\frac{1}{2}SO_4$, or NO_3) with NH_4OH or NaOH gave[205] $Th(OH)_3X$. The electronic spectrum of Bk^{4+} has been discussed[206] and the spin–orbit coupling diagram for f^7 elements presented. The solution chemistry of nobelium has been investigated[207] by solvent extraction and ion-exchange techniques. From the former techniques, the ionic radius of No^{2+}, the principal oxidation state in aqueous acidic solution, was estimated to be 1.1 Å, whereas from the latter techniques, it was found to be 1.0 Å.

Reaction of Pu metal with 1:1 mixtures of CCl_4 and methanol gave[208] $Pu(O_2CH)_3$. In the presence of HF, this reaction afforded PuF_3, and with oxalic acid, $Pu_2(O_2-C_2O_2)_3,xMeOH$. Treatment of CCl_4-MeOH solutions of Pu^{IV} salts led to formation of PuF_4,H_2O, and, in the presence of NH_4F, $[NH_4]_2[PuF_6]$. Neutralization by ammonia of nitric–acetic acid solutions of Th^{IV} afforded[209] $Th(OH)_2(OAc)_2,nH_2O$ (n = 2 or 3), $Th(OH)_3(OAc),H_2O$, and $[NH_4]_2[Th(OAc)_6]$. The stability constants

[192] R. D. Baybarz, *J. Inorg. Nuclear Chem.*, 1973, **35**, 4149.
[193] O. K. Tallent and L. M. Ferris, *J. Inorg. Nuclear Chem.*, 1974, **36**, 1277.
[194] E. V. Kuz'micheva, N. I. Komarevtseva, and L. M. Kovba, *Radiokhimiya*, 1973, **15**, 614.
[195] A. H. Le Page and A. G. Fane, *J. Inorg. Nuclear Chem.*, 1974, **36**, 87.
[196] G. C. Allen, J. A. Crofts, M. T. Curtis, P. M. Tucker, D. Chadwick, and P. J. Hampson, *J.C.S. Dalton*, 197-, 1296.
[197] G. C. Allen and P. M. Tucker, *J.C.S. Dalton*, 1973, 470.
[198] M. C. Ball, C. R. G. Birkett, D. S. Brown, and M. J. Jaycock, *J. Inorg. Nuclear Chem.*, 1974, **36**, 1527.
[199] H.-G. Bachmann, H. Z. Dokuzoguz, and H. M. Muller, *J. Inorg. Nuclear Chem.*, 1974, **36**, 795.
[200] T. Sato, *Naturwiss.*, 1961, **48**, 668.
[201] A. S. Giridharan, M. R. Udupa, and G. Aravamudan, *Z. anorg. Chem.*, 1974, **407**, 345.
[202] G. P. Polunina, L. M. Kovba, and E. A. Ippolitova, *Radiokhimiya*, 1973, **15**, 684.
[203] C. V. Gurumurthy, *Indian J. Chem.*, 1974, **12**, 212.
[204] J. Hauck, *J. Inorg. Nuclear Chem.*, 1974, **36**, 2291.
[205] V. V. Sakharov, T. I. Danilevich, V. M. Klyuchnikov, G. N. Voronskaya, and S. S. Korovin, *Radiokhimiya*, 1974, **16**, 74.
[206] L. P. Varga, R. D. Baybarz, and M. J. Reisfeld, *J. Inorg. Nuclear Chem.*, 1973, **35**, 4313.
[207] R. J. Silva, W. J. McDowell, O. L. Keller, jun., and J. R. Tarrant, *Inorg. Chem.*, 1974, **13**, 2233.
[208] L. R. Crisler, *J. Inorg. Nuclear Chem.*, 1973, **35**, 4309.
[209] V. S. Shmidt, V. G. Andryushin, K. A. Rybakov, and E. G. Teterin, *Radiokhimiya*, 1974, **16**, 391.

of Th^{IV} complexes of α-, β-, and γ-hydroxymonocarboxylic acids have been determined;[210] only the α-acids form chelated complexes. In acidic conditions, Pa^{IV} exists as either PaO^{2+} or $Pa(OH)_2^{2+}$, and as such reacts[211] with edta (H_4L) to give $[PaOL]^{2-}$ or $[Pa(OH)_2L]^{2-}$ and possibly also $[PaO(HL)]^-$ or $[Pa(OH)_2(HL)]^-$. In comparison with UO_2^{2+} and PuO^{2+}, complexing of Pa^{IV} by edta is strong. Uranium(IV) oxalate reacts with diethylenetriaminepenta-acetic acid (H_5L) giving $[UL_2]^{6-}$. Each ligand is quadridentate with co-ordination of two N atoms and two O atoms of carboxygroups; the co-ordination polyhedron is square antiprismatic.[212] The formation constants of the citrate complexes, $[M(Hcit)(cit)]^{2-}$ and $[M(cit)_2]^{3-}$ (M = Am, Cm, Cf, Es, and Fm) have been determined,[213] and 1:1 and 1:2 arsenazo M complexes of Np^{IV} have been reported.[214] The tetraoxalates $Q_4[U(O_2C_2O_2)],nH_2O$ (Q = NH_4, K, or Cs) have been prepared[215] and $[U(O_2C_2O_2)_4]^{4-}$ and $[U(O_2C_2O_2)_2L_2]^{2-}$ (H_2L = tartaric acid) have been precipitated using $[Pt(NH_3)_6]^{4+}$ and $[Pt(NH_3)_4]^{2+}$, respectively.

The Na^+ and $[C(NH_2)_3]^+$ salts of $[Th(CO_3)_3]^{2-}$, $[M(CO_3)_4]^{4-}$ (M = Th or Ce), and $[M(CO_3)_5]^{6-}$ (M = Th or Ce) have been isolated.[216] The dialkylamides $M(NR_2)_4$ (M = Th or U, R = Me or Et) reacted[217] with CX_2 (X = O, S, or Se) giving $M(X_2$-$CNR_2)_4$; COS afforded $M(OCSNR_2)_4$. Uranium(IV) β-diketonates $U\{R^1COCHC$-$(O)R^2\}_4$ (R^1 = CF_3, R^2 = Ph; R^1 = C_3F_7, R^2 = Bu^t; R^1 = CF_3, R^2 = Me) formed[218] adducts with pyridine (1:1), amides, sulphoxides, alcohols, phosphates, phosphine oxides, phosphites, and PBu_3. The proton magnetic resonances of the added ligands were shifted but the shifts caused by $U\{CF_3COCHC(O)Ph\}_4$ were of opposite sign to those caused by $Eu(dpm)_3$.

Series of phosphine and arsine oxides of Th^{IV}, U^{IV}, UO_2^{2+}, and Np^{IV} have been described.[219] These included the nitrates $M(NO_3)_4,xR_3PO$ (R = Me, Pr^n, Bu^n, or Ph), $Th(NO_3)_4,2.5ompa$ (ompa = octamethylpyrophosphoramide), $M(NO_3)_4,$-$1.5ompa$ (M = Th, U, or Np), $UO_2(NO_3)_2,2R_3EO$ (E = P, R = Pr or Ph; E = As, R = Ph), $U(NO_3)_4,4HMPA$ (HMPA = hexamethylphosphoramide) and $[U(HMPA)_4$-$(NO_3)_3][BPh_4]$. Similar perchlorate complexes were isolated. The alkylated phosphate complexes $M\{O_2P(O)(OBu^n)\}_2$ (M = Th or Np), $M\{O_2P(OBu^n)_2\}_4$, $UO_2(O_2PEt_2)_2$, and $UO_2(NO_3)(O_2PBu^n_2)(OPBu^n_3)$ have been prepared.[220]

The stability constants of sulphate complexes of quadrivalent actinides increase in the order $Th^{4+} < U^{4+} > Np^{4+} < Pu^{4+}$, thereby following a single electrostatic

[210] L. Magon, A. Bismondo, L. Maresca, G. Tomat, and R. Portanova, *J. Inorg. Nuclear Chem.*, 1973, **35**, 4237.

[211] R. Lundqvist and J. E. Anderson, *Acta Chem. Scand. (A)*, 1974, **28**, 700.

[212] G. I. Petrzhak, G. S. Lozhkina, and S. S. Zelentsov, *Radiokhimiya*, 1974, **16**, 386.

[213] S. Hubert, M. Hussonois, L. Brillard, G. Goby. and R. Guillaumont, *J. Inorg. Nuclear Chem.*, 1974, **36**, 2361.

[214] Yu. P. Novikov, M. N. Margorina, B. F. Myasoedov, and V. A. Mikhailov, *Zhur, analit. Khim.*, 1974, **29**, 698.

[215] S. K. Awasthi, K. L. Chawla, and D. M. Chackraburtty, *J. Inorg. Nuclear Chem.*, 1973, **35**, 3805; A. A. Grinberg, G. I. Petrzhak, and G. S. Lozhkina, *Radiokhimiya*, 1973, **15**, 879.

[216] J. Dervin, J. Faucherre, P. Herpin, and S. Voliotis, *Bull. Soc. chim. France*, 1973, 2634.

[217] K. W. Bagnall and E. Yanir, *J. Inorg. Nuclear Chem.*, 1974, **36**, 777.

[218] G. Fulcher, J. Paris, P. Plurien, P. Rigny, and E. Soullé, *J.C.S. Chem. Comm.*, 1974, 3.

[219] K. W. Bagnall and M. W. Wakerley, *J.C.S. Dalton*, 1974, 889; J. G. H. du Preez and C. P. J. van Vuuren, *J. Inorg. Nuclear Chem.*, 1974, **36**, 81; K. W. Bagnall and M. W. Wakerley, *J. Less-Common Metals*, 1974, **35**, 267.

[220] K. W. Bagnall and M. W. Wakerley, *J. Less-Common Metals*, 1974, **37**, 149.

model, with the reversal at Np^{4+} being attributed to changes in the shape and size of the ion's hydration shell.[221]

In the Np–Te system, the species $NpTe_3$, $NpTe_{3-x}$, and η- and γ-Np_2Te_3 were identified.[222] All are isostructural with their lanthanide and Pu counterparts. It appears that the metal exists as Np^{III} except in $NpTe_3$, where it may occur as Np^{IV}.

Highly pure UF_3 was obtained[223] by the thermolysis of UF_4 at 1020—1050 °C, the tetrafluoride having been prepared by reduction of either UO_2 or U_3O_8 with a mixture of HF and H_2. A high-temperature trigonal modification of CfF_3 has been prepared,[224] and reduction of this at high temperature gave ^{249}Cf; ^{244}Cm and ^{248}Cm were obtained similarly from CmF_3. UF_4 reacted with Zr and ZrF_4 giving $UZrF_7$,[225] and in the system $UZrF_7$–ZrF_4, UZr_2F_{11} was also detected; both compounds contain uranium(III). From the system NF_4–NH_4F–water, three phases, corresponding to $4NH_4F,UF_4$, $2NH_4F,UF_4$, and $7NH_4F,6UF_4$ have been identified.[226] The octahedral $[MF_6]^{2-}$ (M $\equiv$ Pa, U, Np, and Pu) have been prepared[227] as their Et_4N^+ salts and their electronic spectra discussed. UF_5 has been synthesized[228] from UF_6 by reduction using H_2 catalysed by HF in the presence of Pt, by u.v. irradiation in the presence of H_2, or with Si powder in an anhydrous HF slurry or by reaction with dissolved UF_4 in LiF–BeF_2 mixtures. UF_6 reacted with iodine giving IF_5 and U_2F_9, and the general oxidizing power of metal hexafluorides towards halogens decreased[229] in the order $PtF_6 > PuF_6 > UF_6 > MoF_6 > WF_6$.

A laboratory scale preparation of UCl_4, from UO_2 and CCl_4, has been described,[230] and the stretching frequencies and force constants of MCl_4 (M = Th, Pa, U, and Np) have been determined.[231] The valence force constants increase as the ionic radius of the metal ion decreases. Photolysis (254 nm) of $[UCl_6]^{2-}$ in non-deaerated solvents gave[232] $[UO_2Cl_4]^{2-}$, and in the presence of HCl, $[UCl_6]^-$. The Raman spectra of $QUCl_6$ (Q = Li $\rightarrow$ Cs, Tl) have been interpreted[233] in terms of octahedral symmetry for $[UCl_6]^-$, and the spectrum of solid UCl_5,PCl_5 is consistent with the formulation $[PCl_4]^+[UCl_6]^-$. Spectrophotometric and conductometric evidence has been obtained[234] for the existence of $[UCl_5L]^-$, and the donor-atom strength of L decreased in the order $Cl^- \geqslant$ phosphine oxides > sulphoxides, amides > Br^- > acetone, MeCN > I. Evidence was found for the existence of $[ThCl_5,L_3]^-$ and $[ThCl_8]^{4-}$ in non-aqueous solvents. Ethers formed[235] 1:1 and 2:1 adducts with UX_4 (X = Cl

[221] J. J. Fardy and J. M. Pearson, *J. Inorg. Nuclear Chem.,* 1974, **36**, 671.
[222] D. Damien, *J. Inorg. Nuclear Chem.,* 1974. **36**. 307.
[223] U. Berndt and B. Erdmann, *Radiochim. Acta,* 1973, **19**, 45.
[224] J. N. Stevenson, *Chem. Abs.,* 1974, **80**, 90 432.
[225] G. Fonteneau and J. Lucas, *J. Inorg. Nuclear Chem.,* 1974, **36**, 1515.
[226] L. G. Konev and L. L. Borin, *Zhur. neorg. Khim.,* 1973, **18**, 2850.
[227] D. Brown, B. Whittaker, and N. Edelstein, *Inorg. Chem.,* 1974, **13**, 1805; J. L. Ryan, J. M. Cleveland, and G. H. Bryan, *ibid.,* p. 214.
[228] J. H. Levy and P. W. Wilson, *Austral. J. Chem.,* 1973, **26**, 2711; L. B. Asprey and R. T. Paine, *J.C.S. Chem. Comm.,* 1973, 920; M. R. Bennett and L. M. Ferris, *J. Inorg. Nuclear Chem.,* 1974, **36**, 1285.
[229] N. S. Nikolaev, A. T. Sadikova, and V. F. Sukhoverkhov, *Zhur. neorg. Khim.,* 1973, **18**, 1418.
[230] P. W. Wilson, *Syn. Inorg. Metal-Org. Chem.,* 1973, **3**, 381.
[231] E. W. Bohres, W. Krasser, H.-J. Schenk, and K. Schwochau, *J. Inorg. Nuclear Chem.,* 1974, **36**, 809.
[232] G. Condorelli, L. L. Costanzo, S. Pistara, and E. Tondello, *Inorg. Chim. Acta,* 1974, **10**, 115.
[233] E. Stumpp and G. Piltz, *Z. anorg. Chem.,* 1974, **409**, 53.
[234] J. G. H. du Preez, R. A. Edge, M. L. Gibson, R. E. Rohwer, and C. P. J. van Vuuren, *Inorg. Chim. Acta,* 1974, **10**, 27.
[235] J. D. Ortego and D. L. Perry, *J. Inorg. Nuclear Chem.,* 1974, **36**, 1179.

or Br), and while amides having bulky substituents formed 2:1 adducts,[236] those with small substituents formed 4:1 adducts with UX_4 (X = Cl or Br). Amides also gave 2:1 adducts with UO_2Cl_2, and MeCONHEt afford a 3:1 adduct. Electronic spectral studies of the reaction of UX_4 with amide (L) indicated that UX_4L_2 was formed initially even in the presence of an excess of L. UX_4L_4 reacted with halide ion in non-aqueous solvents giving $[UX_5L]^-$ and $[UX_6]^{2-}$. Picolinic, nicotinic, and iso-nicotinic acid N-oxides formed[237] 1:1 and 2:1 adducts with UCl_4, and the ligands were bonded *via* the O atom attached to nitrogen. A series of sulphoxide complexes $MX_4 \cdot nR_2SO$ (M = Th, U, Np, or Pu; R = alkyl, Ph, or 1-naphthyl; X = Cl or Br; $x = 2$—6), $[UCl_4(Me_2SO)_6][ClO_4]_2$, $[UCl_3(Me_2SO)_5][ClO_4]$, $UBr_4(Me_2SO)_8$ {perhaps $[U(Me_2SO)_8]^{4+}?$}, and $UBr_5(Et_2SO)_5$ has been described.[238] Although many complexes of the aforementioned stoicheiometries have been isolated, a large number of species apparently exist simultaneously in equilibrium in acetone solution. Anthrones (L) formed[239] paramagnetic (one unpaired electron) 1:1 and 2:1 adducts with UCl_5. UCl_4 reacted with NN'-ethylenebis(salicylideneimine) (salenH$_2$) in THF giving[240] U(salenH$_2$)Cl$_4$,THF which, on heating in CHCl$_3$ or pyridine, afforded U(salen)Cl, (μ = 2.85 BM). Recrystallization from THF of this dichloride, [also produced by reaction of UCl_4 with U(salen)$_2$] afforded U(salen)Cl$_2$(THF)$_2$, preliminary X-ray diffraction studies of which reveal eight-co-ordinate uranium. Related monomeric species U(salen)X$_2$ (X = acac$^-$, dpm$^-$, [PhCOCHCOPh]$^-$, or N-methyl-salicylideneiminate) have also been prepared. In the presence of bipyridyl, UCl_4 reacted with U(salen)$_2$ giving U(salen)(bipy)Cl$_2$; Ni(salen),UCl_4 (μ = 3.04 BM at room temperature) was obtained using Ni(salen).

The standard heat of formation of UBr_5 is -197.5 ± 1.3 kcal mol^{-1}, but only below 80 °C could an apparent equilibrium $UBr_4(s) + \frac{1}{2}Br_2(g) \rightleftharpoons UBr_5(s)$ be established. At higher temperatures, irreversible decomposition to UBr_4 and Br_2 occurred, possibly because of a change in crystal structure of UBr_5. The vapour pressure of $PaBr_5$ has been measured.[242] The lactams (L) 2-piperidone (pip), 1-azacyclo-octan-2-one (aco), and ε-caprolactam formed[243] the complexes $UBr_4(pip)_6$, $[U(pip)_6Br_2][ClO_4]_2$, $ThCl_4(aco)_4$, UCl_4L_4, ThI_4L_8, and UI_4L_8. In acetone (Q), the following equilibria were established:

$$UCl_4 \cdot 3Q \underset{2L}{\overset{2L}{\rightleftharpoons}} UCl_4 \cdot 2L \overset{2L}{\underset{2L}{\rightleftharpoons}} \begin{array}{c} UCl_4 \cdot 4L \\ \| 2L \\ UCl_4 \cdot 2L \cdot 2Q \end{array}$$

$$UBr_4 \cdot nQ \overset{2L}{\rightleftharpoons} UBr_4 \cdot 2L \overset{}{\underset{2L}{\searrow}} \begin{array}{c} UBr_4 \cdot 2L \cdot 2Q \\ \\ UBr_4 \cdot 4L \overset{L}{\rightleftharpoons} [UBr_3L_5]^+Br^- \overset{L}{\rightleftharpoons} [UBr_2L_6]^{2+}[Br^-]_2 \end{array}$$

[236] K. W. Bagnall, J. G. H. du Preez, J. Bajorek, L. Bonner, H. Cooper, and G. Segel, *J.C.S. Dalton*, 1973, 2682.
[237] Kh. R. Rakhimov, T. A. Khamraev, and A. G. Muftakhov, *Chem. Abs.*, 1973, **79**, 61053.
[238] P. J. Alvey, K. W. Bagnall, D. Brown, and J. Edwards, *J.C.S. Dalton*, 1973, 2308; J. G. H. du Preez and M. L. Gibson, *J. Inorg. Nuclear Chem.*, 1974, **36**, 1795.
[239] M. Singh, G. Singh, and R. C. Paul, *Inorg. Chim. Acta*, 1974, **10**, 225.
[240] F. Calderazzo, M. Pasquali, and N. Corsi, *J.C.S. Chem. Comm.*, 1973, 784; F. Calderazzo, M. Pasquali and T. Salvatori, *J.C.S. Dalton*, 1974, 1102.
[241] A. Blair and H. Ihle, *J. Inorg. Nuclear Chem.*, 1973, **35**, 3795.
[242] F. Weigel, G. Hoffman, V. Wishnevsky, and D. Brown, *J.C.S. Dalton*, 1974, 1473.
[243] J. G. H. du Preez, M. L. Gibson, and P. J. Steenkamp, *J. Inorg. Nuclear Chem.*, 1974, **36**, 579.

The results indicated that the donor strength of L and the size of L played important roles in determining the nature of the complexes formed.

Reaction of PuI_3 with sodium in liquid ammonia afforded[244] PuN. This may have been produced *via* finely divided Pu metal, which reacted with NH_3 giving the nitride.

Uranyl and Related Compounds.—*Structural Studies.* In oxydiacetatodioxouranium(VI), the UO_2 group is equatorially surrounded by four carboxylato O atoms and one ether O atom, forming an irregular pentagonal bipyramid. Each ligand is shared between three UO_2 groups so that the overall structure is a three-dimensional network of cross-linked U–ligand chains.[245] In the pyridine-2,6-dicarboxylate (pdc^{2-}) complex, $[Ph_4As]_2[UO_2(pdc)_2]$,$6H_2O$, the U atom is eight-co-ordinate,[246] the overall co-ordination polyhedron being an irregular hexagonal bipyramid. In NN'-ethylenebis(salicylideneiminato)methanoldioxouranium, the four donor atoms of the Schiff base, and the O atom of the methanol, are co-ordinated[247] in a plane forming an irregular pentagon. The N—C—C—N group of the base is in a near *gauche* conformation (torsion angle 51.8°) and the complex has an overall 'stepped' geometry. One of the two Schiff bases in the bis-(N-ethylenedimethylaminesalicylaldimine) complex (13) is only bidentate[248] with respect to the metal; the co-ordination polyhedron forms an irregular pentagonal bipyramid and the complex has an overall 'stepped' geometry.

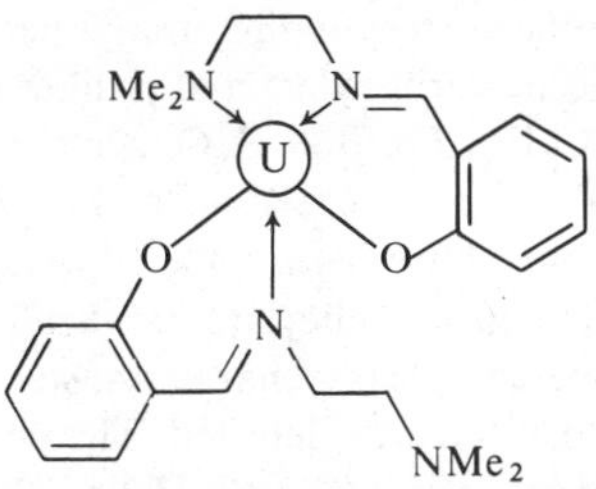

(13) $\textcircled{U} \equiv O=U=O$

Pentagonal bipyramidal co-ordination occurs[249] in β-UOF_4, the (terminal) O atom occupying an equatorial site and the four F atoms forming asymmetric bridges to neighbouring U atoms. The compound is isostructural with β-UF_5, and the structure of α-UOF_4 has also been recently described.[250] The structure of $K_5[(UO_2)_2F_9]$ is derived[251] by the condensation of two $[UO_2F_5]^{3-}$ ions which share one common F atom. The condensation causes torsion between the two planes defined by the

[244] J. M. Cleveland, G. H. Bryan, C. R. Heiple, and R. J. Sironen, *J. Amer. Chem. Soc.*, 1974, **96**, 2285.
[245] G. Bombieri, U. Croatto, R. Graziani, E. Forsellini, and L. Magon, *Acta Cryst.*, 1974, **B30**, 407.
[246] G. Marangoni, S. Degetto, R. Graziani, G. Bombieri, and E. Forsellini, *J. Inorg. Nuclear Chem.*, 1974, **36**, 1787.
[247] G. Bandoli, D. A. Clemente, U. Croatto, M. Vidali, and P. A. Vigato, *J.C.S. Dalton*, 1973, 2331.
[248] D. A. Clemente, G. Bandoli, F. Benetollo, M. Vidali, P. A. Vigato, and U. Casellato, *J. Inorg. Nuclear Chem.*, 1974, **36**, 1999.
[249] J. C. Taylor and P. W. Wilson, *J.C.S. Chem. Comm.*, 1974, 232; *Acta Cryst.*, 1974, **B30**, 1701.
[250] P. W. Wilson, *J.C.S. Chem. Comm.*, 1972, 1241.
[251] H. Brusset, N. Q. Dao, and S. Chourou, *Acta Cryst.*, 1974, **B30**, 768.

sets of F atoms, the angle between these planes being 120°. Neutron-diffraction studies of UO_2Cl_2,H_2O revealed[252] that the metal exhibits pentagonal-bipyramidal co-ordination, with four Cl atoms and a water molecule in the equatorial plane. $UOCl_2$ is isostructural[253] with $PaOCl_2$,[254] there being three independent U atoms. One has dodecahedral co-ordination (three O and five Cl atoms), another tricapped trigonal prismatic (four-fold faces) (four O and five Cl atoms), and the third an approximate dodecahedron with one apex missing (three O and four Cl atoms). The three types of polyhedron are linked together in chains by Cl atoms. As expected, the anion in $[Et_3NH]_2[UO_2Cl_4]$ has octahedral (D_{4d}) symmetry.[255]

Chemical Studies. Photoreduction of uranyl salts in sulphonic acid containing ethanol gave[256] uranium(IV) species with a quantum yield (at $\lambda \simeq 5000$ Å) of *ca.* 0.6. Similar quantum yields were obtained even when the only way for the UO_2^{2+} species to absorb light was by associated water molecules. The electronic and magnetic circular dichroism spectra of di-, tri-, and tetra-nitrato-uranyl complexes have been interpreted[257] in terms of axial symmetry perturbation of $D_{\infty h}$ states. From a temperature-dependent ^1H n.m.r. spectral study of NpO_2^{2+} in aqueous acetone, the hydration number of the metal ion was established[258] as six. Thermal decomposition of single crystals of β-$UO_2(OH)_2$ afforded[259] α-UO_3 and δ-UO_3, which subsequently decomposed into $UO_{2.9}$ and then U_3O_8. Treatment of Np^V salts with H_2O_2 afforded[260] the peroxides $(NpO_2)_2(O_2)$, $Na[(NpO_2)(O_2)]$, and $Na_5[(NpO_2)(O_2)_3]$; these compounds were less stable than the corresponding uranyl peroxides.

A series (14)—(17) of ferrocene-carboxylate and β-diketonate complexes of UO_2^{2+} has been reported [$Fc = (\eta-C_5H_4)_2Fe$; $B = H_2O$, acetone, pyridine or its *N*-oxide; R_3EO, where $R = Bu^n$ or Ph, $E = O$ or $R = Ph$, $E = As$; DMSO].[261] Spectral studies (i.r., n.m.r., and electronic) have been made[262] of a series of 1,2-disubstituted benzene complexes of UO_2^{2+} (*e.g.* salicylate, anthranilate, *o*-aminophenolate). Except in the case of phthalate complexes, charge transfer from the *p*π benzene ring orbitals to the $5f$ and/or $6d$ orbitals was detected. The conformation of the malate ligand in the dimeric malic acid complex $(UO_2)_2(C_4H_4O_5)(C_4H_3O_5)$ was established[263] by ^1H m.n.r. spectroscopy. Photochemical reduction of uranyl citrate afforded[264] uranium(IV) species, acetone, and CO_2, but, in the presence of Tl^+, photoreduction of uranyl lactates is inhibited. The uranyl pamoate complex in DMSO exhibited[265] two reduction waves corresponding to the reduction of a uranyl DMSO cation

[252] J. C. Taylor and P. W. Wilson, *Acta Cryst.* , 1974, **B30**, 169.
[253] J. C. Taylor and P. W. Wilson, *Acta Cryst.*, 1974, **B30**, 175.
[254] R. P. Dodge, G. S. Smith, Q. Johnson, and R. E. Elson, *Acta Cryst.*, 1968, **B24**, 304.
[255] H. Brusset, N. Q. Dao, and F. Haffner, *J. Inorg. Nuclear Chem.*, 1974, **36**, 791; W. Jensen, D. Dickerson, and Q. Johnson, *Acta Cryst.*, 1974, **B30**, 840.
[256] J. T. Bell and S. R. Buxton, *J. Inorg. Nuclear Chem.*, 1974, **36**, 1575.
[257] P. Brint and A. J. McCaffrey, *J.C.S. Dalton*, 1974, 51.
[258] V. A. Shcherbakov, E. V. Iorga, and L. G. Mashirov, *Radiokhimiya*, 1974, 281.
[259] M. J. Bannister, *J. Inorg. Nuclear Chem.*, 1974, **36**, 1991.
[260] C. Musikas, *J. Chim. phys.* 1974, **71**, 197.
[261] P. A. Vigato, U. Casellato, D. A. Clemente, G. Bandoli, and M. Vidali, *J. Inorg. Nuclear Chem.*, 1973, **35**, 4131.
[262] B.-I. Kim, C. Miyake, and S. Imoto, *J. Inorg. Nuclear Chem.*, 1974, **36**, 2015.
[263] J. D. Pedrosa and U. M. S. Gil, *J. Inorg. Nuclear Chem.*, 1974, **36**, 1803.
[264] A. Ohyoshi and K. Ueno, *J. Inorg. Nuclear Chem.*, 1974, **36**, 379; Y. Yokoyama, M. Moriyasu, and S. Ikeda, *ibid.*, p. 385.
[265] T.-T. Lai and W.-Y. Kuu, *J. Inorg. Nuclear Chem.*, 1974, **36**, 631.

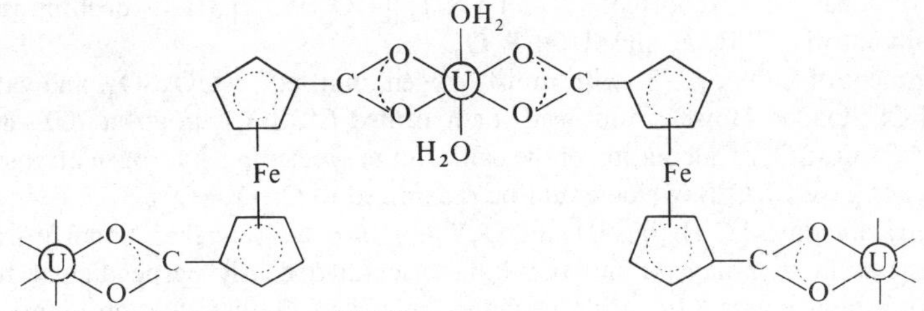

(14) $\textcircled{U} \equiv O=U=O$

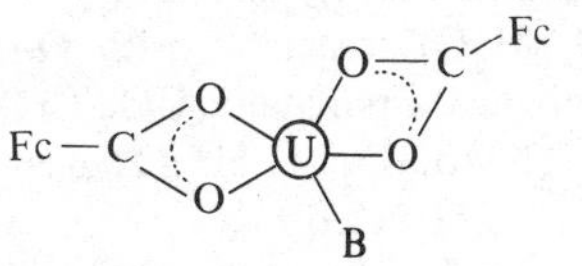

(15)

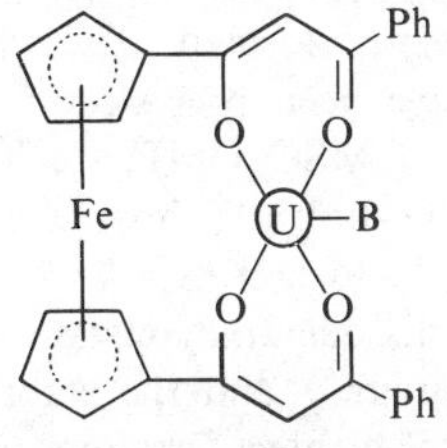

(16)

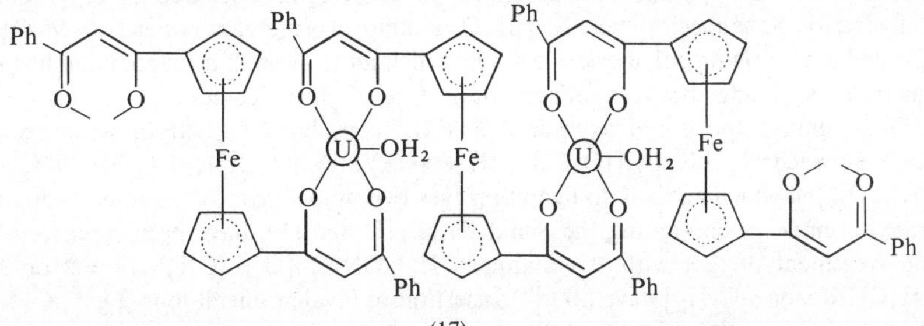

(17)

followed by the reduction of a hydroxy-species. Uranyl oxalate reacted[266] with quaternary ammonium hydroxides giving $[R_4N]_2[UO_2(O_2C_2O_2)(OH)_2]$ or $[R_4N][(UO_2)_2(O_2C_2O_2)_2(OH)]$. Treatment of the latter with $[R_4N]_2[O_2C_2O_2]$ in benzene gave $[R_4N]_3[(UO_2)_2(O_2C_2O_2)_3(OH)]$, and $[R_4N]_5[(UO_2)_2(O_2C_2O_2)_4(OH)]$ and $[R_4N][UO_2(O_2C_2O_2)(OH)]$ were also prepared. The uranyl complex of 5-chloro-2-hydroxy-4-methylpropiophenone oxime has been described.[267] From the Raman and i.r. spectra of $UO_2(NO_3)_2,6H_2O$ it was shown[268] that the nitrate should be formulated as $UO_2(NO_3)_2(H_2O)_2,4H_2O$, in agreement with the structure established by neutron diffraction.[269] The vibrational spectra of $UO_2Cl_2,3H_2O$ could not be interpreted in terms of its known structure. The guanidinium salt of $[UO_2(O_2C_2O_2)_2$-

[266] M. G. Kuzina and A. A. Lipovskii, *Zhur. neorg. Khim.*, 1973, **18**, 1625.
[267] R. N. Saksena and K. K. Panday, *J. Indian Chem. Soc.*, 1973, **50**, 609.
[268] C. Caville and H. Poulet, *J. Inorg. Nuclear Chem.*, 1974, **36**, 1581.
[269] J. C. Taylor and M. H. Mueller, *Acta Cryst.*, 1965, **B19**, 536.

$(ONO)]^{3-}$ has been reported,[270] and $[NH_4][UO_2(PO_4)],3H_2O$ decomposed[271] thermally into $U_2O_3P_2O_7$ and $(UO)_2P_2O_7$.

Treatment of UOS or US_2 with moist oxygen afforded[272] UO_2SO_4, and calcination of Cf^{III}-loaded Dowex resin beads or hydrated $Cf_2(SO_4)_3$ in air at 700—800 °C gave[273] $Cf_2O_2(SO_4)$. Calcination of the californium species in a hydrogen atmosphere or *in vacuo* gave Cf_2O_2S, which could be reoxidized to $Cf_2O_2(SO_4)$.

Uranyl vanadates $[C_nH_{2n+1}NH_3][UO_2V_3O_9]$ ($n = 6$—18) have[274] poorly ordered alkyl chains in monolayers, and these are orientated nearly perpendicular to the metal oxyanion layers. On cation exchange at pH 7—9, intercalation compounds containing up to one additional alkylamine are formed, and these have close-packed well-ordered alkyl monolayers. With excess of amine or alcohol, 1:3 intercalation compounds are formed. The chromates $Q_2[(UO_2)_2(CrO_4)_3],6H_2O$ (Q = NH_4, K → Cs) have been prepared[275] and $UO_2(CrO_4)(H_2O)_3$ reacts[276] with O-donor ligands (L = DMSO, DMF, $MeCONH_2$, or urea) giving the polymeric $UO_2(CrO_4)L_2$. A uranyl 12-molybdophosphate, $(UO_2)_{1.39}H_{1.36}PMo_{12}O_{40},11.5H_2O$, has been described,[277] and $UO_2(S_2PAr_2)_2$ is six-co-ordinate, as expected.[278]

The acid dissociation constants for a series of uranyl β-keto-ester complexes have been measured,[279] and the stability of these UO_2^{2+} compounds is much greater than those of bivalent first-row transition metals. Tropolone ($C_9H_6O_2$) reacted[280] with UO_2^{2+} salts giving $UO_2(C_7H_5O_2)_2$ and $UO_2(C_7H_5O_2)_2,xH_2O$. The former afforded adducts, $UO_2(C_7H_5O_2)_2L$, with O-donor ligands, and the structure of the species where L = py may be described as an irregular pentagonal bipyramid. Dehydration of the species where L = H_2O, or removal of methanol when L = MeOH, afforded a dimeric or polymeric species in which an O atom of a tropolonate ligand functions as a bridge between adjacent $UO_2(C_7H_5O_2)_2$ molecules.

Treatment of the 8-hydroxyquinoline (HL) complex $UO_2L_2(HL)$ with oxalic acid afforded[281] $[H_2L]_2[(O_2C_2O_2)(H_2O)(HL)(UO_2)(O_2C_2O_2)(UO_2)(HL)(H_2O)-(O_2C_2O_2)]$ in which an oxalato ligand bridges two uranyl groups. Seven-co-ordinate uranyl complexes containing the Schiff bases (18) and (19) have been reported,[282] and treatment of (20) with the diamines H_2NQNH_2 [Q = $(CH_2)_n$ ($n = 2$ or 3), CH_2CHMe, or o-C_6H_4] gave (21).[283] Insertion of bivalent metal ions (Fe^{2+}, Co^{2+}, Ni^{2+}, Cu^{2+}) into the vacant N_2O_2 donor-atom site was also reported. Uranyl salts reacted with CuL $[H_2L = NN'$-1,3-propylenediaminebis(salicylideneimine)], giving O-bridged di- and tri-nuclear species, $(CuL)UO_2X_2$ (X = Cl, NO_3, NCS, or OAc) and

[270] R. N. Shchelokov, I. M. Orlova, and G. V. Podnebesnova, *Zhur. neorg. Khim.*, 1973, **18**, 1886.
[271] J. M. Schaekers, *J. Therm. Analysis*, 1974, **6**, 145.
[272] M. Yu. Koval'chuk, K. M. Dunaeva, E. A. Ippolitova, A. V. Dubrovin, and A. I. Zhirov, *Chem. Abs.*, 1974, **80**, 43604.
[273] R. D. Baybarz, J. A. Fahey, and R. G. Haire, *J. Inorg. Nuclear Chem.*, 1974, **36**, 2023.
[274] K. Beneke, U. Grosse-Brauckmann, G. Lagaly, and A. Weiss, *Z. Naturforsch.*, 1973, **28b**, 408.
[275] I. J. Martin, *Ion (Madrid)*, 1973, **33**, 136.
[276] R. N. Shchelokov, I. M. Orlova, and G. V. Podnebesnova, *Zhur. neorg. Khim.*, 1974, **19**, 1581.
[277] A. N. Martra, *J. Indian. Chem. Soc.*, 1974, **51**, 370.
[278] R. N. Mukherjee, S. Shanbag, M. S. Venkateshan, and M. D. Zingde, *Indian J. Chem.*, 1973, **11**, 1066.
[279] N. S. Al-Niaimi and B. M. Al-Saadi, *J. Inorg. Nuclear Chem.*, 1973, **35**, 4207.
[280] S. Degetto, G. Marangoni, G. Bombieri, E. Forsellini, L. Baracco, and R. Gaziani, *J.C.S. Dalton*, 1974, 1933.
[281] E. N. In'lova and V. P. Arsent'eva, *Zhur. neorg. Khim.*, 1974, **19**, 1364.
[282] E. D. McKenzie, R. E. Paine, and S. E. Selvey, *Inorg. Chim. Acta*, 1974, **10**, 41.
[283] U. Casellato, M. Vidali, and P. A. Vigato, *Inorg. Nuclear Chem. Letters*, 1974, **10**, 437.

$[(CuL)_2UO_2][ClO_4]_2$.[284] The hydrazine ligands in $UO_2(RCONHNH_2)_3X_2,nH_2O$ (R = Me or aryl; X = Cl_4 or NO_3) and $UO_2(p-NO_2C_6H_4CONHNH_2)_2,nH_2O$ were bidentate (ON co-ordination).[285]

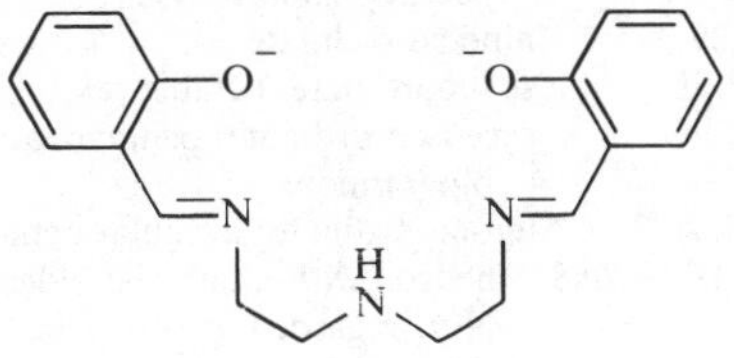

(18)

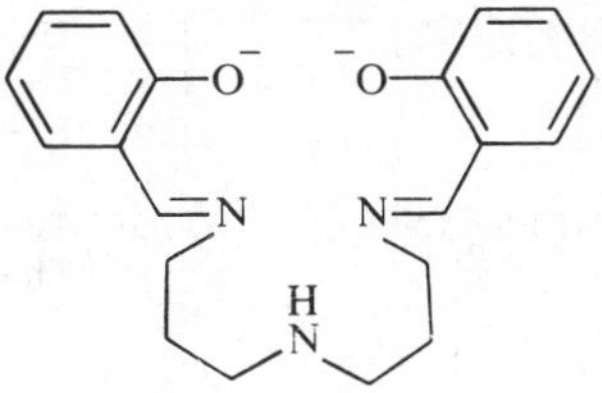

(19)

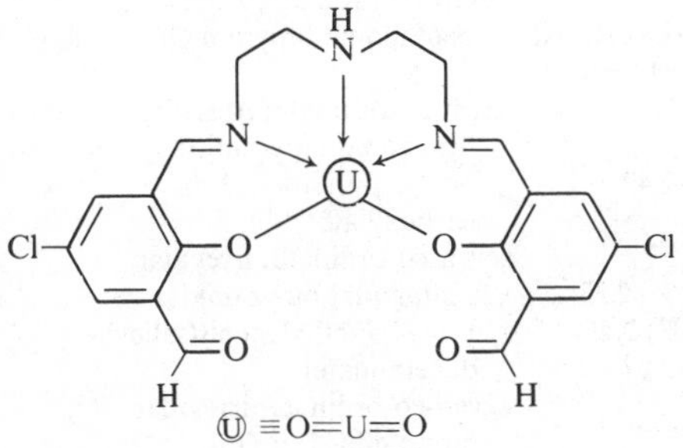

(20)

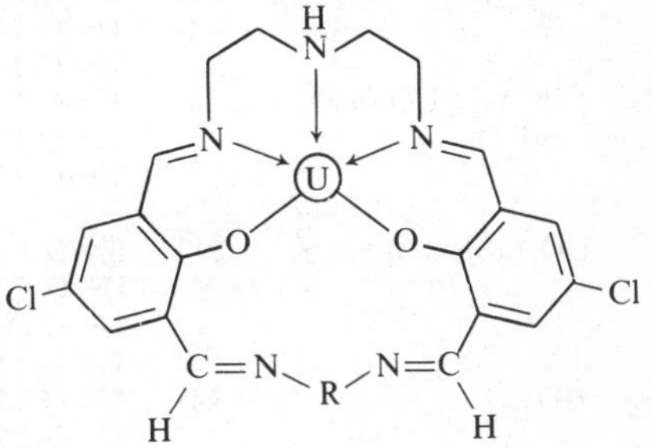

$\textcircled{U} \equiv O{=}U{=}O$

(21)

Reaction of UO_2F_2 with alkali-metal fluorides (QF) afforded[286] QUO_2F_3,nH_2O (Q = Na, n = 0; Q = Rb, n = 1; and the species contains bridging and terminal F; Q = Cs, n = 1), $Q_2UO_2F_4,H_2O$ (Q = Cs or Rb), $NaUO_2F_5$, and $Na(UO_2)_2F_5$. Hydrolysis of UF_6 suspended in an HF slurry led[287] to formation of UOF_4, which decomposed on heating into UO_2F_2 and UF_6. Attempts to prepare UOF_4 by fluorination of U_5O_8 or UO_2Cl_2 gave only UO_2F_2, and reaction of UO_2F_2 with SeF_4 gave an uncharacterized compound containing U, O, Se, and F. Thermal decomposition of UOF_4 at moderate temperatures gave $U_2O_3F_6$ which, on the basis of i.r. spectral studies, contains two types of U=O bond (the compound is not a mixture of UO_2F_2 and UOF_4).[288] From concentrated acetone solutions containing QF (Q = K or Rb) and UO_2F_2, salts of $[(UO_2)_2F_9]^{5-}$, $[(UO_2)_2F_5]^-$, and $[UO_2F_5]^{3-}$ were obtained.[289] Treatment of UO_2F_2 with MF_2 (M = Co, Ni, Cu, Zn, or Cd) in

[284] G. Condorelli, I. Fragala, and S. Giuffrida, *Chem. Letters*, 1974, **7**, 683.

[285] Yu. Ya. Kharitonov, R. N. Shchelokov, R. I. Machkhoshvili, and N. G. Generalova, *Zhur. neorg. Khim.*, 1974, **19**, 858.

[286] M. L. Kotsar, Z. B. Mukhametshina, V. P. Seleznev, B. N. Sudankov, A. A. Tsvetkov, and B. V. Gromov, *Chem. Abs.*, 1974, **81**, 57 697; R. L. Davidovich, Yu. A. Buslaev, V. I. Sergienko, S. B. Ivanov, V. V. Peshkov, and Yu. N. Mikhailov, *Doklady Akad. Nauk S.S.S.R.*, 1974, **214**, 332.

[287] E. Jacobs and W. Polligkert, *Z. Naturforsch.*, 1973, **28b**, 120; P. W. Wilson, *J. Inorg. Nuclear Chem.*, 1974, **36**, 303.

[288] P. W. Wilson, *J. Inorg. Nuclear Chem.*, 1974, **36**, 1783.

[289] Z. B. Mukhatshina, M. L. Kotsar, V. P. Seleznev, A. A. Tsvetko, B. N. Sudarikov, and B. V. Gromov, *Zhur. neorg. Khim.*, 1973, **18**, 1317.

Table 3 *Structures of actinide and uranyl complexes*

Compound	Ref.	Bond lengths/Å	Remarks
β-Th(NO₃)₄,3DMSO	176		eleven-co-ordinate; monocapped pentagonal antiprism
β-[NH₄][UF₅]	177	U—F 2.20—2.39	nine-co-ordinate
UCl₆	178	U—Cl 2.41—2.51	six-coordinate: octahedral
UBr₄	180	U—Br 2.78—2.95	seven-co-ordinate; pentagonal bipyramidal
Th(OPPh₃)₂(NO₃)₄	176a	Th—O (P) 2.33, 2.37 Th—O (NO₃) 2.52—2.58	ten-co-ordinate; irregular octahedron with equatorial bidentate NO₃ regarded as occupying a single co-ordination site
Na₃BeTh₁₀F₄₅	177a		nine-co-ordinate; tricapped trigonal prismatic
[UO₂{(O₂CCH₂)₂O}]ₙ	245	U=O 1.79 U—O 2.37—2.42 (CO₂) U—O 2.42 (—O—)	seven-co-ordinate; irregular pentagonal bipyramidal
[Ph₄As]₂[UO₂(pdc)₂],·6H₂O	246	U=O 1.76 U—N 2.73 U—O 2.37—2.42	eight-co-ordinate; irregular hexagonal bipyramid; pdc = pyridine-2,6-dicarboxylate
UO₂L(MeOH)	247	U=O 1.76—1.77 U—O (L) 2.25—2.33 U—O (MeOH) 2.45 U—N 2.54—2.57	seven-co-ordinate; irregular pentagonal bipyramid; L = NN′-ethylenebis(salicylideneiminate)
UO₂L₂	248	U=O 1.78 U—O 2.24 U—N 2.57—2.69	seven-co-ordinate; irregular bipyramid; L = (13); one N atom of L not bonded
β-UOF₄	249	U—F 2.18—2.34 (bridging) U—F 1.92 (terminal) U—O 1.87	seven-co-ordinate; pentagonal bipyramidal
K₂[(UO₂)₂F₉]	251	U—O 1.79 U—F 2.29	seven-co-ordinate; pentagonal bipyramidal
UO₂Cl₂,H₂O	252	U=O 1.70—1.74 U—O (H₂O) 2.46 U—Cl 2.75—2.80	seven-co-ordinate; pentagonal bipyramidal
UOCl₂	253		U(1): eight-co-ordinate; dodecahedral U(2): nine-co-ordinate; tricapped trigonal prismatic U(3): seven-co-ordinate; approximately dodecahedral with one apex missing
[Et₃HN]₂[UO₂Cl₄]	255	U=O 1.94 U—Cl 2.67—2.70	octahedral, D_{4h}
[Me₄N]₂[UO₂Br₄]	255	U=O 1.72 U—Br 2.79—2.86	octahedral

HF containing NH_4F afforded[290] $[NH_4]_2[M(UO_2F_4)_2]$,$6H_2O$, and $Q_2M^IO_2F_4$ (Q = K, Rb, or Cs; M^1 = Mo^{VI} or W^{VI}) reacted[291] with M^2O_2 (M^2 = U or Ce) giving $Q_{12}M^IM_3^2O_8F_{18}$; $Q_3UO_2F_5$ (Q = Rb or Cs) reacted with $Q_2M^IO_2F_4$, ClF, and M^2O_2 affording $Q_{12}M^2U_{3-x}M_x^IO_8F_{18}$.

290 R. L. Davidovich, V. V. Peshkov, and Yu. A. Buslaev, *Doklady Akad. Nauk S.S.S.R.*, 1973, **212**, 1114.
291 J. Fleckenstein, S. Kemmler-Sack, and W. Ruedorff, *Z. Naturforsch.*, 1974, **29b**, 124.

The methods of preparation of UO_2Cl_2, UO_2Cl_2,H_2O, and $UO_2Cl_2,3H_2O$ have been critically reviewed,[292] and the heats of formation of UO_2Cl_2, its hydrates, and $UO_2(OH)Cl,2H_2O$ were determined. Anhydrous UO_2Cl_2 decomposed to UO_2 *via* $(UO_2)_2Cl_3$. Both UO_2Cl_2 and $Th(NO_3)_4$ formed[293] 1:1 charge-transfer complexes with hexamethylmelamine in anhydrous ethyl acetate, and the i.r. and Raman spectra of $[UO_2Cl_4]^{2-}$ and $[UO_2Br_4]^{2-}$ were assigned.[294] Photolysis of $UO_2(py)_2Cl_2$ in dry ethanol for 10 h afforded[295] $[pyH]_2[UOCl_5]$, but after 50 h, only $U(OEt)_5$ was isolated. Treatment of UO_2Cl_2,nH_2O with $SOCl_2$ in the presence of $[Ph_4P]Cl$ gave $[Ph_4P]$ $[UOCl_5]$.[296] The stability constants for the reaction of Np^V and NpO_2^{2+} salts with Cl^- have been re-evaluated.[297]

[292] G. Prins, *Reactor Cent. Ned. Reports, no.* 186, 1973, p. 115; E. H. P. Cordfunke and G. Prins, *J. Inorg. Nuclear Chem.*, 1974, **36**, 1291.
[293] N. Gunduz, *Chem. Abs.*, 1974, **80**, 103343.
[294] A. Marzotto, *Inorg. Nuclear Chem. Letters*, 1974, **10**, 915.
[295] O. Traverso, R. Portanova, and V. Carassiti, *Inorg. Nuclear Chem. Letters*, 1974, **10**, 771.
[296] K. W. Bagnall and J. G. H. du Preez, *J.C.S. Chem. Comm.*, 1973, 820.
[297] P. R. Danesi, R. Chiariza, G. Scibona, and G. D'Alessandro, *J. Inorg. Nuclear Chem.*, 1974, **36**, 2396.

Author Index

Aarflot, K., 429
Abaeva, S. D., 145
Abbadi, A. R. M., 336
Abbott, E. H., 126, 204
Abd-Elmageed, M. E., 232
Abdelaziz, T., 81
Abdullaeva, Kh. S., 139
Abe, J., 124
Abel, E. W., 169
Ablov, A. V., 56, 98, 197, 250, 365, 393, 451
Abraham, K. M., 165, 219
Abrahams, S. C., 95
Abramowitz, S., 122
Abu-Eittah, R., 326
Achlama, A. M., 96
Ackerman, J. F., 93
Ackermann, G., 27, 83, 216
Ackermann, R. J., 33, 488
Acquista, N., 122
Adachi, G., 479
Adamczak, H., 50
Adams, D. M., 66, 87, 350, 358, 401, 423, 424
Adams, J. M., 400
Adams, R. D., 90, 126, 184, 185, 187
Adams, W. W., 255
Adamson, A. W., 338
Adamson, G. W., 365
Addington, J. W., 338
Addison, A. W., 364, 366
Addison, C. C., 154
Aderjan, R., 373
Adeyemi, S. A., 337
Adler, A. D., 469
Adnitt, S. E., 10
Adrianova, O. N., 412, 422
Afana'ev, T. V., 17
Agambar, C. A., 226
Aganiyazov, K., 142, 480
Agarwal, M. M., 478
Agarwala, U., 331, 342, 363
Aggarwal, A., 27, 177, 240
Aggarwal, B. S., 55
Aggarwal, R. C., 239, 481
Agrawal, D. R., 326
Agrawal, Y. K., 215, 326, 446
Agrinskaya, N. A., 138, 139
Ahakhova, N. F., 343

Ahlberg, I., 461
Ahlborn, E., 107
Ahluwalia, S. C., 26
Ahmad, N., 12, 14
Ahmed, S., 91
Ahn, S., 146
Ahuja, I. S., 226, 439, 440, 461
Aiba, H., 216, 326
Aihara, M., 215, 224, 451
Aijaz Beg, M., 39
Airoldi, C., 482
Aitkin, G. B., 175, 235, 439
Aivazov, M. I., 5
Åkeson, A., 464
Akhmadullina, A. G., 24
Akhtan, S. S., 340
Akimochkina, T. A., 110
Akimoto, K., 236
Akimoto, M., 112
Akitt, J. W., 425
Akl, N. S., 383, 402
Aklan, G., 221
Alam, A. M. S., 97
Alario-Franco, M. A., 91
Albano, V. G., 192, 218, 352, 422
Albertsson, J., 474
Alberti, G., 38
Alchangyan, S. V., 65
Alcock, N. W., 25
Aleksandrovskii, S. V., 10
Aleksanyan, V. T., 87, 440
Alekseev, F. P., 142
Alekseev, N. V., 152
Alekseeva, Z. M., 156
Aléonard, S., 474
Aleskovskii, V. B., 140
Alessandrini, S., 416
Alexander, L. E., 138
Alexander, M. D., 252
Algavish, G., 451
Ali, L. H., 189
Ali, M. A., 217, 258, 301, 328
Ali, S. F. M., 481
Aliev, R. Y., 225, 440, 476
Alimarin, I. P., 343
Alimov, N. S., 40
Aliwi, S. M., 54
Al-Karaghouli, A. R., 54, 473
Allcock, H. R., 146

Allen, D. M., 189
Allen, E. A., 401, 440
Allen, G. C., 489
Allen, G. L., 341
Allen, J. D., 435
Allman, R., 146, 148, 460
Allpress, J. G., 96
Allsop, R. T., 31
Allulli, S., 38
Al-Niaimi, N. S., 54, 327, 445, 465, 496
Alper, H., 191
Al-Saadi, B. M., 445, 465, 496
Altanasio, D., 313, 324
Alter, E., 9, 248, 303
Altybaeva, D. T., 225
Alvarez-Bartolome, M. L., 103
Alves, O. L., 482
Alvey, P. J., 487, 492
Alyamovskii, S. I., 8, 33, 36, 65
Alybakov, A. A., 106
Ameria, R. L., 234, 287, 324, 446
Amestoy, P., 81
Amin, M. I., 482
Amma, E. L., 308, 422
Amos, L. W., 69, 135
Amosov, V. M., 142
Amosova, K. S., 123
Amosova, S. V., 390
Anagnostopoulos, A., 61, 173, 227, 478
Anakawa, K., 446
Anand, S. P., 151
Anan'in, A. V., 81
Anantaraman, R., 26
Andal, R. K., 343
Ander, L. W., 111
Anderegg, G., 326
Andersen, P., 99, 370, 380
Anderson, C. P., 323
Anderson, D. N., 310, 321
Anderson, D. W. W., 424
Anderson, J. E., 490
Anderson, J. S., 79
Anderson, L. B., 231
Anderson, P., 94, 252
Anderson, S., 68
Anderson, S. J., 358
Anderson, T. J., 471

Anderson, W. P., 44, 169
Andreeva, M. I., 82
Andreeva, N. I., 5
Andreeva, T. A., 483
Andrelczyk, B., 97
Andrianov, K. A., 19
Andrianov, M. A., 142
Andriessen, W. T. M., 96
Andrushkevich, M. M., 141
Andryushin, V. G., 489
Angapova, L. E., 479
Angelici, R. J., 87, 182, 291, 321, 326, 327, 382, 400, 467
Angelin, L., 125
Angew, N. H., 225
Angoletta, M., 380
Angyal, S. J., 486
Anisimov, K. N., 71, 157, 220
Anjaneyulu, Y., 69
Anker, M. W., 382
Annopolskii, V. F., 36, 139
Annuar, B., 301
Anson, F. C., 366
Anstey, P., 226
Antimonova, L. N., 158
Antolini, L., 288
Antonovskaya, E. I., 5
Antonescu, L., 135
Antonova, A. B., 71
Antsygina, N. N., 155
Antti, B. M., 175
Aotake, M., 116
Aplincourt, M., 327
Appleton, D. W., 463
Apraksin, I. A., 38, 82, 471
Apte, K. P., 104
Arabi, M. S., 221
Arakelyan, O. I., 4
Araneo, A., 192
Aravamudan, G., 140, 149, 224, 292, 390, 392, 489
Arbuzov, M. P., 7
Archer, R. D., 119
Ardon, M., 112
Arkhangel'skaya, O. I., 408
Armit, P. W., 330, 331
Armor, J. N., 333, 334
Armstrong, J. R., 104
Armstrong, L. G., 294
Armytage, D., 73
Arnaiz, F. J., 485
Arnaud-Neu, F., 136
Arnos, D. W., 19
Arnott, R. J., 248
Arribas, J. S., 103
Arsent'eva, V. P., 496
Ase, K., 429
Ashaev, L. G., 39
Ashby, E. C., 438
Ashley, K. R., 102
Asplund, R. O., 213
Asprey, L. B., 122, 491
Asso, M., 215

Astakhova, A. S., 357, 387
Asting, N., 66
Astolfi, R., 97, 206
Atkinson, S. J., 111
Atohda, T., 7
Attali, S., 221
Attanasio, D., 442
Atwood, J. L., 471
Au, F. H. F., 470
Aubert, J., 52
Aubke, F., 21
Auld, D. S., 462
Aurivillius, K., 456, 457
Averill, B. A., 210, 211
Avery, N. R., 123
Aver'yanova, L. N., 19, 36
Avignant, D., 485
Awasthi, S. K., 490
Aymonino, P. J., 65, 107, 202
Azizov, T. A., 54

Baba, S., 386
Babaev, S. Kh., 99
Babel, D., 17, 94, 201
Babenko, N. L., 48
Babu, R., 18
Babundgi, R. H., 243
Babykutty, P. V., 26
Babynina, N. A., 479
Bacci, M., 199
Bach-Dragutinovic, B., 446
Bachelier, J., 18
Bachelor, F. W., 107
Bachlaus, M. K., 215
Bachmann, H.-G., 489
Badalova, R. I., 484
Badaud, J. P., 141
Baddley, W. H., 356, 362, 398, 405
Baenziger, N. C., 288, 433
Baev, A. K., 32, 71
Baffier, N., 95
Bagchi, M., 134
Bagger, S., 202, 410
Baggio, R. F., 433
Baggio, S., 433
Bagnall, K. W., 487, 490, 492, 499
Bagreev, V. V., 138, 139
Bailar, J. C., 57, 242, 252, 264, 324, 423
Bailey, J., 313, 425
Bailey, N. A., 117, 274
Bailey, W. E., 202
Bajdala, P., 480
Bajorek, J., 492
Bajui, S. P., 260
Bakac, A., 101
Baker, D. 252
Baker, R. W., 471
Balabaeva, R. F., 65
Balaev, A. V., 19
Balahura, R. J., 103, 249

Balan, V. T., 54
Bala-Pala, J., 19
Balashov, V. A., 142
Balch, A. L., 339
Balchin, A. A., 6, 33
Baldea, I., 109
Baldwin, B. A., 117
Baldwin, D. A., 196
Baldwin, J. F., 214
Baldwin, H. W., 94
Baldwin, W. H., 473
Ball, M. C., 489
Ball, P. W., 228
Ballard, R. E., 299
Ballentine, T. A., 9
Balundge, R. H., 296
Balzani, V., 94, 338
Ban, A., 488
Bancroft, G. M., 188
Banditelli, G., 431
Bandoli, G., 355, 459, 493, 494
Bandyopadhyay, P., 117, 155
Banerjee, P., 465
Banerji, S. N., 327
Banks, A. J., 445
Banks, E., 92
Banks, R. L., 334
Bannerjee, S. P., 144
Bannister, M. J., 494
Baour, C., 93
Bapzargashieva, S. D., 407, 408
Baracco, L., 283, 496
Baral, S., 323
Baran, E. J., 65, 100, 107, 150
Baranovskii, I. B., 377
Baratossy, J., 1
Barbeau, C., 168, 170
Barber, M., 87
Barberi, P., 201
Barbier, J. P., 176
Barbieri, R., 297
Barbucci, R., 217, 323, 327, 441
Bard, A. J., 338
Bardin, M. B., 381
Barefield, C. G., 422
Barefield, E. K., 217, 275, 428, 443
Barenkova, Ya G., 39
Barinskii, R. L., 35
Barker, M. G., 49, 66, 79
Barnard, P. F. B., 226
Barnes, D. S., 122
Barnes, J. A., 323
Barnes, J. C., 175, 234
Barnes, R. G., 47
Barnett, G. H., 113
Barnett, K. W., 283
Barr, D. W., 10
Barr, R. M., 385, 454, 459, 460
Barraclough, K. G., 99
Barral, R., 150
Barrett, P. B., 23
Barrow, M. J., 189

Barry, C. D., 486
Bars, O., 124
Barskaya, I. B., 35
Barstow, T. J., 308
Bartecki, A. 68, 144, 150
Bartell, L. S., 111
Barteri, B. M., 263
Bartlett, K. G., 81
Bartlett, N., 345, 370, 400, 433
Bartunik, H. D., 374
Bartusek, M., 50, 144
Barvinok, M., 446
Basato, M., 220
Basco, N., 425
Basile, L. J., 302
Basolo, F., 244, 367, 377
Basov, A. S., 138
Bassett, D. W., 125
Bassett, J. M., 112
Bassi, G., 243, 320, 474
Basson, W. D., 313
Bates, G., 206
Baticle, A. M., 147
Batiste, J. L. H., 384
Batsanov, S. S., 4
Batsanova, L. R., 472
Batson, D. A., 159
Battaglia, L. P., 255
Battistuzzi, R., 384, 458, 459
Batyr, D. G., 54
Bau, R., 162
Baudler, M., 188
Bauer, D. P., 188
Bauer, W. H. 267
Baumann, D., 243
Baumann, E. W., 327
Baumann, H., 128
Baxter, A. C., 176, 325, 465
Baybarz, R. D., 489, 496
Bazhin, N. M., 19
Bazuev, G. V., 480
Beach, R. T., 356
Beadle, E. M., 291, 326, 327, 467
Beal, A. R., 7
Bear, C. A., 321
Beattie, I. R., 138
Beattie, J. K., 265, 336
Beauchamp, A. L., 457
Bechgaard, K., 300
Beck, J. P., 101
Beck, W., 359, 361, 362, 373, 374, 382, 394, 397, 398, 414, 438
Becker, C. A. L., 302
Beckert, W. F., 470
Beckett, R., 238
Beckwith, P. M., 148
Bee, M. W., 334
Beech, G., 48
Beecroft, B., 293
Beer, D. C., 247
Beers, C., 461
Beg, M. A., 118

Begalieva, D. U., 65
Behal, F. J., 463
Behera, P , 240
Behrens, H., 88, 181
Beiles, R. G., 138
Beisekeeva, L. I., 204, 205
Bekker, P. van Z., 363
Bekturov, A., 65
Belen'kaya, I. R., 451
Belford, R. C. E., 322
Belinskii, M. I., 98
Bell, A. P., 34
Bell, J. T., 494
Bellitto, C., 289
Bellon, P. L., 192, 380, 433
Belluco, U., 355, 398
Belokon, A. T., 81
Belon, P. L., 352
Belopotapova, T. S., 332, 353, 357
Beloshitskii, N. V., 216
Belov, N. N., 19
Belov, N. V., 38, 65, 107, 143
Belov, S. F., 485
Belova, I. L., 478
Belozerov, V. V., 81
Bel'skaya, T. I., 140
Belyaev, A. V., 410
Belyaev, E. K., 20, 36, 139
Belyaev, I. N., 19, 36
Belyaev, V. K., 133
Belyaeva, I. I., 19, 36
Belyaeva, K. L., 484
Belyakov, Yu. M., 47
Bemier, J. C., 81
Bencze, L., 117, 353
Benderskaya, L. P., 478
Ben-Dor, L., 123
Benedicenti, C., 380
Beneke, K., 66, 496
Benetollo, F., 493
Benfield, F. W. S., 115, 121, 132
Bengtsson, G., 69
Benko, A., 256
Benlian, D., 215
Benner, G. S., 259
Bennett, C. R., 29
Bennett, L. E., 217, 337
Bennett, M. A., 93, 179, 233, 355, 360, 368
Bennett, M. J., 42, 130, 153
Bennett, M. R., 491
Benno, H., 327
Benno, R. H., 429
Bensoam, J., 122
Bentley, K. W., 253
Bercaw, J. E., 3, 218
Berch, M. L., 332
Berdnikov, V. M., 19
Bereman, R. D., 51, 178, 237, 242, 323, 324, 325, 443
Beresnev, T. I., 35
Berg, T., 99

Bergman, R. G., 218
Berjoan, R., 108
Berkooz, O., 92
Bernal, I., 210
Bernard, M. B., 291
Bernard, S., 81
Bernat, J. G. R., 479
Bernauer, K., 275
Berndt, U., 491
Bernier, J. C., 52, 95
Bernstein, J. L., 95
Berretz, M., 99
Bertaut, E. F., 444
Berthon, G., 451
Bertini, I., 216, 328
Bertoncelj, M., 78
Bertram, K. H., 91
Bertrand, J. A., 208, 209, 258, 321, 322, 325, 469
Beskrovnaya, R. A., 472
Bessarabenko, L. K., 279
Besse, J. P., 78
Bessonov, V. I., 413
Bethuel, L., 52
Beurskens, P. T., 255, 459
Bezman, S., 309
Bhadmaeva, H., 461
Bhaduri, S., 222, 377
Bhalerao, P. D., 95
Bhandari, C. S., 446
Bharaca, P. C., 21
Bhat, A. N., 24
Bhatia, P., 461
Bhatia, V. K., 483
Bhattacharya, A. K., 104
Bhattacharya, G. C., 148
Bhattacharya, P. K., 326, 445, 468
Bhattacharyya, R. G., 39
Bibikova, V. I., 156
Bichler, R. E. J., 165
Biedermann, H. G., 198, 325, 441
Bigoli, F., 441
Bigotto, A., 295, 324
Biloen, P., 112
Bilyk, I. I., 8
Binder, A., 256
Binder, H., 172, 256
Binenboym, J., 161
Biner, F., 17
Bink, T. T., 326
Biradar, N. S., 27, 44, 84
Birayava, V. S., 7
Birchall, T., 192
Bird, P., 371, 405
Bird, P. H., 167, 218
Bird, S. R. A., 220
Birgesson, B., 461
Birkenhäger, G., 222
Birker, P. J. W. L., 255, 265
Birkett, C. R. G,. 489
Birou, M., 19
Biryulina, V. N., 479

Bismondo, A., 490
Bispisa, B., 263
Bissell, E. C., 146
Bite, J., 95
Bizot, D., 78
Black, J. D., 114, 188
Blackmore, M. W., 227
Blair, A., 492
Blake, A. B., 228
Blasse, G., 140, 142, 143
Blauenstein, P., 326
Bleikher, Ya. I., 23
Blessing, R. H., 143
Bliznakov, G. M., 81
Block, B. P., 100, 103
Blokh, M. Sh., 48
Blues, B., 135
Blues, E. T., 427
Blum, P. R., 242, 391, 442
Blume, H., 270
Blyholder, G., 111
Boas, J. F., 476
Bobrik, M. A., 211
Bobrov, Yu. A., 71
Bocard, C., 150
Bochland, H., 18
Bock, H., 373, 394
Bock, M., 188
Bode, J. H. G., 142
Bode, M., 161
Bodeiko, M. V., 49
Bodek, I., 379
Bodner, G. M., 89
Bodrova, L. G., 125
Bodyu, V. G., 209, 240
Böschen, I., 100, 107, 143, 146, 150
Bogatyreva, I. D., 96
Bogdanov, G. A., 65, 144
Boggess, G. W., 435
Boggess, R. K., 215, 468
Bogomolov, G. D., 8
Boguslavskaya, G. I., 141
Bohra, R., 82, 85
Bohres, E. W., 491
Boichinova, E. S., 36, 39, 148
Bois, N., 291
Boiwe, T., 464
Bole, A., 173
Bolen, D. W., 464
Bolletta, F., 94, 338
Bologa, D. A., 365
Bol'shakov, K. A., 158, 347
Bolte, M., 78
Bombieri, G., 296, 394, 493
Bommer, J. C., 308
Bonamico, M., 60
Bonati, F., 216, 431
Bond, A., 359
Bond, A. M., 60, 88
Bondarenko, G. I., 477
Bondarenko, T. N., 5
Bondarenko, V. P., 7, 47, 125, 156

Bonner, L., 492
Bonnet, J. J., 383
Bonnin, A., 108
Bonniol, 103, 104
Bonomo, R. P., 195
Bontchev, P. R., 106
Boorman, P. M., 130, 133
Bor, G., 152, 153, 218, 356
Borden, S. R., 17
Bordignon, E., 221, 432
Bordner, J., 271
Borer, L. L., 286
Borg, D. C., 469
Borin, L. L., 491
Borisov, A. K., 20
Borisov, S. V., 143
Borisova, G. M., 485
Borisova, L. V., 159, 160
Borkett, N. F., 359
Borod'ko, Yu. G., 14, 15, 89
Borokin, G. V., 440
Boronskaya, G. N., 140
Borshagovskii, B. V., 194
Borsheh, A. N., 140
Borzo, M., 446
Bos, K. D., 166
Bos, W. G., 482
Boschi, T., 398
Boschmann, E., 317
Bose, A. K., 4, 294
Bosman, W. P. J. H., 459
Bosnich, B., 259
Bosnyák-Ilcsik, I., 353
Boston, D. R., 264
Bosworth, Y. M., 159
Botar, A. V., 149
Bothner-By, A. A., 446
Bottomley, F., 334, 349
Boucher, L. J., 178, 260
Boudraux, E. A., 94
Bougon, R., 63
Bounthakna, T., 29
Bour, J. J., 255, 265
Boutonnet, R., 485
Bowden, F. L., 14
Bowden, J. A., 88
Bowden, W. L., 335
Bowen, A. R., 127
Bower, B. K., 62, 106
Bowmaker, G. A., 305
Boyd, G. E., 31
Boyd, P. D. W., 228
Boyd, T. E., 220
Boylan, M. J., 188
Bracken, D. E., 94
Braddock, J. N., 337
Braddock, P. D., 408
Bradley, D. C., 29, 60, 473, 488
Bradley, G. F., 219
Braibanti, A., 446
Brai-Koshits, M. A., 43
Brainina, E. M., 40, 41
Braithwaite, A. C., 441
Branca, M., 263

Brand, B. H., 93
Branden, C. I., 464
Brandt, A. O., 446
Brassy, C., 237, 322
Braterman, P. S., 114, 188
Bratibanti, A., 441
Brattus, L., 33
Brauer, D. J., 48
Brauer, G., 305
Braunstein, P., 395, 431
Bravard, D. C., 136
Breece, J. L., 208, 209
Brégeault, J. M., 321
Breitinger, D., 455
Bremard, C., 343
Brencic, J. V., 101, 127
Brennan, J. P., 418
Bresciani, N., 244
Breslin, J. T., 76
Bresler, L. S., 28
Bretthauer, E. W., 470
Breunig, H. J., 268
Breür, H., 87
Breusov, O. N., 81
Brewer, G. J., 467
Brezeanu, M., 313
Brezina, F., 476
Briat, B., 150
Brice, M., 126, 184
Bridgart, G. J., 202
Bright, D., 170
Brill, T. B., 452
Brillard, L., 490
Brindley, P. B., 128
Brint, P., 494
Brintzinger, H. H., 214
Brisdon, B. J., 150, 236
Brisdon, M. E., 227
Britton, D., 429
Brixner, L. H., 142
Brnicevic, N., 83
Brockless, L. D., 305
Brodersen, K., 459
Bronger, W., 387
Brookes, P. R., 454
Brooks, D. W., 209
Brooks, W. V. F., 334
Broszkiewicz, R. K., 376
Brotherton, P. D., 320, 458
Browall, K. W., 384, 422
Brower, W. S., 140
Brown, C. H., 445
Brown, D., 487, 488, 491, 492
Brown, D. G., 217
Brown, D. H., 135, 144, 238, 445
Brown, D. S., 489
Brown, G. M., 320, 337
Brown, I. D., 18, 94, 158
Brown, R. K., 476
Brown, T. L., 163, 220, 256
Brownstein, S., 63, 77, 130
Bruce, M. I., 306, 359
Bruckner, S., 255

Bruder, A. H., 481
Bruice, T. C., 265
Brulet, C. R., 116, 212, 333
Brun, G., 224
Brundle, C. R., 111
Brunette, J. P., 61, 132
Brunner, H., 164, 165
Bruntov, G., 499
Brunvoll, J., 89
Brushmiller, J. G., 262, 263
Buchler, J. W., 277
Brusset, H., 493, 494
Bryce-Smith, D., 427
Brygalina, G. P., 477
Bryan, G. H., 491, 493
Bryan, R. F., 335
Bryan, S. E., 469
Bryukhova, E. V., 40
Bubnova, L. A., 148
Buchanan, D. N. E., 6
Buckler, J. W., 102, 204, 348, 351, 471
Buckingham, D. A., 253
Budarin, L. I., 43
Budd, D. E., 354
Budu, G. V., 451
Buerger, H., 26
Bues, W., 72
Buhikhanov, V. L., 96
Bui Huy, T., 63
Bukin, V. I., 485
Bukovec, P., 472
Bukovszky, A., 138
Bukoyazova, L. I., 216
Bulatov, M. I., 140
Bulatova, A. A., 140
Bulka, G. R., 107
Bullock, J. I., 18, 147
Bulska, H., 119
Bulten, E. J., 166
Bundy. F. P., 384
Bundu, G. V., 430
Burchalova, G. V., 87, 153
Burdett, J. K., 88, 181
Burdykina, R. I., 103
Burg, A. B., 268
Burger, K., 256, 325, 326
Burgess, J., 122, 138
Burkert, P. W., 156
Burlamacchi, L., 112
Burlitch, J. M., 219
Burmeister, J. L., 272, 388, 389, 406, 431
Burmistrova, E. V., 485
Burmistrova, N. P., 108, 425
Burmistrova, T. M., 9
Burns, J. H., 473
Burroughs, P., 87
Bursill, L. A., 91
Burton, C. A., 169
Buryak, N. I., 93, 172
Burykina, A. L., 7
Burylev, B. P., 93
Busch, D. H., 231, 254, 276

Busch, M. A., 151, 217
Busetto, C., 244
Busetto, L., 165, 192
Busev, A. I., 48, 50, 136
Bushnell, G. W., 322, 411
Bushuev, N. N., 142
Buslaev, Yu. A., 77, 133, 145, 149, 155, 497, 498
Buss, B., 146
Butler, I. S., 167, 170, 171
Butler, W., 114
Butterworth, B. E., 468
Buxbaum, P., 325
Buxton, S. R., 494
Buxtorf, R., 229, 275, 324
Buyanov, R. A., 141
Buzina, N. A., 358
Buzuev, G. V., 81
Bycroft, B. M., 394
Byr'ko, V. M., 50, 136
Bystrek, R., 191

Cable, H. D., 312
Cabral, J. de. O., 225
Cabral, M. F., 225
Cadet, A., 63
Caglio, G., 380
Caglioti, L., 394
Cahen, D., 422
Caillet, P., 141
Caira, M. R., 320
Cairns, M. A., 223
Calabrese, A., 104
Calabrese, J. C., 190, 267, 401, 422
Calder, G. V., 269
Calder, V., 476
Calderazzo, F., 1, 46, 492
Calderon, J. L., 90, 192
Calhoun, H. P., 310, 320
Callaham, K. P., 247
Calligaris, M., 244, 257, 297
Calves, J. Y., 81, 144, 145, 201, 239, 443
Calvo, C., 65, 243
Cameron, T. S., 321
Camp, M. J., 32
Campbell, D. H., 441
Campbell, M. J. M., 293, 323, 324
Campero, C. A., 147
Camus, A., 307, 360
Canelli, R. A., 264
Cania, M. R., 310
Cannas, M., 319
Cannon, R. D., 87, 249
Canty, A. J., 389
Canziani, F., 283
Capitan, F., 461
Caraco, R., 318, 327
Carassiti, V., 499

Cardwell, T. J., 98, 365, 393
Carey, J. H., 209
Cariati, F., 244
Carlisle, G. O., 324
Carlson, T. A., 70, 113
Carlyle, D. W., 94
Carmack, M., 317
Carney, H. M., 49
Carpenter, L. J., 258
Carpy, A., 81
Caro, P., 478
Carraher, C. E., 22, 31, 42
Carré, D., 485
Carrell, H. L., 265
Carroll, W. E., 193, 194
Carstons, D. H. W., 73
Carta, G., 319
Carter, F. L., 427
Carter, M. J., 244
Carturan, G., 355
Carty, A. J., 268, 269, 422, 466
Cartz, L., 444
Casabo, J., 96
Casalone, G., 422
Casellato, U., 493, 494, 496
Casey, A. T., 59, 60
Casey, S., 175
Catsikis, B. D., 236, 378
Cattalini, L., 394, 432
Cattermole, P. E., 371
Caughlan, C. N., 473
Caughman, H. D., 305, 312, 320
Caulder, S. M., 427
Caulton, K. G., 114, 193, 222, 330, 334
Cauquis, G., 209
Cavalieri, A., 267, 401
Caville, C., 495
Cefalu, R., 297
Celap, M. B., 261, 265
Celsi, S., 66
Cenini, S., 155, 220, 359, 394
Centineo, G., 324, 393
Cernia, E., 488
Cervona, E., 93
Cesarotti, E., 244
Cetina, R., 326
Cetinkaya, E., 253, 356
Cevales, G., 125
Chackraburtty, D. M., 490
Chadaeva, N. A., 89
Chadwick, B. M., 104
Chadwick, D., 489
Chakladar, J. K., 109
Chakravarty, A. S., 132
Chakravorti, M. C., 132, 138, 150, 160, 161
Chakravorty, A., 217, 243, 296, 323
Chakravorty, S., 134
Chakrawarti, P. B., 11
Chalkov, G. I., 150
Chamberland, B. L., 105

Chaminade, J. P., 78
Chan, A. S. K., 191
Chan, L. Y. Y., 153
Chan, S., 246
Chan, T. T. D., 250
Chandler, G. S., 306
Chandra, M., 365, 393
Chandrasekharan, M., 390, 392
Chandry, S. C., 445
Chang, C. H., 251
Chang, S. W. Y., 74
Chant, R., 100, 261
Chapman, B. E., 176
Charambous, J., 294
Charlu, T. V., 47, 123
Charpin, P., 63
Chase, L. L., 65
Chastain, R. V., 319
Chatt, J., 111, 116, 153, 212, 348, 355, 396
Chatterjee, K. K., 57, 422
Chaturvedi, D. N., 467
Chaturvedi, G. K., 301
Chaudhuri, M. K., 160, 161
Chauhan, S. D., 234, 287, 324
Chaurasia, M., 461
Chawla, K. L., 490
Cheam, V., 327
Chebotarev, N. T., 142, 158
Chedini, M., 394
Chelowski, A., 99
Cheetham, A. K., 20
Chel'tsov, P. A., 412
Chelyanova, D. P., 68
Chen, K. S., 121
Chen, L., 484
Chen, T., 140, 484
Chen, Y. W., 136
Cheng, C. L., 451
Cheng, S. T., 464
Chentsov, V. N., 142
Chepiga, M. V., 156
Cherednichenko, I. F., 20
Cheremisina, I. M., 100, 366, 401, 413, 423
Cheriyan, U. O., 107
Cherkasova, T. G., 358
Chernaya, N. V., 55
Chernikov, S. S., 395
Chernov, D. B., 124
Chernov, R. V., 4, 17
Chernova, V. I., 153
Chernyaev, I. I., 368
Chernyaev, I. Z., 422
Cherry, W., 309
Cherstvova, V. B., 485
Chertkov, A. A., 48
Cherwinski, W. J., 406
Chetverina, R. B., 148
Chevrel, R., 124
Chevrier, J., 137
Chiacchierini, E., 97, 103
Chiariza, R., 499
Chidamboram, R., 320

Chieh, P. C., 422, 466
Chien, J. C. W., 106
Chihara, K., 19
Chikuma, M., 298
Chimura, Y., 9
Chini, P., 218, 267, 283, 352, 401
Chirkin, G. K., 71
Cirkov, A. K., 47, 65
Chisholm, M. H., 112
Chistyakov, V. F., 84
Chiswell, B., 173, 241, 302
Chiu, F. L., 484
Chiukova, A. I., 89
Chivers, T., 15, 130
Chizhikov, D. M., 123
Chjo, K. H., 130
Choarizia, R., 205
Chobanyan, P. M., 140
Chock, P. B., 469
Choi, J. S., 248
Choi, Q. W., 21
Choi, S. N., 94
Chojnacki, J., 146
Chondhury, P., 362
Choppin, G. R., 488
Chopra, S. L., 446, 461
Chosa, M., 302
Choudhar, C. B., 37
Choudhurg, P., 398
Chourou, S., 493
Chow, B. C., 366
Chow, C., 310, 320
Chow, K. K., 234, 282, 374, 396
Chow, S. T., 291, 466
Chow, Y. M., 429
Christensen, J. J., 216
Christoph, G. C., 197, 202, 443
Christopher, R. E., 424
Christophliemk, P., 100, 237, 260, 443, 445, 461
Chu, C. T. W., 192
Chu, Y. H., 464
Chubar, Yu. D., 149
Chubarova, V. K., 343
Chuchelina, G. Ya., 483
Chupakhina, R. A., 482, 483
Chupin, A. I., 20
Church, M. J., 415
Churchill, M. R., 74, 169, 191, 247, 306, 309, 443
Chudik-Major, L., 137
Chum, H. L., 197
Chumakova, G. M., 74
Chuvaev, V. F., 143, 147, 149
Ciampolini, M., 441, 459
Ciani, G., 192, 218, 380, 422
Cieslak, M., 144
Citrin, P. H., 337
Clardy, J., 321, 382, 400
Clare, B. W., 66
Clark, H. C., 165, 406
Clark, N. J., 476

Clark, R. J. H., 29, 81, 127, 159, 217, 431, 452
Clark, R. P., 108
Clarke, J. F., 44
Clarkson, S. G., 334
Clase, H. J., 266
Cleare, M. J., 217, 284
Clearfield, A., 38, 143
Clegg, W., 163, 447, 448
Cleland, A. J., 266
Cleland, W. W., 103
Clemente, D. A., 355, 459, 493, 494
Cleveland, J. M., 491, 493
Clifford, P. R., 454
Coccanari, P., 97
Cocchieri, R., 103
Cocker, D., 236
Cocksey, B. G., 435
Coe, C. G., 260
Coffari, E., 156
Coffin, D. L., 198, 233
Coffery, M. L., 288
Cohen, A. H., 21
Cohen, I. A., 305, 366
Colaitis, D., 155
Colburn, T. J., 56, 324
Cole, G. M., 93, 216
Cole-Hamilton, D. J., 333
Coleman, J. E., 462
Coles, B. R., 253
Coles, M. A., 29
Collamati, I., 97, 206, 313, 324, 416, 442
Collin, G., 179
Collin, R. J., 225, 350
Collin, R. L., 174
Collins, B. M., 140
Collins, D. E., 90
Collins, D. M., 126
Collins, L. W., 250
Collins, M. J., 286
Collman, J. P., 214, 346, 352
Colquitt, K., 179
Colsmann, G., 214
Colton, D. F., 228
Colton, R., 88, 383
Colville, N. J., 167, 171
Conder, H. L., 163, 188
Condorelli, G., 324, 393, 491, 497
Connelly, N. G., 221
Conner, J. A., 168
Connolly, N. G., 117
Connor, J. A., 87, 89
Conroy, L. E., 123
Constant, G., 61, 201
Contant, R., 147
Conti, C., 93
Contreras, R., 326
Cooke, C. G., 162, 221
Cooke, R. R., 389
Cookson, D. J., 12
Cooper, H., 492

Cooper, J. N., 109
Cooper, M. K., 113
Cooper, N. J., 111, 120
Copeland, D. A., 106
Coppens, P., 206
Corbett, M., 255
Corbin, J. L., 136
Corden, B. J., 330, 345
Cordes, A. W., 114
Corenzurit, E., 128
Corey, E. J., 109
Cordfunke, E. H. P., 499
Cornwell, A. B., 220
Coronas, J. M., 96
Corradi, A. B., 255, 296
Correck, P. W., 294
Corsi, N., 492
Corsini, A., 472
Corsmit, A. F., 140, 143
Costa, G., 224
Costantine, U., 38
Costanzo, L. L., 491
Cotton, F. A., 88, 90, 111, 112,
    120, 126, 152, 180, 182,
    184, 185, 187, 190, 353,
    394
Couch, D. A., 307, 338, 339,
    397
Couch, L. S., 46
Coucouvanis, D., 100, 209,
    210, 288, 308, 365, 426
Countryman, R. M., 321, 325
Courbion, G., 94
Court, T. L., 303
Cousseau, J., 33
Cousseins, J. C., 485
Coutts, R. S. P., 14
Coutures, J. P., 108, 478
Covey, W. D., 468
Cowie, M., 42, 130
Cowman, C. D., 127, 298, 401
Cox, A., 189
Cox, D. E., 161
Coyle, C. L., 204
Crabtree, R. H., 110
Craese, A. E., 186
Cragel, J., 263
Craig, D. C., 452
Cramer, R. E., 272, 486
Cras, J. A., 133, 318, 459
Crawford, V. H., 323
Crease, A. E., 126, 167, 182
Creaser, E. H., 253
Creel, R. B., 47
Creswell, P. J., 253
Creternet, J. C., 49
Creutz, C., 342
Creyf, H. S., 451
Crisler, L. R., 488, 489
Cristian, D. F., 244
Cristini, A., 96, 319, 459
Croatto, U., 221, 493
Crociani, B., 398
Crofts, J. A., 489

Cronin, D. L., 88
Cross, R. J., 393
Cruickschank, D. W. J., 352
Crumbliss, A. L., 256
Csontos, G., 354, 372
Csopak, H., 462
Cudennec, Y., 108
Cullen, D. L., 277
Cullen, W. R., 169, 185, 354
Cummings, S. C., 242, 297,
    391, 442
Cundy, C. S., 183, 269
Cunningham, D. E., 466
Cunningham, J. A., 253, 277
Curran, R. T., 323
Curtis, M. D., 165, 219
Curtis, M. T., 489
Curtis, N. F., 254, 274, 275,
    278, 443
Cusachs, L. C., 344, 435
Cusick, J. P., 201
Cutforth, B. D., 452
Cyriac, M. U., 300
Czakis-Sulikowska, D. M., 451

Daganello, G., 355
Dahan, F., 107
Dahl, G. H., 100
Dahl, L. F., 52, 192, 221, 267,
    401
Daireaux, M., 291
Dakternieks, D. R., 295, 298
Dale, B. W., 404
D'Alessandro, G., 499
Dall'asta, G., 145
Dallavalle, F., 325, 446
Dalmonego, H., 67
Daly, J. J., 365, 373
Dalziel, J. R., 21
D'Amello, M. J., 275
Damien, D., 491
Dammann, C. B., 379
da Mota, M. M., 432
D'Amour, H., 146, 148
Danby, C. J., 435
Dance, I. G., 207, 237, 300,
    325
Dance, J. M., 92
Danesi, P. R., 205, 499
Danilevich, T. I., 489
Danilin, V. N., 93
Danilova, T. V., 451
Danot, M., 33
Dantzer, P., 305
Dao, N. Q., 493, 494
Daoudi, A., 105, 480
Dapporto, P., 283
Daran, J. C., 61, 201
Darensbourg, D. J., 87, 154,
    163, 188
Darlington, C. N. W., 81

Darriet, B., 65, 137
Darriet, J., 81, 345
Dart, J. W., 361, 362
Dartinguenave, M., 233, 281
Dartinguenave, Y., 233, 281
Das, M., 200, 298, 365, 391
Das, S. K., 446
Dash, A. C., 215
Dash, S., 467
Datt, I. D., 107
Dave, L. D., 300
Davey, G., 181, 219
David, F., 203
David, P. G., 203
David-Auslaender, J., 161
Davidovich, R. L., 78, 497, 498
Davidson, J. L., 115, 126, 170,
    290
Davidson, M., 255
Davies, B. W., 422
Davies, C. G., 452
Davies, G., 151
Davies, J. E., 131, 177, 208
Davis, A. R., 471
Davis, D. G., 428
Davis, M. H., 42, 103
Davis, R., 154, 155
Davison, A., 2, 332, 362, 418
Davitashvili, E. G., 484
Dawson, J. W., 281
Dawson, P. T., 124
Day, P., 107
Day, R. O., 177
Day, V. W., 90, 111, 177, 192
Dazzi, E., 430
De, R. L., 257
De Alti, G., 295, 324
Dean, P. A. W., 16, 17, 452,
    453
Dean, W. K., 190
Deane, M. E., 116
De Angelis, G., 97
Debaerdemaeker, T., 132
De Beer, A., 189
Debelle, V., 108
De Boer, B. G., 191, 247
De Boer, E. A., 199, 265
De Boer, J. J., 140
De Bolster, M. W. G., 175, 234
Deeney, F. A., 194
De Filippo, D., 200, 233, 301,
    313
Degetto, S., 493, 496
De Haas, K. S., 341, 389
Dehand, J., 395
De Hayes, L. J., 254
Dehnicke, K., 55, 146, 195,
    432
De Jong, I. G., 152
De Jovine, J. M., 101
De Kock, C. W., 92
De Kock, R. L., 217
Delbeke, F. T., 268
Delettre, J., 237, 322

De Liefde Meijer, H. J., 1, 15
De Lima, W. N., 472
Dellesbaugh, G. G., 249
Delluca, R. J., 253
Deloume, J. P., 448
Delphin, W. H., 127
Deluca, J. A., 92
De Maeyer, L., 62, 338
De Maria, G., 156
Demas, J. N., 338, 377
Demazeau, G., 318
Dembicka, D., 150
de Meester, P., 320
Demiray, F., 72
Demkin, A. M., 159
De Mourgues, I., 112
Dem'yanenko, V. P., 81
Dem'yanets, L. N., 38, 485
Denisov, N. T., 16, 49
Denisova, L. A., 35
Denisova, N. E., 39, 148
Dennis, D., 280
Denton, D. L., 16
De Pamphilis, B. V., 211
De Pamphilis, M. L., 103
De Paoli, G., 415
De Pape, R., 94
Dermott, T. E., 31
Dernier, P. D., 128
Derouane, E. G., 198
De Rumi, V. B., 96
Dervin, J., 490
Desai, I. De V. P., 132
Desgardin, G., 81
De Simone, R. E., 333
De Smedt, H., 62, 338
Dessy, G., 60
De Stefano, N. J., 282, 397
Dettingmeijer, J. H., 139
Deutsch, E., 100
Devillanova, F., 200, 301
de Vore, T. C., 476
De Vries, J. L. K. F., 199
Devyatykh, G. G., 46
Dewan, J. C., 69
Dey, K., 57, 257
Dhindsa, K. S., 26
Dias, S. A., 367, 398
Diaz, A., 459
Diaz-Colon, F. A., 81
Dichmann, K. S., 168
Dick, Y. P., 43
Dickens, P. G., 137
Dickerson, D., 494
Dickinson, R. C., 198, 233
Dickinson, R. J., 404
Dickson, I. E., 314
Dieck, R. L., 476, 477
Diemann, E., 150
Diemente, D., 338
Dietzsch, W., 150, 323
Dilebaeva, A. A., 482
Dill, E. D., 114
Dillin, D. R., 422

Dilworth, J. R., 111, 153, 410
Dimante, A., 95
Din, K. U., 118
Dingle, T. W., 424
Di Salvo, F. J., 6, 73, 76
Dittemore, K. M., 433
Dittmer, G., 122
Dixit, L., 32, 159
Dixon, F. L., 367
Dixon, K. R., 372, 389, 390, 406, 411, 415, 424
Djordjevic, C., 83
Djuric, G., 425
Doadrio, A., 151
Dobbie, R. C., 189, 221
Dobcnik, D., 127
Dobizynski, E., 382
Dobson, C. M., 486
Dodge, R. P., 494
Doedens, R. J., 322
Doemeny, P. A., 116
Doi, K., 20
Doichinova, H. I., 144
Dokuzoguz, H. Z., 489
Dolcetti, G., 165, 192
Dollimore, L. S., 255
Dolozoplosk, B. A., 30
Dolphin, D., 469
Dombek, B. D., 87, 182
Domingos, A. M., 107
Dominguez, R. J. G., 52
Dominick, T. L., 319
Donahue, J., 268
Donaldson, J. D., 220
Donato, H., 484
Donohue, P. C., 159
Doonan, D. J., 339
Dopico Vivero, M. T., 9
Dorfman, L. P., 129
Dorfmann, Ya, A., 49
Dori, Z., 112, 216, 244, 306
Dorman, W. C., 128
Dormond, A., 30, 31
Dorn, W. L., 193
Dorokhov, Yu. G., 140
Dorrity, I. A., 228
Dorrity, J., 271
Dosser, R. S., 229
Douglas, B. E., 249, 260, 263, 264
Douglas, P. G., 332
Douglas, W. M., 88, 188
Dove, M. F. A., 303
Dovgei, V. V., 48
Dovlyatshina, R. A., 158
Downs, A. W., 367, 398
Downs, H. H., 97, 239
Doyle, J. R., 433
Doyle, M., 106
Drago, R. S., 232, 243
Dragoo, A. L., 70
Drake, R. F., 323
Drakenberg, T., 461, 466
Drapier, J., 245

Drbalek, Z., 91
Dremin, A. N., 81
Drew, D., 163
Drew, M. G. B., 83, 85, 86, 128, 134, 139
Drew, R. E., 68
Dreyer, R., 446
Drickamer, H. G., 217
Driessen, W. L., 234
Drillon, M., 91
Driss, D., 81
Drobot, D. V., 122, 129
Drobotenko, V. V., 62
Drobyshev, V. N., 81
Drogunova, G. I., 10
Drozdov, Yu, N., 65
Drum, D. A., 119
Dubey, B. L., 18
Dubey, K. P., 461
Dubey, R. J., 168, 170
Dubler, E., 236, 310, 320
Dubois, R., 486
Dubovitsky, V. A., 30
Dubrov, Yu. I., 60
Dubrovin, A. V., 496
Duda, J., 27, 42
Dudeck, E. P., 296
Dudeck, G. O., 296
Dudley, R. J., 318, 323
Dufaux, M., 112
Duff, J. M., 397
Duffy, D. J., 206, 261, 366
Duffy, N. V., 405
Duggan, D. M., 274, 310, 321, 322, 324
Dul'banova, V. G., 408
Dunaeva, K. M., 496
Dunbar, R. C., 88
Duncan, C. S., 175, 234
Dunn, M. F., 463
du Preez, A. L., 313
du Preez, J. G. H., 490, 491, 492, 499
Durif, A., 321
Durkin, T. R., 405
Dustin, D. F., 246, 247, 248
Dutrizac, J. E., 123
Dutta, R. L., 57
Duvansoy, Y., 31
Duyckaerts, G., 1, 488
Dvorina, L. A., 7, 156
Dwyer, M., 253
Dwyer, P. N., 232, 277
D'yachenko, S. A., 451
Dyaltlova, N. M., 239, 326
Dyke, J. M., 278, 324, 325
Dymock, K., 268
D'Yvoire, F., 95, 205
Dzeganovskii, V. P., 5
Dzyadykevich, Yu. V., 125

Eaborn, C., 332, 417, 418
Eady, C. R., 329

Eager, R. L., 340
Earley, J. E., 345
Earp, C. D., 305
Eaton, G. R., 471
Eaton, S. S., 471
Eatough, D. J., 216
Ebinger, H. M., 170
Ebner, J. R., 152
Ebsworth, E. A. V., 112, 424
Echigova, E., 112
Echols, H. F., 280
Eckberg, R. P., 98, 235
Eckel, M. F., 156
Eckelmann, U., 131
Edelstein, N., 488, 491
Edgar, B. L., 206, 261, 366
Edge, R. A., 491
Edgell, W. F., 219
Edmiston, D. J., 76
Edmundson, A. B., 466
Edwards, A. J., 63, 107, 344, 422
Edwards, D. A., 150, 307
Edwards, J., 492
Edwards, W. T., 88
Eeasteal, A. J., 18
Eeckhaut, Z., 268, 427
Efimov, A. I., 48
Efimova, E. M., 379
Efraty, A., 191
Efremov, V. A., 141, 485
Eggerding, D., 288
Eggerman, W. G., 488
Egginton, G. M., 134
Eggli, R., 279
Egorov, A. S., 27
Ehrhardt, H., 312
Ehrl, W., 190, 221, 268
Eick, H. A., 422
Einsphano, H., 268
Einstein, F. W. B., 68, 169, 186, 471
Eisenberg, R., 330, 345
Eisenstein, J. C., 481
Ekland, H., 464
Eland, J. H. D., 435
Elder, R. C., 100, 171, 248, 261, 272
El-Gad, U., 161
Eliezer, I., 451
Eliseev, S. S., 145, 161
Eller, P. G., 208, 321, 325, 469
Ellgen, P. C., 188, 191
Ellis, D. E., 6
Ellis, J., 245
Ellis, J. D., 11
Ellis, V. M., 227, 272
Ellison, R. D., 453
Elmakabaty, S., 326
El'manova, N. A., 161
Elmitt, K., 112
Elphingstone, E. A., 115, 343
El-Sayed, L., 239, 243
El-Sharkawy, G. A. M., 341

Elson, C. M., 116, 212
Elson, R. E., 494
El-Touky, A., 243
Ely, K. R., 466
Ely, N., 476
Emel'yanova, E. N., 485
Emil'yanova, G. N., 362
Emschwiller, G., 194
Endes, H., 302
Endicott, J. F., 229, 254, 368
Endo, H., 50
Endress, H., 243, 384
Enemark, J. H., 328
Englehardt, L. M., 276, 279
Engelsman, F. M. R., 99
Enghag, P., 72
Ennever, J. F., 88
Enomoto, Y., 266
Epov, D. G., 472
Epperlein, B. W., 436
Epps, L. A., 256
Epstein, J. M., 322, 422
Epstein, M., 468
Erametsa, O., 479
Ercolani, C., 97, 206, 282, 313, 324, 416, 442
Erdmann, B., 491
Eremenko, L. A., 81, 480
Eremenko, V. N., 92
Eremin, M. V., 98
Eremin, Yu. G., 477, 483
Ergeshbaev, Z., 446
Erickson, N. E., 111
Eriksson, B., 473
Ermakov, A. N., 60, 61, 160, 428
Ermolaev, M. I., 159, 488
Ermolaeva, T. I., 31
Ermolenko, I. M., 17
Erokhin, E. V., 139
Eroshin, P. K., 39
Escard, J., 224, 324, 435
Eschbach, C. S., 218
Eschchenko, L. S., 205
Esel'sen, B. M., 68
Espenson, J. H., 105, 470
Estes, E. D., 319, 324, 422
Etienne, J., 485
Ettinger, M. J., 323, 325
Eulenberger, G., 215
Evans, D. F., 485
Evans, G. O., 356
Evans, G. W., 104
Evans, H. T. jun., 148
Evans, I. P., 199, 238, 363, 389
Evans, J., 353
Evans, R. S., 94
Evans, S., 1
Evans, W. J., 246, 247, 248
Evarestov, R. A., 156
Evchenko, T. I., 96
Evdokimov, A. A., 142
Everett, G. W., 251, 262, 263, 333, 364

Evreev, V. N., 265
Evtimova, B., 216
Evtushenko, N. P., 438
Ewall, R. X., 337
Eyraud, C., 137
Ezerskaya, N. A., 365
Ezhoo, A. I., 427

Fabretti, A. C., 433, 442
Fabrizzi, L., 217, 327, 441
Fachinetti, G., 2, 29, 46, 257
Fachinetti, L. M., 12
Fackler, J. P., 68, 88, 288, 323, 400
Fadeeva, N. V., 160
Fahey, J. A., 496
Fajer, J., 469
Falconer, W. E., 107, 344, 422
Falicheva, A. I., 103
Falk, M., 324
Fanikan, J. A., 242, 324
Fal'kengof, A. T., 127
Fallon, G. D., 103
Falqui, M. T., 430
Falth, L., 457
Fane, A. G., 489
Fantucci, P., 283
Faraone, F., 371, 374
Farber, M., 91
Fardy, J. J., 491
Fares, V., 60
Farhangi, Y., 452
Farkas, E., 327
Farmer, R. L., 57
Farmery, K., 231
Farona, M. F., 115
Farooq, O., 14
Farooqi, A. S., 174, 229
Farr, J. D., 73
Farrall. M. J., 77
Farrell, M. O., 178
Farthing, R. H., 383
Faucherre, J., 490
Faulkner, T. W., 255
Fava, J., 480
Fawcett, J. P., 167, 220
Fay, R. C., 22, 24
Fayos, J., 382
Fazakerley, G. V., 310, 315, 320
Fedorov, G. G., 72
Fedorov, P. M., 142
Fedorov, V. A., 447, 451
Fedorov, V. B., 124
Fedorov, V. N., 106
Fedorova, A. V., 447
Fedosov, A. P., 38
Fedotovskikh, N. G., 173
Fedulova, L. G., 71
Feigleson, R. S., 65

Feit, B. A., 305, 426
Feitknecht, W., 202
Felden, J. J., 44
Felkin, H., 199
Felps, S., 344
Feltham, R. D., 328
Felton, R. H., 203, 469
Felty, W. L., 468
Fender, B. E. F., 73, 90. 140
Fenerty, J., 91
Fenster, A. E., 171, 354
Fenton, D. E., 322
Ferey, T., 201
Ferguson, B. J., 17
Ferguson, D., 14
Ferguson, J. A., 337
Fergusson, J. E., 343, 459
Fermor, J. H., 425
Fernando, Q., 286, 307, 308
Fernando, W. S., 87
Fernandopulle, M. E., 226
Ferraris, G., 352
Ferraro, J. R., 233, 242, 280, 302
Ferraudi, G., 368
Ferris, L. M., 489, 491
Ferro, R., 70
Fesenko, E. G., 36
Fiat, D., 96
Field, J. S., 169, 222
Fieldhouse, S. A., 162, 218
Figgis, B. N., 322
Filatov, L. Ya., 36
Filatova, N. A., 33
Filipczuk, S. W., 200
Filip'ev, V. S., 36
Filipovic, I., 446
Filippenke, N. V., 33
Filippo, J. S., 363
Filla, A. C., 260
Findlay, M., 106
Fine, D. A., 237
Fink, F. H., 473
Finney, K. S., 262, 263
Finster, J., 150
Finta, Z., 256
Fischer, D. W., 6
Fischer, J., 147, 440
Fisher, D. R., 194
Fisher, K. J., 439
Fishwick, J. P., 324
Fitzpatrick, N. J., 89
Fitzsimmons, B. W., 199
Fiveiskii, E. V. F., 7
Flack, H. D., 433
Flack, P. H., 380
Flamini, A., 289
Fleckenstein, J., 498
Fleet, G. W. J., 109
Fleischer, E. B., 277, 311, 322, 346, 352, 367
Fleischman, W. D., 446
Fleming, C. A., 97
Flengas, S. N., 35, 172

Fletcher, S. R., 151, 158, 320, 352
Flint, C. D., 98
Flood, M. T., 345
Flook, R. J., 467
Floren, S., 385
Florian, L. R., 100, 261
Florian, R., 322, 367
Floriani, C., 2, 29, 46, 257
Flotow, H. E., 143
Fluck, E., 256, 455
Flynn, B. R., 357
Flynn, C. M., 54, 67, 148, 377
Foester, C., 221
Folkesson, B., 97
Fomenko, V. N., 92
Fomichev, E. N., 7, 47, 125, 156
Fomichev, V. A., 125
Fomicheva, K. K., 144
Fontana, S., 90, 192
Fonteneau, G., 491
Fooney, S. W., 266
Ford, P. C., 11, 337
Forder, R. A., 2, 112, 121
Foris, C. M., 142
Forman, A., 469
Forsellini, E., 394, 493, 496
Forster, D., 365
Foster, S., 344
Fotiev, A. A., 49, 65, 472, 479
Fournier, M., 96
Foust, R. D., 11
Fowles, G. W. A., 44, 61, 86, 134, 139
Fragala, I., 195, 324, 393, 497
Francesconi, L., 277, 469
Francis, J. N., 246
Franck, R., 81
Frankel, R. B., 210
Franks, M. L., 127
Franson, G., 321
Frantsuzova, T. E., 141, 179
Franz, K., 10
Franzen, H. F., 6
Fraser, A. J. F., 345
Fraser, C. J. W., 138
Fraser, I. W., 50, 98
Frauendorfer, E., 90, 192
Frazer, A. R., 218
Frederick, C. G., 105
Freeman, A. J., 6
Freeman, F., 107
Freeman, H. C., 194, 467
Freni, M., 155
Frenkel, V. Ya., 35
Frenz, B. A., 112, 120, 152, 182, 185, 187, 190
Fresco, J., 401
Freser, J., 298
Frey, J. E., 378
Friebel, C., 303, 309, 325
Fridman, Va. D., 150, 451
Friec, B., 78
Fried, K., 50

Friedrich, C., 194
Frisch, P. D., 117, 191, 238
Frit, B., 65, 140
Fritchie, C. J., 429
Fritz, H. P., 18, 176, 440, 446, 470
Fritze, K., 174
Fritzer, H. P., 95
Fronczek, F. R., 245, 246, 266
Fruchart, J. M., 147
Frunze, M. F., 42
Frye, H., 169, 387
Fuchs, G., 100
Fuchs, J., 79
Fuger, J., 488
Fuggle, J. C., 78
Fuhr, B. J., 465
Fuhrhop, J. H., 217, 328
Fujimori, K., 321
Fujinami, S., 266
Fujino, T., 68
Fujita, E., 310, 321
Fujita, I., 338
Fujita, J., 251
Fujita, T., 139
Fujita, Y., 59
Fujiwara, I., 239, 445
Fujiwaru, S., 251, 398
Fukuda, Y., 310
Fukumoto, T., 220
Fukunaga, O., 139
Fulcher, G., 490
Fulea, C., 385
Fullam, B. W., 162, 218, 454
Fulrath, R. M., 36
Funabiki, T., 266
Fung, K. W., 453
Funke, A., 107
Furlani, C., 289
Furman, A. A., 4
Furtne, L. A., 136
Furuhashi, A., 289
Furuseth, S., 33
Fusu, I. L., 241

Gabay, A., 129
Gadet, A., 229
Gafner, G., 370
Gafney, H. D., 367
Gaglani, A., 66
Gagne, R. R., 214
Gahan, L. R., 343
Gaiduk, V. V., 40
Gaines, D. F., 152
Gainsford, G. R., 253
Gainulin, I. F., 106
Gaivoronskii, P. E., 46, 88, 89
Gal, D., 236
Gal, I. J., 425, 437

Galambos, G., 62
Galbraith, A. R., 158
Galesic, N., 83
Galinos, A. G., 49, 97
Galitskii, N. V., 46, 95
Galizzioli, D., 244
Galkina, I. P., 478
Gall, C., 26
Gall, R. S., 192, 221
Galliart, G., 43
Gallotti, G., 327
Galsbol, F., 94, 252, 370, 380
Galy, J., 9, 52, 65, 81, 95, 137
Gancarczyk, T., 445
Gancedo, R., 113
Ganelina, E. Sh., 148
Ganescu, I., 102
Gantmakher, A. R., 30
Gaponenko, V. A., 98
Gapotchenko, N. I., 152
Garaj, J., 311
Garcia, F. Z. A., 479
Garcia-Bao, C., 103
Garcia-Blanco, S., 302
Gardhammar, G., 484
Gardiner, J., 249
Gardner, I. R., 66
Gardner, S. A., 44
Gardner, W. E., 92
Garla, L. C., 483
Garner, A. W. B., 411
Garner, C. D., 87, 133, 166, 170, 185
Garner, C. S., 101
Garner, C. W., 463
Garnett, J. L., 50, 98
Garnett, P. J., 263
Garni, M. S. J., 20
Garnier, E., 108
Garrett, B. B., 94, 101
Garrou, P. E., 355, 361
Gasparrini, F., 394
Gasperin, M., 81
Gastaldi, L., 289
Gatehouse, B. M., 103, 131, 146, 177, 208
Gatillov, Yu. V., 19
Gattegno, D., 299, 443
Gattegno, O., 368
Gatteschi, D., 272
Gaughan, A. P., 306, 330, 367
Gaur. J. N., 451
Gauss, P. L., 256
Gavrilova, I. M., 471
Gavrilova, I. V., 424
Gavrish, A. M., 36, 141, 142
Gawande, M. P., 481
Gazo, J., 311, 325
Geaman, J. A., 191
Gearon, I. E., 98
Geary, W. J., 228
Gebala, A. E., 327
Geiger, W. E., 243
Geis-Blazekova, M., 11

Gel, P. V., 7, 92
Gel'bshtein, A. I., 65, 123
Gel'd, P. V., 33, 36
Gel'fman, M. I., 424
Gellings, P. J., 20, 23
Genchev, M., 459
Generalova, N. B., 240, 294, 446, 497
Gennis, M., 345
Gentile, P. S., 234
Geoffrey, G. L., 152, 361, 369
George, T. A., 115
Gerasimova, S. O., 451
Gerasimyuk, G. I., 478
Gerault, Y., 108
Gerbeleu, N. V., 56, 209, 240
Gergely, A., 327
Gerlach, D. H., 269, 381
Gerlach, J. N., 191
Gerloch, M., 228
Germain, J. E., 385
Gerstein, B. C., 113, 286
Gervasio, G., 352
Get'man, E. I., 96, 140, 142
Gheorghiu, C., 102, 135
Ghilardi, C. A., 223
Ghoneim, M. M., 442
Ghose, S., 320
Ghosh, J. C., 430
Ghosh, N., 26
Ghosh, S. N., 436
Ghosh, S. P., 195
Ghotra, J. S., 473, 484, 488
Gianguzza, A., 297
Gibart, P., 99
Gibler, D. D., 138, 345
Gibson, D. M., 87
Gibson, E. K., 250
Gibson, J. F., 75, 152
Gibson, J. G., 274
Gibson, K., 202
Gibson, M. L., 491, 492
Gidney, P. M., 366
Gieger, B., 473
Giesbrecht, E., 195, 479
Gifford, S., 309
Gil, U. M. S., 494
Gil-Arnao, F., 32
Gill, J. S., 144
Gillard, R. D., 255, 263, 264, 336, 364, 366, 411
Gilles, J. C., 20
Gillespie, R. J., 452
Gilli, G., 222, 352
Gillman, H. D., 93, 100, 284
Gillson, J. L., 105
Gingerich, K. A., 340, 417
Ginsberg, A. P., 373
Giordano, R. S., 323, 324, 325
Giovanoli, R., 96, 202
Giovetti, G., 301
Girichev, G. V., 4, 35
Giricheva, N. I., 4, 33
Giridharan, A. S., 489

Girling, R. L., 422, 466
Giuffrida, S., 497
Gizhinksii, A. R., 142
Gladkaya, N. S., 422
Gladkikh, N. T., 123
Gladneva, A. F., 485
Gladyshevskii, E. I., 34
Gladysz, J. A., 116, 212
Glaizer, F., 325, 326
Glass, W. K., 173, 225, 384
Glavic, P., 42, 173
Glazyrin, M. P., 65
Glebov, L. A., 95
Gleitzer, C., 248
Gleizes, A., 33
Glemser, O., 169, 170
Glick, M. D., 114, 229, 286
Glockling, F., 403
Glukhov, I. A., 72, 145, 161
Glushkova, M. A., 155
Glusker, J. P., 265
Gluvchinsley, P., 324
Gnare, G. P., 245
Gnatyshin, O. M., 23, 40
Goby, G., 490
Goddard, W. A., 1
Godina, N. A., 65, 81
Godwin, H. A., 241
Goebel, H., 99
Goedken, V. L., 197, 202, 224, 443
Goeminne, A. M., 427
Goff, H., 257
Goffart, J. I., 488
Goggin, P. L., 368, 383, 397, 403, 407, 419
Gogolin, R., 193
Gokhale, K.V.G., 37
Goldberg, D. E., 272, 310
Goldfield, S. A., 181
Golding, B. T., 265
Gol'dshtein, N. D., 65, 123
Goldstein, M., 385, 454, 459, 460
Golkowska, A., 147
Golic, L., 145
Goll, W., 46
Golobtsova, Z. G., 136
Golosov, V. V., 408
Golovaneva, I. F., 422
Golovanova, A. I., 76
Golovkin, B. G., 65
Golub, A. M., 17, 39, 133, 134, 140, 142, 160, 224, 271, 472, 479, 480
Golub, Ya. S., 7
Golubev, A. A., 10
Golubev, A. M., 149
Golubnichaya, M. A., 422
Gomes, J., 7
Gomez, L. A., 485
Gomez-Lara, J., 326
Gomm, P. S., 200, 228, 433
Goncharik, V. P., 215

Goncharuk, L. V., 92
Gong, H., 246
Gonschorek, W., 140, 149
Gonzalez-Garcia, S., 103
Gonzalez-Vilchez, F., 106
Good, M. L., 210, 236, 378
Good, R., 82
Goodall, B. L., 245
Goodenough, B., 81
Goodenough, J. B., 92, 318
Goodfellow, R. J., 368, 397, 403, 407, 419
Goodwin, H. A., 196, 201, 228, 280
Goodyear, J., 172
Gopienko, V. G., 4, 92
Gopinathan, C., 21, 22
Gordienko, L. N., 479
Gordienko, V. I., 216
Gordon, G., 55
Gordon, H. B., 28, 44
Gore, E. S., 231, 276
Goreaud, M., 108
Gorenbein, E. Y., 438
Gorlova, T. M., 84, 158, 379
Goroschenko, Ya. G., 3, 19, 96
Gorshkova, L. V., 124
Gorter, S., 173
Gorz, G. 459
Gosser, L. W., 223, 269
Gothlicher, S., 440
Goto, M., 81
Goudar, T. R., 84
Gouger, S., 205
Gould, E. S., 145
Gould, R. O., 345
Gouterman, M., 367
Govorov, A. A., 142, 480
Govtsema, F. P., 72
Gowda, N. M. N., 368
Grabaric, B., 446
Grachev, M. K., 133
Graddon, D. P., 286, 295, 298, 325, 452
Graf, W., 198, 325, 441
Graham, L. R., 261
Graham, L. W., 267
Graham, M. A., 88
Graham, R. D., 326, 385
Graham, W. A. G., 153, 183, 219, 330
Gramlich, V., 105, 282
Grandjean, D., 124, 129, 214
Granifo, J., 191
Grankina, Z. Ya., 483
Grannec, J., 481
Gransden, S. E., 68
Grant, W. B., 107
Grassi, R. L., 12
Gravereau, P., 108
Gray, H. B., 35, 127, 152, 176, 195, 210, 281, 298, 345, 361, 369, 401
Grayson, R. L., 127

Graziani, M., 390
Graziani, R., 394, 493, 496
Grdenic, D., 145, 460
Greatrix, R., 189
Grebenova, B., 144, 146
Grechnikhina, V. A., 134
Greco, A., 145
Green, C. R., 400
Green, D. W., 111
Green, J. C., 1, 121
Green, M., 245, 373, 402
Green, M. A., 255
Green, M. L. H., 15, 111, 112, 115, 120, 121, 132
Green, T. H., 393
Greenough, P., 98
Greeve, D. L., 442
Greeves, D., 486
Gregor, I. K., 50, 98
Gregorczyk, Z., 437
Gregory, N. W., 90
Gregson, A. K., 92
Grenthe, I., 484
Gress, M. E., 182
Grey, I. E., 96
Grgas-Kuznar, B., 326, 467
Gribenyuk, A. A., 365
Griessl, E., 325, 441
Griffith, T. J., 81
Griffith, W. P., 75, 106, 343, 350, 352
Griffiths, J. E., 344, 422
Grigorenko, F. F., 37
Grigor'ev, A. I., 485
Grigor'eva, A. S., 51
Grigoryan, M. K., 357
Grim, S. O., 304, 408, 455
Grimes, C. G., 281
Grimes, R. N., 247
Grimm, F. A., 435
Grinberg, A. A., 490
Grinberg, A. N., 36
Grineva, N. P., 216
Grischenko, V. F., 144
Grishina, O. N., 100, 133
Grobe, J., 180
Grobelny, R., 206
Grodneva, M. M., 39
Groen, A., 462
Groenenboom, C. J., 75
Groeneveld, W. L., 175, 234, 235, 323, 446, 451
Grofe, T., 139
Gromov, B. V., 63, 497
Gross, M., 153, 157
Grosse, J., 396, 422
Grosse-Brauckmann, U., 66, 496
Groult, D., 79, 141
Gruba, A. I., 140
Grubanov, V. A., 65
Grubbs, R. H., 422
Gruber, L. I., 447
Gruen, D. M., 111

Gruenenboom, C. J., 1
Grunthaner, F. J., 210
Grunzinger, R. E., 248
Gruz, R., 147
Grynkewich, G. W., 88, 185
Grzeskowiak, R., 293, 324
Guastalla, G., 94
Guastini, C., 260
Gubaidullin, Z. Kh., 106
Gubanova, V. A., 106
Gudel, H. U., 104
Guerchais, J. E., 19, 63, 81, 144, 145, 201, 224, 239, 324, 435, 443
Guerin, H., 205
Guerrera, J. J., 195
Guggenberger, L. J., 2, 70, 121, 471
Guidotti, R. A., 72
Guidry, R. M., 232
Guilbert, L. J., 463
Guiliani, A. M., 299, 443
Guillaumont, R., 490
Guillevic, J., 124, 129, 214
Guitel, J. C., 243, 320
Guka, S., 291
Gukova, Yu. Ya., 159
Gukasyan, Zh. G., 151
Gulens, J., 366
Gul'ko, N. V., 36, 141, 142
Gullotti, M., 244
Gundorina, A. A., 461
Gunduz, N., 106, 133, 134, 499
Gunsulus, R. P., 470
Gunz, H. P., 153
Guo, J. M., 325
Gupta, C. M., 467
Gupta, J., 21, 22
Gupta, M. P., 177
Gupta, R. N., 74
Gupta, R. K., 325, 445
Gupta, S. P., 481
Gupta, S. R., 43
Gupta, V. D., 21
Gupta, V. K., 18
Guraj, J., 325
Guran, C., 102
Gurumurthy, C. V., 489
Gur'yanova, E. N., 24
Gusanova, N. K., 390
Gusch, R., 256
Guseinov, L. I., 485
Guseinov, M. N., 440
Gusev, A. I., 70, 71, 75, 357
Gushchin, N. V., 138
Gushchin, V. I., 418
Gushikem, Y., 482
Gus'kova, V. P., 472
Gustomesova, N. N., 108
Guy, A. L., 469
Guzinski, J. A., 203
Gvozdev, R. I., 116
Gwost, D., 193, 222

Haaland, A., 89
Haase, W., 302, 322
Habenschuss, M., 286
Haberey, F., 20, 214
Hackey, J. M., 109
Haddad, S., 234
Haddock, S. R., 419
Hadenfeldt, C., 473
Haeberle, N., 485
Häfler, M., 222
Haegele, R., 94
Haffner, F., 494
Haga, M., 375
Hagedorn, H. F., 131
Hagen, A. P., 115, 343
Hagenmuller, P., 78, 81, 91, 92, 137, 318, 474, 480
Hagihara, N., 382
Hahn, C. J., 104
Hahn, H., 215
Hahn, Th., 140, 149
Haigh, I., 122
Haight, G. P., jun., 109, 119, 203
Haines, R. J., 189
Haire, R. G., 496
Hairs, N. D., 385, 454
Hajdu, I. P., 236
Halfpenny, M. T., 283
Hall, J. L., 314
Hall, J. R., 424
Hall, M. B., 87
Hall, S. R., 215
Hallgren, J. E., 218
Halloran, L. J., 263, 264
Halpern, J., 245
Halstead, T. K., 137
Ham, C., 464
Hambright, P., 469
Hamer, A. D., 128, 472
Hamilton, D. S., 356, 405
Hamm, D. J., 271
Hamm, K., 464
Hammond, G. S., 152, 361, 369
Hampson, P. J., 489
Hampton, A. F., 91
Hanada, R., 73
Hanaoka, I., 287
Hanby, I., 150
Hancock, R. G. V., 174
Handlovic, M., 12
Handy, R. F., 98
Hangortner, M., 326
Hanic, F., 12
Hansen, R. S., 111
Hanson, J. C., 367
Hanson, J. R., 92
Hanson, L. K., 367
Hansson, M., 37, 108
Hanzlik, J., 202, 266
Happe, J. A., 436
Haque, M. S., 326
Hara, K., 173
Hara, Y., 215

Harada, K., 196
Harady, S., 326
Harakawa, M., 186
Hardner, V., 236
Hardy, A., 108, 201
Hardy, A. M., 201
Hardy, K. J., 469
Hargis, L. G., 101
Harihan, M., 242
Harkema, S., 23
Harlow, R. L., 310, 320, 400
Harman, M., 150
Harnung, S. E., 94, 251, 252, 370, 380
Harris, E. W., 338
Harris, G. M., 364
Harris, L. E., 94
Harris, M. B., 476
Harris, R. L., 272
Harris, R. O., 345
Harrison, B., 379, 422
Harrison, P. G., 220
Harrison, P. M., 469
Harrod, J. F., 371, 405
Harrowfield, J. M., 265
Hart, F. A., 473, 488
Hartley, F. R., 404
Hartmann, H., 92
Hartsuiker, J. G., 16
Hartung, H., 224, 271
Hartwell, G. E., 355, 361
Hartzog, L. D., 149
Harvey, R. A., 464
Hasan, F., 106, 109
Haschke, J. M., 478
Hasebe, K., 245
Hashimoto, K., 198
Hassan, M. K., 104
Hassbrock, F. J., 70
Hassel, R. L., 272, 388, 406
Hastie, J. H., 31
Hasty, E. F., 56, 324
Hatfield, W. E., 98, 235, 319, 323, 324
Hathaway, B. J., 318, 323
Hatterer, A., 105, 129
Hauck, J., 489
Hauge, R. H., 31
Haupt, H. J., 166
Hauser, P. J., 94
Hawkins, C. J., 249
Hawthorne, M. F., 105, 246, 247, 248
Hay, R. W., 251, 253
Hayes, R. G., 104
Haymore, B. L., 360, 367
Hays, J. W., 200
Healy, P. C., 324, 458
Heath, G. A., 110
Heaton, B. T., 352, 366, 418
Heckley, P. R., 482
Hedman, B., 146
Hedvig, G. R., 327
Heger, G., 201

Hegenauer, J., 206
Heinsen, H. H., 150, 238
Heiple, C. R., 493
Heitner, H. I., 400
Hejmo, E., 118
Helm, F. T., 258, 322
Hemingway, R. E., 338
Hempel, H., 46
Hempel, J. C., 98
Hencken, G., 30
Hendrickson, A. R., 100, 222, 261, 289, 392
Hendrickson, D. N., 56, 274, 310, 321, 322, 324
Henneike, H. F., 286
Hennings, D., 81
Henrici-Olivé, G., 217
Henry, P. M., 386
Henry, R. L., 345
Henzler, T. E., 34
Herak, R., 261
Herber, R. H., 191
Herberhold, M., 171
Herbstein, F. H., 179
Herceg, M. 440
Herd, Q., 87
Hergovich, E., 221
Herman, B., 147
Hermann, A. M., 271
Herpin, P., 490
Herrington, D. R., 178, 260
Herskovitz, T., 210, 211
Hertli, L., 326
Herve, G., 147
Hervieu, M., 18
Hetflejs, J., 332
Heveldt, P. F., 343
Heyn, B., 121
Hibdon, D., 324
Hidai, M., 194, 386
Hidaka, J., 250, 262
Hierl, C., 93
Higashiyama, T., 424
Higginson, B. R., 87
Higginson, W. C. E., 202, 266
Higson, B. M., 294
Hildenbrand, D. L., 122
Hill, D. R., 324
Hill, H. A. O., 324
Hill, J. O., 301
Hill, N. L., 287
Hill, R. E. E., 336
Hill, W. E., 234
Hillier, I. H., 87
Hindriks, F. R., 462
Hinkley, C. C., 9
Hinze, F. P., 274
Hirai, S., 251
Hiraki, S., 342
Hiroshima, Y., 341
Hirota, E., 91
Hirota, M., 398
Hirota, R., 326
Hitchman, M. A., 323

Ho, B. Y. K., 88, 185
Ho, C. Y., 248
Ho, R. K. Y., 222, 261
Hoard, J. L., 232, 253
Hochberg, E., 126
Hocks, L., 1
Hodges, K. C., 166
Hodgson, D. J., 98, 319, 324, 422
Hodgson, J. C., 128
Hodgson, P. G., 318, 323
Hodorowicz, S., 146
Höfler, M., 168, 270
Hoekstra, H. R., 143
Hoel, E. L., 247
Hoerz, G., 47
Hoeschle, J., 284
Hoffman, B. M., 21
Hoffman, C. A., 264
Hoffman, G., 492
Hoffman, P. R., 334
Hogue, F., 387
Holah, D. G., 157, 223, 330, 354
Holding, C., 220
Holguin, Q. S., 147
Holland, G., 5
Hollander, F. J., 210, 288, 308
Holloway, J. M., 78
Holm, R. H., 207, 210, 211, 231, 471
Holman, R. L., 36
Holmberg, B., 424
Holmes, R. R., 46
Holsboer, F., 374, 414
Holsk, G., 126
Holt, E. M., 120, 213
Holt, S. L., 90, 93, 99, 120, 213, 216
Holtzclaw, H. F., 312
Hong, H. Y.-P., 473
Hooper, A. J., 49, 66, 79
Hooper, M. A. 87
Hope, H., 287
Hopf, F. R., 337
Hopkinson, M. J., 189
Hoppe, R., 201, 214, 248, 271, 303, 309
Hopper, R., 9
Hora, C. J., 46
Horiba, T., 173
Horike, M., 400
Horlin, T., 52
Horn, R. R., 333
Hornung, N. J., 482
Horrigan, P., 264
Horrocks, W. de. W., 486
Horsler, A. D. J., 324
Horsly, S. E., 38
Horvath, E., 99, 100
Hoskins, B. F., 238, 255, 298
Houlihan, J. F., 5
House, D. A., 249, 272
House, J. E., 394, 395

Howard, J., 422
Howe, J. J., 41
Howe, R. F., 112
Hoy, E. F., 386
Hoy, T. G., 469
Hoyer, E., 119, 136, 150, 323, 325
Hrung, C. P., 153
Hsu Che-Hsiung, 76
Htoon, S., 440
Huang, M. H. A., 191
Huang, T. J., 109
Huang, W. L., 366, 377
Hubbard, A. T., 403
Hubbard, W. N., 46
Huber, H., 152, 381
Huber, M., 95
Hubert, A. J., 198, 245
Hubert, P. H., 142
Hubert, S., 490
Hubert-Pfalzraf, L. G., 82
Huchital, D. M., 244
Hudson, G. A., 168
Huff, J., 214
Hugel, R., 176, 233, 305, 327
Hugel, R. P., 233
Hughes, A. N., 157, 223, 330, 354
Hughes, B., 166, 170
Hughes, H. P., 123
Hughes, J., 26, 102
Hughes, J. G., 261, 265
Hughes, W. B., 117
Hughley, J. L., 379
Hui, B. C., 223, 330, 331, 354
Hull, G. W., 76
Hunt, G. W., 308
Hunter, B. K., 165
Hunter, G., 341
Hursthouse, M. B., 1, 45, 444, 473
Husain, S. M., 197
Hush, N. S., 255, 278, 324, 325
Hussein, F. M., 410
Hussonois, M., 490
Hutchings, R. O., 12, 100, 236
Hutchinson, B. B., 88
Hutchinson, J. L., 79
Hutchinson, J. S., 463
Hutta, P. J., 87
Huttel. R., 433

Iball, J., 474
Ibbott, D. G., 453
Ibers, D. V., 232
Ibers, J. A., 90, 172, 204, 211, 232, 253, 306, 330, 345, 360, 367, 383, 422, 435
Ibraeva, T. D., 23
Ibrahim, E. D., 15
Ibusuki, I., 455
Ichihara, M., 33
Igi, K., 264

Ihle, H., 492
Iijima, S., 81
Ijdo, D. J. W., 20
Ikawa, S., 245
Ikawa, T., 136
Ikeda, S., 111, 172, 494
Ikeda, T., 141
Ikenaga, H., 468
Ikonitskii, I. V., 12
Il'in, E. G., 77
Il'ina, L. A., 100, 366, 401, 423
Il'yashenko, V. S., 158
Il'yasova, A. K., 65
Ilyukhin, V. V., 38
Imanakunov, B. I., 225
Imoto, S., 484
Inagaki, K., 445
Indrasenan, P., 26
Ingen-Schenan, A. D. V., 323
Ingle, D. B., 55
In'lova, E. N., 496
Inomata, T., 387
Inoue, E., 136
Inoue, H., 193, 339, 369
Inoue, M., 173
Inoue, Y., 468
Interrante, L. V., 384, 422
Ioganson, A. A., 157
Iordanov, N., 158
Iorga, E. V., 494
Ippolitova, E. A., 489, 496
Ireland, P. R., 452
Irving, H. M. N. H., 461
Irving, R. J., 98, 177
Irwin, J. G., 427
Isakova, R. A., 123
Isci, H., 417
Ishihara, M., 479
Ishii, E., 477
Ishii, M., 65
Ishii, Y., 383
Ishimori, T., 266, 424
Ishimura, Y., 314
Ishizawa, Y., 72
Isied, S. S., 333
Iskander, M. F., 239, 243
Iske, S. D. A., 192
Iskhakova, L. D., 479
Islam, N., 224
Ismael-Milanovic, A., 112
Isobe, K., 177
Isobe, T., 323
Issa, I. M., 442
Issa, R. M., 442
Issleib, K., 31
Issopoulos, P. B., 446
Isupova, E. N., 65
Ito, H., 251
Itou, T., 98, 111, 266, 276
Ittel, S. D., 422
Iurai, S., 140
Ivakin, A. A., 56, 65, 67, 480
Ivan, E., 484
Ivannikova, N. V., 424

Ivanov, A. A., 139
Ivanov, A. B., 483
Ivanov, S. B., 497
Ivanov, V. A., 216
Ivanov, V. I., 485
Ivanova, E. I., 51
Ivanova, L. A., 78, 81
Ivanova, M. S., 81
Ivanova, N. E., 79
Ivanova, N. N., 145
Ivanova, V. Y., 451
Ivanov-Emin, B. N., 427
Ivantsov, A. E., 89
Ivleva, I. N., 14, 244
Ivnitskaya, R. B., 74
Iwamoto, R. T., 131, 302, 431
Iwasaki, H., 461
Iwasaki, N., 452
Iwase, H., 193
Izatt, R. M., 216

Jacoboni, C., 94
Jacobs, E., 497
Jacobs, H., 473
Jacobs, J. W., 464
Jacobson, A. J., 90, 140, 182
Jacobson, R. A., 252
Jacobson, S., 389, 422
Jackson, S. E., 1
Jackson, W. G. J., 259
Jaecker, J. A., 152, 159
Jagner, S., 62
Jain, A. K., 306
Jain, M. C., 65, 479
Jain, P. C., 104, 144, 177, 321
Jalhoom, M. G., 54
Jamard, A., 158
James, A. D., 202
James, B. R., 204, 232, 330, 331, 332, 354
James, C. W., 62
James, D. W., 401, 424
Janik, B., 445
Janjic, T. J., 265
Jansen, M., 248
Jaouen, G., 87
Jarzyna, T., 119
Jasiewicz, B., 98
Jasim, F., 83
Javora, P. H., 281
Jayasooriya, V. A., 299, 401, 423
Jaycock, M. J., 489
Jeanmaier, D., 464
Jeannin, Y., 33, 61, 201
Jeannot, F., 248
Jeans, D. B., 107
Jecmen, J., 309
Jefferson, I., 112
Jeffery, J. W., 471, 499
Jejurkar, C. R., 326, 468
Jellinek, F., 1, 99
Jendrek, E. F., 81

Jenseen, P. W., 104
Jensen, J. B., 451
Jensen, W., 494
Jesperson, N. D., 146
Jesse, R., 201, 248, 309
Jesson, J. P., 70, 223, 269, 281, 355, 361, 381, 396
Jeter, D. Y., 98, 322
Jetz, W., 183
Jezowska-Trzebiatowska, B., 8, 61, 98, 134, 135, 159, 206, 244
Jicha, D. C., 379
Jilani, I., 313
Job, R. C., 165, 219, 265
Johannes, W., 312
Johannesen, R. B., 188
Johnson, A., 251, 364
Johnson, A. W., 253, 356
Johnson, B. F. G., 117, 329, 347, 353, 377, 486
Johnson, C. W., 344
Johnson, D. A., 477
Johnson, D. K., 46, 58, 282, 397
Johnson, D. W., 99
Johnson, E. C., 337
Johnson, J. R., 90, 163
Johnson, K., 46
Johnson, K. A., 272, 388
Johnson, K. H., 384
Johnson, N. A., 145
Johnson, Q., 494
Johnson, S. M., 288
Johnson, V., 91
Jolly, W. L., 88, 217, 304, 309
Jones, A. G., 431
Jones, C. E., 414
Jones, C. J., 246
Jones, D. W., 104, 423
Jones, G. I. L., 200, 228
Jones, J., 350
Jones, J. G., 197
Jones, L. H., 369, 423
Jones, J. P., 68
Jones, L. H., 104
Jones, M. M., 445
Jones, P. J., 138
Jones, R., 356
Jones, R. D. G., 186
Jordan, B. D., 65
Jorgensen, C. K., 100, 260, 443
Jorgesen, D. R., 47
Jorsey, R. S., 81
Joseph, P. J., 21
Joshi, J. D., 445
Joshi, N. D., 65
Jotham, R. W., 324
Jourdan, G., 224
Joyner, R. D., 114
Judkins, R. R., 251
Judos, F., 62
Jukes, A. E., 328
Jung, H., 270

Jung, W., 42
Jungermann, K., 110
Jungst, R. G., 322
Junsoo So, H., 54
Juza, R., 42, 476

Kabankova, N. N., 63
Kabanova, N. G., 141
Kabassonov, K., 106
Kabesova, M., 311
Kabir-Ud-Din, 39
Kachkar, L. S., 40
Kachkovskaya, E. T., 7
Kaczurba, E., 61
Kaden, T. A., 229, 275, 326, 327
Kadish, K. M., 428
Kaduk, J. A., 253
Kaempfe, L. A., 283
Kafalas, A., 81
Kageyama, H., 194
Kahl, W., 31
Kai, F., 484
Kai, Y., 400
Kaidalova, T. A., 78
Kaitner, B., 460
Kaizer, J., 305
Kakkar, S. N., 60, 446
Kakutani, T., 232, 325
Kalashnik, A. A., 47, 125
Kalbacher, B. J., 237, 242, 443
Kalck, P., 354
Kalieva, K., 479
Kalinichenko, A. M., 19, 142, 478, 480
Kalinichenko, N. B., 60, 61
Kalinina, S. S., 40
Kalinnikov, V. T., 75
Kallesøe, S., 251
Kalontarov, I. Ya., 133
Kalra, K. L., 191, 306
Kalsotra, B. L., 476
Kamata, K., 81
Kamenar, B., 460
Kamijgo, N., 281
Kaminaris, D., 97
Kaminskii, B. T., 5, 6
Kan, H., 326
Kanaeva, O. A., 476
Kanamaru, F., 91, 202, 303
Kanas, A., 118
Kanellakopulos-Drossopulos, W., 182
Kang Kun Wu, 18
Kanie, S., 416
Kan-Nan, Y., 321
Kanth, K. M., 104
Kapatsina, G. A., 204
Kapila, V. P., 21
Kapoor, R. C., 55
Kappelmann, F. A., 484
Kappenstein, C., 216, 305
Kapur, S., 476

Karaban, A. A., 418
Karadakov, B., 27
Karapet'yants, M. Kh., 108
Karaseva, E. V., 478
Karaulov, A. G., 36, 141, 142
Karayannis, N. M., 12, 100, 101, 175, 176, 199, 209, 217, 236, 328, 439, 472
Karitonova, L. K., 472
Karlysheva, F. F., 31
Karmanova, L. F., 38
Karnaukhov, A. S., 108
Karpov, O. P., 144
Karpov, V. N., 140, 142
Karsten, J. H. M., 70
Karwecka, Z., 50, 51
Karweik, D., 428
Kasai, N., 400
Kasai, P. H., 54
Kasaoka, S., 266
Kasenally, A. S., 410
Kashaev, A. A., 82
Kashivazaki, T., 19
Kashiwabara, K., 452
Kashman, J. P., 264
Kasimov, G. G., 20, 81, 214, 480
Kasowski, R. V., 123
Kasper, J. S., 422
Kasper, M., 51
Kasube, Y., 321
Kasuga, K., 98
Katakis, D., 48
Kato, H., 236, 295
Kato, K., 20, 172
Katovic, V., 84, 231
Katsnel'son, L. M., 36
Katsura, T., 50
Katz, L., 81
Kauer, J. C., 468
Kaufmann, J., 180
Kawada, I., 20, 65
Kawaguchi, H., 250, 462
Kawaguchi, S., 177, 250, 386
Kawakami, K., 375, 377, 409
Kawamura, T., 20
Kawasaki, H., 342
Kawase, A., 446
Kawashima, C., 38
Kawashima, K., 321
Kawata, Y., 268
Kawayuchi, S., 320
Kawazura, H., 383
Kay, A. G., 62
Kay, M. S. L., 142
Kaye, L. A., 155
Kazan, V. M., 140
Kazain, A. A., 17
Kazanskii, L. P., 149, 480
Kazantsev, V. V., 20
Kazenas, E. K., 123
Kazuko, Y., 147
Keat, R., 393
Keene, F. R., 252

Keeton, D. P., 267
Keiter, R. L., 455
Kekre, M. G., 234, 287, 324
Keler, E. K., 65, 81
Kelk, N. B., 482
Keller, H. J., 243, 302, 373, 384
Keller, O. L. jun., 489
Kellerman, R., 87
Kelly, T. L., 102
Kelm, H., 92
Kemmitt, R. D. W., 358
Kemmler-Sack, S., 498
Kemp, T. J., 189
Kempny, H. P., 183
Kendall-Torry, A., 277
Kendrow, C. H., 452
Kennedy, B. P., 323
Kennedy, D. J., 172
Kenny, A. J., 463
Kensett, M. J., 294
Kepert, D. L., 66, 69, 128, 443, 456, 457, 474
Kerby, R. C., 96
Kerfoot, D. G. E., 375
Kergoat, R., 145, 224, 324, 435
Kerimov, I. G., 99
Kerr, M. A., 463
Kerridge, D. H., 96, 436, 437
Kertes, A. S., 161
Kerwin, C. M., 297
Kessler, H., 105, 129
Kesterke, D. G., 72
Keton, A. N., 35, 485
Ketrush, P. M., 381
Ketteringham, A. P., 157, 419
Kettle, S. F. A., 324, 334
Khachaturov, A. S., 28
Khadikar, P. V., 234, 287, 324, 326, 446
Khaenko, B. V., 7
Khain, V. S., 155
Khair, A., 117
Khaldoyanidi, K. A.,
Khalifa, M. A., 239
Khalilov, L. M., 54
Khalimov, F. B., 4
Khamar, M. M., 48
Khamraev, T. A., 492
Khan, A. A., 39, 118, 327
Khan, M. A., 322, 411
Khan, M. M., 12, 389
Khan, S. M., 350
Khandozhko, T. Ya., 142
Khandozhow, V. N., 220
Khanolkar, D. D., 55
Kharchenko, L. Yu., 142
Khariton, Kh. Sh., 56, 393
Kharitonov, Yu. Ya., 107, 194, 240, 294, 446, 497
Kharitonova, G. S., 42
Kharitonova, R. I., 446, 451
Kharlamov, I. P., 133, 159
Kharlamova, E. N., 24
Khattak, C. P., 161

Khavtasi, N. S., 451
Khekimov, Yu., 142
Kher, V. G., 95
Khidekel, M. L., 357, 387
Khitrova, O. M., 71
Khivaja, M. A., 309
Khlebodarov, V. G., 77
Khodeev, Yu. S., 32
Khodos, M. Ya., 472, 479
Khodzhaev, O. F., 54
Khokhlova, L. I., 22
Khorasani, S. S. M. A., 97
Khoul, M. K., 461
Khramov, V. P., 483
Khripin, L. A., 478
Khrushch, A. P., 110
Khuller, I. P., 342
Khutoretskaya, I. Sh., 146
Kiba, N., 146
Kida, S., 287, 295, 314, 323
Kidani, Y., 260, 445, 446
Kidwell, S., 257
Kiefer, G. W., 116
Kiernan, P. M., 75
Kiesel, R. F., 51
Kim, B.-I., 494
Kim, J. C., 464
Kim, N. E., 183
Kimizuka, N., 65
Kimoshita, S., 215
Kimura, E., 253
Kimura, M., 20, 245
Kimura, S., 81
Kinard, W. F., 488
Kindred, I. J., 249
King, A. W., 322
King, C. M., 304
King, J. P., 100, 103
King, M. G., 406
King, R. B., 88, 166, 188, 359, 482
King, R. M., 94
King, R. U., 151
King, T. J., 133, 320
Kingston, J. V., 336
Kini, M. R., 140
Kinoshuta, S., 224, 451
Kiprin, L. I., 451
Kircher, C. J., 47
Kirchner, R. M., 90, 172
Kirienko, T. I., 446
Kirillou, N. S., 485
Kirillova, N. I., 70, 71, 75
Kirin, I. S., 488
Kiriyama, H., 460
Kirk, A. D., 102
Kirkaldy, J. S., 91
Kirmse, R., 136, 323, 325
Kirpichinkov, P. A., 24
Kirtley, S. W., 162
Kisel, S. G., 20, 140
Kiselev, Yu. M., 485
Kiseleva, I. N., 365
Kiseleva, N. V., 418, 424

Kishida, Y., 124
Kislyakov, I. P., 65, 141, 151
Kiso, Y., 272
Kiss, A. B., 137
Kiss, J., 96
Kistenmacher, T. J., 251, 258, 261
Kita, W. G., 111, 117
Kitahama, K., 460
Kiyama, M., 202
Kjekshus, A., 33, 248, 425
Klaaijsen, F. W., 272
Klassen, K. L., 405
Klaui, W., 199
Klein, R. L. jun., 101
Kleinberg, J., 121, 131, 431
Kleinert, P., 107
Klenus, V. G., 20, 81, 480
Kleppa, O. J., 47, 123, 305
Klesova, G. M., 57
Klett, R. D., 21
Klevtsov, P. V., 140, 141, 142
Klevtsova, R. F., 140, 141, 142
Klier, K., 87
Klimenko, E. P., 54
Klingen, W., 215
Klissurski, D. G., 81
Klopfer, A., 122
Klotzbücher, W., 381
Klyachkina, K. N., 56
Klyuchnikov, V. M., 82, 489
Klyuchnikova, E. F., 4, 92, 95, 129
Knetsch, D., 175, 446, 451
Knigavko, I. P., 139
Knight, G. F., 203
Knight, J. R., 368, 397, 419
Knobloch, A., 99
Knodo, H., 58
Knop, O., 324
Knowles, P. J., 199
Knutsen, C., 467
Kobayashi, A., 251, 253
Kobayashi, H., 253, 338
Kobayashi, M., 12
Kober, F., 180
Kobets, L. V., 49, 61
Koch, S., 27, 83, 216
Kocheregin, S. B., 36
Kocherzhinskii, Yu. A., 125
Kochetkova, A. P., 365
Kochi, J. K., 259
Kochi, S., 231
Kochler, H., 31
Koden, T. A., 324
Kodess, B. N., 47
Koeguka, H., 398
Koehl, P., 141, 143
Köhler, H., 224, 271
Kogame, T., 326
Kogan, A. V., 27
Kogan, Ya. D., 124
Kohan, J., 117
Kohl, J., 34

Kohl, P., 287
Kohout, J., 325
Kohut, L. N., 271
Koike, H., 260, 445, 446
Koiso, T., 101
Koizumi, M., 91, 203, 303
Kojic-Prodic, B., 145
Kokado, H., 136
Kokarovtseva, I. G., 140
Kokasza, G., 324
Koknat, F. W., 73
Kokot, E., 205
Kokubo, T., 111
Kokunov, Yu. V., 149
Kolari, H. J., 127
Kolbin, N. I., 161
Kolenda, M., 92
Kolesnikov, I. M., 19
Kolesova, R. V., 36
Kolich, C., 9
Kolli, I. D., 96
Kolobova, N. E., 71, 157, 220
Kolodyazhnyi, Yu. V., 79
Kolomnikov, I. S., 29, 353, 357
Kolowos, I., 99
Kolyadin, A. B., 488
Kolyago, E. K., 483
Kolychev, V. B., 216
Komarevtseva, N. I., 489
Komatsu, M., 19
Komissarova, L. N., 40, 472
Komiyama, Y., 52, 136
Konaka, S., 452
Konarev, M. I., 158
Kondrashenkov, A. A., 38
Kondrat'eva, O. I., 89
Kondratov, P. I., 99
Kondratova, T. S., 99
Konev, L. G., 491
Konev, V. A., 446
Kong, P. C., 405, 412
Konietzny, A., 433
Konopka, D., 99
Konovalenko, L. I., 101
Konovalova, G. I., 133
Konstantinova, A. I., 474
Konstants, Z., 95
Konunova, Ts. B., 40, 42
Konya, K., 260, 303
Kopa, M. V., 472
Kopanica, M., 326
Kopanv, V. D., 77
Kopf, H., 31
Kopf, J., 193
Kopperl, S. J., 68
Koppikar, U. V., 1
Koptenko, E. P., 451
Korabel'nikova, T. P., 472
Koran, D., 272
Korant, B. D., 468
Korbut, T. K., 216
Korczynski, A., 458
Kordosky, G., 259
Korenev, Yu. M., 35

Kornilov, I. I., 4, 5
Kornilov, M. Yu., 487
Korobova, Z. N., 133, 159
Korol'kov, D. V., 90, 128, 156, 157
Korosi, G., 62
Korosteleva, A. I., 395
Korotchenko, N. A., 144
Korotkevich, I. B., 140
Korovin, S. S., 38, 79, 82, 489
Korshunov, B. G., 33
Koshechko, L. G., 107, 138
Koseeva, L. P., 142
Koshcheev, G. G., 142
Koshkina, G. N., 59
Koskenlinna, M., 477
Kosman, D. J., 323, 325
Kosmodem'yanskaya, G. V., 147
Kosorukov, A. A., 35
Kossiakoff, A. A., 30
Kost, M., 76
Kosta, M. E., 474
Kostelnik, R. J., 446
Kosteruk, V. P., 7
Kostromina, N. A., 216
Kostrova, A. M., 35
Kosynev, B. M., 58
Kotegov, K. V., 160
Kotel'nikov, A. I., 116
Kotel'nikov, V. P., 410
Kotsar, M. L., 497
Kovac, I., 461
Koval'chenko, M. S., 125
Koval'chuk, M. Yu., 496
Kovalevskii, O. I., 144
Kovba, L. M., 140, 142, 143, 489
Kozeeva, L. P., 141, 142
Kozhemyakin, V. A., 33
Kozhina, I. I., 129
Kozlov, V. D., 478
Kozlov, V. G., 145
Kozlova, N. V., 30
Kozlova, V. A., 311
Kozlova, V. K., 39
Kozlowska, I., 99
Kozlowska-Rog, A., 95
Koz'min, P. A., 156, 157
Krachak, A. N., 108, 480
Kraihanzel, C. S., 217
Krainova, Z. I., 20, 36, 479
Kramer, A. V., 372
Krannich, R., 325
Krankovits, E. M., 358, 390
Krasneninnikova, A. A., 4
Krasnopol'skaya, M. B., 148
Krasnopol'skaya, S. M., 3
Krasnov, B. A., 418
Krasnov, K. S., 4, 33, 35
Krasser, W., 491
Kratsmor-Smogrovic, J., 325
Krause, J., 105
Krauss, H. L., 93

Kravchenko, A. I., 422
Kravchenko, M. D., 478
Krebs, B., 143, 146
Kreisel, G., 51
Krinitskaya, L. V., 472
Krishna, Rao. V., 69
Krishnamachari, N., 243
Krishnamoorthy, C. R., 215, 326, 465
Krishnamurthy, M. V., 54
Kristal'nyi, E. V., 30
Kristoff, J. S., 90, 172, 183, 330
Krivobok, V. I., 140
Krivokhizhiva, L. A., 461
Krivorotenko, A. D., 7, 47, 156
Krivospitskii, A. D., 71
Kriza, A., 27
Kroese, C. J., 309, 321
Kroneck, P., 135
Kroon, A. M., 462
Krot'ko, N. P., 478
Kruck, T., 46, 87, 222, 268, 270
Krueger, C., 48, 187
Krüger, H., 436, 447
Kruger, G. J., 260
Kruglov, A. A., 158
Kruhak, I., 446
Kruk, T. P. A., 326
Krumholz, P., 197
Kruszyna, H. G., 379
Kruzynski, L., 190
Krylov, E. I., 20, 81, 96, 173, 214, 440
Krylov, V. D., 124
Krylov, V. K., 390
Krylov, V. S., 478
Krylova, L. F., 410
Krym, V., 424
Krzewska, S., 216
Kuan, T. S., 363
Kubo, M., 173
Kucharski, A. S., 35, 172, 437
Kudo, K., 386
Kudritskaya, S. A., 27
Kudryavtsev, V. I., 95
Kühr, H., 193
Kuge, Y., 316
Kuhl, K. N., 243
Kuhn, M., 190
Kuijt, H. R., 142
Kukushkin, Y. N., 160, 390, 403, 407, 408, 410, 418
Kukushkina, I. A., 39
Kulakova, T. V., 141
Kuleshov, I. M., 343, 345
Kuliev, A. D., 225
Kulik, O. G., 125
Kulikova, I. M., 35
Kulkarni, D. K., 95
Kulkarni, V. H., 27, 44, 84
Kullberg, L. H., 38
Kumada, M., 272

Kumar, A. 451
Kumari, V., 301
Kume, S., 91
Kumok, V. N., 483
Kundig, E. P., 88, 152, 267, 402
Kunnamann, W., 95
Kunst, H., 7
Kupcik, V., 320
Kuperman, O. M., 142
Kuprina, R. V., 19
Kura, G. 446
Kurahashi, M., 293
Kuran, W., 382
Kuratashvili, Z. A., 343
Kurauchi, T., 320
Kurbatov, L. D., 67
Kuroda, K., 264
Kuroda, R., 251, 302
Kurosaki, T., 356
Kuroya, H., 320
Kurras, E., 105
Kurtak, C. R., 149
Kurtisin, V. B., 47
Kuska, H. A., 53
Kustin, K., 145
Kuszaj, J. M., 286
Kutal, C., 98
Kutoglu, A., 30, 132
Kutolin, S. A., 20, 36
Kutschabsky, L., 298
Kutuzova, M. Y., 451
Kuu, W.-Y., 494
Kuzina, M. G., 495
Kuzina, T. A., 404
Kuzina, T. I., 108, 480
Kuz'ma, Yu. B., 92, 124, 125, 156
Kuz'micheva, E. V., 489
Kuz'micheva, V. P., 148
Kuz'min, E. A., 65, 107
Kuz'mina, N. N., 404
Kuz'mina, N. P., 484
Kuz'mina, T. A., 50
Kuznetsov, A. K., 36
Kuznetsov, N. T., 476
Kuznetsov, V. G., 156
Kuznetsov, V. Ya., 82
Kuznetsova, A. A., 133, 145
Kuzyakov, Yu. Ya., 63
Kvashina, E. F., 15
Kwan, T., 59
Kydynov, M. K., 482
Kyuno, E., 250, 279
Kyrki, J. R., 479

Labes, M. M., 76, 175, 236
Labinger, J. A., 75, 372
Lacey, M. J., 294
Lachenal, D., 209
La Cour, T., 120
Ladd, M. F. C., 440
Lagaly, G., 66, 496

La Ginestra, A., 38
Lahey, M. A. J. Th., 142
Lai, C. N., 403
Lai, T.-T., 494
Laier, T., 94, 252, 370, 380
Laing, K. R., 335, 412
Laing, M., 167, 172
Laitinen, H. A., 144
Lake, R. E., 100, 261
Lakhtin, Yu. M., 124
Lalor, G. C., 260
Lalor, F. J., 116, 182, 193, 194, 359
L'am, C. T., 376
Lam, M. 464
Lamba, R. S., 102
Lambrecht, V. G. jun., 99
La Monica, G., 155, 359, 394
Lancaster, J. C., 226
Landgrebe, J. A., 121
Landia, M. V., 479
Lane, B. C., 377
Lane, R. W., 231
Laney, N. W., 323
Langford, C. H., 209
Langhout, J. P., 331, 346
Lantratov, V. M., 6, 123
Laplace, G., 49
Lappert, M. F., 28, 183, 253, 356, 402
Lappus, M., 164
Lapsina, A., 145
Larin, G. M., 75, 98, 133
Larin, N. V., 46, 89
Larina, T. B., 157
Larionov, S. V., 100, 366, 401, 424
Larkworthy, L. F., 48, 199, 225
La Rosa, R. A., 256
Larsen, E. M., 32, 34
Larsen, K. P., 310, 321
Larson, D. B., 344
Larsson, L. O., 473
Larsson, R., 97
Laruelle, P., 179, 485
Lassocinska, A., 96
Latina, Z. I., 4
Latremouille, G., 63
Latyaeva, V. N., 62
Lau, P. S., 394, 395
Laube, L. G., 39
Laüght, M., 243, 320, 321
Lauher, J. W., 232, 253, 257
Launay, S., 65
Laurence, G. S., 92
Laurie, S. H., 263
Lauzon, G., 464
Lavallec, D. K., 327
Lavielle, L., 105
Lavrenko, V. A., 95
Lawrence, G. A., 249
Lawstron, I. W., 427
Lazarev, V. B., 365
Lazareva, V. A., 4

Lazarus, M. S., 304
Le, A. F., 220
Leary, J. J., 41
Leary, K., 400, 433
Lebas, D., 155
Lebedev, B. G., 95
Lebedev, G. N., 46
Lebedev, V. N., 36
Lebedeva, G. V., 124
Lebedeva, L. I., 136, 147
Le Bihan, M. T., 172
Leblanc, J. C., 29
Leblanc-Soreau, A., 33
Lecaille, C., 155
Lecerf, A., 108
Leciejewicz, J., 92
Lecomte, C., 31
Lecomte, F., 132
Ledner, P. W., 402
Lee, C. W., 20, 95
Lee, H. S., 21, 22
Lee, K. W., 241
Lee, M. R., 65
Leelamani, E. G., 379
Lefelholz, J. P., 23
Le Flem, G., 105, 137
Le Flem, H., 480
Le Fur, Y., 474
Legg, J. I., 263, 264
Legzdins, P., 126, 167, 182, 485
Leigh, G. J., 153, 355, 396
Leigh, M. J., 107
Leipoldt, J. G., 206
Leite, M. T. P., 225
Leith, I. R., 112
Le Marouille, J. Y., 214, 219
Lemoine, P., 153, 157
Lenenko, V. S., 373
Leng, B., 4
Lenhert, P. G., 255
Lenoir, J., 137
Lentz, A., 141
Leonard, A. J., 123
Leonard, J. J., 324
Leong, J., 103
Le Page, A. H., 489
Leporati, E., 325, 446
Leroy, M. J. F., 61, 132
Lesne, J. P., 141
Lester, J. E., 377
Leu, A. D., 325
Leupold, A. A., 76
Leussing, D. L., 468
Levason, W., 175, 216, 234, 238, 328, 415, 427
Levchishina, T. F., 78
Levenson, R. A., 51, 52
Lever, A. B. P., 323
Lever, F. M., 335, 349
Levin, V. A., 141
Levine, B. A., 484
Levitzki, A., 466, 468
Levitskii, M. M., 19

Levitskii, V. A., 95, 139, 142
Levy, H. A., 453
Levy, J. H., 491
Lewin, A. H., 305
Lewis, D. F., 24
Lewis, D. L., 319, 324
Lewis, D. W., 487
Lewis, J., 320, 329, 353, 486
Lewton, D. A., 294
L'Haridon, P., 172
Li, M., 388
Li, N. C., 240, 325
Li, S. N., 484
Li, T., 93, 260, 308
Liang, W. Y., 7, 123
Libbey, E. T., 188
Libby, L. K., 379
Licheri, G., 96
Liefde Meijer, H. J., 75
Liegois, C., 22
Liehr, G., 154
Lienhard, G. E., 56
Likhenshtein, G. I., 116
Lim, H. S., 245
Lim, J. C., 273, 388
Lima, B. J. C., 203
Limar, T. F., 20
Limura, M., 452
Lin, K. K. G., 169
Linarte-Lazcano, R., 385
Lin Ching Lu, 33
Linck, M. H., 125
Lincoln, F. J., 79
Lincoln, S. F., 226, 323
Linde, G., 476
Linder, P. W., 310, 315, 320
Lindman, B., 462
Lindmark, A. F., 24
Lindner, D. L., 355
Lindner, E., 88, 154, 170, 181, 234
Lindoy, L. F., 294, 295, 296
Lindquist, R. N., 56
Lindsell, W. E., 373
Lineva, A. N., 62
Linn, W. J., 304
Linsky, J. P., 422
Linton, C., 5
Lintonbon, R., 49
Lintvedt, R. L., 98, 286
Lip, H. C., 296, 436
Lipovich, V. G., 3
Lipovskii, A. A., 495
Lippard, S. J., 119, 217, 260, 308, 319, 400
Liptoy, G., 439
Lishko, T. P., 39, 224
Lisityna, E. S., 427
Lisson, V. N., 65
Little, L. H., 427
Little, R. G., 232, 322, 345
Little, W. F., 335
Litvin, A. L., 95
Lityinchuk, V. M., 119

Liu, S. T., 145
Livingston, D. M., 462
Livingstone, S. E., 200, 216, 258, 298, 301, 328, 365, 391
Lloyd, D. R., 87, 217
Lloyd, M. K., 90, 117, 191, 361, 362
Lo, J., 125
Lo, S. T. D., 256
Lobachev, A. N., 81
Lobeeva, T. S., 29
Lock, C. J. L., 107, 158
Lock, G. A., 23
Locker, A. L., 44
Lockyer, T. N., 392
Lodzinska, A., 193
Lofgren, P., 108
Logan, N., 154, 379, 422
Logan, P. T., 11
Loh, C. S., 459
Loiseleur, H., 448
Lokio, A. H. J., 479
Lokshin, B. V., 10, 40, 87
Long, G. J., 198, 233, 280
Longfelderova, H., 325
Longo, F. R., 324
Longo, J. M., 159
Longoni, G., 267, 283, 401
Loose, D. J., 323
Lopato, L. M., 478
Lorcher, K. P., 433
Lorenz, D. R., 58
Loriers, J., 477
Loshkareva, N. I., 141
Losee, D. B., 235, 234
Lott, A. L., 233, 440
Lott, J. W., 152
Loub, J., 34
Louis, R., 301, 400
Louw, W. J., 403, 404
Lovecchio, F. V., 231, 276
Low, J. N., 474
Low, W., 129
Lowry, R. N., 22
Loyd, M. K., 238
Lozhkina, G. S., 490
Lozovskaya, P. F., 39
Lubkoll, D., 79
Lucas, C. R., 15
Lucas, J., 491
Lucchini, E., 214
Lucco Borlera, M., 92
Luchian, N., 387
Luchinskii, G. P., 4
Ludi, A., 104
Ludmann-Obier, M. F., 281
Ludwig, W., 279
Lugovoi, S. V., 148
Lugovskaya, E. S., 95
Lui, P. J., 455
Lukas, W., 161
Lukashenko, G. M., 92
Lukes, M. J., 251

Lule, A., 271
Lumme, P., 99
Lunck, Kh. I., 96
Lund, J. S., 87
Lundberg, B. K. S., 321
Lundqvist, R., 490
Lundstrom, T., 380
Luoma, J. R., 201
Lupenko, E., 48
Lusztyk, J., 89
Lutz, H. D., 91
Lutz, O., 436, 447
Lyapilina, M. G., 96
Lyashenko, V. I., 438
Lyban, Yu. P., 37
Lykova, L. N., 140
Lynn, J. L., 56
Lyons, J. R., 411
Lysenko, Yu. A., 22
Lysyak, T. V., 353, 357
Lyubchenko, Yu. A., 119
Lyubimov, G. D., 8

Maaskant, W. J. A., 309, 321
Mabbs, F. E., 133, 323
McAllister, R. M., 243, 258
Macara, I. G., 469
McArdle, J. V., 345
McArdle, P., 486
Macarovici, C. G., 19, 20, 103
Macaskova, L., 311
McAuliffe, C. A., 98, 175, 216, 234, 238, 242, 282, 283, 284, 291, 302, 328, 374, 375, 392, 396, 415, 427, 466
McBreen, J. O., 173, 225, 384
McBride, L., 90
McBride, T., 403
McBryde, W. A. E., 215, 327
McCaffery, A. J., 341, 494
McCaffrey, D. J. A., 352
McCarley, R. E., 73, 128
McCarthy, A. E., 177
McCarthy, P. J., 178
McCleverty, J. A., 90, 111, 117, 191, 238, 361, 362, 432
McClung, R. E. D., 54, 59
McClure, G. L., 227
McConnell, J. F., 294
McConway, J. C., 345
McCormick, B. J., 442
McCormick, I. R. N., 443
McCormick, J., 53
McCormick, M. J., 383
McCourt, J. L., 327
McCreasy, M. D., 487
MacDermott, T. E., 176
Macdonald, C. G., 294
McDonald, I. R., 452
McDonald, J. W., 136
McDonald, R. L., 201

McDonough, P. S., 98, 365, 393
McDowell, R. S., 122
McDowell, W. J., 489
Macek, J., 17
McFadden, D. L., 255
McFarland, J. T., 464
McFarlane, J. C., 470
McFarlane, W., 424
McGinnety, J. A., 435
McGlynn, S. P., 344, 452
MacGregor, A. C., 169, 186
McGregor, K. T., 319, 323, 324
McGuire, G. E., 70, 113
Mach, J., 483
Macheteau, Y., 201
Machkhoshvili, R. I., 240, 294, 446, 497
Maciel, G. E., 446
McIntosh, D., 267
MacKay, A. G., 476
McKean, L., 149, 480
McKenzie, E. D., 255, 274, 294, 496
McKinzie, H. L., 248
McKnight, G. F., 119
McLaren, J. W., 259
McLaughlin, G. M., 253, 356
McLellan, B. E., 101
McLennan, T. J., 281
McLeod, D., 54
McMillan, R., 356
McMullin, S. L., 188
McPartlin, M., 113, 294, 321, 385, 454
McPhail, A. T., 255, 474
McPherson, G. L., 93, 271, 323
McPherson, R., 20
McQuillan, G. P., 175, 235, 406, 439
McWhinnie, W. R., 200, 226, 227, 308, 335, 367, 398
Madaras, A. I., 472
Maddock, A. G., 113
Madey, T. E., 111
Madrazo, L. A. G., 479
Madrid, D., 154
Madura, P., 232
Märkl, G., 199
Maestri, M., 94, 338
Magee, R. J., 227, 242, 301, 309, 358, 390
Magon, L., 153, 218, 490, 493
Magunov, R. L., 478
Mahadev, K. N., 130
Mahale, V. B., 27
Mahe, P., 65
Mahoney, D. F., 336
Maier, A. A., 141, 142
Maier, T. O., 243
Maiezcurrena, R. A., 422
Maillot, F., 20, 65
Maines, R. G., 73

Maiorova, A. G., 362
Maiskaya, T. Z., 32, 39
Maissonnat, A., 221, 354
Maitra, A., 148
Majeste, R. J., 320
Majid, A., 138, 150
Majumdar, A. K., 39
Majumder, M. N., 105, 109
Maker, G. K. R., 325, 326
Makhija, R. C., 457, 458
Makhkamov, K. M., 133
Maki, A. H., 266, 432
Maksimovskaya, R. I., 112
Maksin, V. I., 142, 480
Makurin, Yu. N., 20
Malakha, I. N., 101
Malavasi, W., 301
Malhotra, K. C., 445
Malik, A. U., 14
Malik, K. M. A., 499
Malik, M. A., 299
Malik, W. U., 151, 241, 317, 387
Malin, J. M., 195, 326, 337, 446
Malinina, A. T., 479
Malinovski, A., 106
Malinovskii, T. I., 484
Malisch, W., 190
Mal'kova, T. B., 26
Malouf, G., 337
Mal'tsev, V. T., 140
Malyuk, Yu. I., 81
Mamedov, K. P., 225
Manassero, M., 192, 380, 433
Manchar, H., 400
Mande, C., 151
Manenkov, M. I., 122
Manesis, E. D., 49
Manfredotti, A. G., 260
Mangia, A., 229, 322
Manhas, B. S., 483
Maniv, S., 129
Mann, B. E., 111, 113, 117, 397, 424
Mann, K. R., 322
Mann, R. H., 334
Manning, A. R., 220
Manning, P. E., 465
Manning, P. G., 216
Manoharan, P. T., 343
Manojlovic-Muir, L., 261, 322, 422
Mansour, B., 95
Mansy, S., 466
Marais, P. G., 70
Marangoni, G., 493, 496
Marashkov, S. V., 265
Marcati, F., 356
Marcec, R., 101
Marchenko, V. A., 47
Marcotrigiano, G., 226, 384, 459
Marcus, Y., 447

Marecek, V., 18
Maresca, L., 490
Marezio, M., 128
Margerum, D. W., 213
Margorina, M. N., 490
Margrave, J. L., 31
Margulis, E. V., 204, 205
Marianelli, R. S., 177
Maricondi, C. W., 264
Marina, L. K., 63
Mark, H. B., 272
Mark, W., 37, 108
Markham, L. D., 330, 331, 332
Marko, L., 218, 221, 356, 358
Markov, G. S., 35
Marks, J. A., 324
Marks, L., 117
Marks, R. L., 119
Marks, T. J., 88, 90, 172, 185
Markwell, A. J., 385, 454
Marongiu, G., 319
Marov, I. N., 60, 61, 133, 428
Mars, E., 147
Marsden, C. J., 138
Marsh, R. E., 245
Marshall, R. E., 128
Marsich, N., 307
Martell, A. E., 58, 209, 239,
    244, 291, 324, 446
Martens, H., 156
Martin, B. W., 54
Martin, C., 199
Martin, D. W., 321
Martin, G. W., 65
Martin, I. J., 108, 480, 496
Martin, J. L., 42, 130
Martin, L. Y., 254
Martin, R. B., 215, 468, 484
Martin, R. L., 14, 100, 210,
    222, 261, 289, 392
Martin, W. H., 109
Martinengo, S., 267, 352, 401
Martinez, J., 151
Martinez-Ripoll, M., 302
Martino, G., 112, 115, 127
Martra, A. N., 496
Marumo, F., 140, 251, 253, 261
Martynenko, L. I., 12, 43, 69,
    484, 485
Martyshova, T. I., 483
Marvel, G., 224
Marx, G., 105
Maryanoff, B. E., 12, 100, 236
Marzilli, L. G., 222, 251, 253,
    256, 258
Marzilli, P. A., 258
Marzotto, A., 499
Marzowski, J., 101
Masaguer, J. R., 82
Masaki, N., 488
Masayoshi, K., 446
Mase, H., 146
Mashieu, F., 239
Mashirov, L. G., 494

Maslen, E. N., 69, 226, 276
Maslen, H. S., 178
Maslennikov, I. A., 139
Maslov, L. P., 148
Maslowska, J., 27, 42
Mason, P. R., 189
Mason, R., 110, 158, 181, 194,
    361, 362, 402, 422
Mason, S. F., 113
Mason, W. R., 101, 417
Massart, R., 96
Masset, A., 33
Massey, R. C., 341
Massucoi, M. A., 38
Mastin, S. H., 424
Mastropaolo, D., 321
Masuda, I., 176
Mathe, E., 103
Mather, D. W., 196, 201, 241,
    280
Mather, G. G., 131, 424
Mather, I. H., 255
Matheson, D. S., 248
Matheson, T., 329
Matheson, T. W., 353
Mathew, M., 126, 388
Mathews, N. J., 89
Mathey, F., 122
Mathias, B. T., 128
Mathieu, R., 353
Mathur, H. B., 225, 326
Mathur, P. K., 426
Matienzo, L. J., 304, 408
Matsubayashi, G. E., 398, 410
Matsuda, T., 181, 262
Matsuda, Y., 196, 324, 446
Matsui, M., 320
Matsumoto, A., 476
Matsumoto, M., 266, 382
Matsumoto, S., 124
Matsumura, Y., 186, 220
Matsushita, T., 176
Matsuzaki, R., 122
Matsuzaki, T., 19
Mattes, R., 63, 69, 138, 461
Matveeva, F. A., 37
Matyashov, V. G., 55
Maueler, G., 270
Mauguen, J. M., 145
Mauriello, L., 324
Mavani, I. P., 326, 445, 468
Mavel, G., 324, 435
Mavrodin-Tarabic, M., 38
Mawby, R. J., 375, 402
Maxwell, I. E., 253
Mayerle, J. J., 211, 308
Mays, M. J., 162, 199, 221,
    313, 411, 415, 425
Mazdiyasni, K. S., 1, 31
Mazepa, A. V., 426
Mazo, G. Y., 377
Mazurek, W., 242
Mazurier, A., 485
Mazza, M. C., 400

Mazzei, A., 488
Mazzi, U., 221
Meakin, P., 70, 223, 281, 355,
    361, 396
Mealli, C., 270, 272
Medvedeva, O. A., 7
Medvednikova, E. T., 142
Meek, D. W., 259, 367, 388,
    435
Meerschaut, A., 6, 91
Megaw, H. D., 81
Mègnamisi-Bélombè, M., 302,
    384
Mehra, S. C., 151
Mehrotra, A., 478
Mehrotra, P. K., 343
Mehrota, R. C., 21, 41, 82, 85,
    478
Mehta, R. K., 243, 325, 339,
    445
Meiklejohn, P. T., 136
Meinders, B., 139
Meisal, A., 150
Melamud, E., 244
Mel'chenko, G. G., 479, 485
Mel'gunova, L. G., 390
Mellier, A., 108
Melmed, K. M., 308
Melnik, M., 325
Mel'nik, Ya. I., 100
Mel'nikov, L. B., 161
Mel'nikova, T. Y., 452
Melovski, L., 425
Melpolder, J. B., 431
Melson, G. A., 277, 297, 471
Memering, M. N., 423
Menabue, L., 265, 442
Menaria, K. L., 215
Menchetti, S., 174
Mennenga, H., 105
Menyailova, G. A., 9, 140
Merbach, A. E., 82
Mercer, M., 110
Mercurio, J., 474
Mercurio-Lavaud, D., 65
Meredith, M. K., 101
Meredith, W. N. E., 87
Merehes, M., 27
Merkulov, A. A., 19
Merle, A., 281
Merrithew, P. B., 195
Mertis, K., 152
Messmer, R. P., 384
Mestroni, C., 360
Methan, T. M., 417
Metz, B., 301, 400
Meunier, B., 199
Mews, R., 169, 170
Meyer, D., 250
Meyer, E. F., 277
Meyer, T. J., 335, 337, 379
Meyers, E. A., 97
Meyers, W. H., 367
M'Foundou, E., 451

Michaelis, W., 202
Michejda, C. J., 441
Michel, A., 52
Michel, C., 79, 141
Michel, P., 142
Michelson, K., 101, 252
Michl, R. J., 305
Michnik, M. A., 79
Mido, Y., 239, 445
Midollini, S., 199, 270
Miezis, A., 203, 224, 271, 401
Migal, P. K., 216, 430, 451
Migeon, H. N., 248
Mihara, Y., 91
Mihichuk, L., 169
Mikhailov, M. A., 472
Mikhailov, V. A., 490
Mikhailov, Yu. N., 497
Mikhailova, A. S., 32
Mikhailova, N. S., 84
Mikhailyuk, L. Ya., 150
Mikhal'chenko, T. K., 407
Mikhalevich, K. N., 48, 119
Mikheeva, V. I., 48, 76, 474
Miki, E., 266, 424
Mikulec, Z., 446
Mikulenok, V. V., 46, 107, 138
Mikulski, C. M., 12, 100, 175,
   217, 236, 328, 472
Milburn, G. H. W., 194
Milburn, H., 321
Milburn, R. M., 379
Mildvan, A. S., 464
Milestone, N. B., 274
Mill, B. V., 66
Millar, T. R., 300
Milledge, H. J., 177
Miller, E. M., 372
Miller, F. J., 337
Miller, G. A., 54, 59
Miller, J. D., 335
Miller, J. S., 114, 334
Miller, N. C., 101
Miller, P. T., 109
Miller, R. E., 469
Miller, T. R., 207, 237
Miller, V. R., 247
Milligan, W. O., 202, 422
Mills, O. S., 170
Milne, R., 406
Milovidova, N. D., 84
Minacheva, M. Kh., 41
Minematsu, H., 382
Minghetti, G., 216, 431
Mingos, D. M. P., 218
Minichelli, D., 214
Mininkov, N. E., 478
Miniscloux, C., 115, 127
Mink, J., 383
Minkjewicz, J. V., 232
Minkova, M. M., 140
Minomura, S., 215
Mironov, K. E., 477, 478
Mironov, V. E., 451

Miroshinichenko, O. Ya., 123
Misawa, T., 198
Mishchenko, Yu. A., 65, 123
Mishra, L. K., 195
Miskin, B. K., 146
Misono, A., 194
Misra, G. N., 40
Missner, Kh., 90, 157
Misumi, S., 215, 224, 323, 451
Mitchell, P. R., 264
Mitchell, T. R. B., 368
Mitewa, M., 106
Mitra, G., 161
Mitra, S., 217
Mitrofanov, B. F., 8
Mitrofanova, N. D., 484
Mitschler, A., 199
Mitsuhashi, T., 33
Mitsuko, N., 65
Miyake, C., 494
Miyake, Y., 477
Miyoshi, H., 174
Miyoshi, T., 302
Mizota, T., 19
Mizumachi, K., 266, 424
Mocella, M. T., 217, 275, 428
Mochal'nik., I. A., 480
Mochizuki, A., 478
Mockler, G. M., 205, 295, 324
Modebadze, M. E., 484
Modestino, A. J., 195
Modraz, B., 206
Moeller, T., 476, 477
Moers, F. G., 331, 346, 355
Moggi, L., 94, 338
Moghissi, A. A., 470
Mogilevkina, M. F., 413
Mohai, B., 117
Mohan, M., 241
Mohan, N., 150
Mohandas, P. N., 326, 446
Mohanty, J. G., 323
Mohapatra, B. K., 313, 446
Mohrmann, L. E. jun., 101
Moigue, F., 31
Moise, C., 29
Moiseev, B. M., 84
Mokhosoev, M. V., 65, 96, 140,
   141, 142, 144, 179
Mokudu, J. A. A., 400
Molander, G. A., 470
Molin-Case, J., 197
Mollet, H. F., 113
Molls, W., 91, 268
Monacelli, F., 368
Moody, W. E., 296
Moonen, J. H. E., 133
Moore, C. J., 467
Moore, F. W., 87
Moore, G. W., 250
Moore, J. H., 137
Moraerum, D. W., 274
Morancho, R., 61
Moras, D., 400

Morassi, R., 216, 328
Morazzoni, F., 244
Morcotrigiano, G. O., 458
Moreau, J. M., 433
Moreland, J. A., 322
Mori, G., 325, 446
Mori, K., 19
Morikawa, H., 140
Morioka, Y., 369
Moriwaki, T., 387
Moriyama, J., 124
Moriyasu, M., 494
Mornon, J. P., 237, 322
Moro-Garcia, R., 103
Moroni, W., 302, 384
Morosin, B., 72, 474
Morozov, E. V., 33
Morozov, V. V., 8
Morozova, A. S., 368
Morozova, S. V., 79, 140
Morrell, B. K., 345, 370
Morrell, W., 455
Morris, A., 320
Morris, G. E., 121
Morrison, D. F. C., 438
Morse, K. W., 308
Morton, A. P., 1
Mosbaek, H., 192
Moseeva, E. M., 89
Moser, G. A., 121
Moshnenko, V. M., 4
Mosina, L. V., 57
Moskvitina, E. N., 63
Moskovits, M., 267, 402
Moss, J. H., 4
Moss, K. C., 134, 372, 389,
   390, 406, 415
Motekaitis, R. J., 209, 239, 291
Motooka, I., 12
Motov, D. L., 39
Motzny, H., 206, 241
Mountcastle, W. R. jun., 109
Mounts, R. D., 286, 307
Moustakali, I., 322
Mowat, W., 112
Moyer, J. W., 32
Mozer, T. J., 425
Mrikhina, R. I., 18
Mrozinski, J., 159
Mudretsov, A., 55
Mudrolyubova, L. P., 20
Müller, A., 100, 107, 149, 150,
   237, 238, 260, 322, 443,
   445, 461
Müller, G., 122, 138
Müller, H., 191
Mueller, J., 28, 46, 92
Müller, M., 202
Mueller, M. H., 422, 495
Müller, P., 150
Muenze, R., 158, 446
Müsel, N., 150
Muetterties, E. L., 42, 151, 362,
   471

Muftakhov, A. G., 492
Mui, K. T., 215
Muir, K. W., 253, 261, 356, 422
Muir, M. M., 366, 377
Muir, W. R., 372
Mukaida, M., 310
Mukhametshina, Z. B., 497
Mukherjee, R. N., 58, 237, 289, 496
Mukherjee, S. K., 106
Mukhopadhyay, R. K., 109
Mukmeneva, N. A., 24
Mulay, L. N., 5
Mulcahy, J. K., 318
Muller, H. M., 489
Muller, R. C., 83
Muller-Buschbaum, H. K., 81
Multani, R. K., 134, 476
Mundorf, T., 432
Murakami, T., 250
Murakami, Y., 58, 196, 324, 446
Muramatsu, K., 20
Muranaka, S., 6
Murmann, R. K., 257, 322
Muranovich, A. Kh., 5
Murase, I., 38
Muratova, A. A., 24
Muresan, V., 385, 409
Muriithi, N., 129
Murmann, R. K., 160
Murphy, D. J., 137
Murphy, D. W., 346, 352
Murray, A. J., 117
Murray, K. S., 177, 217
Murray, R. S., 202, 266
Murray, S. G., 238, 374
Murray-Rust, J., 265
Murray-Rust, P., 265, 305
Murtha, D. P., 286, 311, 428
Murty, A. S. R., 69
Murzubraimov, B., 446
Musco, A., 382
Musikas, C., 494
Musker, W. K., 225, 251
Musorin, V. A., 471
Musso, H., 460
Mustaev, A. K., 479
Musumeci, S., 195
Muto, M., 101
Muzhar-ul-Haque., 473
Myagkova, A. A., 95
Myasoedov, B. F., 79
Myers, W. A., 111
Mynott, R. J., 375
Myron, H. W., 6
Mysoedov, B. F., 490
Mzarenlishvili, N. V., 478

Nabilsi, A. H., 461
Naboka, M. N., 123

Nabors, J. B. jun., 109
Nadler, H. G., 195
Nag, K., 465
Naga, S., 260
Nagai, S. I., 485
Nagakura, S., 215
Nagaro, R., 251
Nagashima, N., 478
Nagle, R. J. jun., 111
Nagorsnik, E., 119
Nagypal, I., 327
Nahigian, H., 248
Naik, D. V., 389
Nair, C. G. R., 26
Nakagawa, I., 266
Nakahara, A., 291, 326, 327
Nakahara, M., 318
Nakai, H., 193
Nakajima, S., 207
Nakamoto, K., 150, 302, 303
Nakamura, A., 116, 121
Nakamura, T., 142
Nakamura, Y., 177
Nakao, Y., 291, 326
Nakaraura, Y., 320
Nakatsu, K., 382
Nakayama, K., 91
Nakihara, M., 65
Nakon, R., 291, 326, 327, 467
Nakrasova, V. V., 79
Naksin, V. I., 142
Naldelli, M., 296
Namazov, Z. M., 99
Nametkin, N. S., 440
Nanamatsu, S., 20
Nanda, R. K., 215
Nandi, A. K., 98
Nannelli, P., 100
Nanot, M., 20
Nappier, T. E., 367
Narang, K. K., 27, 177, 239, 240
Nardelli, M., 229, 255, 260, 322, 458
Nardin, G., 244, 307
Narita, T., 91, 99
Narnov, G. A., 479
Narula, S. P., 26
Narzikulova, R. M., 145
Naslain, R. N., 474
Nasonova, S. N., 479
Nassau, K., 95
Nassimbeni, L. R., 125, 205, 310, 315, 320, 441
Natarajan, M., 107
Natarajan, S., 444
Nath, A., 231
Nath, P., 215
Nathur, P. K., 102
Natidze, V. P., 478
Natile, G., 432
Natkaniec, L., 135
Nauer, R., 264
Nauman, C. F., 327

Naumchik, A. N., 32
Navon, G., 305, 426
Nazarova, L. A., 362, 368
Nazarova, L. V., 430, 451
Nechurova, N. I., 69
Nedil'ko, S. A., 479
Nefedov, V. I., 42, 194, 373, 422
Negas, T., 140
Negishi, E., 309
Negoiu, D., 27, 385, 397, 401, 409
Nekrasov, M. M., 81, 480
Neilson, G. W., 162, 218
Nekhamkin, L. G., 39
Nekrasov, Yu. S., 30
Nelson, H. C., 323
Nelson, J. H., 324, 396
Nel'son, K. V., 12
Nelson, N. J., 183
Nelson, T. R., 3
Nelson, W. H., 66
Nemeth, A., 439
Nemets, A. M., 111
Nenova, P., 27
Neshpor, V. S., 125
Nesmeyanov, A. N., 10, 30, 71, 157, 220
Nesterova, L. A., 472
Neuman, M. A., 471
Neumann, F., 166
Neville, G. A., 466
Nevskaya, Yu. A., 23
Newkome, G. R., 227
Newland, M. J., 266
Newman, D. J., 474
Newman, P. W. G., 279
Newnham, R. E., 81
Newton, G., 305
Newton, M. G., 312, 320
Newton, W. E., 136, 334
Neyer, A., 99
Ng, C. S., 106, 109
Ng, T. W., 269
Nguyen, D., 149, 480
Nguyen, H. C., 127
Nguyen, N. C., 129
Nguyen, V. C., 148
Nibert, J. H., 110
Nicholls, D., 10, 52, 61
Nicholson, B. K., 165
Nicolic, R. M., 437
Nieboer, E., 215
Nield, D. J., 137
Nieuwenhuijse, B., 175
Nieuwpoort, A., 131, 133, 173
Nigam, H. L., 11, 96, 104, 234, 392
Niinisto, L., 473, 477, 479
Niklewski, T., 52
Nikolaev, G. I., 111
Nikolaev, N. S., 491
Nikolaev, P. N., 89
Nikonenko, E. A., 440

Nikonova, L. A., 49
Nims, J. L., 350
Nisel'son, L. A., 64, 156
Nishida, K., 91, 99
Nishida, Y., 314
Nishimura, N., 424
Nitanteva, G. G., 447
Nixon, J. F., 89, 223, 353, 354
Nockolds, C. E., 194
Noel, H., 69, 91
Noel, S., 174
Noeth, H., 10
Nöthe, D., 302, 384
Nogina, O. V., 10, 30
Nolan, K. B., 251, 253
Nolan, M. J., 401, 423, 424
Nolte, M. L., 370
Noltes, J. G., 166
Nomiya, K., 253
Nomura, T., 289
Nonaka, Y., 314, 323, 382
Nonayama, K., 310, 317
Nonova, D., 216
Nonoyama, M., 240, 287, 292, 378
Norbury, A. N., 358
Nord, A. G., 243
Nordin, R. J., 42
Nordstrom, B., 464
Norkus, P. K., 344
Norman, J. G. jun., 127, 402
Normanton, A. S., 32
Norne, J. E., 462
Norris, A. R., 249
Norton, J. R., 353
Norton, M. G., 368, 397
Nosvren, N. A., 38
Novak, D. P., 89
Novak. J., 18, 135
Novakovskii, M. S., 43, 279
Novgorodtseva, N. A., 141, 142
Novikov, G. I., 4, 93
Novikov, Yu. P., 490
Novikova, E. M., 141, 142
Novitskaya, G. N., 156
Novolortsev, V. M., 75
Novoselov, G. P., 142
Novoselov, R. I., 432
Novoselova, A. V., 35, 485
Novosel'tseva, L. A., 472
Novotny, M., 119
Novotortseva, V. M., 98
Nowell, D. V., 38
Nowogrocki, G., 343
Nowotny, H., 92
Nunziante, C. S., 156
Nyburg, S. C., 345
Nyein, T. T., 286
Nygren, M., 52, 65, 91

Obol'yaninova, O. A., 123
Oboznenko, Y. V., 239

Obozova, L. A., 446
O'Bryan, N. B., 243
Ochsenreiter, P., 440
O'Connor, M. J., 242, 261, 265, 267, 343, 358, 390
Odell, K., 332
Odenbrett, C., 222
Odintsov, V. V., 7
O'Donnell, S., 67, 148
Odud, Z. Z., 148
Oehme, G., 105
Oelshlegel, F. J., 467
Oetker, C., 361, 362, 397, 398, 438
Oeye, H. A., 72
Offerson, T., 234
Ogden, J. S., 115
Ogino, H., 251
Ogoshi, H., 356
Ogura, T., 307, 308, 386
O'Hare, P. A. G., 143
Ohashi, S., 446
Ohlsonn, I., 464
Ohnesorge, W. E., 378
Ohrimenko, R. F., 39
Ohsaki, K., 194
Ohtsuki, S., 124
Ohyoshi, A., 341, 342, 494
Ojima, H., 310, 317
Okac, A., 50, 144
Okada, K., 140
Okamoto, K., 262
Okamoto, M. S., 217
Okamura, H., 266
Okamura, M., 314
Okano, H., 424
Okawa, H., 287, 295
Okawara, R., 186, 220
Oki, H., 102
Okransinski, S., 161
Okuue, Y., 326
Olabe, J. A., 202
Olah, G. A., 454
Oldham, C., 157, 419
Olejnik, Z., 206
Oliinyk, L. G., 49
Olive, S., 217
Oliver, J. P., 114
Olovsson, I., 108
Olsen, J. P., 162
Ol'Shevskaya, M. M., 63
Olson, L. W., 193
Olszanski, D. J., 471
Omaly, J., 141
Omori, M., 215
Omura, T., 356
Onaca, I., 38
Onaka, S., 165
O'Neill, D. R., 227
Ong, W. K., 325
Ono, A., 37
Ooi, S., 320
Oonishi, I., 252
Oonishi, T., 261

Opalovskii, A. A., 132, 138
Orbin, N. A., 161
Orchard, A. F., 87
Orlandini, A., 270
O'Reilly, E. J., 173
Orekhov, V. T., 122
Orel, B., 17
Orio, A. A., 221
Orioli, P. L., 436
Orlova, I. M., 496
Orrell, K. G., 226, 228, 271, 371
Ortego, J. D., 227, 491
Orth, H., 6
Orthanovic, M., 101
Ortner, H. M., 67
Osachov, V. P., 38
Osaka, S., 99
Osborn, J. A., 217, 372
Osborn, R. S., 422
Osborne, A. G., 371
Osborne, D. W., 143
Oshida, Z., 356
Osipov, O. A., 27, 79
Oskarsson, A., 38, 474
Ostazewski, A. P. P., 306
Österberg, R., 328
Osterheld, R. K., 425
Ostfield, D., 217
O'Sullivan, D. J., 359
O'Sullivan, W. J., 176
Oswald, H. R., 310, 320
Otake, M., 52, 136
Otaki, T., 52, 136
Otsubo, Y., 108, 116, 121, 223, 382
Ottersen, T., 309
Ottewill, R. H., 427
Ouahes, R., 93
Ouhadi, T., 198
Ovchinnikov, I. V., 59
Ovchinnikov, K. V., 157, 161
Ovchinnikov, T. V., 89
Ovchinnikova, N. A., 155
Owens, J. P., 6
Oyama, O., 291
Ozaki, Y., 377, 409
Ozerov, R. P., 107
Ozin, G. A., 88, 152, 216, 267, 353, 381, 402

Paderno, Yu. B., 7
Padurets, L. N., 48
Paez-Pedrosa, M., 385
Paik, H. N., 268, 269
Paine, R. E., 496
Paine, R. T., 122, 160, 491
Pajdowskii, L., 50, 51
Pak, V. N., 128
Pakhomov, V. I., 78, 142
Pakkanen, T., 479
Palant, A. A., 140
Palatnik, L. S., 81, 123

Palavit, G., 174
Palazzi, A., 165, 192
Palazzotto, M. C., 206, 261, 366
Palenik, G. J., 126, 200, 268, 388, 389, 444
Palkin, V. A., 404
Palmer, B. N., 327
Palmer, D. A., 364
Palmieri, G. G., 255, 296
Pal'shin, E. S., 78, 79, 138
Palyi, G., 221, 353, 356, 358
Panasenko, N. M., 20
Panchenkov, G. M., 19
Panday, K. K., 495
Pande, C. S., 40
Pandey, A. N., 18
Pandeya, K. B., 11, 96, 104, 234, 293, 392
Pandit, S. C., 132, 138
Pandu, R., 69
Pandy, R. N., 386
Pangratz, W. R., 242
Pani, R., 240
Pania, S. L., 243, 325, 445
Panin, A. V., 446
Pankratova, L. N., 42
Pannan, C. D., 298
Pannetier, G., 334
Pannu, B. S., 461
Panov, A. S., 7
Panov, V. B., 15
Panova, T. I., 81
Paoletti, P., 217, 327, 323, 441
Paolucci, G., 394
Papatheodorou, G. N., 385, 404
Papp, S., 353
Papp-Molnar, E., 325, 326
Paques-Ledent, M. Th., 65
Paramonova, V. I., 216
Parant, H., 95
Pardo, J., 302
Parent, H., 81
Parentich, A., 427
Parikh, P. G., 326
Paris, J., 490
Paris, R. A., 20, 65
Parish, R. V. D., 404
Park, E., 146
Parker, H. S., 81, 140
Parlenko, L. I., 271
Parmar, S. S., 446
Parmelee, R., 101
Paromova, M. V., 140
Parpiev, N. A., 54, 139, 176
Parrett, F. W., 18
Parris, G. E., 95
Parshall, G. W., 223
Parsons, J. A., 73
Parsons, T. C., 488
Partenheimer, W., 386
Parthe, E., 433
Paruta, L., 397, 401

Pasini, S., 244
Pasquali, M., 492
Pasternak, R. F., 253, 277, 469
Pastorek, R., 96, 483
Pasynskii, A. A., 70, 71, 75
Patel, K. C., 48, 272, 310
Patel, K. S., 57, 242, 324
Patel, R. N., 426
Patel, R. P., 442, 446
Patel, R. R., 199
Pathak, R., 147
Pathmanaban, S., 60
Patil, J. N., 480
Patil, R. N., 37
Patmore, D. J., 185
Patokin, A. P., 125
Paton, R., 215
Patterson, H. H., 350
Patterson, G. S., 207
Patton, R. D., 208
Paul, I. C., 243
Paul, R. C., 21, 26, 484, 492
Paul, S., 325, 341, 386
Pauley, C. R., 325
Paulson, H., 312
Pausewang, G., 17, 69, 78, 94
Pavlenko, E. S., 483
Pavlova, M., 158, 159
Pavlova, S. A., 140
Pavlyuchenko, M. M., 480
Pawson, D., 343, 350, 352
Payne, D. H., 169
Payne, D. S., 400
Payne, N. C., 422
Payne, S. J., 350, 401, 423
Payne, Z. A., 169
Pazdernik, L., 457, 458
Peacock, R. D., 113, 122, 138
Pearson, J. M., 491
Pearson, R. G., 281, 372, 403
Pebler, A., 139
Pebler, J., 195
Pechkovskii, V. V., 49, 61, 205, 214
Pechurova, N. I., 12
Pedelty, R., 210
Pederson, E., 101
Pedley, J. B., 28
Pedregosa, J. C., 65
Pedrosa, J. D., 494
Pelizzi, C., 296
Pellinghelli, M. A., 441, 458
Pellizzi, G., 229, 322
Pell, S., 333, 334
Pellacani, G. C., 226, 265, 301, 433, 442
Pelliccioni, M., 38
Pellerito, L., 297
Pendharkar, A. V., 151
Penfield, J. D., 107
Peng, S. M., 224
Peng, Y. K., 124
Pennella, F., 334
Penneman, R. A., 487

Percy, G. C., 442
Pereira, M. S., 326, 337, 446
Perelyaev, N. A., 47
Perepelitsa, A. P., 142, 480
Perez, G., 140
Perkins, P. G., 135, 144
Pernick, A., 112
Pernoll, I., 461
Perrier, M., 482, 483
Perrotey, J., 174, 444
Perry, A. M., 52
Perry, D. L., 491
Perry, W. B., 88
Perry, W. D., 98, 242
Pershikova, N. I., 49
Pershin, S. V., 42, 81
Persoons, A., 62, 338
Perte, E., 19, 20
Pertikatsis, B., 234
Perumareddi, J. R., 101
Perutz, R. N., 88
Pesavento, M., 327
Peschenko, N. D., 100
Peshchevitskii, B. I., 461
Peshev, P. D., 81
Peshkov, V. V., 497, 498
Pestunovich, V. A., 145
Petersen, J. L., 52
Petersen, R. L., 221
Pethig, R., 302, 393
Petillon, F. Y., 201, 239, 443
Petit, R. H., 150
Petkovic, M., 252
Petrakis, L., 112
Petras, P., 215
Petrinkim, V. E., 265
Petrov, K. I., 151, 485
Petrova, N. V., 84
Petrukhin, O. M., 428
Petrzhak, G. I., 490
Pets, L., 79
Pettifer, M. E., 61
Petz, W., 26
Petzel, T., 485
Peyronel, G., 265, 288, 301, 384, 433, 442, 452, 458, 459
Peytavin, S., 140
Pfeffer, M., 395
Pfefferkorn, K., 105
Pfeiffer, R. M., 196
Philips, C. V., 209
Philippot, E., 140
Phillip, A. T., 242
Phillips, D. J., 48, 258, 299, 301
Phillips, F. L., 345, 352
Phillips, W. D., 211
Philp, D. K., 91
Phipps, D. A., 263
Piccaluga, G., 96
Pick, M. A., 73
Pickard, F. H., 345
Pickerell, M. E., 97

Pickett, C. J., 87
Pickles, V. A., 486
Pidcock, A., 131, 332, 417, 424
Pierce-Butler, M., 25
Pierens, R. K., 451
Pierpont, C. G., 97, 345, 400, 422, 442
Pierrand, J. C., 233
Pietropaolo, D., 165, 371, 374
Pignedoli, A., 288, 301
Pignolet, L. H., 100, 177, 206, 207, 261, 366
Pilbrow, J. R., 12, 53, 323
Pilipenko, A. T., 461
Pilipenko, F. S., 141
Piljac, I., 446
Piltingsrud, D., 100, 209, 365, 425
Piltz, G., 491
Pinaev, G. F., 214
Pink, H., 99
Pinkerton, A. A., 485
Pinna, G., 96
Pinnavaia, T. J., 41
Pinsky, L., 147
Piovesana, O., 289
Piraino, P., 371, 374
Pirinoli, F., 145
Pishchai, I. Ya., 3
Piskor, S. R., 323
Pistara, S., 491
Pitsyuga, V. N., 141, 179
Pittman, C. U., 356
Pivnichny, J. V., 214
Pivovarova, M. G., 446, 451
Plane, R. A., 438
Plante, T., 137
Plapp, B. V., 464
Plath, M., 257
Platt, H., 109
Platt, R. H., 113
Platte, C., 99
Plautz, H., 128
Pletcher, D., 87
Pletnov, R. N., 65, 480
Plewett, G. W., 19
Ploae, K. I., 430
Plostinaru, S., 483
Plotinskii, G. L., 347
Plurien, P., 490
Pluth, J. J., 245
Plygunov, A. S., 5, 6
Plyushchev, V. E., 20, 108, 140, 142, 479, 480
Pocker, Y., 463
Poddar, R. K., 331, 342, 363
Podder, N. G., 326
Poddubnyi, I. Ya., 28
Podlaha, J., 135, 215
Podnebesnova, G. V., 496
Podogomaya, I. V., 56
Podolsky, G., 75
Podozerskaya, E. A., 82

Podzolko, Yu. G., 145
Poe, A. J., 167, 220
Poilblanc, R., 221, 354
Poirier, Y. M., 201, 239
Poix, P., 81, 95
Pokhodenko, A. P., 39
Poliakoff, M., 181
Polinskaya, M. B., 50, 136
Pollard, E. R., 72
Pollert, E., 65
Polligkert, W., 497
Pollock, R., 375
Pollock, R. J. I., 403
Polotebnova, N. A., 136, 148, 149
Poltavtsev, N. S., 125
Polunina, G. P., 489
Polyachenok, L. D., 4
Polyachenok, O. G., 4
Polyakov, V. A., 56
Polyanskaya, T. M., 143
Polynova, T. N., 43
Pomerants, G. B., 430
Pont, L. O., 304
Ponticelli, G., 459
Poon, C. K., 249, 254
Poonia, N. S., 446
Pope, M. T., 54, 67, 136, 148, 149, 480
Popelis, J., 145
Popescu, I., 393
Popesen, I., 387
Popielski, S. E., 119
Popolito, V. I., 81
Popov, A. F., 68
Popov, A. P., 478
Popov, B. N., 144
Popov, M. S., 40
Popov, S. G., 95
Popova, R. A., 82
Popovic, S., 83
Porai-Koshits, M. A., 373, 484
Porta, F., 359, 394
Porta, P., 289
Portanova, R., 490, 499
Porter, J. K., 323
Porter, R. F., 139
Porter, S. K., 321, 400
Potenza, J., 321
Portier, J., 92
Porubstky, I., 439
Poshkute, Z. A., 344
Post, M. L., 448, 450
Postel, M., 83
Postoenko, G. E., 82
Potanina, N. A., 123
Potashnikov, Yu. M., 144
Potoff, A. D., 81
Potsov, A. E., 39
Potts, G. T., 112
Potvin, C., 334
Pouchard, M., 78, 318
Poulet, H., 495
Poulsen, K. G., 192

Pourier, Y. M., 443
Powell, A. R., 335, 349
Powell, D. B., 299, 334, 401, 423, 431
Powell, H. K. J., 327, 479
Power, P. C., 323
Powers, D. A., 35
Pozhidaev, A. I., 43
Pozdin, I. A., 20
Pradelli, G., 92
Pradilla-Sorzano, J., 323
Prado, J. C., 482
Prados, R., 484
Prados, R. A., 136
Praliaud, H., 112
Pramanik, P., 351
Prasad, A. S., 467
Prasolova, O. D., 160
Prather, J. W., 49
Pratt, J. M., 245
Prelesnik, B., 261
Premovic, P. I., 10
Prescott, B. E., 95
Preti, C., 103, 226, 233, 301, 313
Preuss, F., 68
Pribula, C. D., 163
Pribytkova, T. A., 93
Price, D. C., 263
Price, E. R., 100
Priest, P., 307
Prigent, J., 124
Prijis, B., 318
Prijo, B., 327
Prince, R. H., 174, 243, 277, 463
Prins, G., 499
Prokof'eva, G. N., 5
Prokopchik, A. Y., 344
Pronin, V. A., 390
Propach, V., 303
Propof'eva, R. E., 68
Propst, R. C., 477
Prosenko, A. E., 413
Proshina, N. N., 250
Pross, L., 159
Protas, J., 31
Protasova, V. I., 142
Proteasa, M., 102
Prout, K., 2, 112, 121, 321
Provotorov, M. V., 142
Prozorovskaya, Z. N., 40
Pruchnik, F., 228, 259
Pryanchikov, E. N., 125
Pryde, A. J., 371, 414
Przybylinski, J. L., 231
Pszonicki, L., 147
Pudovik, A. N., 24
Pugach, E. A., 7, 48
Pulova, L. A., 29
Pupp, C., 139
Puppe, L., 102
Purik, L. A., 40
Push-Karna, S. K., 446

Puxeddu, A., 224
Puzanowska-Tarasiewicj, H., 134
Pytlewski, L. L., 12, 100, 101, 175, 176, 198, 209, 217, 236, 328, 439, 472

Quagliano, J. V., 281
Quastlerova-Hvastilijova, M., 325
Que, L., 100, 177, 206, 211, 261, 366
Quemeneur, E., 18
Queyroux, F., 20
Quilley, P. J., 217
Quinn, J. D., 322
Quist, A. S., 34

Rabeneck, H., 157
Rabenstein, D. L., 465
Rabinovich, I. B., 87, 89, 153
Raczko, M., 144
Radford, D. V., 200, 451
Radhakrishna, S., 107
Radkowsky, A. E., 109
Radwojsa, P. N., 265
Raethlein, K. H., 455
Raff, D., 277, 469
Rafson, P. A., 204
Rahman, S. M. F., 12
Rai, A. K., 41, 82, 85
Raithby, P. R., 473
Rajabalee, F. J. M., 176, 234, 291, 446
Rajaram, J., 403
Rajender, S., 464
Rake, A. T., 85, 165
Rakhimov, Kh. R., 492
Rakita, P. E., 379
Rakkar, S. N., 326
Rakke, T., 248
Rakov, E. G., 46, 63, 72, 107, 138, 472
Rakova, N. N., 146
Rakshit, S., 117, 155
Ramakrishna, R. S., 60
Ramamoorthy, S., 216, 465
Ramamurthy, P., 325
Rambold, W., 165
Ramesh, B. R., 360
Rana, V. B., 241
Randaccio, L., 244, 255, 297, 307
Randin, J. P., 137
Ranga, V. P., 62
Rankin, D. W. H., 424
Rao, D. W. R., 426
Rao, G. N., 57, 239, 291, 442
Rao, G. V. S., 76, 124
Rao, S. B., 215, 325
Rao, V. R., 481
Rao, V. V. K., 461

Rapsey, G. J. N., 131, 424
Rasmussen, S. E., 305, 422
Rasoul, H. A. A., 327
Rastogi, D. K., 241, 317, 387, 481
Raston, C. L., 69, 226, 279, 287, 306, 458
Rastorgi, S. C., 239, 291
Ratcliff, B., 220, 417
Ratiani, D. D., 95
Rat'kovskii, I. A., 93
Ratnasamy, P., 123
Ratner, E. I., 110
Rattanaphani, V., 200, 227, 308
Rau, H., 28
Rauch, F. C., 88
Rauchfuss, T. B., 361
Rauh, E. G., 33, 488
Rausch, M. D., 28, 44
Raveau, B., 79, 81, 141, 142
Ravez, J., 81, 92
Ray, H. L., 293
Ray, M. P., 376
Ray, S., 234, 321
Raymond, K. N., 103, 181
Rayner-Canham, G. W., 194
Raynor, H. B., 379
Raynor, J. B., 422
Razavi, A., 171
Razgon, E. S., 142, 480
Razumova, E. P., 440
Razumovskii, V. V., 424
Readnour, M., 309
Rebenstorf, B., 93
Rechani, P. R., 291, 327
Rechmann, H., 1
Reddy, G. K. N., 360, 368, 379
Reddy, P. R., 479
Redfield, D. A., 396
Redhouse, A. D., 191
Redman, M. J., 149
Reed, C. A., 214, 374
Reed, F. J. S., 407, 419
Reed, J., 345
Reed, J. L., 367
Reed, T. B., 72
Reedijk, J., 173, 175, 235, 272, 323
Regen, S. L., 109
Rehder, D., 46
Reichardt, W., 105
Reichembach, G., 222
Reichert, W., 112
Reichgott, D. W., 196
Reid, A. F., 96
Reidler, J., 478
Reiff, W. M., 203, 210
Reikhofel'd, V. O., 12
Reilley, C. N., 272
Reilly, J. J., 8
Reimann, R. H., 153, 163, 164, 167, 172
Reinen, D., 303, 309

Reinhard, G. 446
Reinhardt, F. W., 108
Reis, A. H., 443
Reisenhofer, E. P., 224
Reisfeld, M. J., 489
Relevantiev, N. I., 461
Remeika, J. P., 128
Rempel, G. L., 331
Renard, A., 322
Rendall, I. F., 60
Renner, J., 206, 213
Reshetnikova, L. P., 485
Rettig, S. J., 121
Reuben, J., 468
Reuter, B., 214
Revel, A., 148
Revenko, L. V., 357
Revenko, M. D., 56
Reynes, J., 52
Reynhardt, E. C., 260
Reynolds, T. G., 137
Reznichenko, V. A., 4, 9, 26, 140
Ribas, J. G., 485
Ricard, A., 103
Ricard, L., 131, 147, 168, 199
Riccardi, R., 205
Rice, D. A., 44, 61, 86, 134, 139, 341
Richard, P., 444
Richards, J. A., 220
Richards, R., 307
Richards, R. L., 110, 116, 212, 348
Richer, J. C., 109
Rieber, W., 359, 394
Rieck, G. D., 140
Rieck, H., 214
Riedel, E., 99, 100, 214
Rieskamp, H., 63, 69
Riess, J. G., 82, 83, 86
Rietz, R. R., 247
Riggs, W. M., 304
Rigny, P., 147, 490
Rigo, P., 216, 327
Riley, B., 415
Riley, J. T., 214
Riley, P. E., 248
Rillema, D. P., 244
Rimbault, J., 233
Rimmington, H. P. B., 6, 33
Ringenbach, C., 129
Rinke, K., 157
Rinmer, G. D., 358
Riolo, C. B., 327
Rippon, D. M., 452
Risen, W. M., 90, 163
Ritchie, G. L. D., 451
River, J. E., 217
Rivest, R., 212, 457, 458
Rivu, A., 108
Rizzarelli, E., 195
Robb, W., 363
Robbins, D. J., 53

Robbins, M., 99
Roberts, J. D., 3
Roberts, M. W., 111
Robertson, G. B., 113, 194, 210, 265
Robertson, N. J., 437
Robinson, B. A., 76
Robinson, S. D., 307, 332, 335, 338, 339, 397, 412
Robinson, W., 281
Robinson, W. R., 5, 152, 159, 219
Robinson, W. T., 253, 321, 325
Robson, R., 16, 152, 314
Rocchiccioli-Deltcheff, C., 81, 147
Roček, J., 106, 109
Roche, T. S., 254
Rockefeller, H. A., 481
Roddy, J. W., 488
Roder, R., 76
Rodgers, A., 441
Rodrigue, L., 123
Rodriguez, G. J., 461
Roebber, I. L., 452
Roenker, K. P., 48
Roessler, K., 159
Rog, G., 95
Rogachev, B. G., 357, 387
Rogers, A. L., 320
Rogers, C. A., 364
Rogers, D., 422
Rogers, M. T., 78
Rogl, P., 92, 474
Rohbock, K., 102, 348
Rohde, N. M., 100, 210, 261
Rohmer, R., 169
Rohwer, H. E., 491
Rolfe, M., 134
Roman, L., 393
Romeo, R., 408
Romers, C., 284
Romm, I. P., 24
Ronami, G. N., 142
Ronis, M., 95, 205
Ronniger, G., 66
Ronova, I. A., 152
Roper, W. R., 374
Ros, R., 390
Rose, N. J., 196
Rosen, E., 312
Rosenberg, E. H., 3
Rosenblum, C., 90, 216
Rosenblum, M., 217
Rosenthal, U., 105
Rosenthall, I., 263
Rosenzweig, E., 487
Ross, D. S., 122
Ross, P. F., 322
Ross, R. A., 81
Rossi, M., 223
Rossotti, F. J. C., 321
Rostchakovskaya, N. Y., 311
Roth, R. S., 81, 140

Rothschild, B. L., 483
Rothwarf, F., 76
Rotillo, G., 313
Roundhill, D. M., 361, 405
Rousson, R., 63
Roux, M. T., 474
Rouxel, J., 6, 33, 91
Rowbottom, J. F., 154
Rowe, M. D., 341
Rowland, J. F., 215
Rowley, D. A., 286
Roy, S. K., 148
Royer, D. J., 251
Royston, G. H. D., 348
Rozanov, I. G., 480
Rozbianskaya, A. A., 132
Rozhenko, S. P., 17
Rozmanova, Z. E., 138
Rozovskii, G. I., 344
Ruben, H., 108
Rubin, J. A., 81
Rubinchik, Ya. S., 480
Ruchkin, E. D., 4
Rudelle, R., 147
Rudolf, H., 477
Rudolf, M. F., 98, 134, 135
Rudy, E., 92
Ruedorff, W., 69, 498
Ruff, J. K., 88, 188
Ruiz-Ramirez, L., 331
Rukavina, T. G., 97
Rumfeldt, R., 384
Runov, N. N., 108
Rusheed, A., 147
Rusholme, G. A., 194
Rusina, M. N., 326
Russell, R. L. C., 260
Rustagi, S. C., 57, 442
Rustamov, P. G., 485
Ruthardt, R., 6
Ruzic-Toros, Z., 145
Ryabenko, E. A., 5, 26
Ryabov, E. N., 10, 129
Ryabushko, D. P., 461
Ryan, J. L., 491
Ryan, R. R., 104, 487
Ryan, T. A., 10
Rybakov, A. G., 122
Rybakov, K. A., 489
Rybakov, V. K., 142
Rycheck, M. R., 334
Rykl, D., 91

Saarinen, H., 483
Sabadie, J., 385
Sabat, H., 134, 135
Sabelli, C., 174
Sabrowsky, H., 99, 387
Saburi, M., 262
Sacconi, L., 199, 216, 223, 242, 270, 272, 283, 328, 441
Sacerdoti, M., 222

Sadakane, Y., 484
Sadavoy, L. S., 345
Sadikova, A. T., 491
Sadkov, A. P., 116
Sadler, P. J., 324
Sadler, W. A., 272
Sadykova, M. M., 147
Saeki, M., 65
Saeki, Y., 122
Safina, R. G., 39, 148
Safonov, V. A., 89
Safonov, V. V., 74
Safronov, E. K., 23, 32
Sagisawa, K., 460
Saha, A. K., 105
Saha, B. C., 134
Saha, H. K., 134
Sahira, Y., 194
Sahoo, B., 229, 316
Sahotra, S. S., 461
Saibova, M. T., 54
St. Denis, J., 114
St. Krustev, V., 459
St. Nikolov, G., 94
Saito, K., 251
Saito, M., 124
Saito, Y., 251, 455
Saji, T., 45
Sajus, L., 115, 127, 150
Sakaguchi, U., 251
Sakamoto, T., 295
Sakata, K., 324, 446
Sakavov, I. E., 479
Sake Gowda, D. S., 444
Sakharov, V. V., 38, 79, 471, 489
Sakharova, N. N., 485
Sakharova, Yu. G., 485
Sakai, N., 215
Sakata, K., 196
Sakovich, V. N., 36, 479
Saksena, R. N., 495
Sala-Pala, J., 63
Salentine, C. G., 248
Sales, K. D., 60
Salienko, S. I., 14
Salinas, F., 461
Salmon, J. E., 147
Salmon, R., 137, 480
Saloman, C., 282
Saltmann, P., 206, 213
Salvatori, T., 492
Salyn, J. E., 373
Sammartano, L., 195
Samoilenko, V. M., 438
Samoilova, S. A., 110
Samorodova, V. S., 138
Samotus, A., 118, 119
Sams, J. R., 185
Samsonov, G. V., 7, 8
Samsonova, N. P., 451
Samus, N. M., 365
Samvelyan, S. Kh., 87
Sandhu, S. S., 24, 191, 339

Sandler, R. A., 10, 129
San Filippo, J. jun., 126, 127
Sansoni M., 352, 380, 433
Santini-Scampucci, C., 86, 87
Sanyal, N. K., 32, 159
Sanz, F., 373
Sapranova, E. A., 122, 129
Saraceno, A. J., 100
Sarbaev, D., 451
Sargeson, A. M., 251, 253, 265
Sarishvili, L. P., 272, 447
Sarkar, B., 326, 463
Sarkar, S., 106
Sarneski, J. E., 104, 272
Sartori, G., 93
Sarukhanov, M. A., 176
Sasada, Y., 370
Sasaki, Y., 129, 135, 195, 302
Sasse, H. E., 114
Sastri, V. S., 94
Sastry, V. V. R., 62
Sata, T., 142
Sato, M., 31, 59
Sato, S., 251, 252
Sato, T., 489
Sato, Y., 102, 251, 252, 253, 261
Satpathy, S., 229
Saval'eva, M. V., 142
Savchenko, L. A., 204, 205
Savel'era, Z. A., 432
Savin, V. D., 7
Savino, P. C., 443
Savitsku, A. V., 244
Savory, C. G., 117
Sawada, K., 224
Sawai, T., 170
Sawamoto, H., 105
Sawatzky, G., 1
Sawyer, D. T., 69, 135
Saxena, R. S., 60, 65, 479, 485
Saxena, U. B., 41
Sayle, B. J., 291, 466
Sbrignadello, G., 152, 153, 218
Scamellotti, A., 289
Scapivelli, B., 433
Scaramuzza, L., 60
Scarborough, J. P., 109
Scaringe, R., 324
Scatturin, A., 432
Scenery, C. G., 227, 390
Schaefer, H., 128, 139, 157, 161
Schäfer, M., 242
Schäffer, C. E., 251
Schaeffer, R., 418
Schaeffer, W. P., 245, 246, 266
Schaekers, J. M., 496
Schafer, M. W., 124
Schastnev, P. V., 19
Schegrov, L. N., 205
Scheidt, W. R., 232, 253, 277
Schenk, H. J., 159, 491
Scherbakova, M. N., 77

Scherle, J., 139, 144
Scheunemann, K., 81
Schiffer, M., 466
Schlaepfer, C. W., 150
Schlemper, E. O., 160, 257, 280, 322
Schlodder, R., 362, 373, 394, 398
Schmid, G., 183, 398
Schmidbauer, A., 455
Schmidgen, G., 270
Schmidt, J., 46, 193
Schmidt, K. H., 107
Schmidt, V., 69
Schmitz, D., 387
Schmitz-DuMont, O., 271
Schmonsees, W. G., 229
Schmulbach, C. D., 9
Schmutzler, R., 396, 422
Schneehage, H. H., 102, 204, 471
Schneider, M., 46
Schneider, M. L., 167
Schneider, S., 174
Schnidler, J. W., 201
Schnitker, U., 110
Schnitzler, M. S., 168
Schoellhorn, R., 6, 76
Schön, G., 425
Schoen, S., 109
Schoenberg, A. R., 169
Schoenberger, R. J., 47
Schoomaker, E. B., 467
Schorpp, K., 359, 361, 362, 382, 397, 398
Schram, E. P., 51, 405
Schrauzer, G. N., 116
Schreiner, A. F., 94, 271
Schriver, D. F., 330
Schroecke, H., 81
Schröder, F. A., 139, 144, 305
Schröder, J., 122
Schub, J. M., 157
Schug, H., 68
Schugar, H. J., 321
Schultz, A. J., 345
Schulz, H., 139
Schulz, R. A., 98, 177
Schulze, H., 107, 150
Schumann, H., 268
Schunn, R. A., 167, 269, 281, 374, 381, 396
Schupp, R. L., 432
Schussler, D. P., 219
Schwall, R., 73
Schwartz, J., 75
Schwartz, L. H., 76
Schwarzenbach, D., 400
Schweitzer, G. K., 70, 113, 435
Schweizer, A., 195
Schwenk, A., 436, 447
Schwing-Weill, J. M., 136
Schwochau, K., 159, 491
Scibona, G., 205, 499

Scott, D., 134
Scott, J. C., 185
Scott, K. L., 249, 266
Scott, R. A., 109
Scovell, W. M., 213
Screpel, M., 205
Scroggie, J. G., 42, 103
Scudder, M. L., 467
Sculfort, J. L., 147
Searcy, A. W., 155
Searle, G. H., 252
Searle, G. W., 404
Searles, J. E., 136, 334
Sears, P. L., 199
Secco, E. A., 107, 325
Secco, F., 66
Seddon, D., 111, 117, 191, 238
Seddon, K. R., 10, 52
Sedej, B., 17
Seebach, G. L., 58
Seel, F., 250
Seematter, D. J., 262, 263
Seff, K., 174, 234, 248, 309, 486
Sefton, G. L., 205
Segal, G., 492
Segal, J. A., 347
Segall, M. G., 470
Segedin, P., 127
Segel, S. L., 47
Sehaaf, T. F., 88
Seibold, C. D., 115
Seida, M., 232, 325
Seidel, H., 312
Seidel, W., 51, 105
Seidel, W. C., 269
Seidl, V., 91
Seifer, G. B., 194
Seiler, G. J., 101
Sejekan, B. G., 292
Sekido, E., 239, 445
Sekhon, B. S., 446, 461
Sekizaki, M., 321
Selbin, J., 110
Seleznev, V. P., 497
Selig, H., 161
Sellmann, D., 172
Selvey, S. E., 496
Semenishin, D. I., 119, 426
Semenov, E. V., 100, 133
Semenyakova, L. V., 140
Semikinn, L. E., 311
Semochkin, S. Ya., 472
Sen, B., 87
Sen, B. K., 74, 155
Sen, D., 351, 412
Sen, D. N., 480
Sen, S. K., 57
Senegas, J., 9, 81, 95
Sengupta, A. K., 98
Sengupta, G. P., 57
Sen Gupta, K. K., 109
Senior, R. G., 87, 185

Senkowski, T., 119
Senoff, C. V., 376
Sequeira, A., 210
Seralya, V. I., 311
Serebrennikov, V. V., 40, 472, 479, 482, 483, 485
Serebryakova, T. I., 125
Seressova, V., 325
Serezhkin, V. N., 142, 143
Sergeeva, A. N., 48, 271, 426, 461
Sergent, M., 124
Sergienko, V. I., 472, 497
Serree De Roch, I., 150
Serukieva, D. B., 225
Seryakova, G. E., 479
Seshadri, K. S., 112
Seshagiri, V., 215, 325
Seth, A., 6
Sevast'yanova, R. E., 377
Seyferth, D., 218
Seyferth, K., 117
Sgamellotti, A., 375
Sgibnev, E. V., 5
Shadikar, P. V., 60
Shafer, M. W., 76
Shafranski, V. N., 241
Shagisultanova, G. A., 418
Shah, B. F., 235
Shah, J. R., 442, 446
Shaimuradov, I. B., 485
Shakhno, I. V., 108, 142, 480
Shakhova, N. F., 345
Shakirova, D. M., 108, 425
Shalukhina, L. M., 72
Shammasova, A. E., 20, 36
Shamrai, F. I., 124
Shanbhag, S., 58, 496
Shandles, R. S., 160
Shannon, J. S., 294
Shannon, R. D., 91
Shapley, J. R., 217
Shaplygin, I. S., 364, 365
Shapoval, V. I., 144
Sharipov, A., 72
Sharipov, D., 24
Sharma, D. K., 18
Sharma, H. N., 11
Sharma, P. K., 461
Sharma, R. C., 301
Sharma, S. K., 21
Sharov, V. A., 96, 173, 440
Sharova, A. K., 81
Sharp, D. W. A., 78, 115, 126, 150, 170, 290
Sharp, G., 28
Shastina, Z. N., 390
Shatalova, G. E., 36
Shaulov, Yu. Kh., 5
Shaw, B. L., 158, 332, 371, 372, 390, 397, 414, 424, 454
Shawl, E. T., 146
Shchekina, N. S., 451

Shchelokov, R. N., 79, 446, 496, 497
Shcherbakov, V. A., 494
Shcherbina, K. G., 140
Shchetinina, G. P., 472
Shearer, H. M. M., 129
Sheats, J. E., 22
Sheelwant, S. S., 60, 485
Shehadeh, M. A., 336
Sheiman, M. S., 87, 153
Sheinkihan, A. I., 38
Sheka, I. A., 31, 216
Sheldrick, G. M., 107
Sheldrick, J. M., 228
Sheldrick, W. S., 26, 422
Shelest, L. N., 32, 34
Sheliya, N. G., 484
Shelton, E., 469
Shelton, R. A., 32
Shepel'kov, S. V., 158
Sheridan, P. S., 364, 366
Sherrill, H. J., 110
Sherwood, R. C., 95, 99
Shevchenko, A. V., 95
Shevchenko, N. N., 36, 140
Shevchenko, T. M., 483
Shevchuk, I. A., 101
Shewchuk, E., 397
Shibahara, T., 245
Shibasaki, Y., 91
Shibata, M., 261, 262, 266
Shichko, V. A., 36
Shieh, C. F., 90
Shiever, J. W., 95
Shigematsu, T., 195
Shigeru, A., 45
Shihada, A. F., 55, 146
Shilov, A. E., 15, 16, 49, 110
Shilova, A. K., 14
Shimada, M., 303
Shimizu, S., 141
Shimodaira, S., 198
Shimura, A., 310
Shimura, Y., 250, 262
Shin, S., 81
Shindler, Yu. M., 49
Shinik, G. M., 149
Shinike, T., 479
Shinkarev, A. N., 5
Shinmei, M., 485
Shinoda, F., 194
Shinra, K., 176
Shinyaev, A. Ya., 124
Shiokawa, J., 476, 479
Shirotani, I., 215
Shishakov, N. A., 345
Shishkin, E. A., 125
Shishkov, D. A., 144
Shivahare, G. S., 65
Shive, L., 112
Shkol'nikov, S. N., 92, 122
Shlenskaya, V. I., 343
Shmidt, F. K., 3
Shmidt, V. S., 489

Shnaiderman, S. Ya., 23, 54
Shock, J. R., 78
Shokanov, A. K., 145
Shokarev, M. M., 204, 205
Shome, S. C., 376
Shono, T., 176
Shopenko, V. V., 51
Shores, R. D., 114
Shorikov, Yu. S., 379
Short, E. L., 438
Shriver, D. F., 121, 183
Shtrambrand, Yu. M., 5
Shubochkina, E. F., 418, 422
Shubochkina, L. K., 418, 422
Shugal, N. F., 239
Shukla, I. P. 446
Shukla, J. P., 326
Shukla, J. S., 461
Shukla, U. P., 26, 287
Shulakov, A. S., 125
Shul'ga, Yu. M., 15
Shulman, V. M., 432
Shumilin, E. N., 36
Shur, V. B., 373
Shuster, Ya. A., 311
Shuvalova, N. I., 16, 49
Shuydka, L. F., 351
Shvalova, T. G., 479
Shveikin, G. P., 8, 33, 47, 480
Shvelashvili, A. E., 273, 440, 447
Shvetsov, Y. A., 357
Siapkas, D., 81
Sibirskaya, V. V., 407, 408
Sick, E., 76
Siclet, G., 137
Sidorenko, V. I., 216
Sidorko, V. R., 92
Siebert, H., 249, 267
Siedle, A. R., 37, 304
Siegal, B. G., 319
Siegel, J. A., 286
Sieler, J., 99
Sievers, R., 98, 277
Sievert, W., 149, 150, 322
Siftar, J., 472
Sigel, H., 318, 327
Siiman, O., 176, 210, 298
Sikirica, M., 460
Siklos, P., 62
Sikorskaya, E. K., 96
Silber, H. B., 478
Silva, R. J., 489
Silvani, A., 382
Silver, B. L., 244
Silverstein, A. J., 323
Silverton, J. V., 146
Silvon, M. P., 111
Sim, G. A., 111, 189
Simanova, S. A., 390
Simeon, V., 326, 467
Simkin, D., 26
Simono, Y. A., 250
Simonova, T. N., 101

Simonsen, O., 322
Simonsen, S. H., 310, 320, 400
Simpson, G. D., 324
Simpson, J., 165
Sims, A. L., 247
Sims, R., 294
Sindellari, L., 459
Sinel'nikova, V. S., 7
Sing, K. S. W., 91
Singh, B. P., 18
Singh, B., 241, 365, 393
Singh, C. B., 229, 316
Singh G., 484, 492
Singh, H., 24
Singh, H. S., 18
Singh, J. P., 106
Singh, J., 109
Singh, M., 492
Singh, P. P., 26, 241, 280, 287
Singh, R., 109, 226, 365, 393, 439, 440, 461
Singh, R. P., 293, 323, 327
Singh, S. P., 55
Singhi, V. C., 445
Singleton, E., 153, 163, 164, 167, 172
Sinha, D. P., 106
Sinitsyn, N. M., 151, 158, 347, 379
Sinitsyna, E. D., 478
Sinitsyna, S. M., 77, 84
Sinn, E., 216, 321, 324, 325
Sinyakova, S. I., 159
Sipe, J. P., 486
Sironen, R. J., 493
Sirota, A., 298
Site, A. D., 484
Sivak, M., 65
Sizoi, V. F., 30
Sjoberg, S., 327
Sjoborn, K., 146
Skapski, A. C., 151, 158, 320, 345, 352, 370
Skell, P. S., 111
Skinner, H. A., 89, 122
Skinner, H. B., 155
Sklar, L., 277
Sklyarenko, Yu. S., 478
Skoglund, U., 473
Skolis, Yu. Ya., 142
Skopenko, V. V., 234
Skorobogat'ko, E. P., 438
Skudlarski, K., 161
Slade, R. M., 355, 396
Slater, J. L., 476
Sledziewska, E., 275
Sleight, A. W., 141, 142
Sletten, E., 320
Slim, D. R., 63
Slivnik, J., 17, 42, 90, 105, 173
Sljukic, M., 83
Sloan, D. L., 464
Slobodchikov, A. M., 108
Slobodin, B. V., 65, 141, 480

Sloccari, G., 214
Slyudkin, O. P., 412
Slyusar, N. P., 7, 47, 156
Smail, W. R., 323
Smedal, H. S., 362, 398
Smeltzer, W. W., 91
Smirnov, K. P., 123
Smirnov, S. V., 478
Smirnov, V. V., 483
Smirnova, I. N., 141
Smith, A. J., 306
Smith, D. W., 324
Smith, F. E., 196, 201, 228, 241
Smith, G. D., 473
Smith, G. S., 494
Smith, J., 112
Smith, J. D., 323
Smith, J. H., 321, 325
Smith, K. D., 471
Smith, M. A. R., 372, 389, 390, 406, 415
Smith, P. D., 351
Smith, R. T., 142
Smith, T. D., 12, 53
Smits, J. M. M., 255
Snaidoch, H. J., 126, 127, 363
Sneddon, L. G., 247
Snow, M. R., 252
Snyder, R. V., 291
So., H., 136, 148
So, S. P., 72
Sobezak, R., 36
Sobolev, A. S., 125
Sobolova, L. G., 447
Sobota, P., 8, 61
Soderlund, G., 464
Söderbach, E., 305
Sogani, N. C., 446
Sohn, Y. S., 21, 22
Soignet, D. M., 101
Soklosa, H. J., 53
Sokol, V. I., 365
Sokolenko, A. I., 92
Sokolenko, V. I., 7
Sokolenko, V. V., 92
Sokolik, J., 325
Sokolov, D. N., 49
Sokolov, V. I., 391
Sokolov, V. P., 27
Sokolova, G. V., 82
Sokol'skii, D. V., 49
Soldi, T. F., 327
Solntseva, L. S., 84
Solomakha, V. N., 37, 142, 480
Solov'ev, K. N., 62
Solov'eva, A. E., 36
Solovykh, T. P., 365
Solozhenkin, P. M., 100, 133
Solstad, P. J., 237, 325
Solymosi, F., 96
Sommer, B. A., 213
Somova, I. I., 144
Somova, I. K., 178

Sone, K., 310
Songstad, J., 358
Soos, Z. G., 323
Sopueva, A. A., 482
Sordo, J., 82
Sorokin, P. Z., 68
Sorokina, O. V., 140, 142
Sorrell, T. N., 162
Soubeyran, D. L., 229
Souchay, P., 103, 147
Soullé, E., 490
Sovayo, I., 327
Sowerby, J. D., 129
Spacibenko, T. P., 479
Spacu, P., 483, 484
Spaulding, L. D., 469
Speakman, J. C., 400
Speca, A. N., 101, 175, 176, 199, 209, 236, 439
Speier, G., 221
Spek, A. L., 166
Spence, J. T., 135
Spencer, A., 99, 177, 259, 340, 364
Spenling, R., 466
Spevak, V. N., 410
Spevak, V. V., 408
Speyer, I., 256
Spillert, C. R., 252
Spinko, R. I., 36
Spiridonov, V. P., 4, 35
Spiro, E. G., 253, 277, 469
Spiro, T. G., 153, 165, 206, 213
Spitsyn, I., 69
Spitsyn, V. I., 12, 40, 57, 96, 127, 147, 149, 157, 345, 419, 480, 484, 485
Spivack, B., 112
Spodine, E., 311
Spoliti, M., 156
Springer, C. S. jun., 481
Sprirdinov, V. P., 139
Sproul, G., 304
Sprout, O. S. jun., 100
Sprowles, J. C., 466
Sramko, T., 298
Sreckovic, M., 91
Srinivasan, T. K. K., 272, 323
Srinivasan, V., 106
Srinvasan, P. R., 4
Srivasta, K. K., 37
Srivasta, P. C., 11
Srivastava, A. K., 104, 177, 241, 280
Srivastava, B. C., 106
Srivastava, P. C., 96, 234
Srivastava, V. K., 151
Srivastava, R. D., 91
Srivastava, S. K., 225, 326, 392
Srivastava, T. S., 153, 231
Sryvalin, I. T., 93
Stadelman, W., 96
Stadnicka, K., 118
Stadtherr, L. G., 484

Stadtman, E. R., 469
Staffansson, L. I., 72
Stafford, F. E., 160
Staikov, D., 158
Stakhov, D. A., 234
Stalhandske, C., 456
Stambolija, L., 84
Stanislowski, A. G., 120
Starks, D. F., 488
Starowieyski, K. B., 89
Stasicka, Z., 119
Stearus, C. A., 34
Steel, D. F., 391
Steele, C. S., 227
Steele, D., 321
Steenkamp, P. J., 492
Steer, M. L., 466
Stefanov, S. Y., 81
Stefanovic, D., 84
Steffen, E. D., 225
Steffen, W. L., 388
Steger, J., 248
Steggerda, J. J., 355
Steinberg, I. Z., 466
Steinborn, D., 28
Steinecke, H., 18
Steinmann, W., 229, 324
Stelzer, O., 233
Stenger, C. G. F., 20
Stepanets, M. P., 472
Stepaniak, R. F., 422
Stepanov, P. I., 63
Stepanova, G. Yu., 23
Stephens, E. D., 225
Stephens, F. A., 345, 370
Stephens, F. S., 181, 218, 219, 319
Stephenson, N. C., 452
Stephenson, T. A., 330, 331, 333, 391
Stern, K. H., 427
Sternik, B. A., 133
Sternik, B. O., 160
Stetsenko, A. I., 408, 411, 413
Stevenson, J. N., 491
Stewart, D. T., 38
Steward, D. W., 323
Stewart, J. T. P., 135
Stewart, R. C., 256
Stewart, R. P. jun, 117
Stibr, B., 246
Stieffel, E. I., 60
Still, H., 121
Stillman, M. J., 53
Stobart, S. R., 219
Stoeckli-Evans, H., 28
Stoffer, D. A., 277
Stoklosa, H. J., 46, 53, 58, 59
Stone, F. G. A., 115
Stoppioni, P., 199
Stork, W., 461
Storr, A., 121
Stothers, J. B., 453
Stotter, D. A., 174, 243

Stout, E. W.. jun., 94
Straehle, J., 73
Strafford, K. N., 91
Strahle, J., 84, 433
Strajblova, J., 18
Strajescu, M., 19, 20
Strandberg, R., 147, 148
Streitwieser, A. jun., 488
Strekas, T. C., 153, 165
Strelin, S. G., 413
Strelyaev, A. E., 4
Stretkova, Z. V., 479
Stricker, G., 447
Strickson, J. A., 107
Stringer, K. A. J., 424
Strizhev, E. F., 418
Strocko, M. J., 175, 236
Stroganova, N., 478
Strohmeier, W., 354, 372
Strom, D. R., 470
Strom, K. A., 309
Strommen, D. P., 438
Strope, D., 121
Stroud, J. E., 81
Strousse, C. E., 247
Struchkov, Yu. T., 70, 71, 75, 152, 357
Strumolo, D., 352
Stucki, H., 104
Stucky, G. D., 93, 304, 322
Studer, F., 142
Stuehr, J., 205
Stults, B. R., 177
Stumpp, E., 491
Stunzi, H., 326
Sturton, I. A., 96, 436, 437
Stynes, D. V., 204
Stynes, H. C., 204, 232
Suarez-Cardeso, J. M., 103
Subbarao, E. C., 37
Subbotina, N. A., 127, 157
Suchkova, R. V., 43
Sudarikov, B. N., 46, 72, 107, 138, 497
Suetaka, W., 198
Sugimoto, M., 289
Sugiura, C., 90
Sugiyama, I., 1
Sugiyama, K., 323
Sukhanova, N. A., 411
Sukhotin, A. M., 5
Sukhoverkhov, V. F., 491
Sukhushina, I. S., 65
Sulaimankulov, K., 446, 482
Suleimanov, Z. I., 225
Suman, O., 401
Sumarokova, T. N., 23
Sunar, O. P., 326, 446, 484
Sundberg, R. J., 335
Sunder, W. A., 107, 344, 422
Suponitskii, Yu. L., 108
Suquet, H., 93
Sur, K., 376
Suraer, V. V., 3

Surazhaskaya, M. D., 157
Suredo, R. J., 106
Surpina, L. V., 79
Susheelamma, C. H., 360
Sushko, V. A., 438
Sutter, J. H., 179
Sutton, D., 116, 194
Suvorov, A. V., 24, 146
Suzuki, K., 416
Suzuki, S., 318
Svajlenova, O., 325
Svarichevskaya, S. I., 92, 125
Sveshnikova, L. B., 365
Svoboda, P., 332
Swaddle, T. W., 94
Swain, J. R., 354
Swami, M. P., 104, 177
Swank, D. D., 322
Swann, D. A., 275, 322
Swanson, B. I., 104, 369
Swanson, R., 65
Swarez, T. E., 431
Swartz, W. E., 304, 408
Sweigart, D. A., 207
Swift, D., 346, 352
Switkes, E. S., 331
Swoboda, P., 438
Syal, U. K., 321
Syamal, A., 57
Sych, A. M., 18, 20, 81, 480
Sygusch, J., 19
Sykes, A. G., 11, 129, 135, 267
Sykora, V., 91
Symon, D. A., 181, 182
Symons, M. C. R., 162, 218, 454
Syrkin, V. G., 87, 153
Syrtsova, G. P., 256
Szeimies, G., 106
Szternberg, L., 159
Szymansh-Buzar, T., 259
Szytula, A., 92

Ta, N., 351
Tabunchenko, O. Ya., 142
Tachez, M., 49
Tagawa, H., 488
Taha, F. I. M., 104
Tak, S., 484
Takac, M., 488
Takada, T., 6, 202
Takahashi, S., 382
Takahashi, T., 99, 101
Takahushi, M., 464
Takamatsu, T., 336
Takamizawa, M., 8
Takats, J., 42, 130
Takaoka, Y., 40
Takebayashi, N., 341
Takeda, K., 177
Takeda, Y., 303
Takemoto, G., 251
Takenaka, A., 370

Takeuchi, A., 294
Takeuchi, T., 146
Takezhanova, D. F., 65
Takvoryan, N., 231
Tal, N., 466
Talanti, A. M., 235
Tallent, O. K., 489
Tamao, K., 272
Tamura, T., 295
Tanaka, H., 124, 216, 298, 326
Tanaka, K., 268, 298, 398
Tanaka, M., 111, 224, 287, 295
Tanaka, N., 101
Tanaka, T., 207, 268, 298, 375, 377, 393, 398, 409, 410
Tananaev, A. N., 478
Tananaev, I. V., 479
Tandon, J. P., 43, 55
Tandon, S. G., 144, 215, 326
Tandon, S. N., 144
Tang, R., 259
Tang, S. C., 231
Tani, M. E. V., 255
Taniguchi, M., 50
Tanimura, T., 251
Tanny, S. R., 481
Tano, K., 116
Tao, L. J., 76, 124
Taqui Khan, M. M., 215, 326, 343, 465, 479
Taragin, M. F., 481
Tarama, K., 19
Taranets, N. A., 141
Tarantelli, T., 290
Taranukha, N. M., 36, 141, 142
Tarasiwicj, M., 134
Tardy, M., 321
Tarrant, J. R., 489
Tarulli, S. H., 12
Taruma, K., 266
Tasker, P. A., 277
Tasset, E. L., 177
Tatsuke, U., 33
Taube, H., 127, 333, 334, 335
Taube, R., 28, 117
Tayim, H. A., 383, 402
Taylor, B. F., 397, 419
Taylor, D., 69, 128, 210, 456, 457
Taylor, F. B., 277
Taylor, G. A., 260
Taylor, I. F., 305, 335
Taylor, J. C., 122, 304, 487, 488, 493, 494, 495
Taylor, K. R., 22
Taylor, L. T., 208
Taylor, N. J., 18
Taylor, P., 122, 138
Taylor, R. C., 305, 312, 320
Taylor, S. H., 373
Teach, M., 253
Tebbe, F. N., 2, 70
Tedesco, P. H., 96
Teets, R. E., 41

Telegus, V. S., 124
Temerk, Y. M., 442
Temme, F. P., 91
Temple, R. B., 179
Templet, P., 452
Templeton, D. H., 108, 195, 234, 321, 352, 400
Temurova, M. T., 161
Ten Hoedt, R. W. M., 331, 346
Tenhunnen, A., 216, 446
Teotia, M. P., 241, 317, 387
Terada, Y., 91
Terekhina, G. G., 16
Terekhova, L. I., 41
Tereshin, G. S., 427, 472
Tereshkina, R. I., 144
Tergenius, L. E., 380
Terzis, A., 153, 165, 212
Teterin, E. G., 489
Teuben, J. H., 15, 16
Teuerstein, A., 305, 426
Textor, M., 310, 320
Teyssie, P., 198
Teze, A., 147
Thachenko, J. I., 271
Thackeray, D. P. C., 33
Thackeray, J. R., 59, 60
Thackeray, M. M., 205
Thakur, K. P., 241, 365, 393
Than, K. A., 371, 405
Tharwat, M., 438
Thasitis, A., 234
Thauer, R. K., 110
Theobald, F., 49, 52
Theophonides, T., 412
Theriot, L. J., 325
Theyson, T. W., 219
Thick, J. A., 321
Thickett, G. W., 179
Thiele, G., 154, 459
Thiele, K. H., 28
Thierry, J. C., 400
Thomas, B. S., 121
Thomas, G. H., 324
Thomas, J. M., 91
Thomas, K. M., 110, 158
Thomas, P., 119
Thomson, A. J., 53
Thompson, D. W., 23
Thompson, J. S., 395
Thompson, L., 476
Thompson, M., 472
Thoret, J., 65
Thornhill, D. J., 220
Thornton, A. T., 92
Thornton, D. A., 97, 442
Thornton, J. M., 484
Thouvenot, R., 147
Thozet, A., 142
Tice, D. R., 125
Tidwell, S., 49
Tiedt, F., 440
Tikhonova, L. P., 215
Tikhonova, L. S., 411

Tikula, M., 248
Tillack, J., 130
Tilloca, G., 79
Timmons, K. J., 418
Timofeeva, N. I., 7, 20, 36, 479
Timofeeva, V. R., 43
Tipping, L. R. H., 364
Tiripicchio, A., 441
Tirouflet, J., 30, 31
Tishchenko, G. V., 488
Tishchenko, R. P., 69
Tisley, D. G., 151, 472
Titova, Z. M., 81
Titus, D. D., 195, 298, 401
Tkach, E. F., 148
Tobe, M. L., 408
Tobias, R. S., 466
Toda, N., 112
Todd, L. J., 88, 89
Toftlund, H., 101
Tokai, T., 323
Tokel-Takvoryan, N. E., 338
Tokes, B., 461
Tokmergenova, L. A., 482
Toledano, P., 140, 147
Tolman, C. A., 269, 304, 355, 381
Tolstov, L. K., 479
Toma, E. I., 43
Toma, H. E., 195
Tomaja, D. L., 88, 185
Tomar, O. P., 109
Tomat, G., 153, 218, 490
tom Dieck, H., 324
Tomenko, V. M., 20
Tomiyasu, H., 55
Tomlinson, A. A. G., 289, 318
Tomooka, J., 452
Tondello, E., 324, 393, 459, 491
Toneguzzo, F., 472
Tong, D. A., 78
Tong, H. W., 249, 254
Tong, S. B., 334, 349
Topich, J. A., 242
Toptygina, G. M., 35
Torchenkova, E. A., 149, 480
Tordjman, I., 243, 320, 474
Torgashev, P. D., 71
Toriuchi, M., 410
Toryanik, V. V., 124
Tosi, G., 103, 233, 313
Tossidis, I., 100, 237, 260, 443, 445
Toth, L. M., 34
Totsuka, S., 232, 325
Toubol, M., 140
Tourne, C., 148
Tourne, G., 148
Tourneau, J. E., 474
Tourneur, D., 81
Tovey, B. S., 243
Towns, L. R., 51
Townsend, J. M., 75

Toy, A. D., 53, 323
Toyoda, K., 452
Trachevskii, V. V., 134
Traficante, D. D., 362, 418
Traggeim, E. N., 79
Trainor, G., 4
Tran Qui, D., 444
Traverse, J. P., 108
Traverso, O., 499
Travkin, V. F., 158, 347
Trefonas, L. M., 320
Treichel, P. M., 151, 190, 422
Trenkel, M., 139, 161
Treptow, W., 18
Tressaud, A., 92, 370
Tresvyatskii, S. G., 478
Tretnikova, M. G., 95
Tret'yakov, B. N., 47
Tret'yakova, K. V., 64
Treuil, K. L., 160
Triantafillopoulov, I., 97
Tribalat, S., 158
Trichet, L., 33
Tridot, G., 343
Trifiro, F., 112
Tripathi, S. C., 325, 341, 386
Tripp, C., 81
Trivedi, C. P., 326, 446, 484
Trofimov, B. A., 390
Trogler, W. C., 222, 256
Trogu, E. F., 375
Troitskaya, A. D., 59, 89
Trooster, J. M., 199
Trotimchuk, A. K., 438
Trotter, J., 121, 310, 320, 448, 450
Troup, J. M., 126, 180, 187, 190
Troup, G. J., 476
Troyanov, S. I., 32
Truchanowicz, T., 95
Trumble, W. R., 358
Trunov, V. K., 141, 142, 143
Trusov, V. I., 146
Trusova, K. M., 408
Truter, M. R., 321, 322
Tryashin, A. S., 234
Tsailingol'd, A. L., 141
Tsay, Y. H., 146
Tschang, P.-S. W., 474
Tselinskii, Yu. K., 144
Tseryuta, Yu. S., 138, 139
Tsigdinos, G. A., 87
Tsirel'nikov, V. I., 32, 33, 35
Tsrigunov, A. N., 36
Tsuberblat, V. J., 98
Tsuchiya, R., 102, 250, 279
Tsuhako, M., 12
Tsui, C. S., 464
Tsui, Y. H., 464
Tsuji, K., 341
Tsukerblat, B. S., 98
Tsukihara, T., 321
Tsurik, L. A., 40

Tsutsui, M., 153, 217, 476
Tsveniashvili, V. S., 451
Tsvetkov, A. A., 72, 497
Tsvetkov, N. A., 148
Tsvetkov, Yu. V., 123
Tsvirko, M. P., 62
Tsyrenova, S. B., 108
Tu, K. N., 47
Tucker, P. A., 319
Tucker, P. M., 489
Tucker, W. F., 213
Tudo, J., 49
Tulinsky, A., 322
Tullberg, A., 202
Tulyupa, F. M., 351
Tumanova, N. Kh., 93
Tummavuori, J., 99
Tun, K. M., 192
Tune, D. J., 418
Tuney, T. W., 360
Turan, T. S., 326
Turco, A., 216, 327
Turnbull, K. R., 253
Turner, J. J., 88, 181
Turney, T. W., 355
Turov, A. V., 487
Tursunova, G. A., 150
Turtle, B. L., 397, 414
Twigg, M. V., 167, 203, 329
Tytko, K. H., 146
Tyvoll, J. L., 309

Ubozhenko, O. D., 14
Uchida, Y., 194, 386
Udagawa, S., 81
Udovento, V. V., 430
Udupa, M. R., 140, 149, 292, 390, 392, 489
Udy, D. J., 18
Ueda, H., 112
Uehara, A., 279
Ueki, T., 195, 352
Uemara, S., 99, 177, 259, 364
Ueno, K., 494
Ugajin, M., 124
Ugnivenko, T. A., 142
Ugo, R., 244
Ugryumova, L. E., 123
Uguagliati, P., 355
Uh, Y. S., 21, 22
Uhlemann, E., 131, 206
Uhlig, D., 88, 181
Uhlig, E., 242, 325
Uhlemann, E., 241, 257
Ukai, T., 383
Ulicka, L., 65
Ul'ko, N. V., 133, 134, 160
Umanskii, Ya. S., 124
Umetsu, Y., 146
Umilin, V. A., 46
Ummat, P. K., 452

Underhill, A. E., 200, 228, 229, 302, 393, 433
Ungermann, C. B., 334
Ungvary, F., 218
Unsworth, W. D., 459
Urbach, F. L., 57, 242
Urinovich, E. M., 239
Urry, G., 165, 219
Usatsenko, Y. I., 351
Ushatinskii, I. N., 155
Usol'tseva, M. V., 390
Ustimovich, A. B., 205
Ustinov, O. A., 142
Ust'yantsev, V. M., 95, 140
Usucheva, G. M., 89
Uttley, M. F., 332, 335, 412

Vacca, A., 327, 441
Vagg, R. S., 227, 272
Vagina, N. S., 477
Vahrenkamp, H., 167, 190, 221, 268
Vaidya, R. S., 327
Vainshtein, M. N., 438
Vaisbein, Z. Y., 393
Vaivads, A., 95
Val'dman, S. S., 176
Valencia, C. M., 350
Valentine, J. S., 217, 323
Valentova, M., 446
Valiev, L. I., 99
Valikhanova, N. Kh., 124
Vallarino, L. M., 282
Vallee, B. L., 462
Van Baalen, A., 75
Vance, T. B., 174
Vancea, L., 183, 219, 330
Vanchagova, V. K., 46
Vanchikova, E. V., 147
Van Dam, E. M., 111
van den Bergen, A., 258
Van den Elzen, A. F., 140
Van der Kelen, G. P., 268
Van der Linden, W. E., 461
Van Der Planken, J., 70
Vander voet, A., 381
Van de Velde, G. M. H., 20, 23
Vandrish, G., 150, 238
Van Gaal, H. L. M., 355
Van Ingen Schenau, A. D., 173, 235, 284
Van Kemenade, J. T. C., 106
Van Kerkwijk, C. J., 234
Van Laar, B., 99
Van Leirsburg, D. A., 92
Vannerberg, N. G., 62, 202
Van Oven, H. O., 1
Vanquickenborne, L. G., 342
Van Schalkwyk, G. J., 138
Van Tamelen, E. E., 116, 212
van Vuuren, C. P. J., 490, 491
Van Wyk, J. A., 189

Varadi, V., 469
Varadi, Z. B., 131
Varfolomeev, M. B., 418
Varga, L. P., 489
Vargova, C., 65
Varhelyi, C., 102, 256
Varshavsky, Y. S., 358
Vashakidze, R. M., 272
Vasile, M. J., 344, 422
Vasil'ev, V. P., 216
Vasil'eva, A. G., 156
Vasil'eva, I. A., 65
Vasil'eva, I. G., 477
Vasil'eva, V. N., 216
Vasilevskaya, I. I., 156
Vasil'kova, I. V., 129
Vaska, L., 193, 357
Vasvari, G., 236
Vaughan, D. H., 366
Vaughn, J. W., 101
Vauk, C. G., 30
Vavilova, V. V., 5
Veal, J. T., 98
Vedrine, A., 485
Vegh, L., 469
Vekshina, N. V., 6, 123
Veksler, S. F., 139
Velicescu, M., 20, 214
Venanzi, L. M., 282, 375, 397
Vendrell, M., 148
Venevtse, Y. N., 81
Venkappayya, D., 224, 238,
    445
Venkateshan, M. S., 58, 237,
    289, 496
Vennereau, P., 147
Venturini, M., 66
Verani, G., 200, 233, 301, 313
Verdonck, E., 342
Veremei, T. P., 480
Verendyakina, N. A., 194
Vergnon, P., 132
Verkade, G. J., 269
Verkouw, H. T., 1
Verschoor, G. C., 20, 173, 284,
    309, 321
Vershinina, F. I., 205
Vesely, N., 43
Veville, G. A., 461
Vezzosi, I. M., 452
Vicentini, G., 482, 483
Vickery, B. L., 321
Vidal, A., 81
Vidali, M., 493, 494, 496
Vidyarthi, S. K., 425
Viebahn, W., 201
Vielen, P., 103
Vigato, P. A., 394, 493, 494,
    496
Villa, J. F., 323
Villeneuve, G., 91
Vinarov, I. V., 36
Vincentini, G., 472
Vinogradov, I. V., 158

Vinogradova, S. M., 89
Vinokurov, V. A., 140
Vinokurov, V. M., 107
Vinter, I. K., 148
Viossat, B., 130
Virmani, Y., 89, 122
Visscher, M. O., 114
Visser, C. J., 70
Visser, J. W., 174, 444
Vizi-Orasz, A., 221, 356, 358
Vlasse, M., 78
Vlček, A., 222
Vlcek, A. A., 202, 222, 266
Voet, A. V., 216, 353
Vogt, A., 244
Voiculescu, N., 387, 393
Voitko, I. I., 17
Voitovich, R. F., 48
Volchenskova, I. I., 384
Vol'dman, G. M., 146
Volhardt, K. P. C., 218
Voliotis, S., 490
Volkov, J. V., 438
Volkov, S. V., 172
Volkov, V. L., 49, 140
Volkov, V. M., 395
Volkov, S. V., 93
Volkova, V. P., 214
Volodina, G. E., 451
Vol'pin, M. E., 29, 353, 357,
    373
Volponi, L., 415
Volshtein, L. M., 410, 412, 413
Von Dreele, R. B., 20
Vongvusharintra, A., 73
Von Schnering, H. G., 93
Von Stetten, O., 176
Voorhoeve, J. M., 76
Vorob'ev, N. I., 49, 61
Vorob'eva, N. A., 65
Voronina, N., 56
Voronkov, M. G., 145
Voronova, E. M., 67
Voronskaya, G. N., 489
Vorontsova, T. A., 110
Vorotinova, B. G., 32
Votava, H. J., 104
Votoyarov, M., 479
Vovkotrub, E. G., 81, 214, 480
Vrachnov-Astra, E., 48
Vrchlabsky, M., 144, 146
Vuitel, L., 19
Vulikh, A. I., 20, 36
Vyalykh, E. P., 390
Vyshinskii, N. N., 31

Wada, F., 181
Wada, G., 194
Wadas, R., 95
Waddan, D. Y., 335
Waddington, T. C., 91, 181,
    182
Wagner, F., 275, 443

Wagner, K. P., 422
Wailes, P. C., 14, 45
Wainer, I., 464
Wakerley, M. W., 490
Wakihara, M., 50
Wakita, H., 478
Waldmann, J., 222
Walker, D. C., 425
Walker, G. W., 306
Walker, M. D., 465
Walker, N. W. J., 117
Walker, R., 422
Walker, W. R., 227, 240, 325
Walks, P., 126
Wall, F., 94, 201
Wall, J. E., 271
Wall, L. S., 81
Wallace, W. J., 188
Wallis, W. N., 297
Walter, D., 269
Walter, R. H., 117
Walter-Levy, L., 108, 174, 444
Walters, J. M., 321
Walters, K. L., 221
Walton, D. R. M., 418
Walton, E. C., 227, 272
Walton, R. A., 128, 151, 152,
    159, 311, 428, 472
Waltz, W. L., 340
Wan, C., 158, 320
Wang, A. H. J., 275
Wang, B. C., 240
Wang, D. K. W., 332
Wang, F., 367
Wang, F. F. Y., 161
Wang, H. C., 136
Wang, L. P., 484
Wanlass, D. R., 137
Ward, G. A., 106
Waring, J. L., 81
Warner, L. G., 234, 309
Warren, K. D., 1, 89, 341
Warren, L. F., 93, 179, 233
Wasson, J. R., 46, 53, 58, 59,
    100, 175
Waszczak, J. V., 73, 76
Watanabe, E., 356
Watanabe, K., 318
Watanabe, T., 281
Watanabe, Y., 108
Watenabe, O., 5
Waters, D. D., 227
Waters, D. N., 438
Waters, K. L., 245
Waters, T. A., 322
Waters, T. N., 178, 275, 321,
    441, 443
Waters, W. A., 202
Watkins, D. M., 302, 393
Watkins, J. J., 438
Watson, D. G., 284
Watson, K. J., 120, 213
Watt, G. W., 281, 310, 320,
    395

Watton, E. C., 215
Watts, D. W., 66, 256, 263
Watts, J. B., 335
Wayland, B. B., 193, 232
Weakley, T. J. R., 149, 474
Weaver, T. J., 302, 379
Webb, G. A., 217
Webb, M. J., 153
Webb, T. R., 126
Weber, J. H., 243, 258
Weber, L., 398
Weber, O. A., 326, 327, 467
Webster, D., 200
Wedd, A. G., 134
Weeks, B., 371, 414
Wei, R. M. C., 242, 391, 442
Weich, C. F., 263, 264
Weigel, F., 492
Weighardt, K., 267
Weigold, H., 34
Weiher, J. F., 207
Weinstein, G. N., 231
Weinstock, L. M., 317
Weinstock, N., 150
Weiss, A., 6, 76, 496
Weiss, E., 30, 46
Weiss, E. R., 199
Weiss, J., 243, 400
Weiss, K., 452
Weiss, R., 131, 239, 247, 301, 400, 429
Weissman, A., 179
Weitzel, H., 81
Welch, A. J., 115
Wells, W. J., 310, 320
Wendlandt, W. W., 250
Wendt, R. C., 304
Wentworth, R. A. D., 127
Werle, P., 237, 445
Wermeille, M., 275
Werner, H., 199, 236
Wernick, D. L., 487
Wertheim, G. K., 6
Wertz, D. L., 309
Weschler, C. J., 100
West, A. R., 18
West, B. O., 258, 364
West, P. R., 10
West, R., 288
West, R. J., 323
Wester, D., 444
Westland, A. D., 129
Weulersse, J. M., 147
Whan, D. A., 112
Wheatley, P. J., 163, 222, 447, 448
Whei-Lu-Kwik., 60
Whimp, P. O., 265
White, A. H., 69, 226, 276, 279, 287, 306, 320, 322, 422, 456, 458
White, A. J., 182, 190
White, J., 234
White, J. F., 115

White, H., 400
Whitesides, G. M., 109, 487
Whitfield, H. J., 308
Whiting, R., 305
Whitlow, S. H., 320, 457
Whitney, J. F., 142
Whittaker, B., 491
Whittaker, D., 221
Whitten, D. G., 337
Whittingham, M. S., 76
Whyman, R., 381
Wiberg, K. B., 106
Wichmann, K., 325
Wicholas, M., 311
Wiegel, K., 26
Wiegers, G. A., 99
Wiersema, R. J., 105, 246, 247
Wiest, R., 429
Wieting, T. J., 123
Wijnhoven, J. G., 133
Wild, S. B., 259, 397, 422
Wilde, H. J., 104
Wilde, R. E., 272, 323
Wiles, D. R., 152, 182
Wilke, G., 206, 241
Wilkins, J. D., 61, 83, 85, 86, 87, 128, 134
Wilkins, R. G., 245
Wilkinson, G., 87, 99, 126, 154, 162, 177, 199, 238, 259, 340, 345, 363, 364, 389
Wilkinson, W., 401, 440
Wille, G., 31
Willett, R. D., 310, 320, 321, 322
Willey, G. L., 26
Willey, G. R., 25, 102
Williams, A. F., 113
Williams, D. A., 105, 470
Williams, D. H., 214
Williams, D. R., 176, 325, 326, 385, 465
Williams, L. R., 215
Williams, M. L., 278, 325
Williams, P. M., 76
Williams, R. J., 202, 422
Williams, R. J. P., 484, 486
Williams, W. E., 182
Williamson, D. H., 87, 126
Willing, R. I., 34
Willis, A. C., 322, 422
Willis, C. J., 244
Wilson, C. J., 402
Wilson, I. R., 202
Wilson, J. W., 96
Wilson, J. W. L., 249
Wilson, P. W., 122, 304, 487, 488, 491, 493, 494, 497
Wimmer, J. M., 81
Winarko, S., 162
Windisch, S., 92
Winfield, J. M., 78, 138, 150
Wingert, L. A., 262, 263

Wingfield, J. N., 339, 397
Winograd, N., 428
Winter, J. J., 76
Wiseman, P. J., 137
Wishnevsky, V., 492
Wismer, R. K., 252
Wiswall, R. H., 8
Witteveen, H. T., 272
Wittingham, M. S., 137
Wittke, G., 249, 267
Wnuk, M., 464
Woitschach, J., 68
Wold, A., 137, 248
Wolfe, R. W., 81
Wolford, T., 311
Wolterman, G. M., 175
Wong, Y. S., 422, 466
Wood, D. J., 79
Wood, J. M., 470
Wood, J. S., 473
Wood, T. E., 466
Woodhouse, D. I., 111
Woodward, P., 422
Woolley, P. R., 277, 463
Woolsey, I. S., 255, 278, 325
Work, R. A., 201
Worthington, T. M., 274
Wreford, S. S., 2, 362, 418
Wright, C. M., 471
Wright, D. A., 322
Wright, K., 157, 223, 330
Wright, W. B., 97
Wrighton, M. S., 361, 369
Wu, K. K., 94
Wyatt, M., 485

Xavier, A. V., 486

Yablokov, Yu. V., 57, 98
Yacynych, A. M., 100, 261
Yakima, S., 215
Yakovleva, E. G., 12
Yakushin, V. K., 148
Yajima, F., 251
Yamada, O., 99
Yamada, S., 294, 316
Yamada, T., 81
Yamagishi, T., 194
Yamamoto, A., 111, 172
Yamamoto, H., 416
Yamamoto, K., 111
Yamamoto, N., 303
Yamamoto, Y., 98
Yamamura, K., 147
Yamanaka, S., 203
Yamanari, K., 250
Yamaoka, S., 139
Yamasaka, K., 251
Yamasaki, A., 251, 455
Yamasaki, K., 240, 253, 292
Yamase, T., 136

Yamauchi, O., 326, 327
Yamayoshi, T., 316
Yamdagni, R., 139
Yampol'skaya, V. U., 485
Yan, L. Y., 288
Yanagida, R. Y., 174
Yanagisawa, S., 193, 339, 369
Yang, C. H., 267
Yang, P. H., 53
Yang, Y., 136
Yanir, E., 490
Yankina, L. F., 133, 145
Yannopoulos, L. N., 95
Yanushkevich, T. M., 140
Yarbrough, L. W., 269
Yarino, T., 176
Yarkova, E. G., 24
Yarmolovich, A. K., 122
Yarmolyuk, Ya. P., 34
Yarovoi, A. F., 93
Yashioka, H., 382
Yasunaga, T., 326
Yasuoka, M., 102
Yasuoka, N., 400
Yates, J. T. jun., 111
Yatsenko, A. P., 65
Yatsimirskii, K. B., 43, 215, 384
Yawnay, D. B. W., 322
Yeats, P. A., 21
Yellin, N., 447
Yerkess, J., 104
Yin, L. I., 304
Yocom, P. N., 142
Yoika, Y., 251
Yokayama, A., 326
Yokokawa, T., 485
Yokoi, H., 323
Yokoyama, A., 216, 298
Yokoyama, Y., 494
Yolovets, I. A., 56
Yoneda, H., 251
Yonemura, M., 279
Yoshida, N., 476
Yoshida, S., 19
Yoshida, T., 31, 309
Yoshida, Y., 382
Yoshikawa, S., 262
Yoshikawa, Y., 253
Yoshikuni, K., 341
Yoshimura, M., 142
Yoshizumi, F. K., 462
Young, D., 115
Young, D. J., 91
Young, R. J., 345
Ytsma, D., 16
Yukawa, H., 326

Yun, P. T., 225
Yunusov, N. B., 106
Yunusova, N. V., 38
Yupko, L. M., 125
Yuranova, L. I., 40
Yurinov, V. A., 410
Yuzuri, M., 91
Yvon, K., 125

Zabolotskaya, E. V., 30
Zachariasen, W. H., 488
Zaev, E. E., 54
Zagal, J., 311
Zagrebina, R. F., 150
Zagryazkin, V. N., 7
Zaharescu, M., 38
Zahhartseva, A. S., 68
Zaidi, S. A. A., 174, 229
Zaidi, Z. H., 18
Zainulin, Yu. G., 8, 33, 36
Zaitsev, B. E., 427
Zaitsev, L. M., 39, 471
Zaitseva, L. L., 158
Zaitseva, V. A., 427
Zaitseva, Z. A., 95, 478
Zakharov, M. A., 79, 82
Zakharova, I. A., 133
Zalkin, A., 108, 195, 234, 321, 345, 352, 370, 400, 433
Zamanskii, V. Y., 216
Zamoli, A. F., 452
Zamudio, W., 311
Zamyantia, O. V., 65
Zanella, R., 390
Zanger, M., 161
Zanne, M., 248
Zapletnyak, V. M., 68
Zaripova, R. K., 145
Zarli, B., 415
Zarubitskaya, L. I., 144
Zasorin, E. Z., 4, 35
Zassinovich, G., 360
Zastavker, L. A., 81, 480
Zatko, D. A., 49
Zauella, R., 398
Zayats, M. N., 141, 179
Zayan, S. E., 239
Zdunneck, P., 28
Zedelashvili, E. N., 478
Zegarski, W. J., 288, 400
Zeinalov, V. Z., 225
Zelbst, E. A., 82
Zelentsov, S. S., 490
Zelentsov, V. V., 57, 98, 127, 157, 178

Zelikman, A. N., 146
Zelinka, J., 50, 144
Zemlicha, M., 325
Zemlyanskii, N. I., 100
Zemva, B., 90, 105
Zeppezauer, E., 464
Zhantalai, B. P., 311
Zharkikh, A. A., 14
Zhaveronkov, N. M., 34
Zhdanov, A. A., 19
Zhdenovskikh, T. M., 173
Zheligooskaya, N. N., 419
Zhemchuzhnikova, T. A., 98
Zhigach, A. F., 68
Zhigubina, N. S., 142
Zhikhareva, T. N., 98
Zhirov, A. I., 496
Zhukov, V. V., 428
Zhukovskii, V. M., 139, 140
Zhurakovskii, E. A., 5, 7, 26, 92
Zhuravlev, E. F., 216, 478
Zhuraleva, M. D., 482, 483
Ziegenbalg, S., 156
Ziegler, J. F., 47
Ziegler, M. L., 114
Ziegler, R. J., 90, 163
Zimkina, T. M., 125
Zimmerman, D. N., 239, 314
Zimmermann, M. L., 479
Zimtseva, G. P., 419
Zingde, M. D., 58, 237, 289, 496
Zinner, L. B., 482, 483
Zinoune, M., 233
Zioli, R. F., 216, 345
Ziolkowski, J. J., 259
Zmii, V. I., 7
Znamenskaya, A. S., 472
Zolotov, Yu. A., 138, 139
Zolotukhin, V. K., 23, 40
Zoltobrocki, M., 464
Zompa, L. J., 254
Zoz, E. I., 36
Zozlenko, A. A., 136
Zubarev, V. N., 197
Zuberbühler, A. D., 327
Zubieta, J. A., 158, 181, 422
Zubritskaya, D. I., 119
Zuckerman, J. J., 88, 185
Zuech, E. A., 95
Zupan, J., 105
Zurkova, L., 65
Zuznetsova, Z. M., 138
Zvoleiko, P. T., 65
Zwickel, A. M., 342
Zyrnicki, W., 5